33rd EUROPEAN SYMPOSIUM ON COMPUTER AIDED PROCESS ENGINEERING

VOLUME 2

COMPUTER-AIDED CHEMICAL ENGINEERING, 52

33rd EUROPEAN SYMPOSIUM ON COMPUTER AIDED PROCESS ENGINEERING

VOLUME 2

Edited by

Prof. Antonios C. Kokossis
National Technical University of Athens
School of Chemical Engineering
Athens, Greece

Prof. Michael C. Georgiadis
Aristotle University of Thessaloniki
School of Engineering
Department of Chemical Engineering
Greece

Prof. Efstratios Pistikopoulos
Chemical Engineering
Texas A&M Energy Institute
TX, USA

ELSEVIER

Amsterdam – Boston – Heidelberg – London – New York – Oxford
Paris – San Diego – San Francisco – Singapore – Sydney – Tokyo

Elsevier
Radarweg 29, PO Box 211, 1000 AE Amsterdam, Netherlands
The Boulevard, Langford Lane, Kidlington, Oxford OX5 1GB, UK
50 Hampshire Street, 5th Floor, Cambridge, MA 02139, USA

British Library Cataloguing in Publication Data
A catalogue record for this book is available from the British Library

Library of Congress Cataloging-in-Publication Data
A catalog record for this book is available from the Library of Congress

ISBN (Volume 2): 978-0-443-23554-2
ISBN (Set) : 978-0-443-15274-0
ISSN: 1570-7946

For information on all Elsevier publications visit our website at https://www.elsevier.com/

Publisher: Candice Janco
Acquisition Editor: Anita Koch
Editorial Project Manager: Lena Sparks
Production Project Manager: Paul Prasad Chandramohan
Designer: Greg Harris

Typeset by STRAIVE

Contents

T2: Control, scheduling, and operability at the process and enterprise-level

Antonis Kokossis, Michael C. Georgiadis, Efstratios N. Pistikopoulos (Eds.)
PROCEEDINGS OF THE 33rd European Symposium on Computer Aided Process Engineering (ESCAPE33), June 18-21, 2023, Athens, Greece
 http://dx.doi.org/10.1016/B978-0-443-15274-0.50141-4

Optimal design of a microbubble cap to enhance mass transfer efficiency using image processing

Yup Y.,[a,b] Jonghun L.,[a,b] Chonghyo J.,[a,b] Seongbin G.,[a] Sunghyun C.,[a] Jaewon L.,[a] Junghwan K.,[a,b] Hyungtae C.,[a,*]

[a]*Green Materials and Processes R&D Group, Korea Institute of Industrial Technology, 55, Jonga-ro, Ulsan 44413, Korea*

[b]*Department of Chemical and Biomolecular Engineering, Yonsei University, 50, Yonsei-ro, Seoul 03722, Korea*

htcho@kitech.re.kr.82

Abstract

Microbubble technology has been used in various industries to improve the mass transfer efficiency of applications. Although conventional bubble generators using venturi and swilling nozzles are useful, their low gas throughput is insufficient for industrial pollutant reduction. Hence, this study developed an optimal microbubble cap structure with a high gas throughput and microbubble generation rate using an image processing based overall mass transfer coefficient (OMTC) calculation model. Micro bubble cap with brim structure was developed, and the bubble generation pattern by varying the brim size was analyzed in a lab-scale microbubble system. Bubble diameters were measured using image processing and OMTC was calculated using bubble size. The observed bubble size range was approximately 0–250 μm, and the bubble generation ratio was the highest at a 5-mm-brim size. When calculating the OMTC of H_2S in water of 50 μm or less, the value increased by more than 50% compared to other brim-size cases. Therefore, at brim sizes of 5 mm, the dispersion and production of microbubbles were active, and they were expected to endow high mass transfer efficiency.

Keywords: microbubble, vortex, overall mass transfer coefficient, bubble cap, image processing

1. Introduction

Microbubbles are used in various fields, such as water treatment, reactors, nuclear power, and air pollutant reduction (Yup Y. et al., 2021). In the various industries, they are used either for the formation of a large interfacial area between the gas and liquid phases, the generation of radicals, or long interaction time in flow field. The principle of utilization of microbubble is that it creates an environment that enables gas–liquid interaction. Particularly, the gas–liquid interaction is a significant factor in the field of air pollutant reduction. Microbubbles have recently been used to remove air pollutants and increase reduction efficiency using bubble scrubber system (Hyundo P. et al., 2021).
Microbubbles have a role in absorbing air pollutants in a liquid state. Therefore, it is important to generate many small-sized bubbles and create an environment for enhancing mass transfer efficiency by expanding the interfacial area (Al-Mashhadani. et al., 2012).

Microbubbles are generated mainly using a small venturi structure or a sparger-type pressurized generation method (Ali, M. et al., 2013). Although this method is suitable for generating microbubbles, it is difficult to use in the air pollutant reduction field to quickly purify considerable gas owing to low gas throughput. Therefore, a novel microbubble generator is needed.

Confirmation of the bubble size must be an essential step to design the bubble generator. Because it is possible to calculate overall mass transfer coefficient (OMTC) from the bubble size data, which is a significant value in air pollutants removal process (Yup Y. et al., 2023). Nevertheless, previous studies cannot calculate OMTC because they cannot determine the volume of the flow field, which is three-dimensional data, due to limitations in image data. Therefore, we presented a OMTC calculation model that determines bubble size and volume of the flow field by using concept of camera depth of the field and image processing method. This study focuses on an optimal structure for a bubble cap is determined using an OMTC calculation model.

2. Methodology

2.1. Setup of lab-scale microbubble system

This study proposed a novel structure that induces bubble breakage by adding a brim structure that prevents large bubbles from rising. Bubble caps with various brim widths (0, 5, 10, 15, and 20 mm) were fabricated using 3D-printing to investigate the tendency of bubble generation pattern and OMTC to vary depending on the brim width.

A simple experimental device was created to confirm the bubble generation pattern by brim structure. The device has a rectangular structure with an acrylic body, and its width and height are 300 and 500 mm, respectively. Additionally, a vacuum suction blower draws air from the top of the rectangular structure. The induced airflow passes through the air velocity indicator and generates bubbles in the bubble cap installed in the center of the bottom of the experimental device. Water was filled to a height of 10 cm from the bottom of the experimental device. A detailed airflow diagram and component descriptions are shown in Fig. 1.

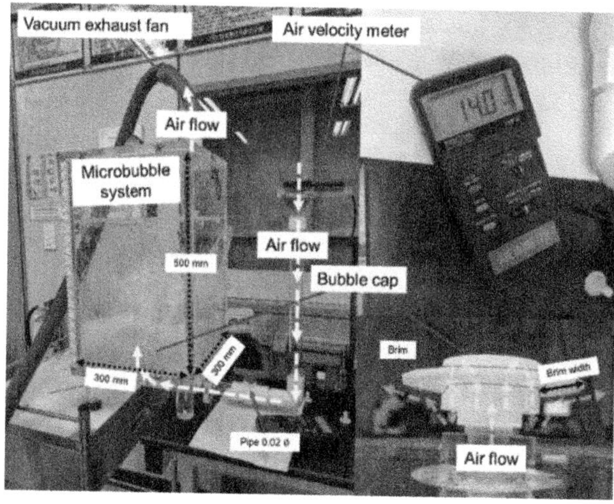

Fig. 1. Bubble generating system experimental setup

2.2. Image processing based overall mass transfer coefficient calculation model

A DSLR camera (SONY A7M3) and a 24 mm-Laowa macro-lens, were used to extract images of the bubbles generated. Pictures were taken using the camera at a height of 1 cm from the bottom of the experimental device. The photographed images were then processed to measure the bubble size. Image processing was performed using OpenCV-Python, and the shape of the bubble was detected using Gaussian blur, Canny edge detection, and morphology transformation functions. Figure 2 shows a sample image generated by the image processing process for bubble size measurement.

Gaussian blur step is a role in suppressing noise in the image, and the Canny edge detection step determines the bubble edge based on the gradient value of brightness. Then, in the morphology transformation step, the closed curve converts to a contour by filling the curve and determining the contour area to calculate the bubble's size. Finally, the extracted bubble size data calculates the number of bubbles and the average bubble size and displays it as the resulting image in Fig. 2.

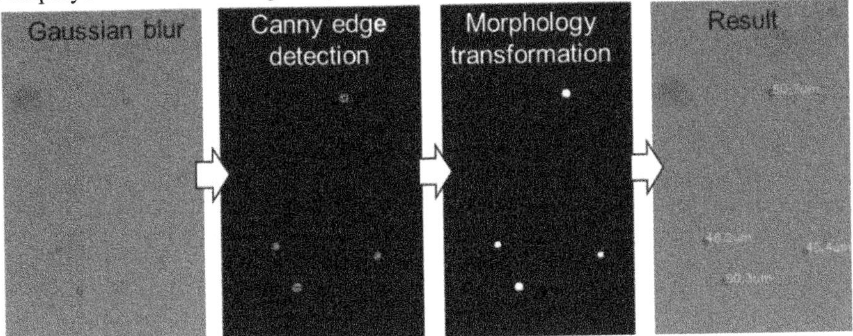

Fig. 2. Sample image of the image processing process for bubble size measurement

The mathematical model for determining OMTC to confirm the mass transfer efficiency due to the formation of microbubbles begins with the following equation.

$$\frac{\partial C_{\mathrm{A}}}{\partial t}+\frac{\partial (uC_{\mathrm{A}})}{\partial z}=c\cdot A_{\mathrm{surf}}(C_{\mathrm{A}}^{*}-C_{\mathrm{A}})=k_{\mathrm{ov}}(C_{\mathrm{A}}^{*}-C_{\mathrm{A}}) \tag{1}$$

$$A_{\mathrm{surf}}=\frac{6\emptyset_{\mathrm{gas}}}{d_{\mathrm{sm}}} \tag{2}$$

where C_{A} represents the concentration of gas component A in a gaseous state, C_{A}^{*} represents the concentration of A in equilibrium with the current liquid, u represents the gas velocity, and A_{surf} represents the interfacial area concentration. Additionally, $\emptyset_{\mathrm{gas}}$ is the volume ratio of the bubble in the flow field, and d_{sm} is the Sauter mean meter of bubbles. Equations (1) and (2) can be used to extract bubble size data from several bubble images in a short time and determine OMTC.

$$\emptyset_{\mathrm{gas}}=\frac{V_{\mathrm{tb}}}{V_{\mathrm{ff}}} \tag{3}$$

where V_{tb} is the sum of bubble volume in the flow field, and V_{ff} is the volume of flow field. V_{ff} is determined from the camera the depth of field.

$$V_{\mathrm{ff}}=A_{\mathrm{ip}}\,C_{dof} \tag{4}$$

A_{ip} is the area where bubbles are detected in the image through image processing, and C_{dof} is the depth of field.

3. Results and Discussions

3.1. Bubble size distribution analysis

The best way to analyze the bubble formation pattern is to determine the bubble size distribution (BSD). The BSD analysis method is a common method used for multi-phase processing. This study also checked the bubble size distribution to confirm the bubble formation pattern according to the bubble cap's brim width. The confirmed result is shown in Fig. 3.

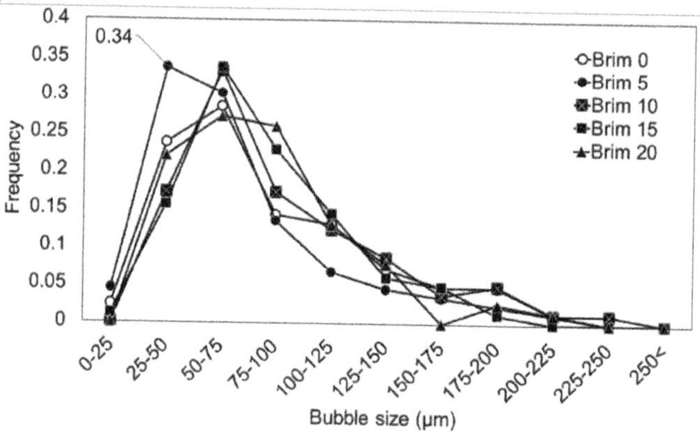

Fig. 3. Bubble size distribution graph according to the brim width of the bubble cap

As can be observed from the figure, when the brim width is 5 mm, bubbles between 25 and 50 μm are generated the most. This shows a frequency (that is, 0.34), which is approximately 40% higher than that of bubble caps with other brim width structure, and the bubble generation ratio of 50 μm or less was approximately 38%. Additionally, in the brim widths of 0, 10, 15, and 20 mm, bubbles between 50 and 75 μm were generated with the highest frequency, and the frequency of bubbles between 75 and100 μm tended to increase as the brim width increased. The benefits of microbubbles with a size less than 50 μm are numerous. Particularly, the interaction time between phases can be maintained for a long time because the buoyancy of small bubbles enables them to stay in the flow field for a long time without rising and disappearing.

3.2. Statistical analysis of bubble generating pattern

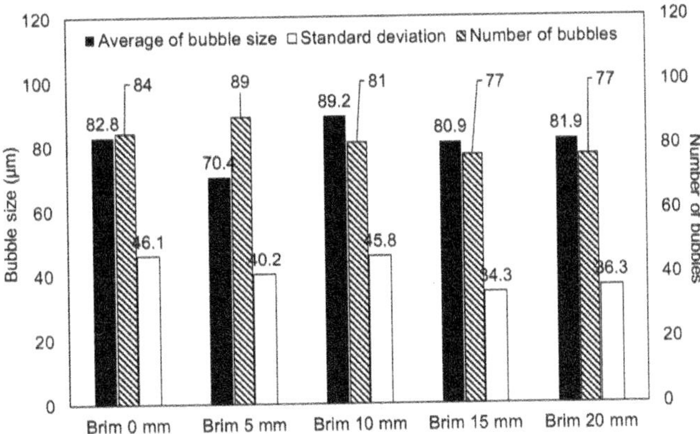

Fig. 4. Results of average bubble size, standard deviation, number of bubbles analysis for generated bubbles

Since the generation of bubbles is accompanied by complex mechanisms and microbubble is generated by various sizes, it is challenging to identify bubble generation patterns visually; hence, identifying the pattern of bubble formation using a statistical analysis method is essential. Therefore, the standard deviation for the average bubble size and number of bubbles was determined and is shown in Fig. 4. The average bubble size was 70.4 μm, which was the smallest in the 5-mm-brim structure. Additionally, 89 bubbles were generated in the observed flow field, which was confirmed to be higher than for other conditions. Where a brim structure creates bubbles of small average size, the standard deviation is lower than when the brim width is 0 and 10 mm. Hence, when the brim width is 5 mm, microbubbles can be well dispersed. Therefore, it is considered to have a characteristic of actively forming turbulence so that many bubbles can be generated.

3.3. Result of overall mass transfer coefficient

Identifying and determining the OMTC in the air pollutant removal process can be an index for selecting process conditions in which actual mass transfer is active. Thus, this index can be used to determine the mass transfer efficiency and improve the process's efficiency under various process conditions. Particularly, the mass transfer effect for the bubbles generated using the novel brim structure must be evaluated because the generation of bubbles of a size of 50 μm or less has a positive effect on the mass transfer efficiency. Figure 5 shows the OMTC for H_2S, a representative air pollutant, which was determined for bubbles less than 50 μm.

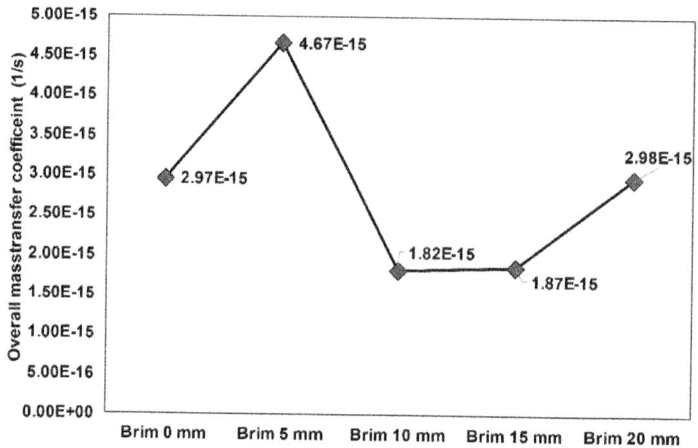

Fig. 5. OMTC calculation results for bubbles less than 50 μm

This outcome demonstrated that the OMTC was at its highest when the brim width was 5 mm. It showed an approximately 50% higher value than the other conditions.

4. Conclusion

In this study, a bubble cap structure that generates a large number of microbubbles with high gas throughput is developed, and the optimal structure of the bubble cap is determined using image processing based OMTC calculation model. Bubble cap structure was then used to confirm the bubble generation pattern according to the width of the additional brim structure that prevents air from rising. A case study was conducted according to the bubble cap brim width (0, 5, 10, 15, and 20 mm) inside a bubble generator system made of acrylic.

Under the condition of brim 5 mm, the rate of formation of bubbles less than 50 μm, which shows the microbubble characteristics, was as high as 38%. Additionally, the highest amount of bubble formation was observed under the 5-mm-brim condition, and the average bubble size was also determined to be the lowest at 70.4 μm. Furthermore, the OMTC value was established to validate the effect of bubbles less than 50 μm on H_2S, a representative air pollutant. OMTC, 4.67E-15 (s^{-1}), was the highest in the 5-mm-brim condition, where bubbles of less than 50 μm were frequently generated. According to these results, the 5-mm-rim bubble cap, in which bubbles of less than 50 μm were induced by the dispersion and destruction of bubbles and appropriate location of turbulent flow, is considered the most suitable structure for removing air pollutants. The structure can only be used to create bubbles of specific size. Additionally, in future studies, we expect to enhance accuracy of OMTC calculation model through gas removal experiments.

Acknowledgment

This work was supported by the Korean Institute of Industrial Technology within the framework of the project "Development and application of AI based microbubble-scrubber system for simultaneous removal of air pollutants [grant number KM-22-0015]."

References

Ali, M. et al., 2013, Iodine removal efficiency in non-submerged and submerged self-priming venturi scrubber, Nuclear engineering and technology, 45, 2, 203-210.

Al-Mashhadani., et al., 2012, CO2 mass transfer induced through an airlift loop by a microbubble cloud generated by fluidic oscillation, Industrial & Engineering Chemistry Research, 51, 4, 1864-1877.

Hyundo P., et al., 2021, Computational fluid dynamic modelling of optimal water level in low-pressure microbubbles scrubbers, Chemical Engineering Transactions, 86, 613-618.

Yup Y., et al., 2021, Method for determining optimum operational conditions of microbubble scrubber using image processing, Journal of Environmental Informatics, 38, 2, 83-92.

Yup Y, et al., 2023, Method for measuring bubble size under low-light conditions for mass transfer enhancement in industrial-scale systems, International Communications in Heat and Mass Transfer, 140, 106525.

Antonis Kokossis, Michael C. Georgiadis, Efstratios N. Pistikopoulos (Eds.)
PROCEEDINGS OF THE 33rd European Symposium on Computer Aided Process Engineering (ESCAPE33), June 18-21, 2023, Athens, Greece
 http://dx.doi.org/10.1016/B978-0-443-15274-0.50142-6

Design of a heterogeneous azeotropic distillation processes using stochastic optimization

Josué J. Herrera Velázquez,[a] Julián Cabrera Ruiz,[a] J. Rafael Alcántara Avila,[b]

[a]*Departamento de Ingeniería Química, Universidad de Guanajuato, Noria Alta S/N, Guanajuato, 36050, Mexico.*
[b]*Department of Chemical Engineering, Kyoto University, Kyoto 615-8510, Japan.*

Abstract

Heterogeneous azeotrope distillation is often preferred industrially over homogeneous azeotrope distillation due to the ease of recovery of the entrainer and the transition across a distillation boundary in the decanter. In a previous study, the synthesis problem for separating an Ethanol and Water mixture by using benzene as an entrainer was solved using a low aggregation superstructure comprised of functional distillation and decanter modules where the utility cost was minimized (Okunishi, 2019). The solution was interpreted to obtain a realistic process. This work optimizes the interpreted solution using a modified Simulated Annealing Algorithm. The new solution makes it possible to reduce the overall energy consumption and the total annual cost of the process. Also, an interface between Python and Aspen Plus® is proposed, and it allows the reduction of convergence problems caused by the complexity of the optimized scheme.

Keywords: Stochastic Optimization, Azeotropic Distillation, Process Simulation, Aspen Plus – Python interface, modified Simulated Annealing Algorithm.

1. Introduction

A heterogeneous azeotropic mixture creates a new azeotropic point by adding a specific third component called entrainer to the binary system. Thus, in heterogeneous azeotropic distillation, a mixture having a composition that exists in the two liquid phase region is separated into an organic phase and an aqueous phase by a decanter. This work proposes the post-optimization of a synthesis problem that departs from the concept of unit operations and proposes functional modules. Hasebe et al. (2015) proposed distillation modules to separate a ternary mixture in previous work. The objective function was the utility cost minimization, and the optimal process structure was interpreted near the minimum reflux condition. Later, the methodology for the design of the separation scheme of the ethanol/water mixture using benzene as an entrainer that was implemented by Okunishi (2019) was taken through the solution of a superstructure. Since the interpreted solution was done assuming a large number of stages in each column section, it can be improved if the equipment cost is included to find a better compromise between equipment cost and operation cost. Therefore, the previous study is complemented with a rigorous simulation of the process that makes the process even more realistic, that is, to a real reflux relationship with finite dimensions.

Three distillation columns formed the distillation process. The first column was a thermally coupled distillation column with a side rectifier, and the ethanol product was obtained at the bottom. Then, the top vapor entered the decanter, and part of the organic stream was recycled to the first column. The aqueous stream was fed to the second distillation column, where water was concentrated at the bottom. In the third column, the water-rich stream was mixed with some of the organic stream to take all possible ethanol and benzene to concentrate water towards the column's bottom. For rigorous model simulation, this work uses an interface between the free access programming language Python and the chemical process simulation software Aspen Plus V12®. A modified Simulated Annealing Algorithm (m-SAA) stochastic algorithm was proposed by Cabrera et al. (2021). The problem was solved by minimizing the Total Annual Cost (TAC).

2. Methodology

2.1. Case Study

The scheme of the base case is shown in Figure 1. It was designed using Aspen Plus® software with the NRTL thermodynamic model. An ethanol/water mixture of 100 kmol/h (89.4% mol of ethanol, 10.6% mol of water) is fed to the system in the main column (MAIN-C). 3 kmol/h of fresh benzene is fed into the system. Optimization is done using the RADFRAC module for the distillation columns. The pressure of all columns is 1 atm, and the pressure drop was neglected. In MAIN-C, ethanol is obtained at a purity of 95 mol%, and in C3, water is obtained with the same purity.

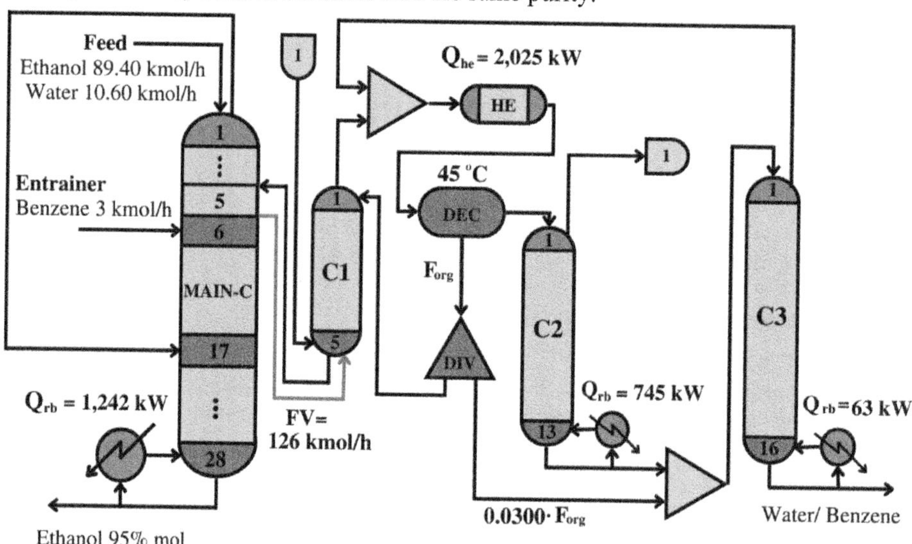

Figure 1. Heterogeneous azeotropic distillation processes (Base case)

MAIN-C is the main column where the ethanol/water and benzene mixtures are feed. Here, the ethanol is also recovered in the bottom. C1 is the rectifier of MAIN-C. HE is the condenser of C1. DEC is the decanter after the C1 condenser that recovers the benzene and sends it to C2, and DIV is a splitter of the organic phase between C1 and C3. The reported value of the splitter is the fraction of flow that goes to the mixer that connects to C3. C3 is the column where water is removed from the process, and a part of the benzene is recovered. Figure 1 shows the solution with process optimal structure after interpreting the synthesis problem for separating mixtures forming heterogeneous azeotropes. Although the process structure is optimal, the number of stages, feed stages, or flows between columns have not been optimized but simulated at near minimum reflux conditions. Therefore, this work solves a post-optimization problem that finds the best structural and operating conditions that minimize the Total Annual Cost (TAC).

In Fig. 1, the variables proposed to optimize the design problem are shown in red. In general, they represent a direct impact on the thermal load of the reboilers of each distillation column since the number of stages of each one and the conditions and amount of organic solution entering these will have a direct impact on these values. The design was proposed based on the number of modules present in the solution by Okunishi (2019).

2.2. Mathematical treatment

The case study was optimized according to the objective function shown in Eq. 1.

$$Min(TAC) = f(N_{MAINC}, N_{C1}, N_{C2}, N_{C3}, Div_{DIV}, T_{DEC}, FV_{MAINC}, N_R, N_F) \quad 1$$

$$s.t. \quad R_i(x) \geq 99.5\ \% \quad \& \quad CC = 1$$

$R_i(x)$ is the vector associated with the recovery percentage (for ethanol and water) that was 99.5 %. Since it is quite a complex simulation, placing a design specification on it will increase the number of iterations to reach convergence. For that reason, a penalty is proposed to the TAC in case the fixed purity of 95% mol is not met for the ethanol in MAIN-C. For the calculation of the cost (FOB) Eq. 2 is used.

$$FOB\left(\frac{USD}{year}\right) = OC + \frac{\left(\frac{CEPCI\ in\ 2019}{CEPCI\ in\ 2003}\right)\left(C_{Shell} + C_{Plates} + C_{Comp} + \sum_{j \in CS}^{i \in HS} C_{Exchanger\ i,j}\right)}{Recovery\ time\ (Year)} \quad 2$$

The FOB modules were calculated using the Guthrie method (Turton et al., 2018) with a payback time of 5 years using the CEPCI value corresponding to the one for 2019. The adjustment is used under the F_C criterion that relates to the purity (In mole fraction) of the system, as shown in Eq. 3.

$$F_C = 1 + [0.95 - Purity_{EtOH}] \quad 3$$

This correction factor is multiplied by the FOB, so the TAC calculation is adjusted, as shown in Eq. 4.

$$TAC\left(\frac{USD}{year}\right) = FOB \cdot F_C \quad 4$$

This correction factor F_C allows us to adjust the TAC to the desired purity so that if the purity is higher, the technology will be less expensive. If the purity is low, the technology will be more expensive than the one calculated. The CC refers to a convergence criterion by the simulation in Aspen Plus V12®, which can take two values as described in Eq. 5.

$$CC = \begin{cases} 0 & then\ TAC = 2{,}000{,}000\ \$/year \\ 1 & then\ \ TAC\ calculated\ by\ Ec. 4 \end{cases} \quad 5$$

The value of CC equal to zero indicates that the simulation does not converge for the optimization variables proposed by the algorithm in that iteration. Therefore, it is penalized with a large TAC (i.e., 2,000,000 \$/yr.). The penalty value in Eq. 5 was determined by sensitivity analysis sufficiently large. The bounds on the variables for optimization are presented in Table 1.

Table 1. Optimization Variables

Variable	Description	Optimization bounds [lower, upper]	Variable type
N_{MAINC}	MAIN-C Stages	[20, 37]	Discrete
N_{C1}	C1 Stages	[3, 10]	Discrete
N_{C2}	C2 Stages	[6, 20]	Discrete
N_{C3}	C3 Stages	[8, 25]	Discrete
Div_{DIV}	Divider	[0.0150, 0.2000]	Continuous
T_{DEC}	Decanter Temperature (°C)	[40, 65]	Continuous
FV_{MAINC}	FV Flow (kmol/h)	[90, 135]	Continuous
F_R	Fresh Benzene Feed Stage	[7, N_{MAINC}-3]	Discrete
F_F	MAINC Dome Vapor Recirculation Feed Stage	[4, N_R-2]	Discrete

The optimization variables followed analogous to the work reported by Alcántara-Maciel et al. (2022) as shown in Eq. 6.

$$L_{Ni} = (L_{Upper} - L_i)/(L_{Upper} - L_{Lower}) \quad 6$$

L_{Ni} is the value of the normalized variable L_i, L_{Upper} is the upper limit of variable i and L_{Lower} is the lower limit of variable i. The algorithm works with random values from 0 to 1. For discrete variables, a rounding function is used. This technique ensures that the search range of the variables in the optimization does not go outside the limits defined in Table 1. Concerning dependent variables, this technique allows us to define some limits based on previous variables so that we have dynamic limits that do not limit the search space. The algorithm determines the variables in the descending order shown in Table 1.

The optimization algorithm used is the modified Simulated Annealing Algorithm (m-SAA) proposed by Cabrera et al. (2021), which is capable of manipulating integer variables more effectively than the typical continuous relaxation of the method. Four annealing of 225 iterations each were used, having a total of 900 iterations that ensured at least 100 iterations per variable to be optimized.

The Spyder IDE 5.2.2 (Spyder, 2022) was used to develop and execute the Python code to run the interface with Aspen Plus. The overall implemented interface is shown in the schematic presented in Figure 2. The computer used for this optimization has an Intel(R) Xeon(R) W-3235 CPU at 3.30 GHz with 48 GB of RAM.

Figure 2. Optimization-simulation software interface

3. Results and Discussion

The post-optimization results are shown in Table 2 and compared with the base case.

Table 2. Design variables and optimal operating specifications for the different cases.

Variable	Base Case	Proposed solution by the post-optimization
N_{MAINC}	28	25
N_{C1}	5	5
N_{C2}	13	6
N_{C3}	16	19
Div_{DIV}	0.0300	0.1082
T_{DEC} (°C)	45.00	53.52
FV_{MAINC} (kmol/h)	126.00	91.84
F_R	17	20
F_F	6	4
TAC (USD/year)	572,551	462,636
Total heat duty (Reboilers only) (kW)	2,050	1,466
Condenser duty (kW)	2,025	1,488

Figure 3 shows the optimal scheme in post-optimization.

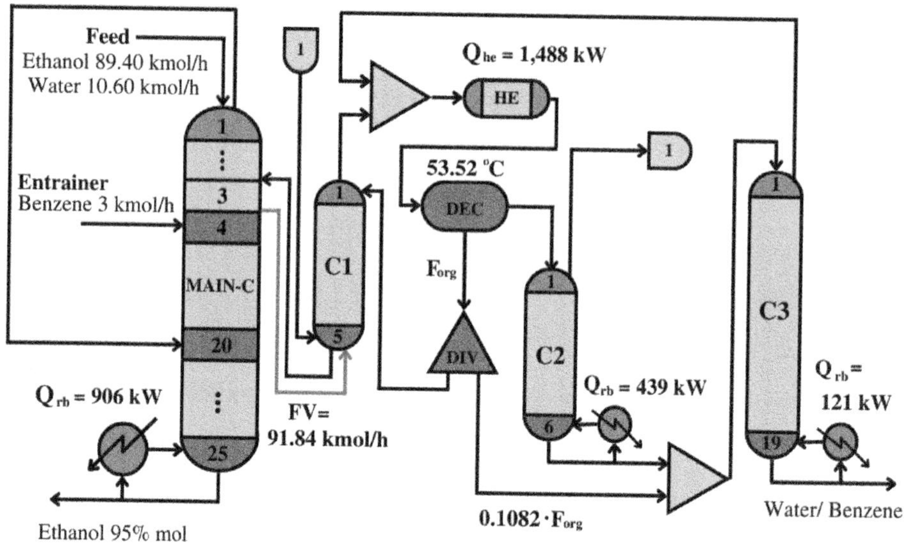

Figure 3. Optimal solution given by m-SAA

In the last solution, the number of stages in the C1 and C3 columns were similar, but the C2 column needed only a few stages. In the base case, the decanter temperature was assumed to be 45℃, but in the best case, a higher temperature (53.5℃) was better because decanters, in general, decrease their separation efficiency when they are operated at low temperatures, this is confirmed by the studies carried out by Skiborowski (2015). In our system, the recirculation flow (Organic) to C1 is less due to the division reflected by the splitter, and this generates lower heat duty in the reboiler of the main column (MAIN-C).The vapor interconnection flow of the first column was 27% lower than that obtained in the base case, which resulted in the reboiler duty reduction in columns MAIN-C and C2. Thus, the overall energy consumption of the process was reduced.

The recycles flows of column C3 (which is rich in organic phase), the one between column C2 and C1 and the recirculation of vapor in MAIN-C reduce the heat of the reboilers in MAIN-C in the same way and C2, the same occurs with the FV interconnection flow. This energy reduction directly reduce the area of the reboilers, and consequently minimize the cost of the TAC in post-optimization. In the case of the C3 column, the reboiler load increases because it receives a greater amount of organic phase flow, but this increase is small compared to all the energy saved in MAIN-C and C2.

4. Conclusions

The proposed problem was solved using an interface with the Python programming language and the Aspen Plus V12® process simulator. The TAC base case was USD 572,551/year. The TAC of the optimized process using the m-SAA stochastic optimization algorithm was 462,636 USD/year, which implies a saving of 109,914 USD/year, which is a saving of approximately 20% for the process obtained at near minimum reflux condition. Using the m-SAA algorithm and rigorous models allows us to have a robust design, which in the case of this work, also implies a saving in its TAC. Using the proposed interface between Python and Aspen Plus V12® reduces the convergence problems caused by the complexity of the optimized scheme. The use of an optimization algorithm as post-optimization was effectively used to improve the operating and structural conditions from the original design proposed by synthesis optimization.

References

F. D. Alcántara-Maciel, V. E. Casillas-Céspedes, J. A. López-García, J. Cabrera-Ruiz, C. Ramiez Marquez, J. R. Alcántara-Ávila, 2022, Economic optimization of a reactive distillation column with multiple reactive sections for silane production, Computer Aided Chemical Engineering, 51, 475–480.

J. Cabrera Ruiz, S. Hasebe, J. R. Alcantara Avila, 2021, Improvement of the Optimal Design Procedure Using Randomized Algorithm and Process Simulators, MATEC Web Conf 2021, 333, 06004.

R. Turton, J. Shaeiwitz, D. Bhattacharyya, W. Whiting, 2018, Analysis, Synthesis, and Design of Chemical Processes, 5th Edition. Boston, USA: Pearson.

M. Okunishi, 2019, Development of systematic design method for azeotropic distillation process using decanter, Kyoto University Master Thesis (in Japanese).

Spyder, 2021, https://www.spyder-ide.org/ (accessed on 27 November 2022).

H. Takase, S. Hasebe, 2015, Optimal Structure Synthesis of Ternary Distillation System, Computer Aided Chemical Engineering, 37, 1097–1102.

M. Skiborowski, A. Harwardt, W. Marquardt, 2015, Efficient optimization-based design for the separation of heterogeneous azeotropic mixtures, Computers and Chemical Engineering, 72, 34-51.

Antonis Kokossis, Michael C. Georgiadis, Efstratios N. Pistikopoulos (Eds.)
PROCEEDINGS OF THE 33rd European Symposium on Computer Aided Process Engineering
(ESCAPE33), June 18-21, 2023, Athens, Greece
 http://dx.doi.org/10.1016/B978-0-443-15274-0.50143-8

Single and multi-objective optimization of heat integrated distillation columns using PYMOO

Josué J. Herrera Velázquez,[a] Julián Cabrera Ruiz,[a] J. Rafael Alcántara Avila,[b] Salvador Hernández Castro,[a]

[a]*Departamento de Ingeniería Química, Universidad de Guanajuato, Noria Alta S/N, Guanajuato, 36050, Mexico.*
[b]*Department of Chemical Engineering, Kyoto University, Kyoto 615-8510, Japan.*

Abstract

Python has become an open-source interpreted programming language that has recently gained popularity. The implementation of interface codes for process optimization between process simulators and closed-source programming languages has been widely reported. This paper presents an interface between the commercial chemical process simulation software Aspen Plus and the open-source programming language Python using the PYMOO (Multi-Objective Optimization in Python) package developed by Blank et al. (2020). This analysis was performed on a A Heat-Integrated Distillation Column (HIDiC) previously optimized by Herrera-Velázquez et al. (2022) was taken as a case study. Multi-objective (NSGA-II) and single-objective (GA and PSO) optimization algorithms were used to minimize economic and energy criteria. Mathematical handling techniques such as normalization of variables and pseudo-normalizations of the objective functions were implemented to improve the solutions and to reduce the computation time during optimization.

Keywords: Global Optimization, Stochastic, PYMOO, Aspen Plus-Python Interface, HIDiC.

1. Introduction

Interfaces between process simulation and optimization software are increasingly becoming more utilized for developing new processes. Stochastic algorithms are particularly interesting because they can link process simulators, which contain material and energy balances represented by highly non-linear equations. The Python programming language has gained popularity among scientists and engineers to solve various mathematical problems. Nevertheless, its capabilities have not been exploited for solving chemical processes.

This work proposes an interface framework between the commercial chemical process simulation software Aspen Plus V12® and the open-source programming language Python. The PYMOO (Multi-Objective Optimization in Python) package developed by Blank et al. (2022) was used to optimize the studied process. This work takes the optimization of a Heat Integrated Distillation Column (HIDiC) to separate a binary mixture of benzene/toluene from a previous work (Herrera-Velázquez et al., 2022). For single optimization, the minimization of an economic criterion, the Total Annual Cost (TAC), was considered. The minimization of energy consumption was also considered for multi-objective optimization because there is a trade-off between energy reduction resulting from heat integration and cost increase due to expensive electricity and compressor costs. Particle Swarm Optimization (PSO) and Genetic Algorithm (GA) were used for single optimization. In contrast, the Non-Dominated Sorting Genetic Algorithm (NSGA-II) was used for multi-objective optimization. The NSGA-II algorithm, which is integrated with the PYMOO package, follows the general scheme of a genetic algorithm and contains a modified survival selection and binary mating, where each individual is compared by rank and crowding distance (Deb et al., 2002). The PSO (Particle Swarm Optimization) algorithm, which is also included in the PYMOO package, uses the concept of particle swarming to guide its search, each particle possessing a velocity and strongly influenced by the best solutions found, both locally and globally (Kennedy et al., 1995).

2. Methodology

2.1. Case Study

The process described in Figure 1 is simulated using Aspen Plus V12® software and the Peng-Robinson thermodynamic model was used. The feed is 50/50 kmol/h of benzene/toluene. By design specifications, each product must be obtained with a purity of 99.9% mol for both components. Optimization is done using the RADFRAC module for the distillation columns. The pressure in the STR column is 1 atm, while for the REC column, the pressure is PR times the STR pressure.

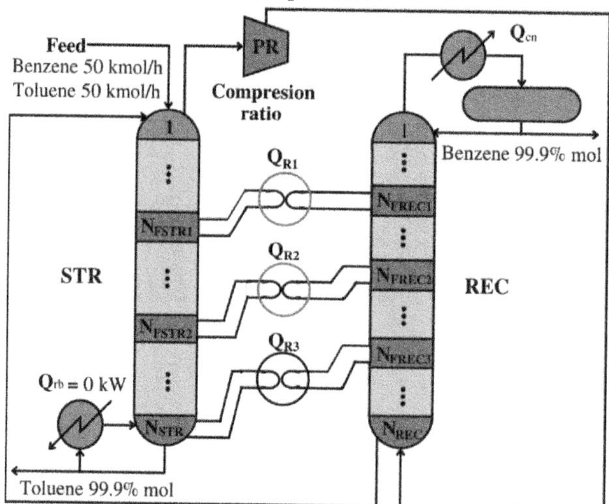

Figure 1. Heat Integrated Distillation Column

In the figure, N_{STR} and N_{REC} are the total stages of the stripping and rectification columns, respectively. PR is the compression ratio (i.e., the pressure in the rectification column divided by the pressure in the stripping column), Q_{R1} and Q_{R2} are the loads of the exchangers that perform the energy integration between the column zones, N_{FSTRj} and N_{FRECi} are the stages in which the integration is carried out for each exchanger. In the case of N_{FREC3}, it is the stage that connects with the exchanger that acts as a reboiler in STR.

2.2. Mathematical treatment

The case study was optimized according to the objective function shown in Eq. 1.

$$Min(TAC, Q_{COMP}) = f(N_{STR}, N_{REC}, PR, Q_{R1}, Q_{R2}, N_{FSTR1}, N_{FSTR2}, N_{FREC1}, N_{FREC2}, N_{FREC3}) \quad 1$$

$$s.t. \quad P_i(x) \geq 99.9\,\%\,mol, \quad R_i(x) \geq 99.5\,\% \quad \& \quad CC = 1$$

$P_i(x)$ and $R_i(x)$ are the constraint vectors associated with percent mol purity (99.9% mol for benzene and toluene) and percent recovery (99.5% for benzene and toluene), respectively. These constraints are integrated into the simulation as design specifications. The TAC objective function is given in Eq. 2:

$$TAC\left(\frac{USD}{year}\right) = OC + \frac{\left(\frac{CEPCI\ in\ 2019}{CEPCI\ in\ 2003}\right)\left(C_{Shell}+C_{Plates}+C_{Comp}+\sum_{j\in CS}^{i\in HS} C_{Exchanger\ i,j}\right)}{Recovery\ time\ (Year)} \qquad 2$$

OC is the operating cost. The TAC modules were calculated using the Guthrie method (Turton et al., 2018) with a payback time of 3 years using the 2019 CEPCI. Q_{COMP} is the value calculated by Aspen Plus of the module COMP (Compressor). Guthrie's method, and Q_{COMP} are a function of the variables presented in Eq. 1. The CC refers to a convergence status by the simulator Aspen Plus®, which can take two values as described in Eq. 3 and modify the Total Annual Cost (TAC) for non-converged simulations.

$$CC = \begin{cases} 0 & then\ TAC = 2{,}000{,}000\ \$/year\ \&\ Q_{COMP} = 300\ kW \\ 1 & then\ TAC\ \&\ Q_{COMP}\ calculated\ by\ algorithm \end{cases} \qquad 3$$

The value of CC equal to zero indicates that the simulation does not converge for the optimization variables proposed in the iteration by the algorithm. The values of TAC and Q_{COMP} given in Eq. 3 are penalty values determined by sensitivity analysis, such that they are sufficiently larger than the values expected by the simulations that converge. The bounds on the variables for optimization are presented in Table 1.

Table 1. Optimization Variables

Variable	Optimization bounds [lower, upper]	Variable type	Units
N_{STR}	[15, 30]	Discrete	--
N_{REC}	[15, 30]	Discrete	--
PR	[2, 3]	Continuous	--
Q_{R1}	[100, 600]	Continuous	kW
Q_{R2}	[100, 600]	Continuous	kW
N_{FSTR1}	[2, N_{STR}]	Discrete	--
N_{FSTR2}	[N_{FSTR1}+1, N_{STR}-1]	Discrete	--
N_{FREC1}	[2, N_{STR}-3]	Discrete	--
N_{FREC2}	[N_{FREC2}, N_{STR}-2]	Discrete	--
N_{FREC3}	[N_{FREC3}, N_{STR}-1]	Discrete	--

The variables have been normalized similarly to Herrera-Velázquez et al. (2022), as shown in Eq. 4, to compare the multi-objective and single-objective results with respect to the previous study and to determine the advantages and disadvantages of the efficiency and effectiveness of the proposed platform

$$L_{Ni} = \left(L_{Upper} - L_i\right)/\left(L_{Upper} - L_{Lower}\right) \qquad 4$$

where L_{Ni} is the value of the normalized variable L_i, L_{Upper} is the upper limit of variable i, and L_{Lower} is the lower limit of variable i. The algorithm works with random values between 0 and 1, and a rounding function is used for discrete variables. The use of dynamic bounds ensures that the search range of the variables in the optimization does not go outside the limits defined in Table 1, which was done similarly to Alcántara-Maciel et al. (2022). This technique considers the bound of dependent variables. The algorithm defines the variables in the order shown in Table 1.

A case study was proposed in which a pseudo-normalization is performed, such that the TAC is intended to be left in the same order of magnitude as Q_{COMP}. For this, the first run of 100 iterations is made, and the rounded value of the mean TAC over Q_{COMP} is taken, as shown in Eq. 5.

$$FACTOR = (1/100)\sum_{i=1}^{100}(TAC_i/Q_{COMP,i}) \quad 5$$

For this case study, the $FACTOR$ value is 5,000. The new objective function results from dividing Eq. 4 by the obtained $FACTOR$ value.

The NSGA-II method was used for multi-objective optimization for 30 generations with 100 individuals with the Augmented Scaling Feature (ASF) decomposition method, a well-known metric in the multi-objective optimization with weights of 0.2, 0.4, 0.5, 0.6 and 0.8 (Wierzbicki., 1980). This study for this method has 3,000 iterations. The ASF method works since the Pareto front is already generated and does not interfere with the NSGA-II method. Therefore, the results from the weights are determined once the algorithm obtains the Pareto front.

The single-objective optimization was performed also using the PSO and GA method included in the PYMOO package for the same optimization variables, but only using the TAC objective function and, subsequently, the Q_{COMP} objective function. For the GA case, 30 generations of 100 individuals are used to obtain a sweep of 3,000 iterations. In the case of the PSO algorithm, freedom was given to run the optimization and terminate it if there was no improvement in the objective function after 500 iterations.

The Spyder IDE 5.2.2 (Spyder, 2022) was used to develop and execute the Python code to run the interface with Aspen Plus V12. The overall implemented interface is shown in the schematic presented in Figure 2. The computer used for this optimization has an Intel(R) Xeon(R) W-3235 CPU at 3.30 GHz with 48 GB of RAM.

Figure 2. Optimization-simulation software interface

3. Results and Discussion

Table 2 summarizes the time and convergence percentage of executed simulations during the TAC minimization using Eq. 2 and the results calculating the TAC using the proposed pseudo-normalization. For the sake or comparison, the results for the base case by Herrera-Velázquez et al. (2022) are also reported.

Table 2. Summary of efficiency results for the TAC pseudo-normalization proposal.

Optimization Algorithm	Use of pseudo-normalization of TAC	Execution Time (Hours)	Convergence Percentage
Herrera-Velázquez et al. (2022) - SA	NO	3.87	99.99%
PYMOO GA	NO	1.42	99.40%
	YES	1.18	99.30%
PYMOO PSO	NO	0.68	99.22%
	YES	0.69	99.22%
PYMOO NSGA-II	NO	1.19	98.70%
	YES	1.16	99.16%

For both GA and NSGA-II, the pseudo normalization reduces the optimization execution time of 16.90% and 2.52%, respectively. However, for PSO, the time increases 1.47%. The results reported in Table 3 for the TAC are those obtained by the pseudo-normalized TAC, the same for those reported in NSGA-II. In the case of the NSGA-II, only the best result between the values of the ASF weights is reported. Figure 3 shows the obtained Pareto Front, as well as each of the solutions obtained by each method.

Table 3. Design variables and optimal operating specifications for the different cases.

	Herrera-Velázquez et al. (2022) SA	**PYMOO PSO**		**PYMOO GA**		**PYMOO NSGA-II**
Variable	**TAC**	**TAC**	**COMP**	**TAC**	**COMP**	**ASF (0.5,0.5)**
N_{STR}	25	23	30	23	30	27
N_{REC}	22	21	30	21	30	24
PR	2.0014	2.0000	2.0000	2.0001	2.0000	2.0000
Q_{R1}	121.37	210.94	227.50	232.34	377.16	148.09
Q_{R2}	102.89	231.39	101.34	298.05	280.51	199.77
N_{FSTR1}	14	20	27	20	27	23
N_{FSTR2}	17	22	29	22	29	26
N_{FREC1}	4	2	2	2	2	2
N_{FREC2}	14	3	3	3	3	3
N_{FREC3}	15	4	4	4	9	4
TAC (USD/year)	548,786	512,196	528,674	512,188	527,952	517,249
Q_{COMP} *(kW)*	122.94	118.10	115.69	118.17	115.73	116.23
Iteration for optimum	231 of 3,000	901 of 1,675	2,478 of 3,000	2,729 of 3,000	2,976 of 3,000	2,646 of 3,000
Time (Hours)	3.87	0.69	1.34	1.18	1.28	1.17

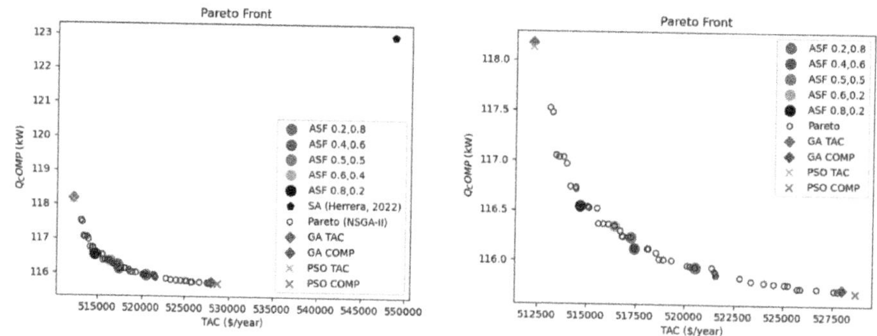

Figure 3. Pareto front solution for SA included (right) and SA not included (left)

4. Conclusions

It was possible to successfully implement the Python-Aspen Plus interface, applying the PYMOO package for a single-objective (PSO and GA) and multi-objective (NSGA-II) optimization. The pseudo-normalization of the TAC improves, on average, from 2% to 17% of the computation time for the GA and NSGA-II algorithms. The best algorithm for the single-objective optimization is the PSO, while the best solution for the multi-objective problem was the ASF(0.5,0.5). The advantage of PYMOO was the reduction of computation times. They are reduced by more than 50% to those used with SA, which makes it an attractive option for its application to complex processes. The SA algorithm can be used for complex processes where convergence is the priority. This work can be extended without loss of generality to solve intensified chemical processes. It can be of interest for improving current interfaces used in Process Systems Engineering.

References

F. D. Alcántara-Maciel, V. E. Casillas-Céspedes, J. A. López-García, J. Cabrera-Ruiz, C. Ramírez-Marquez, J. R. Alcántara-Ávila, 2022, Economic optimization of a reactive distillation column with multiple reactive sections for silane production, Computer Aided Chemical Engineering, 51, 475–480.

J. Blank, K. Deb, 2020, Pymoo: Multi-Objective Optimization in Python, IEEE Access, 8, 89497-89509.

K. Deb, A. Pratap, S. Agarwal, T. Meyarivan, 2002,A Fast and Elitist Multiobjective Genetic Algorithm: NSGA-II, IEEE Transactions on Evolutionary Computation, 2, 182-197.

J. J. Herrera-Velázquez, F. M. Zavala-Durán, L. A. Chávez-Díaz, J. Cabrera-Ruiz, J. R. Alcántara-Avila, 2022, Hybrid two-step optimization of internally heat-integrated distillation columns, Journal of the Taiwan Institute of Chemical Engineers, 130, 1-9.

J. Kennedy, R. Eberhart, 1995, Particle swarm optimization, Proceedings of ICNN'95 - International Conference on Neural Networks, 4, 1942-1948.

R. Turton, J. Shaeiwitz, D. Bhattacharyya, W. Whiting, 2018, Analysis, Synthesis, anD Design of Chemical Processes, 5th Edition. Boston, USA: Pearson.

A. Wierzbicki, 1980, The Use of Reference Objectives in Multi-objective Optimization. Multiple Criteria Decision Making Theory and Application, 177, 468-486.

Spyder, 2021, https://www.spyder-ide.org/ (accessed on 26 November 2022).

Antonis Kokossis, Michael C. Georgiadis, Efstratios N. Pistikopoulos (Eds.)
PROCEEDINGS OF THE 33rd European Symposium on Computer Aided Process Engineering (ESCAPE33), June 18-21, 2023, Athens, Greece
 http://dx.doi.org/10.1016/B978-0-443-15274-0.50144-X

Design of multi-component gradient SMBs

Jesper Frandsen[a], Jakob Kjøbsted Huusom, Krist V. Gernaey[a], Jens Abildskov[a]

[a]*Dept. of Chemical and Biochemical Engineering, Technical University of Denmark, Søltofts Plads, Building 288 A, 2800 Kgs. Lyngby, Denmark.*
ja@kt.dtu.dk

Abstract

In this work, a generic design method has been developed for the design of isocratic and gradient SMBs for multi-component systems. Based on the general method by (Migliorini et al., 2000b), the method is simplified and extended to enable multi-component systems, gradient SMB operation and any isotherm. The method has been verified through two case studies. For the first case study, an analytical solution is possible and the error using this method is insignificant. For the second case study, the system is solved through numerical simulation and the method is verified. With this method, the design of gradient SMBs can be performed with any isotherm and multi-component systems in a simple way.

Keywords: Design methods, Mechanistic Modelling, Gradient SMB

1. Introduction

In the biopharmaceutical industry, chromatography is a workhorse unit operation. Chromatography is typically operated in batch mode, although there are many benefits when transitioning to continuous operation such as reduced capital expenditure, higher capacity utilization, a better-controlled process, reduced utility consumption, higher purity and higher yield (Schmidt-Traub et al., 2020). For batch chromatography, applying a gradient can increase productivity and enable complete separation. This concept can also be transferred to a simulated moving bed (SMB), called gradient SMB. The corresponding True Moving Bed (TMB) is shown in Figure 1. The concept is that in section 1 and section 2, where desorption takes place, a low-affinity zone is created to lower the affinity of the components to the resin, whereas in section 3 and 4, a high-affinity zone is created. For liquid chromatography, this can be done by altering the solvent composition, salt concentration, pH or temperature. The most common gradient operation in SMB is based on altering the modifier concentration in the feed and desorbent. Comparing

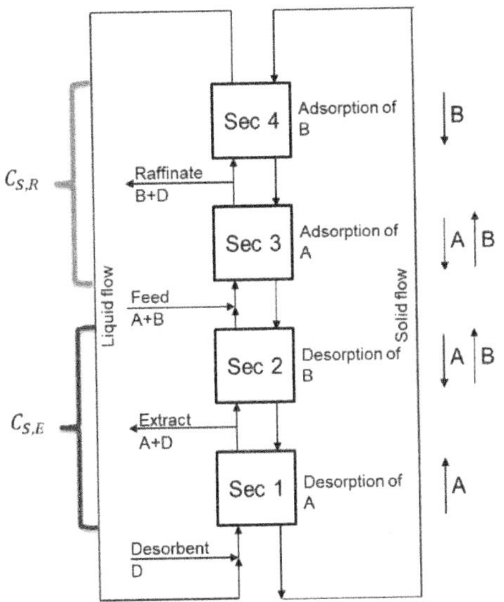

Figure 1: Schematic of a gradient TMB for a binary separation with component A and B and desorbent D. Sec refers to the section, $C_{S,E}$ and $C_{S,R}$ are the modifier concentration in extract zone (sections 1-2) and raffinate zone (sections 3-4), respectively.

gradient SMB to isocratic SMB, there are many advantages such as increased productivity, reduced eluent consumption, reduced production cost and increased product concentration both in closed-loop configuration and open-loop configuration (Kim et al., 2017). A particular challenge with implementing continuous chromatography such as gradient SMB is the design of the operation. Modelling chromatography and SMB processes is very complex as there are many phenomena taking place simultaneously which results in a set of complex coupled partial differential equations (Schmidt-Traub et al., 2020). Furthermore, there are multiple design variables which can make the optimal design computationally expensive. For the gradient SMB, there are two additional design variables i.e., the modifier concentration in the feed and desorbent. Therefore, the design of a SMB is typically performed by designing an ideal TMB, shown in Figure 1. A TMB has steady-state solutions whereas a SMB has cyclic steady-state solutions. The ideal TMB steady-state solutions to the quasi-linear first-order partial differential mass balance equations are solved, assuming isotherm equilibrium. Constraints are set up for each section of the TMB. Solving these equations results in operational conditions for each section of the TMB. For the mid-zone (sections 2 and 3 in Figure 1) where feed is added, the operational range for the internal flowrate ratios results in a 'triangular diagram' i.e., a region of complete separation, composed of three lines: w-b line, w-r line and r-a line. The feasible flow rate ratios can then be converted to SMB flow rates (Migliorini et al., 2000). Migliorini et al. developed a design method that works for any isotherm, but it is not stated explicitly how the method should be applied for neither gradient SMB nor multi-component systems (Migliorini et al., 2000b). Abel et al. developed a design method for gradient SMB for binary systems with Langmuir isotherms (Abel et al., 2004). Lee et al. developed a SMB design method for multi-component systems with any isotherm, but it is not specified explicitly how to incorporate gradients (Lee et al., 2021). Frederiksen developed a design method for any isotherm, and the method can be applied to gradient SMB but only for binary systems (Frederiksen, 2004). With this, there is a need for a design method that can be applied to multi-component systems, gradient SMB with any isotherm. This work proposes a design method for gradient SMB for multi-component systems and any isotherm. The method is verified with SMB simulations in CADET (Leweke & von Lieres, 2018).

2. Methodology

2.1 Design method

For designing the SMB, the dimensionless flowrate ratio, m_j, is often used. It describes the flow rate of the liquid phase per solid flow rate based on the TMB. The dimensionless flow rate ratio is given in eq. (1) (Migliorini et al., 2000b).

$$m_j = \frac{Q_j^* t^* - \epsilon_t V_c}{(1-\epsilon_t) V_c} \tag{1}$$

Where subscript j refers to the section, Q^* is the SMB flowrate, t^* is the switching time for the SMB, ϵ_t is the total porosity and V_c is the volume of the column.

For multi-component systems, the total number of solutes, U, are separated into raffinate components, 1 to n, and extract components, p=n+1 to U.

For each section, there are constraints on the flow rate ratios that ensure complete separation. For section 1, the constraint on m_1 should ensure that the solid phase is completely regenerated i.e., the most adsorbed component is completely desorbed, mathematically given in eq. (2) (Migliorini et al., 2000b).

$$m_1 \geq m_{1,min} = \frac{\partial q_U}{\partial C_U}|_{C_U=0} \tag{2}$$

Where q is the solid phase concentration and C is the mobile phase concentration. The constraint for section 4 ensures that the least retained component does not overflow into section 1. The constraint is given in eq. (3) (Migliorini et al., 2000b).

$$\frac{\epsilon_p}{\epsilon_p - 1} \leq m_4 \leq m_{4,max} = \frac{q_C(C_{1,R})}{C_{1,R}} \tag{3}$$

Where subscript R refers to the raffinate concentration which can be obtained from an overall mass balance (Migliorini et al., 2000b). If having multiple components in the raffinate stream, determining $m_{4,min}$ in eq. (3) requires setting up overall mass balances and isotherm equations for each component in the raffinate stream. The constraints for sections 2 and 3 ensure that the net flow of extract components, components p to U, is towards the extract outlet and the net flow of raffinate components, components 1 to n, is towards the raffinate outlet. Along the w-b line in the separation region, the minimum m_2 required to desorb the most retained raffinate component, component n, is investigated. Hence, the w-b line is drawn by solving the mass balance over section 2 for component p to U and using the minimum m_2 (Migliorini et al., 2000b), given in equation 4.

$$m_2 = m_{2,min} = \frac{\partial q_n}{\partial C_n}|_{C_n=0} \tag{4a}$$

$$(m_3 - m_2)C_{i,F} = q_{i,2} - m_2 C_{i,2} \ for\ i \in p\ to\ U \tag{4b}$$

Where subscripts F and 2 refer to feed and section 2, respectively. In the determination of q_n in eq. (4), only the extract components and modifier concentration in section 2 ($C_{S,E}$), are considered. Along the r-a line, the maximum m_2 to desorb the least retained extract component, component p, is determined (Migliorini et al., 2000b). Combined with mass balances over section 2 for components p to U, the r-a line is drawn by solving equation 5.

$$m_2 = m_{2,max} = \frac{\partial q_p}{\partial C_p} \tag{5a}$$

$$(m_3 - m_2)C_{i,F} = q_{i,2} - m_2 C_{i,2} \ for\ i \in p\ to\ U \tag{5b}$$

where only the extract components and modifier concentration in section 2 are considered in the determination of q_p in eq. (5). The determination of the w-r line is more complicated as it describes the interaction between sections 2 and 3. Along the w-r line, the maximum m_3 to let the least retained extract component, component p, overflow in section 3 is determined (Migliorini et al., 2000b). This is combined with a mixing balance at the feed node as well as with mass balances for extract components in section 2 and raffinate components in section 3 (Frederiksen, 2004). Whereas Frederiksen used this consideration for binary systems, it is here extended to multi-component systems. The equations for the w-r line are given in equation 6.

$$C_{p,\beta} = \frac{C_{p,2} m_2 + (m_3 - m_2) C_{p,F}}{m_3} \tag{6a}$$

$$m_3 = \frac{q_{p,3}}{C_{p,\beta}} \tag{6b}$$

$$(m_3 - m_2)C_{i,F} = q_{i,2} - m_2 C_{i,2} \; for \; i \in p \; to \; U \tag{6c}$$

$$(m_3 - m_2)C_{j,F} = m_3 C_{j,3} - q_{j,3} \; for \; j \in 1 \; to \; n \tag{6d}$$

Where subscript β refers to the point just after the feed node. For the w-r line, this method differs from the method of (Migliorini et al., 2000b) as this method does not involve a numeric integration. The solute concentrations and modifier concentrations used for the determination of the isotherms must be considered carefully and are given in Table 1.

2.2 Numeric SMB simulations

To verify the method, SMB simulations were set up in CADET (Leweke & von Lieres, 2018). A SMB setup with four columns in a 1:1:1:1 configuration was considered. The ideal mass balance for the SMB process for each component was used (Schmidt-Traub et al., 2020). To convert the flow rate ratios to SMB flow rates, equation 1 was used together with mass balances over each section. The column parameters ϵ_t and V_C were set to 0.83 and 7.854e-5 m^3, and a switching time of 180 seconds was used. The values for m_1 and m_4 were set conservatively such that $m_1 = 1.5\, m_{1,min}$ and $m_4 = 0.5\, m_{4,max}$. Simulations were run until cyclic steady state was achieved. Complete separation was defined as when purity ≥ 0.995.

Table 1: Concentrations to use in the isotherms for the determination of the w-r line.

Isotherm	Solute Concentrations	Modifier concentration
$q_{p,3}$	$C_{p+1..U,3} = 0$ $C_{p,3} = C_{p,\beta}$ $C_{1..n} = C_{1..n}$	$C_s = C_{s,3}$
$q_{i,2}$	$C_{p..U,3} = C_{p..U,3}$ $C_{1..n} = 0$	$C_s = C_{s,2}$
$q_{i,3}$	$C_{p..U,3} = 0$ $C_{1..n} = C_{1..n}$	$C_s = C_{s,3}$

2.3 Isotherms

To demonstrate that the method is not restricted to a single isotherm, the two most used isotherms, the Langmuir and Steric Mass Action (SMA), were considered (Kumar & Lenhoff, 2020). Assuming equilibrium, the Langmuir and the SMA isotherms are shown in eq. (7) and eq. (8), respectively (Brooks & Cramer, 1992; Schmidt-Traub et al., 2020).

$$q_i = \frac{H_i C_i}{1 + \sum_{h=1}^{U} K_{L,h} C_h} \tag{7}$$

$$q_i = C_i A_i \left(1 - \sum_{h=1}^{U} \frac{q_h}{q_{max,h}}\right)^{\nu_i} \tag{8}$$

Where H is the Henry's constant, K_L is the Langmuir equilibrium constant, $A = K_{eq}\left(\frac{\Lambda}{C_S}\right)^{\nu_i}$ is the initial isotherm slope, $q_{max} = \frac{\Lambda}{\sigma+\nu}$ is the maximum binding capacity, K_{eq} is the SMA equilibrium constant, Λ is the ionic capacity, C_S is the modifier concentration, ν is the characteristic charge and σ is the shielding factor.

3. Results and discussion

The design method was applied to two case studies. First, the method was verified by determining the separation regions for a multi-component isocratic SMB with the multi-component Langmuir isotherm. For this case, there is a design method based on the analytical solution of the TMB which the proposed design method was compared to (Migliorini et al., 2000a). The comparison of the two methods for determining the separation regions is shown in Figure 1 with parameters shown in the figure. The comparison shows that the proposed design method determines almost the same separation region as the method of (Migliorini et al., 2000a) with insignificant differences. This verifies that the design method can be used for multi-component systems with the competitive Langmuir isotherm. For the second case study, the separation region is determined for separating lysozyme from ribonuclease A and cytochrome c using gradient SMB with $NaCl$ as modifier and parameters from (Meyer et al., 2020). The section 1-2 $NaCl$ concentration was set to 200 mol/m^3 and the section 3-4 $NaCl$ concentration was set to 300 mol/m^3. The separation region and the numeric simulation results are shown in Figure 2 where the green dots indicate a complete separation and red pentagons indicate incomplete separation. Figure 2 shows that the separation region determined with the design method agrees well with the numerical SMB simulations. Some of the numerical simulations show complete separation outside the separation region. The deviation between the simulation results and the separation region from the design method is probably because the modifier concentrations fluctuate due to cyclic nature of the SMB, which has previously observed experimentally (Li et al., 2007) and not accounted for in the design method. Additionally, the assumption made by Frederiksen which avoids the numeric integration compared to Migliorini also leads to a small deviation, but this should not be significant (Frederiksen, 2004; Migliorini et al., 2000b). Despite the minor deviation, this verifies that the design method can be used to make an initial design of multi-component gradient SMBs using the SMA isotherm, though the design method is not limited to the SMA or Langmuir isotherms. Design methods based on the ideal mass balance form an initial step towards an in-depth design and optimization. In preparative chromatography, non-ideal phenomena such as mass

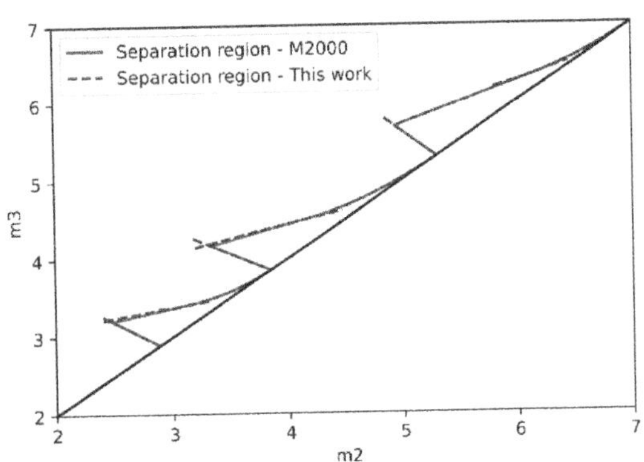

Figure 2: Separation region for m_2 and m_3 with parameters from (Lee et al., 2021) where H = [7.062, 5.297, 3.850, 2.888], K_L = [0.0321, 0.0321, 0.0175, 0.0175] L/g. M2000 refers to the method in (Migliorini et al., 2000a).

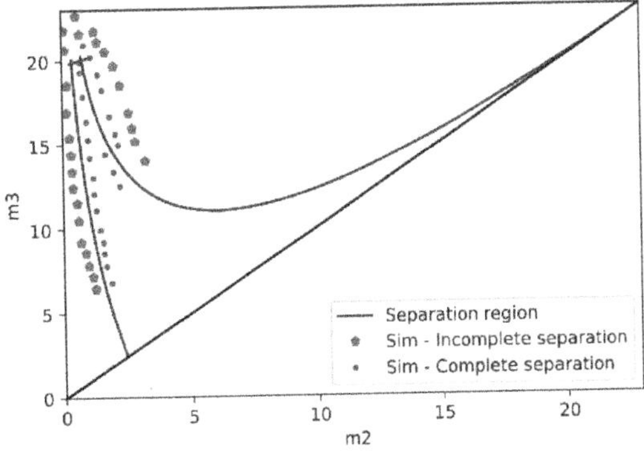

Figure 3: Separation region for m_2 and m_3 where the parameters $\Lambda = 1200$ mol/m^3, K_{eq}= [35.5e-3, 1.59e-3, 7.70e-3], ν = [4.7, 5.29, 3.7], $\sigma = [11.83, 10.6, 10.00]$ and $C_{F,i} = 1$ mol/m^3 were used (Meyer et al., 2020).

transfer resistance and axial dispersion can be significant and can cause a spreading of the components, thus complicating the separation (Schmidt-Traub et al., 2020). For an in-depth design and optimization, a more complex model that takes these non-ideal phenomena into account is required. Here, this design method can provide good initial guesses for optimizing the operational flowrates.

4. Conclusion

In this paper, a straightforward and versatile design method for isocratic SMBs and gradient SMBs was proposed. The design method can be applied to any isotherm and multi-component systems. Through two case studies, the design method was verified for the SMA and the Langmuir isotherms. This method can serve as an initial step in designing operational flowrates. For a more in-depth design and optimization, a complex model that takes phenomena such as axial dispersion and mass transfer resistance into account is required.

Acknowledgements
This study was financially funded by the Technical University of Denmark.

References

S. Abel, M. Mazzotti, M. Morbidelli, 2004, Solvent gradient operation of simulated moving beds - 2. Langmuir isotherms. J. Chromatogr. A, 1026(1–2), 47–55
C. A. Brooks, S. M. Cramer, 1992, Steric mass-action ion exchange: Displacement profiles and induced salt gradients. AIChE Journal, 38(12), 1969–1978
S. S. Frederiksen, 2004, Computer Aided Development and Optimization of Chromatographic Separations, PhD thesis
K. M. Kim, M. Lee, S. Kim, F. V. Santos da Silva, A. Siedel-Morgenstern, C. H. Lee, 2017, Advanced Operating Strategies to Extend the Applications of Simulated Moving Bed Chromatography, Chem. Eng. Technol. 40(12), 2163–2178
V. Kumar, A. M. Lenhoff, 2020, Mechanistic Modeling of Preparative Column Chromatography for Biotherapeutics. Annu. Rev. Chem. Biomol. Eng., 11, 235–255
J. W. Lee, A. Kienle, A. Seidel-Morgenstern, 2021, Numerical Short-Cut Design of Simulated Moving Bed Chromatography for Multicomponent Nonlinear Adsorption Isotherms: Nonstoichiometric Langmuir Model. Ind. Eng. Chem. Res., 60(29), 10753–10763
S. Leweke, E. von Lieres, 2018, Chromatography Analysis and Design Toolkit (CADET). Comput. Chem. Eng., 113, 274–294
P. Li, G. Xiu, A. E. Rodrigues, 2007, Proteins Separation and Purification by Salt Gradient Ion-Exchange SMB. AIChE Journal, 59(4), 215–228
K. Meyer, S. Leweke, E. von Lieres, J. K. Huusom, J. Abildskov, 2020, ChromaTech: A discontinuous Galerkin spectral element simulator for preparative liquid chromatography. Comput. Chem. Eng., 141, 107012
C. Migliorini, M. Mazzotti, M. Morbidelli, 2000a, Design of simulated moving bed multicomponent separations : Langmuir systems. 20, 79–96
C. Migliorini, M. Mazzotti, M. Morbidelli, 2000b, Robust design of countercurrent adsorption separation processes: 5. Nonconstant selectivity. AIChE Journal, 46(7), 1384–1399
H. Schmidt-Traub, M. Schulte, A. Seidel-Morgenstern, 2020, Preparative Chromatography for Separation of Proteins. In John Wiley & Sons (3rd edition), John Wiley & Sons

Antonis Kokossis, Michael C. Georgiadis, Efstratios N. Pistikopoulos (Eds.)
PROCEEDINGS OF THE 33rd European Symposium on Computer Aided Process Engineering (ESCAPE33), June 18-21, 2023, Athens, Greece
 http://dx.doi.org/10.1016/B978-0-443-15274-0.50145-1

A Method for Finding a Design Space as Linear Combinations of Parameter Ranges for Biopharmaceutical Development

Thomas Oberleitner,[a] Thomas Zahel,[b] Christoph Herwig,[b,c]

[a]*Competence Center CHASE GmbH, Ghegastraße 3, Top 3.2, 1030 Vienna, Austria*
[b]*Körber Pharma Austria GmbH, PAS-X Savvy, Mariahilferstr. 88A/1/9, 1070 Vienna, Austria*
[c]*TU WIEN Research Area Biochemcial Engineering, Getreidemarkt 9, 1060 Vienna, Austria*
christoph.herwig@tuwien.ac.at

Abstract

A biopharmaceutical manufacturer might submit a DS definition as part of the regulatory approval application, in which case process parameter (PP) deviations within this space are not considered a change and do not trigger a regulatory post approval procedure. While several methods exist for finding a DS described by non-linear PP ranges, here we introduce a novel method for finding linear combinations – a DS definition often preferred due to its operational simplicity. We use a numeric optimizer to find the design space with the largest parameter space volume that results in critical quality attribute (CQA) boundaries within acceptance criteria, predicted by a regression model. A precomputed approximation of tolerance intervals is used in inequality constraints to facilitate fast evaluations of this boundary using a single matrix multiplication. Correctness of the method was validated against different ground truths with known design spaces.

Keywords: design space; biopharmaceutical development; ICH Q8; numeric optimization; parameter space

1. Introduction

The ICH Q8 guideline for pharmaceutical development defines the design space (DS) as "the multidimensional combination and interaction of input variables (e.g., material attributes) and process parameters that have been demonstrated to provide assurance of quality" (ICH, 2017). The process parameters (PP) described here are generally identified in the risk assessment or process development phases and are considered critical when sufficient evidence was found that they affect the output of a unit operation, i.e., a critical quality attribute (CQA). A design space is comprised of the ranges of these critical process parameters that result in CQA values within acceptable limits. For the biopharmaceutical manufacturer a DS definition can be submitted as part of the regulatory approval application, in which case PP deviations within this space are not considered a change and therefore do not trigger a regulatory post approval procedure. For operators, the DS constitutes a valuable guideline document for controlling a process. While ICH Q8 does not recommend a specific form or method for describing a DS, it provides examples for non-linear and linear combinations of parameter ranges in the form of contour plots (appendix 2c in guide). Non-linear combinations describe the DS as a set of rules, or parameter ranges conditioned on other parameters, e.g., "PP1 is allowed to move between -1 and 1 if PP2 is lower than 0.5". Linear combinations of parameter ranges on the other hand are independent of each other. While the former description generally represents a larger space to operate in and methods for computing it can be found several publications (Kim & Kwak, 2002) (Kusumo, et al., 2020), the latter might be preferred due to its operational simplicity and is the subject of this contribution. Figure 1 shows the different types of design space graphically.

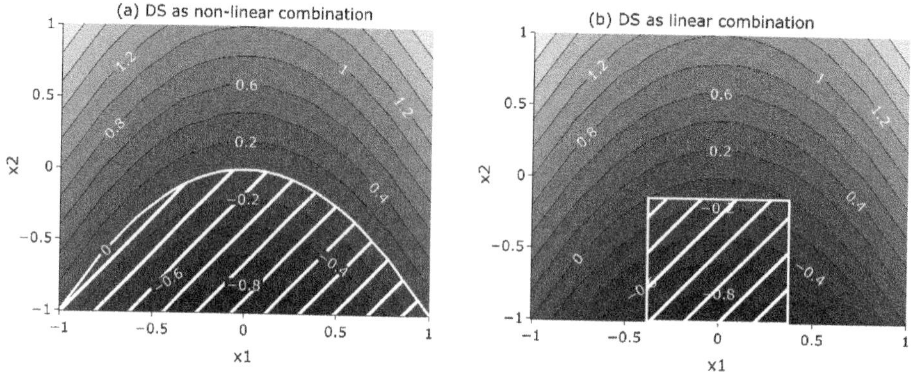

Figure 1: The design space for the function $f(x_1, x_2) = x_1^2 + x_2$, where $f(x_1, x_2) \leq 0$, shown in the contour plat as a non-linear (a) and linear (b) combination of input parameters x_1 and x_2, as defined in (ICH, 2017).

So far, definition of a design space does not incorporate any measure of statistical uncertainty, i.e., the contour shown in Figure 1 directly represents the predicted mean CQA values from the model. In the context of biopharmaceutical process validation, we suggest a more conservative approach. To accurately quantify uncertainty inherent in the regression model due to analytical and process variability, we replace predicted CQA values with the upper and lower boundary of a tolerance interval that incorporates nominal levels of both confidence and coverage. These boundaries are then used to validate acceptance limits in a conservative manner.

In this contribution we present a method for finding a design space comprised of linearly independent parameter ranges, or hyper-rectangles when considered geometrically, while treating CQA predictions conservatively.

2. Method

2.1. Tolerance Interval Approximation

The method proposed here uses a series of numeric optimizers, inequality constraints and approximations specific to polynomial regression models commonly found in process characterization and validation in the biopharmaceutical domain. Such models regress CQAs on PPs and material attributes while considering parameter interactions and quadratic effects. These higher-order effects are relevant to the shape of the response surface and therefore to the optimization problem. While regression model predictions can be computed simply by multiplying the design matrix X by the vector of learned coefficients $\hat{\beta}$, the computation of tolerance intervals can be complex and generally involves multiple evaluations of probability density functions (PDF). For example, one of the simplest equations for calculating TI boundaries around predictions is the one used in ordinary least-squares (OLS) models:

$$\hat{y} \pm \sigma \sqrt{\frac{(n-p)\chi^2_{1;\psi}\left(\frac{1}{n_i^*}\right)}{\chi^2_{n-p;\alpha}}} \tag{1}$$

Like all tolerance intervals, this formulation contains nominal parameters for the proportion of the population to be included in the interval, ψ, as well as the confidence level α. In addition, critical values for the χ^2 distribution are calculated as well as the "effective number of observations" vector $n_i^* = \frac{\hat{\sigma}^2}{se(\hat{y}_i)^2}$. As the evaluation of TI boundaries is used in several of the inequality constraints described in the following sections, this is considered a bottleneck and is replaced by an approximation. To that end, a central composite design (CCD) containing all possible 3-level combinations PPs (min, max, center point) is created and its columns extended to include all interaction and quadratic effects. A variable selection step eliminates all unnecessary effects, and we are left with a parsimonious model describing the relationship between PPs and TI. Consequently, interval boundaries can be computed by multiplying observations with the vector of TI model coefficients, here denoted as τ_l and τ_u to distinguish them from the coefficients of the original model, $\hat{\beta}$. Note that all steps involved in forming this approximation can be vectorized efficiently using linear algebra libraries, which is why the proposed method is much faster than computing the true TI in each iteration of the optimizer. To illustrate this, results of a performance evaluation are presented in section 3.

2.2. Optimization Objective and Constraints

With the TI approximation in place, we can now define the optimization objective:

$$\text{maximize} \qquad f(x) = \prod_{i=1}^{p} x_{p+i} - x_i, \qquad x \in \mathbb{R}^{2p} \tag{2}$$

$$\text{subject to} \qquad c_1(x), \ldots, c_{2^{p+1}+5p+2}(x) \geq 0 \tag{3}$$

Where x is the vector of parameters to be optimized, comprised of the lower parameter boundaries in the first p elements and the upper boundaries in the next p elements, with p being the number of factors in the model. One can see that (2) simply maximizes the volume of the hypercube spanned by lower and upper parameter ranges.

To meet all requirements for a valid design space, a total of $2^{p+1} + 5p + 2$ inequality constraints is defined. First, the parameter space to be searched is constrained by ensuring that a parameter's lower range boundary is smaller than its upper boundary.

$$c_i(x) = x_{p+i} - x_i, \qquad i = 1, 2, \ldots, p \tag{4}$$

Oftentimes the design space is required to contain each parameter's setpoint, so that $x_i \leq s_i \leq x_{p+1}$. This is expressed as the inequality constrains

$$c_i(x) = s_i - x_i, \qquad i = 1, 2, \ldots, p \tag{5}$$

$$c_i(x) = x_i - s_i, \qquad i = p + 1, \ldots, 2p \tag{6}$$

Similarly, optimization of the parameter space should only be performed within the screening range boundaries b_l and b_u. As our main optimization algorithm does not support natural boundaries, this is implemented as inequality constraints:

$$c_i(x) = x_i - b_{l,i}, \qquad i = 1, 2, \ldots, p \tag{7}$$

$$c_i(x) = b_{u,i} - x_i, \qquad i = p + 1, \ldots, 2p \tag{8}$$

The remaining constraints address the evaluation of the TI. To that end, the approximation generated in the pre-optimization step is used to calculate boundaries around CQA predictions that capture model uncertainty. These boundaries are then compared against the upper and lower acceptance limits a_l and a_u in each of the 2^p corner points of the hyperrectangle:

$$c_i(x) = x_* \tau_l - a_l, \qquad i = 1, 2, \ldots, 2^p \tag{9}$$

$$c_i(x) = a_u - x_* \tau_u, \qquad i = 1, 2, \ldots, 2^p \tag{10}$$

x_* denotes a corner point, taken from all possible combinations of upper and lower parameter ranges in x. As the TI approximation is generally not a linear one but at least a second-order polynomial, x is expanded to x_* to contain the same quadratic and interaction effects as τ_l and τ_u.

Evaluating the TI at corners alone does not guarantee a valid design space, as curvature in the response surface might lead to parameter ranges between corner points that exceed acceptance limits. To resolve this problem, a final inequality constraint uses a nested optimization step to find minima and maxima of the TI boundaries inside the hypercube:

$$c(x) = argmin_x(x_* \tau_l)\tau_l - a_l \tag{11}$$

$$c(x) = a_u - argmax_x(x_* \tau_u)\tau_u \tag{12}$$

While these last constraints might seem to make (9) and (10) redundant, having both TI checks in place can improve convergence of the optimizer in certain scenarios. Furthermore, one or the other can be deactivated in practice, depending on the type of optimization problem.

2.3. Optimization Algorithms

From the problem described in section 2.2 one can infer the prerequisites for the optimization algorithm. To find a design space, we want to minimize a single, scalar-valued function that is subject to inequality constraints as well as boundaries, though this last requirement can also be implemented as inequality constraints, see (7) and (8). COBYLA (Constrained Optimization BY Linear Approximation) was chosen as the main optimization algorithm, as it meets those requirements and, as a gradient-free method, shows reasonable robustness against converging in local minima (Powell, 1994).

For the nested optimization problem in (11) and (12), where boundaries and no inequality constraints are needed, a fast L-BFGS-B algorithm was used to find minima and maxima withing the hyperrectangle (Byrd, Lu, Nocedal, & Zhu, 1995).

Due to minor inaccuracies introduced by the TI approximation as well as COBYLA convergence conditions in certain cases, we recommend a final "refinement step" where

COBYLA results are passed to a gradient-based SLSQP optimizer (Kraft, 1988). Here, the same constraints are used as in the main optimization step, except that the true TI calculation is used, as opposed to an approximation. Although iterations are much slower, only a few are needed for convergence as we expect the input to be already near the ideal DS.

3. Results

To evaluate efficiency and effectiveness of the proposed method we compared results with a known design space, calculated by a grid-based method that discretizes the problem space and evaluates the TI at each point. This is illustrated in Figure 2.

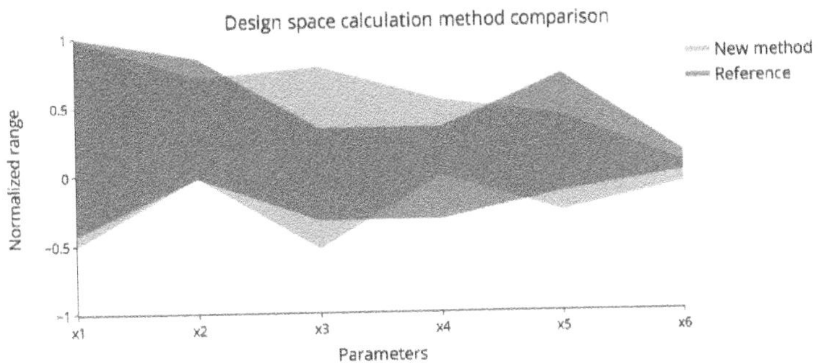

Figure 2: Parameter ranges calculated by both the proposed method and the reference method.

As is to be expected, the volume of the DS is larger in the method based on numeric optimization because it is not constrained by discrete grid points with finite limited resolution. Furthermore, due to parameter interactions, a change in one range can affect other ranges, which overall results in the substantial deviations of parameter ranges shown in in Figure 2. However, the main benefit of the proposed method is improved computing time. Performance evaluation results of an implementation in Python are reported in Figure 3. Due to the exponential increase of the parameter space over the number of parameters added, computing time of a discretization method also grows exponentially, which is not the case with the method based on numerical optimization.

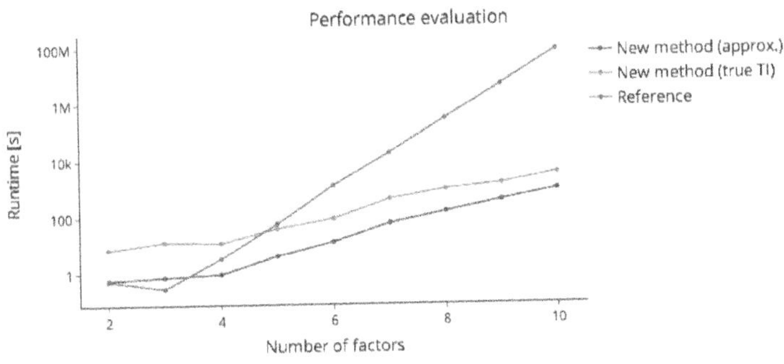

Figure 3: Performance evaluation of the proposed method using both the TI approximation and the true TI, as well as the reference method.

4. Conclusion

In this contribution we outlined a novel method for finding a design space comprised of linear combinations of PP ranges. The relationship between CQA and PPs is represented as a polynomial regression model and its prediction is used to evaluate whether CQAs meet acceptance criteria. Conservative estimation methods are vital in the biopharmaceutical domain, which is why the boundaries of tolerance intervals are used for evaluation of ACs instead of the predicted CQA. As the TI calculation can be complex, an approximation is generated in a pre-optimization step that can be used to calculate the TI by performing a single matrix multiplication. COBYLA is used for the minimization of the main objective, and we suggest refining results with a SLSQP optimizer that uses the true TI instead of an approximation. Performance evaluations show that the proposed method results in design spaces with a larger volume when compared to PP space discretization methods, and that they can be calculated in a fraction of the time. We believe that this approach will facilitate a robust definition of the design space for biopharmaceutical development that reduces patient risk by employing conservative estimators while allowing manufacturers to maximize control ranges. Future work will include a case study where performance is evaluated in a real-world scenario.

5. Acknowledgements

This work was conducted within the COMET Centre CHASE, funded within the COMET – Competence Centers for Excellent Technologies program by the BMK, the BMDW and the Federal Provinces of Upper Austria and Vienna. The COMET program is managed by the Austrian Research Promotion Agency (FFG).
The authors acknowledge TU Wien Bibliothek for financial support through its Open Access Funding Program.

References

Byrd, R., Lu, P., Nocedal, J., & Zhu, C. (1995). A limited memory algorithm for bound constrained optimization. *SIAM Journal on scientific computing*, 1190-1208.

ICH. (2017). ICH guideline Q8 (R2) on pharmaceutical development. EMA.

Kim, I., & Kwak, B. (2002). Design space optimization using a numerical design continuation method. *International Journal for Numerical Methods in Engineering*, 1979-2002.

Kraft, D. (1988). A software package for sequential quadratic programming. *Forschungsbericht-Deutsche Forschungs- und Versuchsanstalt fur Luft- und Raumfahrt.*

Kusumo, K., Gomoescu, L., Paulen, R., García-Muñoz, S., Pantelides, C., Shah, N., & Chachuat, B. (2020). Nested Sampling Strategy for Bayesian Design Space Characterization. *Computer Aided Chemical Engineering*, 1957-1962.

Powell, M. (1994). A direct search optimization method that models the objective and constraint functions by linear interpolation. In *Advances in optimization and numerical analysis* (pp. 51-67). Springer.

UCLA Statistical Consulting Group. (2021). *R Library Contrast Coding Systems for categorical variables.* Retrieved from UCLA Statistical Methods and Data Analysis: https://stats.oarc.ucla.edu/r/library/r-library-contrast-coding-systems-for-categorical-variables/

Antonis Kokossis, Michael C. Georgiadis, Efstratios N. Pistikopoulos (Eds.)
PROCEEDINGS OF THE 33rd European Symposium on Computer Aided Process Engineering (ESCAPE33), June 18-21, 2023, Athens, Greece
 http://dx.doi.org/10.1016/B978-0-443-15274-0.50146-3

Optimal operation of large gas networks: MILP model and decomposition algorithm

Lavinia Marina Paola Ghilardi[a], Francesco Casella[b], Daniele Barbati[c], Roberto Palazzo[c], Emanuele Martelli[a,*]

[a]*Politecnico di Milano, Department of Energy, Via Lambruschini 4, Milano, 20154, Italy*
[b]*Politecnico di Milano, Dipartimento di Elettronica, Informazione e Bioingegneria, Via Ponzio 34/5, Milano, 20133, Italy*
[c]*Snam, San Donato Milanese (MI), 20097, Italy*
[*]*emanuele.martelli@polimi.it*

Abstract

As of today, natural gas is one of the most widely deployed energy sources and its transport relies on large-scale infrastructures managed by the expertise of transmission system operators. In this framework, this paper proposes a Mixed-Integer-Linear-Programming model and a decomposition algorithm to optimize the operation of gas networks. The model aims to minimize the CO_2 emissions from compressor stations by optimizing the unit commitment of the gas-turbine-driven compressors, their loads and the dynamic operation of the network. The formulation includes the detailed linearization of the performance maps of the machines, their technical limitations, the dynamic conservation equations of the pipes, and the operating constraints of control valves. On top of this, the algorithm is capable of handling flow reversals in pipes, being suitable also for networks with cyclic topology. The resulting large scale MILP is extremely challenging to solve, and therefore we developed a decomposition algorithm to speed up the computational time. The decomposition algorithm is composed by two MILP levels, where the master level contains a simplified model of the compression stations and of pipe friction, while the lower level is the actual detailed model. The aim of the master level is to remove unused compression stations and to fix the flow direction in pipes in the lower level, thus reducing the size of this second problem. At the end of each bilevel iteration, an integer cut is added to the master problem to explore different combinations of committed stations. The algorithm is effectively tested on the Italian gas network case study, featuring 112 pipes and 9 compressor stations, each composed by 3-5 units.

Keywords: Gas network, MILP, decomposition, compression stations, Optimal operation.

1. Introduction

Natural gas is currently one of the most widely deployed energy sources and its transport in ensured by large-scale infrastructures, composed by pipelines, valves, and compressor stations. Nowadays, the management of these networks relies on the expertise achieved by the operators. Nevertheless, as the energy transition moves forward, transmission system operators (TSOs) are required to cope with new trends, making the role of optimization tools crucial, as they could help TSOs in their upcoming decisions.

The current literature on gas network optimization covers different optimization purposes, such as minimization of gas compression cost tackled by Liu et al. (2020) and Zlotnik et al. (2015), the stability of operation addressed by Hoppmann-Baum et al. (2021), or gas quality satisfaction. The optimization models can be divided in steady-state and dynamic, where only dynamic models are capable of capturing transient behaviors and considering the pipeline inventory. Nevertheless, very few dynamic approaches allow optimizing the on/off commitment of turbo-compressors, and for large networks they use heuristic rules to solve the optimization problem.

This work addresses the optimal dynamic operation of large gas network with the optimization of on/off commitment of turbo-compressors. The problem is formulated as a Mixed-Integer Non-Linear Program (MINLP), and then it is linearized into a Mixed-Integer Linear Program (MILP). Since also the MILP turns out to be extremely large to be solved in reasonable computational time, we developed a decomposition algorithm capable of finding solution in much shorter time.

2. Problem statement

The dynamic model addresses the optimal operation of the gas network so as to minimize the CO_2 emissions related to compression work. The formulation is as follows.

"Given:

- the topology of the network;
- the performance maps of compressors and of the gas turbines driving them;
- import and consumption profiles across the network;
- air temperature forecasts;

Determine:

- the operating mode of the compressor stations;
- the unit commitment and the operating point of each gas-turbine-driven compressor (GTDC);
- pressures and mass flow rates across the network;

so as to minimize the CO_2 emissions and comply with the following constraints:

- conservation laws in pipes;
- technical limits of GT and natural gas compressors;
- compressor stations and control valves discrete operating modes;
- pressure bounds across the network."

3. Optimization model

The MINLP formulation and the linearized MILP are described in Ghilardi et al. (2022). We report hereafter the main of features of the model.

3.1. Pipes and control valves

The gas dynamics within the pipes are governed by a system of differential equations depending on time t and space x (monodimensional flow). Mass (1), momentum (2) and energy conservations, and gas equation of state (3) depends on gas properties (density ρ, speed u, composition c, temperature T), and pipe features (diameter D and elevation z).

$$\frac{\partial \rho}{\partial t} + \frac{\partial(\rho u)}{\partial x} = 0 \quad (1)$$

$$\frac{\partial(\rho u)}{\partial t} + \frac{\partial(\rho u^2)}{\partial x} + \frac{\partial p}{\partial x} + \rho g \frac{\partial z}{\partial x} + \frac{c_f}{2D}\rho|u|u = 0 \quad (2)$$

$$p = Z(p, T, c) \cdot \rho \frac{R}{MM} T \quad (3)$$

The energy conservation equation can be omitted with the hypothesis of isothermal flow, while the kinetic and the inertial term of gas momentum equation can be neglected due to low speed of gas and large time discretization of the optimization (hourly basis). The differential system is discretized in space with the staggered grid method of Patankar (1980), and in time with backward Euler. As part of the linearization for the MILP model, we adopted an average $\bar{Z}$ in (3), an average gas density for the friction term in (2), and a piecewise linear (PWL) approximation of the function $|u|u$. This second step is responsible for the introduction of the binary variable y_D related to flow direction.

Control valves (CV) are directed elements that continuously adjust the outlet pressure according to setpoints. They are commonly arranged in parallel with a bypass valve which, on the other hand, is undirected (backward flow can occur), and is discretely regulated, e.g. open/closed. The operation of this subsystem (control-bypass valves) can be mathematically modelled with three discrete modes: "regulating", "closed" or "reverse-flow". In "regulating" mode, the gas flows from the inlet to the outlet of the CV reducing its pressure to meet the setpoint; in "closed" mode both the bypass and the CV are closed, and the pressures at their boundaries are thus decoupled; in "reverse-flow" mode the gas flows backward in the bypass valve with a negligible pressure drop.

3.2. Compressor stations

Compressor stations consist of a set of GTDCs arranged in parallel, and, by assumption, they can compress NG along one direction only. The station can be "active" if there is at least one running machine, or, vice versa, it can run in "closed" or "bypass" mode.

Each GTDC is characterized by an operating point with a rotational speed N, inlet pressure p_{in}, mass flow rate w, outlet pressure p_{out}, and fuel consumption w_{fuel}. The variables p_{out} (4) and w_{fuel} (5) are functions of the other variables according the polynomials f_1 and f_2, resulting from the best-fit of technical maps. It must be noted that w_{fuel} depends also on ambient temperature T_0, due to its effect on GT efficiency.

$$p_{out} = f_1(w, N, p_{in}) \tag{4}$$

$$w_{fuel} = f_2(w, N, p_{in}, T_0) \tag{5}$$

These non-convex polynomials are linearized following the convex-hull approach of Makkonen and Lahdelma (2006), discretizing the operating region into binary sub-regions. Moreover, other linearized constraints are added: the choking and the surge curve for the centrifugal compressor, and the load range for the gas turbine, as a function of N.

3.3. Objective function

The objective function to be minimized is the CO_2 emitted by gas-turbines driving the NG compressors, defined by NG emission factor f_{CO2} and fuel consumption w_{fuel} of each machine m at time t. Moreover, a penalty term is added to prevent frequent start-ups and account for the direct and indirect CO_2 emissions emitted during start-up.

$$min \sum_{t \in \mathcal{T}} \left[\sum_{m \in \mathcal{M}} f_{CO2} \cdot w_{fuel}(t, m) + \sum_{m \in \mathcal{M}} f_{start} \cdot y_{start}(t, m) \right] \tag{6}$$

4. Decomposition algorithm

The presented MILP problem for the operation of half-day of the whole Italian gas network, with a time resolution of 1 hour and a spatial resolution of pipe finite volumes of 200 km features 4108 binary variables, 24453 continuous variables, 74992 constraints. Gurobi, one of the best available MILP solver, fails at finding a feasible binary solution in 10 hours. To address this issue, we developed a decomposition technique composed by two levels (Figure 1), where the master problem (MP) and the detailed lower-level (DP) are solved over the iterations k to progressively refine the solution. The MP presents a simplified model of compressor stations and of pipe friction, and it is launched to determine the set of inactive stations S_0^k and the direction of the flow y_D^k within the pipes p. Once these specifications are established, the DP can be solved in the resulting reduced space. Meanwhile, an integer cut is added to the MP to discard the combinations of active stations already explored from the next solutions. After a sufficient number of iterations $\bar{k}$, the algorithm is terminated and the best incumbent solution is maintained.

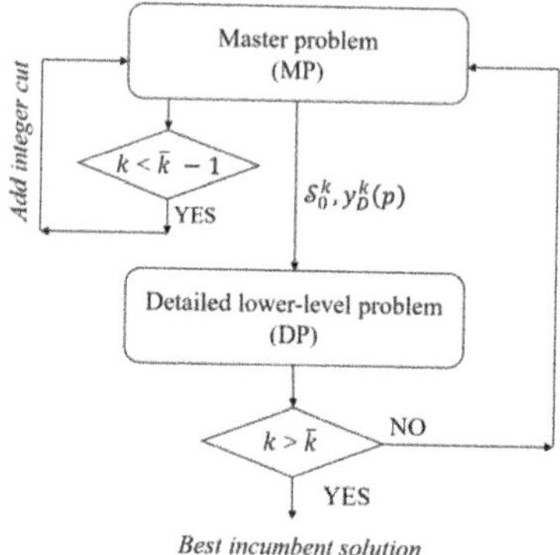

Figure 1. Block diagram of the bilevel decomposition algorithm

4.1. Master problem

The master level MP is a simplification of the original MILP (1)-(6), with regard to the compressor station and the pipe friction model. These two simplifications are intended to reduce the complexity of the problem and especially to decrease the binary variables.
Examining first the pipe model, the PWL representation of $|u|u$ in the friction term is replaced by a straight line $\bar{u}\,u$ with a constant slope $\bar{u}$ for each pipe. This procedure eliminates the binary variable related to flow direction.
As far as the compressor station model is concerned, we defined for the problem MP a single convex hull (CH) for each station operating region (Figure 2), thus removing all individual GTDCs variables (unit commitment and operating point). In this way, the only remaining stations binary variables are those related to their configuration (see 3.3).

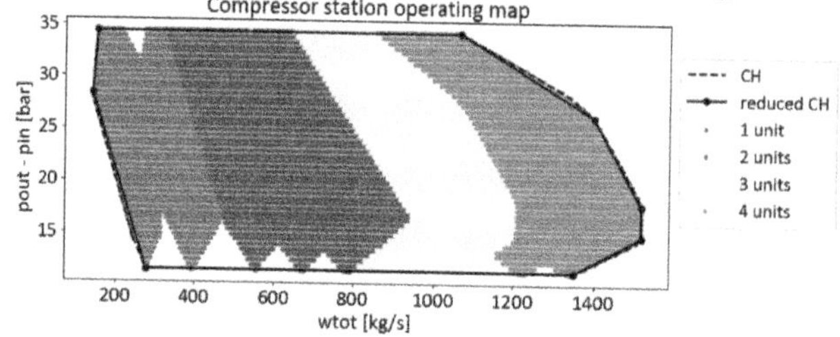

Figure 2. Operating map of a compressor station featuring 4 units projected into space (wtot, pout-pin) with its original convex hull (dashed), and reduced vertices convex hull (continuous).

The above cited convex hull is computed off-line by a series of (CHP) subproblems, one for each value of ambient temperature T_0 and station operating point (p_{in}, p_{out}, w_{tot}). Each subproblem (CHP) is obtained by applying the original MILP (4)-(6) to a single station, and by imposing, as boundary conditions, the coordinates of the operating point at stake. The (CHP) subproblem establishes whether the analyzed operating point is feasible and, if the response is positive, it computes the minimal fuel consumption $w_{fuel,tot}$. Therefore, the convex hull of the feasible points only is established and the number of vertices of the resulting polyhedron is reduced to strike a balance between the number of variables and the accuracy the approximation. The graphical representation of this procedure into the bidimensional space (w_{tot}, $p_{out} - p_{in}$) is provided by Figure 2 for a fixed p_{in} of 50 bar. Once the final CH of the operating region of each station is assessed, the compressor stations variables in MP – when the station is deemed as "active" – are computed as a convex combination of the vertices v of the polyhedron.

4.2. Detailed lower-level problem and integer cuts

The lower level detailed problem (DP) exhibits all the constraints and variables described in (1)-(6), but it has a reduced size compared to the original problem. The DP defines the optimal unit commitment and load of the GTDCs within the resulting reduced space by the MP. Nevertheless, due to the approximations of this MP model, the induced choice of committable stations on DP might be sub-optimal. To mitigate this effect, at the end of each iteration k, an integer cut is progressively added to the MP to exclude the previous explored solutions. This cut (7) guarantees that, in the next iterations, at least one of the previously uncommitted stations ($s \in \mathcal{S}_0^k$) is activated within the time horizon $\mathcal{T}$.
For the practical reasons, given the long computational time, the iterative algorithm stops when the best found solution is not improved for three consecutive iterations.

5. Results

The presented algorithm is applied to a simplified version of the Italian network (Figure 3), still featuring a very complex topology with many loops, 112 pipes, 20 GTDCs arranged in 9 stations, and 16 control valves.

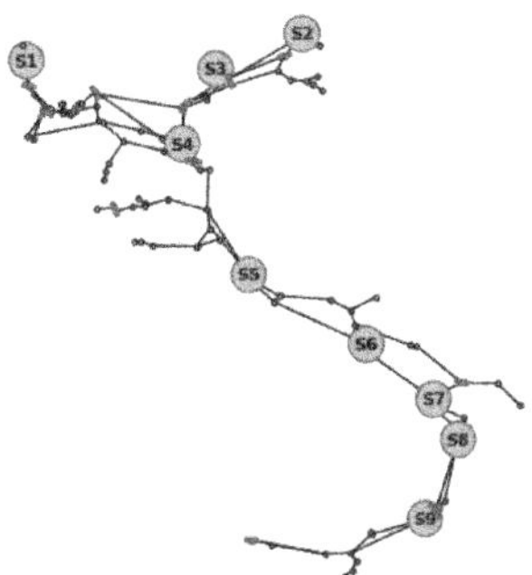

Figure 3. Italian network case study: compressor stations (blue), control valves (red) and pipes (black)

The MILP has been modelled with Pyomo and solved with Gurobi till a gap of 5% on a laptop computer with Intel i7 CPU and 32 GB RAM. The model is applied for the operation of 12 hours in a test scenario featuring a large number of uncommitted stations and backward flow in many pipes. For the practical implementation of the algorithm, the computational time must be limited to about 6 hours. Thus, the iterative MP-DP algorithm stops when the solution is not improved for 3 consecutive iterations.
With a time resolution of 1 hour, the first MP features 14781 continuous variables, 442 binary variables, 18927 constraints, while the DP features 8436, 696 and 74992, respectively. Table 1 depicts for each iteration k the computational time, the stations $\mathcal{S}_1^k$ activated by the MP, and the objective function of the DP.
In all the iterations, the compressor station S2 (at the North-East boundary), the station S9 (near the South border), and S4 are always activated, while different choices of stations along the South-Central pipeline are explored using the integer cut. Among these

solutions, the optimal one is found in the first iteration, as shown in Table 1. Further investigations are needed to define the optimal number of iterations across different scenarios, finding a compromise between optimality and computational load. Nevertheless, it must be underlined that the overall computational time can be reduced in any case by adopting a coarser time grid in MP (without impacting the time resolution of the final solution), and by solving each MP^{k+1} in parallel with DP^k (Figure 1).

k	MP time	DP time	S_1^k	Objective function
1	1135 s	2125 s	S2, S4, S7, S9	586
2	1094 s	2580 s	S2, S4, S8, S9	646
3	1830 s	5811 s	S2, S4, S6, S9	659
4	1910 s	4489 s	S2, S4, S6, S8, S9	614

Table 1. Results of the Italian network case study. The table shows the computational time required to the master problem (MP), the detailed-lower level problem (DP), the activated stations S_1^k by the MP, the optimal CO_2 emissions [$tonCO_2$].

6. Conclusions

This work presents a Mixed Integer Linear Programming algorithmic framework for optimizing the operation of large gas networks including commitment decisions of the turbo-compressors. The model is characterized by the dynamic equations of the network and the performance maps of gas-turbine-driven-compressors. Since the detailed MILP model is too large also for the best available MILP solvers, a decomposition algorithm has been developed. The simplified master problem is firstly launched to define the set of inactive stations and the flow direction within the pipes. Then, the detailed lower-level problem is solved in the resulting reduced space, providing a more accurate solution. Integer cuts are progressively added to the master problem to explore different combinations of active stations, and mitigate the possible sub-optimality of the decomposition. The proposed algorithm is applied to half-day operation of the Italian gas network, which consists of 112 pipes, 9 compressor stations, each one containing 3-5 gas-turbine-driven compressors. For the case study, the first iteration of the bilevel model is solved in less than 1 hour on a laptop computer, and proved to be the best solution across all explored iterations. These results suggest that the algorithm could be applied in decision support systems of transmission system operators. Further development of the research can apply the algorithm to different case studies to assess the optimal number of iterations to be adopted over different scenarios, and to develop some heuristics to speed up the solution.

References

L.M.P. Ghilardi et al., 2022, A MILP approach for the operational optimization of gas networks, IFAC-PapersOnLine, 55, 321–326.

K. Hoppmann-Baum et al. 2021., Optimal Operation of Transient Gas Transport Networks, Optim. Eng., 22, 735–781.

K. Liu et al., 2020, Dynamic optimization of natural gas pipeline networks with demand and composition uncertainty. Chem. Eng. Sci., 215, 115449.

S. Makkonen, R. Lahdelma, 2006, Non-convex power plant modelling in energy optimisation, Eur. J. Oper. Res. 171, 1113–1126.

S. V. Patankar, 1980. Numerical Heat Transfer and Fluid Flow, 1st ed. CRC Press .

A. Zlotnik et al., 2015, Optimal control of transient flow in natural gas networks, Proc. IEEE Conf. Decis. Control 54rd IEEE, 4563–4570.

GUROBI. http://www.gurobi.com.

Antonis Kokossis, Michael C. Georgiadis, Efstratios N. Pistikopoulos (Eds.)
PROCEEDINGS OF THE 33rd European Symposium on Computer Aided Process Engineering (ESCAPE33), June 18-21, 2023, Athens, Greece
 http://dx.doi.org/10.1016/B978-0-443-15274-0.50147-5

An Integrated Framework for the Geometric Optimization of Bio-catalytic Microreactor

Yueheng Han[a], Runze Liang[a], Zhihong Yuan[a,*]

[a]*Department of Chemical Engineering, Tsinghua University, Beijing100084, China*
Corresponding authors: zhihongyuan@mail.tsinghua.edu.cn

Abstract

Biodiesel is an ideal substitute for traditional diesel under the requirements of sustainable and low-carbon issues because of its low-temperature chamber gas emission ratio and good adaptability to the characteristics of existing vehicle engines. However, biodiesel has yet to be fully industrialized due to the traditional batch production mode. Based on the rapid development of microreactor technology that can be used for continuous production, this paper studies the influence of shape parameters of a T-shaped microreactor on biodiesel yield through the enzymatic synthesis of biodiesel from biodiesel castor oil and methanol to prepare fatty acid methyl ester. An automatic simulation optimization platform is thus established based on computational fluid dynamics simulation and derivative-free optimization algorithms. The optimization work makes biodiesel yield increase by 86% compared with the initial result.

Keywords: Derivative free trust region algorithm, Microreactor, Geometric optimization, Automatic CFD simulation

1. Introduction

With the rapid global increase of market competition along with the requirements on safety and environmental issues, the search for sustainable chemical production systems is becoming increasingly important. Biodiesel has been widely regarded as a green alternative to fossil counterparts because of its low-temperature chamber gas emission ratio and good adaptability to the characteristics of existing vehicle engines. However, traditional biodiesel production is mainly carried out in the batch reactor, which takes a long time and consumes much energy and thus hinders its further development and potential large-scale industrial applications.

Benefiting from additive manufacturing and flow chemistry, the new continuous biodiesel production equipment represented by microreactors has started getting attention from academia and industry because of its excellent mass and heat transfer characteristics. However, the selection of microreactor geometric parameters is still determined by experience, lacking a systematic optimization approach (Liang, 2020). Based on the Derivative-Free Trust Region (DFTR) algorithm (Charles A et al., 2017) and the Computational Fluid Dynamics (CFD) simulation, this paper proposes an integrated framework for optimally discovering the geometric parameters of the T-shaped microchannel reactor. In general, the proposed framework can automatically run the CFD simulator based on the offered parameters by DFTR. Accordingly, the information from the CFD simulator feeds DFTR for seeking next-round parameters.

The proposed framework is applied to a biodiesel production example where triglyceride is chosen as the feedstock (Zarejousheghani et al., 2016). Facing the mixing problem caused by the two-phase laminar flow boundary in the reactor channel and the low diffusion coefficient of reactants, this example assumes that the laminar flow boundary can be broken by adding a groove to improve the mixing effect. The main objective is to maximize the yield of biodiesel. The results indicate the feasibility and effectiveness of the proposed framework. The yield of biodiesel associated with the designed microreactor is increased by 86% compared to the existing ones. The designed microreactor is now being fabricated to get further experimentally validated.

2. Reaction

As shown in Fig.1, the production method of enzyme-catalyzed reaction of triglyceride (castor oil) and methanol in a microreactor was selected, and the amount of fatty acid methyl ester obtained from the reaction was regarded as the final biodiesel production.

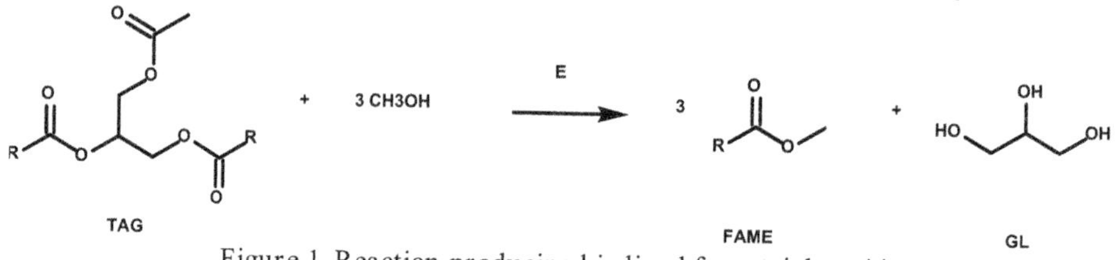

TAG + 3 CH3OH —E→ 3 FAME + GL

Figure 1. Reaction-producing biodiesel from triglyceride

Fig.2 illustrates that the reaction process can be explained by a ping-pong mechanism (Zarejousheghani et al., 2016): (1) The enzyme attacks triglycerides, producing and activating enzyme triglyceride conjugates. (2) The conjugates rearrange to form the enzyme and free fatty acid chain complex and release glycerol. (3) Alcohols link with enzymes in the activated state to release the final product, fatty acid methyl ester, and free enzymes.

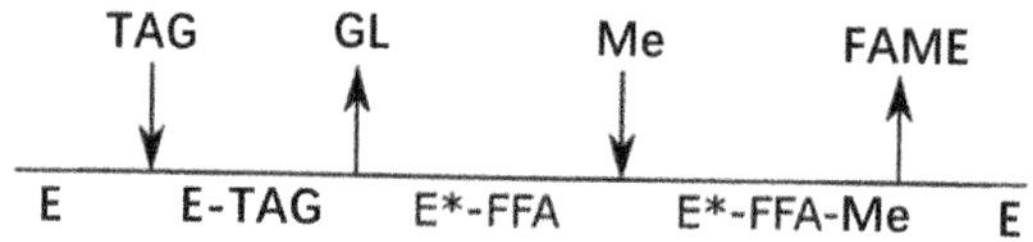

Figure 2. Reaction mechanism

Experimental data (Zarejousheghani et al., 2016) shows that the inhibition effect of methanol in the production process can be ignored so that the reaction mechanism can be characterized approximately by the ping-pong mechanism without inhibition.

$$r = \frac{k[E_0][TAG][Me]}{K_{mTG}[Me] + K_{mA}[TAG] + [TG][Me]} \tag{1}$$

$[E_0]$ represents the enzyme concentration in the reaction system, $[TAG]$ and $[Me]$ are the molar concentrations of triglyceride and methanol respectively. The meanings of other parameters and their estimated values are shown in Table 1.

Table 1. Reaction kinetics parameters

Parameter	Meaning	Value

$k(mmol \cdot ml^{-1} \cdot s^{-1})$	Maximum reaction rate	0.0208
$K_{mTAG}\ (mmol \cdot ml^{-1})$	Michaelis constant	10.54
$K_{mMe}\ (mmol \cdot ml^{-1})$	Michaelis constant	10900.42
$K_{iMe}\ (mmol \cdot ml^{-1})$	Inhibition constant	3976.48

In addition to reaction kinetics, the diffusion coefficient between reactants in the reaction system is also an indispensable parameter in the actual modeling process. In this work, the diffusion coefficient of alcohol in the oil phase is estimated by Wilke Chang's empirical formula. The diffusion coefficient of methanol to triglyceride is approximately $1.8 \times 10^{-10}\ m^2/s$, the diffusion coefficient between lipase and other phases is approximately $5 \times 10^{-11}\ m^2/s$, the diffusion coefficient between other substances is approximately $10^{-9}\ m^2/s$.

3. Reactor

3.1. Initial reactor

The traditional T-type microreactor is selected as the reactor for biodiesel synthesis. The raw materials enter from both ends of the reactor respectively, and the reaction and mixing are carried out in the reaction channel.

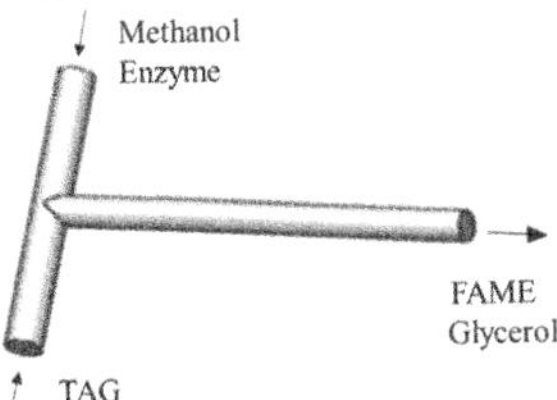

Figure 3. Initial T-shaped Microreactor

Table 2. Geometric parameters and feeding conditions of initial T-type reactor

Variable	Value
Pipe diameter	0.5 mm
Pipe length	10 mm
Branch length before the two-phase intersection	3 mm
Triglyceride feed flow rate	0.1 mm/s
Methanol feed flow rate	0.6 mm/s
The mass fraction of enzyme in the methanol feed solution	1%

3.2. Reactor optimization

Due to the low interphase diffusion coefficient, the contact of reactants forms a laminar boundary, which hinders the diffusion and reaction process of substances. It is feasible to break the original laminar flow by adding grooves to improve the mixing condition. The optimization problem in this work is established by taking the geometric parameters related to the grooves as optimization variables.

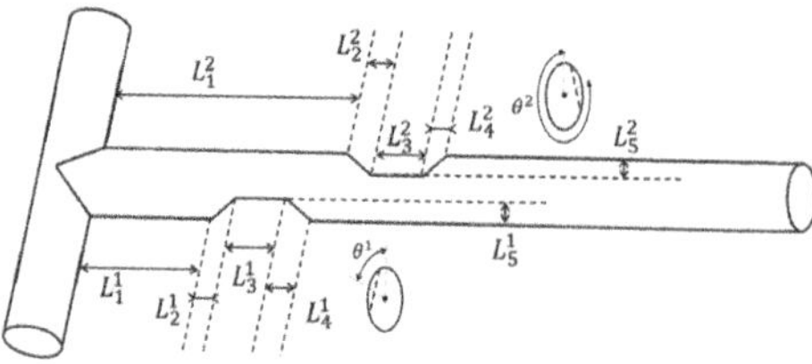

Figure 4. Geometric parameters of the groove of T-type reactor

Constrain the range of each variable: (1)The starting position of the groove is required to be 0.5 mm away from the intersection of two phases; (2)The length of the inclined front section, horizontal section, and rear inclined section of the groove shall be between 0.01 mm and 2 mm; (3)The depth of the two grooves is required to be between 0.05 mm and 0.45 mm to avoid the appearance of too small flow passage cross-section; (4)The deviation angle of the groove centerline is set between 0 and 2π; (5) To simplify the model, the second groove is required to be generated after the first groove.

$$\begin{cases} 0.5 \le L_1^1 \le 10, \quad 0.5 \le L_1^2 \le 10, \quad 0.01 \le L_2^1 \le 2, \quad 0.01 \le L_2^2 \le 2 \\ 0.01 \le L_3^1 \le 2, \quad 0.01 \le L_3^2 \le 2, \quad 0.01 \le L_4^1 \le 2, \quad 0.01 \le L_4^2 \le 2 \\ 0.05 \le L_5^1 \le 0.45, \quad 0.05 \le L_5^1 \le 0.45, \quad 0 < \theta^1 \le 2\pi, \quad 0 < \theta^2 \le 2\pi \\ L_1^2 + L_2^2 + L_3^2 + L_4^2 \le 10, \quad L_1^1 + L_2^1 + L_3^1 + L_4^1 < L_1^2 \end{cases} \tag{2}$$

Table 3. Geometric parameters of the groove of T-type reactor

Variable	Meaning
L_1^1/mm	Distance from the beginning of groove 1 to the entrance
L_2^1/mm	Length of the front section of the inclined section of groove 1
L_3^1/mm	Length of the horizontal section of groove 1
L_4^1/mm	Length of rear section inclined section of groove 1
L_5^1/mm	Depth of groove 1
θ^1	Deviation angle of groove 1 centerline
L_1^2/mm	Distance from the beginning of groove 2 to the entrance
L_2^2/mm	Length of the front section of the inclined section of groove 2
L_3^2/mm	Length of the horizontal section of groove 2
L_4^2/mm	Length of rear section inclined section of groove 1
L_5^2/mm	Depth of groove 2
θ^2	Deviation angle of groove 2 centerline

4. Integrated framework

4.1. Computational fluid dynamics modeling

Empirical formulas in the macro process cannot be directly applied to a micro-scale reactor. The numerical simulation method of solving the fundamental equations in fluid mechanics, computational fluid dynamics (CFD), is used to model the reaction process in the microreactor. Ansys series software was used to complete the modeling process of geometric modeling, mesh generation, and computational fluid dynamics.

4.2. Derivative-free trust-region algorithm

CFD can only obtain the output results under given input conditions, so the optimization process needs an optimization method that does not rely on derivative information. We use a derivative-free trust-region algorithm to solve the problem.

Under the variables and constraints given in section 3.2 to be optimized, the constraint problem $\max f(x)$ is constructed to maximize biodiesel yield. Table 4 shows the algorithm framework:

Table 4. The framework of the derivative-free trust-region algorithm

1. Initial point x_0, trust region radius Δ_0
2. Establish approximate function model $q(x_k + s) = f(x_k) + g^T s + \frac{1}{2} s^T B_k s$
3. Solve the problem $max\ q(x_k + s)$ to obtain s
4. Calculate the ratio of the actual objective function descent to the approximate function model descent
5. Update the trust region radius according to the ratio
6. Output the final result if the termination conditions are met. If not, return to step 2.

4.3. Framework

A fully automatic simulation optimization platform is established based on the modeling and optimization method. The framework is shown in Figure 5.

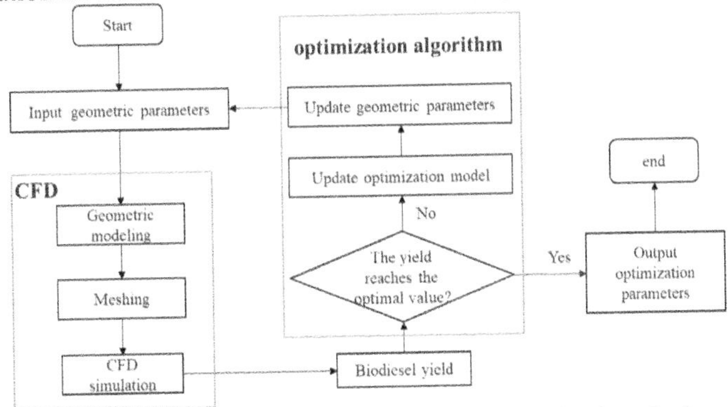

Figure 5. Operation framework of simulation optimization platform

5. Analysis and discussion

With the goal of maximizing the yield of biodiesel, this work calculates the conversion rate of triglyceride at the outlet interface obtained by CFD simulation as the objective function:

$$\max\ f = \frac{\sum c_{FAME}}{\sum c_{FAME} + 3\sum c_{TAG}} \tag{3}$$

Forty feasible groove shapes were selected by Monte Carlo random sampling, with the corresponding diesel yield obtained through CFD simulations. The quality of grids generated in the automatic CFD simulation process is higher than 0.4, the quantity of grids is about 500000, and the calculation accuracy error is about 2%. The groove shape with the highest biodiesel yield was selected as the initial point, and the biodiesel yield was 4.86%.

Figure 6. The product distribution of the initial reactor

After the normalization of variables, the initial trust region radius is selected as 0.05, and the trust region radius less than 0.001 is taken as the termination condition of the trust region algorithm.

As shown in Figure 5 above, according to the geometric parameters and trust region, select the points in the trust region to carry out CFD simulation to establish the proxy model, calculate the proxy model to obtain the best geometric parameters, and carry out CFD simulation to get the calculation results. If the results do not meet the convergence conditions, update the radius of the trust region and the proxy model, and repeat the process. If the result meets the convergence condition, output the optimized geometric parameters and terminate the operation. Each iteration requires about 91 function value calculations. After about 1000 times function value calculation, iterative convergence is achieved. The final yield of biodiesel is 9.03%.

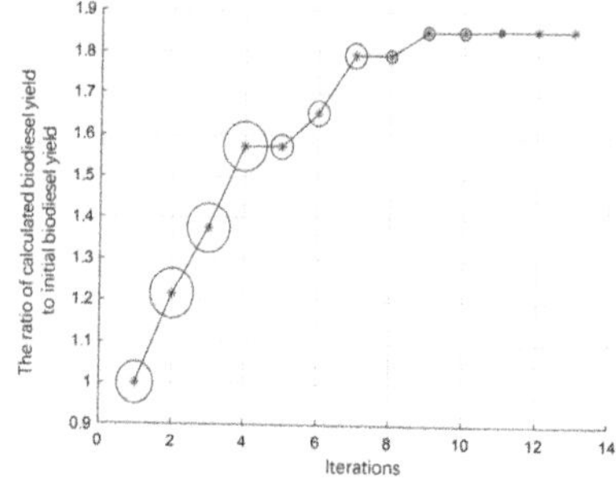

Figure 7. The ratio of optimization result to initial result varies with the number of iterations

The circle size represents the trust region radius size.

Figure 8 shows the geometry of the microreactor under the final optimization results and the product distribution in the reactor at this time. The simulation results show that the optimization process can effectively improve the two-phase mixing condition and promote the reaction.

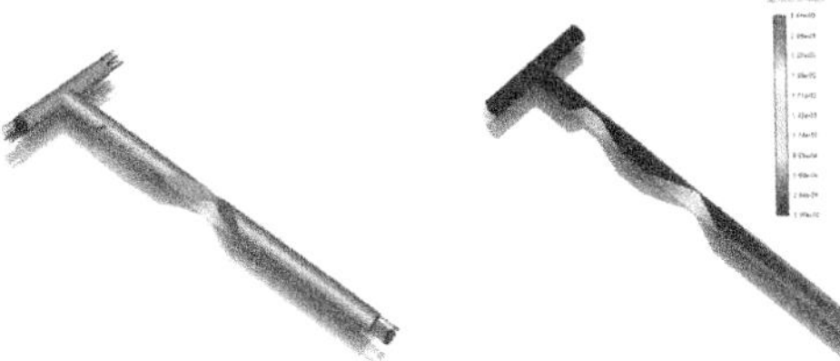

Figure 8. The geometric shape and product distribution of the optimized microreactor

Table 5. Geometric parameters of the optimized microreactor

Parameter	Value	Parameter	Value
L_1^1/mm	1.266	L_1^2/mm	3.754
L_2^1/mm	0.328	L_2^2/mm	1.168
L_3^1/mm	0.437	L_3^2/mm	0.187
L_4^1/mm	1.371	L_4^2/mm	0.998
L_5^1/mm	0.297	L_5^2/mm	0.450
θ^1	141.5°	θ^2	257.7°

6. Summary

An automatic simulation optimization platform has been established, and the biodiesel yield has been successfully improved using a derivative-free trust-region algorithm and CFD. The optimization platform can be extended to more extensive geometric shape optimization research in the future.

References

Liang RZ and Yuan ZH, (2020). Computational Shape Optimization of Microreactors based on CFD Simulation and Surrogate Model-driven Optimization[C]. The 30th European Symposium on Computer Aided Process Engineering, PTS A-C, 48:925-930

Zarejousheghani F, Kariminia HR, Khorasheh F (2016). Kinetic modeling of enzymatic biodiesel production from castor oil: Temperature dependence of the Ping Pong parameters [J]. Canadian Journal of Chemical Engineering, 94(3):512-7

Charles Audet, Warren Hare, et al., (2017). Derivative-Free and Blackbox Optimization[M]. Springer International Publishing.

Antonis Kokossis, Michael C. Georgiadis, Efstratios N. Pistikopoulos (Eds.)
PROCEEDINGS OF THE 33rd European Symposium on Computer Aided Process Engineering (ESCAPE33), June 18-21, 2023, Athens, Greece
 http://dx.doi.org/10.1016/B978-0-443-15274-0.50148-7

A genetic algorithm-active learning approach in energy systems optimization

Rafael Castro Amoedo,[a] Julia Granacher,[a] Yi Zhao,[a] and François Maréchal,[a]

[a] *Industrial Process and Energy Systems Engineering group, École Polytechnique Fédérale de Lausanne, Rue de l'Indutrie 17, 1950 Sion, Switzerland*
rafael.amoedo@epfl.ch

Abstract

Recently, active learning methods have been used to solve large and complex problems. The underlying principle is the construction of computationally cheap surrogate models, which are able to mimic the original problem with a high-level of accuracy. In complex energy systems, (non-linear) simulation models (e.g. generated using Aspen Plus) have grown larger. In the field of process system engineering, these models need to be repeatedly used to optimize operating conditions. The use of surrogate models reduces the required optimization time. However, the quality and the way to obtain these surrogates is crucial. In this work an active learning approach is coupled with a genetic algorithm, that ensures the selection of points that are relevant from the optimization perspective. Nine different surrogate methods are tested. Our results show that artificial neural networks and random forest are the best performing algorithms, with mean square error up to two orders of magnitude better than other more simplistic approaches. A biogas separation case-study is used to validate the approach.

Keywords: Active Learning, Surrogate models, Genetic algorithm, optimization, superstructure

1. Introduction and state-of-the-art

In the field of Process system engineering (PSE), superstructure-based optimization is a common approach for both synthesis and design stages[1]. The backbone of a superstructure is a set of interconnected units (or blocks), that combined are able to generate solutions that meet desired specifications and constraints. Such units are often modeled as black-boxes (BB); thus their accuracy is critical for generating insightful results. These BB models try to capture, via a mathematical formulation, the relation between inputs and outputs. This can be a challenging task, especially if complex mathematical models (e.g., CFD) or large flowsheeting diagrams (e.g. Aspen or Belsim Vali) are required to describe conversion processes. Lengthy computational times and converge issues are among the main concerns with such modelling strategies.

Surrogate models (or metamodels) are a way to by-pass the limitations of complex simulations, while keeping a high-level of accuracy. This type of supervised machine learning is able to explore features between input and output data in a computationally cheap way. They can then be used in superstructure-based optimization, allowing one to obtain fast and reliable results. In the field of PSE, the most used surrogate strategies are Kriging models (or Gaussian processes) and Artificial Neural Networks (ANN)[2], justified by the built-in error measure of the former, and the adaptability to practically any type of data structure of the latter. Examples of such applications can be found for large scale superstructure-based optimization[3] or even to replace state-of-the-art solvers in deterministic global optimization[4].

Building surrogates can be a simple task if large amounts of useful and reliable data are available. Unfortunately, this is often not the case. One technique is to run the simulation model for an extremely large number of points. This is the domain of Deep Learning (DL). Provided enough time and results, DL can build a very accurate (and cheap to run) surrogate model. However, obtaining such data can be (computationally) expensive. Active learning (AL) is another domain of machine learning that leverages the capacity of an algorithm to learn and 'smartly' decide which sample points need to be simulated. Points are selected in regions for which the surrogate level of accuracy is the lowest, and therefore running the expensive simulation program will bring the largest amount of information. The algorithm decides based on the evaluation of a specific function (called learning function)[5].

An AL approach is by itself a good improvement compared with greedy DL strategies. For optimization purposes, one would like to have an extremely accurate surrogate model in the points that are assessed during the optimization stage - rather than a good overall method, that underperforms in the most critical points. In this regard, a way to cover the most 'interesting' space domains for optimization is by using a genetic algorithm (GA) - a nature-inspired meta-heuristic. Such evolutionary strategy has been extensively used for optimization. A review on its properties, advantages and limitations is given in[6].

This work departs from the literature by coupling two 'smart' approaches with the goal of building surrogate models. A GA complements the use of an AL approach, particularly by guiding the surrogate building strategy in the direction that is most interesting from the optimization perspective. The method was tested and validated in a superstructure-based optimization approach for the chemical-absorption separation of biogas.

2. Methodology

2.1. Mathematical formulation and overview

The goal of the GA-AL approach is to create the function **f** that is able to accurately predict the output(s) of interest. An overview of the methodology is shown in Figure 1. There are four macro blocks, explored below.

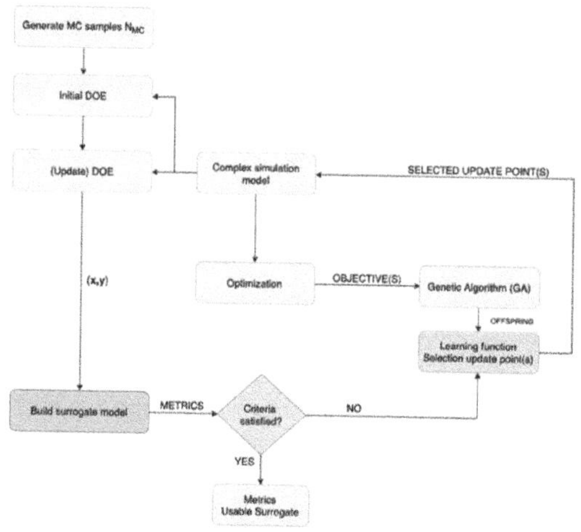

Figure 1 – Simplified block diagram of the GA-AL approach.

2.1.1. Design of experiments (DOE)

The DOE is the section responsible for the initial number of (expensive) simulation runs and the connection between the surrogate building framework and the simulation code. The initial number of samples is a topic of debate. We have adopted the max{n_s +2, 12}, in which n_s is the number of parameters (also known as dimensions), following the suggestion of Hong et al[7] .

2.1.2. Surrogate methods

Table 1 summarizes the list of methods used, comprising simple ones (such as least squares and inverse distance weighting), Kringing-based approaches and more advanced ones such as ANN or Random forests. For this stage, ML python libraries were used, together with the python package 'smt' based on the work of Bouhlel et al.[8], where the pros and cons of each strategy are also discussed and the complete mathematical formulation provided.

Table 1 - Surrogate methods used

Acronym	Surrogate method
LS	Least squares
KRG	Kriging
KPLS	Kriging + partial least squares (PLS)
KPLSK	2-step Kriging + PLS
GEKPLS	Gradient-enhanced Kriging + PLS
IDW	Inverse distance weightings
RBF	Radial basis function
ANN	Artificial Neural Network
RF	Random Forest

2.1.3. Learning function

The learning function decides which point **(x)** (which is in reality a n_s-dimension vector) needs to be integrated in the DOE set. An adaptation of the potential risk function (PRF) as described in Hong et al.[9] is used (eq. 1). It consists of an exploration (e_r) and exploitation (e_t in blue) approach, that is weighted by the parameter τ. The numerator is the distance between any future point and the DOE points, whereas the denominator is the maximum distance within the current DOE points (of size m). For the exploitation, $y(x_i)$ is the real simulation result on the DOE points, ξ is the limit value (equivalent to the limit-state value) and r(x) is given by eq. 2, where $\hat{y}$ represents the surrogate model answer.

$$e_{PFR}(\boldsymbol{x}) = e_r(\boldsymbol{x}) \cdot e_t(\boldsymbol{x}) = \frac{\min_{x_i \in \mathbf{x}} |\boldsymbol{x}_i - \boldsymbol{x}|}{\max_{x_i, x_j \in \mathbf{x}} |\boldsymbol{x}_i - \boldsymbol{x}_j|} \cdot \exp\left(-\tau \frac{r(x)}{\max_{x_i \in x} y(\boldsymbol{x}_i) - \xi}\right), i,j = 1,2,\dots,m, \tau > 0 \quad (1)$$

$$r(x) = \begin{cases} \hat{y}(\boldsymbol{x}) - \xi, \hat{y}(\boldsymbol{x}) \geq \xi \\ \xi - \hat{y}(\boldsymbol{x}), \hat{y}(\boldsymbol{x}) < \xi \end{cases} \quad (2)$$

2.1.4. Stopping criteria

The stopping criteria (eq.3) defines the level at which the surrogate model is good enough to be used without further refinement. The criteria is a convergence based one, adapted from Hong et al.[9]; $\hat{\mu}_i$ and $\hat{\sigma}_i$ are, respectively, the estimated expected value and standard deviation, obtained using the surrogate model, in the GA offspring points. The threshold (ε_{th}) is set to 10^{-5}. In addition, a stopping criterion on the number of runs (100 runs) is included to avoid extremely lengthy surrogate training.

$$\varepsilon = max\left(\frac{\hat{\sigma}_i}{\hat{\mu}_i}\right) \leq \varepsilon_{th} \quad \forall i \in Offspring \quad (3)$$

2.2. Case-study

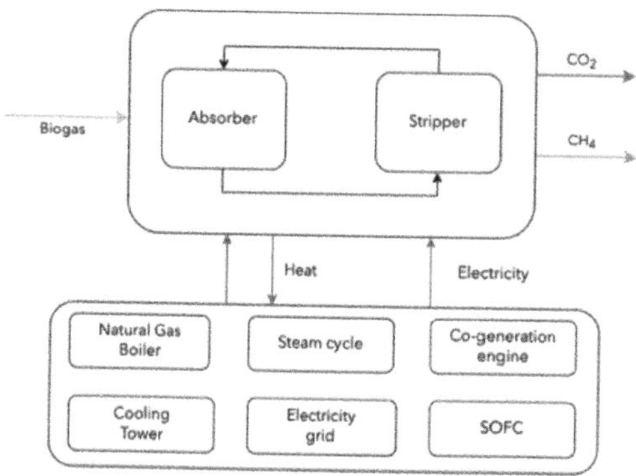

Figure 2 – Superstructure-based approach for biogas separation. The set of technologies was modeled according to Castro-Amoedo et al.[10,11] The average annual (2021) energy price was queried using the approach described in Santecchia et al.[12]

The GA-AL method is applied to the total cost minimization of a biogas (60% CO_2, 40% CH_4) chemical absorption separation (Figure 2). This chemical separation was modeled according to Sharma et al.[13] ;the parameters used for surrogate building are described in Table 2, and the total heat duty is the surrogate output. The superstructure problem is formulated as a mixed-integer linear programming (MILP), following the approach described in Castro-Amoedo et al. [14] .

2.2.1. Performance evaluation

The evaluation of the GA-AL approach is based on **four** criteria. The first is the number of total (expensive) model evaluation per surrogate method (N_{mtd}^{eval}); the second is the beta approximation (β_{rank}) (eq. 4), which measures how close a given method (mth) is from the reference expected value (β_{ref}), which was obtained using a Monte Carlo simulation with 10^5 points. The third (eq. 5) is a trade-off between the first and second criteria, with N_{med} being the median number of model evaluations. Lastly, MSE (eq. 6) represents the mean square error. To avoid the bias introduced by the set of initial points, twenty replications were performed for each surrogate method.

Table 2 – Parameters, range and output

Inputs	Range
Absorber temperature	40-60°C
Stripper temperature	90-120°C
Absorber diameter	15-20 m
Stripper diameter	5-12 m
Absorber tray space	0.6-1.2 m
Stripper tray space	0.6 –1.8m

$$\beta_{rank} = \left|\frac{\widehat{\beta_{mtd}} - \beta_{\text{ref}}}{\beta_{\text{ref}}}\right|, with\ \hat{\beta} = \frac{\hat{\mu}_{mth}}{\hat{\sigma}_{mth}} \tag{4}$$

$$\Delta = \beta_{rank} \cdot \frac{N_{mtd}^{\text{eval}}}{N_{\text{med}}} \tag{5}$$

$$\text{MSE} = \sum_{i=1}^{N_m^{\text{eval}}} \left(y(x_{\text{test},i}) - \hat{y}(x_{\text{test},i})\right)^2 \tag{6}$$

3. Results and Discussion

The systematic evaluation of each approach is based on evaluating the performance criteria for the different strategies. Figure 3 (a) summarizes the results for the set of surrogate models tested. As suggested, random forest and artificial neural networks algorithms outperform the others, scoring the lowest MSE and being close to the reference expected value (β_{ref}). This demonstrates the capacity of such algorithms to adapt to any kind of mathematical structure, even with limited amount of data.

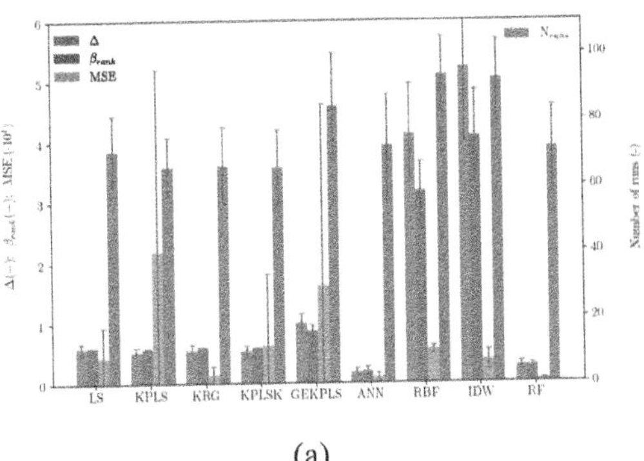

(a)

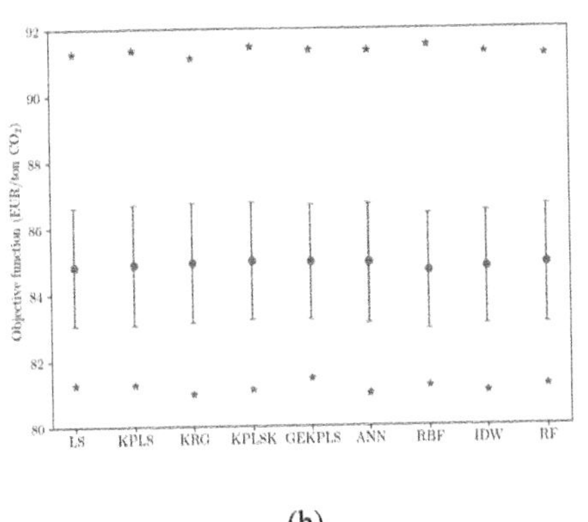

(b)

Figure 3 – (a) surrogate models performance based on the four criteria performance criteria; (b) MILP objective function range

On the other hand, RBF and IDW underperform, due to the combination of a large number of runs required but also a poor approximation to the reference. Their MSE is one order of magnitude greater than the best performing methods, despite the larger number of samples. This might be due to the rather simplistic approach of both methods.

The standard Kriging (KRG) is, after ANN and RF, the best performing algorithm. It brings advantage compared to the least squares (i.e. linear regression) approach. However, it is worth noting that additional features, such as enhanced-gradient (GEKPLS) or the use of partial least squares (KPLS) do not bring advantage. Indeed, for the same average number of runs, an increase (by one order of magnitude) in MSE is recorded.

The surrogate model is part of a superstructure that minimizes the total cost of separating CO_2 from a biogas mixture. At the level of the MILP, the quality of the surrogate influences the GA choice of points and consequently the range of costs. This is visible in Figure 3 (b). Regardless of the surrogate model, the fine tuning of the operating parameters bound the separation process between 81 and 92 EUR/ton of CO_2.

4. Conclusions

In this work, a genetic algorithm is coupled with an active learning methodology to generate surrogate models. Such combined approach harvests the best of both methods:

the genetic algorithm guides the active learning search to the points that are more relevant to the optimization problem at hand, whereas the active learning ensures finding the most interesting point to be simulated and then executed at the MILP optimization interface. Nine surrogate methods were tested in a biogas separation case study. Artificial neural networks and random forest are the best performing algorithms, followed by a standard Kriging formulation. On the other hand, the radial basis function and the inverse distance weighting algorithms have shown the worst performance. Moreover, the coupling of Kriging metamodel with additional features such as enhanced-gradient or partial least squares did not prove advantageous. The cost of separating CO_2 and CH_4 from a biogas mixture was simultaneously assessed, with costs ranging between 81 and 92 EUR/ton of CO_2.

The approach here presented is modular and can accommodate any type of surrogate method, learning function or stopping criteria. Future work will address a more comprehensive set of stopping criteria and their effect on surrogate accuracy, as well as a systematic ranking system. In addition, high-dimensional problems (with more than twenty parameters) will be tested, show-casing the use of this framework for any type of PSE problem.

References

1.Granacher, J., Nguyen, T.-V., Castro-Amoedo, R. & Maréchal, F. Overcoming decision paralysis—A digital twin for decision making in energy system design. *Applied Energy* **306**, 117954 (2022).

2.Cozad, A., Sahinidis, N. V. & Miller, D. C. Learning surrogate models for simulation-based optimization. *AIChE Journal* **60**, 2211–2227 (2014).

3.Granacher, J., Kantor, I. D. & Maréchal, F. Increasing Superstructure Optimization Capacity Through Self-Learning Surrogate Models. *Frontiers in Chemical Engineering* **3**, (2021).

4.Schweidtmann, A. M. & Mitsos, A. Deterministic Global Optimization with Artificial Neural Networks Embedded. *J Optim Theory Appl* **180**, 925–948 (2019).

5.Moustapha, M., Marelli, S. & Sudret, B. Active learning for structural reliability: Survey, general framework and benchmark. *Structural Safety* **96**, 102174 (2022).

6.Lee, S., Kim, J., Kang, H., Kang, D.-Y. & Park, J. Genetic Algorithm Based Deep Learning Neural Network Structure and Hyperparameter Optimization. *Applied Sciences* **11**, 744 (2021).

7.Hong, L., Li, H. & Peng, K. A combined radial basis function and adaptive sequential sampling method for structural reliability analysis. *Applied Mathematical Modelling* **90**, 375–393 (2021).

8.Bouhlel, M. A. *et al.* A Python surrogate modeling framework with derivatives. *Advances in Engineering Software* **135**, 102662 (2019).

9.Hong, L., Li, H. & Fu, J. A novel surrogate-model based active learning method for structural reliability analysis. *Computer Methods in Applied Mechanics and Engineering* **394**, 114835 (2022).

10.Castro-Amoedo, R., Damartzis, T., Granacher, J. & Marechal, F. M. A. System Design and Performance Evaluation of Wastewater Treatment Plants Coupled With Hydrothermal Liquefaction and Gasification. *Front. Energy Res.* **8**, (2020).

11.Castro-Amoedo, R., Morisod, N., Granacher, J. & Maréchal, F. The Role of Biowaste: A Multi-Objective Optimization Platform for Combined Heat, Power and Fuel. *Frontiers in Energy Research* **9**, 417 (2021).

12.Santecchia, A., Kantor, I., Castro-Amoedo, R. & Maréchal, F. Industrial Flexibility as Demand Side Response for Electrical Grid Stability. *Frontiers in Energy Research* **10**, (2022).

13.Sharma, S., Castro-Amoedo, R., Hemrle, J. & Maréchal, F. Assessment of Carbon Capture Technologies for Waste-to-Energy System. in *Computer Aided Chemical Engineering* (eds. Montastruc, L. & Negny, S.) vol. 51 871–876 (Elsevier, 2022).

14.Castro-Amoedo, R., Dahmen, A., Barbosa-Povoa, A. & Maréchal, F. Network design optimization of waste management systems: the case of plastics. in *Computer Aided Chemical Engineering* (eds. Türkay, M. & Gani, R.) vol. 50 185–190 (Elsevier, 2021).

Antonis Kokossis, Michael C. Georgiadis, Efstratios N. Pistikopoulos (Eds.)
PROCEEDINGS OF THE 33rd European Symposium on Computer Aided Process Engineering (ESCAPE33), June 18-21, 2023, Athens, Greece
 http://dx.doi.org/10.1016/B978-0-443-15274-0.50149-9

Data-driven Distributionally Robust Joint Chance-Constrained Optimization for Industrial Utility systems under Uncertainty

Hanxiu Li, Liang Zhao*

Key Laboratory of Smart Manufacturing in Energy Chemical Process, Ministry of Education, East China University of Science and Technology, Shanghai 200237, China

Abstract

The motivation to improve the energy efficiency of utility systems is resonating more intensely since the issue of carbon neutrality was put forward. However, the energy demand uncertainty brings challenges for utility systems modeling and optimization. This paper proposed a multi-objective optimization framework for minimizing operating cost and environmental impact simultaneously. The environmental impact is defined using the life cycle assessment method, and the operating cost consists of the resource consumption of the utility system. A data-driven distributionally robust joint chance-constrained (DRJCC) approach is proposed to deal with the uncertainty of steam demands. The data-driven ambiguity set is constructed based on Wasserstein distance, which contains the empirical distribution and is incorporated into the DRJCC framework. The dual representation for the worst-case probability transforms the problem into a deterministic non-convex reformulation. The proposed method is verified by a case study from an industrial ethylene plant. The results demonstrate that the total cost and environmental impact reduce as the confidence coefficient increases. In addition, the Pareto frontier enables us to select the optimal solutions flexibly, which may minimize the operational cost while the environmental standard is satisfied.

Keywords: industrial utility system, Wasserstein ambiguity set, distributionally robust joint chance-constrained program, life cycle assessment, multi-objective optimization.

1. Background

Emission reduction and energy saving have become the main concerns for the petrochemical industry in the carbon-neutral era. New technology and method applied to improve efficiency and decrease environmental burden is necessary for sustainable development in the next few decades. Process modeling, optimization, and environmental assessment have been focused on for this issue. An improved simulation and optimization method was presented for the sustainable development of utility systems (Varbanov et al., 2020). Ioannou et al. reviewed the role of process systems engineering tools in assessing and optimizing production patterns for alternative chemicals based on renewable resources (Ioannou et al., 2021). A consequential life cycle optimization framework was developed to determine the optimal chemical recycling technology (Zhao and You, 2021). Čuček et al. reviewed the methods of footprints which are defined as indicators and used to measure sustainability (Čuček et al., 2012). With continuous research on system optimization under uncertainty in recent years, stochastic and robust optimization methods were proposed to improve the system's robustness. As a combination of robust and stochastic programming, the distributionally robust chance-constrained program provides a strategy that can adjust the relationship between system reliability and the optimal cost (Ding et al., 2022; Xie and Ahmed, 2016).

As one of the indispensable roles in the process industry, the utility system is generally responsible for providing power and heat for the industrial manufacturing process. Therefore, the deterministic multi-objective model based on life cycle assessment is constructed first. Then, the uncertain parameters of steam demand are collected from the historical data. An improved data-driven DRJCC framework with a Wasserstein ambiguity set is presented, and the deterministic model is reformulated in the dual form for the utility system. Finally, a case study of the utility system is presented to verify the effectiveness of the proposed method.

2. Deterministic Multi-objective Model Based on Life Cycle Assessment

A deterministic multi-objective framework for getting the optimal solutions with sustainability based on economic and environmental factor is proposed in this section.

2.1 Objective function

The operating cost and environmental indicator based on LCA method are defined as sustainability objectives. According to the list of the energy and material consumed in the system, we can calculate operating cost as:

$$COST = \sum_{bo}\sum_{i} p_{i,bo}^{Fuel} \cdot M_{i,bo}^{Fuel} + p^{Ele} \cdot \left(\sum_{mo} M_{mo}^{Ele} + \sum_{m} M_{m}^{Ele} \right) + p^{Water} \cdot \left(\sum_{lv} M_{lv}^{Water} + \sum_{m} M_{m}^{Water} \right) + \sum_{m} p_{m}^{De} \cdot M_{m}^{De} \quad (1)$$

$$i = \{Coal, NG\}, m = \{FGD, SCR\}$$

The life cycle assessment method of CML 2001 was applied for evaluating the environmental load and helped to design the environment-friendly system (Guinée et al., 2001). The total emissions produced in chemical production are composed of direct emissions and indirect emissions. Direct emissions are mainly from the burning of fossil fuels. Equation (2) states the emission of CO_2 and equation (3) indicates the emissions of SO_2 and NO_x under controlled conditions of the flue gas treatment process. The emissions generated by the exploitation, processing, transportation of upstream resources and power production are the indirect emission, which is presented in (4). By weighting the four impact factors, a single environmental assessment index is finally obtained in (5).

$$DE_{CO_2} = \sum_{i} F_{CO_2,i} \cdot M_{i}^{Fuel} + F_{CO_2,FGD} \cdot M_{FGD}^{De} \quad (2)$$

$$DE_k = Produce_k \cdot (1-\theta_k), \quad \theta_k \in [0,1], k = 1,2,...,12 \quad (3)$$

$$IE_k = \sum_{j} LCI_{k,j} \cdot M_j, \quad j = \{Coal, NG, Ele, Water, Lim, Amm\} \quad (4)$$

$$IMPCT = \sum_{n} \frac{\sum_{k} LCIA_{k,n} \cdot (IE_k + DE_k)}{F_{n}^{Normalization}}, \quad n = \{ACI, GWP, EUT, HTP\} \quad (5)$$

2.2 Constraints

The steam demand constraints, power demand constraints and variables range constraints of the proposed utility system are given in equation (6)-(8)(Zhao and You, 2019).

$$W_t^E + W_t^{ST} \geq W_t^{Demand}, \forall t \tag{6}$$

$$\sum_i f_{le,i}^{in} - \sum_j f_{le,j}^{out} \geq d_{le}, le \in \{HS, MS, LS\} \tag{7}$$

$$x^{\min} \leq x \leq x^{\max} \tag{8}$$

3. Data-Driven DRJCC Model for the Utility System

3.1 Data-driven uncertainty sets for DRJCC

The chance-constrained approach can be applied to control the system reliability to a predefined level directly, and help to decide the optimal solutions. The distributionally robust chance-constrained approach over Wasserstein ambiguity set, defined in equations (9) and (10), is presented (Ho-Nguyen et al., 2021) to handle this problem. The ambiguity sets $F_N(\theta)$ are sets of probability measures and defined as the θ-radius Wasserstein ball of distributions around the empirical distribution $\hat{\mathbb{Q}}_N$. It ensures that all probability distributions of uncertain constraints satisfy with a given probability threshold within the selected Wasserstein distance from the empirical distribution.

$$d_W(\mathbb{Q}, \mathbb{Q}') := \inf_{\Pi} \left\{ \begin{array}{l} \mathbb{E}_{(\tilde{\mathbf{u}}, \tilde{\mathbf{u}}') \sim \Pi}[\|\tilde{\mathbf{u}} - \tilde{\mathbf{u}}'\|_*], \\ \Pi \text{ has marginal distributions } \mathbb{Q}, \mathbb{Q}' \end{array} \right\} \tag{9}$$

$$F_N(\theta) := \left\{ \mathbb{Q} : d_W\left(\hat{\mathbb{Q}}_N, \mathbb{Q}\right) \leq \theta \right\} \tag{10}$$

However, the uncertainty involves three levels of steam demand constraints, which can be considered as a joint chance constraint. The joint chance constraint with right hand side uncertainty is reformulated as an individual chance constraint by enforcing the worst-case CVaR approximation method (Chen et al., 2022). The set (11) with demand uncertainty is reformulated as (12).

$$S(\mathbf{x}) := \left\{ \mathbf{x} : \mathbf{b}_p^T \tilde{\mathbf{u}}_i + d_p - \mathbf{a}_p^T \mathbf{x} > 0, p \in [P] \right\} \tag{11}$$

$$\chi_{DR}(S) := \left\{ \mathbf{x} \in \chi : \begin{array}{c} \exists t \geq 0, \mathbf{r} \geq \mathbf{0}, \\ \max\left\{ 0, \min\limits_{p \in [P]} \dfrac{\mathbf{b}_p^T \tilde{\mathbf{u}}_i + d_p - \mathbf{a}_p^T \mathbf{x}}{\|\mathbf{b}_p\|_*} \right\} \geq t - r_i,\ i \in [N], \\ \varepsilon t \geq \theta + \dfrac{1}{N} \sum\limits_{i \in [N]} r_i \end{array} \right\} \tag{12}$$

where $\mathbf{a}_p \in \mathbb{R}^K$, $\mathbf{b}_p \in \mathbb{R}^L$, $d_p \in \mathbb{R}$ for all $p \in [P]$, and $\|\bullet\|_*$ is the dual norm (Boyd and Vandenberghe, 2006). θ is used to adjust the size of Wasserstein ball. ε is the desired safety factor between 0 and 1, which requires a group of m inequalities affected by uncertainty jointly satisfied the probability of at least $1-\varepsilon$.

3.2 Multi-objective optimization with DRJCC framework

We use the weighted-sum method to find the Pareto frontier (Deb and Deb, 2014), the operating cost and environment impact objectives are formulated as follows,

$$\min \ \lambda\text{COST}+(1-\lambda)\text{IMPACT} \tag{13}$$

where λ is the weighting coefficient and satisfying $\lambda \in (0,1)$. The solutions of the multi-objective problem will strongly depend on the changes of weighting coefficient.

We present the DRJCC formulation considering uncertain steam demands distributions (Chen et al., 2022; Xie, 2021). All constraints involving demands $\tilde{d}$ are reformulated in the dual form of DRJCC and shown as follows.

$$\varepsilon t \geq \theta + \frac{1}{N}\sum_{i\in[N]} r_i, \tag{14}$$

$$M(1-z_i) \geq t - r_i, i \in [N], \tag{15}$$

$$\sum_{i\in[N]} z_i \leq \lfloor \varepsilon N \rfloor, \tag{16}$$

$$\left(\sum_t a_{p,t}^{in} f_t^{in} + a_{p,t}^{ext} f_t^{ext}) + \sum_j a_{p,j}^{M}(1-y_j)F_j^S + \sum_{lv}(a_{p,lv}^{in} f_{lv}^{in} + a_{p,lv}^{out} f_{lv}^{out}) - d_{p,i}\right) \Big/ \|m_p\|_* + \left(d_{p,i} - q_p\right) \Big/ \|m_p\|_* \geq t - r_i, \ \text{i} \in [\text{N}]_p, p \in [P], \tag{17}$$

$$\left(\sum_t a_{p,t}^{in} f_t^{in} + a_{p,t}^{ext} f_t^{ext}) + \sum_j a_{p,j}^{M}(1-y_j)F_j^S + \sum_{lv}(a_{p,lv}^{in} f_{lv}^{in} + a_{p,lv}^{out} f_{lv}^{out})\right) \Big/ \|m_p\|_* \geq t, \quad p \in [P], \tag{18}$$

$$f_t^{in,\min} \leq f_t^{in} \leq m_t^{in,\max}, \ \forall t \in T, \tag{19}$$

$$f_t^{exh,\min} \leq f_t^{in} - f_t^{ext} \leq f_t^{exh,\max}, \ \forall t \in T \tag{20}$$

$$\mathbf{z} \in \{0,1\}^N, t \geq 0, \mathbf{r} \geq 0. \tag{21}$$

Then, the multi-objective problem with joint demand uncertainties is converted into a deterministic MINLP problem, which can be solved by the existing solvers easily.

4. Case study

The case of the utility system with four steam grades (SS, HS, MS, LS) for an ethylene plant is presented (Zhao et al., 2014). There are three boilers, four extractions turbines, sixteen backpressure turbines with backup motors, two let-down valves in the given utility system. The demand parameters are collected from the process historical database. The data is not uniformly distributed over a specific range of operating conditions, which can be regarded as a multi-modal, interrelated and asymmetric uncertain parameter set. The superstructure of the utility system and uncertain parameters are given in figure 1.

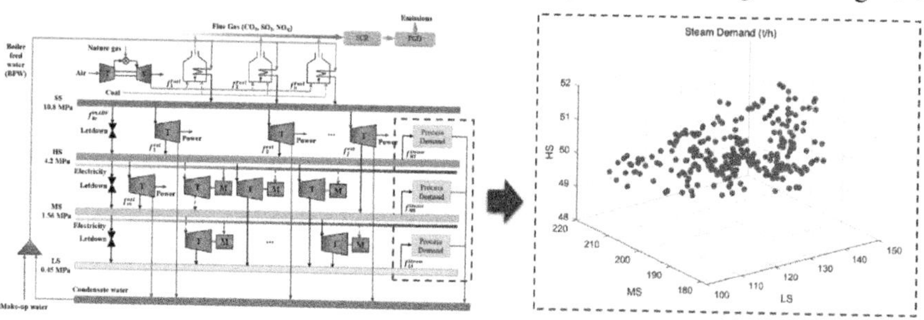

Figure 1. The uncertainty parameters from historical operation data.

Resources consumed by the system and their prices are shown in Table 1. The environmental impact categories and indicators are obtained from the CLCD and openLCA (Ciroth, 2007; Liu et al., 2010).

Table 1 The resources consumed in the system.

	Coal (ton)	Nature gas (ton)	Electricity (kWh)	Water (ton)	Limestone (ton)	Ammonia (ton)
Price (CNY)	800	2700	0.55	0.16	350	3500

We test four joint risk tolerance ε which are 0.05, 0.1, 0.15 and 0.2 and set the Wasserstein radius to $\theta \in \{0.05, 0.1, 0.15, 0.2\}$. All problems are solved with BARON solver. Figure 2 shows the trend of cost and environmental impact with changes of ε and θ.

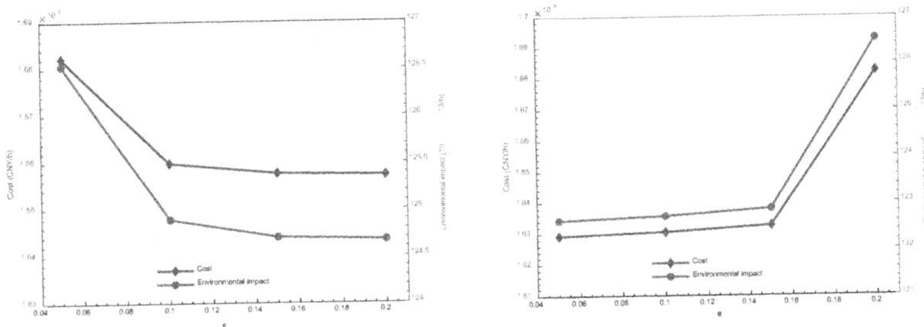

Figure 2. The trend of cost and environmental impact.

The operating cost and environmental impact increase along with the system reliability requirements when the value of risk tolerance decreased from 0.1 to 0.05. On the other hands, extending the Wasserstein radius also increases the economic and environmental burden, which means that the conservation of the solution increases.

Figure 3 shows the Pareto frontier of the multi-objective solutions when ε=0.05 and θ=0.05. Points A and C represent the minimal operating cost environmental impact solutions. Point B is the optimal solution which is supported by the TOPSIS method.

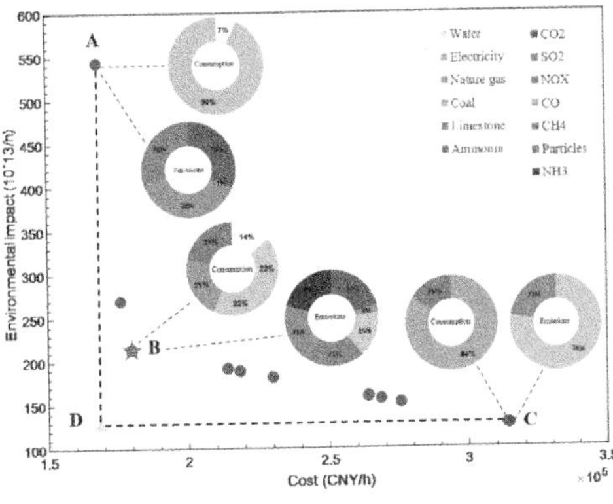

Figure 3. The Pareto frontier of operating cost and environmental impact.

Furthermore, the resources consumption and emissions of different solutions are shown in the pie charts. The total cost ranges from 168,220 to 314,176 CNY per hour, and environmental impact ranges from 126.53 to 543.42 units per hour.

5. Conclusion

An improved DRJCC method with Wasserstein ambiguity set framework was proposed for the utility system optimization based on LCA. Under steam demand uncertainties, it sets economic and environmental objectives and optimizes chance constraints to obtain a trade-off between operating cost, and environmental impact. A case study of utility system in ethylene plant is presented to verify the effectiveness of the proposed method.

Acknowledgment

This work was supported by National Key Research & Development Program - Intergovernmental International Science and Technology Innovation Cooperation Project (2021YFE0112800), National Natural Science Foundation of China (Key Program: 62136003, 22178103, 62173144) and Shanghai AI Lab.

References

S.P. Boyd, L. Vandenberghe, 2006. Convex Optimization. IEEE Transactions on Automatic Control 51, 1859-1859.

Z. Chen, D. Kuhn, W. Wiesemann, 2022. Data-Driven Chance Constrained Programs over Wasserstein Balls. Operations Research.

A. Ciroth, 2007. ICT for environment in life cycle applications openLCA — A new open source software for life cycle assessment. The International Journal of Life Cycle Assessment 12(4), 209.

L. Čuček, J.J. Klemeš, Z. Kravanja, 2012. A Review of Footprint analysis tools for monitoring impacts on sustainability. Journal of Cleaner Production 34, 9-20.

K. Deb, K. Deb, 2014. Multi-objective Optimization, in: Burke, E.K., Kendall, G. (Eds.), Search Methodologies: Introductory Tutorials in Optimization and Decision Support Techniques. Springer US, Boston, MA, pp. 403-449.

Y. Ding, T. Morstyn, M.D. McCulloch, 2022. Distributionally Robust Joint Chance-Constrained Optimization for Networked Microgrids Considering Contingencies and Renewable Uncertainty. IEEE Transactions on Smart Grid 13(3), 2467-2478.

J.B. Guinée, R. Heijungs, G. Huppes, A.d. Koning, L. Oers, A.W. Sleeswijk, U.d. Haes, R.v. Duin, E. Lindeijer, 2001. Life cycle assessment An operational guide to the ISO standards.

Ho-Nguyen, N., Kılınç-Karzan, F., Küçükyavuz, S., Lee, D., 2021. Distributionally robust chance-constrained programs with right-hand side uncertainty under Wasserstein ambiguity. Mathematical Programming.

I. Ioannou, S.C. D'Angelo, Á. Galán-Martín, C. Pozo, J. Pérez-Ramírez, G. Guillén-Gosálbez, 2021. Process modelling and life cycle assessment coupled with experimental work to shape the future sustainable production of chemicals and fuels. Reaction Chemistry & Engineering 6(7), 1179-1194.

X. Liu, H. Wang, J. Chen, Q. He, P. Hou, 2010. Method and basic model for development of Chinese reference life cycle database. Acta Scientiae Circumstantiae 30(10), 2136-2144.

P.S. Varbanov, J. Škorpík, J. Pospíšil, J.J. Klemeš, 2020. Sustainable Utility Systems: Modelling and Optimisation. De Gruyter.

W. Xie, 2021. On distributionally robust chance constrained programs with Wasserstein distance. Mathematical Programming 186(1), 115-155.

W. Xie, S. Ahmed, 2016. On Deterministic Reformulations of Distributionally Robust Joint Chance Constrained Optimization Problems. SIAM Journal on Optimization 28.

H. Zhao, Y. Feng, X. Dong, G. Rong, 2014. Integration Optimization of Production and Utility System for Refinery-wide Planning. IFAC Proceedings Volumes 47(3), 9599-9604.

L. Zhao, F. You, 2019. A data-driven approach for industrial utility systems optimization under uncertainty. Energy 182(SEP.1), 559-569.

X. Zhao, F. You, 2021. Consequential Life Cycle Assessment and Optimization of High-Density Polyethylene Plastic Waste Chemical Recycling. ACS Sustainable Chemistry & Engineering 9(36), 12167-12184.

Antonis Kokossis, Michael C. Georgiadis, Efstratios N. Pistikopoulos (Eds.)
PROCEEDINGS OF THE 33rd European Symposium on Computer Aided Process Engineering (ESCAPE33), June 18-21, 2023, Athens, Greece
 http://dx.doi.org/10.1016/B978-0-443-15274-0.50150-5

Online real-time multi-parameter optimization solution based on parallel EGO algorithm

Xuerong Gu[a], Siyu Yang[a, b*]

[a] *School of Chemistry and Chemical Engineering, South China University of Technology, Guangzhou, 510640, PR China*

[b] *Guangdong Key Laboratory of Green Chemical Products Technology, South China University of Technology, Guangzhou, 510640, PR China*

Email address: cesyyang@scut.edu.cn

Abstract

Chemical process simulation and optimization have the characteristics of high-dimensional and non-linear, which make the simulation higher computational cost or even more difficult to converge. Using surrogate models to replace mechanistic models is an effective way to reduce computational complexity and ensure the accuracy of results. Though the Kriging surrogate model has a stronger nonlinear approximation, it is still difficult to deal with high-dimensional problems. Therefore, this paper investigates the parallel EGO algorithm integrated with the surrogate model, and applies the model to the typical chemical process. Taking eight multi-peak test functions as test cases, it is found that the accuracy and convergence speed of the parallel EGO algorithm is obviously improved. Finally, it is applied to the study case of the two-stage ammonia absorption refrigeration process. The results show that the simulation error is less than 0.01%, and the optimization time is reduced from 9846 s to 3705 s.

Keywords: surrogate model; parallel efficient global optimization algorithm; Multi-parameter optimization; process simulation.

1. Introduction

Process simulation and optimization are the basis for achieving energy saving and emission reduction in the chemical field. Due to the complex engineering design problems with high-dimensional and nonlinear characteristics, the optimization of operating parameters using just traditional process simulation software is often ineffective. To effectively solve this problem, some researchers [1-2] proposed building surrogate models relying on data-driven. The computationally fast surrogate model was used to replace the computationally time-consuming mechanism model, thus greatly reducing the number of computations of time-consuming objectives and constraints. The surrogate models [3-7] commonly used in engineering are polynomial response surface (PRS), radial basis function (RBF), artificial neural network (ANN), support vector regression (SVR), and Kriging surrogate model. Rosma et al [8-9] established PRS and RBF surrogate models for turbines and complex fitness functions and used particle swarm optimization algorithms to dynamically update the models. Yang et al [10] used the ANN surrogate model to optimize the process operation parameters of the crude oil distillation unit to maximize the oil production rate. Singh et al [11] used three surrogate models (SVR, RBF, Kriging) combined with an efficient multi-objective optimization algorithm to identify the Pareto frontier of cyclone separator design with the minimum number of simulations. Compared with the traditional optimization methods, the above-mentioned surrogate-based optimization algorithms can reduce the time-consuming simulation computation in the optimization process. However, this optimization algorithm is too dependent on the sample sets and often requires a large number of samples to be fitted or trained to obtain more reliable results.

To solve the shortcomings of the above optimization algorithms, Jones et al [12] proposed the EGO (Efficient Global Optimization) algorithm. This approach uses fewer sample points to train the Kriging surrogate model and continuously selects the point with the largest value of the EI (Expected Improvement) function as the update point during each iteration. The Kriging surrogate model is continuously updated to find the global optimal solution. This approach can be used to solve the black-box problem but also to speed up the convergence and obtain higher accuracy and computing efficiency.

Based on the superiority of the EGO algorithm, the parallel EGO (Pseudo Efficient Global Optimization) algorithm is proposed based on it in this paper. The algorithm uses parallel computing techniques to select multiple update points for computation during each iteration, thus speeding up the optimization speed of the EGO algorithm. It was applied to 8 multi-peak test functions and the two-stage ammonia absorption refrigeration operation optimization case. The results show that the approach can optimize all variables online in real-time with high accuracy and efficiency, which has a greater value for engineering applications.

2. Parallel EGO algorithm

2.1 Kriging surrogate model

The Kriging model [13] consisting of m samples has a total of 2m+2 parameters, which can be obtained by the maximum likelihood estimation [14]. Finally, the predicted value and variance of any point x can be obtained:

$$\begin{cases} \hat{y} = \hat{\mu} + r^T R^{-1}(y - \hat{\mu}) \\ s^2(x) = \hat{\sigma}^2[1 - r^T R^{-1} r + \dfrac{(1 - R^{-1} r)^2}{R^{-1}}] \end{cases} \tag{1}$$

2.2 EI criterion

For the unknown point, it can be regarded as a normally distributed random variable with mean $\hat{y}(x)$ and standard deviation $s(x)$, it can be expressed as $Y(x) \sim N(\hat{y}(x), s(x))$. If the minimum objective function value in the current sample point is $f_{\min}$, then the improvement of the unknown point to the current optimal solution can also be regarded as a random variable:

$$I(x) = \max(f_{\min} - Y(x), 0) \tag{2}$$

Solving the mathematical expectation of $I(x)$ can get the expression of the EI function as:

$$EI(x) = (f_{\min} - \hat{y}(x))\Phi\left(\frac{f_{\min} - \hat{y}(x)}{s}\right) + s(x)\phi\left(\frac{f_{\min} - \hat{y}(x)}{s}\right) \tag{3}$$

where ϕ and Φ represent the probability density function and the cumulative distribution function of the standard normal distribution. It can be found from equation (3) that the EGO algorithm mainly searches for the minimum value $\hat{y}(x)$ or maximum value $s(x)$ as an update point by maximizing the EI function in each iteration, and finds the optimal solution after several iterations.

2.3 PEI criterion

The main role of the influence function is to approximate the changes brought to the EI function when iteratively updating the Kriging surrogate model, and its equation is as follows:

$$IF(x, x^{(\mu)}) = 1 - Corr[\varepsilon(x), \varepsilon(x^{(\mu)})] \tag{4}$$

where x is an arbitrary unknown point and $x^{(\mu)}$ is an update point. The PEI (Pseudo Expected Improvement) function is constructed by multiplying the initial EI function by the impact function of the update point, and its equation is as follows:

$$\begin{aligned} PEI(x,q-1) &= EI(x)\cdot IF\left(x,x^{(m+1)}\right)\cdot IF\left(x,x^{(m+2)}\right)\cdots IF\left(x,x^{(m+q-1)}\right) \\ &= EI(x)\cdot\prod_{i=1}^{q-1}\left[1-\mathrm{Corr}\left[\varepsilon(x),\varepsilon\left(x^{(m+i)}\right)\right]\right] \\ &= EI(x)\cdot\prod_{i=1}^{q-1}\left[1-\exp\left(-\sum_{k=1}^{d}\theta_k\left|x_k-x_k^{(m+i)}\right|^{p_k}\right)\right] \end{aligned} \tag{5}$$

From the above derivation, it is clear that the parallel EGO algorithm can achieve parallel computation by maximizing the PEI function to get q update points in each iteration. This provides a new breakthrough for solving time-consuming chemical simulation optimization problems and improves optimization efficiency.

3. Numerical experiments

The proposed parallel EGO algorithm is compared with the EGO algorithm. The number of iterations required to find the global optimal solution for eight representative classical multi-peaked test functions is shown in Figure 1. Each data point in the figure was obtained by averaging each point by testing 10 times using 10 different initial designs.

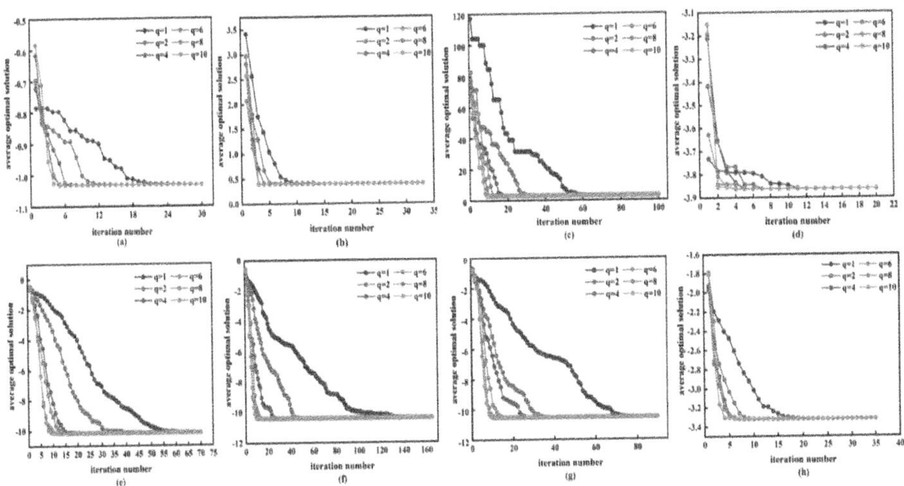

Fig.1 Iteration history curves of different optimization algorithms on test functions. (a) Six-hump. (b) Branin. (c) GoldPrice. (d) Hartman3. (e) Shekekl5. (f) Shekekl7. (g) Shekekl10. (h) Hartman6.

From Figure 1, it can be found that the proposed parallel EGO algorithm performs significantly better than the EGO algorithm no matter how many update points are selected in each iteration. Meanwhile, the convergence speed is significantly faster with the number of update points increasing.

It can be found from the time variation curves in Figure 2 that with guaranteed accuracy, the proposed parallel EGO algorithm is higher efficiency and stronger adaptability when more points are added in each circulation.

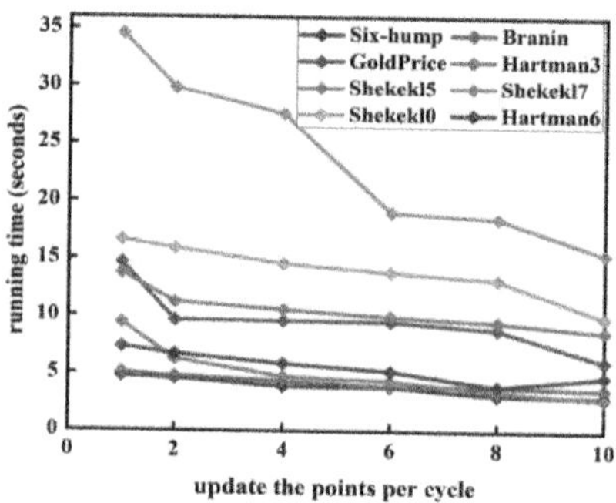

Fig.2 Running time of different multi-peak test functions

4. Multi-parameter optimization of two-stage ammonia absorption refrigeration

4.1 process description

The two-stage ammonia absorption refrigeration unit is an important part of the low-temperature methanol washing waste heat and cooling power co-generation system. Based on this, Aspen Plus is used to model and simulate it. The flowchart of its is shown in Figure 3.

Based on the analysis of key parameters by Liu et al [15], 15 decision variables to be optimized were determined. And the coefficient of performance COP (the ratio of evaporator load to the sum of flasher and distillation tower load) was chosen as the optimization objective for optimization.

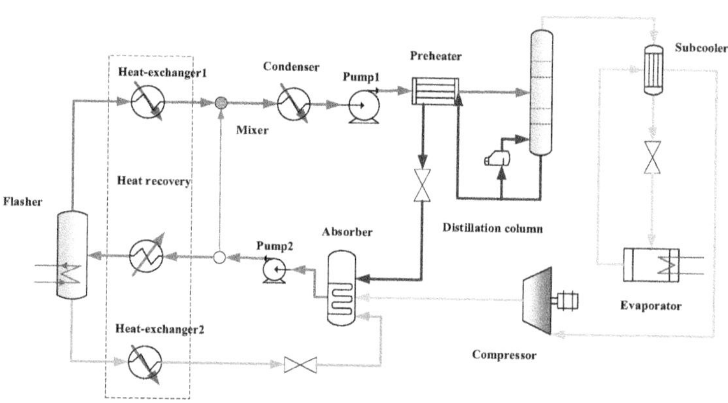

Fig.3 Schematic diagram of the two-stage ammonia absorption refrigeration process

4.2 Results and Discussions

The predicted values using the Kriging surrogate model are compared with the actual values as shown in Figure 4. The mean square error of the model is 2.35×10^{-3}, indicating that the constructed Kriging surrogate model predicts very accurately and can replace the original two-ammonia absorption refrigeration simulation model. Meanwhile, the time required for the numerical calculation of the original simulation model is 21013s, while Kriging only requires 0.383s, so the calculation time is significantly reduced.

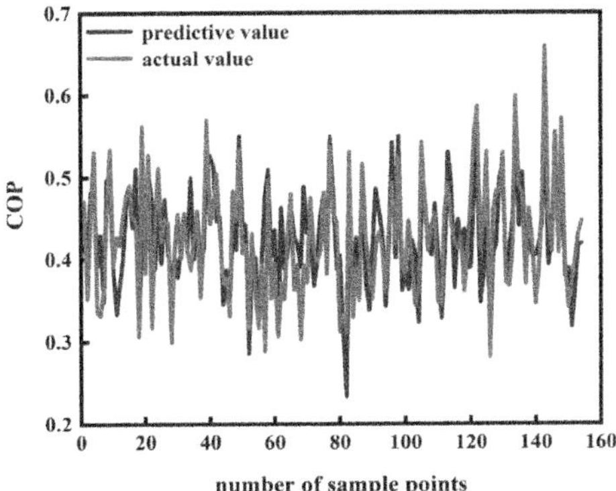

Fig.4 Comparison between predicted and actual values of Kriging model

The iteration curves of the standard EGO algorithm and the parallel EGO algorithm are shown in Figure 5. It can be found from the figure that the standard EGO algorithm requires 330 iterations to converge. In contrast, for the parallel EGO algorithm, the number of iterations gradually decreases to 29 as the number of points updated per circulation increases. The optimization convergence speed is improved by 91.2%. The results show that the parallel EGO algorithm converges faster than the standard EGO algorithm; the optimal solution obtained by the proposed parallel EGO algorithm is also better than the standard EGO algorithm when 600 iterations are ended.

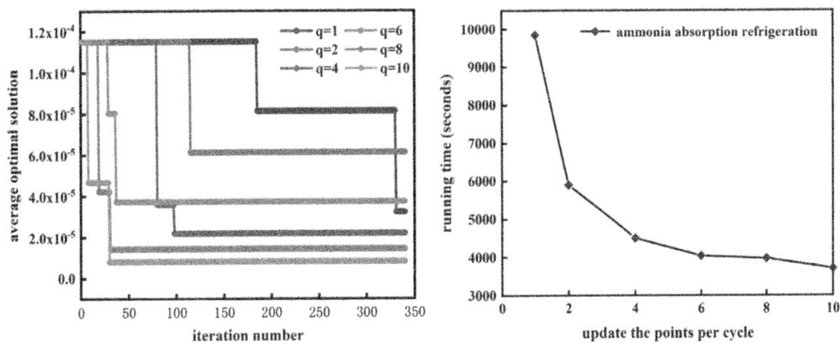

Fig.5 Optimization iteration curves Fig.6 Running time

The relationship that the computation time required by each algorithm relative to the number of update points per iteration is shown in Figure 6. It can be found that under the condition that the maximum number of iterations is set to 600, the time required by the standard EGO algorithm is 9846s, and the time required by the parallel EGO algorithm is 5898-3705s. It is obvious that the parallel EGO algorithm requires less computation time than the standard EGO algorithm, and the operation efficiency is improved by 62.37%.

5. Conclusions

This paper proposed an online real-time parallel EGO optimization algorithm based on Kriging surrogate model for solving computationally intensive and time-consuming optimization problems of chemical process operations. The proposed parallel EGO algorithm uses the PEI criterion to select multiple update points in each iteration for parallel computation. The algorithm is tested on eight test functions of dimensions 2 to 6. The experimental results show that the parallel EGO algorithm proposed in this paper can significantly improve the convergence speed of the standard EGO algorithm; it has higher efficiency and robustness than the standard EGO algorithm in all eight test problems. In addition, when the proposed parallel EGO algorithm is applied to the parameter optimization of two-stage ammonia absorption refrigeration, under the condition of ensuring higher accuracy, using relatively less training data to approach the multi-parameter optimal solution, the obtained solution is also significantly better than the standard EGO algorithm. The convergence speed is improved by 91.2% compared with the standard EGO algorithm.

References

[1] M. Tang, Y. Liu, L. J. Durlofsky, 2020, A deep-learning-based surrogate model for data assimilation in dynamic subsurface flow problems[J]. Journal of Computational Physics, 413: 109-456.

[2] R. M. M. Slot, J. D. Sørensen, B. Sudret, et al, 2020, Surrogate model uncertainty in wind turbine reliability assessment[J]. Renewable Energy, 151: 1150-1162.

[3] S. Wang, J. Xiao, J. Wang, et al, 2018, Application of response surface method and multi-objective genetic algorithm to configuration optimization of Shell-and-tube heat exchanger with fold helical baffles[J]. Applied Thermal Engineering, 129: 512-520.

[4] Y. Deng, X. Zhou, J. Shen, et al, 2021, New methods based on back propagation (BP) and radial basis function (RBF) artificial neural networks (ANNs) for predicting the occurrence of haloketones in tap water[J]. Science of The Total Environment, 772: 145534.

[5] J. Eason, S. Cremaschi, 2014, Adaptive sequential sampling for surrogate model generation with artificial neural networks[J]. Computers & Chemical Engineering, 68: 220-232.

[6] M. W. Ahmad, M. Mourshed, Y. Rezgui, 2018, Tree-based ensemble methods for predicting PV power generation and their comparison with support vector regression[J]. Energy, 164: 465-474.

[7] D. Zhan, J. Qian, Y. Cheng, 2017, Pseudo expected improvement criterion for parallel EGO algorithm[J]. Journal of Global Optimization, 68(3): 641-662.

[8] G. E. González, R. Y. Qassim, P. C. C. Rosman, et al, 2016, Optimization of hydrokinetic turbine array layouts via surrogate modeling [J]. Renewable Energy, 93: 45-57.

[9] H. Yu, Y. Tan, C. Sun, et al, 2016, An adaptive model selection strategy for surrogate-assisted particle swarm optimization algorithm[C]//2016 IEEE symposium series on computational intelligence (SSCI). Los Alamitos: IEEE Computer Society Press.

[10] C. K. Liau, C. K. Yang, M. T. Tsai, 2004, Expert system of a crude oil distillation unit for process optimization using neural networks[J]. Expert Systems with Applications, 26(2): 247-255.

[11] P. Singh, I. Couckuyt, K. Elsayed, et al, 2016, Shape optimization of a cyclone separator using multi-objective surrogate-based optimization[J]. Applied Mathematical Modelling, 40(5-6): 4248-4259.

[12] D. R. Jones, M. Schonlau, W. J. Welch, 1998, Efficient global optimization of expensive black-box functions[J]. Journal of Global Optimization, 13(4): 455-492.

[13] D. G. Krige, 1951, A statistical approach to some basic mine valuation problems on the Witwatersrand[J]. Journal of the Southern African Institute of Mining and Metallurgy, 52(6): 119-139.

[14] S. N. Lophaven, H. B. Nielsen, J. Søndergaard, 2002, DACE: a Matlab kriging toolbox[M]. IMM, Informatics and Mathematical Modelling, The Technical University of Denmark.

[15] S. Liu, Y. Qian, D. Li, et al, 2022, Multi-Scenario scheduling optimization for a novel Double-Stage ammonia absorption refrigeration system incorporating an organic Rankine cycle[J]. Energy Conversion and Management, 270: 116170.

Antonis Kokossis, Michael C. Georgiadis, Efstratios N. Pistikopoulos (Eds.)
PROCEEDINGS OF THE 33rd European Symposium on Computer Aided Process Engineering (ESCAPE33), June 18-21, 2023, Athens, Greece
 http://dx.doi.org/10.1016/B978-0-443-15274-0.50151-7

Optimal design of Water Treatment Networks: effluent and piping disaggregated modelling

Francisco J. G. Patrocínio,[a,b] Hugo M. D. Carabineiro,[c] Henrique A. Matos[b] and Nuno M. C. Oliveira,[a]*

[a]*CIEPQPF, Dep. Chemical Engineering, University of Coimbra, 3030-790 Coimbra, Portugal*
[b]*CERENA, Dep. Chemical Engineering, IST, University of Lisbon, 1049-001 Lisbon, Portugal*
[c]*Petrogal, S.A., 7520-952 Sines, Portugal*
**nuno@eq.uc.pt*

Abstract

Water availability is of crucial importance to industry; therefore, optimal management of this resource has garnered increased importance. This work focuses on industrial water treatment networks (WTN) optimal design models, and can be divided in two sections. The first section addresses the modification of typical mathematical routines to implement efficient and rigorous effluent formulations, using an integer disaggregated formulation. The second section considers network piping expenditures, employing a two-step methodology to account for the discontinuous classes of commercialized piping.

Keywords: WTN optimization, MINLP, effluent modelling, piping, disaggregated model

1. Introduction

WTN models have been the subject of various publications in the past (e.g., Galan and Grossmann, 1998), although the effluent description has been somewhat simplified. Field observation of WTNs brings various additional features: a network can have multiple discharges; these can account for a significant fraction of the network OPEX; and they can be tariffed according to discontinuous concentration classes. Patrocínio *et al.* (2022) considers more detailed modelling of effluents employing indicator variables. This formulation has the downside of introducing new non-linearities in the mathematical model. In this work, a disaggregated formulation is suggested to rigorously model the effluent section of WTNs.

The second section of this work addresses the introduction of piping related expenses in the optimal grassroots design of WTNs, considering the nature of the diameter classes in which pipes are commercialized. Some authors (Alnouri *et al.*, 2014; Caballero and Ravagnani, 2019) already consider this characteristic. In this work, a two-step disaggregated methodology is proposed to model these equipment expenses, according to discontinuous pipe diameter classes.

2. WTN disaggregated model

The water treatment network superstructure (Figure 1) is composed by water flow producer nodes (set i) and consumer nodes (set j). Subsets of wastewater sources ($ww \subset i$), treatment units ($tu \subset i \wedge tu \subset j$) and multiple effluents ($ef \subset j$) are defined. Superstructure connections between the same treatment unit are forbidden (with the set $par_{i,j}$). Eq. (1) models that outlet flow from nodes i (F_i^{nd}) is the sum of flows from i to j

($F_{i,j}$) and eq. (2) models the inlet flow in nodes j. Eq. (3) establishes that that the flow from wastewater sources is defined by the parameter fww_i.

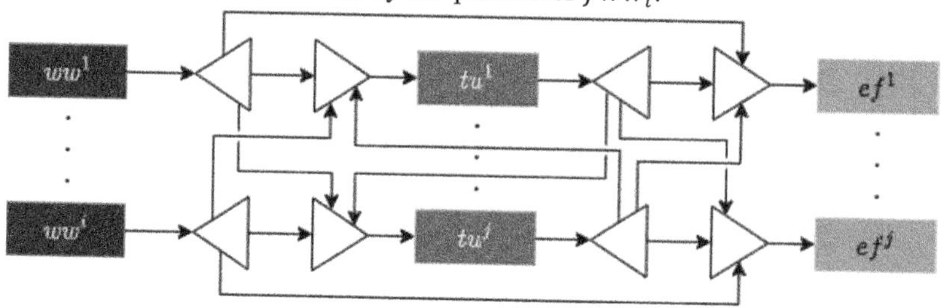

Figure 1 WTN superstructure.

The problem is further specified by a set of network contaminants (set co), with eq. (4) as the mass balance of contaminants in the inlet of consumer nodes ($C^{in}_{j,co}$, $C^{out}_{i,co}$ are the inlet and outlet concentration of nodes j and i). Eq. (4) contains the only non-linearity of the model, as two continuous variables are multiplied. From eq. (5), the outlet concentration of wastewater sources is fixed by the parameter $c^{outWW}_{i,co}$ and eq. (6) defines the removal of contaminants in treatment units, through parameter $rr_{j,co}$. Eq. (7) establishes an upper limit for the inlet concentration of j ($c^{lim}_{j,co}$). Note that each treatment and effluent node may contain inlet concentration limits. Effluents can be further classified into regular or special effluents ($se \subseteq j$). The costs of the former are simply a tariff ($cost^{RE}_j$) multiplied by the effluent flow, and special effluent costs are related to discontinuous contaminants concentration classes (set tcl) of specific contaminants ($tar \subseteq co$), with the highest activated class as the one tariffed. Here a disaggregated formulation for modelling the discharge of special effluents is used, avoiding the non-linearities resulting from the multiplication of integer and continuous variables. The inlet flow and concentration in the special effluent nodes are partitioned ($F^{nd-tcl}_{j,tcl}$, $C^{cl}_{j,co,tcl}$), according with the tariff classes (eqs. (8)-(9)) and only one partition is activated (eq. (10)), introducing the binary variable $Y^{tcl}_{j,tcl}$. Eqs. (11)-(12) guarantee that the concentration for the activated class is lower or equal to the upper bound concentration for this class ($ub^{cl}_{j,co,tcl}$). Eqs. (13)-(14) assign the inlet flow of the node to the activated class, using an upper bound flow for node j (f^{nd-up}_j). The lowest possible class is activated, as higher cost tariff classes are penalized by the objective function (eq. (15)). This function includes the operational costs of the treatment units ($cost^{treat}_j$) multiplied by their inlet flow, and the costs of regular and special effluents.

$$F^{nd}_i = \sum_{j|par_{i,j}} F_{i,j} \ , \forall i \tag{1}$$

$$F^{nd}_j = \sum_{i|par_{i,j}} F_{i,j} \ , \forall j \tag{2}$$

$$F^{nd}_i = fww_i \ , i \in ww \tag{3}$$

$$\sum_{i|par_{i,j}} F_{i,j} C^{out}_{i,co} = F^{nd}_j C^{in}_{j,co}, \forall j, \forall co \tag{4}$$

$$C^{out}_{i,co} = c^{outWW}_{i,co} \ , i \in ww, \forall co \tag{5}$$

$$C^{in}_{j,co} rr_{j,co} = C^{out}_{j,co}, j \in tu, \forall co \tag{6}$$

$$C_{j,co}^{in} \leq c_{j,co}^{lim}, \forall j, \forall co \tag{7}$$

$$C_{j,co}^{in} = \sum_{tcl} C_{j,co,tcl}^{cl}, j \in se, \forall co \tag{8}$$

$$F_j^{nd} = \sum_{tcl} F_{j,tcl}^{nd-tcl}, j \in se \tag{9}$$

$$\sum_{tcl} Y_{j,tcl}^{cl} = 1, j \in se \tag{10}$$

$$C_{j,co,tcl}^{cl} \geq 0, j \in se, co \in tar, \forall tcl \tag{11}$$

$$C_{j,co,tcl}^{cl} \leq ub_{j,co,tcl}^{cl} Y_{j,tcl}^{tcl}, j \in se, co \in tar, \forall tcl \tag{12}$$

$$F_{j,tcl}^{nd-tcl} \geq 0, j \in se, \forall tcl \tag{13}$$

$$F_{j,tcl}^{nd-tcl} \leq f_j^{nd-up} Y_{j,tcl}^{cl}, j \in se, \forall tcl \tag{14}$$

$$Cost^{Network} = \sum_{j \in tu^j} cost_j^{treat} F_j^{nd} + \sum_{\substack{j \notin tu^j \\ j \notin se^j}} cost_j^{RE} F_j^{nd} + \sum_{\substack{j \in se^j \\ tcl}} cost_{j,tcl}^{SE} F_{j,tcl}^{nd-tcl} \tag{15}$$

3. Piping investment and operational expenditures

The grassroots design of WTNs must account for piping expenses, with a cost that depends on the flows and distances between the network nodes. Furthermore, pipes are manufactured according to discontinuous diameter classes. This work proposes a two-stage procedure to incorporate these discontinuous classes in the WTN design model, performing a correspondence between the flows and the pipe classes costs and incorporating that correspondence in a mathematical optimization model, using a disaggregated strategy.

3.1. Pre-processing stage

The pre-processing stage consists in an *a priori* correspondence between flows and the pipes cost. Firstly, the pipe optimal diameter (d_p^{opt} - Peters *et al.*, 2003) is computed (eq. (16)) using a set of p wastewater flows values (F_p). Secondly, the optimal diameter is corresponded with the commercial pipe diameter, with a simple mixed integer linear model – eqs. (17)-(19), where d_p^{opt}t is assigned to a pipe class (set *pcl*), using a binary variable $Z_{p,pcl}^{Pipe-Class}$. Eq. (17) forces the assigned class diameter to be greater-or-equal than the optimal diameter, eq. (18) ensures that one pipe class is corresponded to each flow p, and the objective function (eq. (19)) imposes that the difference between the assigned class diameter and optimal diameter is as small as possible.

$$d_p^{opt} = \left(\frac{F_p}{3600}\right)^{0.45} 0.593, \forall p \tag{16}$$

$$Z_{p,pcl}^{Pipe-Class}(d_{pcl}^{class} - d_p^{opt}) \geq 0, \forall p, \forall pcl \tag{17}$$

$$\sum_{pcl} Z_{p,pcl}^{Pipe-Class} = 1, \forall p \tag{18}$$

$$\min \sum_{p,pcl} Z_{p,pcl}^{Pipe-Class}(d_{pcl}^{class} - d_p^{opt}) \tag{19}$$

The correspondence obtained will be similar to Figure 2, relating individual sch. 40 pipe costs, (per meter length) for several inner diameter classes.

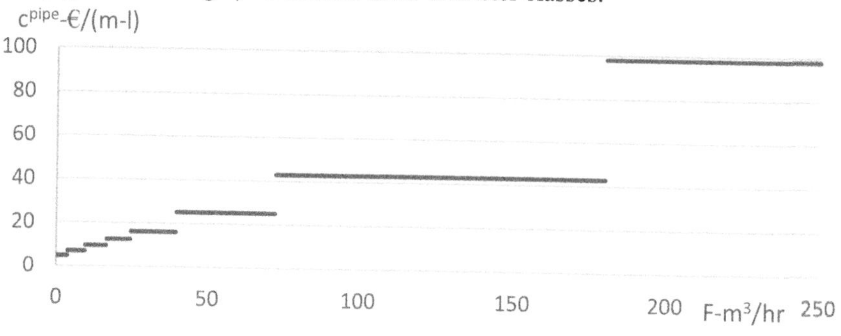

Figure 2 Individual pipe costs as a function of flow.

3.2. WTN disaggregated piping costs model

The proposed disaggregated technique is similar to the one presented for special effluents (section 2). The flow between nodes *i,j* is partitioned ($F_{i,j,pcl}^{pcl}$) regarding the number of piping classes (eq. (20)). A new binary variable ($Y_{i,j,pcl}^{pipe}$) is introduced to assign $F_{i,j,pcl}^{pcl}$ to a class *pcl* (eqs. (21)-(22)), using the lower and upper bounds for each class (f_{pcl}^{lb}, f_{pcl}^{ub}). Eq. (23) assures that, at most, one class is activated. The inequality operator is important as it allows that a null flow does not activate a piping class. Yearly costs of installing a pipe in connection *i,j* ($Cost_{i,j}^{Pipe-install}$) are modelled by eq. (24), considering an annualization factor (af), labor/material cost factors (bm^{Pipe}), distance between nodes ($d_{i,j}^{nd}$) and cost of activated pipe class ($cost_{pcl}$). Eq. (25) expresses the operational costs of pipe between nodes *i,j* ($Cost_{i,j}^{Pipe-oper}$), where a maintenance factor (of^{Pipe}) is multiplied by $Cost_{i,j}^{Pipe-install}$.

$$F_{i,j} = \sum_{pcl} F_{i,j,pcl}^{pcl} \ , \forall i, \forall j \tag{20}$$

$$F_{i,j,pcl}^{pcl} \geq f_{pcl}^{lb} Y_{i,j,pcl}^{pipe} \ , \forall i, \forall j, \forall pcl \tag{21}$$

$$F_{i,j,pcl}^{pcl} \leq f_{pcl}^{ub} Y_{i,j,pcl}^{pipe} \ , \forall i, \forall j, \forall pcl \tag{22}$$

$$\sum_{pcl} Y_{i,j,pcl}^{pipe} \leq 1 \ , \forall i, \forall j \tag{23}$$

$$Cost_{i,j}^{Pipe-install} = af.bm^{Pipe}.d_{i,j}^{nd} \sum_{pcl} Y_{i,j,pcl}^{pipe} \cdot cost_{pcl} \ , \forall i, \forall j \tag{24}$$

$$Cost_{i,j}^{Pipe-oper} = of^{Pipe} Cost_{i,j}^{Pipe-install} \ , \forall i, \forall j \tag{25}$$

$$TCost^{Yearly} = H.Cost^{Network} + \sum_{i,j} \left(Cost_{i,j}^{Pipe-install} + Cost_{i,j}^{Pipe-oper} \right) \tag{26}$$

The original objective function is replaced by eq. (26), and total network cost is computed ($TCost^{Yearly}$) considering the yearly operation of the wastewater network (multiplying the original WTN model objective function by the number of operational hours, in a year – *H*), and adding the piping yearly installation and operational expenditures.

4. Application Examples

The results presented in this section were obtained using an AMD Ryzen 7 4800H computer with 16 GB of RAM, and GAMS 40.3.0.

4.1. Large industrial example of Galan and Grossmann (1998)

The last example of Galan and Grossmann (1998) was used to create the first application example of these techniques. The original problem specification was extended to account for different treatment unit operational expenditures (Table 1) and 2 effluent nodes with different limiting concentrations (Table 2). Effluent 1 is a special effluent with different concentration tariff classes (Table 3). Effluent 2 has a discharge cost of 0.1 €/t.

Table 1 Treatment units operating cost.

Unit	*X1*	*X2*	*X3*	*X4*	*X5*
Cost (€/t)	0.1	0.18	0.25	0.25	0.8

Using the WTN disaggregated model, the optimal network has an hourly operation expenditure of 106.97 €/hr, and the special effluent EF1 represents 53.8% of the total network cost, with 128.1 t/hr and concentrations within Class 1. Effluent EF2 has a flow of 12.9 t/hr, and corresponds to 1.2% of the total network cost. Here, the *ANTIGONE* solver was used, with a time limit of 1 hour, resulting in an optimality gap of 20.8%.
To consider the pipe investment together with the operational expenditures, using the procedure of section 3, distances between nodes were randomly generated within [10;300] meters, for correspondences (ww^{nd};tu^{nd}) and (ww^{nd};ef^{nd}), and [5;50] meters for correspondences (tu^{nd};$tu^{nd'}$) and (tu^{nd};ef^{nd}). Piping classes were constructed based on Figure 2, but aggregating several classes and using median cost values as represented in Table 4. af was assumed as 0.207, bm^{Pipe} as 17 (per recommendation of Woods, 2007), of^{Pipe} as 0.2 (Peters *et al.*, 2003) and H as 8000 operating hours.

Table 2 Effluents concentration limits.

	C1 (ppm)	*C2 (ppm)*	*C3 (ppm)*	*C4 (ppm)*	*C5 (ppm)*	*C6 (ppm)*
ef1	100	100	100	100	100	100
ef2	0	0	200	500	200	0

Table 3 *ef1* tariff classes.

	Cost (€/t)	*C2 (ppm)*	*C3 (ppm)*	*C4 (ppm)*
cl1	0.44	[0;50[	[0;30[	[0;30[
cl2	0.54	[50;85[	[30;55[	[30;55[
cl3	0.72	[85;100]	[55;100]	[55;100]

When computed *a posteriori* for the solution of the WTN disaggregated model, piping investment and operational expenditures correspond to an annual network cost of 975.1 *k*€. Utilising the WTN disaggregated model with piping costs included, the optimal solution found improves 2.75%, with an annual network cost of 948.3 *k*€. The local solver *SBB* was used in this case, with an imposed optimality gap of 0.1%, requiring 11.15 hours of wall clock time for this solution.

Table 4 Pipe classes costs and flow intervals.

	pcl1	*pcl2*	*pcl3*	*pcl4*
Cost (€/m-l)	6.98	12.51	16.13	43.02
F (t/hr)	[0;9.4]	]9.4;24.4]	]24.4;39.5]	]39.5;141]

$d_{i,j}^{nd}$ is a crucial parameter regarding the savings reported by the WTN disaggregated piping costs model (in comparison to the computation of those expenditures *a posteriori* with the WTN disaggregated model), and the computational solving time. When the originally generated distance is decreased by 20% (0.8 $d_{i,j}^{nd}$), the savings and wall clock computation time of the WTN disaggregated piping costs model drop to 1.96% and 3.03 hours. In a 30% decrease (0.7 $d_{i,j}^{nd}$), the drop is to 1.5% and 1.43 hours.

4.2. Sines Refinery wastewater treatment network

The proposed methodologies were applied to an industrial scenario, using data from the Sines Refinery wastewater network, in Portugal, composed by 11 wastewater sources, 5 treatment units, two effluent nodes and 6 contaminants. One of the effluent lines is a special effluent node, with 5 tariff classes, regarding 5 of the contaminants, and the costs associated with that node account for 42% of the WTN expenditures. Applying the optimization model of the first section achieves a 33.6% reduction in the total wastewater network costs. The solver utilized was *ANTIGONE*, with a wall clock time limit of 1 hour, resulting in an optimality gap of 14.2%.

To account for piping expenditures, $d_{i,j}^{nd}$ was obtained with the refinery plot plan and, using a similar procedure to the previous example, 5 pipe classes were considered, as well as the parameters regarding af, bm^{Pipe}, of^{Pipe} and H. When these expenditures are considered, posteriorly and independently of the optimization routine, WTN savings drop to 7.9%. Optimizing the network using the WTN disaggregated piping costs model allows a further increase of WTN savings to 10.4% (with local solver SBB, and an imposed optimality gap of 0.1%, expending 3.04 hours of wall clock time).

Conclusion

This work demonstrates that WTN effluents tariffs cannot be disregarded, in WTN optimization models, focusing in the relevance of discontinuous effluent discharges, as demonstrated by the example of Sines Refinery. Furthermore, piping expenditures are modelled in a more realistic manner, avoiding introducing further non-linearities in the routines. The proposed methodology has the downside of demanding a moderate degree of computational effort, which should be addressed in future work.

Acknowledgments: The authors gratefully acknowledge the financial support provided by Galp Energia and Fundação para a Ciência e Tecnologia, through grant COVID/BDE/152674/2022, and data from INTERFLUIDOS, Lda.

References

Alnouri S. *et al.* 2014. "Water Integration in Industrial Zones: A Spatial Representation with Direct Recycle Applications." Clean Technologies and Environmental Policy 2014 16:8 16 (8): 1637–59.

Caballero J.A. and Ravagnani M.A.S.S. 2019. "Water Distribution Networks Optimization Considering Unknown Flow Directions and Pipe Diameters." Computers & Chemical Engineering 127 (August): 41–48.

Galan B. and Grossmann I.E. 1998. "Optimal Design of Distributed Wastewater Treatment Networks." Industrial and Engineering Chemistry Research 37 (10): 4036-48.

Patrocínio F. *et al.* 2022. "Water Network Optimisation in Chemical Complexes: A Refinery Case Study" Computer Aided Chemical Engineering 51 (January): 817–22.

Peters M. *et al.* 2003. "9. Optimum Design and Design Strategy." In Plant Design and Economics for Chemical Engineers, 3rd ed., 401–6. McGraw-Hill Education.

Woods, D.R. 2007. "Appendix D: Capital Cost Guidelines." In Rules of Thumb in Engineering Practice, 376–436. Weinheim: Wiley-VCH.

Antonis Kokossis, Michael C. Georgiadis, Efstratios N. Pistikopoulos (Eds.)
PROCEEDINGS OF THE 33rd European Symposium on Computer Aided Process Engineering (ESCAPE33), June 18-21, 2023, Athens, Greece
 http://dx.doi.org/10.1016/B978-0-443-15274-0.50152-9

Evaluation of the potential of a deep eutectic solvent for liquid-liquid extraction of furfural using optimization-based process design

Kai Fabian Kruber[a,b], Mariann Kroll[b], Christoph Held[b], Mirko Skiborowski[a,b]

[a]*Hamburg University of Technology, Institute of Process Systems Engineering, Am Schwarzenberg-Campus 4, 21073 Hamburg, Germany*
[b]*TU Dortmund University, Laboratory of Fluid Separations, Emil-Figge-Straße 70, 44227 Dortmund, Germany*
mirko.skiborowski@tuhh.de

Abstract

Reducing our environmental impact is the greatest challenge of our time and affects all areas of the chemical, biochemical, and pharmaceutical industries. The use of sustainable and biodegradable solvents in downstream processes, such as deep eutectic solvents (DESs), is one essential step in this direction. While many studies have focused on the investigation of- the thermo-physical properties of DESs, relatively little attention has been paid to solvent regeneration in the context of a closed-loop process, assuming a simple recovery due to the negligible vapor pressure of the DES. In order to evaluate the suitability of this assumption, the current work performs an optimization-based process design for the recovery of furfural from a diluted fermentation broth by liquid-liquid extraction, comparing a previously screened DES with toluene as an established organic solvent. The comparison reveals important limitations of the DES due to considerable water content in the raffinate and the high boiling point resulting in a more complex and energy-demanding process.

Keywords: process optimization, liquid-liquid extraction, distillation, solvent selection, green solvent.

1. Introduction

While deep eutectic solvents (DESs) were first proposed as an alternative to ionic liquids less than 20 years ago (Abbott et al., 2003), with 200 publications during the first ten years (Smith et al., 2014), publications have rapidly increased to more than 7000 in the last ten years, primarily addressing synthesis and solvent properties. DESs can be considered as molten salts with a eutectic point near or below room temperature due to hydrogen bonding and have been promoted with excellent application potential in the chemical, biochemical, and pharmaceutical industries (Hansen et al., 2021), showing low vapor pressure, production cost, and toxicity, as well as good biodegradability. These characteristics deem DESs environmentally friendly and cost-effective competitors to established organic solvents and ionic liquids. Most studies on fluid separation process applications focus on the investigation of thermodynamic properties, such as vapor pressures, densities, boiling and decomposition temperatures, or partition coefficients and selectivities (Salleh et al., 2017). On the other side, studies on the evaluation of closed-loop process performance, considering solvent recovery, of DES-based processes are still scarce. Larriba et al. (2018) studied the dearomatization of gasoline by liquid-liquid extraction with six different DESs based on choline chloride. Based on experimental investigations paired with simulation studies, they selected a promising sustainable alternative to commonly applied organic solvents. However, the study does not consider the investigation of a complete process flowsheet with solvent recovery and a closed recycle. Han et al. (2022) examined such a closed-loop flowsheet in Aspen Plus for the dehydration of acetonitrile with glycolic acid:choline chloride(3:1) by extractive distillation. They found a substantial cost reduction compared to the benchmark process considering ethylene glycol as solvent. Shang et al. (2019) studied the dehydration of ethanol by extractive distillation using choline chloride:urea(1:2) as mass separating agent. The results based on a closed-loop process indicate a potential reduction in the annualized costs as well as the energy consumption of the process.

While the low vapor pressure implies a simple thermal regeneration of the DES, the high boiling temperature and limitations regarding the degradation temperature are potentially detrimental factors that need to be considered for the closed-loop process, including solvent regeneration. To evaluate the tradeoff of these factors, a rigorous optimization of the closed-loop flowsheet, including solvent recovery, is required. In the current study, hybrid extraction-distillation processes for the purification of furfural from an aqueous fermentation broth using decanoic acid:thymol(1:1) as DES and toluene as established organic solvent are evaluated as optimal design problems. Furfural is widely applied as a herbicide, fungicide, and insecticide in agricultural production and a flavoring agent in the food industry (Yong et al., 2020). Additionally, Furfural is one of the top bio-based chemicals due to its numerous reaction products, like fuel additives and value-added chemicals (Yan et al., 2014).

2. Methodology

As a first step for the targeted evaluation, a suitable DES for the considered separation has to be identified. Thermodynamic properties need to be modeled before a comparison with the established organic solvent can be made based on rigorous model-based process optimization.

2.1. Selection of a Deep Eutectic Solvent

For liquid-liquid extraction, the applicability of a solvent is restricted by the acceptable loss in the raffinate stream, while the economic performance is strongly linked to the amount of solvent required for a feasible separation. Therefore, the distribution coefficient is commonly considered a key performance indicator for extraction performance in an early evaluation step. Thermal recovery of the solvent by means of distillation has to adhere to the boiling and degradation temperature of the DES, which may limit the operating range of a distillation column. Consequently, to prevent degradation and the resulting product pollution and loss of solvent, the DES needs a comparably high degradation temperature to enable an efficient solvent recovery.

For the current example of furfural purification, Dietz et al. (2019c) provide the necessary data for selecting a suitable DES. In an experimental investigation of ten different DESs, decanoic acid:thymol(1:1) (Deca-Thy) was identified as the solvent with the highest distribution coefficient for liquid-liquid extraction of furfural (Dietz, 2019a). Based on investigations of the thermal decomposition of comparable DESs (Dietz, 2019a), a degradation temperature of 190 °C was assumed for Deca-Thy, which is only slightly lower than the highest decomposition temperature reported by Dietz et al. (2019a), for dodecanoic acid:atropine(2:1). However, Deca-Thy provides a distribution coefficient that is about 12 times higher compared to dodecanoic acid:atropine(2:1). Thus, it is selected as most interesting DES for the current application and compared with toluene, as widely applied benchmark organic solvent (Dietz, 2019c).

2.2. Thermodynamic Properties

While the previously described data allows for an initial selection of promising solvent candidates, a variety of thermodynamic properties have to be modeled reliably for a model-based assessment of the respective processes. Besides the pure component ideal gas heat capacity, vapor pressure, and enthalpy of evaporation, which are needed to evaluate the energy balance and thermal regeneration in the hybrid extraction-distillation process, especially the binary interactions for the liquid-liquid and vapor-liquid equilibrium (LLE, VLE) have to be known.

Parameters for pure component properties and activity coefficient models for water, furfural, and toluene are available and exported from Aspen Plus databases DB-PURE32 and APV88 LLE-ASPEN/APV88 VLE-IG, respectively. For modeling the DES and its interactions, similar to other studies (Han and Chen, 2018; Dietz et al., 2019b; Dai et al., 2022), it is assumed that the DES can be modeled as a single pseudo component, for which the ideal gas heat capacity is estimated by the group contribution method by Joback and Reid (1987) for decanoic acid and thymol individually and averaged for the DES. The DIPPR equation for the ideal gas heat capacity is fitted to the obtained data for the process optimization. The vapor pressure is predicted using the perturbed-chain statistical associating fluid theory (PC-SAFT) (Gross and Sadowski, 2001) with the pure component parameters published by Dietz et al. (2019b) and a ratio-specific individual constituent approach. The extended Antoine equation is fitted to the predicted data and used in the optimization-based design. Based on the vapor pressure, the enthalpy of evaporation is assumed to be constant and calculated using the Clausius-Clapeyron equation.

In order to avoid the application of the comparably complex PC-SAFT model in the process optimization, the non-random two-liquid (NRTL) model is considered as a surrogate model to describe the LLE and VLE within the process optimization. Therefore, the NRTL parameters are fitted to predicted equilibrium data determined by the PC-SAFT model. The binary interactions of furfural and Deca-Thy are determined by regressing VLE data at pressures of 0.05, 0.1, and 0.2 bar to ensure a correct phase behavior at vacuum conditions around 0.1 bar. In contrast, the binary interactions of water and Deca-Thy are obtained by regressing LLE data at a pressure of 1 bar. The c_{ij} parameter is fixed at 0.3 and 0.2 for VLE and LLE regression, respectively.

2.3. Rigorous Process Optimization

In order to perform a techno-economical evaluation of the processes with the considered solvents, toluene, and Deca-Thy, a rigorous superstructure optimization based on equilibrium-stage models is performed. The general process configurations do, however, differ due to the different solvent properties. Since toluene is light-boiling compared to furfural, it is obtained as the top product in the solvent recovery column (cf. Figure 1). In contrast, Deca-Thy is heavy boiling and, as such, recovered as the bottoms product in the solvent recovery column (cf. Figure 2). More importantly, furfural is the intermediate boiling component in this mixture. On the basis of the individual process structures, a superstructure model is built, allowing for individual sizing of the extraction column and the rectifying and stripping sections of the distillation column. The respective mathematical models are extended with constraints on product purities as well as sizing and costing models (Biegler et al., 1997), resulting in mixed-integer nonlinear programming (MINLP) problems for the techno-economical designs, which are solved by a series of successively relaxed nonlinear programming (NLP) problems with the respective equilibrium calculations performed via a set of external functions

(Skiborowski et al., 2015). The models and the polylithic solution strategy are implemented in GAMS 34.3.0, and SNOPT is used to solve for local optimality.

3. Process Optimization

For both processes, a feed stream of 5 mol% furfural and 95 mol% water with a flow rate of 100 mol/s (about 7900 kg/h) is assumed. For the wastewater stream comprising the raffinate and the water-rich phase of the decanter, a furfural content $x_{FF} < 0.01$ mol% is defined, and the product stream is constrained to a purity of $x_{FF} > 99.9$ mol%. A depreciation time of ten years with an interest rate of 6% and 8000 working hours per year are set for the calculation of the total annualized costs (TAC). For both solvents, a price of 1000 \$/t (846.5 €/t) is assumed to account for solvent make-up streams. As utilities, cooling water at 25 °C (0.05 €/t), low-pressure steam at 10 bar (14 €/t), and high-pressure steam at 25 bar (16 €/t) are considered. The superstructure consists of a liquid-liquid extraction column with a maximum number of 15 (initial) stages and a distillation column with a maximum number of 80 (initial) stages.

Figure 1 shows the optimization results for the benchmark solvent toluene. The TAC of the process is about 194.4 k€/y and is largely determined by the operating costs, which represent about 69%. The furfural purity in the optimized process even exceeds the desired specification, avoiding unnecessary solvent makeup. Apparently, there is no reduction in the number of stages for the extraction column due to the comparably low investment costs, considering only a column shell with simple sieve tray inlets. In contrast, the cost of potential solvent makeup and increased flow rates in the distillation column would increase the TAC significantly compared to the reduced cost of the extraction column.

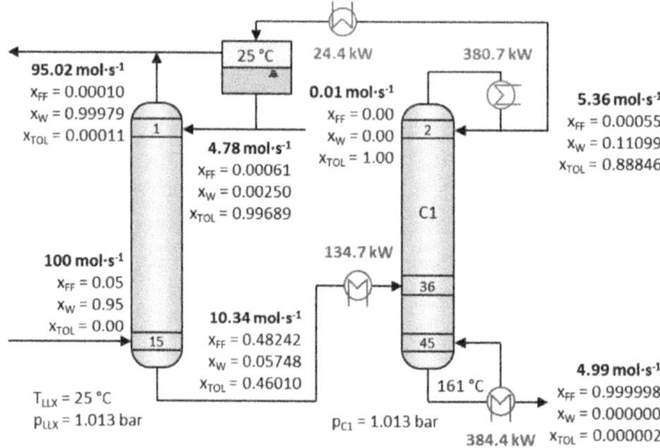

Figure 1: Optimized process flowsheet for the separation of furfural (FF) and water (W) using toluene (TOL) as the solvent in the liquid-liquid extraction (LLX) column

The results for the Deca-Thy-based process are presented in Figure 2. The high boiling temperature of Deca-Thy requires an alternative process design resulting in a three-column setup, including a liquid-liquid extraction, a distillation column for solvent recovery, and a heteroazeotropic distillation column for further purification of furfural. The solvent recovery is carried out in a vacuum column with a pressure of 0.1 bar to avoid the degradation temperature of Deca-Thy. The heteroazeotropic distillation is optimized separately based on the model and solution strategy presented by Kruber and Skiborowski (2022). Due to the furfural losses in the heteroazeotropic distillation, the overall furfural recovery drops slightly from 99.8% to 99.3%, whereas the furfural losses would also add up to the wastewater stream. The TAC of the process is about 246.4 k€/y with almost the same capital expenditures as for the toluene process of about 60.5 k€/y. The increase in the TAC originates exclusively from an increase of 39% in operational expenditures, which is caused by the increased energy demand as well as the use of high-pressure steam (25 bar) in the vacuum column. Notably, the solvent-to-feed (S/F) ratio is significantly lower for the DES, which aligns with the increased distribution coefficient reported by Dietz et al. (2019c). Yet, the thermodynamically derived low S/F ratio should be considered with care, as phase ratios between 1:10 and 10:1 are usually considered for practical application (Goedecke, 2006).

As for toluene, the extraction column is utilized at its maximum number of stages due to the low investment costs. However, the distillation column uses less than half the stages compared to toluene, which is expected based on the extremely low vapor pressure of the DES. The main limitation of the DES is caused by the high cross-solubility of water in the solvent phase, due to which the additional purification in the heteroazeotropic distillation column is required, which counteracts the savings due to the smaller vacuum column. This not only causes a more complex and energy-demanding process but would also complicate the operation of the process.

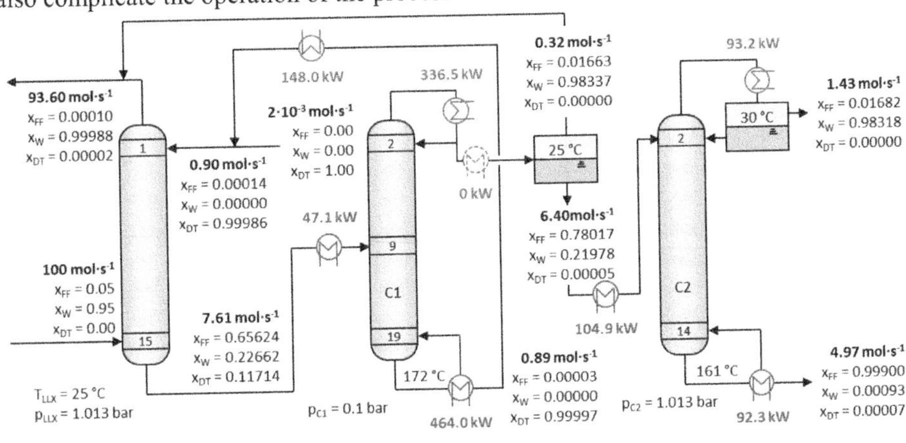

Figure 2: Optimized process flowsheet for the separation of furfural (FF) and water (W) using decanoic acid:thymol (DT) as the solvent in the liquid-liquid extraction (LLX) column

4. Conclusion & Outlook

In order to evaluate the potential benefits of a DES, this work considers the purification of a diluted furfural stream by hybrid extraction-distillation processes by means of a rigorous optimization-based design. DESs are often examined based on simple thermodynamic properties and, therefore, recommended based on their low toxicity, flexibility, and particularly low vapor pressure, which implies an easy thermal regeneration. While all these properties are favorable for solvents, the high boiling temperature can result in a higher energy demand and the necessity for more costly heating utilities. The results illustrate the importance of a holistic consideration of solvent and process design to evaluate if the desired thermodynamic properties of the DES really translate into economic benefits for the overall process. It was shown that thermodynamic properties alone could not directly indicate economic savings. Although the recovery of Deca-Thy from the extract phase is simple, the high cross solubility of water requires further product purification and results in higher TAC of 246.4 k€/y compared to 194.4 k€/y for the toluene-based process. Especially for the purification of high boiling organic compounds in diluted aqueous streams, a volatile solvent may be a favorable choice since the value product will not be an intermediate boiling component that can be separated without additional unit operations. In addition, the assumed degradation temperature for Deca-Thy is rather optimistic, and experimental studies indicate a much lower thermal stability resulting in even higher process cost.

In conclusion, DESs provide many favorable properties which make them interesting solvent candidates, but their suitability should be evaluated based on closed-loop process flowsheets that consider the effort for solvent recovery for each solvent candidate to gain more reliable information with respect to the process performance. In the future, the consideration of rigorous process models during the design and selection of DESs for separation processes should be further investigated, extending the scope to applications in extractive distillation and heteroazeotropic distillation. With the knowledge gained for the furfural purification, alternative DESs can be considered to improve the process performance and find a green alternative to the presented benchmark process and other interesting solvent candidates, such as butyl chloride (Nhien et al., 2017).

References

A.P. Abbott, G. Capper, D.L. Davies, R.K. Rasheed, V. Tambyrajah, 2003, Chem. Comm., 1, 70–71

L.T. Biegler, I.E. Grossmann, A.W.Westerberg, 1997, Prentice-Hall, Upper Saddle River, NJ

Y. Dai, X. Chu, Y. Jiao, Y. Li, F. Shan, S. Zhao, G. Li, Z. Lei, P. Cui, Z. Zhu, Y. Wang, 2022, Chem. Eng. Sci., 264, 118179

C.H.J.T. Dietz, 2019a, Ph.D. Thesis, Chemical Engineering and Chemistry, Eindhoven University of Technology

C.H.J.T. Dietz, J.T. Creemers, M.A. Meuleman, C. Held, G. Sadowski, M. van Sint Annaland, F. Gallucci, M.C. Kroon, 2019b, ACS Sustainable Chem. Eng., 7, 4047–4057

C.H.J.T. Dietz, F. Gallucci, M. van Sint Annaland, C. Held, M.C. Kroon, 2019c, Ind. Eng. Chem. Res., 58 (10), 4240–4247

Goedecke, R., 2006, Fluidverfahrenstechnik, Wiley-VCH, Weinheim

J. Gross, G. Sadowski, 2001, Ind. Eng. Chem. Res., 40 (4), 1244–1260

D. Han, Y. Chen, 2018, Chem. Eng. Process., 131, 203–214

D. Han, Y. Chen, D. Mengru, 2022, Comput. Chem. Eng., 164, 107865

B.B. Hansen, S. Spittle, B. Chen, D. Poe, Y. Zhang, J.M. Klein, A. Horton, L. Adhikari, T. Zelovich, B.W. Doherty, B. Gurkan, E.J. Maginn, A. Ragauskas, M. Dadmun, T.A. Zawodzinski, G.A. Baker, M.E. Tuckerman, R.F. Savinell, J.R. Sangoro, 2021, Chem. Rev., 121, 3, 1232–1285

K.G. Joback, R.C. Reid, 1987, Chem. Eng. Commun., 57 (1-6), 233–243

K.F. Kruber, M. Skiborowski, 2022, 10 (8), 1482

M. Larriba, M. Ayuso, P. Navarro, N. Delgado-Mellado, M. Gonzalez-Miquel, J. García, F. Rodríguez, 2018, ACS Sustainable Chem. Eng., 6 (1), 1039–1047

L.C. Nhien, N. Van Duc Long, S. Kim, M. Lee, 2017, Biotechnol. Biofuels, 10 (81)

Z. Salleh, I. Wazeer, S. Mulyono, L. El-blidi, M.A. Hashim, M.K. Hadj-Kali, 2017, J. Chem. Thermodyn., 104, 33–44

X. Shang, S. Ma, Q. Pan, J. Li, Y. Sun, K. Ji, L. Sun, 2019, Chem. Eng. Res. Des., 148, 298–311

M. Skiborowski, A. Harwardt, W. Marquardt, 2015, Comput. Chem. Eng, 72, 34–51

E.L. Smith, A.P. Abbott, K.S. Ryder, 2014, Chem. Rev., 114, 21, 11060–11082

K. Yan, G. Wu, T. Lafleur, C. Jarvis, 2014, Renewable Sustainable Energy Rev., 38, 663–676

T.L.-K. Yong, K.F. Pa'ee, N. Abd-Talib, N. Mohamad, 2020, in: Inamuddin, A.M., Asiri (Eds.), Advanced Nanotechnology and Application of Supercritical Fluids, Springer International Publishing, Switzerland, 53–73.

Antonis Kokossis, Michael C. Georgiadis, Efstratios N. Pistikopoulos (Eds.)
PROCEEDINGS OF THE 33rd European Symposium on Computer Aided Process Engineering (ESCAPE33), June 18-21, 2023, Athens, Greece
 http://dx.doi.org/10.1016/B978-0-443-15274-0.50153-0

Profit Allocation in Industrial Symbiosis Networks: Utility Exchanges

Fabian Lechtenberg[a], Antonio Espuña[a], Moisès Graells[a*]

[a] *Department of Chemical Engineering, Universitat Politècnica de Catalunya, Barcelona 08019, Spain*
[*]moises.graells@upc.edu

Abstract

Industrial symbiosis networks enable efficient resource sharing between companies and, thereby, generate otherwise unachievable profits. This work studies the multi-actor process integration problem that arises when addressing utility exchange involving different companies. The aim is to find stable profit allocations by combining process integration principles with game theory methods. The profit table provides the necessary information to solve the cooperative game that takes place in the industrial park. For the case study, a simplified utility exchange model is developed and the proposed decision-making framework is applied to it, in order to highlight the importance of knowledge about the potential pitfalls of the employed profit allocation methods.

Keywords: cooperative game theory, process integration, decision-making, steam

Introduction

Collaboration in industrial symbiosis networks potentially leads to profits over standalone operation. These profits may be of any kind (economic, environmental, social ...). While non-cooperative interaction between the actors may be analyzed to find optimal decision-making strategies (Hjaila et al., 2017), it is not ensured that the combined profit of all actors is efficient, i.e. as high as if the actors were selflessly cooperating. Especially, in industrial parks it is reasonable to assume that companies may bargain in a cooperative manner: they can form binding agreements to fairly share the generated profits, regardless of whether an individual member would suffer losses if no allocation occurs. Fair cost and profit allocation was studied before for various case studies (Chin et al., 2021; Hiete et al., 2012) but no study considered utility exchanges via steam transfer so far. To that end, a utility exchange model and the corresponding case study are contributed. Moreover, this work makes a significant step towards a generalized framework for decision-making in general multi-actor process integration problems.

Problem Statement

For the given multi-actor process integration problem (*utility exchanges*) in an industrial park find the highest possible total economic profit and apply allocation methods to fairly distribute the profit among the participants. Check if the allocations are stable, i.e. that an individually rational actor would be likely to accept the proposal.

Methodology

The allocation of profits requires the prior knowledge of the potential profits that can be obtained in the grand coalition and all sub coalitions. To that end, a simplified utility network model is proposed based on works by Galvan-Cara et al. (2022) and Kim et al. (2010). The original model complexity is reduced by omitting the multi-period problem, reducing the number of steam levels, etc. Material, energy and economic balances are maintained. The resulting set of equations is summarized in (1)-(13) and it contains no binary decisions. Thus, the minimization problem is a linear program (LP).

- High-pressure steam (HS) and low-pressure steam (LS) headers:

$$\dot{m}_{HS,i}^{demand} + \dot{m}_{HS,i}^{vent} + \sum_j \dot{m}_{HS,i,j}^{exc} + \dot{m}_{HS,i}^{turbine} + \dot{m}_{HS,i}^{letdown} = \sum_{b_{HS}} \dot{m}_{i,b_{HS}}^{boiler} \quad \forall i \quad (1)$$

$$\dot{m}_{LS,i}^{demand} + \dot{m}_{LS,i}^{vent} + \sum_j \dot{m}_{LS,i,j}^{exc} = \dot{m}_{LS,i}^{turbine} + \dot{m}_{LS,i}^{letdown} + \sum_{b_{LS}} \dot{m}_{i,b_{LS}}^{boiler} \quad \forall i \quad (2)$$

- Boiler mass balance:

$$\dot{m}_{i,b}^{boiler} = Cal_b \cdot \dot{m}_{fuel,i,b}^{purchase} \cdot x_{i,b} \quad \forall i, b \in b_{HS} \cup b_{LS} \quad (3)$$

- Turbine and letdown units:

$$\dot{m}_{HS,i}^{turbine} = \dot{m}_{LS,i}^{turbine} + \dot{m}_i^{cond} \quad \forall i \quad (4)$$

$$\dot{m}_{LS,i}^{turbine} = \text{Split} \cdot \dot{m}_{HS,i}^{turbine} \quad \forall i \quad (5)$$

$$P_i^{turbine} = \text{Yield} \cdot \dot{m}_{HS,i}^{turbine} \quad \forall i \quad (6)$$

$$\dot{m}_{HS,i}^{letdown} = \dot{m}_{LS,i}^{letdown} + \dot{m}_i^{letdown} \quad \forall i \quad (7)$$

$$\dot{m}_i^{letdown} = \text{Fill} \cdot \dot{m}_{HS,i}^{letdown} \quad \forall i \quad (8)$$

- Water units:

$$\dot{m}_{water,i}^{demand} + \sum_b \dot{m}_{i,b}^{boiler} + \dot{m}_i^{letdown} = \dot{m}_{water,i}^{purchase} + \dot{m}_i^{cond} \quad \forall i \quad (9)$$

- Electricity units:

$$P_i^{demand} + \sum_b P_{i,b}^{boiler} = P_i^{turbine} + P_i^{purchase} \quad \forall i \quad (10)$$

- Economic balances:

$$C_i^{raw} = C^{elec} \cdot P_i^{purchase} + C^{water} \cdot \dot{m}_{water,i}^{purchase} + \sum_b C_b^{fuel} \cdot \dot{m}_{fuel,i,b}^{purchase} \quad \forall i \quad (11)$$

$$C_i^{transport} = \frac{1}{2} \cdot \sum_{s \in \{HS,LS\}} \sum_j \beta_s \cdot \left(\dot{m}_{s,i,j}^{exc} + \dot{m}_{s,j,i}^{exc}\right) \cdot L_{i,j} \quad \forall i \quad (12)$$

$$C = \sum_i \left(C_i^{raw} + C_i^{transport}\right) \quad \text{(subject to minimization)} \quad (13)$$

Where i, j is the set of plants and b is the set of boilers. Figure 1 shows the superstructure of the utility system in a plant and visualizes the balances from the model. Utility costs for fuel, water and electricity are considered. Steam exchanges can lead to savings in all utility purchases at the cost of pipeline construction and operation.

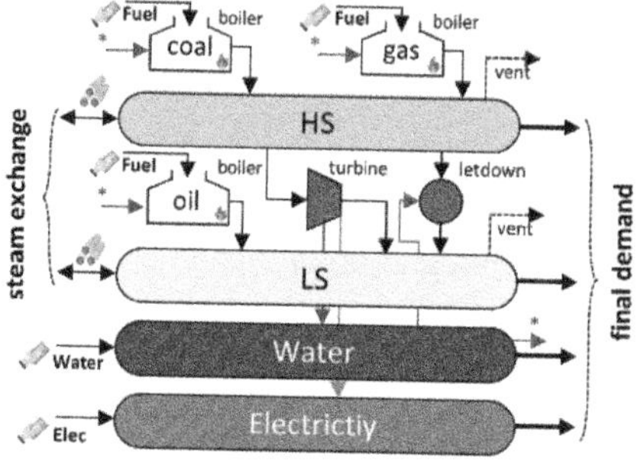

Figure 1. Simplified utility system superstructure for each plant.

Cooperative Game Theory and Profit Allocation:
Multi-actor process integration problems such as the utility-exchange problem presented herein can be adjusted for every existing sub coalition by introducing binary parameters that restrict flows to only allow exchanges between the sub coalition members. The resulting profits are always positive with respect to the standalone case because any participant may decide to not collaborate if no individual profit is in sight. Figure 2 shows the steps followed in this work which is part of a general framework that is to be presented in future work. The profit table in the Figure is an arbitrary example with three players.

The profit table results from solving the process integration problem for the 2^N coalitions, where N is the number of players. Using the information about the worth of each coalition, various solution concepts can be applied to fairly distribute the profits to the members of the coalition. We developed the Python package "***pyCoopGame***" which implements various well-known single-point solution concepts and stability criterion checks for any given profit table. In this work we employ four solution approaches of which the following descriptions briefly outline the underlying intuition.

The core concept (Gillies, 1959) is not a solution approach but a stability criterion that checks if a given profit allocation is efficient and acceptable for individually rational players. That means, an allocation is stable and lies in the core if the total profit equals the profit of the grand coalition and no sub coalition can ensure higher payoff for all members. The Shapley value (Shapley, 1988) makes use of the profit table to determine the marginal contribution of every member to every coalition and calculates a weighted average of the marginal contributions. The Nucleolus (Puerto & Perea, 2013) is the allocation that minimizes the largest dissatisfaction of all coalitions. Dissatisfaction is the difference between the worth of the coalition and the sum of the profits allocated to the coalition members. The τ-value (Tijs & Driessen, 1986) is the efficient allocation that lies between the utopian allocation where each player receives their marginal contribution to the grand coalition and the minimum rights allocation. The least core solution (Maschler et al., 1979) is the profit allocation where the relaxed ε-core set is just composed of a single point.

It should be noted that these four approaches are selected based on their popularity but others may be chosen depending on the use case. Luo et al. (2022) give a good overview of recent developments and applications, as well as current challenges, of cost and profit allocation solution concepts applied to operations management.

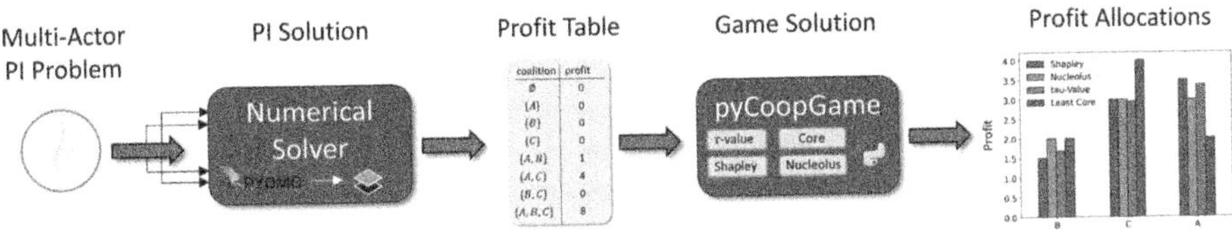

Figure 2. Solving a profit allocation problem in a multi-actor process integration case.

Case Study

The multi-actor process integration problem studied in this work is a utility exchange problem adapted from the Yeosu industrial complex as shown in Figure 3. The game is redefined to include three players (10, 12 and 15) and the utility system is adapted to the previously presented simplified model.

The resulting parameters are summarized in Tables 1, 2 and 3. Demand values for VHS and HS as well as MS and LS are aggregated over the four time periods.

Table 1. Final annual demand values

	HS [kton]	LS [kton]	Water [kton]	Electricity [MWh]
10	5707	0	1908	952
12	0	111	476	396
15	3878	1042	1588	1272

The economic parameters shown in Table 2 are updated to match current market conditions. The fuels are coal (1), methane (2) and fuel oil (3).

Table 2. Economic parameters and calorific content of fuel

Parameter	Value	Unit	Parameter	Price [$/ton]	$Cal_b \left[\frac{kg\ steam}{kg\ fuel}\right]$
Water Price	0.500	$/ton	Fuel 1	205	8.9
Electricity Price	150.0	$/MWh	Fuel 2	290	18.5
Pipe Operation (β)	0.436	$/km/t	Fuel 3	1174	15.9

The distance between plants is kept unchanged.

Table 3. Distance between companies

Plants	Distance	Plants	Distance	Plants	Distance
10 ↔ 12	2.319 km	10 ↔ 15	2.193 km	12 ↔ 15	1.237 km

The utility system of each plant is changed to correspond to the simplified utility model. As can be seen from Figure 3 (right), in scenario 1 plant 10 has a HS coal boiler and a turbine, plant 12 has a LS fuel oil boiler and plant 15 has a HS methane boiler, a turbine and a LS fuel oil boiler. In scenario 2 plant 10 has a methane boiler but no turbine. Results for the whole complex using the original parameters from Galvan-Cara et al. (2022) are also provided.

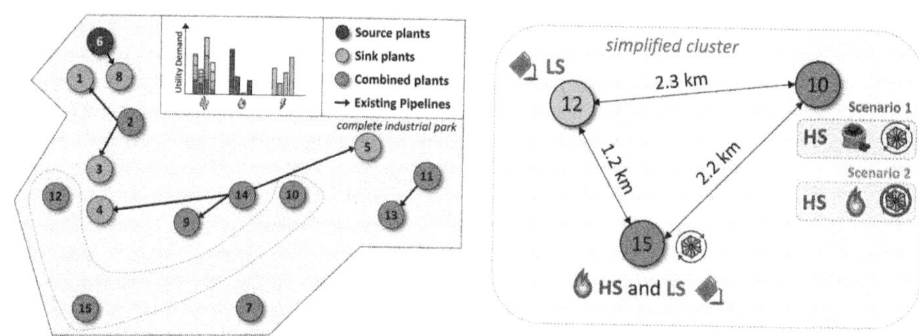

Figure 3. Yeosu industrial complex (Kim et al., 2010) and scenarios.

Results and Discussion

Solving the process integration problem for all coalitions in scenario 1 and 2 leads to the profit tables as shown in Table 4 and 5. Most of the profit in scenario 1 is due to the fuel that plant 10 saves by not using its coal boiler and importing instead HS from plant 12, which has a more cost-efficient methane boiler. Further savings are generated by LS exchange of plant 12 and 15 with 10. In scenario 2 the big savings from plant 10 are not feasible because plant 15 already has a methane boiler. Thus, the only savings come from LS exchange with plant 12, which has a limited demand for this utility.

Applying the single-point solution concepts leads to the profit allocations in Figure 4 and 5. In scenario 1, Shapley, Nucleolus and τ-value yield similar results that follow a simple intuition: plant 10 and 15 both need to be present to generate the large benefits, so both should receive similar payoffs. Plant 12 also has a contribution but its impact is way smaller so it can't claim as much payoff. It is not surprising that these three payoff vectors are indeed stable. However, it is non-intuitive that plant 15 and 10 should receive such asymmetric payoffs according to the least core solution. This outcome follows from the property of the least core method which always yields a solution on the edge of the feasible core allocation set, granted it is non-empty.

In scenario 2 the profit is largely determined by plant 12 accepting LS from either 10 or 15 until its demand is satisfied. Plant 10 is in the most powerful position and can claim the largest payoff. The results of the Nucleolus, τ-value and least core methods are stable and they allocate the payoffs to plants 12 and 15. The latter is preferred over plant 10 due to the slightly closer geographical proximity. Plant 10 is not engaged in the final utility exchange so not assigning payoff to this player is intuitive. However, the Shapley value assigns a significant share of the payoffs to both players 10 and 15 due to their marginal contribution to the sub coalitions. This result is not stable and reveals one of the main weakness of the Shapley value, the most popular approach used in literature. These two scenarios demonstrate that allocation methods should be selected with care, being aware of their potential pitfalls.

Figure 6 shows the results for the whole industrial park. The Nucleolus is not included here as its calculation for games with many players becomes computationally unaffordable. The remaining profit allocations are very different and this is due to the fact that the core for this 15-player game is empty. In this kind of situations, the rules of the game may be adjusted by means of taxation and incentives (Chin et al., 2021).

Table 4. Profit Table (scenario 1)

Coalition	Profit
{10,12}	5,290,864 $/year
{10,15}	36,673,525 $/year
{12,15}	6,127,690 $/year
{10,12,15}	42,804,345 $/year

Table 5. Profit Table (scenario 2)

Coalition	Profit
{10,12}	6,074,326 $/year
{10,15}	0 $/year
{12,15}	6,127,690 $/year
{10,12,15}	6,127,690 $/year

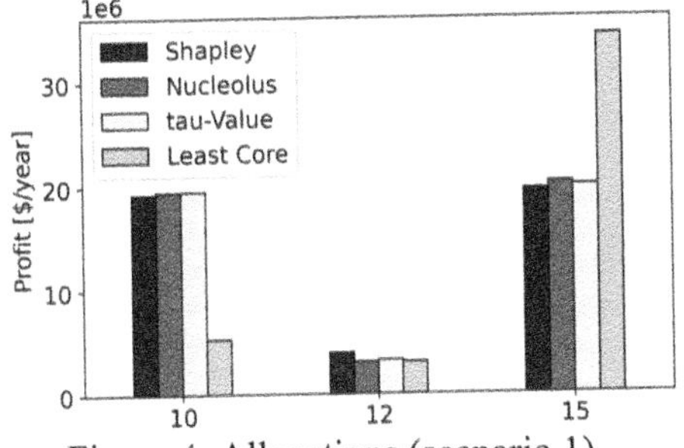

Figure 4. Allocations (scenario 1)

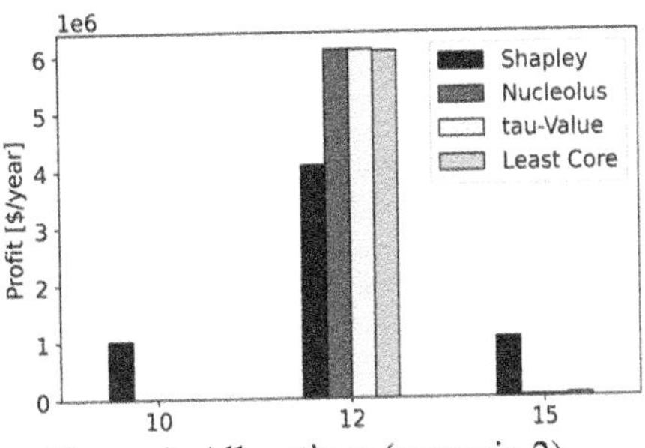

Figure 5. Allocations (scenario 2)

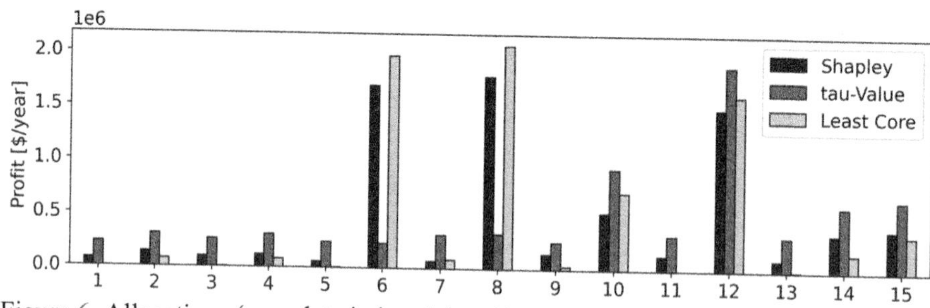

Figure 6. Allocations (complete industrial park). Total profit: 7,157,500 $/year

Conclusions

The application of cooperative game theory methods (and the associated profit allocation) to industrial symbiosis networks facilitates the decision-making and may create trust between the actors by proposing payoff vectors that are intuitive and stable. The presented study demonstrates this on a simplified utility exchange case study. Moreover, a significant step towards the generalization of a profit allocation framework for multi-actor process integration problems is done. Such a framework requires knowledge about strength and weaknesses of the available methods. Future work will further formalize this framework, characterize the allocation methods and demonstrate on a wide variety of process integration problems how decision-making can be systematically improved using cooperative game theory.

Acknowledgements

Financial support received from the Spanish "Ministerio de Ciencia e Innovación" and the ERDF (project CEPI-PID2020-116051RB-I00) is fully acknowledged. Fabian Lechtenberg gratefully acknowledges the "Departament de Recerca i Universitats de la Generalitat de Catalunya" for the financial support of his predoctoral grant FI-2022.

References

A.L. Galvan-Cara, M. Graells, A. Espuña, 2022, Application of Industrial Symbiosis principles to the management of utility networks. Applied Energy, 305

C. Luo, X. Zhou, B. Lev, 2022, Core, shapley value, nucleolus and nash bargaining solution: A Survey of recent developments and applications in operations management. Omega, 110

D.B. Gilies, 1959, Solutions to general non-zero-sum games. In Contributions to the Theory of Games IV, 40

H.H. Chin, P.S. Varbanov, J.J. Klemeš, S. Bandyopadhyay, 2021, Subsidised water symbiosis of eco-industrial parks: A multi-stage game theory approach, Computers & Chemical Engineering, 155

J. Pueto, F. Perea, 2013, Finding the nucleolus of any n-person cooperative game by a single linear program. Computers & Operations Research, 40

K. Hjaila, L. Puigjaner, J.M. Laíney, A. Espuña, 2017, ntegrated game-theory modelling for multi enterprise-wide coordination and collaboration under uncertain competitive environment. Computers & Chemical Engineering, 98

L.S. Shapley, 1988, A value for n-person games, The Shapley Value: Essays in Honor of Lloyd S. Shapley

M. Hiete, J. Ludwig, F. Schultmann, 2012, Intercompany Energy Integration: Adaptation of Thermal Pinch Analysis and Allocation of Savings. Journal of Industrial Ecology, 16

S.H. Kim, S.G. Yoon, S.H. Chae, S.Park, 2010, Economic and environmental optimization of a multi-site utility network for an industrial complex. Journal of Environmental Management, 91

S.H. Tijs, T.S.H. Driessen, 1986, Game Theory and Cost Allocation Problems. Management Science, 32

Antonis Kokossis, Michael C. Georgiadis, Efstratios N. Pistikopoulos (Eds.)
PROCEEDINGS OF THE 33rd European Symposium on Computer Aided Process Engineering (ESCAPE33), June 18-21, 2023, Athens, Greece
 http://dx.doi.org/10.1016/B978-0-443-15274-0.50154-2

Operational optimisation of steam power system under uncertainty using time series prediction optimization model with Markov chain

Kenian Shi, Jingyuan Zheng, Siyu Yang*
School of Chemical Engineering, South China University of Technology, Guangzhou, 510640, China

Abstract

The stochastic programming and robust optimisation methods have been proposed to solve the operational optimisation of industrial energy system under uncertainty. However, these methods can not comprehensive the stability and economy at the same time. This paper proposes a two-stage time series prediction optimization model with Markov chain to solve this problem. The steam demand value is predicted by the method of scenario generation and reduction. Taking the steam power system of a coal-to-gas enterprise as an example, the corresponding optimization model is established, and the predicted steam value is put into the optimization model to solve. The results show that the optimization method in this paper combines the advantages of high economy of stochastic programming and high stability of robust optimization, which provides a new way to solve the uncertain optimization problem of steam power system.

Keywords: Steam demand uncertainty, time series prediction optimization, Markov chain, Scenario generation and reduction.

Introduction

In the operation of steam power system, there are various uncertain factors (uncertain changes of operating temperature, pressure and flow rate), which will eventually lead to uncertain changes of steam demand [1]. The uncertainty of steam demand can lead to the blind production of steam power system, producing more or less steam than the demand. The former will lead to the venting of steam and waste resources, while the latter will affect the safety and stability of the production. Therefore, it is of great theoretical and practical significance to study the operational optimisation of steam power system under uncertainty.

At present, the methods to solve the uncertain optimization problem include stochastic programming [2]and robust optimization [3]. The main difference between these uncertain optimization methods lies in the description of uncertain parameters and solving algorithm [4]. Stochastic programming is only applicable to the uncertain optimization problems whose uncertain parameters can be expressed as probability distribution, but not to the uncertain optimization problems with non-probability distribution [5], and it cannot avoid high risks in the process of system planning [6]. The robust optimization expresses the uncertain parameter as the uncertain parameter set and considers the optimization scheme in the worst case [7].

Each method has its own advantages and disadvantages. Robust optimization has high stability but is too conservative. Stochastic programming has high economy but poor stability. At present, few optimization methods can comprehensive the stability and economy at the same time. This paper proposes a two-stage time series prediction optimization model with Markov chain to solve the coexistence problem of economy and stability.

1. Methodology

The time series prediction of steam demand can be divided into two stages. In the first stage, multiple working conditions are divided based on the distance of three levels of steam. In the second stage, the steam demand is predicted based on the probability of state transition under each working condition, which mainly includes scenario generation and scenario reduction.

1.1. Multiple working condition division

Due to the high volatility of the three levels of steam, it is necessary to classify them to improve the overall efficiency and economy of the steam power system. In this paper, a clustering algorithm is used to classify the annual steam demand [8] The annual demand for three levels of steam can be divided into five operating conditions.

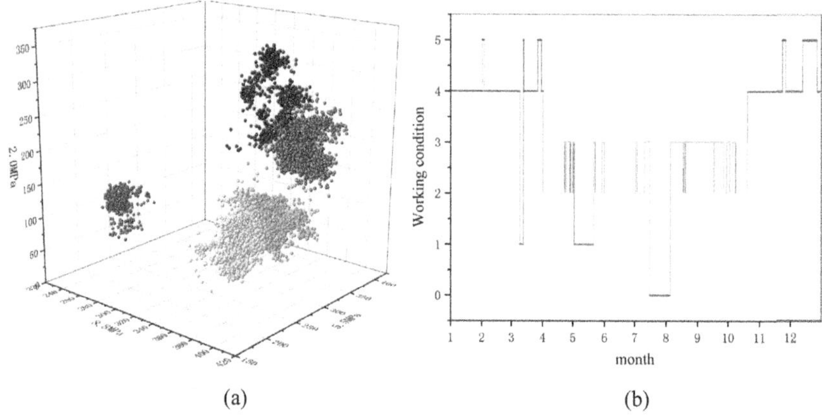

Fig.1 (a)、(b) -Results of steam demand division

The division results of annual steam demand conditions were plotted with time as the abscissa, as shown in Figure 1. There were six operating conditions including parking conditions (operating condition 0).

1.1.2 Scenario generation

The principle of scenario generation is based on the random process of Markov chain, and the probability of the random occurrence is determined by the probability of transition between different states in the research object [9]. Suppose the state of steam demand sequence in the current scenario is...X_{t-2}、X_{t-1}、X_t、X_{t+1}、X_{t+2}..., Then the probability of the steam demand state at time X_{t+1} only depends on the state X_t at the previous moment:

$$P(X_{t+1}| \dots X_{t-2}, X_{t-1}, X_t) = P(X_{t+1}|X_t) \tag{1}$$

In this paper, the steam state is evenly divided into 10 states, the probability transition matrix of three levels of steam can be constructed by statistics of probabilities of transition under different working conditions. According to the state transition probability matrix, the state of the next hour can be predicted continuously until the set prediction time is generated. The prediction time in this paper is 24h.

1.1.3 Scenario reduction

The number of scenarios generated is very large, if all of them are considered, the efficiency of subsequent optimization problems will be seriously affected, and feasible solutions cannot even be obtained [10]. Therefore, the scenario reduction method is needed to eliminate redundant scenarios. The scenario reduction method used in this paper is the simultaneous backward reduction method [11]. The principle of this method is to minimize the probability distance between the initial scenario set and the scenario set saved after reduction, which makes the reduction process easier [12]. The Dr Distance model is shown below:

$$\min D_r(P,Q) = \left\{\sum_{i\in I}\sum_{j\in J} p_i \left\|\varepsilon_i - \tilde{\varepsilon}_j\right\|^r\right\}^{1/r} \tag{3}$$

Where, $\|.\|$ is norm, the Euclidean distance is used here, so r=2.

The number of reserved scenario cuts is generally 10. Figure 2 shows the result of reducing 2000 scenarios generated by the scenario generation to 10. As you can see in Figure 2, the reduced results include all representative cases from minimum to maximum.

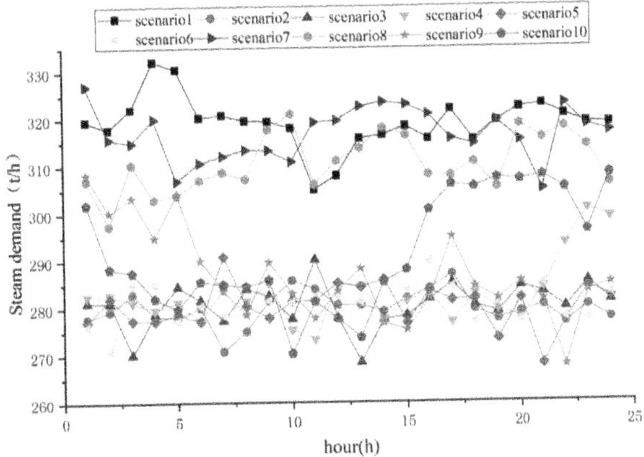

Fig.2 The results of scenario reduction

1.4 Determine the number of initial scenarios Sp

Take the maximum and minimum values of reserved 10 scenarios at each moment to form the upper and lower bounds of the prediction interval. The prediction results of the three levels of steam are shown in Figure 3.

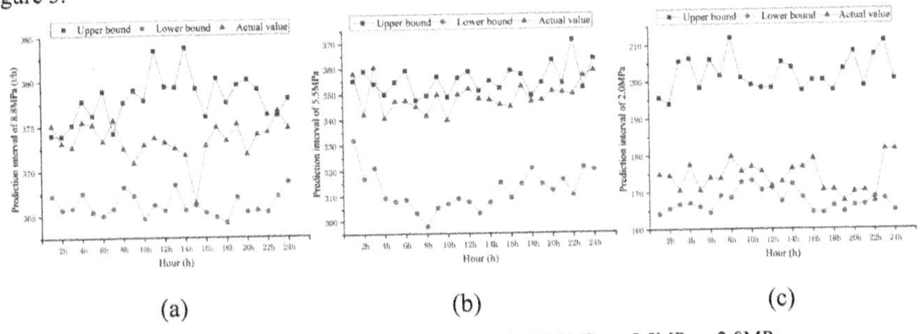

(a) (b) (c)

Fig.3 (a)、(b)、(c)-Prediction interval of 8.8MPa、5.5MPa、2.0MPa

From the prediction intervals of the three level of steam, it can be seen that the actual values of the predicted results of 8.8MPa, 5.5MPa and 2.0MPa are well within the prediction intervals. In order to meet the actual steam demand, the upper bound value is taken as the predicted value of each level of steam.

The upper bound of the prediction interval is related to the number generated in the scenario generation process. The range of the scenario generation number Sp is 500~5000. When the scenario generation number is 5000, the upper bound value tends to be stable. Therefore, by comparing the number of other generated scenarios with the upper bound of Sp=5000, the number of scenarios generated for different levels of steam is determined, as you can see in Figure 4.

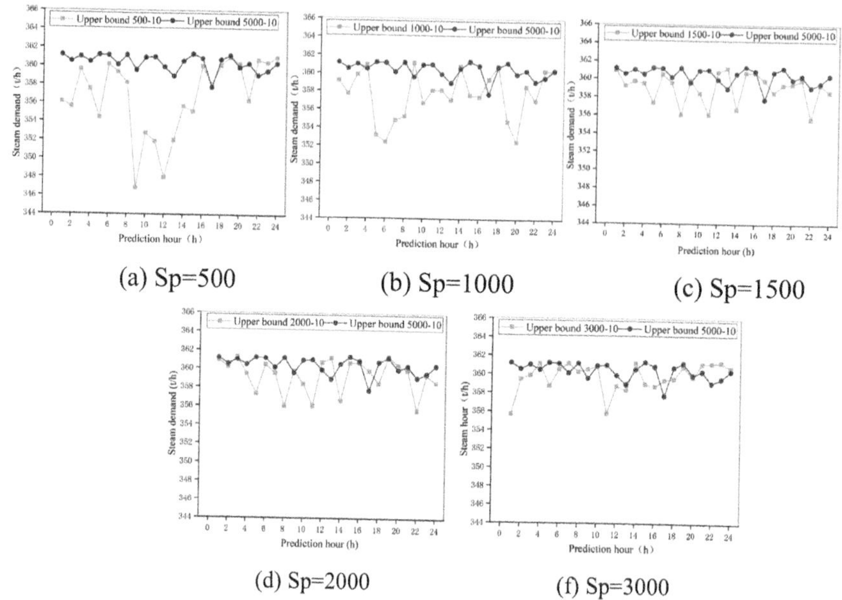

Fig.4 Comparison between the number of different generated scenarios and Sp=5000

For 8.8MPa, the number of scenarios Sp is 500. In the same way, the number of scenarios Sp is 1500 for 5.5MPa, the number scenarios Sp is 1000 for 2.0MPa.

2. Case study

2.1 The prediction results of one month steam demand

The example is the steam power system of a coal-to-gas plant. According to the extracted historical data of steam demand, the prediction results for one month are as follows in Figure 5.

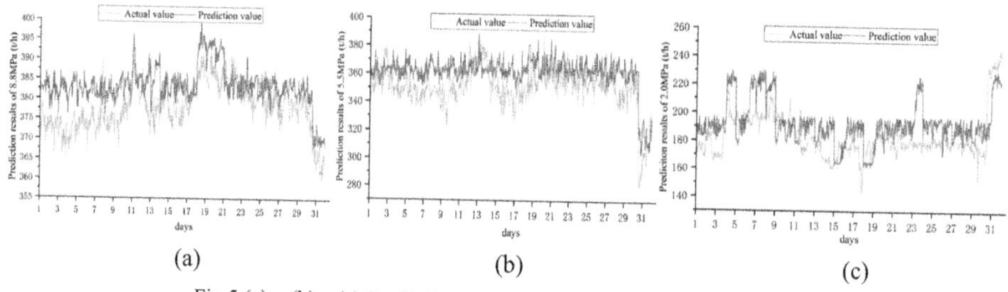

Fig.5 (a)、(b)、(c)-Prediction results of 8.8MPa、5.5MPa and 2.0MPa

As can be seen from Figure 5, the stability (the predicted value is greater than or equal to the actual value) of the three levels of steam is respectively 87.1%、79.4% and 76.2%.

2.2 Steam power system

The flow diagram of steam power system is shown in Figure 6. B1-B4 is the boiler. CN100T is 100MW high pressure extraction type direct air cooled turbine unit, CB30T is 30MW high pressure back pressure turbine unit. Insufficient power and high pressure steam are purchased from outside.

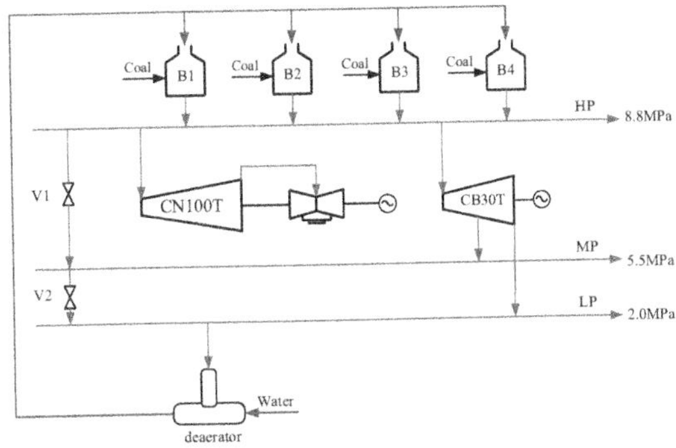

Fig.6 The flow diagram of steam power system

2.3 Analysis of optimization results

In this paper, sequential quadratic programming method was adopted to solve the steam power system model. The objective function is the operating cost, and the monthly steam demand value predicted based two-stage time series prediction model with Markov chain is brought into the optimization model. The comparison of various costs before and after optimization of each optimization method is shown in Table 1.

Table 1 Cost of each optimization method before and after optimization

Cost /10^4¥	Before optimization	This paper optimization	Robust optimization	Stochastic programming
Coal	5871	5721	5828	5694
Water	867	834	859	827
Purchased electricity	0	0	0	0
Purchased steam	0	0	0	0
The total cost	6738	6555	6687	6521
Cost saving	—	183	51	217

As shown in Table 1, the cost savings of the three optimization methods were 1.83 million yuan, 510,000 yuan and 2.17 million yuan respectively.

The optimized stability and economy of the three optimization methods are shown in Figure 7.

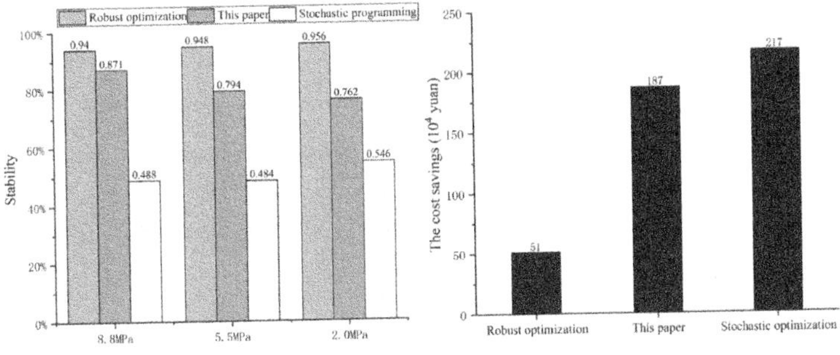

Fig.7 The optimized stability and economy of the three optimization methods

By comparing the stability and economy of the three optimization methods, the two-stage time series prediction optimization model with Markov chain in this paper takes into account the advantages of high economy of stochastic programming and high stability of robust optimization, and its stability and economy are between stochastic programming and robust optimization.

3. Conclusion

In this paper, a two-stage time series prediction optimization model with Markov chain method is proposed. The first stage is based on spatial distance expression to divide multiple operating conditions, and the second stage is based on state transition probability expression to predict the demand value of steam 24h before the day through the method of scenario generation and reduction. This method is used to solve the operation optimization problem of a steam power system of coal-to-gas. The optimization results show that:

(1) The stability of the three levels of steam demand prediction is between that of stochastic programming and robust optimization.

(2) The cost savings of one-month prediction optimization can save 1.83 million yuan, and its economy is between the stochastic programming and robust optimization.

(3) The proposed method combines the advantages of robust optimization and stochastic programming and has better performance.

References

[1] LI S, JIANG X B, HE G H, et al, 2017, Research progress for flexible design and multi-objective optimization of steam power network [J], Chemical Industry and Engineering Progress, 36, 6, 1989-1996.

[2] ALIPOUR M, ZARE K, SEYEDI H , 2018, A multi follower Bi-level stochastic programming approach for energy management of combined heat and power micro-grids [J], Energy, 149, 15, 135-146.

[3] Saeedi M, Moradi M, Hosseini M , Emamifar A , Ghadimi N, 2019, Robust optimization based optimal chiller loading under cooling demand uncertainty[J], Apply Thermal Engineering, 148, 1081−1091.

[4] Sun L, Gai L, Smith R, 2017, Site utility system optimization with operation adjustment under uncertainty[J], Applied Energy, 186, 450-456.

[5] Xie Y L, Huang G H, Li W, et al, 2013, An inexact two-stage stochastic programming model for water resources management in Nansihu Lake Basin, China[J], Journal of environmental management, 127, 188-205.

[6] Niu T, Yin H, Feng E, 2022, An interval two-stage robust stochastic programming approach for steam power systems design and operation optimization under complex uncertainties[J], Chemical Engineering Science, 253, 117533.

[7] Shen F, Zhao L, Du W, et al, 2020, Large-scale industrial energy systems optimization under uncertainty: A data-driven robust optimization approach[J], Applied Energy, 259, 114199.

[8] Sinaga K P, Yang M S, 2020, Unsupervised K-means clustering algorithm[J], IEEE access,8: 80716-80727.
Balzter H. Markov chain models for vegetation dynamics[J]. Ecological modelling, 126, 2-3, 139-154.

[9] Lee D, Baldick R, 2016, Load and wind power scenario generation through the generalized dynamic factor model[J], IEEE Transactions on power Systems, 32, 1, 400-410.

[10] Dupačová J, Gröwe-Kuska N, Römisch W, 2003, Scenario reduction in stochastic programming[J], Mathematical programming, 95, 3, 493-511.

[11] Chen Z, Yan Z, 2018, Scenario tree reduction methods through clustering nodes[J], Computers & Chemical Engineering, 109, 96-111.

Antonis Kokossis, Michael C. Georgiadis, Efstratios N. Pistikopoulos (Eds.)
PROCEEDINGS OF THE 33rd European Symposium on Computer Aided Process Engineering (ESCAPE33), June 18-21, 2023, Athens, Greece
 http://dx.doi.org/10.1016/B978-0-443-15274-0.50155-4

An approach for modelling simultaneous fluid-phase and chemical-reaction equilibria in multicomponent systems via Lagrangian duality: The Reactive HELD algorithm.

Felipe A. Perdomo[a], George Jackson[a], Amparo Galindo[a], Claire S. Adjiman,[a]

[a]*Department of Chemical Engineering, Sargent Centre for Process Systems Engineering, Institute for Molecular Science and Engineering, Imperial College London, South Kensington Campus, London SW7 2AZ, United Kingdom*

Abstract

An approach for the calculation of simultaneous phase and chemical-reaction equilibria and stability that does not require any assumptions on the number of stable phases and is applicable to any number of reactions is presented. It is based on an extension of the dual extremum principle concept for non-reactive systems (HELD) formulated in terms of the Helmholtz energy, where additional constraints are introduced that relate molar amounts and extents of reaction. The extended R-HELD algorithm is applied to reacting mixtures such as the esterification of acetic acid and the formation of methyl tert-butyl ether.

Keywords: Phase stability, reacting mixtures, SAFT-γ Mie equation of state.

1. Introduction

The calculation of equilibria in multiphase, multi-reaction systems is an important subject of study in chemical engineering thermodynamics. The inclusion of chemical reaction equilibria in multiphase equilibria gives rise to significant challenges in terms of thermodynamic consistency and numerical complexity. In reacting systems, stable phases may vanish or emerge as a consequence of the chemical reactions, making it difficult to determine the number and compositions of the phases coexisting at equilibrium. Two main classes of approaches have been presented to overcome these issues (Smith, 1980; Seider and Widagdo, 1996). In the first class, a non-linear programming approach is used to minimize the Gibbs energy function with a fixed number of phases (Castillo and Grossmann, 1981; Lantagne, et al., 1988). In the second class, phases are added (or removed) iteratively and the stability of each solution is tested (Gautam and Seider, 1979). This has led to approaches to solve simultaneously multiphase stability and equilibria in the presence of chemical reactions; (Michelsen, 1989; Gupta et al., 1991Smith et al., 1993; Jiang et al, 1996). These approaches allows for a good compromise between effectiveness and efficiency, but the search for the correct number of stable phases can lead to numerous iterations or convergence to local minima. In our current work, an approach is developed to obtain the stable state comprising one or more phases without making initial assumptions on the number of phases nor restricting the number of components in each phase. To this end, we extend the seminal work of Mitsos and Barton (2007) on the determination of the phase stability of non-reactive multicomponent mixtures via the dual-extremum principle (DEP). We adopt the Helmholtz energy formalism of Pereira et al. (2010; 2012), who recast the DEP in terms of temperature, pressure, and mole numbers as natural variables, a formulation which is better-suited to the use of state-of-the-art equations of state such as SAFT-γ Mie. Building on the DEP, a new phase stability and equilibria algorithm, HELD (Helmholtz energy Lagrangian dual) algorithm is developed, which we now extend to the reactive systems (R-HELD). In Section 2, we outline the extension of the DEP to reacting systems. In Section 3, we embed this stability criterion within the R-HELD algorithm. The performance of the algorithm on reacting systems with one or two phases is illustrated in Section 3.

2. A phase stability criterion for systems at chemical equilibrium

The extension of the DEP of Mitsos and Barton (2007) and of its Helmholtz energy counterpart (Pereira et al. 2010) to reacting systems is presented without proof. We formulate an optimization (dual) problem such that the set of all global solutions admitted by the dual problem is exactly equal to the set of stable phases of the mixture of interest. We consider the (trivial) single-phase minimisation of the Gibbs energy for a reacting system. Given C components, R independent reactions, temperature T_o, pressure P_o and total mole fractions $N_{i,0}$, $i = 1, \dots, C$, the minimisation of interest can be expressed using the molar volume, V, and the vector Ξ of (molar) extents of reaction, ξ_r, $r = 1, \dots, R$, as independent variables and including a material-balance constraint:

$$\begin{aligned} G_V = \quad & \min_{\Xi,V} \quad A(T_0, P_0, \boldsymbol{N}(\Xi)) + P_0 V \\ & s.t. \quad N_i(\Xi) - \left(N_i^0 + \textstyle\sum_{r=1}^{R} v_{i,r}\xi_r\right) = 0, i = 1, \dots, C, \\ & \qquad V \in [\bar{V}, \bar{\bar{V}}], \Xi \in \mathbb{R}_+^R \end{aligned} \tag{1}$$

where A is the Helmholtz energy, N_i is the number of moles of species i, $v_{i,r}$ is the stoichiometric coefficient of component i in reaction r, $\bar{V}$ and $\bar{\bar{V}}$ are lower and upper bounds on the volume, respectively. The application of the first-order optimality conditions for Problem (1) results in

$$-\left(\frac{\partial A(T_o, V^*, \boldsymbol{N}(\Xi^*))}{\partial V}\right)_{T, \xi_{r=1,\dots,R}} = P(T_o, V^*, \boldsymbol{N}(\Xi^*)) = P_o, \tag{2}$$

indicating that the molar volume at the solution (Ξ^*, V^*) corresponds to pressure P_o (Nagarajan et al., 1991), and

$$\textstyle\sum_{i=1}^{C} \mu_i\left(T_o, V^*, \boldsymbol{N}(\Xi^*)\right) v_{i,r} = 0, \quad r = 1, \dots, R, \tag{3}$$

enforcing chemical-reaction equilibria for all R reactions. Thus, these conditions do not need to be stated explicitly as constraints, and Problem (1) is equivalent to the minimisation of the Gibbs energy at $(T_o, P_o, \boldsymbol{N}_0)$ for a phase at chemical equilibria.

We now recast Problem (1) in terms of the set of transformed compositional variables proposed by Ung and Doherty (1995). These $C-R$ compositional variables exhibit useful properties such as independence of the extents of reaction, summation to unity, and thermodynamic consistency. Using the summation to unity property, variable $C - R$ can be eliminated. The resulting vector of $C' = C - R - 1$ variables is denoted by $\widehat{\boldsymbol{x}}$. The primal problem of Pereira et al. (2010) can then be reframed as

$$\begin{aligned} \widehat{G_V^P} = \quad & \min_{\widehat{\boldsymbol{x}},V} \quad A(T_o, V, \widehat{\boldsymbol{x}}) + P_o V \\ & s.t. \quad \widehat{x}_{i,o} - \widehat{x}_i = 0, i = 1, \dots, C' \\ & \qquad \widehat{\boldsymbol{x}} \in \mathcal{X}, V \in [\bar{V}, \bar{\bar{V}}] \end{aligned} \tag{4}$$

where $\mathcal{X}$ defines the domain of the compositional variables. The Lagrange function, corresponding to Problem (5), is given by

$$\widehat{L^V}(\hat{\boldsymbol{x}}, V, \lambda) = A(T_o, V, \hat{\boldsymbol{x}}) + P_o V + \sum_{i=1}^{C'} \lambda_i (\hat{x}_{i,o} - \hat{x}_i), \quad (5)$$

and the dual of Problem (4) is given by

$$\widehat{G^D} = \max_{\lambda \in R^{C'}} \min_{\hat{x} \in \mathcal{X}, V \in [\underline{V}, \overline{\overline{V}}]} \widehat{L^V}(\hat{x}, V, \boldsymbol{\lambda}) \quad (6)$$

Under the assumption that the solutions of the inner problem in Problem (6), i.e., the minimisation of the Lagrange function, lie in the interior of sets $\mathcal{X}$ and $[\underline{V}, \overline{\overline{V}}]$, the solution set of Problem (6) describes the stable state of the reacting mixture at $(T_0, P_0, \boldsymbol{N}_0)$.

3. The reactive HELD (R-HELD) algorithm for reacting systems

Using Problem (6), we extend the implementation of the HELD algorithm (Pereira et. al, 2012) to enable the identification of all stable phases of systems undergoing chemical reactions. The implementation makes use of a multistart approach combined with a SQP solver to solve any nonlinear problem. Thus global optimality cannot be guaranteed.

3.1. Stage 1: Initialization.

Primal problem (4) admits a trivial solution at $\hat{\boldsymbol{x}} = \hat{\boldsymbol{x}}_o$, where $\hat{\boldsymbol{x}}_o$ is found by solving for chemical equilibria at the specified conditions. This yields an upper bound on the minimum Gibbs energy of the system. The tangent-plane distance criterion is applied at $\hat{\boldsymbol{x}}_o$ to determine whether the single phase is stable. If a negative value of the tangent-plane distance is found, the algorithm proceeds to Stage 2. Otherwise, it terminates.

3.2. Stage 2: Identification of candidate stable phases

An iterative search for all global solutions of Dual problem (6) is carried out. At iteration k, (i) given a guess vector $\boldsymbol{\lambda}^k$, the (inner) minimisation of the Lagrange function yields a lower bound (LBD^V) on the minimum Gibbs free energy and a solution $(\hat{\boldsymbol{x}}^k, V^k)$, which is used to construct an approximation of the inner problem by substitution into Eq. (5) and, (ii) an approximate Dual problem, constructed from all solutions to the inner problem up to iteration k is solved with respect to the C' Lagrange multipliers. This outer problem yields an upper bound (UBD^V) on the minimum Gibbs energy and a new guess $\boldsymbol{\lambda}^{k+1}$. The iterations continue until the gap between the lower and upper bounds converges to a specified tolerance, ε, i.e., $|UBD^V - LBD^V| \leq \varepsilon$.

3.3. Stage 3: Material balance

The total Gibbs energy, calculated using all the phases identified so far, is minimised, subject to the overall mass balance. If the problem is found to be feasible, further criteria are applied, including checking the gap between the total Gibbs energy and the upper bound found in Stage 2 and ensuring that the necessary conditions for phase and chemical equilibria are satisfied across all species. This latter check is important to ensure the accuracy of the calculations – the convergence tolerance can be tightened if this is not met and Stage 2 to achieve increased accuracy / identify missing phases.

4. Results

We assess the performance of the extended R-HELD algorithm for several multicomponent reactive systems at conditions where the system is stable as in a single phase or coexists in several phases in the presence of chemical reactions. Our approach only requires knowledge of the stoichiometry of each reaction and the enthalpies and entropies of formation of each species at the selected standard state, here, the ideal gas at 298.15 K and 1.013 bar. All mixtures are modelled with SAFT-γ Mie equation of the state (Papaioannou et al., 2014) with parameters reported in Haslam et al. (2021) and in the supporting information.

4.1. Single *stable phase with reaction equilibria*

We consider the liquid phase synthesis of methyl tert-butyl ether (MTBE) from isobutene (IB) and methanol (MeOH) at 16 bar. The predicted chemical-reaction equilibria surfaces (CES) are presented in Figure 1 for temperatures from 323.15 K to 353 K and initial methanol/isobutene molar ratio from 0.1 to 1.6, with data from Izquierdo et al. (1992).

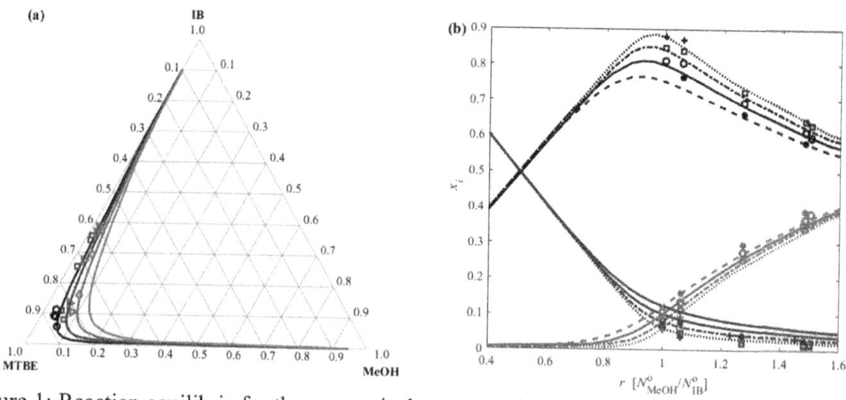

Figure 1: Reaction equilibria for the system isobutene, methanol, and methyl ter-butyl ether at 16 bar: (a) CES at 323.15 K (black circles), 333.15 K (blue squares), 343 K (red triangles), and 353 K (magenta diamonds). (b) Composition profiles for reactants and products as a function of initial MeOH/IB ratio: IB (blue), MeOH (red), and MTBE (black) at 323.15 K (crosses), 333.15 K (squares), 343 K (circles), and 353 K (filled circles). Continuous curves are predictions using SAFT-γ Mie and R-HELD and symbols are experimental data (Izquierdo et. al, 1992)

The quaternary reactive mixture arising from the esterification of acetic acid with ethanol acetic acid [(Ac.Ac) + ethanol (EtOH) $\leftrightharpoons$ ethyl acetate (Et.AcO) + water (H_2O)] is modelled with R-HELD at conditions where one phase is present (1.013 bar and 293.15 K). The predicted CES is presented in Figure 2 as a concentration tetrahedron and is seen to be in excellent agreement with the experimental data of Golikova et al. (2017). The CES is seen to limit feasible compositional space that can be reached.

4.2 Simultaneous phase and chemical reaction equilibria

We present in Figure 3 the calculation at conditions at which phase and chemical equilibria interact for the mixture IB, MeOH, MTBE (5 bar, 342.25 to 378.29 K), together with the molecular simulation data of Lisal et al. (2010). In Figure 3a, the CESs in the vapour and liquid phase (blue) and the bubble and dew curves (black) are seen to intersect at the solutions of the simultaneous phase- and chemical-equilibrium problem.

5. Conclusions and outlook

We have extended the dual extremum principle as a criterion for fluid-phase stability, as proposed by Mitsos and Barton (2007), to include chemical reactions. Specifically, we have extended the Helmholtz energy formalism proposed by Pereira et al. (2010), combined with the compositional transforms of Ung and Doherty (1995). The approach, presented here without proof, is consistent, general, and can be applied to multicomponent and multi-reaction systems coexisting at phase equilibria. We have implemented this within the framework of the HELD algorithm for non-reactive systems (Pereira et al., 2012). The extended R-HELD algorithm has been tested successfully with ternary and quaternary mixtures with one or two stable phases, using the SAFT-γ Mie equation of state. Broadening the scope of the HELD algorithm allows us to consider challenging mixtures that exhibit simultaneous phase and reaction equilibria.

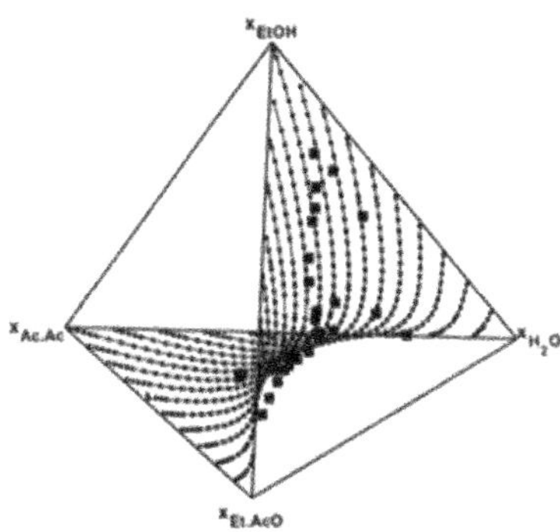

Figure 2 - CES for the esterification reaction of acetic acid 323.15 K and 1.013 bar computed with SAFT-γ Mie and R-HELD (continuous-dotted envelope) and measured experimentally (symbols)

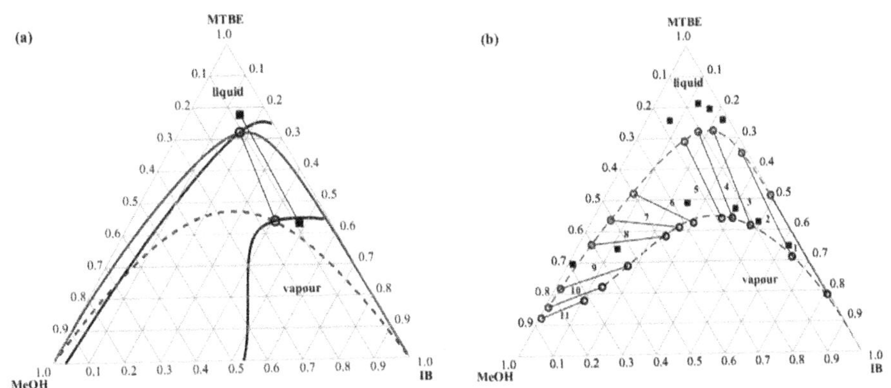

Figure 3: Mixture of IB, MeOH and MTBE at 5 bar (a) CES at 360 K. Black continuous curves indicate non-reactive VLE; continuous and dotted blue curves represent the prediction of the single-phase chemical reaction equilibria (CES) in the liquid and vapour phase respectively. The light solid lines represent reactive tie lines. (b) Complete simultaneous reaction + phase equilibria, with reactive tie-lines (circles connected by solid lines) 1) 342.25 K, 2) 351.8 K, 3) 358.4 K, 4) 360.15 K, 5) 360.65 K, 6) 363.25 K, 7) 365.19 K, 8) 367.15 K, 9) 372.81 K, 10) 376.13 K, 11) 378.29 K. Filled black squares are molecular simulation data (Lisal, et al., 2000).

Acknowledgments

The authors are grateful for financial support from Eli Lilly & Company and the UK Engineering and Physical Sciences Research Council (EP/T005556/1).

Data statement / Supplementary information

All data related to this article and not published elsewhere can be found at doi: 10.5281/zenodo.7418447 and used under the Creative Commons Attribution licence.

References

J. Castillo, and I.E., Grossmann, 1981. Computation of phase and chemical equilibria. Comput. Chem. Eng, 5(2), 99-108.

A. Golikova, A. Samarov, M. Trofimova, S. Rabdano, M. Toikka, O. Pervukhin, and A. Toikka, 2017. Chemical equilibrium for the reacting system acetic acid–ethanol–ethyl acetate–water at 303.15 K, 313.15 K and 323.15 K. J. Solution Chem., 46(2), 374-387.

A.K. Gupta, P.R. Bishnoi, and N. Kalogerakis, 1991. A method for the simultaneous phase equilibria and stability calculations for multiphase reacting and non-reacting systems. Fluid Phase Eq., 63(1-2), 65-89.

A.J. Haslam., A. González-Pérez, S. Di Lecce, S.H. Khalit, F.A. Perdomo, S. Kournopoulos, M. Kohns, T. Lindeboom, M. Wehbe, S. Febra, and G. Jackson, 2020. Expanding the Applications of the SAFT-γ Mie Group-Contribution Equation of State: Prediction of Thermodynamic Properties and Phase Behavior of Mixtures. J. Chem. Eng. Data, 65(12), 5862-5890.

J.F. Izquierdo, F. Cunill, M. Vila, J. Tejero, and M. Iborra, 1992. Equilibrium constants for methyl tert-butyl ether liquid-phase synthesis. J. Chem. Eng. Data, 37(3), 339-343.

Y. Jiang, G.R. Chapman, and W.R. Smith, 1996. On the geometry of chemical reaction and phase equilibria. Fluid Phase Eq., 118(1), 77-102.

G. Lantagne, B. Marcos, and B. Cayrol, 1988. Computation of complex equilibria by nonlinear optimization. Comput. Chem. Eng, 12(6), 589-599.

M. Lísal, W.R. Smith, and I. Nezbeda, 2000. Molecular simulation of multicomponent reaction and phase equilibria in MTBE ternary system. AIChE J., 46(4), 866-875.M.L. Michelsen, 1989. Calculation of multiphase ideal solution chemical equilibrium. Fluid Phase Eq., 53, 73-80.

A. Mitsos, and P.I. Barton, 2007. A dual extremum principle in thermodynamics. AIChE J., 53(8), 2131-2147.

N.R. Nagarajan, A.S. Cullick, and A. Griewank, 1991. New strategy for phase equilibrium and critical point calculations by thermodynamic energy analysis. Part I. Stability analysis and flash. Fluid Phase Eq., 62(3), 191-210.

V. Papaioannou, T. Lafitte, C. Avendaño, C.S. Adjiman, G. Jackson, E.A. Müller, and A. Galindo, 2014. Group contribution methodology based on the statistical associating fluid theory for heteronuclear molecules formed from Mie segments. The J. Chem. Phys., 140(5), 054107.

F.E. Pereira, G. Jackson, A. Galindo, and C.S. Adjiman., 2010. A duality-based optimisation approach for the reliable solution of (P, T) phase equilibrium in volume-composition space. Fluid Phase Eq., 299(1), 1-23.

F.E. Pereira, G. Jackson, A. Galindo, and C.S. Adjiman., 2012. The HELD algorithm for multicomponent, multiphase equilibrium calculations with generic equations of state. Comput. Chem. Eng, 36, 99-118.

W.D. Seider, and S. Widagdo, 1996. Multiphase equilibria of reactive systems. Fluid Phase Eq., 123(1-2), 283-303.

W.R. Smith, 1980. The computation of chemical equilibria in complex systems. Ind. Eng. Chem. Fund., 19(1), 1-10.

J.V. Smith, R.W. Missen, and W.R. Smith, 1993. General optimality criteria for multiphase multireaction chemical equilibrium. AIChE J., 39(4), 707-710.

S. Ung, and M.F. Doherty, 1995. Theory of phase equilibria in multireaction systems. Chem. Eng. Sci., 50(20), 3201-3216.

S. Ung, and M.F. Doherty, 1995. Vapor-liquid phase equilibrium in systems with multiple chemical reactions. Chemical Engineering Science, 50(1), 23-48.

Antonis Kokossis, Michael C. Georgiadis, Efstratios N. Pistikopoulos (Eds.)
PROCEEDINGS OF THE 33rd European Symposium on Computer Aided Process Engineering (ESCAPE33), June 18-21, 2023, Athens, Greece
 http://dx.doi.org/10.1016/B978-0-443-15274-0.50156-6

Optimization of the Biomass Supply Chain for Power Generation with the Steam Rankine Cycle

Monika Dokl, Zdravko Kravanja, Lidija Čuček

Faculty of Chemistry and Chemical Engineering, University of Maribor, Smetanova ulica 17, 2000 Maribor, Slovenia
lidija.cucek@um.si

Abstract

The steam Rankine cycle (SRC) is a mature technology adapted to high temperatures and has a high electrical efficiency. However, the cost-effective and environmentally-sound utilisation of biomass in SRC requires efficient management of the supply chain and overcoming challenges such as complex conversion processes and uncertainties in biomass quality and supply. In this study, the optimisation of biomass utilisation using SRC to produce electricity is conducted considering the entire biomass supply chain. The model is developed as a combined biomass supply chain network and SRC process flowsheet and optimized in GAMS to achieve the most effective power generation while optimising the network economics. The concept of the SRC process-based approach to biomass supply network is evaluated through a demonstrative case study of a relatively small region.

Keywords: supply chain optimisation, mathematical modelling, biomass utilisation, steam Rankine cycle

1. Introduction

The depletion of fossil fuels, their negative impact on the environment and the security of energy supply have stimulated many countries to focus on more extensive use of renewable energy resources to reduce environmental burdens and achieve energy security and fuel diversification. Among renewable energy sources, biomass is becoming increasingly important, as it can be used for heat and power generation and biofuel production. Apart from biofuels, the most widely used technology for biomass utilisation is direct combustion, where the resulting high-temperature gases are typically used to generate high-pressure steam. Power generation using steam cycles such as the steam Rankine cycle (SRC) is suitable for large-scale power plants where thermal efficiencies of over 30 % can be achieved. SRC uses water as working fluid, which has the advantage of being highly available and non-toxic, but it precludes its wide range of applications, especially for low and medium grade heat sources (Abbas et al., 2020).
In many situations, biomass suitable for power generation is available as a feedstock in large quantities with low density and largely distributed production, characterised by its seasonal availability. Achieving competitive electricity generation through biomass utilization requires an efficient management of the biomass supply network, which primarily addresses biomass harvesting, transportation, storage, and utilisation. Biomass supply chain optimisation focuses on supply chain performance and configuration, component sizing and operations scheduling, usually with an objective of minimising cost (Nunes et al., 2020).

In this study a new supply network concept is proposed that considers integrated biomass supply network formulation and detailed optimization of SRC process. The objective of the study is maximising the revenue from selling heat and electricity and subtracting distance-dependent biomass transportation cost and cost of piping network to distribute heat.

2. Methodology

The proposed approach consists of optimizing the integrated biomass supply network and SRC process flowsheet system. First, the proposed integrated design is explained, and further mathematical formulation is briefly presented.

2.1. Biomass Supply Network and Steam Rankine Cycle

An integrated biomass supply network and SRC process, schematically represented in Fig. 1, consists of 3 main layers (L1-L3). The first layer (L1) represents the supply of feedstocks to the plant and accounts for the regional characteristics of each zone (i) and biomass storage at locations s. The second layer (L2) represents the SRC plants which are located at sites j. SRC processes are formulated as a flowsheet structure consisting of two heat exchangers (boiler and condenser) and two pressure changers (turbine and pump). The formulation enables consideration of details, such as optimization of temperatures, pressures, flowrates, specific enthalpies and energy flows. Generated products by SRC plants (heat and electricity) are then distributed to the third layer (L3) at sites d. Road transport is considered as the transportation mode for biomass to SRC utilisation sites, while pipeline is used for delivery of hot water to demand sites.

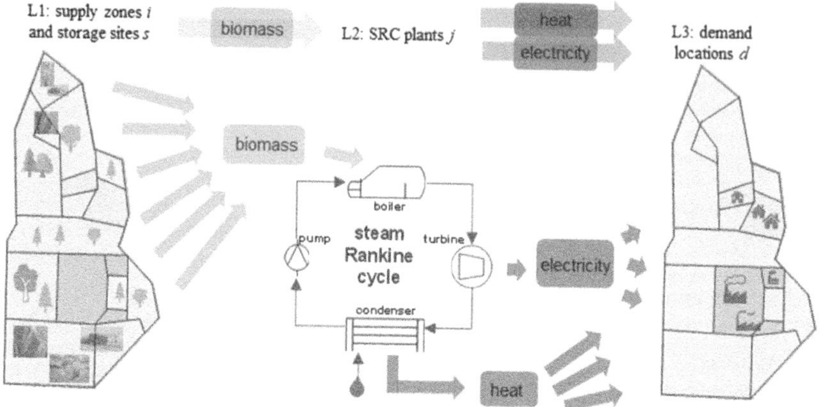

Figure 1: Schematic representation of integrated biomass supply network and SRC

2.2. Model Formulation

The developed model for optimization of biomass supply chain for maximal profit of sold electricity and heat produced by SRC reduced by variable transportation cost comprises material and energy balances and economic constraints (transportation cost, revenue) for the supply and demand layers, interconnections supply - process and process - demand and optimization of SRC which accounts for optimizing the operating conditions of the cycles. The approach for simultaneous consideration of supply network and process flowsheet optimization is based on the work by Kegl et al. (2021) and the approach for optimizing operating conditions in the SRC is adapted from Dokl et al. (2022).

For simplification, the main objective (Z) is to maximize revenue obtained from SRC (R_{SRC_j}) subtracted by transportation cost generated from the supply network (c_{tr}) (Eq. (1)).

$$Z = \max(\sum_{j=1}^{n} R_{\mathrm{SRC}_j} - c_{\mathrm{tr}}) \quad (1)$$

Such objective is selected to identify a simple economic bound of the system (Biegler et al., 1997), while considering only such costs that the selected distances between possible connections will be minimal. Transportation cost is for simplification thus calculated only based on the distance variable costs, and accounts for factors related to return trips and tortuosity factor (Egieya et al., 2019). The obtained revenue of plant j is based on production of electricity and heat (Eq. (2)). Electricity generation in the turbine depends on the fluid flow rate (q_j^{turb}), specific enthalpy difference ($h_j^{\mathrm{turb,out}} - h_j^{\mathrm{turb,in}}$) and turbine efficiency (η_j^{turb}), while the obtained heat in the form of hot water at the condenser outlet is described by the heat duty of the condenser (Q_j^{cond}) reduced by the transportation heat loss ($L_{\mathrm{tr}}^{\mathrm{heat}}$) between plant j and demand site d ($D_{j,d}$).

$$R_j = \left(q_j^{\mathrm{turb}} \cdot \left(h_j^{\mathrm{turb,out}} - h_j^{\mathrm{turb,in}}\right) \cdot \eta_j^{\mathrm{turb}}\right) + \left(Q_j^{\mathrm{cond}} \cdot \left(1 - L_{\mathrm{tr}}^{\mathrm{heat}}\right)^{D_{j,d}}\right) \quad (2)$$

The optimized operating conditions are obtained by correlating thermophysical properties with temperature and/or pressure. Such properties are specific enthalpies of water and steam (Fig. 2), vapor liquid pressure, specific heat capacity at constant pressure and heat capacity ratio.

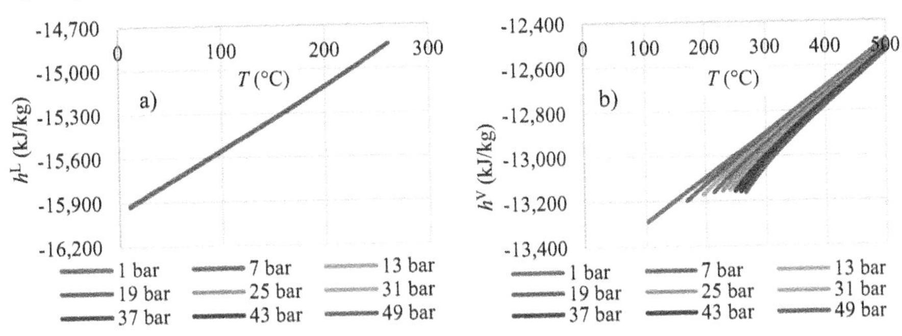

Figure 2: Specific enthalpy of a) water and b) steam, changing with temperature and pressure

3. Case Study and Results

The integrated biomass supply network – SRC flowsheet optimization model is implemented on a demonstration case study of a relatively small region. For simplification, the model assumes single time-period and the average values of parameters in a year. The map of the case study is shown in Fig. 3, with a total area of 906.6 km^2. Up to 17 supply (i_1 - i_{17}), 8 storage (s_1 - s_8), 5 plant (j_1 - j_5) and 5 demand locations (d_1 - d_5) (built-up and industrial areas) could potentially be selected.

The actual areas for growing biomass in zones (forest and agricultural areas) are between 10 and 90 % of the zone areas. The yield of wood is based on MKGP (2011) and of agricultural residues from Jejčič and Verbič (2019). Only the main biomass feedstocks pb are considered, which are wood, corn stover, grain straw and oilseed straw with the yields of 3.21 t/(ha·y) for wood, 5.41 t/(ha·y) for grain straw, 5.59 t/(ha·y) for corn stover and 5.72 t/(ha·y) for oilseed straw. The distribution of area for biomass in the municipality is 87.3 % for wood, 8.6 % for grain straw, 3.9 % for corn stover and 0.2 % for oilseed straw.

For the SRC, all the characteristics of process units and streams were optimized to satisfy the given objective. The only assumptions are minimum temperature difference in condenser, which is 5 °C for all the cycles, 85 % mechanical efficiency of turbines, maximum flow rate of water in the condenser of 200 kg/s (Yıldırım et al., 2006), and temperature of water at condenser inlet (50 °C) and outlet (90 °C) (Čuček et al., 2015). The MINLP model consists of 3852 single equations, 5026 single variables, and 25 binary variables and was solved using DICOPT solver in GAMS (GAMS, 2022). The solutions are provided in around 10 s on a personal computer with an Intel® Core™ i7-10750 H CPU @ 2.60 GHz processor with 16 GB RAM. For the calculation of the revenue from sold electricity and heat, price of electricity of 222.48 €/MWh was assumed as the average day-ahead Slovenia auction trading price of November 2022 (BSP Energy Exchange, 2022) and price of heat of 101.07 €/MWh, which is a recent price for the supply of natural gas for business consumption (Energetika Ljubljana, 2022).

The results from the breakdown of optimal quantities of feedstocks selected and related transportation cost together with optimal conditions of selected SRC plants (j_1, j_3 and j_4) are shown in Table 1. From Table 1 it can be seen that optimal biomass supply chain structure and operating conditions of SRC plant can generate the total annual revenue of 67.89 M€, of which the highest revenue is obtained from heat (29.12 M€/y) and electricity (8.53 M€/y) by SRC plant j_3.

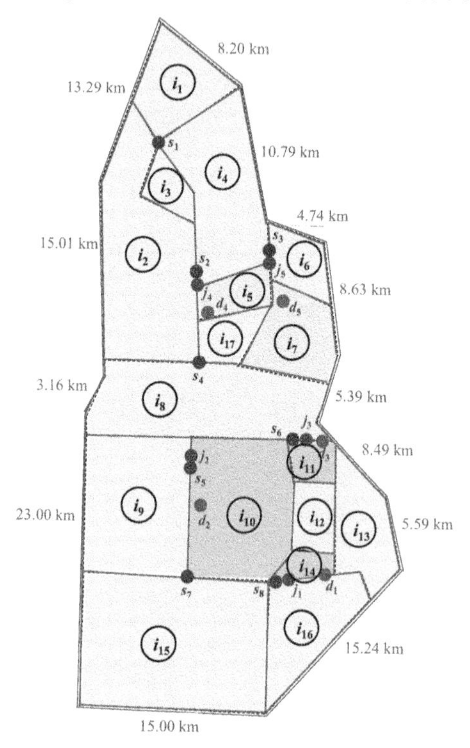

Figure 3: Map of demonstrative case study with areas of zones, storage, plant and demand sites

Optimal turbine working conditions are the same for all selected SRC plants, where steam enters the turbine with temperature of 284.7 °C and pressure of 5.8 bar and passes to the condenser with temperature of 103.9 °C and pressure of 1.1 bar. Higher heat and electricity rates in plant j_3 correspond to higher flows of working fluid and water in the condenser, reaching the upper limit of 200 kg/s. SRC plant j_4 has for 14 M€/y lower revenue than plant j_3, while with plant j_1 only 6.6 M€/y of revenue is obtained. Distance-dependent transportation cost presents 7.98 % of revenue for plant j_1, 2.2 % for plant j_3, and only 0.72 % for j_4. Fig. 4 further shows optimised supply chain superstructure, presenting relation between layers with optimal selection and connections between sites of biomass supply, storage sites, utilisation plants, and demand sites.

Three storage locations are selected for storing feedstocks, where s_2 receives biomass from 8 zones, s_6 from 9 zones, one of which (i_{16}) transport biomass also to s_8. Feedstock is further transported to the selected utilisation plant j_1, j_3, and j_4, delivering produced heat to demand sites d_1, d_3, and d_4.

Table 1: Optimal values of variables for selected SRC plants

Optimal values	j_1	j_3	j_4
Steam Rankine cycle			
Q_{boiler} (kW)	6879.1	38,670.9	24,420.2
$W_{turbine}$ (kW)	778.4	4375.7	2763.2
F_{steam} (kg/s)	2.6	14.8	9.3
F_{water} (kg/s)	35.6	200.0	126.3
$T_{turbine}^{in}$ | $T_{turbine}^{out}$ (°C)		284.7 | 103.9	
$p_{turbine}^{in}$ | $p_{turbine}^{out}$ (bar)		5.8 | 1.1	
Economic parameters			
Road transport cost (M€/y)	0.414	0.771	0.093
Pipeline cost (M€/y)	0.113	0.056	0.077
Revenue from heat (M€/y)	5.08	29.12	18.26
Revenue from electricity (M€/y)	1.52	8.53	5.39
Objective value (M€/y)		66.37	

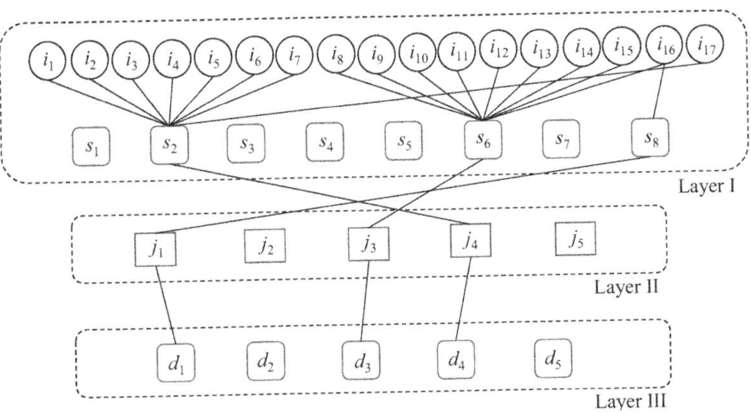

Figure 4: Biomass supply chain superstructure

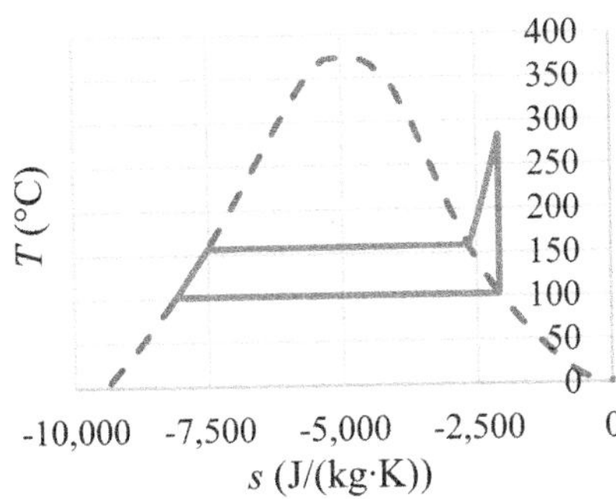

Figure 5: *T-s* diagram for SRC at selected plants

In Fig. 5 temperature vs. entropy (*T-s*) diagram for all the selected SRC plants is shown, representing energy transfer within the cycles. The lower horizontal blue line illustrates steam condensing in the condenser, which is further passed to the boiler, where its temperature is increased to 284.7 °C, extending high above saturation curve (grey dashed curve) as it is being superheated for about 126 °C. The expansion in the turbine is shown by vertical blue line, also closing the cycle by entering the condenser again.

4. Conclusions and Recommendations

In this paper the integration of biomass supply network and SRC flowsheet is performed by optimising the revenue from selling electricity and heat subtracted by transportation cost, and demonstrated on an illustrative case study of a relatively small region. The optimal structure of biomass supply network at optimal operating conditions of selected SRC plants can annually produce almost 8 MW of electricity and about 1300 t/h of hot water for district heating at supply and return temperatures of 90 and 50 °C. Optimal supply chain consists of 17 biomass supply zones, 3 storage locations, 3 SRC plants for biomass utilisation and 3 demand sites of heat consumption. In the future, a multiperiod model could be developed to account for the seasonal characteristics of biomass supply and the complete economic and environmental performance of the studied superstructure could be evaluated.

Acknowledgments

The authors are grateful for funding support from the Slovenian Research Agency (research core fundings No. P2-0412, P2-0421 and P2-0414 and project No. J7 3149).

References

T. Abbas, M. Issa, A. Ilinca, 2020, Biomass cogeneration technologies: A review, Journal of Sustainable Bioenergy Systems, 10(1), 1-15.

L.T. Biegler, I.E. Grossmann, A.W. Westerberg, 1997, Systematic Methods of Chemical Process Design, Prentice Hall PTR, New Jersey, USA

BSP Energy Exchange, 2022, Day-ahead Trading Results: Auction Trading, available at: <bsp-southpool.com/day-ahead-trading-results-si.html> (accessed 29.11.2022)

M. Dokl, R. Gomilšek, L. Čuček, B. Abikoye, Z. Kravanja, 2022, Maximizing the power output and net present value of organic Rankine cycle: Application to aluminium industry, Energy, 239, Part E, 122620

J.M. Egieya, L. Čuček, K. Zirngast, A.J. Iafiade, B. Pahor, Z. Kravanja, 2019, Synthesis of biogas supply networks using various biomass and manure types, Computers and Chemical Engineering, 122, 129-151

L. Čuček, V. Mantelli, J. Y. Yong, P. S. Varbanov, J. J. Klemeš, Z. Kravanja 2015, A procedure for the retrofitting of large-scale heat exchanger networks for fixed and flexible designs applied to existing refinery total site, Chemical Engineering Transactions, 45, 109-114

Energetika Ljubljana, 2022, Price list of public company Energetika Ljubljana d.o.o. for the supply of natural gas for business consumption, available at: <energetika.si/sites/www.jhl.si/files/cenik/datoteke/plin_-_cenik_za_dobavo_-_redni_1._10._2022_poslovni_odjem_0.pdf> (accessed 29.11.2022) (in Slovenian)

GAMS, 2022, available at: <gams.com/39/docs/> (accessed: 8.12.2022)

V. Jejčič, J. Verbič, 2019, Utilization of biomass for energy purposes and potentials to 2050, Overview of biomass potentials from agriculture, available at: <podnebnapot2050.si/wp-content/uploads/2019/03/2b_BIOMASA-IJS-LIFE-CLIMATE-PATH-2050-1-2019.pdf> (accessed: 5.11.2022) (in Slovenian)

T. Kegl, L. Čuček, A. Kovač Kralj, Z. Kravanja, 2021, Conceptual MINLP approach to the development of a CO_2 supply chain network – Simultaneous consideration of capture and utilization process flowsheets, Journal of Cleaner Production, 314, 128008

Ministry of Agriculture, Forestry and Food of Slovenia (MKGP), 2011, Strategy for utilization of biomass from agriculture and forestry for energy purposes, available at: <arhiv2014.skupnostobcin.si/fileadmin/sos/datoteke/pdf/Barbara/PREDLOGI_PREDPISOV/Kmetijstvo/Strategija_biomasa_priloga.pdf> (accessed: 5.11.2022) (in Slovenian)

L.J.R. Nunes, T.P. Causer, D. Ciolkosz, 2020, Biomass for energy: A review on supply chain management models, Renewable and Sustainable Energy Reviews, 120, 109658.

N. Yıldırım, M. Toksoy, G. Gökçen, 2006, District heating system design for a university campus, Energy and buildings, 38(9), 1111-1119.

Antonis Kokossis, Michael C. Georgiadis, Efstratios N. Pistikopoulos (Eds.)
PROCEEDINGS OF THE 33rd European Symposium on Computer Aided Process Engineering (ESCAPE33), June 18-21, 2023, Athens, Greece
 http://dx.doi.org/10.1016/B978-0-443-15274-0.50157-8

Hybrid modelling and data-driven parameterization of monoclonal antibody cultivation processes: Shifts in cell metabolic behavior

Sara Badr,[a] Kota Oishi,[a] Kozue Okamura,[a] Sei Murakami,[b] Hirokazu Sugiyama[a*]

[a]*Department of Chemical System Engineering, The University of Tokyo, 7-3-1, Hongo, Bunkyo-ku, 113-8656, Tokyo, Japan*
[b]*Manufacturing Technology Association of Biologics, 2-6-16, Shinkawa, Chuo-ku, 104-0033, Tokyo, Japan*
sugiyama@chemsys.t.u-tokyo.ac.jp

Abstract

Process developments to achieve more efficient monoclonal antibody (mAb) production processes include cell line modifications to obtain high-performance alternatives. Cell lines with higher growth rates can exhibit higher sensitivities to variations in carbon sources in the cell culture leading to shifts in metabolic behavior. Current kinetic process models often rely on static model parameters like specific cell consumption and production coefficients. Such models can be inadequate for describing process performance in cases of shifts in cell metabolism. Further understanding of the complex biological phenomena is required to develop more robust process models. In this work, pilot-scale experimental results were obtained for a novel high-performance cell line exhibiting switches in metabolic behavior. State estimation-based methods were used to evaluate the model fit over the course of the run and to identify regions where parameter updates are required. Clustering of the underlying experimental conditions was performed to isolate the sources of variations and correlate them with the different regions of model fit. Alternative empirical formulations are accordingly proposed for the varying model parameters. This approach allows for the development of more interpretable hybrid models by using the data-driven insights to categorize and describe the underlying variations in operating parameters and their influence on cell behavior and process performance.

Keywords: Biopharmaceutical production, cell culture, state estimation, hierarchical density-based clustering

1. Introduction

The exponential rise in demand for monoclonal antibodies (mAb) over the past decade (Butler & Meneses-Acosta, 2012; Walsh, 2018) has been driving the search for innovative solutions to increase production efficiency. In addition to exploring different operation modes, e.g., continuous cultivation, and optimizing operating conditions, developments were achieved through modifying the cell lines used (Badr & Sugiyama, 2020). The newly established CHO-MK cell line is a high-performance alternative offering rapid cell growth and higher mAb productivity. This cell line is, however, sensitive to carbon source concentrations, e.g., glucose and lactate, and could exhibit fluctuating behavior in cell metabolism. A shift from lactate production to net consumption has been observed with this cell line. Such shifts have been reported for other cell lines in literature, but the exact mechanism leading to such a switch is still unverified (Hartley et al., 2018). Understanding and manipulating these shifts is important for better process controllability and achieving higher efficiency.

Developing versatile process models to adequately describe the process operation and changes is essential to reach optimal solutions for process design, operation, and control. Kinetic models currently available for describing cell cultivation often rely on static parameters to describe cell-related factors, such as specific productivity or nutrient consumption rates. For stable cell lines, like the standard CHO-K1 cells, these models could show relatively good fit to the experimental and pilot-scale data obtained (Badr et al., 2021). In industrial or experimental applications, cells may temporarily take different metabolic pathways. The reliance on static parameters in the models often leads to discrepancies with industrial applications, especially in the case of sensitive cell lines as the CHO-MK cell line. Discrepancies can be sometimes only partial and not over the course of the entire experimental run. Okamura et al. (2022) have demonstrated the use of a hybrid modelling technique, which inserts a data-driven module to get updated parameter values to better describe shifts to lactate consumption. The input to the data-driven module included the online sensor measurements and the initial predictions of the kinetic model.

In this work, we investigate a data-driven approach to aid the development of more comprehensive and robust mechanistic models for process design and control. This work aimed to isolate the operating conditions correlated to the shifts in metabolic behavior and a change in state. The model fit was analyzed to identify regions where parameter updates are required. Data clustering of the online measurements was used to isolate separate phases of the underlying experimental conditions. The results were combined to suggest new forms for influential model parameters.

2. Methods

2.1 Experimental data

Pilot-scale experimental data were obtained from the Kobe GMP consolidated lab of the Manufacturing Technology Association of Biologics in Japan. CHO-MK 9E-1 cells were cultivated in a 50 L stirred tank in fed-batch mode for 7 days. Supplementary media feeding was started on day 2. Feeding rates were doubled on day 3 and then adjusted on the following days to maintain a glucose concentration of 3 g/L in the solution. Nutrient and metabolite concentrations were measured daily. Online measurements included dissolved oxygen, pH, temperature, pressure, solution weight, air and oxygen sparging, and agitation rate.

2.2 Kinetic model

The model used to describe fed-batch cultivation was based on the one presented by Badr et al. (2021) and adapted by Okamura et al. (2022). The model involves a set of mass balance equations for important system components and Monod expressions for cell growth and death as shown in Eq. (1-7).

$$\frac{d(VX_V)}{dt} = (\mu - \mu_d)VX_V \quad (1)$$

$$\frac{d(VP)}{dt} = Q_P VX_V \quad (2)$$

$$\frac{d(V[GLC])}{dt} = -\left(\frac{\mu}{Y_{X_V/glc}} + m_{glc}\right)VX_V + F_{in}C_{in} \quad (3)$$

$$\frac{d(V[LAC])}{dt} = Q_{lac}VX_V \quad (4)$$

$$\frac{d(V)}{dt} = F_{in} \quad (5)$$

$$\mu = \mu_{\text{max}}\left(\frac{[GLC]}{K_{\text{glc}}+[GLC]}\right)\left(\frac{KI_{\text{lac}}}{KI_{\text{lac}}+[LAC]}\right) \quad (6)$$

$$\mu_{\text{d}} = k_{\text{d}}\left(\frac{[\text{LAC}]}{KD_{\text{lac}}+[LAC]}\right)\left(\frac{KD_{\text{glc}}}{KD_{\text{glc}}+[GLC]}\right) \quad (7)$$

where, X_V is the viable cell density and V is the culture volume. F_{in} is the supplementary feed flows rate. P, $[GLC]$, and $[LAC]$ represent the antibody product, glucose and lactate concentrations, respectively. C_{in} is the glucose concentration in the supplementary feed streams. μ and μ_{d} are cell growth and death rates respectively, while μ_{max} and k_{d} are their corresponding maximum values. Q_P is the specific productivity of mAb, while Q_{lac} is the specific productivity of lactate. $Y_{X_V/glc}$ and m_{glc} represent specific glucose consumption for cell growth and maintenance, respectively. KI_{lac}, K_{glc}, KD_{lac}, KD_{glc} are the Monod model parameters. In the previous works, Q_{lac} [mmol cells^{-1} h^{-1}] has been treated as a constant. However, to account for shifts between lactate production and consumption, Q_{lac} should either be treated as a variable that can vary in magnitude and direction, or an additional term for lactate consumption should be added to Eq. 4.

2.3 Data Analysis

State estimation methods have been extensively used in the chemical industry for control applications where process models are well established (Mohd Ali et al., 2015). Observers can be used to estimate hard to measure state variables. A recent study demonstrated the application of the extended Kalman filter (EKF) with a hybrid model for monitoring and control of mammalian cell cultures with limited mechanistic understanding (Narayanan et al., 2020). In this work, we explored the use of a similar concept for the development of improved process models.

Parameter estimation was first conducted given the measured experimental values of system concentrations. Initial guesses for model parameters were provided and a one step forward solution was obtained. A solution was found to minimize the error between the model estimates and experimental values. The model fit was assessed during different regions of the experimental run. This approach allows us to isolate the regions where parameter updates are required. Further analysis of the experimental conditions enables matching the regions with required parameter updates to the underlying varying conditions thus giving a basis for the generation of modified model equations.

Hierarchical density-based clustering (HDBSCAN) was then applied to the online experimental measurements. This method was chosen due to its resilience to noise in the data and its ability to identify clusters of nonuniform densities. It also requires minimal parameter finetuning or assumptions about the number of clusters in the data. The minimum number of points per cluster was set to 50, which is low compared to the size of the online measurements taken every minute, effectively unrestricting the number of clusters formed. The resulting clusters were compared to the different regions of varying model fit and the underlying varying experimental conditions were accordingly isolated.

Experimental measurements of lactate concentrations were interpolated using the Akima spline method and were used to regress the values for Q_{lac}. Alternative formulations for Q_{lac} were then suggested as a function of the identified varying experimental conditions.

3. Results and discussion

Figure 1 shows an example of the obtained results of the final one step forward simulation using the updated parameters estimated by the implemented algorithm. The figure shows three regions of different model fit to estimate lactate concentrations. Lactate was the component with the highest discrepancies between model estimations and the experimental results relative to all predicted concentrations. There were areas of good model fit observed towards the latter half of the experimental run, areas of partial fit, where model predictions were correct in direction but not in magnitude, and finally areas of poor fit, where the model failed to capture both magnitude and direction of lactate concentration development. Regions of poor fit were those where the cell shifted to net lactate consumption instead of production, this generally coincided with nutrient depletion in the solution.

The results of the clustering step for the operating conditions are shown in **Figure 2**. Three clusters were identified by the algorithm. The strongest separation of the different clusters was observed as a function of the solution weight and the dissolved oxygen (DO) levels as shown in **Figure 2(a)**. Nutrient addition during the fed-batch run leads to changes in the solution weight. The dependency on the solution weight indicates a strong role of nutrient levels in determining the different clusters. In addition, other conditions that showed an impact on cluster formation were the pH and the O_2 sparging levels. **Figure 2(b)** shows a representation of the distribution of the cluster points along the course of the experimental run. The identified clusters were closely related to the different regions of model fit. As a result, it can be concluded that the identified conditions were correlated with the changes in the cell metabolic behavior, resulting in net lactate consumption.

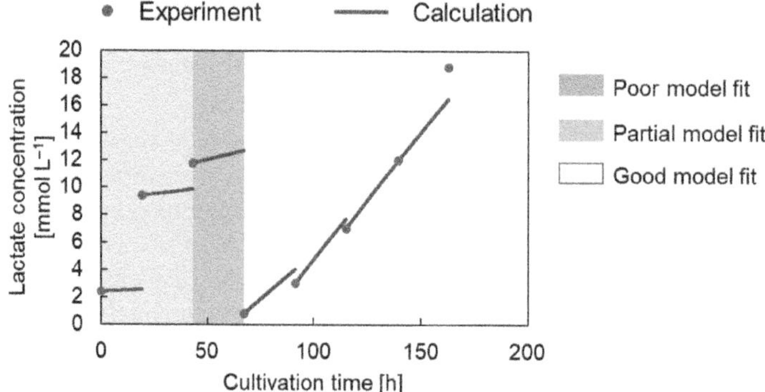

Figure 1. Measured lactate concentrations in comparison to the results of the one step forward simulation using the estimated model parameters to assess the model fit.

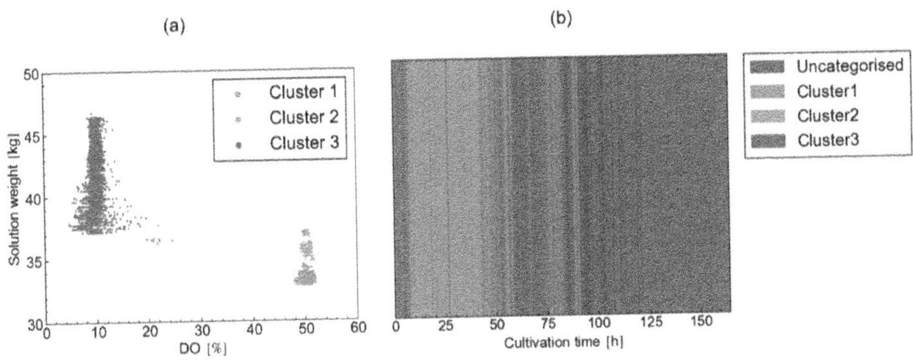

Figure 2. Results of the HDBSCAN clustering. (a) solution weight plotted against dissolved oxygen, (b) the distribution of the clusters with respect to experimental run time.

Regression of the Q_{lac} values from the experimental values and Eq. 4 show that Q_{lac} is only constant in the latter regions of the experimental run (coinciding with clusters 2 and 3). Whereas in the earlier regions (cluster 1), values of Q_{lac} decreased until negative values were reached, indicating net lactate consumption. **Figure 3** shows the regressed values of Q_{lac} along the experimental run. The variations in Q_{lac} values in the earlier regions (cluster 1) showed a strong correlation with variations in the O_2 sparging levels. Therefore, an empirical formula for calculating Q_{lac} in the variable region as a function of the O_2 sparging rate [L min^{-1}] was proposed as follows in Eq. 8. **Figure 3** shows the correspondence between the calculated and the experimentally obtained values of Q_{lac}, where an R^2 value of 0.93 was obtained.

$$Q_{lac} = \frac{k_0}{k_1 + O_2\,\text{sparging rate}} + k_2 \tag{8}$$

The influence of the O_2 sparging rate on the Q_{lac} must still be studied in further detail to draw more concrete mechanistic conclusions. Oxygen sparging is assumed to impact dissolved oxygen levels in the system in addition to affecting the general mixing conditions in the reactor. While the sparging rates varied rapidly in the system, DO levels remained fairly constant in the bulk levels during cluster 1 conditions (**Figure 2 (a)**). This could suggest an impact of sparging levels on local variations in nutrient, metabolite, and

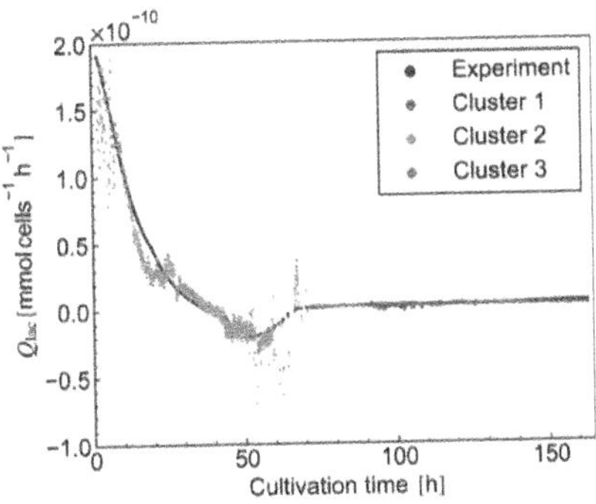

Figure 3. Regressed values of the specific lactate productivity factor (Q_{lac}) from the experimental data and comparison to calculated values for each of the identified clusters.

oxygen concentrations around the cell. Another aspect to be investigated is the impact of changes in pH on Q_{lac} and on the cell metabolic behavior. Lactate concentrations strongly impact pH levels, while concurrently variations in pH have been reported to influence the switch from lactate production to net consumption (Hartley et al., 2018). Further work also includes determining the role of depletion of different nutrients, e.g., glucose and glutamine, on variation of Q_{lac}. Further experimental runs are required to confirm the findings and resolve the uncertainties.

4. Conclusions

This work exploits state estimation concepts to develop an understanding of model fit in cases where shifts in cell metabolic behavior result in a need for model parameter updates. Clustering of experimental conditions is used to isolate the underlying variations in experimental conditions leading to a change of cell metabolic behavior and state. This approach allows the development of more interpretable hybrid models, where the data-driven insights are used to describe correlations between experimental conditions and variations in model parameters. The derived insights in this approach can be applied to develop more flexible and representative mechanistic models which can increase the interpretability of the results and help achieve further understanding of the complex biological phenomena.

Acknowledgements

This work was supported by the Japan Agency for Medical Research and Development (AMED) [grant No. JP21ae0121015, JP21ae0121016].

References

Badr, S., Okamura, K., Takahashi, N., Ubbenjans, V., Shirahata, H., & Sugiyama, H. (2021). Integrated design of biopharmaceutical manufacturing processes: Operation modes and process configurations for monoclonal antibody production. *Computers & Chemical Engineering*, *153*, 107422.

Badr, S., & Sugiyama, H. (2020). A PSE perspective for the efficient production of monoclonal antibodies: integration of process, cell, and product design aspects. *Current Opinion in Chemical Engineering*, *27*, 121–128.

Butler, M., & Meneses-Acosta, A. (2012). Recent advances in technology supporting biopharmaceutical production from mammalian cells. *Applied Microbiology and Biotechnology*, *96*(4), 885–894.

Hartley, F., Walker, T., Chung, V., & Morten, K. (2018). Mechanisms driving the lactate switch in Chinese hamster ovary cells. In *Biotechnology and Bioengineering* (Vol. 115, Issue 8, pp. 1890–1903). John Wiley and Sons Inc.

Mohd Ali, J., Ha Hoang, N., Hussain, M. A., & Dochain, D. (2015). Review and classification of recent observers applied in chemical process systems. In *Computers and Chemical Engineering* (Vol. 76, pp. 27–41). Elsevier Ltd.

Narayanan, H., Behle, L., Luna, M. F., Sokolov, M., Guillén-Gosálbez, G., Morbidelli, M., & Butté, A. (2020). Hybrid-EKF: Hybrid model coupled with extended Kalman filter for real-time monitoring and control of mammalian cell culture. *Biotechnology and Bioengineering*, *117*(9), 2703–2714.

Okamura, K., Badr, S., Murakami, S., & Sugiyama, H. (2022). Hybrid Modeling of CHO Cell Cultivation in Monoclonal Antibody Production with an Impurity Generation Module. *Industrial and Engineering Chemistry Research*, *61*(40), 14898–14909.

Walsh, G. (2018). Biopharmaceutical benchmarks 2018. *Nature Biotechnology*, *36*(12), 1136–1145.

Antonis Kokossis, Michael C. Georgiadis, Efstratios N. Pistikopoulos (Eds.)
PROCEEDINGS OF THE 33rd European Symposium on Computer Aided Process Engineering (ESCAPE33), June 18-21, 2023, Athens, Greece
 http://dx.doi.org/10.1016/B978-0-443-15274-0.50158-X

Optimal Design of Anion-pillared Metal-organic Frameworks for Gas Separation

Xiang Zhang,[a,*] Kai Sundmacher[a,b]

[a]*Department for Process Systems Engineering, Max Planck Institute for Dynamics of Complex Technical Systems, Sandtorstr. 1, D-39106 Magdeburg, Germany*
[b]*Chair of Process Systems Engineering, Otto-von-Guericke University Magdeburg, Universitätsplatz 2, D-39106 Magdeburg, Germany*
[*]*E-mail address of corresponding author: zhangx@mpi-magdeburg.mpg.de*

Abstract

Pressure swing adsorption (PSA) has been widely studied for various gas separation tasks. The efficiency of PSA processes is influenced by the properties of adsorbents and the process operating conditions. As innovative adsorbents, metal-organic frameworks (MOFs) have been designed and synthesized. As special family of MOFs, anion-pillared (AP) MOFs where anion groups (e.g., SiF_6^{2-}, TiF_6^{2-}, AlF_5^{2-}, etc.) are used as linkers have been proposed for the separation of light hydrocarbon. The pore size and framework topology (e.g., interpenetrated and non-interpenetrated) of AP MOFs have a dramatic impact on the adsorptive behavior. Considering the diversity of MOF chemistry and only a limited number of AP MOFs reported, a large design space is worth to be explored. In this work, a computational approach is developed to design new hypothetical AP MOFs. The objective is to find high-performance AP MOFs that can be used in PSA processes for propane/propene separation.

Keywords: metal-organic frameworks, anion-pillared MOF, computational material design, organic linkers, C3 separation

1. Introduction

Pressure swing adsorption (PSA) has been widely applied in the chemical and energy industries for gas separation. It exploits the different affinity of gaseous components to solid adsorbents at different pressures to achieve gas separation. Hence, the adsorbent properties strongly influence the separation efficiency. As well known, multiple types of porous materials can be used as adsorbents such as metal-organic frameworks (MOFs), zeolites, or porous polymers. Among them, MOFs are formed via the self-assembly of different building blocks (i.e., metal nodes and organic linkers) in different topologies. Due to the large variety of available building blocks, MOFs have a nearly infinite design space (Yao et al. 2021). In addition, MOFs have many superior properties such as large porosity, tunable pore geometry, functional pore surface, and importantly modular-ordered. Thus, it is desirable to develop high-performance MOFs that can enhance process efficiency in adsorption-based gas separation (Zhang et al., 2020).

When MOFs are used as adsorbents, four alternative separation mechanisms exist: equilibrium separation, kinetic separation, molecular sieving, and gate-opening separation. Among these, equilibrium separation is the most common method, which enables separation based on the difference of gas equilibrium loadings. So far, an increasing number of MOFs with diverse isotherm characteristics have been synthesized by varying MOF chemistry and structure (e.g., pore geometry and topology). Given certain separation tasks, it has been found that many MOFs fail to achieve the pre-specified separation requirements, not to mention good process performance (Burns et al. 2020). To find promising MOF adsorbents efficiently, various computational screening strategies have been developed. In general, a set of MOF candidates are collected. Their adsorption and diffusion properties are computed via molecular simulations. Then, simple performance metrics (e.g., working capacity and selectivity) are calculated to rank top candidates. Although this approach can be used to screen a large pool of MOFs, using simple performance metrics cannot tell whether specific separation requirements are satisfied, particularly when tough separation specifications (e.g., purity $\geq$ 99.95% in H_2 purification) exist. Thus, a computational methodology is developed in this work to design and screen MOFs that can best serve the process for achieving pre-specified gas separations.

Anion-pillared (AP) MOFs where anion groups (e.g., SiF_6^{2-}, TiF_6^{2-}, AlF_5^{2-}, etc.) are used as linkers have been highlighted for light hydrocarbon separations. The pore sizes of AP MOFs can be fine-tuned by rationally designing their chemistry. It has been found that small changes of pore sizes can have a dramatic impact on the adsorptive behavior of AP MOFs. Moreover, AP MOFs generally have two types of cuboid networks: non-interpenetrated and interpenetrated. The non-interpenetrated network is the standard primitive cubic (i.e., *pcu*) topology, while the interpenetrated one is composed of doubly interpenetrated nets that are isostructural to the non-interpenetrated network (Lin et al. 2011). Evidently, the non-interpenetrated network can form a small pore size with a large organic linker. So far, only a limited number of AP MOFs have been reported and a large design space has not been explored yet considering the diversity of MOF chemistry. This inspired us to design new AP MOFs for efficient gas separation.

As an essential building block for polypropylene production, propene (PE) is a key olefin raw material. It is widely known that polypropylene synthesis requires high-purity propylene (over 99.5%). Thus, the removal of other hydrocarbons, especially propane (PA), from propene is one of the critical gas separation tasks in the modern chemical industry. Currently, this is achieved by the energy-intensive cryogenic distillation. As energy-efficient alternatives, adsorption-based separation methods are of great interest. In this case, the development of advanced adsorbents that can achieve low energy consumption is crucial. Thus, the objective of this work is to use the proposed methodology to design new hypothetical AP MOFs for efficient PA/PE separation.

2. Workflow of Anion-pillared MOF Design

Figure 1 shows the explicit workflow for computational AP MOF design. First, the reported AP MOFs are collected and decomposed to identify the feasible alternatives of metal nodes and anion pillars. Meanwhile, all the MOFs in the 2019 CoRE MOF database are deconstructed. Referring to the common characteristics found in AP MOFs, feasible organic linkers are selectively extracted. Then, based on the topology templates and the collected building blocks, we constructed hypothetical AP MOFs using ToBaCCo that are modified to generate MOFs with two types of organic linkers (i.e., anion pillars and the general organic linker). Afterwards, the single-component adsorption isotherms of hypothetical AP MOFs are predicted using grand canonical Monte Carlo (GCMC) simulations performed in RASPA. To improve the reliability, new Lennard-Jones (LJ) parameters that are suitable for AP MOFs are calibrated based on experimental isotherm data. The obtained single-component isotherms are used to fit the multi-component isotherm models that can be directly applied in a short-cut PSA model. This model can efficiently evaluate whether the new AP MOFs are feasible to satisfy the pre-specified separation requirements. After this feasibility check, rigorous PSA optimization are performed on the feasible AP MOFs to obtain the optimal adsorbent together with optimal PSA process operating conditions.

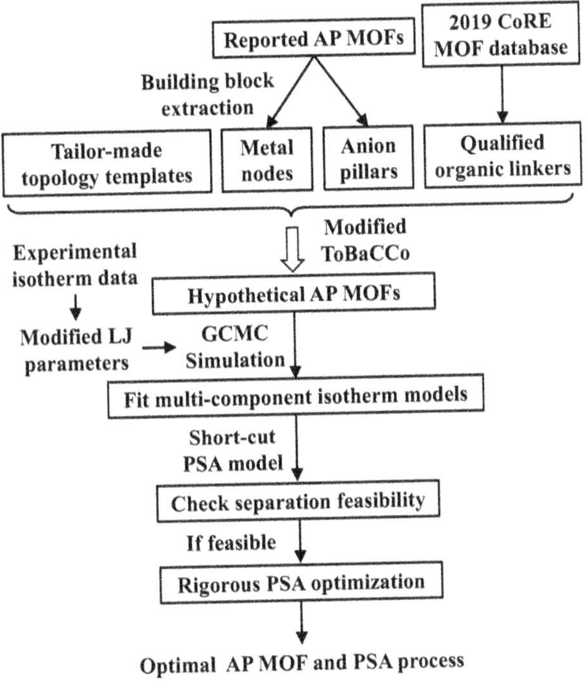

Figure 1. Workflow of computational AP MOF design.

3. Build Blocks Construction

AP MOFs consists of metal nodes and two types of organic linkers (i.e., anion pillars and general organic linkers serving as donor and acceptor, respectively) in certain topologies. For creating sound feasible AP MOFs, proper building blocks and topology templates have to be specified.

3.1. Two topology templates

So far, AP MOFs with limited types of topologies (e.g., *pcu*, *pcu*-i, *mmo*, *pto*, etc.) have been successfully synthesized. Since the non-interpenetrated *pcu* and interpenetrated *pcu*-i are the two most common topologies, only these two types are considered here. The *pcu* template can be found in the original ToBaCCo platform, while the *pcu*-i template is made using Material Studio.

3.2. Sixty-four feasible organic linkers

The general organic linkers that can be used in AP MOFs are extracted from the CoRE MOFs. After removing the CoRE MOFs with structural defects, 9268 CoRE MOFs are left and 14107 organic linkers can be extracted. Then, referring to the most reported AP MOFs, it is found that the general organic linkers should be symmetric and only consist of carbon, nitrogen, and oxygen atoms (Li et al. 2022). Considering these common characteristics, 64 unique symmetric organic linkers are finally obtained.

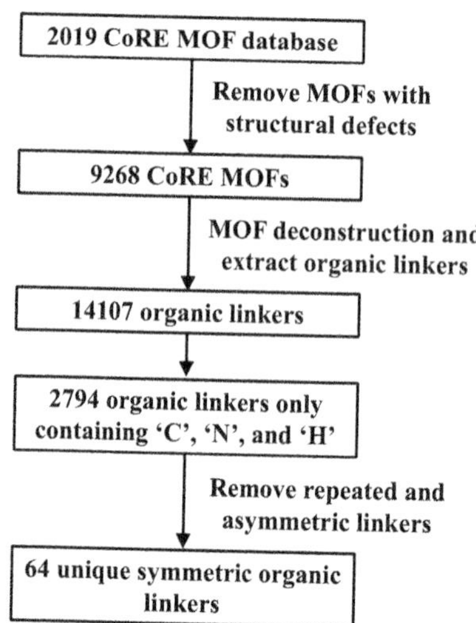

Figure 2. Procedures of extracting general organic linkers used in AP MOFs.

3.3. Five metal nodes and eight anion pillars

After decomposing the reported AP MOFs with *pcu* and *pcu*-i topologies, it is found that the metal nodes are simply single-metal atoms and five metal atoms are commonly involved (i.e., Zn, Fe, Cu, Co, Ni). In addition, 8 kinds of anion pillars are found including SiF_6^{2-}, TiF_6^{2-}, SnF_6^{2-}, GeF_6^{2-}, ZrF_6^{2-}, FeF_5^{2-}, AlF_5^{2-}, and $NbOF_5^{2-}$. With these, the collected AP MOFs in the form of cif files are imported into Material Studio to extract the corresponding metal nodes and anion pillars.

3.4. Bond lengths between building blocks

For MOF construction, building blocks are connected via single-bonds. Given the collected building blocks above, metal-fluorine bonds are formed when connecting metal nodes and anion pillars. Metal-nitrogen bonds are formed when connecting metal nodes and general organic linkers. Since each atom-atom pair has different bond length, the bond lengths between metal-nitrogen and metal-fluorine should be specified explicitly. Based on the reported AP MOFs, Table 1 lists the average bond lengths between metals and two connection atoms. These data can be directly used to adjust the sizes of the five metal nodes.

Table 1. Bond lengths between metals and connection atoms in AP MOFs.

Metal	Metal-fluorine distance (Å)	Metal-nitrogen distance (Å)
Co	2.04	2.15
Cu	2.25	2.04
Fe	1.99	2.19
Ni	1.97	2.08
Zn	2.09	2.13

4. Design of anion-pillared MOFs for PA/PE separation

4.1. ToBaCCo-based AP MOF generation and isotherm prediction

The prepared building blocks and topology templates are imported into the modified ToBaCCo to generate hypothetical AP MOFs. In this process, a series of MOF construction feasibility rules will be verified and the cell sizes will be optimized. After this, over 3000 hypothetical AP MOFs in the form of crystallographic information framework (.cif) files are obtained.

The cif files are directly used to perform GCMC simulations using the open-source software RASPA. The reliability of GCMC simulations relies on the force field parameters (i.e., LJ parameters). It is well-known that generic force fields such as DREIDING force field and Universal Force Field (UFF) are widely used in MOF screening studies. However, for AP MOFs, the PA and PE isotherms cannot be predicted accurately using those general force fields. Thus, a new set of LJ parameters should be specified based on experimental isotherm data. After careful literature review, the single-component PA and PE experimental isotherm data at 298 K of three AP MOFs (i.e., GEFIX-2-Cu-i, SIFSIX-2-Cu-i, and NbOFFIVE-i-Ni) are collected (Wang et al. 2020; Cadiau et al. 2016). On the basis of UFF, the LJ parameters of three atoms (i.e., Ni, Nb, and Ge) and two groups (i.e., CH_3- and CH_2- for PE) should be modified so that the predicted isotherms can match the corresponding experimental data well.

4.2. Process-based AP MOF screening

After specifying the new LJ parameters, the single-component PA and PE adsorption loadings at four pressure levels (i.e., 0.1, 0.5, 1, and 2 bars) and 313 K are calculated for all hypothetical AP MOFs. The obtained GCMC results are used to fit the multi-component dual-site Langmuir isotherm models for the purpose of process-based screening and optimization.

Clearly, it brings great computational burdens to perform rigorous PSA optimization for all the AP MOFs. Thus, for fast screening, a short-cut PSA model is proposed to identify whether an AP MOF can meet the pre-specified separation specifications (i.e., 99.5% PE purity and 65% PE recovery). The short-cut model is based on a one-bed, four-step PSA process that consists of pressurization, adsorption, rinsing, and desorption steps. It is assumed that the column is initially saturated with 99.95% PE at a low pressure (P_{low}). The column goes through the pressurization step where PA/PE gases at a high pressure (P_{high}) are fed into the bottom of the column. The PE mole fraction in the feed gas is set to 85%. The top is closed and the internal pressure gradually increases. When the column's top reaches an intermediate pressure level (i.e., $0.85P_{high}$), the top is opened and the adsorption step begins. The adsorption is stopped when the stream leaving the top is the same as the feed gas (i.e., breakthrough state). Afterwards, based on the mass balance and PE recovery, a certain amount of 99.5% PE gas is fed into the bottom for rinsing. Finally, in the desorption step, a constant low pressure (P_{low}) is enforced at the bottom of the column where purified PE can be collected. If the collected PE stream has a purity above 99.5%, the column state at the end of desorption can presumably be equal to the initial state (i.e., 99.5% PE at the low pressure). This implies that the adsorbent can meet the pre-specified separation requirements. Otherwise, the adsorbent is deemed as infeasible one. Clearly, this short-cut model is a batch PSA process model with two variables (i.e., P_{high} and P_{low}). Grid search on the two variables can be easily performed for a complete feasibility check. Hence, the proposed short-cut model is applied to screen all the generated AP MOFs. It can be found that a set of potential AP MOFs can survive such as NbOFFIVE-15-Fe-i and SIFSIX-2-Co. By minimizing the energy consumption and maximizing the productivity, finally a multi-objective rigorous PSA optimization can be performed on those feasible AP MOFs for final evaluation.

5. Conclusions

This work presented a computational approach to systematically design and screen hypothetical AP MOFs for efficient PA/PE separation in PSA processes. The workflow starts with extracting various building blocks from the reported AP MOFs and the CoRE MOFs. Together with two topology templates, hypothetical AP MOFs are constructed using the modified ToBaCCo. For each generated AP MOF, GCMC simulations are performed using newly calibrated LJ parameters to predict the single-component adsorption loadings of PA and PE. With the predicted loadings, multi-component DSL isotherms are fitted and used in a short-cut PSA model for fast screening. Finally, for the survived candidates, rigorous multi-objective PSA optimization are carried out in order to find the real optimal AP MOFs and the corresponding process operating conditions. Clearly, as demonstrated via the PA/PE separation example, hypothetical AP MOFs leading to better process performance can be computationally synthesized and screened. The major novelty of the present work is the use of modified computational tools and models to design optimal AP MOFs efficiently. Since AP MOFs with specific topologies have similar synthesis routes, this provides great conveniences for the subsequent experimental validation.

References

TD. Burns, KN. Pai, SG. Subraveti, SP. Collins, M. Krykunov, A. Rajendran, TK. Woo, 2020, Prediction of MOF performance in vacuum swing adsorption systems for postcombustion CO_2 capture based on integrated molecular simulations, process optimizations, and machine learning models, Environmental Science & Technology, 54, 4536-4544

A. Cadiau, K. Adil, P.M. Bhatt, Y. Belmabkhout, M. Eddaoudi, 2016, A metal-organic framework–based splitter for separating propylene from propane, Science, 353, 6295

YJ. Colón, DA. Gómez-Gualdrón, RQ. Snurr, 2017, Topologically guided, automated construction of metal–organic frameworks and their evaluation for energy-related applications. Crystal Growth & Design, 17, 11, 5801-5810

X. Li, H. Bian, W. Huang, B. Yan, X. Wang, B. Zhu, 2022, A review on anion-pillared metal–organic frameworks (APMOFs) and their composites with the balance of adsorption capacity and separation selectivity for efficient gas separation, Coordination Chemistry Reviews, 470, 214714

MJ. Lin, A. Jouaiti, N. Kyritsakas, MW. Hosseini, 2011, Molecular tectonics: control of interpenetration in cuboid 3-D coordination networks, CrystEngComm, 13, 776-778

X. Wang, P. Zhang, Z. Zhang, Li. Yang, Q. Ding, X. Cui, J. Wang, H. Bing, 2020, Efficient separation of propene and propane using anion-pillared metal–organic frameworks, Industrial & Engineering Chemistry Research, 59, 3531-3537

Z. Yao, B. Sánchez-Lengeling, NS. Bobbitt, BJ. Bucior, SGH. Kumar, SP. Collins, T. Burns, TK. Woo, OK. Farha, RQ. Snurr, A. Aspuru-Guzik, 2021, Inverse design of nanoporous crystalline reticular materials with deep generative models, Nature Machine Intelligence, 3, 1, 76-86

X. Zhang, T. Zhou, K. Sundmacher, 2021, Integrated MOF and P/VSA process design: Descriptor optimization, AIChE Journal, 68, e17524

Antonis Kokossis, Michael C. Georgiadis, Efstratios N. Pistikopoulos (Eds.)
PROCEEDINGS OF THE 33rd European Symposium on Computer Aided Process Engineering (ESCAPE33), June 18-21, 2023, Athens, Greece
 http://dx.doi.org/10.1016/B978-0-443-15274-0.50159-1

Probabilistic neural networks for mechanical properties and tensile behavior prediction of polymer composites considering uncertainty in tensile testing

Jinkyung Son,[a] Jaewook Lee,[a] and Dongil Shin,[a,b]

[a]*Department of Chemical Engineering, Myongji University, Yongin, Gyeonggido, 17058, Korea*

[b]*Department of Disaster and Safety, Myonji University, Yongin, Gyeonggido, 17058, Korea*

Abstract

Polymer Matrix Composite (PMC) is gaining attention for its excellent modeling processability, light weight, low unit cost, and various mechanical properties by reinforcing the heat resistance and strength of plastics with fillers. However, to this day, PMC development relies on designers' intuition, and the cost of repeated testing is quite high. Therefore, in order to enhance PMC's development competitiveness, an AI model that predicts mechanical properties according to the combination and composition of constituent materials is essential. Since these mechanical properties can be grasped using the Stress-Strain curve (S-S curve) obtained from tests such as tensile, creep, and bend, the S-S curve is a kind of mechanical property evaluation index that can accurately know the mechanical properties.

However, Finite Element Method (FEM)-based simulators or Machine Learning (ML)-based predictive models that calculate S-S curve are deterministic models that provide only one S-S curve and do not have a predictive model that considers the uncertainty (variance) of PMC tensile data. Therefore, in this study, after collecting S-S curve data of PMC, Using the Probabilistic Neural Network (PNN) model, we develop a model that predicts the confidence interval along with the mean value of the tensile S-S curve of the PMC. The performance of the developed model is evaluated on the R^2 basis, and it is confirmed that not only the elastic section, which is relatively easy to predict, but also the S-S curve of the plastic section can be predicted. Based on the results of future research, it is expected that this model can be used to build an AI-based PMC reverse engineering system.

Keywords: Polymer Matrix Composite (PMC), Mechanical behavior, Gaussian Process Regression (GPR), Probabilistic Neural Network (PNN), Feed-Forward Neural Network (FNN)

1. Introduction

The mechanical behavior of the material can be expressed in the S-S curve and not only the mechanical properties such as tensile strength, modulus of elasticity and maximum load are evaluated, but also the deformation from the elastic section to the plastic section. However, polymeric materials have a high uncertainty in the S-S curve due to the inherent nonlinearity properties of the material, which shows a wide distribution. Therefore, in general, when designing polymers, the average value of the data obtained after conducting at least 10 experiments after uniform formulation and manufacturing conditions is used as the S-S curve of the material to consider 'uncertainty'.

However, if the average value of the repeated experiment is used, there is a possibility of over-estimating or under-estimating the physical properties of the composite material because it ignores the variance of the data. Prior studies that directly predicted the S-S curve, which is the result of such a PMC tensile test, G. Chen et al. (2021) In previous studies, the microstructure of composite materials is simulated and analyzed through Convolutional Neural Networks (CNN) models. However, the results of this study produce only one predicted S-S curve and are difficult to apply to composite materials such as PMC with a wide distribution. Therefore, as the first step in the AI-based material reverse engineering system, this study proposes a model that predicts the confidence interval of the S-S curve by reflecting the uncertainty of the PMC tensile S-S curve. First, Domain Knowledge is used to obtain information such as density, porosity, and Poisson's ratio of constituent materials to build a feature set. Next, feature engineering is performed to input a sufficient amount of information to predict the S-S curve of the PMC by the ML-based model and at the same time prevent falling into the dimensional curse. As a feature engineering technique, Feature elimination is used to secure an optimal feature set by selecting only useful features that are most involved in the performance of the model from the feature sets obtained. Using the obtained optimal feature set and tensile test data of PMC, model training is carried out, and a model is designed to predict the S-S curve of PMC using Gaussian Process Regression (GPR), a representative ML model that provides variance of regression results.

2. PMC Tensile Test Data

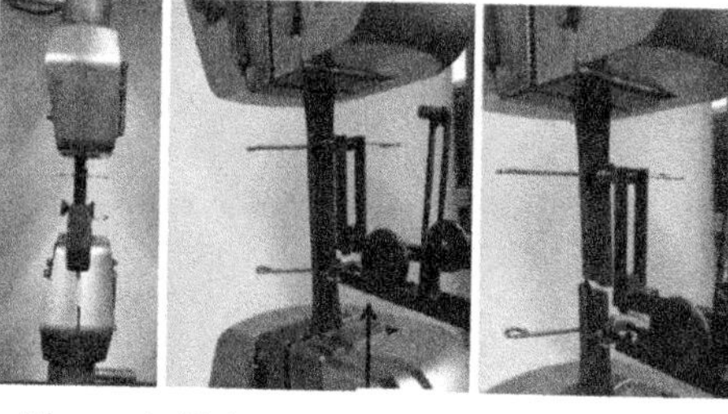

Figure. 1. Universal Testing Machine (UTM)

Matrix	Polypropylene (PP), Polycarbonate (PC), Polyamide 6 (PA6), Polyamide 6,6 (PA6,6)
Filler	Al2O3, Al2O5Si, Boron Nitride (BN), Si3N4
Matrix wt%	90 ~ 40

Figure. 2. Combination and composition of subject to tensile test

2.1. Data collection

A training database of an integrated model for predicting mechanical properties was established by integrating literature material information data, material information bank data, and data factory collection data. As for the tensile test data, the test was conducted with a universal testing machine as shown in Fig. 1. The material to be subjected to the tensile test was manufactured by selecting a PMC, which is widely used as a lightweight material. As shown in Fig. 2, the matrix was set to have Polypropylene (PP), Polycarbonate (PC), Polyamide 6 (PA6), Polyamide 6,6 (PA6,6) and filler were Al2O3, Al2O5Si, Boron Nitride (BN), and Si3N4, and the composition ratio of the polymer and filler was set to between 90% and 40%. According to the combination and composition set in this way, a total of 85 PMC specimens were produced and a tensile test was conducted.

As mentioned earlier, the uncertainty of the tensile test data of the PMC is high due to the high randomness of the polymer itself.

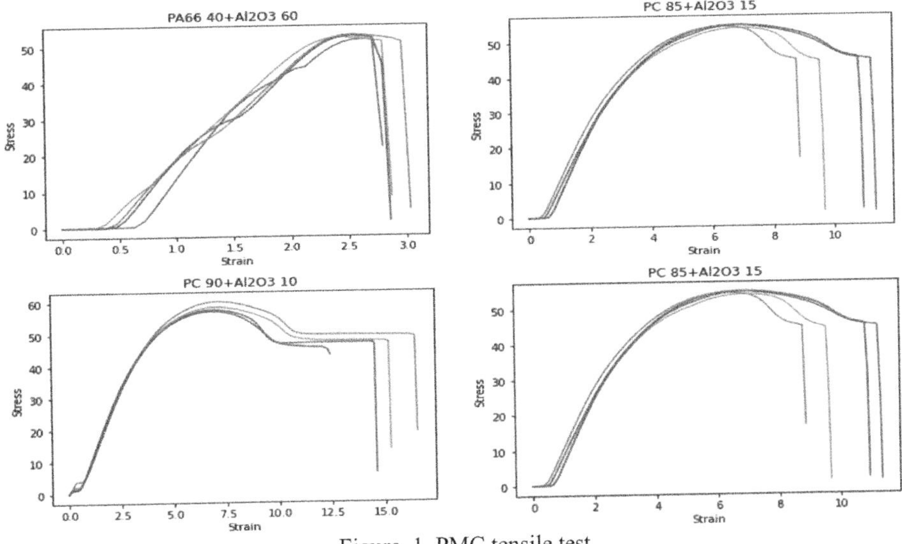

Figure. 1. PMC tensile test

As can be seen from Fig. 3, the high dispersion of the PMC tensile test data shows different S-S curves even when manufactured and tested with the same recipe. Fig. 3 shows the tensile test data of the PMC manufactured with 40wt% PA6,6 and 60wt% Al2O3, 90wt% PC and 10wt% Al2O3. As can be seen from the figures, the tensile test data of each PMC has a high uncertainty of 7% to 24%.

Generally, when designing a PMC, the average value of physical properties obtained after performing five or more repeated experiments is determined and used as the physical properties of the polymer composite water. However, in this study, since the model design considering the uncertainty of the data is the goal, the collected S-S curve raw data was used as it is, not the S-S curve calculated as an average.

2.2. . Data preprocessing

Three processes were conducted for data preprocessing. First, the process of integrating various databases and files that are divided into various categories in an easy way to analyze was conducted. At this time, 58 folders divided by PMC material and target distance and data with an average of 1,400 rows in 1,340 file data were combined to be used as a CSV file with 1,880,000 rows.

Second, we proceeded with data integration. Tensile test data measured in units of 0.01sec consists of an average of 1,400 data per tensile test, and if all 1,340 files are used, the size of the input data becomes too large, reducing learning performance and taking a long time. Therefore, the number of learning data was reduced to 1/10 by adjusting the interval data 0.01sec to 0.1sec through interpolation.

Finally, the data evaluated as outliers by the intuition of the material expert were removed, and only the data with high reliability were selected for learning.

2.3. Train-test data split for evaluation

In the case of general ML related papers, the train-test set is randomly divided at a specific rate. However, if the PMC tensile test data is divided in the same way and the learning is conducted, overfitting the learning data may make it difficult to evaluate the performance of the model. Therefore, we divided all the data corresponding to a specific combination composition into test sets and used the only remaining data for model learning to see if we predict well for the first-time PMC, not simple interpolation fitted to data.

3. Feature Engineering

For prediction through ML-based models, it is very important to discover features that can sufficiently reflect the characteristics of polymers and reinforcement, which are constituent materials of PMC. Therefore, a large amount of features were selected through the domain knowledge of PMC design experts. The most basic feature set is mechanical and chemical property information of polymer and filler, which are constituent materials. The mechanical and chemical property information of polymers and fillers were collected from Polymer Genome and other online databases, and the collected mechanical and chemical property information is shown in Fig. 4 and entanglement of polymers, density to reflect porosity, and tensile strength of polymers that are highly correlated with the tensile strength of composite materials. In addition, MAACCS keys were used to express the presence or absence of functional groups by one-hot encoding to increase the amount of information of the input feature.

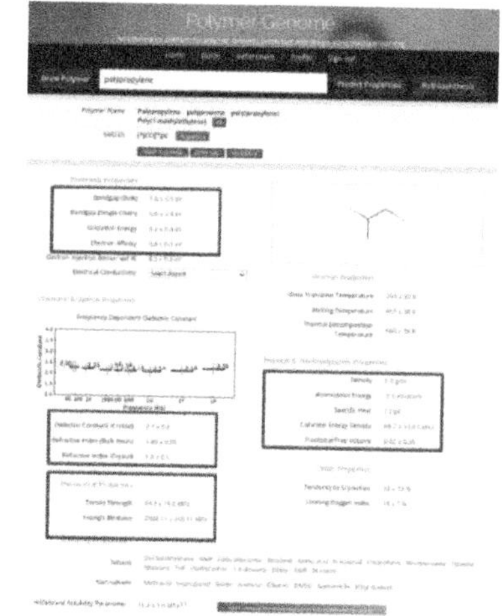

Figure. 2. Mechanical and Chemical properties from the Polymer Genome

4. Design of Prediction Model

In our previous work, we have successfully utilized Deep Neural Network (DNN) to learn the non-linear relationship between the PMC's features and it's S-S curves. However, in this work we utilize Probabilistic Neural Network (PNN) because of two benefits. Firstly, as mentioned above, the deterministic model DNN presents only one numerical value, and data with large variance outputs only one value with minimal loss without considering its variance at all. Therefore, the tensile test data of the PMC with a large dispersion used a PNN model that can estimate the distribution and average value.

Secondly, PNN do not have a separate learning process to determine the weights between neurons compared to CNN, so the calculation time is faster than DNN. The PNN model was designed to learn the mean and distribution of tension-curves according to

PMC's input feature, consisting of a hidden layer, distribution, and distribution layer that calculates the average value of the three layers.

5. Results

In order to accurately determine the performance of the model, we excluded all data of the same blending conditions from the train-test set split. Afterwards, the model excluded from the model evaluation process was verified with PMC test data that was first seen. The performance evaluation was conducted based on the average value of the tension-curve and the average value of the PNN model. R^2 were considered as error metrics to evaluate the performance of this model, and the total R^2 according to PMC's type were shown in Table 1.

PMC's type	R^2	PMC's type	R^2
PP + Al2O3	0.88	PA6 + Al2O3	0.76
PP + Al2O5Si	0.86	PA6 + Al2O5Si	0.74
PP + BN	0.92	PA6 + BN	0.94
PP + Si3N4	0.77	PA6 + Si3N4	0.83
PC + Al2O3	0.82	PA6,6 + Al2O3	0.75
PC + Al2O5Si	0.95	PA6,6 + Al2O5Si	0.87
PC + BN	0.71	PA6,6 + BN	0.91
PC + Si3N4	0.84	PA6,6 + Si3N4	0.79
		Avg. R^2	0.83

Table 1. R^2 of PMC's type

As can be seen from the average result, the average R^2 of the model was 0.83, showing compliant performance. Figs. 5 and 6 are the results of plotting the actual data, the predicted mean, and the confidence interval. The reliability of the model can be confirmed through the fact that most of the tensile test data were included in the predicted confidence interval, which is the core of this study.

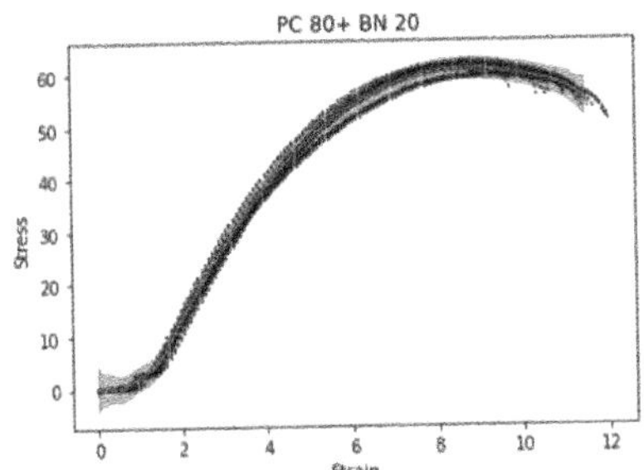

Figure. 4. PMC's prediction results mixed with 80wt% PC and 20wt% BN

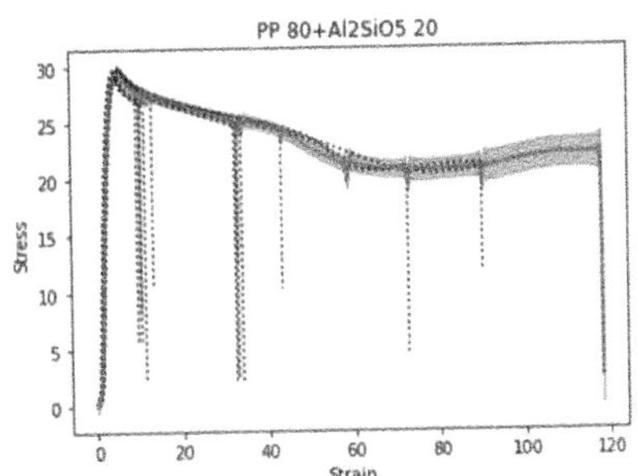

Figure. 4. PMC's prediction results of 80wt% PP and 20wt% Al2SiO5

6. Discussion

GPR made it possible to predict a range of graphs that was previously unpredictable. In order to design PMC, it is necessary to process polymers. At this time, the predicted value of the polymer may vary according to the conditions of each process, and in particular, the predicted value may vary depending on the composite process or the processing type of the polymer. In subsequent studies, considering the various rheological properties of these polymers, we will analyze and predict polymers under various process conditions and compare the results to find representative factors affecting the predicted values.

7. Conclusions

In this research, we designed S-S curve confidence interval prediction model which reflects an uncertainty (variance) of about 7~24% values. This model was processed same conditions such as formulation and environment. After manufacturing the specimen with a composition ratio of 90~40wt% of the polymer and filler, features were secured by feature engineering based PMC's Domain knowledge. These can best describe PMC tensile S-S curve, which is a mixture of polymer and filler. At this time, the performance of the model was about 0.83 based on R^2, and it was confirmed that not only the elastic section, which is relatively easy to predict, but also the S-S curve of the plastic section was well predicted. If only one S-S curve was provided to the user using a rheological model based on the first principle or an ML-based deterministic model to analyze the non-linear behavior of PMC, the PNN-based probabilistic model developed through this study provides material designers with confidence intervals of S-S curve. In addition, it is expected that the developed model can be used to build a PMC reverse engineering system.

References

G.Chen, Z.Song and Z.Qi, 2021, Transformer-convolutional neural network for surface charge density profile prediction: Enabling high-throughput solvent screening with COSMO-SAC, Chemical Engineering Science 246, 117002, https://doi.org/10.1016/j.ces.2021.117002

Wei Zhang, Xiaohui Yang, Yeheng Deng and Anyi Li, 2020, An Inspired Machine-Learning Algorithm with a Hybird Whale Optimization for Power Transformer PHM, Energies,13(12):3143, https://doi.org/10.3390/en13123143

E.R.Davies and Matthew A.Turk, Han Cai, Song Han, Advanced Methods and Deep Learning in Computer Vision, 2022, ISBN 978-0-12-822109-9, https://doi.org/10.116/C2019-0-03221-9

L.Nagyeong, Jaewook Lee, Dongil Shin, 2021, Machine Learning for Predicting Mechanical Behavior of Polymer Matriox Composites for AI-Based Material Design, KIGAS Vol.25, No.2, pp 64~71, https://doi.org/10.7842/kigas.2021.25.264

P.Jihoon,2022, Deep Learning Model for Prediction of Entanglement Molecular Weight of Polymers, polym, korea, vol.46, no.4, pp.515-522 (8 pages), 10.7317/pk.2022.46.4.515

Antonis Kokossis, Michael C. Georgiadis, Efstratios N. Pistikopoulos (Eds.)
PROCEEDINGS OF THE 33rd European Symposium on Computer Aided Process Engineering (ESCAPE33), June 18-21, 2023, Athens, Greece
 http://dx.doi.org/10.1016/B978-0-443-15274-0.50160-8

Data-Driven Indication of Flooding in an Industrial Debutanizer Column

Martin Mojto,[a] Karol Ľubušký,[b] Miroslav Fikar,[a] Radoslav Paulen,[a]

[a]*Slovak University of Technology in Bratislava, Bratislava, Slovakia*
[b]*Slovnaft, a.s., Bratislava, Slovakia*
martin.mojto@stuba.sk

Abstract

The profitability and sustainability of process industries are affected by the performance of each unit involved. A key measure of a unit's performance is based on whether it operates in a desired production window or whether it trips into an abnormal condition. In this contribution, we study flooding of industrial distillation columns. We aim to improve the performance of an industrial debutanizer column by designing a data-driven flooding indicator. The design of the indicator consists of three steps; (a) the data treatment, (b) a priori labeling, and (c) indicator design. The prior knowledge about flooding within the column is used to design a reference indicator. This knowledge is either unused or fully exploited during the design of the indicator. We compare various design methods and show the potential of data-driven approaches for flooding indication.

Keywords: Debutanizer Column, Flooding Indicator, Soft Sensor, Subset Selection

1. Introduction

Flooding is an undesired phenomenon in industrial distillation columns. It occurs when the liquid rises above a tray because of foaming or excessive downcomer fill-up (King, 2016). This state causes a high loss in tray separation efficiency and, hence, plant profitability. Early flooding detection is thus crucial for a profitable and sustainable plant.

Several works dealt with the problem of flooding detection. They considered correlation of the flooding effect with internal process variables, especially the pressure difference (drop) across the column (Peiravan et al., 2020) and the time derivative of the pressure drop (Pihlaja, 2008). Industrial experts use these results and combine them often with an insight into the principal triggering cause of flooding, creating a tailored solution for each column. This requires some effort that could be saved using machine learning (ML).

Several ML approaches (Mojto et al., 2021; Oeing et al., 2021; Fuentes-Cortés et al., 2022) were employed to aid decision making in industrial columns. Unsupervised ML approaches, such as k-means clustering (Forgy, 1965) or principal component analysis (PCA) (Pearson, 1901), consider no knowledge about the model outcome. The supervised ML techniques, e.g., subset (feature) selection (SS) (Smith, 2018) or support vector machine (SVM) (Boser et al., 1992), use knowledge on the desired outcome for training.

This paper investigates the design of data-driven flooding indicators for an industrial debutanizer. Performance of indicators designed via data-driven approaches (unsuper-

Acknowledgments: This research is funded by the Slovak Research and Development Agency (APVV-21-0019), by the Scientific Grant Agency of the Slovak Republic (VEGA 1/0691/21, VEGA 1/0297/22), and by the European Commission (grant no. 101079342, Fostering Opportunities Towards Slovak Excellence in Advanced Control for Smart Industries).

vised and supervised ML) is assessed against the reference indicator (considered as ground truth), which is designed according to the industrial experience about flooding.

2. Problem Statement

Flooding indication is essentially a binary classification problem. We aim to design an indicator **I** that assigns a categorical label $\hat{y}$ via the classification model $f(x)$ as:

$$\hat{y} = \begin{cases} +1 \text{ (flooding)}, & \text{if } f(x) \geq 0, \\ -1 \text{ (normal operation)}, & \text{if } f(x) < 0, \end{cases} \tag{1}$$

where $x \in \mathbb{R}^{n_\mathrm{p}}$ represents a subset (sparse representation, $n > n_\mathrm{p}$) of all online plant measurements $\xi \in \mathbb{R}^n$ at one time instant. We consider a linear classifier in the form:

$$f(x) = w^\top x + w_0, \tag{2}$$

where $w \in \mathbb{R}^{n_\mathrm{p}}$ represents a classifier normal vector and w_0 is a classifier off-set.

2.1. Industrial Debutanizer Column

We study a debutanizer (distillation) column that is a part of the FCC unit of the refinery Slovnaft, a.s. in Bratislava, Slovakia. The column (40 trays) separates the C4/C5 fraction into the C4-fraction-rich distillate product and the C5-fraction-rich bottom product.

The available dataset involves the measurements from January 2019 to April 2021. In this study, we consider 30-minute moving average values of the available minutely data. Overall, the dataset involves 34,297 measurements. The measurements from two plant shutdowns (May – July 2019 and December 2020) are excluded.

The following 41 input variables are directly measured by online sensors at the column:

$$\begin{aligned} \xi = \big(&vo_\mathrm{B}, vo_{\mathrm{D},1-3}, vo_\mathrm{R}, vo_\mathrm{reb,h}, T_{\mathrm{col},1-5}, T_{\mathrm{B},1-2}, T_{\mathrm{D},1-3}, T_\mathrm{F}, T_{\mathrm{reb,h},1-2}, Q_\mathrm{con}, \\ &p_\mathrm{col}, p_{\mathrm{D},1-4}, p_\mathrm{df,col}, p_\mathrm{con}, F_\mathrm{R}, F_\mathrm{B}, F_{\mathrm{D},1-4}, F_\mathrm{F}, F_{\mathrm{reb,h},1-2}, L_{\mathrm{reb},1-2}, L_{\mathrm{con},1-3}\big), \end{aligned} \tag{3}$$

where vo, T, Q, p, and F are a valve opening, temperature, heat input, pressure, and flow rate, respectively. Indices B, D, F, R, col, con, reb, and df represent a bottom section, distillate section, feed section, reflux section, column section, condenser section, reboiler section, and cross-column difference, respectively. Note that exact location of sensors cannot be disclosed due to the confidentiality reasons. The input vector is extended with ratios ($\mathrm{F_R/F_F}$, $\mathrm{F_B/F_F}$, $\mathrm{Q_{con}/F_F}$) and pressure compensated temperatures (PCT_B, PCT_D).

The studied debutanizer column usually operates within the desired operating regime. At times, however, the operating conditions within the unit induce flooding. The envisioned low-cost solution to the flooding problem is to design a reliable indicator. The key aspect of this approach is that the indicator is not only used for monitoring the plant, but it is connected directly to the advanced process controller that can provide a fast response.

The dataset does not contain any direct indication of flooding that could be used to label the data. However, it is possible to attribute flooding occurrence to the increased values of $p_\mathrm{df,col}$, F_R, $F_{\mathrm{reb,h},2}$, and $T_{\mathrm{col},4}$ and decreased values of $T_{\mathrm{reb,h},1}$. We use this knowledge to design the reference indicator to provide the ground truth of the flooding for our study.

3. Methodology

Data-driven indicators are designed using unsupervised ($\mathbf{I}^\mathrm{Uns}$) and supervised ($\mathbf{I}^\mathrm{Sup}$) ML approaches. For training ($\mathbf{I}^\mathrm{Sup}$-type indicator) and testing, the ground truth is provided by the aforementioned reference indicator resulting from industrial expert knowledge.

3.1. Indicator Design

The design procedure of the data-driven indicator consists of three sequential steps:

1. Data processing (data filtering/treatment, distribution to training/testing dataset).
2. A priori labeling of the training dataset (only applied for $\mathbf{I}^{\mathrm{Uns}}$-type indicators).
3. Training of a classifier (fit the $f(x)$ parameters on the labeled training dataset).

After the standardization of the data set (zero mean and unit variance), the aim of the data treatment (the 1st step) is to reduce the number of outliers. Due to the non-ideal (yet close normal) noise distribution within the industrial dataset, minimum covariance determinant (MCD) (Hubert and Debruyne, 2010) is applied. The outlier detection is performed using the F-distribution, retaining data with 99.9999% probability. The high probability follows from the need to eliminate only the most deviated measurements while maintaining the data representing the flooding, which can be otherwise seen as outliers.

It is optional to smoothen the dataset by filtering out the high-frequency noise that does not represent slower effects of flooding. Subsequently, as flooding is characterized by the changes of the variables, we extend the dataset (here, 46 variables) by time differences:

$$\Delta\xi_i(k) = \xi_i(k) - \xi_i(k-1), \qquad \forall_i = \{1,2,\dots,n\}, \tag{4}$$

where k is a time instant. The resulting dataset considers both, the original dataset and time differences, i.e., 92 variables. Effectively, we assign $\xi \leftarrow (\xi, \Delta\xi)$ in this step.

The 2nd step, applied to label the data for $\mathbf{I}^{\mathrm{Uns}}$-type approaches, is performed by k-means clustering (Forgy, 1965) with the elbow method to determine the optimal number of clusters. The clusters with a low cardinality but large distance between the cluster center and the dataset mean are considered to represent the debutanizer flooding.

The training phase needs to choose an appropriate indicator input space ($\mathbb{R}^{n_\mathrm{p}}$) among all the process variables and their time differences. The methods used in this study are:

1. Industrial patent by (Pihlaja, 2008), which uses $\Delta p_{\mathrm{df,col}}$ only (referred to as $\mathbf{I}_{\mathrm{pat}}$).
2. Industrial experience (specific to the studied debutanizer) using $\Delta p_{\mathrm{df,col}}$, ΔF_{R}, $\Delta F_{\mathrm{reb,h,2}}$, $\Delta T_{\mathrm{col,4}}$, and $\Delta T_{\mathrm{reb,h,1}}$ (referred to as $\mathbf{I}_{\mathrm{ref}}$).
3. PCA approach (Pearson, 1901) that chooses a number principal components that explain at least 95% of variance in the dataset (referred to as $\mathbf{I}_{\mathrm{PCA}}$).
4. SS approach (Smith, 2018), which determines the best subset of inputs via cross-validation and comparison of structures with $n_\mathrm{p} = \{1,2,\dots,5\}$ (referred to as $\mathbf{I}_{\mathrm{SS}}$).

In the 3rd step, the linear classifier (see Eq. (2)) is designed according to the chosen inputs ($x \in \mathbb{R}^{n_\mathrm{p}}$). To this end, we use support vector machines (SVM) (Boser et al., 1992).

3.2. Performance Assessment

The outcome of an indicator can fall into four categories: true positive (TP) and false positive (FP), when flooding is indicated correctly and incorrectly, respectively, and, vice versa, true negative (TN) and false negative (FN), for indicating of normal operation. We use some well-known normalized performance criteria for the designed indicators:

$$\mathrm{AC} = \frac{\mathrm{TP}+\mathrm{TN}}{\mathrm{TP}+\mathrm{FP}+\mathrm{TN}+\mathrm{FN}}, \mathrm{PR} = \frac{\mathrm{TP}}{\mathrm{TP}+\mathrm{FP}}, \mathrm{RC} = \frac{\mathrm{TP}}{\mathrm{TP}+\mathrm{FN}}, \mathrm{F1} = \frac{2\times\mathrm{PR}\times\mathrm{RC}}{\mathrm{PR}+\mathrm{RC}}, \tag{5}$$

here AC (accuracy) is a measure of how often the classifier makes the correct prediction, PR (precision or correctness) is a measure of how precisely the true prediction is achieved, RC (recall or sensitivity) is a measure of how actual observations are predicted correctly. F1-score (F1) is a harmonic mean between PR and RC. In the industry, it is important to warn about the potential of flooding and thus low value of FN (high RC) is preferred.

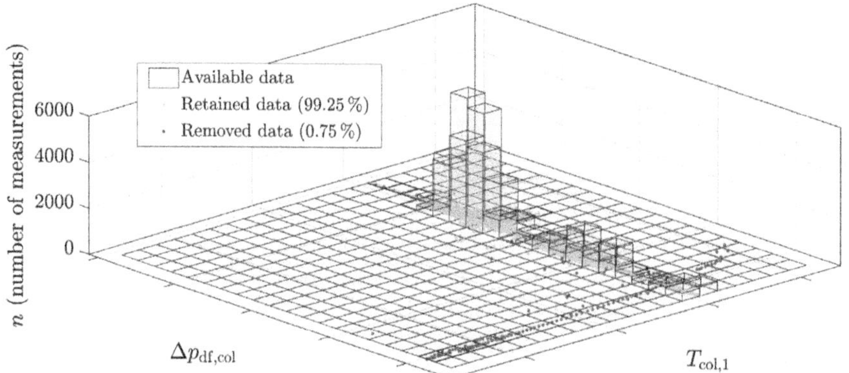

Figure 1: Histogram of two variables from the debutanizer dataset treated by the MCD method.

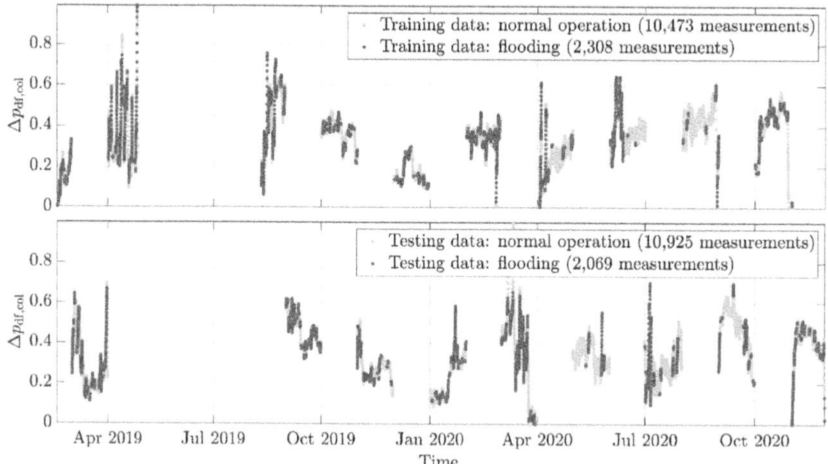

Figure 2: Visualization of training and testing datasets and ground truth labels.

4. Results

The results of data treatment using the MCD method are shown in Fig. 1. The data values are anonymized for confidentiality reasons. Only the most deviated measurements (0.75%) are considered as outliers, and the rest of the measurements (99.25%) is retained. The dataset is further smoothened by filtering using a 10th-order low-pass Butterworth filter with a cut-off frequency of 0.028 mHz (with zero-phase distortion).

For fairness of assessment, we distribute the retained data chronologically on an alternating monthly basis into the training and testing datasets (see Fig. 2). From the entire dataset (25,775 points), 12,781 and 12,994 points are assigned to the training and testing dataset, respectively. Fig. 2 illustrates the training-testing data division together with (ground truth) labels assigned based on industrial experience with the reference indicator.

4.1. Training of Data-Driven Indicators

Design of the data-driven flooding indicators for the debutanizer column is conducted via MATLAB based on the methods from Section 3.1. MATLAB built-in routines for k-means clustering, PCA, and SVM are exploited. We design indicators based on unsupervised ML ($\mathbf{I}_{\text{pat}}^{\text{Uns}}$, $\mathbf{I}_{\text{ref}}^{\text{Uns}}$, $\mathbf{I}_{\text{PCA}}^{\text{Uns}}$, $\mathbf{I}_{\text{SS}}^{\text{Uns}}$) and supervised ML ($\mathbf{I}_{\text{pat}}^{\text{Sup}}$, $\mathbf{I}_{\text{ref}}^{\text{Sup}}$, $\mathbf{I}_{\text{PCA}}^{\text{Sup}}$, $\mathbf{I}_{\text{SS}}^{\text{Sup}}$). A main difference between the approaches is the use of k-means (used for $\mathbf{I}^{\text{Uns}}$ indicators).

Table 1: The comparison of the true positives (TP), false positives (FP), true negatives (TN), false negatives (FN), accuracy (AC), precision (PR), recall (RC), and complexity (no. of input variables n_p, no. of principal components n_{pc}) of the designed data-driven indicators on the testing dataset.

ML method	Unsupervised learning			Supervised learning			
Structure	$\mathbf{I}_{pat}^{Uns}$	$\mathbf{I}_{ref}^{Uns}$	$\mathbf{I}_{PCA}^{Uns}$	$\mathbf{I}_{pat}^{Sup}$	$\mathbf{I}_{ref}^{Sup}$	$\mathbf{I}_{PCA}^{Sup}$	$\mathbf{I}_{SS}^{Sup}$
TP	1,784	1,704	618	1,192	2,031	1,720	2,029
FP	3,828	2,823	3,147	1,358	4	168	0
TN	7,097	8,102	7,778	9,567	10,921	10,757	10,925
FN	285	365	1,451	877	38	349	40
AC	68.3	75.5	64.6	82.8	99.7	96	99.7
PR	31.8	37.6	16.4	46.7	99.8	91.1	100
RC	86.2	82.4	29.9	57.6	98.2	83.1	98.1
F1	46.5	51.7	21.2	51.6	99	86.9	99
n_p/n_{pc}	1	5	17	1	5	17	2

A key to success of unsupervised ML is an appropriate data labeling. The results indicate that, unsurprisingly, the best results are obtained when the k-means clustering is performed on a dataset with reduced dimensionality (e.g., one variable for $\mathbf{I}_{pat}^{Uns}$ indicator or seventeen principal components determined for $\mathbf{I}_{PCA}^{Uns}$), with appropriate input structure. The clustering method reveals 4–5 clusters out of which 1–2 clusters are selected to represent flooding. This result suggest that merging of steps 1 and 2 mentioned in Section 3.1 is a sensible approach to successful indicator design. We can, therefore, expect PCA-based approaches to give inferior performance. We exclude $\mathbf{I}_{SS}^{Uns}$ from the assessment as its performance would suffer from the inaccurate data labelling. Only a more advanced design method (iterating over steps 1–3 from Sec. 3.1) would construct a useful indicator.

The performance assessment of the designed indicators on the testing dataset is shown in Tab. 1, taking into account the so-called confusion matrix elements (i.e., TP, FP, TN, and FN) and performance criteria (i.e., AC, PR, RC, and F1). The complexity of designed indicators is represented by the number of principal components n_{pc} for PCA-based approach and by the number of input variables n_p for the rest of approaches. We can directly see that the supervised ML approaches outperform the unsupervised ones when we compare similar structures. The only exception appear to be the RC criterion when evaluated for $\mathbf{I}_{pat}$ indicator. There are two reasons for this performance drop: 1. RC is given up in training for the AC and precision as the dataset is more populated with data points of normal operation; 2. the industrial data labels indicate flooding based on other variables than pressure (the sole input to $\mathbf{I}_{pat}$ indicator) and thus $\mathbf{I}_{pat}$ indicator falls short in terms of model (input) adequacy (some extra input variables would explain flooding better). Note that, the first reason can be remedied by a modification to SVM objective and some proper tuning, which, however, is beyond the scope of this study.

Among the $\mathbf{I}^{Uns}$-type approaches, it is interesting that, although the structure of the reference indicator is optimal, the highest RC criterion (low FN) is achieved by $\mathbf{I}_{pat}^{Uns}$. Of course, this is paid off by worse accuracy as the classifier mis-indicates flooding (high FP) more often. The PCA-based indicator appears to be the least effective (in all criteria). This is attributed to the aforementioned inappropriate labelling in high dimensions.

Unlike for the unsupervised learning approaches, the performance of the $\mathbf{I}_{\mathrm{PCA}}^{\mathrm{Sup}}$ indicator is sufficient. It also appears that the $\mathbf{I}_{\mathrm{PCA}}^{\mathrm{Sup}}$ is more efficient compared to the $\mathbf{I}_{\mathrm{pat}}^{\mathrm{Sup}}$ indicator viewed by each performance criterion. The highest efficacy among supervised learning approaches is achieved for $\mathbf{I}_{\mathrm{ref}}^{\mathrm{Sup}}$ and $\mathbf{I}_{\mathrm{SS}}^{\mathrm{Sup}}$ indicators. These approaches already consider or can find the best possible input structure. It is noteworthy that $\mathbf{I}_{\mathrm{SS}}^{\mathrm{Sup}}$ achieves the best performance (almost 100% in all performance criteria) using a very simple structure. This effectively tells that the reference structure is overly complicated (some inputs are redundant) and that it is possible to indicate flooding with data from just two sensors. It is also a very interesting result as it allows the industrial practitioners to concentrate efforts regarding sensor maintenance towards smaller subset of online sensors. Surprisingly, pressure is not among the inputs selected for the best indicator. The input structure involves reflux flow ΔF_{R} and the time difference of heating medium flow in the reboiler $\Delta F_{\mathrm{reb,h,2}}$, which are both part of the reference indicator structure. It is possible that the two selected flow rates are measured with better precision and that they do not involve high-frequency fluctuations as pressure measurements do. The results need further validation in an industrial setup to validate the reference indicator (ground truth).

5. Conclusions

This contribution is focused on the design of a data-driven flooding indicator for an industrial debutanizer column. As the ground truth, a reference indicator is used that is designed according to the industrial experience. The effectiveness of unsupervised and supervised machine learning (ML) approaches was evaluated by considering various input structures. The results showed that the unsupervised learning approach can provide sufficient flooding indicators if appropriate input variables are used. The supervised ML approaches achieved higher effectiveness compared to unsupervised ML approaches, resulting from the direct usage of reference indicator labels. The results of supervised learning approaches revealed that the most accurate estimate of the debutanizer flooding is provided by the reflux flow rate and heating medium flow in the reboiler. Future work involves design of indicators for the chosen performance criterion in a multi-objective fashion. It would also be possible to design an unsupervised approach capable of choosing the optimal indicator structure much like the structure achieved with supervised ML.

References

B. E. Boser, I. M. Guyon, V. N. Vapnik, 1992. A training algorithm for optimal margin classifiers. In: Proceedings of the Fifth Annual Workshop on Computational Learning Theory. COLT '92. Association for Computing Machinery, New York, NY, USA, pp. 144–152.

E. Forgy, 1965. Cluster analysis of multivariate data: efficiency versus interpretability of classifications. Biometrics 21, 768–769.

L. F. Fuentes-Cortés, A. Flores-Tlacuahuac, K. D. P. Nigam, 2022. Machine learning algorithms used in pse environments: A didactic approach and critical perspective. Ind. Eng. Chem. Res. 61 (25), 8932–8962.

M. Hubert, M. Debruyne, 2010. Minimum covariance determinant. WIREs Computational Statistics 2, 36–43.

M. King, 2016. Process Control: A Practical Approach. Wiley.

M. Mojto, K. Ľubušký, M. Fikar, R. Paulen, 2021. Data-based design of inferential sensors for petrochemical industry. Computers & Chemical Engineering 153, 107437.

J. Oeing, L. M. Neuendorf, L. Bittorf,W. Krieger, N. Kockmann, 2021. Flooding prevention in distillation and extraction columns with aid of machine learning approaches. Chem. Ing. Tech 93 (12), 1917–1929.

K. Pearson, 1901. LIII. On lines and planes of closest fit to systems of points in space. The London, Edinburgh, and Dublin Philosophical Magazine and Journal of Science 2 (11), 559–572.

H. Peiravan, A. R. Ilkhani, M. J. Sarraf, 2020. Preventing of flooding phenomena on vacuum distillation trays column via controlling coking value factor. SN Applied Sciences 2 (10), 1–11.

R. K. Pihlaja, 2008. Detection of distillation column flooding. U.S. Patent No. US 8.216,429 B2.

G. Smith, 2018. Step away from stepwise. Journal of Big Data 5, 32.

Antonis Kokossis, Michael C. Georgiadis, Efstratios N. Pistikopoulos (Eds.)
PROCEEDINGS OF THE 33rd European Symposium on Computer Aided Process Engineering (ESCAPE33), June 18-21, 2023, Athens, Greece
 http://dx.doi.org/10.1016/B978-0-443-15274-0.50161-X

Can we deploy carbon capture, utilization and storage at a scale?

Ahmed Alhajaj[a,b]

[a]*Research and Innovation center on CO_2 and H_2 (RICH), Khalifa University, PO Box 127788, Abu Dhabi, United Arab Emirates*

[b]*Chemical Engineering Department, Khalifa University, PO Box 127788, Abu Dhabi, United Arab Emirates*

ahmed.alhajaj@ku.ac.ae

Abstract

CO_2 emissions are set to rise in view of continuous growth in economic and supporting infrastructure. This means that the indispensable approach to decarbonize the economy in the medium term is through carbon capture, utilization and storage (CCUS). The main challenges in the large-scale implementation of CCUS network are costs of carbon capture, compression and transportation, energy requirements for CO_2 utilization and conversion, and limited market demand for end products. The development of strategic designs for such CCUS infrastructure at this scale is a non-trivial activity, which must take account of local industry specifics and development plans. A multiscale system modeling and optimization approach is developed to assess alternative CCUS processes and understand transient network behavior. The potential of CCUS in main potential markets in the United Arab Emirates such as simultaneous CO_2 storage with Enhanced Oil Recovery (CO_2-EOR), CO_2 capture using microalgae and mineralization have been examined.

Keywords: CO_2 capture, Optimization, Multiscale Modeling, CCUS, techno-economic.

1. Introduction

The most recent Intergovernmental Panel for Climate Change (IPCC) report indicates that climate change is unequivocal and that without mitigation, average global temperatures may increase by 2-6 ^{O}C by 2100. As a result, most major economies have announced targets for reductions in GHG emissions. UAE is amongst the nations wishing to take a lead in this area and aspires for a flourishing low carbon economy. Sustainable development is a major aspect of Abu Dhabi's Vision 2030, and this is closely linked with achieving significant reductions in carbon emissions, maintaining economic growth and building a circular economy with minimum ecological footprint. Lately, the government of the UAE announced the country's net zero carbon strategic initiative in 2050. This aligns with Paris agreement signed in 2015 by 195 countries including the UAE, which aims to limit the increase in global average temperature to 1.5°C. In addition, the 2nd intended nationally determined contribution of the UAE is to reduce its greenhouse gas (GHG) emissions for the year 2030 by 23.5%, relative to the Business-As-Usual (BAU) scenario (i.e., a reduction of 70 million tons of CO_2 per year in 2030). These ambitious CO_2 reduction targets should go in line with developing technologies and markets for CO_2 capture, utilization and storage (CCUS) along with energy efficiency, renewables, energy storage, and others. CCUS involves CO_2 capture from air or flue gas emitted from power plants and industrial sources using solvents or adsorbents, and then use it as a feedstock in an economically chemical, biological and industrial processes. There are many CO_2 utilization routes that has different market share, temporal storage potential, technological readiness level (TRL) and product value. The main utilization pathways at a scale (Fig. 1) are (1) chemicals; (2) building materials; (3) CO_2-enhanced oil recovery (EOR); (4) soil carbon storage; (5) microalgae; and (6) fuel.

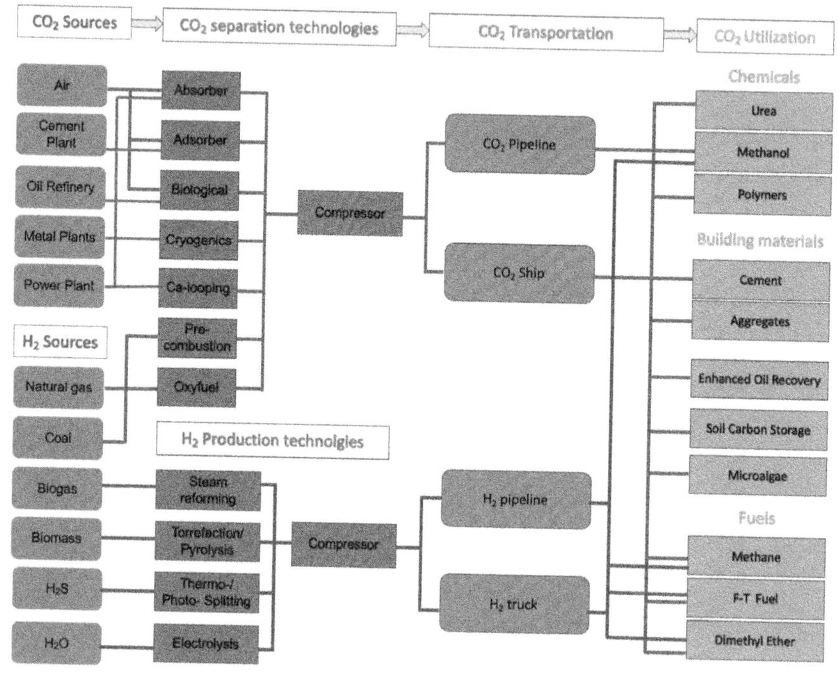

Figure 1: Superstructure of potential CCUS networks

CO_2 capture of around 100 kt yr^{-1} and utilization for urea production has already been implemented commercially in the UAE. Globally, 140 Mt yr^{-1} of CO_2 is used to produce 400 Mt yr^{-1} of urea (Perez-Fortes et al., 2014). Other emerging chemical is polycarbonate polyols that will have a limited market of 10-50 Mt yr^{-1} in 2050 (Hepburn et al., 2019). CO_2 fixation in waste streams from exhausted slag-based and carbon-based sorbents to produce carbonate and silicate aggregates for material construction is a promising long-term utilization route to reduce the overall ecological footprint of CCUS process. This has a large-scale potential in countries climbing the development ladder such as the UAE. One of the main anthropogenic CO_2 utilization route in oil exporting companies such as the UAE is CO_2-Enhanced oil recovery (EOR). EOR is a tertiary approach applied to mature oil reservoirs to improve oil recovery. There are three main temporal phases in oil extraction: primary recovery, at the early stage of oil production, geologic pressure and pumps can push oil from the wellbore to the surface; secondary recovery maintains the pressure of the reservoir and improves oil productivity through injection of water or gas; tertiary recovery, which can recapture 30 to 60% of the original oil in place (OOIP). CO_2 injection decreases the viscosity of the remaining oil, swell it and detach it from the formation. This allows it to move freely in the reservoir reaching the production well, while trapping CO_2 in the reservoir, leading to CO_2-EOR and storage. Part of this CO_2 is produced and recycled back in the operation making it complex to quantify the potential of CO_2 utilization. In this work, a multiscale modeling approach based on system dynamics is demonstrated to examine the potential of combined CO_2-EOR and storage. Another emerging utilization route is soil carbon storage (SCS). Understanding the carbon balance in the soil is essential to evaluate the potential of SCS. Soil stores organic carbon comprising of decaying plants, fungal, bacterial and animal matters. This also enhances the soil structure and reduces the erosion and carbon losses. CO_2 fixation in waste sorbents and carbon rich materials such as char can also play an important role in enhancing the carbon content and water within the soil enhancing the productivity of growing plants. An important utilization route in arid environments such as the UAE is to cultivate microalgae for simultaneous fixation of carbon dioxide (CO_2) and production of high value-added products such as vitamins, antioxidants, and carotenoids for pharmaceutical use. Microalgae are organisms that are able to fix CO_2 with solar energy at higher rates than other plants with no need for arable land and freshwater resources — it can be cultivated using wastewater or salt water medium. Here, a multiscale model is demonstrated to examine the potential and scale of Microalgae cultivation in the UAE. Other utilization routes through recycling CO_2 back into different fuels (Fig. 1) involves production of green hydrogen at a reasonable price as most of the fuel cost comes from hydrogen production (i.e., >85% for methanol (Atsonios et al., 2016)). This is highly linked to the availability of renewables and cost of power. In the UAE, a low power production price was recorded at US$0.015/kWhr. This should go in line with dispatching hydrogen economy at a scale, increasing efficiency and reducing CAPEX of electrolyzers. A price of US$1.0/kg of hydrogen is required to deploy this pathway at a scale. It is worth noting that the fuel generated using this option will be burnt again and emits CO_2, which emphasis an open carbon cycle. This might limit the impact of adoption of power to fuel technologies at a scale while looking at the impact of storage unless carbon is being captured and utilized again in a closed cycle. This work demonstrates the significance of multiscale model as a guide to examine the potential of three major utilization pathways namely CO_2-EOR, microalgae and mineralization.

2. Multiscale modelling methodology

This work builds upon multiscale modelling approach for CCUS system connecting models across different length and time scales (Fig. 2). At picosecond and angstrom scales, molecular dynamics simulations are performed to examine the potential of each sorbent to attract solutes within the available sites in the material. This helps in generating the missing data required to generate isotherm models for CO_2 capture in adsorbents. A continuum model based on partial differential and algebraic equations systems at a scale of seconds and length of meters are used to find the optimal scheduling of different adsorption steps. This helps in establishing the performance of different sorbents through capturing the relationships and trade-offs between structure-property and key performance indicators such as cost, purity, and recovery. Moving to the km scale of reservoir size and decades of EOR operations, a system dynamics model has been developed to quantify the demand of CO_2 for EOR, amount of CO_2 stored and recycled. This integrated multiscale models also applies to other utilization routes such as CO_2 capture using microalgae at the country level. This integrates genomic scale models based on flux balance analysis that describes the change in inter-cellular metabolites with kinetic based models that describes the interactions with extra-cellular environment. The optimal CCUS network can be obtained through Mixed Integer Linear Programming

(MILP) (Alhajaj and Shah, 2020) or Mixed Integer Non-Linear Programming (MINLP) approach.

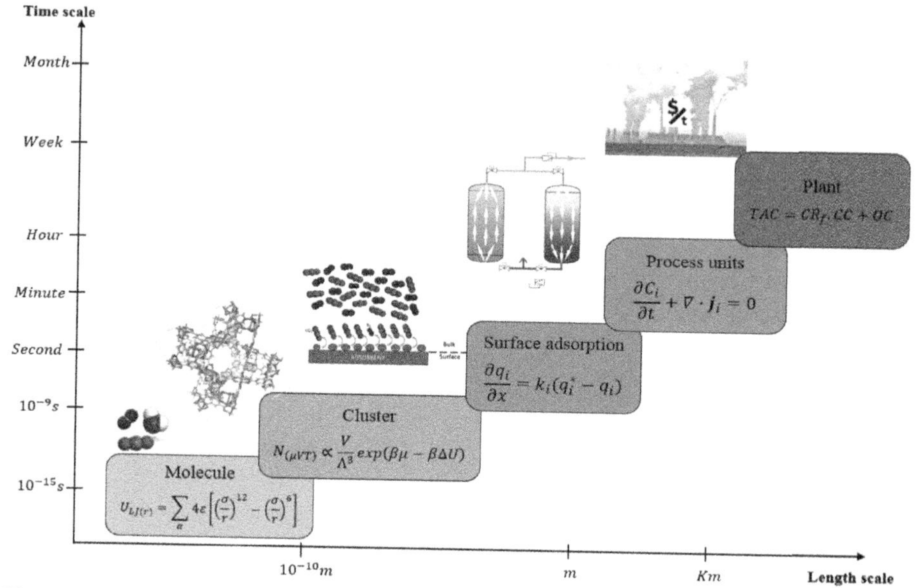

Figure 2: Illustration of multiscale model (Reproduced from (Balogun et al., 2021) with permission of the Royal Society of Chemistry)

2.1. Problem definition

A multiscale model integrating models across different length and time scales have been used to find the potential of CCUS considering main markets in the UAE, namely CO_2-EOR, microalgae and mineralization. The main objective was to examine the potential in meeting the 2nd national reduction target standing at 70 Mt CO_2 yr^{-1}. The multiscale CCUS supply-chain network model was formulated as an MILP problem, similar to our earlier work (Alhajaj and Shah, 2020). The selection of CO_2 sources, main capture technologies with varying configurations, absorbents, and adsorbents in addition to microalgae, transportation modes and CO_2 utilization routes have been considered in this work. The model has been used to find the cost-optimal network that meets a specific CO_2 decarbonization target with a positive long-term environmental impact. The latter is considered through performing life cycle analysis in utilization pathways. This captures the energy requirements for synthesis and upgrading of fuels and materials in addition to the duration of CO_2 storage.

In this work, a multiscale model is used to obtain realistic data for the maximum storage capacity and utilization potential per year, presented as inequality constrains in the MILP problem. This has been used to capture the system dynamics of CO_2-EOR enterprise. The adoption of anthropogenic CO_2 as a solvent to enhance oil recovery in aging reservoirs combined with CO_2 storage to mitigate climate change is a complex process due to the interplay of environmental, technical and market targets. A system dynamics model has been adopted to account for the complex interactions between the dynamics of oil production, climate change mitigation targets, CO_2 sources market, oil production targets, carbon credits, policies, and oil market subsystems (Santos et al., 2021). This has been used to obtain the yearly demand of fresh CO_2 and the potential of CO_2-EOR storage in mitigating greenhouse gas emissions in the UAE and meeting the 2nd national reduction target.

To examine the potential of embracing biological carbon capture and utilization using microalgae at a large scale, bio-process model imbedded with genomic scale model have been used to examine the potential of biological CO_2 capture and utilization from 400 MW combined cycle gas turbine (Alhajaj et al., 2016). This has identified the pond size (length scale) and time scale required for deep decarbonization of flue gas using microalgae. All data obtained from the detailed analysis of utilization pathways and capture technologies are fed into the supply-chain network to obtain realistic and cost-optimal decarbonization pathways.

3. Preliminary Results

The results of the multiscale model have identified CO_2-EOR as a major utilization and storage route in the UAE. This can mitigate 10-20% of the 2nd national reduction target of the UAE (i.e., 13.6-25.6 Mt CO_2 trapped in the reservoir). On contrary, microalgae CO_2 capture and cultivation from combined cycle power plant require a massive land with a size of Dubai Expo (i.e., >10000 hectares) to capture 75% of the flue-gas. A higher capture rate (Fig. 3) is difficult due to the high evaporation rates and emissions to atmosphere resulting from the exponential growth in number of ponds. This demonstrates the complexities in applying microalgae at a scale to mitigate climate change due to the slow process of CO_2 fixation. The main driver to implement microalgae projects should be targeted to small scale high-value and green products.

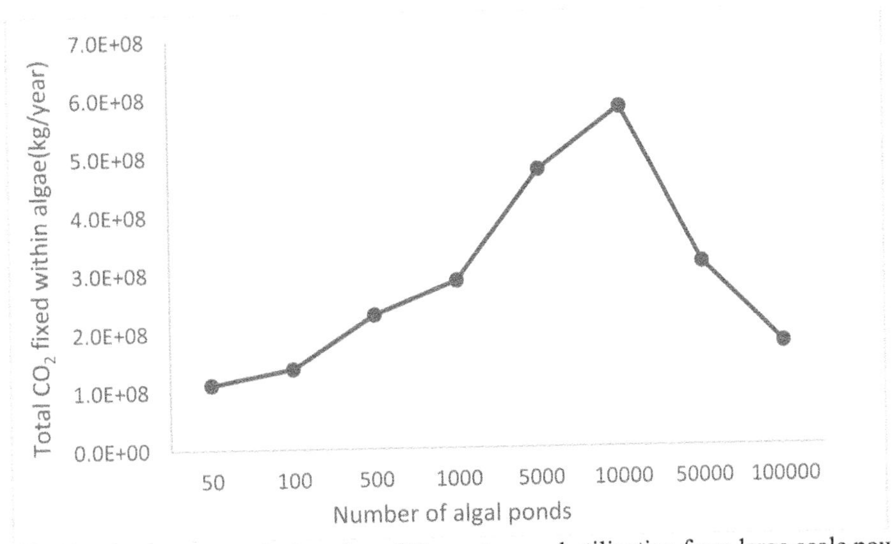

Figure 3: Illustration of microalgae CO_2 capture and utilization from large scale power plant

The results of this study have demonstrated the difficulties of implementing solo-CCUS route at a scale. Therefore, synergized adoption of CCUS with storage should be implemented to mitigate climate change. Emerging markets such as CCUS in building materials and soil carbon storage can complement large scale storage in reservoirs and reactive Basaltic rocks. Other enabling technologies such as hydrogen economy should be supported to create the mass for economies at a scale. This will help in creating new large-scale markets for CO_2-Fuel, especially for aviation.

4. Conclusion

Implementing large scale Carbon Capture, Utilization and Storage (CCUS) projects is required to mitigate climate change and meet national CO_2 reduction targets. In this work, we have developed multiscale models that has helped in examining the potential of large scale CCUS pathways such as CO_2-EOR, mineralization and microalgae in the UAE. The results of this study have highlighted the potential of implementing CCUS at a large scale in the UAE with positive environmental impacts. CO_2-EOR with storage can play an important role in developing CCUS infrastructure, which helps in establishing a network and economies of scale for wider adoption. Emerging markets such as mineralization, soil storage and saline aquifers can help in accelerating the rate and scale of CCUS implementation in the UAE. Microalgae and CO_2-based fuels have an important market such as hard-to-abate aviation but further decarbonization strategies should be implemented to ensure larger net CO_2 reductions. This depends on the availability of enabling resources such as renewables, heat, green hydrogen, and storage sites. These CCUS pathways are region-specific and might not be replicated at a large scale elsewhere. There is also a need for policies and incentives to de-risk the CCUS business and liability, which will assure deep decarbonization of the society. Overall, this study has demonstrated the need for integrated tools and models across different length and time scales to capture the system dynamics using various key performance indicators such as cost and life cycle analysis. This integrated tool can guide policy makers and governments on selecting right incentives and policies that help in wider adoption of CCUS with desired social benefits.

5. Acknowledgements

Funding for this work has been provided by Khalifa University of Science and Technology under projects CIRA-2021-067 and RC2-2019-007. Computational resources from RICH are also acknowledged.

References

A. Alhajaj, N. Shah, , 2020, Multiscale design and anlaysis of CO_2 networks, International Journal of Greenhouse Gas Control, 94, 102925.

A. Alhajaj, N. Mac Dowell, N. Shah, A techno-economic analysis of post-combustion CO_2 capture and compression applied to a combined cycle gas turbine: Part I. A parametric study of the key technical performance indicators, International Journal of Greenhouse Gas Control, 44, 26-41

K. Atsonios, K. Panopoulos, E. Kakaras, 2016, Investigation of technical and economic aspects for methanol production through CO_2 hydrogenation, International Journal of Hydrogen Energy, 41, 4, 2202-2214.

H. Balogun, D. Bahamon, S. AlMenhali, L. Vega, A. Alhajaj, 2021, Are we missing something when evaluating adsorbents at the system level?, Energy & Environmental Science, 14, 12, 6360-6380.

C. Hepburn, E. Adlen, J. Beddington, E. Carter, S. Fuss, N. Mac Dowell, J. Minx, P. Smith, C. Williams, 2019, The technological and economic prospects for CO_2 utilization and removal, Nature, 575, 7781, 87-98.

M. Perez-Fortes, A. Bocin-Dumitriu, E. Tzimas, 2014, CO_2 utilization pathways: Techno-economic assessment and market opportunities, Energy Procedia, 63, 7968–7975.

R. Santos, S. Sgouridis, A. Alhajaj, 2021, Potential of CO_2-enhanced oil recovery coupled with carbon capture and storage in mitigating greenhouse gas emissions in the UAE, International Journal of Greenhous Gas Control, 111, 103485.

Antonis Kokossis, Michael C. Georgiadis, Efstratios N. Pistikopoulos (Eds.)
PROCEEDINGS OF THE 33rd European Symposium on Computer Aided Process Engineering (ESCAPE33), June 18-21, 2023, Athens, Greece
 http://dx.doi.org/10.1016/B978-0-443-15274-0.50162-1

Prediction of capacitance using artificial neural networks for carbon nanofiber-based supercapacitors

Kiran Donthula, Naresh Thota, Sarath Babu Anne, Manohar Kakunuri*

Department of Chemical Engineering, National Institute of Technology Warangal, Telangana, India

manohar@nitw.ac.in

Abstract

Carbon nanofibers (CNF) with different morphologies and microstructures were widely used as electrode material for supercapacitors since they exhibit high electrolyte-electrode interface, flexibility, and tunable physicochemical properties. This CNFs-based electrode electrochemical performance is greatly influenced by their surface area, pore structure, and fiber diameter. No credible physical model is available that can predict the performance of supercapacitors based on the CNFs physicochemical properties. In this work, we have used a data-driven Artificial Neural Network (ANN) model to predict the performance of CNF electrode-based supercapacitors based on electrode material microstructural properties and electrochemical operational parameters. The proposed model demonstrates its feasibility in predicting specific capacitance with a correlation coefficient (R-square) value close to 0.95. A sensitivity analysis was also conducted to understand the effect of independent input parameters on a single output parameter (specific capacitance).

Keywords: ANN, Supercapacitor, Data-driven, Energy storage, Machine learning.

1. Introduction

Recent developments in supercapacitors have attracted great attention due to their superior power density and cyclic stability when compared to conventional batteries (Panda et al., 2020). The use of carbon nanomaterials, such as activated carbons, nanoparticles, nanofibers, etc., as electrode materials for supercapacitors, has been widely studied because of their high surface area, high conductivity, and low production cost (He et al. 2020; Lokhande et al. 2020; Mathis et al. 2019). Compared to other electrode materials, CNFs are unique in that they are both flexible and possess tunable surface area, porosity, and fiber diameter. These characteristics make them ideal electrode materials for flexible supercapacitors. Hence, these CNFs have been extensively studied as electrode materials for supercapacitors (Chandu et al., 2020). However, the improvement in the specific capacitance is inadequate due to charge storage at the electrode-electrolyte interface. Activation and physical etching methods were used to engineer the microstructural properties of these CNFs to enhance their electrochemical performance. The microstructural properties, micro (<2 nm) and mesopore (2nm-50nm) volumes, specific surface area (SSA), and fiber diameter greatly rely on the polymer precursor and carbonization process. In addition to these microstructural properties, supercapacitor performance depends on voltage window, scan rate, cell configuration, and electrolyte characteristics. As the performance of CNF-based supercapacitors depends on these many parameters, there is a significant need to develop a model that correlates these parameters to the performance of the supercapacitors. It is imperative to note that the development of a model does not only assist in predicting the performance of the supercapacitor, but also in planning the experiments to achieve the desired microstructural properties for enhancing its performance. The qualitative deviation of electrochemical performance with electrode SSA, micropore, and mesopore volumes has been examined through experimental studies. However, no physical model has been developed that predicts how these parameters interact to affect electrochemical performance in a synergetic manner. Alternatively, experimental data may be used to develop a data-driven machine-learning model that addresses this problem.

In recent years, various machine learning models have been applied to determine the synergetic effects of multiple variables on the electrochemical performance of energy storage devices. Among different machine-learning models, an ANN was widely used to develop a model using complex and nonlinear data (Himanenmanen et al. 2019; Farouji et al. 2022). In a recent study, Rahimi et al., 2022, developed an ANN-based data-driven model that correlated microstructural characteristics of activated carbon electrodes, electrode synthesizing procedures, and operational conditions that affect the electrode's performance. The developed model accurately predicted the performance of activated carbon electrodes with an R-square value of 0.9. To our knowledge, there are no studies on CNF-based electrodes using data-driven machine-learning models. In this work, we have developed an ANN model to predict the performance of CNF-based electrodes using various independent input variables. Our goal is to examine the effect of input variables independently and synergistically on the prediction of specific capacitance of the CNF-based electrode.

2. Material and methods

2.1 *Experimental data collection and preprocessing*

The ANN model in this work was trained using seven input parameters, which include carbonization temperature (C_t), SSA, total pore volume (V_t), micropore volume (V_{mic}), mesopore volume (V_{meso}), nanofiber diameter (NFD), and voltage window (VW) and a single output parameter (Specific capacitance). Data for both input and output parameters were collected from 70 research articles reporting experimental data,each reference paper we took one single data set. The data was limited to the pristine CNFs-based electrode studies with the aqueous electrolytes and incorrect or outdated data can also impact the accuracy of the model so we considered only recent research articles data and inconsistent formatting or scaling of data can affect the model's accuracy so as a preprocessing step, all the parameters were normalized between 0 and 1. This normalization enables the ANN to learn more quickly since the output is more sensitive to changes in the input (Rafiq et al. 2001). The following function was used to normalize the data in this study:

$$X_{nor} = \frac{X - X_{min}}{X_{max} - X_{min}}$$

where X denotes the variable value and X_{nor} denotes the value of the normalized variable, X_{min} and X_{max} correspond to the minimum and maximum values of the variables, respectively.

2.1.2 *Artificial neural network*

It is possible for an ANN to learn rather than be programmed to make predictions. The ANN consists of input parameters, output parameters, and several hidden layers. Depending on the weight (W_{ij}) and bias (B_i), the hidden layer will augment or cripple the received information. Herein, our goal is to use this ANN and the experimental data collected from the literature to build a data-driven model as shown in Fig. 1 that can predict the specific capacitance performance of a CNF electrode-based supercapacitor. Our model was trained and tested using input and output data collected using the supervised machine learning method and the neural network toolbox in MATLAB 2021a. The artificial neural network consists of one hidden layer with a different number of neurons (11 to 23), a log-sigmoid activation function in a hidden layer, and a purelinear transfer function at the output layer.

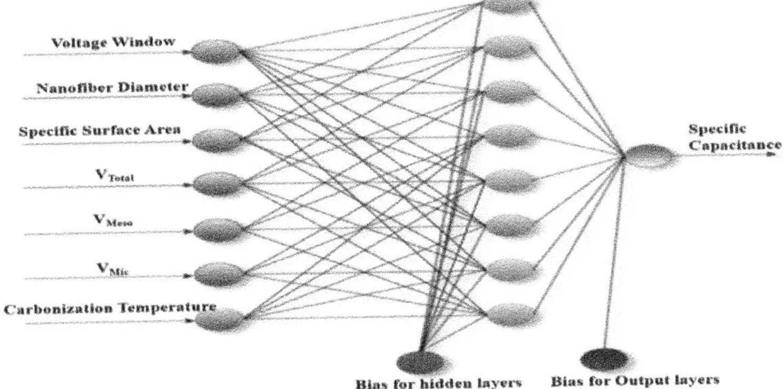

Figure 1. The neural network model of ANN used in this study.

An ANN model was trained using Levernberg–Marquardt with backpropagation of errors (LM-BP)(Willis et al.1991) Levenberg-Marquardt algorithm appears to be the fastest method for training moderate-sized feedforward neural networks (up to several hundred weights). It also has an efficient implementation in MATLAB software, because the solution of the matrix equation is a built-in function. The data was divided into three subsets of 70%, 15%, and 15% for training, testing, and validation, respectively. The schematic illustration of the dataset preparation procedures and modeling implementation is shown in Fig. 2.

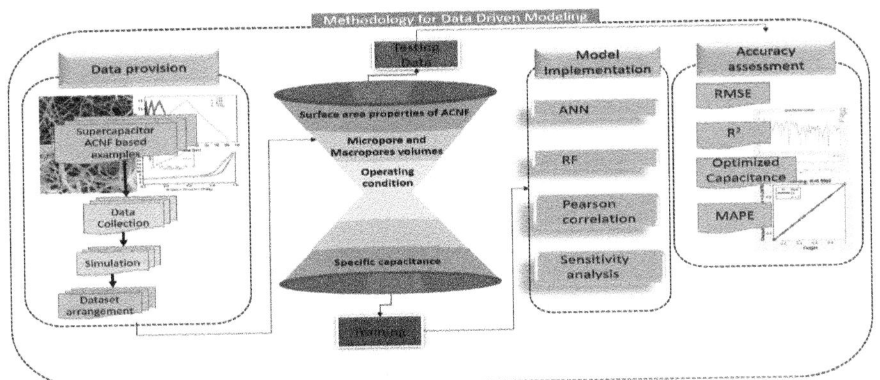

Figure 2. Schematic representation of data-driven modeling methodology

The number of neurons in an ANN model's hidden layer also significantly affects the prediction. To avoid overfitting (a high number of neurons) and offsetting (a low number of neurons), the number of neurons in the ANN model were optimized(Karsoliya, 2012). In this study, the number of neurons in hidden layers (hidden layer size) varied between 11 and 23. After the training, the performance of ANN was evaluated using the test data set.The error of ANN is determined by the mean square error (MSE) and correlation coefficient (R-Square).

3. Results and Discussion

3.1 Result of ANN model

The Pearson correlation coefficient (PCC) was used to investigate the linear relationship and co-occurrence of input and output parameters, as shown in Fig.3. The PCC analysis reveals the dependency of the specific capacitance (output parameter) on all the input parameters. The reliance is in the decreasing order of SSA>V_{meso}>V_t>V_{mic}>C_t>VW. NFD, an independent input variable, showed a negative correlation (-0.16) with specific capacitance in the group of seven input parameters. Due to this, linear/inverse relationships cannot be used accurately to predict capacitance.

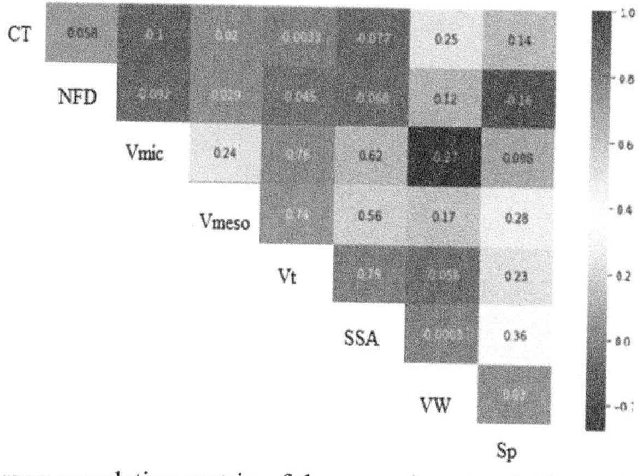

Figure 3. Pearson correlation matrix of the seven input variables and single output variable (specific capacitance).

Table 1 summarizes the effect of hidden layer size on the performance (as MSE & R-square) of the training algorithm LM-BE. Increasing the number of neurons in the hidden layer from 11 to 19 resulted in a moderate decrease in the MSE from 0.0554 to 0.0031, and a further increase in neurons from 19 to 23 resulted in a moderate increase in the MSE to 0.006. As a result, a network with 19 neurons with least MSE value (0.0031) was used for sensitivity analysis studies. Furthermore, the network with 19 neurons demonstrated excellent correlation coefficients, with R-square values exceeding 0.95 for both training and testing achieved performance metrics are consistent across multiple training repetitions.

Table 1: Performance of ANN vs. Number of neurons in hidden layer

Hidden layer size	Training		Testing	
	MSE	R^2	MSE	R^2
11	0.01550	0.786	0.05540	0.918
13	0.00117	0.888	0.00700	0.921
15	0.00392	0.873	0.00310	0.967
17	0.00238	0.878	0.00835	0.910
19	0.00035	0.955	0.00286	0.955
21	0.00703	0.916	0.00160	0.854
23	0.00648	0.867	0.00200	0.866

Based on the results presented in Fig. 4, it can be concluded that substantial agreement exists between the actual and predicted values of specific capacitance during both the training and testing phases. Mostly, ANN models can be compared to other machine learning models such as Decision Trees, Random Forests, Support Vector Machines, etc., so Random Forest (RF) model is used to compare the ANN accuracy results. ANN model showed better accuracy than RF model.

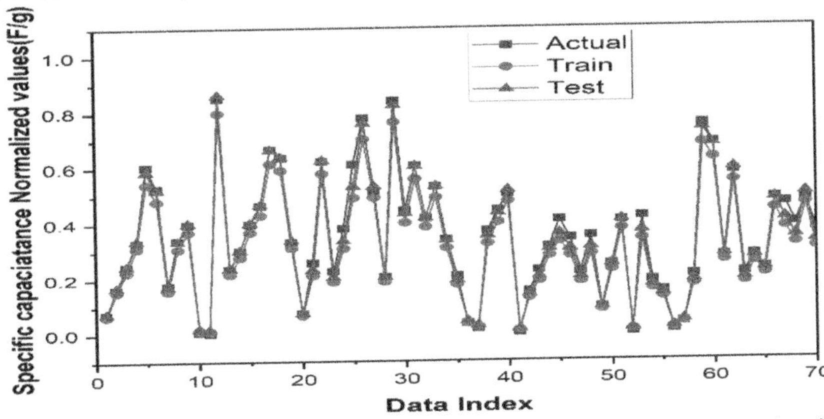

Fig.4 The predicted and actual values of specific capacitance in both test and training, using the ANN.

3.2 Sensitivity analysis

A sensitivity analysis was conducted by selecting input parameters precisely to determine how they affect the performance of CNF-based supercapacitors. The sensitivity analysis is presented in detail in table 2. First, all input variables were considered in this analysis, and then a single input parameter was excluded from other analyses. Microstructural variables appear to have a substantial impact on the electrochemical performance of CNF-based supercapacitors. According to this analysis, specific capacitance is greatly influenced by V_{meso} and SSA, in agreement with experimental observations (Xie et al. 2020). In addition, the mean absolute percentage error (MAPE) was also used to quantify the degree of dependency. The MAPE values for the input set excluding V_{meso} and the input set excluding SSA rose to 4.1% and 4.74%, respectively. This increase in error value indicates the importance of these microstructural properties on the performance of CNF-based supercapactors. The input set without NFD, however, showed lower sensitivity, with MAPE of 1.57.

Table 2 Sensitivity analysis of the ANN model versus independent variables.

Model Input	Input set	MAPE		
		Train	Test	Total
	All	0.56	1.10	0.58
Structural	All exclude Vmic	3.20	3.56	2.93
	All exclude Vmeso	4.63	4.20	4.10
	All exclude SSA	5.14	5.96	4.74
	All exclude Vt	2.10	2.90	2.01
	All exclude NFD	1.90	1.00	1.57
Carbonization	All exclude Ct	2.88	1.69	2.41
Operational	All exclude VW	3.12	3.39	2.84

4. Conclusion

The ANN model predicts the CNF-based electrode performance accurately using microstructural properties and operational parameters as input. In addition, the proposed model with 19 neurons showed a high R-square value of 0.95 and a lower MSE value of 0.003. The results of a sensitivity analysis on seven input parameters suggested that SSA and V_{meso} have a significant impact on electrode performance. In contrast, the input parameter NFD has a relatively small effect. The developed model can be employed to design nanofiber-based electrode materials experimentally, which reduces the need for labor-intensive experimentation to optimize the microstructural properties.

References

P. K. Panda, A. Grigoriev, Y. K. Mishra, and R. Ahuja, 2020, Progress in supercapacitors: roles of two dimensional nanotubular materials Nanoscale Adv., 2, 70–108.

P. E. Lokhande, U. S. Chavan, and A. Pandey, 2020,Materials and Fabrication Methods for Electrochemical Supercapacitors: Overview, Electrochemical Energy Reviews, Springer Singapore, 155–186.

J.He, D. Zhang, Y.Wang, J. Zhang, B. Yang, H.Shi, K. Wang, Y. Wang ,2020, Biomass-derived porous carbons with tailored graphitization degree and pore size distribution for supercapacitors with ultra-high rate capability, Appl. Surf. Sci., 515, 146020.

T. S. Mathis, N. Kurra, X. Wang, D. Pinto, P. Simon,Y.Gogotsi,2019, Energy Storage Data Reporting in Perspective-Guidelines for Interpreting the Performance of Electrochemical Energy Storage Systems ,Adv. Energy Mater., 9.

V.V.Chandu. Muralee Gopia, R.Vinodha, S. Sambasivamb, I. M.Obaidatb,2020,recent progress of advanced energy storage materials for flexible and wearable supercapacitor: From design and development to applications,J. Energy Storage, 27, 101035.

M.S.Farouji, H.V. Thanh , Z. Dai, A. Mehbodniya M. Rahimi , U. Ashraf , A. E. Radwan,2022, Exploring the power of machine learning to predict carbon dioxide trapping efficiency in saline aquifers for carbon geological storage project Journal. Clean. Prod., 372, 133778.

L.Himanen, A.Geurts, A.Foster, P.Rinke, 2019,Data-Driven Materials Science: Status, Challenges, and Perspectives Advanced. Sciences, 6 , 1900808.

M. Rahimi, M. H. Abbaspour-Fard, A. Rohani,2022, synergetic effect of N/O functional groups and microstructures of activated carbon on supercapacitor performance by machine learning,2022, J. Power Sources, 521, 230968.

M. Y. Rafiq, G. Bugmann, D. J. Easterbrook,Neuarl network design forengineering applications,2001, Comput. Struct., 79, 1541–1552.

M. J. Willis, C. Di Massimo, G. A. Montague, M. T. Tham, and A. J. Morris, 1991,Artificial neural networks in process engineering IEE Proc. D Control Theory Appl., 138, 256–266 .

S. Karsoliya, 2012, Approximating Number of Hidden layer neurons in Multiple Hidden Layer BPNN Architecture Int. J. Eng. Trends Technol., 3, 714–717.

Antonis Kokossis, Michael C. Georgiadis, Efstratios N. Pistikopoulos (Eds.)
PROCEEDINGS OF THE 33rd European Symposium on Computer Aided Process Engineering (ESCAPE33), June 18-21, 2023, Athens, Greece
 http://dx.doi.org/10.1016/B978-0-443-15274-0.50163-3

Model size determination using data analysis tools in the context of reaction network identification

Carolina S. Vertis,[a] Belmiro P.M. Duarte,[a,b] Nuno M.C. Oliveira[a,*]

[a]*Univ Coimbra, CIEPQPF, Department of Chemical Engineering, Rua Sílvio Lima — Pólo II, 3030–790 Coimbra, Portugal*

[b]*Polytechnic Institute of Coimbra, ISEC, Department of Chemical & Biological Engineering, Rua Pedro Nunes, 3030–199 Coimbra, Portugal*

[*]*nuno@eq.uc.pt*

Abstract

Data dimensionality analysis or model size determination (identifying the true dimension of the underlying data variant space from experimental data) is a key step for the accurate identification of reaction networks and kinetic modeling of chemical systems. Here, several data analysis tools are compared regarding their relative performance in the identification of the model size from simulated noisy data concerning a challenging case study that presents low reaction extents. Depending on the characteristics of the data used, distinct choices of analytic methods can be suggested. The complementary application of different methods to reinforce the confidence of this analysis is also considered.

Keywords: Data dimension, Model size, Reaction networks, Systematic methods, Statistical approaches.

1. Introduction

Accurately describing chemical reaction systems, including their dynamics, generally requires using systematic frameworks and a detailed set of experimental data. Models under and over-parameterized might present unsatisfactory predictive abilities during the calibration (training) and validation (testing) phases, thus compromising their subsequent applications such as process optimization, design, and control (Bonvin et al., 2016). A common concern to modelers is to maximize the extent of the information explained by the model, while avoiding the addition of model features that can be essentially attributed to noise, thus preventing overfitting.

One crucial step during the modeling task is the identification of the underlying data variant space, whose dimensionality can be described as the model size for the system. The integration of various existing methods within a systematic methodology for chemical reaction model development has been proposed and tested (Vertis, 2022). In this context, the model structure (or the system stoichiometry) can be further identified by determining an abstract (latent) space without using any information about the system kinetics (Bonvin and Rippin, 1990). Here, we compare classical empirical methods such as the scree test (Cattell, 1966), the fractional variance analysis (Bro and Smilde, 2014), and the Kaiser test (Cliff, 1988), in addition to the parametric F-test (Malinowski, 1989) and the procedure for non-parametric cross-validation by eigenvectors (Wise and Ricker, 1991), concerning their ability to identify the "true" model size from noisy reaction data. These data represent cumulative extents corrupted with (white) noise generated from in silico experiments, simulated at different experimental conditions. This type of problem may become challenging if the concentration of some species is of the same order of magnitude as the system noise, leading to various model identifiability issues. As an illustration example, a simulated system including 6 species and 4 reactions is considered. The results allow a comparison of the efficiencies of the methods in capturing the correct model size, and their requirements in terms of the data characteristics. These results also evidence that the choice of the best method to identify the model size depends on the data itself, especially on the characteristics of the underlying noise component.

2. Methods considered

The information from the Singular Value Decomposition (SVD) is very useful to determine the model size R, since it allows the data decomposition in directions ordered by decreasing variance. Thus, many tools for model size identification rely on the analysis of (squared) singular values / eigenvalues. A collection of them is briefly listed next.

2.1. Heuristic approaches

This section introduces three empirical (*ad hoc*) methods, available in the literature, for determining R: the scree test, the Kaiser rule, and the analysis of fractional variances.

2.1.1. Scree test

The scree test assumes that the relevant information can be distinguished from random noise in terms of singular values and that the magnitude of the random noise variation seems to stabilize linearly with the number of components (Cattell, 1966). The common graphical analysis based on heuristic principles requires a visual analysis of the eigenvalues plot and the use of a given threshold.

2.1.2. Kaiser rule

Kaiser's rule assumes that if a component has an eigenvalue greater than one, it explains the variation of more than one original variable (Cliff, 1988). This led to the rule of selecting all components with eigenvalues greater than one; however, recent works demonstrate this rule can be inadequate (Bro and Smilde, 2014).

2.1.3. Fractional variance analysis

Another rule of thumb is to include the first R eigenvectors such that the sum of their variances (represented by eigenvalues) is up to an amount of the total, such as 95% or 99%, which can be seen as an empirical rule to match model accuracy and parsimony.

2.2. Statistical F-test

Malinowski (1989) proposed a parametric test to determine R (within a specified significance level) based on the Fisher variances ratio obtained from pools of samples with normal distribution and i.i.d.

2.3. Cross-validation procedure

The goal of the method selected for cross-validation consists of finding the number of model components that leads to the best overall prediction in the testing phase, established through the minimum PRESS-CV (predicted residual error sum of squares evaluated in a cross-validating approach). The original data matrix $\mathbf{D}[n_o \times n_s]$ is randomly sampled (without replication) into K data files of size $[n_c \times n_s]$ and $[n_v \times n_s]$, respectively, with $n_c + n_v = n_o$. For clarification, let $\mathbf{D}^{-k}[n_c \times n_s]$ and $\mathbf{D}^{k}[n_v \times n_s]$ be the matrices of data used, respectively, for model training and testing in k^{th} iteration ($k \in \{1,\ldots,K\}$). Then, the SVD of $\mathbf{D}^{-k}$ is carried out resulting in $\mathbf{U}[n_c \times n_c]$, $\Sigma[n_c \times n_s]$ and $\mathbf{V}[n_s \times n_s]$ orthogonal matrices where $\mathbf{U}$ and $\mathbf{V}$ contain, respectively, left and right eigenvectors of $\mathbf{D}^{-k}$, and Σ is a rectangular diagonal matrix containing the singular values. This structure is then used (in an internal loop to that of k; the counter is s) to construct a projection matrix $\mathbf{M}$ from $\mathbf{V}_{R*}[(n_s - 1) \times R]$ which, on turn, is obtained from the columns 1 to R of $\mathbf{V}$ by stripping the sth row. For every $s = 1, \dots, n_s$ a column vector $\mathbf{d}_s^k[n_v \times 1]$ containing the variable s is removed from $\mathbf{D}^k$, resulting in a smaller matrix $\mathbf{D}_*^k[n_v \times (n_s - 1)]$. Afterwards, the matrix of scores for data validation, say $\mathbf{T}[n_v \times R]$, is then predicted by projecting $\mathbf{D}_*^k$ onto $\mathbf{M}$. Finally, the variable s in $\mathbf{D}^k$ can be estimated as a linear combination of $\mathbf{T}$ using the excluded row vector $\mathbf{v}_{R,s}[1 \times R]$.

This strategy is commonly designated as *cross-validation by eigenvector* to distinguish it from classic cross-validation where the goal is estimating a model using (partial) least-squares. In contrast, in principal components models (or models from SVD of data) the left-out elements and their predicted values are not independent, resulting in a decreasing PRESS-CV with the number of components (never reaching a minimum) if the traditional (k-fold or leave-one-out) cross-validation approaches are considered (Bro et al., 2008). This result would lead to the incorrect conclusion that all components are significant, thus overestimating the optimal model size. Here, in practice, the problem is handled by predicting the missing variable from the model eigenvectors extracted from the training dataset by excluding one of the columns of the testing data set such that the expectation is maximized (Wise and Ricker, 1991).

3. Case study

The example considered for comparison consists of a catalytic reaction system where 6 chemical species are linked by 4 chemical reactions – the reaction network in Figure 1 (right panel). The reaction kinetics follow the Langmuir-Hinshelwood model (1) where species A, B, and C are adsorbed in the catalyst with a single active site. In this equation, for the reaction $j = 1, \dots, 4$, k_j is the kinetic constant in min^{-1}, C_{reac_j} and α are, respectively, the normalized concentration and the stoichiometric number of moles of the reactant species, and K_s (inverse molar ratio units, dimensionless) and C_s (molar ratio) are, respectively, the adsorption equilibrium constant and the normalized concentration respecting the adsorbed species $s = A, B, C$. Both k and K parameters are temperature-dependent following the Arrhenius (2) and van't Hoff (3) equations, where A_0 (in min^{-1}) is the pre-exponential factor and E_a (in J.mol^{-1}) is activation energy of the reaction, R (in J.mol^{-1}.K^{-1}) is the universal gas constant, T (in K) is the experimental temperature, and, ΔH_{ad} (in J.mol^{-1}) and ΔS_{ad} (in J.mol^{-1}.K^{-1}) are the enthalpy and entropy changes of the species adsorption.

$$r_j = {k_j C_{\mathrm{reac}_j}^{\alpha}} \Big/ {(1 + K_A C_A + K_B C_B + K_C C_C)}, \qquad j = 1, \dots, 4 \tag{1}$$

$$k_j = A_{0_j} \exp\left({-E_{\mathrm{a}_j}} \Big/ {RT} \right), \qquad j = 1, \dots, 4 \tag{2}$$

$$K_s = \exp\left({-\Delta H_{\mathrm{ad}_s}} \Big/ {RT} + {\Delta S_{\mathrm{ad}_s}} \Big/ {R} \right), \qquad s = A, B, C \tag{3}$$

The in-silico experiments were simulated at 50, 70, and 90 °C, respectively, lasting for 30 min with the reaction starting from pure A; thus, three data sets were produced by sampling the process every 2 min. White noise with 0.015 standard deviation was added to the entire set of concentration data. In Figure 1 the model dynamics for a single experiment is presented. As can be seen, the species E is present only in very small concentrations and is strongly affected by noise, leading to potential model identification issues.

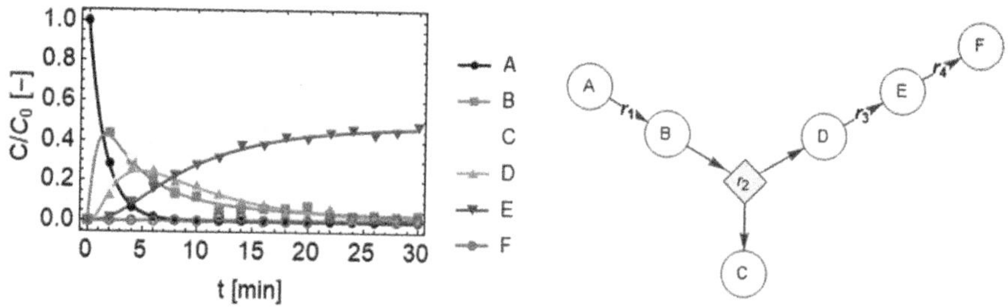

Figure 1 – Model simulation (continuous curves) and noisy experimental data (discrete points) at 70 °C, and model structure (reaction network) on the right-hand side.

The methods described in Section 2 were applied to cumulative molar changes (species extent data) to compare their performance in identifying the correct model size. Note that apart from the measurements, no additional information about the reaction system is used in this task. Also, we considered the cumulative molar changes as they allow minimizing the error in estimation of changes, compared to the estimation of derivatives. In Figure 2 the results obtained for the heuristic approaches are shown. The distribution of the logarithm of the variances in (a) does not present a clear sharp curve, leading to an inconclusive result. Plot (b) shows that the Kaiser test overestimates the model size, indicating a 5-dimensional model. Finally, the plots (c) and (d) for the fractional variance analysis indicate a model with 3 components. However, if 1% of the noise injected was used as a threshold criterion, only a single model component would be selected.

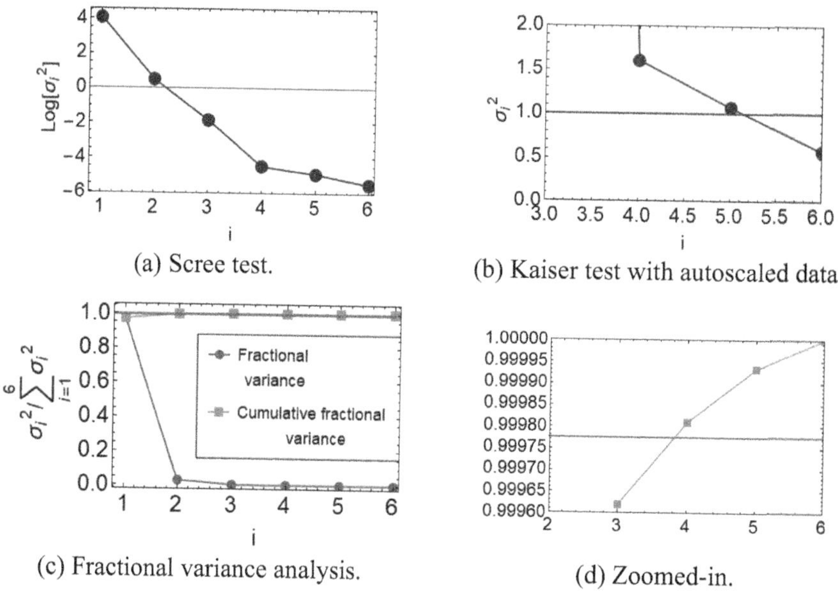

(a) Scree test.

(b) Kaiser test with autoscaled data.

(c) Fractional variance analysis.

(d) Zoomed-in.

Figure 2 – Heuristic methods results based on data variance analysis.

The F-test indicates that 3 model components are statistically significant. The results are shown in Table 1 and Figure 3. The *F*-distribution plots indicate the respective PDF and CDF profiles for different pool sizes. The results demonstrate that $F < F_c$ until the 2-dimensional pool size analysis (where the vertical red dashed line is located to the right of the vertical black dot-dashed line and the horizontal red dashed line is located above the horizontal black dot-dashed line). When the 3rd lowest component is included, the opposite behavior is observed, i.e., $F > F_c$. Therefore, according to these results, the 3rd model component contains structural information related to the data variability that is greater than the noise variability, and hence, it must not be discarded from the pool of noisy components.

Table 1 - F-test results. F_c: critical F value for $\alpha = 5\%$. $\alpha^{\ddagger}$ indicates the percentage of the remain CDF function for each F observed (cutoff value if that F was considered critical).

R	$n_s - R$	$v(R) \times 10^5$	$v(0) \times 10^5$	$F(1, n_s - R)$	$F_c(1, n_s - R)$	$\alpha^{\ddagger}[\%]$	H_0
5	1	8.763	9.904	0.885	161.448	51.95	Failed to reject
4	2	8.856	4.568	1.939	18.513	29.84	Failed to reject
3	3	90.174	2.998	30.077	10.128	1.19	Rejected
2	4	731.095	10.560	69.234	7.709	0.11	Rejected
1	5	21021.1	55.807	376.677	6.608	0.00	Rejected

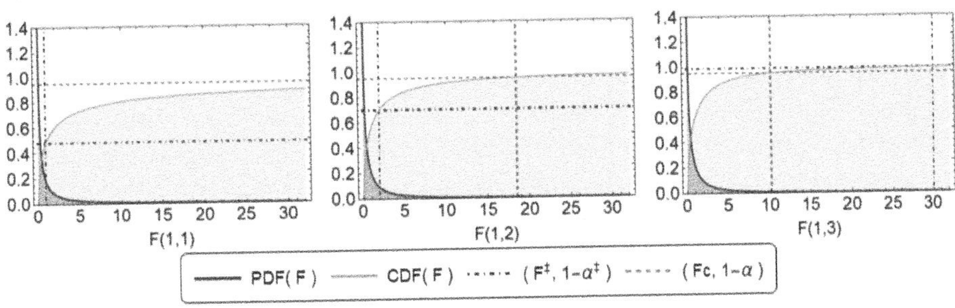

Figure 3 - F-distributions for several sizes of pools. The red dashed lines indicate F_c (x-axis) concerning 1-α = 95% (y-axis). The black dot-dashed lines indicate the observed F (x-axis) and the respective 1-$\alpha^{\ddagger}$ (y-axis).

Figure 4 presents the PRESS-CV for the adopted cross-validation (k-fold) approach. In this plot, a flat behavior of the PRESS-CV curve can be observed between model sizes from 2 to 4, indicating that the predictive ability of the data is similar for models including 2 to 4 components. In a purely data-driven-based approach, 2 components would be selected as it allows maximizing the parsimony. However, only from the fourth dimension the PRESS-CV sharply increases, showing the presence of noise in the fifth component and consequent data overfitting. Although there is no significant improvement in the prediction with the addition of a 3rd and a 4th component to model structure, these are necessary for the complete description of the reaction system through a structure that represents the production/consumption of the entire set of observed species. Thus, from a fundamental perspective where the model needs to be able to describe the underlying phenomena, it is preferable to remain with 4 dimensions in the model, which can be seen as the most plausible model size.

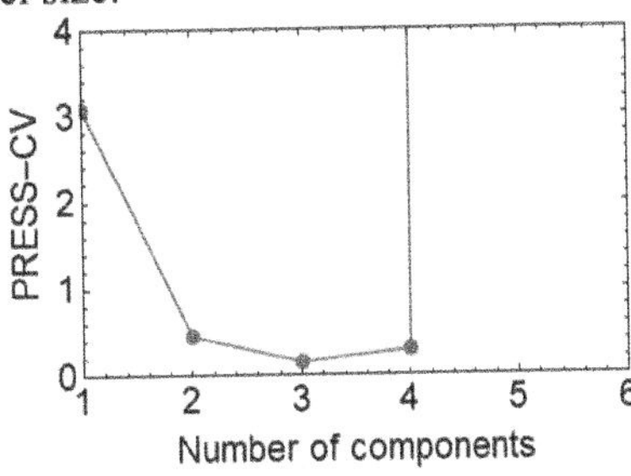

Figure 4 - Cross-validation results for 5 groups of validation datasets.

4. Conclusions

The correct identification of model size from experimental data is important to accurately describe the dynamics of reaction systems. Several methods can be applied, without using any additional system information (like the assumed system kinetics or stoichiometry). The choice of the method must consider the noise, the number of measurements available, the subsequent model application, and the complexity of its structure. When large datasets are considered, the PRESS-CV method tends to perform well and shows a high likelihood of correctly capturing the "true" model size, since the chance of sampling bias is small in this case. The F-test seems preferable to heuristic-based methods, since the former has some theoretical support. When chemical reactions only occur in low extents (i.e., when the signals have variances of the same magnitude as the noise variance) the identification of the "real" size of the model becomes a hard problem. Within this setup, the use of non-heuristic-based methods is advised as they are grounded on theoretically consistent approaches. The results also highlight the advantage of combining different methods to determine the data characteristic space and the complementary space (i.e., the model size and the set of variables included in model representation, and the complementary set of variables). Given the possible synergies of the metrics, the combination of methods can increase the degree of confidence in the identification of the "true" dimensionality of the underlying system. In addition, the results obtained serve as preliminary guidelines for finding new model-based experiments for identifying the real size of the model. Table 2 presents the adequacy of the methods tested (here "+++" indicates a significant preference, "++" an average preference, and "+" a lower preference) and the basic characteristics of the data sets required for its application.

Table 2 - Preference of applicability of the selected methods and respective constraints.

Preference	Method advised	Requirements for successful application
+++	Cross-validation by eigenvectors	Large datasets
++	F-test	---
+	Heuristic approaches*	Low noise content

*When the noise content is known the fractional variance analysis is preferred.

References

D. Bonvin and D. Rippin, 1990, Target factor analysis for the identification of stoichiometric models. Chemical Engineering Science 45, 3417–3426.

D. Bonvin, C. Georgakis, C. Pantelides, M. Barolo, M. Grover, D. Rodrigues, R. Schneider, D. Dochain, 2016, Linking models and experiments, Industrial & Engineering Chemistry Research, 55, 6891–6903.

R. Bro, K. Kjeldahl, A. K. Smilde, H. Kiers, 2008, Cross-validation of component models: a critical look at current methods. Analytical and bioanalytical chemistry 390, 1241–1251

R. Bro and A. K. Smilde, 2014, Principal component analysis, Analytical Methods, 6(9), 2812–2831.

R. B. Cattell, 1966, The scree test for the number of factors, Multivariate behavioral research, 1(2), 245–276.

N. Cliff, 1988, The eigenvalues-greater-than-one rule and the reliability of components, Psychological Bulletin, 103(2), 276.

E. R. Malinowski, 1989, Statistical F-tests for abstract factor analysis and target testing, Journal of Chemometrics, 3(1), 49–60.

C. S. Vertis, 2022, Identification and modeling of chemical reaction networks, Ph.D. thesis, Chemical Engineering Department of University of Coimbra.

B. Wise, and N. Ricker, 1991, Recent advances in multivariate statistical process control: improving robustness and sensitivity, In Proceedings of the IFAC, ADCHEM Symposium, 125–130.

Antonis Kokossis, Michael C. Georgiadis, Efstratios N. Pistikopoulos (Eds.)
PROCEEDINGS OF THE 33rd European Symposium on Computer Aided Process Engineering (ESCAPE33), June 18-21, 2023, Athens, Greece
 http://dx.doi.org/10.1016/B978-0-443-15274-0.50164-5

Modeling of particle formation in pan granulators with sieve-mill recycle

Robert Dürr[a], Mateusz Przywara[b], Eric Otto[c], Dorota Antos[b],Achim Kienle[c,d]

[a]*Magdeburg-Stendal University of Applied Sciences, Breitscheidstraße 2, 39114 Magdeburg, Germany*
[b]*Rzeszow University of Technology, al. Powstańców Warszawy 12 35-959 Rzeszów, Poland*
[c]*Otto von Guericke University, Universitätsplatz 2, 39106 Magdeburg, Germany*
[d]*Max Planck Institute for Dynamics of Complex Technical Systems, Sandtorstraße 1, 39106 Magdeburg, Germany*

Abstract

Agglomeration of powders in pan granulators is a widespread production process, in many industries, e.g., agricultural and pharmaceutical. A continuous operation provides certain desired advantages over batch operation including constant throughput with constant properties of the product particles. An extended setup including a sieve mill recycle can improve the sustainability of overall process. As shown for related process, e.g., fluidized bed granulation, such a setup can exhibit interesting dynamic phenomena like sustained oscillation of the product particle properties, which must be researched in more detail to identify (un-)favorable process regimes. In this contribution, a model for fertilizer production in such a setup is presented. Simulation results indicate the potential benefits not only on increased product particle yield but also manipulation of particle properties.

Keywords: Population balance modeling, pan granulation, numerical solution, digital process twin, fertilizer manufacturing.

1. Introduction

Agglomeration is a particle formation process in which at least two particles are fused. This principle is used in many industries, e.g. fertilizer production. The properties of the formed agglomerates, e.g. size, shape and porosity, significantly affect certain end-use properties, e.g., dissolubility, processability and storeability (*Bück & Tsotsas, 2016*). In particular for fertilizer production and minerals processing, agglomerates are often formed in drums or pans *(Litster & Ennis, 2004*) in continuous operation which may provide constant throughput with constant product quality (in terms of specific agglomerate properties) during steady-state operation. Besides sophisticated experiments, mathematical modeling is also important to come to a deeper understanding and for potential development of model-based process control and intensification. In this report, focus is on population balance modelling of pan granulation which is also known under the term dish or disc granulation. The process setup is given as follows: a powder is fed constantly to an inclined rotating dish into which additional liquid binder is sprayed. Wet granules are sticking together and form larger ones. Flux from the pan does not only contain product particles but also fines and oversized particles. To improve sustainability of the process the non-product fractions can be recycled using a sieve-mill recycle. It is well-known that for size-enlargement processes like pan granulation individual

properties, like characteristic size and porosity, differ from particle to particle. As an alternative to Monte-Carlo modeling approaches, population balance modeling (PBM) *(Ramkrishna & Singh, 2014)* represents an established framework to model such processes. For standard monovariate PBEs, accounting mostly for characteristic particle size or volume, efficient and accurate numerical solution algorithms have been developed, see e.g. *Kumar et al. (2008)*, *Dürr & Bück (2020)*, and the references therein.
In this contribution, a PBM for the continuous pan granulation process with sieve-mill recycle is presented based on a recently reported model for operation without recycle. The model is also extended for a mass controller to guarantee stable process operation. In different scenarios the advantage of recycling is emphasized and also the effects on the product particle characteristics in terms of size distribution is shown.

2. Process Model

2.1. Population Balance Model

Starting from the model for continuous pan agglomeration presented in *Otto et al. (2021)* the model equations must be adapted to account for the screen-mill-recycle. The volume density distribution function of the particles in the pan can be described with the following population balance model

$$\frac{\partial n(t,v)}{\partial t} = +\dot{n}_{\mathrm{feed}}(t,v) + \dot{n}_{\mathrm{rec}}(t,v) - \dot{n}_{\mathrm{out}}(t,v) + \dot{n}_{agg}(t,v)$$

Therein, the expressions on the left hand side represents the accumulation of particles in the pan while the right hand side describes contributions of the powder feed to the pan, the recycled particle stream, particles withdrawn from the pan and new particles generated by agglomeration. It is furthermore assumed that the process parameters are chosen such that particle formation via aggregation is dominant and thus additional effects from breakage of granules is negligible. In the following, all expressions on the right hand side are described in more detail.

2.2. Modeling of Powder Feed and Agglomeration

For the description of the powder feed to the pan the approach for the process without screen-mill-recycle presented in *Otto et al. (2021)* is used.

$$\dot{n}_{feed}(t,v) = \dot{N}_{in}(t)\widetilde{n}_{feed}(v)$$

Here, it is assumed that the powder particle volume characteristics do not change over time and can be represented by the normalized feed powder number density function. Furthermore, denotes the time dependent feed rate.
The change in the particles NDF by formation of a new agglomerate of volume v by binary fusion of two primary particles of volumes *u* and *v-u*, respectively, is described with the following expression

$$\dot{\mathrm{n}}_{\mathrm{agg}}(\mathrm{t},\mathrm{v}) =$$
$$\frac{1}{2}\int_0^v \beta(t,u,v-u)n(t,u)n(t,v-u)\,\mathrm{d}u - \int_0^\infty \beta(t,u,v)n(t,v)n(t,u)\,\mathrm{d}u$$

Generally, the formation of new particles by agglomeration depends on multiple factors, e.g. temperature, binder composition, and particle volume. A common approach to describe such dependencies is a so-called separation approach for the agglomeration kernel function

$$\beta(t,v,u) = \beta_0(t)\beta_v(v,u)$$

with the first factor, the coalescence frequency, accounting for time-dependency due to process conditions or operating parameters and the second for volume dependency. The latter function can either be derived from micro-scale mechanistic modelling, heuristic assumptions or abstract data-driven approaches. In this manuscript, focus is on steady-state operation and thus constant process conditions can be assumed. In consequence,

$$\beta_0(t) = \beta_0 = const.,$$

the so-called Kapur-Kernel is used

$$\beta_v(u,v) = \frac{(u+v)^a}{(u+v)^b}.$$

2.3. Modeling of Outflux Rate

The outflux rate comprises granules smaller than product range, granules in product range and also particles larger than the desired product range.

$$\dot{n_{out}} = \dot{n_{fines}} + \dot{n_{product}} + \dot{n_{over}}.$$

Assuming a sieve with two screens as presented in *Neugebauer et al. (2019)* the rates above can be modeled as

$$\dot{n_{over}} = T_{screen,1}\dot{n_{out}}$$
$$\dot{n_{product}} = (1 - T_{screen,1})T_{screen,2}\dot{n_{out}}$$
$$\dot{n_{fines}} = (1 - T_{screen,1})(1 - T_{screen,2})\dot{n_{out}}.$$

With sieving functions given as

$$T_{screen,i}(L) = \frac{\int_0^L exp\left(-\frac{(\xi - L_{screen,i})^2}{2\sigma^2_{screen,i}}\right) d\xi}{\int_0^\infty exp\left(-\frac{(\xi - L_{screen,i})^2}{2\sigma^2_{screen,i}}\right) d\xi}.$$

In contrast to the previous characterization of granules by volume, the sieving function is described in terms of characteristic particle size and thus reformulation under the assumption of spherical granule shape, i.e.,

$$v = \frac{\pi}{6}L^3$$

When then distribution function is reformulated it must be taken specific care for mass conservation in a subsequent computational implementation, i.e.,

$$n(t,v)\,dv = n\_L(t,L)\,dL$$

2.4. Modeling of recycle rate

The recycle rate includes particles which are out of the desired product particle range with oversized particles being milled before being fed back into the pan

$$\dot{n_{rec}}(t,v) = \dot{n_{fines}}(t,v) + \dot{n_{mill}}(t,v)$$

Assuming an idealized mill, the milled particles follow a normal distribution *(Neugebauer et al., 2019)*

$$\dot{n_{L,mill}}(t,L) = \frac{\int_0^L exp\left(-\frac{(\xi - L_{mill})^2}{2\sigma_{mill}^2}\right) d\,\xi}{\int_0^\infty exp\left(-\frac{(\xi - L_{mill})^2}{2\sigma_{mill}^2}\right) d\,\xi} \dot{n_{L,over}}(t,L)$$

2.5. Powder feed manipulation for mass control

The extension of the process setup for the recycle requires control of the overall volume/mass of particles in the pan to guarantee a constant product stream and omit situations with too much or too few particles in the pan. Different control variables could be used to manipulate the overall mass of the pan content:

- Pan inclination,
- Angular speed of the pan,
- Recycle stream,
- Powder feed rate.

Manipulation of the first two variables are favored for the manipulation of the granule properties and for improved sustainability the goal is to use the full recycle stream in the process. Thus, manipulation of the powder feed rate is supposed to be the best option to obtain a stable mass/volume of the pan content. The change of particle mass in the pan is given as the sum of all the in- and outgoing particle fluxes

$$\dot{m_{pan}} = \dot{m_{feed}} + \dot{m_{rec}} - \dot{m_{out}}$$

One pragmatic choice of control would be to use a two-point controller for manipulation of the feed stream. However, the mass of particles in the pan can not be assessed easily and thus an alternative strategy is proposed: The process is run without recycle at startup until steady-state conditions are obtained for a given mass stream of feed-particles $\dot{m_{feed,nom}}$. Afterwards the recycle is added and the feed is manipulated as follows

$$\dot{m_{feed}} = \dot{m_{out}} - \dot{m_{rec}}\,.$$

3. Results

In the following, simulation scenarios for the setup with recycle are summarized. Feed particle size distribution, separation function and agglomeration kernel parameters are the same as in the non-recycle setup presented in *Otto et al. (2021)*. All additional parameters are summarized in Table 1. Simulations were carried out in MatLab2021a. A detailed description of the simulation algorithm is found in *Otto et al. (2021)* and the references therein. In Fig. 1 – 4 results are shown for four different scenarios: operation without recycle (I), with only milled particles recycled (II), only fines recycled (III) and both recycled (IV).

It can be observed, that recycling of particles which are out of the product specification is favorable in terms increased product particle mass flow. However, the recycling rate also has a clear effect on the characteristic particle size distribution of the product fraction.

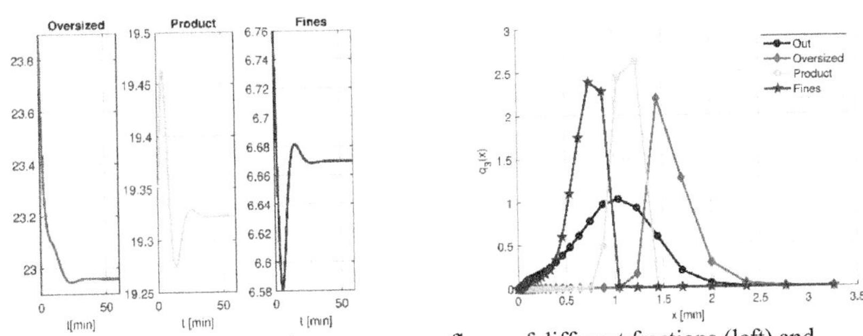

Figure 1: Scenario (I) without recycle stream: mass flows of different fractions (left) and normalized particle size distribution (right)

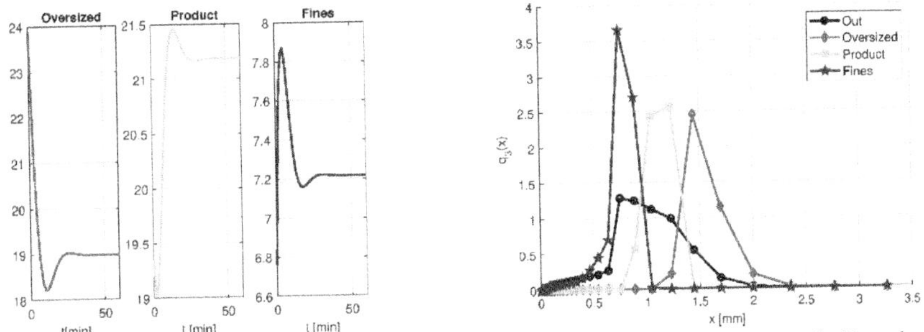

Figure 2: Scenario (II) recycle of milled-particles: mass flows of different fractions (left) and normalized particle size distribution (right)

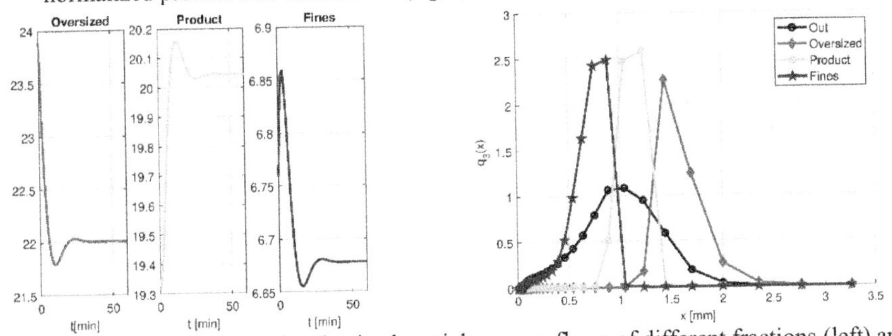

Figure 3: Scenario (III) recycle of undersized particles: mass flows of different fractions (left) and normalized particle size distribution (right)

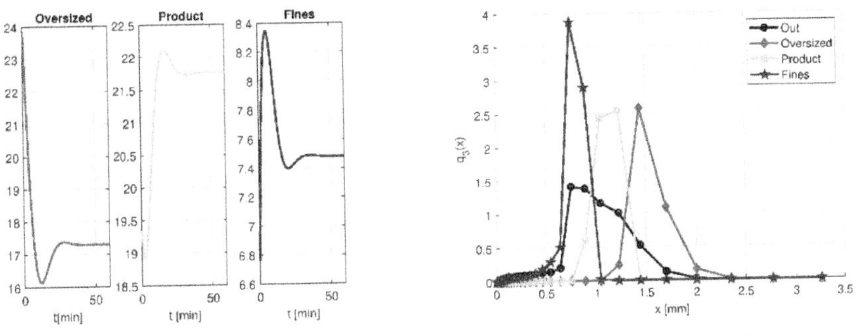

Figure 4: Scenario (IV) recycle of milled and undersized particles: mass flows of different fractions (left) and normalized particle size distribution (right)

4. Summary and Outlook

In this contribution a population balance model for continuous pan granulation with sieve-mill-recycle was presented. Here, it must be mentioned that a mass controller is necessary to guarantee a stable process operation. Simulation results indicate that increased product mass flows are obtained yet also the product particle size distribution and thereby the product properties are affected. Therefore, further research is necessary on this issue as well as identification of (un-)favorable process regimes and interesting dynamic phenomena, such as sustained oscillations, e.g., by bifurcation analysis *(Otto et al., 2022).* Additionally, particle breakage could be taken into account requiring the adaption of the presented concepts. Moreover, the model may be used for the design of model-based sensors or controllers. Of course the previous mentioned research perspectives require model adaption to experimental data (Otto et al., 2021b; Przywara et al., 2021)

References

J. Litster, B. Ennis. The Science and Engineering of Granulation Processes, 2004, Vol. 15 Springer Science & Business Media

C. Neugebauer, E. Diez, A. Bück, S. Palis, S. Heinrich, A. Kienle. On the dynamics and control of continuous fluidized bed layering granulation with screen-mill cycle. 2019, Powder Technology, Vol. 354, p. 765 – 778.

E. Otto, R. Dürr, M. Przywara, D. Antos, and A. Kienle. Population balance modelling of pan granulation processes. 2021, Proceedings to the 31st European Symposium on Computer Aided Process Engineering, p. 965-970

D. Ramkrishna and M. R. Singh. Population balance modeling: Current status and future prospects. Annual Review of Chemical and Biomolecular Engineering, 5:123–146, 2014.

J. Kumar, M. Peglow, G. Warnecke, and S. Heinrich. An efficient numerical technique for solving population balance equation involving aggregation, breakage, growth and nucleation. Powder Technology, 182(1):81 – 104, 2008.

A. Bück and E. Tsotsas. Agglomeration. In Benjamin Caballero, Paul M. Finglas, and Fidel Toldrá, editors, Encyclopedia of Food and Health, pages 73 – 81. Academic Press, Oxford, 2016.

R. Dürr and A. Bück. Approximate Moment Methods for Population Balance Equations in Particulate and Bioengineering Processes, Processes, 8(4), 414, 2020.

E. Otto, R. Dürr, A. Kienle. Bifurcation analysis of combined agglomeration and layering granulation in fluidized bed spray processes, Computer Aided Chemical Engineering 51, 691-696, 2022.

M. Przywara, R. Dürr, E. Otto, A. Kienle, D. Antos. Process behavior and product quality in fertilizer manufacturing using continuous hopper transfer pan granulation—Experimental investigations, Processes 9 (8), 1439, 2021

E. Otto, R. Dürr, G. Strenzke, S. Palis, A. Bück, E. Tsotsas, A. Kienle. Kernel identification in continuous fluidized bed spray agglomeration from steady state data, Advanced Powder Technology 32 (7), 2517-2529, 2021b

Antonis Kokossis, Michael C. Georgiadis, Efstratios N. Pistikopoulos (Eds.)
PROCEEDINGS OF THE 33rd European Symposium on Computer Aided Process Engineering (ESCAPE33), June 18-21, 2023, Athens, Greece
 http://dx.doi.org/10.1016/B978-0-443-15274-0.50165-7

Synthesis and optimization of NGL separation as a complex energy-integrated distillation sequence

Qing Li,[a,b] Adrian J. Finn,[c] Stephen J. Doyle,[b] Robin Smith,[b] Anton A. Kiss [a*]

[a] *Department of Chemical Engineering, Delft University of Technology, Van der Maasweg 9, 2629 HZ, Delft, The Netherlands*

[b] *Centre for Process Integration, Department of Chemical Engineering, The University of Manchester, Sackville Street, Manchester, M13 9PL, United Kingdom*

[c] *Costain, Costain House, 1500 Aviator Way, Manchester Business Park, Manchester, M22 5TG, United Kingdom*

E-mail: tonykiss@gmail.com

Abstract

The synthesis of heat-integrated distillation sequences for energy-efficient separation of zeotropic multicomponent mixtures is complex due to the many interconnected design degrees of freedom. This paper explores the basis on which reliable screening can be carried out. To solve this problem, a screening algorithm has been developed using optimization of a superstructure for the sequence synthesis using shortcut models, in conjunction with a transportation algorithm for the synthesis of the heat integration arrangement. Different approaches for the inclusion of heat integration are explored and compared. Then the best few designs from this screening are evaluated using rigorous simulations. A case study for the separation of NGL is used to compare options. It has been found that separation problems of the type explored can be screened reliably using shortcut distillation models in conjunction with the synthesis of heat exchanger network designs. Unintegrated designs using thermally coupled complex columns show much better performance than the corresponding designs using simple columns. However, once heat integration is included the difference between designs using complex columns and simple columns narrows significantly.

Keywords: Distillation sequencing; energy efficiency; process synthesis and design; process optimization

1. Introduction

Distillation is by far the most widely applied separation technology but is recognized as the most energy intensive operation in the chemical industries due to its inevitable energy degradation. Therefore, to genuinely enhance the energy efficiency of the distillation system, it is necessary that all the degrees of freedom are manipulated simultaneously (Kiss and Smith, 2020), i.e. the basic separation configuration including thermally coupled complex columns, column pressures, reflux ratios, feed conditions, condenser types and heat integration arrangement. For all but the simplest separation problems an exhaustive search is infeasible. The method most often used to address larger problems is to use short-cut distillation models to screen the large number of structural options and determine the best few potential designs, followed by more detailed examination of the best few designs using rigorous simulation. However, this raises many questions regarding the screening procedure. It is not clear whether short cut distillation models have the necessary accuracy for reliable screening, and whether heat integration must be included in the screening. Moreover, if heat integration is included, it is not clear whether the integration can be represented by the energy targets of pinch analysis, or it is necessary to use detailed heat exchanger network designs for screening. This work addresses the problem of multicomponent separation by distillation sequences, by developing an integral systematic optimization method which simultaneously synthesizes sequences, optimizes simple and complex distillation arrangements, heat integration, operating conditions, as well as closed cycle heat pump for refrigeration systems.

The fractionation of natural gas liquids (NGL) is presented here to demonstrate the developed methodology. NGL fractionation, which demands a sequence of several distillation columns, is one of the most energy-demanding processes in the oil and gas industry, producing marketable products such as ethane, propane, butanes, and heavier hydrocarbons. NGL feed is typically 2.5 million tons per year, and approximately 77 GJ per barrel of NGL feed is required, which is equivalent to 4.65 tons of CO_2 emissions per barrel (Manley, 1998). Several attempts based on process intensification principles have been made on the NGL fractionation process to generate more energy efficient distillation sequence design with better separation performance and lower energy use. However, these previous studies made simplifications to reduce model complexities to achieve the challenging optimization: only simple columns are taken into account (Yoo et al., 2016), column pressures are fixed, relative volatilities of the initial feed stream are used for all sequence columns (Ramapriya et al., 2018), or assume only saturated liquid or vapor-phase feed and product streams (Tamuzi et al., 2020). These simplified assumptions on key operating variables lead to loss of accuracy, and none of the previous works considered all degrees of freedom, which risks missing the promising solutions.

2. Problem statement

The optimization of the NGL fractionation system is motivated by the benefits of energy savings in reducing costs and CO_2 emissions. Its design requires many interconnected design degrees of freedom, as well as column arrangement possibilities (simple column or complex column arrangements), which lead to a challenging optimization problem. The present study solves this problem through a novel and systematic optimization approach to evaluate all the complex energy integrated NGL fractionation sequence alternatives simultaneously, eliminating simplifications on key design variables. The best sequences can be ranked based on techno-economic analysis.

3. Methodology and optimization approach

For a given multicomponent mixture, a superstructure is initially created that contains all possible sequences using simple columns, and then the superstructure is extended to merge any two simple columns in series to form different complex column arrangements, including side draw columns, side stripper, side rectifier, prefractionator, Petlyuk and dividing wall column, as illustrated in Figure 1 (a) (Smith, 2016). Therefore, based on the same starting defaults of each simple column sequence, the modified superstructure generates complex column sequences accordingly. Shortcut Fenske-Underwood-Gilliland models are used to design the columns in this task, which can screen the very many options and avoid dramatic computational difficulties created by nonlinear, large tray-based combinations of rigorous design formulation. The sequences screened by shortcut models are then evaluated by Aspen Plus rigorous simulations.

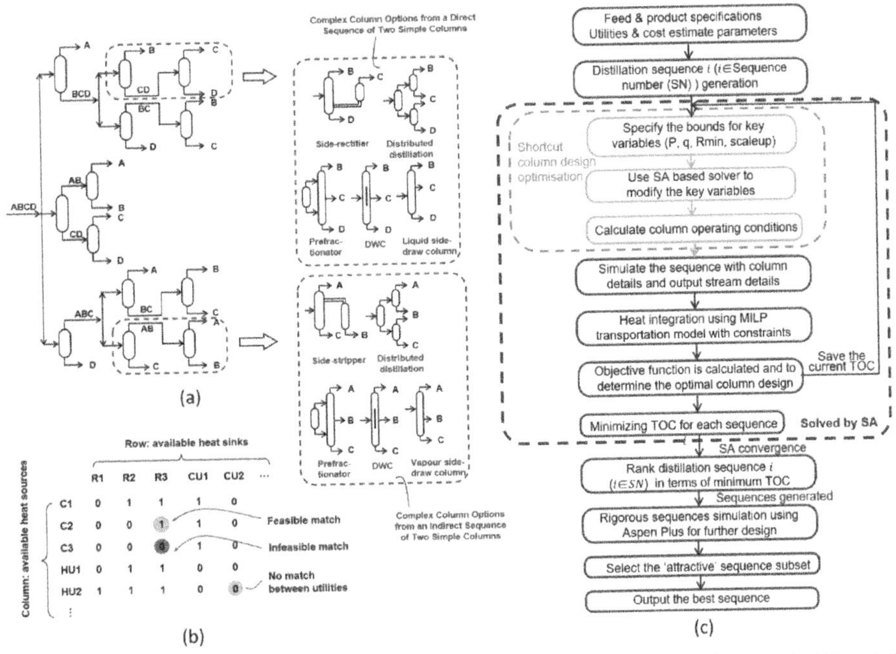

Figure 1 Extended superstructure for sequence synthesis (a), heat integration matrix (b) and the overall methodology flowchart (c)

Heat integration opportunities including feed preheating or precooling and column to column heat recovery are optimized with sequence synthesis and sequence design simultaneously. All heat sinks and heat sources are compiled into an incidence matrix (as shown in Figure 1 (b)) and solved by the MILP transportation method. Binary variables are introduced to make discrete decisions for each match between rows and columns of the matrix. The optimal heat exchanger network structure to minimize operating costs is determined. Liquids and gases are pressurized by pumps and multistage-compressors and depressurized by valves and turbo expanders. The separation is assumed to be sharp split.

A decomposed optimization methodology is created and illustrated by a flowchart shown in Figure 1 (c). The overall model is formed as an MINLP problem and solved by simulated annealing (SA). The shortcut distillation models equations are built into the optimizer, acting as a 'simulation' section to eliminate the non-linearity in the optimization. For a given mixture, the SA creates 'moves' in the optimization that allow for changes in the distillation structure, for example, changing two adjacent simple columns to be a complex column, and the operating variables in their permitted bounds. The values are then sent to the objective function which uses them as inputs for the simulation model. After each move, the sequences are simulated, heat exchanger network designed, and the corresponding total operating cost (TOC) is evaluated by the current column design details and then by the heat integration. The objective value is then returned to the optimizer. The operating variables can subsequently be changed by the optimizer and the process repeated until convergence. This optimization methodology can be carried out within the ColSeq software developed and available at the Centre for Process Integration, The University of Manchester (CPI Suite, 2022).

4. Results and discussion

The conventional configuration of the NGL fractionation uses a train of 4 simple distillation columns: deethanizer (30.98 bar), depropanizer (17.50 bar), debutanizer (5 bar) and deisobutanizer (6.50 bar) to generate 5 products (Manley, 1998; Long et al., 2018). In this work, the NGL fractionation is optimized by applying the developed optimization model to enhance its energy benefits. Multiple utilities considered for heating, cooling and refrigeration and their prices are taken from Turton et al., (2018). The corresponding details of feed and products are taken from Long et al., (2016). For heat exchangers, the minimum temperature difference between hot streams and cold streams (including utilities) is assumed to be 10 K for above ambient heat transfer, whereas for sub-ambient it is assumed to be 4 K (Kiss and Smith, 2020). The optimization is carried out using a desktop with CPU - *Intel(R) Core i7-10700 CPU 2.90 GHz* with 8 cores, and the average time resource is 16 hours.

By optimization, a total of 14 simple sequences using 4 simple columns, and the corresponding 14 complex column sequences are generated. These sequences are developed by two optimization routes for the NGL fractionation: sequences without heat integration and integrated sequences with detailed heat exchanger network design.

4.1. Optimization results evaluation and analysis

Firstly, the developed approach is validated demonstrating by non-heat integrated simple column sequences. Figure 2 (a) shows sequence rankings in terms of the total operating cost. The ranking top 5 sequences (Zone-I) shown in Figure 2, with relatively small differences in TOC (less than 5%), are considered as the most economically attractive sequence family identified by the screening method based on shortcut models. To evaluate the optimized sequences based on shortcut models, the short-cut design details (reflux ratio, trays, condenser temperatures and distillate rate) are extracted as the initialization parameters, and then all optimized sequences are rigorously simulated using Aspen Plus rigorous distillation models. The sequences S7, S6, S4 are the top ranked sequences identified by rigorous simulations in terms of total operating costs, which have all been preselected in the attractive families by the developed pre-screening approach. These results indicate that despite the deviation between shortcut and rigorous models, the developed fast screening method using shortcut models is validated and comparable. Compared to the conventional distillation sequence (Manley, 1998), the optimized operating conditions based on the same separation structures lead to lower column pressure for all four columns without changing the refrigerant level of the deethanizer, and lower condenser and reboiler duties with 2.37 M$/yr TOC savings.

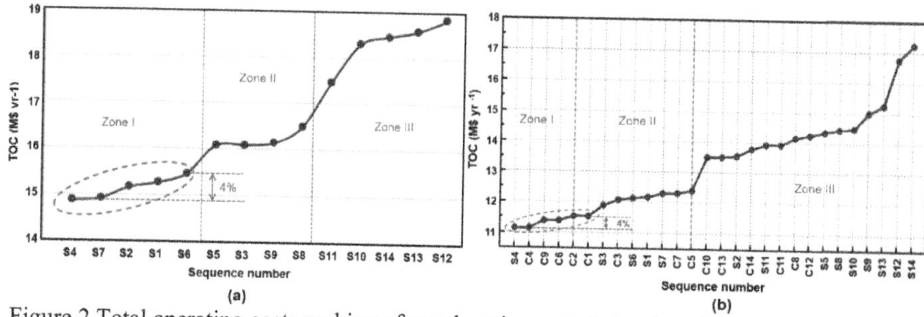

Figure 2 Total operating costs ranking of non-heat integrated simple column sequences (a) and heat integrated sequences (b)

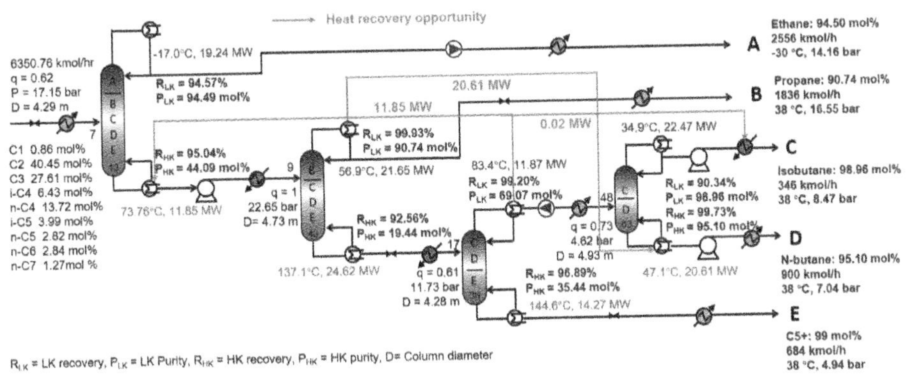

Figure 3. Design details of heat integrated simple column sequence S4

4.2. Energy integrated distillation sequences

Energy intensified sequences using complex columns can bring energy savings. For the non-heat integrated sequences, introducing complex columns can achieve up to 21% operating cost savings compared with the simple column sequences. The conventional sequence (which is the same structure as the simple sequence S4 in this case) that is ranked as the most economical sequence in a simple column sequence, is ranked 11 out of 28 taking into account complex column sequences. The best sequence is now the complex sequence 6, which separates isobutane by a side rectifier, followed by two simple columns separating C2 / C3 and n-butane / C5+, with the TOC of 12.78 M\$/yr.

To achieve further energy savings, heat integration is adopted, achieved by optimizing heat exchanger network with column operating conditions simultaneously for both simple and complex sequences. Figure 2 (b) shows the overall rank of heat integrated simple and complex column sequences. The total operating cost saving brought by heat integration on the optimized non-heat integrated simple columns is up to 26%, and the operating cost saving from optimizing both heat integration and complex columns can achieve up to 38%. The top 6 sequences are considered as the most economically attractive sequences family with a relatively small operating cost difference. Further designs can be taken within this attractive families based on different requirements.

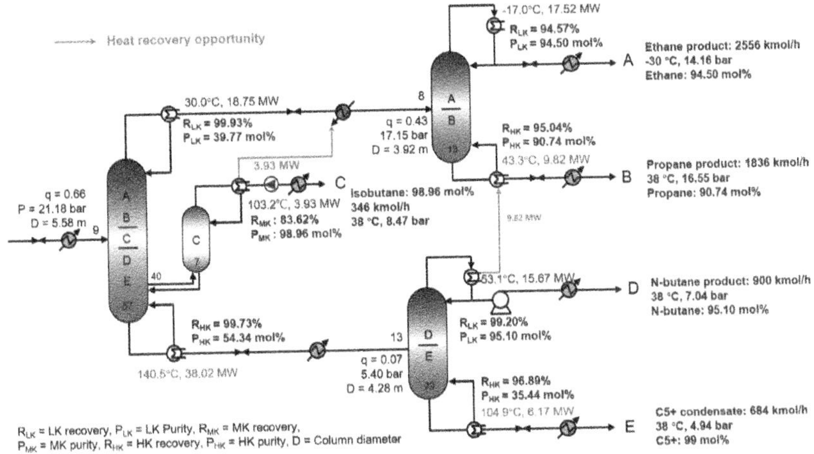

Figure 4. Design details for heat integrated complex column sequence C9

It is shown that when heat integration is adopted, the simple sequence 4 is ranked the best (with the detailed design in Figure 3), followed by the complex sequence C9 (as shown in Figure 4). Simple column sequences benefit to a larger extent from heat integration compared to complex column sequences; simple column sequences can achieve an average energy saving of 16% by introducing heat integration, but complex columns gain an average 11% energy saving by introducing heat integration. This is because complex columns could lose potential heat integration possibilities when constraining the combination of heat sinks and sources during the merging, so that the number of design degrees of freedom of complex columns decrease. The complex sequence C9 may get more benefits when taking into account capital costs due to the reduced equipment sizes. The capital cost is not considered at this stage, as capital costs may not be reflected accurately using approximate capital cost calculation methods.

5. Conclusions

The systematic screening approach developed in this study was successfully used for simultaneously optimizing sequence structures, accounting for all design degrees of freedom and heat integration in the distillation sequences. By optimizing the whole sequencing process, new NGL fractionation sequences have been explored and ranked techno-economically. The generated economically attractive sequences (ranked by the operating costs) are shown to be comparable with those of rigorous simulations using the Aspen Plus RadFrac model. For a given feed composition and product requirements, the ranking of the detail designed energy integrated complex sequences for NGL fractionation are listed, and the most energy efficient and economically attractive sequences family have been selected. These results demonstrate that screening of the search space of energy-efficient distillation sequences, considering synthesis and optimization of the many design degrees of freedom, is achievable with the newly proposed approach.

References

CPI Suite, 2022, The University of Manchester, Centre for Process Integration, Online: https://www.ce.manchester.ac.uk/cpi/research/resources/software/

A.A. Kiss, R. Smith, 2020, Rethinking energy use in distillation processes for a more sustainable chemical industry, Energy, 203, 117788

N.V.D. Long, L.Q. Minh, T.N. Pham, A. Bahadori, M. Lee, 2016, Novel retrofit designs using a modified coordinate descent methodology for improving energy efficiency of natural gas liquid fractionation process. Journal of Natural Gas Science and Engineering, 33, 458e68.

D. Manley, 1998, Thermodynamically efficient distillation: NGL fractionation, Latin American Applied Research, 28, 211-216

G.M. Ramapriya, A. Selvarajah, L.E. Jimenez Cucaita, J. Huff, M. Tawarmalani, R. Agrawal, 2018, Short-cut methods versus rigorous methods for performance-evaluation of distillation configurations, Industrial & Engineering Chemistry Research, 57, 7726-7731

R. Smith, 2016, Chemical process design and integration, Chichester, West Sussex, United Kingdom, John Wiley & Sons, Inc

A. Tamuzi, N. Kasiri, A. Khalili-Garakani, 2020, Design and optimization of distillation column sequencing for NGL fractionation processes, Journal of Natural Gas Science and Engineering, 76, 103180

R. Turton, R.C. Bailie, W.B. Whiting, J.A. Shaeiwitz, 2008, Analysis, synthesis and design of chemical processes, Pearson Education

H. Yoo, M. Binns, M.G. Jang, H. Cho, J.K. Kim, 2016, A design procedure for heat-integrated distillation column sequencing of natural gas liquid fractionation processes, Korean Journal of Chemical Engineering, 33, 2, 405-415

Antonis Kokossis, Michael C. Georgiadis, Efstratios N. Pistikopoulos (Eds.)
PROCEEDINGS OF THE 33rd European Symposium on Computer Aided Process Engineering (ESCAPE33), June 18-21, 2023, Athens, Greece
 http://dx.doi.org/10.1016/B978-0-443-15274-0.50166-9

Multi-scale model of solid oxide fuel cell: enabling microscopic solvers to predict physical features over a macroscopic domain

Hamid Reza Abbasi,[a] Masoud Babaei,[a] Arash Rabbani,[b]

Constantinos Theodoropoulos,[a]

[a]*Department of Chemical Engineering, The University of Manchester, Manchester, M13 9PL, United Kingdom*
[b]*School of Computing, The University of Leeds, UK*
k.theodoropoulos@manchester.ac.uk

Abstract

In this study, a multi-scale model is developed to address the transport of gases, ions, and electrons inside a solid oxide fuel cell. Transport phenomena and electrochemical reactions inside the active layer of anode are simulated at the nanometer scale, while the rest of cell is simulated macroscopically at a coarser resolution. The microstructure of the anode side is reconstructed using pluri-gaussian random fields and the triple phase boundary (TBP), where all the electrochemical reactions are assumed to take place, is captured from the discretized (in voxels) image of the microstructure. Therefore, this multi-scale framework offers a much finer level of description for transport phenomena and for electrochemical reactions, significantly increasing the accuracy of numerical modelling of entire solid oxide fuel cells.

Keywords: Solid Oxide Fuel Cell, Multi-scale model, Triple phase boundary

1. Introduction

Fuel cells, as an alternative power generation device, have shown tremendous potential in clean, small- and large-scale electricity generation in recent decades. Fuel cells convert the chemical energy of a fuel (typically hydrogen) into electricity through electrochemical reactions. The energy conversion (chemical to electrical) is not limited by any thermodynamic law. Theoretically they can have 100% efficiency, whereas the efficiency of e.g., coal-powered stations is limited by the first and second law of thermodynamics and the maximum efficiency attainable is the Carnot efficiency. If hydrogen is to be utilized inside a cell as fuel, the only by-products of the electrochemical reactions would be oxygen gas and water.

Solid oxide fuel cells as the name suggests use a solid ceramic compound as the electrolyte to generate clean electricity (Singh, Zappa, and Comini 2021). SOFCs are well known for their high operating temperatures, which eliminate the need for metallic catalysts. Due to the convoluted interplay between different transport phenomena and chemical/ electrochemical reactions inside the cell, which span different length scales (Grew and Chiu 2012), complex models are needed to capture their behaviour. The electrode microstructure has a substantial effect on the overall electrochemical performance of the cell (dos Santos-Gómez et al. 2021). However, the size of an entire fuel cell is far larger than the microscopic details of the electrodes to be considered in a macroscopic model. It is not, therefore, practically possible to numerically solve the mathematical equations governing the SOFC behaviour over an entire cell while maintaining a fine degree of description of microscopic details. This issue is thoroughly discussed by Grew and Chiu (Grew and Chiu 2012). Hence, most relevant works in the literature are focused either on macroscopic descriptions [4], component-level modelling [5], or on pore-scale modelling [6] of SOFC electrodes. These classes of models make multiple assumptions about the unresolved physics of the "other" scale. To bridge the gap between these two classes of numerical models, an integrated multi-scale numerical framework has been developed in this work by enabling a macroscopic model to obtain information directly from a pore-level model.

The macroscopic model uses the finite-element approach, and its domain covers the entire cell (fuel/air channels, electrolyte, and electrodes, and interconnects). In addition, a microscopic model is developed to simultaneously solve the coupled mass, ion, and electron transport equations, and chemical/electrochemical reactions in a three-dimensional microstructure of the SOFC electrode. Microscopic simulations are only performed on small segments (patches) of the microstructure. A "patch" [7] is a small spatial computational domain extracted from a larger volume/surface of interest, in order to facilitate the efficiency of complex computations. The contribution of this work is that the developed integrated, multi-scale model can resolve the physics linked to the fine microscopic topological features of the electrode microstructure over an entire length of an electrode in a particular fuel cell. This can improve the overall performance prediction of SOFCs, which is highly dependent on both their microscopic and macroscopic features.

2. Methodology

In this section, the two different models that are developed to address the multiscale problem are briefly discussed. The micro model solves the gas/electron/ion transport inside the active electrochemical layer. Using Fick's and Ohm's laws, mass/charge conservation equations can be written as follows. Knudsen diffusivity should also be considered due to small radius of pores. It can be approximated using kinetic theory of gases.

$$\nabla\left(-D_{H_2}\nabla c_{H_2}\right) = -\frac{\rho_{TPB}J}{2F}M_{H_2} \tag{1}$$

$$\nabla(-\sigma_{el}\nabla V_{el}) = -\rho_{TPB}J \tag{2}$$

$$\nabla(-\sigma_{io}\nabla V_{io}) = \rho_{TPB}J \tag{3}$$

Here, c_{H_2}, V_{el}, V_{io} are the hydrogen concentration, electronic potential, and ionic potential, respectively. ρ_{TPB} is the density of the triple phase boundary and J is charge transfer rate. This rate is only non-zero at the TPB. Therefore, all three governing equations (1-3) reduce to the Laplace equation anywhere else in the domain. For the microscopic model, these equations are solved over a representative volume element of a microscopic domain with regular hexahedral mesh. Each grid element represents a pixel in the tomographic

image obtained from 3D imaging. Finite-volume method is employed to discretize equations in 3D space. Boundary conditions for micromodel consist of constant ion and electron potential at microstructure surfaces facing electrolyte and anode, respectively. Hydrogen concentration is also assumed constant at the boundary facing anode side. All other boundary conditions are zero-flux. Three coupled systems of linear equations resulting from discretizing equations 1-3 over the entire microscale domain are then solved until convergence using GMRES algorithm implemented in python.
The macroscopic model solves the mass and charge conservation equations inside an entire cell. The governing equations are the same as mentioned above. Since electrochemical reactions take place on the TPB in microscale, they cannot be resolved with the macroscale model. Instead, these reactions are modeled in the microscale model and then volume-averaged reaction rates are imported into macroscopic model as boundary conditions on the interfaces between electrolyte and porous electrodes. The charge transfer rate is assumed to follow the Butler-Volmer model.

$$J = J_0 \rho_{TPB} \left(\exp\left(\frac{2\beta F}{RT} \eta_{act} \right) - \exp\left(-\frac{2(1-\beta)F}{RT} \eta_{act} \right) \right) \tag{4}$$

Where J_0 $[A/m]$ is the lineal exchange current density taken from data provided by Prokop et al. [8].
A schematic of the macroscopic and the microscopic model and their interactions is shown in Figure 1. The macroscopic domain contains fuel/air channels, porous electrodes, and electrolyte. The microscopic model, on the other hand, is limited to the active electrochemical site between the porous anode and the electrolyte. The microstructure is generated using pluri-gaussian random fields described in [9], and the location of TPB inside the microstructure is calculated with the algorithm discussed by Vivet et al. in [10].

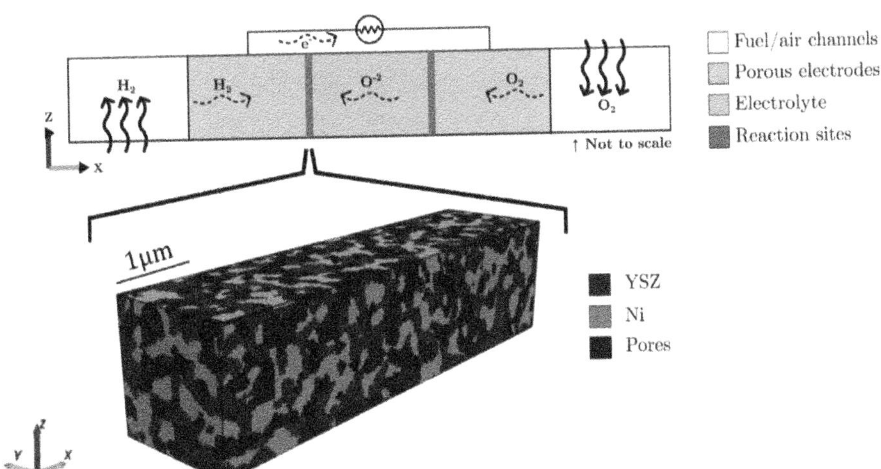

Figure 1 – *Schematic of the multi-scale numerical framework. The macroscopic model spans an entire cell, whereas the microscopic model only covers the electrochemical active layer between electrolyte and electrodes*

The transport equations are solved until convergence both in macro and in micro models. The information that is shared between two models are the physical features of the porous media (porosity, permeability), the effective current density exiting the electrolyte layer, operating voltage, and amount of overpotentials.

Geometrical parameters of the cell and physical properties of Nickel and YSZ are taken from papers by Tseronis et al. [4], and Shearing et al. (Shearing et al. 2010).
The numerical mesh that is used for the microstructure model is essentially the same as that of the voxelized image. Since the voxelized image is the finest description available of the real porous geometry, it has been decided that no smoothing should be performed on the reconstructed image. If the reconstructed geometry is to undergo smoothing, a finer numerical mesh can be constructed, which however would not be useful as its features would have been finer than the finest level of microstructure description available.

3. Results and discussion

In this section, the results of the multi-scale model combining the macro- and micro-scale models are presented and discussed. For the simulations, the operating temperature of the cell is assumed to be 900 °C. The density of the triple phase boundary is measured to be 9.456 μm^{-2}. The variation of the ion potential and current density in the microstructure electrochemical active layer of the anode side is shown in Figure 2. The YSZ phase inside the microstructure is on the left-hand side of the figure with the TPB lines depicted on the outer surface of the YSZ phase as black lines. On the right-hand side of the figure, the variation of the current density is shown on the TPB lines. To visualize the TPB lines, it was assumed that they have an artificial thickness of two pixels on each perpendicular direction. The top interface of the microstructure in this figure is adjacent to the electrolyte and the bottom side faces towards the porous anode layer. Due to the electrochemical reaction taking place on the TPB sites, the ion potential increases from the electrolyte interface (-0.341 V) to the porous anode interface (-0.5 V). The highest charge transfer rate is on the electrolyte interface, and it gradually reduces to zero as the distance from the electrolyte increases.
Changes in hydrogen concentration and electronic potential in the microstructure were insignificant, so they were not shown here. The micromodel is also capable of computing the local activation and concentration overpotential in the microstructure.

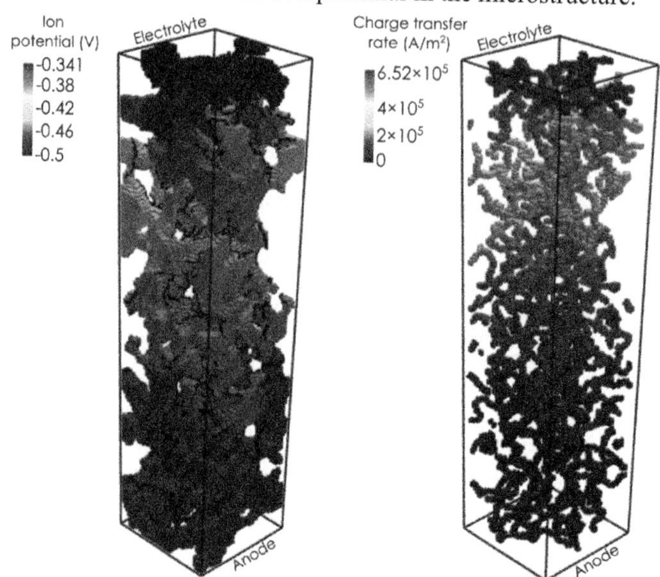

Figure 2 *– spatial variation of ionic potential and current density, respectively in the YSZ phase and on the TPB.*

Results of the multi-scale model are illustrated in Figure 3 below. Figure 3a depicts results from the macroscopic model, which the mass and charge transport through the entire cell, whereas Figure 3b depicts the results of the microscopic model. As mentioned above, the microscopic model solves the same transport equations as of the macroscopic model, with the exception that the micro-scale model also computes the electrochemical reactions on the TPB sites. As shown below, this enables the micro-scale model to capture the rate of charge transfer locally with nanometer resolution. To demonstrate the capability of the multi-scale model, changes of ion potential and current density in the electrolyte and the porous anode are shown in figure 3a, whereas Figure 3b depicts the local variation of the same variables on the interface between electrolyte and porous anode, with much finer (nanoscale) resolution. For comparison, the size of each computational node in the macro-model is roughly 0.1 mm, whereas each computational node of the micro-model is only 25 nm. In Figure 3b, the minimum and maximum value, and the 95% confidence band of ion potential and current density in each slice of the microstructure in x direction (refer to Figure 1) is also shown.

The microscale model covers the electrochemical active layer only. Therefore, the rate of electrochemical reactions (or consequently current density, red curves in Figure 3) approaches zero when the distance from the electrolyte increases in the micromodel. The maximum current density on the electrolyte surface is then transferred to the macroscale model to simulate large-scale mass/charge transport. Potential difference across the micromodel (blue curves in Figure 3) implies the sum of activation and concentration overpotentials in the anode side. This potential difference is forced in the microscale model by macroscale simulation. As a result, until the mass and charge conservation criteria between the micro and macro domains are satisfied, there is an iterative loop between the micro and macro models. This link between micro and macro models allows macroscopic description of mass/charge transport inside a single cell while maintaining a microscopic resolution to explain electrochemical reactions at TPBs. It ultimately creates a framework to investigate the effects of microscopic features on macroscopic performance of the cell in our future studies.

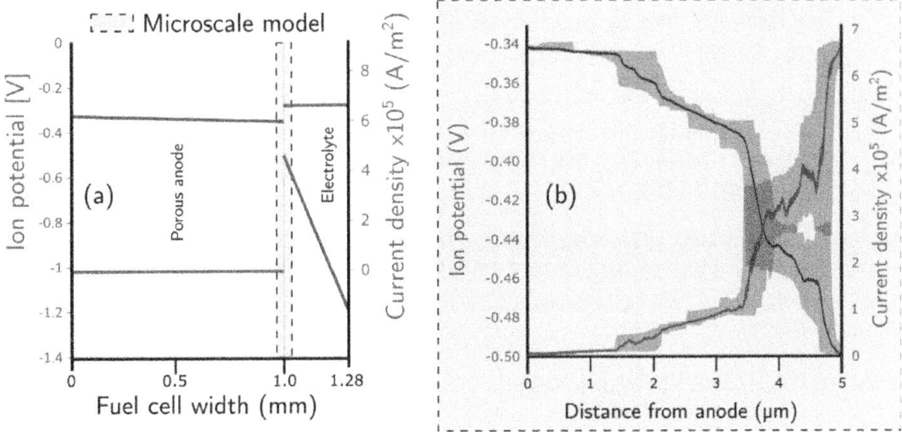

Figure 3 – *Variation of ion potential and current density across the width of the fuel cell both in macro- and micro-models with different numerical resolutions. (a) macroscale model. (b) microscale model*

4. Conclusions

A multi-scale numerical framework has been developed to study the mass/ion/electron transport inside the entire solid oxide fuel cell, from fuel channel to air channel with the capability of resolving the nanometer scale features inside the active electrochemical layer. This can help to achieve a more accurate description of electrochemical reactions taking place on the TPB and also on the double phase boundaries in the case of mixed ionic-electronic conductors, which ultimately increases the accuracy of the numerical model for predicting the performance of the entire cell or even an entire stack of cells. The future research direction is to extend this multi-scale numerical framework to use multiple microstructures in parallel (patches) to capture the nanometric features and rate of electrochemical reactions across the length of the entire fuel cell, and even across multiple cells, significantly expanding the computational capabilities of the multi-scale model.

References

Bertei, A., V. Yufit, F. Tariq, and N.P. Brandon. 2018. 'A Novel Approach for the Quantification of Inhomogeneous 3D Current Distribution in Fuel Cell Electrodes'. *Journal of Power Sources* 396 (August): 246–56. https://doi.org/10.1016/j.jpowsour.2018.06.029.

Grew, Kyle N., and Wilson K.S. Chiu. 2012. 'A Review of Modeling and Simulation Techniques across the Length Scales for the Solid Oxide Fuel Cell'. *Journal of Power Sources* 199 (February): 1–13. https://doi.org/10.1016/j.jpowsour.2011.10.010.

Kevrekidis, Ioannis G, C William Gear, James M Hyman, Panagiotis G Kevrekidis, Olof Runborg, Constantinos Theodoropoulos, and others. 2003. 'Equation-Free, Coarse-Grained Multiscale Computation: Enabling Microscopic Simulators to Perform System-Level Analysis'. *Commun. Math. Sci* 1 (4): 715–62.

Moussaoui, H., J. Laurencin, Y. Gavet, G. Delette, M. Hubert, P. Cloetens, T. Le Bihan, and J. Debayle. 2018. 'Stochastic Geometrical Modeling of Solid Oxide Cells Electrodes Validated on 3D Reconstructions'. *Computational Materials Science* 143 (February): 262–76. https://doi.org/10.1016/j.commatsci.2017.11.015.

Prokop, Tomasz A, Janusz S Szmyd, and Grzegorz Brus. 2018. 'Developing Micro-Scale Heterogeneous Numerical Simulation of a Solid Oxide Fuel Cell Anode'. *Journal of Physics: Conference Series* 1101 (October): 012027. https://doi.org/10.1088/1742-6596/1101/1/012027.

Santos-Gómez, Lucía dos, Javier Zamudio-García, José M Porras-Vázquez, Enrique R Losilla, and David Marrero-López. 2021. 'Recent Progress in Nanostructured Electrodes for Solid Oxide Fuel Cells Deposited by Spray Pyrolysis'. *Journal of Power Sources* 507: 230277.

Shearing, P.R., Q. Cai, J.I. Golbert, V. Yufit, C.S. Adjiman, and N.P. Brandon. 2010. 'Microstructural Analysis of a Solid Oxide Fuel Cell Anode Using Focused Ion Beam Techniques Coupled with Electrochemical Simulation'. *Journal of Power Sources* 195 (15): 4804–10. https://doi.org/10.1016/j.jpowsour.2010.02.047.

Singh, Mandeep, Dario Zappa, and Elisabetta Comini. 2021. 'Solid Oxide Fuel Cell: Decade of Progress, Future Perspectives and Challenges'. *International Journal of Hydrogen Energy* 46 (54): 27643–74.

Tseronis, K., I. Bonis, I.K. Kookos, and C. Theodoropoulos. 2012. 'Parametric and Transient Analysis of Non-Isothermal, Planar Solid Oxide Fuel Cells'. *International Journal of Hydrogen Energy* 37 (1): 530–47. https://doi.org/10.1016/j.ijhydene.2011.09.062.

Tseronis, K., I.S. Fragkopoulos, I. Bonis, and C. Theodoropoulos. 2016. 'Detailed Multi-Dimensional Modeling of Direct Internal Reforming Solid Oxide Fuel Cells'. *Fuel Cells* 16 (3): 294–312. https://doi.org/10.1002/fuce.201500113.

Vivet, N., S. Chupin, E. Estrade, T. Piquero, P.L. Pommier, D. Rochais, and E. Bruneton. 2011. '3D Microstructural Characterization of a Solid Oxide Fuel Cell Anode Reconstructed by Focused Ion Beam Tomography'. *Journal of Power Sources* 196 (18): 7541–49. https://doi.org/10.1016/j.jpowsour.2011.03.060.

Antonis Kokossis, Michael C. Georgiadis, Efstratios N. Pistikopoulos (Eds.)
PROCEEDINGS OF THE 33rd European Symposium on Computer Aided Process Engineering (ESCAPE33), June 18-21, 2023, Athens, Greece
 http://dx.doi.org/10.1016/B978-0-443-15274-0.50167-0

Parameter estimation combined with model reduction techniques for identifiability analysis of biological models

Michael Binns,[a] Alessandro Usai,[b] Constantinos Theodoropoulos,[b]

[a]*Department of Chemical and Biochemical Engineering, Dongguk University-Seoul, 30 Pildong-ro 1-gil, Jung-gu, Seoul 04620, Republic of Korea*
[b]*Department of Chemical Engineering, Biochemical and Bioprocess Engineering Group, The University of Manchester, Manchester M13 9PL, United Kingdom*
k.theodoropoulos@manchester.ac.uk

Abstract

Parameter estimation is typically used as part of model development to determine the values of unknown parameters. However, depending on the model complexity the number of parameters can also vary. High complexity models have large numbers of parameters requiring more computational effort to determine them and are also prone to overfitting. Low complexity models have smaller numbers of parameters but may have reduced accuracy. Based on available experimental data cross-validation can be used to compare different complexity models and determine the most appropriate complexity (James et al., 2013). Alternatively, it is possible to look at the identifiability of parameters based on experimental data which considers the sensitivity and correlation between parameters. Both these types of methodologies can be used to reduce the complexity of models such that insensitive and/or dependent/correlated can be removed or re-estimated and an alternative set of parameters can be computed. In this work both types of methods are explored with examples. Cross-validation combined with a Least Absolute Shrinkage and Selection Operator (LASSO) regularisation method is used to reduce the complexity of linear empirical equations for predicting the performance of downdraft biomass gasification (Binns and Ayub, 2021). Sensitivity and identifiability methods utilizing the Fischer Information Matrix (FIM) are used to reduce the complexity of a nonlinear system of partial integral differential equations describing a population balance model for microalgae cultivation (Usai et al., 2022). Application of these methods allows the number of parameters to be reduced depending on the tolerance and/or accuracy required.

Keywords: Model reduction, Parameter estimation, Optimisation, Identifiability, LASSO.

1. Introduction

Model development typically starts from some knowledge derived from existing experimental data and the equations which have previously been used to model similar systems. Based on this starting point a model might be suggested which includes all known physical variables and parameters which affect the system outputs. This approach might lead to a complex model for which there are a large number of unknown parameters. For this reason model reduction and identifiability methods can be used to reduce the complexity (Baker et al., 2015). In this study we compare two different approaches for model reduction combined with parameter estimation: LASSO based regularization (James et al., 2013; Binns and Ayub, 2021) and a sensitivity and identifiability analysis utilizing the Fischer information matrix and multi-objective optimisation to reduce correlations and improve identifiability.

2. Parameter estimation methods

Parameter estimation involves minimising or reducing the difference between model predictions and experimental values by changing the values of parameters in the models. This is typically achieved by minimising a least-squares type function such as equation 1 where y_z is the experimental value, $\hat{y}_z$ is the corresponding model prediction and N is the number of data points.

$$RSS = \sum_{z=1}^{N} (y_z - \hat{y}_z)^2 \qquad (1)$$

The number of parameters to be estimated depends on the complexity of the terms used in the models to predict $\hat{y}_z$.

2.1. Parameter elimination

Reducing the complexity of models may be possible through the elimination of certain parameters which are either insensitive or do not affect the model outputs. There are potentially three ways by which parameters can be eliminated

- Set parameter to 0
- Set parameter to 1
- Set parameter to a fixed value

In cases where a fixed value is used this is defined using existing knowledge of the system. Alternatively setting values to 0 or 1 will typically allow the model equations to be written in a simpler form.

2.2. LASSO regularisation

$$RSS = \sum_{z=1}^{N} (y_z - \hat{y}_z)^2 + \lambda \sum_{i=1}^{n} |\beta_i| \qquad (2)$$

Regularisation is the methodology where the objective function from equation 1 is modified to include a second set of terms including the sum of the values of n parameters, β, multiplied by a factor, λ. When this function is minimised the values of parameters are also minimised and if the absolute values of β are used (as shown in equation 2) the effect is that a number of parameters are set equal to zero (James et al., 2013). Varying the value of λ will change the number of parameters set to zero. A very small λ value leads to the same solution obtained by minimising equation 1, a very large value λ will lead to a solution where all the parameters are set to zero. Additionally, this value will also affect the model accuracy, so cross-validation is required to find the most appropriate value of λ. In particular cross-validation involves dividing the training set into a number of sections or folds. One of the folds is selected for testing and the remainder are used for training. Sequentially repeating this for every fold will then lead to an overall cross validation mean square error. This cross validation is then repeated for every value of , λ being considered giving a range of different model solutions with varying accuracy and varying numbers of non-zero parameters.

2.3. Sensitivity and identifiability analysis

2.3.1. Sensitivity based parameter reduction

As mentioned in section 2.1 a parameter might be eliminated by setting its value to either zero, one or some pre-defined fixed value. Starting from some existing full model containing possibly a very large set of parameters the change in model output (*y*) resulting from and a small change in each parameter (*P*) will give sensitivity values as shown in equation 3. Then sequentially considering the elimination of each parameter by setting it to some fixed value and checking if the resulting model error exceeds some specified tolerance leads to a reduction of the number of parameters.

$$S_p = \frac{dy}{dP} \tag{3}$$

2.3.2. Identifiability analysis and correlations

After reducing the number of parameters either through the regularisation (as described in section 2.2) or utilizing sensitivities and a sequential removal method (as described in section 2.3.1) it is useful to consider the identifiability of the remaining parameters. Based on the available experimental data identifiability analysis asks which of those parameters can be uniquely identified. Those without a unique solution may have a range of acceptable values or it may not be possible to identify a reasonable value. Starting from the sensitivities it is possible to calculate the Fischer Information Matrix (FIM) which uses the local sensitivities around a fitted set of parameters (S_p) together with the weighted variance ($\hat{\Sigma}$).

$$FIM = \sum_{j=1}^{N_{exp}} \sum_{i=1}^{N_{sample}} \left(S_p(i,j)\right)^T \hat{\Sigma}^{-1} \left(S_p(i,j)\right) \tag{4}$$

The inverse of this FIM gives a lower bound for the covariance matrix which can be used to calculate the correlation matrix (Stoica and Ng, 1998; Baker et al., 2015). The following equations (5-8) are suggested in order to minimise correlations between parameters. From the correlation matrix ($\boldsymbol{\rho}$) it is possible to calculate an overall correlation coefficient as shown in equation 5. Subsequently from this correlation matrix the most highly correlated parameter (with the highest I_j) is chosen as a pivot parameter and other parameters which are strongly correlated with this are defined by linear relations as in equation 6.

$$I_j = \sum_{i=1}^{N_{parameters}} \rho_{ij} \tag{5}$$

$$P_i = C_{ij} P_j \tag{6}$$

$$Z_1 = min \left(\frac{y(C_{ij}P) - y(\hat{P})}{y(\hat{P})} \right)^2 \tag{7}$$

$$Z_2 = min\, I_j(C_{ij}) \tag{8}$$

To reduce the magnitude of correlations present the correlation values can be optimised (subject to any bounds on the final parameter values from the model) by minimising Z_2 as defined in equation 8. A second objective is given in equation 7 which is the deviation in model output between using the original parameters and those modified by changing the correlation values. Hence, this is a multi-objective optimisation which should give a pareto curve of possible solutions. If parameters change without affecting model results this suggests they are not identifiable and can be eliminated or set to constant values.

3. Case studies

3.1. Empirical models for biomass gasification

Biomass gasification models should predict the outlet gas composition based on a number of input variables. In particular the input biomass composition ($Ultimate\ analysis$: $\%C, \%H, \%O, \%Ash$), moisture content (MC), equivalence ratio (ER) and gasification temperature (T_{gas}) are expected to affect the output gas composition. For the gasifier considered here additional inputs are also available including grate rotation speed (Gr), gas fan speed (Fs), Wet bulk biomass density ($bulk$) and biomass void percent ($void$).

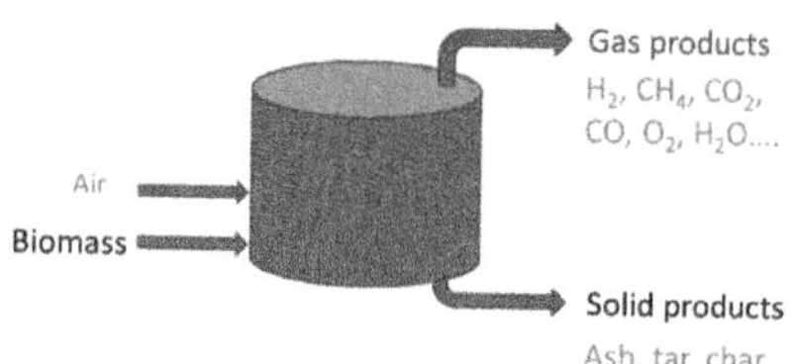

Figure 1. Block diagram representing a biomass gasifier.

Although more complex models are possible, for example based on knowledge of reaction kinetics, in this study simpler empirical models are considered relating the inputs with the outputs through linear and quadratic expressions. In this case the experimental data of Chee (1987) is used including 34 data points with inputs and outputs defined in Table 1. This is the same case investigated by Binns and Ayub (2021) who focused on finding the minimum cross-validation error model for each output. Here we consider the ranges of solutions which can be found with varying λ, the accuracy and simplicity of models obtained. For this purpose we consider the modelling of CO_2 volume percentage in the produced gas with a range of possible solutions shown in table 1 and figure 2. As can be seen from figure 2 and table 2 the model accuracy varies with changing value of λ. For lower values all eleven parameters are non-zero but the cross-validation error is highest. The lowest cross-validation error is achieved in a linear model with 7 non-zero parameters and at high values of λ all the parameters are zero and the model reduces to a constant value with a single fixed value.

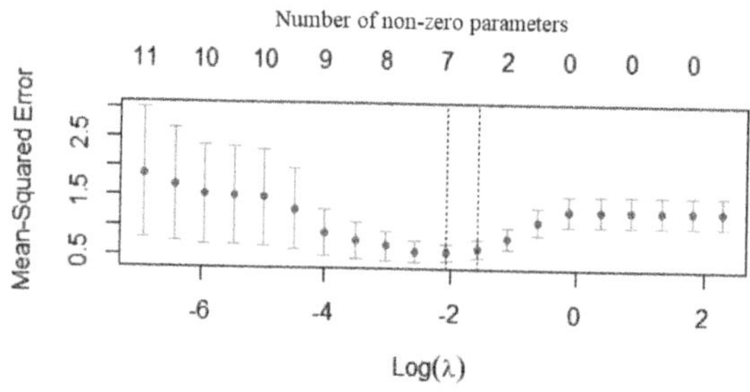

Figure 2. Cross-validation mean square error for CO_2 using a linear model with varying λ. The number of non-zero parameters excludes a constant fixed parameter (β_0).

Table 2. Parameter reduction using LASSO regularisation (Binns and Ayub, 2021)

λ value	Cross validation error	Expression
$\lambda = 0.001$	1.8851	$CO_2(\%) = \beta_0 + \beta_1 Tgas + \beta_2 ER + \beta_3 MC + \beta_4 H + \beta_5 O + \beta_6 C + \beta_7 Ash + \beta_8 Gr + \beta_9 Fs + \beta_{10} Bulk + \beta_{11} void$
$\lambda = 0.2069$	0.6402	$CO_2(\%) = \beta_0 + \beta_2 ER + \beta_3 MC + \beta_5 O + \beta_7 Ash + \beta_8 Gr + \beta_{10} Bulk + \beta_{11} void$
$\lambda = 0.3360$	0.8053	$CO_2(\%) = \beta_0 + \beta_2 ER + \beta_3 MC$
$\lambda = 1.4384$	1.2909	$CO_2(\%) = \beta_0$

3.2. Population balance models for microalgae cultivation

The population balance model considered here is for the growth of the microalgae *Haematococcus pluvialis* from the model developed by Usai et al. (2022). This model describes the growth (G), birth (B), disappearance of mother cells (M) and cell lysis (D) giving the overall population balance in equation 9.

$$\frac{\partial \Psi_V}{\partial t} + G(v) = B(v) - M(v) - D(v) \quad (9)$$

where Ψ_V is the density distribution function of cells in the system.

The full model including expressions for each of the terms in equation 9 requires 34 parameters which are fitted to experimental data (Usai et al. 2022). It can be shown that applying sensitivity based parameter reduction 10 parameters can be eliminated with a tolerance of 0.05. Increasing the tolerance would allow more parameters to be eliminated (similar to the way increasing λ reduces the number of parameters).

Table 3. Parameter reduction using sensitivity approach (Usai et al., 2022b)

Tolerance	Parameters eliminated	Total parameters remaining
0.025	5	29
0.05	10	24
0.1	11	23

After the reduction of parameters, the sensitivity values can also be used to calculate the Fischer information matrix and subsequent covariance and correlation matrices. This identifies a number of strong correlations including existing relations between Monod kinetic parameters. If the objective function in equation 8 is minimised the magnitude of most correlations can be reduced by approximately 35%.

4. Discussion and conclusions

Two methods for model reduction based parameter estimation have been evaluated with two case studies and both are able to reduce complexity of the resulting models. A comparison of the two methods can be seen in table 4. The main advantage of the LASSO regularisation method is that it can start from a very complex model and directly reduce the number of parameters as part of the parameter fitting while the sensitivity and identifiability approach generally starts from a full fitted complex model before reducing model complexity. However, the sensitivity and identifiability approach can more directly control the accuracy of the reduced model through the setting of a tolerance. It also offers more flexibility in terms of how to eliminate parameters, although possibly the LASSO regularisation could be modified so it can eliminate parameters is the same way. The main advantage of the sensitivity and identifiability methods are that they identify potential correlations between parameters and optimisation methods can be used to reduce the correlation values and potentially improve the identifiability of the model.

Table 4. Comparison of model reduction based parameter estimation methods

Issue	LASSO regularisation (Binns and Ayub, 2021)	Sensitivity and Identifiability analysis
Complex initial model	Can identify reduced models starting from a very complex model $\begin{pmatrix} N_{parameters} \gg N_{datapoints} \\ N_{reduced\ parameters} < N_{datapoints} \end{pmatrix}$	Requires full potentially complex model with fitted parameters as starting point
Accuracy vs. simplicity control	Control using λ factor (accuracy or model error must be calculated as an extra step)	Control using model tolerance
Parameter elimination	Set eliminated parameters to zero	Set eliminated parameters to either zero, one, or to specified values
Correlations and identifiability	Does not test for correlations. Reduces some statistical identifiability issues by reducing the number of parameters compared to the number of data points	Identifies potential correlations Can reduce correlation values to improve identifiability

References

S.M. Baker, C.H. Poskar, F. Schreiber, B.H. Junker, 2015, A unified framework for estimating parameters of kinetic biological models, BMC Bioinformatics, 16, 104

M. Binns, H.M.U. Ayub, 2021, Model reduction applied to empirical models for biomass gasification in downdraft gasifiers, Sustainability, 13, 12191

C.S. Chee, 1987, The air gasification of wood chips in a downdraft gasifier, Master's thesis, Kansas State University, Manhattan, USA.

G. James, D. Witten, T. Hastie, R. Tibshirani, An introduction to statistical learning with applications in R, 2nd ed., Springer, New York, 2013, 59-126

P. Stoica, B.C. Ng, 1998, On the Cramer-rao bound under parametric constraints, IEEE Signal Procesing Letters, 5, 177-179.

A. Usai, J.K. Pittman, C. Theodoropoulos, 2022, A multiscale modelling approach for haematococcus pluvvialis cultivation under different environmental conditions, Biotechnology Reports, 36, e00771

Antonis Kokossis, Michael C. Georgiadis, Efstratios N. Pistikopoulos (Eds.)
PROCEEDINGS OF THE 33rd European Symposium on Computer Aided Process Engineering (ESCAPE33), June 18-21, 2023, Athens, Greece
 http://dx.doi.org/10.1016/B978-0-443-15274-0.50168-2

Improving Model Robustness with Transfer Learning for Product Property Models

Per Julian BECKER, Loic IAPTEFF, Benoit CELSE

[a]*IFP Energies nouvelles, Rond-point de l'échangeur de Solaize, BP 3, 69360 Solaize, France*
per.becker@ifpen.fr

Abstract

This paper describes the development of product property models for a new hydrocracking catalyst, developed by IFPEN. The naphtha density is used as an illustrative example. A Bayesian Transfer Learning approach is used to transfer the ordinary least squares (OLS) parameters from a source dataset consisting of 3980 points from industrial units operating previous (n-1) catalyst generations to a target dataset with 96 points from pilot plant tests with the new catalyst (n). Robustness of the new model is greatly improved because information from the larger dataset is included in the new model. The Transfer Learning approach is shown to give good results and lead to a more robust model with respect to feedstock descriptors.

Keywords: Transfer Learning, Modelling, Hydrocracking, Bayesian Statistics

1. Background

Accurate and robust predictive models are essential when commercializing a new chemical process in order to design the industrial units. Models can be derived from first principles or data driven. It is not always possible to do the former, which requires knowledge of all chemical reactions involved, as well as analytical methods for detailed characterization of the feedstocks and products. In this case empirical model must be developed, which require a much larger number of experimental data points in order to achieve the required precision and accuracy. Running pilot-scale experiments is rather expensive and time-consuming, which limits the amount of data which can be obtained in practice. The estimation of physical properties for petroleum refining products is a particularly challenging subject due to the complexity of the feedstocks, which are composed of hundreds of thousands of different hydrocarbon compounds.

1.1. Process Description

Hydroprocessing is a crucial component of a modern petroleum refinery (Jones 2006), which is used to convert heavy fractions obtained by distillation of crude oil (i.e. Vacuum Gas Oil – VGO) into more commercially valuable middle distillate (Diesel & Jet) or naphtha fractions over a fixed-bed catalyst under hydrogen pressure of 80 – 160 bar at temperatures from 360 – 410°C. Naturally occurring nitrogen and sulfur are removed in the Hydrotreatment (HDT) section, wile long-chained hydrocarbons are broken up into smaller compounds in the Hydrocracking (HCK) section(s). The HDT section is necessary because organic nitrogen is a poison to zeolitic HCK catalysts. Ammonia (NH_3) and hydrogen sulfide (H_2S) gasses produced in the HDT section is carried over to the

HCK section, the NH_3 inhibits HCK catalyst activity. The effluent of the reactor is split into the product cuts in a fractionator. It is possible to adjust the desired conversion and yield of product streams via selection of the catalyst(s) or operating conditions, this makes hydrocracking a very versatile process. HCK produces high-quality cracking products.

Higher rates of conversion can be obtained by recycling Unconverted Oil (UCO). In a two-stage process, the UCO-recycle stream is processed in a third HCK reactor, the effluents of the two HCK sections are blended before fractionation. Only trace amounts of NH_3 are present in the second HCK reactor, which therefore requires much lower temperatures to achieve the desired conversion.

In maxi-middle distillate mode, the yields of Diesel and Jet cuts are favored, while maxi-naphtha mode aims to maximize naphtha production. Consequently some, or all, of the produced Jet and Diesel cuts are also recycled in this configuration.

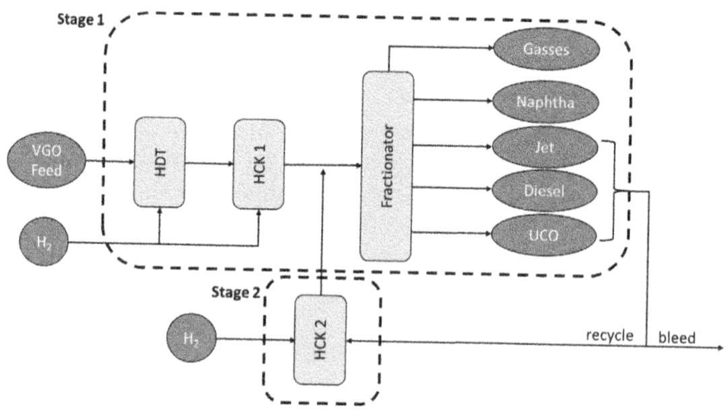

Figure 1 Schematic representation of the two-stage hydrocracking process in maxi-naphtha mode

1.2. Catalysts

IFPEN develops a wide range of HDT and HCK catalysts. The catalyst of generation n-1 is commercialized and used in single-stage maxi-middle distillate production and data from industrial units has been collected over the last 15 years. Catalyst of generation n has been developed for maxi-naphtha production in one- or two-stage configurations and is yet to be commercialized.

1.3. Product & Feedstock Properties

Petroleum feedstocks and products are complex mixtures of long-chained hydrocarbons, which can be roughly classified as paraffins (unbranched chains), iso-paraffins (branched chains), naphthenes (saturated cycles), and aromatics. The detailed characterization of VGO feedstocks, which are composed of 100,000s of individual species remains currently impossible. Petroleum cuts are characterized by Simulated Distillation (ASTM D2887), and density (spgr or d_{154}), as well as the concentration of organic nitrogen and sulfur. Several product properties, such as RON/MON for naphtha or cold-flow properties for middle distillates are also measured – these properties are to be estimated from the macroscopic feedstock and process parameters.

The distillation curves of the products depend on the chosen cut points, which are adjusted to the requirements of each individual refinery, as well as the overlap between cuts, i.e.

efficiency of the fractionation column. The K_{watson} is a calculated feedstock descriptor, often used in petroleum refining, which combines the mean average boiling point (MeABP in °R) and specific gravity (spgr): $K_{watson} = MeABP^{1/3}/spgr$.

Accurate estimation of cut densities is particularly important because it is required to convert mass into volumetric flowrates, and because it is itself used as a descriptor in the models for other product properties.

2. Methodology

2.1. Bayesian Transfer Learning

The linear model for the target dataset is given below, where β_t is the vector of model parameters of size n_p and **X** is the design matrix (a column filled with ones has been added to include the intercept).

$$y = \beta_t \boldsymbol{X} + \epsilon$$

For the transfer approach (Bouveyron 2010, Iapteff 2022, Robert 2007), β_t is considered to be a random variable of prior density $\pi(\beta_t)$. The Bayes theorem gives that the posterior of β_t with respect to the target observations $\boldsymbol{X}_t$ and y_t is:

$$\pi(\beta_t|y_t) = \frac{\pi(\beta_t)f(y_t|\beta_t, \boldsymbol{X}_t)}{f(y_t|\boldsymbol{X}_t)}$$

An improvement of Zellner's prior, also known as g-prior was proposed by L. Iapteff (2022), which acts on prior covariance in addition to prior distribution only:

$$\pi(\beta_t) = \mathcal{N}\left(\hat{\beta}_s, g\sigma_s^2(\boldsymbol{X}_s^T\boldsymbol{X}_s)^{-1}\right)$$

Where $\hat{\beta}_s$ is the maximum likelihood estimator learned on only the source data. The mean of the corresponding posterior distribution is then:

$$\hat{\beta}_t = (\boldsymbol{X}_t^T\boldsymbol{X}_t + \sigma_t^2 g^{-1}\Sigma_s^{-1})^{-1}\left(\boldsymbol{X}_t^T y_t + \sigma_t^2 g^{-1}\Sigma_s^{-1}\hat{\beta}_s\right) \text{ with } \Sigma_s = \sigma_s^2(X_s^T X_s)^{-1}$$

For large values of g the posterior mean tends to $\hat{\beta}_s$, for small values it tends to the prior mean $(\hat{\beta}_t)$. The difficulty lies in determining a suitable value for g. L. Iapteff (2022) showed that for normalized data a good estimate for the optima value of g can be defined as follows, in terms of $\Sigma_s = \sigma_s^2(\boldsymbol{X}_s^T\boldsymbol{X}_s)^{-1}$:

$$g = \left(\frac{1}{n_p}\sum_{j=1}^{n_p}(\Sigma_s)_{j,j}\right)^{-1}$$

2.2. Feature Selection

A limited number of potential features is available due to the limitation described in 1.3. Hydrogen partial pressure and residue conversion (X370+) are retained as process descriptors. Nitrogen, Sulfur, density, and KWatson are retained as feedstock descriptors. Furthermore, the SimDis end-points (5% and 95%) were retained for both, the feedstock and naphtha cuts.

2.3. Data Normalization

The data is normalized according to the median for centering and interquartile range for reduction, based on the source training dataset (Iapteff 2022).

2.4. Process Simulation

In practice, the selected process and feedstock descriptors are always known with high confidence. The SimDis of the HCK effluent is estimated using a Continuous Lumping model and product cut SimDis is calculated using a model for fractionator efficiency (Kister 1992). The ensemble is implemented, together with the various product property correlations in a simulator developed in-house by IFPEN.

3. Datasets

3.1. Source Dataset

Follow-up data was obtained for 31 cycles from 15 industrial plants using HCK catalyst of generation n-1 in maxi-middle distillate, single stage mode, with or without recycle. A total of 3980 data points were retained for model development after expert-filtering and outlier detection (LOF), which was performed on a per-cycle basis. The entire industrial dataset was used to estimate the model parameters, a second dataset with 129 points obtained from IFPEN pilot plant tests was used for validation of the model. Performance indicators for the two datasets are shown in Table 1, the RMSE is in both cases below the acceptability criteria of ± 0.005 g/cm^3.

Table 1 Indicators for source model

	me	*mae*	*rmse*	*+/- 0.01 (%)*	*+/- 0.005 (%)*	*+/- 0.0025 (%)*
Train (Indus)	0.0002	0.0030	0.0040	97.8	83.2	52.5
Test (Pilot)	0.0011	0.0034	0.0042	97.7	72.9	45.7

3.2. Target Dataset

The target dataset for the new catalyst contains 115 experimental points for stage 1 and 24 points for stage 2 from IFPEN pilot plant tests. Seven different feedstocks were available for testing in Stage 1, however, only two feeds were available for stage 2. The feeds were pre-treated in a separate pilot plant with HDT+HCK catalysts and tested in a single HCK reactor. This approach does not exactly reproduce the HCK 2 feeds in an actual recycle configuration but was considered sufficiently close to develop a first model. The principal weakness of this dataset is the lack in feedstock variability, whereas the feeds vary on a day-by-day basis during an industrial cycle.

3.2.1. Gas-Liquid re-combination

An innovative analytical method based on recombination of gas and liquid analysis (chromatography), developed by IFPEN, was used to characterize the cuts below 175°C in terms of family (paraffins, iso-paraffins, naphthenes, and aromatics) and carbon number. Using this data, the distillation curves and densities for arbitrary naphtha cuts can be calculated. Three artificial cuts, 65-150°C, 80-150°C, and 80-175°C where thus generated, greatly increasing the number of points available for model development.

3.2.2. Stage 1 – Changing the Reaction Coordinate

Residue conversion of the residue cut, with boiling points above 370°C (X370+) is used as reaction coordinate in maxi-middle distillate HCK, maxi-naphtha modem on the other hand aims to convert the middle distillate cuts. It is furthermore possible that the feed contains significant amount of Gasoil (250-370°C), the X370+ is less meaningful in this case. The X175+ conversion is used instead. The transformation between X370+ and X370+ is essentially linear in nature.

3.2.3. Stage 2 – Changing the Model Perimeter

The model parameters for stage 1 and stage 2 are shown in Figure 1. The feedstock for stage 1 is the feed of the HDT reactor, which is known. The HDT effluent (i.e. HCK1 reactor feed) is rarely analyzed in industrial units, it is therefore not included in the model. For stage 2 pilot plant tests the HCK2 reactor feed, i.e., the 175+ cut of the blended HCK1 & HCK2 effluents is used in the model. This represents a significant deviation from the stage 1 case, as illustrated in Figure 2.

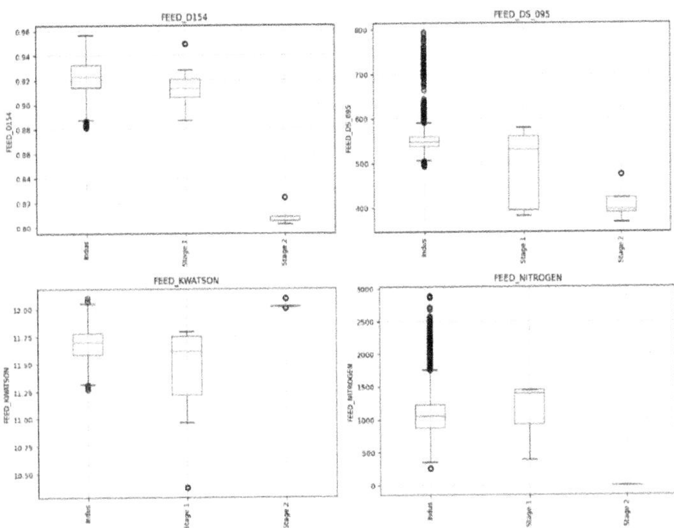

Figure 2 Feedstock characteristics in source (Indus) and target datasets for stage 1 and stage 2

4. Results & Discussion

The model indicators for Stage 1 and Stage 2 with and without transfer from the n-1 dataset are shown in Table 2 and Table 3 respectively.

Table 2 Performance indicators for Stage 1 model with and without transfer

	me	*mae*	*rmse*	*+/- 0.01 (%)*	*+/- 0.005 (%)*	*+/- 0.0025 (%)*
Train without transfer	-0.0008	0.0024	0.0035	97.6	86.0	68.4
Test without transfer	-0.0007	0.0022	0.0034	95.3	88.4	73.3
Train with transfer	-0.0006	0.0025	0.0037	97.9	84.5	64.4
Test with transfer	-0.0005	0.0025	0.0037	95.3	86.0	69.8

Table 3 Performance indicators for Stage 1 model with and without transfer

	me	*mae*	*rmse*	*+/- 0.01 (%)*	*+/- 0.005 (%)*	*+/- 0.0025 (%)*
Train without transfer	0.0011	0.0015	0.0019	100	100	80
Test without transfer	0.0007	0.0014	0.0017	100	100	90.5
Train with transfer	0.0011	0.0016	0.0021	100	97.3	74.7
Test with transfer	0.0008	0.0015	0.0019	100	100	81

For both datasets the results are well within the acceptable range of ±0.05 g/cm³. Model performance with transfer is equivalent to not using transfer learning, the values of the fitted parameter are however rather different. This suggests significant over-fitting when transfer learning is not applied.

An additional test was performed to validate the robustness of the two models. This time the pilot plant was operated in two-stage recycle mode half of the time and in single stage mode the other half. Two, previously unseen feedstocks were tested. The points were simulated using the in-house simulator (see 2.4). Figure 4 shows that the transfer modelling approach leads to more robust model. The model without transfer performs very poorly, confirming the over-fitting of the feedstock related features. The transferred model performs much better because information from the large industrial dataset is retained.

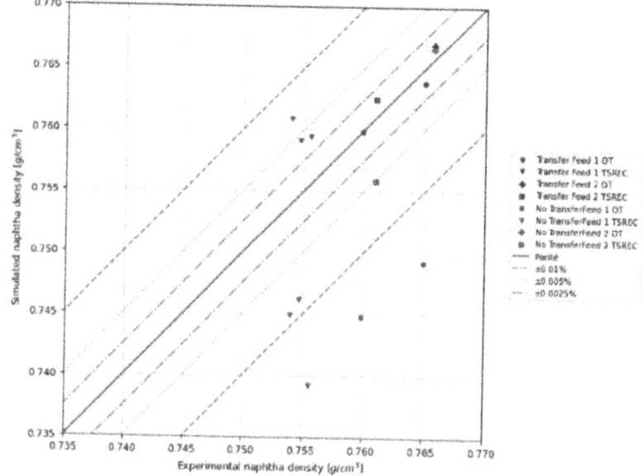

Figure 3 Parity graph for additional experimental points

5. Conclusion

Using transfer learning on the linear model, with an improved prior including the prior mean and covariance, was applied successfully on real-world datasets. Model robustness was shown to be improved when the target dataset is limited.

References

L. Iapteff, 2022, Transfer Learning for Smart Predictive Analytics, PhD thesis

C. Bouveyron and J. Jacques, 2010, Adaptive linear models for regression: improving prediction when population has changed, Pattern Recognition Letters, pages 2237–2247.

P.J. Becker, B. Celse, D. Guillaume, V. Costa, L. Bertier, E. Guillon, and G. Pirngruber, 2016, A continuous lumping model for hydrocracking on a zeolite catalysts: model development and parameter identification, Fuel, pages 73–82.

H.Z. Kister, 1992, Distillation Design, McGraw-Hill, New York, USA, ISBN 0-07-034909-6.

D.S. Jones, and P.P. and Pujado, 2006, Handbook of petroleum processing. Springer Science & Business Media.

L. Iapteff, J. Jacques, M. Rolland, and B. Celse, 2021, Reducing the number of experiments required for modelling the hydrocracking process with kriging through bayesian transfer learning. Journal of the Royal Statistical Society: Series C (Applied Statistics).

C.P. Robert et al., 2007, The Bayesian choice: from decision-theoretic foundations to computational implementation, volume 2. Springer.

Antonis Kokossis, Michael C. Georgiadis, Efstratios N. Pistikopoulos (Eds.)
PROCEEDINGS OF THE 33rd European Symposium on Computer Aided Process Engineering (ESCAPE33), June 18-21, 2023, Athens, Greece
 http://dx.doi.org/10.1016/B978-0-443-15274-0.50169-4

Modeling and simulation approach for modularized hydrogen electrolyzer plants

Kumar Rajan Gopa[b], Isabell Viedt[a], Lucien Beisswenger[c], Leon Urbas[a,b]

[a]*TU Dresden, Process Systems Engineering Group, 01069 Dresden, Germany*
[b]*TU Dresden, Chair of Process Control System, 01069 Dresden, Germany*
[c]*VDMA e.v., P2X4A, 60528 Frankfurt, Germany*
kumar.rajan_gopa@tu-dresden.de

Abstract

The increasing importance of flexible and large-scale green hydrogen production necessitates the accompanying modeling and simulation of these processes to optimize plant flexibility. State-of-the-art monolithic modeling approaches do not address the flexibility and interoperability expected from modular plants, as shown by (Mädler *et al.*, 2022). Therefore, in this work, a new approach to the modeling of modular electrolyzers is proposed. Beginning with a steady-state equation system for the individual Process Equipment Assemblies (PEAs) it is later expanded to a detailed dynamic model. Instead of a single simulation flowsheet for the entire plant, this will result in a set of Functional Mockup Unit ready PEA models which can be combined depending on the plant assembly via co-simulation. This enables the user to generate models for various configurations and test the individual PEAs as well as the composite plant more efficiently. As a first step, a simple case study is presented where two steady state sub-models are developed and connected via co-simulation.

Keywords: Hydrogen Electrolysis, Modular plants, Simulation

1. Introduction

Alkaline water electrolysis (AEL) technology has been present for a long time. Most of the work related to the modeling of these electrolysis plants considered them to be monolithic and focused on the detailed modeling of the electrolyzer or on the entire plant, including the Balance of Plant (BoP). Ulleberg, (2003) presented a detailed parametric model that acts as a baseline for most of the works that followed. The closest work in alkaline electrolysis towards a modular design is done by *Brée et al.*, (2020), where the author defines the electrolysis model as a collection of individual half-cells which could be coupled to form the entire electrolyzer. But this doesn't account for the BoP.
As per VDI 2776 a modular plant is defined as an aggregation of individual process engineering assemblies (PEAs) that offers high flexibility and interoperability within the modules. Oppelt *et al.*, (2015) is one of the earlier works that point towards the use of simulation in the life cycle of a modular plant and describes the future scenarios involving model reuse and co-simulation. A design basis for such plants and their realization on a mini plant, pilot scale, and production scale is discussed by Eilermann *et al.*, (2018). This is clarified by Mädler, Viedt, *et al.*, (2022) in detail, explaining the requirements needed for a digital twin and elaborates the role of modeling and simulation in developing digital twins.

With the eventual goal of developing a digital twin, we try to contextualize this in the case of an alkaline water electrolysis plant, which could be modeled and simulated, taking into account the modular nature of the plant and propose a workflow that could be used for the modeling of different modular electrolysis assemblies. In the first section, the basic principles and functioning of the alkaline water electrolyzers is introduced along with an overview of the modular concept and how this is made part of the simulation process. In the next section a case study and its results are presented where the proposed workflow is applied to a couple of submodels. The paper ends with a conclusion pointing towards the next steps in the workflow.

2. Theory

Large scale and sustainable production of green hydrogen is gaining more and more importance and water electrolysis is one of the major technologies to achieve this goal. In project eModule of the hydrogen lead-project H_2Giga, the focus is on developing modularized electrolysis plants that could be readily assembled and operated to achieve a high degree of flexibility and scaling up. Among the existing technologies, alkaline water electrolysis is the most mature and has been widely used commercially (Brée *et al.*, 2020). Water, being a stable compound, requires a lot of energy for its splitting. With electrolysis, this requires a high potential. In the case of alkaline water electrolysis, this is accomplished with the aid of an alkaline compound. This lowers the potential required for the breakup of water into hydronium and hydroxide ions. In the presence of an electric current hydrogen is produced at the cathode and oxygen is collected from the anode. Typically 20-40% KOH or NaOH solutions are used for electrolysis, and this is depicted in the below figure along with the reactions that occur during electrolysis (Shiva Kumar & Himabindu, 2019).

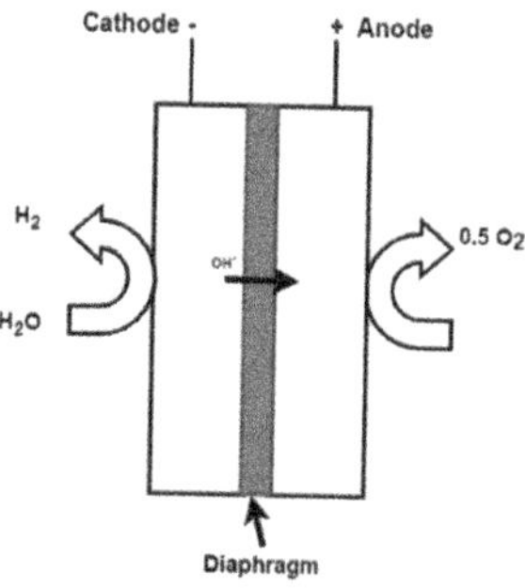

Figure 1: Alkaline Electrolysis (Shiva Kumar & Himabindu, 2019)

Cathode: $$2H_2O + 2e^- \rightarrow H_2 + 2\,OH^- \tag{1}$$

Anode: $$2\,OH^- \rightarrow 0.5\,O_2 + H_2O + 2e^- \tag{2}$$

Overall: $$H_2O \rightarrow 0.5\,O_2 + H_2 \tag{3}$$

For increased flexibility and faster planning, modular approach to design and construction of electrolyzer-plants is assumed (Radatz *et al.*, 2019). With the modular approach, there will be a set of submodels that would have the functional properties corresponding to the individual and independent PEAs. A PEA or process engineering assembly consist of process units that are largely independent, self-sufficient, and can perform a particular process step. It could include single or multiple functional equipment assemblies. These sub-models are brought together in a co-simulation framework where the various sub-models could be coupled together based on the required configuration. The individual PEAs could interact and communicate with each other exchanging information. The assembled model could later be simulated to obtain the overall simulation model of the plant. An overall process flow diagram for the alkaline electrolyzer along with the individual PEAs is shown in Figure 2.

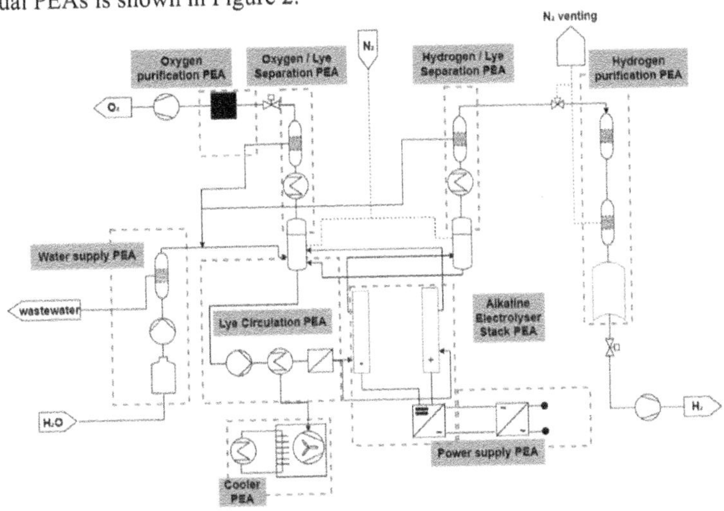

Figure 2: Process flow diagram for the AEL with individual PEAs, Courtesy of the VDMA Power-to-X for Applications platform (Beisswenger & Erdmann, 2022)

Compared to a continuous plant, where these are designated as individual units, here the distinction is made based on the function served. The PEAs present are:

1. Water supply – The function of this PEA is to supply water for the electrolyzer that meets the required specifications.
2. Lye circulation – This PEA assembly mixes the individual lye streams and makes up water, pressurizes the lye stream to the set pressure, heats it to the required temperature specification, and distributes the lye to the Electrolyzer.
3. AEL stack (Electrolyzer) – The water electrolysis occurs here generating hydrogen at the cathode and oxygen at the anode.
4. Cooler – Excess heat from the electrolyzer is removed
5. Power supply – Rectifier arrangement that supplies the necessary power for the electrolyzer
6. Hydrogen Separation - Separates the hydrogen from the gas-liquid lye stream coming from the electrolyzer PEA, cools the outgoing gas to room temperature, and eliminates any fine droplets present in the gas stream.
7. Oxygen separation – Similar to the hydrogen separator, oxygen is separated.
8. Hydrogen purification – removes any oxygen in the stream dries the stream and compress it to the required specification
9. Oxygen purification – Since oxygen is not considered further, this is sent to a combustion chamber and disposed of

Following the iterative modeling process described in Mädler *et al.*, (2021), the first step involves the creation of a simple model. Such a process provides a baseline model for comparison, easier debugging, and testing and, results in models with different levels of complexity. After gathering the necessary preliminary information like the flow diagrams, functional process description, etc. a sub model for each PEA is developed. Since the eModule project consists of several types of electrolyzer modules, individual PEAs are developed in such a way that they can be reused and integrated across multiple electrolyzer technologies with minimal modifications. To do this, each component is developed as an object with its characteristics that can be replicated and reused. They are converted into Functional Mock-up Units (FMU), assembled, and simulated. These steady state models would be used to check the basic performance of the electrolyzer assembly, and it also serve as a baseline with which the results of the more complicated configurations could be compared. In the final step, various tests are performed for the quality of the model (Viedt *et al.*, 2022).
In the second stage, the simple model is extended to include control strategies. Necessary modifications in the individual PEAs as well as the necessary control loops are made during this step. This stage allows for the modular automation of the PEAs. Again, the PEAs are assembled, simulated, and tested. In the final step, the equation system is expanded. If needed, additional equations and correlations for various parameters are supplemented including those for physical properties, thus resulting in a final detailed model that closely resemble a physical plant. Again, the control loops are tested, and the co-simulation is made. The approach is illustrated in the following diagram.

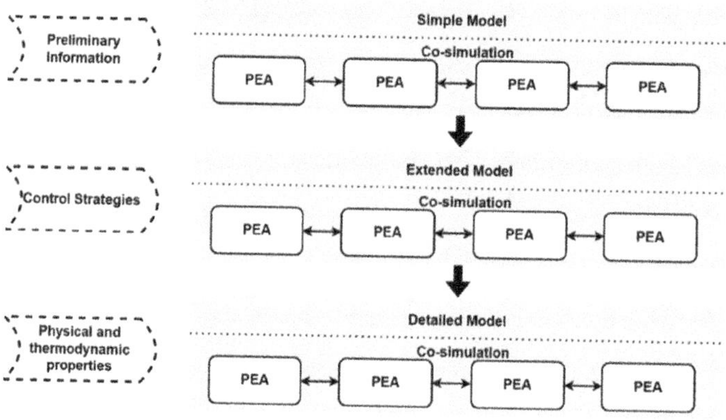

Figure 3: Workflow for Model Development

3. Case study

This is demonstrated by developing a co-simulation model for the PEA of a simple AEL sub-model and combining this with a lye separation sub model. In general, the starting point is a simple steady-state model for each individual PEA. In this phase the goal of the modeling process is to develop a minimal model that could be connected with other PEAs while outputting verified results with a certain degree of accuracy. For developing the initial model for AEL, the model equations presented by Ulleberg (2003) are used. This is a highly parametric model that has been experimentally validated. The main assumptions for this simple steady state model are:

a. KOH is readily ionized, and water is replenished sufficiently to ignore the effect of OH- accumulation.
b. All gases are in ideal condition and hydrogen is freely transported across the membrane (there is no resistance due to bubble formation, diffusion, etc.).
c. A constant Faraday efficiency of 0.86 is assumed and excess heat is transferred to the cooler

For the cooler, the water inlet temperature is assumed constant at 20°C and exits at 30°C and the cooler is considered as an ideal heat exchanger with no pressure drop across it. Also, the physical properties are considered constant within each PEA. The individual PEAs were developed and after verifying them, they were exported as FMUs and simulated. Further details and the simulation models are provided as a GitHub repository(Rajan Gopa, 2022). Thus, totally independent submodels were successfully combined. Input conditions from Sánchez *et al.*, (2020) were used to validate results. For the simulation, 896 kg/h lye flow, Temperature, and pressure at 75 ° C and 7 bar and an electric power supply of 10 kW were used. The results from the co-simulation are:

Results	**eModule model**	**Reference**
Oxygen flow - cathode, kg/h	1.45	1.37
Hydrogen flow - anode, kg/h	0.18 (2.05 Nm^3/h)	0.17
Temperature increase of Lye, °C	1.6	2.2
Current density, A/cm^2	0.4721	0.42
Heat to cooler, W	1500	2392.8
Cooling water, kg/h	-	130

Table 1: Results of the simple model simulation and comparison to Sánchez *et al.* (2020)

The simple model adhered well to the steady state results from Sánchez *et al.*, (2020). The oxygen and hydrogen production are similar and the increase in value could be attributed to the higher current density for the co-simulation model. Only the excess heat that goes to the cooler is comparatively lower than that in the cited work. This could be due to the use of constant values for the physical properties and the underlying assumptions. This is also evident from the difference in the lye temperature increase. Cooling water consumption for this operating point was also obtained but this could not be validated with the referenced work. But the obtained value does correspond to a heat duty in the range of 1500 W. As mentioned before, the overall goal of the simple model is not to produce exact results but to provide a valid and reliable baseline that could be compared to, which the above results justify.

4. Conclusion

In this work, modular workflow to the modeling and simulation of electrolyzers is discussed, and a simple model is implemented for two PEAs of the alkaline electrolyzer plant. These two sub-models are then exported as FMUs and simulated. The results from the co-simulation are compared to a similar work, thus validating this approach. Through this use case the viability of this method is demonstrated. Following this, the other PEAs will be modeled using the same approach to later simulate the entire electrolyzer plant. The deviation in the excess heat could either be due to the simplicity of the model or the absence of any physical property data. This will be incorporated in the detailed models including their individual control strategies so that the final plant configuration could be selected from a combination of these PEAs and various test scenarios could be validated. Different scenarios involving the numbering up and interoperability of different electrolyzers from multiple stakeholders will be tested. To validate the proposed approach for the modular simulation models, the results from one of the co-simulation models will be compared to those obtained from a demonstrator plant.

Acknowledgements

We would like to thank the German Federal Ministry of Education and Research and the Project Management Agency Jülich for their financial support within the framework of the eModule research project (FKZ 03HY116A-E) of the H2Giga lead platform.

References

Beisswenger, L., & Erdmann, D. (2022). Standardized flowsheet for the alkaline electrolysis. VDMA e.V.

Brée, L. C., Wessling, M., & Mitsos, A. (2020). Modular modeling of electrochemical reactors: Comparison of CO2-electolyzers. Computers & Chemical Engineering, 139, 106890. https://doi.org/10.1016/j.compchemeng.2020.106890

Eilermann, M., Post, C., Radatz, H., Bramsiepe, C., & Schembecker, G. (2018). A general approach to module-based plant design. Chemical Engineering Research and Design, 137, 125–140. https://doi.org/10.1016/j.cherd.2018.06.039

Mädler, J., Viedt, I., Lorenz, J., & Urbas, L. (2022). Requirements to a digital twin-centered concept for smart manufacturing in modular plants considering distributed knowledge. In Computer Aided Chemical Engineering (Vol. 49, pp. 1507–1512). Elsevier. https://doi.org/10.1016/B978-0-323-85159-6.50251-7

Mädler, J., Viedt, I., & Urbas, L. (2021). Applying quality assurance concepts from software development to simulation model assessment in smart equipment. In Computer Aided Chemical Engineering (Vol. 50, pp. 813–818). Elsevier. https://doi.org/10.1016/B978-0-323-88506-5.50127-3

Oppelt, M., Wolf, G., & Urbas, L. (2015). Towards an integrated use of simulation within the life-cycle of a process plant. 2015 IEEE 20th Conference on Emerging Technologies & Factory Automation (ETFA), 1–8. https://doi.org/10.1109/ETFA.2015.7301521

Radatz, H., Schröder, M., Becker, C., Bramsiepe, C., & Schembecker, G. (2019a). Selection of equipment modules for a flexible modular production plant by a multi-objective evolutionary algorithm. Computers & Chemical Engineering, 123, 196–221. https://doi.org/10.1016/j.compchemeng.2018.12.009

Rajan Gopa, K. (2022). AEL stack and cooler FMUs (Version 00) [MATLAB/Simulink]. Technische Universität Dresden/P2O Lab. https://github.com/p2o-lab/AEL_stack_cooler

Sánchez, M., Amores, E., Abad, D., Rodríguez, L., & Clemente-Jul, C. (2020). Aspen Plus model of an alkaline electrolysis system for hydrogen production. International Journal of Hydrogen Energy, 45(7), 3916–3929. https://doi.org/10.1016/j.ijhydene.2019.12.027

Shiva Kumar, S., & Himabindu, V. (2019). Hydrogen production by PEM water electrolysis – A review. Materials Science for Energy Technologies, 2(3), 442–454. https://doi.org/10.1016/j.mset.2019.03.002

Ulleberg, O. (2003). Modeling of advanced alkaline electrolyzers: A system simulation approach. International Journal of Hydrogen Energy, 28(1), 21–33. https://doi.org/10.1016/S0360-3199(02)00033-2

VDI 2776—Part 1: Process engineering plants; Modular plants. (2020).

Viedt, I., Mädler, J., & Urbas, L. (2022). Quality assessment for dynamic, hybrid semi-parametric state observers. In Computer Aided Chemical Engineering (Vol. 51, pp. 1255–1260). Elsevier. https://doi.org/10.1016/B978-0-323-95879-0.50210-1

Antonis Kokossis, Michael C. Georgiadis, Efstratios N. Pistikopoulos (Eds.)
PROCEEDINGS OF THE 33rd European Symposium on Computer Aided Process Engineering (ESCAPE33), June 18-21, 2023, Athens, Greece
 http://dx.doi.org/10.1016/B978-0-443-15274-0.50170-0

Design and optimization of a shared heat exchanger network for an integrated rSOC system

Xinyi Wei[a,b], *Shivom Sharma*[a], *Francois Marechal*[a], *Jan Van Herle*[b],
[a]*IPESE, EPFL Valais Wallis, 1950 Sion, Switzerland*
[b]*GEM, EPFL Valais Wallis, 1950 Sion, Switzerland*

xinyi.wei@epfl.ch

Abstract

To limit the global CO_2 emissions, installation of renewables is expected to increase significantly. This would require storage of renewable energy, possibly using an energy carrier such as hydrogen, methane, ammonia, etc. Reversible Solid Oxide Cell (rSOC) is a high temperature modern technology that can be operated in fuel cell or electrolysis mode. The Solid Oxide Electrolysis (SOEC) can be used to convert renewable electricity into hydrogen or carbon-based fuels (Power-to-Fuel). Conversely, Solid Oxide Fuel Cell (SOFC) has high overall efficiency for generating electricity from hydrocarbon fuels (Fuel-to-Power). The SOFC/SOEC technology is underdevelopment and being tested at several hundred kW scales. Generally, there are two different systems, SOFC and its balance of plant (BOP) system; SOE and methanation with their BOP system. The main challenge of rSOC system is to made both systems (SOFC and SOEC) highly integrated. Hence, it must utilize the maximum heat from heat sources and supply maximum heat to heat sinks. In this study, SOFC with oxy-combustion system and its reverse mode (SOEC) with methanation system have been optimized by minimizing the energy requirement and number of heat exchangers for individual system. The heat exchanger network synthesis problem is mixed integer linear programming problem, and it has been implemented in AMPL. For each heat exchanger (i.e., pair of streams), we have computed the heat exchanger area. This study evaluates the operating-flexibility of the shared HEN for rSOC-reformer/methanation system. The results show that heat exchangers can be shared between SOFC and SOEC modes, by manipulating the stream connections.

Keywords: Solid Oxide Cell, Heat Exchanger Network Synthesis, Renewable Energy Storage

1. Introduction

Today, the world's global primary energy supply is still heavily dependent on fossil fuels (up to 84.3% in 2019 [1]). It is well known that burning fossil fuels heavily contributes to climate change through the release of green-house gases (GHG) [2]. Furthermore, fossil fuels have a big impact on human health [3]. With the critical situation regarding climate change, there is a need to decarbonize economical activities with mitigation that occurs through different alternatives.

The scientific community is exploring several viable alternatives to replace the use of fossil fuels with renewables. However, variability, intermittency and storage are the main challenges for the renewable energy usage. Therefore, significant efforts have been invested on the development of efficient energy conversion technologies. Solid oxide fuel cell (SOFC) has attracted huge scientific attentions in the recent years. It is a modern and efficient power production technology that can reduce user's dependency on electricity-grid. SOFC system can generate electricity using liquid [4] or gaseous fuels [5]. It can easily be adjusted for different industry power plant scales. Moreover, SOFC operates at high temperature (600 ^{0}C), and it cogenerates high-quality heat or steam, which can be used as a heat source within the system. As maximum fuel utilization for SOFC is around 85%, a burner is required for the combustion of unconverted fuels. The burner provides additional heat to the SOFC system, at the same time, it generates more CO_2.

In order to meet "net zero" emission target for greenhouse gases in 2050, it is essential to introduce Carbon Capture and Storage (CCS) technology that can retrieve most of the produced CO_2 from emission intensive activities and store it permanently in nature (e.g., underground), leading to an almost carbon neutral activity. The stored CO_2 can also be used in methanation process, to convert H_2 into methane. In SOFC system, it is possible to perform oxy-combustion, by using stoichiometric amount of O_2 inside a catalytic burner. At the outlet of burner, CO_2 and H_2O at high temperature are the only two products. After water condensation, CO_2 can be separated from the mixture automatically.

A reversible SOFC system or solid oxide electrolyzer cell (SOEC) has attracted huge attention for H_2 and fuel production. H_2 and O_2 are produced on the cathodic and anodic side of SOEC, respectively. H_2 can be stored and used in SOFC to produce electricity, whereas O_2 can also be stored and used in catalytic burner. SOEC with methanation is an alternative way of methane production, where excess renewable electricity is converted into methane. SOEC can be considered as a low-cost alternative to address the gap between hydrocarbon fuel/hydrogen supply and demand [6]. The reversibility of the solid oxide cells (rSOC) manages the variability of electrical grids by transforming and storing power through H_2/methane when there is a surplus of electrical power or by generating electrical power through H_2/methane when there is demand. A general system layout has been provided in

Figure 1.

For economic reasons, both systems shall share most of the components, and they should easily be switched. The heat exchange network (HEN) synthesis is very challenging for integrated rSOC system, due to two different operation mode of the system [7]. State of the art of HEN synthesis methodologies have been published over the past few decades, and sequential approach is commonly used [8][9]. In this study, a shared HEN has been designed and optimized for 10 kW SOFC system and 30 kW SOEC system.

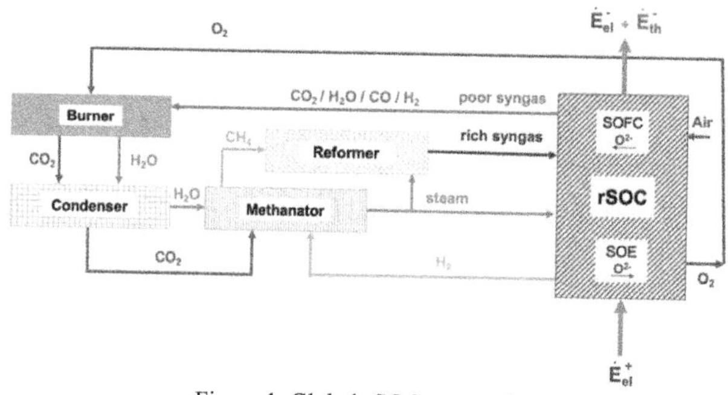

Figure 1. Global rSOC system layout

2. System and methodology description

2.1 SOFC with oxy-combustion

The fuel cell mode is operated when electricity demand cannot be fulfilled by the solar panels. In the fuel cell nominal mode, a mixture of methane and steam is fed into a reformer to convert part of it into syngas (H_2 + CO). The resulting flow is then fed into the anode side of the fuel cell, where rest of the reforming takes place. H_2 and CO are consumed inside the fuel cell, and steam and CO_2 rich syngas comes out of fuel cell. On the cathode side, air is fed to supply oxygen, and oxygen depleted air is released into environment. The unconverted H_2 and CO from fuel cell are combusted using air or pure oxygen. The oxygen combustion ensures absence of N_2 in the exhaust gases, and recovery of product CO_2 that can be used in SOEC/methanation. O_2 has been assumed to be purchased from the market, but this configuration would allow to store and use oxygen produced from electrolyzer mode.

2.2 SOEC with methanation

Electrolyzer mode is operated when excess electricity has to be stored as methane. The excess electricity is fed into the electrolyzer, as well as steam on the cathode side. The cathode side outlet contains H_2 and excess/unreacted steam. On the anode side, an air sweep is typically used to carry product O_2. Alternately, O_2 can be used as sweep, and product O_2 can be stored for further usage. H_2 and CO_2 are consumed inside the methantor to produce methane.

2.3 Heat exchanger network synthesis

The goal of the heat exchanger network synthesis formulation is to define process streams, objective function, heat balance equations, inequalities and potential constraints which will allow to obtain optimal amounts of hot and cold utility, the pairs of process streams which will exchange heat. Here, only essential details are presented, and detailed methodologies can be found in the literature [9].

- *Overall stream heat balance*

$$(T_i^i - T_i^o) \cdot Cp_i = \sum_{k \in ST} \sum_{j \in CS} Q_{ijk} + Q_{CUi} \; i \in HS; \; (T_j^o - T_j^i) \cdot Cp_j = \sum_{k \in ST} \sum_{j \in HS} Q_{ijk} + Q_{HUj} \; j \in CS \quad (1)$$

- *No splitting (avoid splitting of hot stream i and cold stream j)*

$$\sum_{i \in HS} z_{ijk} <= 1 \; j \in CS, k \in ST; \; \sum_{j \in CS} z_{ijk} <= 1 \; i \in HS, k \in ST \quad (2)$$

- *Forbidden matches (avoid heat exchange between hot stream i and cold stream j)*

$$z_{ijk} <= 0 \; k \in ST; \; z_{CUi} <= 0; \; z_{HUj} <= 0 \quad (3)$$

- *Required matches (in the case of a required match, there are two options: data of heat exchange is removed from the streams input data, following constraints are activated)*

$$\sum_{k \in ST} Q_{ijk} - Q_{min,ij} \geq 0; \; Q_{CUi} - Q_{min,CUi} \geq 0; \; Q_{HUj} - Q_{min,HUj} \geq 0 \quad (4)$$

- *Restricted matches (the match is allowed but, in the case, where it exist the heat exchanged is subject to an upper bound)*

$$\sum_{k \in ST} Q_{ijk} - Q_{max,ij} \leq 0; \; Q_{CUi} - Q_{max,CUi} \leq 0; \; Q_{HUj} - Q_{max,HUj} \leq 0 \quad (5)$$

- *Objective functions: There are different options for the objective function. This study uses minimum heat requirement and minimum number of heat exchangers as objective functions.*

$$min\left(\sum_{i \in HS} Q_{CUi} + \sum_{i \in HS} Q_{HUj}\right); \; min\left(\sum_{i \in HS} \sum_{j \in CS} \sum_{k \in ST} z_{ijk} + \sum_{i \in HS} z_{CUi} + \sum_{j \in CS} z_{HUj}\right) \quad (6)$$

The HEN synthesis problem is mixed integer linear programming (MILP) problem, and it has been implemented in AMPL and solved by CPLEX solver. Cooling water (5 to 15 ^{0}C) is used as cold utility. Hot air is used as hot utility (800 to 1000 ^{0}C).

2.4 Heat exchangers area estimation

In the first step, log-mean temperature difference is calculated. This study uses following values of overall heat transfer coefficients (U, $W/m^2/K$): 50 (air to air), 80 (gas to gas) and 160 (gas to water). The heat exchanger areas are computed using heat balance equation.

3. An integrated heat exchanger system design

Typical heat exchanger network of SOFC system has been studied in the previous literature [7]. SOFC system layout with HEN is shown in Figure 2. There are three heaters (H1, H2, H3) and four coolers (C1, C3, C4, C5). In order to improve the practical feasibility of HEN, HEN synthesis has been studied with some constraints. The evaporation of water occurs with direct injection of fuel in H1, therefore, the fuel and water streams are combined together, as shown in Figure 2. In practice, progressive water injection reduces heat requirement at constant temperature while providing a smooth temperature enthalpy profile. This aids in system heat integration. Moreover, in the previous studies, hot air has been used/split, to heat exchange with several streams. In reality, this design brings plant layout complexity, and it will also bring challenges in heat exchanger design, to avoid the potential leakage and explosion. Therefore, in this study, cold air heater (H3) has been restricted to perform heat exchange with hot air cooler (C3).

For SOFC system, the objective functions values for minimum energy requirement and number of heat exchangers are 2.01 kW (cold utility) and six, respectively. As expected, the exhaust gases from burner are used as the only heating source. The outlet stream from burner has been used to heat exchange with the stream before stack (C1-H2), and then it provides heat to the pre-reformer, finally, it passes through H1, to provide the heat for water heating, evaporation and superheating. The heat exchangers design specifications have been presented in Table 1 and Figure 4. There are mainly five heat exchangers: HEX 1, 2, 3, 4 and 5. HEX3 has the largest heat exchange area, where air flow rate is adjusted to balance the heat within the stack (exothermic reaction).

In SOEC system with methanator, the burner will be bypassed, as it is not necessary to burn part of cathode-off gas to provide heat to the system, and it decreases the energy efficiency. H_2 produced from cathode side is pressurized and injected in methanator. Moreover, unlike SOFC exothermic operation mode, SOEC has an endothermic operating mode normally, therefore, additional electrical heating is required. For SOEC system, the objective functions values for minimum energy requirement and number of heat exchangers are 2.97 kW (hot and cold utilities) and six, respectively.

As shown in Figure 3, there are two heaters (H1', H3') and five coolers (C1', C2', C3', C4', C5'). Water flows through methanator as the first heat exchange step. Here, 10% heat loss has been assumed in methanator according to the experiment data. In the following step, steam exchanges heat with SOEC cathode off-gas (HEX6). The cathode off-gas is pressurized to 15 bar, and the temperature increases to 890 ^{0}C (maximum material resistance temperature). Finally, steam is heated up to 754 ^{0}C inside HEX7. Electrical heaters are used, to heat up anode inlet air from 760 to 800 ^{0}C, and steam from 754 to 800 ^{0}C. As shown in Table 1 and Figure 4, there are mainly five heat exchangers. Similar as SOFC system, the biggest heat exchanger is air to air heat exchanger (9.42 m^2).

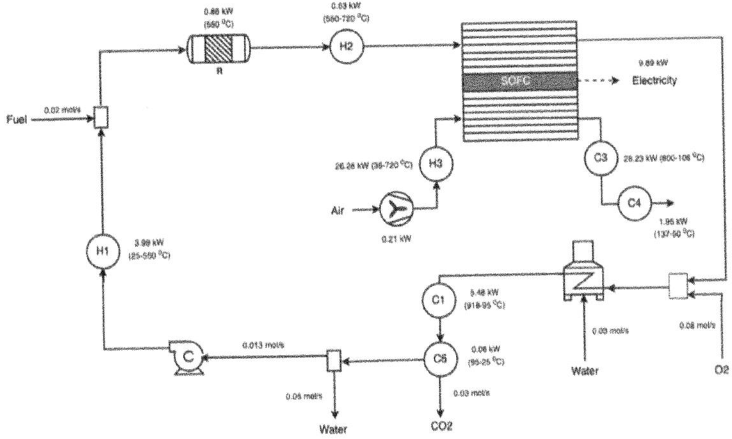

Figure 2. SOFC system with HEN design

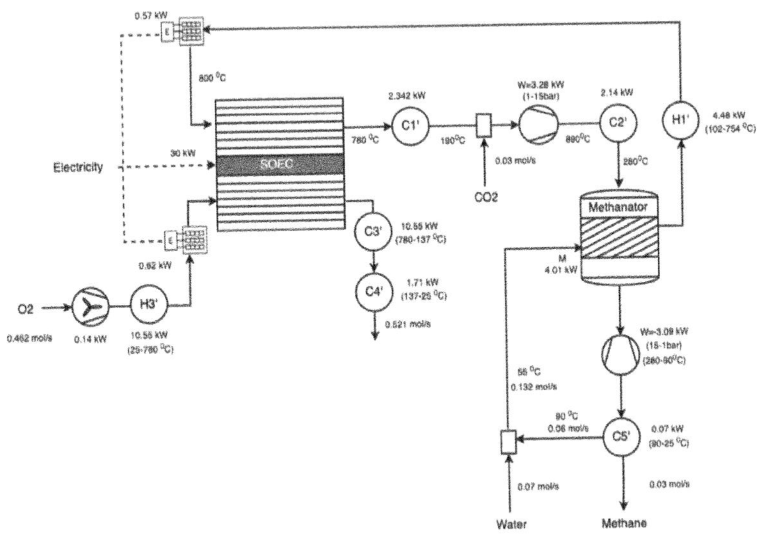

Figure 3. SOEC system with HEN design

By comparing duty and HEX area of two systems, HEX1, HEX2, HEX3, HEX4 and HEX5 from SOFC system can be reused as HEX6, HEX7, HEX9, HEX8 and HEX10 in SOEC system. As a conclusion, five shared heat exchangers are possible in rSOC system.

4. Conclusion

This study presents a shared HEN design for rSOC system, comprising five heat exchangers, reformer and methanator. After studying heat integration, five heat exchangers can be shared. Air to air heat exchanger has largest heat exchanger area. The heat exchangers can be reused by adjusting pipelines connenctions. Moreover, methanator in SOEC system, has also been considered as the heat exchanger, this can be achieved by using adiabatic methanator, it benefits in waste heat usage and tempeature maintainence. Additional electricity is required in electrical heaters and other components. Performing multi-period optimization for two systems simutainiously will be done in future work.

Table 1. SOFC and SOEC system key components specification

	Heater/Cooler	Tin (^{0}C)	Thot (^{0}C)	Duty (kW)	HEX area (m^2)
	SOFC + oxy-combustion				
HEX 1	C1 H2	918 550	829 720	0.63	0.033
C1 - R	C1 R	829 550	695 550.1	0.86	0.053
HEX2	C1 H1	695 25	95 550	3.99	0.242
HEX 3	C3 H3	800 35	50 720	28.23	14.54
HEX 4	C5 CW	95 5	25 15	0.06	0.009
HEX 5	C4 CW	137 5	50 15	1.95	0.158
	SOEC + methanation				
H1' - M	M H1'	280.1 55	280 102	4.01	0.250
HEX 6	C1' H1'	780 102	190 268	2.34	0.061
HEX 7	C2' H1'	890 268	280 754	2.14	0.524
HEX 8	C5' CW	90 5	25 15	0.07	0.011
HEX 9	C3' H3'	780 25	137 760	10.55	9.417
HEX 10	C4' CW	137 5	25 15	1.71	0.189

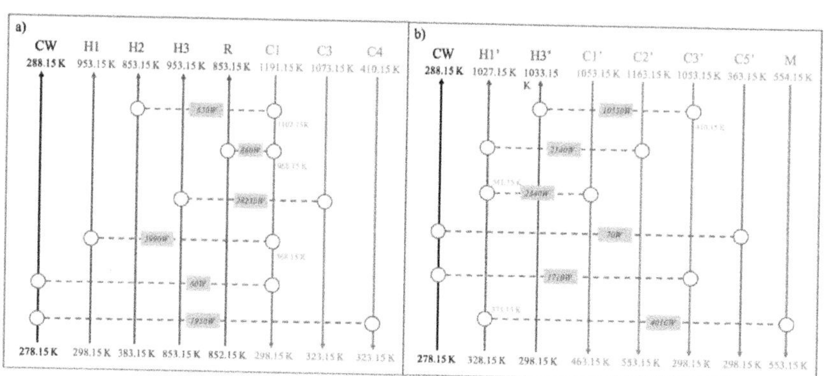

Figure 4. HEN for a) SOFC, b) SOEC systems

References

[1] Hannah Ritchie and Max Roser, "Energy Production and Consumption," *Our World in Data*, 2020. https://ourworldindata.org/energy-production-consumption

[2] A. A. Lacis, G. A. Schmidt, D. Rind, and R. A. Ruedy, "Atmospheric CO $_2$: Principal Control Knob Governing Earth's Temperature," *Science*, vol. 330, no. 6002, pp. 356–359, Oct. 2010, doi: 10.1126/science.1190653.

[3] J. Lelieveld, K. Klingmüller, A. Pozzer, R. T. Burnett, A. Haines, and V. Ramanathan, "Effects of fossil fuel and total anthropogenic emission removal on public health and climate," *Proc. Natl. Acad. Sci.*, vol. 116, no. 15, pp. 7192–7197, Apr. 2019, doi: 10.1073/pnas.1819989116.

[4] M. Santin, A. Traverso, and L. Magistri, "Liquid fuel utilization in SOFC hybrid systems," *Appl. Energy*, vol. 86, no. 10, pp. 2204–2212, Oct. 2009, doi: 10.1016/j.apenergy.2008.12.023.

[5] Y. Kalinci and I. Dincer, "Analysis and performance assessment of NH3 and H2 fed SOFC with proton-conducting electrolyte," *Int. J. Hydrog. Energy*, vol. 43, no. 11, pp. 5795–5807, Mar. 2018, doi: 10.1016/j.ijhydene.2017.07.234.

[6] P. Prabhakaran *et al.*, "Cost optimisation and life cycle analysis of SOEC based Power to Gas systems used for seasonal energy storage in decentral systems," *J. Energy Storage*, vol. 26, p. 100987, Dec. 2019, doi: 10.1016/j.est.2019.100987.

[7] F. Tanozzi, S. Sharma, F. Maréchal, and U. Desideri, "3D design and optimization of heat exchanger network for solid oxide fuel cell-gas turbine in hybrid electric vehicles," *Appl. Therm. Eng.*, vol. 163, p. 114310, Dec. 2019, doi: 10.1016/j.applthermaleng.2019.114310.

[8] S. Sharma and F. Maréchal, "Robust Multi-Objective Optimization of Solid Oxide Fuel Cell–Gas Turbine Hybrid Cycle and Uncertainty Analysis," *J. Electrochem. Energy Convers. Storage*, vol. 15, no. 4, p. 041007, Nov. 2018, doi: 10.1115/1.4039944.

[9] A. Mian, E. Martelli, and F. Maréchal, "Framework for the Multiperiod Sequential Synthesis of Heat Exchanger Networks with Selection, Design, and Scheduling of Multiple Utilities," *Ind. Eng. Chem. Res.*, vol. 55, no. 1, pp. 168–186, Jan. 2016, doi: 10.1021/acs.iecr.5b02104

Antonis Kokossis, Michael C. Georgiadis, Efstratios N. Pistikopoulos (Eds.)
PROCEEDINGS OF THE 33rd European Symposium on Computer Aided Process Engineering (ESCAPE33), June 18-21, 2023, Athens, Greece
 http://dx.doi.org/10.1016/B978-0-443-15274-0.50171-2

Multi-objective optimisation of the algae Arthrospira platensis under uncertainty: a Pareto ellipsoids approach

Wannes Mores, Satyajeet S. Bhonsale, Filip Logist, Jan F.M. Van Impe*

BioTeC+, KU Leuven, Gebroeders De Smetstraat 1, Gent 9000, Belgium

* jan.vanimpe@kuleuven.be

Abstract

Multi-objective optimisation allows a decision maker to evaluate trade-offs between objectives by supplying a set of optimal solutions, for example to find the balance between process performance and sustainability in process design. Accurate prediction of these objectives and, more importantly, safe operating conditions are needed. Thus, we need to account for uncertainty within the model. The Pareto ellipsoids approach ensures efficient use of computation while visualising the effect of parametric uncertainty. The novel Pareto ellipsoids-based algorithm is applied to the algae *Arthrospira platensis* and determines optimal light conditions for both biomass and product yield. Uncertain parameters are selected to investigate propagation of uncertainty from both the light intensity model and the Droop model. Finally, the obtained Pareto ellipsoids' accuracy is verified using a Monte Carlo approach.

Keywords: Multi-objective optimisation, Parametric uncertainty, Droop model

1. Introduction

Multi-objective dynamic optimisation is an important tool for (bio-)process design and operation (Logist et al., 2010). It allows a decision maker (DM) to evaluate different criteria, which are often of a conflicting nature. A typical example of this is the trade-off between an efficient, low-cost process and a high throughput process. To evaluate this trade-off, model-based multi-objective optimisation is often used. Here, a set of optimal solutions is generated, known as the Pareto front. With this set, a DM can decide where the point of operation lies. Pareto fronts can be generated with vectorisation (Bhaskar et al., 2000) and scalarisation approaches (Das and Dennis, 1998).

Model-based optimisation is a useful tool if the model represents the real process well. However, uncertainty on model parameters propagate through the model and influence the performance of the process. This not only leads to sub-optimal performance but can also cause critical errors in the process. Incorporating the effect of parametric uncertainty is important for safe and effective process design and operation (Vallerio et al., 2015).

Incorporating uncertainty into the optimisation is done in two ways. Robust optimisation accounts for the worst-case scenario and is a very conservative approach. Stochastic optimisation uses information on the distribution of the uncertainty and defines confidence on model outputs. Incorporating uncertainty into stochastic optimisation can be done in different ways, an overview is given in Nimmegeers et al. (2016).

In this work, stochastic methods are used to incorporate uncertainty. These methods lead to a significant increase in computational expense, it is therefore important to use computationally efficient stochastic approaches and generate only significant Pareto points. Hashem et al. (2017) developed the D&C algorithm that finds relevant regions of the Pareto front with user-specified filter settings. In this work, the idea behind D&C is expanded by using information obtained from stochastic optimisation approaches. The uncertainty on model parameters will influence the performance regarding the objectives. In the case that the uncertainty on the objective is Gaussian, this can be shown by drawing ellipsoids around the Pareto points. These Pareto ellipsoids are then used to determine the significance of a Pareto point and aid in the spread of points on the Pareto front.

This novel Pareto ellipsoids-based optimisation algorithm is applied to the production of C-Phycocyanin through *Arthrospira platensis.* C-Phycocyanin is commonly used as a natural colorant in the food and cosmetic industry, reducing the need for synthetic colorants with possible toxic effects (Martelli et al., 2014). The use of *Arthrospira platensis* for production is promising as it yields high C-Phycocyanin concentration in cells (Xie et al., 2015). Del Rio Chanona et al. (2015) developed advanced dynamic models for this production process using experimental data from both batch and fed-batch processes. For this process, it was shown that light intensity plays a big role in C-Phycocyanin production. Higher light intensities promote biomass production but inhibit and even decrease intracellular C-Phycocyanin content. The substrate concentration promotes biomass growth but also suppresses C-Phycocyanin accumulation. In this work, a Droop model is used to determine optimal light conditions and substrate level for the trade-off between biomass production and C-Phycocyanin yield.

2. Methodology

In this section, the methodology for multi-objective optimisation under uncertainty is presented. The concept of Pareto ellipsoids is introduced, which is used to define a novel significance criterion in a Pareto ellipsoids-based algorithm. These methods are applied to the Droop model for *Arthrospira platensis*, described in the final part of this section.

2.1. Multi-objective dynamic optimisation

A multi-objective optimisation problem can generally be formulated as:

$$\min \boldsymbol{J} \tag{1}$$

subject to

$$\frac{d\boldsymbol{x}}{dt} = \boldsymbol{f}(\boldsymbol{x}(t), \boldsymbol{u}(t), \boldsymbol{p}, t) \tag{2}$$

$$\boldsymbol{c}(\boldsymbol{x}(t), \boldsymbol{u}(t), \boldsymbol{p}, t) \leq 0 \tag{3}$$

Here, $\boldsymbol{J}$ is the vector of objective functions to be evaluated, $\boldsymbol{x}$ is the state vector, $\boldsymbol{f}$ is the vector of state equations and $\boldsymbol{c}$ is the constraint vector. The vector $\boldsymbol{p}$ contains all the model parameters, including the uncertain parameters. It is assumed that the distribution is known for the uncertain parameters.

The propagation of this uncertainty is quantified using the sigma points method (Julier et al., 1996). This method relies on selecting sample points within the uncertainty distribution, which are propagated to the outputs of the model and combined to obtain the expected value E and variance V of model outputs. Other methodologies for uncertainty propagation exist, an overview is presented in Nimmegeers et al. (2016). The solutions of the multi-objective optimisation problem are obtained through the nominal boundary intersection (NBI) method, presented in Das and Dennis (1998).

2.2. Pareto ellipsoids

The uncertainty on the objective functions is used to define the concept of Pareto ellipsoids. If normally distributed uncertainty on the objective functions is assumed, the uncertainty can be visualised using an ellipsoid. The center of the ellipsoid is the expected value E, and the radii can be defined using the covariance matrix $\mathbf{\Sigma}$ as follows:

$$(x - E[J])^{\top} \mathbf{\Sigma}_J (x - E[J]) = c_{\alpha}^2 \quad (4)$$

Where c_{α}^2 is the upper αth percentile of a χ^2distribution. The value of α determines the level of confidence that the objective value is inside the ellipsoid.

2.3. Pareto ellipsoids significance criterion

Regions of the Pareto front with high uncertainty levels on objective performance are usually undesirable. A novel significance criterion is used to include this effect in the *a priori* smart filter. The Pareto ellipsoids criterion can be defined as follows:

A Pareto point is considered insignificant if the Pareto point falls within the Pareto ellipsoid of a neighboring Pareto point or if one of its neighboring Pareto points falls within the Pareto ellipsoid of the Pareto point.

For two Pareto points $P1$ and $P2$, this can be checked by evaluating 2 inequalities:

$$(E[J_{P1}] - E[J_{P2}])^{\top} \mathbf{\Sigma}_J (E[J_{P1}] - E[J_{P2}])^{\top} \leq c_{\alpha}^2 \quad (5)$$

$$(E[J_{P2}] - E[J_{P1}])^{\top} \mathbf{\Sigma}_J (E[J_{P2}] - E[J_{P1}])^{\top} \leq c_{\alpha}^2 \quad (6)$$

2.4. Pareto ellipsoids-based algorithm

The underlying concept of the D&C algorithm with the information criterion from Hashem et al. (2017) can be extended with this novel criterion to reduce points in regions of high uncertainty. The information criterion settings are specified with $\sigma_{low}^2, \sigma_{high}^2$. The ellipsoid criterion is added when evaluating the information criterion. This requires the expected values and covariances of the objective functions, in this work obtained using the sigma points method. The ellipsoid criterion setting is specified using α.

Algorithm 1 Pareto ellipsoid-based algorithm

Input: Specify α, $\sigma_{low}^2, \sigma_{high}^2$

Output: Pareto (ellipsoid) set with adaptive resolution S.

Step 1: Initialise solution set $S = \{\}$.

Step 2: Construct weight cell C_{in} corresponding with the boundaries of the Pareto set

Step 3: Initialise the weight cell C_w: $C_w = C_{in}$

Step 4: Start the recursive function for weight distribution

for $i = 1{:}n_w$ (n_w = number of entries in C_w)

  if $C_w \in$ CHIM **do**

    Solve an NBI subproblem using $C_w[i]$ to obtain Pareto point P_i

    Compute $E[J_{P_i}]$ and $V[J_{P_i}]$ (using sigma points)

end for

Step 5: Construct a Pareto cell $C_P = [P_1, \ldots, P_{n_{Pareto}}]$, with $n_{Pareto} \leq n_w$

If information criterion and Pareto ellipsoid criterion not activated **do**

  Add C_P to S

  Divide: Construct daughter cells $C_{d1}, \ldots, C_{dn_w}$

  Repeat **Step 4** with $C_w = C_d \in \{C_{d1}, \ldots, C_{dn_w}\}$

else *Conquer:* Exit recursive call and stop exploring current segment.

Step 6: Produce S when all recursive calls are ended

End

2.5. Model description

The model used for optimisation is the Droop model for *Arthrospira platensis*. Droop models use intracellular quota to represent the nutrient storage inside the cell. In total, the model considers 4 states: biomass concentration c_X, nitrogen concentration c_N, intracellular normalised nitrogen quota c_q, and intracellular Phycocyanin quota c_{q_c}. The model also incorporates the Aiba model, where illumination affects the phycocyanin production constant and specific growth rate. For brevity, the model equations are not included, a detailed description can be found in del Rio Chanona et al. (2015).

The light intensity throughout the reactor is determined using the Lambert-Beer law.

$$I(z) = I_0[\exp(-(\tau \cdot X + K_a) \cdot z) + \exp(-(\tau \cdot X + K_a) \cdot (L - z)] \quad (7)$$

Where I_0 is the incident light intensity, τ the cell absorption coefficient, K_a the bubble reflection coefficient, and z the distance from the light source. In this study, the photobioreactor is lit from both sides and has width L.

The light intensity plays a big role in the C-Phycocyanin production process as it promotes biomass growth and inhibits C-Phycocyanin production. To evaluate optimal lighting conditions for the process, two objectives are considered. These are maximisation of biomass growth (J_1) and maximisation of C-Phycocyanin yield (J_2). The control input for the optimisation is the incident light intensity I_0, and contributes to the contradicting aspect of the two objective functions as described previously. As optimisation is usually done by minimisation, the objective functions are negative resulting in maximisation.

$$J_1 = -c_{X,f} \quad (8)$$

$$J_2 = -\frac{c_{q_c X,f}}{c_{N,0}} \quad (9)$$

Constraints are imposed on the model, which include minimal product return, a lower and upper bound on initial substrate concentration, and the initial state of c_X. The constraints are found in equations (9-11) The two parameters considered uncertain are chosen from both the Droop model equations and the nested Aiba model. These are the cell absorption coefficient τ and the cell specific decay rate u_d. Both parameters are assumed to have a normally distributed uncertainty with a standard deviation of 10% of their nominal value.

$$c_X(0) = 0.1\frac{g}{L} \quad (10)$$

$$c_N(0) \geq 1000\frac{mg}{L} \quad (11)$$

$$c_{q_c X}(f) \geq 100\frac{mg}{L} \quad (12)$$

3. Results

The pareto ellipsoids-based algorithm is implemented in the POMODORO toolbox (Bhonsale et al., 2018), available at https://cit.kuleuven.be/biotec/software/pomodoro. It is a python-based toolkit for solving multi-objective optimal control problems. For optimisation, an α of 70% is chosen, which means that the ellipsoids should have a 70% confidence that the true objective value lies within its boundaries. The settings for the information criterion $\sigma^2_{low}, \sigma^2_{high}$ are chosen to be 0.001.

3.1. Pareto ellipsoids set

The multi-objective optimisation study results in a set of Pareto ellipsoids shown in Figure 1. Looking at the control profiles for the anchor points and a chosen trade-off solution, an early spike in light intensity is beneficial for initial biomass growth. Retaining the high light intensity results in efficient biomass production and limits the C-Phycocyanin production, which is seen in the maximised biomass control profile. The spike in control profile is later for the maximised yield, which is explained through the fact that light intensity decreases the intracellular C-Phycocyanin content, and a later spike means less loss of product. The starting nitrate concentration is also lower for maximised yield. High nitrate concentrations suppress intracellular phycocyanin accumulation but promote biomass growth. The trade-off solution shows a mix of the two anchor points.

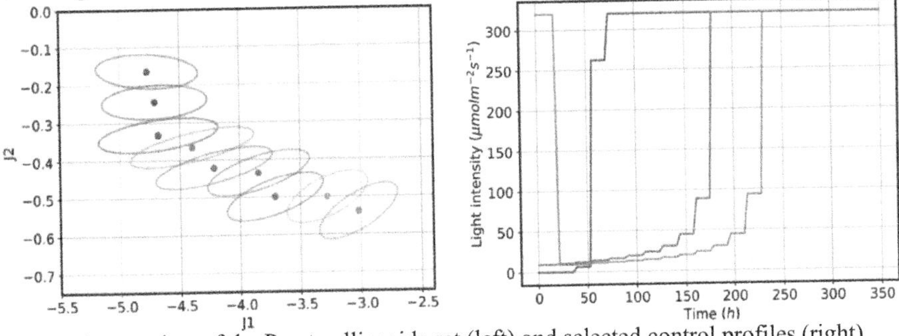

Figure 1: overview of the Pareto ellipsoids set (left) and selected control profiles (right)

Overall, the front of Pareto ellipsoids has relatively constant shape of ellipsoids. There is an increase in uncertainty in the middle of the trade-off zone between the anchor points. When minimising J_1, the ellipsoid seems relatively aligned with the objective axes, when moving towards minimising J_2, the ellipsoids have more rotation. Using this visualisation of uncertainty, a DM can improve his awareness of the uncertainty's effect and make a better decision regarding expected process performance and safety of operation.

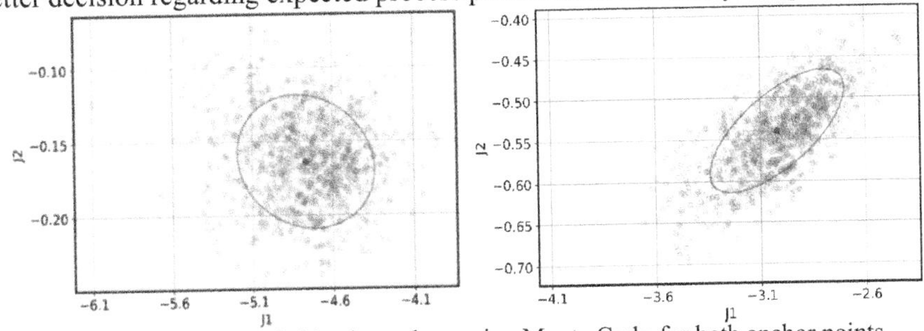

Figure 2: visualisation of objective values using Monte Carlo for both anchor points

3.2. Pareto ellipsoids verification

Monte Carlo simulation is used to verify the accuracy of ellipsoids. 1000 samples from the uncertain parameters' distribution are evaluated and shown in Figure 2. As the setting for the Pareto-ellipsoids based algorithm was set to $\alpha = 70\%$, the expectation is that 70% of all Monte Carlo iterations are within the boundary of the ellipsoid (green points).

Table 1: Sigma points accuracy through Monte Carlo simulation

Ellipsoid:	1	2	3	4	5	6	7	8	9
σ^2_{J1}	97.6%	101.6%	100.2%	99.4%	101.9%	97.9%	98.0%	97.7%	97.9%
σ^2_{J2}	98.2%	97.8%	97.2%	96.8%	96.4%	95.6%	96.1%	95.5%	96.0%
$\mathrm{Cov}(J_1, J_2)$	103.7%	101.3%	99.7%	98.8%	108.7%	95.7%	94.9%	95.6%	95.8%
% green points	70.3%	69.3%	69.3%	70.1%	69.4%	68.8%	68.4%	68.5%	69.0%

Additionally, the covariance matrix obtained through sigma points can be compared to the covariance matrix from the Monte Carlo dataset to determine its accuracy. An overview of the covariance matrix accuracy is found in Table 1. Figure 2 shows the ellipsoid, which is using sigma points, and the Monte Carlo iterations as points.

Overall, the ellipsoids are very accurate. All ellipsoids contain between $68 - 71\%$ of Monte Carlo iterations, which is very close to the desired 70%. The shape of the Monte Carlo dataset also seems to resemble the ellipsoid's shape. The covariance accuracy shows that the variance of objective 2 is underapproximated by around $3 - 4\%$. A more in-depth analysis is done for a range of different case studies in Mores et al. (2023).

4. Conclusion

In this work, a novel Pareto ellipsoids-based algorithm for multi-objective optimisation is presented and applied to a case study of C-Phycocyanin production through *Arthrospira platensis*. The algorithm provides only relevant points, as defined by the user through settings $\sigma_{low}^2, \sigma_{high}^2$ and α. This ensures efficient calculation of the Pareto set, which is important when implementing uncertainty propagation techniques. The Droop model as presented in del Rio Chanona et al. (2015) is used and two parameters are considered uncertain. Application of the novel algorithm results in a set of Pareto ellipsoids, which show how the uncertainty will propagate to the performance of the process. Early increase in light intensity is used to promote biomass growth, while delaying it leads to a more efficient C-Phycocyanin production. The ellipsoids' accuracy is evaluated using a Monte Carlo approach and shows that they represent the uncertainty accurately, allowing a DM to make a more informed decision on how to design and operate the process.

Acknowledgement

This research is co-funded by the Research Foundation Flanders (FWO) through the Senior Research Project G0B4121N ("Pharma4S") and the Strategic Basic Research Project S008121N ("BlueBioChain", H2020 ERA-NET Cofund on Blue Bioeconomy)

References

Bhaskar et al. 2000, "Applications of multiobjective optimization in chemical engineering.", Reviews in Chemical Engineering, 16, 1

Bhonsale et al. 2018, "Pomodoro: A novel toolkit for dynamic (multiobjective) optimization, and model based control and estimation.", IFAC-PapersOnline, 51(2), 719

Das, I., & Dennis, J. 1998, "Normal-boundary intersection: A new method for generating the Pareto surface in nonlinear multicriteria optimization problems.", Siam Journal On Optimization, 8, 631

del Rio-Chanona et al. 2015, "Dynamic simulation and optimization for Arthrospira platensis growth and C-phycocyanin production.", Industrial & Engineering Chemistry Research 54.43, 10606

Hashem et al. 2017, "A novel algorithm for fast representation of a Pareto front with adaptive resolution: Application to multi-objective optimization of a chemical reactor.", Computers & Chemical Engineering, 106, 544

Julier et al. 1996, "A general method for approximating non-linear transformations of probability distributions.", Technical report, Robotics research group University of Oxford 1

Logist et al. 2010, "Fast Pareto set generation for nonlinear optimal control problems with multiple objectives.", Structural and Multidisciplinary Optimization 42.4, 591

Martelli et al. 2014, "Thermal stability improvement of blue colorant C-Phycocyanin from Spirulina platensis for food industry applications.", Process Biochemistry, 49.1,154

Mores et al. 2023, "Multi-objective optimization under parametric uncertainty: A Pareto ellipsoids-based algorithm." Computers & Chemical Engineering, 169: 108099.

Nimmegeers et al. 2016, "Dynamic optimization of biological networks under parametric uncertainty.", BMC Systems Biology, 10(1), 1

Vallerio et al. 2015, "An interactive decision-support system for multi-objective optimization of nonlinear dynamic processes with uncertainty.", Expert Systems with Applications 42.21, 7710

Xie et al. 2015, "Fed-batch strategy for enhancing cell growth and C-phycocyanin production of Arthrospira (Spirulina) platensis under phototrophic cultivation.", Bioresource technology, 180, 281

Antonis Kokossis, Michael C. Georgiadis, Efstratios N. Pistikopoulos (Eds.)
PROCEEDINGS OF THE 33rd European Symposium on Computer Aided Process Engineering (ESCAPE33), June 18-21, 2023, Athens, Greece
 http://dx.doi.org/10.1016/B978-0-443-15274-0.50172-4

Predicting the Evolution of Flammable Gases During Li-ion Battery Thermal Runaway Using Micro-Kinetic Modelling

Dr. Peter Bugryniec,[a] Dr. Sergio Vernuccio,[a] Prof. Solomon Brown[a]

[a]*Department of Chemical & Biological Engineering, The University of Sheffield, Sheffield, S1 3JD, UK*
s.f.brown@sheffield.ac.uk

Abstract

Li-ion batteries are a widely used electrochemical energy storage device. But, catastrophic failure via thermal runaway leads to great flammability and toxicity hazards. As such, there is a need to better understand the thermal runaway process. In doing so, reducing its occurrence and improving predictions of its hazards. To achieve this, we aim to develop a more detailed model of thermal runaway. This is based on fundamental reaction theory. Micro-kinetic modelling techniques are applied to predict the kinetic evolution of the reacting systems on a mechanistic level, based on a detailed analysis of the elementary reaction steps. Using this methodology, we simulate the thermal decomposition of dimethyl carbonate, as a model electrolyte solvent, and predict the product species present in the off-gas. We also investigate the impact of the temperature on the composition of the off-gas and the lower flammability limit. This demonstrates a method for predictive hazard assessments of Li-ion battery failure. For the DMC case study, we show that the LFL increases with increasing the operating temperature due to the large proportion of CO_2 generated. This effectively makes the off-gas safer in terms of explosion hazards. Further work will extend this methodology to construct the reaction systems for a complete Li-ion cell.

Keywords: Reaction network analysis, Hazard prediction, Dimethyl carbonate, Lower explosion limit, Thermal decomposition

1. Introduction

Li-ion batteries (LIBs) have been extensively adopted as energy storage devices. Hence, the safety of LIBs is of great importance. This is because under adverse conditions LIBs can undergo thermal runaway (TR) (Wang *et al.*, 2012) that occurs when the Li-ion cell's component materials exothermically decompose. This leads to an exponential increase in heat generation and cell temperature. Further, the decomposition process leads to the generation of a complex vapour-gas mixture including electrolyte vapours, as well as flammable, toxic, and poisonous gases (Fernandes *et al.*, 2018). The most notable toxic gas is HF. While CO_2 is not poisonous and can cause suffocation in large enough quantities. Flammable gasses also have the potential to cause explosions. LIBs are increasingly used in EVs, and domestic and commercial stationary energy storage. As a result, these hazards are present to more people and methods to improve LIB safety are essential.

To help achieve this goal LIB researchers and developers need to improve their ability to predict the complexities of TR. TR modelling has been used to predict thermal hazards of cells, some including flammability behaviour (Zhang *et al.*, 2022a). However, studies on the flammability of the off-gas are based on a given gas composition determined from experimental studies. This is because typical TR models that simulate the decomposition of the battery materials do not provide a detailed description of the reaction products (for example, Bugryniec *et al.*, 2020). Micro-kinetic modelling can overcome this limitation by including all reactants, e.g. Li-ion cell components, and all possible products, e.g. off-gasses (Bugryniec *et al.*, 2022) in the reaction network as well as each possible elementary step and reaction intermediate. Further, the kinetic parameters of these detailed models are typically estimated based on a solid theoretical basis and no a priori assumptions on the rate-determining steps are needed (Hermes *et al.*, 2019).

An important parameter to quantify the combustion risk is the lower flammability limit (LFL), or the lower explosion limit (LFL) (Ludwig, E., 1991). The LFL is the threshold percentage of gas that is required to be present in the air for it to ignite/explode. Below this threshold, the fuel conditions are too lean for combustion. Complimentary to this is the upper flammability limit (UFL) which indicates the highest fraction of flammable gas at which a flammable mixture is formed. In this work, we focus on the LFL as it is the start of the flammability hazard. The LFL can be determined experimentally or theoretically by the Le Chatelier formula which is discussed in detail in Section 2. The error calculating the LFL by Le Chatelier's formula is relatively small (<7%) (Li et al., 2019). As such it is a widely accepted method in engineering applications.

The LFL has been calculated for the off-gas generated when experimentally abusing Li-ion cells. This can range from 4% to 50% depending on SOC and chemistry. Typically, the higher values of LFL relate to a higher fractional amount of CO_2 being generated (Chen *et al.,* 2020, Li et al., 2019, Zhang *et al.*, 2022b). The LFLs of electrolyte solvents have also been reported. For dimethyl carbonate, diethyl carbonate and ethyl methyl carbonate the LFL are 3.2%-4.2%, 1.4%-1.6%, and 2.1% respectively (Henriksen et al., 2019, IPCS INCHEM, 2000, IPCS INCHEM, 2005). However, the LFL of the off-gas from solvent decomposition has not been calculated, nor has the LFL of Li-ion cell off-gas been predicted. To the authors' knowledge, this is the first work on the prediction of the off-gas flammability behaviour on the topic of LIB safety.

Considering this, we aim to develop a TR model that predicts the generation and composition of LIB off-gas by applying micro-kinetic modelling methodologies. From this, the flammability (LFL) of the gas composition is assessed. In this work, we focus on one subcomponent of the cell, the electrolyte solvent dimethyl carbonate (DMC). For this component, we construct a reaction model in MATLAB® to simulate the generation of gas species under heating. With this, predictions are made on the gas composition during different stages of TR, from which changes in flammability hazard (i.e. LFL) are assessed.

2. Methodology

As stated previously, this work focuses on the reaction network for DMC thermal decomposition in an inert atmosphere. The reaction network applied is presented in Figure 1, reproduced from Fernandes et al., 2019. The reaction network consists of 16 forward reactions and 12 species. The arrows in Figure 1 indicate the reactions proceeding in the forward direction. However, each elementary step included in the network was considered to be reversible. Within this, there are 3 possible pathways leading to the decomposition of DMC: The direct decomposition of DMC to dimethyl ether, and the reaction of DMC with water or hydrogen.

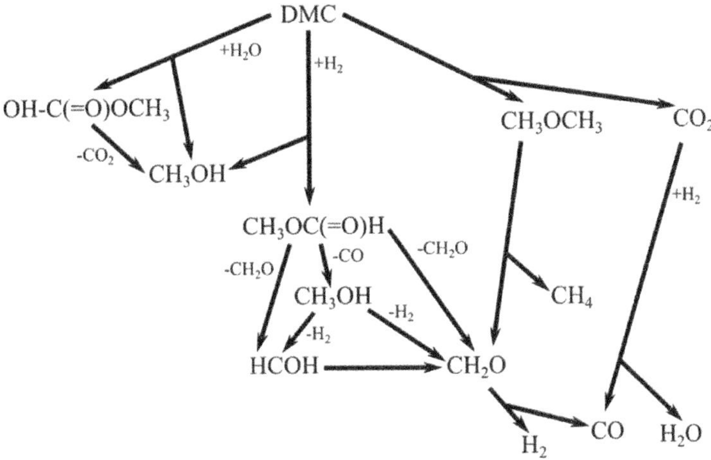

Figure 1. Reaction network for DMC thermal decomposition.

The micro-kinetic model consists of a system of 12 ordinary differential equations (ODEs) describing the change in concentration over time of the 12 species included in the mechanism. Each ODE is expressed as the algebraic sum of all relevant reaction rates (R_i) corresponding to a given reaction step. R_i is governed by an Arrhenius equation of the form $R_i = A_i e^{-E_{a,i}/RT} \prod_j C_j$. In which A_i is the frequency factor, $E_{a,i}$ is the activation energy, R is the gas constant, T is the temperature and C_j is the species concentration. The kinetic parameters of each rate-determining step are determined from transition state theory. The thermochemistry (of the reaction species, complexes, and transition state structures) was determined using density functional theory (DFT). This was carried out in Gaussian (Gaussian 16, 2019) with the method and basis set combination of B3LYP/6-311++G(d,p). The solution to the system of ODES is computed in MATLAB® (MATLAB, 2021).

The initial reactant system is assumed to be a mixture of only DMC and water with concentrations of 11.87 mole/L and 1.30E-3 mole/L, respectively. This is in line with the typical water content of commercially available DMC. The time-varying simulation assumes a constant linear temperature increase of 1°C/min, starting from 20°C and ending at 800°C. This is to approximate the common LIB abuse tests of adiabatic calorimetry. The predicted decomposition species are then used to calculate the temperature dependant LFL of the off gas.

As CO_2 is expected to develop, the off-gas mixture can be considered dilute. The LFL for a dilute mixture, LFL_M (see Eq. 1), is calculated using the extended Le Chatelier's formula (Schroeder, 2016). The LFL_M is determined from the LFL'_i, the modified LFL of each species accounting for the diluting species, i.e. CO_2, (see Eq. 2). In which the LFL'_M is the LFL of only the flammable gas mixture (see Eq. 3). Further, a_i and B_k are the molar fraction of flammable gas i and molar fraction of inert gas k, respectively. $\overline{K}$ is the average value of the nitrogen equivalency K_k, of the inert gases weighted according to their molar fractions. For pure CO_2 $\overline{K} = 1.5$.

$$\mathrm{LFL}_M = \frac{100}{\sum_i^n \frac{a_i}{LFL'_i}} \tag{1}$$

$$LFL'_i = \frac{100 - LFL'_M - \left((1 - \bar{K}) \frac{\sum_{p=1}^{k} B_k}{\sum_{i=1}^{n} a_i} \times LFL'_M \right)}{100 - LFL'_M} \times LFL_i \tag{2}$$

$$LFL'_M = \frac{100}{\sum_i^n \frac{a_i}{LFL_i}} \tag{3}$$

3. Results and Discussion

The predicted decomposition of DMC and generation of the off-gas species, along with the calculated LFL of the off-gas is presented in Figure 2. Only 8 species are presented in Figure 2 as the concentrations of the other 4 species (H_2O, OH-C(=O)$OOCH_3$, CH_3OC(=O)H, and HCOH) included in the reaction network do not reach a significant value. This is due to the low concentration of water in the initial solution which prevents OH-C(=O)$OOCH_3$ from being generated in high concentrations. On the other hand, CH_3OC(=O)H is not observed in high concentrations due to the number of favourable decomposition paths that this species can decompose by.

From Figure 2 it can be seen that DMC begins to decompose at 450°C. At this initial stage, DMC leads to CH_3OCH_3 and CO_2 generation only. At approximately 550°C CH_3OCH_3 starts to decompose leading to CH_4 and CH_2O generation. At even greater temperatures, 650°C, CH_2O decomposes to H_2 and CO. At 800°C the gas produced predominantly consists of CO_2 and CH_4, with H_2 and CO in low quantities.

The LFL_M is calculated only considering the major species present. The LFLs of these individual species, LFL_i, are presented in Table 1. It can be seen from Figure 2 that LFL_M increases compared to the $\mathrm{LFL}_{\mathrm{DMC}}$ as decomposition progresses. This is mainly driven by the diluting effect of CO_2. As even in the early stages of decomposition when CH_3OCH_3 is present with a lower LFL than DMC the LFL_M still rises. This indicates that the greatest flammability risk when cell venting occurs (at ~200°C) would be from the solvent vapour (whilst not considering the electrode decomposition reactions). Further, we can see that generation of CO_2 has a beneficial effect of increasing the LFL_M. However, a large quantity of CO_2 presents a significant suffocation hazard.

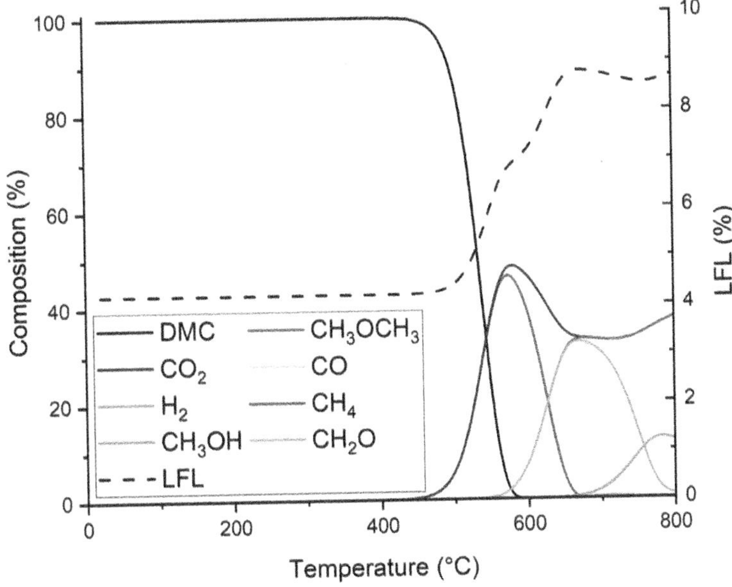

Figure 2. DMC decomposition and predicted off-gas species with calculated LFL of the mixture.

Table 1. LFL of major species present in the off-gas.

Species	LFL (%)	Source
DMC	4.20	(IPCS INCHEM, 2005)
CH_3OCH_3	3.35	(Zlochower, 2012)
CO_2	n/a	(Chen et al., 2020)
CH_3OH	6.00	(PubChem, Methanol)
H_2	4.00	(Chen et al., 2020)
CH_4	5.00	(Chen et al., 2020)
CH_2O	7.00	(PubChem, Formaldehyde)
CO	12.50	(Chen et al., 2020)

4. Conclusion

This work proves that micro-kinetic modelling can predict gas generation during LIB TR. We have shown the flammability risk can be evaluated using the extended Le Chatelier's formula. In doing so, the LFL of the off-gas is calculated. A large amount of CO_2 increases the LFL of the off-gas. This reduces the flammability hazard but increases the suffocation hazard. The composition of off-gas varies over the temperature range studied.

This work will be extended to full Li-ion cell systems to be practical for LIB developers. Including all electrolyte and electrode materials. Also, this method can be extended to calculate the ignition energy and the heat of reaction of the off-gas. Providing further predictive capabilities for risk assessment of LIB energy storage systems.

5. Conflicts of Interest & Acknowledgements

There are no conflicts to declare. This work was supported by the Faraday Institution [grant number FIRG028].

References

P. Bugryniec, J. Davidson, S. Brown, 2020. Advanced abuse modelling of Li-ion cells – A novel description of cell pressurisation and simmering reactions. Journal of Power Sources, 474, 228396.

P. Bugryniec, A. Yeardly, A. Jain, N. Price, S. Vernuccio, S. Brown, 2022. Gaussian-Process based inference of electrolyte decomposition reaction networks in Li-ion battery failure. Computer Aided Chemical Engineering, 51, 157-162.

S. Chen, Z. Wang, J. Wang, X. Tong, W. Yan, 2020. Lower explosion limit of the vented gases from Li-ion batteries thermal runaway in high temperature condition. Journal of Loss Prevention in the Process Industries, 63, 103992.

Y. Fernandes, A. Bry, S. de Persis, 2018. Identification and quantification of gases emitted during abuse tests by overcharge of a commercial Li-ion battery. Journal of Power Sources, 389, 106-119.

Y. Fernandes, A. Bry, S. de Persis, 2019. Thermal degradation analyses of carbonate solvents used in Li-ion batteries. Journal of Power Sources, 414, 250-261.

Gaussian 16, 2019. Revision C.01. M. J. Frisch, G. W. Trucks, H. B. Schlegel, G. E. Scuseria, *et al.*. Gaussian, Inc., Wallingford CT, 2019.

M. Henriksen, K. Vaagsaether, J. Lundberg, S. Forseth, D. Bjerketvedt, 2019. Explosion characteristics for Li-ion battery electrolytes at elevated temperatures. Journal of Hazardous Materials, 371, 1-7.

E. Hermes, A. Janes, J. Schmidt, 2019. Micki: A python-based object-orientaed microkinetic modeling code. The Journal of Chemical Physics, 151 (1), 014112.

IPCS INCHEM, 2000. Diethyl carbonate. International Chemical Safety Card: 1022. https://inchem.org/documents/icsc/icsc/eics1022.htm

IPCS INCHEM, 2005. Diemethyl carbonate. International Chemical Safety Card: 1080. https://inchem.org/documents/icsc/icsc/eics1080.htm

W. Li, H. Weng, Y. Zhang, M. Ouyang, 2019. Flammability characteristics of the battery vent gas: A case of NCA and LFP lithium-ion batteries during external heating abuse. Journal of Energy Storage, 24, 100775.

E. Ludwig, 1999. Process Safety and Pressure-Relieving Devices, in Applied Process Design for Chemical & Petrochemical Plants, 1, 485.

MATLAB, 2021. version 9.10.0 (R2021a), Natick, Massachusetts: The MathWorks Inc.

PubChem, Methanol. https://pubchem.ncbi.nlm.nih.gov/compound/Methanol

PubChem, Formaldehyde. https://pubchem.ncbi.nlm.nih.gov/compound/712

V, Schroeder, 2016. Calculation of Flammability and Lower Flammability Limits of Gas Mixtures for Classification Purposes. Bundesanstalt für Materialforschung und – prüfung (BAM), Berlin.

Q. Wang, P. Ping, X. Zhao, G. Chu, J. Sun, C. Chen, 2012. Thermal runaway caused fire and explosion of lithium ion battery. Journal of Power Sources, 208, 210–224.

Y. Zhang, E. Wang, C. Li, H. Wang, 2022a. 2D Combustion Modeling of Cell Venting Gas in a Lithium-Ion Battery Pack. Energies, 15, 15530.

Q. Zhang, J. Niu, Z. Zhao, Q. Wang, 2022b. Research on the effect of thermal runaway gas components and explosion limits of lithium-ion batteries under different charge states. Journal of Energy Storage, 45, 103759.

I. Zlochower, 2012. Experimental flammability limits and associated theoretical flame temperatures as a tool for predicting the temperature dependence of these limits. J Loss Prev Process Ind, 25(3), 555-560.

Antonis Kokossis, Michael C. Georgiadis, Efstratios N. Pistikopoulos (Eds.)
PROCEEDINGS OF THE 33rd European Symposium on Computer Aided Process Engineering (ESCAPE33), June 18-21, 2023, Athens, Greece
 http://dx.doi.org/10.1016/B978-0-443-15274-0.50173-6

Modeling of carbon offset networks for process systems to achieve net zero emissions

Soo Hyoung Choi

Division of Chemical Engineering, Clean Energy Research Center, Jeonbuk National University, Jeonju, 54896, S. Korea
soochoi@jbnu.ac.kr

Abstract

A carbon cycle model based method is proposed for performance evaluation of carbon offset projects that compensate industrial emissions. As a case study, carbon cycle impact assessment was conducted for five representative approaches to carbon neutrality. The result indicates that fossil to biomass fuel switching is doomed to a high concentration of atmospheric carbon dioxide and large scale deforestation, and that reduction of emissions is more effective than extension of sequestration from an environmental point of view. Finally, from a social point of view, optimization of carbon offset networks in terms of both carbon cycle dynamics and carbon market economics is suggested as future work.

Keywords: carbon neutrality, net zero emissions, carbon offset, carbon credit, carbon cycle impact assessment

1. Introduction

Carbon neutrality is an arithmetic concept that can be achieved by net zero emissions, which only guarantees a steady state for the carbon cycle (Manousiouthakis and Choi, 2021). If emissions and absorptions are equally large, the steady state concentration of carbon dioxide in the atmosphere can still be high even though carbon neutrality is attained. A typical example is harvested plant biomass, which inherently owes a huge carbon debt (Choi and Manousiouthakis, 2020) when combusted, and thus cannot be a proper alternative to fossil fuels (Manousiouthakis and Choi, 2021). Furthermore, emissions can be overcompensated by carbon offsets (Kotchen, 2009) such as renewable energy and energy efficiency projects that only indirectly reduce emissions. Therefore, carbon cycle impact assessment is necessary when a carbon offset project is considered as an approach to carbon neutrality.

Chemical plant industries, mostly carbon based, are on the way to carbon neutrality, reducing emissions by carbon capture and utilization (CCU), carbon capture and storage (CCS), and by growing and processing of biomass (Gabrielli et al., 2020), and earning or purchasing carbon credits (Blaufelder et al., 2021) that can be obtained by carrying out or investing in carbon offset projects such as forestry, renewable energy, energy conservation, and conversion of waste to energy (EIC, 2020). Among these projects, forestry is essential, because it directly controls atmospheric carbon dioxide. Forests should be protected because deforestation is equivalent to carbon emissions (Photopoulos, 2017), and one to one reforestation cannot fully recover the initial state of atmospheric carbon dioxide as long as deforestation continues (Choi and Manousiouthakis, 2020). Besides, as an alternative to forestry, direct air capture (DAC) is receiving attention (Chatelain, 2022) with skepticism on its efficiency (Day, 2021).

Carbon offsetting is criticized, especially strongly by environmentalists, because it does not directly reduce emissions (Ghussain, 2020). This issue could be moderated by corrective evaluation of carbon credits, reflecting the fact that less emissions are better than more sequestration. Therefore, carbon cycle impact assessment is suggested for emissions trading system (ETS) also. In this work, a model is proposed as shown in Fig. 1, as a tool for performance evaluation of carbon offset projects. The solid arrows represent main active flows involved in carbon cycle and carbon offsetting. The dashed arrows represent flows that are much less dynamic than the others. The model indicates that the total amount of carbon credits, i.e., allowances of carbon emissions, should be limited to the total capacity of sequestration by forests in order to achieve net zero emissions. Furthermore, the prices of carbon credits should be sufficiently high in order to drive the industry to substantially reduce emissions (Opanda, 2022).

2. Proposed Method

2.1. Modeling of Carbon Cycle

Carbon cycle impact assessment is required in order to rigorously evaluate carbon offset projects in terms of contribution to climate change mitigation. Proposed carbon cycle model equations are as follows:

$$\dot{C}_a = f_{rp}(C_p) + f_{rs}(C_s) - f_p(C_a, C_p, A_v) - f_a(C_a, C_{so}) + r_v + q(t) \tag{1}$$

$$\dot{C}_p = f_p(C_a, C_p, A_v) - f_{rp}(C_p) - f_l(C_p) - h(t) \tag{2}$$

$$\dot{C}_s = f_l(C_p) - f_{rs}(C_s) - f_t(C_s) \tag{3}$$

$$\dot{C}_{so} = f_a(C_a, C_{so}) + f_t(C_s) + f_{up}(C_{do}) - f_{dw}(C_{so}) \tag{4}$$

$$\dot{C}_{do} = f_{dw}(C_{so}) - f_{up}(C_{do}) - f_s(C_{do}) \tag{5}$$

$$\dot{C}_r = f_s(C_{do}) - r_v \tag{6}$$

$$\dot{A}_v = \left[\int_0^t s(t-\tau)\dot{g}(\tau)\,d\tau - h(t)\right]\frac{A_v}{C_p} \tag{7}$$

where

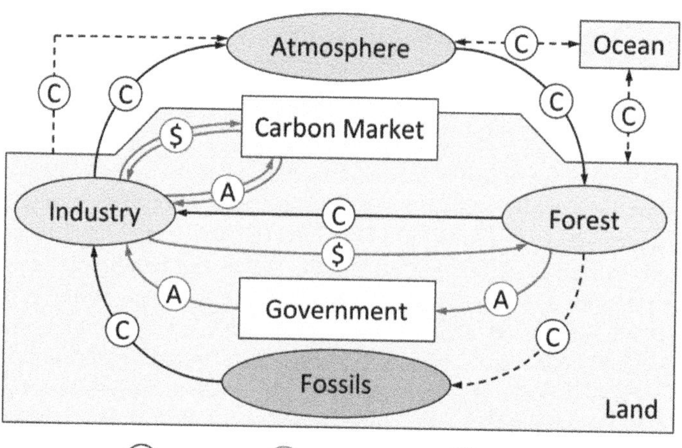

Figure 1. Carbon cycle and emissions trading system.

C_a = mass of carbon in the atmosphere, PgC
C_p = mass of carbon in plants, PgC
C_s = mass of carbon in soils, PgC
C_{so} = mass of carbon in the surface ocean, PgC
C_{do} = mass of carbon in the deep ocean, PgC
C_r = mass of carbon in rocks, PgC
A_v = normalized vegetated land area
f_p = rate of photosynthesis, PgC/y
f_{rp} = rate of plant respiration, PgC/y
f_l = rate of litterfall, PgC/y
f_{rs} = rate of soil respiration, PgC/y
f_t = rate of transfer from soils to the surface ocean by the river flow, PgC/y
f_a = net rate of absorption to the surface ocean, PgC/y
f_{up} = rate of upwelling from the deep ocean, PgC/y
f_{dw} = rate of down-welling from the surface ocean, PgC/y
f_s = rate of sedimentation to the crust, PgC/y
r_v = rate of volcano emission, 0.1 PgC/y
q = rate of carbon emissions by fuel combustion, PgC/y
h = rate of plant harvesting, PgC/y
s = target rate of biomass production by planting seedlings, PgC/y
g = normalized tree mass growth function, $g(0) = 0, g(\infty) = 1$

The unit PgC represents 10^{15} grams of carbon, which is equivalent to GtC. The initial conditions at present are as follows:

$$C_a(0) \approx 1.5C_a^\circ, C_p(0) \approx C_p^\circ, C_s(0) \approx C_s^\circ, C_{so}(0) \approx 1.02C_{so}^\circ, C_{do}(0) \approx C_{do}^\circ,$$
$$C_r(0) \approx C_r^\circ, A_v(0) \approx 0.75$$

where the parameters with degree symbols represent reference values estimated for the time right before the industrial revolution. Detailed expressions of the functions in the above equations and model validation are referred to Choi and Manousiouthakis (2022).

2.2. Simulation of Carbon Offset Networks

Simultaneous implementation of multiple carbon offset projects can be viewed as a carbon offset network. The resulting rate of carbon emissions can be calculated as follows:

$$q(t) = \sum_{i=0}^{n} q_i(t) \tag{8}$$

where $q_i < 0$ represents the reduced rate of emissions by the i-th project, which should be close to the annually earned carbon credit if properly evaluated. Rates of plant harvesting and planting seedlings can be calculated in the same way as follows:

$$h(t) = \sum_{i=0}^{n} h_i(t) \tag{9}$$

$$s(t) = \sum_{i=0}^{n} s_i(t) \tag{10}$$

Let us assume that harvested plant biomass is partially combusted as a fuel, and the rest is stored as a product. Then, carbon neutrality is achieved at $t = t^*$ if the following condition is satisfied:

$$q(t) = h(t),\ t \geq t^* \tag{11}$$

Therefore, the above three functions q, h, and s compose design variables for a carbon offset network to be optimized subject to the following constraints:

$$C_a \leq C_a^* \tag{12}$$

$$A_v \leq A_v^* \tag{13}$$

where C_a^* is set by the maximum allowable increase in the global temperature, and A_v^* by the maximum available land area for forestry.

3. Case Study

3.1. Approaches to Carbon Neutrality

Consider a hypothetical task of achieving carbon neutrality in 25 years with initial emissions of 10 PgC/y, which approximately corresponds to the current situation. Five approaches are summarized in Table 1, and six cases are defined as follows:

Case 0. Fossil fuels
As a base case, it is assumed that fossil fuels are used as ever.

$$q(t) = 10,\ h(t) = s(t) = 0 \tag{14}$$

Case 1. Fossil to biomass with reforestation
Fossil fuels are gradually replaced by harvested plant biomass with 1:1 reforestation.

$$q(t) = 10,\ h(t) = s(t) = \begin{cases} 0.4t, 0 \leq t < 25 \\ 10, t \geq 25 \end{cases} \tag{15}$$

Case 2. Fossil to biomass with afforestation
Fossil fuels are gradually replaced by harvested plant biomass with 1:1 afforestation.

$$q(t) = 10,\ h(t) = \begin{cases} 0.4t, 0 \leq t < 25 \\ 10, t \geq 25 \end{cases},\ s(t) = 10 \tag{16}$$

Table 1. Changes in design variables for different approaches to carbon neutrality

Case	Approach to carbon neutrality	Changes in design variables
1	Fossil to biomass with reforestation	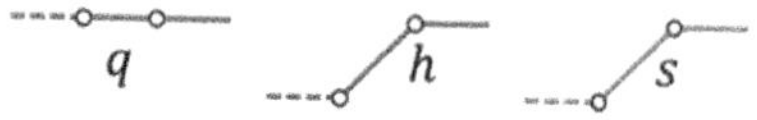 q h s
2	Fossil to biomass with afforestation	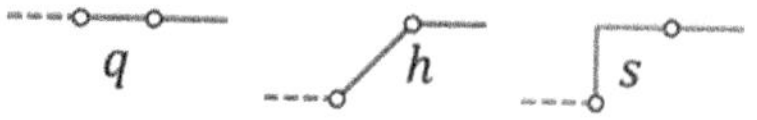 q h s
3	Extension of sequestration	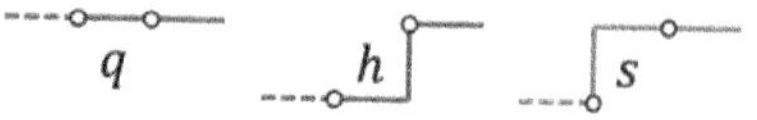 q h s

4	Reduction of emissions	q h s
5	Half reduction and half afforestation	q h s

Case 3. Extension of sequestration
Fossil fuels are used while sequestration is extended by 1:1 afforestation.

$$q(t) = 10,\ h(t) = \begin{cases} 0, 0 \le t < 25 \\ 10, t \ge 25 \end{cases},\ s(t) = 10 \tag{17}$$

Case 4. Reduction of emissions
Fossil fuels are gradually replaced by non-carbon based energy.

$$q(t) = \begin{cases} 10 - 0.4t, 0 \le t < 25 \\ 0, t \ge 25 \end{cases},\ h(t) = s(t) = 0 \tag{18}$$

Case 5. Half reduction and half afforestation
Fossil fuels are gradually reduced to half, and sequestrated by 1:1 afforestation.

$$q(t) = \begin{cases} 10 - 0.2t, 0 \le t < 25 \\ 5, t \ge 25 \end{cases},\ h(t) = \begin{cases} 0,\ 0 \le t < 25 \\ 5,\ t \ge 25 \end{cases},\ s(t) = 5 \tag{19}$$

3.2. Performance Evaluation

Carbon cycle impact assessment was carried out for the above six cases. The results are as shown in Fig. 2. Cases 1 and 2 indicate that fossil to biomass fuel switching is an improper approach to carbon neutrality because of huge carbon debts, i.e., the initial atmospheric carbon dioxide that exceeds that of fossil fuels, and large loss of forests. Cases 3 and 4 show that extension of sequestration cannot replace reduction of emissions in effect. Case 5 is a compromise between cases 3 and 4, which is considered to be a feasible approach whose impacts on the carbon cycle are acceptable.

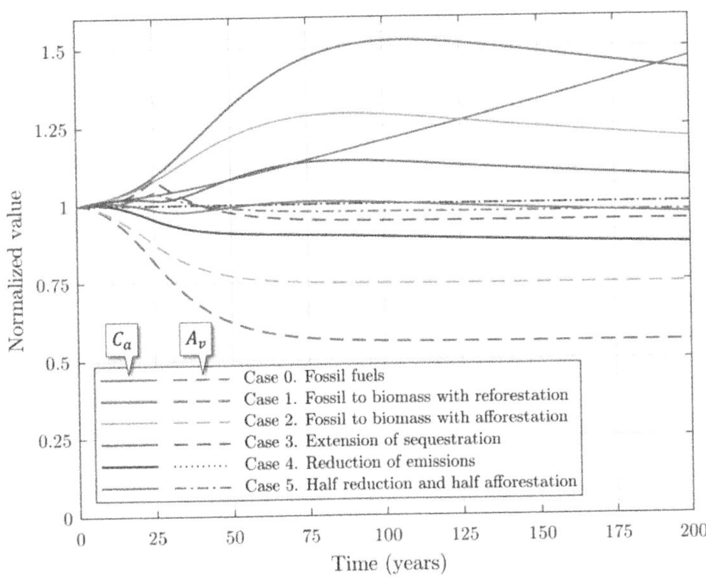

Figure 2. Predicted carbon cycle impacts of different approaches to carbon neutrality.

4. Conclusions

The result of case study shows that net zero emissions can result in various levels of concentration of carbon dioxide in the atmosphere. Therefore, it is suggested that the concept of carbon neutrality should be more conservatively redefined in view of dynamic effects, and the credits for carbon offset projects should be more rigorously revaluated based on carbon cycle impact assessment. Then, the proposed method is expected to be able to contribute to optimal design of carbon offset networks in terms of not only carbon cycle dynamics, but also carbon market economics, for process systems to achieve net zero emissions within a given period of time.

Acknowledgements

This work was supported by National Research Foundation of Korea (NRF) funded by Ministry of Science and ICT (MSIT) (Grant No. 2022R1F1A1062901).

References

C. Blaufelder, C. Levy, P. Mannion, and D. Pinner, 2021, A blueprint for scaling voluntary carbon markets to meet the climate challenge, https://www.mckinsey.com/capabilities/sustainability/our-insights/a-blueprint-for-scaling-voluntary-carbon-markets-to-meet-the-climate-challenge

R. Chatelane, 2022, Companies set sights on removing troublesome CO_2 from air, https://www.ny1.com/nyc/all-boroughs/news/2022/03/30/companies-set-sights-on-removing-troublesome-co2-from-air

S. H. Choi and V. I. Manousiouthakis, 2020, On the carbon cycle impact of combustion of harvested plant biomass vs. fossil carbon resources, Computers and Chem. Eng., 140, 106942. https://doi.org/10.1016/j.compchemeng.2020.106942

S. H. Choi and V. I. Manousiouthakis, 2022, Modeling the carbon cycle dynamics and the greenhouse effect, IFAC-PapersOnLine, 55, 7, 424-428. https://doi.org/10.1016/j.ifacol.2022.07.480

L. Day, 2021, Carbon sequestration as a service doesn't quite add up, https://hackaday.com/2021/10/15/carbon-sequestration-as-a-service-doesnt-quite-add-up/

EIC, 2020, 4 types of carbon offset projects, https://www.eic.co.uk/4-types-of-carbon-offset-projects/

P. Gabrielli, M. Gazzani, and M. Mazzotti, 2020, The role of carbon capture and utilization, carbon capture and storage, and biomass to enable a net-zero-CO_2 emissions chemical industry, Ind. Eng. Chem. Res., 59, 7033-7045. https://doi.org/10.1021/acs.iecr.9b06579

A. A. Ghussain, 2020, The biggest problem with carbon offsetting is that it doesn't really work, https://www.greenpeace.org.uk/news/the-biggest-problem-with-carbon-offsetting-is-that-it-doesnt-really-work/

M. J. Kotchen, 2009, Offsetting green guilt, Stanford Social Innovation Review, 7, 2, 26-31. https://doi.org/10.48558/515j-d290

V. I. Manousiouthakis and S. H. Choi, 2021, A carbon cycle optimization method for fossil and biomass energy utilization, Korean J. Chem. Eng., 38, 10, 2003-2008. https://doi.org/10.1007/s11814-021-0899-9

S. Opanda, 2022, Carbon credit pricing chart: updated 2022, https://8billiontrees.com/carbon-offsets-credits/new-buyers-market-guide/carbon-credit-pricing/

J. Photopoulos, 2017, Counting the carbon cost of forest destruction, https://ec.europa.eu/research-and-innovation/en/horizon-magazine/counting-carbon-cost-forest-destruction

Antonis Kokossis, Michael C. Georgiadis, Efstratios N. Pistikopoulos (Eds.)
PROCEEDINGS OF THE 33rd European Symposium on Computer Aided Process Engineering (ESCAPE33), June 18-21, 2023, Athens, Greece
 http://dx.doi.org/10.1016/B978-0-443-15274-0.50174-8

Identifying first-principles models for bubble column aeration using machine learning

Peter Jul-Rasmussen[a], Arijit Chakraborty[b], Venkat Venkatasubramanian[b], Xiaodong Liang[a], Jakob Kjøbsted Huusom[a]

[a]*Department of Chemical and Biochemical Engineering, Technical University of Denmark, Søltofts Plads 228A, 2800 Kgs. Lyngby, Denmark*
[b]*Complex Resilient Intelligent Systems Laboratory, Department of Chemical Engineering, Columbia University, New York, NY 10027, USA*

Abstract

Mass transfer of oxygen is investigated in this work using a pilot-scale bubble column unit with a two-fluid nozzle for aeration. First-principles models for the bubble column unit are identified by utilizing concepts in artificial intelligence (AI) and machine learning (ML), and applying the same to experimental data. By combining process knowledge with data-driven modeling, we discovered interpretable models for oxygen transport phenomena in bubble columns. By virtue of obtaining symbolic models, it is possible to perform post-hoc analyses on the same in order to gain physical insights into the mechanisms occurring in the system -- a convenience lost when using black-box models such as neural networks. This provides valuable understanding which can be applied when modeling more complex systems such as fermentation processes.

Keywords: Artificial intelligence; Machine Learning; Hybrid AI; Interpretable models; Mechanistic modeling

1. Introduction

With the advent of improved computing power, and the ability to generate and handle copius amounts of data, artificial intelligence (AI) has yielded successes in the domains of computer vision, game-playing, recommender systems, among other applications. It must be noted however, that such domains do not have guiding mathematical equations that determine their existence. In stark contrast, physicochemical phenomena, are governed by constitutive equations and conservation laws, among other functional relationships (Venkatasubramanian 2019). Further, knowledge about these phenomena is known in many cases, and can be leveraged to better describe the same. Consequently, when attempting to derive mathematical models of such systems using data-driven tools, it is imperative that one utilize any prior knowledge available about the system, and incorporate the same into the conventional data-driven modeling framework (Chakraborty et al. 2022). This hybrid approach results in simpler, more interpretable models, and can yield models that can provide mechanistic insights about the system.

The majority of industrial fermentation processes are aerobic i.e. they require oxygen to be supplied for the microbial metabolic processes to take place. Oxygen is supplied through aeration, in which air is bubbled through the fermentation broth, facilitating a mass transfer of oxygen from the bulk gas phase through the gas-liquid interface to the bulk liquid phase (Doran 2013). Typically modeling of the oxygen transfer in fermentation processes is based on the difference between the dissolved oxygen concentration at saturation and the dissolved oxygen concentration in the fermentation broth as driving force, an overall liquid-phase mass transfer coefficient, and the surface area of the gas-liquid interface. Experimentally it is difficult to separate the overall liquid-phase mass transfer coefficient from the surface area of the gas-liquid interface, therefore these two terms are generally combined to a volumetric mass transfer coefficient, which can be determined through methods such as the sulphite oxidation technique and gassing-out techniques (Stanbury et al. 1995). Prediction of the volumetric mass transfer coefficient has typically relied on emperical correlations, while theoretical correlations have been introduced using Higbie's penetration theory and Kolmogoroff's theory of isotropic turbulence (Kawase et al. 1992).

In this work, we attempt to obtain an interpretable model of a bubble column aeration system. The data used for modeling has been experimentally generated under 6 different conditions. We propose a hierarchical modeling approach by broadly classifying the same into 2 operating regimes - with, and without addition of salt (here, $MgSO_4$). This enables us to obtain high fidelity models accounting for the nonidealities that are manifested as a result of (or lack thereof) salt addition. In the remainder of the manuscript, we describe the bubble column aeration system in Section 2, and briefly mention the model discovery engine used to obtain the models in Section 3. We then discuss the results when compared to purely first-principles models, in Section 4. Finally, we conclude our findings and, highlight potential future directions to investigate.

2. Problem Description

The experimental set-up consists of a main column, a recycle loop for circulating liquid in the system, a compressed air supply, and a H_2O_2 solution storage tank (Figure 1). Aeration in the bubble column is introduced using a two-fluid nozzle, in which air and recycled liquid are premixed before entering the column. The recycle flow in the system is enforced using a centrifugal pump, and a H_2O_2 solution is fed to the system either in the bottom of the column, in the top of the column, or through the aeration nozzle using a variable speed dosing pump. Online sensor measurements are accessible through a Supervisory Control And Data Acquisition (SCADA) system and stored in a central data base. Online sensor measurements in the system includes: flow sensors for liquid recirculation (FI_1) and compressed air (FI_2), dissolved oxygen sensors at bottom (AI_1) and top (AI_2) of the column, temperature sensor (TI_1), conductivity sensor (CT_1), level sensor (LI_1), and weight measurements of the H_2O_2 solution storage tank (WT_1). The system is used to investigate oxygen mass transfer and the influence of air flow, liquid recycle flow, and salts on the mass transfer. The hydrogenperoxide-catalase method is employed in the system, in which H_2O_2 is decomposed to water and oxygen in the presence of an enzyme (Eq. 1).

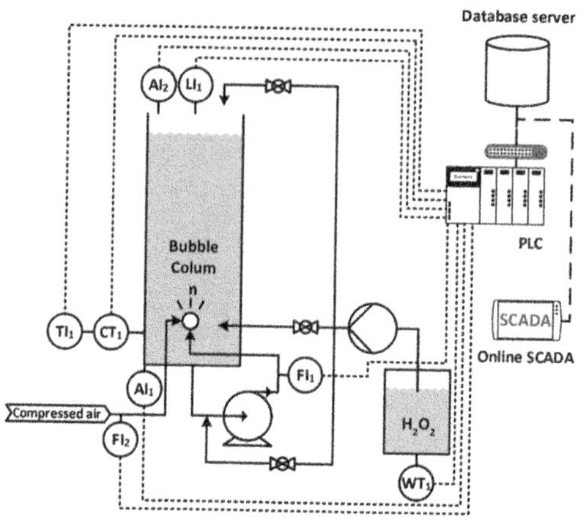

Figure 1 Experimental bubble column set-up with live measurements accessible through SCADA system.

$$H_2O_2 \xrightarrow{catalase} H_2O + \frac{1}{2}O_2 \quad (1)$$

The oxygen liberated in reaction (Eq. 1) will be dissolved in the liquid phase, effectively making the solution supersaturated. The oxygen transfer is thus going to be from the liquid phase to the gas phase. A model for the oxygen transfer in the bubble column is given in (Eq. 2) assuming that the conversion through reaction (Eq. 1) is instantaneous, irreversible, and that the column is ideally mixed.

$$\frac{dO}{dt} = \frac{r}{2} - K_L a(O - O^*) \tag{2}$$

In which O is the dissolved oxygen concentration, O^* is the dissolved oxygen concentration at saturation, K_L is the overall liquid-phase mass transfer coefficient, and a is the surface area of the gas/liquid interface. $r = c_{H_2O_2}m/V$ is the rate of added H_2O_2 per column volume, $c_{H_2O_2}$ is the concentration of H_2O_2, m is the volumetric flowrate of H_2O_2, and V is the volume of the column. Integrating (Eq. 2) from t_0 to t using $O(t_0) = O_0$ yields the analytical solution (Eq. 3).

$$O = O^* + \frac{c_{H_2O_2}}{2K_L aV}m - \frac{c_{H_2O_2}}{2K_L aV}me^{-K_L a(t-t_0)} - O^* e^{-K_L a(t-t_0)} + O_0 e^{-K_L a(t-t_0)} \tag{3}$$

In the system, the mass transfer coefficient and the surface area of the gas/liquid interface is a function of the liquid recirculation flowrate (F_1), the air flowrate (F_2), and the presence of salts in the liquid. A total of six experimental runs were performed for the system using the experimental conditions given in Table (1).

Table 1 Experimental conditions used for the bubble column system

Experiment	$F_1(l/min)$	$F_2(l/min)$	Salt
1	0	300	None
2	0	600	None
3	100	300	None
4	100	600	None
5	0	300	$MgSO_4$
6	100	300	$MgSO_4$

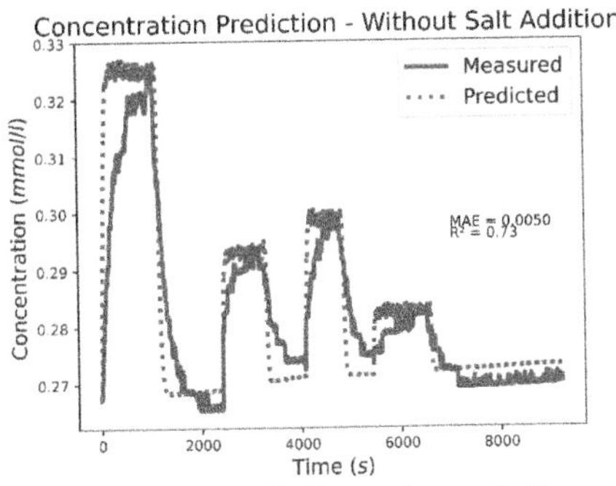

Figure 2 Model predictions using analytic model and measured oxygen concentrations for experiments in which no $MgSO_4$ was added.

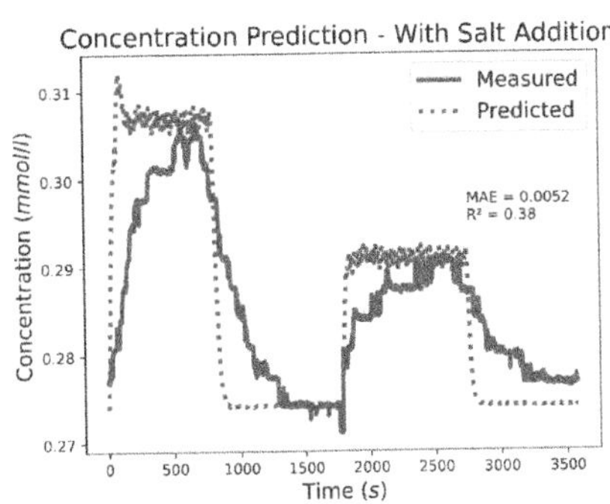

Figure 3 Model predictions using analytic model and measured oxygen concentrations for experiments in which $MgSO_4$ was present in the system.

By performing a post-hoc analysis of the data from the six experiments, steady-state $K_L a$ values can be determined and introduced as a constant in (Eq. 3) to model the dissolved oxygen concentration for each of the six experiments.
Predictions of the experiments using the analytic model reveals that the full system dynamics are not captured in the deterministic model (Figure 2 and 3). This suggests that some of the assumptions in the deterministic model should be relaxed to account for the unmodelled phenomena. However, deriving a deterministic expression for the unmodelled phenomena is a challenge, as it includes complicated enzyme kinetics and mixing phenomena dependent on the configuration and geometry of the bubble column. Due to the complicated nature of the unmodelled phenomena in the system, it is considered to be unknown in this work.

3. Methods

The data obtained from the pilot-plant is used to discover interpretable models that bear some resemblance to the analytical solution, whilst also being able to account for the nonidealities in the system. A genetic algorithm-based approach - titled *AI-DARWIN* (Chakraborty et al. 2020, 2021) - was used for the same. This approach takes a list of permitted tractable function transformations for each of the inputs, and extends the concept of genetic algorithm (Koza 1995) to mathematical models, in an attempt to explore the possible search space of different possible models. The advantage of such an approach is that it searches through the space of potential models with only the user-permitted function transformations. This allows one to curb the complexity of the models obtained, and allows for inclusion of prior knowledge into the modeling exercise.
The bubble column is a complex dynamical system with a multitude of disparate operating regimes. Consequently, a hierarchical modeling approach was used due to the low variation in the experimental conditions. Two different models were determined: one for the system without $MgSO_4$, and one for the system containing $MgSO_4$.

4. Results

The models discovered using *AI-DARWIN* (Table 2 and 3) contain several features that are either a direct analogue, or are similar to features used in the analytic solution (Eq. 3).

Table 2 Models predicting oxygen concentration discovered using *AI-DARWIN* with standardized features

$MgSO_4$	*AI-DARWIN* output	Fitness (MAE and R^2)
No	$O = \beta_0 + (\beta_1(-t)^2 + \beta_2 F_1 + \beta_3 F_2)m + \beta_4 m e^{-t} + \beta_5 (O^*)^2 e^{-t} + \beta_6(-t)O^* + \beta_7(-t)(O^*)^2 + \beta_8 F_2 O^* + \beta_9 F_2^2 (O^*)^2 + \beta_{10} e^{-2t} + \beta_{11} e^{-t}$	0.0016 ($R^2 = 0.97$)
Yes	$O = \beta_0 + (\beta_1 m (O^*)^2 + \beta_2 m)m + \left(\frac{\beta_3(-t)}{O^*} + \beta_4(-t)^2 + \frac{\beta_5(-t)^3}{O^*}\right)O^* + \beta_6 F_1 e^{F_1} + \beta_7 O^* + \beta_8 (O^*)^2 F_2^2 + \beta_9 (O^*)^2(-t)^2 + \beta_{10} e^{-t} + \beta_{11} e^{m}$	0.0011 ($R^2 = 0.98$)

Terms containing the features m and e^{-t} are found in all models, while the Taylor series expansion $(-t + (-t)^2 + (-t)^3)$ is found as a feature in the model for experiments with $MgSO_4$ discovered using *AI-DARWIN*. The intercept (β_0) is approximately the same as the initial value of O^* for both the discovered models. Additional terms correlating the features F_1 and F_2 to the output O appear in the discovered models, a correlation handled in the analytic solution by the K_La values determined from post-hoc analysis. The extra terms found in the discovered models are accounting for the unknown phenomena in the bubble column.

Using the discovered models for predicting the concentration of dissolved oxygen in the bubble column it is found that the dynamic behavior in the system is significantly better represented compared to the analytic solution, as indicated by the improved fitness (Figure 4 and 5). The improved performance of the discovered models indicates that the unknown phenomena is successfully accounted for in the six different experiments in the bubble column system.

Table 2 Parameters for models predicting oxygen concentration discovered using *AI-DARWIN* with standardized features

$MgSO_4$	No	Yes
β_0	0.26	0.29
β_1	-0.014	-0.000
β_2	-0.002	-0.004
β_3	-0.003	0.013
β_4	0.005	0.004
β_5	-0.015	-0.003
β_6	-0.004	-0.000
β_7	0.021	-0.009
β_8	-0.003	0.000
β_9	0.016	-0.001
β_{10}	-0.003	-0.003
β_{11}	0.017	0.003

Table 3 Features in deterministic model and selected features from the models discovered using *AI-DARWIN*

Model	Feature				
Deterministic	O^*	m	$me^{-(t-t_0)}$	$O^*e^{-(t-t_0)}$	$O_0e^{-(t-t_0)}$
AI-DARWIN w/o. $MgSO_4$	β_0	m	me^{-t}	$(O^*)^2e^{-t}$	-
AI-DARWIN w. $MgSO_4$	β_0	m	-	$O^*(-t+(-t)^2+(-t)^3)$	-

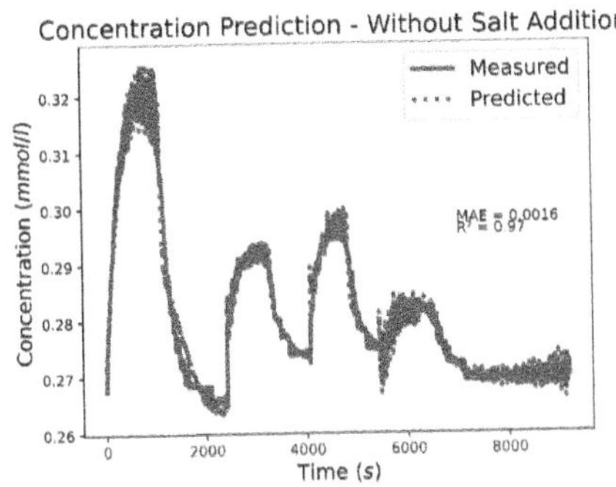

Figure 4 Predicted and measured oxygen concentrations for experiments 1, 2, 3, and 4 in which no $MgSO_4$ was added.

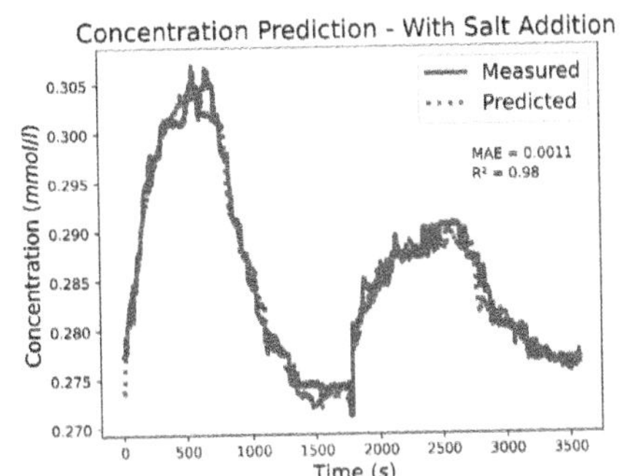

Figure 5 Predicted and measured oxygen concentrations for experiments 5 and 6 in which $MgSO_4$ was present in the system.

An interesting observation is the inclusion of certain terms which exist in the analytical solution, in the models predicted by *AI-DARWIN*. While some complex nonlinear parameterized features are represented through a Taylor Series expansion, other novel terms are included to account for nonidealities in the system. We encounter the interaction effects of the different input variables which yield more complex function transformations, but result in better predictability. While it would be the desired goal to be able to decouple the effects of the ideal conditions as expressed by the analytical solution, and the nonidealities which are accounted for by virtue of the additional terms, it remains an ongoing challenge in the interpretability of ML models in the sciences and engineering.

5. Conclusion

A step forward in the direction of creating a digital-twin for a complex bubble column system has been made, by virtue of an interpretable modeling approach. Such a technique relied on the use of *a priori* knowledge of the system's dynamics, and was able to account for the analytical solution obtained from first-principles. Such a data-driven approach combined with first-principles permits interpretability, which is not possible in purely black-box models (Chakraborty et al. 2022). Due to the high cost of performing experiments, varied set of conditions for some input variables (F_1 and F_2) have not been considered in this work (at most 2 conditions were considered). Future steps include sampling of the data at different operating conditions using appropriate experimental design techniques. Incorporating this into the overall modeling framework can help us obtain more reliable and robust models. Certain key underlying assumptions of the first-principles-based modeling approach include the instantaneity of Reaction 1 and the bubble column being well-mixed -- both of which do not seem to hold upon performing the experiments. As a result, the $K_L a$ values have been assumed to be constant at given conditions when modeling using first-principles. It would be worthwhile to explore the possible function form of the mass transfer coefficient under different operating regimes.

Acknowledgements

This study was financially supported by the Sino-Danish Center for Education and Research (SDC).

References

A. Chakraborty, S. Serneels, H. Claussen, V. Venkatasubramanian, 2022. Hybrid ai models in chemical engineering–a purpose-driven perspective. Computer Aided Chemical Engineering 51, 1507–1512.

A. Chakraborty, A. Sivaram, L. Samavedham, V. Venkatasubramanian, 2020. Mechanism discovery and model identification using genetic feature extraction and statistical testing. Computers & Chemical Engineering 140, 106900.

A. Chakraborty, A. Sivaram, V. Venkatasubramanian, 2021. Ai-darwin: A first principles-based model discovery engine using machine learning. Computers & Chemical Engineering 154, 107470

P. M. Doran, 2013, Chapter 10 - mass transfer, Bioprocess Engineering Principles (Second Edition), Academic Press, London, 379–444.

Y. Kawase, B. Halard, M. Mooyoung,1992. Liquid-phase mass transfer coefficients in bioreactors. Biotechnology and Bioengineering 39 (11), 1133–1140.

J. R. Koza, 1995. Survey of genetic algorithms and genetic programming. In: Wescon conference record. Western Periodicals Company, pp. 589–594.

P. F. Stanbury, A. Whitaker, S. J. Hall, 1995, Chapter 9 - aeration and agitation, Principles of Fermentation Technology (Second Edition), Pergamon, Amsterdam, 243–275.

V. Venkatasubramanian, 2019. The promise of artificial intelligence in chemical engineering: Is it here, finally? AIChE Journal 65 (2), 466–478.

Antonis Kokossis, Michael C. Georgiadis, Efstratios N. Pistikopoulos (Eds.)
PROCEEDINGS OF THE 33rd European Symposium on Computer Aided Process Engineering (ESCAPE33), June 18-21, 2023, Athens, Greece
 http://dx.doi.org/10.1016/B978-0-443-15274-0.50175-X

Optimization framework based on a sensitivity analysis for the identification of the critical design variables

Maria de los A. Villarreal-de-Aquino,[a] Jaime D. Ponce-Rocha,[b] Eduardo S. Perez-Cisneros,[c] Verónica Rodríguez-López,[d] Edgar I. Murillo-Andrade,[c] Divanery Rodriguez-Gomez,[e] Ricardo Morales-Rodriguez[a*]

[a]*Departamento de Ingeniería Química, Universidad de Guanajuato, Noria Alta S/N, Noria Alta, Guanajuato, Guanajuato, 36050, México. *ricardo.morales@ugto.mx*
[b]*Dirección de Ingeniería y construcción de plantas, CIATEQ, A.C. Centro de Tecnología Avanzada, Av. del Retablo 150, Constituyentes Fovissste, Querétaro, Querétaro, 76150, México*
[c]*Departamento de Ingeniería de Procesos e Hidráulica, Universidad Autónoma Metropolitana – Iztapalapa, San Rafael Atlixco No. 186, Col. Vicentina, Iztapalapa, Ciudad de México, 09340, México*
[d]*Facultad de Farmacia, Universidad Autónoma del Estado de Morelos, Av. Universidad 1001 Chamilpa, Cuernavaca, Morelos, 62209, México*
[e]*Coordinación de Ingeniería Bioquímica, Instituto Tecnológico Superior de Irapuato, Carretera Irapuato-Silao Km 12.5, Irapuato, Guanajuato 36821, Guanajuato, México.*

Abstract

In this work, a systematic framework was developed and implemented to optimize large scale processes, through a particular case study. In this process, 32 operating and design variables were found, which can be subject to modification using an optimization process. The identification of high impact variables was carried out through a sensitivity analysis using the method of standardized regression coefficients, together with the Latin hypercube method. Once these variables were identified, the optimization of the process was carried out for the maximization of profits together with the minimization of the operating costs using only the most important identified variables. The solution of the process model was done using Aspen Plus and the optimization was performed using the technique of genetic algorithms, which is available in MATLAB. The results illustrated that was possible to increase the profit by 5.02 % and decrease the energy cost 4.6 %. The framework allowed to reduce the computational time 34.2 % compared when all the manipulating variables were used in the optimization task.

Keywords: Latin hypercube method, sensitivity analysis, phytosterols production.

1. 1. Introduction

The development of new products and their manufacturing processes have demanded innovative manners to solve those challenges (Pistikopoulos et al., 2021). Process optimization has been employed as a paramount strategy to improve the performance in the development of new products and processes, through the introduction of new optimization strategies, advanced methods, combinations of computer-aided tools, etc.

Process optimization most of the time has been categorized as time consuming task, especially when metaheuristic methods are employed, such as genetic algorithms. The implementation of optimization in large and new processes could rise the complexity in the solution, since the number of variables to optimize could be large increasing the time to obtain the optimal solutions (Katok et al., 2021). In a previous work, Ponce-Rocha et al. (2022) proposed a rigorous process design optimization under uncertainty combining Aspen Plus-Matlab, using genetic algorithms (GA) obtaining remarkable results, and showing the functionality of the combination of the computer-aided tools. GA method uses a set of initial solutions (initial populations), randomly selected from the feasible solution space. The fitness function is evaluated for each solution, and the solutions are consequently ranked. Then, the population evolves through several operations, such as reproduction, crossover, and mutation to optimize the fitness function and obtain the final optimal solution. The process is repeated until some criteria are satisfied. This evolutionary algorithm is preferred to classical optimization approaches because it can handle the nonlinear, nonconvex, and nonsmooth optimization problem of the component sizing for the hybrid system (Katok et al., 2021); the nonconvexity of the problem makes it difficult for classical optimization methods to obtain a global optimum. On the other hand, GA globally searches the domain of possible solutions for an optimal solution. Therefore, the objective of this work is to present a systematic framework to optimize large processes, decreasing optimization time, maximizing profits, and decreasing operating costs, where the first step is the identification of the most significant variables of the process, followed by optimization using GA to find the best values of these variables using the combination of Aspen Plus and Matlab. The employed case study was the production of phytosterols and biodiesel from crude vegetable oils.

2. Methodology

The framework includes three main sections applied to one case study of relevant interest.

2.1. Case study

The production of phytosterols (ß-sitosterol and stigmasterol), glycerol and clean biodiesel from crude vegetable oils was taken as a case study. The integrated process consists of two main sections as illustrated in Figure 1: A) Phytosterols production: the separation of triglycerides from phytosterols, phyto-glucoside compounds and fatty acids (FFA) is performed employing supercritical extraction using CO_2 and a mixture of CO_2 – CH_3OH as supercritical fluids and vacuum distillation processes. B) Biodiesel production: this is performed using a reactive distillation column to carry out the transesterification reactions of triglycerides to produce clean biodiesel and glycerol, followed by two distillation columns. It is possible to identify diverse important designs and operating variables in the process.

2.2. Generation of the process diagram in Aspen Plus and the connection with MATLAB.

Firstly, the process flow diagram is generated using a process simulator, in this case Aspen Plus can be used as simulation environment. Then, the connection between MATLAB® and Aspen plus® can be made using COM technology, which allows making the bridge between these two computer-aided tools. For the case study mentioned above, 32 variables were identified for case A-B and for case A only 13 variables were distinguished.

2.3. Sensitivity analysis.

As second step, a sensitivity analysis is performed where the Latin hypercube method can be used to obtain different possible scenarios and analysis the functionality of certain variables using a Monte Carlo approach. One advantage of the Latin hypercube method is that guarantees that each of these components is represented in a completely stratified way (*i.e.*, the population is separated into exclusive segments). For sensitivity analysis, a random sampling of 1,000 samples were generated, which are subject to a 5 % variation for upper and lower limits. Then, the identification of the most significant variables of the system was performed using the standardized regression coefficient method.

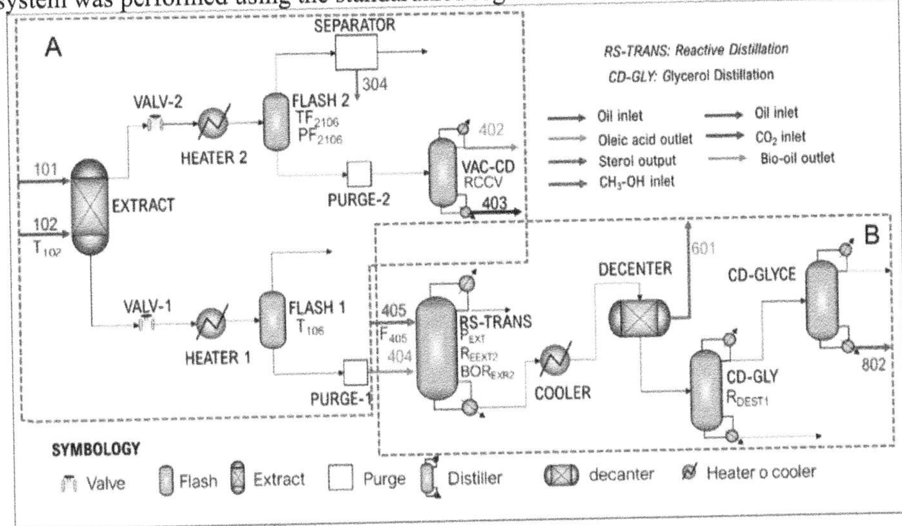

Figure 1. Flowsheet for the production process of phytosterols and biodiesel from palm oil.

2.4. Optimization using multi-objective genetic algorithms.

The identified variables in the sensitivity analysis were used as decisions variables for the optimization step to improve the performance of the process. The optimization was carried out using the Genetic Algorithm Toolbox included in MATLAB. In a previous work Ponce-Rocha et al. (2022) found that the best working condition for genetic algorithms in MATLAB together with Aspen Plus was having a population of 100 and generations of 50.

3. Results

The framework was evaluated using the production of phytosterols and the process to produce phytosterols and biodiesel.

3.1. Case study A: phytosterols production

The identifiability analysis largely depends on the interpretation of the sensitivity function of parameters and variables with respect to model outputs, usually corresponding to the measured variables such as operating costs or gross profit. The sensitivity analysis allowed to identify 6 variables with the highest impact in the operating cost, gross profit and net earnings. Table 1 illustrates the identified variables. For this case the confidence factor was 0.95-0.99 %. Figure 2 shows the significant variables for the operating cost, gross profit and the net earnings defined as the difference between the gross profit and operating costs. The results show that 3 model variables were especially important for the 3 output variables, 1 variable is significant for 2 output variables and 1 variable is important for the operating cost; the identified operating and design variables were employed in the optimization step.

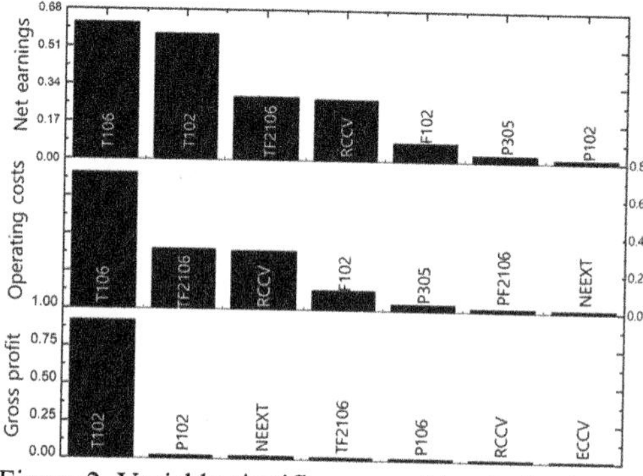

Figure 2. Variable significance ranking (only the first 7 highest ranking variables are shown) for case A model outputs based on the sensitivity measure of gross profit, operating costs and, net earnings.

Table 1. Results of Sensitivity analysis for case study A

Variable	Gross profit, %	Operating costs, %	Net earnings, %
F_{102}	0.93	**11.6**	8.4
T_{102}	**91.9**	1.0	**56.9**
T_{106}	0.17	**79.8**	**61.9**
TF_{2106}	1.72	**35.3**	**28.5**
RC_{CV}	1.28	**34.5**	**27.6**
P_{102}	**2.8**	0.5	1.6

3.2. Case study A-B: phytosterols and biodiesel production.

Figure 3 shows the identified variables with high impact in the output variables for process A-B. Table 2 illustrates the variables with high impact in the operating costs, gross profit and the net earnings (defined above) and, the percentage of relevance in process A-B.

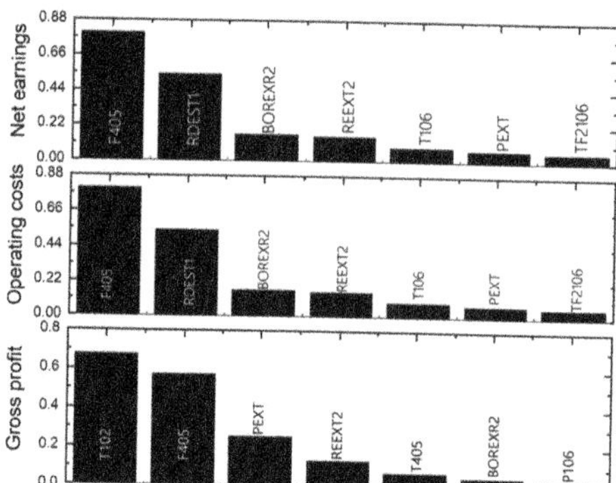

Figure 3. Variable significance ranking (only the first 7 highest ranking variables are shown) for case A-B model outputs based on the sensitivity measure of gross profit, operating costs and, net earnings.

Table 2. Results of sensitivity analysis for case study A-B

Variable	Gross profit, %	Operating cost, %	Net earnings, %
T_{102}	**67.3**	0.14	0.41
F_{405}	**57.2**	**79.9**	**79.8**
R_{EXT2}	**13.1**	**15.1**	**15.1**
BOR_{EXR2}	4.4	**16.2**	**16.3**
P_{EXT}	**25.2**	7.0	7.1
R_{Dest1}	0.98	**53.9**	**53.9**

3.3. Results of the optimization for case study A

Once sensitivity measures have identified the significant variables in the process, the following step is to find out the optimal operating conditions with the aim of reducing the operating costs and increasing the net profit of the system. The variables related and identified with the gross profit were the temperature and pressure in the stream 102 (T_{102}, P_{102}, respectively); for the operating costs, the flowrate for stream 102 (F_{102}), temperature of the flash unit 1 (T_{106}), temperature for flash 2 (TF_{2106}), reflux ratio (RC_{CV}) were related and identified; a finally for the net earnings, the variables identified for the operating costs were also selected adding the temperature of the stream 102 (T_{102}). The purple diamond (right and lower side of the plot) illustrates the value of the objective function for the base case scenario (without optimization). The light blue triangles show the values of the objective function when the most important variables identified with the operating costs in the sensitivity analysis step, (see red and bold values in Table 1) were used as manipulated variables in the optimization task, the results showed a reduction of operating costs of about 4.5 % with respect to the base case, and a small increase in gross profit (about 3.4 %) as illustrated by the star with the same color. Similar case was observed when the optimization was performed using the significant variables for gross profits (see red and bold values in Table 1), here it was possible to see an increase of 4.3 % gross profit and only 3.0 % of reduction in the operating costs as illustrated by green star. In the case of net earnings, a considerable increase in profits was obtained (4.3 %) and a reduction of 4.6 % in operating costs was reached compared to the base case scenario as it is illustrated by the blue star. In the case of the optimization tasks using all the available manipulated variables (11 variables), it resulted in an increase of 3.7 % in the profit and a reduction of 2.1 % in the operating costs (see red star in Figure 4).

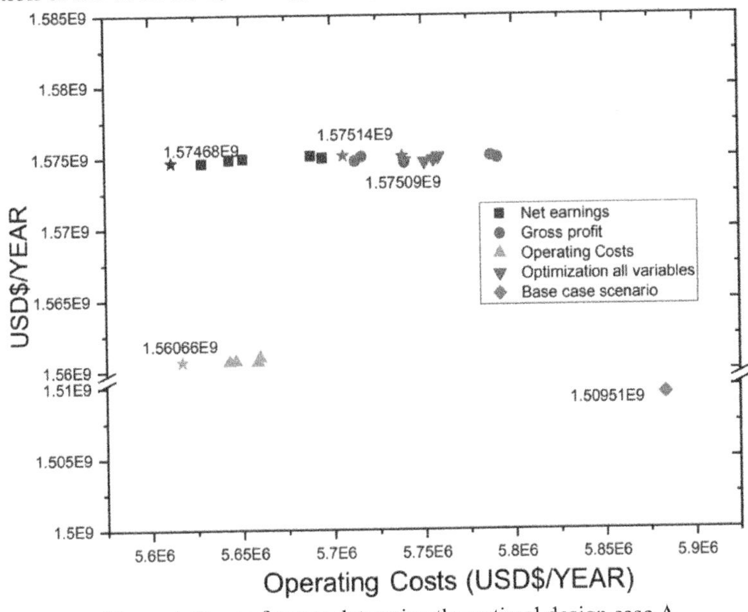

Figure 4. Pareto front to determine the optimal design case A.

3.4. Results of the optimization for case study A-B.

The results of case study A-B are shown in the Pareto front illustrated in Figure 5. The highest improvement was found when the identified variables using the net earnings (see red and bold values in Table 2) were used in the optimization step; the energy cost decreased by 4.4 % and the profit increased 5.02 % compared with base case scenario (purple diamond and blue star, respectively in Figure 5). When the optimization was performed using the 32 manipulated variables of the process A-B, an increase in profit of 3.02 % and a decrease in operating costs of 3.87 % (illustrated with the red star) were found compared with the base case scenario. The optimization results for the variables selected to reduce energy costs (see color red values in Table 2) provided good results because the reduction in operating costs was 4.2 %, but the increase in the profits was only 0.27 % (see light blue star). A similar case was observed when the variables with higher impact for the gross profit (see color red values in Table 2), where it was possible to increase 4.6 % the gross profit and decrease the energy consumption by 0.47 % (see green star).

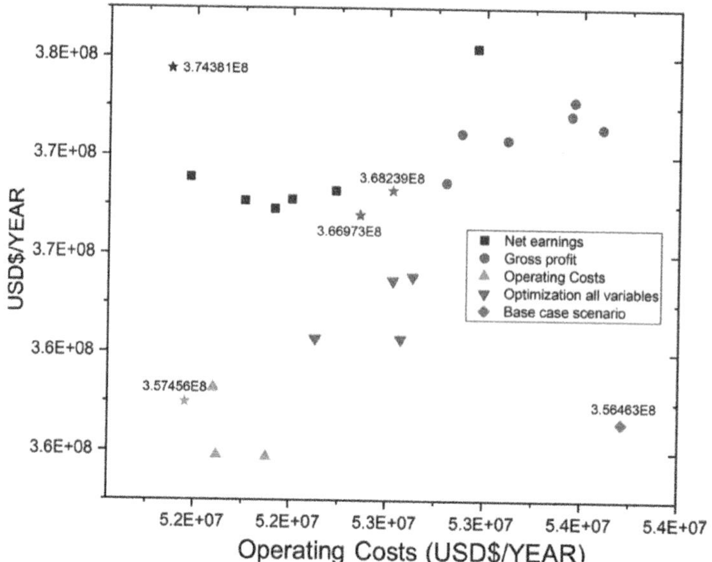

Figure 5. Pareto front to determine the optimal design case A-B.

During the optimization of the 32 variables, the optimization time was 10.56 hours, while the optimization time using only the variable identified in the sensitivity analysis was 6.95 hours, that is, the computational time was reduced 34.2 %, in addition with better results.

4. Conclusion

A framework for a more efficient optimization task was developed. The sensitivity analysis revealed that based on the objectives (operating costs, gross profit and net earnings) only a few variables had an impact in the magnitude of the sensitivity measure. It was possible to observe that the identified variables were the most important in the optimization problem, allowing to reduce computing time and improving the optimal results. The sensitivity analysis generated for the variables provided an accurate picture in the selection of high impact variables with respect to the objective established for each process section and, this can be clearly seen in the optimization results for both case studies. The results also showed that was possible to find a better alternative design and operation of the plant in comparison to the base case scenario.

5. Acknowledgments

M.A. Villarreal-de-Aquino acknowledges the National Council of Science and Technology (CONACyT) for the financial support on the development of this project.

6. References

S. Katoch, S. Chauhan, V. Kumar, 2021, A review on genetic algorithm: past, present, and future, Multimedia Tools and Applications, 80, 5, 8091-8126.

E. N. Pistikopoulos, A. Barbosa-Povoa, J.H. Lee, R. Misener, A. Mitsos, G.V. Reklaitis, V Venkatasubramanian, F. You, R. Gani, 2021, Process systems engineering – The generation next?, Computers & Chemical Engineering, 147, 107252.

J.D. Ponce-Rocha, M. Picón-Núñez, R. Morales-Rodriguez, 2022, A framework for optimal and flexible schemes design under uncertainty & sustainable aspects, Computer Aided Chemical Engineering, 51, 757-762

Antonis Kokossis, Michael C. Georgiadis, Efstratios N. Pistikopoulos (Eds.)
PROCEEDINGS OF THE 33rd European Symposium on Computer Aided Process Engineering (ESCAPE33), June 18-21, 2023, Athens, Greece
 http://dx.doi.org/10.1016/B978-0-443-15274-0.50176-1

Effect of coalescence models on the prediction of the separation of dispersed oil-water pipe flows

Nikola Evripidou,[a] Federico Galvanin,[a] Panagiota Angeli,[a]

[a] *Department of Chemical Engineering, University College London, London WC1E 7JE, UK*

Abstract

The effect of coalescence models on the prediction of the separation of dispersed oil-water pipe flows was assessed using a one-dimensional mechanistic model. The mechanistic model predicts the formation and evolution of four characteristic layers along the pipe: a pure water layer at the bottom, a flotation/sedimentation layer, a dense-packed zone, and a pure oil layer on the top. It was shown that the film drainage coalescence model by Jeelani and Hartland (1994) that considers interfacial mobility produces good predictions at low mixture velocity, but it depends on the flowrate. The asymmetric film drainage coalescence model by Henschke et al. (2002) is independent of the mixture velocity and the dispersed-phase fraction, and produces reasonable predictions. There was small deviation between the model outputs in the presence of the four characteristic layers, but further investigation of the regions where a single dense-packed layer persists is required.

Keywords: coalescence; liquid-liquid; dispersion; separation; modelling

1. Introduction

Liquid-liquid pipe flows are common in the petroleum industry. Deepwater and marginal fields with lower volumes of reserve entail new development challenges, while heavy oils and mature wells require water flooding to enhance production. Increased volumes of water in the pipelines affect the flow, as they can alter the spatial configuration of the two immiscible phases. Considering the high cost associated with oil extraction and separation, as well as the increased demands for further offshore drilling, ensuring a successful and economical flow of oil-water mixtures is essential in optimising transportation and downstream separation of the extracted oil. Additionally, subsea separation facilities are used to reduce the cost and space requirements of remote deepwater operations. The oil-water flows in pipes are often in the dispersed pattern. Models that can predict flow pattern transitions in unstable dispersed pipe flows are essential during both design and operation of industrial facilities.

Henschke et al. (2002) developed a mechanistic model that predicts the evolution of heights of the characteristic layers that develop in separating batch dispersions as well as the average drop size. The model uses coalescence time correlations based on asymmetrical film drainage between the drops. Pereyra et al. (2013) attempted to extend Henschke's model to one-dimensional pipe flows by changing the time scale to a length scale. Evripidou et al. (2019) further modified the model to account for hindered settling of drops in dense dispersions. Another coalescence model was proposed by Jeelani and Hartland (1994) and is based on the interfacial mobility. In this work, we use the mechanistic model as presented in Evripidou et al. (2019). We consider both the *asymmetric film drainage coalescence model* by Henschke et al. (2002) and the *interfacial mobility film drainage coalescence model* by Jeelani and Hartland (1994), to assess their ability to predict the formation and evolution of the characteristic layers in a pipe, for different oil-in-water dispersed flows.

2. Model description

Four characteristic layers may emerge in a separating dispersed pipe flow. In an oil-in-water dispersion, these are a pure water layer at the bottom, a settling layer (SL), a dense-packed layer (DPL), and a pure oil layer at the top. Figure 1 shows schematically these layers. The thickness of each layer depends on the drop settling (flotation/sedimentation) rate and the coalescence rate of drops, with their homophase. The drop size may also change through drop-drop coalescence.

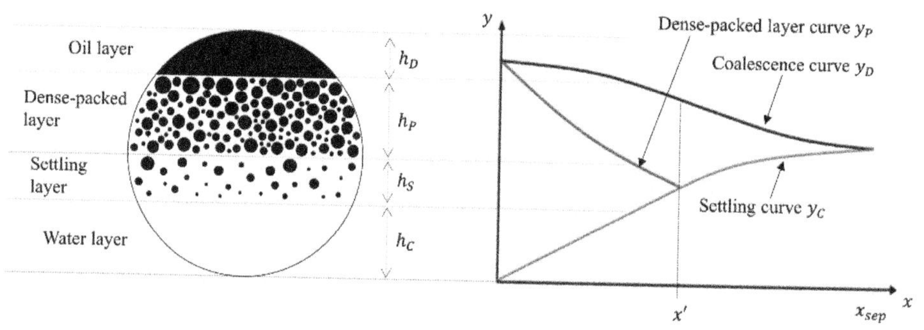

Figure 1: Diagram of the cross-sectional area of the pipe for an oil-in-water dispersion and diagram of the evolution of the characteristic layers.

The mechanistic model is applicable to liquid-liquid pipe flows where the separation is primarily gravity-driven. A constant mixture velocity, u_M, is assumed for both phases – this allows velocity profiles to be ignored, which means that there is no exchange of momentum between layers. For the dispersed layers, monodisperse drop distributions are assumed for simplification, while interfacial tension is considered constant. Lastly, drop break-up and turbulence effects are neglected. In what follows, the model equations associated with coalescence are briefly discussed.

2.1. Coalescence analysis

2.1.1. Drop-interface coalescence

Accumulation of drops near the top of the pipe results in coalescence and the formation of a pure oil layer of thickness h_D. Assuming a monodispersed DPL with drops of diameter $d_{p,I}$,

$$\frac{dh_D}{dx} = \frac{2\varphi_I d_{p,I}}{3\tau_I u_M}, \qquad (1)$$

where τ_I is the drop-interface coalescence time. The oil fraction at the interface φ_I is taken as approximately equal to 1.

2.1.2. Drop-drop coalescence

Drop-drop coalescence is considered only in the DPL. Assuming that in each step, all drops within the DPL are of equal size, $d_{p,I}$,

$$\frac{d(d_{p,I})}{dx} = \frac{d_{P,I}}{6\tau_C u_M}. \qquad (2)$$

where τ_C is the drop-drop coalescence time.

2.1.3. Coalescence time

Two coalescence models are considered. These are outlined in table 1. $d_{p,0}$ is the drop size at the inlet, μ is viscosity, ρ is density, σ is the interfacial tension whereas the subscripts C and D refer to the continuous and dispersed phases respectively; g is the gravitational constant.

Table 1: Coalescence models

Asymmetric film drainage model		*Interfacial mobility film drainage model*	
Drop-interface coalescence time: $\tau_I = \frac{(6\pi)^{\frac{7}{6}}\mu_C r_a^{\frac{7}{3}}}{4\sigma^{\frac{5}{6}} H^{\frac{1}{6}} r_{F,I} r_V^*}$	(3)	Drop-interface coalescence time: $\tau_I = \frac{\tau_{I,0} d_{p,I}}{h_P}$	(9)
Drop-drop coalescence time: $\tau_C = \frac{(6\pi)^{\frac{7}{6}}\mu_C r_a^{\frac{7}{3}}}{4\sigma^{\frac{5}{6}} H^{\frac{1}{6}} r_{F,C} r_V^*}$	(4)	Drop-drop coalescence time: $\tau_{I,0} = \frac{3\pi\mu_C r^4}{4(1+2m)F\delta_r^2}$	(10)
Drop-drop contact radius: $r_{F,C} = 0.3025 d_{p,I}\sqrt{1-\frac{4.7}{La+4.7}}$	(5)	Modified coalescence equation: $\frac{dh_D}{dx} = \frac{2\varphi_I h_P}{3\tau_{I,0} u_M}$	(11)
Drop-interface contact radius: $r_{F,I} = \sqrt{3} r_{F,C}$	(6)	Drop-interface film radius: $r = d_{p,0}^2\sqrt{\frac{\lvert\rho_C-\rho_D\rvert g}{12\sigma}}$	(12)
Channel contour radius: $r_a = 0.5 d_{p,I}\left(1-\sqrt{1-\frac{4.7}{La+4.7}}\right)$	(7)	Force due to gravity: $F = \frac{\pi d_{p,0}^3 \lvert\rho_C-\rho_D\rvert g}{6}$	(13)
Modified Laplace number: $La = \left(\frac{\lvert\rho_C-\rho_D\rvert g}{\sigma}\right)^{0.6} h_P^{\ 0.2} d_{p,I}$	(8)	Critical film thickness: $\delta_r = 0.267\left(\frac{\pi r^4 H^2}{6\sigma F}\right)^{\frac{1}{7}}$	(14)

The left column of table 1 shows the *asymmetric film drainage coalescence model*, which depends on the deformation of the drops. Deformation increases with dense packed layer thickness below the drop considered. Two unknown parameters are present in this model: the Hamaker coefficient, H, and the asymmetry parameter, r_V^*. H is set to 10^{-20} N m as proposed by Henschke et al. (2002) for all systems. r_V^* is system specific and can be obtained experimentally.

The right column of table 1 presents the *interfacial mobility film drainage coalescence model*. This coalescence model does not depend on drop size. Instead, the model allows equation (1) to be simplified into equation (11) through the use of equation (9), making h_D independent of $d_{P,I}$.The fitted parameter m is the interface mobility, i.e. the sum of the mobilities due to induced circulation in the adjacent phases and the interfacial tension gradient and is characteristic of each system. When $m = 0$ the velocity at the interfaces on both sides of the draining film is 0, and the surfaces are deemed immobile; when $m =$ 1.5 the velocity at one of the interfaces is 0 while the velocity gradient at the other surface is 0. Under these conditions, film drainage, and thus the rate of coalescence, is extremely slow. Other values of m are also possible and correspond to different surface velocities and velocity gradients. Values of m larger than 1.5 correspond to more mobile interfaces.

3. Results and discussion

3.1. Experimental methods

The experimental data used to assess the performance of this model were obtained by Voulgaropoulos (2018) in a two-phase liquid-liquid flow facility discussed in detail in Voulgaropoulos et al. (2016). In the experiments tap water and oil (828 kg m^{-3}, 5.5 mPa s) were used as test fluids. The test section comprised of transparent acrylic pipes with an internal diameter of 37 mm and overall length of around 8 m. Partial dispersions of oil in water were generated at the inlet of the test section using a multi-nozzle mixer. High-speed imaging was employed at three locations along the spanwise dimension of the pipe to enable the identification of the flow patterns. A dual-conductance probe was implemented to measure the local volume fractions and the drop size distributions of the dispersions. Measurements were taken every 2 mm, spanning the whole pipe diameter.

3.2. Results and discussion

Three case studies were investigated (c.f. Table 2). For each case, we solved the mechanistic model twice, implementing a different coalescence model each time, using gPROMS ModelBuilder at intervals of 0.1 m pipe length. C_h was taken as 0.01 as suggested by Evripidou et al. (2019) and r_V^* was set to 0.007 as suggested by Pereyra et al. (2013). m was fitted to experimental data obtained at $u_M = 0.52 \text{ m s}^{-1}$ and was found to be 360. The resulting flow profiles were nondimensionalized using the pipe diameter and are presented in Figures 2 and 3.

Table 2: Inlet conditions of the experiments

u_M (m s^{-1})	φ_0
0.52	0.30
	0.45
1.04	0.60

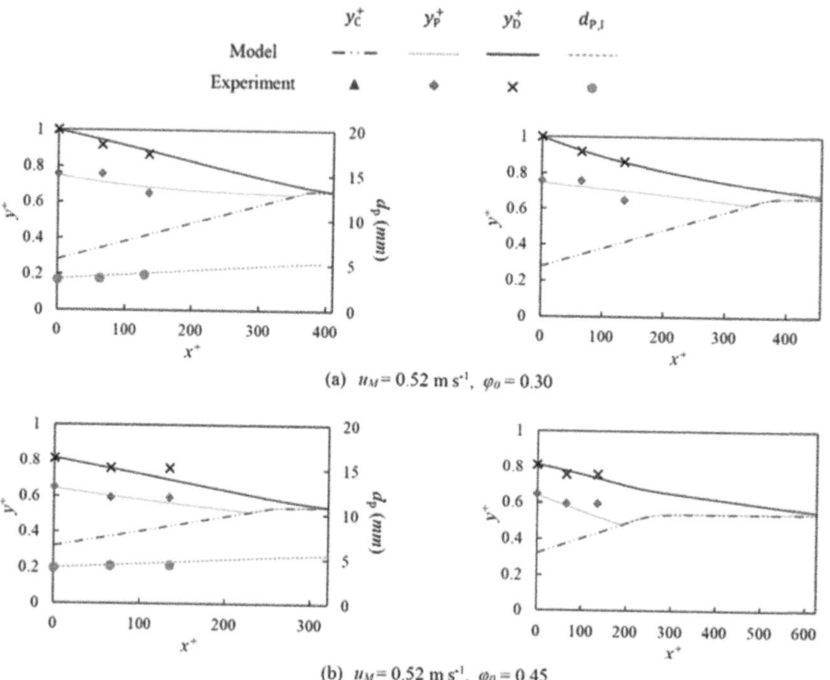

Figure 2: Model predictions of the *asymmetrical film drainage coalescence model* (left) and the *interfacial mobility film drainage coalescence model* (right) for $u_M = 0.52 \text{ m s}^{-1}$.

Figure 2 shows the two case studies at $u_M = 0.52\ \mathrm{m\ s^{-1}}$ and $\varphi = 0.30$ and 0.45. The two mixtures separate in a similar fashion and both coalescence models predict the separation with reasonable accuracy. The rate of drop-settling is large enough to deplete the SL first. On the contrary, the DPL persists throughout the pipe and coalescence controls the rate of separation.

Although both coalescence models show reasonable agreement with experimental data, the *interfacial mobility film drainage model*, where the coalescence rate is a function of the DPL thickness shows a better fit to the experiments. Despite that, the SL is depleted at similar axial lengths for both cases with deviations of 13% or less in x'^{+} between the predictions of the two models. The predictions of the interfacial mobility film drainage model for the total separation length are consistently larger than the predictions of the asymmetric film drainage model. Specifically, for the case with oil fraction of 0.30, the interfacial mobility film drainage model predicts an x^{+}_{sep} of 455, while the prediction of the asymmetric film drainage model is 9% less at 412. This difference is even larger for the case of oil fraction of 0.45. For this case, the asymmetric film drainage model predicts a separation length of 320, while the other coalescence model predicts $x^{+}_{sep} = 624$, a value that is almost twice as large. The above observations suggest that the models behave in a similar manner at the pipe locations where all four characteristic layers are present, and any major deviations between the two arise past the point of depletion of the SL (i.e. at lengths $x > x'$).

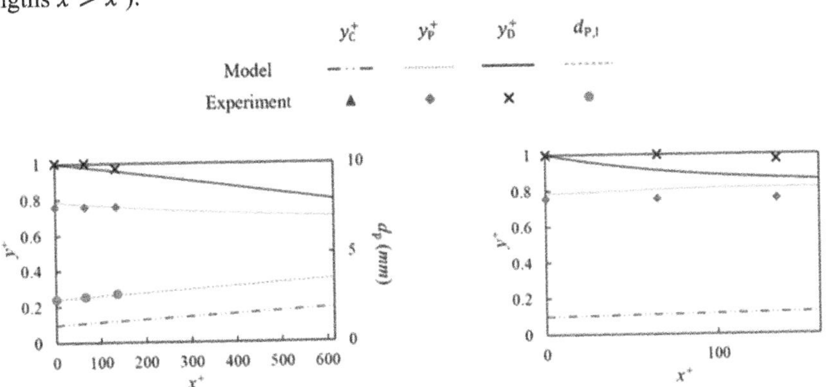

Figure 3: Model predictions of the *asymmetrical film drainage coalescence model* (left) and the *interfacial mobility film drainage coalescence model* (right) for $u_M = 1.04\ \mathrm{m\ s^{-1}}$ and $\varphi_0 = 0.60$.

Figure 3 presents the results obtained at a mixture velocity of 1.04 m s^{-1} and oil fraction of 0.60. The deviations in the predictions of the two models for this case study are significant. The coalescence model by Henschke et al. (2002) produces reasonable results with deviations of 2% or less to experimental measurements. On the other hand, the *interfacial mobility film drainage model* significantly overestimates the coalescence rate of drops with their homophase. As a result, the oil layer acquires a large thickness at the beginning of the pipe. This suggests that the mixture velocity may affect interfacial mobility and that m should be fitted for each mixture velocity.

Nevertheless, both models predict depletion of the DPL, which, according to Evripidou et al. (2022), occurs once the thickness of the DPL becomes smaller than the drop diameter along the interface. At this point, settling becomes the limiting separation mechanism and controls the rate of separation. Predictions of the complete flow profile up to the point of total separation are not possible with the current model, as it is only applicable to regions where a DPL is present and coalescence controls the rate of separation.

4. Conclusions

The paper presents a comparison between two coalescence models that can be used in mechanistic models of separating dispersed pipe flows. At the low mixture velocity, both coalescence models capture the drop-interface coalescence adequately. The *interfacial mobility film drainage model* shows better agreement with experimental data, hence may be preferred over the *asymmetric film drainage coalescence model*. The two models result in similar predictions for the length of depletion of the settling layer (SL), but the *interfacial mobility film drainage model* predicts significantly larger separation lengths x_{sep}. Hence, for the low mixture velocity, we concluded that the models behave similarly in pipe locations where all four layers are present, and any major deviations between the two models occur at pipe lengths greater than x' for which we have no experimental data. Therefore, further experimental studies, especially in the region $x > x'$, are needed to provide the necessary information to differentiate between the two coalescence models.

At the high mixture velocity, the *asymmetric film drainage coalescence model* shows good agreement with the experimental data. The *interfacial mobility film drainage model*, however, significantly overpredicts the rate of drop-interface coalescence. The observations suggests that the interfacial mobility coefficient m varies with u_M, but not with oil fraction φ. To the contrary, r_V^* is specific to the oil-water system but independent to both u_M and φ, hence the *asymmetric film drainage model* may be preferred for cases with variable flowrates. Nevertheless, both coalescence models predict the depletion of the DPL at the high mixture velocity. To account for the depletion of the DPL a different approach must be used such as that described in Evripidou et al. (2022).

Acknowledgements
NE would like to acknowledge Chevron Corporation and UCL for her PhD studentship.

References

N. Evripidou, V. Voulgaropoulos, P. Angeli, 2019, Simplified mechanistic model for the separation of dispersed oil-water horizontal pipe flows, In BHR 19th International Conference on Multiphase Production Technology. OnePetro.

N. Evripidou, C. Avila, P. Angeli, 2022, A mechanistic model for the prediction of flow pattern transitions during separation of liquid-liquid pipe flows, International Journal of Multiphase Flow, p.104172.

M. Henschke, L.H. Schlieper, A. Pfennig, 2002, Determination of a coalescence parameter from batch-settling experiments, Chem Eng J 85, 369-378.

S.A.K. Jeelani, S. Hartland, 1994, Effect of interfacial mobility on thin film drainage, J. Colloid Interface Sci., 164, 296-308.

E. Pereyra, R.S. Mohan, O. Shoham, 2013, A simplified mechanistic model for an oil/water horizontal pipe separator, Oil and Gas Facilities, 2 (03): 40-46.

V. Voulgaropoulos, L. Zhai, C. Ioannou, P. Angeli, 2016. Evolution of unstable liquid-liquid dispersions in horizontal pipes, 10th North American Conference on Multiphase Technology. BHR Group. Banff, 8-10 June.

V. Voulgaropoulos, 2018, Dynamics of spatially evolving dispersed flows, PhD dissertation, University College London, London, UK.

Antonis Kokossis, Michael C. Georgiadis, Efstratios N. Pistikopoulos (Eds.)
PROCEEDINGS OF THE 33rd European Symposium on Computer Aided Process Engineering (ESCAPE33), June 18-21, 2023, Athens, Greece
 http://dx.doi.org/10.1016/B978-0-443-15274-0.50177-3

Learning interpretable multi-output models: Kaizen Programming based symbolic regression for estimating outlet concentrations of a splitter

Jimena Ferreira,[a,b] Martín Pedemonte,[b] Ana I. Torres,[c*]

[a] *Instituto de Ingeniería Química, IIQ, Facultad de Ingeniería, Universidad de la República, Montevideo, 11600, Uruguay*
[b] *Instituto de Computación, INCO, Facultad de Ingeniería, Universidad de la República, Montevideo, 11600, Uruguay*
[c] *Department of Chemical Engineering, Carnegie Mellon University, 5000 Forbes Ave., Doherty Hall, Pittsburgh, PA, 15213, USA*
* *aitorres@cmu.edu*

Abstract

Machine learning techniques for building surrogate models typically generate single-output expressions. However, Chemical Engineering processes are naturally multi-output, with several of the outputs depending on the same physico-chemical phenomena. This work is an extension of our previous work on Symbolic Regression aimed to learn interpretable models from data that reflect the shared physico-chemical phenomena as common bases in the derived expressions. To do so, we consider two multi-output algorithms: one based on multi-output linear squares regression, and the other one on an islands model. As case studies, we use real historical data from a C3/C4 splitter from an oil refinery. The results from the proposed multi-output algorithms are compared to those from the previous single-output algorithm and a multi-output Gaussian Process implementation. We conclude that both multi output strategies are competitive, and in some cases better, than the single-output one although within the same order of magnitude in the error in the validation set. In all cases the multi-output algorithms perform better than Gaussian Process.

Keywords: machine learning, surrogate modeling, symbolic regression, interpretable models, multi-output regression.

1. Introduction

Machine learning techniques for learning models from data are increasingly receiving attention within the Chemical Engineering community. As examples, the work of Eagle and Sahinidis (2022), developed a symbolic regression (SR) algorithm to obtain surrogate models for equations of state; and the work of Chakraborty et al. (2021), considers a machine learning technique to obtain interpretable models based on genetic feature extraction. Our own previous work in Ferreira et al., 2022a, also addresses the problem of learning an explicit interpretable model from data without a priori definition of the functional bases. The techniques used in all these examples are single output, meaning that if one wishes to derive expressions for multiple outputs, the algorithms must be repeatedly executed.

However, for many Chemical Engineering applications it is highly desirable to be able to learn models that result in common (interpretable) terms for those outputs that are known to depend on common phenomena. In Ferreira et al., 2022b, we proposed two multi output

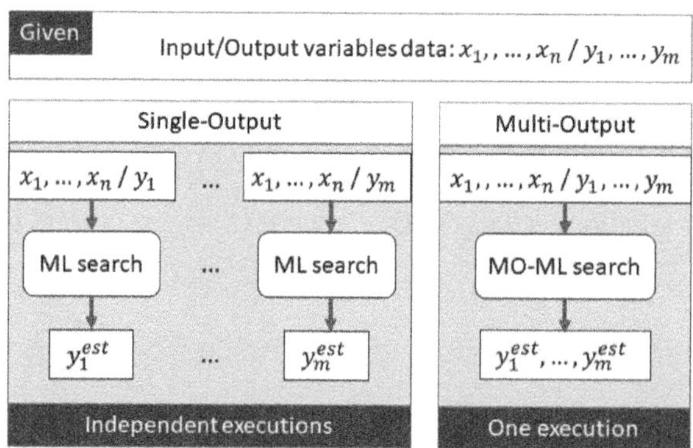

Figure 1 – Scheme of generation of surrogate models for several outputs with the same input data. Left: single-output algorithm executes to learn one model at a time. Right: the multi-output strategy learns all the models in the same run.

algorithms and tested them using benchmark functions or made-up data stemming from reaction engineering problems. In this work, we extend our previous contributions to learn multi-output interpretable models from real data from a distillation column that operates in an oil refinery. Specifically, we consider the data for the C3/C4 splitter in Ferreira et al. (2022a) and apply the two multi-output algorithms. The results from these, are compared with those from the previous single-output algorithm and a multi-output Gaussian Process implementation.

2. Algorithms

Figure 1 shows a scheme of the approaches considered in this work. On the left, we include the strategy used by most ML methods to solve multi-output surrogate modeling. This is, for each input-output dataset (represented by several vectors x, and one of the vectors y), an individual execution is performed for each y_j to learn a model that relates it to the x variables. We named this strategy as the Single-Output (SO) approach. On the right, we consider the multi-output (MO) approaches where models for all y are learned in the same execution. Figure 2 shows the specific algorithms used in this work. In all cases, KP refers to the iterative algorithm that solves SR problems from input-output data. KP searches for models that are linear combinations of nonlinear expressions. These nonlinear expressions are the functional bases f_i^j that are created and modified by operations of Genetic Programming, and the coefficients β_i^j are calculated by Ordinary Least Squares (OLS). The details about KP have been published in Ferreira et al. (2022a). The first multi-output strategy, MO_{MLR}, performs a multi-output linear regression based on OLS in the step of calculation of the coefficients. This change forces term sharing between the y_j models at each iteration. The second multi-output strategy, MO_{IM}, is based on the islands model for parallelization of evolutionary algorithms. MO_{IM} alternates independent search steps with migration steps. In the independent search steps, each output variable y_j is learned as a SO strategy; in the migration step the functional bases of each output variable (that were searched in the independent executions) are shared with the others output variables. Thus, in the next independent search step, each independent search for y_j starts with all the functional bases of the other y. It is important to note that contrary to the other multi-output strategy, MO_{IM} does not force the final expressions of the models to all have the same functional bases but enhances the probability of that happening. More details on the MO algorithms can be found in Ferreira et al. (2022b).

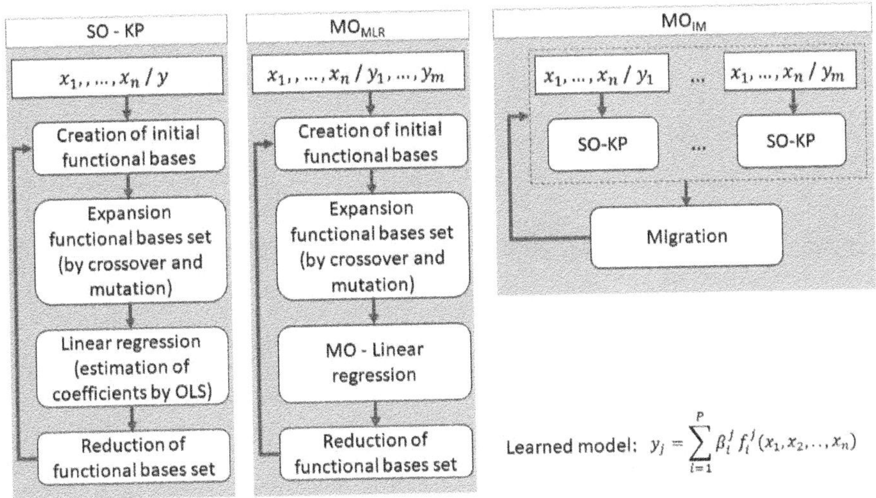

Figure 2 – Scheme of algorithms used in this work.

3. Case studies

As mentioned earlier, we applied the MO algorithms to the C3/C4 oil refinery splitter in Ferreira et al. (2022a). In such splitter, the inlet stream is mainly composed of C3 and C4 hydrocarbons, while other hydrocarbons with 1, 2, 5 and 6 carbons, as well as H_2 and H_2S are the main minor components. The system is continuously monitored through eight online sensors that measure the feed, reboiler, head and distillate temperatures; the reflux, bottoms and distillate flows; and the bottom's pressure. These 8 variables are the input variables to the models, i.e. $x = [T_{feed}, F_{reflux}, F_{bottoms}, F_{distil}, T_{reb}, T_{head}, T_{distil}, P_{bottom}]$.

The quality of the outlet streams is monitored through laboratory experiments (chromatography) that determine the composition of 19 species with daily or bi-daily frequency. As case studies we evaluate the development of models in two instances. One to obtain the models of the volumetric fraction of C4s in the distillate and the volumetric fraction of C3s in the bottoms (reboiler) $y_{case-1} = [C4_{distillate}, C3_{reboiler}]$. These fractions evaluate the quality of the outlet streams and depend on the temperature and pressure of the distillate/reboiler; we do not expect the expressions for the outlets to share these terms. In the second case study, we are interested in developing models for the volumetric fractions of the seven C4 hydrocarbons in the distillate $y_{case-2} = \begin{bmatrix} \text{1,3-butadiene, 1-butene, cis-2-butene, isobutane, isobutene, n-butane,} \\ \text{trans-2-butene} \end{bmatrix}$. This is the case where we expect the output variables to share many terms.

To compare the three algorithms (SO-KP, MO_{MLR} and MO_{IM}), we followed the statistical procedure in Alba and Luque (2005). This is, first, one hundred independent executions were performed for each algorithm; then, the Friedman test was used as an omnibus test to compare the RMSE distributions over validation sets. Finally, a pairwise Wilcoxon signed test to the RMSE distributions was performed in those cases in which differences among the distributions were found through the Friedman test. All the statistical tests were performed with a confidence level of 99%.

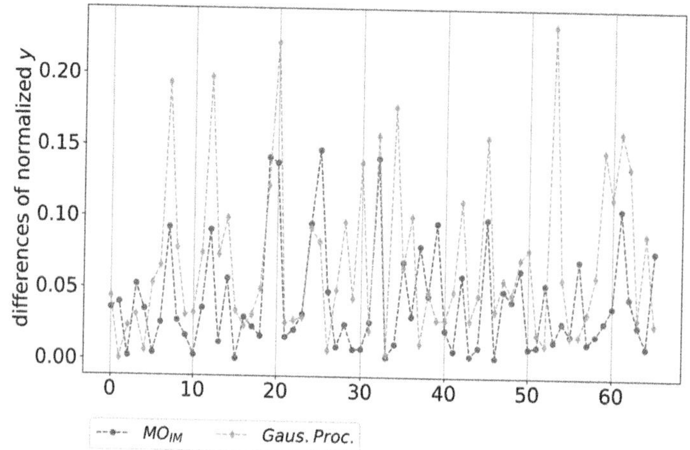

Figure 3 – Absolute differences of the estimation and data point by the best model with MO_{IM} and GP, over validation set.

4. Numerical results and discussion

4.1. Case study 1

Based on Friedman test we found that for y_1 there is statistical difference between the distributions of the RMSE of the models obtained with the three algorithms ($p_{value} = 4.81 \cdot 10^{-12}$), and for y_2 there is not statistical difference ($p_{value} = 0.0105$), with a median of RMSE between 0.062 and 0.067.

From the Wilcoxon signed rank test for y_1, we found that there is no statistical difference between the error's distributions of the models obtained by MO_{IM} and MO_{MRL} ($p_{value} = 0.033$), whereas the models obtained by SO-KP had an overall lower median for RMSE ($p_{value} = 2.51 \cdot 10^{-9}$ for MO_{IM} and SO-KP and $p_{value} = 2.35 \cdot 10^{-7}$for MO_{MRL} and SO-KP). The RMSEs are: 0.051, 0.056 and 0.057, respectively for SO-KP, MO_{MRL} and MO_{IM}.

Figure 3 presents the absolute difference between estimation and data over the validation set, for the second output variable by the best model obtained by MO_{IM}, and the one obtained using the multi-output Gaussian Process algorithm (GP, implemented in the *scikit-learn* package). From the figure, we can see that all estimations obtained with the KP-based algorithm are better than the ones obtained with GP. This is reinforced by the value of RMSE of the model obtained by GP: 0.088. It is worth commenting that all the values presented in these figures were normalized to preserve data confidentiality.

Table 1 shows the best model for each output variable obtained with each KP-based algorithm. It can be seen that the models obtained with MO_{IM} correctly capture the dependence of y_1 on T_{head} and T_{distil}, and of y_2 on T_{reb} and P_{bottom}. On the other hand, those obtained with SO-KP include almost all inlet variables in the mathematical expression. Thus, despite its slightly largest RMSE, we consider the MO_{IM} model to be superior.

4.2. Case study 2

In this case, based on Friedman test, we found that for y_1 there is no statistical difference between the distributions of the RMSE of the models obtained with the three algorithms. For the rest of the output variables there exists statistical difference.

Table 1 – The best learned models for the second output variable with SO-KP, MO_{IM} and MO_{MLR} algorithms. The coefficients are presented as Greek letters in order to preserve confidentiality.

y_1	SO-KP	$\beta_1^1 \frac{F_{reflux}^2}{(F_{reflux} + P_{bottom})T_{head}} + \beta_2^1 \frac{F_{reflux}F_{distil}}{T_{head}P_{bottom}} + \beta_3^1 \frac{T_{reb}}{T_{head}P_{bottom}} + \beta_4^1 \frac{F_{reflux}T_{distil}}{T_{head}P_{bottom}} + \beta_5^1 \frac{F_{reflux}^2}{P_{bottom}} + \beta_6^1 F_{reflux} + \beta_7^1 F_{bottoms} + \beta_8^1 F_{distil} + \beta_9^1 T_{feed} + \beta_{10}^1 T_{head} + \beta_{11}^1 T_{feed}(F_{bottoms} + F_{distil}) + \beta_{12}^1 T_{distil} + \beta_{13}^1 \left(\frac{T_{head}}{T_{reb}}\right)^{\left[(P_{bottom}-F_{distil})\frac{F_{reflux}-F_{bottoms}}{T_{distil}-P_{bottom}}\right]} + \beta_{12}^1 (F_{bottoms} + F_{distil})F_{distil}$
	MO_{IM}	$\gamma_1^1 F_{distil} + \gamma_2^1 P_{bottom} + \gamma_3^1 F_{reflux} + \gamma_4^1 T_{distil} + \gamma_5^1 T_{feed} + \gamma_6^1 T_{reb} + \gamma_7^1 F_{reflux}F_{distil} + \gamma_8^1 F_{reflux}T_{feed} + \gamma_9^1 (F_{reflux}^2 - T_{head} T_{distil}) + \gamma_1^1 T_{reb}T_{head}$
	MO_{MLR}	$\delta_1^1 P_{bottom} + \delta_2^1 T_{head} + \delta_3^1 F_{distil} + \delta_4^1 F_{reflux} + \delta_5^1 T_{feed} + \delta_6^1 F_{reflux}^2 + \delta_7^1 F_{reflux}T_{reb} + \delta_8^1 T_{reb} + \delta_9^1 \frac{T_{feed}F_{bottoms}T_{reb}}{T_{distil}}$
y_2	SO-KP	$\beta_1^2 \left(\frac{F_{reflux}-F_{bottoms}}{F_{distil}+T_{reb}}\right) + \beta_2^2 F_{bottoms} + \beta_3^2 F_{reflux} + \beta_4^2 F_{bottoms} + \beta_5^2 P_{bottom} + \beta_6^2 \frac{F_{bottoms}-F_{reflux}}{2F_{distil} - T_{distil}} + \beta_7^2 T_{reb} + \beta_8^2 F_{distil}$
	MO_{IM}	$\gamma_1^2 \frac{F_{bottoms}}{T_{reb}} + \gamma_2^2 F_{reflux} + \gamma_3^2 F_{bottoms} + \gamma_4^2 P_{bottom} + \gamma_5^2 T_{reb} + \gamma_6^2 P_{bottom}^2 + \gamma_7^2 F_{reflux}T_{reb} + \gamma_8^2 F_{reflux}F_{distil}$
	MO_{MLR}	$\delta_1^2 \left(\frac{T_{feed}}{F_{reflux} - T_{reb}} - \frac{P_{bottom}}{T_{head}}\right) + \frac{\delta_2^2 F_{distil}}{F_{reflux} - T_{reb}} + \frac{\delta_3^2 P_{bottom}}{T_{feed}} + \delta_4^2 F_{reflux} + \delta_5^2 F_{bottoms} + \delta_6^2 P_{bottom} + \delta_7^2 T_{reb} + \delta_8^2 T_{head} + \delta_9^2 (F_{reflux} - T_{feed})F_{bottoms}T_{distil} + \delta_{10}^2 (F_{bottoms} - F_{reflux}) T_{reb}$

For output variables y_2 to y_7, the Wilcoxon signed rank test showed that there is not statistical difference for three output variables, MO_{IM} has lower median RMSE than SO-KP for two output variables, and SO-KP lower than MO_{IM}. in one case. Comparing SO-KP with MO_{MLR}, SO-KP median RMSEs were lower than MO_{MLR} in two cases.
Finally, comparing MO_{MRL} and MO_{IM}, there was no statistical difference in two cases, MO_{IM} had lower RMSEs in three cases, and MO_{MRL} lower than MO_{IM}, in one case.

Then, the models obtained by multi-output algorithms generally perform better in terms of RMSEs. MO models also showed better performance in terms of term sharing which is coherent with the fact that in this case, output variables are strongly related, and the multi-output learning is enriched by incorporation of data from all variables.

5. Conclusions

In this paper we tested the performance of two multi output (MO) strategies to learn an interpretable models (symbolic regression) where the bases functions were learned via genetic programming and the coefficients via least squares regression. The first MO strategy, MO_{MLR}, is based on the use of a multi-output linear regression instead of a single-output linear regression. The second algorithm, MO_{IM}, is based on the islands model for parallelization of evolutionary algorithms, in which steps of independent search alternate migration steps where the best functional bases are exchanged. The models obtained by the multi-output algorithms were compared with the single-output executions for each output variable, and a multi-output Gaussian Process algorithm.
Data from a real C3/C4 splitter from an oil refinery was used in the case studies. We considered a first case where the output variables were weakly related by the physico-chemical phenomena, and a second case, where the output variables were strongly related. We showed that in the first case study, the island based multi-output strategy MO_{IM} was able to identify the variables that were relevant for each output variable model, although having a slightly worse performance in terms of RMSE (in the validation set). In the second case study, multi-output strategies performed better than the single-output one, although within the same order of magnitude in RMSE. The three algorithms performed better than the multi output Gaussian Process one.

Acknowledgements

The authors gratefully acknowledge the collaboration of ANCAP-Uruguay for providing the data used for the case study.
J. Ferreira thanks the Agencia Nacional de Investigación e Innovación of Uruguay for financial support for her graduate studies (Award No. POS_NAC_2018_1_152185).

References

E. Alba and G. Luque, Measuring the Performance of Parallel Metaheuristics. John Wiley & Sons, Ltd, 2005, ch. 2, pp. 43–62.

A. Chakraborty, A. Sivaram and V. Venkatasubramanian, 2021, AI-DARWIN: A first principles-based model discovery engine using machine learning, Computers & Chemical Engineering, 54, 107470.

M.R. Engle and N.V. Sahinidis, 2022, Deterministic symbolic regression with derivative information: General methodology and application to equations of state. AIChE J. 68 (6): e17457.

J. Ferreira, A. I. Torres and M. Pedemonte, 2021, Towards a Multi-Output Kaizen Programming Algorithm, 2021 IEEE Latin American Conference on Computational Intelligence (LA-CCI), Temuco, Chile.

J. Ferreira, M. Pedemonte, and A. I. Torres, 2022a, Development of a machine learning-based soft sensor for an oil refinery's distillation column. Computers & Chemical Engineering, 161, 107756.

J. Ferreira, M. Pedemonte, and A. I. Torres, 2022b, A multi-output machine learning approach for generation of surrogate models in process engineering. Computer Aided Chemical Engineering, 49, 1771-1776.

D.G. Krige, 1951, A statistical approach to some mine valuations and allied problems at the Witwatersrand. Master's thesis of the University of Witwatersrand.

Antonis Kokossis, Michael C. Georgiadis, Efstratios N. Pistikopoulos (Eds.)
PROCEEDINGS OF THE 33rd European Symposium on Computer Aided Process Engineering (ESCAPE33), June 18-21, 2023, Athens, Greece
 http://dx.doi.org/10.1016/B978-0-443-15274-0.50178-5

Decision regression for modelling of supply chain resilience in interdependent networks: LNG case

Adnan Al-Banna,[a,b] Brenno C. Menezes,[b] Mohammed Yaqot,[b] Jeffrey D. Kelly[c]

[b]*Division of Engineering Management and Decision Sciences, College of Science and Engineering, Hamad Bin Khalifa University, Doha, Qatar Foundation, Qatar*
[b]*Department of Logistics and Supply Chain, Milaha, Doha, Qatar*
[c]*Industrial Algorithms Ltd., 15 St. Andrews Road, Toronto M1P 4C3, Canada*
bmenezes@hbku.edu.qa

Abstract

As an application of advanced analytics (AA) in supply chains (SCs), to model supply chain resilience (SCR) of transactions, logistics, operations, etc., of such complex representation of networks, we propose a supervised machine learning approach as a predictive analytics decision regression modelling framework that uses a coefficient setup MIQP (mixed-integer quadratic programming) technique. This can determine optimisable surrogate models to correlate independent X variables (e.g., resistance and recovery of the SC resilience) to the dependent Y variable SCR considering a dynamic behaviour of the SC with lag- and dead-time. A novel methodology to quantify SCR, based on a tree of continuous x and binary y variables of resistance (avoidance and containment) and recovery (stabilisation and return), considers ad hoc relationships of x and y to be part of the SCR predictions in the machine learning MIQP identification method. Such SCR algebraic or analytical formulas obtained in this constrained decision regression approach (a type of predictive analytics) can be used in optimisation and control problems of prescriptive and detective analytics types of AA. The proposed model is applied in an oil and gas case pertaining to liquid natural gas (LNG) and its leaks known as boil-off gas (BOG). The vapors generated by the leaks or venting of BOG in the LNG supply chain reduces the SCR of this commodity. There are losses of materials, environmental impacts, reduction in the calorific value of the LNG to be re-gasified, potential safety issues, to name a few. This research aims to introduce a methodology to model and predict SCR in a general way (by a new design as example), but particularly, covers the digital transformation implementations that can potentially lead to an enhanced resilience in the supply chain of LNG towards the desired level of digital supply chain resilience (DSCR).

Keywords: supply chain, resilience, advanced analytics, surrogate modelling.

1. Introduction

Making determinations on complex design, operation, and control of today's industry may count on recent capabilities of advanced analytics (AA) that is widespread in machine learning techniques and big data novelties and applications. Particularly, supply chain and logistics organisations operate in a data-rich environment, managing huge amount of data about customers' demands, offerings of suppliers, previous and current orders, satisfaction levels and reviews, shipment movements, locations of containers, temperature, humidity, among others. Hence, many supply chain organisations have put data analytics at the top of their strategic priorities. This is driving interest in smart analytics techniques designed to help diagnose operational issues, optimise network planning, and predict future scenarios. As a branch of the so-called smart manufacturing, AA can consider multiple scopes, scales, scenarios, and layers to model and solve decision-making problems as found in the engineering management and decision science literature. It moves from the mature or old-fashion descriptive analytics of past activities to the needed cognitive analytics with self-learning and adaptive structures in both manufacturing systems and supply chains (Menezes et al., 2019a).

The proposed SCR modelling in this work uses a methodology that considers a MIP-based machine learning modeling approach with the objective of identifying the causation and correlation among (1) two parent inputs (resistance and recovery) and (2) their four child inputs (avoidance and containment, and stabilisation and return) of the SCR from Melnyk et al. (2014) as seen in Figure 1. This can guide directions in the investments on new design and digital transformation building blocks for improvement of the liquid natural gas (LNG) supply chain resilience that is impacted by the boil-off gas (BOG) losses during the shipping and its adjacent areas (upstream and downstream).

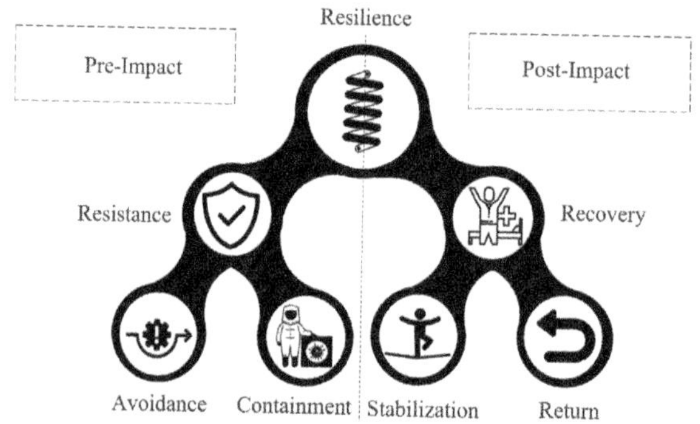

Figure 1: Tree of supply chain resilience (adapted from Melnyk et al., 2014).

2. Boil-off gas in the LNG supply chain

To be stocked and transported in the liquid state, natural gas (NG) requires very low temperatures, therefore it is very susceptible to heat invasion and other driven-forces and starts to boiling at -162 °C. This vaporised LNG represents a substantial energy waste with a negative impact in economics, climate warming, and operational hazard. Henceforth, the LNG losses throughout the supply chain requires to be controlled and managed in the most efficient manner, avoiding the losses of BOG or containing it, if LNG boils. Also, if it is lost, measures to recovery the commodity to its initial product condition in terms of its properties and specifications (after the liquefaction plant).

Figure 2 illustrates potential deployment of internet-of-things (IoT), automated decision-making and controls, artificial intelligence (AI), etc., elements of the Industry 4.0 age on LNG vessels and their associated facilities. Besides the use of technology (more operational related) to increase SCR, other types of investments are related to changes in design, by changes in grassroots projects or retrofit of current assets. Among these investments, there are (1) new vessels propelled by LNG, (2) retrofit in current vessels to install liquefaction systems to liquefy the BOG; and (3) different storage tank inside the vessels considering the two main LNG vessel types, the MOSS and membrane. In the latter, as example, the membrane design is more resilient and advantageous due to many reasons, but mainly by its robust insulation, as it contains an amalgamation of multiple layers of carefully selected materials that is split into two membranes (a primary and a secondary) that spans across almost one meter distance between the ship's inner hull and the LNG cargo.

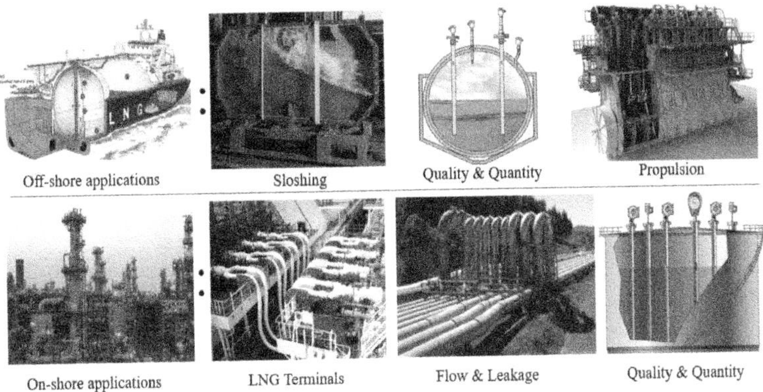

Figure 2: Potential investments to increase supply chain resilience.

3. LNG supply chain resilience stages

In the following, a detailed description of examples of resistance and recovery stages related to the BOG losses in the LNG supply chain is given.

1) Resistance Stage

In the avoidance stage, supply chain resilience (SCR) can be associated to the supply chain vulnerability (SCV) for the BOG generation driven-forces. Some SCVs are the (a) fluctuation in the ambient temperature and tank pressure, and (b) sloshing (LNG fluid excessive movement and friction with its internal reservoir surfaces), which is influenced by excessive rolling of the ship during voyage. Besides, by augmenting the. The insulation between the ambient and the stored LNG layer in LNG ships is under constant development, and it is one of the main reasons the old MOSS type vessels are now progressively replaced with the new membrane type vessel. However, insulation enhancement is extensive work and requires excessive out of operations periods and expensive drydock repairs.

In the same context, controlling the tank pressure is a critical factor to avoid BOG generation, and this could be established with any sort of sensing-calculating-actuating (SCA) capability to proactively regulate the tank pressure (Menezes et at., 2019b). Moreover, route optimisation is an efficient methodology to main the ship voyage smooth and hence significantly reduce the time-period of the friction between the LNG and its tank surfaces, eventually reducing the BOG generation. Such an enhancement can be achieved via deploying artificial intelligence (AI) and big data (BD) enablers with the objective of calculating maritime records about waves, currents, and winds, among others to find optimised routes. From work experience, ship arriving late due to taking longer voyage route could have better economics compared to a ship that charted a shorter route but through rough waves that generated excessive BOG.

In the containment stage, the objective is to explore methodologies to contain and utilise the generated BOG within the LNG ship eco-system instead of losing it to the atmosphere.

Such enhancement can be via deploying re-liquification plants onboard the ship, which receive the generated BOG, re-liquefy it, and eventually direct it to the LNG storage tank. This non-I4.0 enhancement requires excessive time out of operations and drydock repairs due to the extensive pipe work involved. Another enhancement in this stage is directing the generated BOG towards the propulsion system of the ship, however limited percentage of LNG ships worldwide have this capability. Currently, there are two types of two stroke engines that deliver this flexible dual-fuel propulsion, that are MEGI (M-type, Electronically Controlled, Gas Injection), and XDF (low pressure 2-stroke engine). Most of the newly built LNG ships are either MEGI or XDF; however, despite some major marine engine manufacturers are starting to offer retrofitting ships solutions, this option may not be economically viable (Man Energy Solutions, 2022).

2) Recovery Stage: in this stage and its sub-stages of stabilisation and return, the objective is to achieve stabilisation and return post the generation of BOG, where it is not attainable to avoid, nor contain it. An envisaged enhancement in this stage is to deploy AI solutions with the objective to optimise receiving terminals pumps operation algorithms in a manner that assist in the recovery process. Similarly, another enhancement is injecting ethylene ($CH_2{=}CH_2$) in the receiving terminals to enhance the calorific value and heat content of the delivered LNG during its re-gasification stage and injection in pipelines.

The aforementioned discussion is summarised in Table 1.

Table 1: BOG resilience tree.

Avoidance (On the ship), potential I4.0 enabler and desired function		
Tank temperature	Operational (I4.0)	Deploy
	Design (Retrofit)	Install enhanced insulation around the tank
Tank pressure	Operational (I4.0)	Deploy S-C-A to keep the pressure in the tank regulated
Optimise route	Operational (I4.0)	Deploy AI and BD to calculate maritime records about waves, currents, and winds, etc., and find optimised routes
Containment (On the ship), Potential I4.0 enabler and desired function		
Propulsion	Design (Grassroots)	Generated BOG fuelling propulsion systems
Re-liquefaction	Design (Retrofit)	Re-liquefy the BOG
Stabilisation (At receiving terminal), Potential I4.0 enabler and desired function		
Pumping at receiving terminals	Operational (I4.0)	Deploy AI to optimise operation of the pumps in receiving terminal (KBC, 2021)
Return (At the receiving terminal), Potential I4.0 enabler and desired function		
Injecting ethylene ($CH_2{=}CH_2$)	Operational	Inject ethane in the re-gasification stage in the receiving terminal to correct the calorific value and heat content of LNG

From the operational and design related projects as seen in Table 1, with I4.0 content or not, a plot can be sketched for the proposed resilience domains (resistance and recovery) that are function of the investments. This is shown in Figure 3, where two planes are highlighted to exemplify SCR investments more related to design and those with focus on operational investments (supposedly less expensive, but not such efficient in comparison with the design expenditures.

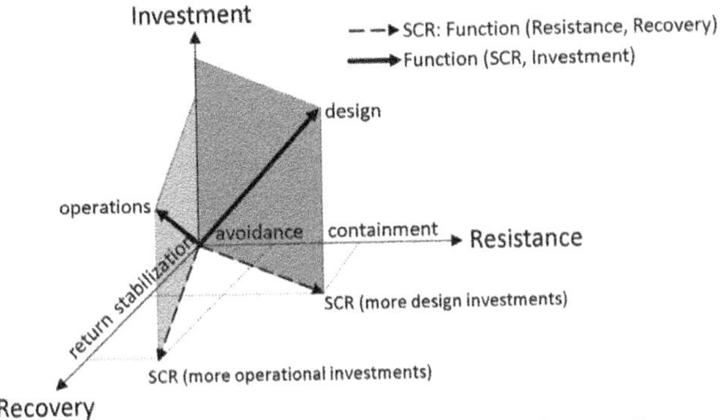

Figure 3. SCR investments in design and operatios.

4. Methodology and results

This research proposes a modeling approach for the supply chain resilience (SCR) concerning the BOG generation in the LNG supply chain by evaluating SCR in two distinct stages as shown in Figure 1: resistance and recovery, and their consequential sub-stages of: avoidance and containment (for resistance) and stabilisation and return (for recovery). The surrogates are built to represent dependent variables of interest Y, such as the SCR. Independent-dependent variations correlate as $Y_i = f(X_j)$. The input variables considered to form the surrogates represent the BOG resilience tree as in Table 1. When building surrogate models, it is important to consider some important aspects concerning the required data for training and testing, functional form of the surrogates and proper evaluation to verify their performance. A simplified framework methodology proposed by Franzoi et al. (2021a), as shown in Figure 4.

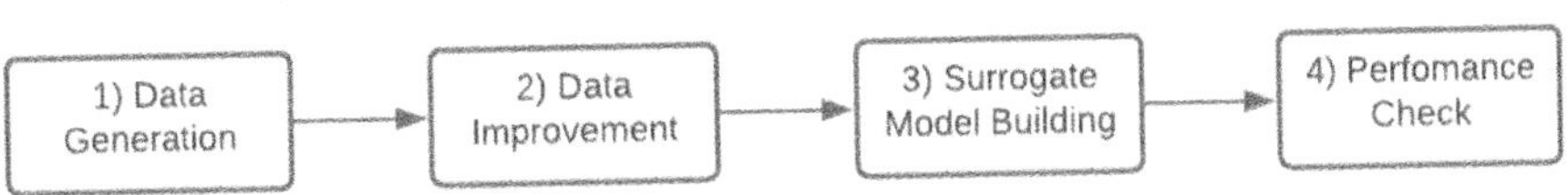

Figure 3: Framework for the surrogate model building strategy.

1) *Surrogate Model Building*: To model the SCR, bilinear surrogate models can be used, as shown in Equation (1) to calculate the dependent variable Y_i (see Franzoi et al., 2020).

$$Y_i = I_i + \sum_{j \in DV} b_{ij} X_j + \sum_{j \in IV} \sum_{k \in IV} c_{ijk} X_j X_k \qquad \forall\ i \in DV,\ j \leq k,\ k \leq n \quad (1)$$

In which I_i is the intercept of each point i within dependent variables (DV), and b_{ij} and c_{ijk}, are the coefficients to be determined or estimated during the building process of the surrogate model by evaluating the accuracy of the input-output data to achieve the target of minimising the prediction error for each independent variable point (*IV*). This optimisation target is presented in Eq.(2) for minimising the least-squares error (LSE).

$$Minimize \sum_{p=1}^{n_{training}} (y_{ip} - Y_{ip})^2 \quad (2)$$

In which y_{ip} is the actual calculated using a simulation of the for the variable i at each point p within the training data set $n_{training}$, whereas Y_{ip} represents the calculated values using the surrogate model estimated coefficients for each independent variable i at each point p.

The dataset to construct the surrogate model is determined by concatenating the operational and design investments from Table 1. As example, if no installation of more efficient temperature insulation by using membrane technology instead of MOSS (independent variable X_2), it is expected to have losses of the calorific value Y due to the no perfect control of the temperature (independent variable X_1). The SCR modelling in this case is considered investment projects (operational- or design-related) that would avoid and contain the BOG or, if in the case of recovery, those measure that would stabilise or recovery to the initial stage of the calorific value Y in the beginning of the supply chain, from liquefaction plants to receiving terminals. Equation (3) is the surrogate for the calorific value Y. Cases of application of this method can be found in chemical engineering for applications on cutpoint optimization (Franzoi et al., 2020).and reactor systems where an adaptive sampling algorithm can converge to parameters of the surrogate formulas for ranged spaces (Franzoi et al., 2021b).

$$\begin{aligned} Y_{CV} = -10X_1 - 5.1X_2 - 3.2X_3 - 10X_4 - 10X_5 + 7.4X_7 - 9.2X_1X_2 + 10X_1X_3 \\ - 10X_1X_4 - 10X_2X_3 + 10X_2X_4 + 7.3X_2X_7 - 10X_3X_4 + 10X_3X_5 \\ + 10X_3X_6 + 10X_3X_7 - 10X_4X_5 - 10X_4X_6 - 1.4X_4X_7 + 5.5X_5X_7 \end{aligned} \quad (3)$$

5. Conclusion

The main purpose of this paper is to build surrogate models that can effectively determine digital supply chain resilience in considering the boil-off gas in the LNG supply chain. The surrogates can be built to represent the resistance and recovery measures to be managed in the proposed modeling. Considering the generic SCR modeling using MIP-based machine learning for the parent and child components as in the Figure 1 tree (parent: resistance and recovery, children: resistance, recovery, containment, stabilisation), a more complex combination will be proposed for the second-order or cross-product of the components. It will cover the following aspects of the CSR in the LNG supply chain such as (1) physical elements of the LNG shipping: pressure, temperature, BOG losses, etc., and (2) trustworthiness levels of the IoT considering communication, timeliness, etc.

References

R. E., Franzoi, B. C., Menezes, J. D., Kelly, J. A., Gut, I. E., Grossmann, 2020, Cutpoint temperature surrogate modeling for distillation yields and properties, Industrial and Engineering Chemistry Research, 59(41), 18616-18628.

R. E., Franzoi, T., Ali, A., Al-Hammadi, B. C, Menezes, 2021a, Surrogate modeling approach for nonlinear blending processes. 1st International Conference on Emerging Smart Technologies and Applications, 1-8. IEEE.

R. E., Franzoi, J. D., Kelly, B. C., Menezes, C. L. E., Swartz, 2021b, An adaptive sampling surrogate model building framework for the optimization of reaction systems, Computers and Chemical Engineering, 152, 107371.

S. A. Melnyk, D. J. Closs, S. E. Griffis, C. W. Zobel, J. R. Macdonald, 2014, Understanding supply chain resilience. Supply Chain Management Review, 18, 34-41.

B. C. Menezes, J. D. Kelly, A. G. Leal, G. C. Le Roux, 2019, Predictive, prescriptive and detective analytics for smart manufacturing in the information age, IFAC-PapersOnline, 52 (1), 568-573.

B. C. Menezes, J. D. Kelly, A. G. Leal, 2019b, Identification and design of Industry 4.0 opportunities in manufacturing: examples from mature industries to laboratory level systems, IFAC-PapersOnLine, 52(13), 2494-2500.

Man Energy Solutions, 2022, Retrofits for two-stroke marine engines, available on: https://www.man-es.com/services/industries/marine/retrofit-upgrade/two-stroke-engines, accessed on: 20 October 2022.

KBC, a Yokogawa Company, 2021, available on: https://www.kbc.global/events/transforming-decision-making-in-scheduling-of-terminal-operations, accessed on 6/10/2022.

Antonis Kokossis, Michael C. Georgiadis, Efstratios N. Pistikopoulos (Eds.)
PROCEEDINGS OF THE 33rd European Symposium on Computer Aided Process Engineering (ESCAPE33), June 18-21, 2023, Athens, Greece
 http://dx.doi.org/10.1016/B978-0-443-15274-0.50179-7

Multi objective decision-making methodologies applied to GREENSCOPE sustainability indicators

Ricardo N. Dias[a], Rui M. Filipe[a,b], Henrique A. Matos[a]

[a]CERENA, Instituto Superior Técnico, Av. Rovisco Pais 1, 1049-001 Lisboa, Portugal
[b]Instituto Superior de Engenharia de Lisboa, Instituto Politécnico de Lisboa, R. Conselheiro Emídio Navarro 1, Lisboa 1959-007, Portugal

ricardo.n.dias@tecnico.ulisboa.pt

Abstract

The current socio-economic and environmental concerns envisage entities to make sustainable investments and improvements. However, sustainable investment selection in a given industry might not be straightforward from the several technological alternatives available. Several methodologies exist to help in the alternatives assessment and help in the decision-making activity. GREENSCOPE (Ruiz-Mercado et al., 2012), used in this work, calculates 139 normalised sustainability indicators from four different classes: efficiency, environment, energy, and economy. Analysing such indicators for investment alternatives is a complex multi-objective decision-making process (MODM) for which many methodologies exist. GREENSCOPE suggests using the Additive Utility Method (AUM) (Smith & Ruiz-Mercado, 2014), which might have some limitations in identifying different alternatives. An alternative method is implemented in this work, the Analytic Hierarchy Process (AHP) (Saaty, 1978). Both methods were tested and tweaked, originating AHPmod and AUMmod. These were applied to three case studies.

Keywords: Sustainability, decision-making, multi-objective, sustainability indicators

1. Introduction

The newest IPCC report, released on April the 4th, 2022, focuses on the mitigation of climate change, calling for an improvement in sustainable development and an increase in efficiency(WGIII IPCC, 2022). Therefore, decision-making must take into consideration, not just economic sustainability but environmental sustainability as well. UN sustainable goals is a set of internationally recognized goals that call for sustainable development (UN General Assembly, 2015). The main challenge when evaluating the sustainability of a given project is the assessment of the environmental impacts since they might have a different importance in different regions. For a region frequently affected by droughts, local industry's water consumption and efficiency might be of higher priority in the agenda against CO_2 emissions mitigation. Therefore, the decision-making process should always consider the socio-economic impacts (Ruiz-Mercado et al., 2012).

GREENSCOPE (GRNS) is a life cycle assessment methodology focused on gate-to-gate analysis that uses a total of 139 indicators. They are spread (not evenly) across 4 metrics: Mass Efficiency, Environment, Energy and Economy.

$$\%G = \frac{|x_i - x_{i,worst}|}{|x_{i,best} - x_{i,worst}|} \qquad (1)$$

The normalization technique used on the GRNS methodology implies an indicator score between 0 % and 100 % (max-min normalisation). It makes for a straightforward interpretation of the results, with the user always aiming for a score as close to 100% as possible. Even though the maximum and minimum values assumed by the methodology for the normalisation step might have some degree of uncertainty and subjectivity, it is considered a well-based methodology. Despite the representation for the scores make for eye-appealing graphs, as per Figure 1, it is not easy to quantify the improvement form alternative to alternative.

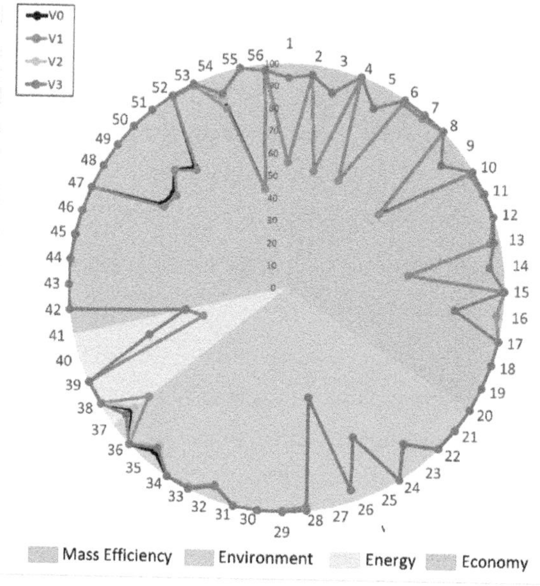

Figure 1 - Graphical representation of acetaldehyde case study GRNS scores.

This work's main goal is to establish a decision-making process based on the indicators calculated using GRNS. Even though the representation of the scores in spider-like graphs (Figure 1) intends to help the decision-maker by showing that a given alternative with scores closer to the outside boundaries is better than other not as close, a quantification method is also needed.

Smith et al. (2013) use marginal rates of substitution (MRS) to determine the relative importance of any indicator, compared to the others, and AUM to decide upon the best alternative. However, the decision-making process is based on a limited number of indicators that are selected subjectively.

In this work, 2 different methods are implemented, AUM and AHP, together with 2 new variations of these methods. An analysis of the results obtained with each method is performed to decide on which is more suited to this type of multi-objective decision-making (MODM) problems.

2. Methodology

In this section, MODM methodologies, such as AUM and AHP, are applied to GRNS indicators, to decide what is the best alternative for each case study.

2.1. Additive Utility Method (AUM)

GRNS uses a scoring between 0 % and 100 % for each indicator, which may lead to accept that a higher sum of scores indicates the best alternative. However, the increase in one indicator score can come at the expense of another (Smith & Ruiz-Mercado, 2014). AUM is a method that uses the summation of indicator scores combined with MRS.

MRS uses the derivative of one indicator to another, with which it is possible to create an interaction matrix with size n (where n is the number of indicators considered for the analysis). From this matrix, one column, to be used as reference indicator, is selected and multiplied by the indicator scoring obtained from GRNS. The resulting vector values are summated, and a final value is obtained. This procedure, using the same interaction matrix from MRS, is repeated for each alternative. Comparison between alternatives is made using the calculated single scores for each alternative.

2.2. Analytic Hierarchy Process (AHP)

AHP, a well-known Multi-Criteria Decision Making (MCDM) method (Saaty, 1994), proposes a systematic methodology to compare multiple possible choices structuring a problem in three parts: goal, possible solutions (alternatives), and criteria (indicators). To determine the best possible solution, one must assign priorities to the various criteria being considered. It can be different for any given problem; Saaty proposes a scale of relative importance from 1 to 9. Using this relative importance is possible to establish a pairwise comparison matrix A, which is a $n \times n$ reciprocal matrix (where n is the number of indicators taken into consideration).

For each indicator a matrix M must be calculated, which is a $m \times m$ matrix (where m is the number of alternatives considered). And, like matrix A, it is a reciprocal matrix that is obtained by dividing the scores of every indicator by each other.

Vectors P and p (obtained from the geometric mean and normalisation of matrices A and M) are then rearranged in a table, where their values are multiplied and summed to obtain a final score, see eq. (2).

$$Prirority = \sum_{i=1}^{n} P''(i) \times p_k''(i) \tag{2}$$

2.3. Additive Utility Method modified (AUMmod)

Since the AUM method uses a reference indicator arbitrarily chosen, whose selection may be arguable, a modified version that considers all the indicators is herein proposed. This method consists in using the vector P'', originated from the interaction matrix A on the AHP method, to balance the scores from GRNS, allowing for a single score.

2.4. Analytic Hierarchy Process modified (AHPmod)

A proposed modification to the AHP method is also presented. An interaction matrix for every alternative with the values of every indicator was calculated, and the same mathematical steps of matrix line geometric mean and normalisation were applied. These two vectors, P'' and new p are to be multiplied and summed, with the highest value being the best performing alternative.

2.5. Implementation

The four methods previously outlined were implemented using MATLAB. A code was written to obtain the indicators scoring from the GRNS file and make the calculations. The implementation of AUM was straightforward since it had already been used for GRNS indicator scoring comparison.

For the AHP and AHPmod methods implementation, a systematic methodology to overcome the subjective score selection (1-9) was carried out. Therefore, matrix A was generated not based on the relative difference attributed by the user, but based on the delta calculated through the difference between maximum and minimum possible indicator values, that corresponds to the denominator of eq. (1).

The indicators used in every method were determined via a set of restrictions outlined to ensure calculations feasibility, since MRS uses delta divisions, which cannot be zero. The set of restrictions imposed on the indicator selection are defined in Table 1, as stages.

Table 1. Indicator selection, elimination stages with description.

Stage	Description
0	139 GRNS indicators
1	Remove $\nabla_i < 0.01$
2	Remove $\%G_i < 0.01$
3	Remove $P_i'' < 0.01$
4	Remove $P_i'' = \infty$
5	Remove $p_i'' = \infty$

In stages 4 and 5, the elimination of infinite values on the P'' and p_i'' vectors, is essential for the implementation of AHP and AUM methods. Stage 6 ensures that the same set of indicators is used on every alternative evaluation.

Finally, to facilitate the interpretation, a ranking was attributed to the scenarios evaluated, so if a scenario has a better final score than any other, it is ranked as the best alternative in a scale from 1st to 4th.

3. Case study

To carry out a sustainability analysis, there must be a reference scenario upon which the comparison can take place; otherwise, the score obtained can't be interpreted. In the context of the present work, 3 case studies were evaluated, for which 4 different scenarios were created. Since decision-making can only take place if there is more than one alternative to choose from, in Table 2 is possible to find a quick overlook of scenarios for the different case studies.

Table 2. Case studies and its scenarios.

Scenario	Description		
	Biodiesel plant	**Acetaldehyde plant**	**Propylene Glycol plant**
V0	Base scenario	Base scenario	Base scenario
V1	Lower price of feed streams	Energy integration (EI)	Energy integration (EI)
V2	Lower price of product streams	Recirculation, no EI	Recirculation, no EI
V3	Lower energy consumption (EI)	Recirculation with EI	Recirculation with EI

3.1. Biodiesel plant

The biodiesel plant was used as case study in GRNS (Ruiz-Mercado et al., 2013). Since the model for this case study was not available, modifications to the original biodiesel demonstration file (V0) were made, as follows: a decrease of 10 % in raw materials cost (V1), a 10 % reduction in products value (V2) and decrease of 10 % in energy consumption (V3). These scenarios were created to better understand the decision-making process based on the indicators calculated.

3.2. Acetaldehyde and propylene glycol plant

Acetaldehyde production is a very well-known academic case study, and so is the propylene glycol. These cases are particularly suited for a sustainability analysis since the assessment outcomes can be clearly validated. As for the biodiesel plant, several alternatives were created: base case scenario with no energy integration or material recirculation (V0), energy integration and no material recirculation (V1), recirculation without energy integration (V2) and material recirculation with energy integration (V3). The V3 alternative is expected to be more sustainable, with higher sustainable scores across the GRNS indicators.

4. Results

The methods previously described (see paragraphs 2.1 to 2.4) were applied to the GRNS scores obtained for every alternative of the case studies. Since the results obtained through the implementation of the MODM methods are not comparable and have very different orders of magnitude, the results obtained are presented in ranking order in Table 3.

Table 3. Rankings obtained to the alternatives of the case studies.

Biodiesel plant case study

	Expected	AUM	AUMmod	AHP	AHPmod
V0	3	4	4	2	**3**
V1	1	3	3	**1**	**1**
V2	4	1	1	3	**4**
V3	2	**2**	**2**	4	**2**
Acetaldehyde plant case study					
V0	4	4	4	2	3
V1	3	1	1	1	4
V2	2	**2**	**2**	3	**2**
V3	1	3	3	4	**1**
Propylene glycol plant case study					
V0	4	2	2	2	3
V1	3	1	1	1	2
V2	2	3	3	3	4
V3	1	4	4	4	**1**

From Table 3 the pattern set by not having material recirculation (V0 and V1) could not be identified by AHP, AUM and AUMmod methods as the worst alternatives. Moreover, those methods also failed in identifying the best and worst performing alternatives.

The AHPmod method ranked the alternatives in the expected order. In the biodiesel case, the ranking corresponds to the expected ranking set. For the acetaldehyde case, two main groups without material recirculation (V0 and V1) are identified as the worst performing alternatives, and ranked 3 and 4, respectively. The order in which V0 and V1 were ranked is not unexpected since energy integration represents capital investment increase, that should be carefully analyzed and justified. It can have a rippling effect across many sustainability indicators. V4 was considered the best-performing one, as expected, since it considers material recirculation and energy integration.

However, in the propylene glycol case, only the best-expected alternative was ranked as anticipated (using the AHPmod). This represents a good accuracy in the AHPmod method for the best-performing cases, even though it can miss rank other alternatives that do not have a very different scoring.

5. Conclusion and future work

Decision-making using sustainability indicators can be a very subjective process. Many methodologies use weights that are determined subjectively to decide on the best and worst-performing alternatives. The work developed and presented in this paper shows that it is possible to use MODM methods to help in sustainability analysis. The proposed methods remove the need for weights and can, with a certain degree of uncertainty, identify a pattern of increasing sustainability. Even though AHP and AUM were not the best performing MODM methods, those are the base for the AHPmod method presented and proposed for further validation. Future work will use TOPSIS method for additional validation of the techniques presented, sensibility analysis of the indicator eliminations steps and Pareto analysis of the vector P'' generated in the MRS reciprocal matrix.

Acknowledgements

The present work was financed by the Portuguese Foundation for Technology and Science (FCT) PTDC/EAM-PEC/32342/2017 and CERENA strategic project FCT-UIDB/04028/2020. This support is gratefully appreciated.

References

Ruiz-Mercado, G. J., Gonzalez, M. A., & Smith, R. L. (2013). Sustainability indicators for chemical processes: III. biodiesel case study. *Industrial and Engineering Chemistry Research*, *52*(20). https://doi.org/10.1021/ie302804x

Ruiz-Mercado, G. J., Smith, R. L., & Gonzalez, M. A. (2012). Sustainability indicators for chemical processes: I. Taxonomy. *Industrial and Engineering Chemistry Research*, *51*(5). https://doi.org/10.1021/ie102116e

Saaty, T. L. (1994). How to make a decision: the analytic hierarchy process. *Interfaces*, *24*(6), 19–43.

Smith, R. L., & Ruiz-Mercado, G. J. (2014). A method for decision making using sustainability indicators. *Clean Technologies and Environmental Policy*, *16*(4). https://doi.org/10.1007/s10098-013-0684-5

UN General Assembly. (2015). UN General Assembly, Transforming our world : the 2030 Agenda for Sustainable Development. In *Resolution adopted by the General Assembly on 25 September 2015* (Vol. 16301, Issue October).

WGIII IPCC. (2022). *IPCC, 2022: Climate Change 2022: Mitigation of Climate Change*.

Antonis Kokossis, Michael C. Georgiadis, Efstratios N. Pistikopoulos (Eds.)
PROCEEDINGS OF THE 33rd European Symposium on Computer Aided Process Engineering (ESCAPE33), June 18-21, 2023, Athens, Greece
 http://dx.doi.org/10.1016/B978-0-443-15274-0.50180-3

Process as a battery: Robust dynamic optimal operation of zeolite crystallization in a COBR with respect to the carbon footprint of electric power

Robin Semrau[a] , Sebastian Engell[a]

[a]*TU Dortmund University, August-Schmidt-Straße 1, 44227 Dortmund, Germany*
Robin.Semrau@tu-dortmund.de

Abstract

The process industries are increasingly electrifying their processes in order to reduce the carbon footprint by the use of "green" electric power. However, the supply of power from renewable sources and the prices of electric energy are volatile which creates an opportunity and a challenge. On the one hand, the volatility of the price creates an opportunity to reduce cost and at the same time reduce the CO_2-footprint as the price and the availability of power from renewables are strongly correlated. On the other hand, the plants must be flexible enough to adapt to the price levels and the operational strategy must become more dynamic. This paper investigates the flexible operation of a continuous tubular reactor for the hydrothermal synthesis of zeolites with the goal to determine the economically optimal operating trajectory in the presence of parametric model-plant mismatch. The robust dynamic optimization strategy is presented and the benefits over the usage of the nominal model are evaluated.

Keywords: Demand response, Dynamic optimization, Robust optimization, Process Intensification, Continuous crystallization

1. Introduction

The industrial sector accounts for 38 % of the total global energy consumption. Therefore, besides increasing the energy efficiency of the production processes, a change of the primary energy sources from fossil to renewable sources in the industrial sector has to take place. In the chemical industry, the direct electrification of processes is one of the main contributors to this transition. The supply of electric power is fluctuating due to the increasing share of renewable power, while there is a (up to now mostly) steady demand for electric energy from the industrial sector. The synchronization of supply and demand via price signals is referred to as Demand Side Management or Demand Response. This adaptation has multiple benefits. The production company is able to use electric energy in periods of lower prices and therefore has an economic benefit. The grid operator on the other hand profits because supply and demand are balanced better. As a result, the share of fluctuating sources in the grid can be increased, while maintaining grid stability. So overall also the CO_2-footprint is reduced. The procurement of electrical energy by large consumers is usually realized by longer-term contracts (e.g. time of use tariffs) and by the participation in different power markets, such as the day-ahead spot market the intraday market and the frequency reserves. The time procurement horizons of these markets range from the next day on the day-ahead spot market to 15 minutes before consumption on the intraday market to 30 seconds on the frequency reserves.

Demand Side Management strategies are frequently formulated as Mixed Integer Linear Problems (Zhang and Pinto, 2022, Dalle Ave et. al., 2019), where the dynamics of the processes are reduced to minimum up and down times or ramp constraints on the change of the production levels, neglecting the real dynamics of the processes. However, for slow processes and fast load changes, these formulations are too coarse and dynamic models of the behavior of the plants have to be included. This has been investigated by several authors for a variety of examples, e.g. Air Separation Units (Pattison et al., 2016) and a continuous zeolite production plant (Semrau et al., 2022). The economic and environmental advantages of demand side management for the zeolite plant were investigated in detail in Semrau and Engell (2023), assuming a perfect plant model. In this work, the effect of stochastic parameter uncertainties on the optimal day-ahead commitment of the electric energy demand is studied. First, the plant and the optimization problem are described. Then the robust dynamic optimization problem is presented and the results are shown for historic data for the prices of electric energy. Finally, the difference to the usage of a nominal model is quantified.

2. Problem statement

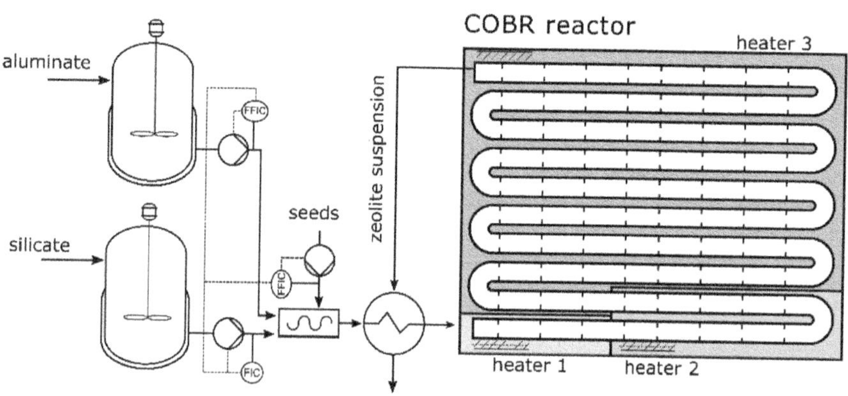

Figure 1: Scheme of the continuous zeolite production process

The process for the continuous production of NaX zeolites is shown in Figure 1. The silicate and the aluminate suspensions are mixed along with seed crystals and fed into the tubular reactor. The reactor is equipped with three heaters along the reactor. In the continuous oscillatory baffled reactor, the amorphous solids are converted to the desired crystalline zeolites. This takes place at elevated temperatures. The outlet stream is used to preheat the entering stream. The inputs of the system are the total flowrate and the heat duty of the three heaters. The zeolite crystal content in the outflow should be higher than 98%. The reaction temperature is limited to 403.5 K, due to the formation of side products. The production process is described in detail in Ramirez Mendoza et al. (2020). The reactor is modelled by the axial dispersion partial differential equation with reaction terms for the dissolution of the amorphous solid and the nucleation and growth of the zeolite particles. To solve the population balance of the zeolite particle formation, the method of moments is used. A general, notation of the model is given by (1) with the states x , the inputs u, and the parameters p.

$$\dot{x} = f(x, u, p) \tag{1}$$

2.1.1. Parametric uncertainty

Models are only simplified description of the plant behavior. Especially, the nucleation and growth of the crystalline particles is a highly complex, stochastic process. Therefore, the prediction of the plant behavior is prone to errors. In this work, a parametric stochastic deviation from the nominal model prediction is assumed, which can be described by a stochastic differential equation (2) with a linear time invariant matrices A, a, B and the Wiener process dW. The evolution of mean value $\bar{p}$ and the covariance matrix P can be described the differential equations (3),(4). The nucleation rate and the heat transfer coefficient of the preheater are assumed to be stochastic. The deterministic description of the system is replaced by a stochastic one.

$$dp(t) = (A\,p(t) + a)dt + BdW \quad (2)$$

$$\dot{P} = AP + PA^T + BB^T \quad (3)$$

$$\dot{\bar{p}}(t) = A\bar{p}(t) + a \quad (4)$$

$$d\begin{pmatrix} x \\ p \end{pmatrix} = \begin{pmatrix} f(x,u,p) \\ Ap + a \end{pmatrix} dt + \begin{pmatrix} 0 \\ B \end{pmatrix} dW \quad (5)$$

2.1.2. Economic cost structure

In the considered case, the energy costs and the revenue are considered. Therefore, the costs can be described by the normalized usage of electric energy multiplied by the normalized price $\gamma(t)$ subtracted by the relative throughput.

$$\phi(t) = \gamma(t)\,\frac{\sum_j^3 \dot{Q}_j}{\sum_j^3 \dot{Q}_{j,ref}} - \frac{F}{F_{ref}} \quad (6)$$

3. Method

In the following, the dynamic optimal schedule based on the day-ahead and the intraday electric energy price markets in the presence of stochastic parametric model-plant mismatch is presented.

3.1. The Unscented Transform

The Unscented Transform (Julier 2002) approximates the propagation of a multivariate Gaussian distribution $\mathcal{N}(m,\Sigma)$ of dimension N through a nonlinear function $g(m)$ with individual samples m, the sigma points. The expected value of the function $g(m)$ can be approximated by a weighted sum of the sigma points. The individual weights w_i are chosen, such that the corresponding moments of the distribution are accurate to the third order.

$$m \sim \mathcal{N}(\bar{m},\Sigma) \quad (7)$$

$$m_i = \bar{m} \pm \sqrt{(N+l)\Sigma}_i \quad (8)$$

$$\mathbb{E}(g(m)) \approx \sum_{i=0}^{2N} w_i\, g(m_i) \quad (9)$$

3.2. Robust dynamic optimization

The optimization problem under uncertainty is formulated in (10). The stage cost is the expected cost of electric energy. For the robustification of the solution, the constraints are satisfied for all sigma points. Similar approaches were used by e.g. Vallerio et.al.(2016). The optimization variables are the extended state vector $X(t)$, which consists of the states x_i for the corresponding sigma points p_i, and the input trajectory $u(t)$ which is the same for all sigma points. The terminal cost is introduced to reduce the terminal sell-off effect, it consists of a linear term with $\lambda_{ss}^{*,T}$ being the Lagrange multiplier at the optimal steady state (Zanon and Faulwasser, 2018). To further reduce the sell-off effect, the optimization problem is solved on two horizons, of which the second one is discarded, similar to model predictive control approaches. The constraints of the optimization are the dynamic model and the bounds on the process variables and the product quality, which need to be satisfied for all sigma points. Furthermore, the mean production is ensured.

$$\min_{X,U,E} \int_0^{tf} \mathbb{E}(\phi(t))dt + \lambda_{ss}^{*,T} X(t_f) \tag{6a}$$

$$s.t.: \dot{x} = f(x_i, u, p_i) \ \ \forall i \in (0,2N) \tag{6b}$$

$$x_i(0) = x_{0,i} \ \ \forall i \in (0,2N) \tag{6d}$$

$$0 \geq h(x_i, u, p_i) \ \ \forall i \in (0,2N) \tag{6e}$$

$$0.98 \leq X_i(t) \ \ \forall i \in (0,2N) \tag{6f}$$

$$F_{fix} = \frac{1}{t_{k+1} - t_k} \int_{t_k}^{t_{k+1}} F(t)\, dt \ \ \forall\, k \in [0,1] \ \forall i \in (0,2N) \tag{6g}$$

Time is discretized with the OCFE method using Radau polynomials of a second order. The optimization problem is implemented CasADi and solved with the NLP solver IPOPT. The length of the first horizon is chosen to be 24 h, the length of the second horizon is 12 h. The resulting nonlinear optimization problem has around 320.000 variables and is solved within 12h.

4. Results

For the selection of the sigma points, the mean and the covariance of the parameters are the steady state solution of the equation (2), (3). The resulting mean value of the parameter, is one, parameter covariance matrix is given by

$$P_{ss} = diag([65.53, 4.08])10^{-6}. \tag{7}$$

The tuning parameter of the Unscented Transform was chosen such that the isoprobability curve on which the Sigma Points lie, covers 95% of the distribution. The electric energy price trajectory for the day-ahead electric energy price is data from the EPEX Spot market from the 13.01.2020.

The results of the optimization are shown in Figure 2 for the nominal optimization, in which only the mean is considered and the robust optimization. The normalized electric energy price is shown in the bottom plot. The resulting optimal trajectories for the temperature (T) and crystallinity (X) at the outlet for the different sigma points.

Additionally, the throughput (F) and the heating powers for individual the heaters ($\dot{Q}_j$) are displayed. The energy usage is shifted in both cases to the low energy price period, while the constraints are satisfied for all sigma points. It is clearly observable, that the inputs vary in a smaller range, and the height of input movement is smaller.

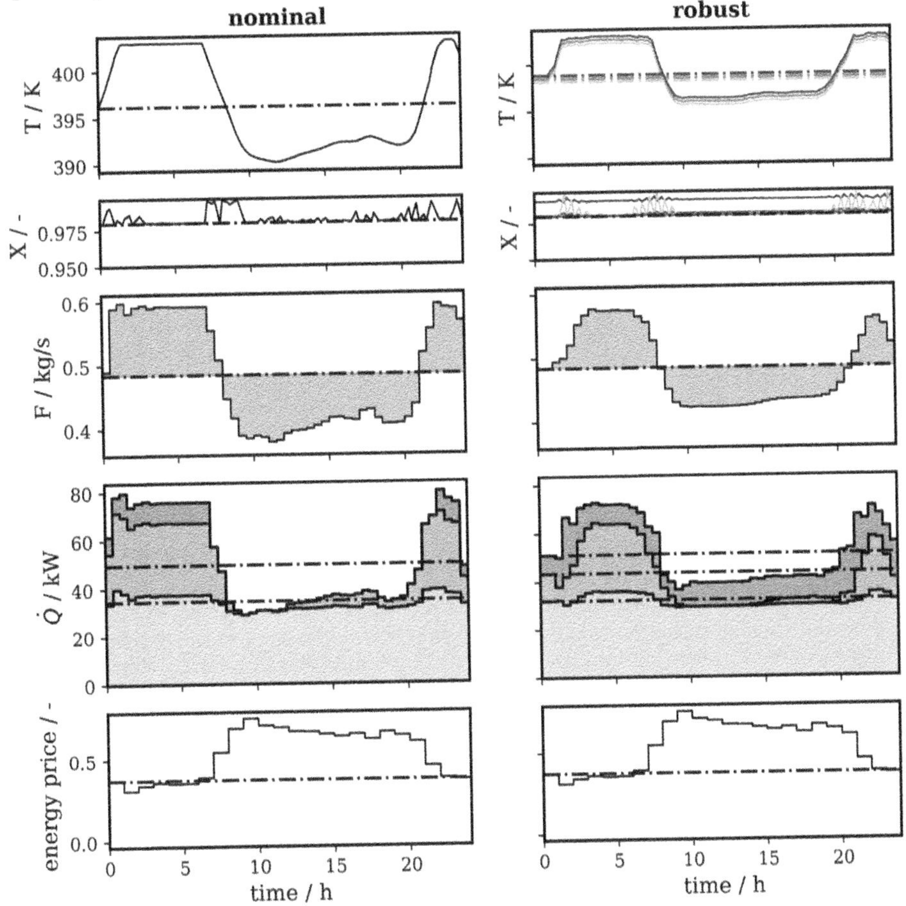

Figure 2: Optimal dynamic operation with energy price data for 13.01.2020 as the result of the robust optimization

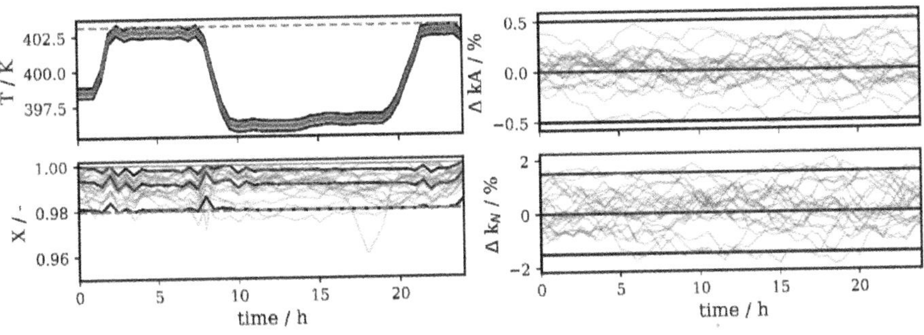

Figure 3: Application of the robust optimal trajectory to the stochastic system

To test the validity of the calculated solution the stochastic differential equation of the model (5) is numerically solved for 20 different realization of the noise the results are shown in Figure 3. In which, the outlet temperature, the crystallinity and the two varying parameters are shown. The grey lines represent the solution of the stochastic differential equation for different realizations, the constraints are marked in red and the considered sigma points in black lines. The robust optimal operation of the process limits the constraint violation to 3.67% instead of 33.26% in the nominal case. However, the economic energy cost savings decrease from 9.29% using the nominal trajectory to 4.13% for the robust optimization.

5. Conclusion

This work presents the extension of a moving horizon day-ahead optimization for the flexible usage of electric energy based on the day-ahead market by the inclusion of stochastic parametric model-plant mismatch. The problem statement and the robust optimization problem are introduced. The optimization problem is solved for a one-day operation using past energy price data. The benefit of using the optimization under uncertainty is demonstrated for multiple realization of the stochastic model description, demonstrating an improvement of the robustness with a negative impact on the calculated cost. Further work will address the estimator design to identify the model parameters and robust economic NMPC to be able to reduce the performance loss, while maintaining the robustness of the solution. Furthermore, the application of two stage stochastic programs will exploited to optimally choose the energy purchase on the different markets.

References

Andersson, J.A.E., Gillis, J., Horn, G., Rawlings, J.B., Diehl, M., 2019. CasADi: a software framework for nonlinear optimization and optimal control. Mathematical Programming Computation 11, 1–36.

Dalle Ave, G., Harjunkoski, I., Engell, S., 2019. A non-uniform grid approach for scheduling considering electricity load tracking and future load prediction. Computers & Chemical Engineering 129, 106506.

Julier, S.J., 2002. The scaled unscented transformation, Proceedings of 2002 American Control Conference, IEEE, Anchorage, AK, USA, pp. 4555–4559 vol.6.

Pattison, R.C., Touretzky, C.R., Johansson, T., Harjunkoski, I., Baldea, M., 2016. Optimal Process Operations in Fast-Changing Electricity Markets: Framework for Scheduling with Low-Order Dynamic Models and an Air Separation Application. Ind. Eng. Chem. Res. 55, 4562–4584.

Ramirez, H., Valdez, M., van Gerven, T., Lutz, C., 2020. Continuous flow synthesis of zeolite FAU in an oscillatory baffled reactor. Journal of Advanced Manufacturing and Processing 2.

Semrau, R., Engell, S., 2023. Process as a battery: Electricity price aware optimal operation of zeolite crystallization in a continuous oscillatory baffled reactor. Computers & Chemical Engineering 108143. https://doi.org/10.1016/j.compchemeng.2023.10814

Semrau, R., Yang, J., Engell, S., 2022. Process as a battery: Electricity price based optimal operation of zeolite crystallization in a COBR, in: Computer Aided Chemical Engineering. Elsevier, pp. 1165–1170

Vallerio, M., Telen, D., Cabianca, L., Manenti, F., Impe, J.V., Logist, F., 2016. Robust multi-objective dynamic optimization of chemical processes using the Sigma Point method. Chemical Engineering Science 140, 201–216.

Zanon, M., Faulwasser, T., 2018. Economic MPC without terminal constraints: Gradient-correcting end penalties enforce asymptotic stability. Journal of Process Control 63, 1–14.

Zhang, Q., Pinto, J.M., 2022. Energy-aware enterprise-wide optimization and clean energy in the industrial gas industry. Computers & Chemical Engineering 165, 107927.

Antonis Kokossis, Michael C. Georgiadis, Efstratios N. Pistikopoulos (Eds.)
PROCEEDINGS OF THE 33rd European Symposium on Computer Aided Process Engineering (ESCAPE33), June 18-21, 2023, Athens, Greece
 http://dx.doi.org/10.1016/B978-0-443-15274-0.50181-5

Machine Learning models to optimize dairy cow culling: the case of the experimental farm UniLaSalle Beauvais

[a,b]Abdifatah OUSELEYE,[a,c] José RODRIGUEZ, [a] Hamilton ARAUJO *

[a] *Institut Polytechnique UniLaSalle, 19 rue Pierre Waguet, Beauvais 60000, France*
[b] *Université de Haute-Alsace, 2 rue des Frères Lumière, 68100 Mulhouse, France*
[c] *Université d'Artois, ULR7519, Beauvais 60026, France*
**hamilton.araujo@unilasalle.fr*

Abstract

In the process of milk production, dairy farmers are confronted with the management of their herd. Indeed, when a group of cattle is in the lactation phase, several genetic and phenotypic factors can impact the milk production (W. L. Weigel, 2017). That way, good herd management can increase production, lower costs, and at the same time reduce the carbon emissions. In this sense, one of the pillars of a good herd management is the renewal rate of the cows (voluntary culling). However, optimize the decision to cull cow is very important. For example, culling a cow with a good milk production potential or keeping a cow in the herd with a non-satisfactory production is not economically profitable. Regarding this, the objective of this work was to process the database of the experimental farm of the UniLaSalle polytechnic institute and try to understand, using machine learning models, the phenotypic and genetic parameters that played an important role in the longevity of cattle cow.

Keywords: cow culling; milk production; ai-based decision tool; machine learning; optimization

1. Introduction

The current global scenario is very unstable and uncertain. In particular, the tensions created by the war in Ukraine, the energy crisis, and the environmental crisis, among other implications, create for instance a pression on the price of raw materials. An important input for dairy cows is soybean, see in the Figure 1 the price evolution (from 208,51 U$/ metric ton at the end of 2002 to 561,79 U$ / metric ton at the end of 2022 or an increase of 169% in 10 years)

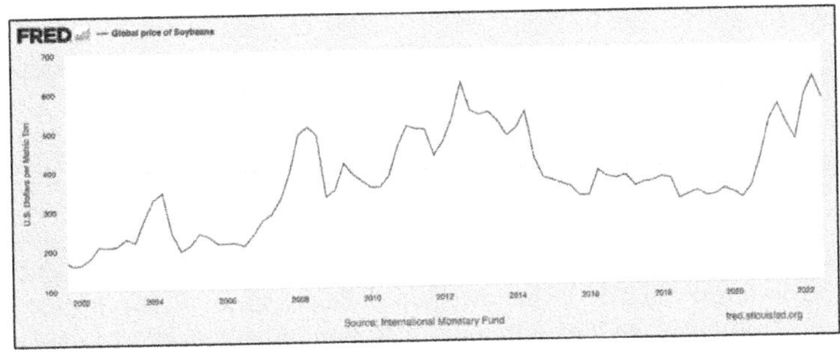

Figure 1: U.S. Dollars per metric ton of soybeans in the last 10 years. Source: https://fred.stlouisfed.org/series/PSOYBUSDQ (International Monetary Fund)

In this context, dairy farms face multiple challenges while remaining efficient and competitive. The trivial way to increase milk production on a farm would be to increase the size of the dairy herd. However, this approach will lead to economic and environmental consequences. In that sense, with the advancement of technology and genetic improvement of the herds, the trend observed in the field is the opposite: the number of daily cattle decreases and productivity increases, as illustrated in Figure 2 (from 2009 to 2020 a 5,40% decrease in the number of cows and a 14,75% increase in milk production in France)

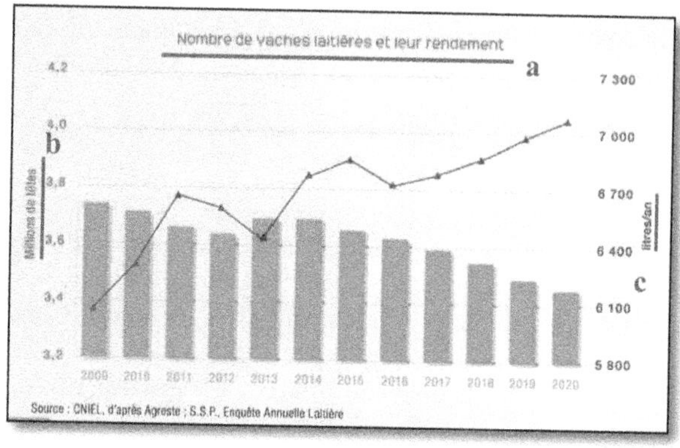

Figure 2: Milk production per year from 2009 to 2020 vs number of cows in France. Source: CNIEL, Agreste. (a: Number of dairy cows and their yield, b: millions of cows and c: liters of milk per year)

Knowing that the trend is to reduce the number of cattle and increase production, mastering herd management becomes very important. We should mention 3 aspects of herd management. First, if we keep productive cattle with good genetic and phenotypic parameters, the production tends to increase. Secondly, if we detect the least productive cattle, and the causes of this weak production, we can act to correct these malfunctions and reduce the costs of breeding. Finally, reduce the number of cows and keep the same production yield reduces the environmental impact and the economic. Accord to with (Lengagne, 2019) the average renewal rate of the herds is around 34% and the objective according to the Regional Chamber of Agriculture in 2022 (French Government) is to be less than 25%.

Milk production depends on several phenotypic and genetic factors. Phenotypic parameters are subject to external interferences (stochastic events) caused, as for example, by breeding conditions. For instance, diseases, artificial inseminations, or milk quality yields. These parameters can be explained, and they are influenced, by genetic indexes. In the sense that, the observations of reality are the expressions of the genetic indexes. Precisely, these genetic indexes tell us about the performance of the cows. On the field, the interactions experienced by animals in the environment will contribute positively or negatively to the milk production and it can disturb the veracity of these genetic variables.

On the other hand, with the advancement of data collection, processing and modeling, the use of phenotypic and genetic parameters is widely used to understand the milk production process (Weigel, 2017 and M. Lopez-Suarez et al, 2018). In that sense, our objective is to understand the longevity of a dairy cow. With genetic and phenotypic data can we explain the factors that have contributed, positively and negatively, in the longevity of a dairy cow?

2. The environment and the database

2.1. UniLaSalle's experimental farm

At *UniLaSalle*, a French engineering school in agriculture, based in Beauvais, northern region of France, "*La Ferme du Bois*" is an experimental farm used by the professors and the engineers-students to develop them projects. The farm is functional and productive to deserve the school's community.

2.2. The database

We have built up a database from our experimental farm with phenotypic data (milk production data, calving data, diseases data etc.) and genetic data (milk indexes, health index etc). See Figure 3. The phenotypic parameters are collected on the farm and referred in the database either by a farm employee or automatically by the milk robot. Genetic parameters are collected directly from the cattle by an external company and after referred it in the database.

	Example of variables in the database
532 cows	calving_calving_interval
	diseases and diseases_types
	artificial_inseminations
	milk_protein_rate
	milk_fat_rate
	milk_production_rate
	number_of_calves
	longevity
	milk_production_index
	milk_quallity_index
	general synthesis index

Figure 3: some variables in the database.

An important part of this work was cleaning up and extracting exploitable datasets for training machine learning models. In total we have 532 cattle already culled in the database, however if we would have, for instance, utilize 10 variables to evaluate their impact on the longevity we need enough culled cattle information to have a model.

3. Methodology and models

In a first step, we have processed this database to verify the veracity of some genetic indexes and its influence in the phenotypic parameters. Secondly, we applied some machine learning models to understand the longevity of already reformed cattle referenced in the database.

For this, we used a python environment, in form of Jupyter Notebook, on Google Colab and we used the basic libraries for manipulation and visualization of tabular data (Pandas, Numpy, Seaborn and Matplotlib). For the predictions we used the models available in Scikit-Learn library (Boyu.Ji, 2022).

3.1. Genetic index phenotypic performance

Initially, we have looked for the veracity of some genetic indexes by making a comparison with the collected phenotypic data. In general, we have noticed 3 different instances between the indexes and the phenotypic parameters that we can exemplify with these 3 examples (see figure 4): The milk index (genetic) is 53% correlated with the milk production (phenotypic). Then the milk index (genetic) is 61% correlated with the first lactation milk production (phenotypic). Finally, the butter index is 84% correlated with the butter index of the first lactation (Gabriel M. Dallago et al., 2022).

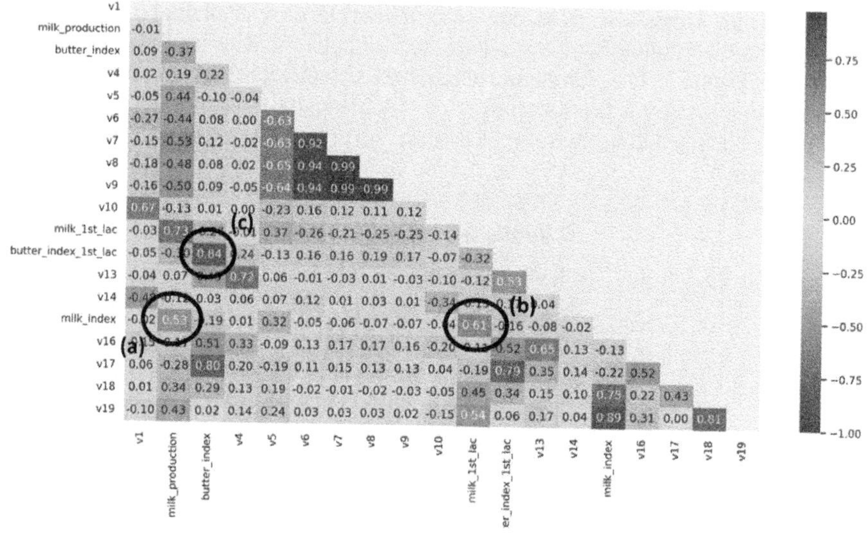

Figure 4: (a) milk_production correlated 53% with the milk_index (b) milk_index correlated 61% with the milk production in the first calving (c) butter_index correlated with 84% the butter_index in the first lactation

3.2. Machine Learning

We have deployed several supervised machine learning models, essentially linear regression, XGBoost and random forrest to predict a phenotypic variable: longevity (A.Essl, 1998). For that, we standardized data by

$$\tilde{x} = \frac{x - \bar{x}}{\sigma}$$

were $\bar{x}$ is the average of each variable and σ the standard deviation. To train our models we also used a 5-cross-validation test with parameters specific to each model. The data in each model was divided in 70% train et 30% test and the metric utilized to compare and to estimate the precision of the model was the R^2-metric. Regardless, we have also looked for the feature-importance of the of each prediction because it is important to understand the variables influence the longevity. In fact, initially all variables were considered and after some models were tested, we selected only the most pertinent variables with the objective of optimizing the models.

4. Results

In some senses, we have succeeded in explaining the longevity of a dairy cow with the genetic and phenotypic data was available. The Figure 5 below shows a visual example of the prediction of one of the trained models. The example shows predictions for the longevity of at least 60 cattle. It means that the dataset used initially had 200 cows with enough information for the selected variables (more than 30 variables). Thus, the model was trained with 140 cows (70% training) and was tested with 60 cows (30% test) as shown in the Figure 5.

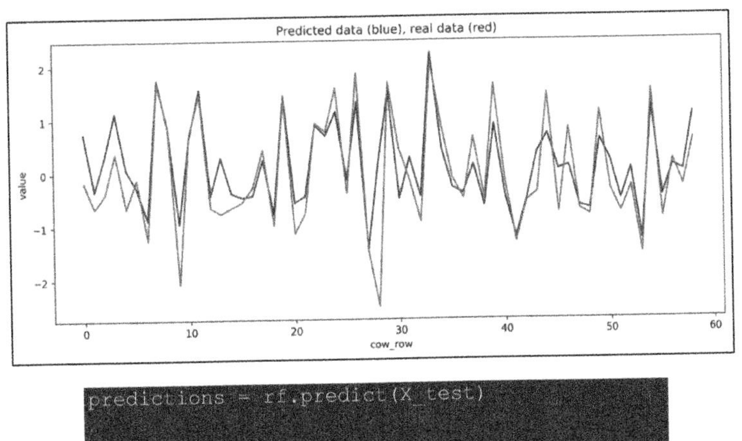


```
predictions = rf.predict(X_test)

print("R2: ", rf.score(X_train, y_train))
```

Figure 5: Random Forrest results: blue predicted values and red reality values

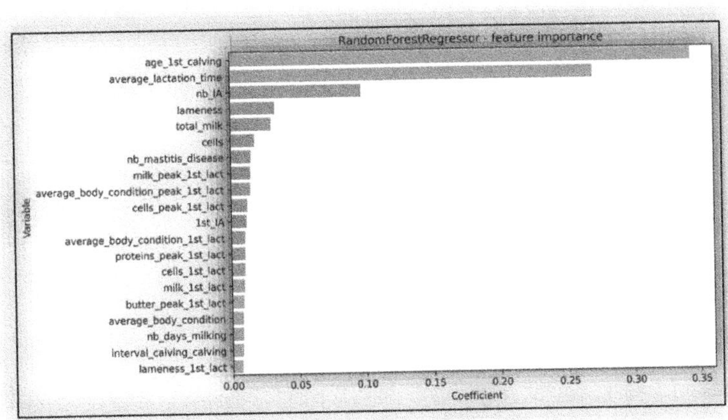

Figure 6: feature importance of the random forest model

In the figure 6 we can see that the age at first calving, the average of the duration of the lactation and the number of artificial inseminations are the most impacting variables concerning longevity. These factors must be evaluated and corrected by the cattle farmer.

5. Discussion and outlook

In terms of improvement, a major difficulty in this work was the data preparation and processing phases. The database has an important number of missing values for several variables. In this case, we reduced the quantity of cows to have more information about them in the models. Then, expanding our database is extremely important for the future. In addition, predicting longevity is important, but also linking milk production in the model is interesting. For instance, the following synthetic variable can be modeled bay the parameters in our database.

$$\frac{m \; x \; 300 \; x \; (l - c_{age})}{l \; x \; 365}$$

were m is the average of the dairy milk production, l is the longevity and c_{age} is the age at first calving. The value 300 is explained by the rest that the cow had after each calving, around two months.

Another axis of improvement is to find a way to use the information from the feature importance to help the cow farmers. It means, not only to have an estimate of the longevity of his cattle, but also what the breeder can act to improve the parameters impacting the longevity. Finally, training a clustering model to classify the cattle to be culled or not is also intended (Esther van der Heide, 2020).

The authors would like to mention that this work was carried out by the first author in the context of his master's thesis.

References

W. L. Weigel, 2017, A 100-Year Review: Methods and impact of genetic selection in dairy cattle—From daughter–dam comparisons to deep learning algorithms, Journal of Dairy Science, 100, 12, 10234-10250.

M. Lopez-Suarez et al, 2018, Using Decision Trees to Extract Patterns for Dairy Culling Management. 14th IFIP International Conference on Artificial Intelligence Applications and Innovations (AIAI), Rhodes, Greece. pp.231-239, https://hal.inria.fr/hal-01821072.

E. M. M. van der Heide, R. F. Veerkamp, M. L. van Pelt, C. Kamphuis, B. J. Ducro, "Predicting survival in dairy cattle by combining genomic breeding values and phenotypic information", Wageningen University & Research Animal Breeding and Genomics, PO Box 338, 6700 AH, Wageningen, the Netherlands, 2020

A.Essl, "Longevity in dairy cattle breeding: a review", Department of Livestock Sciences, University of Agricultural Sciences Vienna, Gregor-Mendel-Strabe 33, A-1180 Vienna, Austria,1998

Boyu.Ji, " A machine learning framework to predict the next month's daily milk yield, milk composition and milking frequency for cows in a robotic dairy farm", 2022

Gabriel M.Dallago, Roger I. Cue, Kevin M. Wade, René Lacroix, Elsa Vasseur, "Birth conditions affect the longevity of Holstein offspring", Department of Animal Science, McGill University, Sainte-Anne-de-Bellevue, Quebec H9X 3V9, Canada , 2022.

Lengagne L., Analyse du coût de renouvellement en élevage bovin lait : Analyse du renouvellement et de la carrière des vaches laitières, coût de production des génisses et identification de marges de progrès. Sciences du Vivant [q-bio]. 2019. ffdumas-02649011

Ousleye A., Rodriguez J., Araujo H., Machine Learning models to optimize dairy cow culling: the case of the experimental farm UniLaSalle Beauvais, preprint, 2022.

Antonis Kokossis, Michael C. Georgiadis, Efstratios N. Pistikopoulos (Eds.)
PROCEEDINGS OF THE 33rd European Symposium on Computer Aided Process Engineering (ESCAPE33), June 18-21, 2023, Athens, Greece
 http://dx.doi.org/10.1016/B978-0-443-15274-0.50182-7

Analyzing the Large-Scale Supply of Low-Carbon Hydrogen in Germany

Paul Sizaire[a], Emre Gençer[a]

[a]*MIT Energy Initiative, Massachusetts Institute of Technology, 400 Main Street, Cambridge, MA 02142, USA*

Abstract

The increasing political momentum advocating for decarbonization efforts, in Europe and elsewhere, has led many governments to unveil national hydrogen strategies. Hydrogen is viewed as a potential enabler of deep decarbonization, notably in hard to abate sectors such as the industry. A linear programming model was developed to assess the procurement of low carbon hydrogen in Germany, a large industrial consumer.
Results show that decarbonizing the current industrial demand (55 TWh/yr) with domestic electrolytic hydrogen would result in a levelized cost of hydrogen totaling ~\$5kg/H_2, requiring 22 GW of electrolyzer, 30 GW of solar and 15 GW of wind, highlighting the requirement to import. Retrofitting natural gas derived production will likely result in a carbon intensity exceeding low carbon requirements unless emissions from the natural gas supply chain are adequately addressed, engendering significant additional investments. Imports from neighboring countries such as Norway can significantly reduce costs and meet carbon intensity requirements if sourced correctly, leveraging the relatively short distance between countries using an undersea pipeline.
This model highlights the likely importance of imports in decarbonizing hydrogen procurement in Germany to address high costs of domestic electrolytic production and high carbon intensity of retrofitted natural gas derived production.

Keywords: hydrogen, supply chain, electrolyzer, linear programming, Germany

1. Introduction

Despite the ratification of the Paris agreement by 189 countries in 2015, carbon emissions have remained fairly constant and above 30 Gt per year (IEA, 2021). This has incited calls for further action, which resulted in several countries pledging to attain carbon neutrality by mid-century. An advocated decarbonization pathway promotes the reduction of carbon emissions from electricity generation and the electrification of several end-uses that currently rely on a variety of incumbent fuels. However, electrification may be impaired by feasibility or cost-effectiveness issues in several applications, notably in the industry sector. Among alternatives, a portfolio of low-carbon fuels such as hydrogen and biofuels are considered. Hydrogen's versatility is deemed to make it suitable to various end-uses such as industrial processes, long-haul transport, or renewable energy storage. This has incited several countries worldwide to release national hydrogen strategies, which highlights the potential future role of hydrogen in the decarbonization effort.

In Europe, the current largest consumer of hydrogen is Germany. It almost entirely consumes it in the chemical industry, predominantly for refining, followed by ammonia and methanol production. Germany produces most of its hydrogen, relying very little on imports, with an annual output reaching ~55 TWh. This production is dominated by a single process – Steam Methane Reforming (SMR), which uses natural gas as a fuel, across 44 plants in the country. The predominance of hydrogen production from SMR is associated with substantial carbon emissions - this process emits ~10 tCO_2/tH_2 (IEA, 2019). To date, Germany possesses less than 200 MW of electrolytic capacity. To comply with the amendment to the Climate Change Act stating a net-zero target by 2045 (Bundestag, 2021), Germany has produced a national hydrogen strategy. The government only considers electrolytic hydrogen produced from renewable electricity to be sustainable in the long term. The strategy focuses on demand applications that are close to commercial viability and those which cannot be decarbonized in other ways, citing the steel and chemical industries. It expects 90-110 TWh of hydrogen demand in 2030. Setting a target of 5 GW of electrolyzers in 2030 aiming to produce 14 TWh of hydrogen using 20 TWh of renewable electricity, Germany is aware that this will not be sufficient to meet the entire demand. As such, Germany views imports as crucial.

The ambitious targets are likely to face several challenges. Current global electrolytic capacity is dwarfed by the sheer scale of envisioned deployment - reaching Germany's target corresponds to a 75% compound annual growth rate until 2030. Electrolytic hydrogen may be expensive in the short- to medium- term which may impair competitiveness in the industrial sector, current incumbent of hydrogen demand. Furthermore, this production requires low-carbon electricity to provide low-carbon hydrogen, triggering an extensive addition of renewable energy in the German electricity grid, which is currently characterized by a relatively large carbon intensity. The creation of a holistic supply chain will also require transmission pipelines and large-scale storage.

To provide insights regarding the least cost scenario for a developed hydrogen supply chain while addressing the various concerns aforementioned, an optimization model has been developed. Using a linear approach, a multi-nodal system representing Germany's regions optimizes for electricity generation, hydrogen generation, transmission, and storage. Hydrogen produced in Norway sent through a pipeline is evaluated as a method of import. Several power generation, hydrogen production, and storage options are considered. This model aims to provide insights regarding the development of a hydrogen supply chain realistically, by implementing several objectives laid out in the German hydrogen strategy.

2. Methodology & Case Study

2.1. Overview

The model presented in this report is a multi-nodal system that seeks to optimize for a hydrogen supply chain comprising power and hydrogen production, transmission, and storage. It is developed from a macroeconomic modeler's perspective, thus optimizing for social welfare by minimizing the overall costs. A perfect foresight is adopted using power generation data, which will be described subsequently.

Each node in the system represents a region of Germany. Within all nodes, capacities of renewable generation from solar and wind, electrolyzers, Auto Thermal Reforming (ATR) and SMR plants, and hydrogen and electricity storage are optimized for. Power generation is assumed to be wholly used for powering electrolyzers and compressors for hydrogen storage and transmission. A portion of the electricity that cannot be used is therefore curtailed, not sent into the grid. Most of the infrastructure considered in this analysis is greenfield. The cost of electricity omits any taxes, according to the stated objective that will see the EEG surcharge non-applicable to electrolyzers (The Federal Government, 2020). The relative share of wind and solar resources to install is optimized for depending on the regions in which this production is located.

Transmission is optimized for both electricity and hydrogen. This model also performs a life cycle analysis, which is accounted for in the objective function via a carbon tax. It includes emissions for all infrastructure installed, which is assumed to linearly scale with increasing capacity. It notably includes carbon emissions not captured by CCS from ATR and SMR. Hydrogen storage is assumed to be provided by salt caverns. The technical potential of salt caverns has been evaluated and reported in several studies (Michalski et al., 2017; Welder et al., 2018). Hydrogen demand is assumed to come exclusively from industrial applications – as such, demand is considered to be flat throughout the year. Regional demand reflects their current relative hydrogen consumption.

2.2. Objective Function, Variables & Constraints

The objective function aims to minimize the annualized costs of the system, which comprises capital and operational expenditures of assets, fuel, and emissions costs. A set of electrical production infrastructure is denoted P_e and comprises solar PV, wind turbines, inverters, and electrical storage (Li-Ion batteries). Another set of hydrogen production/storage infrastructure is denoted P_h and comprises electrolyzers, ATR plants, storage and compressors, while P_{ng} contains SMR plants only. A set T denotes transmission, encompassing electrical cables as well as domestic and international hydrogen transmission. Finally, set E corresponds to electrolyzers, solar, and wind, while C includes SMR, ATR, and Norwegian hydrogen. Cap denotes the capacity of the assets, DF is a detour factor, AF is an annualized factor, and RCC is the retrofit cost. The resulting objective function is composed of several cost components, laid out below:

Capacity:
$$\begin{aligned}\boldsymbol{CC} = &\textstyle\sum_{p_e \in P_e} CapEx_{p_e} \times (AF_{p_e} + OpEx_{p_e}) \times \sum_{n \in N} Cap_{p_e,n} \\ &+ \textstyle\sum_{p_h \in P_h} CapEx_{p_h} \times (AF_{p_h} + OpEx_{p_h}) \times \sum_{n \in N} Cap_{p_h,n} \\ &+ \textstyle\sum_{t \in T} CapEx_t \times (AF_t + OpEx_t) \times \sum_{n_{in} \in N} \sum_{n_{out} \in N} Cap_{t,n_{in},n_{out}} * DF_{n_{in},n_{out}}\end{aligned} \tag{1}$$

Retrofit:
$$\boldsymbol{RC} = \textstyle\sum_{p_{ng} \in P_{ng}} RCC_{p_{ng}} \times AF_{p_{ng}} \times \sum_{n \in N} Cap_{p_{ng},n} \tag{2}$$

Fuel Cost:
$$\boldsymbol{FC} = Production_{ng} * Cost_{ng} + Production_{Norway\, h_2} \times Cost_{Norway\, h_2} \tag{3}$$

Em. Costs:
$$\boldsymbol{EC} = \textstyle\sum_{e \in E} Emissions_e \times Cap_e + \sum_{c \in C} Emissions_c \times Production_c \tag{4}$$

Objective:
$$\boldsymbol{OF} = \boldsymbol{CC} + \boldsymbol{RC} + \boldsymbol{FC} + \boldsymbol{EC} \tag{5}$$

The variables in the model relate to the capacities of all technologies, electricity flows, hydrogen flows, and hydrogen and electricity storage levels. The capacities are optimized for every node, while all other variables are additionally optimized hourly, enabling great insights into the temporal operation of the various flows within the supply chain. Three main constraints dictate the hourly production of electricity, associated flows, and electrolytic production ($CF_{n,h}$ = hourly capacity factor at node n and time h, F = flow, H = hours in a year, EE = electrolyzer efficiency):

Production:
$$Gen_{elec_{n,h}} = Capacity_{wind_n} \times CF_{wind_{n,h}} + Capacity_{solar_n} \times CF_{solar_{n,h}} \quad \forall n \in N, h \in H \tag{6}$$

flows:
$$\begin{aligned}Gen_{elec_{n,h}} = &F_{el \to etlz_{n,h}} + F_{el \to curtail_{n,h}} + F_{el \to storage_{n,h}} \\ &+ \textstyle\sum_{n_{out} \in N} F_{el \to transmission_{n_{in},n_{out},h}}\end{aligned} \quad \forall n \in N, h \in H \tag{7}$$

Electrolyzer:
$$Gen_{h_2,n,h} = \left(F_{el \to etlz_{n,h}} + F_{el_{storage} \to etlz_{n,h}} + \textstyle\sum_{n_{in} \in N} F_{el \to transmission_{n_{in},n_{out},h}}\right) \times EE \quad \forall n \in N, h \in H \tag{8}$$

Resulting electricity and hydrogen flows are subsequently used to model the operation of the supply chain, such as the compression, transmission, storage, and so on. They also set the required capacity variables with a variation of the constraint $flow_{n,h} \leq capacity_n \; \forall \, n \in N, h \in H$.

2.3. Case Study

The system analyzed in this report is Germany, in which 12 regions are considered and modelled as nodes. The focus of this analysis consists of short- to medium-term decarbonization of hydrogen production – as such, most costs are not projections but current values. Similarly, demand was assumed to equal the current demand of 55 TWh/yr, kept instead of the projected 100 TWh/yr due to large uncertainties. This demand was spread across regions according to the industrial hubs using data from Welder et al. (2018). The cost parameters of the model consist of the CapEx, OpEx, and lifetime of all assets considered, gathered from public sources. Emission values were also obtained from publicly available sources. Hourly renewable production data for wind and solar were gathered from the RE-Europe dataset, which provides granular hourly capacity factors for both technologies (Jensen and Pinson, 2017). Finally, a land eligibility analysis was performed to define thresholds for solar, wind, and salt caverns availability regionally.

An initial scenario considered a theoretical edge case in which all production is ensured with domestic electrolyzers, to illustrate the potential required infrastructure deployment and the associated costs entailed. Another more realistic scenario caps the electrolytic production to 14 TWh/yr (as per the German strategy target for 2030), while the remainder of the production comes from either new ATR plants or retrofitted SMR plants. A final scenario allows imports to occur between Norway and Germany.

3. Results

3.1. First Scenario: Self-Sufficiency with Electrolytic Hydrogen Only

In this situation, all hydrogen production stems from electrolyzers. The resulting supply chain as well as the $LOCH_2$ are illustrated in Figure 1. The resulting $LCOH_2$ is 5.12 €/kgH_2.The required renewable energy/electrolyzer capacity oversize ratio is ~2, resulting in a total of 11% of electricity curtailed. Total electrolytic capacity reaches 22 GW, which is more than twice as much as what Germany plans to build by 2040. By 2021, installed solar capacity totaled 58 GW and wind capacity reached 64 GW in Germany (IRENA, 2022). Such a system would therefore monopolize a significant portion of renewable capacity. This edge case highlights the requirement to either import or continue producing domestically using other processes.

Hydrogen transmission accounts for less than 1% of the total costs. While hydrogen storage negligibly participates in the levelized cost, the required capacity is significant. The technical potential, evaluated at ~200 TWh for Germany, heavily relies on assumptions – actual potential might be closer to the order of magnitude of the optimum capacity. The cost of storage is also notably complicated to quantify and bears significant uncertainties. Since the deployment of such assets is unlikely in the short-term, storage may have to be performed by tanks, which would drive up the levelized cost.

3.2. Addition of Natural Gas Derived Hydrogen with New Plants and Retrofits

A further self-sufficiency scenario was run with a cap on electrolytic hydrogen production set to 14 TWh/yr. The remainder of the demand, or 41 TWh/yr, is fulfilled by natural gas derived hydrogen production. Two cases were considered: all fossil production came from new ATR plants, or all fossil production came from SMR retrofits.

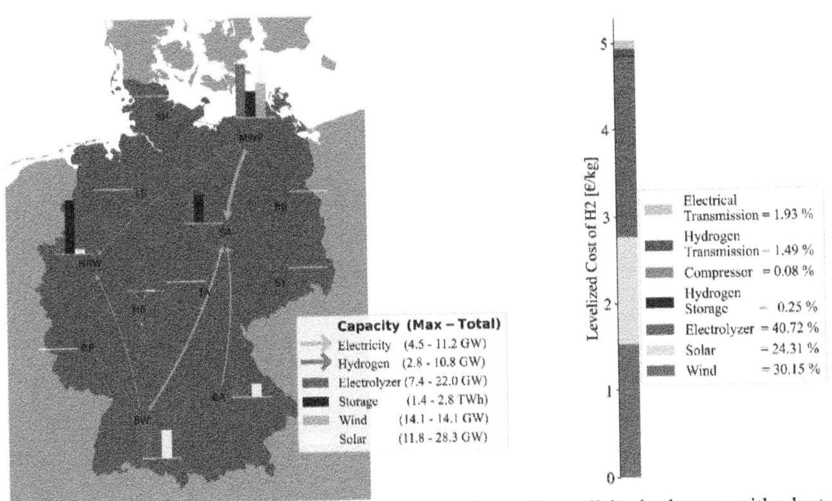

Figure 1: Supply chain required if Germany were to produce all its hydrogen with electrolyzers. "Max" represents maximum capacity between two regions (for electricity and hydrogen transmission) and within a region (for all other technologies), and "Total" represents overall capacity.

The resulting carbon intensities are laid out in the following table:

Table 1: Emission intensity for ATR (91% capture) and SMR (85%)

	ATR [$kgCO_{2,eq}/kgH_2$]	SMR [$kgCO_{2,eq}/kgH_2$]
Process (91% capture for ATR, 85% capture for SMR)	0.62	1.38
Indirect upstream (compression, transport infrastructure and flaring)	1.79	2.99
Electricity	1.44	1.77
Leaks and venting, production and transmission	1.35	2.25
Total	**5.20**	**8.38**
Emissions per Year	**6.39 $MtCO_2/yr$**	**10.31 $MtCO_2/yr$**

This assumes 9 $gCO_{2,eq}/MJ$ of natural gas for indirect upstream emissions (Bauer et al., 2021), 400 $gCO_{2,eq}/kWh$ intensity from the electricity grid, 1.5% methane leakage rate along the supply chain, a natural gas pipeline length of ~2000 km, and a 100 year global warming potential for methane. In contrast, electrolytic hydrogen has a carbon intensity of 0.8 $kgCO_2/kgH_2$. While direct emissions from the process are greatly reduced (down from ~10 $kgCO_2/kgH_2$), these results highlight the need to adequately reduce upstream emissions both from the natural gas supply chain as well as from the electricity.

3.3. Addition of Norwegian Hydrogen

This scenario considers imports of Norwegian hydrogen. This import is considered exogenously – hydrogen is assumed to be available at a constant rate and for a certain price (2€/khH_2). Once again, electrolytic production is capped at 14 TWh/yr, implying 41 TWh/yr of imports. Imports occur through an undersea pipeline. The results show that a pipeline with a 4.7 GW capacity is required, dispatching hydrogen at a constant flow rate. An equal total capacity of domestic pipelines (~10.5 GW) is required in this scenario as would be in the 100% electrolysis case. Storage requirements are greatly decreased, highlighting the benefits of a steady source of hydrogen as opposed to an intermittent production.

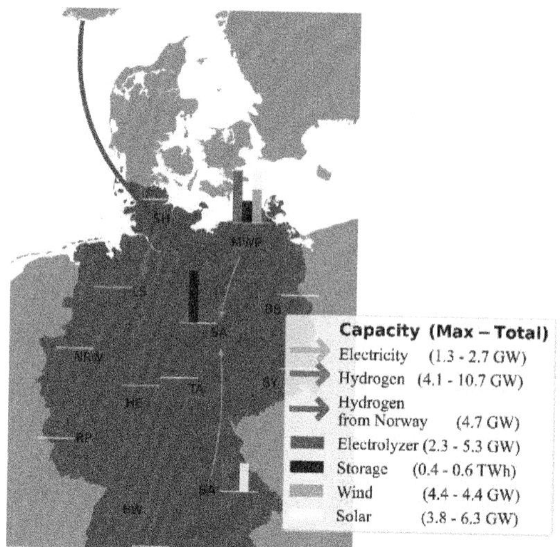

Figure 2: Hydrogen supply chain with 41 TWh/yr of imports from Norway.

Even though the levelized cost of electrolytic remains close to 5 €/kgH_2, sourcing cheaper hydrogen from imports would in this case reduce the average levelized cost to 3.1 €/kgH_2.

4. Conclusion

This paper demonstrated that reaching hydrogen production self-sufficiency in a decarbonization context would face several limitations: cost hurdles, capacity limitation, both for electrolyzers and renewable technologies, and scale of natural gas plant retrofit required. Imports are therefore likely to become central to decarbonize procurement and should be further evaluated.

References

Bauer, C., Treyer, K., Antonini, C., Bergerson, J., Gazzani, M., Gencer, E., Gibbins, J., Mazzotti, M., McCoy, S.T., McKenna, R., Pietzcker, R., Ravikumar, A.P., Romano, M.C., Ueckerdt, F., Vente, J., Spek, M. van der, 2021. On the climate impacts of blue hydrogen production. Sustain. Energy Fuels 6, 66–75.

Bundestag, 2021. Climate Change Act - climate neutrality by 2045.

IEA, 2021. Global Energy Review: CO2 Emissions in 2020. International Energy Agency, Paris.

IEA, 2019. The Future of Hydrogen - Analysis. International Energy Agency, Paris.

IRENA, 2022. Renewable Capacity Statistics 2022. International Renewable Energy Agency.

Jensen, T.V., Pinson, P., 2017. RE-Europe, a large-scale dataset for modeling a highly renewable European electricity system. Sci. Data 4, 170175.

Michalski, J., Bünger, U., Crotogino, F., Donadei, S., Schneider, G.-S., Pregger, T., Cao, K.-K., Heide, D., 2017. Hydrogen generation by electrolysis and storage in salt caverns: Potentials, economics and systems aspects with regard to the German energy transition. Int. J. Hydrog. Energy, Special Issue on The 21st World Hydrogen Energy Conference (WHEC 2016), 13-16 June 2016, Zaragoza, Spain 42, 13427–13443.

The Federal Government, 2020. The National Hydrogen Strategy. Federal Ministry for Economic Affairs and Energy, Berlin.

Welder, L., Ryberg, D.S., Kotzur, L., Grube, T., Robinius, M., Stolten, D., 2018. Spatio-temporal optimization of a future energy system for power-to-hydrogen applications in Germany. Energy 158, 1130–1149.

Antonis Kokossis, Michael C. Georgiadis, Efstratios N. Pistikopoulos (Eds.)
PROCEEDINGS OF THE 33rd European Symposium on Computer Aided Process Engineering (ESCAPE33), June 18-21, 2023, Athens, Greece
 http://dx.doi.org/10.1016/B978-0-443-15274-0.50183-9

Equilibrium sensitivity analysis of Carbon dioxide hydrogenation to methanol reaction system and comparison with conversions achievable in a simplified packed bed reactor model.

Milton D. Cárdenas[a], Camilo Rengifo [a], Manuel Figueredo [a]
[a]*Facultad de Ingeniería, Universidad de la Sabana, Campus del Puente del Común, Km. 7, Autopista Norte de Bogotá. Chía, Cundinamarca, Colombia.*
Contacting email: manuel.figueredo@unisabana.edu.co

Abstract
An equilibrium analysis of the reaction system for the synthesis of methanol based on CO_2 & CO hydrogenation and Water-Gas Shift reaction was performed to define the conditions under which the reaction system favours methanol production. The results were compared with a sensitivity analysis in a simplified Tubular Packed Bed Reactor (TPBR) with commercial $Cu/Zn/Al_2O_3$ catalytic bed. As a result, the highest feedstock conversion to methanol at equilibrium was found with increasing pressure, decreasing temperature, and setting the initial H_2/CO_2 ratio to 2.55. The reactor model showed similar behaviour in pressure and H_2/CO_2 inlet ratio but differed in the temperature effect on feedstock conversion to methanol. Differences between equilibrium analysis and final reactor conversions are briefly discussed.

Keywords: CO_2 hydrogenation, Equilibrium analysis, Tubular packed bed reactor (TPBR) simulation, Sensitivity analysis.

1. Introduction

Methanol is an important commodity commonly used in the chemical industry to produce diverse value-added chemicals. According to the analysis made by Sonthalia (2021), this molecule is the basis for synthesizing diverse important compounds such as alcohols, dimethyl ether (DME), formaldehyde, methyl tertiary butyl ether (MTBE), acetic acid, and methyl formate among others. This molecule is not only used not only as a feedstock but also as a fuel in internal combustion engines with great environmental advantages since it can be produced more easily from renewable energies than commonly used fuels (Verhelst et al., 2019). In the work made by Khojasteh-Salkuyeh (2021) are established some used methanol production pathways in the industry such as natural gas conventional reforming, dry-reforming, tri-reforming or direct carbon dioxide hydrogenation. Among them, implementing direct CO_2 hydrogenation can carry out some advantages compared with natural gas reforming (widely used production method), like providing a more controlled reaction process due to a lower temperature in the reaction and a favourable environmental analysis of net CO_2 emissions when using green H_2. The direct CO_2 hydrogenation to methanol is a reaction system highly dependent on equilibrium conditions, thus, this study aims to state the implication of temperature, pressure, and inlet feedstock ratio on the methanol production from an equilibrium approach with actualized data and relate these results with a sensitivity analysis in a simplified tubular packed bed reactor (TPBR).

2. Model Formulation

2.1 Model for species concentration at equilibrium

The carbon dioxide hydrogenation to methanol is an exothermic reaction system $\Delta H_r = -49.4\ kJ/mol$. Defined by Graaf (1986) as:

Reaction	Name	Abbreviature	$\Delta H_{i@298K}$
$CO + 2H_2 \leftrightarrow CH_3OH$	CO Hydrogenation	r_1	-90.5 kJ/mol
$CO_2 + H_2 \leftrightarrow CO + H_2O$	Water Gas Shift (WGS)	r_2	41.1 kJ/mol
$CO_2 + 3H_2 \leftrightarrow CH_3OH + H_2O$	CO_2 Hydrogenation	r_3	-49.4 kJ/mol

Table 1 – Reaction system for the synthesis of methanol from carbon dioxide.

The molar fraction of a component in a gas system when the equilibrium is reached $(y_i)_{eq}$ depends on the equilibrium constants $(K_{eq,j})$, the pressure (P) and temperature (T) according to the implicit relation in eq.1.

$$0 = K_{\phi,j}(T, P, y_i(\epsilon_j))_{eq} \cdot K_{y,j}\left(y_i(\epsilon_j)\right)_{eq} \cdot P^{\bar{\alpha}_j} - K_{eq,j}(T) \qquad \text{eq.1}$$

The equilibrium constants can be calculated with a polynomial temperature-dependent expression as presented in eq.2, the constants for this equation were taken from Graaf (2016).

$$\ln K_{eq,j}(T) = (1/RT) \cdot (a_1 + a_2T + a_3T^2 + a_4T^3 + a_5T^4 + a_6T^5 + a_7T \ln T)_j \qquad \text{eq.2}$$

The non-ideal behavior of gas species was corrected by calculating the fugacity coefficients $(\widehat{\Phi}_i)$ from the modified Soave-Redlich-Kwong equation of state (SRK-EoS) proposed by Van Bennekom (2012) and replacing these values in eq.3. $K_{y,j}$ was calculated with eq.4.

$$K_{\phi,j} = \prod_{i=1}^{n} \left(\widehat{\Phi}_i(T, P, y_i(\epsilon_j))\right)_{eq}^{\alpha_{ij}} \qquad \text{eq.3}$$

$$K_{y,j} = \prod_{i=1}^{n} \left(y_i(\epsilon_j)\right)_{eq}^{\alpha_{ij}} \qquad \text{eq.4}$$

Note that the composition is expressed as a function of the reaction extent $y_i(\epsilon_j)$ and this is useful to avoid extra degrees of freedom since each variation in species concentration follows a stoichiometric relation. Eq.5-6.

$$y_i(\epsilon_j) = \frac{n_{i0} + \sum_{j=1}^{k} \alpha_{ij}\epsilon_j}{n_T} \qquad \text{eq.5}$$

$$n_T(\epsilon_j) = n_0 + \sum_{i=1}^{n} \sum_{j=i}^{k} \alpha_{ij}\epsilon_j \qquad \text{eq.6}$$

Finally, by finding reaction extents values that make true the equality of eq.1, is possible to calculate the equilibrium compositions of each specie when temperature, pressure, and initial compositions changes.

2.2 Model for the tubular packed bed reactor

It is important to highlight that all variables affecting the concentration at equilibrium in eq.1 (T, P, y_i) can only be manipulated indirectly by affecting conveniently the reactor design variables and input conditions. To compare this ideal equilibrium approach to a more accurate analysis, it was proposed a simplified model for a tubular packed bed reactor (TPBR) and the mathematical model is summarized in Table 2.

Mass Balance		$\frac{dF_i}{dW} = r_i' = \sum_{j=1}^{2} c_{ij} \cdot r_j'$	eq.7
Energy Balance		$\frac{dT_R}{dW} = \frac{UA_s(T_j - T_R) + \sum_{j=1}^{2} -\Delta H_j \cdot r_j}{F_{tot} \cdot Cp_m}$	eq.8
	Co-Current Jacket Configuration	$\frac{dT_J}{dW} = \frac{UA(T_R - T_J)}{\dot{m}_c C_p}$	eq.9
	Counter-Current Jacket Configuration	$\frac{dT_J}{dW} = \frac{-UA(T_R - T_J)}{\dot{m}_J C_{pJ}}$	eq.10
Momentum Balance		$\frac{dP_R}{dW} = -\frac{G}{\rho_g g_c D_p}\left(\frac{1-e}{e^3}\right)\left[\frac{150(1-e)\mu}{D_p} + 1.75G\right] * \frac{1}{(1-e)A_c\, \rho_c n_t}$	eq.11

Table 2 – Mathematical model for a Tubular Packed Bed Reactor.

2.4 Model for properties estimation

The properties of the gas mixture (gas density, volumetric flow, heat capacity and viscosity) were calculated or estimated under ideal conditions considering the temperature and pressure in which the reactor simulation was developed (175°C - 255 °C and 25 bar – 60 bar respectively).

2.5 Kinetic model

The kinetic model was taken from Graaf (1988) and is summarized in Table 3. This kinetic model is based on widely used $Cu/Zn/Al_2O_3$ catalysts. This kinetic study implies no mass diffusion limitations inside the catalyst particle and no mass transfer limitation in the solid-gas interface.

r_1	=	$k_1 b_{CO} \cdot \dfrac{\left(P_{CO}P_{H_2}^{3/2} - \dfrac{P_{CH_3OH}}{P_{H_2}^{1/2}K_1}\right)}{\left(1 + b_{CO}P_{CO} + b_{CO_2}P_{CO_2}\right)\cdot\left(P_{H_2}^{1/2} + b_{H_2O}/b_{H_2}^{1/2}P_{H_2O}\right)}$	eq.12
r_2	=	$k_2 b_{CO_2} \cdot \dfrac{P_{CO_2}P_{H_2} - P_{CO}P_{H_2O}/K_2}{\left(1 + b_{CO}P_{CO} + b_{CO_2}P_{CO_2}\right)\cdot\left(P_{H_2}^{1/2} + b_{H_2O}/b_{H_2}^{1/2}P_{H_2O}\right)}$	eq.13
r_3	=	$k_3 b_{CO_2} \cdot \dfrac{\left(P_{CO}P_{H_2}^{3/2} - \dfrac{P_{CH_3OH}P_{H_2O}}{P_{H_2}^{3/2}K_3}\right)}{\left(1 + b_{CO}P_{CO} + b_{CO_2}P_{CO_2}\right)\cdot\left(P_{H_2}^{1/2} + b_{H_2O}/b_{H_2}^{1/2}P_{H_2O}\right)}$	eq.14

Table 3 – Kinetic model of the reaction system for CO_2 hydrogenation to methanol.

3. Computational and simulation aspects

For solving the reaction extents that make true the equilibrium relation in eq.1, it was used a trust-region algorithm with finite central numerical evaluation provided by Matlab integrated function *fsolve*. In this numerical process is present a high instability since $K_{\phi,j}(T,P,\epsilon_j)$ and $K_{y,j}(\epsilon_j)$ (eq.3-4) are defined by a product operator, and it is necessary to rearrange these equations to avoid singularities in the numerical evaluation. The reactor model was simulated with the Matlab function *ode45* and the optimization of both equilibrium and reactor was performed by *fmincon* Matlab function.

The variable to evaluate the effect of inlet conditions (temperature, pressure, and initial H_2/CO_2 ratio) on methanol concentration when equilibrium is reached was the feedstock conversion to methanol (X_C). The meaning of this response variable is the total conversion of hydrogen and carbon dioxide to methanol when the equilibrium is reached compared to the stoichiometric quantities (eq.15).

$$X_C = \frac{\left(n_{CH_3OH}\right)_{eq}}{\left(\left(n_{CO_2}\right)_0 + \left(n_{H_2}\right)_0\right)} \cdot \left(\frac{\alpha_{CH_3OH}}{\alpha_{CO_2} + \alpha_{H_2}}\right)^{-1} \quad \text{eq.15}$$

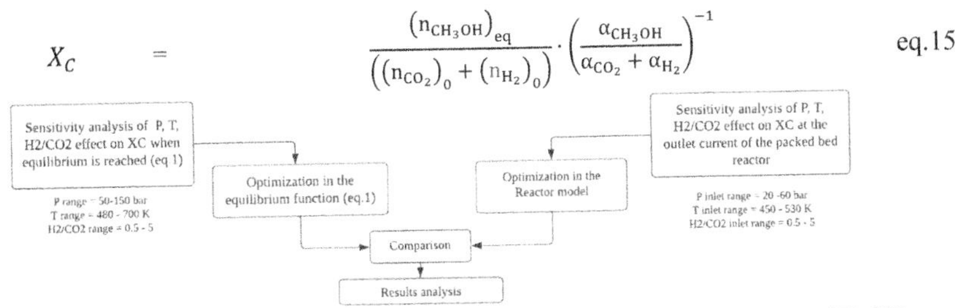

Figure 1 – General diagram to compare the effect of inlet temperature, pressure and H_2/CO_2 on both equilibrium and the reactor final current.

4. Results

4.1 Sensitivity analysis and optimization of feedstock conversion at equilibrium

The temperature and pressure effect on X_C were evaluated according to values in Table 4. As a result, it was evidenced a higher conversion of carbon dioxide and hydrogen to methanol at lower temperatures and higher pressures (Figure 2a). When evaluating the inlet H_2/CO_2 ratio effect on X_C, it was evidenced a maximum value of feedstock conversion to methanol in a H_2/CO_2 ratio equal to 2.55 (Figure 2b).

Constrained variables	Evaluation range		Optimal Values
	Min	Max	
Temperature	480 K	700 K	480 K
Pressure	50 bar	150 bar	150 bar
H_2/CO_2 ratio	0.5	5	2.55
			$X_c = 0.79$

Table 4 – Optimal temperature, pressure and H_2/CO_2 ratio to maximize feedstock conversion to methanol at equilibrium in the constrained numerical optimization of eq.1.

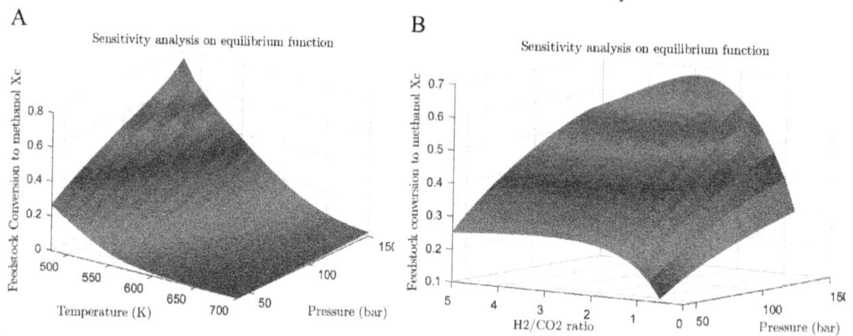

Figure 2 – Temperature, pressure and initial H_2/CO_2 effect on feedstock conversion to methanol. a) T and P effect on X_c. b) P and inlet H_2/CO_2 ratio effect on X_c.

4.2 Sensitivity analysis and optimization of feedstock conversion in the TPBR

Only were evaluated temperature, pressure, and inlet H_2/CO_2 effect, and other reactor-related variables was considered as parameters. When the sensitivity analysis was performed in the reactor model it was found a similar effect of inlet pressure and H_2/CO_2 ratio on X_C as in equilibrium function (Figure 3a-b) but different in the temperature analysis (Figure 3c). An increasement of the inlet temperature of feedstock current or in the liquid at the different jacket configurations increased X_c until limit value in which the trend was reversed and finally followed the same tendency as in equilibrium analysis. Note that the equilibrium is not reached in the TPBR at the end of the process and the temperature and pressure ranges differs in both equilibrium and reactor sensitivity analysis; this is why all values of X_C in the reactor model are lower than the direct analysis of the equilibrium expression (see Figure 3d). The optimization constrains of the TPBR are in Table 5.

	Evaluation range			Optimal Values
	Min		Max	
Temperature	Feedstock	480 K	530 K	508 K
	Jacket - Countercurrent	480 K	530 K	530 K
	Jacket – Co-current	480 K	530 K	530 K
Pressure	20 bar		60 bar	60 bar
H_2/CO_2 ratio	0.5		5	4
				$X_c = 0.165$

Table 5 – Optimal temperature, pressure and H_2/CO_2 ratio to maximize feedstock conversion to methanol at the outlet of the TPBR in the constrained numerical optimization of EDO in Table 2.

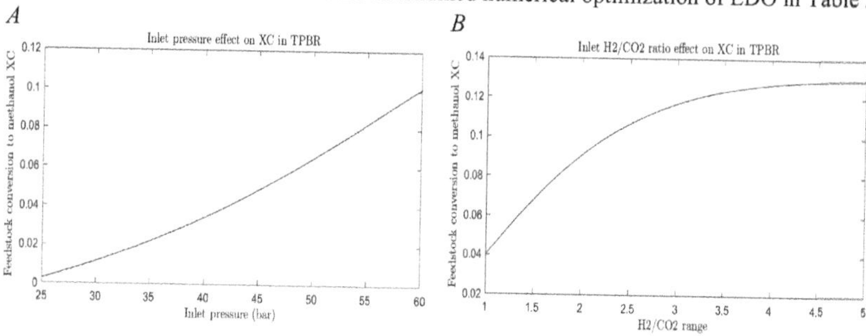

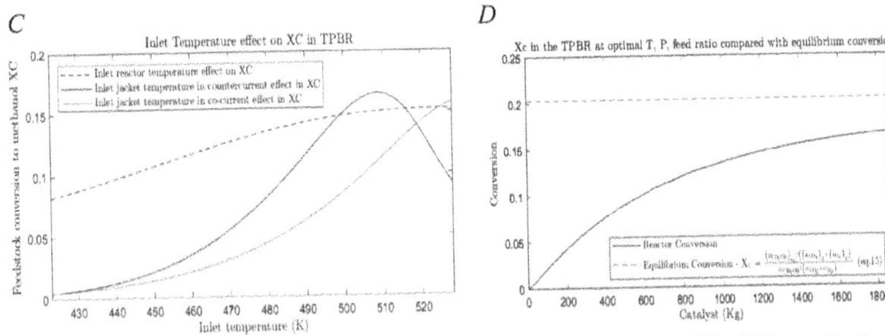

Figure 3 – a) Effect of inlet pressure on X_C in the TPBR. b) Effect of inlet H_2/CO_2 on X_C the TPBR. c) Effect of inlet temperature (Inlet species temperature vs inlet Jacket temperature) on X_C in the TPBR. d) Direct comparison of equilibrium analysis and de conversions achievable in the proposed TPBR.

5. Comparison and brief results discussion

5.1 Temperature

Temperature affects mainly the equilibrium constants ($K_{eq,j}$) and the fugacity coefficients of the components in the mixture ($\widehat{\phi}_i$). The evaluation of eq.1 in Figure 2a shows that lower temperatures favor methanol synthesis. As higher the temperature, higher the methanol oxidation to carbon dioxide and higher the production of undesirable products like CO and H_2O (WGS reaction). When the temperature was evaluated in the TPBR, the tendency was different, among others, because the activation energies in Graaf (1988) kinetic model. For instance, in Figure 4 the reaction rate of methanol is favored with temperature despite the different tendency in equilibrium sensitivity analysis. Then, it is necessary to optimize temperature to be high enough to reach a good react rate and be low enough to achieve good feedstock conversion to methanol. Finally, the optimal inlet temperature is different depending on the heat dissipation system properties throughout the catalytic bed weight. In the proposed model, the countercurrent configuration reached a higher feedstock conversion to methanol faster than co-current configuration.

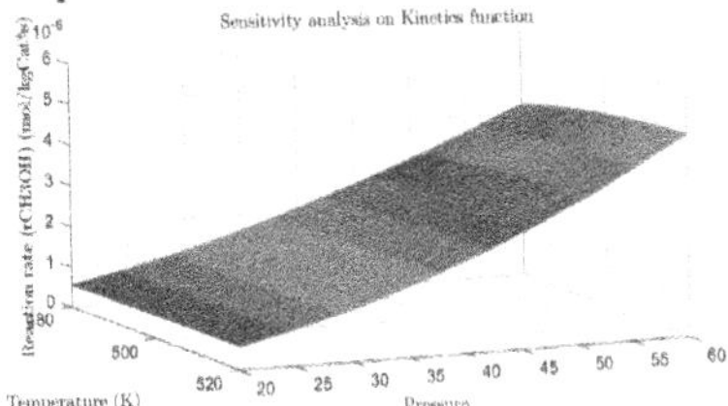

Figure 4 – Temperature and pressure effect on methanol reaction rate in Graaf (1988) kinetic model at the inlet conditions of the TPBR reactor.

5.2 Pressure and H_2/CO_2 ratio

The equilibrium composition of methanol increases with pressure (Figure 2, 3a). Pressure is directly related with concentration of moles in the gas phase and the species activity. When the H_2/CO_2 ratio effect is analyzed (Figure 2b), the feedstock conversion to methanol reaches a maximum value when inlet feedstock ratio is 2.55. This value represents an optimal feedstock input and is a quantity close to the stoichiometric relation in the reaction.

6. Conclusions & further recommendations

As lower the values of temperature and higher the pressure in the gas-phase system, the feedstock conversion to methanol increases, but the temperature should be enough to promote the methanol reaction rate to achieve considerable feedstock conversion in the reactor geometry. When temperature raises, the system shifts towards carbon monoxide and water production. For a constrained optimization of equilibrium expression, the optimal values of temperature, pressure and H_2/CO_2 ratio values were 480 K, 150 bar and 2.55 respectively. The temperature control in the TPBR model depends on the jacket fluid configuration and the characteristics of the heat dissipation system. For this specific set of conditions, it was found a jacket liquid temperature in countercurrent to maximize X_c of 508K, an inlet pressure of 60 bar and inlet H_2/CO_2 ratio of 4. Also, is secure to affirm that an increasement of the pressure is an excellent form to increase feedstock conversion to methanol due to its positive effect in both equilibrium expression and TPBR model analysis. The most substantial differences between the analysis of the equilibrium expression and the reactor model may lie in the fact that this reactor model is limited and cannot reach a closer conversion to equilibrium concentrations with the specified reactor design parameters. Another limitation is the small range of temperatures and pressures over which the reactor model is valid. Further analysis should robust the reactor model and compare all the temperatures and pressures ranges in the reactor with the equilibrium expression. But, with the basis relations founded in this simple reactor model is possible to understand the fundamentals of the reaction behavior and design more complex reactor models.

7. Notation table

T_R	Reactor Temperature (K)	U	Global Heat Transfer coefficient (J/m^2sK)
T_J	Jacket Temperature (K)	A_s	Heat Transfer Area ($m^2/kgCat$)
P_R	Reactor pressure (bar)	$\widehat{\phi_i}$	Fugacity coefficient of the components in mixture (bar)
F	Total molar flow (mol/s)	y_i	Molar fraction in gas phase (-)
F_i	Species molar flow (mol/s)	ϵ_j	Reaction extent
H_2/CO_2	Inlet Hydrogen vs Carbon dioxide relation (-)	P, T	Pressure (bar) – Temperature (K) for equilibrium function analysis.
ρ_g	Gas mixture density (kg/m^3)	α_{ij}	Stoichometric coefficents
μ_g	Gas mixture viscosity ($Pa \cdot s$)	$\overline{\alpha_j}$	Sum of stoichiometric coefficients for j reaction.
G_g	Gas mass Flux density ($k_g m^{-2} s^{-1}$)	K_ϕ	Equilibrium constant based in fugacity coefficients
Cp_g	Molar gas mixture heat capacity (J/molK)	K_y	Equilibrium constant based on molar fractions
Cp_J	Molar Jacket liquid heat capacity (J/molK)	K_p^0	Equilibrium constant based in partial pressures
e	Catalyst void fraction (-)	X_c	Feedstock conversion to methanol (-)
ρ_{cat}	Catalyst density (kg_{cat}/m^3)	$\Delta H_{r,j}^0$	Enthalpy change of formation (J/mol)
D_p	Particle diameter (m)	r'	Reaction rate per weight of catalyst (mol/skg_{cat})
n_t	Tubes number (-)	Subscripts	
D_t	Tubes diameter (m)	i	Refers to species $n = \{1,2,3,4,5\}$
A_c	Tube cross-sectional area (m^2)	j	Refers to reactions k = {1, 2, 3}
g_c	Conversion factor (SI=1)	0	Refers Initial state
b	Adsorption constant	eq	Refers to equilibrium

References

Graaf, G. H., Stamhuis, E. J., & Beenackersz, A. A. C. M. (1988). Kinetics Of Low-Pressure Methanol Synthesis. In *Chemical Engineering Science* (Vol. 43, Issue 12).

Khojasteh-Salkuyeh, Y., Ashrafi, O., Mostafavi, E., & Navarri, P. (2021). CO2 utilization for methanol production; Part I: Process design and life cycle GHG assessment of different pathways. *Journal of CO2 Utilization, 50*. https://doi.org/10.1016/j.jcou.2021.101608

Sonthalia, A., Kumar, N., Tomar, M., Edwin Geo, V., Thiyagarajan, S., & Pugazhendhi, A. (2021). Moving ahead from hydrogen to methanol economy: scope and challenges. *Clean Technologies and Environmental Policy, 1*, 1–25. https://doi.org/10.1007/S10098-021-02193-X/TABLES/7

van Bennekom, J. G., Winkelman, J. G. M., Venderbosch, R. H., Nieland, S. D. G. B., & Heeres, H. J. (2012). Modeling and experimental studies on phase and chemical equilibria in high-pressure methanol synthesis. *Industrial and Engineering Chemistry Research, 51*(38), 12233–12243. https://doi.org/10.1021/ie3017362

Verhelst, S., Turner, J. W., Sileghem, L., & Vancoillie, J. (2019). Methanol as a fuel for internal combustion engines. *Progress in Energy and Combustion Science, 70*, 43–88. https://doi.org/10.1016/J.PECS.2018.10.001

Antonis Kokossis, Michael C. Georgiadis, Efstratios N. Pistikopoulos (Eds.)
PROCEEDINGS OF THE 33rd European Symposium on Computer Aided Process Engineering (ESCAPE33), June 18-21, 2023, Athens, Greece
 http://dx.doi.org/10.1016/B978-0-443-15274-0.50184-0

Improving Transferable Force-Fields for Describing Crystal Structures Containing Hydrogen-Bonds

Benjamin I. Tan[a], David H. Bowskill[b], Adam Keates[b], Constantinos C. Pantelides[a] and Claire S. Adjiman[a]

[a] *Department of Chemical Engineering, Sargent Centre for Process Systems Engineering and Institute for Molecular Science and Engineering, Imperial College London, London SW7 2AZ, United Kingdom*

[b] *Process Studies Group, Syngenta, Jealott's Hill International Research Centre, Bracknell, Berkshire, RG42 6EY, United Kingdom*

Abstract

In Crystal Structure Prediction, transferable force-fields (FF) offer an efficient means to construct cheap surrogates of more accurate models. FF development, however, demands large volumes of high-quality reference data which are scarce in experimental databases. In this work, we use periodic DFT-D generated reference data to parameterize a FF apt for predicting properties of crystals in which molecules interact via hydrogen-bonds. Tang-Toennies damping is applied to the underlying Buckingham potential, and several fitting schemes are tested. Our FF achieves good agreement with the energy and geometry training data, for structures with and without hydrogen-bonds. When applied to a validation set, the quality of property predictions are largely preserved, evidence of FF transferability. Testing the popular FIT FF on our data fails to achieve the same accuracy, likely since it was derived from different reference data and lacks the damping function.

Keywords: Crystal Structure Prediction, Transferable Force-field, Parameter Estimation

1. Introduction

Many industrial process intermediates and commercial products exist in crystalline form. To predict the physicochemical properties of such materials requires knowledge of the three-dimensional arrangement of molecules within their crystal lattice. This, however, is complicated by the fact that most molecular crystals can exist in geometries other than their thermodynamically most stable form (Cruz-Cabeza *et al.*, 2015). Failure to identify these metastable polymorphs can have severe financial, legal, and social repercussions (Bučar *et al.*, 2015), especially since the interconversion to a more stable form is spontaneous yet highly unpredictable if the metastable form is kinetically trapped.

The central goal of Crystal Structure Prediction (CSP) is thus to characterize the structure and thermodynamic stability of all putative crystal forms from a description of only molecular connectivity (Pantelides *et al.*, 2014). Most CSP methodologies seek to accomplish this by minimizing the crystal static lattice-energy (U_{latt}) relative to its geometry at 0 K and 0 Pa (Bowskill, 2021). By minimizing an ample number of crystalline geometries, an adequate survey of the crystal structure landscape may be achieved. Energetically low-lying minima, within 10 kJ/mol of the global minimum (Nyman & Day, 2015), are then regarded as potentially observable experimental forms.

The validity of any CSP study is thus dependent on both the breadth of the search and the accuracy of the energy model used. Hybrid energy models which use a combination of *ab initio* calculations and empirical force-fields (FFs) offer a good balance between accuracy and computational cost. Whilst the *ab initio* components of these models have been studied extensively, the accompanying empirical FFs have not. Transferable FFs like FIT (Coombes *et al.*, 1996) and W99rev (Pyzer-Knapp *et al.*, 2016) are widely used despite deficiencies in the estimation of FF parameters (Gatsiou *et al.*, 2018). For instance, reliance on experimental heat of sublimation (ΔH_{sub}) data to provide "experimental" values of U_{latt} limits FF validity for two reasons. First, the relationship used to derive U_{latt} from ΔH_{sub} is approximate. Second, experimental ΔH_{sub} values are scarce, resulting in FF parameters biased towards geometry prediction. Finally, both FFs mentioned were derived using multipole (MP) descriptions of electrostatics. This may invalidate their use in the initial stages of CSP, with point-charge (PC) electrostatics.

In the larger context of CSP, dispersion-corrected periodic density-functional theory (DFT-D) is regarded as the gold-standard of accuracy (Brandenburg & Grimme, 2016) and is usually applied at end of CSP studies to refine results obtained with cheaper models. In the spirit of a blind CSP study, it is thus more sensible to design hybrid models that act as surrogates for the DFT-D landscape since the true experimental landscape remains unknown. For these reasons, in this work, we seek to develop a robust FF compatible with PC electrostatics, fit to high-quality DFT-D generated reference data. Details of the energy model and fitting procedure are described in Section 2. In Section 3, the performance of the newly parameterized FF is then explored.

2. Methodology

2.1. Lattice-Energy Model

Within our energy model, U_{latt} comprises intra- and intermolecular contributions,

$$U_{\text{latt}}(\boldsymbol{X},\boldsymbol{p}) = \Delta U_{\text{intra}}(\boldsymbol{X}) + U_{\text{inter}}(\boldsymbol{X},\boldsymbol{p}) = \Delta U_{\text{intra}}(\boldsymbol{X}) + U_{\text{elec}}(\boldsymbol{X}) + U_{\text{FF}}(\boldsymbol{X},\boldsymbol{p}), \quad (1)$$

where $\boldsymbol{X}$ is the vector describing the crystal geometry (cell lengths and angles, and atomic positions). U_{latt} contains the energy cost of distorting a molecule's conformation relative to its gas-phase conformation (ΔU_{intra}), and stabilizing packing forces (U_{inter}). U_{inter} comprises intermolecular charge-charge interactions (U_{elec}), electron-correlation, and exchange-repulsion. The latter two effects are described by the pairwise summation of interactions between atoms i and j separated by distance r_{ij}, interacting via a Buckingham potential (BHP) with model parameters $\boldsymbol{p} = \{A_{ij}, B_{ij}, C_{ij}\}$,

$$U_{\text{FF}}(\boldsymbol{X},\boldsymbol{p}) = \sum_i \sum_j A_{ij} \cdot \exp\left(-\frac{r_{ij}}{B_{ij}}\right) - f_6(\beta_{ij}, r_{ij}) \cdot \frac{C_{ij}}{r_{ij}^6}, \quad (2)$$

$$f_6(\beta_{ij}, r_{ij}) = 1 - \left[\sum_{k=0}^{6} \frac{(\beta_{ij} \cdot r_{ij})^k}{k!}\right] \cdot \exp(-\beta_{ij} \cdot r_{ij}) \quad \text{where } \beta_{ij} = \frac{1}{B_{ij}}. \quad (3)$$

In its native form ($f_6 = 1$) the BHP potential exhibits unphysical short-range attractions. When A_{ij}/B_{ij} is small and/or C_{ij} is large, this unphysical region extends to longer ranges and eventually the potential can become fully attractive. To mitigate this occurring in highly attractive hydrogen-bond (HB) interactions, the Tang-Toennies (Tang & Toennies, 1984) damping function (f_6) is applied within our potential. Its damping factor (β_{ij}) is correlated to the B_{ij} repulsion term to avoid overfitting. Finally, U_{elec} and ΔU_{intra} are determined from *ab initio* calculations performed on the isolated molecules of interest, while the $\boldsymbol{p}$ in U_{FF} are found empirically by fitting to crystal data. As a result, U_{FF} absorbs systematic errors from the *ab initio* components of the model (Pantelides *et al.*, 2014). For instance, the inductive effects molecules exert on each other in the crystal lattice are neglected by isolated molecule calculations and must be captured by U_{FF}.

2.2. Reference Data

DFT-D reference data are generated in the Vienna Ab initio Simulation Package (VASP v5.4.4, Kresse & Joubert, 1999) using the Tao-Perdew-Staroverov-Scuseria (Tao *et al.*, 2003) functional with D3 dispersion-correction (Grimme *et al.*, 2010). Experimental structures from the Cambridge Structural Database are first relaxed to obtain reference geometries. A single-point calculation is done on isolated molecule m at its crystalline conformation to obtain its monomer energy ($U_{\mathrm{iso},m}$). Using the number of molecules in the unit cell (Z) and asymmetric unit cell (Z'), the reference energy ($U_{\mathrm{inter,ref}}$) is found from

$$U_{\mathrm{inter,ref}} = \frac{U_{\mathrm{total,vasp}}}{Z} - \frac{\sum_{m=1}^{Z'} U_{\mathrm{iso},m}}{Z'}. \tag{4}$$

The training set comprises 313 non-HB (NHB) crystals and 132 HB crystals, while an additional 28 structures (12 NHB and 16 HB) are used in the validation set. More details about the computational methods and training structures are described in Bowskill (2021).

2.3. Parameter Estimation Algorithm

The parameter estimation is conducted using CrystalEstimator (v1.4) as described in Bowskill (2021). During the parameter estimation, the following problem is solved,

$$\min_{\boldsymbol{p}} \frac{1}{2}\sum_{i=1}^{N_c}\left\{\sum_{j=1}^{N_{X,i}} \frac{\omega_g\left[X_{ij}^{\mathrm{ref}} - X_{ij}^{*}(\boldsymbol{p})\right]^2}{N_{X,i}\cdot\left[X_{ij}^{\mathrm{ref}}\right]^2} + \sum_{j=1}^{N_{R,i}} \frac{\omega_g\left[R_{ij}^{\mathrm{ref}} - R_{ij}^{*}(\boldsymbol{p})\right]^2}{3(N_{R,i}-1)} + \frac{\omega_e\left[U_{\mathrm{inter},i}^{\mathrm{ref}} - U_{\mathrm{inter},i}^{*}(\boldsymbol{p})\right]^2}{\left[U_{\mathrm{inter},i}^{\mathrm{ref}}\right]^2}\right\}, \tag{5}$$

where X_{ij} is the jth lattice cell parameter (lengths and angles) for crystal i, R_{ij} its jth atomic fractional coordinate, and $U_{\mathrm{inter},i}$ its intermolecular energy respectively. The number of crystals is N_c, and the dimensions of X_i and R_i are crystal-specific ($N_{X,i}$ and $N_{R,i}$). ω_g and ω_e are the respective weights for the geometry and energy residuals. The superscripts 'ref' and '*' respectively indicate the reference values and the values calculated by minimizing U_{latt} at a given $\boldsymbol{p}$. The parameter estimation is thus a bilevel optimization since at each trial $\boldsymbol{p}$, all reference crystals must have their U_{latt} minimized first. In this minimization, molecules are kept rigid at their reference conformation ($\Delta U_{\mathrm{intra}} = 0$, $U_{\mathrm{latt}} = U_{\mathrm{inter}}$) and U_{elec} is calculated from PCs derived in Gaussian09 using the HLYGAt scheme with the PBE0 functional and 6-311G(d,p) basis set. Due to the multiplicity of local minima on the $\boldsymbol{p}$-solution landscape, multi-start initialization using low-discrepancy Sobol' sequences is used to generate the initial $\boldsymbol{p}$ vectors. The Sobol' point which yields the best objective function value (OF) is taken to be the final solution. At the solution, we evaluate the RMSD_{15} value between optimized and reference structures, determined with the Cambridge Crystallographic Data Centre's COMPACK tool (distance and angle tolerances are set to 0.4 and 40°), and deviations in U_{inter} ($dU = U_{\mathrm{inter}}^{*} - U_{\mathrm{inter}}^{\mathrm{ref}}$).

2.4. Lattice-Energy Model

To select a suitable number of parameters which capture the underlying physics without overfitting, several fitting schemes are tested. We opt to parameterize all heteroatomic interactions ($i \neq j$) rather than using combining rules. Analysis of our training set and experimental crystal structures (Gavezzotti & Filippini, 1994) indicates that Hs bonded to O (H_o) and N (H_n) exhibit distinct HB geometries, hence we distinguish the two. It was previously asserted that polarized Hs would be stripped of electron density and exhibit no dispersive attractions ($C_{ij} = 0$ for $i = H_{o/n}$) (Pyzer-Knapp *et al.*, 2016). In practice, other parts of the model may not fully capture the HB effects and require the FF to compensate, hence, we test setting $C_{\mathrm{Ho\ or\ Hn}/j} = 0$ and varying $C_{\mathrm{Ho\ or\ Hn}/j}$. Finally, A_{ij} and B_{ij} are highly correlated, thus it is common to fix B_{ij} and fit only A_{ij} and C_{ij}. B_{ij} were initially fixed to the FIT FF values, but subsequently some $B_{\mathrm{Ho\ or\ Hn}/j}$ were also allowed to vary.

Table 1: Performance of each FF in describing the training set. Only FF1, FF2, and FF3 are damped. The average $RMSD_{15}$, average dU, and average absolute dU ($|dU|$) are presented for the entire training set (NHB+HB), NHB, and HB crystals. The number of structures which failed to give a valid $RMSD_{15}$ are given in parenthesis. Parameters are obtained either by combining rules (CR), varied in the optimization (VR), or fixed (FX).

FF	$i \neq j$	B_{ij}	C_{ij}	OF	Data	$RMSD_{15}$ (Å)	dU (kJ/mol)	$\|dU\|$ (kJ/mol)
					NHB+HB	0.2701 (-)	12.2450	14.1433
FIT	CR	FX	VR	-	NHB	0.2566 (-)	7.2671	9.6378
					HB	0.3021 (-)	24.0486	24.8268
					NHB+HB	0.2406 (32)	0.8560	4.7473
FF1	VR	FX	0	0.692	NHB	0.2168 (-)	-0.4978	3.6014
					HB	0.3154 (32)	4.0662	7.4643
					NHB+HB	0.2214 (1)	0.3874	4.1879
FF2	VR	FX	VR	0.538	NHB	0.2136 (-)	0.1417	3.4808
					HB	0.2401 (1)	0.9701	5.8645
					NHB+HB	0.2203 (-)	0.2888	4.0604
FF3	VR	VR	VR	0.524	NHB	0.2129 (-)	0.1730	3.4290
					HB	0.2377 (-)	0.5635	5.5573

3. Results

Three FFs are obtained using the fitting schemes described in Section 2. Their performance in describing the training set is compared against the FIT FF in Table 1. In some structures, failure to match 15 molecules results in no $RMSD_{15}$ being generated.

Over the training set, all three newly parameterized FFs present with similarly small errors in average dU (< 5 kJ/mol) and $|dU|$ (< 10 kJ/mol). In a previous work which used the same training set (Bowskill, 2021), the average dU and $|dU|$ of the HB structures were 22.31 kJ/mol and 22.48 kJ/mol respectively, indicating that the new FFs exhibit a marked reduction in the energy errors for these structures. In the previous work, energy data for the HB structures was omitted due to numerical issues associated with the optimization. The inclusion of the damping function in this work appears to have mitigated these problems. However, to match the HB energies well, the geometry predictions worsen significantly, particularly for the HB structures. Using FF1 and FF2, not only do the average $RMSD_{15}$ values for HB structures increase over the previous work (0.238 Å), but more critically, some structures fail to produce any $RMSD_{15}$ values at all.

This tradeoff between energy and geometry may be explained as follows. In FF1, fixing $C_{ij} = 0$ for the $H_{o/n}$ interactions prevents these interactions from contributing to the deficit in U_{inter}. The only recourse is for the optimizer to set A_{ij} in these interactions to 0 (or close to 0), allowing greater short-range charge-charge interactions and other atoms to interact deeper within their energy-wells, at the expense of the fidelity to geometry. Allowing non-zero C_{ij} within FF2 does somewhat alleviate this issue, but there remain some interactions (e.g., O/H_o) where $A_{ij} = 0$, resulting in fully attractive potentials. The absence of repulsive walls in FF1 and FF2 (Figure 1, left) permit unnaturally close contact distances that prevent the assessment of $RMSD_{15}$ and cast doubt on the transferability of these FFs. In contrast, by allowing some of the B_{ij} involving $H_{o/n}$ to vary (FF3), we attain potentials which *all* have repulsive walls. As the only 'physically sensible' FF, the transferability of FF3 is tested on an independent validation set. Overall, the performance of FF3 does not deteriorate radically from the training set (Table 2), with a similar quality of energy prediction. While the average $RMSD_{15}$ for HB structures increases, it is still within a cutoff that most CSP practitioners would consider acceptable (< 0.5 Å).

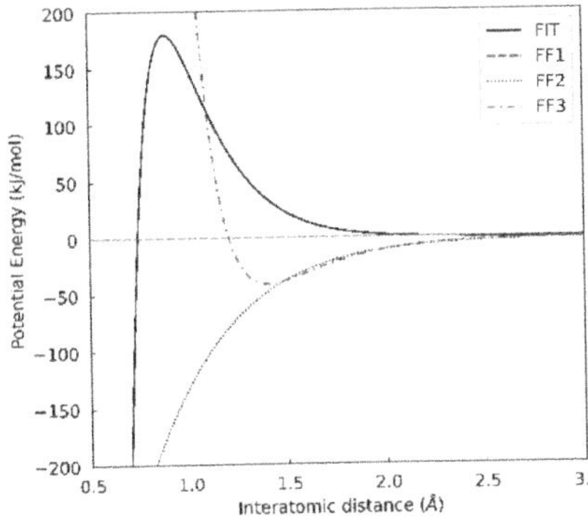

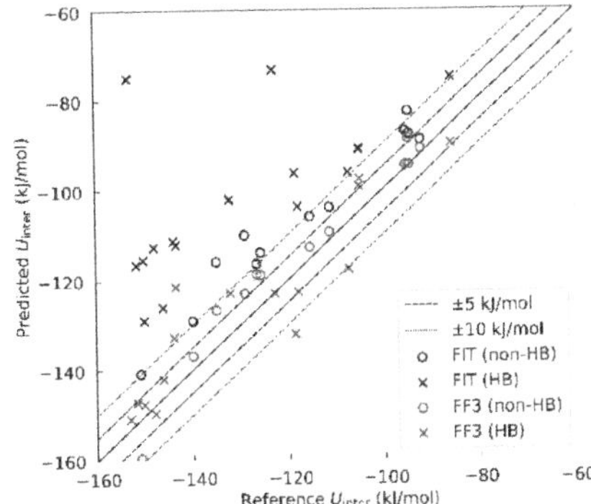

Figure 1: (Left) Comparison of the H_o/O potential for the four FFs. (Right) Parity plot of the validation set U_{inter}, comparing the performance using the FIT FF and FF3.

As mentioned before, the goal of modern CSP FFs is to act as proxies for the DFT-D landscape. To further benchmark our FFs, we compare against the FIT FF. Within both the training and validation set, whilst the FIT FF offers good geometry predictions, energies are substantially underbound, especially for HB structures. We note FIT is being used well beyond the data to which it was parameterized. Correcting from their ΔH_{sub} to our DFT-D $U_{inter,ref}$ is approximately given by $-\Delta H_{sub} = U_{inter} + \Delta U_{intra} + 2RT$. In deriving FIT, it was assumed that $\Delta U_{intra} \approx 0$, while our VASP calculations predict ΔU_{intra} as large as 20 kJ/mol in HB structures even with small conformational changes. Moreover, our $U_{inter,ref}$ are effectively at 0 K, while the room-temperature data used in FIT would result in a positive $2RT$, but this was ignored in the original fitting. Overall, the FIT reference energies underpredict the DFT-D $U_{inter,ref}$, explaining the large, positive average dU observed when applying FIT here. This is also apparent when one considers the shape of the potentials (Figure 1, left): FF3 has deeper energy-wells and shorter-range repulsive walls, while FIT has a repulsive wall at distances exceeding the typical HB lengths (O-H···O distances ≈ 1.6-1.9 Å, Gavezzotti & Filippini, 1994), and shallower energy-wells.

4. Conclusion

Electronic-structure methods like DFT-D offer our 'best guess' of the polymorphic landscape and are used in the final stages of a CSP study. With this in mind, we construct hybrid energy models to emulate the DFT-D landscape as opposed to the (unknown) experimental landscape. In parameterizing a FF using DFT-D generated training data, commonly used model-simplifications were found to yield unphysical results. By altering the functional form of the FF and avoiding setting model parameters to arbitrary values,

Table 2: Performance of FF3 and FIT in describing the validation set. The average $RMSD_{15}$, average dU, and average absolute dU ($|dU|$) are presented for the full dataset, NHB, and HB crystals. All structures gave valid $RMSD_{15}$ with both FFs. The change in each metric from the training set is given in parenthesis for FF3.

FF	Data	$RMSD_{15}$ (Å)	dU (kJ/mol)	$\lvert dU \rvert$ (kJ/mol)
FIT	NHB+HB	0.2811	20.8127	20.8127
	NHB	0.2097	10.7082	10.7082
	HB	0.3346	28.3911	28.3911
FF3	NHB+HB	0.2380 (+0.0177)	1.6665 (+1.3777)	6.3550 (+2.2946)
	NHB	0.1549 (-0.0580)	2.9774 (+2.8044)	4.4317 (+1.0027)
	HB	0.3003 (+0.0626)	0.6833 (+0.1198)	7.7975 (+2.2402)

we were able to obtain a FF that makes physical sense and is effective in describing crystal properties. In both the training and validation sets, our new FF has an average $|dU| < 10$ kJ/mol, well within the threshold for identifying putative structures. Conversely, a FF like FIT that was parameterized to experimental data is unsuited to predict the DFT-D landscape. Ultimately, using computationally generated reference data will allow us to circumvent the limitations of scarce experimental data and explore new avenues in CSP FF development, such as the expansion of atom-typing.

Acknowledgements

We are grateful to the UK Materials and Molecular Modelling Hub for computational resources, which is partially funded by EPSRC (EP/P020194/1 and EP/T022213/1). The authors also gratefully acknowledge funding provided from Syngenta, and computational resources provided by Imperial College London's High-Performance Computing Cluster.

References

D. H. Bowskill, 2021, Reliable and Efficient Parameter Estimation Methodologies for Crystal Structure Prediction, PhD Thesis, Imperial College London

J. G. Brandenburg, S. Grimme, 2016, Organic crystal polymorphism: a benchmark for dispersion-corrected mean-field electronic structure methods, *Acta Crystallographica Section B, Structural Science, Crystal Engineering and Materials*, 72, 4, 502-513

D. K. Bučar, R.W. Lancaster, J. Bernstein, 2015, Disappearing Polymorphs Revisited, *Angewandte Chemie International Edition*, 54, 24, 6972-6993

D. S. Coombes, S. L. Price, D. J. Willock, M. Leslie, 1996, Role of Electrostatic Interactions in Determining the Crystal Structures of Polar Organic Molecules. A Distributed Multipole Study, *The Journal of Physical Chemistry*, 100, 18, 7352-7360

A. J. Cruz-Cabeza, S.M. Reutzel-Edens, J. Bernstein, 2015, Facts and fictions about polymorphism, *Chemical Society Reviews*, 44, 23, 8619-8635

C. A. Gatsiou, C. S. Adjiman, C. C. Pantelides, 2018, Repulsion–dispersion parameters for the modelling of organic molecular crystals containing N, O, S and Cl, *Faraday Discussions*, 211, 297-323

A. Gavezzotti, G. Filippini, 1994, Geometry of the Intermolecular X-H···Y (X, Y = N, O) Hydrogen Bond and the Calibration of Empirical Hydrogen-Bond Potentials, *The Journal of Physical Chemistry*, 98, 18, 4831-4837

S. Grimme, J. Antony, S. Ehrlich, H. Krieg, 2010, A consistent and accurate ab initio parametrization of density functional dispersion correction (DFT-D) for the 94 elements H-Pu, *The Journal of Chemical Physics*, 132, 15, 154104

G. Kresse, D. Joubert, 1999, From ultrasoft pseudopotentials to the projector augmented-wave method, *Physics Review B*, 59, 3, 1758-1775

J. Nyman, G. M. Day, 2015, Static and lattice vibrational energy differences between polymorphs, *Crystal Engineering Communications*, 17, 28, 5154-5165

C. C. Pantelides, C. S. Adjiman, A.V. Kazantsev, 2014, General computational algorithms for ab initio crystal structure prediction for organic molecules, *Prediction and Calculation of Crystal Structures: Methods and Applications*, Springer, 25-58

E. O. Pyzer-Knapp, H. P. G. Thompson, G. M. Day, 2016, An optimized intermolecular force field for hydrogen-bonded organic molecular crystals using atomic multipole electrostatics, *Acta Crystallographica Section B, Structural Science, Crystal Engineering and Materials*, 72, 4, 477-487

K. T. Tang, P. J. Toennies, 1984, An improved simple model for the van der Waals potential based on universal damping functions for the dispersion coefficients, *The Journal of Chemical Physics*, 80, 8, 3726-3741

J. Tao, J. P. Perdew, V. N. Staroverov, G. E. Scuseria, 2003, Climbing the Density Functional Ladder: Nonempirical Meta--Generalized Gradient Approximation Designed for Molecules and Solids, *Physics Review Letter*, 91, 14, 146401

Antonis Kokossis, Michael C. Georgiadis, Efstratios N. Pistikopoulos (Eds.)
PROCEEDINGS OF THE 33rd European Symposium on Computer Aided Process Engineering (ESCAPE33), June 18-21, 2023, Athens, Greece
 http://dx.doi.org/10.1016/B978-0-443-15274-0.50185-2

A data-driven uncertainty modelling and reduction approach for energy optimisation problems

Julien Vaes, Vassilis M. Charitopoulos

Departement of Chemical Engineering, The Sargent Centre for Process Systems Engineering, University College London, Torrington Place, London WC1E 7JE, UK.

Abstract

Taking uncertainty into account is crucial when making strategic decisions. To guard against the risk of adverse scenarios, traditional optimisation techniques incorporate uncertainty based on prior knowledge on its distribution. In this paper, we show how, based on limited historical data, we can generate from a low-dimensional space the underlying structure of uncertainty. To this end, we first exploit the correlation between the sources of uncertainty through a principal component analysis (PCA) to reduce dimensionality. Next, we perform clustering to reveal the typical uncertainty patterns, and finally we generate polyhedral uncertainty sets based on a kernel density estimation (KDE) of marginal probability functions.

Keywords: Polyhedral Uncertainty Set, Robust Optimisation (RO), PCA, Kernel density estimation (KDE), Dimensionality Reduction, Data scarcity.

1. Introduction

Optimisation under uncertainty aims to find optimal solutions given uncertain outcomes. This is typically done through methods like chance-constrained optimisation, robust optimisation, or stochastic optimisation, which require prior knowledge of an uncertainty set or probability distribution. Robust optimisation (RO) is for instance a suitable method in situations where insufficient historical data is available to accurately estimate distributions (Roald et al., 2023). In this method, the solution hedges against a range of possible realisations called uncertainty set with emphasis on the worst-case to guarantee feasibility of the policy derived. The definition of such uncertainty set is crucial: the larger it is, the less informative it becomes and the more conservative is the solution, which could render it impractical (Ben-Tal et al., 2009).

We mention here some methods and their limitation that have been recently proposed in the literature to generate uncertainty sets. First, Bianchi et al. (2018) employ statistical hypothesis tests to derive uncertainty sets' sizes. Nonetheless, these tests become impractical in cases of data scarcity. In Shang et al. (2017); Ning and You (2018), a single polyhedral uncertainty set (PUS) is obtained using respectively kernel learning (support vector clustering) or a principal component analysis (PCA) coupled with kernel smoothing. However, this assumes continuous data and cannot handle uncertainty formed as a collection of independent states. To circumvent this limitation, Ning and You (2017, 2019) define the uncertainty set as the union of PUS, which are generated using a Dirichlet process mixture model. However, the latent uncertainty used to generate the PUSs is of same dimension as the original uncertainty, leading to computational intractability for high dimensional uncertainty sets when used in RO (Ben-Tal et al. 2009).

There is consequently the need for a method that generates polyhedral uncertainty sets in the case of data sparsity where uncertainty is highly dimensional and consists of independent sets of realisations; this is the objective of the present paper.

The rest of this paper is structured as follows: In Section 2, we present the method which derives polyhedral uncertainty sets from a low-dimensional space based on historical data. We then show in Section 3 how it can be applied to the energy data in the United Kingdom. In Section 4 a discussion of the key findings is presented and finally in Section 5 conclusions on the main contributions of this work are drawn.

2. Method for deriving polyhedral uncertainty sets

To motivate the importance of building polyhedral uncertainty sets, we consider the following scenario-based linear adaptive robust optimisation (ARO) problem, which is recurrently used to formulate energy problems (e.g. unit commitment):

$$\underset{x}{\text{minimise}}\, \boldsymbol{c}^T\mathbf{x} + \sum_{k\in\mathcal{K}} p_k \left(\max_{\boldsymbol{v}_k \in \mathcal{V}_k} \min_{\boldsymbol{y}_k} \boldsymbol{b}_k^T \boldsymbol{y}_k \right) \tag{1a}$$

$$\text{such that } \boldsymbol{Ax} \leq \boldsymbol{d} \text{ and } \forall k \in \mathcal{K}: \boldsymbol{h}_k - \boldsymbol{T}_k\boldsymbol{x} - \boldsymbol{M}_k\boldsymbol{v}_k \leq \boldsymbol{W}_k\boldsymbol{y}_k \tag{1b}$$

where $\boldsymbol{x}$ and $\boldsymbol{y}_k$ with $k \in \mathcal{K}$ are respectively the first and second stage variables, $\mathcal{K}$ is the set of distinct operation conditions/scenarios, $\mathcal{V}_k$ is the uncertainty set associated to the uncertain parameters for scenario $k \in \mathcal{K}$, and p_k is its weight (or probability of occurrence). Such problem can be solved either in a monolithic fashion or by using variants of Benders' decomposition. If each set $\mathcal{V}_k$ is a polyhedral uncertainty set, then each set can be expressed as linear constraints, which facilitates the resolution of the ARO problem by exploiting the dual formulation associated to a linear program (Birge and Louveaux, 2011). We now show how to derive these PUSs $\mathcal{V}_k$ based on limited historical data and how to generate them from lower dimensional spaces. The method is here illustrated for energy uncertainty data but is not restricted to such application.

Let $n \in N$ be the number of uncertain attributes (*e.g.* electricity demand, temperature). Let $\boldsymbol{v}^{(i)} \in \mathbb{R}^{24n}$, $i \in [\![1,m]\!]$, be m historical data points, where the vector $\boldsymbol{v}^{(i)}$ contains the standardised daily profiles (hourly values) of all attributes on a given day $i \in [\![1,m]\!]$. We assume that the data is standardised in the sense that the sample mean and standard deviation of each component over all m data points is respectively 0 and 1. This allows for attributing an even weight to any uncertain component. Given a data point $\boldsymbol{v}^{(i)} \in \mathbb{R}^{24n}$, let $\boldsymbol{V}^{(i)} \in \mathbb{R}^{n\times 24}$ be the matrix with the daily profiles of each attribute on its rows.

As an initial step, based a correlation analysis, we define a set $\mathcal{A}$ of disjoint groups of attributes such that the data inside a group is strongly correlated while being weakly correlated with any other group. We then assume that these groups of attributes are independent and separately generate a polyhedral uncertainty set for each one of them. The (weak) correlation between the groups will partly be reinstated on a later stage when defining the scenarios. For any group of attributes $a \in \mathcal{A}$, we denote by $\boldsymbol{v}_a^{(i)} \in \mathbb{R}^{24n_a}$ the data related to the n_a attributes associated to a for a day $i \in [\![1,m]\!]$.

The following steps allow for deriving a PUS based on a lower dimensional space for each group $a \in \mathcal{A}$: First, we perform a principal component analysis (PCA) and express the data in the PCA basis: this allows for finding the directions along which the data has greatest variance. Mathematically, there exists an orthogonal matrix $\boldsymbol{P}_a \in \mathbb{R}^{24n_a \times 24n_a}$

such that, any data point $\boldsymbol{v}_a^{(i)}$, $i \in [\![1,m]\!]$, expressed in the original basis can be expressed in the PCA basis using the linear transformation: $\boldsymbol{w}_a^{(i)} = \boldsymbol{P}_a \boldsymbol{v}_a^{(i)}$. As $\boldsymbol{P}_a$ is orthogonal, the linear transformation from the PCA basis to the original basis is given by $\boldsymbol{v}_a^{(i)} = \boldsymbol{P}_a^T \boldsymbol{w}_a^{(i)}$. Given the data in the PCA basis, we perform dimensionality reduction by retaining only the first r_a components that explain the most the variability inside the data. As the data is strongly correlated inside any group $a \in \mathcal{A}$, we expect to have $r_a \ll 24n_a$. Let $\overline{\boldsymbol{w}}_a^{(i)} \in \mathbb{R}^{r_a}$ denote the data points in the truncated PCA basis, where only the first r_a components are kept.

To define the typical operational conditions associated to a group $a \in \mathcal{A}$ of attributes, we perform clustering (e.g. K-means) on the data points expressed in the truncated PCA basis. Clustering in the PCA basis rather than in the original basis allows to give relatively more importance to the directions explaining the most the variance and obtain more meaningful clusters (Ding and He, 2004). We denote by $\mathcal{K}_a$ the set of clusters related to $a \in \mathcal{A}$, call each cluster $k_a \in \mathcal{K}_a$ a *group-scenario* and define its probability of occurrence p_{k_a} as the proportion of data points attributed to this cluster.

We now follow a similar approach as Ning and You (2018) to construct a polyhedral uncertainty set for each group-scenario $k_a \in \mathcal{K}_a$. To this end, we estimate the marginal probability density function $\hat{f}_{a,k_a,r}$, $r \in [\![1, r_a]\!]$, along each truncated principal component direction (with a kernel smoothing method such as KDE for instance (Chen, 2017)). We denote by $\hat{F}_{a,k_a,r}$ the associated cdf, and by $\hat{F}_{a,k_a,r}^{-1}$ the associated quantile function. We now define the PUS related to the group of attributes $a \in \mathcal{A}$ and the cluster $k_a \in \mathcal{K}_a$, which we call a *group-scenario PUS*, as follows:

$$\overline{\mathcal{W}}_{a,k_a}^{pol} := \left\{ \overline{\boldsymbol{w}} \in \mathbb{R}^{r_a} \;\middle\|\; \left\{ \begin{array}{c} \mathbf{0} \leq \mathbf{z}^-, \mathbf{z}^+ \leq \mathbf{1} \\ \mathbf{1}^T(\mathbf{z}^- + \mathbf{z}^+\} \leq \Phi_{a,k_a} \\ \forall i,j \in [\![1, s_{a,k_a}]\!]: z_i^- + z_i^+ + z_j^- + z_j^+ \leq \Psi_{a,k_a} \\ \boldsymbol{\xi}_{a,k_a}^{lb} = \left[\hat{F}_{a,1,k_a}^{-1}(\alpha_{a,k_a}), \dots, \hat{F}_{a,r_a,k_a}^{-1}(\alpha_{a,k_a})\right]^T \\ \boldsymbol{\xi}_{a,k_a}^{ub} = \left[\hat{F}_{a,1,k_a}^{-1}(1-\alpha_{a,k_a}), \dots, \hat{F}_{a,r_a,k_a}^{-1}(1-\alpha_{a,k_a})\right]^T \\ \boldsymbol{\lambda} = \frac{1}{2}(\mathbf{z}^+ - \mathbf{z}^- + \mathbf{1}) \\ \overline{\boldsymbol{w}} = \boldsymbol{\xi}_{a,k_a}^{lb} \circ (\mathbf{1} - \boldsymbol{\lambda}) + \boldsymbol{\xi}_{a,k_a}^{ub} \circ \boldsymbol{\lambda} \end{array} \right. \right\} \tag{2}$$

where $\mathbf{0}$ and $\mathbf{1}$ are vectors of respectively zeros and ones of size r_a, where $\circ$ defines the Hadamard product, and where vectors inequalities must be understood componentwise. This set is similar to the budgeted uncertainty set proposed by (Bertsimas and Sim, 2004), which enables to control the degree of conservatism: the larger the volume of the PUS, the more conservative is the resulting robust Optimisation Problem (1). The uncertainty set is parametrised by α_{a,k_a}, Φ_{a,k_a} and Ψ_{a,k_a}. First, α_{a,k_a} is used to exclude both tails of the marginal pdf along each principal component axis. Then, Φ_{a,k_a} limits the cumulative dispersion from the nominal value along all retained PCA axes. Finally, the parameter Ψ_{a,k_a} additionally limits the pairwise dispersion along the first s_{a,k_a} PCA components. This parameter is important to further exclude unlikely data points from the uncertainty that would otherwise not be cut with the general budget constraint parametrised by Φ_{a,k_a}.

We have so far generated a PUS for each group-scenario $k_a \in \mathcal{K}_a$ for each $a \in \mathcal{A}$. Let $\mathcal{K}^\times := \times_{a \in \mathcal{A}} \mathcal{K}_a$ denote the Cartesian product of the sets of group-scenarios. The probability $\hat{p}_k$ associated to $k \in \mathcal{K}^\times$ is estimated as the proportion of the data points such that the uncertainty part related to a is attributed to cluster k_a for all $a \in \mathcal{A}$. To take into account the weak correlation between groups of attributes (so far assumed to be independent), we retain only the most probable combinations of group-scenarios $k := \{k_a \in \mathcal{K}_a, a \in \mathcal{A}\} \in \mathcal{K}^\times$, keeping those with associated probability greater than a threshold probability $\tilde{p}$. The more conservative we desire to be, the greater the acceptance probability threshold $\tilde{p}$ should be. The set of scenarios $\mathcal{K}$ is then constituted of the combinations in $\mathcal{K}^\times$ satisfying the probability threshold $\tilde{p}$. The probability of occurrence p_k of each scenario $k \in \mathcal{K}$ used in (1a) is then defined as the scaled probability when the scenarios that do not satisfy the acceptance threshold are excluded, *i.e.* $p_k = \hat{p}_k / \sum_{k \in \mathcal{K}} \hat{p}_k$. Finally, the PUS in the original basis related to a scenario $k = \{k_a \in \mathcal{K}_a, a \in \mathcal{A}\} \in \mathcal{K}$ is then defined as follows:

$$\mathcal{V}_k := \left\{ \boldsymbol{v} \in \mathbb{R}^{24n} \middle| \forall a \in \mathcal{A}: \begin{cases} \overline{\boldsymbol{w}}_a \in \overline{\mathcal{W}}_{a,k_a}^{pol} \\ \boldsymbol{v}_a = \boldsymbol{P}_a^T \begin{bmatrix} \overline{\boldsymbol{w}}_a \\ \boldsymbol{0}_{n_a - r_a} \end{bmatrix} \end{cases} \right\} \tag{3}$$

where $\boldsymbol{0}_{n_a - r_a}$ is a vector of zeros of size $n_a - r_a$. The reduction in dimension appears clearly: the uncertain part $\boldsymbol{v}_a \in \mathbb{R}^{24 n_a}$ related to $a \in \mathcal{A}$ is generated from elements of $\overline{\mathcal{W}}_{a,k_a}^{pol}$, which is in the lower-dimensional space $\mathbb{R}^{r_a}$.

In some applications (e.g. unit commitment), we are interested in taking into account the uncertainty over several days. Given the succession of N daily scenarios $k_{1\to N} := (k_1, \dots, k_N) \in \mathcal{K} \times \dots \times \mathcal{K}$, we can define an associated PUS based on the definition of $\mathcal{V}_k$ in (3) as follows:

$$\mathcal{V}_{k_{1\to N}} := \left\{ \boldsymbol{v} \in \mathbb{R}^{24nN} \middle| \begin{cases} \boldsymbol{v} = [\boldsymbol{v}_1, \dots, \boldsymbol{v}_N] \\ \forall i \in [\![1, N]\!]: \boldsymbol{v}_i \in \mathcal{V}_{k_i} \\ \forall i \in [\![1, N-1]\!]: |(\boldsymbol{V}_{i+1}[:,1] - \boldsymbol{V}_i[:,24]) - \boldsymbol{\mu}_\Delta| \circ \frac{1}{\boldsymbol{\sigma}_\Delta} \leq c \end{cases} \right\} \tag{4}$$

where $1/\boldsymbol{\sigma}_\Delta$ must be understood componentwise, where $\boldsymbol{V}_i \in \mathbb{R}^{n \times 24}$ is the matrix representation of $\boldsymbol{v}_i$, and where $\boldsymbol{\mu}_\Delta$ and $\boldsymbol{\sigma}_\Delta$ are vectors in $\mathbb{R}^n$ that correspond respectively to the sample mean and standard deviation of the step size between successive profiles for all n attributes. The third constraint in (4) enforces continuity between the daily profiles by limiting the step size with the parameter $c > 0$.

3. Case study: uncertainty related to energy data in the UK

We apply the method presented in the previous section to derive uncertainty sets for quantifiable sources of uncertainty related to power generation based on historical data for the UK in 2015. This includes national electricity demand, as well as regional data on gas demand, temperature, solar availability, offshore and onshore wind availability (we refer to the paper Charitopoulos et al. (2022) for more details). Each data point corresponds to the collection of hourly daily profiles of the different attributes. The size of the uncertainty space is therefore high dimensional, which motivates the need for a reduction technique to generate uncertainty sets. Based on the correlation plot (see Figure 1), three groups of attributes have strong intra-correlation and weak inter-

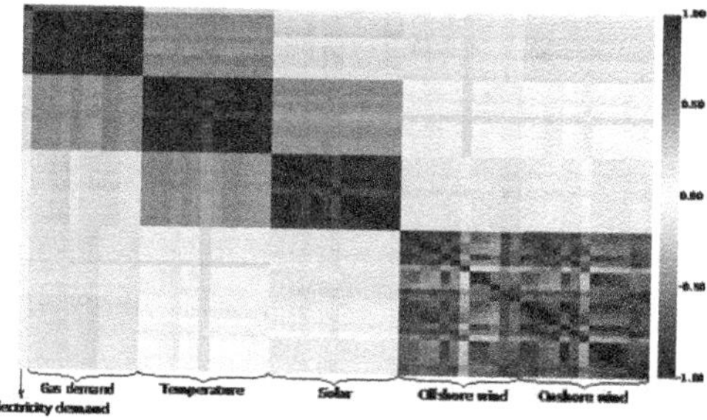

Figure 1: Correlation between all the attributes

correlation: *seasonal* data (electricity/gas demand and temperature), *solar* data and *wind* data (off/onshore wind), i.e. $\mathcal{A} = \{seasonal, solar, wind\}$.

Figure 3 illustrates for each month the proportions of days that are allocated to each cluster (4 clusters are generated for each group of attributes $a \in \mathcal{A}$). The correlation between the solar and seasonality data is clearly apparent: the simultaneous occurrence of the blue wind cluster with the red solar cluster is for instance highly improbable. Now, if we focus on the clustering from April to October, we observe that the clusters proportions in the wind data are roughly the same for each month. This translates the idea that the seasonal data and wind data are independent for this period of the year. As a consequence of these observations, we do not define the scenarios as the Cartesian product of group-scenarios since all combinations are not as probable. Figure 2 depicts how the number of scenarios retained varies with the acceptance threshold $\tilde{p}$. It seems that adopting a probability threshold of $\tilde{p} = 3\%$, which results in keeping 14 scenarios, is in line with the trade-off between accuracy and computational tractability: a higher value would lead to too few scenarios (6 or less), while a lower value would lead to too many scenarios and therefore would not allow for the pruning of the least likely realisations.

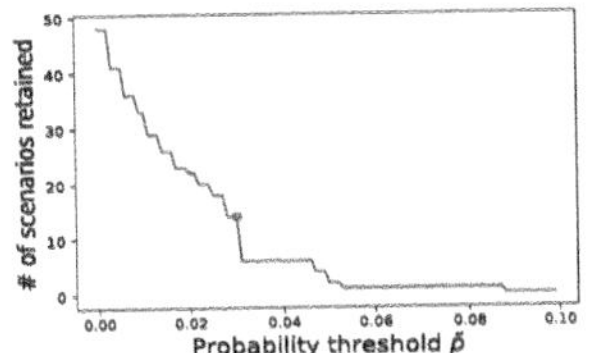

Figure 2: Number of scenarios given the probability threshold $\tilde{p}$

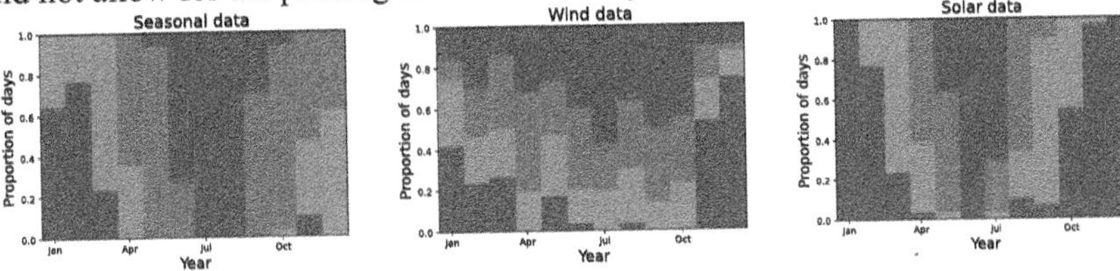

Figure 3: Proportion of days allocated to each cluster for all groups of attributes

Finally, Figure 4 depicts data generated with (3) for the parameters values: $a = seasonal$, $k_a = blue$, $r_a = 24$, $\alpha_{a,k_a} = 5\%$, $\Phi_{a,k_a} = 5$, $s_{a,k_a} = 24$ and $\Psi_{a,k_a} = 1.5$. We observe that the daily outliers in the historical data are ignored and, as desired, that the synthetic realisations are similar to the 90% of cases around the median. The size reduction is as follows: $n_a = 40$ attributes are associated with the group $a = seasonal$, which implies that the seasonal data is of size $24n_a = 960$ and is generated on the basis of $r_a = 24$ PCA components; this corresponds to a size reduction factor of 40. Figure 5 depicts realisations generated with (4) for the seasonal group of attributes for the succession of five days, where $k_{1\to5} = \{blue, blue, orange, orange, orange\}$ and where the continuity constraint parameter is equal to $c = 2.5$.

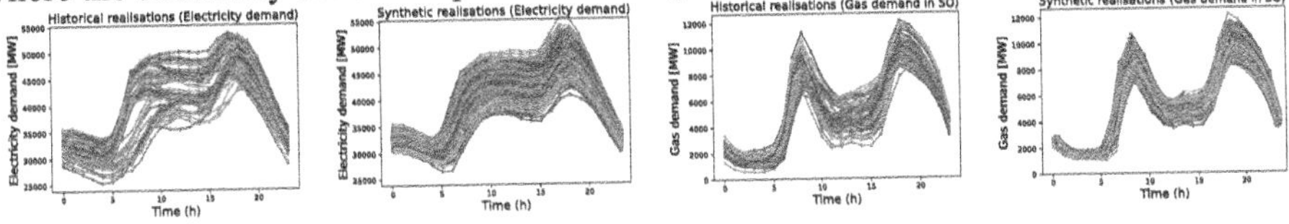

Figure 4: Comparison of historical and synthetic realisations for the blue cluster.

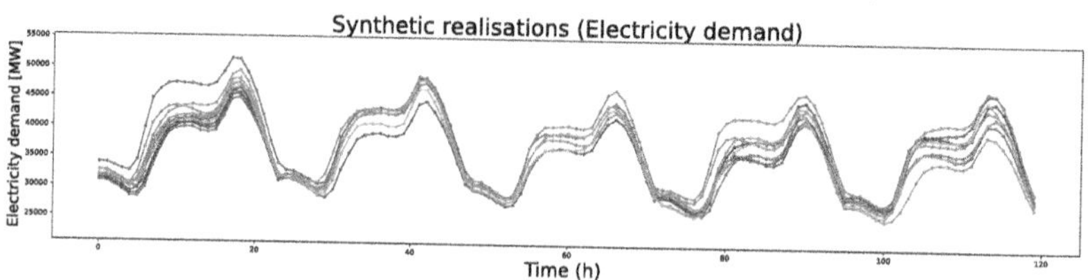

Figure 5: Synthetic realisations for $k_{1\to5} = \{blue, blue, orange, orange, orange\}$.

4. Results and discussion

We first point out why we did not start by clustering the data and then, for each cluster, generate a PUS, as it would implicitly take into account the (weak) correlation between attributes and avoid generating clusters based on a correlation analysis. The answer lies in two main points: Firstly, the scarcity of data. Indeed, in the case study, if we were to generate the 14 scenarios by clustering over the 365 historical data points, then each of them would have on average 26 data points. In contrast, in the method presented in this paper, only 4 clusters are generated per group of attributes. The quantiles of each cluster are then derived on the basis of around 90 data points. The PUS derived with our method are thus less subject to statistical estimate errors. Secondly, establishing groups of uncorrelated attributes allows for associating to each one of them a uncertainty dispersion budget Φ_{a,k_a}. In this way, we do not constrain the realisations of independent attributes together.

5. Concluding remarks

In conclusion, the advantages of our method are the following: First, it considerably reduces the size of the original uncertainty set as it can be expressed as a linear transformation of a truncated basis using PCA. Secondly, the PUS generated are statistically meaningful and can exploit the possibly independence of uncertain attributes. Thirdly, it generates consistent uncertainty daily profiles for which the continuity from hour to hour is guaranteed. Finally, it derives polyhedral uncertainty sets that can then be used in a robust optimisation (RO) type of problem.

Reproducibility: All the graphs of this article can be reproduced with the code made available on GitHub, where we have also shared the 2015 UK energy uncertainty dataset.

Acknowledgements: Financial support from the EPSRC (EP/T022930/1) is gratefully acknowledged.

References

A. Ben-Tal, L. El Ghaoui, and A. Nemirovski. Robust Optimization. Princeton University Press, 2009. doi:10.1515/9781400831050.

D. Bertsimas and M. Sim. The Price of Robustness. Oper. Res., 52(1):35–53, 2004. doi:10.1287/opre.1030.0065.

M. Bianchi, V. Gupta, and N. Kallus. Data-driven robust optimization. Math. Program., 167(2):235–292, 2018. doi:10.1007/s10107-017-1125-8.

J. R. Birge and F. Louveaux. Introduction to Stochastic Programming. Springer New York, 2011. doi:10.1007/978-1-4614-0237-4.

V. M. Charitopoulos, M. Fajardy, C. K. Chyong, and D. Reiner. The case of 100% electrification of domestic heat in Great Britain. 2022. doi:10.17863/CAM.81913.

Y. C. Chen. A tutorial on kernel density estimation and recent advances. Biostat. Epidemiol., 1(1):161–187, 2017. doi:10.1080/24709360.2017.1396742.

C. Ding and X. He. K -means clustering via principal component analysis. In Twenty-first Int. Conf. Mach. Learn. - ICML '04, page 29. ACM Press, 2004. doi:10.1145/1015330.1015408.

C. Ning and F. You. Data-driven adaptive nested robust optimization: General modeling framework and efficient computational algorithm for decision making under uncertainty. AIChE J., 63(9):3790–3817, 2017. doi:10.1002/aic.15717.

C. Ning and F. You. Data-driven decision making under uncertainty integrating robust optimization with principal component analysis and kernel smoothing methods. Comput. Chem. Eng., 112:190–210, 2018. doi:10.1016/j.compchemeng.2018.02.007.

C. Ning and F. You. Data-Driven Adaptive Robust Unit Commitment Under Wind Power Uncertainty: A Bayesian Nonparametric Approach. IEEE Trans. Power Syst., 34(3):2409–2418, 2019. doi:10.1109/TPWRS.2019.2891057.

L. A. Roald, D. Pozo, A. Papavasiliou, D. K. Molzahn, J. Kazempour, and A. Conejo. Power systems optimization under uncertainty: A review of methods and applications. Electr. Power Syst. Res., 214:108725, 2023. doi:10.1016/j.epsr.2022.108725.

C. Shang, X. Huang, and F. You. Data-driven robust optimization based on kernel learning. Comput. Chem. Eng., 106:464–479, 2017. doi:10.1016/j.compchemeng.2017.07.004.

Antonis Kokossis, Michael C. Georgiadis, Efstratios N. Pistikopoulos (Eds.)
PROCEEDINGS OF THE 33rd European Symposium on Computer Aided Process Engineering (ESCAPE33), June 18-21, 2023, Athens, Greece
 http://dx.doi.org/10.1016/B978-0-443-15274-0.50186-4

Moisture and Throughput Control in an Integrated Pharmaceutical Purification Platform using PharmaPy

Inyoung Hur[a], Daniel Casas-Orozco[a], Gintaras Reklaitis[a], Zoltan K. Nagy[a]*

[a]*Davidson School of Chemical Engineering, Purdue University, West Lafayette, IN 47907 USA.*

* *Corresponding author email address: zknagy@purdue.edu*

Abstract

A novel integrated continuous filter-drying carousel unit has as an enabling technology to handle filtration and drying of crystallization slurries within an end-to-end continuous pharmaceutical manufacturing train. However, due to the complexity of available contactless moisture sensing technologies, the challenges lie in developing an online monitoring system, which can accurately predict and control moisture content of the crystal product. In this work, we develop a hybrid monitoring framework to enhance the operation of this continuous filtration-drying unit by incorporating a knowledge-driven process model implemented in the object-oriented modeling platform, PharmaPy. The model is iteratively recalibrated from online measurements using a moving horizon state estimation methodology. The digital twin is shown to effectively track the offline measured moisture content. The integrated quality-by-control (QbC) framework achieves a substantial increase in process productivity and product quality.

Keywords: Continuous pharmaceutical manufacturing, Continuous filtration, Quality-by-Control, Digital Twin, Process Control

1. Introduction

Integrated filtration-drying has become an important emerging technology for continuous drug purification to produce high-purity small molecule oral drugs. [1] These steps serve as vital operations to achieve purity requirements of the final drug product [2]. Hence, process intensification (PI) of the filtration technology has attracted increased attention in the pharmaceutical sector due to its potential to achieve the critical quality attributes (CQAs) and target purity with improved economic and environmental outcomes [3]. The semi-continuous filtration carousel (CFC) has become a crucial technology by adopting separated small-scale batch process within continuous manufacturing (CM) mode. However, the CFC process is highly nonlinear due to; 1) the dynamic operation of the filtration-drying batch steps; 2) the uncertainty in the model parameters of these steps, and 3) the occurrence of unmodeled disturbances in the system, e.g., drifting critical process parameters (CPPs) due to fouling phenomena.

To overcome the challenges posed by input variability and model parametric uncertainty, quality-by-design (QbD) initiatives have been developed in the last years by the US Food and Drug Administration (FDA). QbD seeks to define the viable operating region within which the process can be operated while meeting target CQAs for given sets of critical material attributes (CMAs) and CPPs. With the increasing availability of process

analytical tools (PAT) and digital design tools, QbD has been used in various open-loop studies that create and use process digital twins to optimize process metrics (CQAs, costs, waste generation) with CPPs as decision variables [4]. For the pharmaceutical isolation step, crystal properties and final product impurities of regulated materials are the main CQAs. Since impurities remaining in the solid at the end of the drying step will be directed to the downstream process, failure to reach the drug's targeted purity will lead to off-spec products under the stringent regulatory requirements imposed on the CQAs [5]. To improve the estimation and monitoring of the CQAs, it is critical to incorporate the decision-making framework developed from the design space (DS) into the closed-loop control of the product quality to achieve a higher level of assurance.
In this work, we develop a digital twin of the physical CFC system, which combines knowledge-driven model predictions calculated from dynamic flowsheet simulations with actual data retrieved in real-time from sensor measurements. The flowsheet model is simulated in the Python-based, object-oriented modeling platform PharmaPy [6]. The PharmaPy simulator is a first-principles model of the process, which includes dynamic mass, energy, and momentum balances for the multi-phase component system [7]. A moving horizon estimation (MHE) framework that relies on state estimation of average moisture content of active pharmaceutical ingredients (API) filtration cakes is applied to the flowsheet which simulates the pharmaceutical filtration-drying step of a real continuous drug substance isolation operation in CFC. This paper describes how the resulting digital twin accurately predicts the main CQAs, namely, cake moisture content (MC) and improves productivity, while minimizing plant-model mismatch.

2. Digital Twin Development on the Continuous Filtration Carousel (CFC)

2.1. Continuous CFC platform and Real-time Data Communication

In this study, we incorporate a prototype version of carousel designed and manufactured from the Alconbury Weston Ltd (UK) as a platform for performing continuous solid liquid separation of crystal slurries. The flowsheet of the process installed at Purdue University CryPTSys lab is shown in Fig 1. The main body of the carousel consists of 5 different stations in which the material goes through one of the filtration and drying steps. The slurry with API crystals is intermittently introduced into a charge vessel and transferred to the first station where filtration step occurs driven by the pressure gradient developed across the cake and filter mesh. Filtration is automatically followed by a deliquoring step (DL01-DL02) in station 3. Finally, the remaining mother liquor inside the cake is reduced by introducing a hot stream of air at DR01 (4th station). After the given rotation time, the final dried product is ejected from the system by a mechanical piston in 5th station.
Real-time data acquisition is enabled via OPC (OLE (Object Linking and Embedding) for Process Control) protocol, using the existing CFC process sensors and a mass spectrometer. A vacuum pump connected to the filtrate receiver creates a pressure gradient at the bottom of the processing stations, and pressure sensor (PC) measures the pressure drop in the system. Moreover, a flow sensor measures the flowrate of the gas coming from the DR01. The thermocouples (TCIN, TCOUT) measure the temperature of the inlet and outlet of gas streams of the drying port respectively. A weighing scale, WI, monitors the mass of the filtrate collected in the filtrate receiver and a mass spectrometer is used to determine the composition of the drying gas. The data acquisition is implemented in the programmable logic control (PLC) of the unit by read/write functions and fed as the inputs to the PharmaPy simulator, which calculates the residual moisture content of the final dry product, acting as a soft sensor.

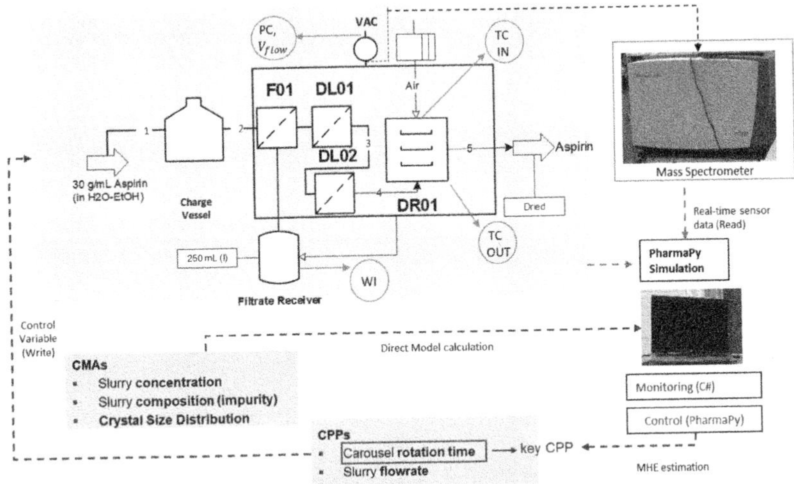

Fig 1. Schematic of CFC flowsheet and sensor network for a continuous purification process

2.2. Hybrid State Estimation Framework

A hybrid state estimation approach is presented to account for the plant-model mismatch under process disturbances. The purpose of the methodology is developing an online monitoring framework, which accurately captures the measurements from the physical sensor, hence it is used to construct controllers to control the MCs of the dry product by adapting cycle duration (t_{dry}). First, the model calculates the process outputs by solving the mechanistic equations given real-time input information (Eq. 2). A moving horizon estimation (MHE) approach iterates this to solve optimization problem by adapting model parameters to the fluctuations of the CFC system (Eq. 1).

$$\min_{\theta_p} \quad J(t, x, y, u, \theta_p) \tag{1}$$

$$s.t. \quad \dot{x} = PharmaPy(x, u, \theta_p), \;\; x_0 = x(t_0) \tag{2}$$

$$y = h(x, u, \theta_p) + v \tag{3}$$

$$\theta_p^{lb} < \theta_p < \theta_p^{ub} \tag{4}$$

The variable J is an estimation function to determine the state disturbances, $x = [T_{node}, x_{liq}, y_{gas}]$ is the state matrix comprised of vectors of the dynamic gas temperature, the species composition of the liquid phase, and the species composition of the gas phase from the drying port, $y = \left[T_{out}^T, \sum_{i=0}^{k} \frac{dy_{gas}(t_i)}{dt}\right]$ is the measurement vector, $u = [t_{dry}]$ is the process input, $\theta_p = [S_{var}, k_y, h_T]$ is the vector of model parameters, *PharmaPy* represents the vector function of the dynamic state equations, h is the vector function of the measurement equations, v is measurement noise vector, k is the sampling instance within a cycle window.

To compensate for the performance degradation from the presence of process uncertainties, the digital twin process model should be continuously updated using the new measurements obtained at each batch cycle (k). Here, the online state estimation framework is depicted in Fig 2. A MHE is exploited for state estimation in which the state variables are estimated by solving two consecutive optimization problems. First, the

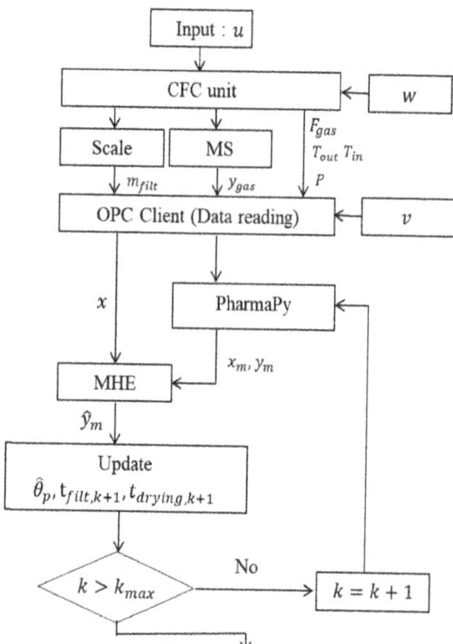

Fig 2. Block diagram of the framework

distance between simulated f_{sim,t_i} and measured f_{p,t_i} gas composition is minimized by varying S_{var} (cake saturation after deliquoring) over a respective batch time frame (Eq. 5) from the best fitting line on the parity plot at i^{th} sampling point within the estimation horizon (N_{past}) of the kth cycle window.

$$\min_{S_{var}} \mathrm{J} = \sum_{t_i=(k-1)N_{past}}^{kN_{past}} f_{p,t_i} - f_{sim,t_i} \tag{5}$$

Next, heat and mass transfer coefficients (h_T, k_y) are estimated by solving Eq. 6. The variable W is the weighting matrix, y is vector of measurements and $\hat{y}$ is vector of predicted measurement. The first optimization problem (Eq. 5) is first solved to maximize the correlation of the dynamic trend of the EtOH composition in the gas phase. The purpose of maximizing the correlation is to inform the model with an intermediate saturation value after the final deliquoring step based on different drying rate period MS data. The output of the decision variable is utilized to reinitialize the initial saturation of the deliquored

$$\min_{h_T,k_y} \mathrm{J} = \sum_{t_i=(k-1)N_{past}}^{kN_{past}} \left(y_{t_i} - \hat{y}_{t_i}\right)^T W_E \left(y_{t_i} - \hat{y}_{t_i}\right) \tag{6}$$

cake on the second problem (Eq. 6), and the problem is sequentially solved.

3. Result and Discussion

3.1. Case study: CFC experiment on Aspirin-H2O/EtOH slurry

The monitoring framework is tested on a continuous purification process of Aspirin from solvent mixture of water-ethanol under the presence of disturbances, e.g., uncertain model parameters and control variables. The performance on the final MC prediction is examined by comparing with off-line measured MC of dry product from each batch cycle via a gravimetric method.

Fig 3 demonstrates the trajectory tracking control of the monitoring framework on the outlet temperature of the drying gas and 1st derivative of ethanol composition read from the MS (a) by adapting a set of uncertain model parameters per every cycle window (b). The soft sensor on MC of the dry product is validated against the off-line measured MC as shown in Fig 3-c. Overall, the soft sensor performs well in terms of following the trend of the measurements throughout the whole cycle.

Moreover, the adaptive model parameter enables accurate capture of the relative performances of the subsequent drying batch compared to the fixed parameter model and shows good comparability to the offline measured residual MC. This proves our hybrid monitoring system allows the informative and efficient solution of state estimation problems to be real-time feasible within an online optimization framework.

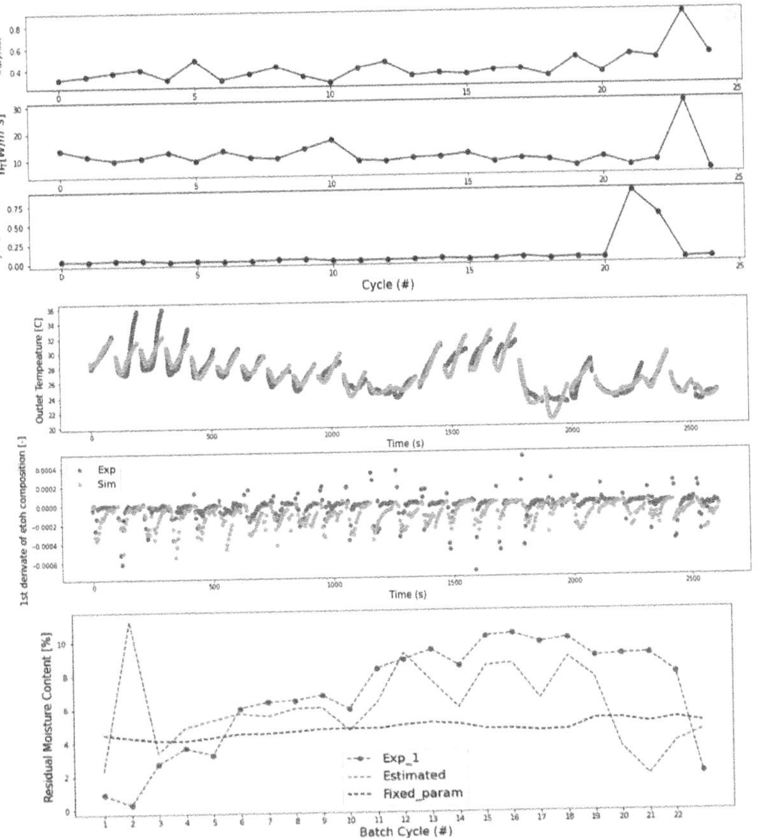

Fig 3. Result on online state estimation; top (a): decision variables from the problem; middle (b): comparison between prediction and measurement variables; bottom (c): comparison of moisture content (MC) from the model and measurement

The monitoring framework is further coupled with a proportional-integral (PI) controller to update the control variables (t_{drying}) based on the differences of the MC between prediction and the target MC (5%) (Eq. 8).

$$u(t) = u_{bias} + K_c e_i(t) + \frac{K_c}{\tau_I}\sum_{i=1}^{k} e_i(t)\Delta t \quad (7)$$

$$e_i(t) = \widehat{MC}_{est}(t_i) - MC_{target}\ ,\ MC_{target} = 5\ (\%)$$

The initial t_{drying} of 60 s is chosen for the base control variables (u_{bias}). The response of the control system to attain the target moisture content and resulting residual MC measurement is depicted in Fig 4. With initial cycle time, the batches produced in Cycle 1-2 was above 20% which do not meet the desired quality. The controller increases the cycle time to allow a longer operation based on the abnormal MC on the Cycle 3 and we see the MC to be regulated under the target MC. The controller starts decreasing t_{drying} by reducing the excessive operation duration, which is not needed anymore as batches produced in Cycle 3-14 are maintained either below or slightly above the target MC, hence effectively increasing the throughput. At Cycle 16, however, controller again increases the cycle time as the residual MC shows a consistent increment of its value due to the fouling of the process and simulator predicts that the CQA does not meet the quality threshold.

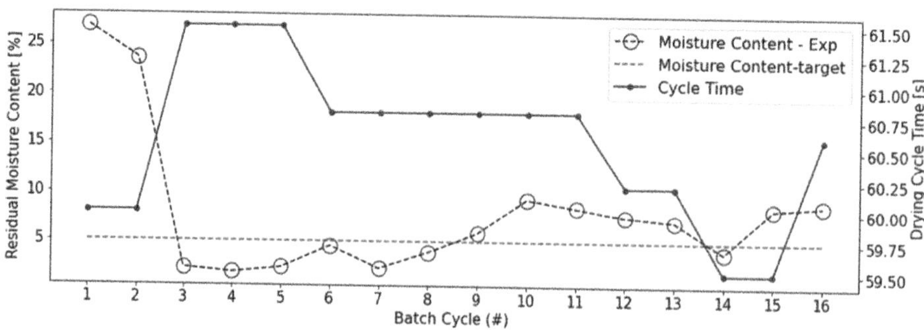

Fig 4. Controller responses and resulting measured MC of the dried product from experimetation

4. Conclusions

A digital twin is developed and tested for the continuous filtration and drying operation of Aspirin and shown to produce sufficiently accurate estimates of the residual MC of the dried crystals with fluctuation in absolute errors within 4%. The estimated outputs are fed to a PI controller that adjusts MC and throughput of the dried product by manipulating drying cycle time. The result showcases a successful application of a real-time hybrid monitoring framework that offers the advantages of both knowledge-driven process modeling and optimization that rely on real-time data acquisition and analysis of the continuous filtration and drying process.

5. Acknowledgments

This project was supported by the United States Food and Drug Administration through grant U01FD006738. The views expressed by the authors do not necessarily reflect official policies of the Department of Health and Human Services; nor does any mention of trade names, commercial practices, or organization imply endorsement by the United States Government.

References

[1] Price, C. J et al., (2020). CHAPTER 13. Continuous Isolation of Active Pharmaceutical Ingredients. In The Handbook of Continuous Crystallization. The Royal Society of Chemistry

[2] ICH Quality Implementation Working Group. Point to Consider: ICH-endorsed Guide for ICH Q8/Q9/Q10 Implementation. 2011.

[3] Van Gerven, T. (2014). In Workshop on Process Intensification 2014. The Royal Society of Chemistry.

[4] Su, Q et al., (2019). Data reconciliation in the Quality-by-Design (QbD) implementation of pharmaceutical continuous tablet manufacturing. International Journal of Pharmaceutics, 563(April), 259–272

[5] International Council for Harmonisation (2016), Impurities: guideline for residual solvents Q3C(R6)

[6] Casas-Orozco, D et al., (2021) PharmaPy: An object-oriented tool for the development of hybrid pharmaceutical flowsheets. Computers and Chemical Engineering, 153, 107408.

[7] Hur, I et al., Dynamic Flowsheet Simulation and Application of Soft Sensors on an Intensified and Integrated Purification Step for Pharmaceutical Upstream Manufacturing, AIChE Annual Meeting, 2021

Antonis Kokossis, Michael C. Georgiadis, Efstratios N. Pistikopoulos (Eds.)
PROCEEDINGS OF THE 33rd European Symposium on Computer Aided Process Engineering (ESCAPE33), June 18-21, 2023, Athens, Greece
 http://dx.doi.org/10.1016/B978-0-443-15274-0.50187-6

Parameter estimation and dynamic optimization of an industrial fed-batch reactor

Jan G. Rittig,[a] Jan C. Schulze,[a] Lars Henrichfreise,[a] Sebastian Recker,[b] Rolf Feller,[b] Alexander Mitsos,[a,c,d] Adel Mhamdi,[a,*]

[a]*RWTH Aachen University, Process Systems Engineering (AVT.SVT), Aachen, Germany*
[b]*LANXESS Deutschland GmbH, Cologne, Germany*
[c]*Forschungszentrum Jülich GmbH, Institute for Energy and Climate Research IEK-10: Energy Systems Engineering, Jülich, Germany*
[d]*JARA-ENERGY, Aachen, Germany*
**Corresponding author, email: adel.mhamdi@avt.rwth-aachen.de*

Abstract

Modeling and optimization of fed-batch reactors with several multi-step reaction pathways is challenging due to the nonlinear dynamic system behavior and large number of kinetic parameters. We showcase the model-based optimization of an industrial (20 m^3) fed-batch reactor by using our open-source dynamic optimization software DyOS. First, we build a detailed mechanistic model of the fed-batch reactor. Second, we conduct parameter estimation of the mechanistic model with 25 states and 44 fitting parameters using historic time-series industrial production data. Third, we perform dynamic multi-stage optimization and identify optimal feeding profiles for the operation, targeting improvements in economic profit over the established experience-based production routine. We demonstrate substantial economic improvement: The optimized production recipe can save up to 10% of raw material at the same yield of main product. Our findings underline the strong capabilities of model-based process optimization and its application to industrial challenges in process design and operation.

Keywords: Dynamic process optimization, Fed-batch reactor, Industrial application, Reaction optimization.

1. Introduction

The determination of feeding strategies is of paramount importance in the design and control of fed-batch reactors. Finding optimal feeding strategies is typically based on the dynamic optimization of a reactor model. However, modeling reaction processes with several multi-step reaction pathways can be challenging due to nonlinear system behavior that needs to be expressed by many equations and parameters, and a lack of mechanistic knowledge about the considered reaction, e.g., reaction kinetics and corresponding parameters. When experimental data of the reactor is available, kinetic parameters can be estimated using dynamic optimization. Subsequently, dynamic optimization can be employed for determining optimal feeding profiles with respect to a defined objective, e.g., reaction yield, product quality, and economic targets. Dynamic optimization of fed-batch reactors has been performed successfully for several applications including the production of (bio-)polymers (Zavala et al, 2005; Lopez et al., 2010, Jung et al. 2015), yeast (Hjersted & Henson, 2006) and drugs (Banga et al., 2005). We have also recently optimized fed-batch reactors for microgel (Jung et al., 2019) and polymer (Faust et al.

2021) production. To this end, we have employed our dynamic optimization software DyOS (Caspari et al., 2019), which is an open-source framework for the optimization of large-scale differential algebraic equation (DAE) systems.
Herein, we use DyOS for the optimization of feeding strategies for a fed-batch reactor of industrial scale. Specifically, we conduct three steps with the overall goal to identify optimal feeding strategies: (i) mechanistic modeling of the fed-batch reactor with the reaction kinetics, (ii) estimation of kinetic reaction and heat transfer parameters, and (iii) dynamic multi-stage optimization of raw material feeding profiles. In the first step, we develop together with the plant operators a mechanistic reactor model in the Modelica modeling language (Modelica, 2022). The parameter estimation in the second step includes the collection of historic industrial production data of the currently established experienced-based feeding profiles that we then use to fit model parameters with DyOS. In the third step, we start from the established feeding profiles and apply DyOS to identify economically optimized feeding strategies. Our in-silico results suggest that the optimized feeding strategies can lead to substantial economic improvements and savings in raw materials.
Due to confidentiality reasons, we cannot provide detailed insights on the model, the corresponding data, and the literature sources we used to collect information about the considered reaction process. We thus focus on the methodological approach of using dynamic optimization for fitting model parameters and optimizing feeding profiles. In the following, we schematically describe the considered fed-batch reactor, explain the three steps (i – iii) in detail, and present the modeling and optimization results.

2. Fed-Batch Reactor Model Development

We consider an industrial stirred-tank fed-batch reactor producing one main product in a multi-step reaction. The reactor has a volume of 20 m^3. A schematic illustration of the reactor is shown in Figure 1. The inputs to the reactor are composed of four raw materials (RM1, RM2, RM3, RM4), water (H2O), and a catalyst (CAT). All input components are fed in liquid form, either in pure species or in aqueous solutions. RM2, H2O, and CAT are filled into the reactor before the reactions start. When RM1 and RM3 are introduced to the reactor, several multi-step reactions involving multiple intermediate products (IPs) are initiated. RM1 and RM3 are dynamically fed during the reaction process and can be used to influence the reaction rates. The production goal is the main product (MP). Additionally, several undesired byproducts (BPs) occur in the reaction competing with the MP production. The reactor temperature during operation is controlled with cooling water (CW) that flows through the cooling jacket of the reactor and enters with ambient temperature. To stop the reaction process an inhibitor (RM4) is fed to the reactor. After feeding RM4, the species concentrations remain approximately constant and the reactor can be drained.

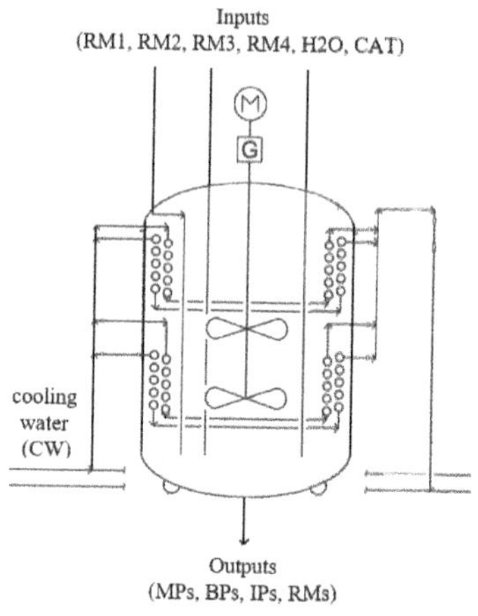

Figure 1. Schematic illustration of fed-batch reactor with input, output, and cooling water flows.

2.1. Mechanistic Modeling

We develop a dynamic model of the reactor and reaction system. The model includes 25 differential equations with 44 fitting parameters corresponding to the mass balance with the reaction kinetics and to the energy balance. The reaction kinetics involve 21 reactions for which reaction constants are calculated using the Arrhenius equation, i.e., two parameters in the form of the pre-exponential factor and the activation energy. Note that 3 side reactions have the same rate constant such that the reaction kinetics effectively have 38 fitting parameters in total. The energy balance accounts for the enthalpy of reactions, enthalpy flows, and cooling heat transfer rate affecting the temperature of the reaction mixture. Five reaction enthalpy parameters need to be estimated as several side reactions are modeled by the same enthalpy parameter. For the cooling, the heat transfer coefficient represents another fitting parameter, whereas we calculate the heat exchange surface and the cooling-water hold up based on the reactor geometry. For all other model parameters, e.g., as part of density and heat capacity correlations, we use standard values from the literature. In addition, we neglect other energy flows, e.g., heat loss over the reactor surface. Next to the differential equations, several algebraic equations, e.g., for pH, density, heat capacity, and yield calculation, are part of the model.

For modeling the dynamic input streams, we discretize the feed flows of RM1, RM3, RM4, and CW in one-minute intervals. Since RM2, H2O, and CAT are loaded to the reactor without any main reactions occurring, we use the respective total amount loaded to the reactor as initial values.

2.2. Implementation

We implement the reactor model in the modeling language Modelica (Modelica, 2022). Further, we apply our dynamic optimization framework DyOS (Caspari et al., 2019) with integrator NIXE (Hannemann et al., 2010) and NLP solver SNOPT (Gill et al., 2005). We use the Python interface of DyOS to implement and run the case study in Python. We use a Microsoft Windows Server with an Intel(R) Xeon(R) E5-2630 v2 processor running at 2.6 GHz and 128 GB RAM; simulating the model with given parameters took ca. 2 sec.

3. Parameter estimation

The model constitutes overall 44 fitting parameters that we initialize by data obtained from laboratory experiments and literature. However, as the simulation results when using these parametric values did not match the measured temperature profiles and final concentrations available from historical reactor data, we performed parameter estimation. To this end, we measured concentrations of some main components and the reactor temperature profile for one fed-batch run. In Figure 2, the feeding strategy for this run for (a) RM1, RM3, RM4, and (b) CW is illustrated. Subsequently, we used this experimental data for fitting the parameters of the model.

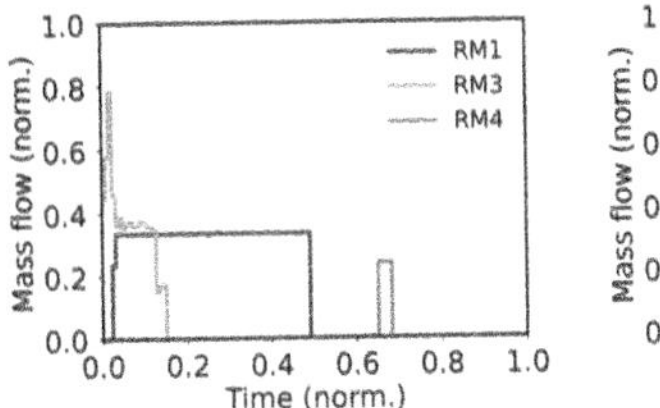

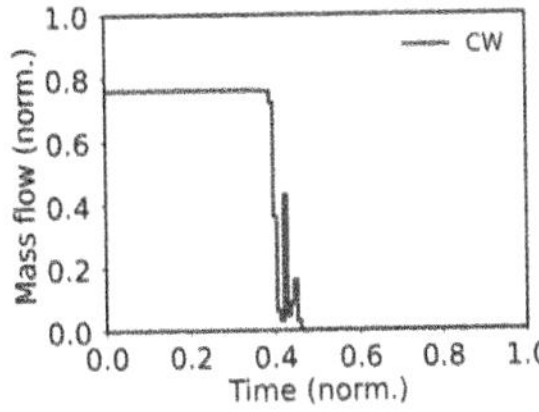

Figure 2. Feeding profiles of one fed-batch run for (a) RM1, RM3, RM4 and (b) CW.

In the parameter estimation process, we utilized DyOS to fit the 44 model parameters with the objective to minimize the mean squared error (MSE) of the model deviation to the measured time-series of the concentration profiles and the reactor temperature profile. We solve the optimization problem using direct single shooting with optimality and feasibility tolerances of 10^{-3}. We apply absolute and relative integrator tolerances of 10^{-4}. Figure 3 shows the simulation results with the initial and fitted parameters in comparison to the experimental data for the temperature profile and the concentration profiles of the MP and RM2. In contrast to the initial model, the results from the fitted model match the temperature profile with slight deviations. The fitted model also predicts the concentration measurements of MP and RM2 with high accuracy. We stress that due to the large model size and large number of parameters, we did not perform a parameter identifiability study. Other parameter values might also yield a good fit but were not explored since we find a sufficient match of the experimental data using the fitted model.

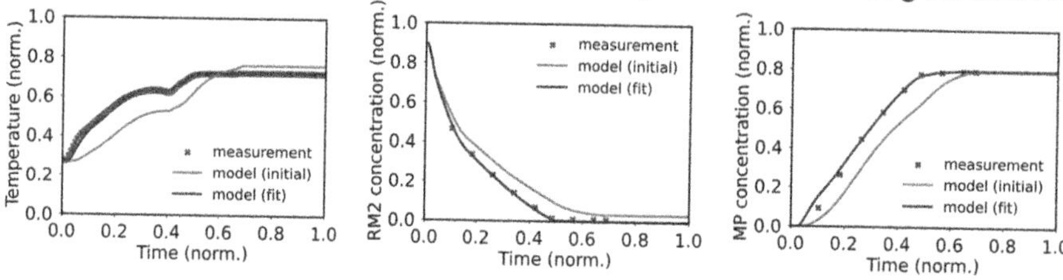

Figure 3. Comparison of the experimental results from batch run used for fitting with the simulation results obtained with the initial and the fitted model: Temperature (left) and concentrations for RM2 (middle) and MP (right).

For validation of the fitted model parameters, we conducted another experimental run of the industrial reactor at similar conditions. Figure 4 illustrates the temperature profiles and concentration measurements against the simulation results of the fitted model. We observe that the model matches the profiles closely. Some deviations regarding the MP concentration are, however, noticeable.

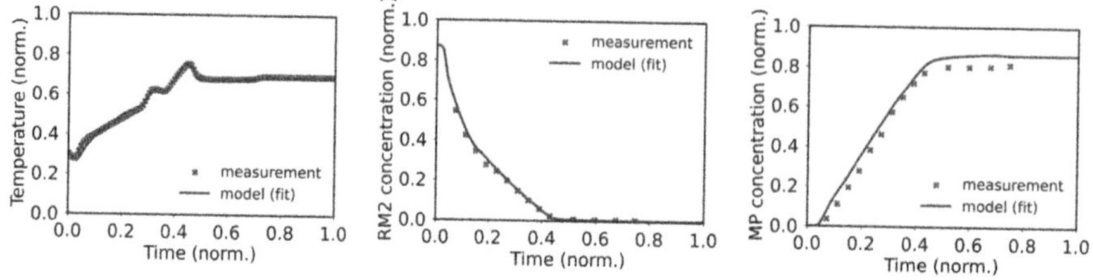

Figure 4. Comparison of the experimental results from validation batch run with the simulation results obtained with the fitted model: Temperature (left) and concentrations for RM2 (middle) and MP (right).

We also tested the model for other batch runs for which historical data on the temperature profile and the end concentrations but not all initial conditions, e.g., temperatures and impurities of raw material feed flows, was available. Here, we also observe a generally good match of the fitted model results and the experimental data, but also find some deviations, especially in the concentrations of some byproducts. We attribute the deviations to uncertainty factors in the initial conditions including BP impurities in the feed flows, limitations of the mechanistic model in capturing all characteristics of the multi-step reaction, and presumably slight overfitting of the parameters. In comparison to the initial model, the parameter fitting improves the prediction accuracy.

Overall, we conclude that the fitted model correctly captures the trends of temperature profiles and concentration measurements. In the next step, we will thus use the model with the fitted parameters for optimization of the feed profiles.

4. Dynamic optimization of the reactor feeding profiles

In the optimization of the feeding profiles, we aim at identifying changes in the currently established feeding strategy to produce the same amount of MP while using fewer raw materials per batch. As a starting point for the optimization, we herein consider the feeding profiles from the fed-batch run for which we collected concentration measurements and fitted the model parameters. The objective of the optimization is to decrease the total costs of RM1 and RM3 fed to the reactor, which we formulate as

$$\begin{aligned} \min_{\dot{m}_i(t)} \quad & m_{RM1} \cdot c_{RM1} + m_{RM3} \cdot c_{RM3} \\ s.t. \quad & m_{MP,cur}(t_{end}) \leq m_{MP}(t_{end}), \\ & T_{R,min} \leq T_R \leq T_{R,max}, \\ & V_{L,min} \leq V_L \leq V_{L,max}, \\ & 0 \leq \dot{m}_i(t) \leq \dot{m}_{i,max}, \\ & c_{BP,j} \leq c_{BP,j,max}, \\ & i \in \{RM1, RM3, CW\}, j \in BPs, \end{aligned} \tag{1}$$

where $c_{RM1}$and c_{RM3} are cost factors and the mass flows, $\dot{m}_i\,(t)$, of the raw materials RM1 and RM3 and the CW are the control variables, i.e., degrees of freedom. To enforce a minimal MP production, we add a constraint for the final produced amount of MP, $m_{MP}(t_{end})$, to at least matching the currently established strategy (cur), $m_{MP,cur}(t_{end})$. We consider additional constraints that include limits on the reactor temperature, T_R, the reactor level V_L, the mass flows, $\dot{m}_i(t)$, and concentrations of BPs, $c_{BP,j}$. The upper and lower limits on the reactor temperature and level arise from safety considerations, the upper limits on the mass flows are based on plant equipment limits, and the upper limit on the BPs' concentrations results from process downstream quality limits. Furthermore, the process control system only allows for a maximum number of four step changes in the mass flow of RM1 which we formulated as a multi-stage problem with free stage duration. For solving the optimization problem (Eq. 1), we apply DyOS with the same integrator and optimizer settings as for parameter estimation (cf. Section 3).

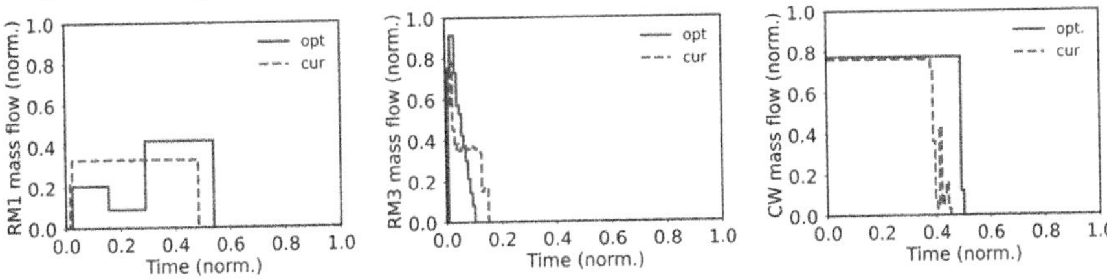

Figure 5: Optimized vs. currently established feeding strategy for RM1, RM3, and CW.

Figure 5 illustrates the currently established (dashed line) and optimized (full line) feed. According to the optimization results, RM1 should be fed in three steps with different flow rates instead one constant rate. The optimized RM3 feed is at a higher rate from the start of the batch and is then reduced almost linearly instead of in three steps. The cooling water feed is also at the maximum possible rate and changes only slightly compared to the established routine; specifically, the reactor is cooled longer. With the optimized feeding strategy, we find the produced MP amount being at the lower bound, i.e., matching the MP amount produced with the established feed profiles, which is expected as it is enforced by a constraint in the problem formulation. At the same time, the amounts of RM1 and RM3 can be reduced by 6% and 17%, respectively. This corresponds to about 10% cost savings as the same MP amount is produced by using less raw material. The main reasons for the reduced amount of required raw materials are higher main product yield, less byproduct production, and less overfeeding of RM3. We are currently working on the experimental validation of these results.

5. Conclusions

We present the dynamic modeling and optimization of an industrial 20 m^3 fed-batch reactor. We first develop a mechanistic reactor model and then collect experimental data on the reaction process and perform parameter estimation using our in-house dynamic optimization software DyOS. Based on the developed model, we apply DyOS to identify promising feeding strategies for which the model suggests improved reaction yield and higher economic profit by 10 %. The experimental validation of the identified feeding strategies is planned as the next step with the overall goal to establish a new feeding routine for the industrial operation of the reactor.

Future work could include further model validation and potentially improvements of the model, e.g., by collecting additional experimental data that can be used for parameter fitting. Next to model refinements, more fine-grained feed profiles could be investigated to incentivize changes to the current process control system limitations. Furthermore, the presented method could be transferred to other processes with fed-batch reactors.

Acknowledgments: This work was funded by LANXESS Deutschland GmbH. The authors also gratefully acknowledge the financial support of the Kopernikus project SynErgie 2 by the German Federal Ministry of Education and Research (BMBF) and the project supervision by the project management organization Projektträger Jülich (PtJ). This project was further supported by the Deutsche Forschungsgemeinschaft (DFG, German Research Foundation) - 466417970 - within the Priority Programme 'SPP 2331: Machine Learning in Chemical Engineering'.

References

J. R. Banga, E. Balsa-Canto, C. G. Moles, & A. A. Alonso, 2005. Dynamic optimization of bioprocesses: Efficient and robust numerical strategies. J. Biotechnol., 117(4), 407-419.

A. Caspari, A. M. Bremen, J. M. M. Faust, F. Jung, C. D. Kappatou, S. Sass, Y. Vaupel, R. Hannemann-Tamás, A. Mhamdi, & A. Mitsos, 2019. Dyos-a framework for optimization of large-scale differential algebraic equation systems. Comput. Aided Chem. Eng., 46, 619-624.

J. M. M. Faust, S. Hamzehlou, J. R. Leiza, J. M. Asua, A. Mhamdi, & A. Mitsos, 2021. Closed-loop in-silico control of a two-stage emulsion polymerization to obtain desired particle morphologies. Chem. Eng. J., 414, 128808.

P. E. Gill, W. Murray, & M. A. Saunders, 2005. SNOPT: An SQP algorithm for large-scale constrained optimization. SIAM review, 47(1), 99-131.

R. Hannemann, W. Marquardt, U. Naumann, & B. Gendler, 2010. Discrete first-and second-order adjoints and automatic differentiation for the sensitivity analysis of dynamic models. Procedia Comput. Sci., 1(1), 297-305.

J. L. Hjersted & M.A Henson, 2006. Optimization of fed-batch Saccharomyces cerevisiae fermentation using dynamic flux balance models. Biotechnol. Prog., 22(5), 1239-1248.

F. Jung, F. A. Janssen, A. Caspari, H. Spütz, L. Kröger, K. Leonhard, A. Mhamdi, & A. Mitsos, 2019. Dynamic optimization of a fed-batch microgel synthesis. IFAC-PapersOnLine, 52(1), 394-399.

T. Y. Jung, Y. Nie, J. H. Lee, & L. T. Biegler, 2015. Model-based on-line optimization framework for semi-batch polymerization reactors. IFAC-PapersOnLine, 48(8), 164-169.

J. A. López, V. Bucalá, & M. A. Villar (2010). Application of dynamic optimization techniques for poly (β-hydroxybutyrate) production in a fed-batch bioreactor. Ind. Eng. Chem. Res., 49(4), 1762-1769.

Modelica, 2022. https://modelica.org/ (accessed on 31/10/2022).

V. M. Zavala, A. Flores-Tlacuahuac, & E. Vivaldo-Lima, 2005. Dynamic optimization of a semi-batch reactor for polyurethane production. Chemical Engineering Science, 60(11), 3061-3079.

Antonis Kokossis, Michael C. Georgiadis, Efstratios N. Pistikopoulos (Eds.)
PROCEEDINGS OF THE 33rd European Symposium on Computer Aided Process Engineering (ESCAPE33), June 18-21, 2023, Athens, Greece

Optimization of large-scale Direct Air Capture (DAC) Model using SCR algorithm

So-mang Kim,[a] Syed Ali Zaryab,[b] Salar Fakhraddinfakhriazar,[a] Emanuele Martelli,[b] Grégoire Léonard[a]

[a]*Chemical Engineering, University of Liège, B6a Sart-Tilman, 4000, Liège, Belgium*
[b]*Politecnico di Milano, Department of Energy, Via Lambruschini 4, Milan 20156, Italy*

Abstract

This paper describes optimization studies of a large-scale fixed bed reactor model using Aspen Adsorption cycle simulations. The model uses Lewatit® VP OC 1065 amine-functionalized adsorbents and temperature vacuum swing adsorption (TVSA) cycles to capture CO_2 from the ambient air. Building a comprehensive direct air capture (DAC) model to optimize process design and cost is crucial to assess the feasibility of DAC technologies and therefore, a novel surrogate-based derivative free global optimization algorithm, referred to as SCR is implemented to evaluate a large-scale DAC system. Lastly, in order to achieve zero emissions and assess the viability of deploying large-scale DAC systems, assessments involving emission factors of various energy sources in different countries' electricity energy grid systems are studied. The optimization study showed a reduction in capture cost by 45% compared to the base case.

Keywords: Direct air capture, TVSA, process modelling, surrogate-based optimization

1. Introduction

Negative emission technologies (NETs) with a cumulative CO_2 capture of 450 to 1000 Gt are necessary to meet the strict 1.5 °C temperature increase target between 2010 and 2100 (Rogelj et al., 2015). One of the potential NETs is Direct Air Capture (DAC), and the likelihood of technological progress could significantly reduce DAC's energy usage, making it a practical long-term mitigation strategy (Realmonte et al., 2020). DAC was first proposed by Lackner and coauthors in 1999 to reduce climate change and the captured CO_2 may be used in various ways, including enhanced oil recovery, construction materials, and production of goods like chemicals, fuels, cement, etc. The CO_2 can also be permanently stored in deep geological formations (IEA, 2022).
Negative emission technologies (NETs) with a cumulative CO_2 capture of 450 to 1000 Gt are necessary to meet the strict 1.5 °C temperature increase target between 2010 and 2100 (Rogelj et al., 2015). One of the potential NETs is Direct Air Capture (DAC), and the likelihood of technological progress could significantly reduce DAC's energy usage, making it a practical long-term mitigation strategy (Realmonte et al., 2020). DAC was first proposed by Lackner and coauthors in 1999 to reduce climate change and the captured CO_2 may be used in various ways, including enhanced oil recovery, construction materials, and production of goods like chemicals, fuels, cement, etc. The CO_2 can also be permanently stored in deep geological formations (IEA, 2022).
There are several DAC techniques, including membrane, chemical, and electro-swing adsorption (ESA) methods (IEA, 2022, Voskian and Hatton, 2019, Fujikawa et al., 2021). The chemical methods can be further classified as a liquid solvent and solid sorbent

techniques which are the furthest technologies in development. In the case of the liquid solvent approach, an aqueous KOH solution is used as the capture medium and a calcium caustic loop is used as the recovery mechanism at high temperatures between 300-900°C (Keith et al., 2018). Solid sorbents technologies usually employ amine groups on the internal surface of the sorbent to react with CO_2 from the ambient air at low pressure, followed by regeneration at medium temperature of 80 – 120°C (McQueen et al., 2021). There are 18 operating DAC plants that capture almost 0.01 $MtCO_2$ annually according to the IEA report (2022) however, the majority of current DAC plants operate on a small scale for testing and demonstration purposes. Also, few studies for DAC have been conducted on process scales (Schellevis et al., 2021). Therefore, in order to address the aforementioned gaps, this paper presents a large-scale fixed bed reactor system of Kim and Léonard (2022), to capture CO_2 from ambient air via Temperature Vacuum Swing Adsorption (TVSA) using Lewatit® VP OC 1065 amine-functionalized adsorbents. Moreover, important operating parameters impacting the energy and cost of DAC systems are identified and CO_2 capture cost is optimized via SCR algorithm. Lastly, the viability of deploying DAC systems based on various electricity grid systems in different countries is explored to achieve net zero emissions.

2. Modelling

A large-scale one-dimensional fixed bed reactor model for gas adsorption (at 400 ppm CO_2) presented in the work of Kim and Léonard (2022b) is studied further to find feasible operating conditions under TVSA cyclic operation (adsorption, evacuation, desorption and cooling). The large-scale model was scaled up in the aforementioned work from a kg-scale DAC model where this small-scaled model was validated with experimental results presented in the works of Bos et al. (2019) and Yu et al. (2017). The validations of equilibrium loadings of CO_2 onto the solid sorbent, breakthrough curves and model assumptions are presented in Kim and Léonard (2022a&b).

3. Optimization Methodology

In this work, the DAC unit is being optimized using a recently published optimization algorithm called SCR which was presented by Zaryab et. al. (2022). SCR is a surrogate - based derivative free optimization algorithm; this algorithm was chosen because the simulation model of DAC has a very high computational cost, but the use of a surrogate model helps reduce it. SCR considers the DAC simulation code as a black box function and creates Kriging-based surrogate models of this function and its constraints. SCR then finds the region where the global minimum of this surrogate model lies and then calls the black box function within this region to look for the optimum. Using this optimization strategy helps to greatly reduce the number of functional evaluations required to reach the optimum thus saving significant time. For the optimization a total of five optimization variables have been selected, these include: the adsorption time (t_{ads}), depressurization time (t_{eva}), desorption time (t_{des}), cooling time (t_{cool}) and the vacuum pressure (P_L). In the case of the desorption temperature (T_{des}), Aspen Adsorption model was not robust enough to allow T_{des} to be a part of the optimization variables within the time limitation and therefore, two optimization runs were carried out at two fixed desorption temperatures (T_{des}) of 352.15 K and 360.15 K. The goal of the optimization is to minimize the specific cost of the DAC unit (€/ton_{CO2}). The constraints of the optimization are that the recovery should be greater than 90% and the purity should be greater than 85%.

4. Results and Discussions

4.1. Optimization of large-scale DAC systems

A benchmark study of Climeworks Hinwil plant (Climeworks, 2017) capturing 2460 kg/day with 18 DAC units was selected in the work of Kim and Léonard (2022b), where a single large-scale DAC unit model was developed. This model could capture 127 kg of CO_2 per day via 4 cycles and this led to a total capture of 2286 kg/day using 18 DAC units which is similar to the Climeworks Hinwil plant's capture scale. Based on this result, a CO_2 capture cost of 620.7 €/ton$_{CO2}$ was reported in the aforementioned work. In the current paper, the DAC costs of Kim and Léonard (2022b) are re-evaluated with updated equipment and utility costs (case A) and then optimized using SCR algorithm of Zaryab et al. (2022) at two T_{des} of 352.15 K (case B) and 360.15 K (case C). Case A's revised capture cost was found to be 552.78 €/ton$_{CO2}$ and then TVSA step times (t_{ads}, t_{eva}, t_{de}, t_{cool}) and P_{vac} were optimized in the cases B and C using SCR by minimizing the capture cost (€/ton$_{CO2}$). Variables before and after optimization are presented in Table 1 below.

Table 1. Summary of variables before and after optimization

	Before optimization	**After optimization**	
	Case A	Case B	Case C
T_{des} **[K]**	352.15	352.15	360.15
P_{vac} **[bar]**	0.33	0.37	0.46
t_{ads} **[s]**	6500	6467	6664
t_{eva} **[s]**	30	72	56
t_{des} **[s]**	6700	7539	5561
t_{cool} **[s]**	6300	1535	1000
Captured CO_2 per cycle [kg/cycle]	31.66	34.57	34.75
Capture cost [€, 2021/tonCO_2]	552.78	367.23	301.35

It was found in both cases B and C that t_{cool} has been shortened significantly. Before optimization, the adsorption bed temperature was allowed to cool down to ambient temperature while in the optimized cases, due to the shortened cooling time, the bed temperature is not fully cooled down to its initial temperature. However, as air is fed into the bed during the adsorption step, the bed rapidly cools down to its initial state. This may seem highly ideal but the temperature profile of the simulation can be further refined through experimental data to validate the cooling steps. Based on the simulation results, the drastic decrease in cooling time allows more cycles to be achieved per day (grey line) and this allows more CO_2 to be captured which reduces capture cost per ton of CO_2 (orange line) for Case B and C as depicted in Figure 1.

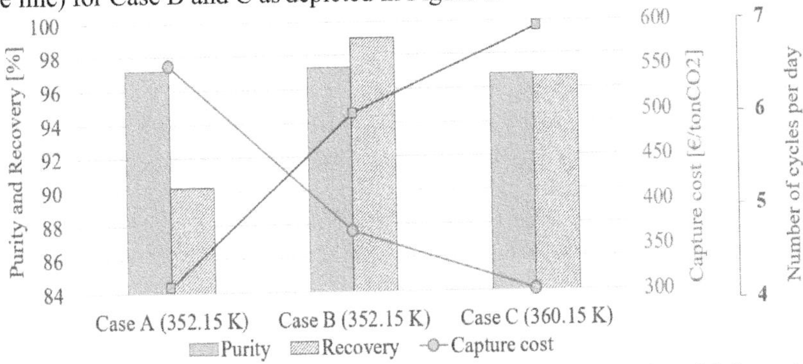

Figure 1. Purity and recovery of the system with capture cost and number of daily cycles

The purity of the different cases stayed around 97% while recoveries of case B increased to 99% from 90% (case A). This is due to an increase in desorption time in case B which allows improved utilization of the adsorption bed. Case C shows desorption at a higher temperature of 360.15 K with longer t_{ads} (6664 s) and the shortest t_{des} of 5561 s among cases. Although the recovery may be slightly lower (97%) than case B (99%), this indicates that it is more advantageous to operate at a higher T_{des} as even at a shorter desorption time, a large recovery per cycle can be achieved.
Specific energy duties of the cases are also calculated and presented in Figure 2 with the annualized CAPEX and OPEX. The electrical energy includes feed (air) compression, vacuum compression of purge streams and air blowers (in hatched column graph sections) while the thermal energy comprises the heat of reaction of CO_2 /H_2O, sensible and latent heat of purge streams as well as sensible heat of sorbents and reactors (in solid column graph sections). Note that water adsorption isotherm is not included in this study and therefore, a theoretical heat of reaction of H_2O (the energy required for H_2O desorption) is calculated using the vaporization enthalpy (Veneman et al., 2015).

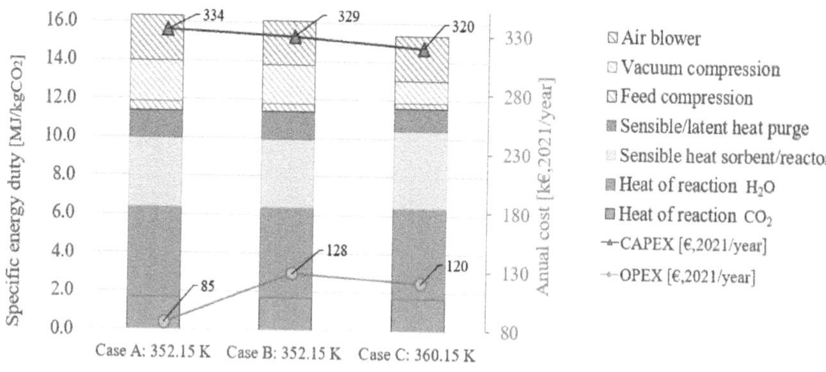

Figure 2. Specific energy duties (column graphs) with annualized costs (line graphs)

There are small changes in specific energy duties in the case of A and B (before and after optimization at 352.15 K respectively). For the optimized results at 360.15 K (case C), due to its higher T_{des}, there is an increase in sensible heat of sorbent/reactor but operating at higher T_{des} is beneficial since this reduces compression duties and sensible/latent heat of purge streams (less steam is required) which makes the overall specific energy duty to decrease. Climeworks process needs between 1.8 and 2.6 MJ/kg of electrical energy and between 5.4 and 11.9 MJ/kg of thermal energy (Deutz and Bardow, 2021), where the minimum value represents the target value for the future and the maximum value depicts the energy requirements at present. For all cases, the thermal energy is in the agreement with the reported range in literature while the electrical energies including air blowing obtained in this study are in the same magnitude but higher in both cases B and C. However, in case C, an electrical duty of 3.76 MJ/kg was estimated which is 31% higher than the upper boundary from the literature. More detailed optimization can be performed to further reduce the electrical duties.
In terms of economics, the annualized CAPEX throughout the cases showed a small change since the dimensions of the columns and the air feed flow rates are kept constant. However, due to the increase in the number of cycles per day in cases B and C, higher operating costs are expected. Nevertheless, the capture costs after the optimization showed a reduction of 33.6% and 45.5% for case B and C respectively as compared to case A since more CO_2 can be captured with a small increase in costs.

4.2. Achieving net-zero emission with DAC

As it is known, the energy source for DAC systems is a key factor in achieving negative emissions and, low-cost and low-carbon resources are crucial to obtain this goal of the DAC process. To point out this issue, the assessment of integrating different renewable resources into DAC systems has been suggested in the literature (Zolfaghari et al., 2022). Moreover, using currently available electricity sources for DAC is considered the most cost-effective approach for this technology (Hanna et al., 2021). One way to decrease the amount of emitted CO_2 by the DAC process itself is the utilization of electricity not only for electrical equipment (e.g. fans and pumps) but also for heat generation. Therefore, in this section, the total energy requirement, which is acquired through the optimization of the DAC model, is used to calculate the amount of CO_2 produced from DAC operations. The energy mix of electricity grids presented in Figure 3 (Global change data lab, 2022) and the emission factor of different sources from Bruckner et al. (2014) are used to assess the possibility of achieving negative emission in the case of a fully electrified DAC process.

Figure 3. Share of energy sources in electricity production of different countries.

As it is depicted in Figure 4, due to the lower CO_2 emission from electricity grid systems in countries with a higher share of renewables or nuclear energy, the usage of grid electricity for DAC is effective to achieve negative emissions in countries like Norway, Sweden and France, producing about 0.15-0.3 $kg_{emitted,CO2}/kg_{captured,CO2}$. On the other hand, the application of grid electricity for carbon removal from ambient air in countries that have a higher share of fossil fuels in power productions (e.g. China, India, and the USA) results in producing more carbon dioxide than captured. Nevertheless, to be able to compare DAC processes with other capture technologies and minimize the side effects of these systems, a detailed Life Cycle Assessment (LCA) is essential.

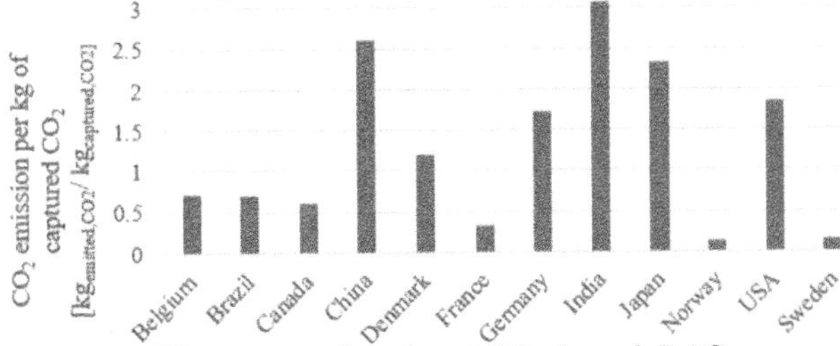

Figure 4. The amount of produced CO_2 through DAC process.

5. Conclusion and perspectives

In this paper, a large-scale DAC model of Kim and Léonard (2022b) is studied. The capture cost (€,2021/tonCO_2) of DAC systems was minimized by varying cyclic operating parameters (adsorption time, evacuation time, desorption time, cooling time and vacuum pressure) and using the SCR algorithm of Zaryab et. al. (2022) at two fixed

desorption temperatures of 352.15 K and 360.15 K. The optimized results showed that there are desirable step times which can yield a higher capture rate and there is a significant reduction in the cooling time (assuming ideal cooling profiles) allowing more CO_2 to be captured per day, increasing its performance. In terms of the specific energy duties and CAPEX of the systems, small changes were observed before and after the optimization but larger operating costs were observed due to more cycles per day. The increase in the number of cycles per day and the higher cyclic capture rate [kg/cycle] however greatly decreased the capture cost by 33.6% and 45.5% for optimized case B and C respectively. The thermal energy duties of the considered cases agree with Climeworks' benchmark unit. The higher desorption temperature reduces electrical duties but further studies involving different bed configurations, dimensions, different sorbents and air/purge flows are required to further reduce the electrical (as well as thermal) energies of the current model. In next steps, more robust optimization studies including desorption temperature, steam flow rates and cooling cost would be needed.
Also, a location for implementing a fully electrified DAC system is important in achieving net-zero emissions since depending on the country's electricity grid systems, the same DAC system may or may not be a negative emission option. However, a detailed LCA is required to obtain more insights into the relevance and potential deployment of DAC technology. Lastly, the co-adsorption of CO_2 and H_2O, not considered in the present study, can be included to fully assess the impacts of humidity on the performances of DAC units. Overall, this study shows a complex model with a high computational cost can benefit from SCR algorithm which is a surrogate-based derivative free optimization method. The results from the optimization also suggest the potential for DAC improvement as well as its promising potential as a negative emission technology.

Acknowledgement

The authors are grateful to the Belgian Federal Public Service Economy and to the Belgian Energy Transition Fund which support this research within the project PROCURA.

References

M. J. Bos, S. Pietersen and D.W.F. Brilman, 2019, *Chem. Eng. Sci: X, 2*, p.100020.
T. Bruckner et al., 2014. In *Clim. Change*, (pp. 1329-1356). Cambridge University Press.
Climeworks, 2017, Climeworks makes history with world's first commercial DAC plant
S. Deutz and A. Bardow, 2021, *Nat. Energy* 6, 203–213.
S. Fujikawa, R. Selyanchyn and T. Kunitake, 2021, *Polym. J., 53*(1), pp.111-119.
Global change data lab, 2022, Published online at OurWorldInData.org.
R. Hanna, A. Abdulla, Y. Xu and D.G. Victor, 2021, Nat. Commun., 12(1), 1-13.
IEA, 2022, Direct Air Capture, IEA, Paris https://www.iea.org/reports/direct-air-capture
D.W. Keith, G. Holmes, D. St.Angelo, K. Heidel, Joule, 2018. 2(8): p. 1573-1594.
S. Kim and G. Léonard, 2022a, *Comput. Aided Chem. Eng* (Vol. 51, pp. 265-270). Elsevier.
S. Kim and G. Léonard, 2022b, SSRN, GHGT-16.
K. Lackner, H. J. Ziock and P. Grimes, 1999, (No. LA-UR-99-583), LANL
N. McQueen et al., 2021, *Prog. Energy, 3*(3), p.032001.
G. Realmonte et al., 2019, *Nat. Commun., 10.*
J. Rogelj et al., 2015, *Nat. Clim. Change, 5*(6), pp.519-527.
H.M. Schellevis et al., 2021, *Int. J. Greenh. Gas Control, 110*, p.103431.
Q. Yu and D.W.F. Brilman, 2017, *Energy Procedia, 114*, pp.6102-6114.
R. Veneman et al., 2015, *Int. J. Greenh. Gas Control,* 41, 268–275
S. Voskian and T.A. Hatton, 2019, *Energy Environ. Science.* 12 3530–47
S. Zaryab, A. Manno, E. Martelli, 2022, *Comput. Aided Chem. Eng.* 51, 1213-1218, Elsevier.
Z. Zolfaghari, A. Aslani, A. Moshari, and M. Malekli, 2022, *Int. J. Energy. Res.*, 46(1), 383-396.

Antonis Kokossis, Michael C. Georgiadis, Efstratios N. Pistikopoulos (Eds.)
PROCEEDINGS OF THE 33rd European Symposium on Computer Aided Process Engineering (ESCAPE33), June 18-21, 2023, Athens, Greece
 http://dx.doi.org/10.1016/B978-0-443-15274-0.50189-X

Simulation of biodiesel production from algae

Suresh Kumar Jayaraman,

AVEVA Group Plc, Lake Forest, California 92630, United States of America

Abstract

Increasing oil demand and depletion of fossil fuel acts as a motivator for alternate fuel research. Biofuels, derived from renewable resources, have the potential to replace their petroleum counterparts, providing a domestic, and carbon-neutral fuel. This study analyzes a large-scale alga dewatering technique and production of biodiesel from algae oil. The dewatering stage extracts algae from water with liquefied propane or butane, which is part of a refrigeration process, is found to be cost effective compared to other dewatering methods. Dry algae are processed into algal oil, and the oil is then trans esterified to Fatty Acid Methyl Ester (FAME), a common biodiesel. The glycerol byproduct is converted to methanol, a more profitable product. This article involves developing a detailed steady state simulation of production of biodiesel from algal oil using first principles unit-op models and rigorous thermodynamics on a next generation commercial simulation tool, AVEVA Process Simulation.

Keywords: dewatering, algae oil, biodiesel, process simulation

1 Introduction

Biofuels, derived from renewable resources, have the potential to replace their petroleum counterparts, providing a domestic, and carbon-neutral fuel. Common sources of biofuel include sunflower, safflower, canola, soy, and corn. In order to address the issue of food crops being used for fuel production, researchers have turned their focus to non-food resources like algae [1]. Algae are ubiquitous photosynthetic organisms, ranging from single-celled forms to complex multicellular colonies [2]. Algae require sunlight, carbon dioxide, and nutrients like nitrates and phosphates to grow.

The major four processing steps in the production of biofuels are cultivation, Dewatering, Algae extraction, and Biodiesel production. Nutrients are the most cost-involving requirements of algae growth, so optimizing the nutrients requirement can reduce the associated production cost [3]. Out of various algae species, members of the marine green algae such as *Nannochloropsis* are the best species for biodiesel production, because of high lipid content and biomass productivity [4]. The fully grown concentration of algae in an aqueous culture medium is usually 0.1 w/w%, and this concentration is enhanced up to 20 w/w% through a traditional method like drying using heat or electricity [5]. Given the high latent heat of water, algae dewatering by evaporation is an energy intense process [6], alternative approaches are necessary to improve the energy efficiency of algae dewatering process. Mechanical disruption techniques and hexane extraction are often used to extract algae oil from algae biomass [7]. The algae oil is mostly composed of triglycerides, which is converted to biodiesel by the transesterification reaction[8]. This paper proposes a novel dewatering technique using propane as solvent, as part of a plant-wide refrigeration system.

2 Process Simulation

Process simulations are used to assess the commercial feasibilities of the proposed processes. The Next generation process simulation software, AVEVA™ Process Simulation developed by AVEVA, is used in this work. Microalgae consists of carbohydrates, proteins, and lipids. Here, algae are assumed to be of 50 % lipids, 25 % proteins, 25 % carbohydrates. Glucose is used to model carbohydrates, and poly lysine is used to model protein, and both are simulated as solid components. For simplicity just triolein ($C_{57}H_{104}O_6$) is used to represent lipid content of the algae in the simulation of dewatering process [2]. Triolein (glycerol esterified with a free fatty acid called oleic acid commonly found in various oil sources) is used to represent triglycerides and Methyl Oleate (MOLEATE) is used to represent FAME. All the components are added from DIPPR (Design Institute for Physical Properties) data bank. Non-Random Two-Liquid (NRTL) and Ideal Gas Law thermodynamic properties are used in this simulation. Missing thermodynamic properties are entered manually in the fluid.

3 Algae-dewatering process:

Algae-dewatering has been identified as one of the bottlenecks limiting the overall energy efficiency of algal biofuel technology[9], which currently inhibits the large-scale processing of algae into useable products. Centrifuges are usually used to concentrate the algae and remove lipids, but this processing step is energy intensive and is currently an area of significant research. Algae biomass production rates can vary based on the species chosen, cultivation technique, nutrients, and pond conditions. It is suggested that algal biomass production of 30 g m^{-2} d^{-1} (109. 6 t ha^{-1} y^{-1}) is possible[10]. The cultivation land of 1,500 acres, wet algae flowrate of 760,000 Kg/hr, and 1 w/w % of algae concentration is assumed in the simulation. Figure 1 shows the algae-dewatering process using liquid propane to reduce the moisture content from 99 % to 0.02%. This process stands out because a chemical manufacturer (such as fertilizer plant) generates waste (such as N, CO_2 and heat) and needs refrigeration. The proposed propane extraction idea uses the dewatering energy to supplement the plant refrigeration need and integrating the algae production with the plant converts waste to needed raw material. The condenser is used in the process to cool the compressed propane gas to liquid propane. Since propane and water are immiscible, they can be separated by decanting. Then propane is separated from algae biomass using flash separators

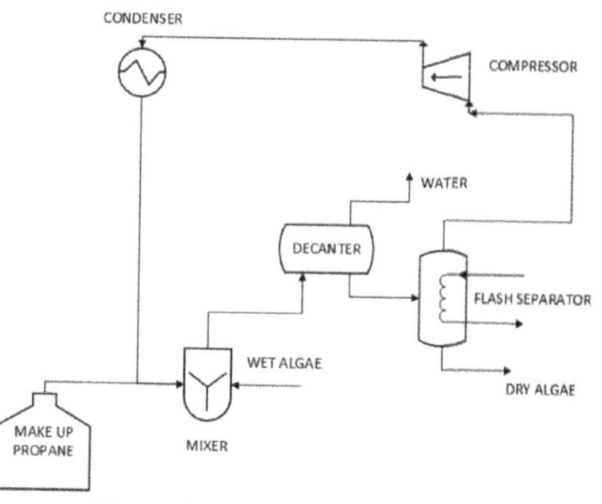

Figure 1: Algae dewatering process

and recycled to the fresh feed. The water from the decanter can be recycled back to the pond, thereby minimizing the wastage of water. In this process the water is removed, not by heating or drying but by a much less energy intensive process. The dried algae from this process are sent to an extraction unit to extract algae oil. The propane is evaporated in the flash separator, leaving the algal paste to be removed mechanically. The gaseous propane is compressed, constituting a refrigeration cycle, thereby supplementing an energy need for an adjacent plant. Butane can be used as an alternative solvent choice in the dewatering process. Though Coefficient of performance (COP) of butane is better than propane, the process economics suggest that propane dewatering technique would be a cost-effective way to dewater algae. However, the interaction of algae with propane, and ways to remove algal paste from the separator remain unknown.

4 Algae Oil Extraction Process:

After dewatering, the algae biomass undergoes a pre-treatment process to enhance the efficiency of lipid extraction process. The pre-treatment is either performed in a single step or multiple steps. If pre-treatment is not performed, the algae biomass concentrate is directly processed for lipid extraction. The algae biomass concentrate is either exposed to a cell disruption technique or dried and then milled into fine powders[11]. The pre-treated algae (disrupted concentrate or dried powder) is then exposed to an eluting extraction solvent which extracts the lipids out of the cellular matrixes[11]. Due to limited data on oil extraction process, a simple conversion model is used in this simulation study to convert 50% of the dry algae to algae oil/lipid. Under nutrient stress during the cultivation stage, the increase in total lipid primarily consists of triglycerides [12 13].

5 Biodiesel Production Process:

Component information and thermodynamic models are set up and the process steady-state simulation is simulated in a next generation simulation tool, AVEVA Process Simulation. Process flowsheet of production of biodiesel is shown in Figure 2. The oil content of algae is assumed to be 50% in the simulation. The Biodiesel production process in this work consists of following sections:

5.1 Transesterification reactions

The most common way to produce biodiesel is by stepwise transesterification, which is a catalyzed chemical reaction between algae oil and an alcohol to yield fatty acid alkyl esters (biodiesel) and glycerol as shown below.

Triglyceride + Methanol ←→ Diglyceride + FAME

Diglyceride + Methanol ←→ Monoglyceride + FAME

Monoglyceride + Methanol ←→ Glycerol + FAME

When triglycerides react with alcohol (e.g., methanol), the three fatty acid chains are released from the glycerol skeleton and combine with the alcohol to yield fatty acid alkyl esters [10] and glycerol as a by-product. Methanol is the most commonly used alcohol and Transesterification reactions can be alkali-catalyzed, acid-catalyzed, or enzyme-catalyzed [10]. Based on several studies in the literature [10, 14], the transesterification reaction is carried out at a temperature about 60°C and a 6:1 molar ratio of methanol to algae oil is fed to the Plugflow reactor, which helps in shifting the equilibrium to the product side. The reaction submodel is developed to define the reaction kinetics of the transesterification reactions [15]. Fresh make up methanol, recycled methanol, and sodium hydroxide with a concentration of 1.0 wt% of the algae oil are mixed prior to being pumped into stoichiometric reactor by a pump. For the study, 97% conversion of triglycerides in transesterification reaction is assumed in the reactor.

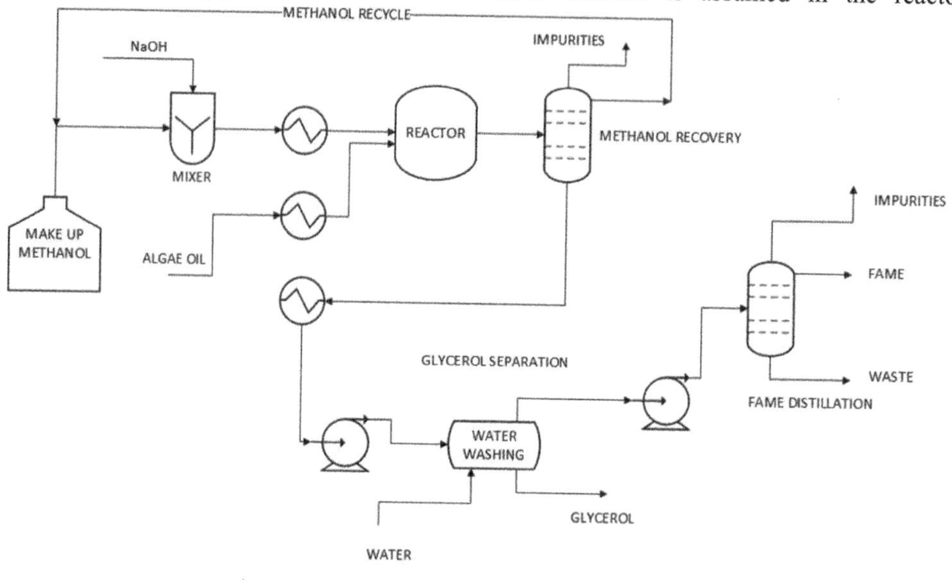

Figure 2: Biodiesel production process

5.2 *Methanol recovery*

The distillation column used for MeOH recovery has four theoretical stages, a reflux ratio of 2, total condenser, and kettle reboiler for a good separation of methanol from other components. Pressure in the range of 20 kPa is used to maintain the temperature of bottom stream under 150°C. Methanol is collected in the distillate stream with a purity of 96%. Methanol from distillate stream is recycled to the fresh make up methanol and then fed to the stoichiometric reactor. Bottom stream is sent to decanter to separate glycerol and FAME. Carbon-dioxide and oxygen are removed as impurities in this section.

5.3 *FAME and glycerol separation*

FAME is separated from glycerol, methanol and catalyst using a water washing column in this study. The FAME is separated from glycerol, methanol, and catalyst by adding 110 kg/hr water. The unreacted triglycerides, methanol and water in the top stream are less than 6 % and glycerol is collected in the bottom stream. Feed water amount is determined by performing the sensitivity analysis to achieve complete separation between FAME and glycerol.

5.4 *FAME purification*

Biodiesel purity of 99.65% is required to meet ASTM D 6751 of biodiesel specification [22a]. To obtain a final biodiesel product adhering to specifications, FAME distillation column with four theoretical stages, a reflux ratio of 1, total condenser, and kettle reboiler is used. Distillation column is operated under vacuum to keep temperatures low to avoid degradation of the FAME. Methanol, water, and unreacted oil are collected in the column overhead using a total condenser. FAME is obtained as distillate with the purity of 99.8 %.

5.5 *Methanol production from glycerol*

For every 9 kg of biodiesel produced, about 1 kg of a crude glycerol is formed as a by-product. The crude natural glycerol could be converted to a useful chemical, to increase the profitability of the biodiesel production plant. Glycerol can be converted to methanol, which can be used as fresh make up solvent to the transesterification reaction. Glycerol is converted to methanol and CO_2 as shown below

3 Glycerol + 2 Water → 7 Methanol + 2 Carbon-dioxide

Glycerol from the water washing column is sent to the conversion reactor along with water, and 99% of glycerol is assumed to be converted to methanol and CO_2. The reaction takes place at 470 °C and 5 MPa to produce methanol and CO_2. Separation of methanol

and CO_2 is done in a distillation column with 5 theoretical stages and a reflux ratio of 1.5. Pure CO_2 is collected as a distillate, containing 99% of the total CO_2 and pure methanol is collected as a bottom product, containing 95% of the total methanol. Combining this process along with the biodiesel production, methanol can be used as fresh make up solvent for the transesterification reaction and CO_2 can be used as a nutrient for the growth of algae.

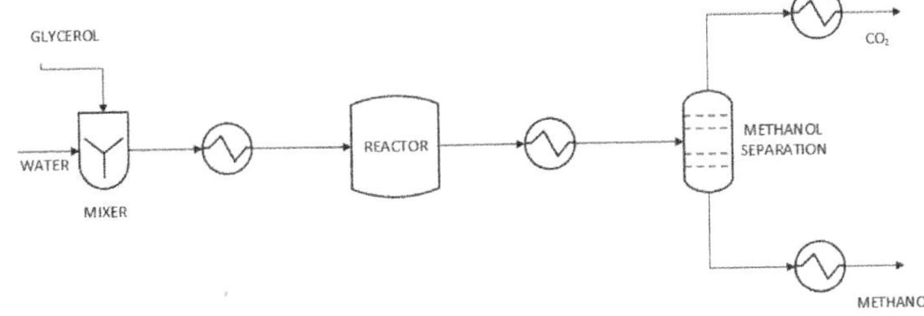

Figure 3: Methanol production process

6 Conclusion:

In this work, a new dewatering technique using propane is discussed and simulated using AVEVA Process Simulation. This work simulated a biodiesel production plant using the simulation tool to produce 33000 tons of biodiesel per year, with algae species of 50% oil content. The results show that the production unit is profitable only with 50% and 70% oil content, suggesting that the choice of algae species plays a vital role in the production of biodiesel. Algae species with 70% oil content is not achievable, so this work concludes that the production plant is profitable with algae species of 50% oil content. So future research should focus on cultivation and pond construction to make this biodiesel production more profitable. Since the production of glycerol exceeds the world demand, the byproduct glycerol is converted to methanol and CO_2, which can be used as make up solvents and nutrients for algae respectively. However, the present study shows that the biodiesel can be produced at a reasonable cost, thereby making algae a potential source for biofuels.

References:

1. Chisti, Y., Biodiesel from microalgae. *Biotechnology Advances* **2007,** *25* (3), 294-306.
2. Brennan, L.; Owende, P., Biofuels from microalgae—A review of technologies for production, processing, and extractions of biofuels and co-products. *Renewable and Sustainable Energy Reviews* **2010,** *14* (2), 557-577.
3. Blair, M. F.; Kokabian, B.; Gude, V. G., Light and growth medium effect on Chlorella vulgaris biomass production. *Journal of Environmental Chemical Engineering* **2014,** *2* (1), 665-674.
4. Sani, Y. M.; Daud, W. M. A. W.; Abdul Aziz, A. R., Solid acid-catalyzed biodiesel production from microalgal oil—The dual advantage. *Journal of Environmental Chemical Engineering* **2013,** *1* (3), 113-121.
5. Lardon, L.; Hélias, A.; Sialve, B.; Steyer, J.-P.; Bernard, O., Life-Cycle Assessment of Biodiesel Production from Microalgae. *Environmental Science & Technology* **2009,** *43* (17), 6475-6481.
6. Shao, P.; Darcovich, K.; McCracken, T.; Ordorica-Garcia, G.; Reith, M.; O'Leary, S., Algae-dewatering using rotary drum vacuum filters: Process modeling, simulation and techno-economics. *Chemical Engineering Journal* **2015,** *268* (0), 67-75.
7. Geciova, J.; Bury, D.; Jelen, P., Methods for disruption of microbial cells for potential use in the dairy industry—a review. *International Dairy Journal* **2002,** *12* (6), 541-553.
8. Vyas, A. P.; Verma, J. L.; Subrahmanyam, N., A review on FAME production processes. *Fuel* **2010,** *89* (1), 1-9.
9. Rawat, I.; Ranjith Kumar, R.; Mutanda, T.; Bux, F., Biodiesel from microalgae: A critical evaluation from laboratory to large scale production. *Applied Energy* **2013,** *103* (0), 444-467.
10. Benemann, J. R.; Oswald, W. J. *Systems and economic analysis of microalgae ponds for conversion of CO2 to biomass. Final report.*
11. Silva, C.; Soliman, E.; Cameron, G.; Fabiano, L. A.; Seider, W. D.; Dunlop, E. H.; Coaldrake, A. K., Commercial-Scale Biodiesel Production from Algae. *Industrial & Engineering Chemistry Research* **2014,** *53* (13), 5311-5324.
12. Halim, R.; Danquah, M. K.; Webley, P. A., Extraction of oil from microalgae for biodiesel production: A review. *Biotechnology Advances* **2012,** *30* (3), 709-732.
13. Hu, Q.; Sommerfeld, M.; Jarvis, E.; Ghirardi, M.; Posewitz, M.; Seibert, M.; Darzins, A., Microalgal triacylglycerols as feedstocks for biofuel production: perspectives and advances. *The Plant Journal* **2008,** *54* (4), 621-639.
14. (a) Pankaj Kumar, M. R. S., Kiran Toppo, Physico-Chemical Characterization of Algal oil: a Potential Biofuel. *Asian Journal of Experimental Biology* **2011,** *2* (3); (b) Demirbas, A., Progress and recent trends in biodiesel fuels. *Energy Conversion and Management* **2009,** *50* (1), 14-34.
15. Integrated Process Modeling and Product Design of Biodiesel Manufacturing, Ai-Fu Chang and Y. A. Liu, Industrial & Engineering Chemistry Research 2010 49 (3), 1197-1213

Antonis Kokossis, Michael C. Georgiadis, Efstratios N. Pistikopoulos (Eds.)
PROCEEDINGS OF THE 33rd European Symposium on Computer Aided Process Engineering (ESCAPE33), June 18-21, 2023, Athens, Greece
 http://dx.doi.org/10.1016/B978-0-443-15274-0.50190-6

Different methods of model-based process optimization presented on real world examples from water industry

Dr. Ewa Bozek[a], Dr. Corinna Busse[a], Delia Gietmann[a], Dr. Bernd-Markus Pfeiffer[a]
[a]Siemens AG, Siemenspromenade 3, Erlangen 91058, Germany
ewa.bozek@siemens.com

Abstract

With the digitalization of industry and further development of computational methods, different ways to control and optimize production processes are possible. Depending on which automation level (i.e. process, field or control level) is chosen for optimization, a variety of operator decision support systems exists. Using the example of water and wastewater industry this contribution explores existing optimization methods: high-fidelity first-principle, data-driven and linear dynamic black-box models. All described solutions can connect online to distributed control systems, but use case, engineering and deployment differ significantly for each of these applications.

Keywords: process simulation, optimization, model-predictive control

1. Introduction

Saving resources, securing sustainable supply, lowering maintenance and increasing market volatility are some of many challenges that the water industry is facing. A variety of model-based methods for process optimization exists to advice on optimal operating setpoints in an increasingly complex environment while meeting internal and external constraints. The following methods are described below in more details:

- Process level: high-fidelity first-principles models, optimization problem can be formulated in an arbitrary way, additional value provided by e.g. soft sensors, steady state or dynamic optimization for a specified time horizon (gPROMS & gDAP example: reverse osmosis desalination plant)
- Field level: Data-driven models, optimization problem is predefined: minimize switching operations or energy costs for a time horizon of 24 h. (SIWA Optim example: drinking water supply network)
- Control level: linear dynamic black-box models, dynamic optimization problem is essentially to minimize future deviations from specified setpoints in a specified prediction horizon for any multivariable feedback control problem (PCS 7 embedded MPC (Model Predictive Control) Example: wastewater treatment plant).

2. Real-time optimization based on first principles model

First principles models rely on fundamental laws of physics, chemistry and biology. The knowledge captured in these models is used since a long time for offline simulation and optimization. Using such models for online applications is a fairly new use case.

2.1. General Application Use Case

The provision of desalinated water by reverse osmosis is an energy intensive task. Due to the high osmotic pressure of seawater, significant energy is required to generate water of a certain quality (salt content). The energy consumption of the process is influenced by various decisions operators take based on process experience. To support operators' decisions, real-time optimization (RTO) based on first principles models is explored. In RTO, the optimum values of setpoints for underlying controllers are re-calculated periodically (e.g., every hour or every day). These repetitive calculations involve solving a constrained steady-state optimization problem. Therefore, a steady-state process model, a performance index (e.g., energy consumption) and constraints are required.

2.2. Process Model

The process model of the reverse osmosis section of a desalination plant was developed in gPROMS PROCESS using the library for membranes including a detailed description of the mass transfer (see [1.]). The model also includes dynamics like the gradually increasing fouling layer on the membranes. Overall, in gPROMS a differential-algebraic system of equations (DAE) is created, which can be used for a variety of use cases: simulation, global system analysis, soft-sensing and optimization.

2.2.1. Optimization problem

The optimization problem was configured within gPROMS PROCESS itself. A major advantage of a first principles model in the context of optimization is the flexibility to define an optimization problem: any calculated variable can be chosen as objective function, e.g., summation of all energy consumers. Similarly, all assigned variables can be decision variables and constraints can be specified according to process requirements. gPROMS also allows the use of discrete decision variables leading to mixed-integer optimization problems. Finally, constraints like product quality can be configured based on all calculated variables. The time horizon can be set-up as steady-state or dynamic.

2.2.2. Engineering Workflow

The engineering workflow of an RTO application from a first principles model follows four steps: 1. Development of a model based on design data like tank size, pump curves, membrane specifics etc. 2. Model validation based on historical process data to determine unknown parameters like mass transfer coefficients or fouling functions. 3. The calibrated model are then used to determine optimal operating points offline. 4. If the robustness of the model is proven, the optimization task can be transferred into an online application.

2.2.3. Deployment for Online Operation

For the online deployment the gPROMS digital application platform gDAP is used. The core component is the computational module required for the optimization based on the offline optimization task. Additionally, online connections are configured like e.g., OPC-UA, SQL Server, Historian. Measurement data coming from the real world like flowrates, pressure, quality measurements are validated before using in the computational module. Optimization results are then validated as well before sending them back to the real world.

2.2.4. Example from Water Industry

A high-fidelity process model including a detailed model of the membranes used for the separation of retentate (high salt loading) and permeate (lower salt content) with a detailed description of the mass transfer incl. fouling function depending on the load is developed in gPROMS PROCESS [1.] . This model was used for offline optimization studies to determine the optimization potential in typical operation conditions, as described by [1.]. As objective function, the energy consumption of a defined time horizon is chosen. As decision variables the following assigned variables are chosen: split fraction between 2nd stage and product tank, setpoints for the flowrate of the booster pumps in the 1st stage and setpoints for the flowrate of the high-pressure pump. As constraints, salt content in the permeate and overall production capacity are considered. Based on the offline optimization study, a significant potential for energy savings is determined. The developed set up is ready for online deployment with the dedicated framework gDAP.

3. Demand driven Optimization

The economic operation of pumps is one of the biggest challenges in the operation of water networks due to different energy tariffs and the increasing number of renewable energies. Since pumps are the largest energy consumer in the network they should be operated in a smooth and efficient way. This is a special use case of RTO, that can be solved by a Siemens software SIWA Optim dedicated to this optimization task.

3.1. General Application Use Case

The goal of RTO for drinking water networks is calculating a schedule for the most efficient pump operation, ensuring water supply at all times, taking all important parameters such as tank level, energy prices, consumer demands or pump availabilities into consideration. SIWA Optim assists in daily operation by continuously calculating schedules in individually set time intervals or supporting maintenance and operation planning with manually started what-if scenarios and calculations [2.].

3.2. Process Model

The water network is modeled in a flowsheet with realistic tank sizes and linearized pump models describing the relationship between transported volumes and required energy in the given environment. Model parameters are derived from process data. The simulation uses a linear, flow-based model (in contrast to a hydraulic model that consider pressure calculation). Besides filling of tanks, no other dynamic effects are described in the model.

3.2.1. Optimization problem

The optimization problem is solved by a Mixed Integer Linear Program Solver (MILP). It can be executed in a manual or cyclic way. In a manual mode it is possible to change the parameters and study the system in more detail, mainly for operation or maintenance purpose. Cyclic calculations are running autonomously in the background in a defined time interval between 1-24 h. Pump status and tank levels are updated before every cyclic optimization. Short intervals between the optimizations can increase operation safety since unexpected events like damaged pumps or low tank levels can be recognized sooner and taken into consideration for the next pump schedule. The calculated pump schedule can also be sent back to the DCS where a dedicated function block is receiving it and transfers the flow setpoints per station or pump to the control unit of the pump group.

3.2.2. Engineering Workflow

The water network is modeled as a flow based, abstract distribution network. It includes consumer forecast demands, water suppliers feeding fixed quantities of water or providing water in a specific range. Tanks with minimum and maximum level restrictions, pumping stations with limited allowed combinatorics of pumps as well as energy suppliers which calculate the working price for the specific stations and limit power consumption are also included into the model.

This data is used to calculate an optimal schedule for the upcoming 24 hours. Data like pump status and current tank level can be imported from the control system via an OPC UA connection, which allows reading independently of control system provider. The model is fine-tuned by validating the results with the real plant data.

3.2.3. Deployment

The application is running in Linux dockers within a virtual machine hosted by Hyper-V Manager. The user interface is accessible via Google Chrome. The network engineering for the application is done in an MS Visio based engineering tool.

4. DCS embedded Model Predictive Control (MPC)

An optimization problem for a dynamic process model on a finite time horizon, where the performance index includes deviations of model output variables from specified setpoints (or ranges) can be classified as an MPC problem, if it is solved repeatedly in a fixed sample time and the time horizon is shifted accordingly ("moving horizon").

4.1. General Application Use Case

The general use case is multivariable feedback control, namely model predictive control. In some cases, the aim is to reduce the demand for manual operator interactions and increase the degree of automation. In most cases the goal is to optimize process operation with respect to resource consumption, throughput or yield. This is achieved in two steps: 1. Reduce variances of process variables by improved feedback control. 2. This allows to push process operation towards constraints without the risk to violate them. Constraint pushing means to exploit the full potential of the process. As an example from water industry, aeration control in a waste water treatment plant will be described.

4.2. Process Model

All MPC algorithms exploit the IMC principle (Internal Model Control [3.]): a dynamic process model is not only used for offline controller design but constitutes an integral part of the online control algorithm, where it is used to predict future process behavior along a specified finite prediction horizon. Most industrial applications of MPC rely on linear dynamic process models that are identified from active experiments (like step testing) in the plant. The typical models are multivariable FIR (Finite Impulse Response) or FSR (Finite Step Response) models as they allow simulation of model response with minimal computing effort. Nonlinear MPC is out of the scope here but described elsewhere [4.] .

4.2.1. Optimization problem

The second central idea of MPC is to formulate the feedback control problem as an optimization problem: minimize the squared sum of "control deviations" and other "costs" of control like MV (Manipulated Variables) moves over a finite time horizon. The specific formulation of the objective function for optimization offers many degrees of freedom for controller design. Control deviations are differences between predicted process outputs and future setpoints. Different controlled variables (CVs) can be assigned weights in the performance index to reflect their relative importance to process operation. Move penalties for manipulated variables in the objective function restrict controller moves to be more or less aggressive. Hard constraints like MV limits and soft constraints for model output variables are considered in the optimization problem. The time horizon for optimization is user-defined and in typical applications set for a few hours. Special version of Quadratic Programming solvers are applied inside linear MPC algorithms.

4.2.2. Engineering Workflow

Linear MPC models are typically not derived from first principles but from process data. Active experiments in the plant like step tests around the typical plant operating point are required to create process data with a sufficient information content on process dynamics. From such data, the linear dynamic process model is created automatically, using one of the existing MPC software packages. Simulations of process model are compared to real process data to verify model quality. Definition of optimization problem reduces to specification of MPC design parameters like priority of controlled variables and move penalties for manipulated variables. Offline simulations of closed MPC control loop with linear dynamic process model inside MPC software allow fine-tuning of parameters.

4.2.3. Deployment for Online Operation

There are two deployment types of MPC in context of DCS in process industry: 1. deployment on a Windows PC interfaced to DCS via OPC. 2. DCS embedded MPC, where the online control algorithm is a function block similar to a PID controller. This is currently only available from two DCS vendors: Emerson DeltaV and Simatic PCS 7.

4.2.4. Example from Water Industry

An aeration control in a wastewater treatment plant is reviewed: [5.], [6.]. Closed-loop control of the biological operations in the activated sludge tank is a big challenge for sewage treatment plant automation, because aeration is the most sensitive and energy-consuming part of the WWTP. O_2-concentration in aeration tank is controlled by PID controllers which manipulate air blower speed. However, the decisive KPIs for water cleaning performance are NH_4 and NO_x outlet concentrations. Instead of leaving those KPIs to the plant operator, closed loop feedback control by MPC can be established. Manipulated variables of MPC are setpoints for existing PID controllers for recirculation rate and O_2-concentation. Fluctuations in WWTP feed (feed rate and pollution load) are a challenge for aeration control. Thus, they are included as measurable disturbance variables in the MPC concept, improving prediction quality and providing better compensation of the disturbances effects on NH_4 and NO_x concentration. WWTP plant operation by MPC assures the fulfillment of the legal specifications for purified wastewater. Test runs in a large WWTP operated by Hansewasser in Germany confirm that energy savings of 3-10% compared to manual operation with constant O_2-setpoint and heuristic reflux can be achieved.

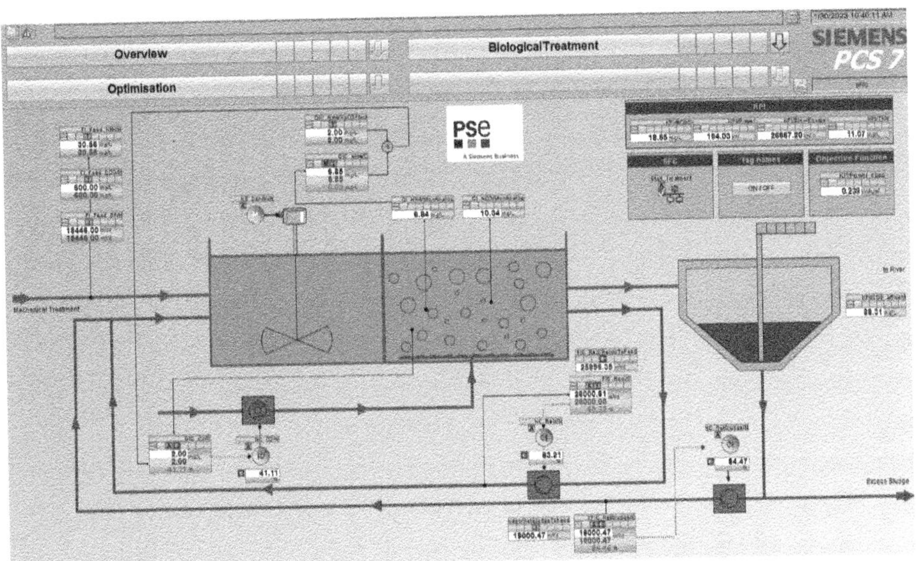

Figure 1: Simatic PCS 7 Unit template for aeration control in wastewater treatment plant [7.]

5. Conclusion

If a dedicated software solution for a given optimization problem (like SIWA Optim) is available in context of existing automation hardware and software architecture of a process plant, it is always recommended to exploit this offering. Otherwise, a very general approach like RTO can be applied. In some cases the term "optimization" in requirement specifications is not used in the narrow sense of mathematics. It has to be carefully checked if there is actually a feedback control problem in the plant, or rather potential for RTO. The typical cycle time for feedback control in process plants is seconds or minutes, whereas steady state RTO is calculated daily or hourly. In general, apply feedback control if target values (or target ranges) for the most important process variables are known and can be used as setpoints for MPC. Apply RTO if the focus is not on stabilizing plant operation at specified setpoints, but rather driving the plant in a desired direction of optimization, e.g. minimize energy consumption. The decision support workflow is shown in Figure 2. It is recommended to discuss it with the experts, to consider all factors.

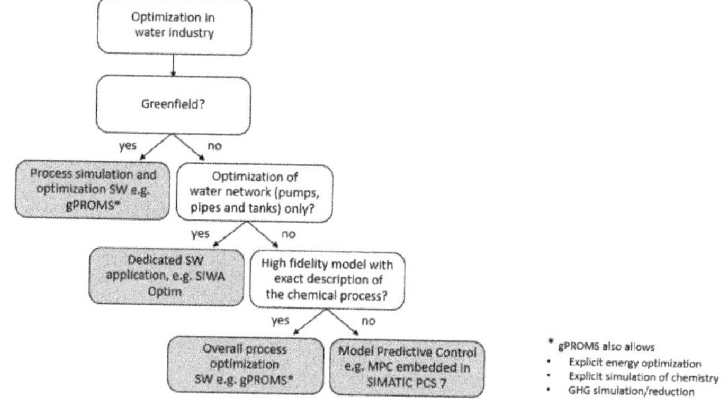

Figure 2: Decision support workflow for optimization in the water industry.

References

[1.] Pfeiffer, B-M., Lawal, A., Oppelt, M., Ruminski, L., Lade, M.: Plantwide Dynamic Optimization of Seawater Desalination with a Digital Process Twin. Industry Talks, Automation 2020, Baden-Baden (virtual), Jun. 2020.

[2.] Siemens AG, Flyer: SIWA Optim – Operate your water supply in a resource-saving and efficient way, 2018. https://assets.new.siemens.com/siemens/assets/api/uuid:c2befa03-0f72-468c-b03e-b9d92698203c/flyer-siwa-optim-en.pdf

[3.] Garcia, C.E., Prett, M., Morari, M. (1989): Model predictive control: theory and practice – a survey. Automatica 25 Nr. 3, pp. 335-348.

[4.] Pfeiffer, B-M., Oppelt, M., Leingang, Ch., Pantelides, C., Pereira, F.: Nonlinear Model Predictive Control based on Existing Mechanistic Models of Polymerisation Reactors. IFAC World Congress 2020, Berlin (virtual), Jul. 2020.

[5.] Oppermann, J., Radtke, N., Haber-Quebe, J., Pirsing, A., Pfeiffer, B-M., Labisch, D.: Einsatz von Advanced Process Control auf kommunalen Kläranlagen. 12. Aachener Tagung Wassertechnologie, Okt. 2017, Aachen.

[6.] Siemens AG, Whitepaper: Optimization of Sewage Treatment Plants by Advanced Process Control, March 2015. https://silo.tips/download/optimization-of-sewage-treatment-plants-by-advanced-process-control

[7.] Simatic PCS 7 Water Unit Template - Control of Biological Stage of a Wastewater Treatment Plant with upstream Denitrification. https://support.industry.siemens.com/cs/de/en/view/109485916

Antonis Kokossis, Michael C. Georgiadis, Efstratios N. Pistikopoulos (Eds.)
PROCEEDINGS OF THE 33rd European Symposium on Computer Aided Process Engineering (ESCAPE33), June 18-21, 2023, Athens, Greece
 http://dx.doi.org/10.1016/B978-0-443-15274-0.50191-8

A Reduced Order Model for the Prediction of the Dynamics of a Methane Reactor

Enrico A. Cutillo,[a*] Erasmo Mancusi,[a] Katarzyna Bizon,[b] Piero Bareschino,[a,c] Gaetano Continillo,[a]

[a]*Dipartimento di Ingegneria, Università degli Studi del Sannio, Benevento 82100, Italy*

[b]*Faculty of Chemical Engineering and Technology, Cracow University of Technology, ul. Warszawska 24, 31-155 Kraków, Poland*

[c]*Istituto di Scienze e Tecnologie per l'Energia e la Mobilità Sostenibili - Consiglio Nazionale delle Ricerche, Italy*

**cutillo@unisannio.it*

Abstract

The present work studies fixed bed methanation reactor for the upgrading of biogas, which is expected to provide chemical energy storage for non-programmable sources. The dynamics of such systems can exhibit complex behaviors, that could make the numerical solution computationally expensive. In this work the influence of both the inlet temperature and the recycle ratio on the reactor dynamics are studied employing an empirically based reduced-order model allowing a significant reduction of computational time. Moreover, an assessment of the accuracy of the proposed reduced-order model enables its use within real-time control applications, or in speeding up the resolution of time-consuming optimization.

Keywords: reduced models, POD, methanation, step response, Power-to-Methane.

1. Introduction

Numerical simulation of high-fidelity models of complex systems can take hours or even days. Some applications, such as parametric studies or optimization, often require thousands of those simulations, and for this reason, overall CPU time rapidly increases. Anyhow, simulations enable to optimize process parameters, to reduce costs, or maximize the productivity of the studied process (Bizon and Continillo, 2021) without the need for time-consuming and expensive experimental work but, most importantly, permit to predict unwanted behaviors and adopt adequate control strategies (Bareschino et al., 2022). In this work, the studied process is related to sustainable energy conversion and reduction of CO_2 emissions, which are challenges that cannot be procrastinated any further, in line with the core principles of the European Green Deal and Power-to-Gas concept. Particularly, the production of biofuels as green energy resources is very promising (Mancusi et al., 2021). Among biofuels, biogas from the anaerobic digestion of organic waste stands out as an attractive way to reduce landfilling and to produce energy. However, the CH_4 content in biogas usually never reaches values above 70%, so it needs to be purified (removal of trace components) and improved (removal of CO_2) before it can be used. One of the most widely studied upgrading processes involves the methanation of CO_2 (Bareschino et al., 2021). Recently, modeling studies on the methanation process have addressed specific aspects such as the dynamical characterization of methanation reactors (Zimmermann et al., 2022). In this work, the simulations of a dynamical system that describes the methanation of biogas in an adiabatic fixed-bed reactor with a nickel-based catalyst and a recycle loop are conducted by applying classical proper orthogonal decomposition (POD) for the model-order reduction. This methodology enables the use of the previously derived robust numerical model (Bareschino et al., 2022) for more computationally intensive applications. POD is a popular model reduction technique that, combined with Galërkin projection, is widely applied in the field of chemical engineering and many other fields, to alleviate the computational expense required for the solution of very high-dimensional systems (Sahyoun and Djouadi, 2013). Both methanation (Bremer et al., 2017) and fixed-bed reactors with recycle have been simulated successfully with this methodology (Bizon and Continillo, 2012).

To understand if the proposed POD/Galërkin-based reduced-order model (ROM) can correctly predict the dynamics of the methanation reactor, in this work a step response analysis is performed in a region of the space of the parameters where periodic solutions are encountered, and in its vicinity. Temperature oscillations inside the reactor must generally be avoided, as they can lead to permanent deactivation of the catalyst. Outlet concentration variations are also generally inconvenient. The ROM developed here can be useful if incorporated into the design of a control system.

2. Mathematical model and computational methods

2.1. Mathematical model of an adiabatic fixed bed reactor for biogas upgrading

The methanation reactor investigated is a fixed-bed adiabatic reactor with mass recycle. The catalyst considered is a nickel-based catalyst and the kinetic model employed is the one proposed by Xu and Froment (1989). The chosen model is one-dimensional pseudo-homogeneous with axial dispersion (Rönsch et al., 2016). Gas recycle is necessary to keep the temperature inside the reactor within an acceptable limit, determined by catalyst deactivation and carbon formation. The model is mathematically formalized in species and heat balances, Eq.(1) and (2) respectively.

$$\varepsilon_g \frac{\partial c_i}{\partial t} = D_{ax} \frac{\partial^2 c_i}{\partial z^2} - u_{sg} \frac{\partial c_i}{\partial z} - (1-\varepsilon_g)\rho_c r_i \tag{1}$$

$$\left(\varepsilon_g \rho_g c_{pg} + (1-\varepsilon_g) c_{pc}\right)\frac{\partial T}{\partial t} = \lambda_{eff} \frac{\partial^2 T}{\partial z^2} - u_{sg}\rho_g c_{pg} \frac{\partial T}{\partial z} - \rho_c \sum_{j=1}^{3} \Delta H_{Rj} R_j \tag{2}$$

where z is the axial position along the reactor, ε_g represents the bed porosity, c (kmol·m^{-3}) the gas concentration, ρ_c (kg·m^{-3}) the packed-bed density, and r_i (kmol·kgcat^{-1}·s^{-1}) the rate of consumption or formation of the ith species (i=CH_4, CO, CO_2, H_2, H_2O); the latter is determined by summing up the reaction rates of that species in all the reactions, R_j (Xu and Froment, 1989) according to the stoichiometric coefficient, as follows:

$$r_i = \sum_{j=1}^{3} \nu_{i,j} R_j \, . \tag{3}$$

The boundary conditions are:

$$c_i(0,t) = c_{i,in}(t); \; T(0,t) = T_{in} \tag{4}$$

where

$$c_{i,in}(0,t) = \frac{c_{i,0}}{1+\mathrm{R}} + \mathrm{R}\frac{c_{i,out}(t)}{1+\mathrm{R}} \tag{5}$$

In Eq.(5) R represents the recycle ratio. To solve the above system of partial differential equations, the method of lines was initially employed: spatial derivatives were approximated using finite differences over a uniform grid of 200 discrete nodes, yielding a system of 1200 ordinary differential equations (ODE) that was solved with MATLAB ode15s, a solver for stiff ODEs. To simplify the description of the reduction method, and to efficiently code the system of ODEs in MATLAB, the following matrix notation is employed:

$$\frac{d}{dt}\mathbf{c}_i = \mathbf{A}_c\mathbf{c}_i + \mathbf{B}_c\mathbf{c}_i \odot \mathbf{u} + \mathbf{F}_{\mathbf{c}_i}(\mathbf{c}_i, \mathbf{T}) \tag{6}$$

$$\frac{d}{dt}\mathbf{T} = \left(\mathbf{A}_T\mathbf{T} + \mathbf{B}_T\mathbf{T} \odot \mathbf{u}\right) \odot \frac{1}{\varepsilon_g \rho_g(\mathbf{c}_i, \mathbf{T})c_{pg} + \rho_c c_{pc}\left(1-\varepsilon_g\right)} + \mathbf{F}_T(\mathbf{c}_i, \mathbf{T}) \tag{7}$$

where **A** and **B** are the coefficient matrices discretizing spatial derivatives which include also the terms resulting from the outlet boundary conditions, whereas the nonlinear functions $\mathbf{F}_{\mathbf{c}_i}$ and $\mathbf{F}_T$ incorporate the inlet boundary conditions (Eq. (4) and Eq. (5)). The discretized system above is hereafter referred to as the full-order model (FOM).

2.2. Reduced order model formulation.

The POD method allows to extract a set of empirical basis functions from observations originating from experiments or numerical simulation (Holmes et al. 1996). The observations, often referred to as snapshots, are collected in the snapshots matrix **Y**. The choice of the most relevant snapshots is often driven by the knowledge of the spatiotemporal complexity of the system and its variability when changing key parameters of the model. In this work, the most straightforward strategy (uniform sampling) is employed (Bizon et al., 2012). After collecting the snapshots, the POD basis is determined by solving the eigenvalue problem:

$$\mathbf{C\Phi} = \mathbf{\Lambda\Phi} \tag{8}$$

In Eq. (8) **C** is the autocorrelation matrix defined as $\mathbf{YY}^T/M$, where M is the number of observations used for the construction of the basis, while $\mathbf{\Phi}$ and $\mathbf{\Lambda}$ are, respectively, the matrix containing basis functions and the relevant eigenvalues. The magnitude of each eigenvalue indicates the contribution of the corresponding POD mode to the "information" content of the original data. Using the POD basis, the generic state variable vector **y** can be expressed in truncated form as:

$$\mathbf{y} \approx \tilde{\mathbf{y}} = \sum_{k=1}^{K} \phi_k x_k = \mathbf{\Phi}_K \mathbf{x}_K \tag{9}$$

where $K \leq N$ is the dimension of the reduced model, and $\mathbf{\Phi}_K \in \mathbf{R}^{N\times K}$ is a matrix composed of the first K columns of matrix $\mathbf{\Phi}$. In Eq. (9), the unknowns are the K time-dependent coefficients $\mathbf{x}_K$. They can be determined by resolving the following ROM obtained after substitution of Eq. (9) into Eq. (6)-(7), followed by Galërkin projection onto $\mathbf{\Phi}_K$:

$$\frac{d}{dt}\mathbf{x}_K = \mathbf{\Phi}_K^T\mathbf{A}\mathbf{\Phi}_K\mathbf{x}_K + \mathbf{\Phi}_K^T\mathbf{B}\mathbf{\Phi}_K\mathbf{x}_K \odot \mathbf{u} + \mathbf{\Phi}_K^T\mathbf{F}(\mathbf{\Phi}_K\mathbf{x}_K) \tag{10}$$

The state variables vector has been divided into six sets of variables, associated with the species concentrations and temperature. Each one of these sets has been reduced with its own POD basis.

3. Results and discussion

The values of the main operating conditions and parameters characterizing the reactor geometry used in both FOM and ROM simulations are reported in Table 1. Fig. 1 reports a schematic of the system (a) and the grid of parameters explored (b). Before starting the construction and validation of the reduced-order model, several simulations of FOM were performed for the different values of the three parameters analyzed in this work, namely: T_{in} – the inlet temperature assumed to calculate the steady state, ΔT_{in} – the amplitude of the step forcing, and R – the value of the recycle ratio used for both steady-state and step response calculations.

Table 1. Main model parameters

Parameter	Value	Parameter	Value
L_r [m]	7	d_r [m]	1.2
P [atm]	20	ρ_c [kg·m^{-3}]	2350
c_c [J·kg^{-1}·k^{-1}]	1107	$y_{CH4,in}$	0.271
Q_{in} [Nm^3·s^{-1}]	1	$y_{CO2,in}$	0.146
D_{ax} [m^2·s^{-1}]	10^{-2}	$y_{H2,in}$	0.583

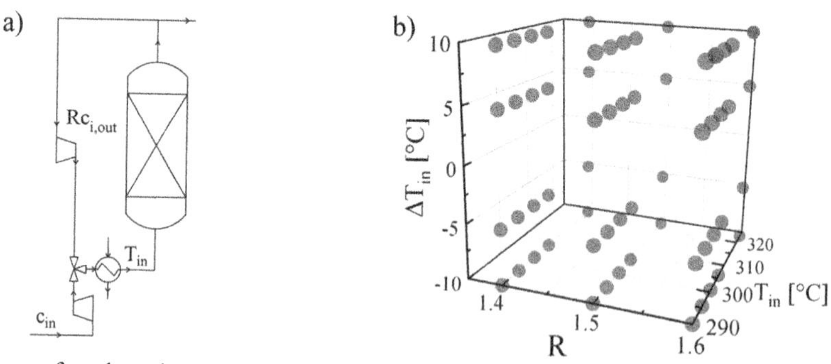

Figure 1. Scheme of methanation reactor (a) and values of the parameters used in simulations performed with the FOM (b). The two darker dots correspond to the set of parameter values employed in the construction of the ROM.

The simulations were conducted in two steps: first, the steady state (ss) solution is determined for different values of T_{in} and R; then, a step change ΔT_{in} of the inlet temperature is introduced, and the response of the reactor is simulated. Different values of ΔT_{in} are considered. The whole set of FOM simulations performed is represented inside the space of the studied parameters (Fig. 1b). The snapshots matrix is built with the spatial profiles of the step response of the FOM simulations, sampled uniformly in time. To build the snapshots matrix, different sampling strategies in the parameter space have been evaluated, but only the one that has given the best results in terms of ROM accuracy is reported in this work. Particularly, the snapshots matrix is built using only two FOM step response solutions. Those solutions were obtained for the values of the parameters highlighted in Fig. 1b, that is T_{in}= [295.3, 323] °C, ΔT_{in}=10 °C, and R=1.6. After computing the set of basis functions, the truncation order was set to K=35 for each state variable. Temperature time series and spatial profiles obtained using FOM and ROM are shown in Fig. 2(a) and (b), respectively. Before the inlet temperature step change, the system was simulated only using FOM, once the step change is imposed both FOM and ROM solutions are determined and reported in Figs. (2). This is because the ROM was built to reproduce only the step response of the system. As an interesting case study, to assess the ROM accuracy, an inlet temperature step change was applied to the operating conditions for which the oscillatory regime (or) was encountered. The parameters used in the simulation are T_{in}=306 °C and R=1.4. As reported in a previous study (Bareschino et al., 2022), this set of parameters is inside a region of periodic regimes. To avoid these undesired behaviors, the inlet temperature can be reduced, for instance, by 10 °C. Forcing the system with a step of $\Delta T_{in}=-10$ °C results in a steady-state solution. As can be seen from the overlapping curves shown in Fig. 2a, the ROM predicts even the steepest transient oscillations quite accurately.

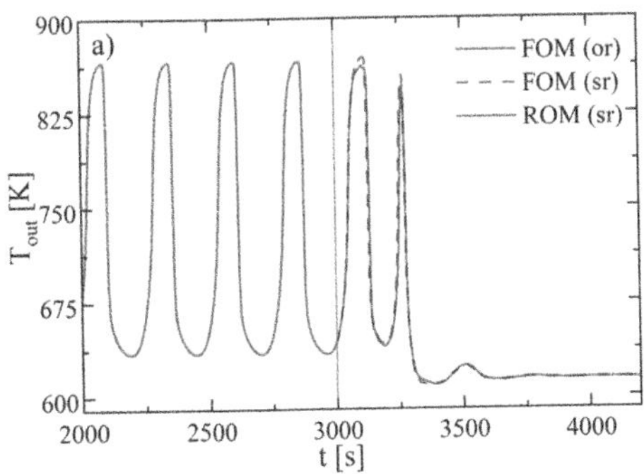

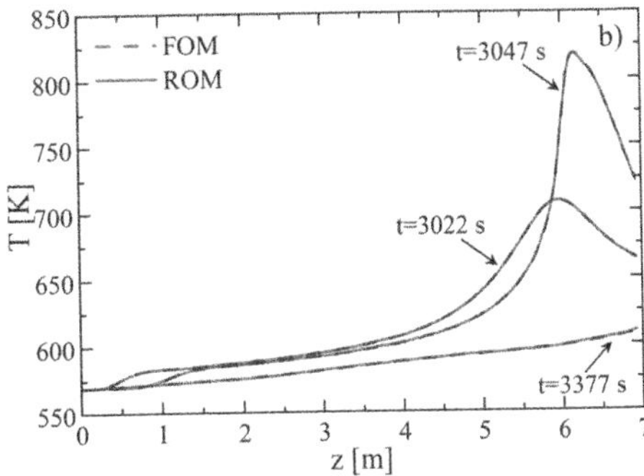

Figure 2. Time series (a) of the outlet temperature as computed with FOM before the step forcing (in the legend, or: oscillatory regime), and with both FOM and ROM after the step forcing (in the legend, sr: step response); (b) spatial profile of the temperature at different time instants confirming the accuracy of the ROM solution inside the reactor.

The choice of using only 35 POD modes was made to achieve a sufficient reduction of CPU time and to maintain suitable accuracy of the obtained ROM. The resulting ROM has 210 degrees of freedom, which is one order of magnitude lower than the corresponding FOM. The simulation of the step response toward a stable steady state is less computationally expensive, and in this case, the speed up (the ratio of CPU time of ROM simulation over FOM simulation) of the analyzed ROM is 15. For the chosen approximation, the computed relative root mean squared error (RMSE) between FOM and ROM (calculated for the solutions described in Fig. 2) is of the order of 10^{-3} with a maximum value of the error in correspondence of the peak value of T (at 3130 s) of $1.9\cdot10^{-2}$ and a steady state error of $6.5\cdot10^{-5}$. Although the ROM was built without taking into account periodic regime solutions, it can predict such regimes accurately.

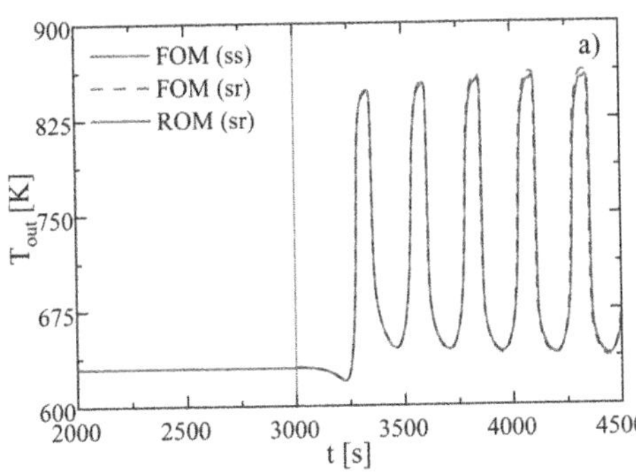

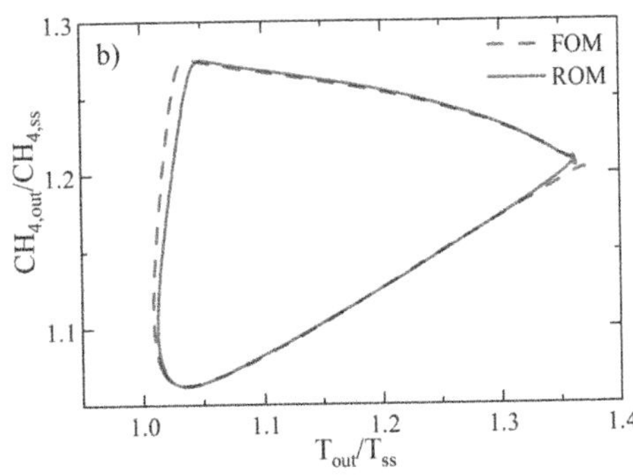

Figure 3. Time series as computed with FOM before the step forcing (in the legend, ss: steady state), and with both FOM and ROM after the step forcing – after the vertical line – (a); and phase space to highlight the accuracy of the ROM predicting the oscillatory regime (b).

This is demonstrated in Fig. 3: as it can be seen, if the system is forced to go from a stable steady state (ss) regime towards an oscillatory regime, the ROM correctly predicts periodic oscillations. The parameter values used in the simulation are T_{in}=300.7, R=1.6, ΔT_{in}=10 °C. In Fig. 3b, to highlight the accuracy of the ROM, the limit cycle obtained using both models are reported. In this case, the simulation is more expensive in terms of CPU time both for FOM and ROM because the solution is characterized by steep gradients that slow the convergence of the employed stiff-stable solver. In this case, the ROM achieves a speed up of 5 with a computed RMSE of $4 \cdot 10^{-3}$.

Conclusions

The simulation of the complex system that describes the methanation of biogas in a fixed bed reactor with mass recycle has been conducted by means of a purposedly developed efficient reduced-order model, and its application is critically discussed. For the case analyzed in this study, it has been shown that sufficiently reliable POD basis can be extracted using two solutions obtained from FOM simulations. Despite the nonlinearities, which might have made ROM perform worse due to the need of recomputing the nonlinear terms in the FOM space at each step, the chosen approach resulted in a significant reduction in calculation time, with speedup between 5 and 15 for periodic and stationary regimes, respectively. The dimension of the system has been reduced by one order of magnitude, preserving accuracy both of stationary and periodic solutions. Having assessed the performance of the proposed ROM, a more efficient optimization or a real-time control application can be developed. A considerably larger benefit in terms of reduction of CPU time is expected if the methodology is applied to reduce more complex models such as those considering the radial distribution of temperature and concentration.

References

P. Bareschino, A.E. Cutillo, C. Tregambi, F. Pepe, G. Continillo, E. Mancusi, 2022, Periodic Oscillations in Methane Reactor: Effects of the Main Operating Parameters, Comput. Aided Chem. Eng. 51 1–6.

P. Bareschino, E. Mancusi, C. Tregambi, F. Pepe, M. Urciuolo, P. Brachi, G. Ruoppolo, 2021, Integration of biomasses gasification and renewable-energies-driven water electrolysis for methane production, Energy, 230, 120863

K. Bizon, G. Continillo, 2021, Efficient optimization of a multifunctional catalytic fixed-bed reactor via reduced-order modelling approach, Chem. Eng. Res. Des. 165 214–229.

K. Bizon, G. Continillo, M. Berezowski, J. Smuła-Ostaszewska, Optimal model reduction by empirical spectral methods via sampling of chaotic orbits, 2012, Physica D, 241, 1441-1449.

K. Bizon, G. Continillo, 2012, Reduced order modelling of chemical reactors with recycle by means of POD-penalty method, Comput. Chem. Eng. 39 22–32.

J. Bremer, P. Goyal, L. Feng, P. Benner, K. Sundmacher, 2017, POD-DEIM for efficient reduction of a dynamic 2D catalytic reactor model, Comput. Chem. Eng. 106 777–784.

E. Mancusi, P. Bareschino, P. Brachi, A. Coppola, G. Ruoppolo, M. Urciuolo, F. Pepe, 2021, Feasibility of an integrated biomass-based CLC combustion and a renewable-energy-based methanol production systems, Renew. Energy. 179 29–36.

S. Rönsch, J. Schneider, S. Matthischke, M. Schlüter, M. Götz, J. Lefebvre, P. Prabhakaran, S. Bajohr, 2016, Review on methanation - From fundamentals to current projects, Fuel. 166 276–296.

S. Sahyoun, S. Djouadi, 2013, Local Proper Orthogonal Decomposition based on space vectors clustering, 2013 3rd Int. Conf. Syst. Control. ICSC 2013. 665–670.

J. Xu, G.F. Froment, 1989, Methane steam reforming, methanation and water-gas shift: I. Intrinsic kinetics, AIChE J. 35 88–96.

R. T. Zimmermann, J. Bremer, K. Sundmacher, 2022, Load-flexible fixed-bed reactors by multi-period design optimization, Chem. Eng. J. 428 130771.

Antonis Kokossis, Michael C. Georgiadis, Efstratios N. Pistikopoulos (Eds.)
PROCEEDINGS OF THE 33rd European Symposium on Computer Aided Process Engineering (ESCAPE33), June 18-21, 2023, Athens, Greece
 http://dx.doi.org/10.1016/B978-0-443-15274-0.50192-X

A Novel Approach for Cost-effective and Sustainable Capacity Expansion Utilizing Process Intensification in Chemical Process Industries

Chinmoy B. Mukta, Selen Cremaschi, Mario R. Eden*

Department of Chemical Engineering, Auburn University, Auburn, AL 36849, USA
**edenmar@auburn.edu*

Abstract

Sustainable and cost-effective capacity expansion in chemical process industries is essential. In this work, a systematic capacity expansion procedure is developed based on reactor intensification, which can help reliably formulate the expansion horizon and investment plans to be determined from a large solution space. Using the developed approach, the case study showed the impact of ethylene oxide reactor intensification on various capacity expansion modes. With a base case of 100 kton/yr, the Brownfield process is the cheapest capacity expansion option up to 112 kton/yr. Subsequently, the retrofitted mode yielded a minimum cost until a capacity of 157 kton/yr, and above that, we would need green field capacity expansion. On the other hand, in terms of utility carbon emission, Greenfield was the lowest carbon producer among the investment alternatives.

Keywords: Process Intensification; Capacity expansion; Greenfield process; Brownfield process; Process Retrofitting

1. Introduction

Widespread use of modern amenities by human society requires continued expansion of the capacity of the chemical process industry. As the commodity industry contributes about 32% of US emissions stemming from energy usage (Brueske *et al.*, 2015), the question remains: What are the most cost-effective and sustainable ways to expand the capacity of chemical processes? In this regard, various process intensification technologies might play a pivotal role in sustainable capacity expansions. Process intensification can provide innovative approaches to process and equipment design, which can bring significant benefits in process efficiency and lower capital and operating expenses (Demirel *et al.*, 2019).

New reactor intensification technologies have shown significant advancement, such as enhanced heat transfer reactors (Sheng *et al.*, 2011; Rye *et al.*, 2017). Reactor intensification has a profound impact on downstream separation and as such the overall process. We analyzed the keywords of 1800 research journal papers highlighting "Process Intensification". A high number of journal papers indicate significant interest in process intensification and their distribution in various fields is shown in Figure 1 by the size of the circle representing the relative research concentration in each field. That also emphasizes the need for considering separation process and capability upon reactor intensification.

Figure 1. Research areas within Process Intensification (*Supercritical extraction; **Precipitation disk; r.: reactors.)

Traditional ways of capacity expansion mostly hinge on ultimately building new plants (Greenfield). Increasing the capacity of existing plants (Brownfield) through retrofitting sometimes occurs without effectively considering alternative technologies. We have shown the potential capacity expansion option advantages in Figure 2. First, it can be assumed that potential process intensification options can be readily applied to the green field mode in all parts of the process as we are building the process plant from the ground up. For retrofit, the process intensification options can only be applied to a particular (smaller) section of the plant that we are retrofitting, as we might be using rest of the existing process flowsheet and equipment configuration. For Brownfield, assuming no retrofitting is done in this mode, the scope of process intensification gets reduced even further. It is also evident that for Greenfield the execution time and capital investment might be higher than the Brownfield and retrofitted processes.

It is important to find a suitable intensification technology candidate which can capture the effects of the different capacity expansion modes that we have discussed. Most of the intensification technologies might only affect certain unit operations rather than the overall process. Among the intensification technologies discussed, reactor intensification may have a profound impact on both feed conditioning and downstream separations. Micro fibrous Entrapped Catalyst (MFEC), a reactor intensification technique, can be an ideal candidate in this case (Cheng *et al.*, 2022). MFEC is a microfibrous metal mesh of either Nickel, steel, or Copper. Moreover, MFEC can remove the reaction heats fast from the catalyst-filled tubes through a conductive heat transfer mechanism instead of traditional convective heat transfer. Moreover, the smaller particles used in MFEC-based reactors, in comparison to larger particles (>1mm) in the traditional packed bed reactor, provide better control of the reactor temperature and thus quench the heat. MFEC has the potential to reduce the intra-particle and inter-particle transport resistances considerably. All these properties of MFEC are advantageous and affect the downstream process and overall capacity expansion plans.

2. Methodology

This section expands on cost calculation methodologies that we discussed in our previous work for a different capacity expansion scenario (Mukta *et al.*, 2022). In this work we are applying the methodology to reactor intensification by Microfibrous Entrapped Catalyst (MFEC), which is a highly conductive catalyst support that intensifies reactors through conductive rather than convective heat transport and large surface area. This technology is an ideal candidate to test our methodology for the impact of reactor intensification on various capacity expansion modes as MFEC intensified reactors have the advantage of positively impacting reaction rates, mass transfer, and heat transfer inside the reactor. This can significantly affect the downstream separation process i.e. capacity expansion plan.

2.1. Capacity Expansion Options

Greenfield cost calculation followed the conventional process plant installation procedure. Different formulations considering the lack of land development requirements, module cost, and utility requirements should be considered for Brownfield and retrofitting cases.

Table 1: Capacity expansion options

Greenfield	Retrofit 1	Retrofit 2	Brownfield
A new plant on a newly developed land	Replacing reactor or other equipment	Replacing reactor	Replacing the catalyst in the existing reactor with MFEC
Capital cost calculations based on grass root equipment cost	Capital cost calculations based on modular equipment cost		
New utility systems	Existing utility systems and new utility systems can be installed		Existing utility systems

2.2. Systematic Procedure for Capacity Expansion

It is important to utilize existing infrastructure from both economic and sustainability perspectives. A capacity expansion plan in process industries considering existing facility utilization entails large number of decisions. Identification of the optimum capacity expansion may require significant computational resources. A systematic procedure is necessary to improve the tractability of this potentially large problem. That is why in addition to traditional Greenfield, we analyzed and compared process feasibility and economic performance of other viable capacity expansion modes such as Brownfield and retrofitted processes. The developed systematic procedure is illustrated in Figure 2, where a step-by-step approach is used to selectively reduce the solution space as we expand the capacity of process plants. First, we determine process flowsheet alternatives of green field mode upon reactor intensification using ProCAFD (Tula *et al.*, 2019). Subsequently, the Brownfield alternative is analyzed to see if equipment configurations can be optimized further to achieve capacity expansion. Next, the retrofitted process is analyzed as it has a significantly higher number of decision variables/constraints compared to the Greenfield and Brownfield options.

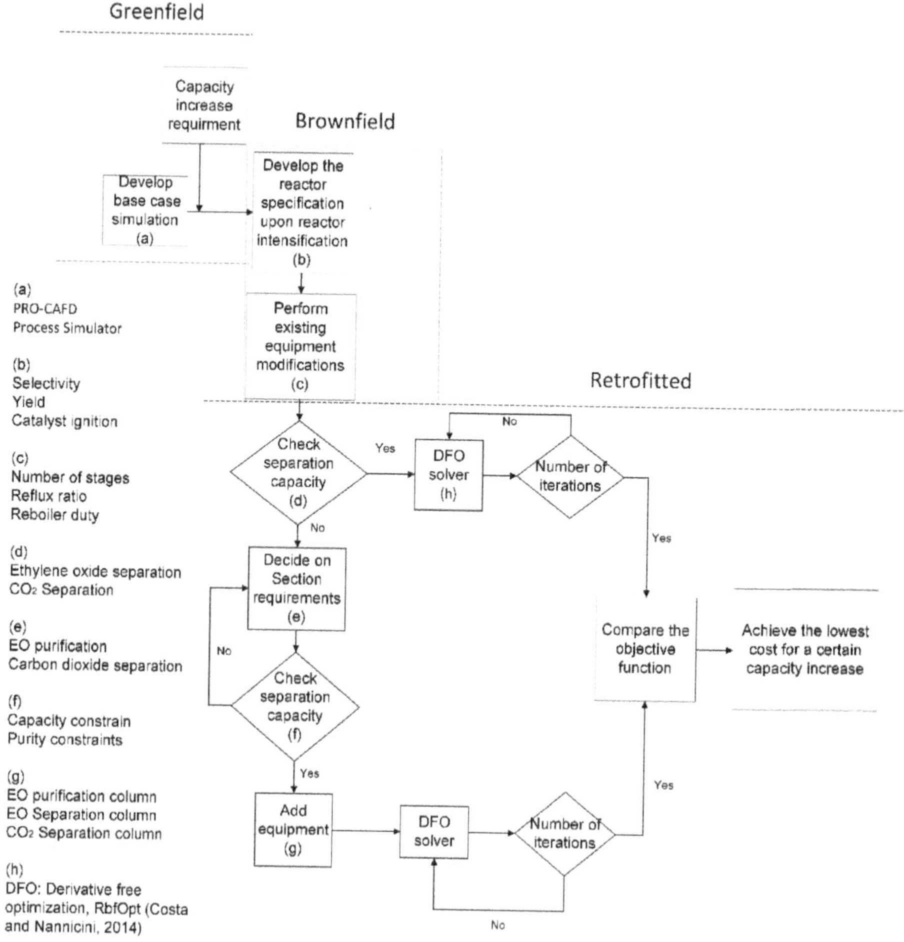

Figure 2. Flowchart showing capacity expansion procedure.

3. Case Study

A baseline simulation model of the conventional Ethylene oxide (EO) process has been developed using Aspen Plus^{TM} and was described in our previous work (Mukta *et al.*, 2020). Next, standard coefficient-based costing equations were used to calculate capital and operating costs. Then the cost was annualized to form our objective function for the derivative-free optimization algorithm.

Brownfield process achieved the highest cost reduction, but above 112 kton/yr, the separation capacity was insufficient to handle excess rector capacity. So the framework suggests the next favorable process is the retrofitted process. The retrofitted process-1 achieved a considerable decrease in production cost, but above 130 kton/yr, production costs increased dramatically with additional columns for ethylene oxide separation. Greenfield process expansion becomes cost-effective only after 160 kton/yr. The Greenfield process achieved the lowest carbon emission. The retrofitted process had a moderate increase in carbon emissions. The Brownfield process showed significant increase in carbon emissions upon an increase in capacity as the separation column operates at a higher capacity than it was designed for thus resulting in suboptimal conditions. It can be seen from the utility carbon emission graph that the Greenfield mode produces the minimum utility carbon among all capacity expansion scenarios that we checked. Carbon emission due to construction was not considered for the new equipment for retrofitted process or Greenfield, which might affect the overall carbon emissions of the process plant.

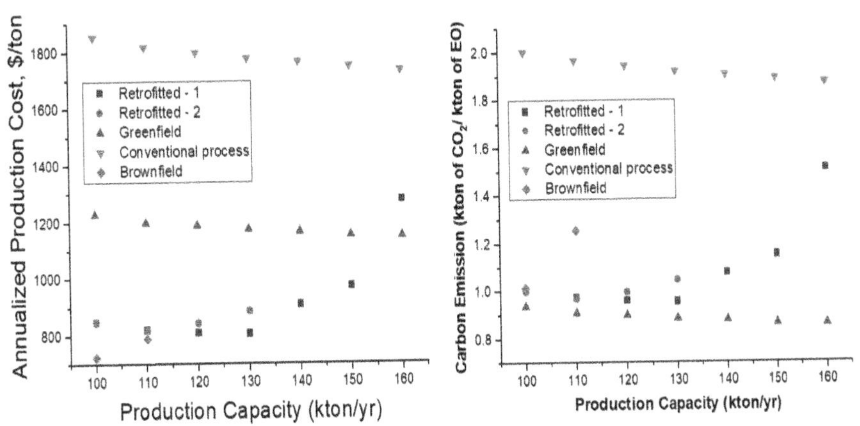

Figure 3: Techno-economic capacity expansion results and associated utility carbon emission.

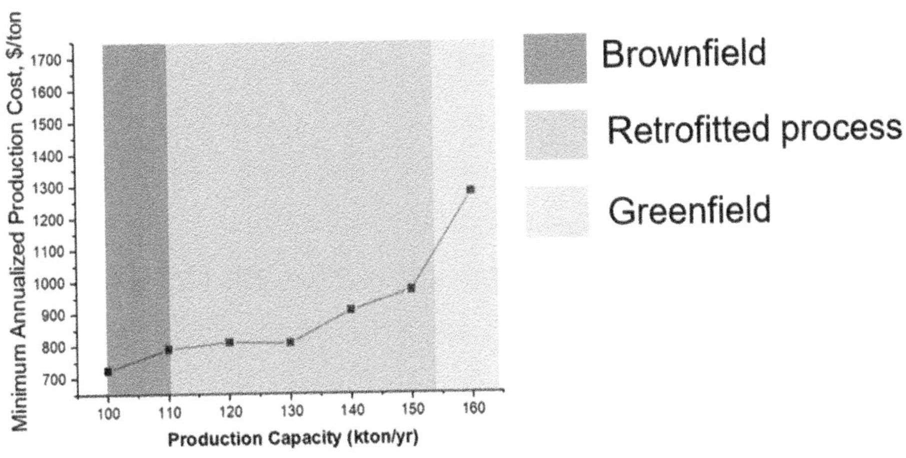

Figure 4: Various capacity expansion modes regions.

Conclusion

A systematic procedure of capacity expansion upon reactor intensification for ethylene oxide production has been developed that might be applicable to other chemical processes. The Brownfield process is found to be the cheapest capacity expansion option up to 112 kton/yr. Subsequently, the Retrofitted mode yielded a minimum cost until 157 kton/yr capacity, and above that, we would need Greenfield expansion. But in terms of utility carbon emission, Greenfield was the lowest carbon producer among the investment alternatives.

References

Brueske, S., Kramer, C., & Fisher, A. (2015). Bandwidth Study on Energy Use and Potential Energy Saving Opportunities in US Chemical Manufacturing (No. DOE/EE-1229). Energetics.

Cheng, X., Yang, H., & Tatarchuk, B. J. (2016). Microfibrous entrapped hybrid iron-based catalysts for Fischer–Tropsch synthesis. Catalysis Today, 273, 62-71.

Costa, A., Nannicini, G., 2014. RBFOpt : an open-source library for black-box optimization with costly function evaluations. Optim. online no.4538.Demirel, S. E., Li, J., & Hasan, M. F. (2019). Systematic process intensification. Current Opinion in Chemical Engineering, 25, 108-113.

Lutze, P., Babi, D. K., Woodley, J. M., & Gani, R. (2013). Phenomena-based methodology for process synthesis incorporating process intensification. Industrial & Engineering Chemistry Research, 52(22), 7127-714

Mukta, C. B., Rayaprolu, N. R., Cremaschi, S., Eden, M. R., & Tatarchuk, B. J. (2022). Techno-Economic Study of Intensified Ethylene Oxide Production Using High Thermal Conductivity Microfibrous Entrapped Catalyst. In Computer Aided Chemical Engineering (Vol. 49, pp. 697-702). Elsevier.

Ryu, J. H., Lee, K. Y., La, H., Kim, H. J., Yang, J. I., & Jung, H. (2007). Ni catalyst wash-coated on metal monolith with enhanced heat-transfer capability for steam reforming. Journal of Power Sources, 171(2), 499-505.

Sheng, M., Yang, H., Cahela, D. R., & Tatarchuk, B. J. (2011). Novel catalyst structures with enhanced heat transfer characteristics. Journal of Catalysis, 281(2), 254-262.

Tula, A. K., Eden, M. R., & Gani, R. (2019). ProCAFD: Computer-aided Tool for Sustainable Process Synthesis, Intensification and Hybrid solutions. In Computer Aided Chemical Engineering (Vol. 46, pp. 481-486). Elsevier.

Antonis Kokossis, Michael C. Georgiadis, Efstratios N. Pistikopoulos (Eds.)
PROCEEDINGS OF THE 33rd European Symposium on Computer Aided Process Engineering (ESCAPE33), June 18-21, 2023, Athens, Greece
 http://dx.doi.org/10.1016/B978-0-443-15274-0.50193-1

Determining the performance and network properties of petrochemical clusters

Michael D. Tan,[a*] Paola Ibarra-Gonzalez,[a] Igor Nikolic,[b] Andrea Ramirez [a]

[a] *Department of Engineering Systems and Services, Faculty of Technology, Policy and Management, Delft University of Technology, Jaffalaan 5, 2628BX Delft, Netherlands*
[b]*Department of Multi-Actor Systems, Faculty of Technology, Policy and Management, Delft University of Technology, Jaffalaan 5, 2628BX Delft, Netherlands*
m.d.tan@tudelft.nl

Abstract

The reliance of the petrochemical industry on fossil-based sources will need to be reduced by the introduction of Alternative carbon sources (ACS). Introducing ACS in a petrochemical cluster will require existing processes to be modified or replaced, potentially affecting other chemical processes within the cluster due to existing material and energy interconnections. Therefore, it is important to understand the current level of interconnections, functioning, and performance of the petrochemical cluster before introducing ACS. In this work, a representative cluster model based on the petrochemical cluster of the Port of Rotterdam was developed and considered as a case study. This model was analyzed using complex network analysis and environmental and technical key performance indicators. The selected key performance indicators (KPIs) provide insight into the performance of a petrochemical cluster, while the network properties give an understanding of the exchange of material and energy in an industrial cluster.

Keywords: petrochemical cluster, key performance indicators, complex network properties, material transition, alternative carbon sources

1. Introduction

The petrochemical industry is reliant on fossil-based sources as an energy and carbon source, and therefore it will have to transition to alternative carbon sources to reach the CO_2 reduction goals. The use of new carbon feedstocks (CO_2, biomass, waste) will require the modification or removal of existing chemical processes. Given the high level of interconnections in existing industrial clusters, these changes are likely to impact existing material and energy connections between processes and among companies in petrochemical clusters. Before these impacts can be assessed, it is needed first to evaluate the complexity and interdependency of petrochemical clusters in terms of material and energy connections and quantify their performance so they can be used as a point of departure.

Network analysis is a method used to study complex systems such as industrial clusters. It allows the interdependencies of processes or companies within a cluster to be assessed by quantifying the exchange of material, energy, or knowledge between processes and or companies. Prior studies (Domenech and Davies, 2011; Song et al., 2018) focused on the occurrence of a link between processes and companies, and assumed that each link is equally important. However, processes are diverse, not only in terms of products but also

in terms of production capacities, and the magnitude of their incoming and outgoing materials and energy flows will also vary as a result. Therefore, it is important that the relative importance of the interconnections is identified and included in the analysis.
In addition to understanding the interdependencies of processes within a petrochemical cluster, additional metrics are required to determine the environmental and technical performance of a petrochemical cluster. Key performance indicators (KPIs) that have been used to assess the performance of industrial clusters are, for instance, CO_2 emissions (Yu et al., 2015) and energy consumption (Sokka et al., 2011).

In this work, a new framework for assessing the complex network properties and performance of a petrochemical cluster is presented. The framework was applied to a case study based on the petrochemical cluster in the Port of Rotterdam (PoR).

2. Petrochemical Cluster Model Framework

The methodological framework used to characterize and evaluate the performance of the representative petrochemical cluster (RPC) is presented in Figure 1. The first step is to select the petrochemical processes and utility units that are part of the RPC. For instance, processes producing chemical building blocks (CBBs), intermediate chemicals (ICs), and end-of-value chain chemicals (EVCs) were selected based on the PoR cluster. Additionally, utility production processes that provide the required steam, electricity, and auxiliary chemicals for the chemical processes need to be selected.

In the next step, detailed process models based on publicly available data were built for each selected chemical and utility generation process using Aspen Plus. Each process model was modeled according to the production capacity and process technology of its counterpart in the PoR cluster. These process models provided detailed material and energy balances for each process and utility generation unit. Furthermore, the bare equipment costs and land footprint were determined for each process using Aspen Process Economics Analyzer.

In the third step, the material and energy connections between the processes in the RPC were mapped. This was done by collecting the results of the models and publicly available data on infrastructure and connections of the PoR and using it to match the material and energy requirements within the processes in the RPC. The resulting mapping of exchanges contains all connections between the processes within the cluster and the processes' connections to the outside world. For example, natural gas imported into the cluster, wastewater sent to wastewater treatment plants, and emissions emitted to the environment.

Next, detailed material and energy stream data from the Aspen Plus process models is automatically extracted using an in-house developed Python module. The material and

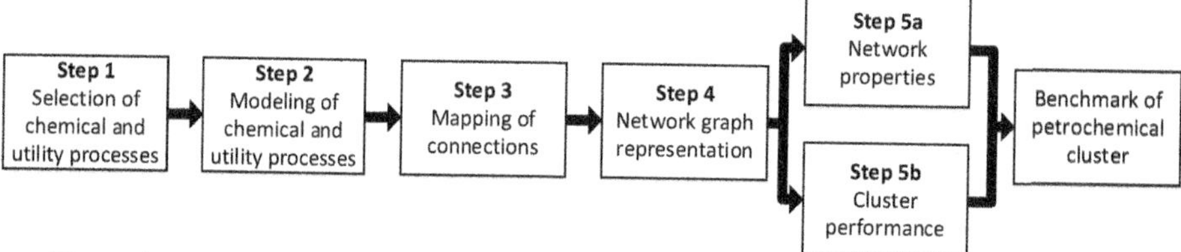

Figure 1: Methodological framework for modeling and assessing a petrochemical cluster

energy data is used to construct a complex network representation in Python using Py3plex (Škrlj et al., 2019). This complex network representation uses graph theory to describe the petrochemical cluster, where the material and energy connections are represented as links, and the chemical and utility generation processes are depicted as nodes. In this work, we developed a multiplex graph of the cluster, where each layer represents a different type of interaction between the nodes. For instance, the first layer represents all the material interactions, and the second layer depicts all the energy exchanges between processes. The only type of interlayer connections are with the counterpart of each node present in the other layers.

Py3plex is based on the complex network analysis Python module NetworkX by Hagberg et al. (2008), thereby allowing the complex network of the graph to be calculated. The number of connections a node has k_i, also known as the degree of node i, was determined by:

$$k_i = \sum_{j \in N} a_{ij} \tag{1}$$

where N is the number of nodes, and a_{ij} is zero if there is no direct link between nodes i and j and one if there is. Alternatively, the degree centrality $C_{D,i}$ was calculated by:

$$C_{D,i} = \frac{\sum_{j \in N} a_{ij}}{N-1} \tag{2}$$

The degree centrality only considers whether a connection between processes is present, and the magnitude of the material or energy exchange is not considered in its calculation. It determines the importance of a process in the petrochemical cluster by considering the number of connections it has to other processes in the cluster. To consider the magnitude of the material or energy exchange, the strength of a node s_i was calculated by:

$$s_i = \sum_{j \in N} w_{ij} \tag{3}$$

where w_{ij} is the weight of the link between nodes i and j. For instance, in the context of the transformation of petrochemical clusters, the mass of carbon in a link could be considered to calculate the weight of a link as it allows the most import nodes in terms of carbon flows between processes to be identified.

The final step consisted of determining the environmental and technical performance of a petrochemical cluster. The graph representation and list of connections were used to calculate the key performance indicators carbon efficiency, CO_2 emissions, and total energy usage within the boundaries of the cluster. The carbon efficiency η_{Carbon} of the cluster is calculated by:

$$\eta_{Carbon} = \frac{\sum m_{Carbon,p}^{Product}}{\sum m_{Carbon,f}^{Feed}} \tag{4}$$

where $m_{Carbon,p}^{Product}$ is the mass of carbon present in stream p leaving the petrochemical cluster, while $m_{Carbon,f}^{Feed}$ is the total mass of carbon present in the material feed stream f entering the cluster. The total CO_2 emissions of the petrochemical cluster are determined by the mass of CO_2 present in the streams being emitted to the environment:

$$m_{CO_2} = \sum_{i=1}^{N_{streams}} m_{i,CO_2}^{Environment} \quad (5)$$

Where, $m_{i,CO_2}^{Environment}$ is the mass flow rate of CO_2 in stream i emitted to the environment, and $N_{streams}$ is the total amount of streams emitted to the environment.

As described before, the framework presented was implemented using an RPC based on the PoR cluster, and the results of its implementation are discussed in the next section.

3. Results and Discussion

Nine utility generation processes and 33 petrochemical processes producing 52 chemicals were selected and modeled in Aspen Plus. Based on these models, a complete mapping of all the material and energy streams of the cluster was created and used to construct a complex network representation of the petrochemical cluster. A two-layer multiplex graph was created containing 49 nodes, where the first layer represented the material exchanges and contained 64 links, while the second layer represented the exchange of energy and contained 48 links. In figure 2, a multiplex graph representation of a small section of the cluster consisting of four chemical processes and three utility generation units is shown.

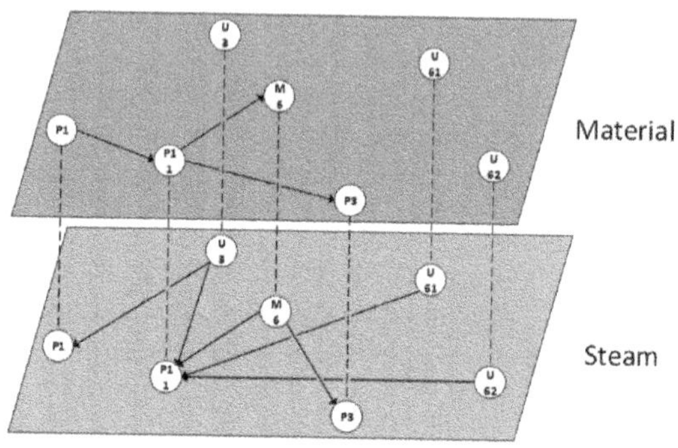

Figure 2: Multiplex graph representation.

The degree centrality and strength of the nodes of the material layer are presented in Table 1. It shows that the EDC/VCM plant and the SMR plant are the most interconnected processes of the cluster, followed by the aromatics and olefins plants. The EDC/VCM plant is part of the chlorine sub-cluster, which contains many interconnections. The SMR plant supplies H_2, which is used as auxiliary material in the production of several chemicals, while the aromatics and olefins plant produce the CBBs that are either directly or indirectly used by the other processes within the cluster. Therefore a high level of interconnectivity is expected for these processes. When considering the strength of the nodes, the aromatics plant appears as the most critical process in terms of carbon flows, followed by the ethylbenzene (EB) and olefins plants, respectively.

Table 1: Degree centrality and strength of the most interconnected processes on the material layer

Process	Degree centrality	Process	Strength (ktonne of carbon per year)
Ethylene dichloride/Vinyl chloride monomer (EDC/VCM)	0,19	Aromatics	1757,19
SMR	0,19	Ethylbenzene (EB)	1430,39
Olefins	0,17	Olefins	1110,50
Aromatics	0,17	Propylene oxide/Tert-butyl alcohol (PO/TBA)	978,34
Chlorine	0,13	Propylene oxide/Styrene monomer (PO/SM)	866,78

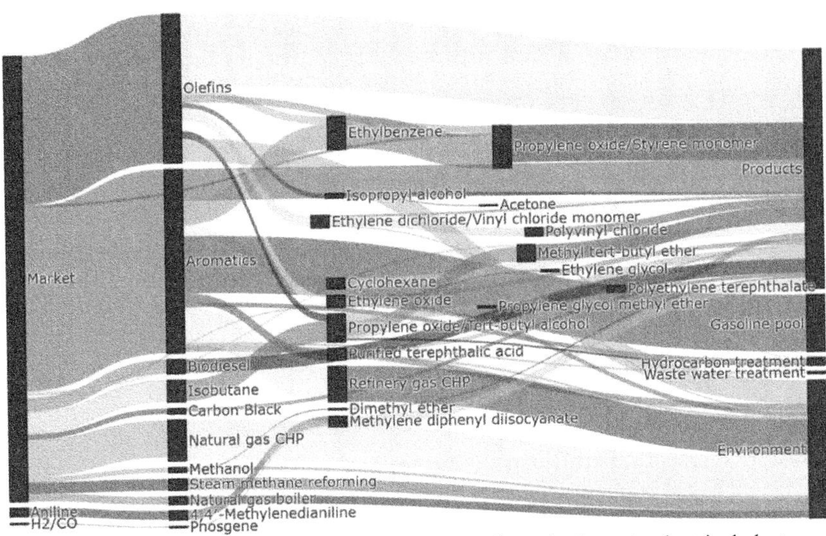

Figure 3: Sankey diagram of all the carbon mass flows in the petrochemical cluster.

In terms of performance, carbon mass flows of the petrochemical cluster are shown in a Sankey diagram in Figure 3. In this diagram, all the chemical processes, the mass flows of carbon between the processes in the cluster, and the carbon flows from and to the outside world are shown. The market represents any material flows being imported into the cluster, while the products represent all material being exported out of the cluster, and the environment represents any CO_2 directly being emitted to the environment. The petrochemical cluster has an overall carbon efficiency of 58.5%. Nearly all the fossil-based carbon feedstock imported into the cluster is sent to the aromatics and olefins plants, with 44.3% and 35. 9% of the carbon feedstock, respectively. These plants transform the carbon feedstock into CBB, and these are distributed across the different value chains that make up the cluster. Therefore, replacing these fossil-based processes with alternative carbon source processes such as CO_2, biomass, or waste will most likely have the most significant impact on the transformation of the petrochemical industry. Additionally, not all chemicals produced by the olefins plant and the aromatics plants are used by the downstream chemical processes in the petrochemical cluster and instead exported out of the cluster. Compared to the strength of the nodes, the importance of the olefins and aromatics is more clearly defined in the Sankey diagram. This is due to the

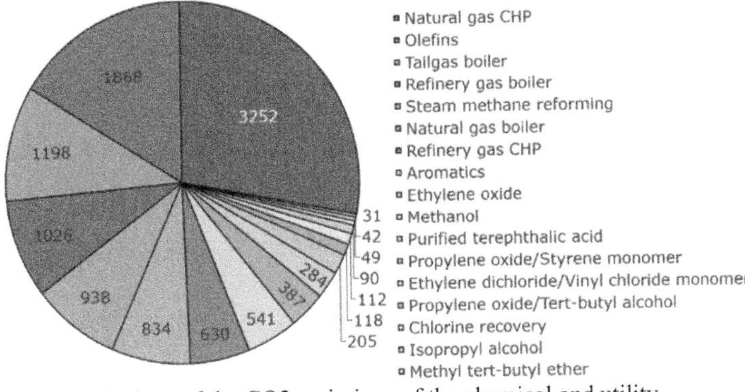

Figure 4: Pie chart of the CO2 emissions of the chemical and utility generation processes in the cluster in ktonne of CO2 per year.

manner the strength of a node is calculated, with only links between processes inside the cluster being considered and not the stream entering or leaving the cluster.

In this case study, the petrochemical cluster emits 11.647 Ktonne of CO_2 every year. The distribution of direct CO_2 emission for all chemical and utility processes emitting at least 25 ktonne per year is shown in Figure 4. A major part of all CO_2 emissions in the cluster is from the utility units that provide the required steam, electricity, and auxiliary chemicals by the chemical processes. The olefins plant has the highest direct CO_2 emissions, as it burns byproducts to provide the heat required for the process. However, new alternative carbon source processes could result in a shift of CO_2 from inside the cluster to outside the cluster. Therefore, an LCA cradle-to-gate approach should be implemented. This will be investigated in further research.

4. Conclusions

The transformation of a petrochemical cluster to more sustainable carbon sources can impact the exchange of material and energy between processes and the performance of the cluster. Thus, network properties and performance indicators are required to assess the current petrochemical cluster and future configurations. In this work, a framework was presented for modeling a petrochemical cluster from the bottom up based on the cluster in the Port of Rotterdam. This model was analyzed using complex network analysis and environmental and technical key performance indicators. In future work, the network properties and performance of a modified cluster, including alternative carbon source processes, will be compared with the benchmark developed in this work, allowing the potential impacts of the transformation of petrochemical cluster to be identified.

Acknowledgements

This publication is part of the project Unravelling the impacts of using alternative raw materials in industrial clusters (with project number VI.C.183.010 of the research programme Vici DO which is (partly) financed by the Dutch Research Council (NWO).

References

Domenech, T., Davies, M., 2011. Structure and morphology of industrial symbiosis networks: The case of Kalundborg. Procedia - Soc. Behav. Sci. 10, 79–89. https://doi.org/10.1016/j.sbspro.2011.01.011

Hagberg, A.A., Schult, D.A., Swart, P.J., 2008. Exploring network structure, dynamics, and function using NetworkX. 7th Python Sci. Conf. (SciPy 2008) 11–15.

Škrlj, B., Kralj, J., Lavrač, N., 2019. Py3plex toolkit for visualization and analysis of multilayer networks. Appl. Netw. Sci. 4, 94. https://doi.org/10.1007/s41109-019-0203-7

Sokka, L., Pakarinen, S., Melanen, M., 2011. Industrial symbiosis contributing to more sustainable energy use - An example from the forest industry in Kymenlaakso, Finland. J. Clean. Prod. 19, 285–293. https://doi.org/10.1016/j.jclepro.2009.08.014

Song, X., Geng, Y., Dong, H., Chen, W., 2018. Social network analysis on industrial symbiosis: A case of Gujiao eco-industrial park. J. Clean. Prod. 193, 414–423. https://doi.org/10.1016/j.jclepro.2018.05.058

Yu, F., Han, F., Cui, Z., 2015. Reducing carbon emissions through industrial symbiosis: A case study of a large enterprise group in China. J. Clean. Prod. 103, 811–818. https://doi.org/10.1016/j.jclepro.2014.05.038

Antonis Kokossis, Michael C. Georgiadis, Efstratios N. Pistikopoulos (Eds.)
PROCEEDINGS OF THE 33rd European Symposium on Computer Aided Process Engineering (ESCAPE33), June 18-21, 2023, Athens, Greece
 http://dx.doi.org/10.1016/B978-0-443-15274-0.50194-3

Quality assessment of partial models for co-simulation of modular electrolysis plants

Isabell Viedt[a], Kumar Rajan Gopa[a], Jonathan Mädler[a], Leon Urbas[a,b]

[a]*TU Dresden, Process Systems Engineering Group, 01069 Dresden, Germany*
[b]*TU Dresden, Chair of Process Control System, 01069 Dresden, Germany*

Abstract

Co-simulation of modular process plants combines partial models for each Process Equipment Assembly (PEA) from different manufacturers. Considering the manufacturer's intellectual property (IP), the co-simulation of the partial models must be realized via a standardized interface using Functional Mock-up Units (FMU). This approach also requires a flexible approach to model quality assessment. In this paper, a workflow for the quality assurance of modularized simulations models for co-simulation is presented via the example of a modular electrolysis plant. The first step is an analysis of the required partial models to derive requirement specifications, the resulting quality models, and the test specifications. These specifications lay the groundwork for the test implementation for the modular electrolysis plant. This use case addresses eighteen partial models of varying modeling depth, including models provided solely as a FMU.

Keywords: Model Quality Assurance, Modular Process Plants, Modular Electrolysis.

1. Introduction

Modular process plants according to VDI 2776 consist of individual Process Equipment Assemblies (PEA) from different manufacturers (VDI, 2020). With these PEAs envisioned to be equipped with simulations models and a digital twin (Mädler et al., 2022), the simulation of the entire modular process plants must also combine partial models from the individual manufacturers which are potentially confidential. Integrated into varying composite models, co-simulation of different plant configurations can not only be used to determine the best plant setup, but also to test out different scale-up scenarios. As most manufacturers are assumed to be reluctant to share their internal model structure because of IP concerns (Mädler et al., 2022), the co-simulation of the partial models must be realized via a standardized interface, the Functional Mock-up Interface (FMI), using Functional Mock-up Units (FMUs).

Since green hydrogen production will play an important role in the decarbonization of the process industry, scale-up of the hydrogen production is necessary (Buttler & Spliethoff, 2018; IEA, 2021). One approach is the transfer of the concepts from VDI 2776 to electrolysis plants (VDI, 2020). Simulation models will be vital in the testing of the necessary scale-up concepts. In this paper, the workflow for quality assurance of simulations models for co-simulation for modular hydrogen production is presented. The case study consists of eighteen partial models of varying modeling depth and approach, including models provided solely as a FMU. The models are implemented in MATLAB/Simulink and tested via the test-driven modeling workflow.

The remainder of this paper is structured as follows. Section 2 introduces current methods for quality assessment and the methodology for model quality assurance approach regarding modular plants. In section 3 the case study for the quality assurance of a modular electrolysis plant is presented. The first step is an analysis of the required partial models to derive the requirement specifications, the resulting quality models, and the test specifications. With this, necessary quality factors and criteria such as interoperability, flexibility, and security are discussed as well as functional suitability and reliability. These specifications lay the groundwork for the test implementation and modeling of the electrolysis sample plant. Section 4 provides a conclusion and further research potential.

2. Methodology for model quality assurance

2.1. Quality assessment of simulation models

In literature, different strategies for the quality assessment of simulation models exist (Murray-Smith, 2015). The most widely known and used strategies are the methods for verification and validation (V&V) by Balci and Sargent (Sargent & Balci, 2017). Although they provide a wide spectrum of assessment methods, the main focus remains on the models' accuracy. Within the meta analysis of Sargent and Balci (2017) it was found that many simulation studies do not consequently apply V&V methods or even disregard them altogether. Mädler et al. (2021) and Viedt et al. (2022) discuss the applicability of quality assurance methods from software development to simulation models. For this approach, quality assessment strategies from software development such as test-driven development and quality models to define quality features in form of FCM (factor–criteria–metrics) models are applied to ensure test-driven modeling.

2.2. Quality assurance methodology for modular process plant simulation models

With the increasing importance of model-based approaches for plant optimization and scaling concepts, the quality of simulation models also gains importance. Especially for large-scale co-simulations, where partial models from different manufacturers are combined to form a process simulation of the entire modular plant, quality aspects for the interaction of those models must be considered more directly (Viedt et al., 2022). To ensure high quality co-simulation, quality assurance moves into focus. Thus, quality needs to be a driving factor for the implementation of the models. A shift from 'tagged-on' quality assessment to a quality driven-modeling approach is necessary.

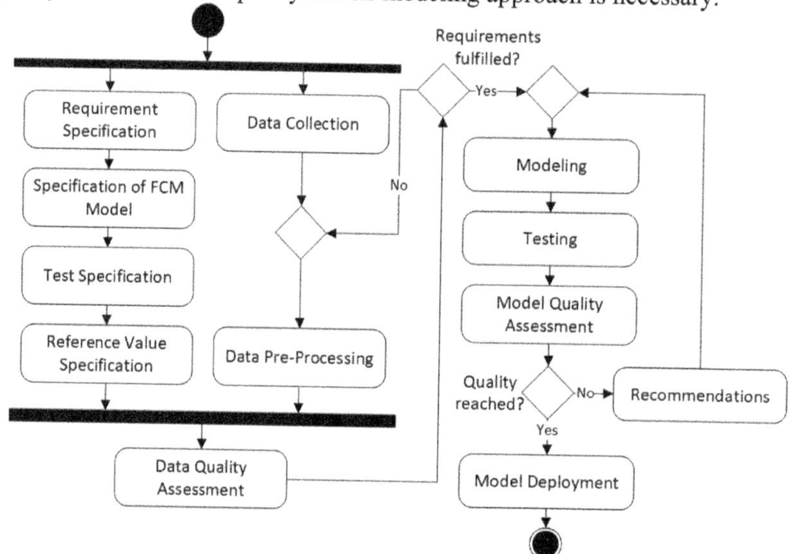

Figure 1: Quality assurance workflow for simulation models

As a result, the pre-definition of requirements, quality features, and tests prior to the start of the actual modeling is of paramount importance. With this, the authors propose a workflow for the quality assurance of simulation models which is shown in Figure 1. The workflow for model quality assurance considers two distinct quality iteration loop: data quality assurance and model quality assurance. Data quality assurance will be necessary in almost all modeling cases. This is because even if data is not used during the actual implementation of the model for a data-driven modeling approach, it will at least be utilized for model validation as well as for parameter identification (Mädler et al., 2022).

3. Case study modular electrolysis plant

3.1. Modular electrolysis sample plant

In the research project eModule, standardized plant concepts consisting of PEAs were developed for all three major electrolysis technologies: alkaline electrolysis (AEL), proton exchange membrane electrolysis (PEMEL) and high-temperature electrolysis (HTEL) (Buttler & Spliethoff, 2018). Fig. 2 shows the exemplary flowsheet for the AEL.

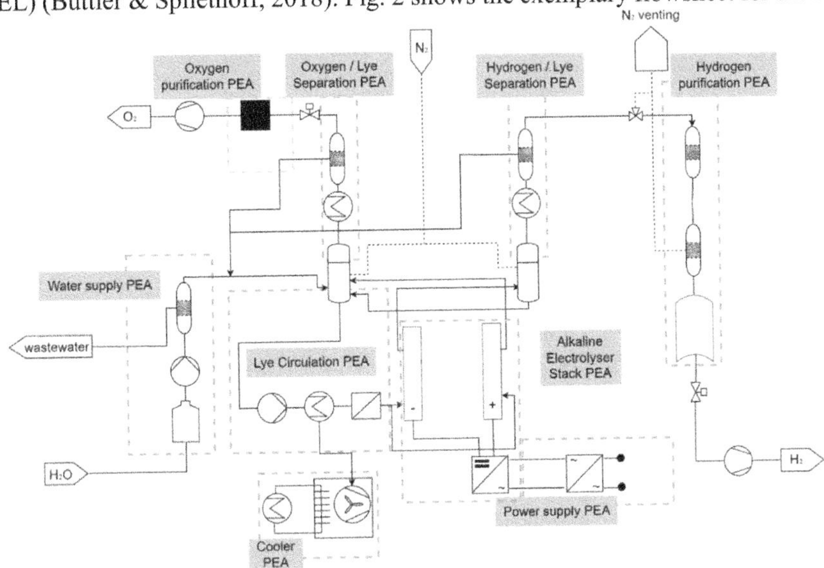

Figure 2: Flowsheet for the AEL - Courtesy of the VDMA Power-to-X for Applications platform (Beisswenger and Erdmann. 2022).

3.2. Partial models and requirements definition

The first step in the test-driven modeling workflow is the specification of the model goal and scope. For the modular electrolysis plant shown in Fig. 2 for the AEL, eight independent partial models were defined: water supply, cooler, lye circulation, AEL stack, oxygen/lye separation, hydrogen/lye separation, oxygen purification and hydrogen purification. In accordance with the project scope, the power supply was not implemented as an individual partial model. For all three considered technologies, AEL, PEMEL, and HTEL, eighteen partial models are defined as necessary according to the requirements specification. This specification defines the necessary functions e.g. interfaces, components, and control functions of each individual model as well as the required modeling depth and approach e.g. a data-driven approach. Once the requirement specification for a model is derived from the modeling and simulation goal, a corresponding FCM model is derived to structure the quality features of the models.

Every metric is allocated a tag that is later used to assign the necessary test specification during the test execution. This tag consists of the tag for the partial model, e.g. WS for water supply or H2LS for the hydrogen/lye separation, and the tag for the specific metric in the FCM model. An excerpt is shown in Table 1.

Table 1: Excerpt of the FCM model for the water supply partial model.

Factor	Criterion	Metric	Tag
Functional Suitability	Functional Correctness	Target conductivity	WS-F1C1M1
	Functional Appropriateness	Degradation rate	WS-F1C2M1
		Conversion efficiency	WS-F1C2M2
	Functional Completeness	Implemented components	WS-F1C3M1
		Implemented control functions	WS-F1C3M2
Reliability	Fault Tolerance	Fault injection tolerance	WS-F2C1M1
Compatibility	Interoperability	Interface standard adherence	WS-F3C1M1

3.3. *Test specification*

The next step in the test-driven modeling workflow is the test specification which also includes the corresponding reference values for the tests. These test specifications are compiled from the aggregated information from the requirements specification, process flow diagram (PFD), the piping and instrumentation diagram (P&ID), the formalized process description, and manufacturer information. Test specification in an early model development phase allows a more comprehensive testing approach, which also makes it easier to detect missing information at an earlier development stage. An excerpt of an exemplary test specification for the AEL stack model is shown in Table 2.

Table 2: Excerpt of an exemplary test specification for the AEL stack partial model.

Tag	Test
AEL-F1C1M1	If the AEL is supplied with power, H_2O is converted into H_2 and O_2.
AEL-F1C2M2	If the AEL is not supplied with power, H_2O is not converted into H_2 and O_2.
AEL-F3C1M1	The model must adhere to the FMI standard.

3.4. *Modeling and FMU preparation*

In the modeling step the pre-defined partial models are implemented in the chosen modeling environment, here MATLAB/Simulink 2022b. In Fig. 3 the model for the hydrogen/lye separation is shown in a first iteration. In the next step, FMUs must be created for all implemented models. The corresponding FMUs will have the same inlets and outlets as the non-FMU model shown in Fig. 3.

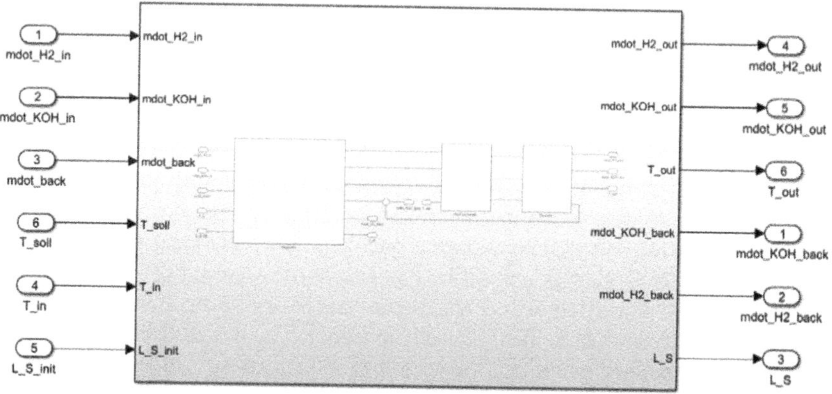

Figure 3: Partial model for the hydrogen/lye separation for AEL.

To enable co-simulation at a later stage, the model and system configurations, which describe the connections between the individual FMUs and the system structure as a whole, must be created. These configurations are compiled from the aggregated information from the requirements specification, block flow diagram (BFD), process flow diagram (PFD), piping and instrumentation diagram (P&ID), and manufacturer information.

3.5. Testing and model quality assessment

Once the first modeling iteration step is concluded, the models are tested against the requirements specification and the test specification. For this case study, the tests are implemented and automated via the Simulink Test Manager for both the models implemented in MATLAB/Simulink and the provided FMUs. Since the FMUs represent the partial models provided by the different PEA manufacturers, the internal structure and modeling approach is not known. Therefore, to ensure trust in them, the testing of these partial models must be even more rigorous with more tests and a higher test coverage.

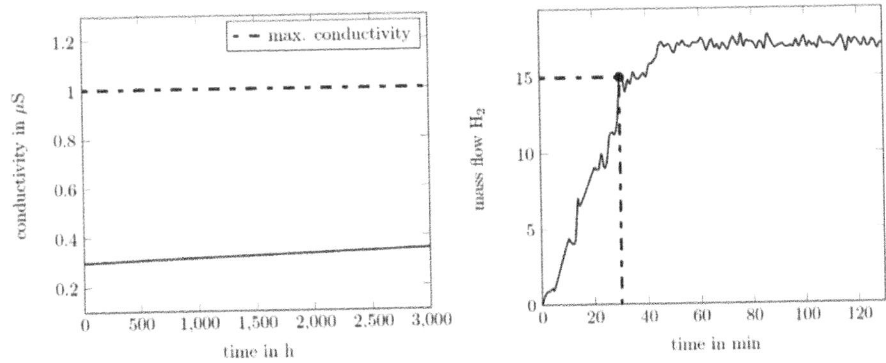

Figure 4: Excerpt of the test results for water supply (left) and hydrogen/lye separation (right).

For the modular electrolysis case study all partial models are tested against their test specification and the requirements specification. Figure 4 shows an excerpt of these test results for the tests tagged WS-F1C1M1 for the water supply and H2LS-F1C1M2 for the hydrogen/lye separation. Both partial models reach the required quality for the two shown metrics of target conductivity and target hydrogen production and can therefore be deployed for the intended purpose. If a partial model does not reach the desired quality, as per the quality assurance workflow (cf. Table 1) another modeling iteration must follow to adjust the model to improve quality. One example of this is the behavior of the hydrogen/lye separation model. In the first modeling iteration, the hydrogen purity metric, H2LS-F1C1M1, did not meet the specification. A corresponding recommendation was the adjustment of the assumptions for vapor pressure and gas solubility. The requirements specification and the FCM model also include further quality factors such as compatibility and efficiency, which are only tested qualitatively. For co-simulation in particular, quality features which do not describe the explicit model behavior are of paramount importance as runtime efficiency and model compatibility will enable successful execution.

4. Conclusion and outlook

With this case study it was shown that the presented quality assurance workflow allows for an easy integration of additional requirements and tests to cover multiple modular plant configurations while enabling test-driven modeling and co-simulation. Especially for a large number of models, this approach allows for a test-driven and structured

assurance process which enables better result tracking. Since the partial models are compiled into a co-simulation framework, quality features outside of the models behavior must also be considered to assure successful model interaction. One future challenge will be the rigorous assessment of FMUs and composite FMUs. Since the person conducting quality assurance is not the model builder, acceptable quality of the individual models does not automatically lead to sufficient quality of the composite models if the modeling approach etc. is not known. Therefore, there is a need for an exchange of context information between manufacturer and operator of the co-simulation to ensure compatibility in addition to standard adherence like the modeling goal and some information about the modeling approach (Mädler et al., 2022).
Another challenge for the quality assurance of the co-simulation and the model process will be the definition and creation of the co-simulation configuration. As of now, the model assurance workflow does not include the quality assessment of the configurations content besides standard adherence itself.

Acknowledgements

We would like to thank the German Federal Ministry of Education and Research and the Project Management Agency Jülich for their financial support within the framework of the eModule research project (FKZ 03HY116A) of the H2Giga lead platform.

References

Beisswenger, L., Erdmann, D., 2022. Standardized flowsheet for the alkaline electrolysis.

Buttler, A., Spliethoff, H., 2018. Current status of water electrolysis for energy storage, grid balancing and sector coupling via power-to-gas and power-to-liquids: A review. Renewable and Sustainable Energy Reviews 82, 2440–2454. https://doi.org/10.1016/j.rser.2017.09.003

International Energy Agency (IEA), 2021. Global Hydrogen Review 2021. OECD. https://doi.org/10.1787/39351842-en

Mädler, J., Viedt, I., Lorenz, J., Urbas, L., 2022. Requirements to a digital twin-centered concept for smart manufacturing in modular plants considering distributed knowledge, in: Yamashita, Y., Kano, M. (Eds.), Computer Aided Chemical Engineering, 14 International Symposium on Process Systems Engineering. Elsevier, pp. 1507–1512. https://doi.org/10.1016/B978-0-323-85159-6.50251-7

Mädler, J., Viedt, I., Urbas, L., 2021. Applying quality assurance concepts from software development to simulation model assessment in smart equipment, in: Türkay, M., Gani, R. (Eds.), Computer Aided Chemical Engineering. Elsevier, pp. 813–818. https://doi.org/10.1016/B978-0-323-88506-5.50127-3

Murray-Smith, D.J., 2015. Testing and Validation of Computer Simulation Models, Simulation Foundations, Methods and Applications. Springer International Publishing, Cham. https://doi.org/10.1007/978-3-319-15099-4

Sargent, R.G., 2010. Verification and validation of simulation models, in: Proceedings of the 2010 Winter Simulation Conference. Presented at the 2010 Winter Simulation Conference - (WSC 2010), IEEE, Baltimore, MD, USA, pp. 166–183. https://doi.org/10.1109/WSC.2010.5679166

VDI 2776-1:2020, 2020. Verfahrenstechnische Anlagen - Modulare Anlagen - Grundlagen und Planung modularer Anlagen.

Viedt, I., Mädler, J., Lorenz, J., Urbas, L., 2022a. Requirements for the quality assessment of virtual commissioning models for modular process plants, in: Computer Aided Chemical Engineering. Elsevier, pp. 805–810. https://doi.org/10.1016/B978-0-323-85159-6.50134-2

Viedt, I., Mädler, J., Urbas, L., 2022b. Quality assessment for dynamic, hybrid semi-parametric state observers, in: Computer Aided Chemical Engineering. Elsevier, pp. 1255–1260. https://doi.org/10.1016/B978-0-323-95879-0.50210-1

Antonis Kokossis, Michael C. Georgiadis, Efstratios N. Pistikopoulos (Eds.)
PROCEEDINGS OF THE 33rd European Symposium on Computer Aided Process Engineering (ESCAPE33), June 18-21, 2023, Athens, Greece
 http://dx.doi.org/10.1016/B978-0-443-15274-0.50195-5

Mechanistic modeling of industrial fermentation processes for antibiotic production

Atli F. Magnússon,[a] Stuart M. Stocks,[b] Jari P. Pajander,[b] Gürkan Sin,[a]

[a]Technical University of Denmark, Søltofts Plads, 227, 2800 Kgs. Lyngby, Denmark
[b]LEO Pharma A/S, Industriparken 55, 2750 Ballerup, Denmark
afrmag@kt.dtu.dk

Abstract

An unstructured mechanistic model is proposed to describe the industrial-scale production of Fusidic Acid in fed-batch cultivations. The model accounts for differences in dead and viable biomass and the effect of the primary carbon source and oxygen on cell growth and production. The model parameter is calibrated, and performance is tested using experimental data obtained from an operating industrial production in Denmark. The model predicted the main product concentration with a relative mean error of 7%. These successful implementations open up opportunities for soft sensor implementations for key state variables but also set a good foundation for further model extensions, such as hybrid modeling.

Keywords: Fermentation, Process Modelling, Pharmaceuticals

1. Introduction

Industrial microbiology has been one of the most common methods for the mass preproduction of antibiotics since the widescale production of penicillin. In a very similar sense, Fusidic Acid is a secondary metabolite that is commercially produced using a filamentous microorganism. At the same time, there is extensive literature on the modeling of penicillin production with varying degrees of complexity (Goldrick et al. 2015), but there is no published research on the mechanistic modeling of the Fucidin process. The bioprocess industry has seen rapid advancements toward digitalization. This has led to increasing interest in industrial applications of digital twins. These technologies require good-quality process models to act as the foundation. Mechanistic models are the gold standard as they incorporate current knowledge into a first principles mathematical description of the system of interest (Gernaey et al. 2010). While more challenging to develop, these model types are notably beneficial for application in the fermentation industry since they can be better extrapolated to new scenarios than machine learning algorithms and artificial intelligence.

A mechanistic model describes the dynamic behavior of a system with a series of mathematical formulas, typically ordinary differential equations (ODEs). These models are based on prior knowledge of the system phenomena like mass, energy, and moment balances.

The mechanistic model of Bajpai and Reuss (Bajpaj and Reuss 1980) is a good starting point when designing a model structure for a poorly researched organism. It's unstructured, has low complexity, few parameters, and can be calibrated based on experimentally measured growth profiles. Furthermore, it was made to model the secondary metabolites of a filamentous organism and showed excellent agreement with experimental data. Extensions and changes will be proposed based on observed phenomena on industrial and lab scales to improve predictive qualities and applications. This study aims to design and evaluate a mechanistic model to quantitatively describe the behavior of an industrial-scale Fusidic Acid process. To this end, a mechanistic model structure is inspired by unstructured penicillin models with extensions or modifications to account for observable phenomena in the production setting and during lab experiments performed at LEO Pharma A/S. Data from industrial-scale fed-batch fermentation is collected to calibrate and validate model parameters and structure.

2. Materials and Methods

A model is developed to describe a filamentous fermentation fed-batch process operated at LEO Pharma A/S. The traditional fed-batch equation is utilized for the overall concentration balance with an extension to account for the effects of evaporation.

$$\frac{dC_i}{dt} = q_i + \frac{E}{M}C_i + \frac{F_{feed}}{M}(C_{i,f} - C_i) + k_l a_i (C_i^* - C_i)$$

Where the first term qi denotes the biochemical kinetic rate of component i, the second term accounts for increased concentration due to mass loss from water evaporation and offgas balance, where E describes the broth mass change due to evaporation and offgas balance $E = F_{evap} + F_{OUR} - F_{CER}$. It is assumed that all relevant components are nonvolatile and thus are not present in the offgas. The third term describes the dilution due to feed where F_{feed} is the feed rate, M is the broth weight, and $C_{i,f}$ is the concentration of component i in the feed. The final term describes the mass transfer of component i from the gas phase to the liquid phase, where $k_l a_i$ is the mass transfer coefficient of component i and C_i^* is the equilibrium solubility of component i. For simplifications, it is assumed that $k_l a$ is 0 for all components except oxygen.

As the model is unstructured, it is assumed that all the biochemical kinetics can be explained with a specific growth rate term, which can be either positive, indicating growth/production, or negative, indicating consumption.

$$\boldsymbol{q} = \boldsymbol{\mu} X$$

Where X is the concentration of viable biomass and $\boldsymbol{\mu}$ is a vector containing the specific rates of each species calculated with an unstructured model equation structure. Note that X is viable biomass and not the total dry weight, which accounts for the accumulation of cell debris and precipitated main product.

$$X_{TDW} = X + X_D + P_{precipitated}$$

The Contois model kinetics is utilized for growth due to the excellent agreement with experimental data for multiple microbial systems (Alqahtani et al. 2015). The overall system of equations describing the biochemical model of the main components are

$$\mu_X = \mu_{X,max}\frac{S}{K_{SX}X+S}\frac{DO}{K_{OX}X+DO}\left(1-\exp\left(\frac{-t}{t_{lag}}\right)\right)-k_d$$

$$\mu_P = \mu_{P,max}\frac{S}{K_{SP}X+S}$$

$$\mu_S = \frac{\mu_X}{Y_{SX}}+\frac{\mu_P}{Y_{SP}}+m_S\frac{S}{K_{SS}X+S}$$

The changes in broth mass are modelled by considering all components entering and leaving the fermenter

$$\frac{dM}{dt} = F_{feed} - F_{evap} + F_{OUR} - F_{CER}$$

Feedrate and offgas measurements are available. To estimate the evaporation, we consider the amount of water vapors in the process air and offgas.

$$F_{evap} = Q_{air}\left(\frac{\phi_{out}p^*_{H_2O,out}}{RT_{out}} - \frac{\phi_{in}p^*_{H_2O,in}}{RT_{in}}\right)$$

Where ϕ is the relative humidity and $p^*_{H_2O}$ is the saturation pressure of water in the air. Relative humidity of process air is available from batch measurements ,a nd it is assumed that the offgas is fully saturated with water ,i.e., $\phi_{out} = 1$. The saturation pressure of water can be estimated using a steam table or the Antoine equation.

The model equation structure is implemented into MATLAB R2021B and solved using the stiff solver *ode15s*. Measured online data for Dissolved Oxygen (DO), pH and offgas F_{OUR} and F_{CER} are used as inputs when model is calibrated. The model contains 11 growth parameters that need to be estimated, no previous nominal values exist for the fungus and are all estimated simultaneously using *lsqnonlin* function in MATLAB.

3. Results

Experimental data collected from sampling industrial production are used for parameter estimation. An initial fit for the validation batch can be seen in Figure 1. The prior estimation showed a generally satisfactory fit with experimental data and can accurately describe the concentration profiles of viable biomass and main product. It also captured the concentration of the primary carbon source reasonably.

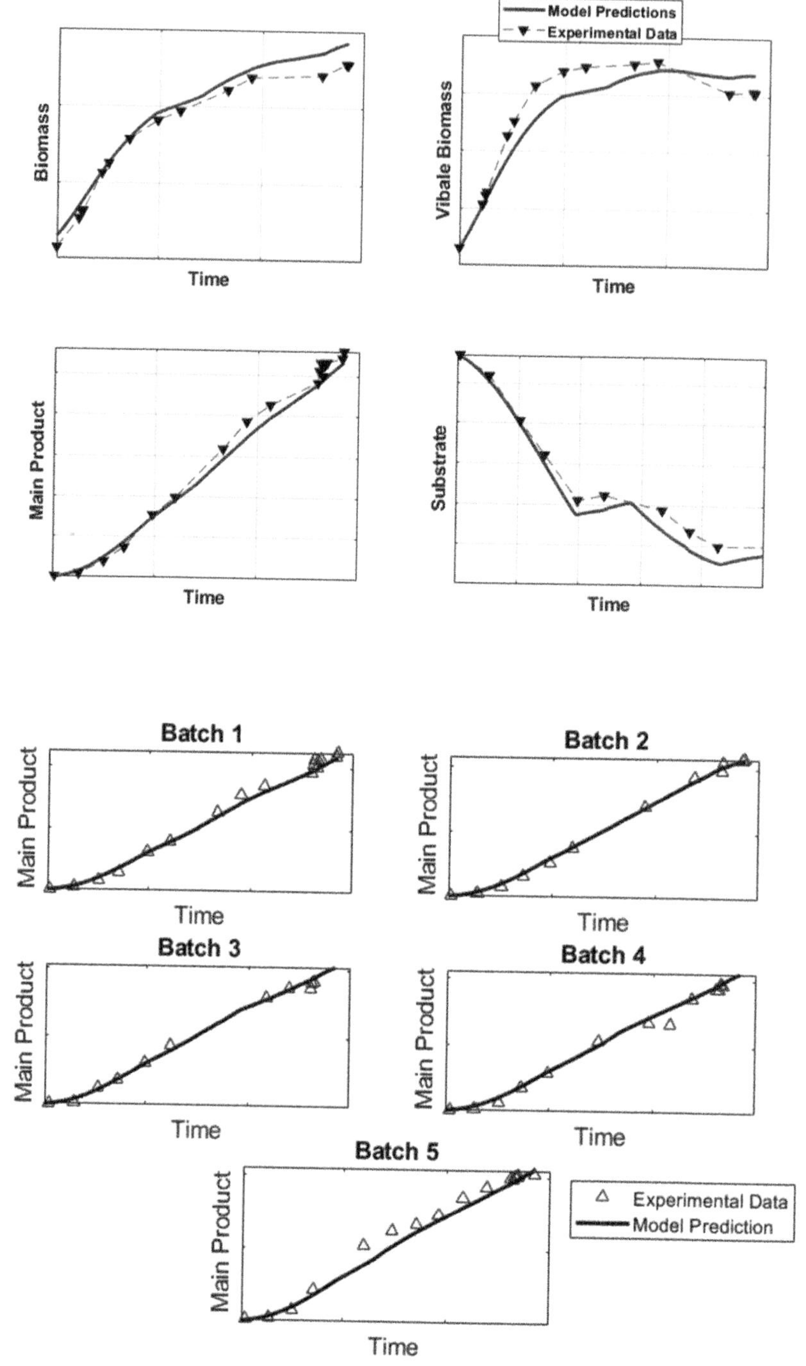

However, for engineering purposes, we are primarily interested in the main product concentration, which is the main economic driver of the entire process. Therefore, the comparison of measured and simulated growth profiles of all sampled batches is shown in Figure 2. The overall fit to the main product considering all experimental values is depicted in figure 3 with an overall prediction error of 6.6%. This Model shows excellent fits for the main product.

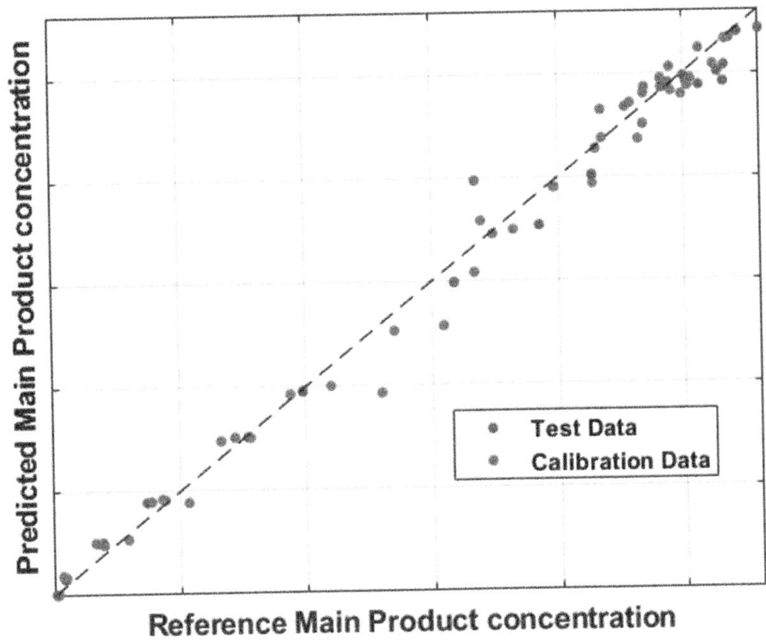

Model quality was assessed with the root mean sum of squared errors (RMSSE) but is reported here as a scaled percentage deviation from the experimental measured mean value to preserve confidentiality of process scales of industrial sponsor.

$$RMSSE(\%) = 100 * \frac{\sqrt{\frac{1}{n}\sum_{i=1}^{N}\left(y_{meas,i} - y_{pred,i}\right)^2}}{\bar{y}}$$

Process Variable	RMSSE (%)	R^2
Total Dry Weight	12.75	0.98
Viable Biomass	12.99	0.91
Main Product	6.60	0.99
Substrate	19.52	0.98

Initial parameter estimation shows that the Model has the most trouble predicting main carbon source concentrations. If all measurements are included, there is an error of approximately 20%. It's hypothesized that the complex media blend gives rise to multiple available carbon sources. At the same time, the Model focuses only on a singular primary carbon source. This may lead to different consumption rates depending on the variation in the initial media blending. Both biomass measurements show a similar error of approximately 13% which is considered a good fit. Similarly, with the substrate predictions, it may be possible to improve fits by considering the effects of potential alternative carbon sources, but this is a highly complicated data collection and modeling process. One of the main drawbacks of using the current iteration for batch planning or as the basis for digital twins is that the model cannot run as an independent simulation. This is because the model relies on Oxygen Uptake and Carbon Evolution data to estimate evaporation accurately. Without this information, any optimization that relies on model outputs will probably underestimate final product concentrations while overestimating the current broth weight. In the current state, the mechanistic model uses batch measurements to calculate estimates of key process variables that describe the batch states that are not observable meaning it has potential applications in soft-sensor technology.

4. Conclusions

A mechanistic model structure was proposed to describe an antibiotic production process of a novel filamentous fungi strain. The model can be used to simulate growth profiles on an industrial scale. It can simulate biomass growth, main product synthesis, and substrate uptake. Due to the lack of sophisticated hardware sensor technology, these key process variables are not commonly measured online. The predictions are supported by readily available online measurements such as offgas mass spectrometry to evaluate carbon evolution and oxygen uptake. Future work will use this model as a foundation for integrating machine learning techniques to expand the model capabilities to obtain readily available predictions of the missing key state variables that went unexplained in this work. We also wish to study the model's reliability in the framework of good modeling practices (Sin et al. 2009) by statistically analyzing the effects of uncertainty of the model structure and parameters.

References

S. Goldrick, A. Ştefan, D. Lovett, 2015, The development of an industrial-scale fed-batch fermentation simulation, Journal of Biotechnology, 193, 70-92

K. V. Gernaey, A.E. Lantz, P. Tufvesson, J. M. Woodley, G. Sin, 2010, Application of Mechanistic Models to Fermentation and Biocatalysis for next-Generation Processes, Trends in Biotechnology, 28, 7, 346-354

R. K. Bajpai, M. Reuss, 1980, Mechanistic Model for Penicillin Production, Journal of chemical technology and biotechnology, 30, 6, 332-344

R. T. Alqahtani, M. I. Nelson, A. L. Worthy, 2015, A biological treatment of industrial wastewaters: Contois kinetics, Anziam Jounral, 55, 4, 397-415

G. Sin, K. V. Gernaey, A. Elliasson Lantz, 2009, Good Modeling Practice for PAT Applications: Propagation of Input Uncertainty and Sensitivity Analysis, Biotechnology Progress, 25, 4, 1043-1053

Antonis Kokossis, Michael C. Georgiadis, Efstratios N. Pistikopoulos (Eds.)
PROCEEDINGS OF THE 33rd European Symposium on Computer Aided Process Engineering (ESCAPE33), June 18-21, 2023, Athens, Greece
 http://dx.doi.org/10.1016/B978-0-443-15274-0.50196-7

A new framework and online solution engines for multiparametric Model Predictive Control

Diogo A.C. Narciso[a,b*], Dustin Kenefake[c,d], Sahithi Srijana Akundi[c,d], F.G. Martins[a,b] and Efstratios N. Pistikopoulos[c,d]

[a]*LEPABE - Laboratory for Process Engineering, Environment, Biotechnology and Energy, Faculty of Engineering, University of Porto, Rua Dr. Roberto Frias, 4200-465 Porto, Portugal*

[b]*ALiCE - Associate Laboratory in Chemical Engineering, Faculty of Engineering, University of Porto, Rua Dr. Roberto Frias, 4200-465 Porto, Portugal*

[c]*Texas A&M Energy Institute, Texas A&M University, College Station, TX 77843, USA*

[d]*Artie McFerrin Department of Chemical Engineering, Texas A&M University, College Station, TX77843, USA*

dnarciso@fe.up.pt

Abstract

In this paper, we combine the multiparametric programming approach proposed in Pistikopoulos et al., 2002 for quadratic Model Predictive Control (mp-QP/MPC) problems, with the solution strategy proposed in Narciso et al. 2022 for mp-QP problems. The problem statements in these two contributions are fully compatible, and we show firstly how they can be combined to deliver critical regions and the corresponding control laws of MPC problems in the offline stage. Then, a new online architecture is proposed for the calculation of optimal control actions. To this end we explore the convenient solution structure presented in Narciso et al. 2022 with the specific aim to calculate them as fast as possible via parallel processing, and where a restricted collection of critical regions is kept and continuously updated in time.

Keywords: Multiparametric programming, Quadratic Programming, Model Predictive Control.

1. Introduction

Model Predictive Control (MPC) enables an optimization-based approach to control problems, whereby a cost function driving the system' states to a predefined set point and a set of operational constraints are defined (Pistikopoulos et al. 2020). As the size of MPC problems increase, so does the total time required to solve them. This is particularly critical when the total times required to solve MPC problems exceed the required response times, which renders the optimization-based route ineffective at best (Pappas et al. 2020).

Multiparametric programming (mp) has been shown to mitigate this limitation (Pistikopoulos et al., 2002): the expensive optimization calculations are moved to the *offline* stage, where a collection of explicit critical regions and matching control laws are obtained, and thus requiring a set of inexpensive function evaluations during the *online* calculation of the optimal control actions. The drawback of mp is that as MPC problems grow very large, so do the corresponding explicit solutions, which in turn generally slow down the calculation of optimal controls in the online stage.

Several strategies have been proposed to deliver faster online calculations for MPC problems (Pappas et al. 2020). In this work, we focus on the class of mp-QP/MPC (quadratic programming) problems presented in Pistikopoulos et al., 2002, and use the mp-QP algorithm presented Narciso et al. 2022 to deliver critical regions and control laws in compact form during the offline stage. We explore this convenient solution structure and propose a new architecture for the fast calculation of controls during the online stage.

2. Offline calculation of optimal control laws

We consider the following class of QP/MPC problems, where the state, output, and control vectors are denoted as $x \in \mathbb{R}^n$, $y \in \mathbb{R}^p$ and $u \in \mathbb{R}^m$ (Pistikopoulos et al., 2002):

$$
\begin{aligned}
\min_{U} \quad & x'_{t+N_y|t} P\, x_{t+N_y|t} + \sum_{k=0}^{N_y-1} \left[x'_{t+k|t} Q x_{t+k|t} + u'_{t+k} R u_{t+k} \right] \\
s.t. \quad & y_{min} \le y_{t+k|t} \le y_{max}, k = 1, \dots, N_c \\
& u_{min} \le u_{t+k|t} \le u_{max}, k = 1, \dots, N_c \\
& x_{t|t} = x(t) \\
& x_{t+k+1|t} = A x_{t+k|t} + B u_{t+k}, k \ge 0 \\
& y_{t+k|t} = C x_{t+k|t}, k \ge 0 \\
& u_{t+k|t} = K x_{t+k|t}, N_u \le k \le N_y
\end{aligned}
\qquad \text{(Eq. 1)}
$$

where $\mathrm{U} = \left[\mathrm{u_t}', \mathrm{u_{t+1}}', \dots, \mathrm{u_{t+N_u-1}}'\right]' \in \mathbb{R}^s$, with $s = mN_u$. $P = P' \succcurlyeq 0$, $Q = Q' \succcurlyeq 0$ and $R \succcurlyeq 0$ are constant cost matrices. A, B and C are constant matrices defining the state-space model, and subscripts min and max denote the lower and upper bounds, respectively, on y and u. K is a constant feedback gain matrix, and $N_y \ge N_u$. Using $x_{t+k|t} = A^k x(t) + \sum_{j=0}^{k-1} A^j B u_{t+k-1-j}$, an equivalent QP/MPC is defined:

$$
\begin{aligned}
\min_{U} \quad & 1/2\, U' H\, U + x(t)' F U + 1/2\, x(t)'\, Y\, x(t) \\
s.t. \quad & G U \le W + E x(t)
\end{aligned}
\qquad \text{(Eq. 2)}
$$

where $H \succcurlyeq 0$, and all additional matrices are obtained from Equation 1 and $x_{t+k|t} = A^k x(t) + \sum_{j=0}^{k-1} A^j B u_{t+k-1-j}$. Using the variable transform, $\gamma = U + H^{-1} F' x(t)$, and defining the allowable space of state variables (space of independent parameters), a mp-QP/MPC problem is obtained accordingly:

$$
\begin{aligned}
\min_{\gamma} \quad & 1/2\, \gamma' Q\, \gamma \\
s.t. \quad & G\gamma \le W + S x(t) \\
& P_A x(t) \le P_b
\end{aligned}
\qquad \text{(Eq. 3)}
$$

where $\gamma \in \mathbb{R}^s$ and $S = E + GH^{-1}F'$. Multiple routes are available to solve Eq. 3, namely via the explicit calculation of critical regions of the state space and the corresponding control laws (Pappas et al. 2020). In this work we consider the solution strategy for mp-QP problems presented in Narciso et al. 2022, where a final recast step is required. Note that a new transformed vector of parameters $z = Sx(t)$ is defined, and no bounds are enforced on the corresponding space of parameters Z:

$$\min_{\gamma} \ 1/2\,\gamma' Q\,\gamma$$
$$s.t. \quad G\gamma \leq W + z \qquad \text{(Eq. 4)}$$

In this convenient format, it is possible to deliver the map of all critical regions (Eq. 5) and the corresponding optimal control laws (Eq. 6) in compact form for all active sets, as follows:

$$z = z^* + V_i^z\left((\alpha^{full} - \alpha)l\right) + V_a^z(\alpha l), \alpha \in \{0,1\}, l \geq 0 \qquad \text{(Eq. 5)}$$

$$\gamma^{opt} = \gamma^* + V^{\gamma}(\alpha l) \qquad \text{(Eq. 6)}$$

where α is the binary active set vector of Eq. 4 with size $q = (p+m)N_c$, and such that 0/1 denote inactive/active constraints, respectively. α^{full} denotes the full active ($\alpha^{full} = [1, \dots, 1]'$). Optimizers are denoted as γ^{opt}, and l is a non-negative multiplier vector of size q. All vectors (z^*, γ^*) and matrices (V_i^z, V_a^z, V^{γ}) in Eqs 5 and 6 are obtained from the Karush-Kuhn-Tucker (KKT) conditions on all the q single-constraint active sets.

An algorithm and discussion on the significance of these vectors/matrices is presented in Narciso et al. 2022. For any given active set α, it suffices to set it directly in Eqs. 5 and 6 to obtain the corresponding critical region in the transformed parameter space and optimal control law, respectively; the explicit calculation of critical regions may also optionally be enabled, where $V_{\{\alpha\}}^z$ includes the relevant column vectors from V_i^z and V_a^z for any α:

$$z = z^* + V_{\{\alpha\}}^z l, l \geq 0 \qquad \text{(Eq. 7)}$$

3. Online calculation of optimal control actions

To obtain the optimal control actions for any $x(t)$, 4 steps are required:

1) Compute $z^{input} = Sx(t)$;
2) Solve Eq. 5 for $z = z^{input}$ to obtain (α, l);
3) Compute γ^{opt} from Eq. 6;
4) Compute the first set of control actions (u_t) from $U = \gamma^{opt} - H^{-1}F'x(t)$.

Steps 1, 3 and 4 are inexpensive matrix/vector operations. Step 2, on the other hand, is a less trivial problem: when solving Eq. 5 for a given z, such statement defines in fact a Linear Complementarity Problem (LCP) (Murty and Yu, 2010). Using one of the state-of-the art LCP solvers is the obvious choice for this step, but in fact the convenient structure of Eq. 5 enables other alternative routes for its solution in the context of mp-QP/MPC, namely via: (i) the explicit calculation of critical regions/control laws (as is in fact the common practice in the field), or (ii) recast Eq. 5 as an NLP or MILP problem.

All routes above are equivalent with respect to the solutions obtained from them. In a control context, however, not only are we concerned with finding the optimal control actions, but also to deliver them as fast as possible to avoid any undesirable delays and the consequent loss of control efficiency (Pistikopoulos et al. 2020). We propose the following procedure to calculate (α, l) via step 2 as depicted in Figure 1.

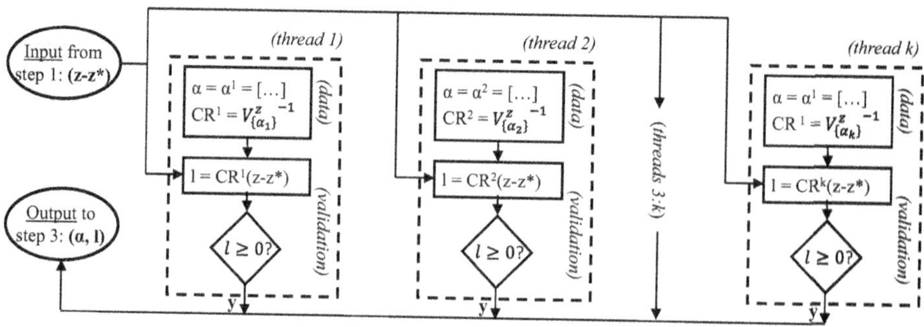

Figure 1: Online calculation of (α, l) via parallel processing of all critical regions.

From Eq. 7, critical regions are obtained for all $k \leq 2^q$ relevant active sets of Eq. 4 as $V_{\{\alpha\}}^{z\ -1}$, which we refer to as the critical region *identifier* for α. In this convenient solution format, it suffices to compute $l = V_{\{\alpha\}}^{z\ -1}(z - z^*)$, and check which critical region satisfies $l \geq 0$, to identify α for any given $z = z^{input}$. Since all identifiers are tested in *parallel* in Figure 1, this amounts in fact (and in theory) to a very fast calculation of (α, l).

In MPC problems containing many constraints, the scheme depicted in Figure 1 may become prohibitive, since the online controller requires that up to a total of 2^q threads are created to store and process all active sets. A modified architecture is proposed where $k \ll 2^q$; this avoids the large memory/processing requirements, while preserving the speed and effectiveness of the online controller. This architecture is shown below:

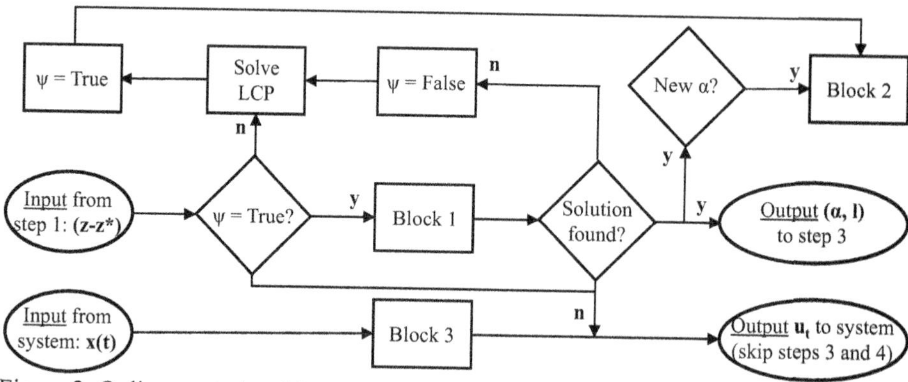

Figure 2: Online control architecture (step 2).

The middle row of tasks in Figure 2 captures the *nominal* mode of operation, where the pair (α, l) is calculated via Block 1: it shares the architecture depicted in Figure 1, but where a much smaller number of *candidate* active sets and their identifiers are kept in memory (details below). Whenever a new α is found between consecutive sampling times, an update of the k threads is executed in Block 2 to ensure the candidate active sets and their identifiers remain relevant as the vector of states is steered towards the set-point. ψ is a Boolean variable providing information if the current set of identifiers kept in memory are up to date and managing the calculation of controls accordingly.

Since only a small fraction of all active sets is kept in memory, it is possible that the matching active set for a given $x(t)$ is not found for a particular $t = t$. In this case, a *remedy* control action is implemented via Block 3, using a conventional feedback controller (e.g. PID). When this occurs, Block 1 becomes unavailable until the optimal solution is obtained, which requires the solution of a more expensive LCP problem.

The structure of active sets and their identifiers kept in the k threads are as follows. As a minimum, a single thread must be defined: if an active set α is the optimal solution at $t = t$, the unique defined thread includes α and the matching identifier at $t = t + 1$. Using a single thread/active set, however, would most likely frequently trigger solving the LCP problem and using the feedback controller, as it is expected that the optimal active set does not remain constant. To strengthen the predictive capability of Block 1, additional threads are defined: this includes all active sets matching α, except for up to r coordinates, where r is a design parameter. The total number of threads is given by the number of combinations $C(q, 0) + C(q, 1) + \cdots + C(q, r) = k$, where $k \geq 1$ (when $r = 1$) and $k \leq 2^q$ (when $r = q$). When a new α is found, Block 2 updates the full set of active sets in all threads, taking α as a reference and updating selectively their coordinates:

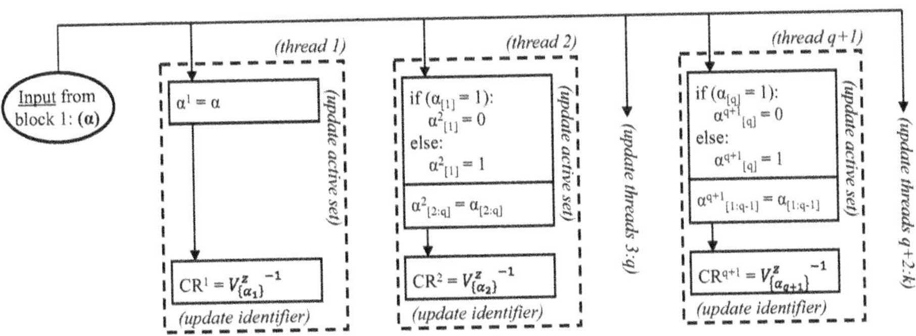

Figure 3: Threads' update of in step 2 (Block 2).

Parameter r defines a *compromise* between the computational complexity (number of active sets and identifiers kept in memory) and the predictive capability of Block 1. In stable systems, it is expected that the vector of states undergoes a smooth trajectory, and where the optimal active set frequently remains constant between consecutive times, or where a restricted number of constraints change status. Therefore, a careful tuning of r facilitates an efficient route for step 2, with a minimal fraction of time spent on solving LCPs. The most frequent or nominal control tasks are highlighted below:

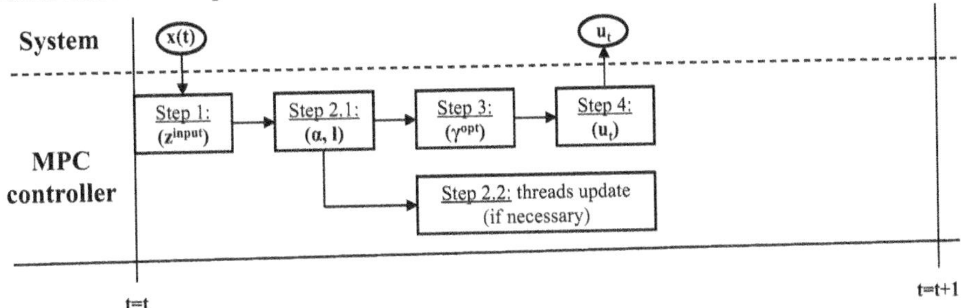

Figure 4: Control tasks in the nominal operation mode for any given $t = t$.

Steps 1-4 are all executed in sequence. Step 2 is broken down in two stages where steps 2.1 and 2.2 correspond to Blocks 1 and 2, respectively. Note that the threads update via step 2.2 is launched in parallel (when necessary) with steps 3 and 4 to speed up this task.

4. Discussion and future work

We have shown in Section 2 that combining the mp-QP/MPC problem formulation in Pistikopoulos et al. 2002 with the mp-QP solution strategy in Narciso et al. 2022 is a trivial task, to deliver critical regions and control laws in compact form.

While this is a significant gain from the offline perspective, even more important is the ability to calculate optimal control actions very fast. With this goal in mind, a new architecture for online calculations was proposed. It makes use of the convenient structure of Eq. 5, to keep a restricted collection of active sets and their critical region identifiers. Parallel computing is used to compute the optimal control actions from the states vector via a sequence of inexpensive matrix/vector multiplications.

Several improvements to this architecture have been identified for future work, and which we discuss briefly here. Firstly, we remark that this architecture includes no provisions to deal with infeasible solutions: failing to obtain a solution from Block 1 is not necessarily a result of the identifiers stored in memory not being sufficient, but simply because no feasible solution exists for the MPC problem with a given $x(t)$. Additionally, to the explicit calculation of identifiers, the framework in Narciso et al. 2022 also allows the calculation of infeasible regions, which can be used for a more efficient control strategy.

A maximum of 2^q critical region identifiers may be obtained from Eq. 5 since the parameter space (Z) for $x(t)$ is defined unbounded. This choice is made for the generality of the solution obtained in Narciso et al. 2022. In the present context, where the explicit identifiers are kept in memory, an additional check of Z may simplify the online control problem: checking for instance which constraints are always inactive or pre-calculating subspaces in Z where a constraint can be active reduces the online complexity.

More broadly, this research aims at implementing the fastest online strategy for MPC problems. Additional developments on the solution of LCPs in this specific context is of great significance. Two LCP solvers are currently under development for this class of problem. An assessment of the performance of all the viable routes on computing control actions for MPC problems of varying sizes is a key research objective. It is expected that this work can facilitate establishing several guidelines for practical applications.

References

E. N. Pistikopoulos, N. A. Diangelakis and R. Oberdieck, 2020, Multi-parametric Optimization and Control, ISBN: 978-1-119-26518-4.

I. Pappas, D. Kenefake, B. Burnak, S. Avraamidou, H. S. Ganesh, J. Katz, N. A. Diangelakis and E. N. Pistikopoulos, 2021, Multiparametric Programming in Process Systems Engineering: Recent Developments and Path Forward, Front. Chem. Eng., 21.

E. N. Pistikopoulos, V. Dua, N. Bozinis, A. Bemporad, M. Morari, 2002, On-line optimization via off-line parametric optimization tools, Comput. Chem. Eng., 26, 175-185.

D. A. C. Narciso, I. Pappas, F. G. Martins and E. N. Pistikopoulos, 2022, A new solution strategy for multiparametric quadratic programming, Comput. Chem. Eng., 164, 107882.

K. Murty, F. Yu, 2010. Linear Complementarity, Linear and Non Linear Programming. http://www-personal.umich.edu/murty/books/linear complementarity webbook/.

Antonis Kokossis, Michael C. Georgiadis, Efstratios N. Pistikopoulos (Eds.)
PROCEEDINGS OF THE 33rd European Symposium on Computer Aided Process Engineering (ESCAPE33), June 18-21, 2023, Athens, Greece
 http://dx.doi.org/10.1016/B978-0-443-15274-0.50197-9

Comparison of regression techniques for generating surrogate models to predict the thermodynamic behavior of biomass gasification systems

Meire Ellen Gorete Ribeiro Domingos,[a,b] Daniel Flórez-Orrego,[b] Julia Granacher,[b] Marie Jones,[b] Moisés Teles dos Santos,[a] François Maréchal[b]

[a]*Polytechnic School, University of Sao Paulo, Department of Chemical Engineering, Sao Paulo, Brazil.*
[b]*École Polytechnique Fédérale de Lausanne, Valais, Sion, Switzerland*
meire.ribeirodomingos@epfl.ch

Abstract

Biomass resources can play an important role in the energy transition, being a dynamic feedstock that can be transformed either into solid, liquid or gaseous fuels. Biomass gasification is a versatile way to convert waste into energy. In this work, the modeling and simulation of two different gasification processes using wood and black liquor as feedstock are performed using Aspen Plus®. The surrogate models for these biomass-based gasification systems are generated considering different techniques (e.g. artificial neural networks, random forest, and Gaussian process regression) using an Active Learning Artificial Intelligence approach. These techniques are compared in terms of their capabilities for predicting the thermodynamic behavior of the gasification systems for the different biomass resources. As a result, the surrogate models developed were able to estimate the process design and operating conditions, and the Gaussian process regression outperformed the artificial neural networks and random forest techniques. The generated models could be helpful to be further used for replacing the simulation systems in other applications, such as multi-objective optimization, at expense of lower computational requirements.

Keywords: Surrogate models, Biomass, Gasification, Artificial intelligence.

1. Introduction

The gasification process poses modeling complexities involving several stages that occur simultaneously, such as drying process, and pyrolysis, gasification and combustion reactions. These reactions are highly endothermic and the energy balance becomes a function of the combustion and the reduction steps. In addition, the production rate, composition and properties of the syngas produced strongly vary according to the biomass properties, the gasification agent used (type and flow), the equivalence ratio, the type of reactor used (fixed bed, fluidized bed or indirect gasifier), the operating conditions (temperature, pressure, residence time) and the downstream processes considered for gas cleaning and conditioning. Numerous models have been reported, aiming at understanding and describing the behavior of such systems, using commercial simulation software. However, if integrating these complex processes into industrial systems and performing optimization on the system level, the direct simulation approach entails high computational expenses. Surrogate models emerge as a robust solution that may help accelerate the convergence of the optimization problem (Granacher et al., 2021). This work proposes the generation of surrogate models for two gasification systems, considering wood and black liquor as feedstock, using different regression techniques applying the Active Learning Artificial Intelligence (ALAI) approach (Granacher et al., 2021). The models are compared to demonstrate their capability in predicting the processes behavior.

2. Methodology

2.1. Modeling and simulation of the gasification systems

Figure 1 illustrates the wood and black liquor gasification setups based on (Domingos et al., 2022, Flórez-Orrego et al., 2019). The simulations are performed in the Aspen Plus® software (Aspentech, 2015), using the Peng-Robinson EoS with Boston-Mathias modifications as thermodynamic model.

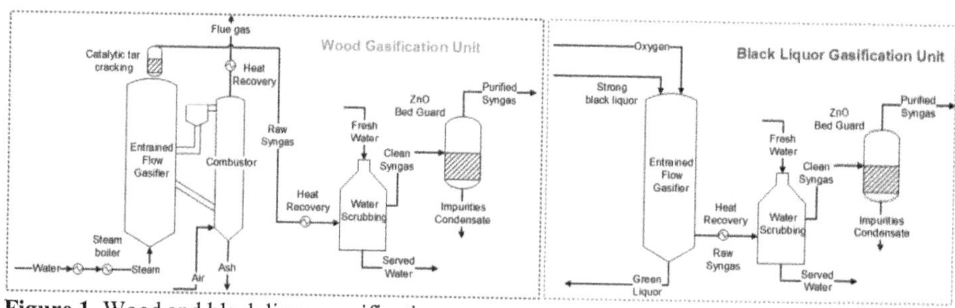

Figure 1. Wood and black liquor gasification systems.

The ultimate mass-based wood composition is set to 48.2%C, 5.8% H, 45% O, 0.2% N, 0.02% S and 0.8% ashes, whereas the mass-based proximate analysis is considered as 10% moisture (after drying), 16.4% fixed carbon, 82.8% volatiles, and ash in balance (Telmo et al., 2010). The Battelle Columbus Laboratory (BCL) indirect gasifier operating at atmospheric pressure and using steam as gasification medium is used for modeling the wood gasification system shown in Fig. 1. This system avoids the dilution with nitrogen of the syngas produced, as the combustion and gasification processes occur in a separate double column system (Kinchin, 2009). After the produced syngas leaves the gasifier, a thermal catalytic cracking of the produced tar is performed.

The black liquor ultimate composition (mass) is set as 29.86%C, 3.27% H, 29.05% O, 0.1% N, 4.09% S, 0.90%Cl and 32.73% ashes, whereas proximate analysis (mass) is considered as 15% moisture (after drying), 10.21% fixed carbon, 57.06% volatiles, and ash in balance. The pressurized entrained flow high temperature black liquor gasification (PEHT-BLG) available in the market as Chemrec® gasifier (Consonni et al., 2009) is used for modeling the black liquor gasification system. This technology has the advantage of including tar-free and uniform temperature gasification at turbulent conditions which results in a high conversion efficiency and high reaction rate. In the gasification step, the carbonaceous materials in the BL are converted into syngas rich in CO, H_2, H_2O and CO_2 using oxygen as agent, while a smelt containing mainly Na_2CO_3 and Na_2S is formed in the bottom when a quench cooler separates the regenerated salts that will form the green liquor (Darmawan et al., 2017).

In both systems, a fraction of the char produced in the pyrolysis step is combusted to supply the heat required by the endothermic drying, pyrolysis and reduction reactions.

After the produced syngas leaves the gasifier, it is cooled down and scrubbed with water, in order to remove the impurities that may affect the downstream equipment.

2.2. Surrogate model design

The algorithm used to generate and test the surrogate models considering the ALAI approach (Granacher et al., 2021) is presented in Fig. 2. The simulation systems developed in Aspen Plus® (Aspentech, 2015) are used to retrieve the dataset comprising simulation inputs and outputs, taking into account the decision variables selected. The quasi-random Latin-Hypercube sampling (McKay et al., 1979) algorithm is applied for sampling the design space *D*. The dataset is then used to generate and train the surrogate models, where the uncertainty with which a prediction is made is quantified. The active learning is responsible for continuously enriching the database, and new data points are selected from a pool of unlabeled data based on the predicted uncertainty. This strategy results in a fast improvement of the surrogates' quality. When the desired quality is met following the defined performance metrics, the surrogate model is saved (Granacher et al., 2021).

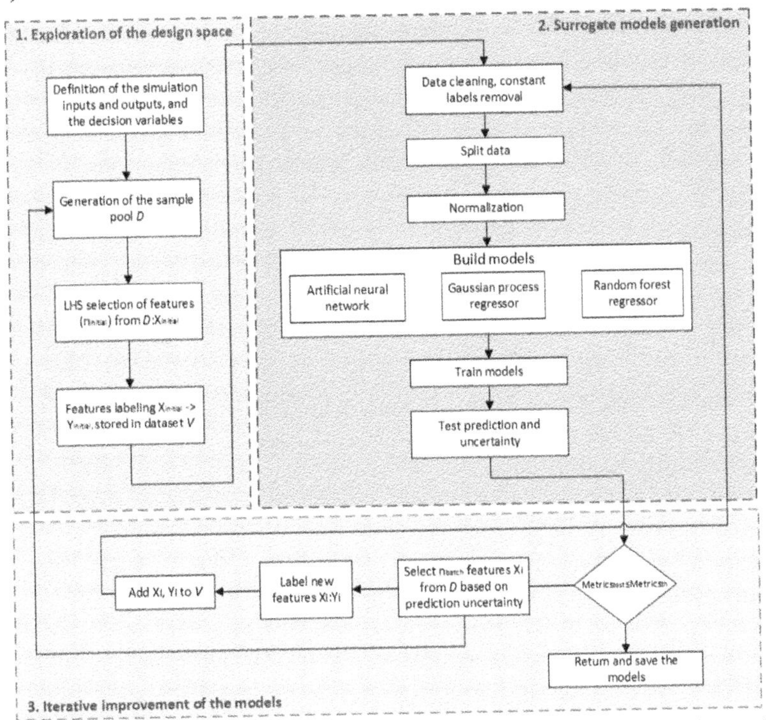

Figure 2. Algorithm for generation of the surrogate models, adapted from (Granacher et al., 2021).

In this work, three regression techniques, namely Gaussian process regression, Artificial neural networks (ANN) and Random Forest regression, are assessed and compared to generate surrogate models for the wood and black liquor gasification systems. In brief, the Gaussian process regression is a nonparametric, Bayesian approach to regression that makes predictions with uncertainty (Rasmussen and Williams, 2006). This method calculates the probability distribution over all admissible functions that fit the data instead of returning only exact predictions for each. The ANN is inspired by the biological nervous system and it consists of several layers connected by neurons that perform the input data transformations aiming to obtain the required outputs (Silva et al., 2016).

Finally, Random Forest regression is a supervised learning algorithm that combines predictions from multiple algorithms to make a more accurate prediction compared to a single model (Breiman, 2001).

For each gasification system, three parameters were chosen to be varied based on the operating conditions reported in the literature for these setups (Larsson et al., 2006, Puig-Arnavat et al., 2012). The parameters and their respective range are reported in Table 1. All the important simulation inputs, outputs, and operating conditions were set to be extracted to compose the dataset for building the surrogate modes.

Table 1. Operating conditions varied for wood and black liquor gasification systems.

Wood gasification		**Black liquor gasification**	
Parameter	Range	Parameter	Range
Gasification temperature (°C)	[760,960]	Gasification temperature (°C)	[900,1100]
Air preheating (°C)	[300,500]	Gasification pressure (bar)	[25,32]
Steam-to-biomass ratio	[0.75±30%]	Oxygen-to-fuel ratio	[0.2,0.5]

In order to measure the quality of the surrogate model, a random set of additional points is generated for testing. The new simulation points are compared against the surrogate model's response using error-based performance metrics. The mean squared error (MSE) (Eq.1) measures the average squared difference between the estimated values coming from the response of the surrogate models and the actual simulation value.

$$MSE = \frac{\sum_{i=1}^{n}(x_{simul,i} - x_{surr,i})^2}{n} \tag{1}$$

where $x_{simul,i}$ is the simulation output and $x_{surr,i}$ is the surrogate response for each parameter i, and n is the number of points in the data set. In order to facilitate the comparison between the dataset with different scales, the simulation data is normalized.

3. Results and discussion

Table 2 presents the mean squared errors in absolute dimensions for the surrogate models generated considering the three regression techniques. It can be seen that the Gaussian Process regression outperformed the artificial neural networks and the random forest approaches, since it presented the lowest MSE.

Table 2. Mean squared error (MSE_{test}) considering 100 data points, absolute dimensions.

Regression model	**MSE_{test} - wood gasification**	**MSE_{test} - black liquor gasification**
Gaussian Process regression	0.02459	0.01498
ANN	0.02696	0.06296
Random forest	0.03061	0.05464

In Fig. 3, the predictions are displayed against the true values, with their respective uncertainty (see the orange bars in Fig. 3), for the hydrogen mole fraction in the purified syngas stream for the two gasification systems. The results demonstrate that the Gaussian process regression technique resulted in lower uncertainty, related to the prediction capability of the surrogate model considering the syngas composition. The same behavior was observed for the prediction of the heat exchanger duty related to the 'Raw Syngas' stream compared to the true values retrieved from the simulation (see Fig. 4).

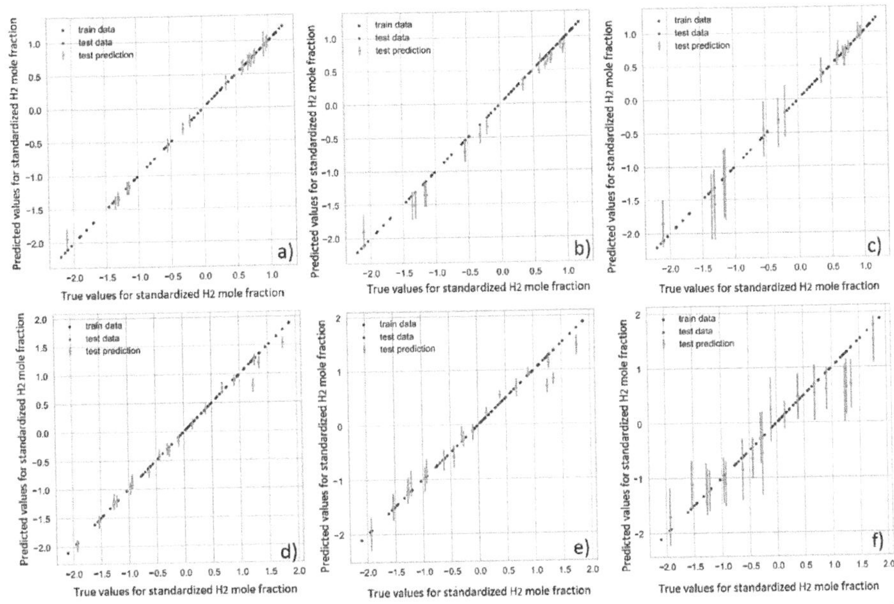

Figure 3. Test predictions and uncertainty considering 100 data points for hydrogen mole fraction in the syngas from wood gasification using a) Gaussian regression, b) ANN, c) random forest, and from black liquor gasification using d) Gaussian regression, e) ANN, f) random forest.

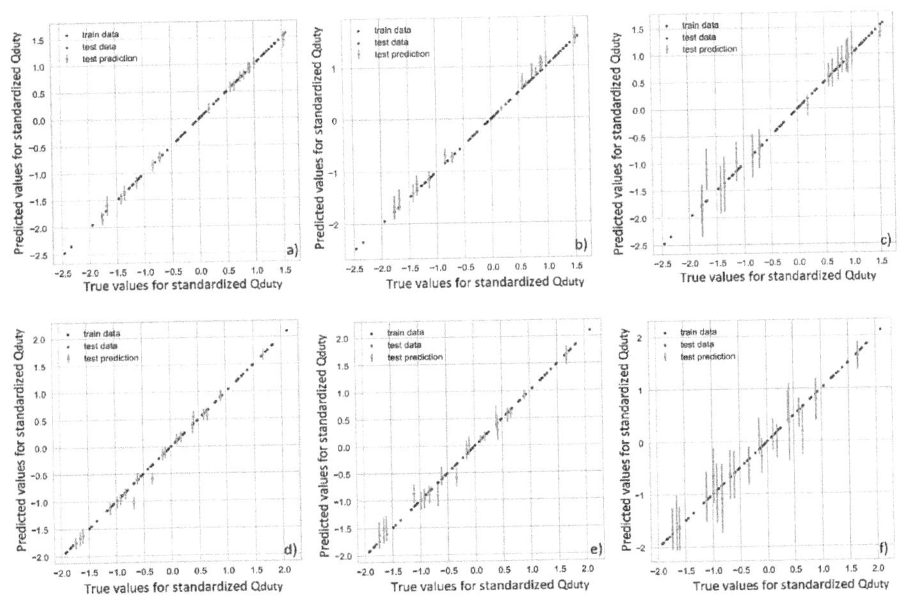

Figure 4. Test predictions and uncertainty considering 100 data points for the heat recovery heat exchanger in the 'Raw Syngas' stream for wood gasification using a) Gaussian regression, b) ANN, c) random forest, and for black liquor gasification using d) Gaussian regression, e) ANN, f) random forest.

The comparison between the predictions of the surrogate models against the simulation results were also performed for the other components of the streams as well as all other process design and operating conditions. The results pointed towards a good capability of the surrogate models in predicting the process design and operating conditions for the studied gasification systems.

4. Conclusion

In this work, three regression techniques were compared to generate surrogate models applying the ALAI approach for wood and black liquor gasification systems simulated in the Aspen Plus software. The Gaussian process regression outperformed the artificial neural networks and random forest techniques for both analyzed processes, achieving lower mean squared error and associated uncertainties. Overall, the surrogate models were able to accurately predict the thermodynamic performance and the high nonlinear behavior associated to both gasification systems. Additionally, the derived surrogate models are computationally less expensive and can be further used to explore the tradeoffs between conflicting objectives, such as operating costs and syngas composition.

5. Acknowledgements

MD and MJ thank European Union's Horizon Europe Research and Innovation programme under Grant Agreement No. 101084288 and No. 945363.

References

Aspentech. 2015. Aspen Plus V8.8. Bedford, United States: Aspen technology Inc.

J. Granacher, I. D. Kantor, F. Maréchal. 2021. Increasing Superstructure Optimization Capacity Through Self-Learning Surrogate Models. Frontiers in Chemical Engineering, 3.

M. E. G. R. Domingos, D. Florez-Orrego, M. Teles dos Santos, S. de Oliveira Jr., F. Maréchal. 2022. Techno-economic and environmental analysis of methanol and dimethyl ether production from syngas in a kraft pulp process. Computers & Chemical Engineering, 163, 107810.

D. Florez-Orrego, F. Maréchal, S. de Oliveira Jr. 2019. Comparative exergy and economic assessment of fossil and biomass-based routes for ammonia production. Energy Conversion and Management, 194, 22-36.

C. Telmo, J. Lousada, N. Moreira. 2010. Proximate analysis, backwards stepwise regression between gross calorific value, ultimate and chemical analysis of wood. Bioresource technology, 101, 3808-3815.

C. M. Kinchin, R. L. Bain. 2009. Hydrogen Production from Biomass via Indirect Gasification: The Impact of NREL Process Development Unit Gasifier Correlations, Technical Report NREL/TP-510-44868, May 2009. National Renewable Energy Laboratory.

S. Consonni, R. E. Katofsky, E. D. Larson. 2009. A gasification-based biorefinery for the pulp and paper industry. Chemical Engineering Research and Design, 87, 1293-1317.

A. Darmawan, F. Hardi, K. Yoshikawa, M. Aziz, K. Tokimatsu. 2017. Enhanced Process Integration of Entrained Flow Gasification and Combined Cycle: Modeling and Simulation Using Aspen Plus. Energy Procedia, 105, 303-308.

McKay, M. D. et al. (1979). Comparison of Three Methods for Selecting Values of Input Variables in the Analysis of Output from a Computer Code. Technometrics 21, 239–245.

C. E. Rasmussen, C. Williams, Gaussian Processes for Machine Learning. 2006. MIT Press. ISBN 026218253X.

I. N. Silva, D. Hernane Spatti, R. Andrade Flauzino, L. Liboni, S. dos Reis Alves. 2016. Artificial Neural Networks. New York, NY: SpringerBerlin Heidelberg.

L. Breiman. (2001). Random Forests. Machine Learn. 45, 5–32.

A. Larsson, A. Nordin, R. Backman, B. Warnqvist, G. Eriksson. 2006. Influence of black liquor variability, combustion, and gasification process variables and inaccuracies in thermochemical data on equilibrium modeling results. Energy and Fuels, 20, 359-363.

M. Puig-Arnavat, J. C. Bruno, A. Coronas. 2012. Modified thermodynamic equilibrium model for biomass gasification: a study of the influence of operating conditions. Energy & Fuels, 26, 1385-1394.

Antonis Kokossis, Michael C. Georgiadis, Efstratios N. Pistikopoulos (Eds.)
PROCEEDINGS OF THE 33rd European Symposium on Computer Aided Process Engineering (ESCAPE33), June 18-21, 2023, Athens, Greece
 http://dx.doi.org/10.1016/B978-0-443-15274-0.50198-0

Global optimization of symbolic surrogate process models based on Bayesian learning

Tim Forster[a], Daniel Vázquez[a], Gonzalo Guillén-Gosálbez[a,*]
[a] *Department of Chemistry and Applied Biosciences, Institute for Chemical and Bioengineering, ETH Zurich, Vladimir-Prelog-Weg 1, 8093 Zurich, Switzerland*
[*] *gonzalo.guillen.gosalbez@chem.ethz.ch*

Abstract

In this work, we address the global optimization of process surrogates using Bayesian symbolic regression and deterministic global optimization algorithms. In contrast to other surrogates of process models that are hard to (globally) optimize, e.g., artificial neural networks or Gaussian processes, symbolic regression leads to a closed-form mathematical expression describing the observed data that can subsequently be globally optimized using off-the-shelf deterministic solvers. After providing an introductory example, we show the capabilities of our approach in the optimization of a methanol production plant. We further discuss the model accuracy, CPU times for model building and optimization, and outline the advantages and limitations of the proposed strategy.

Keywords: Global optimization, Symbolic Regression, Surrogate Modelling.

1. Introduction

Optimizing process flowsheets is a challenging problem in Process Systems Engineering, for which several approaches have been proposed to date. Process optimization often relies on mechanistic models based on first principles (Haydary, 2019) and closed-form expressions that enable the direct application of deterministic optimization algorithms, including global optimization methods. However, building such first principles is a challenging task, potentially leading to convergence issues during their optimization. Alternatively, when there is limited knowledge about the process, data-driven models can be the method of choice to build a mathematical expression that can be used for optimization purposes. Specifically, state-of-the-art machine learning algorithms (i.e., neural networks or support vector machines) can be applied to build a data-driven process model. However, optimizing classical machine learning models to global optimality is not trivial due to their intrinsic complexity and nonlinearities (Mitsos et al., 2009; Schweidtmann and Mitsos, 2019). Recently, symbolic regression algorithms have emerged to build algebraic models from data without assuming any specific structure (Cozad et al., 2014; Guimerà et al., 2020). This work explores the use of symbolic regression to build analytical surrogates of process flowsheets that can then be optimized with state-of-the-art global optimization (GO) algorithms. A priori assumptions about the mathematical structure of the model are avoided, and well-established global optimization algorithms are applied to the algebraic surrogate without the need for tailored strategies. Notably, after building a well-fitting analytical model using Bayesian symbolic regression, BARON (Sahinidis, 1996) is used to identify its global optimum. We deploy the proposed approach in an introductory example. Additionally, we then demonstrate the performance of the approach with a complex flowsheet of an industrial process.

2. Methodology

2.1. Problem Statement

In this work, we consider a given process in steady state and a set of input variables x_i, with $i \in I$ being the set of dependent variables, i.e., degrees of freedom, of the flowhseet. A target objective is defined by $f(x)$, which can be, for example, the total cost of a plant or the conversion of a reactor. In the first step, a suitable surrogate model $F(x)$ that approximates the studied process target $f(x)$ appropriately is identified. Second, a model-based global optimization is performed to find the global optimum x^*, as follows:

$$\begin{aligned} F^* = & \min_x F(x) \\ s.t. \quad & g_n(x) \leq 0, \forall n \in N \\ & h_m(x) = 0, \forall m \in M \\ & \underline{x} \leq x \leq \overline{x} \\ & x \in \mathbb{R}^{|I|} \end{aligned} \tag{1}$$

Where F^* corresponds to the model-based optimum of the process approximation $F(x)$, and $g_n(x)$ and $h_m(x)$ represent inequality and equality constraints, respectively. Herein, we aim to identify an accurate surrogate model without assuming a pre-defined model structure. In addition, we apply state-of-the-art deterministic GO solvers to model (1). Due to the sake of simplicity, the purpose of this work is a proof-of-concept, rather than a performance comparison.

2.2. Model Building

We seek to identify an accurate surrogate process model $F(x)$ that maps the input data x to the corresponding objectives $f(x)$. In addition, to apply deterministic optimization approaches, $F(x)$ is required to be in an algebraic form. To this end, in this work, we use the algorithm developed by Guimerà et al. (2020), the Bayesian machine scientist (BMS). It implements symbolic regression that uses a Markov chain Monte Carlo (MCMC) approach to explore the space E of closed-form mathematical expressions $z_e, e \in E$. During this explorative search, the algorithm minimizes the description length $\mathcal{L}(z_e)$, which can be approximated by the Bayesian Information Criterion (BIC) and an initial prior knowledge about a given expression z_e, where Bayesian learning is incorporated:

$$\mathcal{L}(z_e) \approx \frac{BIC(z_e)}{2} - \log(p(z_e)) \tag{2}$$

For simplicity, only the approximation of the description length is given here. For a detailed derivation, the reader is referred to the original work by Guimerà et al. (2020).

2.2.1. Model-Based Global Optimization

With the obtained algebraic surrogate model $F(x)$ at hand, the model-based optimization problem in (1) can be solved to identify the optimal decision variables values x^*, providing the best value of the surrogate $F^* = F(x^*)$ and which can be evaluated in the original process model as $f^* = f(x^*)$. This observed response f^* can finally be compared to the model-based optimum F^* to measure the mismatch between the surrogate and the process model in the optimum of the surrogate x^*.

3. Case Studies

3.1. Software Implementation

The calculations were carried out on an AMD Ryzen 5 3600 CPU (introductory example) and on an Intel®Core™ i7-8700 CPU (methanol production plant), both with 16 GB of RAM. Python v3.9 (introductory example) and v3.8 (methanol production plant) were used to implement the algorithm. The flowsheet was simulated in Aspen HYSYS v11. Python and HYSYS were connected through a COM interface. The algorithm provided by Guimerà et al. (2020) was used to train the BMS. The obtained algebraic expression was implemented using the General Algebraic Modeling System (GAMS Software GmbH, 2022) v40.2 and optimized using BARON 20.4 (Sahinidis, 1996).

3.2. Introductory Example

We illustrate the approach using first a simple two-dimensional unconstrained global optimization problem, the Rosenbrock function (Rosenbrock, 1960):

$$f(x) = (x_1 - a)^2 + b(x_2 - x_1^2)^2 \quad (3)$$

There are two dependent variables x_1 and x_2, and two parameters a and b. For the subsequent illustrations, parameters a and b are chosen to be 1 and 10, respectively. A graphical representation of the resulting contour plot bound to $\underline{x} = [-2, -1]$ and $\overline{x} = [2,3]$, is given in Figure 1.

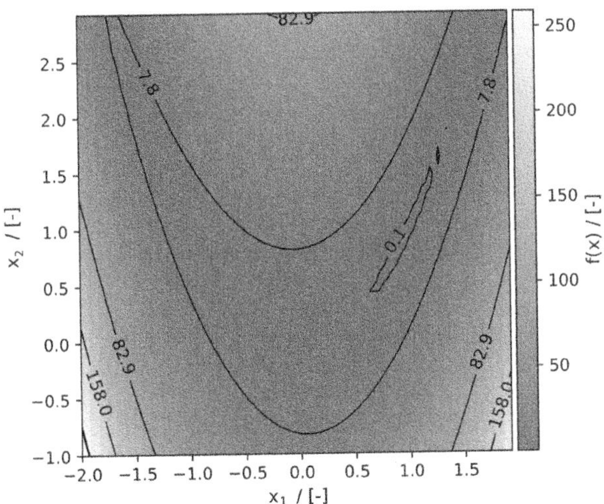

Figure 1: Contour of the Rosenbrock function. The color gradient refers to the value of $f(x)$. The contour levels are indicated by the black lines with the corresponding values of $f(x)$.

To evaluate $f(x)$ according to (3), 225 equally distant data points were generated within the bounds. The global optimum (minimization) is given by $f^* = 0$ at $x^* = [1,1]$. The obtained samples ($N = 225$) were randomly split into training ($N_{tr} = 157$) and test set ($N_{te} = 68$). Details of the BMS training are shown in Table 1.

3.3. Methanol Production Plant

A methanol (MeOH) production plant model is investigated next (Figure 2), which was adapted from Vázquez and Guillén-Gosálbez (2021). The feed consists of carbon dioxide and hydrogen, whose pressure and temperature are adjusted to the desired conditions before being sent to a plug-flow reactor. The reactor outlet is sent through two flash drums and into a distillation column. Methanol is collected in the distillate, while water is the main product at the bottom. The vapor streams of the flash drums are partly recycled (from the first flash) and purged (from both drums). For a detailed description, the reader is referred to the work by Vázquez and Guillén-Gosálbez (2021).

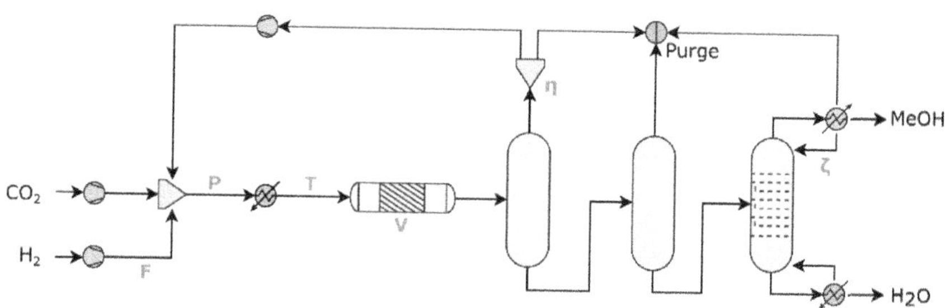

Figure 2: Simplified schematic representation of the MeOH plant under study adapted from Vázquez and Guillén-Gosálbez (2021), with the decision variables highlighted in green.

The degrees of freedom $x = [T, P, \eta, V, F, \zeta]$ to be optimized (in green in Figure 2) consist of the reaction temperature (T), the reaction pressure (P), the purge ratio of the splitter (η), the reactor volume (V), the hydrogen flow rate (F), and the reflux ratio of the distillation column (ζ). The objective is to minimize the unitary cost (UC) of methanol. The carbon dioxide inlet flow is fixed. Vázquez and Guillén-Gosálbez (2021) provide the exact calculation of UC and a more detailed description of the simulation.
To determine the optimum heat exchanger network (HEN), and, therefore, the cost of the HEN, the mixed-integer nonlinear programming model developed by Yee and Grossmann (1990) (SYNHEAT) is solved for each sampling point. The obtained samples ($N = 1000$) were randomly split into training ($N_{tr} = 800$) and test set ($N_{te} = 200$). The bounds indicated in Table 1 were used to generate these samples. The applied settings and hyperparameters for the BMS training are shown in Table 1.

Table 1: The chosen BMS hyperparameters (MCMC steps and allowed operations) are given, whereas otherwise, the default settings chosen by the authors (Guimerà et al., 2020) were applied. Lastly, used relative optimality gap ϵ_R of BARON is indicated for both examples.

Case Study	BMS MCMC steps	Allowed Operations	Optimizer Settings
Rosenbrock	$5 \cdot 10^3$	$\exp(x), \log(x), x^2, x^3, \sqrt{x}, +, -, \div, x^a$, and $\|x\|$.	$\epsilon_R = 0$
Methanol	$20 \cdot 10^3$		

4. Results and Discussion

4.1. Introductory Example

A summary of the results is given in Table 2. The BMS could identify the true equation, given by the following expression:

$$F(x) = (x_1 - \theta_1)^2 + \theta_2(x_2 - x_1^2)^2 \quad (4)$$

The simultaneously identified parameters were found to be $\theta = [10,\ 1]$. The model training required 788 s. These results can also be visually compared by considering the true contour plot in Figure 1, and the predictions given in Figure 3 (b).

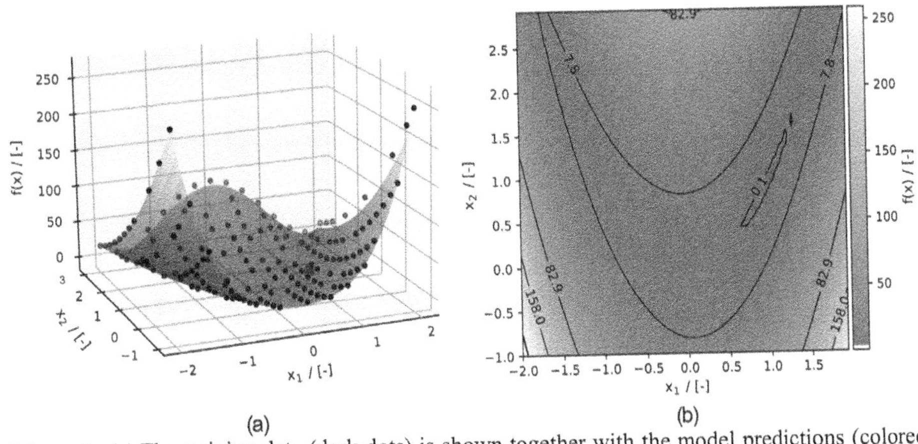

Figure 3: *(a)* The training data (dark dots) is shown together with the model predictions (colored surface). The identified optimum is depicted as a red dot. *(b)* The contour of the BMS model predictions. The identified model-based global optimum F^* is again shown as a red dot.

Figure 3 (a) provides the training data together with the model predictions, where Figure 3 (b) represents the obtained contour plot of the model predictions. Optimizing the equation in GAMS leads to the optimum solution given by $F^* = 0$ with $x^* = [1, 1]$. The optimization with BARON was carried out in 0.2 s. Implementing x^* into $f(x)$ results in $f^* = 0$, which is also the true global optimum of the Rosenbrock function.

Table 2: Training and optimization results summary for the introductory example. The CPU times for training and optimization (minimization) are shown together with the calculated root mean squared error (RMSE) for the training and testing set. In addition, the obtained optimum x^* with the corresponding solution F^* is shown. Lastly, the process output f^* is shown for the corresponding optimum x^*.

CPU Training	RMSE	CPU Optimization	x^*	F^*	f^*
788 s	0 [-] (Train) 0 [-] (Test)	0.2 s	$x_1 = 1$ [-] $x_2 = 1$ [-]	0	0

4.2. Methanol Production Plant

The goal of this six-dimensional case study is to minimize the unitary cost, which is a nonlinear programming problem (NLP). After training the BMS, an algebraic model could be obtained and then optimized in GAMS. The training and optimization steps took 40'600 s and 1 s, respectively. Inserting the identified model-based optimum x^* into HYSYS again led to an optimum solution $f^* = 0.73$ \$/kg, which was very close to the prediction made with the algebraic surrogate at its optimum of $F^* = 0.72$ \$/kg. The training time of the BMS is substantial (>10 hours), yet this allows for a fast and easy optimization. However, more advanced symbolic regression algorithms could reduce the computational time of the training phase. Moreover, an additional advantage of having an algebraic surrogate expression at hand is that it could be used for calculations that need to solve optimization models iteratively, such as in multi-objective optimization or in real-time optimization, since the optimization step is greatly simplified using the algebraic surrogate.

5. Conclusions

In this work, we explored the use of algebraic surrogate process models constructed using symbolic regression. Accurate process models were built using symbolic regression based on Bayesian learning and then solved in GAMS using an off-the-shelf global optimization algorithm (BARON). We showed the capabilities of the approach in two examples. The high computational training times for symbolic regression still represent a computational bottleneck, yet they might be reduced with the future emergence of faster regression algorithms. Moreover, analytical surrogate process models could be used to simplify the calculations in applications requiring solving multiple optimization problems iteratively.

References

A. Cozad, N.V. Sahinidis, D.C. Miller, 2014, Learning surrogate models for simulation-based optimization, AIChE Journal, 60, 2211–2227.

GAMS Software GmbH, 2022, GAMS Development Corporation. General Algebraic Modeling System (GAMS) Release 40.2.0, 2021.

R. Guimerà, I. Reichardt, A. Aguilar-Mogas, F.A. Massucci, M. Miranda, J. Pallarès, M. Sales-Pardo, 2020, A Bayesian machine scientist to aid in the solution of challenging scientific problems, Science Advances, 6.

J. Haydary, 2019, Chemical Process Design and Simulation: Aspen Plus and Aspen Hysys Applications, Wiley, Haydary2019.

A. Mitsos, B. Chachuat, P.I. Barton, 2009, McCormick-Based Relaxations of Algorithms, SIAM Journal on Optimization, 20, 573–601.

H.H. Rosenbrock, 1960, An Automatic Method for Finding the Greatest or Least Value of a Function, The Computer Journal, 3, 175–184.

N.V. Sahinidis, 1996, BARON: A general purpose global optimization software package, Journal of Global Optimization, 8, 201–205.

A.M. Schweidtmann, A. Mitsos, 2019, Deterministic Global Optimization with Artificial Neural Networks Embedded, Journal of Optimization Theory and Applications, 180, 925–948.

D. Vázquez, G. Guillén-Gosálbez, 2021, Process design within planetary boundaries: Application to CO2 based methanol production, Chemical Engineering Science, 246, 116891.

T.F. Yee, I.E. Grossmann, 1990, Simultaneous optimization models for heat integration—II. Heat exchanger network synthesis, Computers & Chemical Engineering, 14, 1165–1184.

Antonis Kokossis, Michael C. Georgiadis, Efstratios N. Pistikopoulos (Eds.)
PROCEEDINGS OF THE 33rd European Symposium on Computer Aided Process Engineering (ESCAPE33), June 18-21, 2023, Athens, Greece
 http://dx.doi.org/10.1016/B978-0-443-15274-0.50199-2

Molecular Property Targeting for Optimal Solvent Design in Extractive Distillation Processes

Zihao Wang,[a] Teng Zhou,[b,#] Kai Sundmacher[a,c,*]

[a] *Department for Process Systems Engineering, Max Planck Institute for Dynamics of Complex Technical Systems, Sandtorstraße 1, D-39106 Magdeburg, Germany*

[b] *Sustainable Energy and Environment Thrust, The Hong Kong University of Science and Technology (Guangzhou), Guangzhou, China*

[c] *Chair of Process Systems Engineering, Otto-von-Guericke University Magdeburg, Universitätsplatz 2, D-39106 Magdeburg, Germany*

[*] *sundmacher@mpi-magdeburg.mpg.de (K. Sundmacher),* [#] *tengzhou@ust.hk (T. Zhou)*

Abstract

In extractive distillation (ED), a suitable solvent is added to increase the relative volatility of azeotropic or close-boiling mixtures, thereby facilitating the separation. The viability and efficiency of this separation are largely dependent on the selection of solvent. In this work, solvents are optimally designed to improve the performance of ED processes by a molecular property targeting method. Specifically, optimal solvents are identified by directly targeting desirable molecular properties obtained from process optimization. First, data-driven process models are established to estimate key performance indicators of the ED process with the most important process-relevant physical properties of the solvents. Based on the established models, multi-objective optimization is then performed to maximize the product purity while minimizing energy cost, through which optimal molecular properties are obtained. Hypothetical target molecules featuring the desirable properties are thereby generated. Subsequently, several real solvents that approximate the optimal property profiles of the hypothetical molecules are selected from a large solvent database. In a final assessment step, the performance of these real solvent candidates is evaluated via detailed process simulations. The proposed molecular-property-targeting solvent design methodology is demonstrated by the separation of 1-butene and 1,3-butadiene. For this separation task, three real solvents are finally identified, allowing for a higher product purity and lower energy consumption compared to the benchmark solvent n-methyl-2-pyrrolidone (NMP).

Keywords: solvent design, artificial neural network, data-driven modeling, continuous molecular targeting, extractive distillation

1. Introduction

Extractive distillation (ED) processes enable the separation of azeotropic or close-boiling mixtures, where a suitable solvent is introduced to increase the relative volatility of the mixture (Gerbaud et al., 2019). The viability and efficiency of this separation largely depends on the selection of the solvent. In the past, most solvents used in chemical industries were primarily chosen based on chemical knowledge and have not been systematically designed (Chemmangattuvalappil, 2020). They may not be the real optimal candidates to meet the separation requirements. Therefore, discovering better alternatives to replace current solvents following a systematic solvent screening or design strategy can improve the performance of the ED processes.

Computer-aided molecular design (CAMD) provides a promising route to systematically design solvents fulfilling a set of target molecular attributes or process performance indicators (Gani, 2004). Taking the advantages of modern molecular property models and process models, CAMD methods have been extensively used to design solvents for various applications such as absorption (Burger et al., 2015; Zhang et al., 2021; Zhou et al., 2021), extraction (Song et al., 2018; Ten et al., 2021), and reaction (Struebing et al., 2013; Zhou et al., 2015; Zhou et al., 2016). So far, the most commonly used approach for representing molecular structures in CAMD is the decomposition into functional groups (Marrero & Gani, 2001). By this method, solvents are represented by discrete variables (e.g., the number of molecular building blocks) which are optimized under certain molecular structural constraints. To circumvent discrete molecular decisions in CAMD, Bardow et al. (2010) proposed a continuous-molecular-targeting (CoMT) approach by considering a continuous molecular structure space in terms of molecular property-related parameters. The optimal CAMD solution is then used to identify existing molecules, the properties of which approximate the optimal parameter values best. The CoMT approach has been applied to design solvents for CO_2 absorption (Stavrou et al., 2014) and working fluids for the organic Rankine cycle (Schilling et al., 2017).

In this contribution, the CoMT approach is adopted for the optimal design of solvents in ED processes. To achieve higher product purity and lower energy consumption, solvents are designed by directly targeting desirable solvent properties obtained from process optimization. First, data-driven process models are established to estimate key performance indicators of the ED process from the most important process-relevant physical properties of the solvent. Subsequently, solvent design is performed in two steps, namely molecular property targeting and molecular mapping. In the first step, the optimal molecular parameters are obtained by model-based optimization. Thereby, hypothetical target molecules featuring the desirable molecular properties are generated. In the subsequent step, real solvents that approximate the optimal property profiles of the hypothetical molecules are selected from a large solvent database. Finally, the performance of these solvent candidates is evaluated via detailed process simulations. The proposed molecular-property-targeting solvent design methodology is illustrated using an industrially relevant case, i.e., the separation of the close-boiling mixture 1-butene/1,3-butadiene (C_4H_8/C_4H_6).

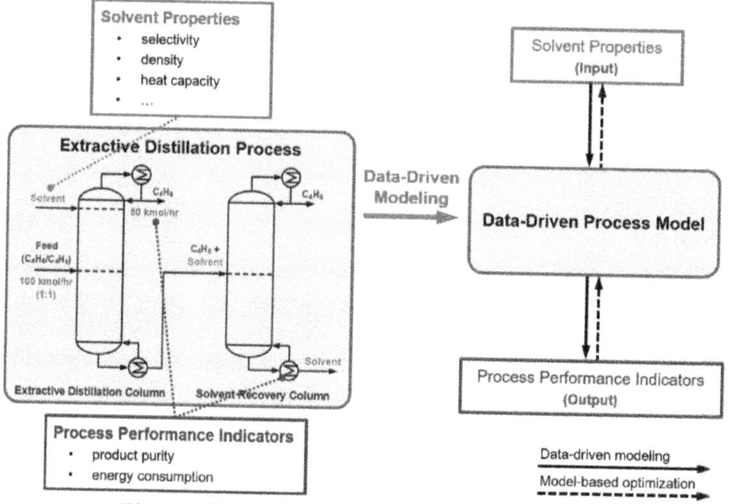

Figure 1. Data-driven modeling for the ED process.

2. Data-Driven Process Modeling

Process models are required to guide the design of solvents that meet the separation requirements. In a data-driven manner, process models are built where solvents described by molecular properties are directly linked to their corresponding process performance (Figure 1). Such models can direct the design of solvents by identifying the optimal values of molecular properties to maximize the process performance. We perform the solvent design under the operating conditions pre-determined by process optimization using n-methyl-2-pyrrolidone (NMP) as the solvent. Thus, the present work can be considered as the search for potentially better alternatives (e.g., showing lower energy consumption in extractive distillation) to replace the industrially used solvent NMP.

The dataset for data-driven modeling contains input-output pairs for 126 solvents. The inputs are five process-relevant physical properties of the solvent, including selectivity of 1-butene/1,3-butadiene at the infinite dilution condition, molecular weight, density, molar heat capacity, and viscosity. The outputs are the key performance indicators of the ED column, including product purity (C_4H_8 purity in the distillate) and energy consumption (described by the reboiler heat duty for simplicity). The dataset is generated through rigorous process simulation in Aspen Plus based on the UNIFAC thermodynamic model. It is randomly split into two sets, i.e., a training set (80%) for model development and a test set (20%) for model evaluation. Feature scaling is adopted to standardize the model input.

Feedforward neural network (FNN), the most straightforward type of artificial neural networks, is used to build the data-driven process model using PyTorch (Paszke et al., 2019). To constrain model complexity and reduce overfitting, the FNN has up to two hidden layers with a maximum of four neurons in each layer during the hyperparameter optimization. Different types of non-linear activation functions are considered.

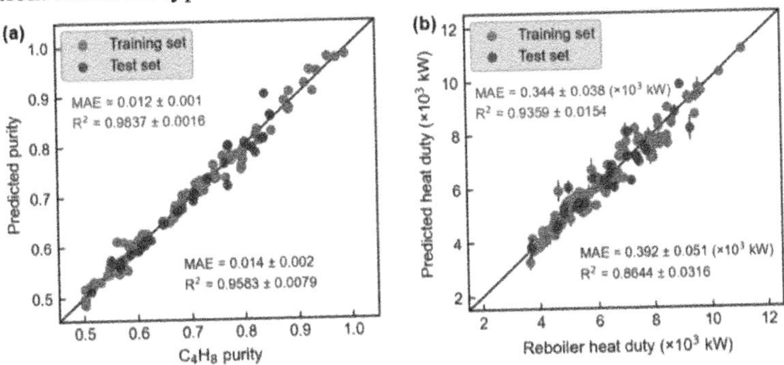

Figure 2. Performance of the data-driven models in predicting (a) C_4H_8 purity and (b) reboiler heat duty. The corresponding process performance for each solvent is predicted using five models derived from the five-fold cross-validation, and the point and error bar show the average and standard deviation of these predictions, respectively.

After identifying the optimal FNN configurations (hyperparameters) using five-fold cross-validation, the models are evaluated on the training and test sets. As they present satisfactory accuracy in the prediction of C_4H_8 purity and heat duty (Figure 2), the data-driven models are subsequently used for solvent design.

3. Molecular Property Targeting

Based on the data-driven models, a multi-objective optimization problem aiming at maximizing the C_4H_8 purity, while minimizing the heat duty is formulated as follows:

$$\min_{x \in X} (1 - f_1(x), f_2(x))$$

$$s.t. \quad X_i^L \leq X_i \leq X_i^U$$

$$X_i \subseteq \mathbb{R}$$

where $f_1(x)$ and $f_2(x)$ are the data-driven models for the estimation of C_4H_8 purity and reboiler heat duty, respectively. X is the molecular property space, in which each variable is standardized individually. L and U denotes the lower and upper bounds. Box constraints are set on each variable based on the corresponding minimum and maximum values from the dataset.

The multi-objective optimization problem is solved using a genetic algorithm implemented in Pymoo (Blank & Deb, 2020). It converges in 100 generations, obtaining a set of Pareto-optimal solutions (i.e., hypothetical target solvent molecules). The hypothetical molecules are considered to be the optimal solutions found in the design space closest to the ideal point. Figure 3 depicts the objective function values for the hypothetical target molecules featuring optimal molecular properties.

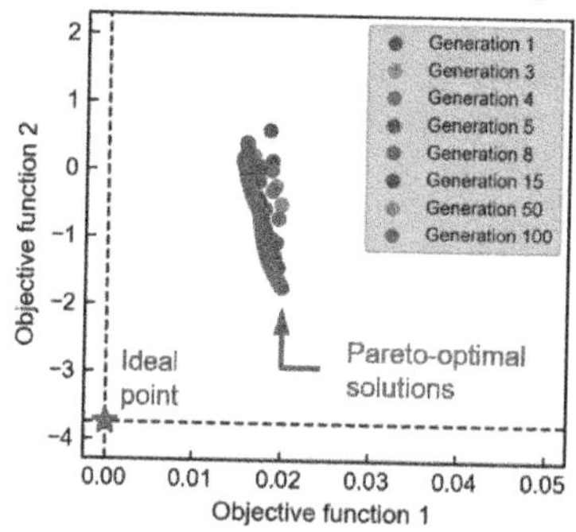

Figure 3. Multi-objective optimization visualized in the space of objective functions. Standardization is applied to the heat duty (objective function 2) in the optimization so that the magnitudes of two objective functions are comparable.

4. Molecular Mapping

The industrially used solvent NMP is set as the benchmark for comparison. Our target is to find better solvents that allow for a higher C_4H_8 purity and a lower heat duty under the given operating conditions. Figure 4a shows the estimated process performance for hypothetical molecules (in gray) and benchmark solvent NMP (in green). The hypothetical molecules are closer to the ideal point than NMP, demonstrating the success of the molecular property targeting step.

In the molecular mapping step, the hypothetical target molecules obtained by molecular property targeting are mapped into real solvents. The molecular mapping is performed by searching a large database consisting of 1,259 real solvents. A preliminary criterion to find the real solvents closest to the hypothetical target molecules is based on the Euclidean distance in the molecular property space. Thereby, optimal real solvents that approximate the optimal property values are identified.

Nineteen solvent candidates are obtained from the molecular mapping. Their estimated process performance is also presented in Figure 4a. It is observed that two solvent candidates are closer to the ideal point than the hypothetical molecules. Therefore, they could in principle show better process performance than the hypothetical molecules.

A detailed simulation of the ED process on the 19 solvent candidates proves that 16 of them are technically viable to achieve the separation of C_4H_8/C_4H_6. Among them, nine solvents present both a lower heat duty and a lower C_4H_8 purity compared to NMP under the same operating conditions. Besides, three solvents, methyl cyanoacetate, glutaronitrile, and 1,4-dicyano-2-butene (blue dots in the green area of Figure 4b), are better alternatives to the benchmark solvent NMP, because they allow for a higher product purity and lower energy consumption under the specified operating conditions. Moreover, their process performance can be further improved by performing rigorous process optimization.

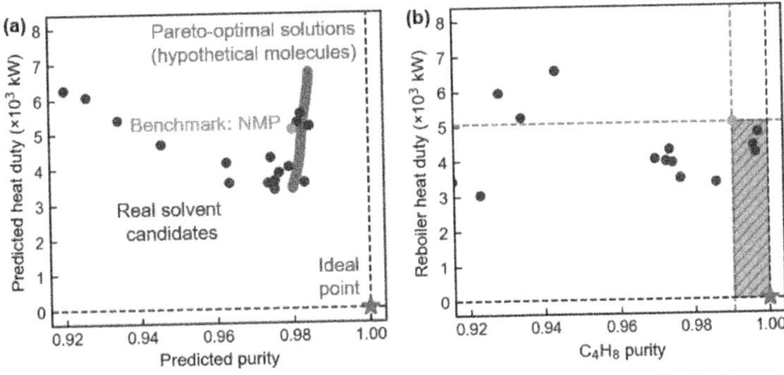

Figure 4. (a) Process performance estimated by the data-driven models for hypothetical molecules, NMP, and real solvent candidates identified by molecular mapping. (b) Process performance evaluated via rigorous simulation of the ED process. Vertical and horizontal lines in green represent the C_4H_8 purity and heat duty of the benchmark process.

For the three solvent candidates, their molecular properties and corresponding process performance are summarized in Table 1. These solvent candidates have similar molecular properties (except viscosity) to the benchmark solvent NMP. A correlation analysis performed on the dataset indicates that a solvent with a higher selectivity at infinite dilution could lead to a higher C_4H_8 purity. This can also be inferred from Table 1. All the three solvents have higher selectivity and higher C_4H_8 purity than NMP. Therefore, the infinite dilution selectivity of the solvent may have played a vital role in designing solvents for a better separation of C_4H_8 and C_4H_6.

Table 1. Molecular properties and the corresponding process performance of the solvents.

Solvent	NMP	Methyl cyanoacetate	Glutaronitrile	1,4-Dicyano-2-butene
CAS number	872-50-4	105-34-0	544-13-8	18715-38-3
Molecular formula	C_5H_9NO	$C_4H_5NO_2$	$C_5H_6N_2$	$C_6H_6N_2$
Selectivity at infinite dilution	1.642	1.753	1.666	1.689
Molecular weight (g/mol)	99.13	99.09	94.12	106.13
Density (kg/m³)	1027	1117	983	1002
Molar heat capacity (J/mol·K)	161.7	192.5	191.0	200.9
Viscosity (cP)	1.89	2.82	6.17	7.13
C_4H_8 purity	0.9904	0.9965	0.9960	0.9972
Reboiler heat duty (kW)	5059	4198	4394	4761

5. Conclusion

This work presents a molecular property targeting approach for the design of solvents in ED processes. Based on data-driven process models, optimal molecular properties are obtained, thereby generating a set of hypothetical target molecules. Nineteen real solvent candidates are identified from a large solvent database by molecular mapping. Their performance on the separation of 1-butene and 1,3-butadiene is evaluated via detailed process simulations. It turns out that three solvents allow for higher product purity and lower energy consumption under the given operating conditions, compared to the benchmark solvent NMP. Further process optimization is necessary to find the optimal ED process for these solvent candidates respectively. Future work will also account for the integrated design of the solvent and the ED process incorporating the molecular property targeting approach, where environmental, health, and safety impacts of solvents can be considered.

Acknowledgement

This research work is supported by the International Max Planck Research School for Advanced Methods in Process and Systems Engineering (IMPRS ProEng), Magdeburg, Germany.

References

A. Bardow, K. Steur, J. Gross, 2010, Continuous-molecular targeting for integrated solvent and process design, Industrial & Engineering Chemistry Research, 49, 6, 2834-2840.

J. Blank, K. Deb, 2020, Pymoo: Multi-objective optimization in python, IEEE Access, 8, 89497-89509.

J. Burger, V. Papaioannou, S. Gopinath, G. Jackson, A. Galindo, C.S. Adjiman, 2015, A hierarchical method to integrated solvent and process design of physical CO_2 absorption using the SAFT-γ Mie approach, AIChE Journal, 61, 10, 3249-3269.

N.G. Chemmangattuvalappil, 2020, Development of solvent design methodologies using computer-aided molecular design tools, Current Opinion in Chemical Engineering, 27, 51-59.

R. Gani, 2004, Computer-aided methods and tools for chemical product design, Chemical Engineering Research and Design, 82, 11, 1494-1504.

V. Gerbaud, I. Rodriguez-Donis, L. Hegely, P. Lang, F. Denes, X. You, 2019, Review of extractive distillation. Process design, operation, optimization and control, Chemical Engineering Research and Design, 141, 229-271.

J. Marrero, R. Gani, 2001, Group-contribution based estimation of pure component properties, Fluid Phase Equilibria, 183, 183-208.

A. Paszke, S. Gross, F. Massa, A. Lerer, J. Bradbury, G. Chanan, T. Killeen, Z. Lin, N. Gimelshein, L. Antiga, 2019, Pytorch: An imperative style, high-performance deep learning library, Advances in Neural Information Processing Systems, 32.

J. Schilling, M. Lampe, J. Gross, A. Bardow, 2017, 1-stage CoMT-CAMD: An approach for integrated design of ORC process and working fluid using PC-SAFT, Chemical Engineering Science, 159, 217-230.

Z. Song, C. Zhang, Z. Qi, T. Zhou, K. Sundmacher, 2018, Computer-aided design of ionic liquids as solvents for extractive desulfurization, AIChE Journal, 64, 3, 1013-1025.

M. Stavrou, M. Lampe, A. Bardow, J. Gross, 2014, Continuous molecular targeting-computer-aided molecular design (CoMT-CAMD) for simultaneous process and solvent design for CO_2 capture, Industrial & Engineering Chemistry Research, 53, 46, 18029-18041.

H. Struebing, Z. Ganase, P.G. Karamertzanis, E. Siougkrou, P. Haycock, P.M. Piccione, A. Armstrong, A. Galindo, C.S. Adjiman, 2013, Computer-aided molecular design of solvents for accelerated reaction kinetics, Nature Chemistry, 5, 11, 952-957.

J.Y. Ten, Z.H. Liew, X.Y. Oh, M.H. Hassim, N. Chemmangattuvalappil, 2021, Computer-aided molecular design of optimal sustainable solvent for liquid-liquid extraction, Process Integration and Optimization for Sustainability, 5, 2, 269-284.

X. Zhang, X. Ding, Z. Song, T. Zhou, K. Sundmacher, 2021, Integrated ionic liquid and rate-based absorption process design for gas separation: Global optimization using hybrid models, AIChE Journal, 67, 10, e17340.

T. Zhou, K. McBride, X. Zhang, Z. Qi, K. Sundmacher, 2015, Integrated solvent and process design exemplified for a Diels-Alder reaction, AIChE Journal, 61, 1, 147-158.

T. Zhou, H. Shi, X. Ding, Y. Zhou, 2021, Thermodynamic modeling and rational design of ionic liquids for pre-combustion carbon capture, Chemical Engineering Science, 229, 116076.

T. Zhou, J. Wang, K. McBride, K. Sundmacher, 2016, Optimal design of solvents for extractive reaction processes, AIChE Journal, 62, 9, 3238-3249.

Antonis Kokossis, Michael C. Georgiadis, Efstratios N. Pistikopoulos (Eds.)
PROCEEDINGS OF THE 33rd European Symposium on Computer Aided Process Engineering (ESCAPE33), June 18-21, 2023, Athens, Greece
 http://dx.doi.org/10.1016/B978-0-443-15274-0.50200-6

A Bayesian-based screening framework for optimal development of safe-by-design nanomaterials

Kostas Blekos, Effie Marcoulaki

System Reliability and Industrial Safety Laboratory, National Centre for Scientific Research "Demokritos", P.O. Box 60037, Agia Paraskevi 15310, Greece
emarcoulaki@ipta.demokritos.gr

Abstract

In this article we propose a novel screening approach for the development of new nanomaterials (NM) which are safe(r)-by-design. An integral part of our approach involves the use of recently proposed notation standards and extensions, which facilitate the representation and processing of data related to NMs. In particular, we utilize a general extendable representation for NMs which is built in line with InChI extensions for NMs. This representation enables us to link the composition, structure, and experimental conditions of a NM to its toxicity profile, using descriptor information obtained from available datasets. The processed data are then complemented with predictions from QSAR models and fed to a Bayesian-based procedure to assess the safety of new NM designs. This comprehensive approach can be used to build a local screening tool to search for less hazardous NM configurations.

Keywords: nanomaterial, toxicity, safe-by-design, Bayesian, nanoinformatics.

1. Introduction

During the last few decades, nanotechnology and nanomaterial (NM) applications have emerged as key enabling technologies for the global industry, since the unique properties of NMs and nano-enabled products allow for a very wide range of applications with significant potential. The risks of engineered nanomaterials (ENM) are largely unknown, however their size and composition make them inherently threatening to the human health and the environment. In effect, WHO (2017) published precautionary guidelines for their handling and use, and OECD developed appropriate Test Guidelines to identify, assess and manage potential emerging risks related to NMs throughout the development chain (OECD, 2020) towards a Safe(r) Innovation Approach (Shandilya et al., 2020).

Early identification, assessment and management of the potential health, safety and environmental risks related to ENMs throughout their life-cycle is extremely important. Risk assessment of ENMs is challenging because of the diversity in their physical and chemical composition, their surface functionalization and structural properties, and the variety of alternative configurations. Significant public and private investment over the past decades enabled the development of advanced experimental methods for generation of extensive nanotoxicity data, and supported the development of informatics tools for data management / processing and for the prediction of physicochemical, exposure and (eco)toxicological properties of novel NMs to enable in silico risk assessment (Marcoulaki et al., 2021). Among these, the NanoInformaTIX H2020 project develops a database comprising the significant amounts of available data, and a set of tools for modeling materials, exposure and dose-response. These tools will become available on a user-friendly web-based platform for ENM risk management, to support sustainable manufacturing of ENM-based products.

The development of optimization tools for designing ENMs with defined properties is a very new research field. Preliminary works applied Generative Adversarial Networks (GAN) to optimize functionality (So and Rho, 2019; Kim et al., 2020). This work proposes an incremental procedure to develop a screening framework for safe-by-design ENMs. We start from novel representation schemes, build extensions to improve the data analysis potential, propose an assessment procedure using similarity functions built on top of the representation. All these finally enable the generation and assessment of an ensemble of alternative ENM designs, starting from a given ENM design.

2. Representation methodology

In a recent work we proposed a general, extendable representation for NMs from the point of view of computational applications and we set qualifications for a computational-friendly representation (Blekos et al., 2022). Our proposal was in line with the recently proposed InChI for NMs (Lynch et al., 2020) and its realization greatly facilitates similarity assessments between components which, in turn, is a crucial part of calculating NM affinities. Our work identified key challenges for a computational-friendly NM representation and proposed strategies to address them. Through the resulting representation qualifications, we created a proposal of 5 extensions for the InChI for NMs that aimed to address the identified challenges.

These extensions enable the computationally-friendly representation of important and commonly found features of NMs. They are designed to facilitate standardization, curation, and comparison of data on NMs. This can help improve the sharing and management of data on NMs, and can assist researchers in identifying potential safety concerns and making informed decisions about the use of these materials. The extensions are: a) an extension to describe distributions, which can be applied anywhere a real number needs to be represented; b) an extension to describe shapes; c) an extension for the representation of mixtures and interfaces using logical operators that apply to other layers already defined, building on the mixture-InChI; d) an extension for structured comments that allows for consistent and uniform capturing of auxiliary information and facilitates database curation; and e) an extension for the definition of macros and shortcuts that allows for the creation of reusable, parametrizable, and human-readable components.

3. Data preprocessing

The representation and extensions described in Section 2 provide a consistent and comprehensive framework for describing and analyzing the properties of NMs. In particular, the proposed extensions for distributions and shape/morphology, provide new capabilities for data-driven tools, and enables safety assessment, which is crucial for the design of materials with improved safety profiles.

3.1. Data preprocessing

Starting from a given dataset, the proposed screening framework uses a bottom-up and a top-down stage to generate two outputs that will be fed to the next stages. The two data preprocessing stages are to be conducted simultaneously and they are complementary, as shown in Figure 1 and described below.

In the *top-down stage*, we access available NM databases to obtain descriptors that provide information about the composition, structure, and other characteristics of NMs. Descriptor information is taken from available datasets, linking composition, structure and experimental conditions information to the NM toxicity profile. These descriptors can be further analyzed and encoded through QSAR and/or other similar models, to explore the properties of NMs and identify safety and toxicity profiles.

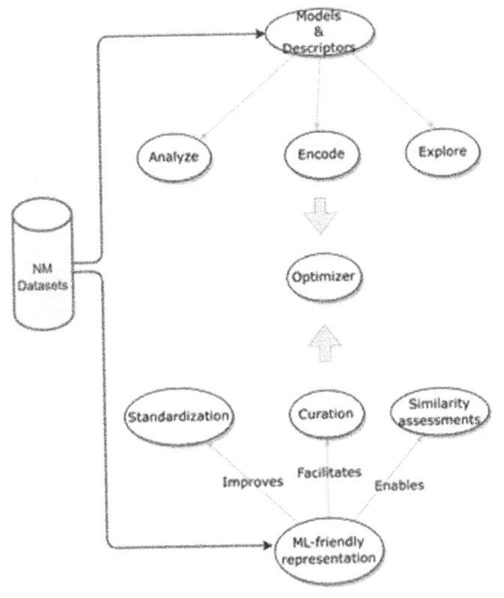

Figure 1: Two stage data preprocessing

In the *bottom-up stage*, we convert the NM data into the computational-friendly representation outlined in Section 2. The representation and the proposed extensions provide a comprehensive and consistent framework for describing and analyzing the properties of NMs, including their chemical composition, morphology, and interactions with biological systems. This allows the more detailed and accurate evaluation of the comparative properties (e.g. toxicity) of NMs.

Additionally, the proposed extensions to the representation, such as the ability to describe probability distribution functions (pdfs) and bypass categorical shape descriptors, provide new capabilities for analyzing and improving the safety of NMs. For example, the extension for describing distributions can be used to better characterize the distribution of particles within a NM, which can have important implications for its safety and toxicity. Similarly, the extension for bypassing categorical shape descriptors allows for more flexible and comprehensive analysis of the toxicity of the NMs, which can be useful for identifying novel materials with improved safety profiles.

3.2. Similarity and toxicity matrices

In the next stage, the outputs of data preprocessing are used to create a matrix of toxicity values, with each toxicity model providing a separate value for each NM (Figure 2). We can additionally conduct similarity assessments based on the proposed representation model. This provides information on the similarity between different NMs, based e.g. on their chemical composition, structure, zeta potential and other characteristics etc. (Figure 3). The obtained information is used to create a matrix of similarity values for each NM and each axis of comparison (Figure 4).

4. Bayesian-based screening

The toxicity and similarity matrices of Section 3 are utilized as input data to create and train a Bayesian network (BN). The BN is a probabilistic graphical model that represents the conditional dependencies between different variables in a system, such as the likelihood of a given NM being toxic given its similarity to other NMs. Once the BN is obtained, it can be used to perform probabilistic inference and assess the toxicity of NMs options which are not included in the dataset.

As shown in Figure 4, the BN constitutes the core function of a local search procedure towards less toxic NMs. Note that, in principle we can consider functionality as an additional objective. Otherwise, we assume that small changes to the design descriptors are unlikely to significantly affect the functionality of the NM.

The search starts from an existing NM with known safety characteristics and the desired functionality for a particular application. A local search is performed in the vicinity of similar designs, to identify design options featuring lower toxicity. The search uses the similarity functions and the Bayesian network, to alter the NM descriptors and evaluate the new toxicity, respectively. Iterative application of this procedure on the new NMs, progressively generates design alternatives with the desired safety profile.

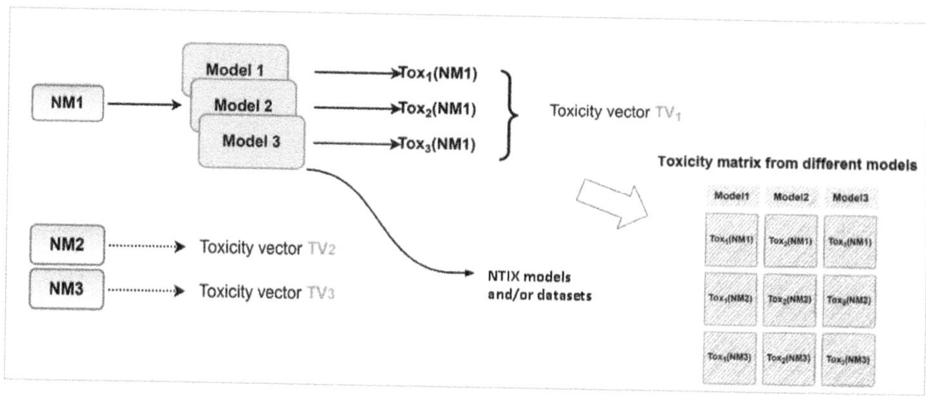

Figure 2: Toxicity matrix using externally available QSAR models

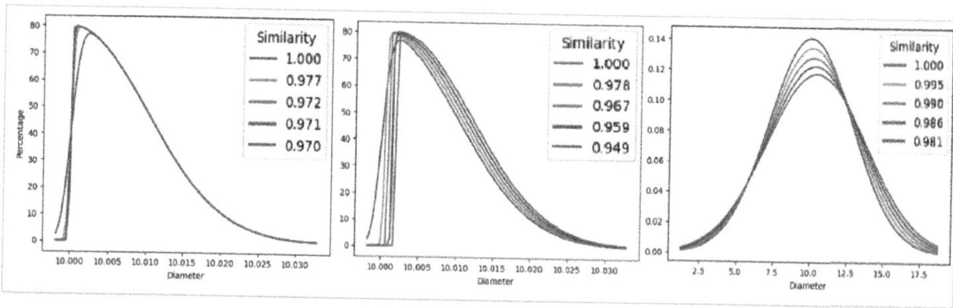

Figure 3: Results of similarity calculations for different pdfs of the NM major diameter

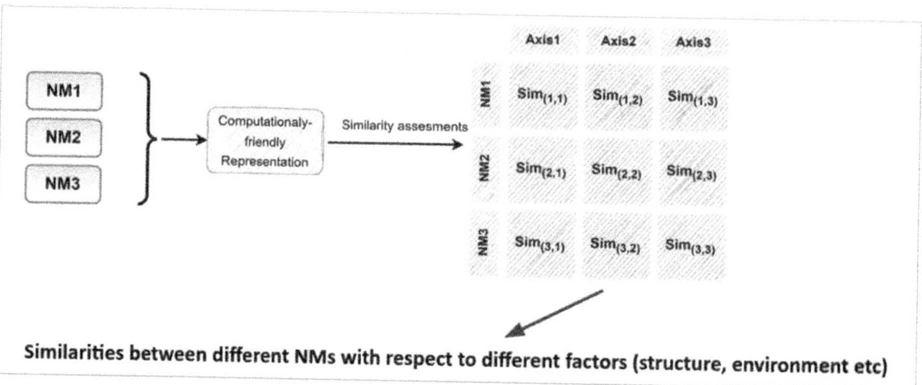

Figure 4: Similarity matrix obtained from the computational-friendly representation

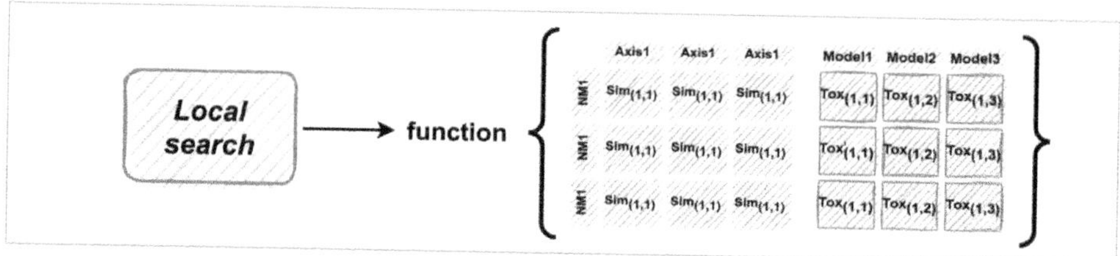

Figure 5: Bayesian-based screening using the Toxicity and Similarity matrices as inputs.

The success of the proposed framework relies heavily on the availability, the accuracy and the range of descriptor information to generate reliable inputs for the BN. The data can come from relevant datasets and/or from QSAR predictions as described in section 2.

QSAR models allow for cost-effective and efficient assessment of the potential risks of NMs without the need for extensive in vivo experiments (Huang et al., 2021). Latest trends use machine learning (ML) algorithms to augment or develop QSARs (Furxhi et al., 2020; Balraadjsing et al., 2022; Singh et al., 2020). There are however many challenges in the development and use of these models for nano-toxicity predictions. A major challenge is the limited number of experimental data, making it difficult to train models that require large and diverse datasets to achieve acceptable prediction accuracy. There is also need to connect and harmonize different data sets, for example by using read across and other methods of filling data gaps. Finally, the complexity and diversity of NMs hinder the identification of the most relevant structural and/or physicochemical properties that affect their toxicity. The above can lead to the development of ML models that have low predictive power and/or a limited domain of applicability. On a different perspective, all the predictive toxicology models should be validated in a manner that is acceptable to regulatory agencies. This requires the development of transparent and reproducible validation procedures, to ensure that these models are suited to support regulatory decision-making in the field of nanotoxicity (Winkler, 2020).

To avoid these concerns, we are currently implementing a simplified version of the proposed framework, relying only on an extended and well curated dataset. This version will soon be integrated on the NanoInformaTIX project platform. Advanced QSARs are also under development in NanoInformaTIX and will be available on the platform, along with data sources and other models to enable risk assessment and property predictions. These will enable to fully implement the proposed SbD framework in the future.

5. Discussion and conclusions

The physical, chemical and biological behavior of the materials at nanoscale is very different from their conventional behavior at macroscale, due to a major shift in the dominant phenomena. This creates enormous opportunities towards novel advanced solutions in a wide range of application areas. It also raises concerns on possible risks on the human health and safety and the environment, during the entire life cycle of ENMs and nano-enabled products.

The above concerns need to be taken into account ex-ante, during the early stages of material development, under the safe-by-design (SbD) principles. This work proposes a framework for SbD, using a **general and extendable representation for NMs** suitable for computational applications, and a **data-driven assessment method using Bayesian networks**. These are herein the **essential components for the systematic and efficient screening for new NM designs with improved safety profiles**.

The representation is based on the InChI for NMs, and facilitates the comparison of NM components, which is crucial for calculating NM affinities. A set of extensions to NM InChI aim to address key challenges in computational-friendly NM representation, including the ability to describe statistical distributions and bypass categorical shape descriptors typically used to describe NMs (Blekos et al., 2022). Note that, the authors are part of the NM InChI Working Group and many aspects of these extensions will be included in the proposal for a new InChI standard (InChI Trust, 2022).

The screening framework proposed here depends on the availability of many and reliable data and appropriate QSAR models to fil gaps in nanotoxicity predictions. The platform developed under the NanoInformaTIX H2020 project will contain a variety of suitable models integrated into workflows for human and environmental risk assessment, providing the perfect playground to evaluate the proposed approach. A simplified version of the proposed framework, without dependency on predictions from external models, will soon be available at the NanoInformaTIX platform.

The proposed framework has the potential to be used in other or extended applications, including the design of new NMs with improved sustainability and functionality profiles. By taking into account these additional factors, we can better formulate the NM design problem and consider a variety of NMs, their production processes and applications.

Funding

This work is funded via the EU H2020 project "NanoInformaTIX: Development and Implementation of a Sustainable Modelling Platform for NanoInformatics" (grant agreement No. 814426).

References

Balraadjsing, S., Peijnenburg, W.J.G.M., Vijver, M.G., 2022. Exploring the potential of in silico machine learning tools for the prediction of acute Daphnia magna nanotoxicity. *Chemosphere* 307, 135930.

Blekos, K., Chairetakis, K., Lynch, I., Marcoulaki, E., 2022. Principles for nanomaterial representations to facilitate machine processing and cooperation with nanoinformatics tools, *Journal of Cheminformatics*, accepted

Furxhi, I., Murphy, F., Mullins, M., Arvanitis, A., Poland, C.A., 2020. Practices and Trends of Machine Learning Application in Nanotoxicology. *Nanomaterials* 10, 116.

Huang, H.-J., Lee, Y.-H., Hsu, Y.-H., Liao, C.-T., Lin, Y.-F., Chiu, H.-W., 2021. Current Strategies in Assessment of Nanotoxicity: Alternatives to In Vivo Animal Testing. *International Journal of Molecular Sciences* 22, 4216.

InChI Trust, 2022. Nanomaterials Working Group. https://www.inchi-trust.org/nanomaterials/

Kim, I., Viswanathan, K., Kasi, G., Sadeghi, K., Thanakkasaranee, S., Seo, J., 2020. Preparation and characterization of positively surface charged zinc oxide nanoparticles against bacterial pathogens. *Microb Pathog* 149, 104290.

Lynch, I., Afantitis, A., Exner, T., Himly, M., Lobaskin, V., Doganis, P., et al., 2020. Can an InChI for nano address the need for a simplified representation of complex nanomaterials across experimental and nanoinformatics studies?. *Nanomaterials*, 10(12), p.2493.

Marcoulaki, E., de Ipiña, J.M.L., Vercauteren, S., Bouillard, J., Himly, M., Lynch, I., e. al., 2021. Blueprint for a self-sustained European Centre for service provision in safe and sustainable innovation for nanotechnology. *NanoImpact*, 23, p.100337.

NanoInformaTIX: Development and Implementation of a Sustainable Modelling Platform for NanoInformatics (H2020-NMBP-TO-IND https://cordis.europa.eu/project/id/814426

OECD, 2020. Moving Towards a Safe(r) Innovation Approach (SIA) for more sustainable nanomaterials and nano-enabled products, ENV/JM/MONO(2020)36/REV1.

Shandilya, N., Marcoulaki, E., Barruetabena, L., Llopis, I.R., Noorlander, C., Jiménez, A.S., et al., 2020. Perspective on a risk-based roadmap towards the implementation of the safe innovation approach for industry. *NanoImpact*, 20, p.100258.

Singh, A.V., Ansari, M.H.D., Rosenkranz, D., Maharjan, R.S., Kriegel, F.L., Gandhi, K., Kanase, A., Singh, R., Laux, P., Luch, A., 2020. Artificial Intelligence and Machine Learning in Computational Nanotoxicology: Unlocking and Empowering Nanomedicine. *Advanced Healthcare Materials* 9, 1901862.

So, S., Rho, J., 2019. Designing nanophotonic structures using conditional deep convolutional generative adversarial networks. *Nanophotonics* 8, 1255–1261.

WHO guidelines on protecting workers from potential risks of manufactured nanomaterials, 2017

Winkler, D.A., 2020. Role of Artificial Intelligence and Machine Learning in Nanosafety. *Small* 16, 2001883.

Antonis Kokossis, Michael C. Georgiadis, Efstratios N. Pistikopoulos (Eds.)
PROCEEDINGS OF THE 33rd European Symposium on Computer Aided Process Engineering (ESCAPE33), June 18-21, 2023, Athens, Greece
 http://dx.doi.org/10.1016/B978-0-443-15274-0.50201-8

Artificial neural network modelling for the prediction of the deactivation of CaO-based adsorbents in the calcium looping process for CO_2 capture

Rui M. Filipe,[a,b] Sérgio González-De-La-Cruz,[c] Adrian Bonilla-Petriciolet,[c] Carla C. Pinheiro[d,e]

[a]*Instituto Superior de Engenharia de Lisboa, Instituto Politécnico de Lisboa, R. Conselheiro Emídio Navarro 1, 1959-007 Lisboa, Portugal*

[b]*Centro de Recursos Naturais e Ambiente, Instituto Superior Técnico, Universidade de Lisboa, Av. Rovisco Pais 1, 1049-001 Lisboa, Portugal*

[c]*Instituto Tecnológico de Aguascalientes, Aguascalientes 20256, Mexico*

[d]*Instituto Superior Técnico, Universidade de Lisboa, Av. Rovisco Pais 1, 1049-001 Lisboa, Portugal*

[e]*Centro de Química Estrutural, Institute of Molecular Sciences, Instituto Superior Técnico, Universidade de Lisboa, Av. Rovisco Pais 1, 1049-001 Lisboa, Portugal*

Abstract

Calcium-looping (CaL) is a promising cyclic process for CO_2 capture based on the reversible chemical reaction between CaO-based adsorbents and CO_2 to form $CaCO_3$. Currently, the main CaL challenge is the decay of the adsorbents carrying capacity with increasing number of cycles of calcination-carbonation due to the particles sintering and pores blockage.

In this work, different types of CaO-based adsorbents are investigated in a cyclic laboratory scale CaL unit for CO_2 capture using a real flue gas from a cement plant and a synthetic gas mixture, both with same CO_2 content.

An artificial neural network (ANN) model is developed to predict the adsorbents' deactivation along the cycles, based on their composition and on the type of gas used. A good agreement between the experimental and predicted values was obtained, demonstrating that ANN can be successfully used to predict the adsorbents' deactivation based on the materials composition and type of gas.

Keywords: Calcium looping, CaO-based adsorbents, CaO deactivation, neural networks.

1. Introduction

The calcium looping (CaL) process is a promising CO_2 capture technology for industrial plants, based on the reversible chemical reaction between CaO and CO_2 to form $CaCO_3$, which has received great attention in recent years. CaL process is based on the cyclic calcination-carbonation of CaO-based adsorbents, usually natural limestones ($CaCO_3$), to capture CO_2 from a flue gas. The adsorbents can be reused for several cycles, but their CO_2 carrying capacity decreases along the cycles due to the particles sintering and pores blockage (Abreu et al. 2020; Teixeira et al., 2020). CaL is currently a potential sustainable option for emissions' reduction in the cement industry because the purged deactivated CaO-based materials can be conveniently reused and integrated in the clinker production.

Despite the synergy between the CaL process and the cement production plants, the decay of the adsorbents' CO_2 carrying capacity is a problem that needs to be tackled for improving the operation and efficiency of the CaL process. A fresh adsorbent make-up stream may be used to maintain the adsorbent activity and the desired CO_2 capture level. The degree of deactivation of each type of adsorbent material depends, among other factors, on its chemical composition as well as on the flue gas composition, and it may be assessed in the laboratory through a sequence of several calcination-carbonation cycles. A tool to predict the deactivation along the cycles is thus urgent and valuable for the CaL process at industrial scale, as it would avoid the laboratory testing burden and it would also allow the estimation of the flowrate of a purge stream of the deactivated CaO-based adsorbent, as well as a make-up stream of the fresh adsorbent.

CaO-based adsorbents are usually non expensive materials such as natural geological or waste-based materials precursors. In this work, the use of different samples of both types of CaO-based materials in a cyclic laboratory scale CaL unit for CO_2 capture (Marques et al., 2022) is investigated, using a real flue gas from a cement plant and a synthetic gas mixture, both with the same CO_2 content.

An artificial neural network (ANN) is developed to obtain a model that predicts the adsorbents' deactivation along the cycles, based on their composition and on the type of gas used.

2. Methodology

2.1. Adsorbent deactivation data

The data reported by Marques et al. (2022) for six different samples of Ca-adsorbents comprising four natural limestones (LA, LB, LC, and LD) and two waste marble powders (WMA, and WMB), was used in this work. The oxide content of the adsorbents is shown in Table 1. The samples were tested in a fluidized bed reactor using two different gases for the carbonation step, a real flue gas (FG) from a cement plant chimney (pressurized to a bottle) and a synthetic gas (SG) mixture of nitrogen and air, that replicates a real carbonation atmosphere, both with 15 % of CO_2. The calcination was always performed with a SG mixture (70 % CO_2 balanced in air) to replicate realistic calcination conditions. The chemical composition of the gases and a simplified diagram of the experimental set-up may be found in Marques et al. (2022).

The deactivation of the adsorbent was evaluated using the amount of CO_2 captured by unit mass of adsorbent as indicator, the carrying capacity (CC), over ten cycles of calcination and carbonation. The characterization of the gases used, and the experimental procedure details are given by Marques et al. (2022) and will not be reproduced here.

Table 1. Adsorbents chemical composition (wt. %, adapted from Marques et al. (2022))

	SiO_2	Al_2O_3	Fe_2O_3	CaO	MgO	K_2O
LA	9.00	3.69	1.72	45.52	0.99	0.72
LB	14.78	3.94	1.80	40.57	2.02	0.92
LC	0.49	0.26	0.11	54.90	0.46	0.06
LD	2.31	0.85	0.60	45.89	0.78	0.12

WMA	0.60	0.02	0.05	55.33	0.25	0.02
WMB	3.38	0.60	0.28	52.22	0.79	0.30

A set of 120 experimental results is available. After a careful analysis of the data, two points were considered as outliers and removed from the data set, resulting in a total of 118 points available for training, testing and validation of the ANN.

2.2. Model development

This work aims at developing an ANN to correlate the adsorbent characteristics and its working conditions with the capacity of capturing CO_2. Eight inputs were selected for the ANN: adsorbent oxide composition (SiO_2, Al_2O_3, Fe_2O_3, CaO, MgO, and K_2O), number of the cycle, and type of gas used. The output of the ANN is the carrying capacity.

A feedforward backpropagation neural network model using Levenberg-Marquardt backpropagation and a hyperbolic tangent sigmoid activation function were used. The data was divided for training (70 %), testing (15 %), and validation (15 %) using interleaved indices, with the same sets being used to train, test, and validate each network.

As there are no fixed rules for the selection of the ANN size and configuration, a preliminary analysis was performed to identify the simplest architecture giving adequate results. Different ANN architectures were created, by varying the number of neurons and hidden layers. The number of internal layers tested was 1 and 2, with the number of neurons varying between 4 and 20 for layer 1, and between 1 and 3 for layer 2.

The different architectures were tested using MATLAB Deep Learning Toolbox. To increase the probabilities of achieving better performance the model was trained 500 times for each architecture, and the performance evaluated using mean squared error (MSE) as criteria. The best performing case for each architecture was then used to evaluate the influence of the number of neurons and layers in ANN performance, while also checking for an adequate description of the process. The criteria used to select the final ANN architecture was the minimization of the number of layers and neurons, while having a low value of MSE and providing a proper description of the process.

Statistical parameters were calculated to assess ANN model predictions, and included the coefficient of correlation (R) and the fitting error (e_i (%)) for each data point i, defined as:

$$e_i = \frac{(y_i^p - y_i)}{y_i} 100 \tag{1}$$

where y_i^p is the value predicted by the ANN model and y_i the experimental value.

3. Results and discussion

3.1. ANN architecture

Several tests were made to identify the proper architecture to be used, varying the number of layers, and the number of neurons in each layer. Although different results were obtained depending on the conditions used to train the ANN, as the number of times the network is trained, or the sets used for each task, for example, only a small improvement was obtained when using two layers instead of one. Furthermore, no significative improvement was observed in the performance indicator (MSE) when the number of neurons is increased over a certain number.

As an illustrative example, Figure 1 depicts the variation of the minimum value of MSE obtained for each ANN architecture, with a varying number of neurons and internal layers. The small improvements observed, particularly for more than nine neurons, indicate that a simpler ANN, with only one layer, can still provide adequate results, while requiring a minimum number of parameters to be estimated. In view of this, an architecture with only one layer and nine neurons was selected for this problem, resulting in an ANN with eight inputs, nine neurons and one output (8:9:1).

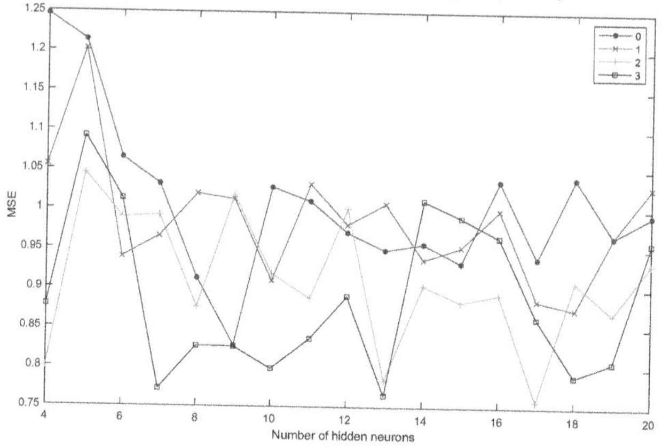

Figure 1. Mean squared error for different ANN architecture and number of hidden neurons in the first layer (the legend indicates the number of neurons in the second layer).

3.2. ANN performance

Figure 2 shows the fitting performance for the three subsets used, individually and for all the data used, for the architecture 8:9:1. An overall good behavior is observed, with R values greater than 0.99 for all the cases. The histogram in Figure 3 shows a normal distribution, with most of the estimated points having a deviation lower than 0.34. The calculated overall deviations are comprised in the interval [-3.2, 2.8] and the error (e_i) in the interval [-18, 14] %.

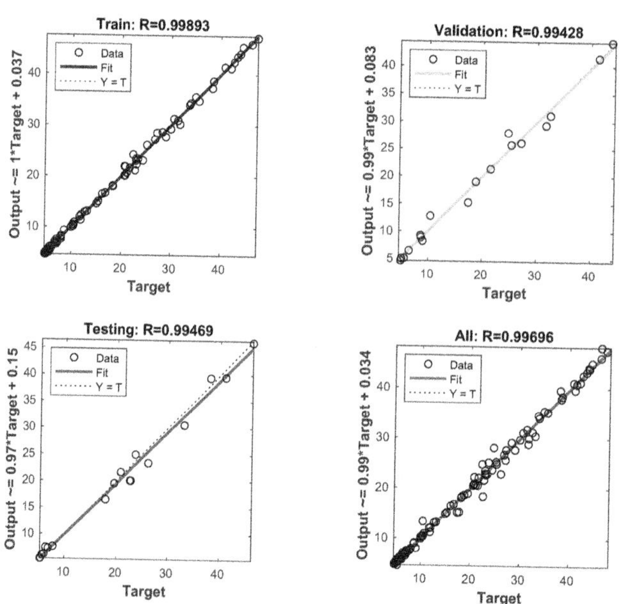

Figure 2. Fitting performance for the sets used.

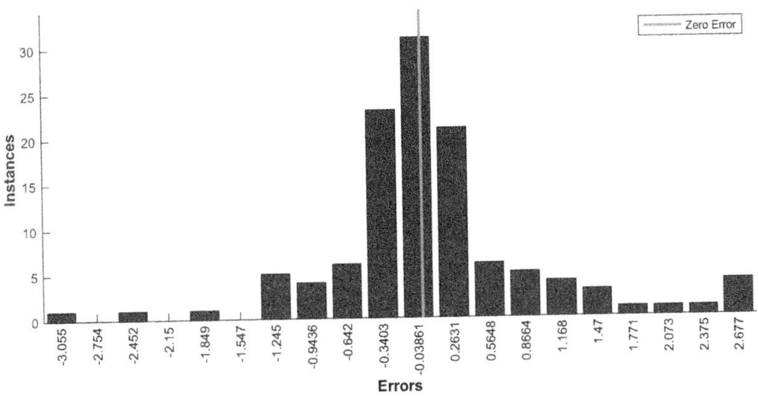

Figure 3. Histogram of error obtained in fitting.

Figure 4 and Figure 5 depict the evolution of the adsorbents carrying capacity along the carbonation-calcination cycles. As expected, a decay over the cycles is observed, which is justified by adsorbent particles sintering and pores blockage. It is also observed that each adsorbent has different initial and final carrying capacities, and deactivation profile depending on the adsorbent composition and type of gas used.

The figures show a good agreement between the target measured values (y) and the predicted values (y^p) for most of the cases, whit no apparent signs of overfitting. Some of the major discrepancies observed may be justified by the inclusion of some experimental data that would be considered as outliers, in a stricter data validation process.

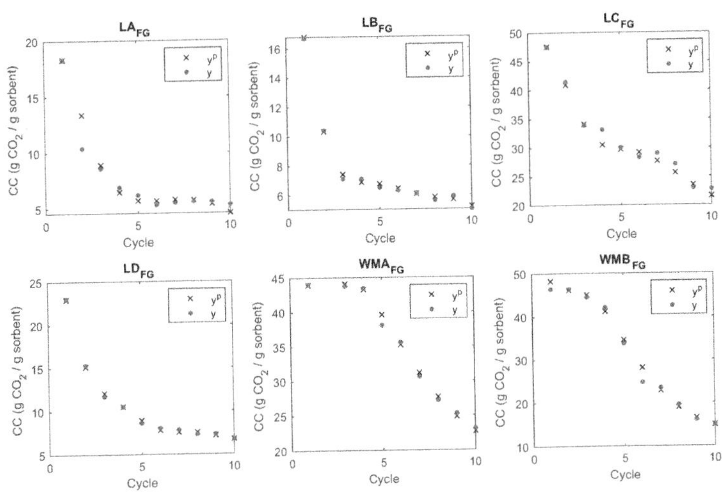

Figure 4. Experimental (y) and predicted (y^p) carrying capacity (CC) for each sorbent as a function of the number of cycles using real flue gas.

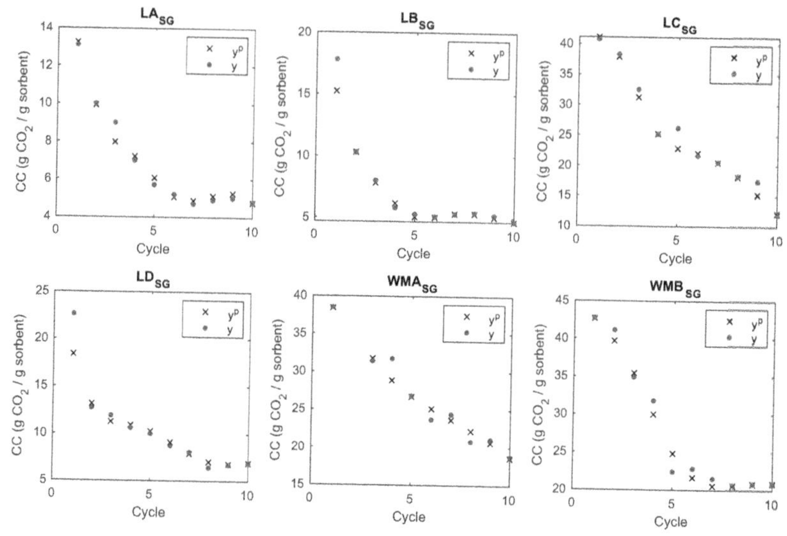

Figure 5. Experimental (y) and predicted (y^p) carrying capacity (CC) for each sorbent as a function of the number of cycles using synthetic gas.

4. Conclusions

An ANN was developed to predict CaO-based adsorbents deactivation along the Ca-looping cycles. A study on the architecture of the ANN was performed, with no significative improvements observed above a minimum of nine neurons, even when more complex configurations were used. A good agreement between the experimental and the predicted values of the adsorbents' CO_2 carrying capacity was obtained for the tested materials, demonstrating that the ANN is sensible to the input variables, and can differentiate among samples composition, type of gas used, and number of cycles applied to each sorbent. The ANN model developed is a powerful tool, that can now be used to assist the selection of new adsorbents by predicting the behavior over several cycles. It may also provide useful information for models used in process design, optimization, and operation. Future work will investigate how extra information supplied to the ANN affects performance.

Acknowledgments

The financial support from FCT under projects UIDB/00100/2020, UIDP/00100/2020, UIDB/04028/2020, UIDP/04028/2020, and PTDC/EAM-PEC/32342/2017, and the European Union's Horizon 2020 research and innovation programme under the Marie Sklodowska-Curie grant agreement No. 778168, is gratefully acknowledged.

References

M. Abreu, P. Teixeira, R. Filipe, L. Domingues, C.I.C. Pinheiro, H. Matos, 2020, Modeling the deactivation of CaO-based sorbents during multiple Ca-looping cycles for CO_2 post-combustion capture, Computers & Chemical Engineering, 134, 106679.

L.M. Marques, S.M. Mota, P. Teixeira, C.I.C. Pinheiro, H.A. Matos, (2022), Ca-looping process using wastes of marble powders and limestones for CO_2 capture from real flue gas in the cement industry, Journal of CO_2 Utilization, (submitted).

P. Teixeira, I. Mohamed, A. Fernandes, J. Silva, F. Ribeiro, C.I.C. Pinheiro, 2020, Enhancement of sintering resistance of CaO-based sorbents using industrial waste resources for Ca-looping in the cement industry, Separation and Purification Technology, 235, 116190.

Antonis Kokossis, Michael C. Georgiadis, Efstratios N. Pistikopoulos (Eds.)
PROCEEDINGS OF THE 33rd European Symposium on Computer Aided Process Engineering (ESCAPE33), June 18-21, 2023, Athens, Greece
 http://dx.doi.org/10.1016/B978-0-443-15274-0.50202-X

Dynamic operation for the effective use of green hydrogen in Power-to-X value chains

Michael Mock,[a] Hannes Lange,[a] Isabell Viedt,[a] Kumar Rajan Gopa,[a] Jonathan Mädler,[a] Leon Urbas[a]

[a] *TU Dresden, Chair of Process Control System and Process Systems Engineering Group, Helmholtzstr. 14, 01069 Dresden, Germany*
michael.mock@tu-dresden.de

Abstract

In the upcoming decades, the integration of renewable energies into our continuous process routes, by producing and processing green hydrogen, is of paramount importance to combat climate change. To include green hydrogen in conventional process routes large hydrogen storages or dynamic process control strategies for hydrogen processing are needed. In this work a dynamic process control strategy for a Power-to-Ammonia (PtA) value chain is studied. A PtA process plant and respective process control strategies to adjust the plant load are conceptualized. To study the feasibility, a Haber-Bosch reactor system is modeled and implemented in MATLAB 2022b. Simulation case studies are carried out to emulate load changes. The simulation shows that the conceptualized process enables an intermittent production of ammonia based on green hydrogen. Further studies with more detailed models and refined process control strategies are planned for a detailed analysis to optimize the efficiency and flexibility of the production.

Keywords: Power-to-X, Power-to-Ammonia, Requirements Analysis, Dynamic Simulation, Renewable Energies

1. Introduction

The integration of renewable energies, like wind energy or photovoltaics, into our existing chemical production lines is necessary to combat climate change and reduce our economic dependency on fossil resources (IEA, 2021). Power-to-X (PtX) technologies offer the possibility to integrate fluctuating energy into our value chains, by producing and processing green hydrogen. To integrate intermittent energy in conventional production lines large electricity storages or large hydrogen storages are necessary to decouple the subsystems. Since both electrical storage and hydrogen storage are expensive and exhibit increasing losses over time, flexible PtX value chains through dynamic process control is a promising approach towards integrating renewable energies in the chemical industry (Moradi, 2019). The aim of this work is to elaborate a plant and control concept for the dynamic control of a PtX value chain. The implementation and simulation of the developed concepts will be used to study whether the PtX value chain can adapt to the load changes of sustainable energy sources without large-scale electricity or large-scale intermediate hydrogen storage. Chapter 2 will shortly describe the methodology used in this work. The developed requirements, concepts and simulation will be specified in chapter 3. The conclusion of the work can be found in chapter 4.

2. Methodology

The plant and process control concept are developed for a selected PtX use case, namely Power-to-Ammonia (PtA), based on the available literature and a preceding requirements analysis. The concept primarily includes conventional applications to study the possibility to use established technologies for PtX value chains. Subsequently, simplified concepts are implemented and simulated in MATLAB 2022b. In this work, the focus is on the simulation of a Haber-Bosch reactor system. Necessary parameters for the reactors are taken from Khademi et al. (2017) to create models for dynamic tubular Haber-Bosch reactors. The resulting PDE system of the reactors models are discretized to derive an ODE system and are solved using the ode23t solver.

Case studies are simulated in which different load change rates are simulated and the dynamic response of the reactor system is considered. The simulation results, such as temperature and concentration profiles and molar flow rates, are validated by those from literature, but due to missing data, validation with real plant data is omitted. The time-dependent temperature and concentration profiles under load variations are evaluated to assess whether dynamic operation is feasible with the input of sustainable energy. The focus here is on whether the plant concept can maintain the permitted operating parameters even under fluctuating input conditions and different load change rates. In later iterations the complete PtA value chain with added electrolyzers, a small-scale intermediate hydrogen tank and auxiliary components shall be modeled and simulated. The determined range of load change rates can subsequently be used in the dynamic PtA value chain.

3. Methodology

Since PtX technologies include a wide range of production routes a specific value chain must be chosen. Ammonia production accounts for about three-quarters of the hydrogen consumption in chemical production lines (IEA, 2021). Therefore, the Power-to-Ammonia (PtA) value chain is chosen as an example to conceptualize and evaluate the flexible operation of PtX technologies through dynamic process control.

3.1. Use-case for main energy source, secondary energy source and plant location

The process plant design and process control strategies have to fit the chosen use-case of the PtA value chain. In the first step a use-case is defined, considering the main energy source, a potential secondary energy source and the location of the plant.

Wind energy is chosen as the main energy source, since it is the most widely used energy source for PtX applications and offers a high number of nominal operational hours with 1400-5000 h/a (Chehade, 2019). A grid connection is chosen as a secondary energy source. This enables the use of a stand-by operation, which is necessary for safe and economical operation with fluctuating energy supply without large electricity or hydrogen storage. The grid connection makes the use of a stand-by operation during times of low wind energy availability possible. The main disadvantage is that this can lead to the use of electricity from fossil resources to maintain the stand-by operation.

Finally, the process plant location is chosen to be next to the energy provider. While a positioning next to ammonia consumers or existing ammonia production facilities offer advantages like existing infrastructure, ammonia is far more suitable for long distance transport than hydrogen (IEA, 2021).

3.2. Plant requirements for a dynamic Power-to-Ammonia value chain

The plant, especially the electrolyzer, must be able to adjust to the load gradient of a wind turbine in order to respond to incoming energy fluctuations. It must have a large partial load range in which the plant can be safely operated efficiently and economically. To minimize the potential use of fossil fuels, the energy consumption in stand-by operation should be as low as possible. The generated heat should be integrated in the plant. The CAPEX should be kept as low as possible, as this is a significant cost factor for the ammonia produced in small- and medium-scale plants (Chehade, 2021). The hydrogen tank necessary to act as a buffer for the different system dynamics should be as small as possible to reduce CAPEX and storage losses.

3.3. Process control requirements for a dynamic Power-to-Ammonia chain

The operating conditions have to be kept within the permitted value range to prevent the degradation of plant equipment and catalyst materials. The maximum warm start-up time for minimal operation of the ammonia synthesis should be less than 1 hour. If the available energy falls below minimal load, the plant transitions to stand-by operation automatically and holds the operating conditions with inert gas. The stand-by operation can be upheld for a maximum of 20 hours. Energy forecast data is used to decide, whether the plant will be kept in stand-by operation or is shut down.

3.4. Concept of the Power-to-Ammonia value chain using dynamic process control

The PtA plant and their respective process control strategies for the developed use-case are conceptualized, modeled and simulated. The goal is to study the plant and the respective process control concept in terms of flexibility and efficiency.

3.4.1. The Power-to-Ammonia plant concept

In simplified form, the plant concept consists of a technology selection for a water electrolysis, a nitrogen recovery and a reactor concept for the Haber-Bosch synthesis for ammonia production. The concept for the process control strategies consists of strategies to hold the operating conditions in the permitted value ranges as well as strategies to change the plant load during operation.

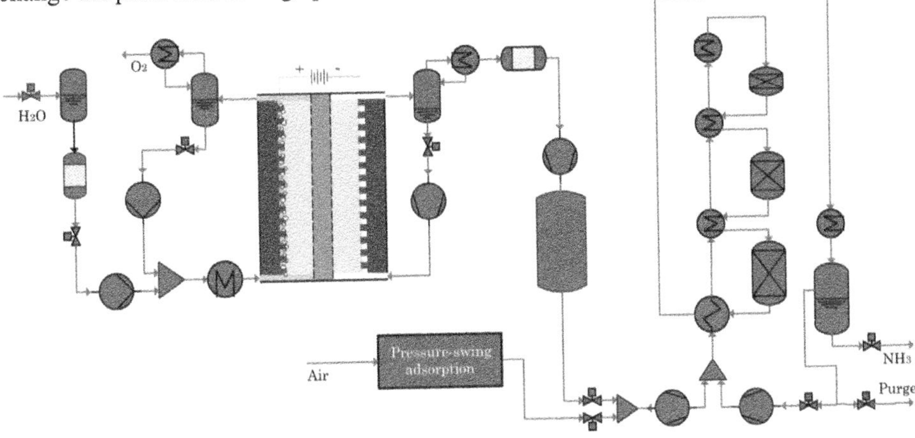

Figure 1: Simplified process flow diagram of the PtA plant concept

The Polymer Membrane Electrolysis (PEMEL) was chosen for the hydrogen production. The advantage of the PEMEL is the flexibility of the plant operation. Both the cold start with 10 15 min and the warm start with <10 s allow the PEMEL to start up quickly. A heating curve of 0.3 1 K/min can be achieved (Buttler, 2018; Hermesmann, 2021). PEMEL systems achieve load gradients of 10 90 %/s with load flexibility of 0 160 % if the power electronics are also designed for it (Grigoriev, 2020; Milanzi, 2018; IEA, 2019). Likewise, PEMELs are more compact than other electrolysis systems for the same hydrogen output (Buttler, 2018). On the other hand, the PEM has higher requirements for

water purity and is also more expensive to procure (1390 – 2320 €/kWel) than an alkaline electrolysis (AEL) with a procurement price of 620 – 1200 €/kWel (Buttler, 2018; Hermesmann, 2021).
For nitrogen production a pressure swing adsorption is used. A minimum load range of 30 %, fast start-up behavior and efficient dynamic operation makes the compact PSA attractive for operating small- and medium-scale PtA systems (Bañares Alcántara; 2015). The ammonia synthesis takes place within a reactor cascade of three tubular reactors with interstage cooling. The Haber-Bosch process is catalyzed through an iron based catalyst. While a ruthenium based catalyst is better suited for a dynamic process due to the lower operating pressure and temperature, an iron based catalyst is chosen due to the lower cost (Chehade, 2021).

3.4.2. The Power-to-Ammonia dynamic process control concept

The process control focuses on the control strategy for the reactor load, because the reactors are the production speed determining components. Nitrogen production is not considered in detail and a constant nitrogen supply is assumed.
The electrolyzers are started up within 10 minutes and begin to feed the produced hydrogen into the buffer tank. Temperature in the reactor system is continuously ramped up with 20 K/h to the operating conditions of about 400 °C and the reactor is pressurized to 250 bar while the electrolyzers are starting up and already producing. The produced hydrogen is temporarily stored in the buffer tank until the operating conditions of the ammonia synthesis with 690 K reactor inlet temperature and 286 bar reactor inlet pressure are reached. The load of the Haber-Bosch reactors is controlled by adjusting the hydrogen input, since hydrogen production accounts for about 85 – 95 % of the total energy consumption (Khademi, 2017). With the assumed constant nitrogen supply a feedforward control system is used, which controls the hydrogen input to the reactor system based on the available power, the fill level of the hydrogen buffer tank and the hydrogen content in the recirculation. Adjusting the stoichiometry of the reaction to the load change through hydrogen input has been established as the most efficient method to control the Haber Bosch reaction (Khademi, 2017). The nominal ramp-up rate of the hydrogen input is set to 50 % of nominal load per hour (Viedt, 2022). The reactor pressure is maintained by the recycle compressor. In stand-by operation the temperature is maintained by the heating element upstream of the reactor.

3.5. Simulation of the Power-to-Ammonia value chain

The concept is modeled and simulated in MATLAB 2022b. Due to the fast dynamics of the PEM electrolysis, the electrolyzer models are simplified by assuming constant material parameters and quasi-stationary conditions in the cells, while neglecting the influence of water vaporization, gas cross-permeation and the heat capacity of the membrane and electrodes. The reactors are modeled as dynamic tubular flow reactors. Assuming constant material parameters, the one-dimensional material and energy balances of the adiabatic reactors result in:

$$\frac{dc_i}{dt} = v\frac{dc_i}{dz} + c_i\frac{dv}{dz} + R_i \quad \text{for } i = H_2, N_2, NH_3 \tag{1}$$

$$\frac{dc_{Ar}}{dt} = v\frac{dc_{Ar}}{dz} + c_{Ar}\frac{dv}{dz} \tag{2}$$

$$\frac{dT_R}{dt} = v\frac{dT_R}{dz} + T_R\frac{dv}{dz} + \frac{r(\text{-}\Delta h_R)}{\rho c_p} \tag{3}$$

The reactors are discretized in equidistant z-locations to convert the PDE system into an ODE system. A simplified convergence study shows that the temporal behavior changes negligibly with a number above 40 equidistant discretization points. The load change rate is varied from 50 500 % per hour to study the influence of the load change rate on the dynamic operation of the Haber-Bosch reactors.

3.5.1. Simulation results

The simulation of the reactor system provides results for the temperature and concentration curves in the Haber-Bosch reactors. For model validation, the local concentration and temperature profiles of the steady-state operating points were compared with the steady-state simulation of different Haber Bosch reactor systems by Khademi et al. (2017). The continuous operation delivers satisfactory results in this respect. The dynamic response of temperature and concentrations is difficult to compare with other literature sources due to low availability of data and different design and model assumptions. However, due to the simplifying assumptions, it is expected, that the simulated plant has a significantly faster dynamic response than real plants or more complex simulations. Figure 2 shows the temperature curve and ammonia concentration curve for 50, 200 and 500 % load change rate per hour of the first reactor during transition from zero load to nominal load.

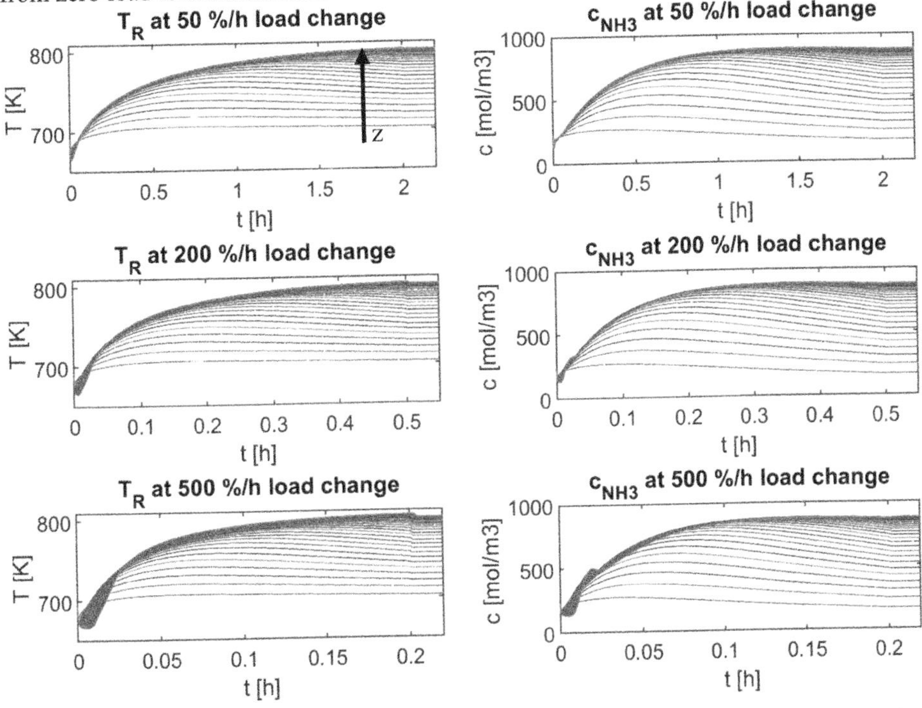

Figure 2: Temperature profile and ammonia concentration profile of the first reactor during transition from nominal load to minimal load with 50, 200 and 500 %/h load change rate

It can be seen, that the maximum reactor temperature slightly increases with increasing load change rate. At load change rates above 200 % per hour the temperature peak in the reactor exceeds the allowed maximum operation temperature of 805 K.

4. Conclusion

In summary, the operation of conventional Haber-Bosch reactors for PtA value chains can be done with load change rates of up to 200 % of nominal load per hour. This information will be used in further simulations of the entire plant and the dynamic control strategies to enable the integration of sustainable energy sources into chemical production chains even without large-scale energy storage or intermediate hydrogen storage. The simplified simulation study showed that the dynamic operation of Haber-Bosch reactors is possible without exceeding or falling below the permitted temperature range The simulation study is thus only applicable for a limited value range. The assumptions of adiabatic operation and the neglect of loss terms and thermal masses in the plant simulation provide faster simulated system dynamics than reality and need to be included in more advanced simulations. Complex models must be created and real plant data must be used for simulation validation to provide a more reliable simulation.
Another important aspect is the choice of concept. The choice of an iron-based catalyst for ammonia synthesis represents the conventional technology well, but is not well suited for dynamic operation, due to the high operating temperature and pressure required. The potential use of new catalyst and reactor systems, such as ruthenium-based catalysts and adsorption reactors would increase the usability of the plant concept for dynamic operation, however at the cost of an increased CAPEX and a more complex control system.

Acknowledgements:

We would like to thank the German Federal Ministry of Education and Research and the Project Management Agency Jülich for their financial support within the framework of the eModule research project (FKZ 03HY116A) of the H2Giga lead platform.

References

R. Bañares-Alcántara, 2015, Analysis of islanded ammonia-based energy storage systems, University of Oxford

A. Buttler, 2018, Current status of water electrolysis for energy storage, grid balancing and sector coupling via power-to-gas and power-to-liquids: A review, Renewable and Sustainable Energy Reviews 82

G. Chehade, 2021, Progress in green ammonia production as potential carbon-free fuel, Fuel 299

D.S. Falcão, 2020, A review on PEM electrolyzer modelling: Guidelines for beginners, Journal of Cleaner Production 261

R. Fang, 2019, Control strategy of electrolyzer in a wind-hydrogen system considering the constraints of switching times, Internatinal Yournal of Hydrogen Energy

M. Hermesmann, 2021, Promising pathways: The geographic and energetic potential of power-to-x technologies based on regeneratively obtained hydrogen, Renewable and Sustainable Energy Reviews 138

International Energy Agency, 2019, The future of hydrogen

International Energy Agency, 2021, Global Hydrogen Review 2021

M.H. Khademi, 2017, Comparison between three types of ammonia synthesis reactor configurations in terms of cooling methods, Chemical Engineering Research and Design 128

H. Kim, 2013, One-dimensional dynamic modeling of a high-pressure water electrolysis system for hydrogen production, International Journal of Hydrogen Energy 38

S. Milanzi, 2018, Technischer Stand und Flexibilität des Power-to-Gas-Verfahrens

R. Moradi, 2019, Hydrogen storage and delivery: Review of the state of the art technologies and risk and reliability analysis, International Journal of Hydrogen Energy 44.23

I. Viedt, 2022, Anforderungen aus der dynamischen Nutzung von grünem Wasserstoff in Power-to-X-Wertschöpfungsketten, Automation 2022

Antonis Kokossis, Michael C. Georgiadis, Efstratios N. Pistikopoulos (Eds.)
PROCEEDINGS OF THE 33rd European Symposium on Computer Aided Process Engineering (ESCAPE33), June 18-21, 2023, Athens, Greece
 http://dx.doi.org/10.1016/B978-0-443-15274-0.50203-1

Evaluation of two-stage stochastic programming applied to the optimization of crude oil operations scheduling

Tomas Garcia Garcia-Verdier [a,b], Gloria Gutierrez [a,b], Carlos Mendez [c], Cesar de Prada [a,b]

[a]*Dpt. of Systems Engineering and Automatic Control, University of Valladolid, c/ Prado de la Magdalena 3-5, Sede Mergelina EII, 47011 Valladolid, Spain*
[b]*Institute of Sustainable Processes, Dr. Mergelina s/n, 47011 Valladolid, Spainw*
[c]*Center for Advanced Process Systems Engineering (CAPSE), INTEC (UNL - CONICET), Industrial Engineering Dpt. (FIQ-UNL), Güemes 3450, 3000 Santa Fe, Argentina*

Abstract

This paper addresses the optimization of crude oil operations scheduling in a marine-access refinery, considering uncertainty in the arrival date of the ships that supply the crudes. Furthermore, we evaluate the performance of a two-stage stochastic mixed-integer nonlinear programming (MINLP) model based on continuous-time representation. For this purpose, we calculate the Expected Value of Perfect Information (EVPI) and the Value of the Stochastic Solution (VSS), which allow us to assess and compare the solution of the stochastic model against solutions obtained from deterministic models. Finally, we consider the Conditional Value-at-Risk (CVaR) measure as the objective function and evaluate the solutions obtained for different risk levels.

Keywords: Stochastic optimization; Continuous-time representation; Crude oil scheduling; Expected Value of Perfect Information; Conditional Value-at-Risk.

1. Introduction

In this paper, the problem under study is the optimization of crude oil operations scheduling in a refinery with marine access. In addition, we address two different topics related by means of stochastic programming.

First, we evaluate the solution obtained from a two-stage stochastic MINLP model based on continuous-time representation by comparing it with the one obtained from a deterministic model. We are interested in finding out if the deterministic solution is close to the stochastic optimum or, on the contrary, if it is worth using stochastic optimization in spite of the increase in computational effort. Another question that arises is how much we are willing to pay to obtain more information about the uncertainties. To answer these inquiries, there are two measures: the value of the stochastic solution (VSS) and the expected value of perfect information (EVPI), which are explained throughout the article.

Second, we analyze the result of including risk management, that is, the effect of deviations from average values on the quality of the solution. For this purpose, we employ the Conditional Value-at-Risk (CVaR) measure as the objective function and evaluate the solutions obtained for different risk levels.

The rest of the paper is structured as follows. The definition of the problem is given in Section 2. The mathematical formulation is described in Section 3. The concepts EVPI and VSS are defined in Section 4. The risk management method is described in Section 5. The solution strategy is mentioned in Section 6. Then, problem instances and computational results are reported in Section 7. Finally, conclusions are drawn in Section 8.

2. Problem definition

As stated before, we tackle the optimization of crude oil operations scheduling in a refinery, which is supplied with crude oil by ships. Before beginning with the description of the problem, it is convenient to clarify the meaning of "crude operations" and "operations scheduling".

The term "crude operations" refers to the operations involved in the supply, storage, and processing of crude oil. Then, when we talk about operations scheduling, we refer to the process of allocating resources to operations and sequencing their execution to comply with the production plan.

Considering these definitions, we can determine that the optimization of crude operations scheduling consists of deciding the best way to operate the terminal-refinery system, considering the management of the arrival and unloading of vessels, crude oil inventory in tanks, and loading of CDUs. Moreover, in this case, the crude oil supply availability is subject to uncertainty due to the impact of weather conditions on ship arrival dates.

Figure 1 shows a typical scheme of a refinery with a marine terminal. The crude oil is received through this terminal and stored in the refinery tanks; these two areas (terminal and tanks) are connected by a pipeline. Then, from the stored crude, the feed blends are formed, which must meet certain quality specifications. Finally, the blends obtained are loaded into the crude distillation units (CDUs) through a mixing pipeline system.

In this article, we address the problem by means of a two-stage stochastic programming with recourse model (Birge & Louveaux, 2011). In these types of models, we must define two kinds of decision variables: first-stage variables ("here and now") which have to be implemented now and affect all future decisions, and second-stage variables that would be implemented later on when more information about the process becomes available (recourse, "wait and see" variables).

For this case, the first-stage variables are: allocation of tanks to CDUs, total volumes transferred to CDUs, and start-time, end-time, and duration of slots. With respect to the second-stage ones, the following were defined: variables related to the activities carried out at the marine terminal, inventory level in tanks, and amount of each type of crude oil

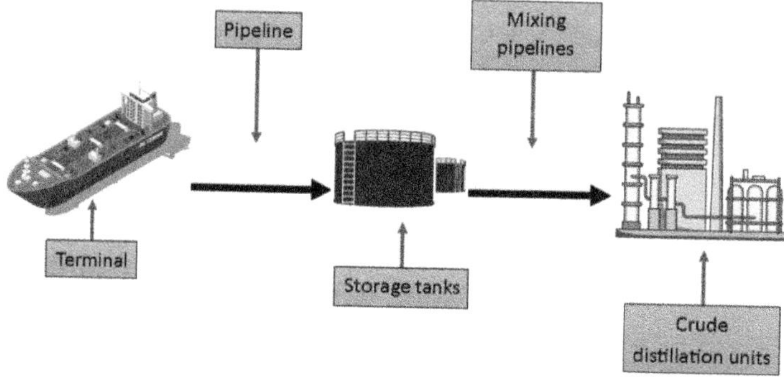

transferred from tanks to CDUs. Finally, the uncertainty in the supply is represented by a discrete set of scenarios that contemplate different arrival dates.

3. Model formulation

We use the model developed in (García-Verdier et al., 2022), which consists of a two-stage stochastic MINLP model with continuous-time representation. In this model, the first and second-stage variables are solved jointly, i.e., the deterministic equivalent program is solved. In order not to exceed the allowed page limit, we do not show the complete mathematical formulation.
A relevant feature of this model is that the precedence between ships does not depend on the order of the elements of the set "Vessels" (Cerdá et al., 2015), nor is there a pre-assignment of slots for each ship (Reddy et al., 2004). For this case, it has been implemented the concept of predefined precedence, which consists in considering the set of slots preordered, and then, the optimization algorithm is in charge of assigning each ship to some of those slots (Gómez Palacín, 2020; Palacín et al., 2019).

3.1. Objective function

To define the objective function, the cost associated with each scenario is calculated by using (1). The first summation comprises the costs due to the difference between processed volume and required demand in each CDU and constitutes the first-stage cost. The second summation involves the costs due to demurrage and departure tardiness of vessels and represents the second-stage cost.

$$ze_e = \sum_u (COP_u * op_u + CSP_u * sp_u) + \sum_b (CDMG_b * dmg_{b,e} + CTDN_b * tdn_{b,e}) \quad \forall e \in E \tag{1}$$

Where,
COP_u = cost due to positive difference between processed volume and required demand by a CDU.
CSP_u = cost due to negative difference between processed volume and required demand by a CDU.
$CDMG_b$ = demurrage or sea waiting cost.
$CTDN_b$ = departure tardiness cost.
op_u = positive difference between processed volume and required demand by CDU u.
sp_u = negative difference between processed volume and required demand by CDU u.
$dmg_{b,e}$ = demurrage of vessel b under scenario e.
$tdn_{b,e}$ = departure tardiness of vessel b under scenario e.
Finally, the objective function consists of minimizing the cost associated with the first stage and the expected cost of the second stage, considering all scenarios (2). The parameter π_e represents the probability of occurrence of scenario e.

$$MIN \sum_e \pi_e * ze_e \tag{2}$$

4. Assessment of the stochastic solution

As mentioned above, one of the objectives of the paper is to evaluate whether two-stage stochastic programming offers any advantage over simpler deterministic approaches. To this end, (Birge & Louveaux, 2011) proposed the value of the stochastic solution (VSS)

and the expected value of perfect information (EVPI) but before defining both measures it is necessary to explain some concepts.
First, the two-stage stochastic programming model is known as the recourse problem (RP), and its solution as the here-and-now solution. Second, the expected value (EV) problem corresponds to the model obtained by substituting all random variables by their expected values, and the EEV value refers to the expected result of fixing the first-stage variables according to the solution of the EV problem. Finally, if we have perfect information, we can apply the optimal solution corresponding to each scenario. Then, the expected long-term cost, known as the "wait and see" (WS) solution, will be equal to the sum of the costs of each scenario weighted by their probabilities of occurrence. In summary, the WS value represents the expected long-term cost of using the optimal solution for each scenario.
Once these concepts are established, we continue with the definition of the value of the stochastic solution (VSS) and the expected value of perfect information (EVPI). The VSS (3) quantifies the improvement obtained in the objective function when considering the randomness of the uncertainty (RP), versus its weighted average (EEV). The EVPI (4) measures the maximum amount that the decision maker would be willing to pay in exchange for complete and accurate information about the future, i.e., for perfect information.

$$VSS = EEV - RP \tag{3}$$

$$EVPI = RP - WS \tag{4}$$

5. Risk management

The approach described above does not evaluate the risk associated with the objective function, i.e., it is risk-neutral. The purpose of this approach is to minimize the expected value in the long run without considering the probability distribution of the objective function. However, it is often important to consider this distribution to reduce the risk that the solution obtained takes extreme values in the most unfavorable scenarios. For this purpose, there are two popular risk measures, Value-at-Risk (VaR) and Conditional Value-at-Risk (CVaR). On the one hand, VaR at confidence level 1-α determines the minimum value ω^* such that the probability of the objective function taking a value less than ω^* is greater than 1-α. On the other hand, CVaR at confidence level 1-α represents the average value of the tail of the distribution, above $\text{VaR}_{1-\alpha}$. The latter is more useful in optimization problems since it is simple to calculate, and is convex.
The stochastic programming model is reformulated by incorporating constraints (5) and (6), where ϕ_e is an auxiliary variable to evaluate CVaR and the scalar α corresponds to the CVaR significance level.

$$ze_e - var \leq \phi_e \ \forall e \in E \tag{5}$$

$$cvar = var + (1/\alpha) * (\sum_e \pi_e * \phi_e) \tag{6}$$

Finally, the new objective function (7) minimizes CVaR for a given confidence level.

$$MIN \ cvar \tag{7}$$

6. Solution strategy

The stochastic programming problem is solved from its deterministic equivalent, following a solution strategy. The solution strategy for the two-stage stochastic MINLP model consists of two steps. A more detailed description of the strategy is given in (García-Verdier et al., 2022).

7. Results

In this section, the resolution of an example is carried out. It consists of a 120-hour scheduling horizon, 5 storage tanks, 2 crude distillation units, and 5 classes of crude characterized by a single property. The arrival of 2 ships is expected. The arrival dates and probabilities for each scenario are detailed in Table 1. The expected departure date is 12 hours after the arrival. The demand for CDU 1 is 100,000 m^3 and for CDU 2 is 65,000 m^3. The example has been solved using GAMS 39.2.1 software, OsiGurobi for MILPs, and CONOPT 4.19 for NLPs on a computer with Intel Core i7-10510U 2.30 GHz processor and 16 GB RAM.

Table 1. Arrival times and probabilities.

Scenarios	Probabilities	Arrival time (h)	
		Ship 1	Ship 2
1	0.01	10	40
2	0.05	50	40
3	0.01	90	40
4	0.18	10	70
5	0.5	50	70
6	0.18	90	70
7	0.01	10	100
8	0.05	50	100
9	0.01	90	100

7.1. EVPI and VSS

The values obtained for RP, EEV, and WS are shown in Table 2. The RP problem involves 9,345 continuous variables, 1,272 binary variables, and 19,700 constraints; it was solved in 51.58 seconds. In addition, EVPI and VSS values are presented in the same table to analyze the effect of considering uncertainty. Figure 2 (a) compares the costs of each scenario for both the two-stage stochastic programming model and the deterministic EV model.

Table 2. Expected costs associated with proposed models, and values of EVPI and VSS.

RP	WS	EEV	EVPI	VSS
		(x10³ €)		
24.42	0.74	74.79	23.68	50.37

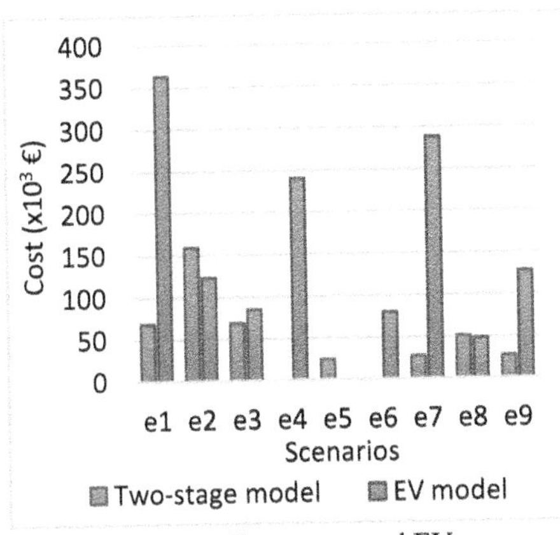

(a) Two-stage and EV

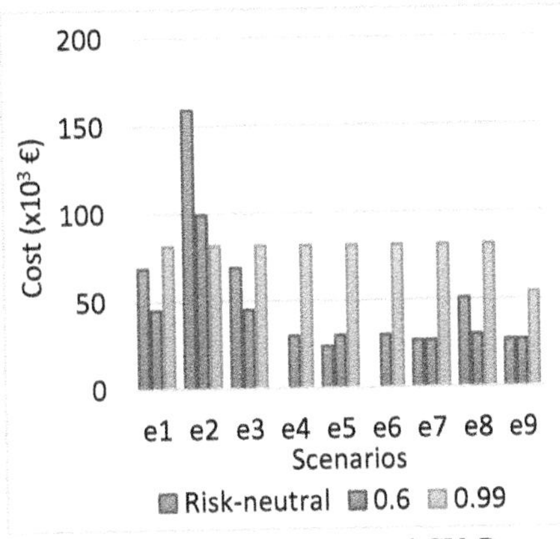

(b) Risk-neutral, $CVaR_{0.6}$, and $CVaR_{0.99}$

Figure 2. Cost of scenarios.

7.2. CVaR

The results obtained for the two-stage stochastic programming model with risk management are shown below. Table 3 shows the VaR and CVaR values at confidence level 0.99 and 0.6. In addition, the expected costs and resolution time for both cases, including the risk-neutral approach, are shown. Figure 2 (b) compares the costs of each scenario in each of the previously mentioned approaches.

Table 3. VaR and CVaR values with confidence levels of 99% and 60%.

$1\text{-}\alpha$	$VaR_{1-\alpha}$ (x10³€)	$CVaR_{1-\alpha}$ (x10³ €)	Expected cost (x10³ €)	Time (s)
0.99	81.56	81.56	81.28	93.67
0.6	30	39.38	33.69	131.06
Risk-neutral (RP)	NA	NA	24.42	51.58

8. Conclusions

If we have access to perfect information, the RP solution will improve, and we would be willing to pay up to 23.68 k€.

The VSS indicates that it is worth using the two-stage stochastic optimization since the expected cost, when using the mean values of the uncertain parameters, increases by 50.37 k€.

Based on Figure 2 (a), we can see that, although there are some scenarios where the EV solution is better than the RP solution, the performance of the RP solution is superior. There are scenarios where the EV solution incurs very high costs (e1, e4, and e7). Therefore, it is preferable to apply the RP solution.

Finally, when including risk in the two-stage stochastic model, it is observed that the expected cost increases with increasing risk aversion. However, from Figure 2 (b) we notice that the solutions obtained by considering risk are more "stable" than the risk-neutral solution. Specifically, in scenario 2, the cost of the risk-neutral solution is twice the cost of CVaR at confidence level 0.99.

Acknowledgements

Financial support received from the Spanish Government with projects a-CIDiT (PID2021-123654OB-C31) and InCo4In (PGC 2018-099312-B-C31), and from European Social Fund.

References

Birge, J. R., & Louveaux, F. (2011). Introduction to Stochastic Programming. In Springer Series in Operations Research and Financial Engineering.

Cerdá, J., Pautasso, P. C., & Cafaro, D. C. (2015). Efficient approach for scheduling crude oil operations in marine- Access refineries. Industrial and Engineering Chemistry Research, 54(33), 8219–8238.

García-Verdier, T. G., Gutiérrez, G., Méndez, C., Palacín, C. G., & de Prada, C. (2022). Minimizing risk in the scheduling of crudes in an oil refinery. IFAC-PapersOnLine, 55(7), 809–814.

Gómez Palacín, C. (2020). Efficient scheduling of batch processes in continuous processing lines. Universidad de Valladolid.

Palacín, C. G., Méndez, C. A., & de Prada, C. (2019). Slots start-up synchronization with shared resources dependency. Chemical Engineering Transactions, 74, 1321–1326.

Reddy, P. C. P., Karimi, I. A., & Srinivasan, R. (2004). A new continuous-time formulation for scheduling crude oil operations. Chemical Engineering Science, 59(6), 1325–1341.

Antonis Kokossis, Michael C. Georgiadis, Efstratios N. Pistikopoulos (Eds.)
PROCEEDINGS OF THE 33rd European Symposium on Computer Aided Process Engineering (ESCAPE33), June 18-21, 2023, Athens, Greece
 http://dx.doi.org/10.1016/B978-0-443-15274-0.50204-3

Generalized First-Principles Suite for Dynamic Modeling of Circular Solid Waste Thermal Treatments: a Sewage Sludge Oxy-Gasification Case Study

Francesco Negri[a,c], Anna Nova[a,b], Andrea Galeazzi[a,b], Francesco Gallo[c], Flavio Manenti[a,b,*]

[a]*Politecnico di Milano, CMIC Dept. "Giulio Natta", Piazza Leonardo da Vinci 32, Milan 20133, Italy*

[b]*Consorzio Interuniversitario Nazionale per la Scienza e Tecnologia dei Materiali, Via Giusti 9, Firenze 50121, Italy*

[c]*Itelyum Regeneration spa, Via Tavernelle 19, Pieve Fissiraga 26854, Lodi, Italy*

**flavio.manenti@polimi.it*

Abstract

It is estimated that 2000 Mt/y of Municipal Solid Waste (MSW) are currently produced worldwide. A strategy for disposing of the non-recyclable fraction of MSW through chemical recycling based on thermal treatments is proposed, leading to the conversion of waste into high-value chemicals. A first-principles suite for the dynamic modeling of thermal conversion processes has been developed to reliably predict the product distribution starting from several types of waste. The GasDS suite models pyrolysis, gas-solid interactions, and gas-phase reactions by dividing the domain into finite volumes, with the possibility to choose among different fluid-dynamic configurations for a given unit. A case study based on a steam-moderated oxy-gasification of sewage-sludge-derived hydrochar is presented in this work. Detailed results showing the syngas productivity and the H_2/CO molar ratio as functions of the Equivalence Ratio (ER) and the Oxygen content of the gasifying medium (O_2 mol%) are provided.

Keywords: Circularity, First-principles, GasDS, Municipal Solid Waste, Sewage Sludge

1. Introduction

Solid waste management is a crucial aspect of every socio-economic system worldwide. The option of conventional recycling is crucial, since it is the one that allows to accelerate the transition towards a circular economy in which waste is transformed back into a new product. In the last years, the topic of chemical recycling has been growing in importance, since it allows to valorize the non-recyclable fraction of waste which would be otherwise incinerated and/or landfilled. Chemical recycling has been shown to have a great potential in reducing CO_2 emissions from the treatment of residual municipal solid waste compared with incineration, especially when the electric grid has a high percentage of renewables (Voss et al., 2021). Some chemical recycling routes start from biological processes and are then followed by appropriate chemical syntheses, such as the production of biofuels from organic-waste-derived biogas (Fedeli et al., 2022; Negri et al., 2022a). Other pathways may start from the thermal degradation of solid waste and then lead to the synthesis of biofuels and biosolvents. Some examples of said processes are based on air-based gasification (Prifti et al., 2021a), and oxy-gasification (Negri et al., 2022b; Prifti et al., 2021b). Chemical recycling processes based on thermal degradation of solid matter need a proper description of this reactive step. The most common approach typically involves a thermodynamic analysis, which has several limitations since it cannot analyze in detail complex phenomena such as chemical kinetics, intra-particle and inter-particle mass and heat transfer. Moreover, it cannot analyze the feasibility of the start-up policies proposed for the unit. This work shows an innovative approach to this problem that exploits the GasDS package, a generalized first-principle suite for the dynamic modeling of gas-solid reactive systems, already validated for biomasses (Ranzi et al., 2014), and coal (Corbetta et al., 2015). The structure of the suite is shown in its entirety, along with a case study that shows a detailed analysis of a gasification chamber working with sewage-sludge-derived hydrochar, thus opening the possibility for design and optimization of innovative thermochemical recycling processes.

2. Methods

2.1. Waste Characterization

The GasDS suite includes a rich and expandable database. Lignin, cellulose, and hemicellulose can be used to describe organic wastes, while several types of coal and tar can be used to describe char-like substances. These species are characterized in detail inside the GasDS database, which contains basic thermochemistry data and properties related to mass and energy transfer. Any general waste can be expressed as a linear combination of said species to perform a simulation. Higher precision can be obtained by entering new tailored species in the database. For example, this is the case for plastic waste, where it is advisable to add entries related to pure polymers like PE, PP, PS, which are the main components of the solid matrix. This paper analyzes the gasification of hydrochar, a waste obtained from the hydrothermal carbonization of sewage sludge. It is possible to express the composition of hydrochar as a linear combination of two types of coal, given its similarity with this substance (Knötig et al., 2021). The sample is almost a perfect linear combination of the species named COAL1 and COAL3, but it is necessary to add a dummy species to close the atomic balance on carbon and hydrogen, here represented by methane and named GCH4. Of course, this provides accurate results only if the final amount of the dummy species is relatively small on a dry-and-ash-free basis like in this case, as shown in Figure 1.

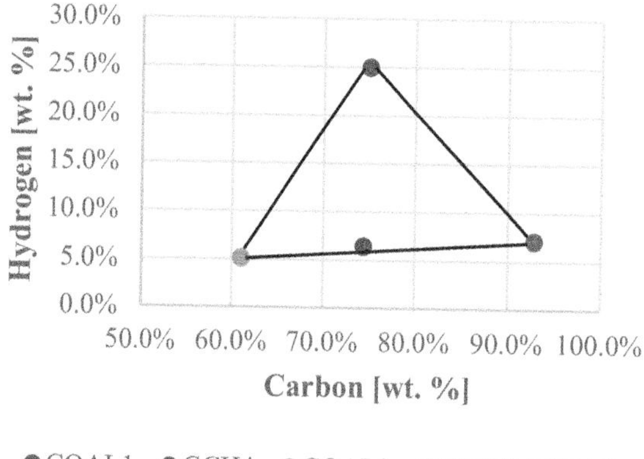

Figure 1. Visual representation of the linear combination for the characterization of hydrochar

2.2. Detailed Kinetic and Mass/Energy Transfer Modeling

The kinetic model implemented in the GasDS suite describes the pyrolysis of the solid matrix, heterogeneous gas-solid interactions, and secondary gas-phase reactions. Pyrolysis models from the literature are already implemented for biomass (Ranzi et al., 2008) and coal (Vascellari et al., 2015), and it is possible to add new reaction sets with the aim of analyzing different type of feedstocks. For example, one may include models relative to the most common polymers (PE, PP, PS, etc.) to describe the pyrolysis of plastic waste (Kaminsky et al., 2004). Gas-solid heterogeneous interactions make use of kinetic models related to the gasification and combustion of char generated by the previous pyrolysis step (Groeneveld and van Swaaij, 1980). The heterogeneous gas-solid system is modeled in detail for what concerns intra-particle mass and energy transfer taking place in parallel with the reactions. The migration of the generated gaseous species to the homogeneous gas phase is modeled by considering appropriate inter-particle transport phenomena. Finally, the homogeneous gas phase is modeled with a detailed kinetic scheme that describes the secondary gas-phase reactions occurring in said environment (Ranzi et al., 2012). The multi-scale approach used to model the reactive system is based on two crucial assumptions: both the solid particle and the reaction environment are discretized into a finite number of volumes, which are then modeled by employing zero-dimensional dynamic mass and energy balances relative to the ith component and the jth sector. The equations are shown in Table 1 (Corbetta et al., 2015; Ranzi et al., 2014).

Table 1. Balance equations used in the GasDS suite (Corbetta et al., 2015; Ranzi et al., 2014)

EQUATION	DESCRIPTION	(#)
$\frac{dm_{j,i}}{dt}=V_jR_{j,i}$	Mass balance of the solid phase	(1)
$\frac{dm_{j,i}}{dt}=J_{j-1,i}S_{j-1}-J_{j,i}S_j+V_jR_{j,i}$	Mass balance of the gas phase in the solid particle	(2)
$\frac{d\sum_{i=1}^{NCP} m_{j,i}h_{j,i}}{dt}=JC_{j-1}S_{j-1}-JC_jS_j+S_{j-1}\sum_{i=1}^{NCG} J_{j-1,i}h_{j-1,i}-S_j\sum_{i=1}^{NCG} J_{j,i}h_{j,i}+V_jHR_j$	Energy balance of the solid particle	(3)
$J_{j,i}=-D_{j,i}^{eff}MW_i\frac{dc_{j,i}}{dr}\Big\vert_{rj}-\frac{Da_j}{\mu_j}\frac{dP_j}{dr}\Big\vert_{rj}c_{j,i}MW_i$	Diffusive mass and heat flux in the particle	(4)
$JC_j=-k_j^{eff}\frac{dT_j}{dr}\Big\vert_{rj}$	Heat flux in the particle by conduction	(5)
$\frac{dg_i}{dt}=G_{in,i}-G_{out,i}+J_{N,i}S_N\eta+V_RR_{g,i}$	Gas-phase mass balance for the reactor	(6)
$\frac{d\sum_{i=1}^{NCG} g_ih_{gi}}{dt}=\sum_{i=1}^{NCG} G_{in,i}h_{gin,i}-\sum_{i=1}^{NCG} G_{out,i}h_{g,i}+\sum_{i=1}^{NCG} J_{N,i}h_{N,i}S_N\eta+h_{ext}(T_N-T^{bulk})S_N\eta+V_RHR_g$	Gas-phase energy balance for the reactor	(7)

The discretization of the domain makes the GasDS suite extremely flexible and adaptable to several system configurations. Dividing the particle in sectors is useful to describe thermally thick particles, which are defined as those having a significant temperature gradient between the core and the surface, a situation that may show up whenever the particle cannot be reduced to a small enough size (Ranzi et al., 2014). The discretization of the reaction chamber allows to describe a lot of reactor morphologies. A single reaction volume can be used to describe fluidized-bed systems, in which the approximation of perfect mixing typically holds. More than one volume may be used to describe fixed/moving bed configurations, both in co-current and counter-current. The gasification chamber modeled in this work is a fixed-bed, countercurrent, cylindrical gasifier that processes 100 kg/h of hydrochar by using a binary H_2O/O_2 mixture as gasifying agent. The unit is shown in Figure 2 and the following specifications are implemented:

- Reactor height/diameter: 2 m / 1 m
- Temperature of gasification agent at start-up/steady-state: 930°C / 230°C
- Particle average diameter: 0.01 m
- Number of discrete volumes for the analysis of the solid particles/gasifier: 2/3

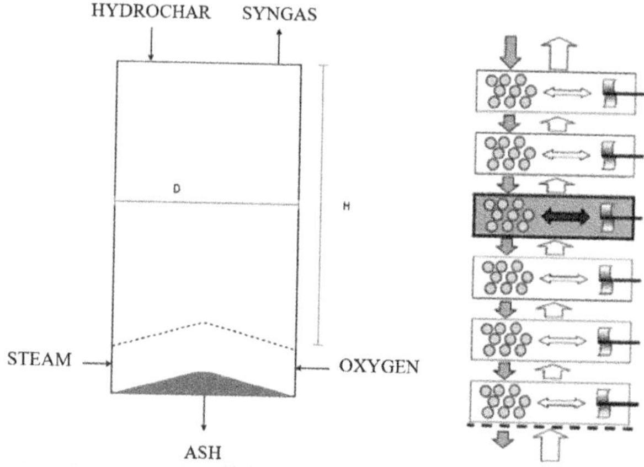

Figure 2. Simplified scheme of the gasification chamber modeled in GasDS

3. Results and Discussion

The system described beforehand still has two degrees of freedom that will be the pivots of the detailed analysis. The first one is the *Equivalence Ratio* (ER), which is here defined as per equation 8, with ER < 1 for fuel-lean mixtures, and ER > 1 for fuel-rich ones.

$$ER = \frac{O_2|_{REAL}}{O_2|_{STOICH}} \tag{8}$$

The second degree of freedom of the system is the oxygen content of the binary gasifying mixture, which is here expressed on a molar basis. It is thus possible to evaluate the behavior of the gasification chamber in a two-dimensional domain identified by these two independent variables. Figure 3 shows the first KPIs of the analysis, the syngas productivity on the left (accounting only for CO and H_2), and the H_2/CO molar ratio on the right. The productivity shows an optimal value of ER for any given O_2 mol%, with a monotonically increasing behaviour as the Oxygen content increases. Looking at the H_2/CO ratio, the value remains relatively constant with respect to ER at a given O_2 mol%, and the curve is monotonically decreasing as the Oxygen content increases. Discovering this kind of behaviour is crucial for a proper optimization of the operating conditions of the gasifier. For example, if the user aims at producing an hydrogen-rich molecule such as methanol and/or dimethyl ether starting from this syngas, it will be necessary to find an operating point which is a trade-off between productivity and H_2/CO ratio, which have opposite behaviour with respect to the Oxygen content of the gasifying mixture (Negri et al., 2022b). Moreover, the GasDS suite allows to highlight deviations from engineering "best practices", which is a typical situation when dealing with unconventional feedstocks such as solid waste. Specifically, the gasification of carbonaceous material typically uses an ER of around 0.30 – 0.33 as rule of thumb for optimal operations (Jangsawang et al., 2015); however, this study highlights that the maximum syngas productivity is systematically achieved for higher values of ER, around 0.35 – 0.45. Finally, GasDS shows an interesting behaviour of the gasifier for low values of ER, which would be lost in a thermodynamic analysis. For these extremely fuel-lean mixtures, the system cannot start-up properly with the procedure described in the previous paragraph. The system is too fuel-poor to sustain the reaction environment during the switch from the hot gasifying agent used for starting-up the chamber, to the colder steady-state current. Trying to operate the unit in this region would bring the system to a premature shut-down.

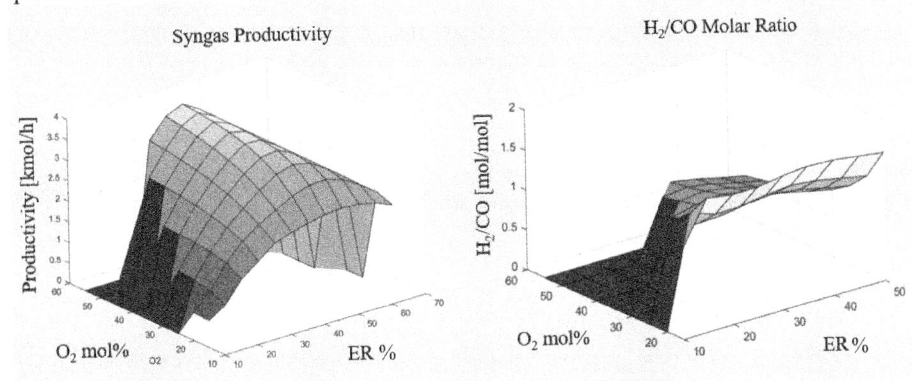

Figure 3. Syngas productivity in kmol/h (left) and H_2/CO molar ratio (right)

4. Conclusions

A generalized first-principles suite for the dynamic modeling of thermal treatments of carbonaceous feedstocks has been introduced. The steps of a GasDS simulation have been shown in detail, starting from the characterization of the raw material to be processed, continuing with the balance equations that make up the core of the suite, and finishing with the set-up of a case study based on a steam-moderated oxy-gasification of hydrochar. The results include a detailed analysis of the syngas productivity and the H_2/CO ratio, which are fundamental KPIs for the steps following the thermal degradation. These parameters are shown as a function of the Equivalence Ratio (ER) and the oxygen content of the gasifying agent (O_2 mol%). The GasDS suite allows to understand functional dependencies between the analyzed variables, deviations from engineering "best practices" shown by the system due to the unconventional feedstock, and even unfeasible operating conditions which would lead to a premature shut-down of the system. Thus, the suite can be used to model and optimize thermo-chemical waste recycling processes.

Acknowledgments

The authors acknowledge HBI Group for providing data on hydrochar composition.

References

Corbetta, M., Bassani, A., Manenti, F., Pirola, C., Maggio, E., Pettinau, A., Deiana, P., Pierucci, S., Ranzi, E., 2015. Multi-scale kinetic modeling and experimental investigation of syngas production from coal gasification in updraft gasifiers. Energy and Fuels 29, 3972–3984. https://doi.org/10.1021/acs.energyfuels.5b00648

Fedeli, M., Negri, F., Manenti, F., 2022. Biogas to advanced biofuels: Techno-economic analysis of one-step dimethyl ether synthesis. J. Clean. Prod. 376. https://doi.org/10.1016/j.jclepro.2022.134076

Groeneveld, M.J., van Swaaij, W.P.M., 1980. 39 Gasification of char particles with CO2 AND H2O. Chem. Eng. Sci. 35, 307–313. https://doi.org/https://doi.org/10.1016/0009-2509(80)80101-1

Jangsawang, W., Laohalidanond, K., Kerdsuwan, S., 2015. Optimum Equivalence Ratio of Biomass Gasification Process Based on Thermodynamic Equilibrium Model. Energy Procedia 79, 520–527. https://doi.org/https://doi.org/10.1016/j.egypro.2015.11.528

Kaminsky, W., Predel, M., Sadiki, A., 2004. Feedstock recycling of polymers by pyrolysis in a fluidised bed. Polym. Degrad. Stab. 85, 1045–1050. https://doi.org/https://doi.org/10.1016/j.polymdegradstab.2003.05.002

Knötig, P., Etzold, H., Wirth, B., 2021. Model-based evaluation of hydrothermal treatment for the energy efficient dewatering and drying of sewage sludge. Processes 9. https://doi.org/10.3390/pr9081346

Negri, F., Fedeli, M., Barbieri, M., Manenti, F., 2022a. A versatile modular plant for converting biogas into advanced biofuels. Invent. Discl. 2, 100008. https://doi.org/https://doi.org/10.1016/j.inv.2022.100008

Negri, F., Nova, A., Basso, D., Manenti, F., 2022b. Techno-Economic Analysis of Dimethyl Ether Biofuel Production Plant Based on Sewage Sludge Gasification. Chem. Eng. Trans. 94, 1087–1092. https://doi.org/10.3303/CET2294181

Prifti, K., Basso, D., Pavanetto, R., Manenti, F., 2021a. Improving hydrothermal carbonization (HTC) processes by hydrochar gasification. Chem. Eng. Trans. 86, 925–930. https://doi.org/10.3303/CET2186155

Prifti, K., Galeazzi, A., Margarita, I., Papale, A., Miele, S., Bargiacchi, E., Barbieri, M., Petea, M., Manenti, F., 2021b. Converting end-of-life plastic waste into methanol: The gasiforming™ process as new, efficient and circular pathway. Environ. Eng. Manag. J. 20, 1629–1636. https://doi.org/10.30638/eemj.2021.151

Ranzi, E., Corbetta, M., Manenti, F., Pierucci, S., 2014. Kinetic modeling of the thermal degradation and combustion of biomass. Chem. Eng. Sci. 110, 2–12. https://doi.org/10.1016/j.ces.2013.08.014

Ranzi, E., Cuoci, A., Faravelli, T., Frassoldati, A., Migliavacca, G., Pierucci, S., Sommariva, S., 2008. Chemical Kinetics of Biomass Pyrolysis. Energy & Fuels 22, 4292–4300. https://doi.org/10.1021/ef800551t

Ranzi, E., Frassoldati, A., Grana, R., Cuoci, A., Faravelli, T., Kelley, A.P., Law, C.K., 2012. Hierarchical and comparative kinetic modeling of laminar flame speeds of hydrocarbon and oxygenated fuels. Prog. Energy Combust. Sci. 38, 468–501. https://doi.org/https://doi.org/10.1016/j.pecs.2012.03.004

Vascellari, M., Roberts, D.G., Hla, S.S., Harris, D.J., Hasse, C., 2015. From laboratory-scale experiments to industrial-scale CFD simulations of entrained flow coal gasification. Fuel 152, 58–73. https://doi.org/https://doi.org/10.1016/j.fuel.2015.01.038

Voss, R., Lee, R.P., Seidl, L., Keller, F., Fröhling, M., 2021. Global warming potential and economic performance of gasification-based chemical recycling and incineration pathways for residual municipal solid waste treatment in Germany. Waste Manag. 134, 206–219. https://doi.org/https://doi.org/10.1016/j.wasman.2021.07.040

Antonis Kokossis, Michael C. Georgiadis, Efstratios N. Pistikopoulos (Eds.)
PROCEEDINGS OF THE 33rd European Symposium on Computer Aided Process Engineering (ESCAPE33), June 18-21, 2023, Athens, Greece
 http://dx.doi.org/10.1016/B978-0-443-15274-0.50205-5

A Discrete-Event Simulation Approach to the Design and Planning of Biomass Supply Chains considering Technological Learning

Helena Paulo[a,b], Miguel Vieira[a,c,d], Bruno S. Gonçalves[e], Susana Relvas[a], Tânia Pinto-Varela[a] and Ana P. Barbosa-Povoa[a]

[a]*CEG-IST, Instituto Superior Técnico, Universidade de Lisboa, Lisboa, Portugal*
[b]*ISEL - Instituto Superior de Engenharia de Lisboa, IPL, Lisboa, Portugal*
[c]*EIGeS, Universidade Lusófona, Lisboa, Portugal*
[d]*Univ Coimbra, CEMMPRE, Dept Mech Engn, Coimbra, Portugal*
[e]*ESTG, Politécnico de Leiria, Leiria, Portugal*

Abstract

The present work intends to contribute to the supply chain design and planning, by addressing the technological learning due to non-mature technologies at biorefineries. The methodological approach combines optimization with discrete-event simulation, assessing its investment risk through the Conditional Value at Risk. Through the optimization model, the optimal strategic variables for the biomass supply chain under deterministic conditions are obtained. In contrast, the simulation model allows to analyze the technological uncertainty supported by the learning curve concept to evaluate the technologies' performance. The results highlight the advantage of exploring this combined approach to consider variability parameters for strategic and tactical solutions and the particular influence of technological learning in supply chain design and planning.

Keywords: Learning curves, Discrete-event simulation, Biomass, Supply chain design.

1. Introduction

Most decisions related to the design and planning of supply chains take place in an increasingly dynamic environment where variability is present at multiple decisional levels. The impact of uncertainty in decision-making has long been recognized, although modelling and solution approaches still call for improvements. For example, the complexity of strategic decisions is greatly influenced by the uncertainty on availability, quality, product demand, and process operations. Moreover, considering technological process parameters represent an additional challenge in their design and planning to ensure long-term financial viability. The general learning curve concept can be set to model the uncertain technological performance in adopting new technology. The mathematical relation reflects the technical and economic progress of the technology as users gain experience with time (Weiss et al., 2010). It is assumed that the learning factor can be translated into the technological conversion of the main process (e.g. the transesterification conversion factor from biomass to biodiesel), which becomes an uncertain parameter. In this study, we propose the development of an approach capable of modelling the impact of an expected technological development based on the learning curve concept. The approach is applied to the design and planning of integrated biorefineries supply chains. Integrated biorefineries are a promising alternative to bioenergy and biomaterials production, becoming an emerging industry where the technology performance involving biomass conversion at an industrial scale has yet to be fully known. Therefore, the design and planning of integrated biorefineries involve a significant degree of uncertainty associated with the low maturity of technological learning, with correlated influence in the processing capacity and bioproducts availability or demand. In this scope of multilevel decision-support, discrete-event simulation (DES) can provide a useful approach to model uncertainty and evaluate a solution for the supply chain under different scenarios of the network.

The Portuguese case of a biomass supply chain is used as a real case study to illustrate the application of the proposed methodology. By combining optimization-based simulation, a detailed DES model allows to incorporate of the dynamics and uncertain conditions in the supply chain structure, evaluating a solution with the optimal number of integrated biorefineries, their location, capacities, and technology, while considering the assessment of their financial viability using Conditional Value-at-Risk (CVaR).

2. Methodology

The problem scope consists of assessing the biomass-to-biofuels supply chain design and planning through measuring the financial risk, with particular relevance to the technological uncertainty at integrated biorefineries. The learning curve is introduced to model the process conversion efficiency and its impact on the overall supply chain decisions. The methodology proposes an evaluation approach combining an optimization model which generates an initial deterministic solution, then simulated under the conditions of a detailed DES model with variable technological learning. Considering this overall perspective based on strategic long-term investment, the Conditional Value at Risk (CVaR) is used as a risk measure to evaluate the likelihood that a specific loss will exceed a certain value at risk. The hybridization of optimization with simulation excels in the advantages of having an analytical model that formulates a solution under ideal conditions, with a more precise simulation model that mimics how the system reacts under realistic outcomes and hard-to-model constraints (Figueira et al., 2014). The approach allows to explore alternative network solutions with reduced computational hindrance by accounting for the variability provided by uncertain operational parameters.

2.1. Supply chain simulation-optimization modeling

The approach combines an optimization model and a simulation model. A Mixed Integer Linear Programming model, based on Paulo et al. (2015), is implemented in GAMS to consider the supply chain superstructure with all possible locations of biomass sources (with the corresponding availability per type), biorefineries (set of available technologies and capacities), demand centers and transportation modes with the distances between all points. The objective is to minimize supply chain total cost, given by Eq.1, identifying the number, location, technology, and capacity of biorefineries to install, with the corresponding amounts produced and demand centers to serve, as well as the amount of biomass type collected at each source and transportation flows.

$$\min Total\ Cost = \sum (Biomass\ Cost + Fixed/Variable\ Operating\ Cost\ at\ Biorefiner \\ + Transportantion\ Costs + Annualized\ Investment\ Cost) \qquad (1)$$

The DES model aims to represent the real-world environment (as is the Portuguese case) to evaluate the optimized solution under variable conditions. Through the SIMIO platform, the model computes multiple output scenario statistics involving the occurrence of events (e.g. a new bioproduct order) using probability distributions of parameters. The main library of objects available in the SIMIO software is used, which, in general, considers model entities (e.g., products) processed at servers (e.g., biorefinery), and then departs from the system. Suitable to the case level of detail, the focus is set on the impact of the technological learning curve in the design capacity of the facilities to accommodate an expected increase in bioproduct demand. For simplification, the availability of biomass is considered non-limiting. Additional uncertainty in overall cost and prices forecast are introduced as sources of variability, being considered 1000 replications to guarantee statistical significance. The interaction of the two models aim to evaluate a supply chain configuration with a risk measure of the techno-economic investment viability. To account for the time value of money, the simulation model computes the net present value (NPV) of cashflow operations (Eq.2), for n time periods and CF_k as the annualized value of cashflows at time period k with i discount rate.

$$\mathrm{NPV} = \sum_{k=0}^{n} \frac{CF_k}{(1+i)^k} \tag{2}$$

2.2. Learning Curves and Technological Conversion Efficiency

Empirical assessments can demonstrate that costs decrease with the gain of experience from the use of a particular technology, which usually translates in a performance improvement of the production process. There are different types of learning curves: the one-factor learning curve represents learning by accumulated production; the two-factor formulates learning by doing and by searching effects; and other multi-factor learning curves are explained by other factors. Although a detailed characterization of the learning process seems appealing, multiple factors are often difficult to quantify, therefore the one-factor learning curve is chosen for the development of the present work. Mapping the integrated biorefineries and the information on their specifications allow to find the trend of the process conversion efficiency for a given biomass-to-biofuel technology. And by defining these equations, it is possible to calculate values of conversion that reflects the learning-by-doing performance of each technology over the time horizon, which is expected to stabilize once reached a plateau (Weiss et al., 2010). The learning curve effect on costs can be generalized as a power law function given by (Eq.3), with CC_{cum} the unit cost of production in monetary units, CP the cumulated production amounts, and CC_{ref} and CP_{ref} the initial conditions at an arbitrary starting point.

$$CC_{cum} = CC_{ref} \left(\frac{CP}{CP_{ref}} \right)^{-\varepsilon} \tag{3}$$

The factor $(CP/CP_{ref})^{-\varepsilon}$ represents the decrease ratio in the unit cost of production, with ε the learning coefficient related to the technology. Considering the conversion efficiency μ as the ratio of the amount of bioproduct by the amount of biomass feedstock CB used to process it, the equation can be rewritten as (Eq.4).

$$CC_{cum} = CC_{ref} \left(\frac{\mu . CB}{\mu_{ref} . CB_{ref}} \right)^{-\varepsilon} \tag{4}$$

From (Eq.4), the relationship between the technological conversion efficiency and the learning-by-doing performance can be obtained. As higher the values of conversion, higher the accumulated production quantities obtained from the same feedstock, and therefore, lower unit production costs.

2.3. Financial risk assessment

Attracting investment to the biomass-based industry represents a substantial challenge, whereas making the supply chain efficient under uncertain technological process parameters increases the impact in its design and planning to assure a long-term viability. The assessment of uncertainty appeals for risk management studies and the research community has been demonstrated their importance. The CVaR is a widely used risk metric to quantify the tail risk, where conclusions can be drawn to either downside losses or upside gains for an industrial investment, as demonstrated by Vieira et al. (2021). Given a discretized probability distribution function of NPV values, CVaR measures the weighted average outcome of far-edge tail events beyond the VaR for a specified confidence level (α). The VaR for downside losses is set as the outcome value for the (1-α)-quantile of the distribution. The CVaR is given in (Eq.5), multiplying the probability of scenario s occurrence Π_s by the positive variable δ_s calculated in (Eq.6), which measures the positive difference between the VaR and the NPV of each shortfall scenarios.

$$CVaR = VaR - \frac{1}{1-\alpha} \sum_{s} \Pi_s \, \delta_s \tag{5}$$

$$\delta_s = VaR - NPV_s, \delta_s \geq 0 \tag{6}$$

3. Case Study and Results

The described methodology has been applied to an illustrative case study of biomass supply chain of bioethanol and biodiesel production in Portugal. Supply chain characteristics and data considered in this study is summarized on Fig.1, considering two technologies – fermentation for bioethanol and transesterification for biodiesel.

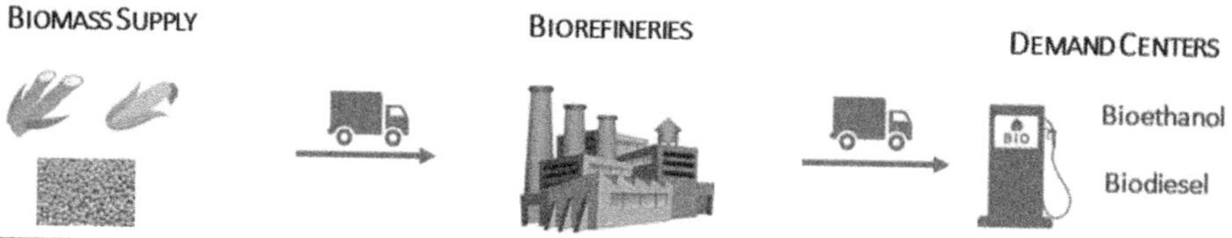

BIOMASS SUPPLY | BIOREFINERIES | DEMAND CENTERS

Bioethanol

Biodiesel

278 candidate locations biomass supply (i)	28 candidate locations to biorefineries installation (k)		18 demand centers (v)
Biomass type (b)	**Technology (m)**	**Capacity (q)**	**Products (p)**
Starch/Sugar (b1)	Transesterification (m1)	75 (q1) / 250 (q2) kton/year	Bioethanol (p1)
Seed/Animal fats (b2)	Fermentation (m2)	50(q1) / 200(q2) kton/year	Biodiesel (p2)
Biomass acquisition cost	**Investment cost**		**Selling price**
b1: 185.33€/ton	m1q1: 36415 k€ m1q2: 74991k€		p1: 904.09€/ton
b2: 411.85€/ton	m2q1: 12164 k€ m2q2: 27945k€		p2: 947.72€/ton
	Fixed annual costs	**Variable operating costs**	
	m1q1: 2798k€ m1q2: 9328k€	p1: 318.32 €/ton	
	m2q1: 1190k€ m2q2: 4763k€	p2: 198.79 €/ton	

Figure 1. Case study data

The initial solution provided by the optimization model defines the supply chain superstructure, identifying 4 biorefineries to install. Lisboa, Pombal and Vila Real are the selected locations to install biorefineries using fermentation technology, with a processing capacity of 250 000 tons and 75 000 tons of bioethanol, respectively.

Santarém is selected to install a transesterification biorefinery of 50 000 tons of production capacity for biodiesel. The optimal solution also defines the biomass flows from each collecting point to each biorefinery, as well as the product flows to each demand center. This information is used to compile the simulation model details. With regard to the learning curve integration, it is considered that the two technologies selected have dissimilar characteristics in its learning-by-doing progress. From the work by Seabra (2021) based on real data, the equations used to calculate the decreasing costs for each level n of accumulated production for the fermentation (m1) and transesterification ($m2$) processes were, respectively (Eq.7) and (Eq.8).

$$CC_{n\,m1} = 534.43\,(CP_{n\,m1}/10000)^{-0.3219} \tag{7}$$

$$CC_{n\,m2} = 282.10\,(CP_{n\,m2}/10000)^{-0.1520} \tag{8}$$

The derived conversion efficiency trends are translated by (Eq.9) and (Eq.10), whereas the values pondered for the first time period were set as $\mu_{\text{fermentation}}$=25.58% and $\mu_{\text{transesterification}}$= 27.51%.

$$\mu_{fermentation} = 0.055(\mathrm{t}-1) + 0.2558 \qquad \forall \mathrm{t} > 1 \tag{9}$$

$$\mu_{transesterification} = 0.035(\mathrm{t}-1) + 0.2751 \qquad \forall \mathrm{t} > 1 \tag{10}$$

It should be noted that the biomass conversion efficiency progress is expected to stabilize, which the study by Sharew et al. (2022) suggested the interval [46%-56%] for these technologies, so it is assumed a plateau value of 51%. This is added to the characteristic of each corresponding biorefinery in the model, which computes the statistical simulation for a time horizon of n=10 years, and the NPV considering the defined parameters distributions, such as cost and price variability, with a discount factor i of 10%. Assuming the scenario where the transition to biofuels is fostered, the increase in the demand follows a uniform distribution between 2-4%. The results displayed in Fig. 2 demonstrates the output simulation replications, represented by a probability distribution, which enable the calculation of the CVaR by discretizing the scenarios under the VaR for a confidence level of 95% (13,6.10⁶€). The value of CVaR means that the proposed supply chain configuration and scenario conditions guarantees at least a NPV of 12,1.10⁶€ with a 95% of probability.

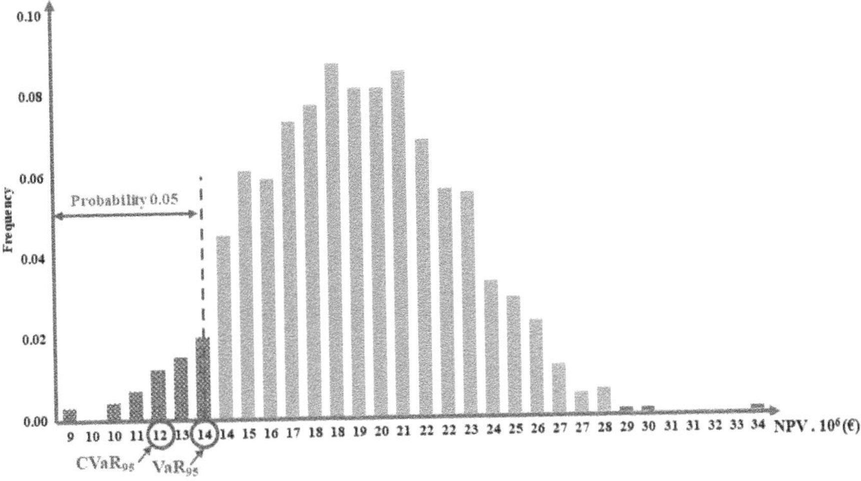

Figure 2 – Risk evaluation results: NPV VaR and CVaR representation

With Fig. 3 it is possible to analyze the learning curve effect in the increase of overall capacity of the biorefineries, as well as its utilized capacity while accommodating the demand increase. For example, Lisboa and Vila Real show a higher utilization of its installed capacity, while the remaining recommend to evaluate an alternative scenario of a reduced capacity or network configuration. Likewise, sensitivity analysis to learning curve progress can be simulated, for example the CVaR obtained, for a 95% confidence level, with an increase of 0,5% in the learning rate (faster learning) rises to 20,1.10⁶€, while a decrease of 0,5% (slower learning) generates a probable downside loss of 3,9.10⁶€.

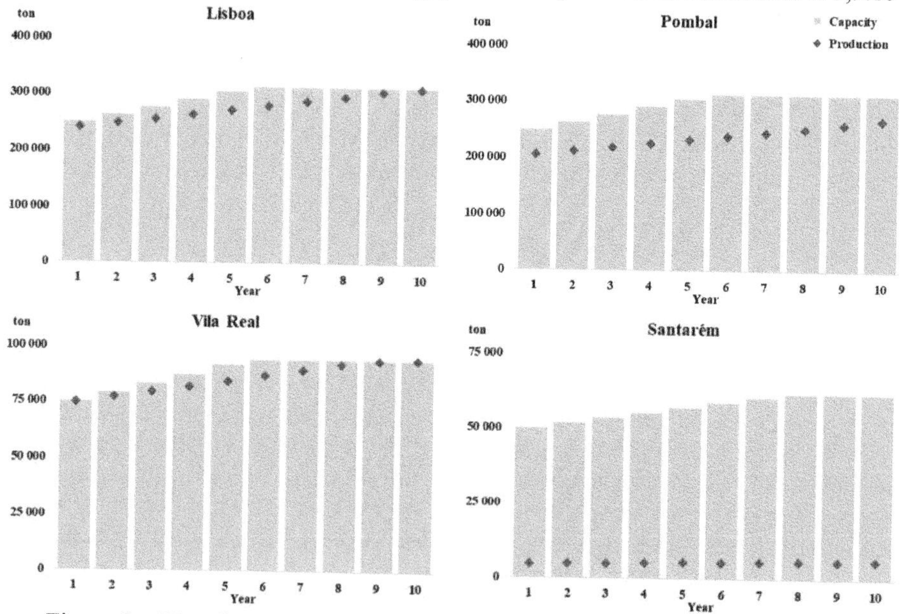

Figure 3 – Biorefineries installed capacity (bars) and production (dots) by year

4. Conclusions

This work explores the effect of learning-by-doing in non-mature technologies over the biomass supply chain performance. The methodology measures the investor risk by combining simulation and optimization techniques, allowing to explore an optimal-based solution through the simulation of variable dynamic conditions, parameters distributions, and technological learning conversion. By using CVaR to measure NPV as an investment risk, the decision maker is able to have more robust information on its viability, as well as to evaluate alterative scenarios. The results demonstrate the impact of different learning rates and uphold the simulation of alterative configurations to the supply chain design, pursuing the assessment of the multiple uncertainty factors affecting these decisions.

Acknowledgments

UE/FEDER through COMPETE and FCT projects UIDB/00097/2020 and UIDB/00285/2020.

References

M. Weiss, et al., 2010, A review of experience curve analyses for energy demand technologies. Technological Forecasting and Social Change, 77(3), 411–428.

G. Figueira. and B. Almada-Lobo, 2014, Hybrid simulation–optimization methods: A taxonomy and discussion. Simulation Modelling Practice and Theory, 46:118-134.

H. Paulo et al., 2015, Supply chain optimization of residual forestry biomass for bioenergy production: The case study of Portugal. Biomass and Bioenergy, 83, 245-256

M. Seabra, 2021, Modelling uncertainty in the technologies' conversion efficiency for the design and planning of biomass supply chains, Master in Industrial Engineering and Management, Master Thesis, Instituto Superior Técnico, Universidade de Lisboa.

M. Vieira et al., 2021, Assessment of financial risk in the design and scheduling of multipurpose plants under demand uncertainty. International Journal Production Research, 59(20), 6125–6145

S. Sharew et al., 2022, Alternative Energy Potential and Conversion Efficiency of Biomass into Target Biofuels: A Case Study in Ethiopian Sugar IndustryWonji-Shoa. Biomass,2(4), 279–298.

Antonis Kokossis, Michael C. Georgiadis, Efstratios N. Pistikopoulos (Eds.)
PROCEEDINGS OF THE 33rd European Symposium on Computer Aided Process Engineering (ESCAPE33), June 18-21, 2023, Athens, Greece
 http://dx.doi.org/10.1016/B978-0-443-15274-0.50206-7

Data-driven modeling to predict the rate of Boil-off Gas (BOG) generation in an industrial LNG storage tank

Suraj Prakash Singh,[a] Rajagopalan Srinivasan,[a,b*] I. A. Karimi[c]

[a] *Department of Chemical Engineering, Indian Institute of Technology Madras, Chennai 600036, India*

[b] *American Express Lab for Data Analytics, Risk and Technology, Indian Institute of Technologuy Madras, Chennai 600036, India*

[c] *Department of Chemical and Biomolecular Engineering, National University of Singapore, Singapore*

*raj@iitm.ac.in

Abstract

Natural gas is stored as Liquefied Natural Gas (LNG) under cryogenic conditions in a storage tank. The tank is highly insulated; nevertheless, there is heat ingress from the surrounding, causing the generation of boil-off-gas (BOG). BOG generation over-pressurizes the tank and can lead to tank failure, so proper BOG management is critical to plant safety. This paper seeks to develop a data-driven model for a real industrial LNG terminal. Modeling was performed on historical real-time LNG data using different machine learning algorithms - Linear regression, Random Forest, and XGBoost. The performance of algorithms is analyzed based on R^2, Mean-Absolute-Error (MAE %), and Root-mean-square error (RMSE %) values. The critical input features are calculated based on the Shapley additive explanation method (SHAP value); less important features were removed to decrease the model complexity. Our studies show that Random Forest outperforms the other two algorithms in terms of accuracy. The developed model can help plant operators make decisions quickly with better confidence.

Keywords: Data-driven modeling, Liquefied Natural Gas, Boil-Off-Gas, BOG management, Machine learning.

1. Introduction

Natural gas (NG) is considered one of the cleanest fossil fuels, with a variety of end-uses, including fertilizer production, power generation, along with industrial and residential heating. With the global annual consumption of NG being 4.04 Tm3 (Statistical Review of World Energy, 2021), it is likely to become the second most widely used fossil fuel by 2035. It is globally traded in its liquefied form (LNG) via specialized carriers that transport LNG from suppliers to receiving terminals, where it is unloaded into storage tanks. Storing LNG under cryogenic conditions enables safe, convenient, and cost-effective distribution. However, continuous heat ingress from the surrounding environment causes the vaporization of LNG, known as boil-off gas (BOG). When LNG of two different densities is stored in the same storage tank, rather than forming a homogeneously mixed liquid, it can result in stratification. This is because when the different layers are mixed, a significant amount of BOG is generated due to interface disruptions, and the tank pressure builds up. This heterogeneous mixing phenomenon is termed rollover. In this work, the broad objective is to prevent or alleviate the adverse consequences of rollover.

Rollover is a common problem across storage units in LNG supplier facilities, FSRUs, LNG carriers, and regasification terminals, with a higher frequency of occurrence at the latter (Energy report, 2018). Rollover can have catastrophic repercussions; for example, the incidents at La Spezia (Italy, 1971) and Partington (the UK, 1993) are prominent examples of how BOG generation due to rollover can have devastating impacts on the regasification terminal, causing significant economic losses. According to the GIIGNL 2019, more than 24 rollover incidents have been reported over the last 60 years. Such events are continuously increasing due to the ever-increasing use and trade of LNG. The amount of BOG generation during a rollover is enormous; nevertheless, the mechanism is, until now, not fully understood. Therefore, it is essential to develop a model to predict the rate of BOG generation in the LNG regasification terminal.

The basics of rollover and the rate of BOG generation have been the subject of several studies. These have a theoretical, experimental, or numerical foundation. Theoretical studies concentrated on how many intermediate stages were assumed to better comprehend rollover events, such as two-stage (Heestand, 1983), three-stage (Koyama, 2007), and four-stage (Li, 2015). Experimental studies focused on the effects of LNG composition, initial concentration difference (Munakata, 1994), visual observations of boundary layers, and formation and decomposition of evolved LNG layers (Bates, 1997). Numerical studies involved computations based on different geometries, modes, and angles of LNG filling (Hubert, 2019; Saleem, 2020) that can include initial conditions and boundary conditions (Zakaria, 2017), interface instability (Wei, 2018), among other factors. However, all of them are computationally expensive and difficult to model. Making industrial decisions based on these models during critical situations is difficult. We aim to address this issue in the current work. In this paper, we develop a data-driven model based on collected historical data to predict the rate of BOG generation in an LNG tank in a regasification terminal. The modelling was performed by using a simple linear model (Linear regression (LR)) as well as more complex nonlinear models (Random Forest (RF) and XGBoost). Further, the performance of all three models is analyzed to select the best model. Finally, the model is used to perform various case studies based on industrial problem scenarios.

2. Process description

In the LNG regasification terminal, LNG is imported through ships and unloaded into the highly insulated storage tank. This stored LNG is continuously supplied to the customers, and whenever the tank level reaches the minimum value, a new ship arrives for fresh LNG unloading. The LNG in the tank is stored nearly at its saturation temperature under atmospheric pressure. The continuous heat ingress from the surroundings to the tanks and pipeline increases the temperature of LNG. Due to this, LNG evaporates and forms BOG, which leads to pressure build-up in the tank. For the safety of the tank and regasification terminal, the generated BOG is continuously removed from the tank through a compressor. The generated BOG contains nearly 99% methane, which has the highest calorific value, and its removal reduces the overall calorific value of LNG. The BOG can be managed in various ways within the plant, such as (1) directly supplied to the customer, (2) recondensed back to the tank, and (3) flared in of the plant. These methods are either not economically suitable or permissible due to environmental concerns. So, it is crucial to understand the behavior and BOG generation rate under various conditions and operate the plant so as to reduce BOG formation. A data-driven model is developed for this purpose.

2.1. Data collection

Industries continuously monitor the pressure inside the tank and accordingly estimate the BOG generation. For process control and operations, various other variables like temperature and density of LNG, tank vapor pressure, LNG level, compressors operation, and LNG outflow are also continuously monitored. These data are continuously monitored and saved on an hourly basis between two consecutive LNG ship unloading.

This research aims to predict the rate of BOG generation under various conditions and under the influence of various parameters. Thus, an accurate relationship has to be established between the input and output variables. The model input variables are LNG level (x_1), LNG density at the top (x_2), LNG density at the bottom (x_3), LNG temperature at the top (x_4), LNG temperature at the bottom (x_5), vapor pressure (x_6), LNG outlet flowrate (x_7) and operation mode of two compressors (x_8 and x_9). The model output variable is the rate of generation of BOG (y) in the LNG storage tank.

In our work, we have used historical real-time data of all the input and output variables collected over nearly one month (934 datasets) for model development. These data are collected every second but saved at every hour interval (sampling time) for future study purposes. The data consists of various missing values and outliers, so proper data pre-processing is necessary. In our case, as the number of missing values (1.60%) and outliers (0.96 %) are small, we removed those samples from the datasets. The whole dataset of every month is segregated into different segments based on the mode of operation in the regasification terminal. The first mode considered is the unloading operation, where LNG is unloaded from an LNG ship to the storage tank. The second mode is normal operations when there is no input to the tank, but LNG is sent to the regasification process, and the gas is sent-out to customers. In this paper, we consider the latter, i.e., the normal operations mode. The data from normal operations is split into training and testing data sets in the ratio of 4:1. The training datasets are used to develop various models based on different machine learning algorithms.

3. Results and discussion

The approach considered to establish a relationship between input and output features includes (1) data collection, (2) data pre-processing, (3) designing a model, (4) training with suitable algorithms, and (5) model validation. Initially, the nine input features mentioned above were used to design the model without considering the effect of time (i.e., a steady-state model was developed). We start from the simple LR model and then compare its performance against two nonlinear models – RF and XGBoost. The model's performance was further enhanced by removing outliers. Finally, a complete comparison of all three models with and without outliers is considered in the analysis, and results are reported accordingly in the results and discussion section.

3.1. Linear Regression

In this section, the model is trained using multivariate linear regression, and the results are analyzed. The model was initially trained by considering the outliers and then checked the performance based on both training and testing datasets. The results clearly show that the presence of outliers in the training and testing datasets decreases the model's accuracy. The accuracy of the model was further improved by removing outliers from both training (0.64 %) and testing (0.32 %) datasets. The performance of both results is compared by calculating the R-Square value (R^2), Mean-Absolute-Error (MAE %), and Root-mean-square error (RMSE %). After removing the outliers, the R^2 value improved from 0.84 to

0.95 in training and from 0.82 to 0.96 in the testing datasets. Similarly, MAE (%) enhanced from 3.66% to 2.21% in training and 4.23% to 2.29% in testing datasets, and RMSE (%) improved from 5.5% to 3% in training and 6.3% to 2.9% in testing datasets. The mode of compressor operation, tank pressure, bottom layer temperature and LNG level are considered as important variables in the linear regression modelling.

In order to improve accuracy, two nonlinear machine learning algorithms Random Forest (RF) and XGBoost were also studied.

3.2. Random Forest

In this section, the model is trained and tested based on Random Forest (nonlinear algorithm. The value of R^2 remains 0.99 in both cases in training datasets, while it increases from 0.88 to 0.99 in testing datasets. Similarly, the value of MAE (%) and RMSE (%) remain the same at 0.49% and 1.23% with the training datasets, while it improves from 1.29% to 0.65% and 5.03% to 1.02% in testing datasets, respectively. The model indicates mode of compressor operation, LNG level and LNG density plays crucial role in the generation of BOG in the storage tank.

3.3. XGBoost

The performance of XGBoost algorithm is also good but lesser than the RF algorithm. The value of R^2 remains 0.98, which is unchanged with or without considering the outliers in training datasets, while it improves from 0.88 to 0.99 in testing datasets. Similarly, the MAE (%) and RMSE (%) value remains the same at 1.15% and 1.83% in training datasets, while it improves from 1.68% to 1% and 5.82% to 1.32% in the testing dataset, respectively. This algorithm suggested mode of compressor operation, LNG level, LNG outflow and LNG temperature are important input variables.

Table 1. *Comparison of the performance of LR, RF and XGBoost, based on R^2, RMSE (%) and MAE (%)*

		R^2		RMSE (%)		MAE (%)	
Models		*Training datasets*	*Testing datasets*	*Training datasets*	*Testing datasets*	*Training datasets*	*Testing datasets*
Linear regression	*With outliers*	0.84	0.82	5.50	6.30	3.66	4.23
	Without outliers	0.95	0.96	3.00	2.90	2.21	2.29
Random Forest	*With outliers*	0.99	0.88	1.23	5.03	0.49	1.29
	Without outliers	0.99	0.99	1.23	1.02	0.49	0.65
XGBoost	*With outliers*	0.98	0.88	1.83	5.82	1.15	1.68
	Without outliers	0.98	0.99	1.83	1.32	1.15	1.00

The comparison of all three models revealed that the mode of compressor operation has a significant impact on the rate of BOG generation. The compressor operation at lower capacity increases vapor pressure inside the tank, resulting in a lower rate of BOG generation. Another important variable suggested by the models are LNG level in the tank. The LNG level provides direct information about the volume of LNG in the storage tank. Thus, higher volume results more generation of BOG.

3.4. Performance of models

The performance of the three models (LR, RF and XGBoost) are compared on training and testing data sets with and without considering the outliers. The performance of the Linear regression model is the least as compared to the other two nonlinear models. Among the two nonlinear models, the RF has better performance than XGBoost. The performance of all three models can easily be visualized in Fig. 1 and table 1 based on R^2, MAE and RMSE values.

3.5. Sensitivity analysis

The sensitivity analysis of the model helps to calculate the contribution of each individual feature in the output features. Initially, all the measurable input features (9 features) are used for modelling, but it may be possible that the model is independent of some of the input features. These extra independent features increase the complexity of the model and are responsible for both overfit and increased simulation time for large datasets. To avoid these possibilities, a sensitivity analysis was performed and analyzed the features really crucial for the modelling by using Shapley additive explanation method (SHAP value). By considering all nine input features, the SHAP value is calculated and reported in Fig. 2. (a). The lowest SHAP value input features have to be removed, and checked the performance of the model. Similarly, one by one, input features have to be removed and reviewed the performance of the model until a significant decline is observed.

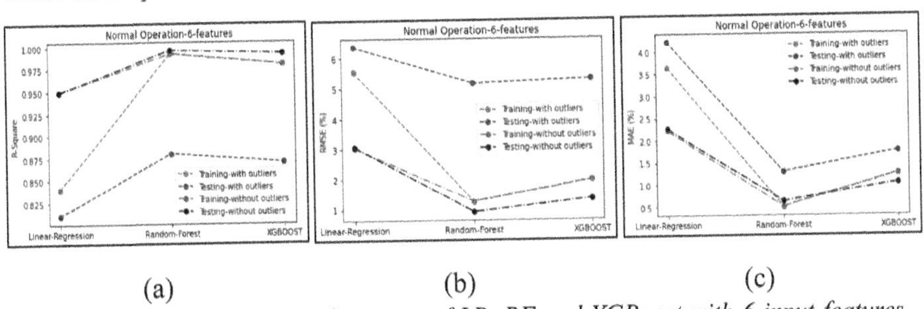

(a) (b) (c)

Fig. 1. Comparison of the performance of LR, RF and XGBoost with 6 input features, based on (a) R^2, (b) RMSE (%) and (c) MAE (%) with and without considering the outliers in both training and testing dataset

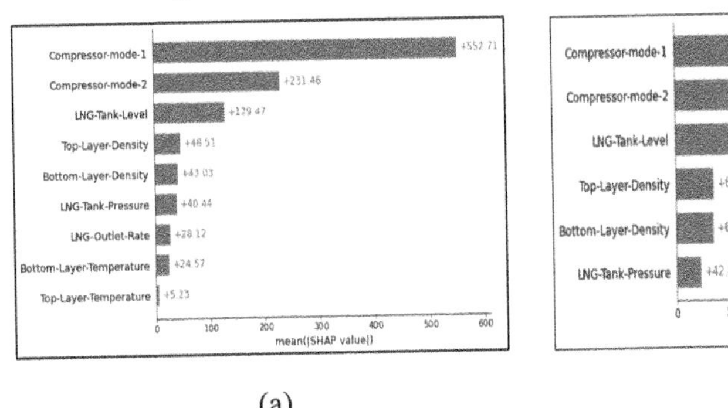

(a) (b)

Fig. 2. Comparison of the mean (/SHAP/) value by considering (a) 9 input features, and (b) 6 input features in the model

In this modelling, we would not have observed a significant change in the model's performance after removing the three lowest SHAP value (Fig. 2 (b)) input features (top-temperature, bottom-temperature and LNG flowrate). Thus, six input features have to be considered in the model and the performance reported in Fig. 1 (a), (b), and (c).

4. Conclusions

A data-driven model is used to predict the rate of boil-off-gas (BOR) generation during normal operation in an industrial LNG storage tank based on different linear and nonlinear machine learning (ML) algorithms. The three different ML algorithms (Linear regression, Random Forest and XGBOOST) are used to model the data by considering nine input features and checking the accuracy of the model. The performance of all three algorithms is analyzed based on R^2, MAE (%) and RMSE (%) values. Out of all three algorithms, the performance of Random Forest is considered to be the best, with R^2 value of 0.99 in training datasets with and without outliers. The performance of R^2 in testing datasets improve from 0.88 with outliers to 0.99 without outliers.

Further, sensitivity analysis is performed using the SHAP value method to find the important input features for the model. The input features with lesser SHAP values are removed one by one from the model and again checked the performance of the model. After removing top layer temperature, bottom layer temperature and LNG outlet flow rate, the performance of the model remains almost the same in both the training and testing datasets. This analysis indicates that the remaining six input features – compressor operation of 1 and 2, level of LNG, bottom LNG density, top LNG density and tank vapor pressure – play the most important role in BOG generation in the industrial LNG storage tank. Specifically, the mode of compressor operation and the LNG level are the most significant features. While the resulting model has good overall performance, it has several limitations. The dynamic behavior of the system, i.e., the temporal dimension is not considered in this model. The model also does not cover other operating modes such as LNG unloading, compressor maintenance, and special events like rollover. These will be addressed in our future work. Also, other input features like the effect of ambient temperature, and dimensions of the LNG tank would be incorporated into the model.

Acknowledgments

The authors gratefully acknowledge financial support from the Indian Institute of Technology Madras and the Ministry of Human Resource Development, Government of India, under the Scheme for Promotion of Academic and Research Collaboration (SPARC).

References

GIIGNL 2019 Annual Report - https://giignl.org/document/giignl-2019-annual-report

https://www.energy.gov/sites/prod/files/2018/11/f57/draft-eis-0531-port-delfin-lng-app-r-2016-07.pdf

Heestand, J.; Shipman, C.W.; Meader, J.W., 1983, A predictive model for rollover in stratified LNG tanks. AIChE J. 29, 199–207.

Bates, S.; Morrison, D.S., 1997, Modeling the behaviour of stratified liquid natural gas in storage tanks: A study of the rollover phenomenon. Int. J. Heat Mass Transf., 40, 1875–1884.

Saleem, A.; Farooq, S.; Karimi, I.A.; Banerjee, R., 2020, CFD Analysis of Stratification and Rollover Phenomena in an Industrial-Scale LNG Storage Tank. Ind. Eng. Chem. Res., 59, 14126–14144.

Zakaria, Z.; Kamarulzaman, K.; Samsuri, A., 2017, Rollover phenomena in liquefied natural gas storage: Analysis on heat and pressure distribution through CFD simulation. Int. J. Innov. Eng. Technol., 8, 392–400

Antonis Kokossis, Michael C. Georgiadis, Efstratios N. Pistikopoulos (Eds.)
PROCEEDINGS OF THE 33rd European Symposium on Computer Aided Process Engineering (ESCAPE33), June 18-21, 2023, Athens, Greece
 http://dx.doi.org/10.1016/B978-0-443-15274-0.50207-9

Synthesis of process configurations and solvent blend for CO_2 capture from CCGT power plants

Nahyan Arshad,[a,b] Ahmed Al Hajaj,[a,b]

[a] *Department of Chemical Engineering, Khalifa University of Science and Technology, Abu Dhabi, UAE*

[b] *Research and Innovation Center of CO_2 and Hydrogen, Abu Dhabi, UAE*

Abstract

Post-combustion CO_2 capture (PCCC) from flue gas of power plants is a promising route to mitigate the climate change in the near term. However, high energy consumption remains the major challenge of MEA based PCCC process. This study aims to investigate the potential of a hybrid approach by replacing MEA with an energy efficient MDEA/PZ blend and applying process configurations to have the maximum reduction in the regeneration energy. Rigorous rate-based model of a MDEA/PZ based PCCC process integrated with a 750 MW CCGT power plant was developed in Aspen Plus. In addition, the study examined the synergistic effects of combining three process configurations in a single flowsheet on the regeneration energy. The energy analysis results show that AIC+RSS+LVC is the energy-optimal combination as it provides the lowest regeneration energy (2.12 GJ/t_{CO2}). However, considering the significant electricity consumption by the lean vapor compressor, the equivalent energy results highlight that AIC+RSS+IHS is the optimum combination as it results in lowest equivalent energy (0.985 GJ/t_{CO2}) corresponding to an overall energy saving of 8.96%.

Keywords: Post-combustion CO_2 capture; MDEA/PZ; Process configurations; Energy savings

1. Introduction

The significant use of fossil fuels for energy generation has resulted in high concentration of CO_2 in the atmosphere which primarily contributes to the global climate change. Combined Cycle Gas Turbine (CCGT) power plants are rapidly adopted worldwide for energy generation compared to coal power plants due to higher energy efficiency, lower CO_2 emissions rate and flexible power output; hence, they can be used in conjunction with variable renewable energy sources (Cheng et al.,2022). One of the key pathways to minimize the anthropogenic CO_2 emissions in the short term is Post Combustion CO_2 Capture (PCCC) using amines due to its technological maturity and large-scale availability. Monoethanolamine (MEA) is the benchmark solvent in PCCC applications due to its low cost, low volatility, low viscosity, and high absorption kinetics at low CO_2 concentrations as in the case of flue gases from power plants. In the recent years, research experts have proposed various solvents with optimal properties for CO_2 capture such as ammonia, ionic liquids, deep eutectic solvents and water lean solvents. However, the commercial applications of these solvents are still immature and require further testing to evaluate their techno-economic feasibility at industrial scale (Li et al.,2016). One of the major limitations of MEA based PCCC process is the significant thermal energy requirement for CO_2 desorption which accounts for majority of the energy consumption and operational costs of the process (Li et al.,2016). Other limitations of MEA include high corrosion and degradation which increases the solvent makeup costs.

The two practical ways of reducing this high energy demand are the development of energy efficient solvents and synthesis of process configurations. Methyldiethanolamine (MDEA) is a potential solvent to replace MEA as it has attractive properties for CO_2 capture such as higher CO_2 absorption capacity, lower regeneration energy and degradation rate compared to MEA (Ibrahim et al.,2014). The major drawback of MDEA is the low CO_2 absorption rate which increases the capital costs associated with height of the absorber column. This limitation of MDEA can be improved by blending it with an absorption kinetics promoter. Among the amine family, Piperazine (PZ) has the highest CO_2 reaction kinetics as it rapidly forms carbamates with CO_2; hence it can be considered as an efficient absorption kinetics promoter in tertiary amines such as MDEA (Ibrahim et al.,2014). The experimental study by Bishnoi and Rochelle showed the absorption rate of MDEA/PZ blend is comparatively higher than MDEA/MEA and MDEA/DEA blends (2002). Khan et al. (2020) evaluated the CO_2 absorption rate and regeneration energy of MDEA/PZ at varying concentrations of PZ. The Aspen Plus results of the study revealed that the optimum concentration ratio of MDEA/PZ is 30/20 wt%. The other approach to reduce the high energy demand is process configurations which are modifications in PCC plants to reduce the parastic energy consumption of the capture process. The configurations evaluated in this paper include Absorber Inter Cooling (AIC), Rich Solvent Preheating (RSP), Rich Solvent Split (RSS), Rich Solvent Flash (RSF), Lean Vapor Compression (LVC), Rich Vapor Compression (RVC) and Inter Heated Stripper (IHS).

Majority of the studies on process configurations were based on MEA and very few studies have examined the potential of MDEA/PZ and configurations on the energy requirement of PCCC process. Therefore, this paper focuses on a combined approach of replacing MEA with MDEA/PZ blend and applying process configurations to maximize the reduction in the regeneration energy. In addition, previous work on process configurations have considered few individual configurations and single combination of three configurations due to convergence challenges. Hence, there is a need to examine several individual configurations and multiple combinations of three configurations in a single flowsheet to demonstrate the cumulative effect of combining three configurations on the regeneration energy of MDEA/PZ process. This work aims to evaluate the performance of seven process configurations and their combinations on the regeneration energy of MDEA/PZ based PCCC process integrated with a 750 MW CCGT power plant. Rate-based model was developed followed by a detailed energy analysis to identify the optimum combination interms of overall energy savings.

2. Methodology

2.1 Baseline CO_2 Capture Process

A conventional CO_2 capture process is shown below in Fig 1. The hot flue gas (106°C) composed of 74.32% N_2, 12.09% O_2, 8.67% H_2O, 4.04% CO_2 and 0.88% Ar from a 750 MW CCGT power plant was precooled to 40°C in a direct contact cooling (DCC) column. The cooled flue gas then enters the bottom of the absorber column with lean solvent fed from top of the absorber to allow efficient chemical absorption of CO_2 from flue gas into lean solvent. The capture rate is set at 90% after which the lean flue gas leaves from top of absorber to a waterwash unit to condense any escaped solvent vapor. The rich solvent from absorber is preheated in a Lean Rich Heat Exchanger (LRHX) followed by the desorption of CO_2 at elevated temperatures in a stripper column. The hot lean solvent from the reboiler is precooled in the LRHX after which it is recycled back to the top of absorber. The desorbed CO_2 is cooled in an overhead condenser and sent to a compression train which compresses the CO_2 product to 152 bars for storage purposes.

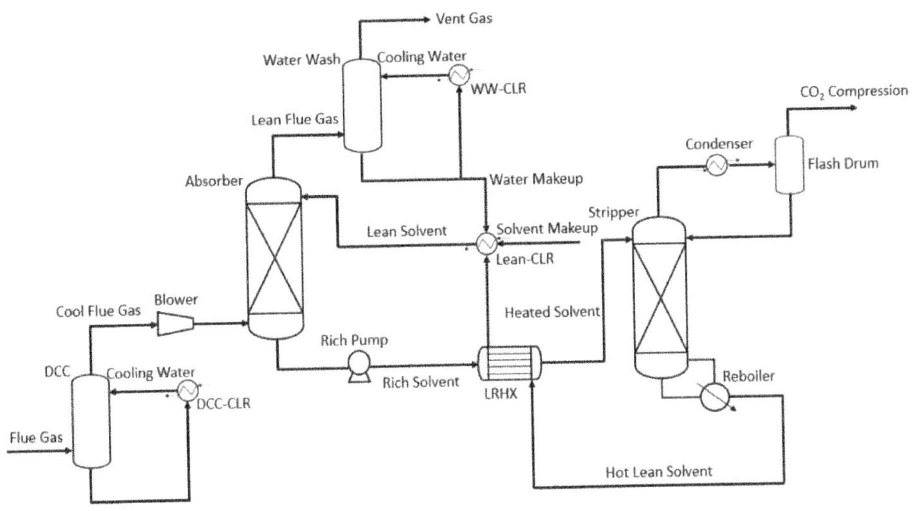

Figure 1: Conventional CO_2 Capture Process

2.2 Thermodynamic and Rate Based Model

The process model for the MDEA/PZ based CO_2 capture was developed in Aspen Plus V10 software. The liquid properties of the ionic species were computed by the Electrolyte Non Random Two Liquid (ENRTL) method while the vapor properties of the species were computed using Redlich Kwong (RK) Equation of State. The equilibrium and kinetic reactions of the MDEA-PZ-CO_2-H_2O system were the same as in the inbuilt MDEA/PZ CO_2 capture model in Aspen Plus.

The absorber and stripper columns were modeled using the rate-based modeling as it considers mass transfer resistance between phases and reaction kinetics assuming equilibrium only at the fluid interface. Hence, rate based modeling provide more reliable results for column design compared to equilibrium based modeling. The initial data for rate-based modeling such as packing specifications, film resistance, mass and heat transfer correlation methods, etc was obtained from the work of Li et al. (2016).

2.3 Equivalent Energy

Equivalent energy is a standard KPI which unifies the thermal and electrical energy consumptions to make a reasonable comparison of overall energy consumption of conventional process with process configurations such as LVC and RVC consuming additional electrical energy. The equation to determine equivalent energy is shown below where the first part represents the equivalent electrical penalty of thermal energy while the second part represents the electrical energy consumed by compressors and pumps in the process.

$$E_{eq} = 0.75Q_{reb}\left(\frac{T_{reb} + 10K - T_{sink}}{T_{reb} + 10K}\right) + E_{elec}$$

2.4 Process Configurations

The descriptions and schematics of process configurations evaluated in this study are summarized in Table 1:

Table 1: Description and Schematics of Process Configurations

Configuration	Description	Schematic
Absorber Inter Cooling	Extraction and cooling of a portion of semi-rich solvent to enhance the driving force for CO_2 absorption	Lean Flue Gas; Absorber; Lean Solvent; AIC; Flue Gas; Rich Solvent
Rich Solvent Preheating	Additional heating of the preheated rich solvent after LRHX to lower the amount of energy required to heat the solvent	CO_2/H_2O; Stripper; RSP; From LRHX; To LRHX; Lean Solvent
Rich Solvent Split	Splitting of rich solvent into cold stream which enters from top of stripper and heated stream which enters from middle to allow pre-stripping of CO_2 from hot vapors	Cold Solvent; Stripper; To Absorber; Heated Solvent; From Absorber; LRHX; Lean Solvent
Rich Solvent Flash	Flashing of the heated rich solvent after which vapor is sent to overhead condenser to reduce the heat required for CO_2 desorption	CO_2; Condenser; Stripper; Flash Drum; From LRHX; To LRHX; Lean Solvent
Lean Vapor Compression	Flashing of the lean solvent followed by the compression and reinjection of vapor into stripper to minimize steam requirement in the reboiler	CO_2/H_2O; Stripper; From LRHX; Compressor; Vapor; Flash drum; To LRHX; Lean Solvent

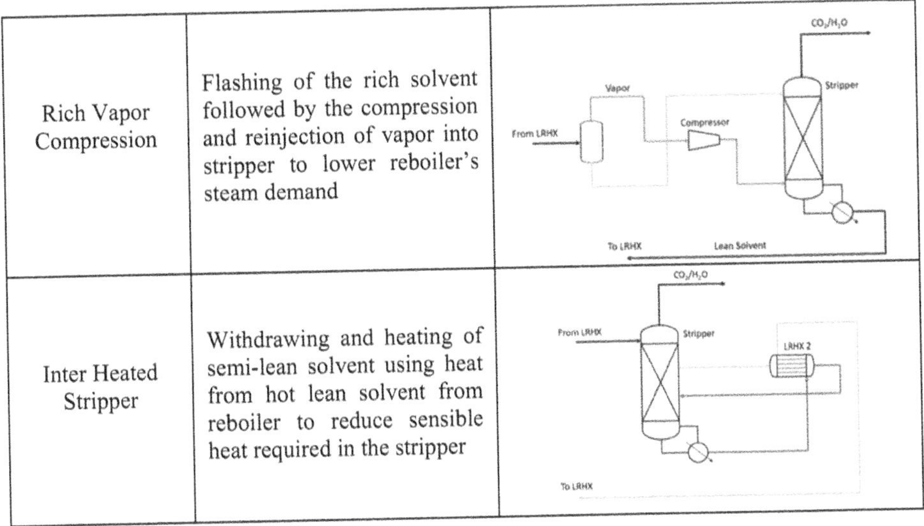

Rich Vapor Compression	Flashing of the rich solvent followed by the compression and reinjection of vapor into stripper to lower reboiler's steam demand	
Inter Heated Stripper	Withdrawing and heating of semi-lean solvent using heat from hot lean solvent from reboiler to reduce sensible heat required in the stripper	

3. Results and Discussion

The summary of regeneration energy, equivalent energy and percentage energy savings for the individual and combined process configurations are presented below in Table 2:

Table 2: Summary of Energy Analysis Results

Configuration	Regeneration Energy (GJ/t_{CO2})	Equivalent Energy (GJ/t_{CO2})	% Savings in Energy
Base Case	2.87	1.082	-
AIC	2.58	1.046	3.33
RSP	2.97	1.102	-1.85
RSS	2.76	1.070	1.12
RSF	2.87	1.083	-0.10
LVC	2.38	1.053	2.68
RVC	3.02	1.135	-4.90
IHS	2.72	1.058	2.22
AIC+RSS+LVC	2.12	0.996	7.95
AIC+RSS+IHS	2.30	0.985	8.96
AIC+IHS+LVC	2.18	1.002	7.40
RSS+IHS+LVC	2.22	1.011	6.56

3.1 Discussion of energy analysis results

The results presented in the above table demonstrate that AIC provided the highest savings in equivalent energy (3.33%) as it enhances the absorption capacity of MDEA/PZ blend, thus reducing the regeneration energy. RSP provided negative energy savings (-1.85%) due to consideration of additional heat consumption by the preheater in the overall energy analysis. The integration of waste heat from flue gas cooling process for the rich

MDEA/PZ preheating process is not possible due to significant difference in specific heat capacities of flue gas and rich MDEA/PZ. RSS resulted in an energy savings of 1.12% as there is less pre-stripping of CO_2 by hot vapors from heated fraction of amine, hence having less impact on the driving force for CO_2 desorption. RSF had negligible energy savings (0.10%) as the reduction in regeneration energy was almost equivalent to an increase in the sensible heat demand in the stripper due to flashing the heated rich solvent. LVC and RVC have similar mechansims; however, the results demonstrate that LVC provided positive energy savings (2.68%) while RVC resulted in negative energy savings (-4.90%). This is because the electrical energy consumption by rich amine compressor and pump is relatively higher than the thermal energy reduction by RVC. IHS provided an energy savings of 2.22% as the heating of semi lean solvent elevates the temperature of the stripper which enhances the CO_2 desorption kinetics, thus lowering the overall energy requirement.
Various combinations of three configurations were proposed, so the advantages of each configuration can be combined to have the maximum reduction in the regeneration energy. The energy analysis results show that all combinations resulted in positive energy savings with AIC+RSS+LVC providing the lowest regeneration energy (2.12 GJ/t_{CO2}). However, LVC consumes substantial amount of electrical energy which reflects in the equivalent energy results concluding that AIC+RSS+IHS as the optimal combination in terms of overall energy savings (8.96%). In comparison to MEA base case (1.202 GJ/t_{CO2}) present in the literature, the hybrid case (MDEA/PZ and AIC+RSS+IHS) provides 18.2% savings in the equivalent energy.

4. Conclusion and Future Work

This study investigated the performance of seven process configurations and their combinations applied to MDEA/PZ based PCC process attached with 750 MW CCGT power plant. In addition, the study demonstrated that combining three configurations showed higher energy savings than individual configurations due to combined advantages of each configuration. The results of the energy analysis conclude that AIC+RSS+IHS is the optimum combination based on the equivalent energy savings. The future work should include a detailed techno-economic analysis to gain deeper insights on the tradeoffs between energy savings and higher capture costs of different configurations in order to identify optimum combination in terms of energy and cost savings.

References

S. Bishnoi and G. Rochelle, 2002, Thermodynamics of Piperazine/Methyldiethanolamine/Water/Carbon Dioxide, Industrial & Engineering Chemistry Research, 41, 3, 604-612

F. Cheng, N. Patankar, S. Chakrabarti and J. Jenkins, 2022, Modeling the operational flexibility of natural gas combined cycle power plants coupled with flexible carbon capture and storage via solvent storage and flexible regeneration, International Journal of Greenhouse Gas Control, 118, 103686

A. Ibrahim, F. Ashour, A. Ghallab and M. Ali, 2014, Effects of piperazine on carbon dioxide removal from natural gas using aqueous methyl diethanol amine, Journal of Natural Gas Science and Engineering, 21, 894-899

B. Khan, A. Ullah, M. Saleem, A. Khan, M. Faiq and M. Haris, 2020, Energy Minimization in Piperazine Promoted MDEA-Based CO_2 Capture Process, Sustainability, 12, 20, 8524

K. Li, A. Cousins, H. Yu, P. Feron, M. Tade, W. Luo and J. Chen, 2016, Systematic study of aqueous monoethanolamine-based CO_2 capture process: model development and process improvement, Energy Science & Engineering, 4, 1, 23-39

Antonis Kokossis, Michael C. Georgiadis, Efstratios N. Pistikopoulos (Eds.)
PROCEEDINGS OF THE 33rd European Symposium on Computer Aided Process Engineering (ESCAPE33), June 18-21, 2023, Athens, Greece
 http://dx.doi.org/10.1016/B978-0-443-15274-0.50208-0

PREDICTION OF BIOMEDICAL WASTE GENERATION IN SANITARY EMERGENCIES FOR URBAN REGIONS USING MULTIVARIATE RECURRENT NEURAL NETWORKS

Nicolas Galvan-Alvarez [a], David Rojas-Casadiego [a], Viatcheslav Kafarov [a], David Romo-Bucheli [a]

[a] *Universidad Industrial de Santander, Bucaramanga, Santander, Colombia.*

Abstract

Biomedical waste (BMW) generation is severely affected by generalized sanitary emergencies such as epidemics, as shown recently during the COVID-19 pandemic. These sanitary emergencies often increase plastic use in personal protection items, single-use plastics, and other healthcare elements. This increase might surpass the capacity of the waste management mechanism of a specific region, leading to a potential increase in its population health risks. Predicting the trends of BMW generation is not straightforward because it depends on several variables associated with the local health system and the health emergency status. However, a substantial amount of work has been done in epidemics modelling. Our main hypothesis is that BMW generation is strongly associated with sanitary emergencies dynamics. We propose a simulation framework that uses historical data from an ongoing sanitary emergency to build a model that can predict BMW generation trends in urban regions of developing countries.

Keywords: Biomedical waste, Simulation, Epidemics, Neural networks, Developing countries.

1. Introduction

Biomedical waste (BMW) is defined as any waste produced during the diagnosis, treatment, or immunization of human or animal research activities or in the production or testing of biological or in health camps. Recently disposal of BMW has become a major problem all over the world due to the influence of health emergencies such as epidemics on the generation of this type of waste. Specifically, the COVID-19 pandemic has exacerbated the problem of BMW management in several regions around the world. For instance, the world health organization (WHO) estimates that between March 2020 and November 2021, approximately 87,000 tons were purchased and are presumed to have ended up as waste (WHO, 2022). Also, according to press releases in March 2021, the generation of medical waste (infectious and non-infectious) increased significantly (+370%) in Hubei Province, with a high proportion of plastics. Additionally, from January 20 to March 31, 2020, accumulated medical waste across China was estimated at 207 kt, and in Wuhan medical waste increased from a normal level of 40 tons per day (t/d) to a peak of 240 t/d, exceeding the maximum incineration capacity in that city of 49 t/d (Klemeš, 2020). The purpose of this work is to develop a prediction model to estimate the amount of BMW during different scenarios associated with sanitary emergencies, such as the COVID-19 pandemic. The main motivation of this type of simulation models would be to inform the government institutions in the selection of strategies for BMW management and avoid potential health risks.

2. Related Work

Several strategies have been used to simulate waste generation and treatment. Kannangara et al. carried out a simulation study for waste management. The purpose was to develop models for accurate prediction of municipal solid waste generation and diversion from demographic and socioeconomic variables (Kannangara, 2018). Additionally, in May 2021, an article was published by Vu et al. in which machine learning models are used for modelling municipal solid waste disposal rates during COVID-19 (Vu, 2021). On the other hand, Yu et al. proposed a reverse logistics network strategy to address the drastic increase in healthcare waste generation during the pandemic (Yu, 2020). This reverse logistics network aims to optimize decisions to establish temporary waste management facilities and transportation strategies at different stages. Similarly, in 2022 Chowdhury et al. carried out a simulation study by calculating the amount of waste generated per patient. A full estimate of the healthcare waste generated from March 2020 to May 2021, during the COVID-19 pandemic in the period, depended on whether the patient is infected, isolated, deceased, ICU or in quarantine (Chowdhury, 2022). The main contribution of this work is the development of a simulation strategy aiming to predict the amount of BMW in a specific geographical region based on epidemics variables. The simulation integrates a "Susceptible Exposed Infectious Recovered Deceased" (SEIR-D) compartmental model and a multivariate RNN (LSTM) model. The integrated strategy leverages the SEIRD model, broadly known in epidemics modelling, to analyze different sanitary emergency scenarios (pessimistic, neutral, and optimistic). The SEIR-D model generates synthetic data associated with the number of cases and deaths. These variables are then fed to the LSTM to simulate BMW generation rates. The simulation tool allows to study the influence of the sanitary emergency dynamics and its environmental impact.

3. Data Collection

3.1. India's COVID-19 dynamics and India Biomedical Waste Dataset

For this study, a dataset was constructed with daily values of BMW generation for five states of India. Monthly reports available from the central pollution control board (CPCB) of India's ministry of environment, forest, and climate change, with measurements of BMW during COVID-19 emergency (from June 2020 to December 2021) were retrieved (CPCB, 2022). For the COVID-19 dynamics, we added daily data on infected and deceased individuals for each of the states and territorial unions of India. The data used come from daily worldwide reports from the COVID-19 Data Repository by the Center for Systems Science and Engineering (CSSE) at Johns Hopkins University (Dong, 2020). Additionally, data associated with changes in residential mobility, workplaces, and public transportation from Google's Local Mobility reports were also included (Google, 2021).

3.2. COVID-19 Colombian Dataset and AMB Biomedical Waste Dataset

Data on COVID-19 cases and deaths in Bucaramanga Metropolitan Area (AMB in Spanish), were extracted from a dataset designed by the Colombian government, which includes COVID-19 cases and deaths for each municipality in Colombia. On the other hand, BMW data was obtained from a waste collection company in AMB. Moreover, data associated with changes in residential mobility, workplaces, and public transportation from Google's Local Mobility reports on COVID-19 were also included (Google, 2021).

4. Methods

The simulation model consists of two integrated components: a health emergency dynamic modelling component (COVID-19) and a BMW generation estimation component (Figure 1a).

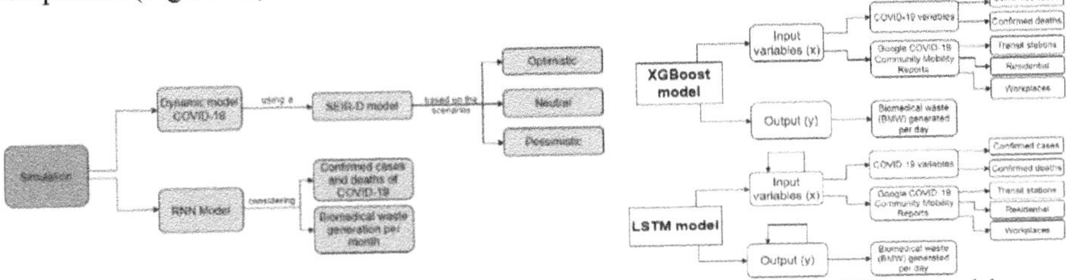

Figure 1. Left (a): Schematic of the model. Upper right (b): Inputs and outputs of the XGBoost model. Predictions for x days consider only the input variables of these x days. Down right (c): Inputs and outputs of the LSTM model. The network previous state is used to make a new prediction.

4.1. XGBoost Model

For the BMW estimation component, the XGBoost algorithm (Figure 1b) was implemented. The model is fed with variables such as confirmed COVID-19 cases and deaths per day, and mobility indices (residential, workplaces and transit stations). The task is to predict the BMW. It should be clarified that the model predicts data for the same day on which it is predicted, i.e., the data of the 5 variables mentioned for a certain day are entered, and the amount of BMW is predicted for that same day.

4.2. Multivariate LSTM Model

Subsequently, to develop a much more complete model, a recurrent neural network (RNN) was implemented for the prediction of such BMW, which is a type of neural network that is naturally adapted to the processing of time series data and other sequential data. To predict biomedical residuals based on COVID and mobility variables, a type of RNN called "*Long short-term memory*" (LSTM) was used (Figure 2). Introduced by Cho et al. in 2014, LSTM aims to solve the leakage gradient problem that comes with a standard recurrent neural network (Cho, 2014).

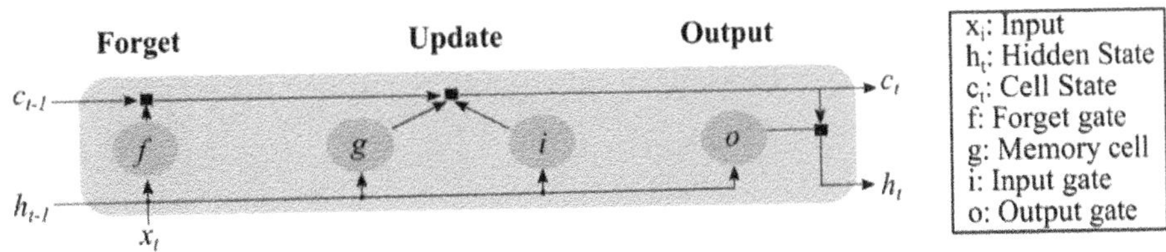

Figure 2. Structure of the LSTM cell (Adapted from Matlab, 2023).

Similar to the XGBoost model, the LSTM (Figure 1c) is fed with variables such as confirmed cases, deaths per day, and some mobility indices calculated during the pandemic (residential, workplaces, and transit stations). The model aims to predict BMW in a specific time window. However, the main difference with respect to the XGBoost is that it takes 7 days of input as a reference for the initialization of the model, and from then on it is fed back with the data it is predicting. This behavior is more convenient, since it allows us to make longer predictions with a smaller amount of data.

Several LSTM models for each region were trained with data from: new COVID-19 cases and deaths, mobility indexes, and amount of BMW from June 11st 2020 to June 29th, 2021. First, data is normalized by the region population. Then a windowing process is performed, which consists of dividing the data into windows of 7 days. The LSTM model is trained to predict the amount of BMW on the 8th day.

4.3. Epidemics Model (SEIR-D)

For the dynamics component (COVID-19) a SEIR-D compartmental epidemiological model was implemented, which is applied for the mathematical modelling of infectious diseases and allows dividing the individuals of a population N into five compartments: Susceptible (S), Exposed (E), Infected (I), Recovered (R) and Deceased (D), which are governed by the following set of ordinary differential equations:

$$\frac{dS}{dt} = -\frac{\beta SI}{N} \qquad \frac{dE}{dt} = \frac{\beta SI}{N} - \sigma E \qquad \frac{dD}{dt} = \mu I$$

$$\frac{dI}{dt} = \sigma E - \gamma I - \mu I \qquad \frac{dR}{dt} = \gamma I \qquad N = S + E + I + R + D$$

Equation 2.

Where β is the rate at which infected individuals interact with other infected individuals, σ is the estimated incubation period of the disease, γ is the duration of illness or recovery rate, μ is the mortality rate due to the disease, and R_0 is the transmission of the disease or basic reproduction number. Considering the variability in the trends and the different COVID-19 dynamics, pessimistic, neutral, and optimistic scenarios were proposed based on historical data from the AMB. The parameters for each of the scenarios were defined by optimizing the least squares of the residual data between the initial simulations and the historical data depending on whether there was a high number of cases (pessimistic scenario), a medium number of cases (neutral scenario), or a low number of cases (optimistic scenario). Table 1 shows the optimized parameters for each scenario.

Table 1. Parameters defined for the scenarios.

Scenario	R_0	β	σ	γ	μ
Optimistic	1.4	0.121	0.001	0.083	2.21E-10
Neutral	1.99	0.173	0.002	0.043	1.39E-04
Pessimistic	1.07	6.755	0.815	6.274	-2.28E-04

4.4. Performance metrics for predictive models

The mean squared error (MSE) was used to evaluate the models. MSE represents the mean squared difference between the estimated values and the true value, and can be expressed as $MSE = \frac{1}{n}\sum_{i=1}^{n}(x_i - y_i)^2$, where x_i are the estimated values, and y_i are the actual values of the corresponding vectors.

5. Results

This section will explain the results by analyzing the predictive capability of the LSTM model and subsequently integrating it with the full model.

5.1. Results of multivariate predictive models

When evaluating the prediction of BMW in the AMB, the XGBoost obtained a mean square error of 0.2551 for the 3 months predicted, obtaining the results presented in Figure 2a. As can be seen, the performance of this model is poor, which may be since in this prediction period there is a drop in BMW never seen before in the training window. On the other hand, Table 3 shows the results of the LSTM model based on its MSE. For the study, 5 states of India whose population numbers are similar to the AMB were taken. As can be seen, the best and worst performing region was Nagaland and the AMB respectively. The associated error for the Nagaland state is extremely low because in the predicted period its BMW and COVID variables remained somewhat constant (probably related to bad BMW data collection). On the other hand, the main cause of the poor results in the AMB region is that the model has not seen such a steep drop in the amount of BMW during the training phase, so it is very difficult to predict.

Table 2. MSE results for the LSTM Model.

Region	window-size		
	7	14	28
Puducherry	0.025368	0.140410	0.152633
Goa	0.030012	0.025989	0.007791
Manipur	0.018458	0.012636	0.016037
Nagaland	5.403496E-05	8.69878E-06	2.12439E-07
Mizoram	0.001639	0.000436	0.000371

More specifically, the results obtained in the AMB case study are presented in Figure 2b, obtaining a mean square error of 0.1186 tons on average for the 3 months predicted, using the model with a 7-day window.

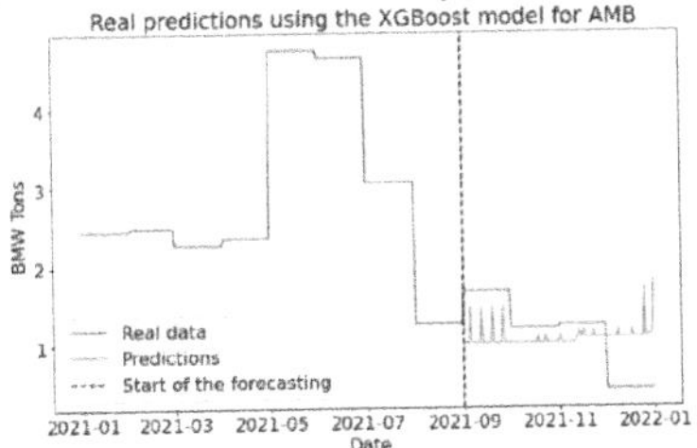

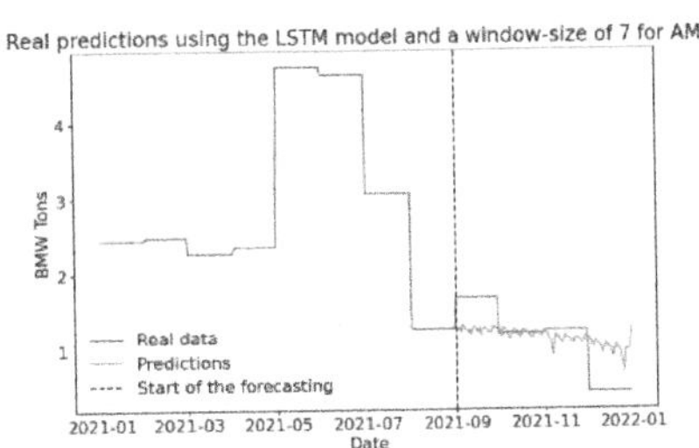

Figure 3. Right (a): LSTM model results for the AMB. Left (b): XGBoost model results for the AMB.

5.2. SEIR-D integrated model

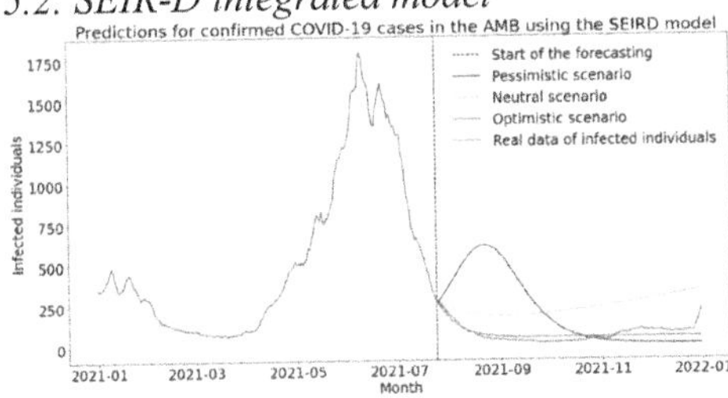

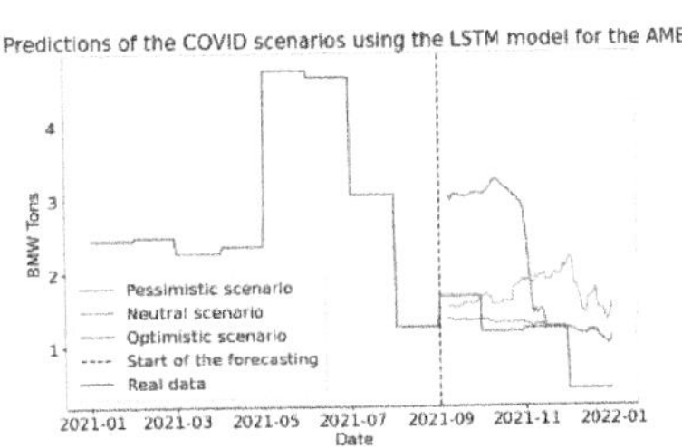

Figure 4. Left (a): SEIR-D model predictions for confirmed COVID-19 deaths in the AMB. Right (b): Results of the LSTM model for simulating BMW generation for the three COVID-19 scenarios in the AMB.

Once the parameters of Table 1 were defined, predictions were made with the SEIR-D model, and the results collected (Figure 3a). The pessimistic scenario has a peak of infected persons in September 2021. The cases, and deaths of the neutral scenario increase over time, and both variables remain constant for the optimistic scenario. The cases and deaths predicted by the SEIR-D model were used as input for the LSTM network (Figure 1c), and for the mobility indicators the actual data was used. Figure 3b shows the predictions for the proposed scenarios: the pessimistic scenario produces the highest amount of BMW at the peak of COVID cases in September 2021. The neutral scenario has the second largest BMW increase. Finally, for the optimistic scenario, the output is fairly constant and has the lowest amount of BMW. The simulation results are difficult to evaluate, but the general tendency of each scenario is coherent. Additionally, it is important to note that the closest scenario to the actual COVID19 data (optimistic scenario), also obtained the smallest error in terms of BMW generation (Table 3).

Table 3. MSE results for the simulated scenarios in the integrated model.

Scenario	COVID-19 cases	BMW
Pessimistic	67513.363322	1.516234
Neutral	29823.396586	0.700147
Optimistic	***1013.564280***	***0.172478***

6. Conclusions

In this work, a widely used epidemic model for modelling the transmission of diseases (SEIR-D) was successfully developed as an initial stage to simulate the process of plastic BMW generation. The SEIR-D model simulates different scenarios (pessimistic, neutral, and optimistic) for the development of a health emergency. The output of the SEIR-D compartmental model is then fed to multivariate recurrent models that can estimate BMW generation rates that are consistent with historical values observed in a specific geographic region. Simulation results demonstrate that the recurrent neural networks successfully capture the characteristics associated with the healthcare system of a particular location during a health emergency. We have integrated data from different domains (i.e, epidemiological (number of cases and deaths) and mobility indexes (based on Google mobility report)) to estimate the biomedical waste generation via a recurrent neural network. The model trained on historical data, could be useful afterwards to simulate different developing scenarios for sanitary emergencies. The simulation outputs are consistent with expected tendencies, showing that non-linear recurrent models using multimodal information might become an important modelling approach for complex scenarios.

Acknowledgments

Acknowledgements to the Universidad Industrial de Santander, the Vice Rector's Office for Research and Extension (VIE) and to the internal call for research "UIS contributions to the COVID-19 pandemic".

References

Cho, K., van Merrienboer, B., Gulcehre, C., Bahdanau, D., Bougares, F., Schwenk, H., & Bengio, Y. (2014). Learning Phrase Representations using RNN Encoder-Decoder for Statistical Machine Translation. ArXiv:1406.1078 [Cs, Stat]. http://arxiv.org/abs/1406.1078

COVID-19 biomedical waste management status. "CPCB | Central Pollution Control Board." https://cpcb.nic.in/covid-waste-management/ (accessed Jan. 20, 2022).

E. Dong, H. Du, and L. Gardner, "An interactive web-based dashboard to track COVID-19 in real time," The Lancet Infectious Diseases, vol. 20, no. 5, pp. 533–534, May 2020, doi: 10.1016/S1473-3099(20)30120-1.

Google, "COVID-19 Community Mobility Report". https://www.google.com/covid19/mobility?hl=es (accessed Jun. 21, 2021).

H. L. Vu, K. T. W. Ng, A. Richter, N. Karimi, and G. Kabir, "Modeling of municipal waste disposal rates during COVID-19 using separated waste fraction models," Science of The Total Environment, vol. 789, p. 148024, Oct. 2021, doi: 10.1016/j.scitotenv.2021.148024.

H. Yu, X. Sun, W. D. Solvang, and X. Zhao, "Reverse Logistics Network Design for Effective Management of Medical Waste in Epidemic Outbreaks: Insights from the Coronavirus Disease 2019 (COVID-19) Outbreak in Wuhan (China)," IJERPH, vol. 17, no. 5, p. 1770, Mar. 2020, doi: 10.3390/ijerph17051770.

J. J. Klemeš, Y. V. Fan, R. R. Tan, and P. Jiang, "Minimising the present and future plastic waste, energy and environmental footprints related to COVID-19," Renewable and Sustainable Energy Reviews, vol. 127, p. 109883, Jul. 2020, doi: 10.1016/j.rser.2020.109883.

Matlab, "Long Short-Term Memory (LSTM) Networks". https://es.mathworks.com/discovery/lstm.html (accessed Feb. 6, 2023).

M. Kannangara, R. Dua, L. Ahmadi, and F. Bensebaa, "Modeling and prediction of regional municipal solid waste generation and diversion in Canada using machine learning approaches," Waste Management, vol. 74, pp. 3–15, Apr. 2018, doi: 10.1016/j.wasman.2017.11.057.

T. Chowdhury, H. Chowdhury, M. S. Rahman, N. Hossain, A. Ahmed, and S. M. Sait, "Estimation of the healthcare waste generation during COVID-19 pandemic in Bangladesh," Science of The Total Environment, vol. 811, p. 152295, Mar. 2022, doi: 10.1016/j.scitotenv.2021.152295.

World Health Organization, "Tonnes of COVID-19 health care waste expose urgent need to improve waste management systems." https://www.who.int/news/item/01-02-2022-tonnes-of-covid-19-health-care-waste-expose-urgent-need-to-improve-waste-management-systems (accessed Nov. 15, 2022).

Antonis Kokossis, Michael C. Georgiadis, Efstratios N. Pistikopoulos (Eds.)
PROCEEDINGS OF THE 33rd European Symposium on Computer Aided Process Engineering (ESCAPE33), June 18-21, 2023, Athens, Greece
 http://dx.doi.org/10.1016/B978-0-443-15274-0.50209-2

Hybrid Modeling and Multi-Fidelity Approaches for Data-Driven Branch-and-Bound Optimization

Suryateja Ravutla[a], Jianyuan Zhai[b], Fani Boukouvala[a]

[a]*Department of Chemical and BiomolecularEngineering, Georgia Institute of Technology, Atlanta, GA 30332 USA*
[b]*Engineering & Data Sciences, Cargill Inc, Shanghai, 200031, China*
Email: sravutla3@gatech.edu

Abstract

High-Fidelity (HF) simulations are essential in quantitative analysis and decision making in engineering. In cases where explicit equations and/or derivatives are unavailable, or in the form of intractable nonlinear formulations, simulation-based optimization methods are used. We recently proposed a data-driven equivalent of spatial branch-and-bound that constructs underestimators of high-fidelity simulation data. Within this framework, low-fidelity surrogate data can also be used to inform underestimators. In this work, we utilize the recent advances in hybrid multifidelity surrogate modeling techniques to improve the validity of our underestimators, which leads to better bounds and incumbent optima with lower sampling requirements. Specifically, we show that by modeling the error between the high-fidelity and low-fidelity data, the surrogates learn more about the underlying function with less sampling requirements.

Keywords: Data-driven Optimization, Branch-and-bound, Convex underestimators, Multifidelity surrogate models, neural networks, support vector regression, hybrid modeling.

1. Introduction

In many engineering fields, it is desirable to simulate complex processes and use these quantifications of system performance for decision-making. Many such cases exist for chemical engineers, including molecular simulations, flowsheet simulations, computational fluid dynamic models, agent-based models, and more. These models are referred to as High-Fidelity (HF) simulations. But often these HF function evaluations are computationally expensive, and this implies that the number of function evaluations is limited by cost or time. Adding to this difficulty, in most cases, the objective functions and the constraints are only available as black-box evaluation function outputs (Fisher, Watson et al. 2020, van de Berg, Savage et al. 2022, Zhai and Boukouvala 2022). As a result, optimization of such systems becomes increasingly prohibitive. The absence of information on the system calls in for derivative-free optimization (DFO) or simulation-based optimization techniques. DFO techniques can be broadly classified into Sampling-based and Model-based methods. Numerous algorithms have been proposed in the recent years (Rios and Sahinidis 2013) for both cases. Sampling-based methods rely on comparing the function values directly and utilize this information to further sample adaptively. On the other hand, Model-based methods rely on constructing Machine Learning (ML) surrogate models as approximations of HF simulations with the aim to expedite the optimization process (Kim and Boukouvala 2020, Li, Dong et al. 2021). A disadvantage of using sampling-based methods is that they require too many samples to infer the optimum solution, while in surrogate model-based approaches, constructing an accurate surrogate can be challenging and computationally expensive. A solution that we have recently proposed combines advantages of both the methods in the form of a data-driven equivalent of the spatial branch-and-bound algorithm (DDSBB) (Zhai and Boukouvala 2022). The key idea of DDSBB is based on constructing convex underestimators of simulated data. A schematic of spatial branch-and-bound and its data-driven equivalent are shown is Figure 1.

These underestimators serve as relaxations and are convex, so they can be efficiently optimized, circumventing the task of directly optimizing nonconvex fitted surrogates. Samples drawn from the HF simulation serve as upper bounds (UB) of the global optimum and the minimum of the convex underestimator serve as lower bounds (LB). The search space is then progressively partitioned by using branching, node selection and pruning rules and adaptively sampling in the non-pruned subspaces. To build underestimators, using N samples, the underlying formulation is shown in Equation 1.

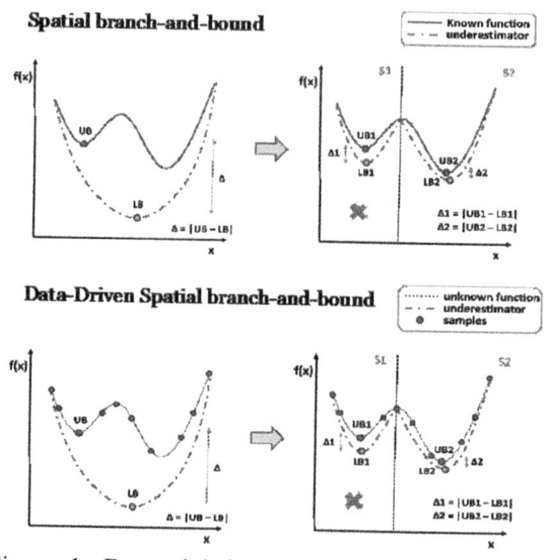

Figure 1: Deterministic Spatial Branch-and-Bound, and Data-Driven Spatial Branch-and-Bound, and the process of branching and bounding in both variants

$$\begin{aligned} & min_{a,b,c}\textstyle\sum_i^N \left(f(\boldsymbol{x}_i) - f_{lb}(\boldsymbol{x}_i)\right) \\ & s.t. f(\boldsymbol{x}_i) - f_{lb}(\boldsymbol{x}_i) \geq 0 \quad \forall\, i = 1 \; to \, N \\ & f_{lb}(\boldsymbol{x}_i) = \boldsymbol{a}\boldsymbol{x}_i^2 + \boldsymbol{b}\boldsymbol{x}_i + c \quad \forall\, i = 1 \; to \, N \\ & \boldsymbol{a} \geq 0, \qquad \boldsymbol{a}, \boldsymbol{b} \in \mathbb{R}^D, \; c \in \mathbb{R} \end{aligned} \tag{1}$$

In contrast to the conventional sample-based and surrogate-based approaches, DDSBB employs some of the positive aspects of both. Like the sample-based approaches, it adaptively samples in the search space and uses this information to improve the bounding of the original function, by constructing the relaxations and pruning the subspaces that are not promising. It also utilizes surrogates to construct Low fidelity (LF) data but does not directly optimize the surrogates or rely on a single surrogate prediction. This LF data can be utilized along with the HF data to build the convex relaxations. In our recent work (Zhai and Boukouvala 2022), we have shown that by jointly using LF and HF data (multifidelity MF), we can optimize a higher fraction of benchmark problems

While the MF approach employed in DDSBB showed promising performance, a fraction of benchmark studies was still not optimized given limits on sampling requirements. In this current work, we improve the performance of our framework by incorporating more advanced hybrid multifidelity modeling techniques. Specifically, using the same amount of HF data as before, we attempt to learn more about our underlying black-box problem by modeling the error between the HF and LF data. The hypothesis is that this will overall improve our underestimators, and consequently the overall efficiency of the DDSBB approach, without increasing sampling requirements. Recent studies show that there is an increasing amount of focus in constructing surrogate models that combine data with different fidelity (Meng and Karniadakis 2020, Bradley, Kim et al. 2022). These Multifidelity Surrogate Models (MFSMs) exploit the relation between LF data and HF data. In this study we integrate these MFSMs into the DDSBB architecture and quantify their performance by benchmarking the method on 2-10 dimensional black-box optimization problems.

2. Overview and explored methods

2.1.1. *Multifidelity approach in DDSBB*

The overview of DDSBB is shown in Figure 2. Initially, an input design is generated in the variable search domain by employing Latin Hypercube Sampling (LHS) to generate HF samples. This HF samples are then used for constructing quadratic underestimators by employing the formulation shown in Equation (1). This is shown in the figure using a solid red line. Subsequently, a set of branching, node selection and pruning rules are used to adaptively add samples in the non-pruned subspaces until convergence. Alternatively, as shown in red dotted line, if the MF approach is selected, these HF samples are used to generate LF surrogate models and LF samples. These LF samples are then used along with the HF samples to construct the underestimators. Currently, DDSBB has the capability to use Support Vector Regression (SVR), Neural Networks (NN) and Gaussian Process Regression (GPR) as the surrogate options. For the rest of the study, we utilize SVR as the surrogate option.

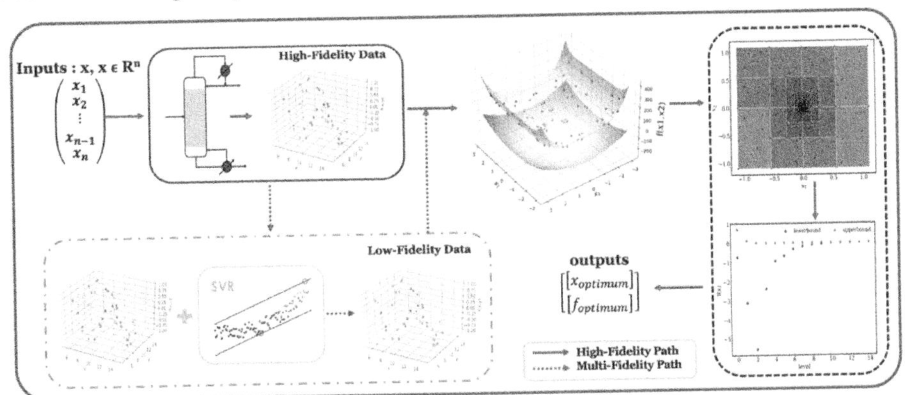

Figure 2: Overview of DDSBB. High-fidelity approach is shown in solid red line and multi-fidelity approach is shown in dotted red line.

2.1.2. *Multifidelity Surrogate models to improve LF data*

There is an inherent trade-off between the number of HF samples that are used to construct LF surrogates, and the accuracy of the constructed surrogates. Also, adding the LF samples makes the convex underestimators more conservative with respect to bounding the function. We have found that in certain cases this leads to improving their validity, and as a result it leads to locating the global optimum of challenging benchmarks (Zhai and Boukouvala 2022). At the same time, LF surrogate predictions can make the underestimators overly conservative, thus leading to large sampling requirements for convergence. In this work, our hypothesis is that, MFSMs can be used to improve the accuracy of the LF data, which leads to improvement in underestimator validity. A widely used correlation to build MFSMs is: $y_H = \rho(x)y_L + \delta(x)$ where y_L, y_H represent the low and high-fidelity data respectively, $\rho(x)$ is multiplicative correlation surrogate and $\delta(x)$ is the additive surrogate. In a more general way, we can re-write it as $y_H = F(x, y_L)$. To establish a correlation between the HF and LF data, one would need a HF and a LF model to generate that data (Meng and Karniadakis 2020, Bradley, Kim et al. 2022). In case an LF model is available, it can be directly used. In cases where the LF model is not available, we propose a framework (Workflow 1) shown below to create a LF model using a fraction of available data.

Workflow 1: Constructing MFSMs

Let the data set $[\boldsymbol{x}_{HF}, y_{HF}]_{tot}$ represent the complete HF data.
Generate a training set data $[\boldsymbol{x}_{HF}, y_{HF}] = 75\%[\boldsymbol{x}_{HF}, y_{HF}]_{tot}$

TRAIN SVR

Set $\boldsymbol{x}_{HF} \leftarrow$ *input* and $y_{HF} \leftarrow$ *output,*
While termination criteria **not true:**
Calculate $MSE_{SVR} = \frac{1}{N_{LF}}\sum_{i=1}^{N_{LF}}(|y_{LF} - y_{HF}|^2) + \beta_1 \parallel \phi_{SVR} \parallel_2$
Tune SVR parameters

Generate LF dataset $[\boldsymbol{x}_{HF}, y_{LF}]_{tot}$ using HF input $[\boldsymbol{x}_{HF}]_{tot}$
Utilize the correlation $y_H = F(x, y_L)$ to model error between LF and HF outputs $[y_{LF}]_{tot}$ and $[y_{HF}]_{tot}$ respectively using a NN

TRAIN NN

Set $[\boldsymbol{x}_{HF}, y_{LF}]_{tot} \leftarrow$ *input* and $y_{HF} \leftarrow$ *output,*
While termination criteria **not true:**
Calculate $MSE_{NN} = \frac{1}{N_{HF}}\sum_{i=1}^{N_{HF}}(|y_{LF}^* - y_{HF}|^2) + \beta_2 \parallel \phi_{NN} \parallel_2$
Tune NN parameters

In the Workflow 1, y_{LF}, y_{LF}^* represent output from SVR and NN. N_{LF}, N_{HF} is the number of training data points and total HF data points. β_1, β_2 represent the regularization weights and ϕ_{SVR}, ϕ_{NN} represent the associated parameters with SVR and NN respectively. A schematic for the Workflow 1 is shown in Figure 3.

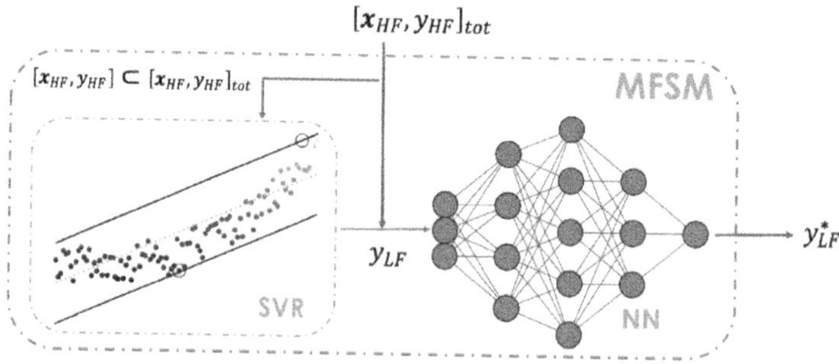

Figure 3: A schematic of the workflow for fitting the MFSMs. Part of the total HF data set (shown in blue) is used to train LF SVR model and generate LF data y_{LF}

3. Results and discussion

To understand and visualize the effect of MF and MFSM approach on underestimator construction, we first take a 1- dimensional case study. Let us consider that the function $f(x) = sin(x) + sin(10x/3)$ is available as a black-box function, with $x \in [0, 9]$ and known global solution at $xopt = 5.14574$. Figure 4 shows underestimators constructed using HF, MF and MFSM approaches. We can see among all three approaches, the optimum solution identified is more accurate in MFSM approach and LF data in case of MFSM approach is more accurate in comparison to MF approach, without increasing the number of HF data collected.

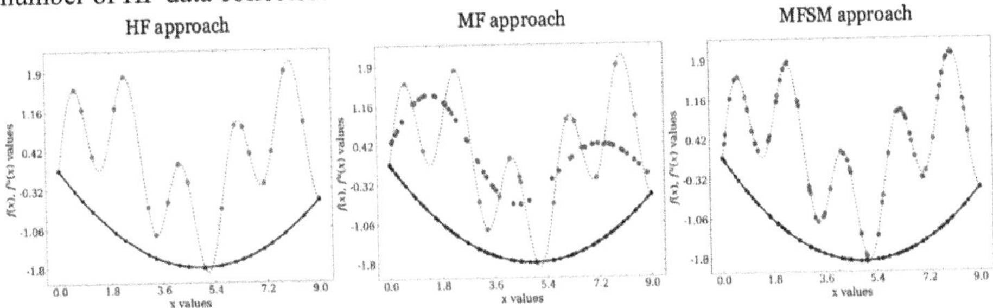

Figure 4: Comparison of constructed underestimators and LF data using HF, MF and MFSM approaches. Black-box function is shown in red dotted line, HF samples in red circles and LF samples in green circles. Underestimator is shown in solid black line.

Next, we test the performance of all three approaches on a large set of continuous box-constrained benchmark problems with known global solutions (BARON 2022). The benchmark problems were divided into two groups based on their dimensionality: lower dimensional group containing 118 problems with 2-3 variables and higher dimensional group containing 63 problems with 4-10 variables. All three approaches were initialized with 10*dimension+1 initial samples and converge when the $|UB - LB| \leq 0.05$ or $|UB - LB|/|LB| \leq 0.001$. For the performance analysis, we use the criterion $f_{best} \leq max(f^* + 0.01, (1.01)f^*)$ to allow a tolerance limit towards the optimal solution found. $f^*$and f_{best} represent the known global solution and the solution reported by DDSBB as optimum. We study the fractions of problems solved with sampling and CPU requirements based on pre and post convergence solutions. *Pre-convergence* and *post-convergence* solutions represent the optimum solution found by the algorithm *before* and *after* closing the UB-LB gap respectively. The performance curves for all three approaches are shown in Figure 5. In Figure 5A, we can see that the MFSM approach solves higher fraction of benchmark problems, and the pre-convergence solution quality is better in both low and high dimensions groups. In Figure 5B, we show the CPU requirements for the three approaches. The HF approach does not involve any surrogate fitting, so the CPU time largely corresponds to HF sampling. On the other hand, CPU time in MF and MFSM approaches also include surrogate modeling costs. As expected MFSM approach takes higher CPU time for fitting the complex MFSM structure. Thus, the advantage of this method is expected to be even greater when sampling cost increases.

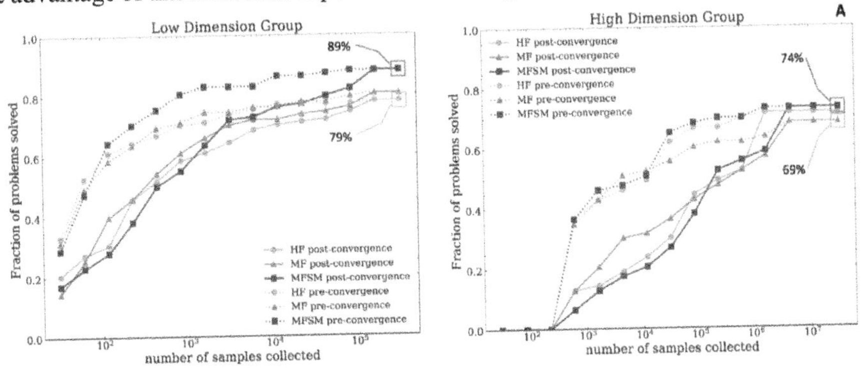

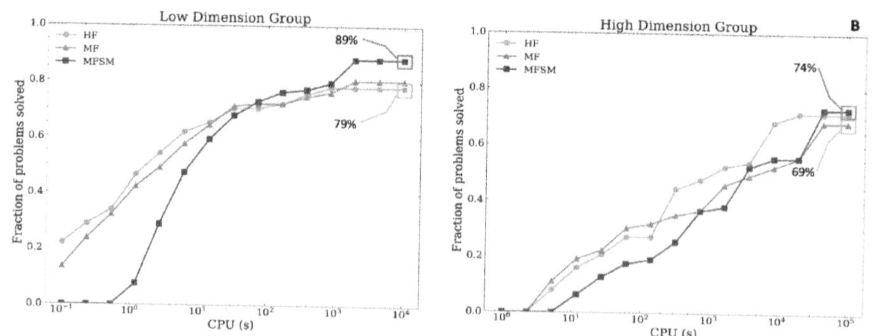

Figure 5: Performance curves of HF, MF, MFSM approaches. A) Fraction of problems solved vs sampling requirement, compared with pre and post convergence solutions reported by the algorithm. B) Fraction of problems solved vs CPU requirement.

4. Conclusions

In this work, we utilized multifidelity surrogate models as composite structures to model the error between HF and LF data to improve the validity of constructed data-driven underestimators embedded within a branch-and-bound framework. Results show that using composite/hybrid multifidelity models for surrogate-based optimization is promising, because it leads to more accurate surrogates with the same sampling cost but requires additional CPU time for training.

5. Acknowledgements

The authors acknowledge support from the National Science Foundation (NSF-1944678)

References

BARON (2022). Bound-constrained programs.

Bradley, W., J. Kim, Z. Kilwein, L. Blakely, M. Eydenberg, J. Jalvin, C. Laird and F. Boukouvala (2022). "Perspectives on the integration between first-principles and data-driven modeling." Computers & Chemical Engineering **166**: 107898.

Fisher, O. J., N. J. Watson, J. E. Escrig, R. Witt, L. Porcu, D. Bacon, M. Rigley and R. L. Gomes (2020). "Considerations, challenges and opportunities when developing data-driven models for process manufacturing systems." Computers & Chemical Engineering **140**: 106881-106881.

Kim, S. H. and F. Boukouvala (2020). "Surrogate-based optimization for mixed-integer nonlinear problems." Computers & Chemical Engineering **140**: 106847-106847.

Li, Z., Z. Dong, Z. Liang and Z. Ding (2021). "Surrogate-based distributed optimisation for expensive black-box functions." Automatica **125**: 109407.

Meng, X. and G. E. Karniadakis (2020). "A composite neural network that learns from multi-fidelity data: Application to function approximation and inverse PDE problems." Journal of Computational Physics **401**: 109020-109020.

Rios, L. M. and N. V. Sahinidis (2013). "Derivative-free optimization: a review of algorithms and comparison of software implementations." Journal of Global Optimization **56**(3): 1247-1293.

van de Berg, D., T. Savage, P. Petsagkourakis, D. Zhang, N. Shah and E. A. del Rio-Chanona (2022). "Data-driven optimization for process systems engineering applications." Chemical Engineering Science **248**: 117135-117135.

Zhai, J. and F. Boukouvala (2022). "Data-driven spatial branch-and-bound algorithms for box-constrained simulation-based optimization." Journal of Global Optimization **82**(1): 21-50.

Antonis Kokossis, Michael C. Georgiadis, Efstratios N. Pistikopoulos (Eds.)
PROCEEDINGS OF THE 33rd European Symposium on Computer Aided Process Engineering (ESCAPE33), June 18-21, 2023, Athens, Greece
 http://dx.doi.org/10.1016/B978-0-443-15274-0.50210-9

Intensified Process for the Production of Biojet Fuel from Mexican Biomass

Rosa Angelica Morales-Gutierrez[a], Eduardo Sánchez-Ramírez[a], Brenda Huerta-Rosas[a], Juan José Quíroz-Ramírez[b], Gabriel Contreras-Zarazua[b], Juan Gabriel Segovia-Hernández[a]

[a] *Universidad de Guanajuato, Noria Alta S/N, Guanajuato 36050, México*
[b] *CONACyT – CIATEC A.C. Centro de Innovación Aplicada en Tecnologías Competitivas, Omega 201 Col. Industrial Delta, León 37545, México*

Abstract

The use of fossil fuels has direct environmental impacts, including the use of jet fuel. An alternative to jet fuel is biojet fuel, which can be produced from biomass. Recently the ATJ (alcohol to jet) process has been promoted as an alternative that may offer better biojet production performance compared to some other process routes. Although there are several proposals for jet fuel production using the ATJ route, no intensified alternative has been reported to increase the performance, and decrease the economic and environmental impact. In this proposal, an intensified ATJ production process is presented, considering the intensification of the pretreatment stage using a saccharification-fermentation reactor, ethanol purification considering a divided wall column, and a distillation column with two reaction zones for biojet fuel production. Once the intensified proposal was evaluated in a multi-objective optimization framework, a saving of 18% in the environmental impact was obtained, and a saving of 12% in the environmental impact.

Keywords: biojet production, ATJ process, process intensification.

1. Introduction

Despite the COVID-19 pandemic's reduced jet fuel demand, growth in the services provided by the industry is anticipated once more. This poses a difficulty, as does the fuel's intrinsic environmental impact throughout its life cycle, in an economic climate that is still dependent on finite oil. Given this, using biojet fuel as a renewable aviation fuel is a possible alternative. The production of biojet can be done by several routes, for example, from the conversion of oils. There are some other routes such as hydroprocessing of fatty acids and esters (HEFA), catalytic hydro thermolysis (CH) and hydroprocessing to depolymerized cellulosic biojet (HDCJ). These processes are not very feasible despite their low-cost processing. There are other routes to obtain it, for example, starting from alcohol (Alcohol-to-jet, ATJ) that was recently certified in April 2016 by ASTM. The ATJ process consists of 4 stages starting from ethanol, i) dehydration, ii) oligomerization, iii) hydrogenation, and iv) purification. The ATJ route has bioethanol as a raw material, which has lignocellulosic agroindustrial waste as a precursor raw material that does not require additional water or a long time to obtain it. However, one of the disadvantages is that it is a relatively expensive process and is still not sufficiently competitive in the market, compared to the conventional jet fuel production process. The process of obtaining biojet fuel has been previously studied by several authors; however, it is a process with a wide field for improvement.

Process intensification, on the other hand, is an improvement strategy that can involve the reduction of equipment size, the development of several operations in a single piece of equipment, the reduction of process times, etc. An example of an intensified process is the reactive distillation column, where a chemical reaction and a separation process based on liquid-vapor equilibrium can be carried out. Taking into consideration that the process of obtaining biojet fuel has already been previously studied and applied through the ATJ route; however, it is a process with a wide field of improvements where process intensification strategies can be implemented. Thus, the goal of this work was to design an intensified process to produce biojet fuel from Mexican lignocellulosic biomass (corn stover, and cane bagasse) having alcohols as intermediates. The process was designed and modeled considering an intensified process for pretreatment/hydrolysis/fermentation/purification for the biomass-ethanol process.

2. Description of the intensified process

The creation of intermediate alcohols is necessary for the whole process of producing biojet fuel from lignocellulosic biomass. The general process for producing alcohols from lignocellulosic agricultural waste includes the following steps: (1) biomass pretreatment (2) enzymatic hydrolysis (3) fermentation of sugars (4) and purification of the alcohols produced. In this manner, Figure 1 shows how the conventional turns into an intensified process in some sections of the ATJ process.

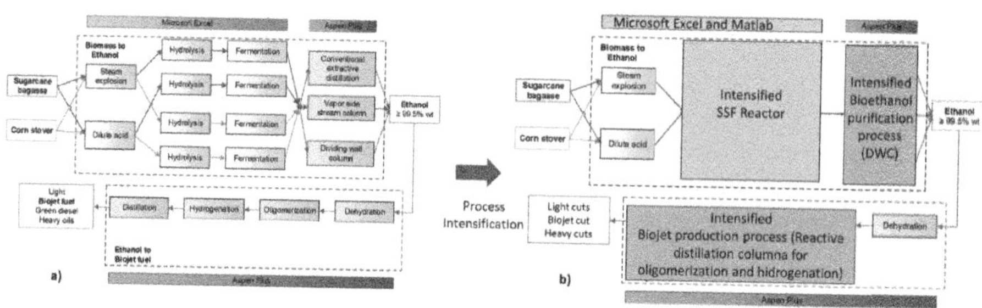

Figure 1. Description of conventional (a) and intensified (b) biojet fuel production.

The main difference between the conventional process and the intensified process is the use of the intensified fermentation-saccharification (SSF) reactor instead of separate process steps. Additionally, an intensified process is used for bioethanol separation. Finally, in the final part of the process, the oligomerization, hydrogenation, and purification stages are carried out simultaneously using a double-reactive distillation column.

3. Case study

According to SENER (2017), Mexico needs to replace at least 5.5% of its current demand for conventional jet fuel with biojet fuel, or 258 million liters, to develop a viable biojet fuel market. Two pretreatment alternatives, dilute acid, and steam explosion, are considered in this proposal. For the hydrolysis and fermentation section, an intensified saccharification-fermentation (SSF) reactor will be considered. Once an effluent of ethanol, water, etc. has been produced from the reactor, it will be purified in an intensified separation scheme (divided wall column). The alcohol is dehydrated in a dehydration reactor to obtain ethylene. Finally, starting from ethylene, the biojet fuel stream is obtained.

4. Methodology and Process Modeling

The two feedstocks and pretreatments were combined to create the biomass-to-ethanol process, which was created using the superstructure scheme. Kinetic data were gathered from the literature to describe the pretreatment. The modeling of the intensified saccharification-fermentation reactor was carried out in the Matlab computational package considering the kinetics reported by Kadam et al. 2004. The S. cerevisiae strain was considered for fermentation. Aspen Plus was used to simulate the ethanol purification zone and the module ethanol-to-biojet fuel using the NRTL and ENRTL thermodynamic models, respectively. A column sequence with a vapor side stream and a column sequence of dividing wall (DWC) was proposed as enhanced methods because the expense of separation represents a well-known area of opportunity. For the dehydration reaction, a performance reactor as reported by Rivas-Interian (2022) was considered. For the replacement of the oligomerization and hydrogenation process by a reactive double distillation column, the approach previously proposed by Heydenrych et al. was considered (Heydenrych et al., 2001). The complexity of the system (alkenes starting from C2) is reflected in the following chemical equations.

$2\,(C_2H_4) \rightarrow C_4H_8$ (1) $C_4H_8 + H_2 \rightarrow C_4H_{10}$ (9)

$C_2H_4 + C_4H_8 \rightarrow C_6H_{12}$ (2) $C_6H_{12} + H_2 \rightarrow C_6H_{14}$ (10)

$C_2H_4 + C_6H_{12} \rightarrow C_8H_{16}$ (3) $C_8H_{16} + H_2 \rightarrow C_8H_{18}$ (11)

$C_2H_4 + C_8H_{16} \rightarrow C_{10}H_{20}$ (4) $C_{10}H_{20} + H_2 \rightarrow C_{10}H_{22}$ (12)

$C_2H_4 + C_{10}H_{20} \rightarrow C_{12}H_{24}$ (5) $C_{12}H_{24} + H_2 \rightarrow C_{12}H_{26}$ (13)

$C_2H_4 + C_{12}H_{24} \rightarrow C_{14}H_{28}$ (6) $C_{14}H_{28} + H_2 \rightarrow C_{14}H_{30}$ (14)

$C_2H_4 + C_{14}H_{28} \rightarrow C_{16}H_{32}$ (7) $C_{16}H_{32} + H_2 \rightarrow C_{16}H_{34}$ (15)

$2(C_4H_8) \rightarrow C_8H_{16}$ (8) $C_8H_{16} + H_2 \rightarrow C_8H_{18}$ (16)

The estimated Kinect parameters of power law are shown in Table 1.

Table 1. Estimated kinetic parameters of the power-law model in equations 1-16

Oligomerization-Hydrogenation production

Reaction Number	**k**	**E (kJ/kmol)**	**Reaction Number**	**k**	**E (kJ/kmol)**
1 (C2)	38728.5907	4.623	9 (C2)	6146.3487	5.1584
2 (C4)	74825.7294	14.934	10 (C4)	7150.2255	2.4686
3 (C6)	48935.42	164.8806	11 (C6)	6409.9637	5.5759
4 (C8)	110614.018	16.2112	12 (C8)	2465.1438	2.8607
5 (C10)	83400.8644	17.5098	13 (C10)	6052.2691	7.6072
6 (C12)	116517.955	21.4748	14 (C12)	4027.1747	7.9935
7 (C14)	27140.8737	19.7567	15 (C14)	1723.794	6.935
8 (C16)	3349.1216	158.5374	16 (C16)	746.1161	6.0205

5. Process Optimization

An optimization effort is implied by the quest for the design and operation parameters that maximize savings and minimize the environmental effect. The following describes these goals as well as the optimization method.

5.1. Objective functions

The total annual cost (TAC) and eco-indicator-99 (EI99) were chosen as the objective functions to evaluate the process' sustainability. Equations 17 and 18 were used to calculate them, respectively:

$$TAC = \frac{Capital\ cost}{Recovery\ time} + Operation\ cost \quad (17) \quad EI99 = \sum_b \sum_d \sum_{k \in K} \delta_d \omega_d \beta_b \alpha_{b,k} \quad (18)$$

Where β_b is the total amount of chemical b released per unit of reference flow due to direct emissions, $\alpha_{b,k}$ is the damage caused in category k per unit of chemical b released to the environment, ω_d is a weighting factor for damage in categories d, and δ_d is the normalization factor for damage of category d. Guthrie's approach was used to compute the capital cost. The annual production of alcohol was also factored into the ethanol production process to meet Mexico's demand for biojet fuel, which accounts for 5.5% of the country's demand for conventional jet fuel (SENER, 2017). Following is a description of the objective function:

$$\begin{gathered}\text{Min}(TAC, EI99, IR) = f(X) \\ \text{Subject to } \vec{x_m} > \vec{y_m} \qquad (19)\end{gathered}$$

Being the variables involved in the whole modeling process can serve as degrees of freedom of the process. The Differential Evolution with Tabu List (DETL) stochastic approach was used to optimize the case of research. The best design parameters and the best design were chosen using this methodology. It was carried out on a hybrid platform that combined Matlab, Aspen Plus, and Microsoft Excel. The following criteria were employed: 120 individuals, 1000 generations, 50% of the entire population on the tabu list, 0.0001 tabu radius, and probability of crossover and mutation of 0.9 and 0.3, respectively.

6. Results

In the following section, the results obtained will be shown. The results will be shown in three sections: general pretreatment section, ethanol purification, and biojet production.

6.1.1. Pretreatment section

As mentioned in the methodology section, two scenarios were analyzed considering two types of feedstock. As shown in Figure 2, the use of an intensified strategy (SSF reactor) considerably reduces costs compared to a conventional process in both pretreatments and with the two feedstocks considered. Figure 2 shows the optimization results of all costs associated with the entire reaction section for both processes (conventional and intensified) when considering various pretreatments. Regarding the ethanol purification process, a direct comparison with the conventional process was not performed, since in the previous work of Rivas-Interian et al. (2022) a substantial saving of the intensified process versus the conventional one was observed. Thus, Figure 3a) shows the Pareto front of the optimization results schemes of all alternatives that accomplish recovery and purity restrictions considering economic and environmental objectives for ethanol purification. On the other hand, Figure 3b) shows similar information for the biojet fuel section after being fed with ethane to be converted by oligomerization and hydrogenation.

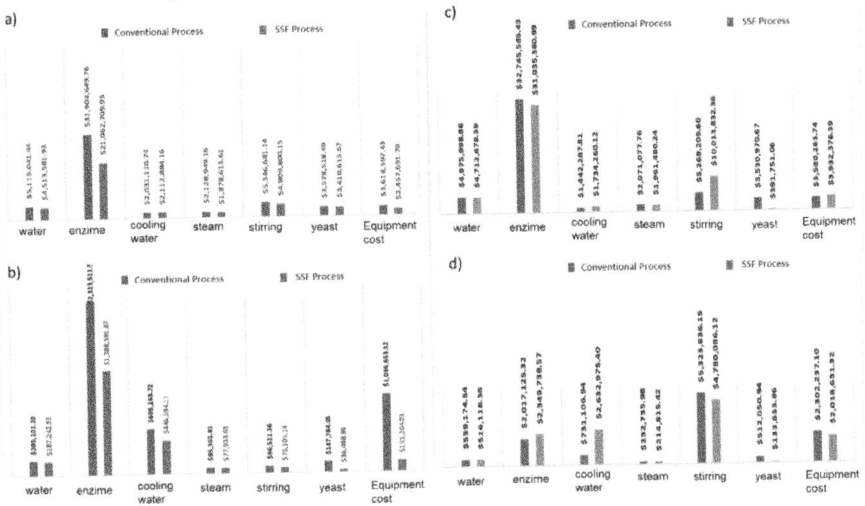

Figure 2. Comparison of pretreatment processes cost considering steam explosion a) sugarcane bagasse, b) corn stover; and dilute acid c) sugarcane bagasse, d) corn stover; and dilute acid.

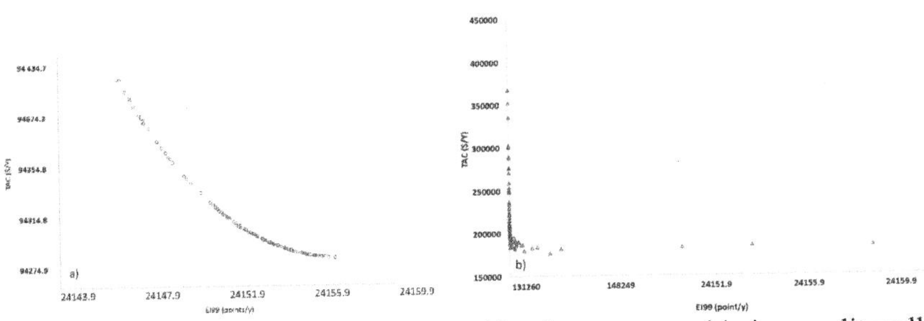

Figure 3. Pareto front for the ethanol purification zone considering a split-wall column (DWC) (a), and for the biojet production zone (b).

Figure 3a) shows that a TAC and EI99. weighted by the number of kilograms produced 1.18USD/kg ethanol and 0.3015 ecopoints/kg ethanol, respectively, were obtained. These values represent a reduction of 10% for the TAC, and close to 40% in the EI99 concerning the conventional design that reports TAC values of 1.29USD/kg ethanol and EI99 is 0.4716 ecopoints/kg ethanol. Finally, Figure 3b) shows the Pareto obtained from the transformation of ethanol to biojet fuel considering the dehydration, oligomerization, and hydrogenation stages. An important aspect to consider for savings in all stages is the reduced use of smaller equipment. In the case of the pretreatment and fermentation section, the use of equipment was reduced, the purification section was carried out in a piece of single process equipment, and in the biojet production zone, the double reactive distillation column performed the work of two reactors and a separation column. These capital cost savings are immediately reflected in the overall cost of the entire process. Table 2 shows the design parameters for the SSF reactor section as well as the double reactive column for biojet production. Data for the ethanol separation section can be found in the work of Rivas-Interian et al. (2022).

Table 2. Design parameter for SSF reactor, and reactive distillation column.

Scheme	Residence time (h)	Volume (m3)	Enzyme (FPU/g cellulose)	Yeast (FPU/g cellulose)	Stirring (RPM)
Sugar Cane/Steam	29.63	20492	9.013	0.9986	150.8353
Corn/Steam	29.6492	6633	18.75	0.9952	200.81
Sugar cane/acid	32.82	62528	10.008	1.0031	150.03
Corn/acid	25.034	16455	10.7791	1.0059	18086

	Oligomerization-Hydrogenation Reactive Column
Number of stages	33
Reflux ratio	69
Feed stage	24 and 30
Side stream stage	33
Reactive stages	12-20, 21-29
Hold Up (l)	3.5, 48
Distillate flowrate (kg h^{-1})	5422.44
Condenser duty (kcal h^{-1})	9.56216 x 10 6
Reboiler duty (kcal h^{-1})	854153
Jet Fuel Production (kg/h)	19935.821

The synergistic effect generated by joining the oligomerization and hydrogenation processes in a column with two reactive zones is quite good. During the parametric optimization process, it was possible to obtain design conditions that allowed obtaining a significant amount of product with better operating conditions.

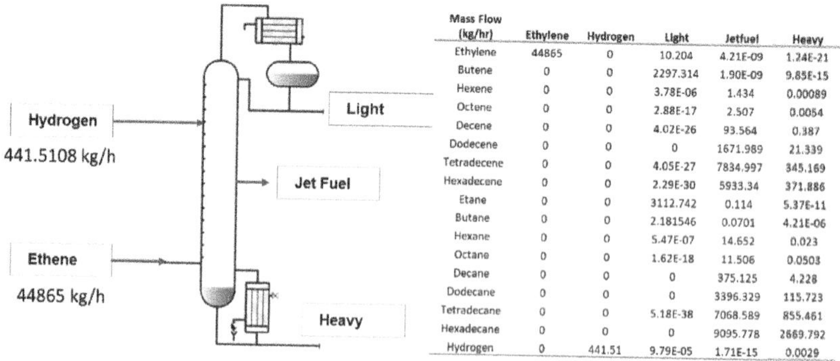

Mass Flow (kg/hr)	Ethylene	Hydrogen	Light	Jetfuel	Heavy
Ethylene	44865	0	10.204	4.21E-09	1.24E-21
Butene	0	0	2297.314	1.90E-09	9.85E-15
Hexene	0	0	3.78E-06	1.434	0.00089
Octene	0	0	2.88E-17	2.507	0.0054
Decene	0	0	4.02E-26	93.564	0.387
Dodecene	0	0	0	1671.989	21.339
Tetradecene	0	0	4.05E-27	7834.997	345.169
Hexadecene	0	0	2.29E-30	5933.34	371.886
Etane	0	0	3112.742	0.114	5.37E-11
Butane	0	0	2.181546	0.0701	4.21E-06
Hexane	0	0	5.47E-07	14.652	0.023
Octane	0	0	1.62E-18	11.506	0.0503
Decane	0	0	0	375.125	4.228
Dodecane	0	0	0	3396.329	115.723
Tetradecane	0	0	5.18E-38	7068.589	855.461
Hexadecane	0	0	0	9095.778	2669.792
Hydrogen	0	441.51	9.79E-05	1.71E-15	0.0029

Figure 4. Mass balance for the reactive distillation column.

Conclusions

In this work, an intensified approach to the production of biojet from biomass was presented. Process intensification strategies were applied in the pretreatment section, ethanol purification, as well as in biojet production area. Once the intensified alternatives were implemented, an improvement was observed in the economic and environmental indicators that were evaluated. In general terms, an improvement of 18% was observed in the economic indicator, and 12% in the environmental indicator when comparing the conventional and the intensified strategy.

References

SENER. Mapa de Ruta Tecnológica Bioturbosina, 2017. https://www.gob.mx/cms/uploads/attachment/file/324219/MRT_Bioturbosina_Final.pdf (accessed Abril 07, 2020).Y. Brown, Year, Article or Chapter Title, etc.

Heydenrych, M. D., et al. (2001). Oligomerization of ethene in a slurry reactor using a nickel(II)-exchanged silica-alumina catalyst. Journal of Catalysis, 197(1), 49–57.

Kadam, K., et al. (2004). Development and Validation of a Kinetic Model for Enzymatic Saccharification of Lignocellulosic Biomass. Biotechnology Process, 20(3), pp.698- 705.

Rivas-Interianet al. (2022). Economic and Environmental Optimization and Feedstock Planning for the Renewable Jet Production Using an Intensified Process. In Computer Aided Chemical Engineering (Vol. 51, pp. 643-648). Elsevier.

Antonis Kokossis, Michael C. Georgiadis, Efstratios N. Pistikopoulos (Eds.)
PROCEEDINGS OF THE 33rd European Symposium on Computer Aided Process Engineering (ESCAPE33), June 18-21, 2023, Athens, Greece
 http://dx.doi.org/10.1016/B978-0-443-15274-0.50211-0

Metrics for Evaluating Machine Learning Models Prediction Accuracy and Uncertainty

Yushi Deng, Mario R. Eden, Selen Cremaschi

Department of Chemical Engineering, Auburn University, Auburn, AL 36830, USA
selen-cremaschi@auburn.edu

Abstract

Models with stochastic outputs are commonly employed to make decisions for design and operation of chemical processes. The accuracy and precision of the stochastic outputs, i.e., model predictions, should be compared to the observation based on the design and operation requirements. The area metric measures the overall mismatch between prediction and observation but does provide precision and accuracy information separately. In this paper, we proposed a new metric called uncertainty width that decomposes area metrics into precision and bias components. This new metric enables the consideration of precision and accuracy simultaneously for model evaluation. We applied this metric to evaluate and compare the performance of three models fitted using three liquid entrainment fraction measurement datasets with different flow orientations, using Gaussian Process (GP) and Bayesian Neural Network (BNN). The results reveal that the overall prediction accuracy of GP model (GPM) is better than that of BNN model. The overestimation of the GPM prediction precision causes the high average area metric.

Keywords: Metrics, Machine Learning, Area Metrics, Accuracy, Precision, Uncertainty

1. Introduction

Design and operations decisions for chemical processes are regularly made considering various uncertainties (Sharifian et al., 2021). Models that provide predictions and their uncertainty over the target domain are essential in the risk assessment of these decisions. Before applying these decisions in chemical engineering analysis and design, the model performance evaluation must be performed. For models with stochastic outputs, the model performance is evaluated by measuring the mismatch between the model output and the true observation. Various metrics exist for this evaluation.

In classical hypothesis testing, a test is constructed under the hypothesis that the observations come from the predicted population. Decisions about rejecting or not rejecting the hypothesis can be made according to whether the test statistic falls within the critical region (Liu et al., 2011). Common hypotheses include the comparison of several moments (e.g., t-test, F-test) and comparison of full statistical distributions (e.g., Kolmogorov-Smirnov test (Haldar and Mahadevan, 2000), Anderson-Darling test (Rebba and Mahadevan, 2008)). The Bayes factor, expressed as the ratio of the relative likelihood of the null hypothesis over the alternative hypothesis, is used as the criteria (Mahadevan and Rebba, 2005). The null hypothesis is not rejected when the Bayes factor is larger than one. The Frequentist Metric evaluates the distance between the estimated mean of the observations and the mean of the predictions (Oberkampf and Trucano, 2002). The confidence bound is generated from the uncertainty of the distance.

The area metric (AM) measures the disagreement area of the cumulative distributions between the prediction and the observation (Ferson and Oberkampf, 2009). The u-pooling method derived from the area metric is proposed to measure the global dispersion of the distributions of observation and prediction. The measures of the cumulative distribution function at each validation site, called the u-value, are aggregated over the entire input space. The evaluation of the overall disagreement between the observation and model prediction is expressed as the area difference between the u-value distribution and the uniform distribution (Ferson et al., 2008). The model acceptance is determined according to a predefined accuracy requirement for the frequentist and area metrics.

For the area metric, the mismatch between the stochastic observation and prediction can be decomposed into precision and accuracy. The precision refers to the variance of multiple predictions at one validation site. Accuracy measures the degree of closeness between the prediction and the observation (Eisenhart, 1969). Generally, models with high prediction accuracy (i.e., low average prediction bias) and precision (i.e., narrow average uncertainty) are good performers. The area metric measures the sum of accuracy and precision. For predictions with high area metric values, it is difficult to determine whether this is due to low accuracy or poor precision. The accuracy and precision should be quantified separately when the decisions depend on both accuracy and precision.

This study proposes uncertainty width to evaluate the prediction precision within the area metric. By separating the area metric into the prediction bias related to accuracy and the uncertainty width related to precision, the performance between various models can be compared, and a model can be selected according to the acceptance criterion. Large bias may be acceptable with narrow uncertainty width when the model is employed to provide a prediction range. High uncertainty may be tolerated with a low bias if a point prediction is adopted for solving the problem. The proposed metric is applied to evaluate the performance of two machine learning models built using Gaussian Process Modeling (GPM) and Bayesian Neural Network (BNN).

The uncertainty width is defined and discussed in Section 2. Section 3 presents the details of the application of the uncertainty width for model performance evaluation. The results are given and discussed in Section 4. Finally, Section 5 summarizes the conclusions and future directions.

2. Methodology

A graphical illustration of the area metric is given in Figure 1. For one sample, the cumulative distribution functions (CDFs) of the model prediction and true observation are plotted as $f_1(y)$ and $f_2(y)$. According to the definition, the area metric is the area between the two curves. For symmetric distributions, the means of the two curves, μ_1 and μ_2, are located at the cumulative probability value equal to 0.5. The bias of the prediction $\mu_2 - \mu_1$, defined as the difference between the two means, is shown as the rectangular area between μ_1 and μ_2. Assume the two curves intersect at $y = \mu_I$.

The area metric is defined as the integral of the difference in area between the two CDFs and its analytical expression is given in Equation 1.

$$\int_{-\infty}^{\infty} |f_1(y) - f_2(y)| dy = \int_{-\infty}^{\mu_I} [f_2(y) - f_1(y)] dy + \int_{\mu_I}^{\infty} [f_1(y) - f_2(y)] dy \qquad (1)$$

The bias between the prediction and observation mean is equal to the integral of the cumulative probability $\int_{\mu_1}^{\mu_2} 1 dx$. By shifting the $f_1(y)$ to the left by $2(\mu_2 - \mu_1)$ and shifting $f_2(y)$ to the right by $2(\mu_2 - \mu_1)$, the bias can be cut into several pieces. The integrals of the resulting pieces are given in Equation 2.

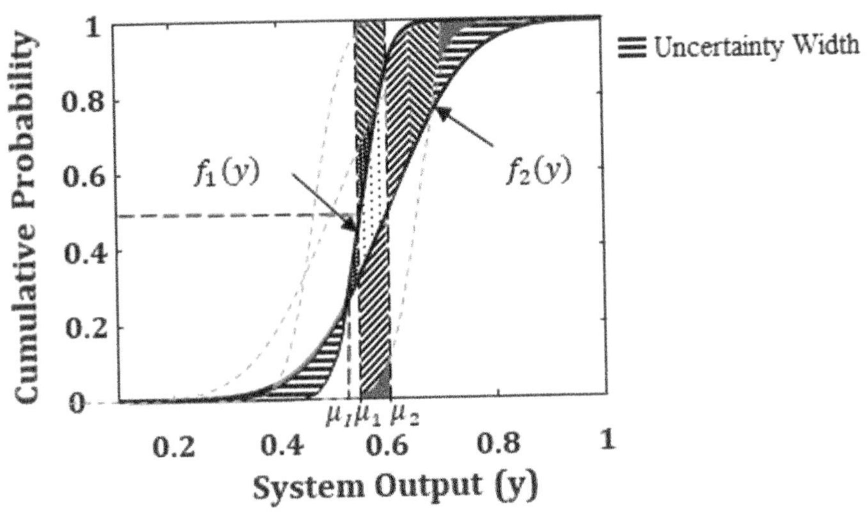

Figure 1 Area Metric Decomposition

$$
\begin{aligned}
\int_{\mu_1}^{\mu_2} 1dy = {} & \int_{\mu_1}^{\mu_2}\left[1 - f_1\left(y + 2(\mu_2 - \mu_1)\right)\right]dy \\
& + \int_{\mu_1}^{2\mu_1-\mu_I}\left[f_2\left(y + 2(\mu_2 - \mu_1)\right) - f_1\left(y + 2(\mu_2 - \mu_1)\right)\right]dy \\
& + \int_{2\mu_1-\mu_I}^{\mu_2}\left[f_1\left(y + 2(\mu_2 - \mu_1)\right) - f_1(y)\right]dy \\
& + \int_{\mu_1}^{\mu_2}\left[f_2\left(y + 2(\mu_2 - \mu_1)\right) - f_1(y)\right]dy + \int_{\mu_1}^{\mu_2}[f_1(y) - f_2(y)]dy \\
& + \int_{\mu_1}^{\mu_2}\left[f_2(y) - f_1\left(y - 2(\mu_2 - \mu_1)\right)\right]dy \\
& + \int_{\mu_1}^{\mu_2}\left[f_1\left(y - 2(\mu_2 - \mu_1)\right)\right]dy
\end{aligned} \tag{2}
$$

The second term in area metric shown in Equation 1 can also be decomposed into integrals given in Equation 3.

$$
\begin{aligned}
\int_{\mu_I}^{\infty}[f_1(y) - f_2(y)]dy = {} & \int_{\mu_I}^{\mu_1}[f_1(y) - f_2(y)]dy + \int_{\mu_1}^{\mu_2}[f_1(y) - f_2(y)]dy \\
& + \int_{\mu_2}^{2\mu_2-\mu_1}[f_1(y) - f_2(y)]dy + \int_{2\mu_2-\mu_1}^{2\mu_2-\mu_I}[f_1(y) - f_2(y)]dy \\
& + \int_{2\mu_2-\mu_I}^{3\mu_2-2\mu_1}\left[f_1(y) - f_1\left(y - 2(\mu_2 - \mu_1)\right)\right]dy \\
& + \int_{3\mu_2-2\mu_1}^{5\mu_2-4\mu_1}\left[f_1(y) - f_1\left(y - 2(\mu_2 - \mu_1)\right)\right]dy \\
& + \int_{3\mu_2-2\mu_1}^{5\mu_2-4\mu_1}\left[f_1\left(y - 2(\mu_2 - \mu_1)\right) - f_2(y)\right]dy \\
& + \int_{5\mu_2-4\mu_1}^{\infty}[f_1(y) - f_2(y)]dy
\end{aligned} \tag{3}
$$

Because $f_2(y)$ is symmetric at $(\mu_2, 0.5)$ and $f_2\left(y + 2(\mu_2 - \mu_1)\right)$ is shifted from $f_2(y)$, the functions $f_2\left(y + 2(\mu_2 - \mu_1)\right)$ and $f_2(y)$ are symmetric at $(\mu_1, 0.5)$. It can also be inferred that the functions $f_1(y)$ and $f_1\left(y - 2(\mu_2 - \mu_1)\right)$ are symmetric at $(\mu_2, 0.5)$. As such, $\int_{\mu_1}^{2\mu_1-\mu_I}\left[f_2\left(y + 2(\mu_2 - \mu_1)\right) - f_1\left(y + 2(\mu_2 - \mu_1)\right)\right]dy = \int_{\mu_I}^{\mu_1}[f_1(y) - f_2(y)]dy$. Other equal integrals of Equation 2 and Equation 3 are given in Equations 4 to 7.

$$\int_{3\mu_2-2\mu_1}^{5\mu_2-4\mu_1}[f_1(y)-f_1(y-2(\mu_2-\mu_1))]dy$$
$$=\int_{\mu_1}^{\mu_2}\left[1-f_1\left(y+2(\mu_2-\mu_1)\right)\right]dy+\int_{\mu_1}^{\mu_2}[f_1(y-2(\mu_2-\mu_1))]dy \quad (4)$$

$$\int_{\mu_1}^{2\mu_1-\mu_I}[f_1(y+2(\mu_2-\mu_1))-f_2(y+2(\mu_2-\mu_1))]dy$$
$$=\int_{2\mu_2-\mu_1}^{2\mu_2-\mu_I}[f_1(y)-f_2(y)]dy$$
$$+\int_{2\mu_2-\mu_I}^{3\mu_2-2\mu_1}[f_1(y)-f_1(y-2(\mu_2-\mu_1))]dy \quad (5)$$

$$\int_{\mu_1}^{2\mu_1-\mu_i}[f_2(y+2(\mu_2-\mu_1))-f_1(y+2(\mu_2-\mu_1))]dy$$
$$=\int_{\mu_I}^{\mu_1}[f_1(y)-f_2(y)]dy \quad (6)$$

$$\int_{\mu_1}^{\mu_2}[f_2(y)-f_1(y-2(\mu_2-\mu_1))]dy=\int_{\mu_2}^{2\mu_2-\mu_1}[f_1(y)-f_2(y)]dy \quad (7)$$

For visualization, the equal integral pairs are shaded with the same patterns in Figure 1. The uncertainty width is defined as the difference between the area metric $\int_{-\infty}^{\infty}|f_1(y)-f_2(y)|dy$ and the prediction bias $\int_{\mu_1}^{\mu_2}1dy$. If Equations 4 to 7 are inserted into the uncertainty width definition, Equation 8 is obtained to calculate the uncertainty width.

$$Uncertainty\ width$$
$$=\int_{-\infty}^{\infty}|f_1(y)-f_2(y)|dy-\int_{\mu_1}^{\mu_2}1dy$$
$$=\int_{-\infty}^{\mu_I}[f_2(y)-f_1(y)]dy+\int_{3\mu_2-2\mu_1}^{5\mu_2-4\mu_1}[f_1(y-2(\mu_2-\mu_1))-f_2(y)]dy$$
$$+\int_{5\mu_2-4\mu_1}^{\infty}[f_1(y)-f_2(y)]dy \quad (8)$$

3. Case Study

3.1. Machine Learning Models

Two Bayesian-based machine learning techniques, GPM and BNN, are adopted to build models in this study. GPM is a supervised learning method with a theoretical basis in statistics. A GPM is characterized by its mean function and covariance function or kernel function (Williams and Rasmussen, 2006). A constant mean function and square exponential kernel function with different length scales for each input dimension are utilized. The GP prior is defined in Equation 9:

$$f\sim \mathcal{GP}(m,k)$$
$$m(x)=a,\ and\ k(x_p,x_q)=\sigma_f^2\exp\left(-\sum_{h=1}^{d}\frac{(x_p-x_q)^2}{2l_h^2}\right)+\sigma_n^2\Delta_{pq} \quad (9)$$

In Equation 9, f represents the underlying function value at the input x, $m(x)$ is the mean function of the GP, and $k(x_p,x_q)$ is the covariance function representing the spatial covariance between any two points (x_p and x_q) at the process. In the covariance function, d is the dimension of input x, l_h represents the characteristic length scale corresponding to the hth dimension of input x, σ_n^2 is the output variance, and the parameter σ_f^2 is output-scale amplitude. Δ_{pq} is Kronecker delta, which is one if p = q and zero otherwise.

BNN applies Bayesian statistics to Neural Networks (NNs) (Goan and Fookes, 2020). Instead of treating the model weights as a fixed value, BNN assumes the weights are random variables. After specifying the prior distribution of weights and likelihood, the Bayes theorem is applied to calculate the posterior distribution. BNN model parameters are optimized using variational learning that minimizes the Kullback-Leibler (KL) divergence to avoid intractable computations (Blundell et al., 2015). The prediction mean and variance are calculated using Monte Carlo sampling.

3.2. Model Performance Evaluation using Uncertainty Width

The uncertainty width is applied to evaluate and compare the performance of the selected machine learning (ML) models from both accuracy and precision perspectives. Three datasets containing liquid entrainment fraction measurements and two machine learning techniques, GPM and BNN, are used to build six ML models. The three datasets for vertical, horizontal, and inclined flow orientations are composed of 1,083, 478, and 100 samples with 9 input features representing the experimental operating conditions. For each sample, the liquid entrainment fraction prediction with its uncertainty is calculated and evaluated using the area metric. The area metric is then decomposed into accuracy (absolute bias) and precision (uncertainty width). The overall model performance is evaluated using the average area metric, average absolute bias (mean absolute error), and average uncertainty width of all the samples.

4. Results and Discussion

The model performance evaluation results are shown in Figure 2. The average area metric, mean absolute error (MAE), and the average uncertainty width of the models built using experimental datasets for vertical, horizontal, and inclined flow orientations are shown in Figures 2 (a), (b), and (c), respectively. The total height of each bar represents the average area metric. The MAE and the average uncertainty width values are separated and marked using different colors and shades within the area metric. While the average area metric shows the overall performance of the model, models with low MAE have better accuracy, while models with low uncertainty have better precision.

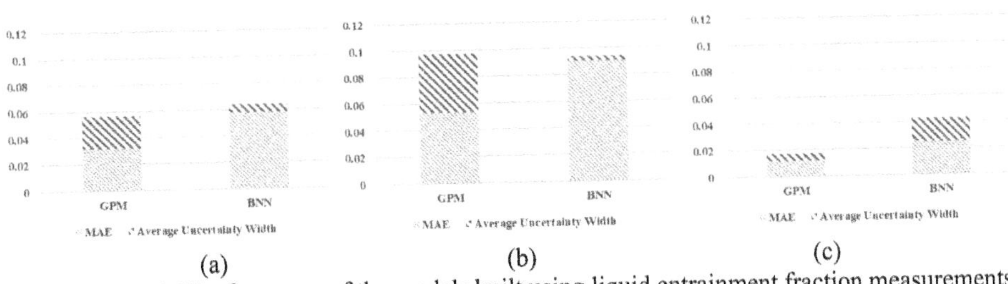

Figure 2 Model Performance of the models built using liquid entrainment fraction measurements for (a) vertical (b) horizontal (c) inclined flow orientations

Considering the overall model performance evaluated using the average area metric, the GPM and BNN fitted using datasets for vertical and horizontal flow orientations have identical performance. This observation suggests that when an entrainment fraction prediction range is used for developing a solution from the model output, the GPM and BNN performances are close. The GPM outperforms the BNN when fitted using the dataset for inclined flow orientation. For the model accuracy, the MAE of GPM, is lower than that of BNN for all three pairs of models. This observation indicates that the accuracy of the GPM is always better than the BNN. The GPM should be selected when entrainment fraction point prediction is employed to make a decision. For models built using the dataset for inclined flow orientation, both MAE and average uncertainty width of BNN are higher than those of GPM, indicating that the GPM outperforms BNN in both accuracy and precision.

5. Conclusions

In this study, a metric called uncertainty width is developed to evaluate the performance of machine learning models with stochastic outputs. The prediction performance at each sample is separated into accuracy and precision by decomposing the area metric into absolute bias and uncertainty width. The average absolute error and average uncertainty bias from all samples are used to simultaneously evaluate the overall model performance, considering accuracy and precision. The new metric is applied to evaluate the performances of six models built using GPM and BNN using three datasets containing liquid entrainment fraction measurements for vertical, horizontal, and inclined flow orientations. The results suggest that GPM performs better than BNN, considering both accuracy and precision using the dataset for the inclined flow orientation. For vertical and horizontal flow orientation datasets, although the overall model performance evaluated using the average area metric are identical, the accuracy of GPM is better than that of BNN. The GPM performs better when an entrainment fraction point prediction is needed for making design and operation decisions using these models. The high area metric of GPM built using vertical and horizontal flow orientation datasets comes from the high uncertainty width, which means that GPM overestimates the prediction uncertainty. Future work will investigate the application of uncertainty width to outputs with skewed distributions.

References

Blundell, C., Cornebise, J., Kavukcuoglu, K., Wierstra, D., 2015. Weight uncertainty in neural networks. 32nd Int. Conf. Mach. Learn. ICML 2015 2, 1613–1622.

Eisenhart, C., 1969. Realistic Evaluation of the Precision and Accuracy. NBS Spec. Publ. 21.

Ferson, S., Oberkampf, W.L., 2009. Validation of imprecise probability models. Int. J. Reliab. Saf. 3, 3–22.

Ferson, S., Oberkampf, W.L., Ginzburg, L., 2008. Model validation and predictive capability for the thermal challenge problem 197, 2408–2430. https://doi.org/10.1016/j.cma.2007.07.030

Goan, E., Fookes, C., 2020. Bayesian Neural Networks: An Introduction and Survey. Lect. Notes Math. 2259, 45–87. https://doi.org/10.1007/978-3-030-42553-1_3

Haldar, A., Mahadevan, S., 2000. Probability, reliability, and statistical methods in engineering design. John Wiley & Sons Incorporated.

Liu, Y., Chen, W., Arendt, P., Huang, H.Z., 2011. Toward a better understanding of model validation metrics. J. Mech. Des. Trans. ASME 133, 1–13. https://doi.org/10.1115/1.4004223

Mahadevan, S., Rebba, R., 2005. Validation of reliability computational models using Bayes networks. Reliab. Eng. Syst. Saf. 87, 223–232.

Oberkampf, W.L., Trucano, T.G., 2002. Verification and validation in computational fluid dynamics. Prog. Aerosp. Sci. 38, 209–272.

Rebba, R., Mahadevan, S., 2008. Computational methods for model reliability assessment. Reliab. Eng. Syst. Saf. 93, 1197–1207.

Sharifian, S., Sotudeh-Gharebagh, R., Zarghami, R., Tanguy, P., Mostoufi, N., 2021. Uncertainty in chemical process systems engineering: a critical review. Reviews in Chemical Engineering 37, 687–714. https://doi.org/doi:10.1515/revce-2018-0067

Williams, C.K.I., Rasmussen, C.E., 2006. Gaussian Processes for Machine Learning. MIT press Cambridge, MA.

Antonis Kokossis, Michael C. Georgiadis, Efstratios N. Pistikopoulos (Eds.)
PROCEEDINGS OF THE 33rd European Symposium on Computer Aided Process Engineering (ESCAPE33), June 18-21, 2023, Athens, Greece
 http://dx.doi.org/10.1016/B978-0-443-15274-0.50212-2

Optimization of Process Families for Deployment of Carbon Capture Processes using Machine Learning Surrogates

Georgia Stinchfield[a], Bashar L. Ammari[a], Joshua C. Morgan[c,d], John D. Siirola[b], Miguel Zamarripa[c,d], Carl D. Laird[a*]

[a]*Carnegie Mellon University, 5000 Forbes Ave., Pittsburgh, PA, 15232*

[b]*Sandia National Laboratories, P.O. Box 5800, Albuquerque, NM, 87185*

[c]*National Energy Technology Laboratory (NETL), 626 Cochran Mill Road, Pittsburgh, PA, 15236*

[d]*NETL Support Contractor, 626 Cochran Mill Road, Pittsburgh, PA, 15236*

**claird@andrew.cmu.edu*

Abstract

Traditional process design approaches focus on exploiting economies of scale but are inefficient when designing a large number of similar processes for decentralized applications. By optimizing each process individually, these approaches do not allow for manufacturing standardization or make use of economies of numbers. In this work, we design a family of processes (i.e., multiple processes) of a carbon capture facility with various performance requirements. We identify sub-components within the process and create a small set of sub-component designs that can be shared across all the processes in the process family. We formulate this optimization problem as a nonlinear generalized disjunctive program (GDP) and, in previous work, developed two approaches for reformulating and solving this problem: one based on full-discretization of the design space (Zhang et al., 2022) and one that used Machine Learning (ML) surrogates to replace the nonlinear process models (Stinchfield et al., 2022). Using ML surrogates to predict required system costs and performance allows us to reformulate the nonlinearities in the GDP to generate an efficient MILP formulation. In this work, we apply the ML surrogate approach to design a family of carbon capture systems to cover a set of different flue gas flow rates and inlet CO_2 concentrations, where we consider the absorber and stripper as sub-component types.

Keywords: Manufacturing, Optimization, Energy Systems, Machine Learning.

1. Introduction

Effectively mitigating climate change requires the broad deployment of green energy and industrial decarbonization processes. When designing *multiple* processes, a conventional design approach produces a set of unique unit designs for each process that can be expensive to manufacture. Modularity, a well-explored design approach, provides significant reductions in manufacturing costs and has proven beneficial in particular when designing multiple instances of a process (Baldea et al., 2017). However, modular designs are typically 'numbered-up' to achieve capacity requirements and often fail to fully exploit economies of scale. Rather than using traditional or modular design, we propose a hybrid approach that applies concepts from product family design to the design of process systems to gain the manufacturing benefits of modularity while still obtaining process customization and economies of scale.

In product family design (PFD), a product family is broadly defined as “a set of products that share one or more common ‘element(s)’ yet target a variety of different market segments” (Simpson et al., 2014). This approach is used extensively in automotive industries, for example, where a wide variety of products are available, but designed and assembled using a smaller, commonly designed platform of sub-components. As process vendors tackle the broad deployment of large numbers of processes for industrial decarbonization, they face similar design goals. We are extending the ideas of PFD into this area of process design. In our work, the “products” are the set of processes we must design. Each individual process is referred to as a process variant; each variant is associated with a set of conditions that the process design must meet. The “common ‘element(s)’” are sub-components shared among the process variants. For our case study, we consider the design of a carbon capture system (CCS). Typically, a CCS is designed for a specific point source capture location (i.e., coal-fired power plants, natural gas combined cycle, cement, refineries, etc.) mainly based on the flue gas conditions and flowrate. PFD has the potential to accelerate the deployment of CCS for industrial processes through the simultaneous design of process families that share common sub-components. Our case study considers 12 process variants. Each process variant in the family is designed for a particular flue gas flow rate and CO_2 concentration. We wish to design each of these process variants using a common platform that includes shared sub-component types, in this case, the absorber and the stripper. We require one of each sub-component for each process variant (i.e., 12 absorbers and 12 strippers in total) but hope to meet all the process variant requirements with a small number of sub-component designs (e.g., here we consider a platform with only two absorber designs and two stripper designs from which to design all 12 process variants).

This approach captures the benefits of manufacturing standardization, akin to modularity, since a large number of process variants share the same common sub-component designs. At the same time, this retains economies of scale as the platform of manufactured sub-components is designed simultaneously with the set of process variants in the family. This approach has proven beneficial in other industries; companies such as Nissan, Toyota, and Boeing have reported substantial cost and time savings by using this approach (Simpson et al., 2014). However, despite documentation of significant success in other industrial manufacturing settings, PFD has largely not been applied to chemical process design. Furthermore, reported approaches have largely been heuristic, and there is a need for rigorous optimization approaches.

Our optimization problem simultaneously determines the individual *designs* of the sub-components in the platform (i.e., if we only offer two absorber designs, what should they be) and *which design* each process variant should use (i.e., for each of 12 process variants, which of the two absorber designs should each variant use). We first formulated this problem as a nonlinear GDP. We solved this problem by reformulating it as a MILP formulation based on full discretization of the design space (Zhang et al., 2022). We extended this approach, proposing another MILP reformulation that avoids discretization by utilizing ReLU activated Neural Network (NN) as piecewise linear surrogate models (Stinchfield et al., 2022, Ceccon et al., 2022). This significantly decreased pre-computation requirements for simulations and allowed us to search for an approximation of the continuous design space. In this paper, we further demonstrate this piecewise linear surrogate approach to optimize a process family of carbon capture systems, using both an ReLU activated NN and a linear model decision tree as surrogates. We design 12 carbon capture systems with different flue gas flow rates and CO_2 concentrations with the absorber and stripper columns as the shared sub-components.

2. Optimization Formulation

The following formulation for optimization of a process family design was first proposed by Stinchfield, et al., (2022). A process variant is represented by v, and is characterized by the design requirements for that variant. The parameter set V represents all process variants. The boundary conditions of a process variant are parameterized in b_v. In our case study, $|V| = 12$ and the boundary conditions for each variant include the flue gas flow rate and CO_2 concentration. The total annualized cost of a process variant is represented by variable c_v. The objection function includes a weight for each process variant, w_v, that captures the expected sales or number of installations of each variant. The variable p_v captures the performance metrics of a particular process variant. In many cases, this will be an indicator function that determines whether a particular design is feasible for the variant. This ensures that an infeasible combination of sub-component designs is not selected for a process variant v. We use trained piecewise linear surrogate models, represented by the system of equations f^c and f^p, to predict the variable cost of the variant, c_v, and the performance indicator, p_v. Piecewise linear surrogates are used because they can be represented exactly as MILPs within the optimization formulation.

The set of common sub-component types considered in the platform is given by K. In our case study, $K = \{\text{absorber, stripper}\}$. The set J_k includes the sub-component designs for a particular sub-component of type k. The cardinality of J_k dictates the number of sub-component designs to consider in the platform for sub-component type k (i.e., if we want two absorber designs, then $J_k = \{1,2\}$ where $k = \text{absorber}$). The variable $\hat{d}_{k,j}$ represents the design variable values for design j of sub-component type k, while variable $d_{v,k}$ represents the design variable values of sub-component type k for process variant v. With this, if design j is chosen for sub-component type k in variant v, then $d_{v,k} = \hat{d}_{k,j}$.

$$min. \sum_{v \in V} w_v c_v \qquad (1a)$$

s.t.

$$c_v = f^c(b_v, d_{v,1}, \ldots, d_{v,k}) \qquad \forall v \in V \qquad (1b)$$

$$p_v = f^p(b_v, d_{v,1}, \ldots, d_{v,k}) \qquad \forall v \in V \qquad (1c)$$

$$\bigvee_{j \in J_k} \begin{bmatrix} Y_{v,k,j} \\ d_{v,k} = \hat{d}_{k,j} \end{bmatrix} \qquad \forall v \in V, k \in K \qquad (1d)$$

$$\hat{d}^L_{k,j} \leq \hat{d}_{k,j} \leq \hat{d}^U_{k,j} \qquad \forall k \in K, j \in J_k \qquad (1e)$$

$$p^L_v \leq p_v \leq p^U_v \qquad \forall v \in V \qquad (1f)$$

$$c_v \geq 0 \qquad \forall v \in V \qquad (1g)$$

$$Y_{v,k,j} \in \{\text{True, False}\} \qquad \forall v \in V, k \in K, j \in J_k \qquad (1h)$$

The objective, Eq. (1a), is to minimize the total weighted annualized cost of all process variants, $v \in V$. Piecewise linear surrogates predict the cost of each variant c_v, Eq. (1b), and the performance indicator p_v, Eq. (1c). Both surrogates are functions of the boundary conditions at a particular variant, b_v, and the design of each sub-component type $k \in K$. The disjunctions in Eq. (1d) select which of the sub-component designs, $j \in J_k$, for a sub-component of type k are selected for process variant v. The number of disjunctions is equal to $|K| \times |V|$ because we make this decision for each sub-component k and each process variant v. The number of disjuncts corresponds to $|J_k|$ for a particular sub-component of type k, because we decide on which sub-component design, $j \in J_k$, to select. Eq. (1e) bounds the design variable of sub-component type k, which also corresponds to the ranges used to train the surrogate. Eq. (1f) captures the performance constraints for each process variant.

3. Numerical Case Study

We demonstrate this formulation by designing a process family of 12 aqueous monoethanolamine (MEA) solvent-based carbon capture variants. Each variant requires a design that captures at least 90% of CO_2 entering the facility. In this case study, each process variant v is described by one of three flue gas flow rates and one of four carbon dioxide concentrations. This led to $3 \times 4 = 12$ possible combinations, representing the boundary conditions for the 12 process variants. Considering the units in the capture facility flowsheet, we chose the common sub-component types to be the absorber and stripper, $K = \{$absorber, stripper$\}$, since they contribute heavily to the overall purchase cost and have large effects on the performance of the system, thereby directly affecting operating costs.

To gather surrogate training data, we discretized the design ranges for the absorber and stripper. For the absorber, we considered 10 diameters in the range of 0.3 m – 1.2 m. For the stripper, we considered 10 diameters in the range of 10 in.–55 in. We performed a simulation of the carbon capture system for all possible combinations of the process variant boundary conditions and sub-component sizes. This led to $12 \times 10 \times 10 = 1{,}200$ simulations. We used an Aspen Plus model of the aqueous MEA system for the simulations (Morgan et al., 2018). Each simulation was marked as *feasible* if the key performance requirements (e.g., CO_2 recovery) were met and *infeasible* otherwise. This set of data and corresponding *feasible, infeasible* labels were used to train a 15-node, single-layer ReLU activated NN to determine the performance indicator, p_v. The NN was imported directly in the formulation, as shown in Eq. (1c), using OMLT: Optimization and Machine Learning Toolkit (Ceccon et al., 2022). Purchase cost was estimated using the IDAES costing framework (Lee et al., 2021). Combined with estimates for annual operating cost, total annualized cost was used to train a linear decision tree (Ammari et al., 2022) and imported using OMLT, represented by Eq. (1b), to predict c_v of process variant v. Note that both surrogates are captured as piecewise linear functions, resulting in an MILP formulation.

For this case study, we considered two absorbers and two strippers for the sub-component designs (i.e., $J_k = \{1,2\}$ for each sub-component type). The problem was formulated using Pyomo (Bynum et al., 2021) and solved using Gurobi (Optimization, LLC Gurobi, et al., 2020). Additionally, we used OMLT and the Pyomo extension, Pyomo.GDP (Chen et al., 2022), to capture the disjunctions. The optimal designs for the two absorbers and two strippers are reported in Table (1), and Figure (1) shows which combinations of the two possible absorber designs and two possible stripper designs were selected for each of the 12 process variants. The optimization determined the sizes for the common sub-components. The larger absorber was used for the upper right corner of the variant process space, while the larger stripper was needed only for the most demanding variant in the upper right-hand corner.

Table 1. Optimal Designs for Common Units

Sub-component k	Design 1 $j_1 \in J_k$	Design 2 $j_2 \in J_k$
Absorber	0.3 m. diameter	0.61 m. diameter
Stripper	0.254 m. diameter	1.075 m. diameter

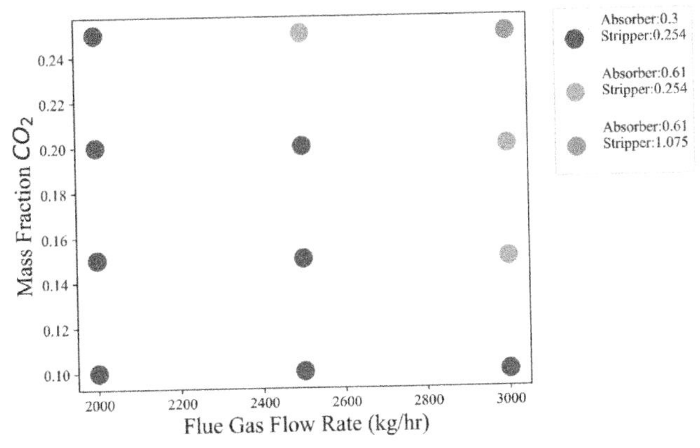

Figure 1. Process Family Design

Since the optimization is performed with surrogate models, we verified this result using the original model to simulate each of the 12 process variant boundary conditions with the corresponding set of assigned absorber and stripper designs. The total annualized verified cost was $\$1.29 \times 10^6$, which is an overall surrogate error of approximately 2%.

Table 2. Optimization Problem Specs

Num. of Binary Variables	191
Num. of Continuous Variables	273
Num. of Constraints	720
Total Annualized Cost (Obj.)	$\$1.26 \times 10^6$

The combination of training time for both the ReLU NN surrogate and linear decision tree was approximately 300 seconds, while the optimization formulation itself took approximately 1 second along with the problem structure as described in Table (2).

4. Conclusions and Future Work

We demonstrated a surrogate-based approach for optimally designing a process family of 12 carbon capture variants. We considered a platform of common sub-component designs that considered two absorbers and two strippers, whereas a conventional design approach would have generated 12 unique absorber designs and 12 unique stripper designs. This approach uses surrogates to predict required system costs and performance. This removes the nonlinearities in the GDP and allows the transformation to an MILP. The machine learning surrogates were easily included in the optimization with OMLT, and the overall formulation solved in under a second, demonstrating tractability in this approach.

For future work, we plan on expanding the sub-component design ranges and incorporating more sub-component types. Furthermore, we plan on incorporating an estimation of savings due to economies of numbers within the optimization formulation, exploring uncertainty in weights, and quantifying the degree of flexibility our designs offer across the design space.

Disclaimer: This project was funded by the United States Department of Energy, National Energy Technology Laboratory an agency of the United States Government, in part, through a support contract. Neither the United States Government nor any agency thereof, nor any of their employees, nor the support contractor, nor any of their employees, makes any warranty, express or implied, or assumes any legal liability or responsibility for the accuracy, completeness, or usefulness of any information, apparatus, product, or process disclosed, or represents that its use would not infringe privately owned rights. Reference herein to any specific commercial product, process, or service by trade name, trademark, manufacturer, or otherwise does not necessarily constitute or imply its endorsement, recommendation, or favoring by the United States Government or any agency thereof. The views and opinions of authors expressed herein do not necessarily state or reflect those of the United States Government or any agency thereof.

SNL is managed and operated by NTESS under DOE NNSA contract DE-NA0003525

Acknowledgments: This effort is part of the U.S. Department of Energy's (DOE) Institute for the Design of Advanced Energy Systems (IDAES) supported by the DOE Office of Fossil Energy and Carbon Management's Simulation-Based Engineering/Crosscutting Research Program.

References

Ammari, B., et al., 2022, Machine Learning Surrogates with OMLT and ideas for Improved Design and Analysis of Energy Systems, INFORMs Annual Meeting 2022, Indianapolis, IL.

Baldea, M., et al., 2017, Modular manufacturing processes: Status, challenges, and opportunities. *AIChE journal 63(10)*, 4262-4272

Bynum, M. L., et al., 2021, Pyomo- Optimization Modeling in Python. *Springer.*

Chen, Q., et al., 2022, Pyomo gdp: an ecosystem for logic based modeling and optimization development. *Optimization and Engineering 23(1)*, 607-642-

Ceccon, F., et al., 2022, Omlt: Optimization & machine learning toolkit, submitted, preprint on arXiv.

Lee, A., et al., 2021, The idaes process modeling framework and model library- flexibility for process simulation and optimization. *Journal of Advanced Manufacturing and Processing 3(3)*

Morgan, J.C., et al., 2018. Development of a rigorous modeling framework for solvent-based CO_2 capture. part 2: steady-state validation and uncertainty quantification with pilot plant data. *Industry & Engineering Chemistry Research 57*, 10464-10481.

Optimization, LLC Gurobi, et al., 2020, Gurobi optimizer reference manual, *URL http://www.gurobi.com12.*

Stinchfield, G., et al., 2022, Optimization-based Approaches for Design of Chemical Process Families Using ReLU Surrogates, submitted to Foundations of Computer Aided Process Operations, 2022.

Wilberforce, T., et al., 2019, Outlook of carbon capture technology and challenges, *Science of the total environment*, 56-72

Zhang, C., et al., 2022, Optimization-based design of product families with common components. *Computer Aided Chemical Engineering, Proceedings of 14th International Symposium on Process Systems Engineering (PSE 2021+)*. Volume 49, pp. 91-96. Elsevier.

Antonis Kokossis, Michael C. Georgiadis, Efstratios N. Pistikopoulos (Eds.)
PROCEEDINGS OF THE 33rd European Symposium on Computer Aided Process Engineering (ESCAPE33), June 18-21, 2023, Athens, Greece
 http://dx.doi.org/10.1016/B978-0-443-15274-0.50213-4

Global Optimal Explainable Models for Biorefining

Jiayang Ren[a], Kaixun Hua[a], Heather Trajano[a], Yankai Cao[a,b]

[a] *Department of Chemical and Biological Engineering, University of British Columbia, 2360 East Mall, Vancouver V6T 1Z3, Canada*

[b] *yankai.cao@ubc.ca*

Abstract

Hemicelluloses are amorphous polymers of various sugar molecules and have been widely utilized in bioenergy, mining, and textile. Through hydrolysis, hemicellulose is transformed into sugar oligomers and monomers. In this paper, we build a global optimal decision tree (GODT) model on an extensive hemicellulose hydrolysis dataset containing 1955 experimental data points from 71 published papers from 1985 to 2019. The GODT model is trained to predict xylose yield from hardwood hemicellulose hydrolysis in batch reactors. Compared with the heuristic method, our global optimal algorithm can obtain an average absolute improvement of 1.54% in test accuracy. Moreover, we demonstrate that the reasoning procedure of predictions is easy to comprehend by human decision-makers, thus contributing an explainable model for biorefining.

Keywords: Biorefining; Global Optimization; Machine learning; Decision Tree

1. Introduction

Hemicelluloses are amorphous polymers of various sugar molecules, such as xyloses, mannoses, arabinoses, glucoses, and galactoses (Scheller & Ulvskov, 2010). It is the second most component in lignocellulosic biomass, ranging from 25 to 35 wt% (Isikgor & Becer, 2015). Hemicellulose has been broadly utilized in the bioenergy, mining, textile, cosmetic and pharmaceutical industries (Spiridon & Popa, 2008). Industrial use of hemicellulose often requires that the polymer be hydrolyzed into constituent oligomers and monomers. Scale-up and commercialization of hemicellulose hydrolysis require robust models. However, the traditional kinetic models can only consider limited operating regimes and species (Fearon et al., 2020). Fortunately, extensive studies on kinetic models have contributed abundant data to enable the development of statistical models. In our previous work (Wang et al., 2022), we mined the literature on hardwood hemicellulose hydrolysis and collected a dataset containing 1955 experimental points on batch hemicellulose hydrolysis from 71 published papers dating from 1985 to 2019. Three machine learning models (i.e., ridge regression, support vector regression, and artificial neural networks (ANN)) were developed to predict the xylose yield of hemicellulose hydrolysis utilizing this dataset. This work also showed the superior performance of ANN in predicting the xylose yield of hemicellulose hydrolysis, compared with kinetic models. However, the black-box nature of ANN makes the interpretability of predictions difficult.

This paper aims to develop an interpretable model based on decision trees to predict the xylose yield of hemicellulose hydrolysis. The decision tree is a powerful machine-learning method for classification and regression tasks. The key advantage of decision tree modeling lies in interpretability, as such models closely resemble human reasoning and are easy to comprehend by human decision-makers (Weiss & Kulikowski, 1991). For example, the operator of a biorefinery plant is more likely to understand and trust a decision tree model over black-box models (e.g., neural network) since the decision tree breaks a complex decision-making process (e.g., the efficiency of the operation) into a collection of more straightforward logical tests (e.g., temperature range).

A fundamental problem is decision tree learning, a task to find the most accurate decision tree model from a dataset. For decades, we have relied on greedy-based heuristics, such as CART (Loh, 2011), ID3 (Quinlan, 1986), and C4.5 (Quinlan, 2014), to obtain suboptimal decision trees. Recent studies based on a comprehensive benchmark have shown that global solutions obtained through mixed-integer optimization (MIO) led to an average absolute improvement in test accuracy of 1 - 5% (Bertsimas & Dunn, 2017). Despite its superior performance, existing MIO-based approaches (Bertsimas & Dunn, 2017; Aghaei et al., 2021; Verwer & Zhang, 2019) rely on all-purpose global optimization solvers and can only address a small dataset (e.g., several hundred samples). Recently, we developed a tailored reduced-space branch and bound algorithm to train optimal decision trees for classification tasks (Hua et al., 2022). The algorithm is based on the observation that decision tree learning can be formulated as a two-stage stochastic programming problem, and thus various approaches proposed in the stochastic programming community can be utilized to exploit the problem structure. This work extended the solvable optimal decision tree problem to 245,000 samples.

In this paper, we apply the global optimal algorithm proposed by Hua et al. to the biorefining dataset and show that the average training and testing accuracy can be improved by 1.70% and 1.54% compared with the heuristic method.

2. Problem Description

2.1. Dataset Description and pre-process

We adopt the hemicellulose hydrolysis dataset from our previous work (Wang et al., 2022). This dataset collected 1955 data points from 71 published works from 1985 to 2019. Each data point has eight numeric variables and two categorical variables. Wang et al. have demonstrated that the two categorical variables (wood and acid species) contribute little in predicting the xylose yield.

Besides, these two categorical variables need one-hot encoding to be utilized in the decision tree model, which will introduce redundant binary variables and make the decision tree modeling unstable. Therefore, we only use eight numeric variables in this paper, as shown in Table 1. Notably, there are some repeating experiments with the same reaction and biomass conditions but resulting in different xylose yields. We unified these data points with the average value of xylose yields and obtained a unique dataset with 1828 data points.

Table 1 Dataset variable list

No.	Variables	Type	Units	Range
1	Total Reaction Time	Numeric	minute	0 - 1155
2	Reactor Temperature	Numeric	Kelvin	313 - 553
3	Liquid Solid Ratio	Numeric	g liquid / g solid	1.53 - 50
4	Initial Proton Concentration	Numeric	mols/L	0-1.22
5	Particle Size	Numeric	mm	0.117 - 50
6	Isothermal Reaction Time	Numeric	minute	0 - 1140
7	Initial Hemicellulosic Xylose in Feedstock	Numeric	weight %	11.932 - 34.9
8	Xylose Yield	Numeric	%	0 – 100

To adapt to the classification decision tree algorithms, we further encoded the variable xylose yield from a numeric variable to a categorical variable with K categories. Specifically, we created two classification datasets with three and five categories. For the first dataset, the yield is either lower [0%, 30%], medium (30%, 75%], or high (75%, 100%]. For the second one, the labels of yield are very low [0, 20%], low (20%, 40%], medium (40%, 60%], high (60%, 80%], or very high (80%, 100%].

2.2. Problem Description

The paper aims to develop a classification decision tree to predict the encoded xylose yield of hemicellulose hydrolysis utilizing variables 1-7 in Table 1. Firstly, we randomly divide the whole dataset into training and testing datasets. Then, let $\mathcal{S} = \{1, \dots, n_{train}\}$ be the index set of the training dataset, $\mathcal{S}_{test} = \{1, \dots, n_{test}\}$ be the index set of the testing dataset, $\mathcal{K} = \{1, \dots, K\}$ be the class set of the encoded xylose yield. For sth data point, we denote the encoded xylose yield as $\boldsymbol{y}_s \in \{0,1\}^K$ and the feature variables as $\boldsymbol{x}_s \in \mathbb{R}^P$. Sequentially, the training dataset can be represented as $\boldsymbol{X} = \{\boldsymbol{x}_s | \boldsymbol{x}_s \in \mathbb{R}^P, s \in \mathcal{S}\}$ with the label set $\boldsymbol{Y} = \{\boldsymbol{y}_s | \boldsymbol{y}_s \in \{0,1\}^K, s \in \mathcal{S}\}$. Then, the problem of this paper can be represented as finding a decision tree $F: \mathbb{R}^P \to \{0,1\}^K$ with maximum depth D to solve the following optimization problem:

$$\min \sum_{s \in \mathcal{S}} E(F(x_s), y_s) + \lambda R(F) \tag{1}$$

where $E(\cdot)$ is the misclassification error between the predicted label of tree $F(x_s)$ and the corresponding ground truth y_s, λ is the complexity parameter to balance the accuracy and tree complexity, $R(\cdot)$ is the complexity function to measure the tree's complexity. After obtaining the decision tree, the labels of testing samples are predicted by $F(x_s), s \in \mathcal{S}_{test}$.

3. Global Optimal Decision Tree (GODT)

Given the maximum depth of a decision tree D, the nodes of the tree are denoted as $t \in \{1, \dots, T = 2^{D+1} - 1\}$. We further define branch nodes as $t \in \mathcal{T}_B = \{1, \dots, \lfloor T/2 \rfloor\}$ and leaf nodes as $t \in \mathcal{T}_L = \{\lfloor T/2 \rfloor + 1, \dots, T\}$. Then, the decision tree problem (1) can be reformulated in the following MIO form:

$$\min_{a,b,c,d,z,L} \frac{1}{\hat{L}} \sum_{s \in \mathcal{S}} L_s + \lambda \sum_{t \in \mathcal{T}_B} d_t \tag{2}$$

$$\begin{aligned}
\text{s.t.} \quad & \frac{1}{2} \sum_{k \in \mathcal{K}} (y_{sk} + c_{kt} - 2y_{sk}c_{kt}) - L_s \leq 1 - z_{st}, \forall t \in \mathcal{T}_L \\
& \sum_{k \in \mathcal{K}} c_{kt} = 1, \forall t \in \mathcal{T}_L \\
& \sum_{t \in \mathcal{T}_L} z_{st} = 1 \\
& \mathbf{a}_m^T (\mathbf{x}_s + \epsilon - \epsilon_{min}) + \epsilon_{min} \leq b_m + (1 + \epsilon_{max})(1 - z_{st}), \forall m \in A_L(t), t \in \mathcal{T}_L \\
& \mathbf{a}_m^T \mathbf{x}_s \geq b_m - (1 - z_{st}), \forall m \in A_R(t), t \in \mathcal{T}_L \\
& \sum_{j=1}^{P} a_{jt} = d_t, \forall t \in \mathcal{T}_B, j \in \{1, \cdots, P\} \\
& 0 \leq b_t \leq d_t, \forall t \in \mathcal{T}_B \\
& d_t \leq d_{p(t)}, \forall t \in \mathcal{T}_B \\
& 0 \leq L_s \leq 1 \\
& a_{jt}, d_t \in \{0,1\}, 0 \leq b_t \leq 1 \forall t \in \mathcal{T}_B, j \in \{1, \cdots, P\} \\
& z_{st}, c_{kt} \in \{0,1\}, \forall t \in \mathcal{T}_L
\end{aligned}$$

where $L_s \in [0,1]$ is the loss of sample s. L_s equals to 1 if sample s is misclassified, otherwise 0. Variables a, b, c, d are the structure variables of the decision tree. Specially, $a_t = [a_{1t}, \dots, a_{Pt}] \in \{0,1\}^P$ represents the split dimension on node t. $b_t \in [0,1]$ is the split value on node t. $c_{kt} \in \{0,1\}$ is the class indicator at the leaf node $t \in \mathcal{T}_L$. $d_t \in \{0,1\}$ is the spilt indicator at the decision node $t \in \mathcal{T}_B$. A more detailed description of this MIO formulation can be found in Hua et al., 2022.

The fundamental idea to solve this MIO problem is to decompose the problem into a two-stage problem, in which structure variables a, b, c, d are the first-stage variables. The two-stage problem is shown as follows:

$$\text{First-stage:} \quad f(M_0) = \min_{m \in M_0} \sum_{s \in \mathcal{S}} Q_s(m) \tag{3}$$

$$\text{Second-Stage:} \quad Q_s(m) = \min_{z_s, L_s} \frac{1}{\hat{L}} L_s + \frac{\lambda}{n} \sum_{t \in \mathcal{T}_B} d_t \tag{4}$$

s.t. Constraints in problem (2)

where $m = (a, b, c, d)$ is the first-stage variables, and $M_0 := [m^l, m^u]$ is the initial region of the first-stage variables. Then we can use the branch and bound scheme to solve the resulting two-stage problem. The key advantage here is that we can prove that branching only on variables denoting the tree structures (a, b, c, d), our branch and bound training algorithm can guarantee the convergence to a global optimal solution with properly designed lower and upper bounds. Moreover, the decomposable nature of our lower and upper bound methods allows us to easily parallelize the branch and bound decision tree training algorithm on thousands of CPU cores. The detailed procure of the tailored reduced-space branch and bound algorithm to solve problem (2) can be found in Hua et al., 2022.

4. Results and Discussions

4.1. Experiment setup

In our study, we randomly split the dataset into a training dataset (75% of the entire dataset) and a testing dataset (25%). We compare the performance of our global optimal decision tree (GODT) with the popular heuristic algorithm CART (Loh, 2011). Different tree depths, including 2 and 3 levels, were considered in our experiments. All experiments were executed using Julia 1.7.0 with 4 hours time limit on a high-performance computing cluster, of which each node contains 40 Intel cores at 2.4 GHz and 202 GB RAM. We executed all the experiments in parallel mode with 40 cores for two-level trees and 80 cores for three-level trees.

To compare the performance of each algorithm, we reported train and test accuracy (%):= $\frac{correct\ classification}{all\ classification}$. We also reported the optimality gap for GODT as $Gap\ (\%) := \frac{upper\ bound - lower\ bound}{upper\ bound} \times 100\%$. This optimality gap is the worst gap of the current solution from the global optimal solution, which is a unique property of the deterministic global optimization algorithms. Heuristic algorithms such as CART cannot provide this worst gap. Besides, to fairly compare the performance, we repeated the experiments five times with five different random splits and reported means and standard deviations of each index in the form of *mean (standard deviation).*

4.2. Comparison results

Table 2 shows the comparison results of CART and GODT on different datasets with 3 and 5 categories. The method with better training and test accuracy is bolded in this table. Generally, both GODT and CART perform better on the dataset with fewer categories and trees with more levels. For the optimality gap, the GODT method can obtain a small and acceptable optimality gap on two-level trees while maintaining a larger optimality gap on three-level trees. Despite the larger optimality gap on three-level trees, GODT still performs better than CART. As for accuracy, GODT can obtain average absolute improvements of 1.70% and 1.54% in train and test accuracies, respectively.

Table 2 Comparison results of CART and GODT. Each index is in the form of *mean (standard deviation).*

Yield Category	Tree Depth	Method	Optimality Gap (%)	Train Accuracy (%)	P value	Test Accuracy (%)	P value
3	2	CART	-	64.62 (0.86)	9.80E-04	64.74 (2.62)	6.98E-04
		GODT	**0.92 (0.04)**	**65.73 (0.92)**		**66.14 (2.72)**	
3	3	CART	-	68.46 (1.45)	1.96E-01	67.98 (2.23)	1.96E-01
		GODT	**88.86 (0.61)**	**68.86 (0.98)**		**68.99 (1.07)**	
5	2	CART	-	44.24 (0.30)	2.18E-05	43.78 (1.00)	4.54E-03
		GODT	**5.27 (0.79)**	**48.61 (0.24)**		**46.51 (0.37)**	
5	3	CART	-	49.37 (0.59)	4.88E-02	47.12 (1.45)	1.40E-01
		GODT	**79.55 (1.70)**	**50.30 (0.83)**		**48.13 (2.91)**	

Remark 1. GODT focuses on optimizing the training accuracy in problem (2) . We have proved that global optimality can be reached on the problem (2). However, there is no guarantee that the test accuracy of GODT can also be globally optimal or consistently exceed CART (considering over-fitting). Nevertheless, extensive numerical results in this paper and previous works have demonstrated that an improvement in the test accuracy can be expected on most datasets for GODT, especially when the number of samples is large (e.g., >=1000) and the tree depth is small (e.g., 2-4), since over-fitting is avoided.

Remark 2. The p-values of one-tailed paired t-tests show that when tree depth is 2, GODT's train and test accuracy is statistically significantly larger than CART. When tree depth is 3, the results are not statistically significant. That is possible because we have not reached a small optimality gap in these experiments. We may obtain a more significant result if we develop a more efficient global algorithm for the decision tree.

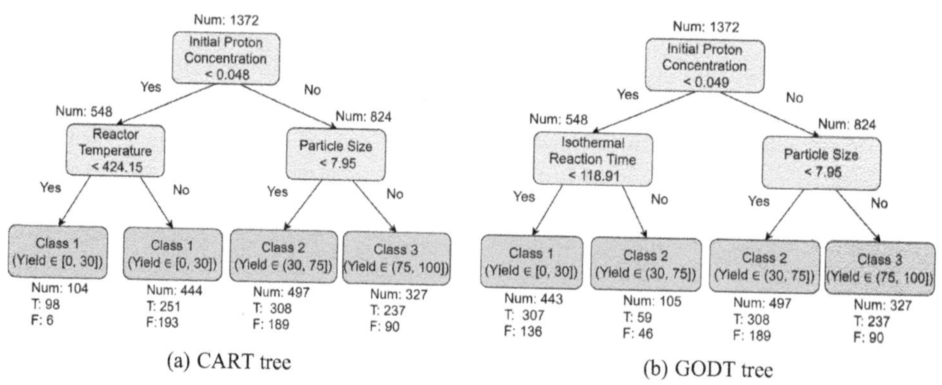

(a) CART tree (b) GODT tree

Figure 1 Decision Trees with two levels

4.3. Interpretability of models

The reasoning procedures of decision trees are similar to human reasoning and easy to comprehend by human decision-makers. Figure 1 shows the detailed reasoning procedures of two-level trees on the three-category training dataset in one experiment. In this experiment, the CART and GODT trees obtained a training accuracy of 65.16% and 66.40%, respectively. Notably, the optimality gap of this GODT tree is 0.92%. The CART tree selected initial proton concentration, reactor temperature, and particle size as the reasoning variables, while the GODT tree selected initial proton concentration, isothermal reaction time, and particle size as the reasoning variables. It should be noted that the CART trees have two leaf nodes with the same parent node and the same predicted class. This kind of split is abnormal since this split does not improve accuracy. The reason for this abnormal split is that CART utilizes the Gini index as the objective function at each node to make its split decisions, which differs from the accuracy we want to optimize. This kind of split may reduce the Gini index but not influence the accuracy. As an outcome, the accuracy of the GODT tree is better than the CART tree.

5. Conclusion

In this paper, we developed a global optimal explainable decision tree model for hemicellulose hydrolysis. This global optimal decision tree algorithm utilizes the MIO form of the decision tree and solves the problem by decomposing the MIO problem into a two-stage stochastic programming problem. Experiments on a hemicellulose hydrolysis dataset show that the global optimal decision tree model can obtain an average absolute improvement of 1.54% in test accuracy compared to the heuristic decision tree model while providing an acceptable optimality gap. Moreover, as a nature of decision tree models, the proposed models closely resemble human reasoning and are easy to comprehend by human decision-makers. In the future, we aim to develop more efficient global optimal explainable decision tree models to achieve higher prediction accuracies and lower computational costs.

References

H. V. Scheller, & P. Ulvskov, 2010, Hemicelluloses, Annual Review of Plant Biology, 61, 263–289

F. H. Isikgor, & C. R. Becer, 2015, Lignocellulosic biomass: a sustainable platform for the production of bio-based chemicals and polymers, Polymer Chemistry, 6(25), 4497-4559.

I. Spiridon, & V. I. Popa, 2008, Hemicelluloses: major sources, properties and applications, Monomers, Polymers and Composites from Renewable Resources, 289-304.

O. Fearon, V. Nykänen, S. Kuitunen, K. Ruuttunen, R. Alén, V. Alopaeus, & T. Vuorinen, 2020, Detailed modeling of the kraft pulping chemistry: carbohydrate reactions, AIChE Journal, 66(8), e16252.

E. Wang, G. Cai, R. Ballachay, Y. Cao, & H. L. Trajano, 2022, Predicting Xylose Yield in Prehydrolysis of Hardwoods: A Machine Learning Approach, Frontiers in Chemical Engineering, 84.

S. M. Weiss, & C. A. Kulikowski, 1991, Computer systems that learn: classification and prediction methods from statistics, neural nets, machine learning, and expert systems, Morgan Kaufmann Publishers Inc..

W. Y. Loh, 2011, Classification and regression trees, Wiley interdisciplinary reviews: data mining and knowledge discovery, 1(1), 14-23.

J. R. Quinlan, 1986, Induction of decision trees, Machine learning, 1(1), 81-106.

J. R. Quinlan, 2014, C4. 5: programs for machine learning, Elsevier.

D. Bertsimas, & J. Dunn, 2017, Optimal classification trees, Machine Learning, 106(7), 1039-1082.

S. Aghaei, A. Gómez, & P. Vayanos, 2021, Strong optimal classification trees, arXiv preprint arXiv:2103.15965.

S. Verwer, & Y. Zhang, 2019, Learning optimal classification trees using a binary linear program formulation, In Proceedings of the AAAI conference on artificial intelligence, 33(1), 1625-1632.

K. Hua, J. Ren, & Yankai Cao, 2022. A Scalable Deterministic Global Optimization Algorithm for Training Optimal Decision Tree. Advances in Neural Information Processing Systems.

Antonis Kokossis, Michael C. Georgiadis, Efstratios N. Pistikopoulos (Eds.)
PROCEEDINGS OF THE 33rd European Symposium on Computer Aided Process Engineering (ESCAPE33), June 18-21, 2023, Athens, Greece
 http://dx.doi.org/10.1016/B978-0-443-15274-0.50214-6

NUMERICAL SIMULATION OF THE L-V EQUILIBRIUM WITHIN A STAGE IN A DISTILLATION COLUMN USING CFD

Perla G. Canchola-López[a], Ariadna E. Vázquez-Hernández[a], Jazmín Cortez-González[a], Rodolfo Murrieta-Dueñas[a], Roberto Gutiérrez-Guerra[b], Carlos E. Alvarado-Rodríguez[c]

[a] *Tecnológico Nacional de México/ Irapuato, Ingeniería Química y Bioquímica,; Carretera Irapuato-Silao Km. 12.5, C.P:36821 Irapuato, Guanajuato, MEXICO;*
[b] *Universidad Tecnológica de Leon, Departamento de Sustentabilidad para el desarrollo. Blvd. Universidad Tecnológica 225, Universidad Tecnologica, San Carlos la Roncha, 37670 León, Gto.*
[c] *Departamento de Ingeniería Química, División de Ciencias Naturales y Exactas, Universidad de Guanajuato, Noria Alta S/N, C.P. 36050, Guanajuato, Guanajuato, México.*

Abstract

This work presents the numerical simulation in L -V equilibrium stage in a plate distillation column using the SPH method, considering Sieve and Bubble cap plates. To perform the simulation of the equilibrium stage, periodic conditions in temperature were established. The sizing of the column was performed in Aspen One considering an equimolar mixture of Benzene-Toluene and an operating pressure that guarantees that the cooling water temperature of the condenser is 120°F, the thermodynamic model used was Chao-Seader. The following information was obtained from the Aspen simulation: liquid and vapor velocity per stage, viscosity and density of the mixture, operating pressure, and column diameter. Data from Aspen simulation was used for the CFD simulations using DualSPHysics. The results allow it to compare the thermal distribution and the velocity fields between plate design and the time to achieve the thermal equilibrium.

Keywords: CFD, Simulation of Distillation, thermal equilibrium.

1. Introduction

Traditionally, the numerical simulation of hydrodynamics and heat transfer in distillation column has been performed with Eulerian methods. These methods use mesh for the discretization of the medium, thus generating averages in the transfer zones and interfaces between fluids. On the other hand, Lagrangian methods allow to see in detail the phenomena at the interface and to discretize the continuous medium through points meshless. In this work we propose to perform a numerical simulation of distillation columns using the SPH method (Lagrangian nature).

Distillation is a highly inefficient unitary operation and its energy consumption is high, so some authors have proposed new alternatives to improve efficiency and reduce energy consumption. Yin et al., (2002) studied the liquid retention in a packed column with Pall rings and the distribution of the fluid over the packing's. They find that the distribution of the liquid retention is not uniform due to the design of the distributor.

Fu et al., 2020 investigated the flow and retention of a liquid in a packed column with pall rings finding that the randomness of the packing generates transitions from viscous flow to turbulent regimes in a Reynolds Number range between 6.7 and 40.2. Jain et al., 2021 developed an Eulerian model to simulate multiphase gas-liquid flow in a wide variety of regimes over a manifold, finding that the discharge patterns, the uniformity and symmetry of the flow lines and the pressure drop across the manifold exhibit significant differences from each other. Pichler et al., 2021, studied heat transfer in packed columns by numerical simulations considering conduction and natural convection suggesting a correction as a function of particle diameter that can reduce up to 75% the error in surface temperature. In this work the numerical simulation of the hydrodynamics and heat transfer in one stage inside a distillation column using the Smoothed Particle Hydrodynamics (SPH) method is performed in order to compare velocity fields, retention times and heat transfer in one stage. The SPH method is a mesh-free method with which it is possible to track the interface between the fluids immersed in a flow in a clear and simple way, thus allowing a better analysis of the phenomena occurring in it.

2. Methodology

In this work a methodology for the simulation of the hydrodynamics and thermal equilibrium of plate distillation columns is proposed. This methodology consists of 3 stages: rigorous design of the distillation column in Aspen Plus, 3D design of the column - sieve plates and hydrodynamic analysis of the columns using the SPH method. The Aspen Plus simulation results were used as input data for SPH. The 3D design was performed in SolidWorks.

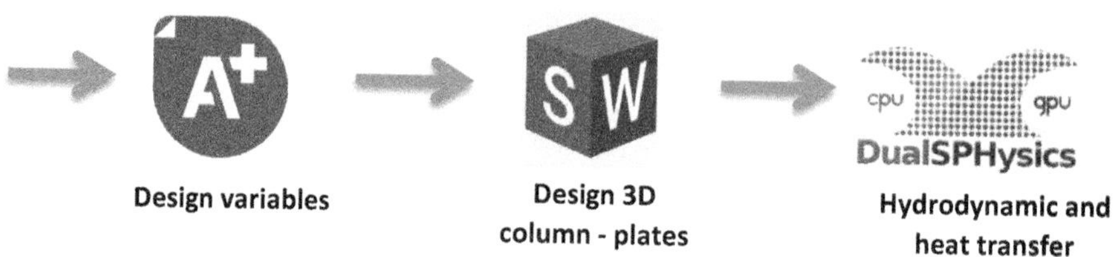

Figure 1. Methodology for the numerical simulation of the hydrodynamics and equilibrium of distillation columns.

2.1. Rigorous design of the distillation column in Aspen Plus

The design of the distillation column was carried out considering a binary equimolar Benzene-Toluene mixture, whose equilibrium is modeled with Chao-Seader. A Rad-Frac model was used for column equilibrium. It was proposed at 120°F to ensure cooling water in the condenser. The fed is 100lbmol/hr. According to the rigorous design, 10 stages were obtained to achieve 98% purity and recovery for each component. Table 1 shows the results obtained from the Aspen One® simulation of stages 6 and 7, which were used as input data for the SPH method.

2.2. Column and sieve plate sizing

The 3D design of the column and plates was carried out in SolidWorks. The design parameters used were: column diameter, effective and downspout areas, obtained in Aspen One. The distance between plates was proposed to be 0.15m. Stainless steel 316L was chosen as construction material. Figure 1 shown in detail the 3D design of the sieve and bubble cap plates.

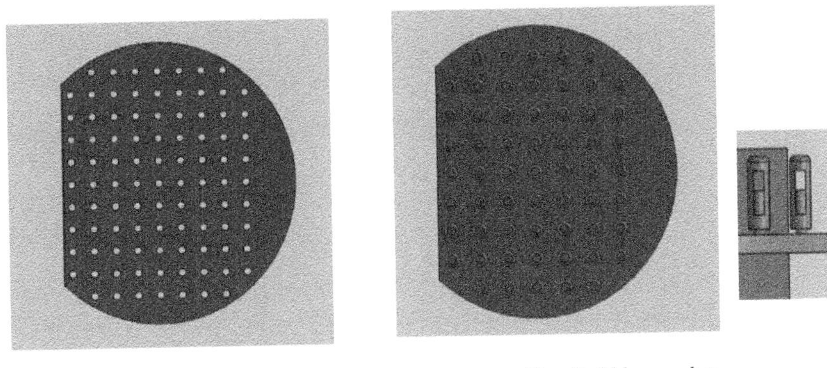

a) Sieve plate *b) Bubblecap plate*

Figure 1. 3D design plates using in distillation columns. a) sieve and b) bubble cap.

Table 1. Results of the rigorous simulation of the distillation column in Aspen Plus.

Parameter	Liquid	Vapor
Density (kg/m$^{3)}$	783.8023	4.0206
Viscosity (Ns/m2)	2.5x10^{-4}	9.5x10^{-6}
Temperature (°C)	110.39	105.99
Mass fraction	0.5886	0.5405
Velocity (m/s)	0.0261	0.0214
Coefficient of thermal diffusivity (m2/s)	7.9247E-8	
Surface tension	0.018 N/m	0.018 N/m

2.3. Hydrodynamics and heat transfer in SPH

SPH is a Lagrangian meshless method with applications in the field of Computational Fluid Dynamics. Originally invented for astrophysics in the 1970s (Monagan J, 1992) it has been applied in many different fields, including fluid dynamics (Alvarado-Rodríguez C.E. et al., 2019) and wastewater treatment (Mokos et al., 2015). The method uses points named particles to represent the continuum and these particles move according to the governing equations in the fluid dynamic. When simulating free-surface flows, no special surface treatment is necessary due to the Lagrangian nature of SPH, making this technique ideal for studying violent free-surface motion. Moreover, the movement of the boundaries can be set easily without the necessity of update the mesh as in the Eulerian methods. The SPH formalism used in the simulations is reported by Dominguez et al. (2021) which is set in the DualSPHysics code, in this work only the continuity equation (Eq. 1), the momentum equation (Eq. 2), the equation of state (Eq. 3) and the energy transport equation (Eq. 4) in the SPH formalism are reported. Mass transport can be added for future work (Chiron et al., 2019)

$$\frac{d\mathbf{v}_a}{dt} = -\sum_b m_b \left(\frac{P_a+P_b}{\rho_a\rho_b} + \sum_b m_b \left(\frac{4v_0 r_{ab}\cdot\nabla_a W_{ab}}{(\rho_a+\rho_b)({r^2}_{ab}+\eta^2)}\right)\mathbf{v}_{ab} + \sum_b m_b \left(\frac{\vec{\tau}^{j}_{ab}}{{\rho_b}^2} + \frac{\vec{\tau}^{i}_{ab}}{{\rho_a}^2}\right)\nabla_a W_{ab}\right)\nabla_a W_{ab} + g\,, \tag{1}$$

$$\frac{d\rho_a}{dt} = -\rho_a \sum_b \frac{m_b}{\rho_b}(\mathrm{v}_b - \mathrm{v}_a)\cdot\nabla_a W_{ab}\,, \tag{2}$$

$$P = B\left[\left(\frac{\rho}{\rho_0}\right)^{\gamma} - 1\right], \tag{3}$$

$$\frac{dT_a}{dt} = -\frac{1}{Cp}\sum_b \frac{m_b(k_a-k_b)(r_a-r_b)\cdot\nabla_a W_{ab}}{\rho_a\rho_b({r^2}_{ab}-\eta)}(T_a - T_b), \tag{4}$$

where the subscripts a and b are denoted for the mean particle "a" and the neighbors particles "b", v is the velocity, t is time, m is mass, P is pressure, ρ is density, υ_0 is the kinematic viscosity, τ is the stress tensor, $B = {c_0}^2\rho_0/\gamma$, c_0 is and artificial sound speed, and $\gamma = 7$, Cp is the specific heat of the fluid, k is the conductivity constant, T is the temperature and W is the kernel function defined in the SPH method. This method was implemented for the numerical analysis of the distillation columns under the following considerations: periodic conditions were used to perform the hydrodynamic and equilibrium analysis. The initial properties of the fluids are shown in Table 1. In all cases, no-slip conditions were considered in the interaction between fluid and boundary particles using the dynamic particle method reported by (Crespo et al., 2007a). A different resolution was used in each case due to the specific characteristics for the geometries in order to save computational time, obtained through a convergence analysis for each phase. SPH results have so far only been compared with those obtained from the Aspen One simulation.

3. Discussion of results

With the results obtained from the SPH numerical simulations, it is possible to analyze the velocity profile and the temperature profile inside the stage, for each plate design. In addition, it is possible to approximate the required time that each stage needs to reach equilibrium under the initial conditions established.

3.1. Bubble cap plate

The simulation was performed with a total of 1,686,948 fluid particles of which 1,042,345 are liquid and 684,936 are vapor. The particles modeling the plate and column are 535,069. We simulated 30 seconds of real-time. The simulation took a total of 10.5h on the ITESI computer with an Nvidia GPU Geforce RTX 3060. Figure 2 shows the velocity field obtained in the simulation from the top and side perspectives. According to the results obtained, it can be deduced that the vapor velocity below the cap does not register fluid movement, however above it, the profile is almost homogeneous. This can be contrasted with the results of the temperature profile shown in Figure 3, for the same stages analyzed. The stagnation of the fluid shown in Figure 2 generates a retard in the homogeneity of the temperature in that zone. This is due to the fact that the temperature variation is dominated by the effect of diffusion, which means that even after 30 seconds, the final temperature corresponds to the initial value.

3.2. Sieve plate

The simulation was performed with a total of 1,433,006 fluid particles of which 962,243 are liquid and 470,763 are vapor. The particles modeling the plate and column are 387,365. 10 seconds of real-time were simulated. The simulation was completed in a total of 2.5h on the same ITESI computer. Figure 4 shows the velocity field obtained in the simulation from the top and side perspectives. As in the bubble cap plate, a higher velocity is obtained in the area where the fluid enters from the downspout, however, the velocity distribution is more homogeneous compared to the bubble cap plate due to the velocity generated by the vapor in the orifices. This homogeneity in the flow also generates homogeneity in the temperature as shown in Figure 5.

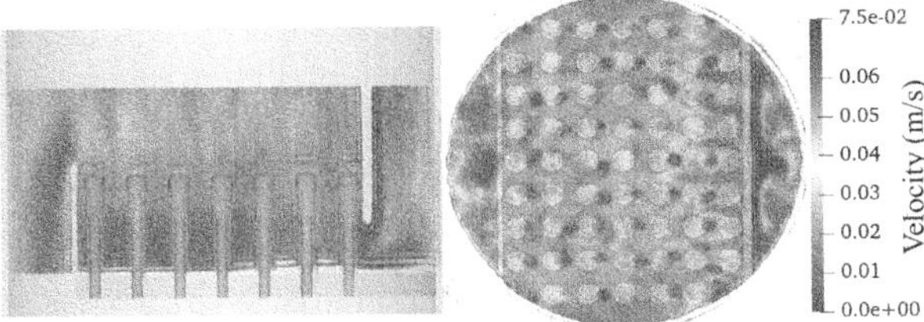

Figure 2. Side and top perspectives of the velocity fields in the stage for the bubble cap plate.

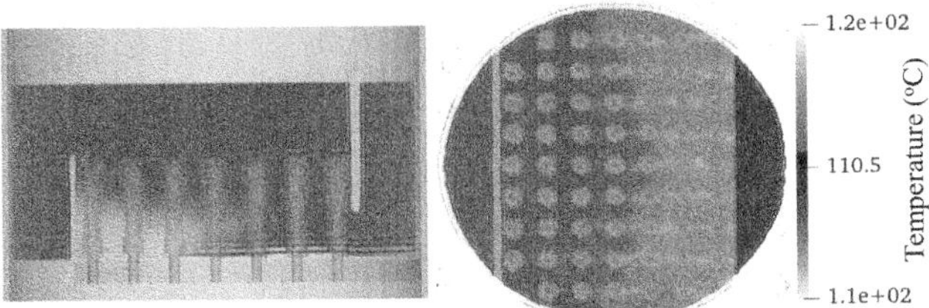

Figure 3. Side and top perspectives of the temperature fields in the stage for the bubble cap plate.

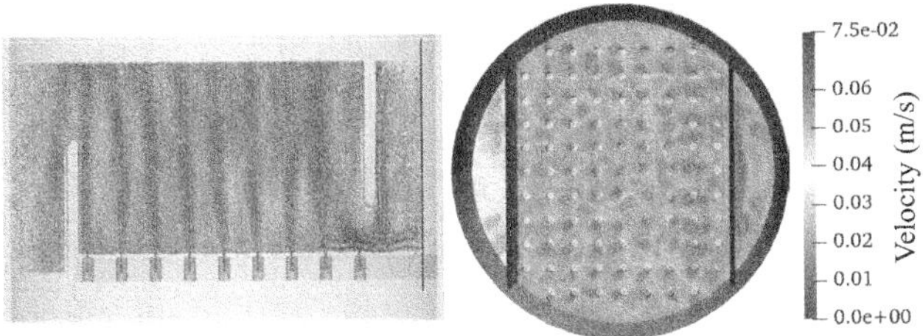

Figure 4. Side and top perspectives of the velocity fields in the stage for the sieve plate.

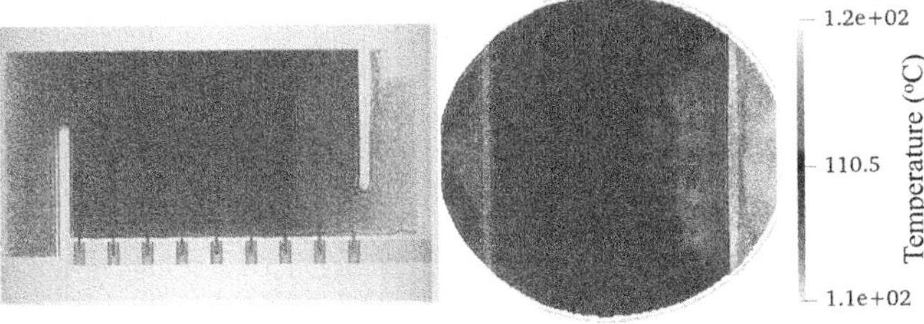

Figure 5. Side and top perspectives of the temperature fields in the stage for the sieve plate.

4. Conclusions

In this work, the simulation of a stage inside a distillation column is presented. Two case studies were developed: bubble cap and sieve plates. The geometry of the column and the plates were made in SolidWorks. Then, the geometries were exported to DualSPHysics to perform the corresponding simulations, based on the parameters obtained with the rigorous simulation in Aspen One. The numerical results indicate that the height of the bubble caps is not adequate since there is no homogeneous temperature distribution in the area near the downspout towards the next stage. A proper configuration in the distribution of the caps or in the height of the caps could benefit a homogeneous distribution of velocity and temperature. In contrast to the bubble cap plate, in the sieve plate system, the velocity distribution is more homogeneous and reaches a constant temperature after about 10 seconds. Based on the methodology proposed in this paper, it is possible to evaluate the hydrodynamic and heat transfer performance of different geometries for packed or plate columns. These analyses will allow obtaining mathematical correlations with which it will be possible to quantify the impact of a design modification on the hydrodynamic performance of the column. This methodology can be extended to packed columns to determine how liquid retention increases or decreases with packing design and distribution.

References

Alvarado-Rodríguez, C. E., Klapp, J., Domínguez, J. M., Uribe-Ramírez, A. R., Ramírez-Minguela, J. J., & Gómez-Gesteira, M. (2019, March). Multiphase Flows Simulation with the Smoothed Particle Hydrodynamics Method. In International Conference on Supercomputing in Mexico (pp. 282-301). Springer, Cham.

Domínguez JM, Fourtakas G, Altomare C, Canelas RB, Tafuni A, García-Feal O, Martínez-Estévez I, Mokos A, Vacondio R, Crespo AJC, Rogers BD, Stansby PK, Gómez-Gesteira M. (2021), DualSPHysics: from fluid dynamics to multiphysics problems. Computational Particle Mechanics.

Fu, Y., Bao, J., Singh, R., Wang, C., & Xu, Z. (2020). Investigation of countercurrent flow profile and liquid holdup in random packed column with local CFD data. Chemical Engineering Science, 221, 115693.Mokos, A., Benedict, D. R., Stansby, K. P., & Dominguez, J. M. (2015). Multi-phase SPH modelling of violent hydrodynamics on GPUs. Journal Computer Physics Communications, 196, 304-31.

Monagan J J, "Smoothed Particles Hydrodynamics", Annu. Rev. Astron. Astrphys., 30:534-74, 1992.

Pichler, M., Haddadi, B., Jordan, C., Norouzi, H., & Harasek, M. (2021). Effect of particle contact point treatment on the CFD simulation of the heat transfer in packed beds. Chemical Engineering Research and Design, 165, 242-253.

Jain, E., Sau, M., & Buwa, V. V. (2021). Eulerian simulations of liquid distribution generated by chimney and bubble cap distributors. Chemical Engineering Journal, 421, 127799.

Yin, F., Afacan, A., Nandakumar, K., & Chuang, K. T. (2002). Liquid holdup distribution in packed columns: gamma ray tomography and CFD simulation. Chemical Engineering and Processing: Process Intensification, 41(5), 473-483.

Chiron, S. Marrone, A. Di Mascio, D. Le Touzé, (2018). Coupled SPH–FV method with net vorticity and mass transfer, Journal of Computational Physics, Volume 364, Pages 111-136, ISSN 0021-9991.

Crespo, A.J.C. ,Gómez-Gesteira, M.,Dalrymple,R.A., Boundary conditions generated by dynamic particles in SPH methods, Computers, Materials & Continua ,5,173–184, 2007a.

Antonis Kokossis, Michael C. Georgiadis, Efstratios N. Pistikopoulos (Eds.)
PROCEEDINGS OF THE 33rd European Symposium on Computer Aided Process Engineering (ESCAPE33), June 18-21, 2023, Athens, Greece
 http://dx.doi.org/10.1016/B978-0-443-15274-0.50215-8

MONITORING OF THE DRILLING REGION IN OIL WELLS USING A CONVOLUTIONAL NEURAL NETWORK

Caroline Dias Grossi[a,b], Vinícius P. Barbosa[c], Rubens Gedraite[c], Maurício B. de Souza Jr.[a], Cláudia Mirian Scheid[b], Luís Américo Calçada[b], Luiz Augusto da Cruz Meleiro[b].

[a]*Federal University of Rio de Janeiro, EPQB/UFRJ, Rio de Janeiro/RJ, 21941-598, Brasil*
[b]*Federal Rural University of Rio de Janeiro, IT/DEQ/UFRRJ, Seropédica/RJ, 23897-000, Brasil*
[c]*Federal University of Uberlândia, DEQ/UFU, Uberlândia/MG, 38400-902, Brasil*
cdiasgrossi@eq.ufrj.br

Abstract

Currently, in the oil well drilling industry, the process is monitored by pressure and temperature sensors in the well and of the drilling fluid. Instabilities can be noticed with the gravel that returns in the mud, demandingthe implementation of monitoring systems. This work proposes applying computer vision techniques to monitor vibrating screens and detect particulate material flow. Experiments were conducted on a pilot screen using a suspension of sand and 0.1% xanthan gum, simulating drilling mud. Each experiment varied operating conditions, obtaining 26 videos. The methodology evaluated the response of U-Net neural networks in semantic segmentation with Python. A database of 11,140 images and corresponding templates was built. The evaluation metrics were accuracy, F1-Score, and MeanIoU. The results show high segmentation ability, with accuracy 97%, F1-Score 92%, and MeanIoU 91%. The results show that the U-Net has good segmentation capabilities and provide a promising alternative for non-invasive process monitoring.

Keywords: computational vision, convolutional neural networks, oil wells drilling.

1. Introduction

Artificial Intelligence is a science that studies the ability of machines to act similarly to humans. For this, it a machine learning step is necessary [1] . Currently, several algorithms can be found in the literature capable of solving problems in various industry segments. In this context, the choice of the appropriate algorithm will depend on the final objective and nature of the data[2] .

Image acquisition of industrial processes is a common practice in facility surveillance. The rise of machine learning models and the increased processing power of computers have allowed the development of models capable of processing increasingly complex data, such as images, which has motivated various industries to explore the information that the so-called computer vision can offer. Such a perspective can bring advantages to the automation of quality control systems, sensing, and safety monitoring, with a high potential of identifying signs of operational accidents before their occurrence[3] .

Shallow learning algorithms, in general, need a feature extraction phase before they can make predictions and classifications. The manual process of feature extraction from images is costly and can generate attributes irrelevant to the proposed objective, leading

to highly complex models that do not solve the problem. When these methods are not able to provide feasible results, deep learning techniques are applied [3][4].
One example is Convolutional Neural Networks (CNN's), traditionally consisting of two modules: general learning and specific learning. The general learning module is responsible for receiving the original image and identifying, from it, information common to a diversity of objects from convolution and pooling operations. The response is the so-called attribute vector, which stores the learned characteristics [4] . The specific learning module consists of a fully connected layer responsible for solving the problem. It takes as input the attribute vector and provides the class to which the image belongs.
Besides the classification of images as a whole, the classification of each image pixel can be performed, a particular feature of Fully Convolutional Neural Networks (FCN). In such a type of FCN, the specific learning module is replaced by another convolutional layer capable of classifying individual pixels. This type of neural network can be trained without the need for feature extraction steps and exhibits a more accurate response regarding the shape of the objects of interest [5].
The U-net (Figure 1), a particular type of FCN, was developed for biomedical image segmentation. Given the difficulty of creating a sizeable medical image bank, U-Net was initially designed to provide high-accuracy responses regarding segmentation with a low number of images available for training[6] .

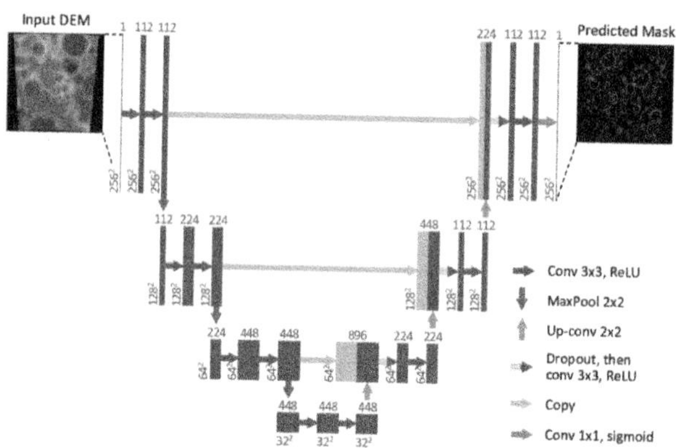

Figure 1. Representation of the U-Net architecture. Source: Silburt et al. 2018.

The U-Net architecture is formed by a sequence of contraction and expansion layers. In the shrinking step occurs dimensionality reduction, generating feature maps that preserve essential information, such as object contours. At the end of the contraction step, the expansion step occurs, which performs the opposite path, increasing the dimensionality of the image from the generated feature maps. One of the main advantages of this network is in the expansion path, which uses the output of the previous unpooling concatenated with the corresponding pooling layer to increase the number of feature maps. This feature improves the high-level feature extraction capability and makes the network learn the content and location of each pixel in the image[7] .
Some relevant uses of the U-Net, far from those cases applied to medical imaging, have been published in recent works with adaptations of this type of FCN, as follows: the detection of manufacturing defects in clothing fabrics[8] ; mapping impervious surfaces from satellite images[9]; the identification of falsification in image splices using dual encoder U-Nets[10]; quantification of gas leakage in the subsea environment using CFD and fully convolutional U-net networks[11].

The process of drilling oil wells is complex, and with advances in-depth, instabilities are more likely to occur[12] . Therefore, it is convenient to monitor all stages of the process, which is already done using sensors installed on the rig. Most of the variables currently monitored are related to the pressure and temperature conditions along the well and the drilling fluid. An interesting approach would be to evaluate the mud that reaches the surface, which can also indicate the beginning of some instabilities, such as the well collapsing and insights from the column trapped.

Thus, this work proposes using computer vision methods to detect the gravel that reaches the surface. The method evaluated the response of U-Net type fully convolutional neural networks in the semantic segmentation of images that simulate the flow in vibrating screens, the first equipment to receive the drilling mud. Another justification for using this type of network is the difficulty in obtaining images of gravel flow on actual screens, in rigs, and the adaptability of U-Net for the scenario of small image banks.

2. Methodology

Due to the sake of industrial secrecy related to obtaining images from the operation of the actual sieves, the behavior of this process equipment was investigated from experiments performed in a pilot-scale sieve using a suspension of sand and xanthan gum. A total of 26 videos with varied operating conditions was employed in the present study. The experiments are detailed in [P].

The tests with the proposed U-Net architecture used 11,140 flow images with their respective templates; 20% of these were used to test the FCN and 80% for the training stage. From the training data bank, 10% was used for the validation stage, which used a cross-entropy loss function. Examples of images from the database can be seen in Figure 2. The templates were obtained in a semi-automated way from the construction of image processing algorithms specific to the color, brightness, and contrast conditions present in each video. Conversion to color spaces other than RGB, histogram equalization, and digital filtering for image quality improvement were explored.

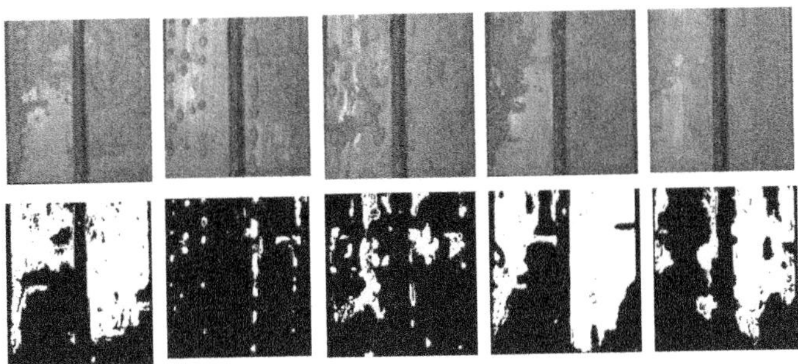

Figure 2. Examples of images from the database.

The U-Net algorithm was developed in Python, using Keras and TensorFlow. The preprocessing step performed the image acquisition in its original form, resizing to size [128, 128] and storing the data in tensors. The procedure was the same for the network input images and the template, with the difference that the input images are in color channels and the templates in binary masks.

Next, the network architecture was built using convolution, pooling, and ReLu activation layers in both the contraction and expansion stages. After input, the data goes through a normalization step that forces the pixels in each channel to assume values between zero and one. Next, the image passes through each layer of the network, which has a depth equal to 4 levels, until it reaches the final layer, which uses a sigmoidal function to perform the pixel-by-pixel classification.

At the end of the model construction, training parameters were defined, such as early stopping criteria, which helps minimize the occurrence of overfitting and the maximum number of epochs. The model was trained, and the network was saved for later use.

3. Results and Discussions

During training, accuracy and loss data were collected for training and validation and used in constructing the graphs in Figure 3. It was noticed that the curves converged quickly to a model with high accuracy and low loss rate, with no indication of overfitting. The final accuracy found for this data set was 98.85% for training and 98.88% for testing.

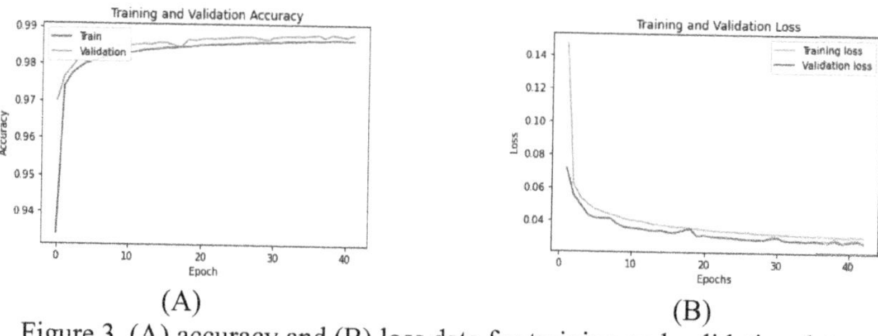

Figure 3. (A) accuracy and (B) loss data for training and validation data.

Other metrics were implemented to evaluate the network response: accuracy, recall, F1-Score, and MeanIoU. Accuracy evaluates the hits in each class considering all positive predictions; recall is the proportion of positives correctly classified as positives. The F1-score is a balancing metric between the response of precision and recall.

The MeanIoU metric was considered the most important for this work since it is the primary metric for semantic segmentation evaluation. It checked the overlap between the prediction and the template, considering the classification value of each pixel and its location. To calculate this metric, the number of pixels that were classified as true positives, which represented the intersection between the template and the prediction, was divided by the sum of all positive pixels in both the template and the prediction, which represented the union between them. The results of the metrics and the confusion matrix can be seen in Figure 4 and Tables 1 and 2.

PREDICT	TARGET	
	SAND	BACKGROUND
SAND	98.35%	0.97%
BACKGROUND	1.65%	99.03%

(A)

PREDICT	TARGET	
	SAND	BACKGROUND
SAND	98.36%	0.95%
BACKGROUND	1.64%	99.05%

(B)

Figure.4 Results for the (A) training and (B) test image set.

Table 1. Metrics by class for the training and test image set.

	TRAINING			TEST		
CLASS	Recall	Precison	F1-score	Recall	Precision	F1-score
SAND	0.983	0.973	0.978	0.983	0.973	0.978
BACKGROUND	0.994	0.990	0.992	0.994	0.990	0.992

Table 2. Global metrics for the training and test image set.

	TRAINING	TEST
Mean Accuracy	0.998	0.998
Mean IoU	0.973	0.942

The results showed high segmentation ability for both training and test data, so it can be concluded that the chosen parameters for the proposed U-Net were adequate for solving this problem. The adequate segmentation performance measured how many pixels of sand suspension and background were in the image and evaluated the location of each. The prediction of the network by image response was qualitatively verified, and the result is shown in Figure 5.

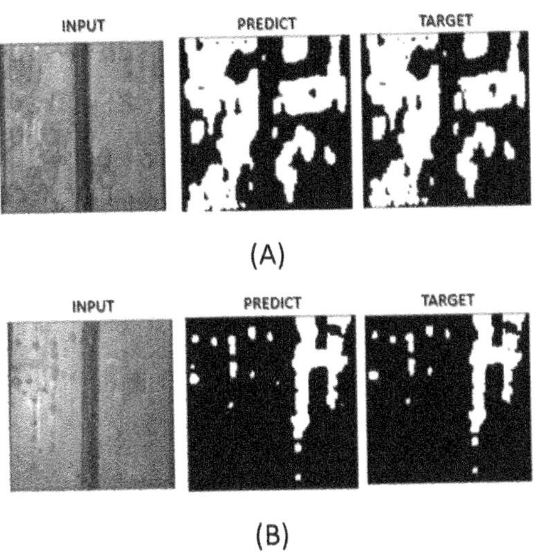

Figure 5. Segmentation examples provided by the trained U-Net for images from the (A) training and (B) test set.

Inputs from the training and test data sets were tested, and the prediction results showed that the U-Net appropriately described the expected template, showing its potential for feature learning in future predictions.

4. Conclusion

It can be concluded that convolutional networks are powerful tools for image processing, presenting great results with minimal pre-processing procedures for pattern corrections. Parameter tuning is an essential step for the best neural network design. In the case of supervised learning, the quality of the templates is essential to ensure the reliability of the predictions. As a proposed sequence of the present work, the authors expect to evaluate the proposed U-Net in analyzing video images from actual industrial sieves operations for estimating useful process variables for monitoring the drilling process.

This study allows the extraction of operational information from the objects of interest. In the case of this application, the calculation of estimates of percentage filling of the cutout analysis and average flow velocity of the agglomerates is under development. These parameters are not currently measured for rig screens but are useful in the online identification of instabilities and well-collapse. Extension to other areas, such as mineral and food processing, is possible with the estimation of other variables, such as flow rate and morphological properties of particles, at macro or micro scales, provided images are available.

Acknowledgments
CAPES (Finance code 001), CENPES/PETROBRAS (Grant No. 4600580875), CNPq (Grant No. 311153/2021-6) and FAPERJ (Grant No. E-26/201.148/2022).

References

1. GONÇALVES, L. B. **Intelligent image classification system of macroscopic rock for the oil and gas industry.** PhD Thesis (In Portuguese). Universidade Federal Fluminense. Niterói – RJ, Brazi. p. 309, 2010.
2. OSAROGIAGBON, A. U., KHAN, F., VENKATESAN, R., GILLARD, P. **Review and analysis of supervised machine learning algorithms for hazardous events in drilling operations**. Process Safety and Environmental Protection, Volume 147, 2021, Pages 367-384, ISSN 0957-5820, https://doi.org/10.1016/j.psep.2020.09.038.
3. MAGANA-MORA, A., AFFLECK, M., IBRAHIM, M., MAKOWSKI, G., KAPOOR, H., OTALVORA, W. C., JAMEA, M. A., UMAIRIN I. S., ZHAN, G., GOONERATNE, C. P. **Well Control Space Out: A Deep-Learning Approach for the Optimization of Drilling Safety Operations**. *IEEE Access*, vol. 9, pág. 76479-76492, 2021, https://doi.org/10.1109/ACCESS.2021.3082661.
4. LECUN, Y., BENGIO, Y. & HINTON, G. **Deep Learning**. Nature 521, 436-444 (2015).
5. IJJEH, A. A., ULLAH, S., KUDELA, P. **Full wavefield processing by using FCN for delamination detection**. Mechanical Systems and Signal Processing, Volume 153, 2021, 107537, ISSN 0888-3270, https://doi.org/10.1016/j.ymssp.2020.107537.
6. JOHN, L. D., ZHANG, C.. **An attention-based U-Net for detecting deforestation within satellite sensor imagery.** International Journal of Applied Earth Observation and Geoinformation, Volume 107, 2022, 102685, ISSN 0303-2434, https://doi.org/10.1016/j.jag.2022.102685.
7. SILBURT, ARI & ALI-DIB, MOHAMAD & ZHU, CHENCHONG & JACKSON, ALAN & VALENCIA, DIANA & KISSIN, YEVGENI & TAMAYO, DANIEL & MENOU, KRISTEN. (2018). Lunar Crater Identification via Deep Learning. Icarus. 317. https://doi.org/10.1016/j.icarus.2018.06.022.
8. JING, J., WANG, Z., RÄTSCH, M., ZHANG, H. **Mobile-Unet: An efficient convolutional neural network for fabric defect detection.** Article Information. Volume: 92 issue: 1-2, page(s): 30-42. Article first published online: May 29, 2020; Issue published: January 1, 2022. https://doi.org/10.1177/0040517520928604
9. MCGLINCHY, B. JOHNSON, B. MULLER, M. JOSEPH AND J. DIAZ. **Application of UNet Fully Convolutional Neural Network to Impervious Surface Segmentation in Urban Environment from High Resolution Satellite Imagery.** *IGARSS 2019 - 2019 IEEE International Geoscience and Remote Sensing Symposium*, 2019, pp. 3915-3918, https://doi.org/10.1109/IGARSS.2019.8900453.
10. BI, X., LIU, Y., XIAO, B., LI, W., PUN, C. M., WANG, G., GAO, X. **D-Unet: A Dual-encoder U-Net for Image Splicing Forgery Detection and Localization.** Computer Vision and Pattern Recognition. https://doi.org/10.48550/arXiv.2012.01821
11. CALDAS, G. L. R., BENTO, T. F. B., MOREIRA, R. M., SOUZA, M. B. Quantifying Subsea Gas Leakages using Machine Learning: a CFD-based study, Editor(s): Yoshiyuki Yamashita, Manabu Kano, Computer Aided Chemical Engineering, Elsevier, Volume 49, 2022, Pages 1345-1350, ISSN 1570-7946, ISBN 9780323851596, https://doi.org/10.1016/B978-0-323-85159-6.50224-4.
12. M. BARBOSA, V.P.; MENEZES, A.L.; GEDRAITE, R.; ATAÍDE, C.H. **Vibration screening: A detailed study using image analysis techniques to characterize the bed behavior in solid–liquid separation.** Minerals Engineering, Volume 154, 2020, 106383, ISSN 0892-6875

Antonis Kokossis, Michael C. Georgiadis, Efstratios N. Pistikopoulos (Eds.)
PROCEEDINGS OF THE 33rd European Symposium on Computer Aided Process Engineering (ESCAPE33), June 18-21, 2023, Athens, Greece
 http://dx.doi.org/10.1016/B978-0-443-15274-0.50216-X

An inverse optimization approach to decision-focused learning

Rishabh Gupta[a], Qi Zhang*[a]

[a]*Department of Chemical Engineering and Materials Science, University of Minnesota, Minneapolis, MN 55455, USA*

* *qizh@umn.edu*

Abstract

Decision-focused learning is an emerging paradigm specifically aimed at improving the data-driven learning of input parameters to optimization models. The main idea is to learn predictive models that result in the best decisions rather than focusing on minimizing the parameter estimation error. Virtually all existing works on decision-focused learning only consider the case where the unknown model parameters merely affect the objective function. In this work, extend the framework to also consider unknown parameters in the constraints, where feasibility becomes a major concern. We address the problem by leveraging recently developed methods in data-driven inverse optimization, specifically applying a penalty-based block coordinate descent algorithm to solve the resulting large-scale bilevel optimization problem. The results from our computational case study demonstrate the effectiveness of the proposed approach and highlight its benefits compared with the conventional predict-then-optimize approach, which treats the prediction and optimization steps separately.

Keywords: Decision-focused learning, inverse optimization, constraint learning.

1. Introduction

In traditional data-driven optimization, we often follow a two-step predict-then-optimize approach, i.e. we first predict the unknown model parameters from data with external features and then solve the optimization problem with those predicted inputs. Here, the learning step focuses on minimizing the parameter estimation error; however, this does not necessarily lead to the best decisions (evaluated with the true parameter values) in the optimization step. In contrast, decision-focused learning (Wilder et al., 2019), also known as smart predict-then-optimize (Elmachtoub and Grigas, 2022), integrates the two steps to explicitly account for the quality of the optimization solution in the learning of the model parameters (i.e. minimize the decision error).

Existing works on decision-focused learning, many of which are based on deep learning with differentiable optimization layers (Amos and Kolter, 2017), have shown that significantly improved solutions can be achieved compared to the traditional predict-then-optimize approach. However, virtually all of them consider the case where the unknown model parameters only affect the objective function, which simplifies the problem considerably since feasibility is not a concern. Yet in many applications, we also need to use data to predict parameters in the constraints; the treatment of this case is in theory possible but difficult using existing methods. In this work, we address this problem by leveraging methods that we recently developed for data-driven inverse optimization (Gupta and Zhang, 2022a), which provide a natural way of incorporating constraints with unknown parameters.

The goal of inverse optimization is to infer an unknown optimization model from decisions that are assumed to be optimal solutions to that optimization problem (Chan et al., 2021). While early works primarily addressed the deterministic setting in which observations are assumed to be exactly optimal solutions of the optimization model, more recent contributions focus on the case with multiple noisy observations (Aswani et al., 2018; Chan et al., 2019; Gupta and Zhang, 2022b). Decision-focused learning can be viewed as a data-driven inverse optimization problem by treating the predictive model for the input parameters as the unknown part of the overall optimization model.

In the remainder of this paper, we first present the mathematical formulation of the decision-focused learning problem where we explicitly incorporate constraints that ensure feasibility of the optimal solutions obtained from the model with the estimated input parameters. To solve the resulting large-scale bilevel optimization problem, we apply our recently proposed penalty-based block coordinate descent algorithm. In a computational case study, we demonstrate the effectiveness of the proposed approach and highlight its benefits compared with the conventional predict-then-optimize approach.

2. Mathematical formulation

We assume that the optimization problem to be solved can be generally formulated in the following compact form:

$$\begin{array}{ll} \text{minimize} & f(x,u) \\ \text{subject to} & g(x,u) \leq 0, \end{array} \tag{1}$$

where $x \in \mathbb{R}^n$ are the decision variables, and the model parameters are denoted by u. In this work, we assume that problem (1) is convex, with f and g being differentiable and convex in x. The model parameters u (or a subset of them) change with some external features r and are not exactly known; hence, they need to be estimated from data. The goal is to construct a predictive model $u = m(r)$ given a set of N data points, where each data point i corresponds to a feature-output pair $(\bar{r}_i, \bar{u}_i)$. Given a new r, problem (1) will then be solved using the predicted values $u = m(r)$.

2.1. Conventional predict-then-optimize approach

In the conventional two-step process, the prediction of the model parameters is carried out independent from the later optimization. To obtain a predictive model m, one typically solves an empirical risk minimization problem of the following form:

$$\text{minimize} \quad \frac{1}{N}\sum_{i\in I} \ell(\bar{u}_i, m(\bar{r}_i)), \tag{2}$$

where $I = \{1, \dots, N\}$ denotes the set of data points, and the loss function ℓ is some measure of the difference between the true output $\bar{u}_i$ and the prediction $m(\bar{r}_i)$. The underlying assumption is that if we minimize the difference between the true and predicted parameter values, this will also lead to optimal solutions to problem (1) that are the closest possible to the solutions we would obtain if we knew the true parameter values.

2.2. Decision-focused learning

In decision-focused learning, we integrate the prediction and optimization steps to construct a predictive model that directly takes the quality of the resulting optimization solution into account. We do so by solving the following problem:

$$\text{minimize} \quad \frac{1}{N}\sum_{i\in I} f(\hat{x}_i, \bar{u}_i) \tag{3a}$$

$$\text{subject to} \quad \hat{x}_i \in \arg\min_{\tilde{x}}\{f(\tilde{x}, u) : g(\tilde{x}, u) \leq 0, u = m(\bar{r}_i)\} \quad \forall\, i \in I \tag{3b}$$

$$g(\hat{x}_i, \bar{u}_i) \leq 0 \quad \forall\, i \in I, \tag{3c}$$

where per constraints (3b), $\hat{x}_i$ is an optimal solution to problem (1) with $u = m(\bar{r}_i)$. The objective is to minimize the true cost averaged over the training set I, i.e. it considers the cost of $\hat{x}_i$ evaluated at the true parameter values $\bar{u}_i$ for each $i \in I$. Importantly, constraints (3c) ensure feasibility of each $\hat{x}_i$ given $\bar{u}_i$. These last set of constraints are omitted in virtually all existing works on decision-focused learning since they consider the case in which only the objective function f depends on u such that feasibility is not an issue.

3. Solution approach

The decision-focused learning problem (3) is a bilevel optimization problem with $|I|$ convex optimization problems in its lower-level. We reformulate (3) into a single-level problem by replacing the lower-level problems with their KKT conditions. This results in a nonconvex nonlinear optimization problem which generally lacks regularization. To address the convergence difficulties of standard nonlinear solvers on this problem, we consider a penalty reformulation and apply an efficient block coordinate descent (BCD) algorithm. We do not provide more details about our solution algorithm here but refer the reader to Gupta and Zhang (2022a) for more details. We end this section by highlighting the fact that our approach is restricted to the case where problem (1) is a strictly convex problem and satisfies Slater's condition.

4. Case study

In this section, we apply the proposed decision-focused learning approach to a (single-period) production planning problem for a small interconnected process network. This network, as depicted in Figure 1, consists of 5 materials and 3 processes. The goal is to determine the optimal quantities of raw materials to purchase and the amounts of products to manufacture to satisfy a given demands. This problem can be formulated as follows:

$$\text{minimize} \quad z = \textstyle\sum_{m\in\mathcal{M}}\left(\sum_{p\in\mathcal{P}} c_p y_p^2 + f_m w_m^2\right) \tag{4a}$$

$$\text{subject to} \quad q_m^{\min} \leq q_m^0 + \left(\sum_{p\in\hat{\mathcal{P}}_m} \mu_{pm} y_p - \sum_{p\in\bar{\mathcal{P}}_m} \mu_{pm} y_p + w_m - d_m\right) \leq q_m^{\max} \quad \forall\, m \in \mathcal{M} \tag{4b}$$

$$0 \leq w_m \leq w_m^{\max} \quad \forall\, m \in \mathcal{M} \tag{4c}$$

$$0 \leq y_p \leq y_p^{\max} \quad \forall\, p \in \mathcal{P}, \tag{4d}$$

where $\mathcal{M}$ and $\mathcal{P}$ are the sets of materials and processes, respectively. Further, the set $\hat{\mathcal{P}}_m$ consists of the processes that can produce material m, and the set $\bar{\mathcal{P}}_m$ contains the processes that consume m. The amount of a reference material produced by process p is denoted by y_p and we use w_m to specify the amount of material m purchased from the market. The conversion factor μ_{pm} determines the amount of a material m produced or consumed by process p for one unit of the reference material. Constraints (4b) restrict the inventory levels while accounting for product demand represented by d_m, (4c) limit the amount of a material that can be acquired from the market, and (4d) set the capacities of processes. The objective is to minimize the total production and material purchasing cost.

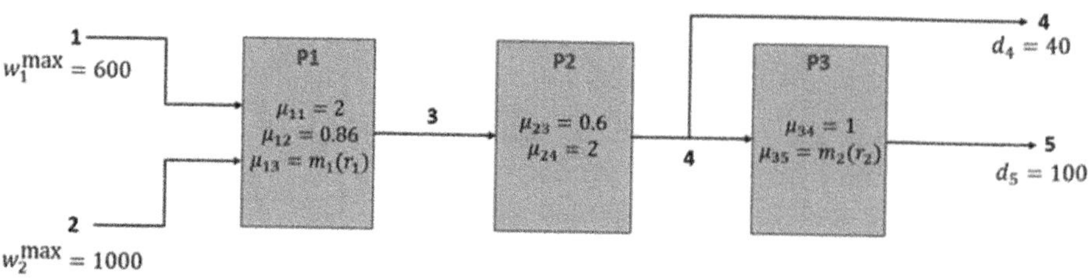

Figure 1 Process network for the production planning problem (4). The minimum and maximum allowed inventory values are 0 and 200, respectively, for all materials. For all processes, the values of y_p^{max} is set to 400.

For this case study, we consider a scenario where the conversion factors vary based on some observable external feature r. For the sake of simplicity, we assume that the change in most of the conversion factors is negligible; only μ_{13} and μ_{35} deviate significantly enough from their nominal values to affect optimal production decisions. Our goal is to build predictive models for these two uncertain parameters using a data set containing observed (r_1, μ_{13}) and (r_2, μ_{35}) values.

4.1. Synthetic data generation

We now describe the process used to generate the training data set for the case study. We start by assigning models (5a) and (5b) to the uncertain parameters. These are the underlying true models which are assumed to be unknown. To obtain the training data, we sample $|I|$ values of the features r_1 and r_2 from the uniform distributions $\mathbb{U}(0.1, 0.45)$ and $\mathbb{U}(-2, 1)$, respectively. Following that, we evaluate the models for μ_{13} and μ_{35} at each of the sampled feature values to complete the training data set.

$$\mu_{13}(r_1) = \frac{1}{10}\left(\sin(20\pi r_1) + 7r_1\right) + 2 \tag{5a}$$

$$\mu_{35}(r_2) = 2 + \tfrac{1}{10}\left((r_2 - 1)\, r_2\, (r_2 + 2)^2\right) \tag{5b}$$

In order to estimate predictive models m_1 and m_2 with the proposed approach, we assume a hypothesis class consisting of cubic polynomials. We solve problem (3) using our BCD algorithm with training data sets of four different sizes: 10, 25, 50, and 100. The quality of the resulting model estimates is evaluated through a test data set of 100 unseen data points, which is generated using the same parameter generation scheme as the training data set. A model is considered good if the produced $\hat{\mu}$ values result in production decisions that are not only close to the true optimal decisions but also feasible for the true model (i.e., problem (4) with the true μ values).

In addition to the decision-focused approach, we also estimate cubic models for m_1 and m_2 using the traditional two-stage predict-then-optimize approach. Specifically, we use least squares regression to fit a cubic polynomial to the observed data. This estimated model is then used to solve problem (4) for the points in the test data set.

4.2. Results and discussion

Here we compare the performance of the models estimated using the decision-focused and two-stage approaches. Figure 2 compares the plots of the true μ models with their estimates obtained using the two approaches. In all cases, we find that the decision-focused approach constructs an underestimator function for the training data points. This happens because in problem (4), an optimal solution will always be such that inventories of all materials are close to or at their minimum values. If the estimated $\hat{\mu}$ values are such that production gets overestimated, then there is a high probability that the inventory will fall below its permissible value when the process is actually run with a lower conversion value, resulting in an infeasible operation. Therefore, decision-focused learning obtains an underestimate of μ to avoid violating the lower bound on the inventory constraint. From Figure 2, we find that as we provide more training data, the proposed approach finds better underestimators. With 50 data points, it is able to find almost perfect underestimators for both μ parameters.

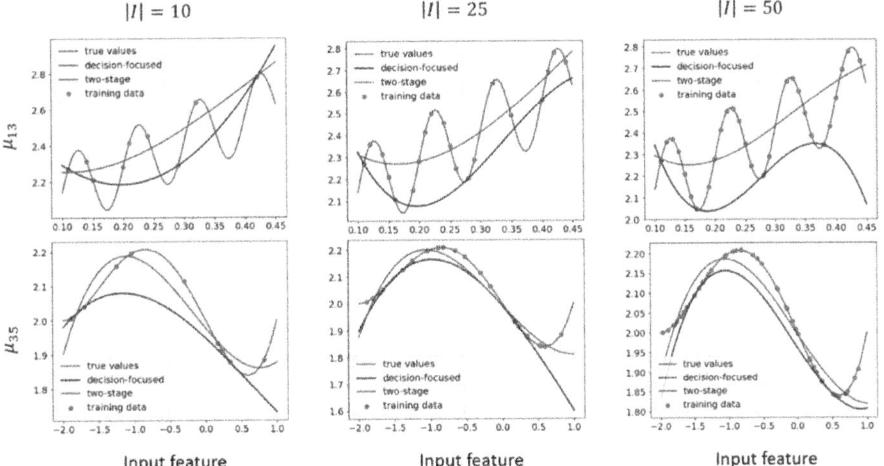

Figure 2 True μ models compared with their approximations estimated using the decision-focused and two-stage learning approaches

In contrast, the goal of the two-stage approach is to build the best approximation of the function itself using the provided data. The estimated model tries to closely mimic the behavior of the actual function to the extent that the assumed hypothesis class allows. This difference in approach leads to differences in performance, as seen in Table 1. The "feasible fraction" column indicates the fraction of the points in the test data set for which the estimated μ models produced a feasible decision. The data in this column shows that the decision-focused approach significantly outperforms the two-stage approach. Moreover, one can see that for the decision-focused case, the better the estimated function underestimates the true function, the higher the fraction of the feasible points. Here since the two-stage approach does not focus on yielding underestimators, the fraction of feasible decisions is very low.

For the test data points that yield feasible decisions, Table 1 also compares the distance of those decisions from the true optimal solutions. While the decision-focused learning generates feasible solutions with a high degree of confidence, these solutions are slightly more different from the true optimal solutions compared to the two-stage approach. However, as can be seen from the last column in the table, which compares the optimality gap of the decisions generated by the two approaches, the mean optimality gap of the decision-focused approach is still less than 10% (compared to ~5% in the two-stage case). This suggests that the decision-focused approach produces high quality decisions while almost guaranteeing their feasibility.

$\|I\|$	feasible fraction		mean $\frac{\|y(\mu)-y(\hat{\mu})\|_1}{\|y(\mu)\|_1}$		mean $\frac{\|w(\mu)-w(\hat{\mu})\|_1}{\|w(\mu)\|_1}$		mean $\frac{\|z(\mu)-z(\hat{\mu})\|}{z(\mu)}$	
	decision -focused	two-stage	decision-focused	two-stage	decision-focused	two-stage	decision-focused	two-stage
10	0.54	0.21	0.04	0.02	0.09	0.06	0.08	0.04
25	0.7	0.32	0.04	0.02	0.11	0.06	0.08	0.05
50	0.9	0.25	0.03	0.03	0.13	0.07	0.07	0.06
100	0.91	0.27	0.03	0.02	0.12	0.06	0.07	0.04

Table 1. A comparison of the performance of the models estimated using the decision-focused and two-stage approaches

5. Conclusions

In this work, we extended the decision-focused learning framework to include cases where the unknown parameters are in the constraints. We used an inverse optimization approach in which the problem is formulated as a bilevel program. Our approach allows inclusion of constraints that, with a high degree of confidence, ensure that the estimated model produces decisions that remain feasible for the true model. We illustrated our approach by applying it on a production planning problem with unknown process parameters. Our results show that the models obtained using decision-focused learning produce feasible decisions at a significantly higher rate compared to traditional two-stage learning without substantially sacrificing the optimality of these decisions.

Acknowledgements

The authors gratefully acknowledge the financial support from the National Science Foundation under Grant #2044077. R.G. acknowledges financial support from a departmental fellowship sponsored by 3M and a Doctoral Dissertation Fellowship from the University of Minnesota.

References

Amos, B. and Kolter, J.Z., 2017. Optnet: Differentiable optimization as a layer in neural networks. *Proceedings of the International Conference on Machine Learning*, pp. 136-145.

Aswani, A., Shen, Z.-J. M., and Siddiq, A., 2018. Inverse optimization with noisy data. *Operations Research, 66(3),* 870–892.

Chan, T. C., Lee, T., and Terekhov, D., 2019. Inverse optimization: Closed-form solutions, geometry, and goodness of fit. *Management Science, 65(3)*, 1115–1135.

Chan, T.C., Mahmood, R., and Zhu, I.Y., 2021. Inverse optimization: Theory and applications. arXiv:2109.03920.

Elmachtoub, A.N. and Grigas, P., 2022. Smart "predict, then optimize". *Management Science*, *68*(1), pp. 9-26.

Gupta, R. and Zhang, Q., 2022a. Efficient learning of decision-making models: A penalty block coordinate descent algorithm for data-driven inverse optimization. arXiv:2210.15393.

Gupta, R. and Zhang, Q., 2022b. Decomposition and adaptive sampling for data-driven inverse linear optimization. *INFORMS Journal on Computing*.

Wilder, B., Dilkina, B., and Tambe, M., 2019. Melding the data-decisions pipeline: Decision-focused learning for combinatorial optimization. *Proceedings of the AAAI Conference on Artificial Intelligence*, pp. 1658-1665.

Antonis Kokossis, Michael C. Georgiadis, Efstratios N. Pistikopoulos (Eds.)
PROCEEDINGS OF THE 33rd European Symposium on Computer Aided Process Engineering (ESCAPE33), June 18-21, 2023, Athens, Greece
 http://dx.doi.org/10.1016/B978-0-443-15274-0.50217-1

Multi-scale modeling and techno-economic analysis of biogas catalytic reforming for hydrogen & power production with CO_2 capture feature

Alessandra-Diana Selejan [a], Simion Dragan [a*], Ana-Maria Cormos [a], Mihaela Dragan [b], Calin-Cristian Cormos [a]

[a] *Faculty of Chemistry and Chemical Engineering, Babes-Bolyai University, 11 Arany Janos street, RO-400028, Cluj-Napoca, Romania*

[b] *Faculty of Economics and Business Administration, Babes-Bolyai University, Teodor Mihali street, No. 58-60, RO-400591, Cluj-Napoca, Romania*

Abstract

Hydrogen is one of the main pillars needed for a future low-carbon economy. The biogas steam reforming process can produce hydrogen without consuming fossil-based (non-renewable) resources. In this work, a dynamic 2D heterogeneous model (at macroscale and microscale) is developed for catalytic hydrogen production, applying the biogas steam reforming process. The model was implemented in Matlab/Simulink and the simulation results were compared with literature data, obtaining good agreement. The temperature profiles and those for the gas composition were obtained, related to the height of the catalyst layer in the reactor, as well as methane profile in the catalyst particle. A special emphasis was placed on highlighting the influence of a high content of carbon dioxide on the reactor design parameters. The dynamic behavior was also studied for ramp input changes on the feed flow. Based on mass and energy balanced derived from whole process simulation, the detailed techno-economic assessment of decarbonized biogas catalytic reforming plant for flexible hydrogen & power generation was assessed to evaluate the overall advantages.

Keywords: Biogas catalytic reforming; Green hydrogen production; Multiscale modeling; Techno-economic assessment; Negative CO_2 emissions.

1. Introduction

Hydrogen is seen as the fuel of the future, because it has the potential to be produced and used with near-zero greenhouse gas emissions. There are several industries using hydrogen, from fertilizer production and refining processes to methanol production and synthesis of organic compounds. Biogas got significant attention as a promising renewable energy source and an energy-efficient way of converting various biowastes into energy carriers. Steam methane reforming is the most used process for hydrogen production. By using biogas to produce hydrogen, the whole process can be considered environmentally friendly (IEA 2022, Lepage et al. 2021, Nahar et al. 2017). The steam reforming process, which uses biogas as feedstock for hydrogen production, takes place in a multi-tubular reactor with side furnaces flanking tubes filled with a packed-bed of catalyst (figure 1). The tubes are made of IN-519 alloy, while the catalyst consists of nickel particles on $MgAl_2O_4$ support. The reactor operates at a temperature range of 700 to 1200 K and a pressure range of 28 to 37 bar. The inlet gas contains one-part biogas and at least three parts steam (Soliman et al. 1988).

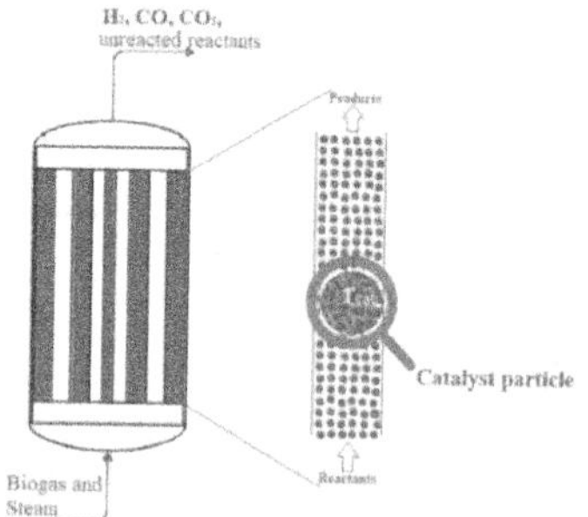

Figure 1. The multi-tubular reforming reactor with catalyst

There are three reactions involved in the endothermic process described above, R_1 and R_2 are parallel reactions, while R_3 follows them (Calistru et al. 1984, Xu and Froment 1989, Ghouse and Adams 2013):

R_1: $CH_4 + H_2O \rightleftarrows CO + 3H_2$ (ΔH_{298K}=+206.3 kJ/mol)
R_2: $CO + H_2O \rightleftarrows CO_2 + H_2$ (ΔH_{298K}=−41.1 kJ/mol)
R_3: $CH_4 + 2H_2O \rightleftarrows CO_2 + 4H_2$ (ΔH_{298K}=+164.9 kJ/mol)

In this work, a dynamic 2D heterogeneous model (at macroscale and microscale) is developed for catalytic hydrogen production, applying the biogas steam reforming process. The spaces between the catalyst particles, inside of reactor' tubes represent the macroscale of the system. The model of the microscale described the pores inside the catalyst particle. A special emphasis was placed on highlighting the influence of a high content of carbon dioxide on the reactor design parameters. The dynamic behavior was also studied for step and ramp input changes on the feed flow. As other relevant research contributions of present work, the detailed technical and economic assessment of decarbonized biogas catalytic reforming process for flexible hydrogen & power generation was assessed.

2. Mathematical model

2.1. Model development

The multi-scale heterogeneous dynamic model of catalytic hydrogen production by steam reforming of biogas is developed based on following assumptions: along the length of the reactor the mass transfer through the catalyst layer takes place by convective and diffusional mechanism, while for the mass transfer through the catalyst pores, just the radial direction is considered, while it is thought that the catalyst particles have a homogenous porous structure. In the case of the catalyst phase, the mass balance equations are functions of time and catalyst particle radius. Any particle at the z-position is equivalent to all other particles at the same position. The partial differential equations describing the model are presented in Table 1 (Ghouse and Adams 2013).

Table 1. Mass and energy balances of the reforming reactor

Observations	Equation and boundary conditions
Mass balance for the gas phase	$\frac{\partial c_i}{\partial t} = -\frac{F}{A}\frac{\partial c_i}{\partial z} - k_{Ti}\left(c_i - c_{ci}\|_{r=R_p}\right) - r_i\rho_{cat}$
Energy balance for the gas phase	$\frac{\partial T}{\partial t} = -\frac{F}{A}\frac{\partial T}{\partial z} - \frac{\sum_{i=1}^{Nc}\Delta H_i k_i (c_i - c_{ci})}{cp_{mix}\rho_g} + \frac{K_T A_T (T_w - T)}{cp_{mix}\rho_g A dz}$ at z=0 and t>0 $c_{i\|z=0} = c_{i,inlet}$ $T_{g\|z=0} = T_{inlet}$
Mass balance for the catalyst phase	$\frac{\partial c_{ci}}{\partial t} = \frac{1}{\theta_c}\left\{\frac{2}{r}D_{eimix}\frac{\partial c_{ci}}{\partial r} + \frac{\partial}{\partial r}\left[D_{eimix}\frac{\partial c_{ci}}{\partial r}\right] - r_i\rho_{cat}\right\}$ at t>0 and any z $\left[\frac{\partial c_{ci}}{\partial r}\right]_{r=0} = 0$ $\left[D_{eimix}\frac{\partial c_{ci}}{\partial r}\right]_{r=Rp} = k_{Ti}\left(c_i - c_{ci}\|_{r=Rp}\right)$

Table 2. Process parameters

Inlet flow, F – 23.271 [kmol/h]
Inlet temperature, T_0 – 723 [K]
Inlet pressure, P – 30 [bar]
Inlet gas composition (mole fraction):
x_{CH_4}: 0.1808
x_{H_2O}: 0.7981
x_{CO} : 0.0049
x_{CO_2} : 0.0061
x_{H_2}: 0.0098
Catalyst particle diameter, D_p – 0.017 [m]
Catalyst void fraction, θ_c – 0.519
Tub diameter, D_t – 0.107 [m]
Tub wall thickness, X_t – 0.015 [m]
Packed-bed length, L_{cat} – 9.18 [m]

The developed mathematical model was implemented in Matlab/Simulink, using process parameters presented in Table 2 (Calistru et al. 1984, Soliman et al. 1988, Ghouse and Adams 2013).
Since all 247 tubes present in the reforming reactor are operated and designed identically, the simulation was performed for a single tube (Calistru et al. 1984, Ghouse and Adams 2013). The connection between the two models (macroscale and microscale) have been made with multiple functions, repetitive and nested structures.

2.2. Results and discussions

Table 3. Model validation

Parameter	Model	Literature
CH_4 conversion	69.10%	71.66%
H_2 fraction	69.00%	70.60%

The predicted values of the developed mathematical model were compared with literature data in terms of CH_4 percentage conversion and H_2 mole fraction at exit in dry basis - Table 3 (Calistru et al. 1984, Ghouse and Adams 2013, Soliman et al. 1988).
Based on simulations results, the profiles for the gas molar concentrations in the macroscale (reported to the length of the reforming reactor) and in the microscale (reported to the catalyst particle radius) were obtained (Figure 2). It can be observed that the hydrogen is produced manly in the first half of the reforming reactor length. In the case of the microscale, close to the interface, due to the mass transfer, the density of reactants is higher than into the particle center.

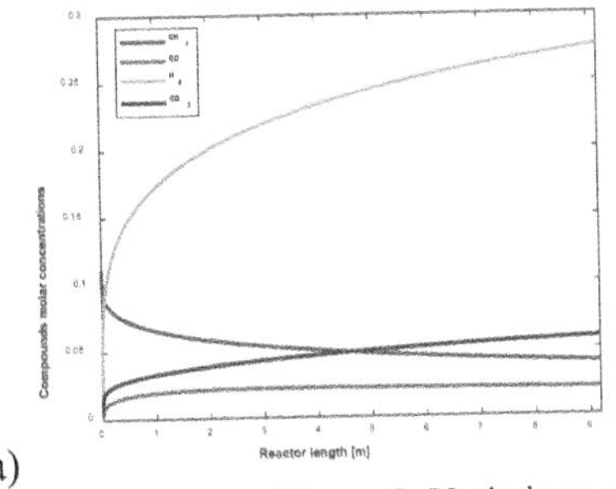

a)

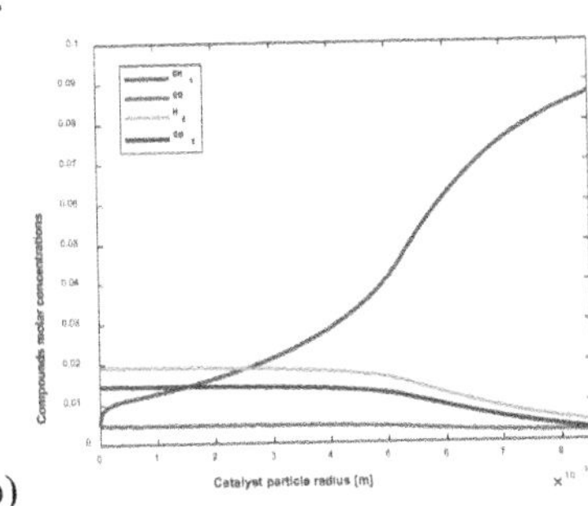

b)

Figure 2. Variations of chemical species concentrations a) macroscale and b) microscale

The biogas high content of carbon dioxide could influence the reactor design parameters. Simulations of the model were performed to observe how the degrees of transformation, and hence the conversions of the reactions are changed at a pressure of 30 bar, different $\dot{x}_{H_2O} = H_2O_{(g)} : CH_4$ and $\dot{x}_{CO_2} = CO_2 : CH_4$ molar ratios in the reactor feed gas, for a temperature between 820-1200 K. The upper temperature limit was chosen according to the limitations imposed by the material from which the reactor tubes are made. Comparing Figures 3a and 3b for the same molar ratio of $H_2O_{(g)} : CH_4$, reaction methane conversion (R_1) is favoured by an excess of CO_2. If in Figure 3a the corresponding yellow line $\dot{x}_{H_2O}$=4.41 barely reaches the value of 0.90 for η_{CH_4}, in

Figure 3b it exceeds it, reaching a value of 0.92. The carbon monoxide conversion (R_2 reaction), it is strongly based by an excess of CO_2, (Figure 3a), for $\dot{x}_{H_2O}$=4.41, η_{CO}=0.35, and in Figure 3b) for the same $\dot{x}_{H_2O}$, η_{CO}<0.1.

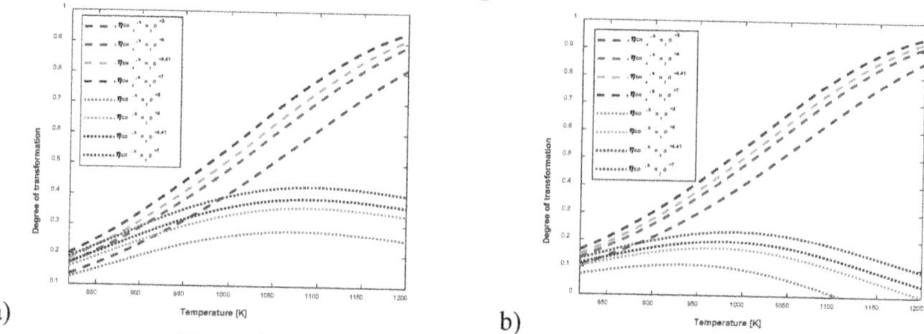

a) b)

Figure 3. Variation of transformation degrees with temperature for a) P=30 [bar] $\dot{x}_{CO_2}$=0.03 and b) P=30 [bar] $\dot{x}_{CO_2}$=0.75

How an excess of CO_2 in the reactor inlet flow requires a change in the height of the catalyst packed-bed to achieve the same results is presented in Figure 4. Although the excess CO_2 is very large, the catalyst layer height increases by a maximum of one meter and the desired reaction is favored.

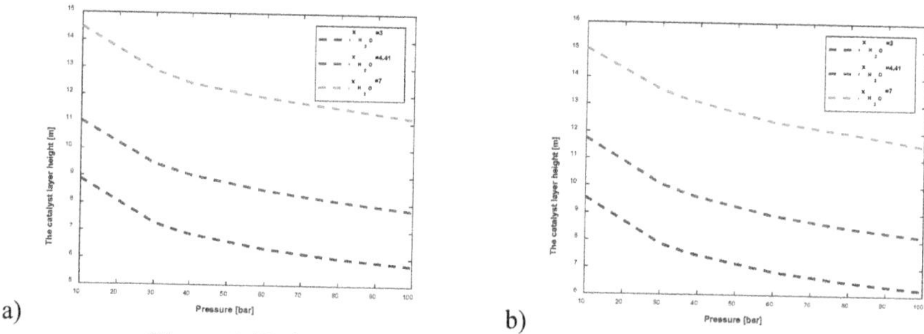

a) b)

Figure 4. Variation of catalyst layer height with pressure at different $\dot{x}_{H_2O}$ a) P=30 [bar] $\dot{x}_{CO_2}$=0.03 and b) P=30 [bar] $\dot{x}_{CO_2}$=0.75

The dynamic behaviour of the process was investigated by implementing a ramp signal disturbance of -20% on the biogas feed flow (Figure 5). The simulations showed that for the ramp signal disturbance, the system needs a long time to reach the new steady state values, as the disturbance propagated with a delay from one end of the reactor to the other. It is also observed that when the biogas feed flow rate is reduced, the amount of methane is reduced and the produced hydrogen quantity increases.

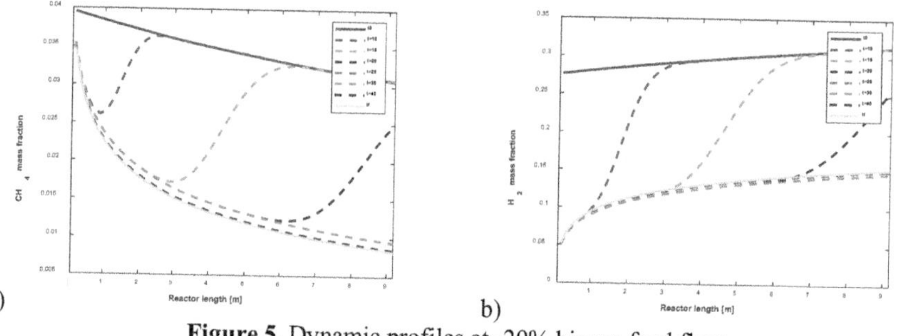

a) b)

Figure 5. Dynamic profiles at -20% biogas feed flow a) Methane and b) Hydrogen

3. Techno-economic evaluation

The biogas catalytic reforming process for hydrogen & power co-generation with CO_2 capture feature was evaluated also in term of overall techno-economic performances. As decarbonization method, the pre-combustion CO_2 capture using reactive gas-liquid absorption with Methyl-D-Ethanol-Amine (MDEA) solvent was used. The overall process was modeled and simulated using ChemCAD software, the mass and energy balances being employed for calculation of main performance indicators. As illustrative concept, 100 MW thermal hydrogen output with 65% CO_2 capture rate were considered. For the economic evaluation methodology, the overall International Energy Agency - Greenhouse Gas R&D Programme method (reported in IEAGHG, 2017) was used. In terms of main economic assumptions, the ones defined in Cormos et al. (2022) was used. For instance, the capital costs were estimated using the cost correlation method with the most recent values of Chemical Engineering Plant Cost Index (CEPCI). The levelized costs of hydrogen and power were calculated based on Net Present Value (NPV) method with an internal rate of return of 8%. The CO_2 removal and avoidance costs were calculated using as benchmark a biogas reforming plant without CO_2 capture feature as reported in Cormos et al. 2022.

Table 4 presents the overall techno-economic and environmental performances. As can be noticed, the cumulative energy efficiency is high enough (66.50%) as well the decarbonization rate (65%). The economic parameters (e.g., hydrogen production cost, CO_2 capture costs) look promising in comparison to relevant studies (IEAGHG, 2017) based on natural gas as fuel input. For instance, the utilization of biogas (renewable source) into the decarbonized reforming plant gives overall negative CO_2 emissions.

Table 4. Techno-economic performance indicators

Parameter	Unit	Value
Biogas input flowrate	t/h	31.22
Lower heating value of biogas	MJ/kg	17.58
Thermal energy of biogas	MW_{th}	152.45
Gross power production	MW_e	8.00
Ancillary electricity consumption	MW_e	6.50
Hydrogen thermal energy	MW_{th}	100.00
Net power production	MW_e	1.50
Hydrogen thermal efficiency	%	65.59
Net power efficiency	%	0.98
Cumulative energy efficiency	%	66.57
CO_2 capture rate	%	65.00
Specific CO_2 emissions	kg/MWh	172.50
Capital investment cost	M€	129.95
Specific capital investment cost	€/kW net	1280.00
Operational & maintenance cost	€/MWh	37.95
Levelized cost of hydrogen	€/MWh	61.42
Levelized cost of electricity	€/MWh	60.60
CO_2 removal cost	€/t	24.12
CO_2 avoidance cost	€/t	28.65

4. Conclusions

The 2D heterogeneous multiscale dynamic model for the biogas steam reforming process presented in this paper has been validated with data available in the literature. Particular attention was paid to how a higher CO_2 concentration influences the reactor design parameters while favoring the methane conversion (R_1). The dynamic response of the system was highlighted by its response to a ramp disturbance of the feed flow.

When the feed flow rate decreases, the amount of methane decreases, but the amount of hydrogen increases in the reactor. In addition, a detailed techno-economic evaluation of hydrogen and power co-generation based on biogas reforming with CO_2 capture feature was done to illustrate its key advantages e.g., high energy efficiency, competitive hydrogen production cost, negative CO_2 emissions (due to CO_2 capture capability) etc.

Acknowledgments

This work was supported by a grant of the Ministry of Research, Innovation and Digitization, CNCS/CCCDI - UEFISCDI, project number PN-III-P4-ID-PCE-2020-0632, within PNCDI III.

Nomenclatures

c_i	concentration of component *i* in the gas phase [kmol/m^3]
c_{ci}	concentration of component *i* in the catalyst phase [kmol/m^3]
k_{Ti}	mass transfer coefficient [mol/m^{2}s]
A	section area [m^2]
r_i	reaction rate of component *i*[kmol/kg h]
ρ_{cat}	catalyst density [kg/m^3]
T	gas phase temperature [K]
T_w	temperature of the tube wall [K]
ΔH_i	enthalpy of reaction [J/mol]
k_i	rate coefficient of reaction *i*
cp_{mix}	specific heat of the gas mixture [J/mol K]
ρ_g	gas mixture density [kg/m^3]
D_{eimix}	effective diffusivity of component *i* inside the catalyst pores [cm^2/s]
$\dot{x}_{H_2O}$	H_2O : CH_4 moles ratio in the inlet stream
$\dot{x}_{CO_2}$	CO_2 : CH_4 moles ratio in the inlet stream

References

C. Calistru, C. Leonte, I. Siminiceanu, C. Hagiu, O. Popa, 1984, Mineral Fertilizer Technology Vol. I -Nitrogen Fertilizers, Bucharest Technical Publishing House, Romania

C.C. Cormos, A.M. Cormos, L. Petrescu, S. Dragan, 2022, Techno-economic assessment of decarbonized biogas catalytic reforming for flexible hydrogen and power production, Appl. Therm. Eng., 207, 118218

G. Nahar, D. Mote, V. Dupont, 2017, Hydrogen production from reforming of biogas: Review of technological advances and an Indian perspective, Renewable and Sustainable Energy Reviews, 76, 1032-52

International Energy Agency - Greenhouse Gas R&D Programme (IEAGHG), 2017, Techno-economic evaluation of SMR based standalone (merchant) hydrogen plant with CCS, Report 2017/02, Cheltenham, UK

J. Xu, G.F. Froment, 1989, Methane Steam Reforming, Methanation and Water-Gas Shift: I. Intrinsic Kinetics, AIChE Journal Jan, 35(1), 88-96

J.H. Ghouse, T.A. Adams II, 2013, A multi-scale dynamic two-dimensional heterogeneous model for catalytic steam methane reforming reactors, International Journal of Hydrogen Energy, 38, 9984-99

M.A. Soliman, S.S.E.H. El-Nashaie, A.S. Al-Ubaid, A. Adris, 1988, Simulation of steam reformers for methane, Chemical Engineering Science, 43(8), 1801-6

T. Lepage, M. Kammoun, Q. Schmetz, A. Richel, 2021, Biomass-to-hydrogen: A review of main routes production, processes evaluation and techno-economical assessment, Biomass and Bioenergy, 144, 105920

The International Energy Agency (IEA), 2022, Hydrogen IEA Paris, https://www.iea.org/reports/hydrogen; [accessed 1 December 2022].

Antonis Kokossis, Michael C. Georgiadis, Efstratios N. Pistikopoulos (Eds.)
PROCEEDINGS OF THE 33rd European Symposium on Computer Aided Process Engineering (ESCAPE33), June 18-21, 2023, Athens, Greece
 http://dx.doi.org/10.1016/B978-0-443-15274-0.50218-3

Modeling hierarchical systems via nested generalized disjunctive programming

Hector D. Perez,[a] Ignacio E. Grossmann[a]

[a]*Carnegie Mellon University, 5000 Forbes Ave, Pittsburgh 15207, USA*

Abstract

Modeling systems with discrete-continuous decisions is traditionally done in algebraic form with mixed-integer programming models, which can be linear or nonlinear in the continuous variables. A more systematic approach to modeling such systems is to use Generalized Disjunctive Programming (GDP), which extends the Disjunctive Programming paradigm proposed by Egon Balas. GDP allows modeling systems from a logic-based level of abstraction that captures the fundamental rules governing such systems via algebraic constraints and logic. Although GDP provides a more general way of modeling systems, it warrants further generalization for systems presenting a hierarchical structure. This work extends the GDP literature to address alternatives for modeling and solving systems with nested (hierarchical) disjunctions. We also provide two theorems on the relaxation tightness of such alternatives, establishing that a explicitly modeling nested disjunctions is superior to the traditional approach for dealing with nested disjunctions.

Keywords: generalized disjunctive programming, hierarchical systems, discrete-continuous optimization.

1. Introduction

Discrete-continuous optimization is one of the main modeling approaches for addressing design, planning, and scheduling problems in Process Systems Engineering (PSE) (Grossmann, 2012). Raman and Grossmann (1994) present a new modeling paradigm that extends the work done by Balas (1985) on disjunctive programming. This new paradigm, called Generalized Disjunctive Programming (GDP), was further developed by others in the PSE community over the years to account for additional features, such as nonlinearities and nonconvexities, in the problems encountered (Grossmann & Trespalacios, 2013). GDP relies on the intersection of disjunctions of convex sets to model the feasible space. Boolean variables are used as the indicator variables for each convex set, meaning that if True, the constraints in the disjunct are enforced. Logical constraints are also included to describe the relationships between the Boolean indicator variables via propositional logic and constraint programming logic.

The present work extends the GDP theory to allow modeling hierarchical systems, which are commonly encountered in PSE, and more particularly in Enterprise-Wide Optimization (EWO) (Grossmann, 2012), and flowsheet superstructure optimization (Türkay & Grossmann, 1996). Multi-level decisions can be concisely modelled via nested disjunctions. However, traditional GDP does not consider such formulations. Existing GDP literature suggests reformulating nested disjunctions into equivalent single-level disjunctions (Vecchietti & Grossmann, 2000). Such an approach requires introducing additional Boolean variables and logical propositions. An alternate approach is used in the work by van den Heever and Grossmann (1999), in which a direct or inside-out reformulation to MI(N)LP is performed. We formalize these two approaches and provide theoretical proofs on the tightness of their continuous relaxations. The model tightness and computational performance of the different approaches are compared. A series of examples are used to show the modeling and computational advantages obtained by explicitly modeling nested disjunctions.

2. GDP Modeling Overview

The classical GDP formulation is given below, where x is the set of continuous variables (bounded between x^{LB} and x^{UB}), $f(x)$ is the objective function, $r(x) \leq 0$ is the set of global constraints, $g_{ij}(x) \leq 0$ is the set of constraints applied when the indicator Boolean Y_{ij} is *True* for disjunct j in disjunction i. $\Omega(Y)$ defines the set of logic constraints, which are described via propositional logic on a subset of Boolean variables. These constraints describe the interactions between the Boolean variables via clauses that contain with one or more of the following logic operators: AND ($\wedge$), OR ($\vee$), implication ($\Rightarrow$), equivalence ($\Leftrightarrow$), and negation ($\neg$). The set of logic constraints may also include cardinality clauses of the form *choose exactly* (or *at least* or *at most*) m Boolean variables from a subset of Booleans to be *True* (Yan & Hooker, 1999). Cardinality clauses are indicated with the notation $(Y_s \ \forall s \in S)_m$, which means that at least m of the Boolean variables Y_s are *True*. An alternative is to use predicate logic to define the following predicates: $\Xi(m, Y_s \ \forall s \in S)$ for *exactly* m are *True*, $\Lambda(m, Y_s \ \forall s \in S)$ for *at least* m are *True*, and $\Gamma(m, Y_s \ \forall s \in S)$ for *at most m are True*. GDP models typically include a cardinality clause to enforce that exactly 1 disjunct is chosen in each disjunction, i.e., $\Xi(1, Y_{ij} \ \forall j \in J_i) \ \forall i \in I$. The GDP literature often uses the exclusive OR operator, $\underline{\vee}$, to define this constraint. However, such an operator is only correct for proper disjunctions (non-overlapping disjuncts). Thus, to avoid any ambiguity, we use the predicate logic notation, $\Xi(1, Y)$, here instead.

$$
\begin{aligned}
\min z = f(x) \qquad & && \text{(GDP)}\\
s.t. \qquad & r(x) \leq 0 && \\
& \bigvee_{j \in J_i} \begin{bmatrix} Y_{ij} \\ g_{ij}(x) \leq 0 \end{bmatrix} && \forall i \in I \\
& \Xi(1, Y_{ij} \ \forall j \in J_i) && \forall i \in I \\
& \Omega(Y) && \\
& x^{LB} \leq x \leq x^{UB} && \\
& x \in \mathbb{R}^n && \\
& Y_{ij} \in \{True, False\} && \forall i \in I, j \in J_i
\end{aligned}
$$

One of the main advantages of modeling discrete-continuous problems using GDP is the collection of methods that are available for optimizing such systems (Grossmann & Trespalacios, 2013; Trespalacios & Grossmann, 2016). These include, 1) reformulating to mixed-integer (non)linear models (MI(N)LP) via either Big-M or Hull reformulations, 2) logic-based decomposition methods such as Logic-based Outer Approximation (LOA), 3) disjunctive branch-and-bound, 4) basic steps, and 5) hybrid cutting planes. The reader is referred to the above references for a detailed understanding of each of these solution methods.

3. Extended formulation for hierarchies

3.1. Hierarchical GDP

We propose extending the GDP paradigm to include multi-level decisions by means of nested disjunctions. Although the notion of nesting disjunctions to represent hierarchical decisions is not new, the limitations in the traditional GDP notation have made it difficult to exploit the benefits of nesting disjunctions. One of the first references to nested disjunctions is found in the work by Vecchietti and Grossmann (2000), which describes the transformations required to conform to the current GDP notation.

Therefore, from a model development point of view, the use of disjunction nesting is shown to add value. However, its implementation has often required breaking the explicit hierarchical structure. An exception is the work by Van Den Heever and Grossmann (1999), which does not transform the nested GDP into a logically equivalent single-level GDP, but rather suggests performing the hull reformulation on the inner disjunction and then reformulating the outer disjunction.

The proposed extension to the classical GDP notation for hierarchical systems is given below for a 2 level GDP (*2L-GDP*), where the upper-level decisions, Y, enforce the constraints $g(x) \leq 0$ and the nested decisions, W, which have constraints $h(x) \leq 0$. Here the cardinality clause of selecting exactly one disjunct from the upper-level decisions, Y, is expressed explicitly, along with a new set of cardinality rules that enforce selecting exactly one of the lower-level decisions, W, if and only if the upper-level decision has been selected, and selecting no lover-level decisions when the upper-level decision is not selected. This constraint is expressed as the conjunction of two cardinality rules: $\left[Y_{ij} \Rightarrow \Xi\left(1, W_{ijkl}\ \forall l \in L_{ijk}\right)\right] \wedge \left[\neg Y_{ij} \Rightarrow \Xi\left(0, W_{ijkl}\ \forall l \in L_{ijk}\right)\right] \forall i \in I, j \in J_i, k \in K_{ij}$. In the GDP literature, this constraint has been traditionally written as $Y_{ij} \Leftrightarrow \underline{\vee}_{l \in L_{ijk}} W_{ijkl}\ \forall i \in I, j \in J_i, k \in K_{ij}$. However, such a logic proposition is incomplete because it would allow the following to occur: $Y_{ij} = False$ and $W_{ijkl} = True$ for more than 1 index $l \in L_{ijk}$. As a result, the cardinality rule $\Gamma\left(1, W_{ijkl}\ \forall l \in L_{ijk}\right) \forall i \in I, j \in J_i, k \in K_{ij}$ should be added to ensure that no more than 1 literal W_{ijkl} can be *True*. A more compact form would be to use the predicate constraint, $\Xi\left(t\left(Y_{ij}\right), W_{ijkl}\ \forall l \in L_{ijk}\right)$, where $t(\cdot)$ is a unary function that maps a Boolean variable to its binary counterpart (i.e., $t(True) = 1$ and $t(False) = 0$). For simplicity, we make a slight abuse of notation, dropping the mapping function $t(\cdot)$, and using the expression $\Xi\left(Y_{ij}, W_{ijkl}\ \forall l \in L_{ijk}\right)$ instead. Such a model can be generalized to a multi-level hierarchical GDP with n levels.

$$
\begin{aligned}
&\min z = f(x) && \text{(2L-GDP)}\\
&s.t.\quad r(x) \leq 0 \\
&\bigvee_{j \in J_i} \begin{bmatrix} Y_{ij} \\ g_{ij}(x) \leq 0 \\ \bigvee_{l \in L_{ijk}} \begin{bmatrix} W_{ijkl} \\ h_{ijkl}(x) \leq 0 \end{bmatrix} \quad \forall k \in K_{ij} \end{bmatrix} && \forall i \in I \\
&\Xi\left(1, Y_{ij}\ \forall j \in J_i\right) && \forall i \in I \\
&\Xi\left(Y_{ij}, W_{ijkl}\ \forall l \in L_{ijk}\right) && \forall i \in I, j \in J_i, k \in K_{ij} \\
&\Omega(Y, W) \\
&x^{LB} \leq x \leq x^{UB}
\end{aligned}
$$

$$
\begin{aligned}
& x \in \mathbb{R}^n \\
& Y_{ij} \in \{True, False\} && \forall i \in I, j \in J_i \\
& W_{ijkl} \in \{True, False\} && \forall i \in I, j \in J_i, k \in K_{ij}, l \in L_{ijk}
\end{aligned}
$$

3.2. Equivalent Single-level GDP

Previous references to GDP with nested disjunctions in literature have proposed transforming the 2L-GDP model into the equivalent single-level GDP (*2E-GDP*) given below (Vecchietti & Grossmann, 2000). Here, the nested disjunction is extracted and a dummy or "slack" disjunct is added to preserve feasibility. Thus, if none of the nested disjuncts is selected, the slack disjunct is selected, which contains the entire feasible set for x. The exclusive cardinality rule on the inner Boolean variables, W, is also augmented to include the slack Boolean variable, W_{ijk0}. This slack variable is, however, not included in the linking logic constraint for the upper and lower-level decisions. This ensures that the nested decisions are only selected if their master Boolean is *True*. This method for transforming a nested disjunction can also be applied to the multi-level system *ML-GDP*.

$$
\begin{aligned}
& \min z = f(x) && \text{(2E-GDP)} \\
& s.t. \quad r(x) \le 0 \\
& \bigvee_{j \in J_i} \begin{bmatrix} Y_{ij} \\ g_{ij}(x) \le 0 \end{bmatrix} && \forall i \in I \\
& \left(\bigvee_{l \in L_{ijk}} \begin{bmatrix} W_{ijkl} \\ h_{ijkl}(x) \le 0 \end{bmatrix} \right) \bigvee \begin{bmatrix} W_{ijk0} \\ x^{LB} \le x \le x^{UB} \end{bmatrix} && \forall i \in I, j \in J_i, k \in K_{ij} \\
& \Xi(1, Y_{ij}\ \forall j \in J_i) && \forall \square \in \square \\
& \Xi(1, W_{ijkl}\ \forall l \in L_{ijk} \cup \{0\}) && \forall i \in I, j \in J_i, k \in K_{ij} \\
& \Xi(Y_{ij}, W_{ijkl}\ \forall l \in L_{ijk}) && \forall i \in I, j \in J_i, k \in K_{ij} \\
& \Omega(Y, W) \\
& x^{LB} \le x \le x^{UB} \\
& x \in \mathbb{R}^n \\
& Y_{ij} \in \{True, False\} && \forall i \in I, j \in J_i \\
& W_{ijkl} \in \{True, False\} && \forall i \in I, j \in J_i, k \in K_{ij}, l \in L_{ijk}
\end{aligned}
$$

Although the above formulation, allows modeling hierarchical systems in the standard GDP notation, it has two major drawbacks: 1) the explicit hierarchical structure is lost, and 2) although the equivalent single-level GDP model is logically equivalent to the nested GDP model, it requires introducing additional disjuncts and Boolean variables. Introducing "slack" disjuncts and Boolean variables results in models whose continuous relaxations are less tight, as given in the below theorem.

3.3. Tightness of Continuous Relaxations

The following two theorems address the advantages of modeling multi-level decisions problems via Hierarchical GDP, rather than the traditional single-level GDP approach. The advantages are shown by discussing the tightness of the continuous relaxations of these two approaches. For proofs of these theorems see Perez and Grossmann (2023).

Theorem 1. Let *rML-GDP-HR* denote the continuous relaxation of the mixed-integer program (MIP) obtained from a multi-level hierarchical GDP via the hull reformulation, and let *rME-GDP-HR* denote the continuous relaxation of the MIP obtained from the equivalent single-level GDP via the hull reformulation. The feasible space of the former is contained within the feasible space of the latter, meaning *rML-GDP-HR* ⊆ *rME-GDP-HR*.

Theorem 2. Let *rML-GDP-BM* denote the continuous relaxation of the mixed-integer program (MIP) obtained from a multi-level hierarchical GDP via the Big-M reformulation, and let *rME-GDP-BM* denote the continuous relaxation of the MIP obtained from the equivalent single-level GDP via the Big-M reformulation. The feasible space of the former is contained within the feasible space of the latter, meaning *rML-GDP-BM* ⊆ *rME-GDP-BM*.

3.4. Graphical Example of Model Tightness

Consider the hierarchical (nested) GDP constraint system given in (1), which can be expressed as the equivalent single-level GDP in (2). Each of these models is reformulated to a MIP using the Big-M reformulation, with both a loose *M* value and a tight *M* value, and the Hull reformulation. Their continuous relaxations are then projected onto the x_1, x_2 plane in *Figure 1*. The projections show that flattening via basic steps is advantageous when the hull reformulation is performed, but not necessarily when the Big-M reformulation is performed with a tight *M* value. Preserving the explicit hierarchical relationship via nested GDP reduces the feasible region of the continuous relaxation more than the equivalent single-level GDP approach in both the Big-M and hull reformulation cases. Furthermore, in this example the Big-M reformulation with a tight *M* on the nested GDP model produces the same relaxation as the hull reformulation on the equivalent GDP model with only a fraction of the model size. It should also be noted that, in this example, the convex hull of the system is obtained when either the hull reformulation is applied to the nested GDP or to the flattened GDP after two basic steps. The continuous relaxation of either of these two formulations will yield the optimum.

$$\begin{bmatrix} Y_1 \\ 1 \le x_1 \le 3 \\ 4 \le x_2 \le 6 \\ \begin{bmatrix} W_1 \\ 1 \le x_1 \le 2 \\ 5 \le x_2 \le 6 \end{bmatrix} \vee \begin{bmatrix} W_2 \\ 2 \le x_1 \le 3 \\ 4 \le x_2 \le 5 \end{bmatrix} \end{bmatrix} \vee \begin{bmatrix} Y_2 \\ 8 \le x_1 \le 9 \\ 1 \le x_2 \le 2 \end{bmatrix} \\ \Xi(1, \{Y_1, Y_2\}) \\ \Xi(Y_1, \{W_1, W_2\}) \tag{1}$$

$$\begin{bmatrix} Y_1 \\ 1 \le x_1 \le 3 \\ 4 \le x_2 \le 6 \end{bmatrix} \vee \begin{bmatrix} Y_2 \\ 8 \le x_1 \le 9 \\ 1 \le x_2 \le 2 \end{bmatrix} \\ \begin{bmatrix} W_1 \\ 1 \le x_1 \le 2 \\ 5 \le x_2 \le 6 \end{bmatrix} \vee \begin{bmatrix} W_2 \\ 2 \le x_1 \le 3 \\ 4 \le x_2 \le 5 \end{bmatrix} \vee \begin{bmatrix} W_3 \\ 1 \le x_1 \le 9 \\ 1 \le x_2 \le 6 \end{bmatrix} \\ \Xi(1, \{Y_1, Y_2\}) \\ \Xi(1, \{W_1, W_2, W_3\}) \\ \Xi(Y_1, \{W_1, W_2\}) \tag{2}$$

4. Conclusions

The contribution of this paper has been to extend GDP for modeling hierarchical systems via nested disjunctions. Such an approach results in more intuitive models, but has not been formalized in the past, as classical GDP does not consider disjunction nesting. The notation and logic constraints for such structures is provided, along with theoretical proofs to the tightness of such models, versus equivalent single-level GDP models. It is shown that mixed-integer programming reformulations of nested GDP models have tighter relaxations than the reformulations of their single-level counterparts in both the hull reformulation, as well as the Big-M reformulation when tight M values are used.

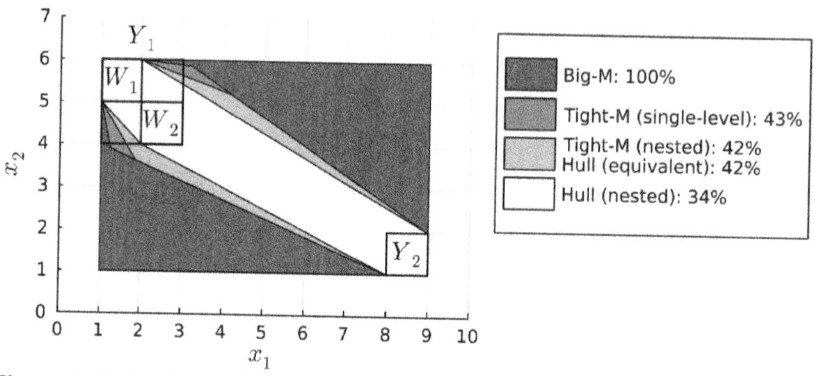

Figure 1. Projections of the continuous relaxations of Example 4.1 for different reformulations (Big-M = Big-M Reformulation, Tight-M = Big-M Reformulation with tight M values, Hull = Hull Reformulation; equivalent = equivalent single-level GDP, nested = multi-level hierarchical GDP). Projection areas, relative to Big-M are indicated in %.

Acknowledgement

The authors gratefully acknowledge the financial support from the Center of Advanced Process Decision-making at Carnegie Mellon University.

References

E. Balas, 1985, Disjunctive Programming and a Hierarchy of Relaxations for Discrete Optimization Problems, SIAM J. Alg. Disc. Meth., 6(3), 466–486.

I.E. Grossmann, 2012, Advances in mathematical programming models for enterprise-wide optimization, Comput. Chem. Eng., 47, 2–18.

I.E. Grossmann, F. Trespalacios, 2013, Systematic modeling of discrete-continuous optimization models through generalized disjunctive programming. AIChE J., 59(9), 3276–3295.

H.D. Perez, I.E. Grossmann, 2023, An Extension to Generalized Disjunctive Programming for Modeling Hierarchical Systems, submitted for publication.

R. Raman, I.E. Grossmann, 1994, Modelling and computational techniques for logic based integer programming, Comput. Chem. Eng., 18(7), 563–578.

F. Trespalacios, I.E. Grossmann, 2016, Cutting Plane Algorithm for Convex Generalized Disjunctive Programs, INFORMS J. Comput., 28(2), 209–222.

M. Türkay, I.E. Grossmann, 1996, Logic-based MINLP algorithms for the optimal synthesis of process networks, Comput. Chem. Eng., 20(8), 959–978.

S.A. Van Den Heever, I.E. Grossmann, 1999, Disjunctive multiperiod optimization methods for design and planning of chemical process systems, Comput. Chem. Eng., 23(8), 1075–1095.

A. Vecchietti, I.E. Grossmann, 2000, Modeling issues and implementation of language for disjunctive programming, Comput. Chem. Eng., 24(9–10), 2143–2155.

H. Yan, J.N. Hooker, 1999, Tight representation of logical constraints as cardinality rules. Math. Program., 85(2), 363–377.

Antonis Kokossis, Michael C. Georgiadis, Efstratios N. Pistikopoulos (Eds.)
PROCEEDINGS OF THE 33rd European Symposium on Computer Aided Process Engineering (ESCAPE33), June 18-21, 2023, Athens, Greece
 http://dx.doi.org/10.1016/B978-0-443-15274-0.50219-5

Fusion and integrated correction of chemometrics and machine learning models based on data reconciliation

Pál P. Hanzelik,[a,b] Alex Kummer,[b] Ádám Ipkovich,[b] János Abonyi[b]

[a] *Dombóvári út 28, Budapest, Hungary*
[b]*Egyetem Street 10, Veszprém, Hungary*

Abstract

Soft-sensors can provide real-time estimates of unmeasured process variables to improve process control and performance. A Multi-Input-Multi-Output system is considered, where the outputs need to satisfy a linear constraint (e.g. balance equations), while the output variables are predicted using machine learning (ML) models. The key idea is that data reconciliation (DR) can be used to modify both the measurements and the model predictions to meet the prescribed constraints. In addition, the model prediction errors can be used as weighting factors for data reconciliation.

The paper summarizes the mathematical derivation and application of the approach to predict the composition of various rocks in an industrial environment. The results show that the proposed, integrated ML+DR approach greatly contributes to the models' usability. Furthermore, we obtained smaller model errors in four cases out of the six properties, and the prediction sum also considers the constraints.

Keywords: machine learning, algorithm development, data reconciliation, chemometrics, model prediction

1. Introduction

This paper proposes a methodology that can improve ML models applied in industrial processes and help optimization based on digitization tools. Moreover, during the Industry 4.0 processes, software sensors' use is becoming more widespread [1]. The motivation of the work is that the use of correlations between process parameters can make the industrial application of machine learning algorithms more efficient and robust. These options include various ML algorithms that can replace traditional quality assurance processes. During the development of ML models of Quality 4.0 techniques, we can estimate only specific parameters with sufficient accuracy in many cases. The key idea is that in many cases the variables to be estimated are not independent of each other, for example they have to satisfy some balance equation, but these constraints are not taken into account when teaching ML models. We recommend a data reconciliation technique to meet the limits, thus increasing the applicability of ML models.

The DR technique has already been used for energy monitoring chemical processes, which considers the balance equations' consistency. In addition to the DR method, principal component analysis (PCA) can also use to forecast operating costs. The advantage of this method is that if we apply the PCA technique, the reliability of the data increases significantly, which increases the accuracy of this estimated cost.[2], furthermore, DR and PCA are also excellent for error detection [3] [4]. Xie et al.

investigated the industrial application of sodium-aluminium solution vaporization. They presented that DR is a key technique in improving the accuracy of measured and predicted data in the case of a chemical process [5]. The DR-based approach results in better data quality and, thus, fewer operational errors [6]. Furthermore, DR is used in the control design of online processes, where the goal is to estimate the valid values of erroneous measurements within limits. Non-linear DR was used in material flow analysis for the systematic and efficient management of material flow and inventory [7] or even for eliminating errors caused by measurement noise in the control of reactor processes [8]. For example, the hierarchical DR provides a high-quality measurement of predicting a natural industrial system's material and heat energy consumption [9]. In addition, DR can be used in 4.0 quality, and there is a large literature on Fourier transform infrared spectroscopy (FTIR) measurement techniques and dielectric spectroscopy techniques for predicting the concentration of different metabolites. DR using FTIR belongs to Dabros et al. according to them, the predictive performance of the primary models was found to be poor, but with DR techniques the problem can be improved by real-time mass and elemental balance [10].
We present a machine learning and data reconciliation-based approach (see Section 2) to predict elemental composition of rocks, where the predictions need to satisfy a linear constraint (a balance equation). Furthermore, since the model predictions are loaded with errors, the standard deviation of the model errors can be used in data reconciliation to modify the predicted values. The accuracy of the ML models with and without using DR are compared (see Section 3).

2. Data reconciliation-based integration of machine learning models

This section introduces the methodology of our approach to integrate ML and DR techniques to improve the quality of ML predictions. The goal is to predict several related variables, where the predictions need to satisfy some constraints, e.g. balance equations. In order to achieve the highest accuracy of the ML predictions, and to fulfil the constraints, we recommend to integrate ML and DR techniques.

We consider a Multi-Input-Multi-Output system with N samples and N_m measurement variables. The measurement outputs of the k-th sample can be written as in Equation 1.

$$\boldsymbol{y}_k = f(\boldsymbol{x}_k, \boldsymbol{\Theta}) + \boldsymbol{\varepsilon}_k \tag{1}$$

where $\boldsymbol{y}_k = \left[\boldsymbol{y}_k^{(1)}, \dots, \boldsymbol{y}_k^{(N_m)}\right]^T$ is the measurement vector of the k-th sample, $\boldsymbol{x}_k$ is the input features of the k-th sample, and $\boldsymbol{\varepsilon}_k$ is the measurement error at the k-th sample. The k-th sample of the i-th output variable can be defined as Equation 2.

$$y_k^{(i)} = f_i\left(\boldsymbol{x}_k^{(i)}, \boldsymbol{\Theta}^{(i)}\right) + \varepsilon_k^{(i)} \tag{2}$$

where $y_k^{(i)}$ is the i-th output variable as the kth sample, f_i is the i-th model, $\boldsymbol{x}_k^{(i)}$ stands for the input features fot the ith model at the kth sample, $\boldsymbol{\Theta}^{(i)}$is the model parameters of the ith mdoel and $\varepsilon_k^{(i)}$ represents the model error for the ith model at the kth sample.

The prediction of the model at the kth sample is represented as $\hat{\mathbf{y}}_k = \left[\hat{y}_k^{(1)}, \ldots, \hat{y}_k^{(N_m)}\right]^T$, and the prediction of the ith model at the kth sample $\left(\hat{y}_k^{(i)}\right)$ can be calculated as presented in Equation 3.

$$\hat{y}_k^{(i)} = f_i\left(\boldsymbol{x}_k^{(i)}, \boldsymbol{\Theta}^{(i)}\right) \tag{3}$$

The measrurement variables are not independent of each other, because they need to satisfy a linear constraint, while the measurements are loaded with errors (e_k), hence this constraint is not fulfilled.

$$g(\mathbf{y}_k) + e_k = 0 \tag{4}$$

where $g()$ is an equality constraint function.

Data reconciliation calculates minimal correction of the predicted variables to make them verify a set of model constraints. It minimizes the difference between the predicted and reconciled values considering the variance of the variables and ensures that reconciled parameters satisfy some equality and inequality constraints. Similarly, the corrected model predictions for the kth sample are represented as $\tilde{\mathbf{y}}_k = \left[\tilde{\mathbf{y}}_k^{(1)}, \ldots, \tilde{\mathbf{y}}_k^{(N_m)}\right]^T$.

$$\min_{\tilde{\mathbf{y}}_k} (\hat{\mathbf{y}}_k - \tilde{\mathbf{y}}_k)^{\mathrm{T}} \mathbf{Q} (\hat{\mathbf{y}}_k - \tilde{\mathbf{y}}_k) \tag{5}$$

$$\text{s.t.} \qquad g(\tilde{\mathbf{y}}_k) = 0 \tag{6}$$

where $\tilde{\mathbf{y}}_k$is the corrected value by data reconciliation, the $\mathbf{Q}$ is the inverse covariance matrix of the measurements/predictions error, the $g(\tilde{\mathbf{y}}_k)$ is a model equality constraint. In the inverse covariance matrix $\mathbf{Q}$ the standard deviation of the measurement errors are presented, as in Equation 7.

$$\mathbf{Q} = \begin{bmatrix} 1/\sigma_1^2 & \cdots & 0 \\ \vdots & \ddots & \vdots \\ 0 & \cdots & 1/\sigma_{N_m}^2 \end{bmatrix} \tag{7}$$

In case of a linear constraint Equation 5 becomes the following:

$$\mathbf{A}\tilde{\mathbf{y}}_k = \mathbf{b} \tag{8}$$

where $\mathbf{A}$ is the incidence matrix, and $\mathbf{b}$ stands for constraint values.
The analytical solution of the system of Equations 5 and 8 is the following:

$$\tilde{\mathbf{y}}_k = (\mathbf{I} - \mathbf{Q}\mathbf{A}^{\mathrm{T}}(\mathbf{A}\mathbf{Q}\mathbf{A}^{\mathrm{T}})^{-1}\mathbf{A})\hat{\mathbf{y}}_k + \mathbf{Q}\mathbf{A}^{\mathrm{T}}(\mathbf{A}\mathbf{Q}\mathbf{A}^{\mathrm{T}})^{-1}\mathbf{b} \tag{9}$$

where $\mathbf{I}$ is the identity matrix, $\mathbf{Q}$ is the inverse covariance matrix, $\mathbf{A}$ is the incidence matrix and $\mathbf{b}$ is the constraint values.

In the developed methodology, industrial data collection in a data store provides input data to the ML models. The essence of the developed methodology is that various DR restrictions are applied after the models are incorporated, which enables more accurate and robust ML models. As a result, we can build more effective multivariate data analysis on our industrial processes. Figure 1 shows in a general diagram how the developed technique can be applied in an industrial environment. The essential part of the methodology is shown with the orange arrows, and the blue dashed line shows the traditional approach in the following figure.

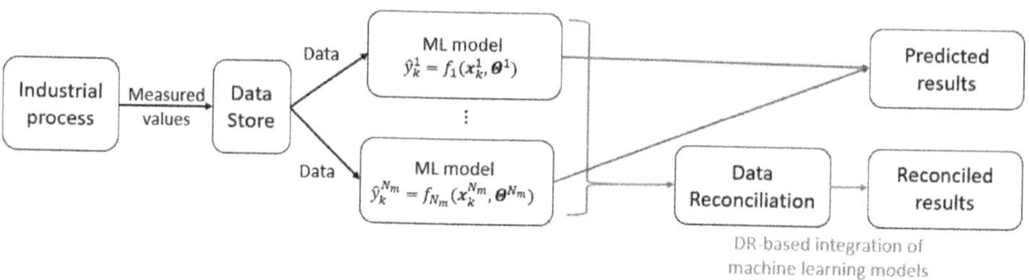

Figure 1: Diagram of the recommended method

3. Data reconciliation-based integration of machine learning models

A mineral composition prediction model was built on 618 rock samples, where the target variables were mass percentage of quartz, clay content without kaolinite, kaolinite, plagioclase with K-feldspar, calcite and dolomite. The reference measurements were performed with an X-ray diffraction (XRD) device. PLS models were trained on pretreated spectra from FTIR measurements, and individual models were trained for each of the six target variables. Optimal number of factors for PLS models were tuned based on the results of 10-fold cross-validation. The ML input spectra of samples were scanned (4 scans co-added) from 1998 to 600 cm^{-1} in attenuated total reflectance (ATR) mode by using a Spectrum 400 spectrometer equipped with a Universal ATR (UATR) accessory, which contains a single reflection diamond/ZnSe composite crystal.
Table 1 summarizes the target values of a chosen sample, the predictions (ML(PLS)) and the reconciled predictions (ML(PLS)+DR).

Table 1. Prediction and reconciled results a chosen sample (weight%)

	Quartz	Clay without kaolinite	Kaolinite	Plagioclase + K-feldspar	Calcite	Dolomite	*sum*
Target	49	16	1	17	10	7	*100*
ML (PLS)	49.39	17.95	1.637	14.32	10.51	8.053	*101.86*
ML(PLS)+DR	49.21	17.55	1.085	14.19	10.36	7.605	*100*

The reconciled predictions for five of the six output variables were better compared to the non-reconciled model predictions. Table 2 summarizes the results obtained for all 618 rock samples, as well as the standard deviation of the reference measurement. The performance indicators of the models were the cross-validation error (*RMSEcv*).

Table 2. Reference measuring standard deviation, optimal factor number of the PLS, $RMSE_{cv}$ (%) results of ML and ML + DR models

	Quartz	Clay without kaolinite	Kaolinite	Plagioclase + K-feldspar	Calcite	Dolomite
Factor number	8	5	9	9	5	8
ML(PLS) $RMSE_{cv}$	5.934	6.265	1.467	3.263	4.173	2.414
ML(PLS) + DR $RMSE_{cv}$	5.924	6.232	1.624	3.260	4.172	2.481

The values of the six target variables ranged from 1.47 to 6.27 with the ML solution and from 1.62 to 6.23 with the ML+DR technique. The accuracy of the ML models constructed from FTIR spectra approaches the uncertainty of the reference measurement in two cases (kaolinite, dolomite), and the ML+DR models give better results when the six properties are considered together. In the latter case, the ML+DR models perform worse for kaolinite and dolomite (Table 2). This can be explained by the fact that these models already performed very close to the accuracy of the reference models.

4. Conclusion

The article proposes a methodology that improves the usability of ML models by reconciling the model predictions. The proposed methodology takes into account the prediction error of the ML models, by which the DR technique corrects the prediction values while satisfying certain constraints. Since in many cases the variables to be estimated are not independent of each other, for example, they correspond to balance equations. However, these constrains should be fulfilled by ML models. We present a methodology on how to integrate machine learning and data-reconciliation, and we tested the approach to estimate the elemental composition of rock samples.

With the inclusion of the constraints, the results showed more accurate outputs for four of the six target variables. The results will help geologists by enabling them to obtain mineral composition data of rocks. Furthermore, the presented methodology allows faster and cheaper access to compositional data, either with a mobile device or with devices installed in a soft sensor.

Our future research is to take constraints into account using DR during the training of the ML models, rather than performing the training and data reconciliation serially.

References

1. P. P. Hanzelik, A. Kummer, J. Abonyi, 2022, Edge-Computing and Machine-Learning-Based Framework for Software Sensor Development, Sensors, 22, 4268.
2. B. Farsang, S. Nemeth, J. Abonyi, 2014, Synergy between data reconciliation and principal component analysis in energy monitoring. Chemical Engineering, 39.
3. T. Amand, G. Heyen, B. Kalitventzeff, 2001, Plant monitoring and fault detection: Synergy between data reconciliation and principal component analysis. Computers & Chemical Engineering, 25(4-6), 501-507.

4. B. Farsang, S. Nemeth, J. Abonyi, 2014, Life-cycle Modelling for Fault Detection–Extraction of PCA Models from Flowsheeting Simulators. In Computer Aided Chemical Engineering, 33, 421-426, Elsevier
5. S. Xie, C. Yang, X. Yuan, X. Wang, Y. Xie, 2018, Layered online data reconciliation strategy with multiple modes for industrial processes. Control Engineering Practice, 77, 63-72.
6. X. Jiang, P. Liu, Z. Li, 2013, A data reconciliation based approach to accuracy enhancement of operational data in power plants. Chemical Engineering Transactions, 35, 1213-1218.
7. J. H. Taylor, R. del Pilar Moreno, 2013, Nonlinear dynamic data reconciliation: In-depth case study. In 2013 IEEE international conference on control applications (CCA) (pp. 746-753). IEEE.
8. M.J. Leibman, T.F. Edgar, L.S. Lasdon, 1992, Efficient data reconciliation and estimation for dynamic processes using nonlinear programming techniques, Computers & Chemical Engineering, 16, 10-11, 963-986
9. S. Xie, H. Wang, J. Peng, X. Liu, X. Yuan, 2020, A hierarchical data reconciliation based on multiple time-delay interval estimation for industrial processes. ISA transactions, 105, 198-209.
10. M. Dabros, M. Amrhein, D. Bonvin, I. W. Marison, U. von Stockar, 2009, Data reconciliation of concentration estimates from mid-infrared and dielectric spectral measurements for improved on-line monitoring of bioprocesses. Biotechnology progress, 25(2), 578-588.

Antonis Kokossis, Michael C. Georgiadis, Efstratios N. Pistikopoulos (Eds.)
PROCEEDINGS OF THE 33rd European Symposium on Computer Aided Process Engineering (ESCAPE33), June 18-21, 2023, Athens, Greece
 http://dx.doi.org/10.1016/B978-0-443-15274-0.50220-1

Model-based Optimisation of Regional Nutrient Flow and Recovery for Resource and Environmental Sustainability

Purusothmn Nair S Bhasker Nair,[a] Nan-Hua Nadja Yang,[a] Wei Zhang,[a] Kok Siew Ng,[a,b] Aidong Yang[a]

[a] *Department of Engineering Science, University of Oxford, Parks Road, Oxford OX1 3PJ, United Kingdom.*

[b] *Department of Chemical Engineering, College of Engineering, Design and Physical Sciences, Brunel University London, Kingston Lane, Uxbridge UB8 3PH, United Kingdom.*

Abstract

The unsustainably managed flow of nutrients combined with the rise in the generation of waste calls for urgent actions involving nutrient recovery. This work develops a mathematical programming model intended to optimise regional nutrient recovery from various waste treatment technologies. Food waste, green waste, and wastewater are evaluated in this work with an option of treating these wastes individually or as mixed streams. Five waste treatment technologies are available, each with a specific profile of nutrient recovery. The results of this work demonstrate the attractiveness of anaerobic digestion as well as the benefit of a mixed waste stream for optimum nutrient recovery.

Keywords: Mathematical Optimisation, Nutrient Recovery, Circular Economy, Climate Change, Sustainable Resource Management

1. Introduction

Nutrients such as nitrogen and phosphorus are vital for food production. Presently, there is a vast usage of inorganic fertilisers derived from ammonia and phosphate (van der Wiel et al., 2020). Phosphorus is solely sourced from phosphate rocks which are finite resources and limited in terms of their geographic availability (Álvarez et al., 2018) while nitrogen in ammonia production represents one of the major industrial energy demands and sources of greenhouse gas emissions (Liu et al., 2020). Furthermore, the existing managed flow of phosphorus and nitrogen from its source (mines) to the ocean or other aquatic environments is not sustainable (Cordell and White, 2015). The current farming and waste management practices create a significant nutrient influx to water streams, resulting in undesirable consequences e.g., eutrophication (Cordell and White, 2015). Aside from that, nutrient depletion due to current practices would induce instability in the food systems, thus driving up the costs of food production. A key solution to these problems lies in innovative waste management. In the United Kingdom, 27 million tonnes of household waste were produced in the year 2018 (DEFRA, 2022). Although close to half of this waste is recycled, much more needs to be done to align with environmental sustainability and thus create a circular economy (DEFRA, 2022). With the rise in population, the total volume of waste generated annually is projected to rise. Nutrients such as nitrogen and phosphorus should be extracted from these waste streams as they may be re-utilised as fertilisers.

In the past, several studies concerning nutrient recovery had been conducted (Mehta et al., 2014; Perera et al., 2019; Sniatala et al., 2023), which were all focused on technological advancement for nutrient recovery. There was a limited focus in terms of modelling and optimising nutrient flows from their sources (water, food, and green waste) to potential sinks e.g., land dispersion, fertilisers etc. Despite some work being done recently for optimising nutrient recovery, much of the focus was associated with transportation costs (Metson et al., 2020) and biogas generation from anaerobic digestion (Feiz et al., 2022). There has been limited emphasis on optimising nutrient recovery from these waste treatment technologies. This work, therefore, seeks to model and optimise the regional nutrient flows and maximise recovery from waste streams into valuable products such as fertilisers. Firstly, the flows of wastewater, food and green wastes are mapped to understand their resource recovery potential in creating a circular economy. Based on the available data, an optimisation model is subsequently built toward determining potential opportunities that may be present for nutrient recovery. Potential technologies such as anaerobic digestion, composting and incineration are considered in the optimisation model. This work makes use of a case study from Leicestershire County (LC) in the UK and is expected to consolidate efforts toward building a circular economy by tapping into the underexplored potentials of nutrients embedded in the waste streams, thus promoting a sustainable practice of nutrient management in a regional setting.

2. Problem Statement

Given a set of waste streams i.e., food waste, green waste and wastewater that may be treated by processing unit (PU) j as either an individual waste (IW) stream i or mixed waste (MW) stream im. The treatment of IW i and MW im by PU j results in the recovery of nutrient l. A superstructural mathematical programming formulation is developed to synthesise a waste treatment network for optimised nutrient recovery.

3. Mathematical Programming Formulation

The total flow of IW i to PU $j \in J$ $(x1_{i,j})$ and MW $im \in IM$ $(x2_{i,im})$ should equate to the total waste available (TW_i), as shown in Eq.(1).

$$\sum_{j} x1_{i,j} + \sum_{im} x2_{i,im} = TW_i \qquad \forall i \tag{1}$$

Next, the total flow of IW $i \in I$ and MW $im \in IM$ $(x3_{im,j})$ to PU j should not exceed the maximum treatment capacity of PU j (M_j), as shown in Eq.(2).

$$\sum_{i} x1_{i,j} + \sum_{im} x3_{im,j} \leq M_j \qquad \forall j \tag{2}$$

Eq.(3) ensures that the flow of MW im to PU j equate the flow summations of IW i to MW im.

$$\sum_{i} x2_{i,im} = \sum_{j} x3_{im,j} \qquad \forall im \tag{3}$$

Following this, the total recovery of nutrient l from PU j treating IW i $(NT1_{i,j,l})$ and MW im $(NT2_{im,j,l})$ are depicted in Eq.(4) and Eq.(5) respectively.

$$NT1_{i,j,l} = x1_{i,j} \times a1_{i,j,l} \qquad \forall i\ \forall j\ \forall l \tag{4}$$

$$NT2_{im,j,l} = x3_{im,j} \times a2_{im,j,l} \qquad \forall im\ \forall j\ \forall l \tag{5}$$

where $a1_{i,j,l}$ and $a2_{im,j,l}$ are the recovery of nutrient l from PU j treating IW i and MW im respectively.

The objective function of this optimisation problem is to maximise the total nutrient recovery, i.e., the sum of nutrients $l \in L$ from PU $j \in J$ treating IW $i \in I$ and MW $im \in IM$, as shown in Eq.(6):

$$\max NTT = \sum_{i}\sum_{j}\sum_{l} NT1_{i,j,l} + \sum_{im}\sum_{j}\sum_{l} NT2_{im,j,l} \tag{6}$$

The linear programming model is implemented in GAMS using the CPLEX solver.

4. Nutrient Flow Data and Operational Constraints

This work makes use of a case study from LC in the UK. Firstly, the type and mass of waste available within the LC region are presented in Table 1.

Table 1: Type and volume of waste

Waste	Waste Mass, t y^{-1}
Food (F)	149,845
Green (G)	61,143
Wastewater sludge (W)	463,696

Based on Table 1, the three types of waste investigated in this work are food waste, green waste, and wastewater sludge. It is estimated that there are 149,845 tonnes of food waste generated annually in LC (WRAP, 2020). Meanwhile, the annual volume of green waste amounted to 61,143 tonnes (WasteDataFlow, 2021). Moving on to the wastewater, it is estimated that the total wastewater generated in LC is 0.7 m^3 d^{-1} $capita^{-1}$, covering both domestic households and small businesses (Severn Trent Plc, 2022). Considering that the population in LC is 712,300 (Richardson, 2022), the total annual volume of wastewater in LC is estimated to be 180.9 million m^3, resulting in 463,696 tonnes of wastewater sludge annually. In this work, five PUs i.e., anaerobic digestion (AD), windrow composting (WC), bin composting (BC), incineration (IN) and the landfill (LF) are considered for treating various waste in LC. Table 2 presents the assumed maximum treatment capacity of each PU.

Table 2: Assumed maximum treatment capacity of PU

Processing Unit	Waste processing capacity, t y^{-1}
AD	700,000
WC	300,000
BC	60,000
IN	700,000
LF	700,000

Both IN and LF are well-established technologies and are assumed to treat a maximum of 700,000 tonnes of waste, as observed in Table 2. Meanwhile, BC which is done within households is assumed to have the least waste processing capacity. In this work, AD is assumed to have a significant annual waste treatment capacity of 700,000 tonnes of organic waste due to the potential for nutrient recovery and biogas generation. Next, the recovery of nutrients l from the PU j treating IW i and MW im is presented in Table 3.

Note that the two nutrients considered in this work i.e., nitrogen (N) and phosphorus (P) may be recovered in solid (NS/PS) and liquid (NL/PL) streams.

Table 3: Nutrient recovery

Waste Stream	Processing Unit	NL, kg t^{-1}	NS, kg t^{-1}	PL, kg t^{-1}	PS, kg t^{-1}
Food [a]	AD	2.28	1.94	0.230	0.61
Water [b]	AD	0	1.80	0	1.47
Food + Green [c, d]	AD	2.15	1.82	0.350	0.92
Green [e]	WC	0	3.22	0	0.61
Food + Green [e]	BC	0.05	2.48	0.010	0.56
Food + Green [f, g]	IN	0	0	0	1.20
Food + Green [f, h]	LF	0.15	0	0.005	0

References: a) (Dimambro, 2015) b) (Havukainen et al., 2022) c) (Banks et al., 2018) d) (Cesaro, 2021) e) (Andersen et al., 2010) f) (Jeswani and Azapagic, 2016) g) (Šyc et al., 2020) h) (Kulikowska and Klimiuk, 2008)

Based on Table 3, AD may be used to treat food wastes, wastewater sludge as well as a mixture of food and green wastes. WC may be used for the treatment of green waste. Besides, the mixture of food and green wastes may be treated using BC, IN or LF. Although the ratio of food and green waste may change in a MW stream, this work assumes a fixed ratio of the food and green waste (9:1) in MW streams to all PUs.

5. Case Study

This work is solved according to Eq.(6) subject to the constraints in Eq.(1) till Eq.(5). Table 4 presents the optimised volumes of waste flows to the PUs. Note that Table 4 also includes the results of two supplementary runs that maximise the recovery of total N or P.

Table 4: Waste Flow Volume to PU *j*

Waste Stream	Waste flow to PU (max NTT/max total N/max total P)				
	AD	WC	BC	IN	LF
Food, t y^{-1}	0/149,845/0	0	0	0	0
Green, t y^{-1}	0	44,493/61,143/44,493	0	0	0
Wastewater, t y^{-1}	463,696	0	0	0	0
Food + Green, t y^{-1}	166,495/0/166,495	0	0	0	0

Based on Table 4, maximising the total nutrient recovery (Eq.(6)) resulted in no food waste being treated via AD as an IW stream. All food waste is combined with approximately 27% of the green waste for treatment via AD. The higher nutrient recovery of the MW stream via AD resulted in it being the primary choice of waste treatment in this optimisation problem (see Table 3). Note that all wastewater sludge is treated via AD for a similar reason. The objective function in this work is maximised at 2,559 tonnes of nutrients, comprising 1,639 tonnes of N and 920 tonnes of P. Moving on, the total recovery of N is maximised at 1,663 tonnes. All wastes are treated via AD and WC as an

IW stream. The greater N recovery from all types of waste being treated as an IW stream led to this solution. The total recovery of P is maximised at 920 tonnes, with similar results observed as to maximising the total nutrient recovery. It is good to note that if incineration is selected, the P recovery from incineration ash would require additional extraction processes before it may be applied to agricultural land or replacing phosphate for fertiliser production (Kalmykova and Karlfeldt Fedje, 2013). The results of this optimisation problem align with the aim of this work in maximising nutrient recovery and exploring the potential offered by these waste streams.

6. Conclusions

An optimal waste treatment plan for maximising nutrient recovery is suggested in this work based on mathematical programming models. This work is expected to aid in value-creation among waste streams that would otherwise remain unutilised. The results of this work demonstrate the importance of AD in maximising nutrient recovery. Additionally, combining the food and green waste streams presents an added value for better use of resources. Future work should include the cost analysis of potential by-products i.e., biogas, and digestate as well as emissions that may arise from the waste treatment. Also, the operational and capital costs associated with the operations of waste treatment plants must be considered for a more realistic waste management scenario.

Acknowledgement

This work is supported by the Natural Environment Research Council (NERC), UK as part of the AGILE initiative (grant number NE/W004976/1). The authors would like to also thank Dr John Ingram and Dr Bhawana Gupta from the Environmental Change Institute at the University of Oxford for the support received throughout this work.

References

Álvarez, J., Roca, M., Valderrama, C., Cortina, J.L., 2018. A Phosphorous Flow Analysis in Spain. Science of The Total Environment 612, 995–1006. https://doi.org/10.1016/J.SCITOTENV.2017.08.299

Andersen, J.K., Boldrin, A., Christensen, T.H., Scheutz, C., 2010. Greenhouse gas emissions from home composting of organic household waste. Waste Management 30, 2475–2482. https://doi.org/10.1016/J.WASMAN.2010.07.004

Banks, C., Heaven, S., Zhang, Y., Baier, U., IEA Bioenergy Programme, 2018. Food waste digestion : Anaerobic digestion of food waste for a circular economy. IEA Bioenergy Task 37 12.

Cesaro, A., 2021. The valorization of the anaerobic digestate from the organic fractions of municipal solid waste: Challenges and perspectives. J Environ Manage 280, 111742. https://doi.org/10.1016/J.JENVMAN.2020.111742

Cordell, D., White, S., 2015. Tracking phosphorus security: indicators of phosphorus vulnerability in the global food system. Food Secur 7, 337–350. https://doi.org/10.1007/S12571-015-0442-0/FIGURES/6

DEFRA, 2022. UK statistics on waste [WWW Document]. Department for Environment, Food & Rural Affairs. URL https://www.gov.uk/government/statistics/uk-waste-data/uk-statistics-on-waste#data-revisions-in-this-update (accessed 9.27.22).

Dimambro, M.E., 2015. Novel Uses for Digestates: Protected Horticulture. 20th European Biosolids & Organic Resources Conference & Exhibition.

Feiz, R., Johansson, M., Lindkvist, E., Moestedt, J., Påledal, S.N., Ometto, F., 2022. The biogas yield, climate impact, energy balance, nutrient recovery, and resource cost of biogas production from household food waste—A comparison of multiple cases from Sweden. J Clean Prod 378, 134536. https://doi.org/10.1016/J.JCLEPRO.2022.134536

Havukainen, J., Saud, A., Astrup, T.F., Peltola, P., Horttanainen, M., 2022. Environmental performance of dewatered sewage sludge digestate utilization based on life cycle assessment. Waste Management 137, 210–221. https://doi.org/10.1016/J.WASMAN.2021.11.005

Jeswani, H.K., Azapagic, A., 2016. Assessing the environmental sustainability of energy recovery from municipal solid waste in the UK. Waste Management 50, 346–363. https://doi.org/10.1016/J.WASMAN.2016.02.010

Kalmykova, Y., Karlfeldt Fedje, K., 2013. Phosphorus recovery from municipal solid waste incineration fly ash. Waste Management 33, 1403–1410. https://doi.org/10.1016/J.WASMAN.2013.01.040

Kulikowska, D., Klimiuk, E., 2008. The effect of landfill age on municipal leachate composition. Bioresour Technol 99, 5981–5985. https://doi.org/10.1016/J.BIORTECH.2007.10.015

Liu, X., Elgowainy, A., Wang, M., 2020. Life cycle energy use and greenhouse gas emissions of ammonia production from renewable resources and industrial by-products. Green Chemistry 22, 5751–5761. https://doi.org/10.1039/D0GC02301A

Mehta, C.M., Khunjar, W.O., Nguyen, V., Tait, S., Batstone, D.J., 2014. Technologies to Recover Nutrients from Waste Streams: A Critical Review. Crit Rev Environ Sci Technol 45, 385–427. https://doi.org/10.1080/10643389.2013.866621

Metson, G.S., Feiz, R., Quttineh, N.H., Tonderski, K., 2020. Optimizing transport to maximize nutrient recycling and green energy recovery. Resources, Conservation & Recycling: X 9–10, 100049. https://doi.org/10.1016/J.RCRX.2021.100049

Perera, M.K., Englehardt, J.D., Dvorak, A.C., 2019. Technologies for Recovering Nutrients from Wastewater: A Critical Review. Environ Eng Sci. https://doi.org/10.1089/ees.2018.0436

Richardson, H., 2022. Leicestershire population rises by more than 100,000 in last ten years - Leicestershire Live [WWW Document]. LeicestershireLive. URL https://www.leicestermercury.co.uk/news/local-news/leicestershire-population-rises-more-100000-7290712 (accessed 11.16.22).

Severn Trent Plc, 2022. Who We Are [WWW Document]. About Us. URL https://www.severntrent.com/about-us/who-we-are/ (accessed 11.16.22).

Sniatala, B., Kurniawan, T.A., Sobotka, D., Makinia, J., Othman, M.H.D., 2023. Macro-nutrients recovery from liquid waste as a sustainable resource for production of recovered mineral fertilizer: Uncovering alternative options to sustain global food security cost-effectively. Science of The Total Environment 856, 159283. https://doi.org/10.1016/J.SCITOTENV.2022.159283

Šyc, M., Simon, F.G., Hykš, J., Braga, R., Biganzoli, L., Costa, G., Funari, V., Grosso, M., 2020. Metal recovery from incineration bottom ash: State-of-the-art and recent developments. J Hazard Mater 393, 122433. https://doi.org/10.1016/J.JHAZMAT.2020.122433

van der Wiel, B.Z., Weijma, J., van Middelaar, C.E., Kleinke, M., Buisman, C.J.N., Wichern, F., 2020. Restoring nutrient circularity: A review of nutrient stock and flow analyses of local agro-food-waste systems. Resour Conserv Recycl 160, 104901. https://doi.org/10.1016/J.RESCONREC.2020.104901

WasteDataFlow, 2021. WasteDataFlow [WWW Document]. What is WasteDataFlow? URL https://www.wastedataflow.org/ (accessed 11.16.22).

WRAP, 2020. UK progress against Courtauld 2025 targets and UN Sustainable Development Goal 12.3 [WWW Document]. Report. URL https://wrap.org.uk/resources/report/uk-progress-against-courtauld-2025-targets-and-un-sustainable-development-goal-123 (accessed 11.16.22).

Antonis Kokossis, Michael C. Georgiadis, Efstratios N. Pistikopoulos (Eds.)
PROCEEDINGS OF THE 33rd European Symposium on Computer Aided Process Engineering (ESCAPE33), June 18-21, 2023, Athens, Greece
 http://dx.doi.org/10.1016/B978-0-443-15274-0.50221-3

Improving results of a Continuous Fluidized Bed Process for the Separation of Enantiomers by applying mathematical optimization

Nadiia Huskova[1,3], Jonathan Gänsch[1], Michael Mangold[2], Heike Lorenz[1,3], Andreas Seidel-Morgenstern[1,3]

[1]*Max Planck Institute for Dynamics of Complex Technical Systems, Sandtorstraße 1, 39106 Magdeburg*

[2]*University of Applied Sciences Bingen, Berlinstraße 109, 55411 Bingen*

[3]*Otto von Guericke University Magdeburg, Universitätsplatz 2, 39106 Magdeburg*

Abstract

The focus of the presented work is on the model-based optimization of a purification process of chemical compounds called enantiomers. As an example, the separation of asparagine monohydrate from the racemic mixture is considered, which has been proven to be effective in experiments. The experimental results indicate the need for a careful selection of operating conditions to provide the desired process performance. To describe the studied phenomena, a validated mathematical model based on population balance equations is used. Numerical optimization finds a set of operating conditions ensuring high product purity, while keeping the product contamination under a defined threshold

Keywords: Separation of Enantiomers, Continuous Crystallization, Population Balance Equations, Mathematical Optimization, Genetic Algorithms, Parallel Computations.

1. Introduction

More than ten million substances have been discovered in organic chemistry over the past two centuries (Dengale et al., 2016). Organic synthesis is becoming more and more prevalent due to the need to solve fundamental problems, such as determining the relationship between a chemical substance and its reactivity. To prevent and treat diseases, there is a need for affordable and effective drugs. To satisfy this need, chemists and pharmaceutical researchers are joining forces.

To reduce the time of drug production, great efforts are made to study already known compounds that are optical or geometric isomers.

One of the potential candidates are enantiomers, which are pairs of optically active isomers differently orientated in space (McConathy and Owens, 2003). A large interest in producing single enantiomers is confirmed by the life science industries, pharmaceuticals, and agrochemical companies. To supply pure enantiomers different approaches can be performed. The focus of this study is on the racemic approach based on separating mixtures of both enantiomers to obtain a significant amount of the desired single enantiomer characterized by high purity.

2. Process description

Chemical industries have increasingly relied on crystallization as a separation and purification technology, especially for those chiral molecules that crystallize in conglomerates, where the two enantiomers are mechanically mixed in the solid phase crystallizing separately as enantiopure crystals. In this kinetically controlled process, different rates of crystallization of each enantiomer are important.

Implementing a fluidized bed crystallization approach, two tubular crystallizers are connected spatially and are capable of crystallizing both enantiomers of chiral molecules selectively out of racemic solutions. Each of the operated crystallizers consists of the cylindrical top and the conical bottom parts. Liquid solvent without crystals enters the crystallizer at the bottom, and a mixture of solvent and formed crystals quits from the crystallizer at the top. The mechanical milling element located at the bottom ensures the continuity of the process and the supply of seed crystals. The volumetric flow rate characterizes the incoming stream; the removal of crystals is carried out periodically, and the outflow rate is determined by the applied withdrawal strategy. Corresponding scheme of the process is given in Figure 1. Special attention within this work is paid to modeling the crystal's withdrawal in a pulse-wise manner how it is performed in a laboratory.

3. Modeling crystallization process in fluidized beds

The fundamentals of the mathematical model applied are proposed in (Huskova, 2022). The model consists of three main components: population balances for the enantiomers in both crystallizers, distributed in one space and one property coordinate; a total mass balance for describing changes in the liquid phase; an energy balance to account for the non-isothermal nature of the process. The model distinguishes between the two types of enantiomers, as they may crystallize at different rates The structural change in the crystal shape is not considered. Instead, the physical parameters of the model are sorted into four groups:

- crystallizer design parameters;
- thermodynamic and kinetic properties of a particular chemical compound;
- operational parameters and parameters related to the product withdrawal strategy;
- set of parameters related to the model assumptions that have been adapted within the validation procedure.

A detailed description of the mathematical model and all the related model equations can be found in (Huskova, 2022). The reference numerical solutions are obtained in Matlab by a time integration with the ODE23 integrator applying Runge–Kutta method.

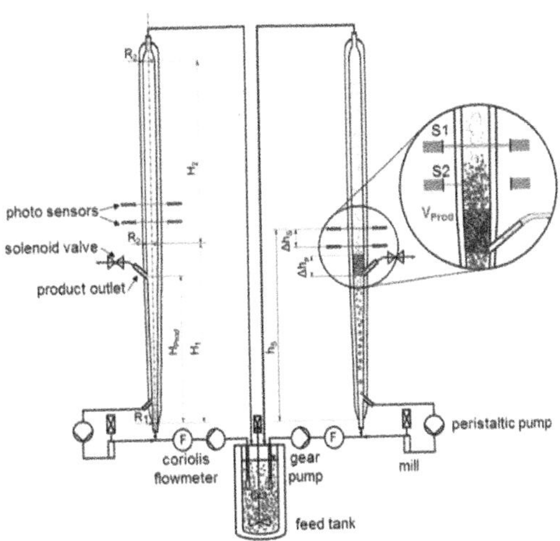

Figure 1. Schematic representation of continuous crystallization in a fluidized bed described in (Gänsch et al., 2021)

4. Model-based optimization

The next step after finding the reference solution and providing a model validation is to determine relevant model parameters to improve the process. The main goal of the optimization is to improve productivity of the target compound, which in general can be expressed as follows:

$$OF = Pr^L \rightarrow max \,,$$

where L stands for a target enantiomer.

A sensitivity analysis presented in (Gänsch et al., 2021) indicates that the operational conditions and geometrical parameters of the crystallizer listed in Table 1 have the biggest leverage.

Table 1. List of optimization variables

v_1	H	Crystallizer height
v_2	H_1	Height of conical section
v_3	R_1	Radius of the bottom
v_4	R_2	Radius of the top
v_5	T_{cryst}	Crystallizer Temperature
v_6	$u_{emptytube}$	Empty tube velocity

Empirical threshold to nucleation assessment

However, while analyzing productivity, it should be noted that when the sensor is in a high position, the risk of product contamination significantly increases, as was experimentally proven in (Gänsch et al., 2021). The occurrence of contamination can be explained by critical values of nucleation factors – supersaturation of the counter-enantiomer, S^D, and asymmetry of the solution, $asym^L$ (from here on, the target enantiomer is indexed L, and the counter-enantiomer D). From previous work, it is known that a low value of $asym^L$ as well as a high value of S^D cause a drop in purity Based on the experimental study, the following restrictions are proposed to ensure an acceptable product contamination:

$$\begin{cases} S^D \leq 1.4 \\ asym_L \geq 0.48 \end{cases} \qquad [\text{Eq.1}]$$

Optimization results

The simulation results shown on the left-hand side of Figure 1 do not include imposed constraints. It demonstrates that the productivity of the process can reach $400 gL^{-1}h^{-1}$, however with low purity and high contamination of the product. When the nucleation factors from [Eq.1] are considered in the optimization, the achieved productivity is lower – see the right-hand side of Figure 2.

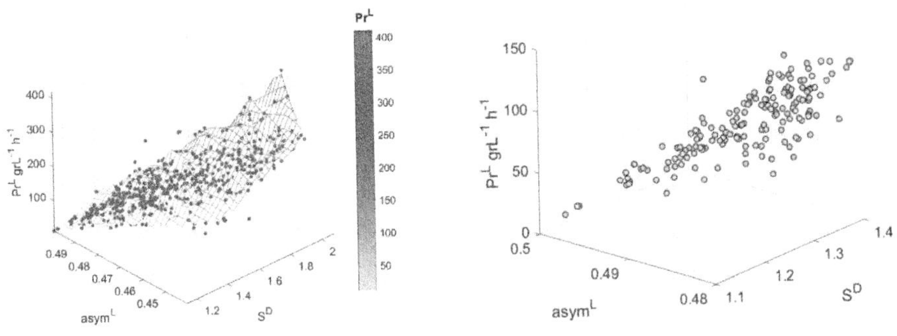

Figure 2. Simulation results achieved ignoring (left) following (right) the imposed nucleation factors constraints.

Also in this case, the obtained values significantly exceed the productivity noted in the available literature. Compared to the reference case study reported in (Gänsch et al., 2021), the increase in productivity reaches about 60%. The proposed approach makes it possible to ensure the minimum acceptable level of product purity and focus on increasing productivity by selecting the previously identified optimization variables.

The use of evolutionary algorithms for finding optimal solutions made it possible to avoid getting stuck in a local minimum, which would have happened with a high probability when using gradient-based approaches due to the non-convexity problem.

Applicability of parallel computing

Another challenge for the mathematical optimization is related to the resource intensity of the numerical solution. To ensure high accuracy of the numerical solution, the number of differential equations resulting from a discretization of the space and property coordinate exceeds 20 thousand. The elapsed time to calculate the transient to a cyclic steady state for one set of parameters exceeds several hours. The size of the developed model and time constraints lead to the inability to use a single processor to obtain optimal parameters and improve the process. In this regard, an attempt is made to directly solve the problem of reducing the computational requirements without compromising the accuracy of the developed mathematical model through parallelization methods. At each iteration, the formed sets of parameters describing the operating and design conditions were distributed among the processors, which made it possible to significantly speed up the search for the optimal solution.

5. Conclusions

The proposed model-based optimization approach is applied to improve the productivity and robustness of the separation of asparagine monohydrate in a fluidized bed reactor. It is demonstrated that the proposed optimal design of the crystallizer setup increases productivity, and additionally, it allows the evaluation of the nucleation factors once the suggested geometry is simulated. These recommendations can be used by experimenters to further achieve the desired purity and productivity of the process.

By shifting the focus away from asparagine monohydrate, the developed mathematical model and optimization procedure can be applied to design the purification of other substances. The study showed that substances for which the proposed fluidized bed approach is applicable should have a certain ratio of the densities of the solid and liquid phases. As an outlook, the approach may also be used to optimize particle characteristics like average particle size.

Acknowledgments

The authors acknowledge the financial support of Deutsche Forschung Gemeinschaft (DFG) within the Priority Program "Dynamic Simulation of Interconnected Solids Processes (SPP 1679)". N.H. is grateful for the support by the "Research Center of Dynamic Systems (CDS)" of Otto von Guericke University Magdeburg.

References

(Dengale et al., 2016). Dengale, S.J., Grohganz, H., Rades, T., Löbmann, K., 2016. Recent advances in co-amorphous drug formulations. Adv. Drug Delivery Rev. 100,116-125.

(Gänsch et al., 2021). Gänsch, J., Huskova, N., Kerst, K., Temmel, E., Lorenz, H., Mangold, M., Janiga, G., Seidel-Morgenstern, A., 2021. Continuous enantioselective crystallization of chiral compounds in coupled fluidized beds. Chem. Eng. J. 422.Gooding, W. B., Pekny, J. F., McCroskey, P. S. (1994). Enumerative Approaches to Parallel Flowshop Scheduling via Problem Transformation. *Comput. Chem. Eng.*, *18*, 909.

(Huskova, 2022). Huskova, N. 2022. Dynamic Modeling and Optimization of a Continuous Fluidized Bed Process for the Separation of Enantiomers by Preferential Crystallization. PhD thesis. Otto von Guericke University, Magdeburg.

(McConathy and Owens, 2003). McConathy, J., Owens, M., 2003. Stereochemistry in Drug Action. J. Clin. Psychiatry 2, 70-73.

Antonis Kokossis, Michael C. Georgiadis, Efstratios N. Pistikopoulos (Eds.)
PROCEEDINGS OF THE 33rd European Symposium on Computer Aided Process Engineering (ESCAPE33), June 18-21, 2023, Athens, Greece
 http://dx.doi.org/10.1016/B978-0-443-15274-0.50222-5

Machine-learning-based optimization of operating conditions of naphtha cracking furnace to maximize plant profit

Chonghyo Joo,[a,b] Hyukwon Kwon,[a,b] Junghwan Kim[a,b], Hyungtae Cho[a], and Jaewon Lee[a,*]

[a]*Green Materials and Processes R&D Group, Korea Institute of Industrial Technology, 55, Jongga-ro, Ulsan 44413, Korea*

[b]*Department of Chemical and Biomolecular Engineering, Yonsei University, 50, Yonsei-ro, Seoul 03722, Korea*

j.lee@kitech.re.kr

Abstract

Naphtha cracking is the primary process for propylene (PL) and ethylene (EL) production and depends on the operating conditions, such as the coil outlet temperature and feedstock composition. The product yields, in turn, determine the profit of the naphtha cracking plant. Therefore, the operating conditions should be optimized to maximize plant profit. However, it is challenging to optimize the conditions using conventional simulation methods since a high-fidelity model is hard to develop owing to the nonlinearity and complexity of the cracking process. In this study, we used machine learning to optimize the operating conditions of a naphtha cracking furnace to maximize plant profit. First, a data-driven model was developed to predict the product yields for different operating conditions using a deep neural network (DNN). The model could predict the PL and EL yields with high accuracy ($R^2 = 0.965$ and 0.900, respectively). Next, a genetic algorithm was used for optimization based on the developed DNN model. Finally, the developed model was used with real-world plant data and product prices for 2020. The plant profit under the optimized operating conditions was 30% higher than that corresponding to the original operating conditions. Thus, the proposed method is suitable for determining the optimal operating conditions of various types of plants in order to maximize profit.

Keywords: naphtha cracking furnace, optimization, machine learning, genetic algorithm

1. Introduction

Growth in the petrochemical industry has increased the demand for ethylene (EL) and propylene (PL), which are the elementary materials for various petrochemical products. To meet this demand, EL and PL are generally produced in a naphtha cracking center (NCC), which breaks down naphtha into light and valuable hydrocarbons.

The NCC consists of four main units: a cracking furnace and quenching, compression, and distillation units, as shown in Figure 1. In the first unit, that is, the cracking furnace, naphtha is passed through the coils of the furnace as the feedstock, where it receives thermal energy for the cracking reaction from the furnace. As a result, it undergoes cracking into light hydrocarbons. In the second unit, that is, the quenching unit, the light hydrocarbons are quenched to prevent the formation of coke and tar in the hydrocarbon mixture and recover the waste heat. In the third unit, that is, the compression unit, the light hydrocarbons receive energy for transportation. Moreover, catalyst poisons and impurities, such as H_2S and CO_2, are removed in this unit. Lastly, in the distillation unit, the light hydrocarbons are fractionated into different products, such as EL and PL.

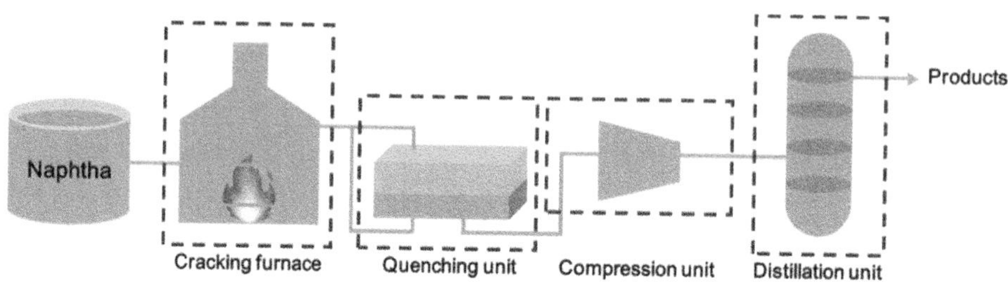

Figure 1. Illustration of NCC for EL and PL production

Of the various units that form the NCC, the cracking furnace has the greatest impact on the product yields, which determine the profit of the naphtha cracking plant because the cracking reaction occurs in the furnace. Figure 2 describes the cracking reaction in the furnace. First, naphtha is preheated before being mixed with diluting steam. After being preheated in the convection section, this mixture is transported to the coil in the radiant section and broken into EL and PL via the cracking reaction. The products are subsequently transported to the quenching unit through the coil outlet.

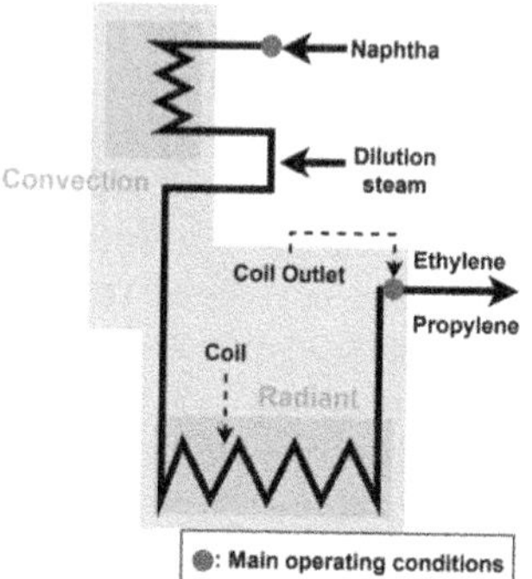

Figure 2. Schematic of cracking furnace showing main operating conditions: naphtha composition and COT

The yields of EL and PL depend on the operating conditions, that is, the naphtha composition and coil outlet temperature (COT). Thus, the operating conditions of the cracking furnace must be optimized to maximize the profit of the naphtha cracking plant. However, it is challenging to optimize the conditions using conventional simulation methods since a high-fidelity model is hard develop owing to the nonlinearity and complexity of the cracking process.

In this study, we use machine learning to optimize the operating conditions in a naphtha cracking furnace in order to maximize the plant profit. Figure 3 provides an overview of the study. First, 784 actual industrial operating data points were collected from an NCC plant. In this dataset, the naphtha composition, which included 25 components, and the

COT are used as the input variables, while the EL and PL yields are used as the output variables for the developed model. Next, a deep neural network (DNN)-based model is developed using the data. The developed model is evaluated based on the R^2 values before optimization. Finally, the operating conditions are optimized for profit maximization using a genetic algorithm. For model evaluation, the optimal conditions and the corresponding profit are compared with the actual plant data for 2020.

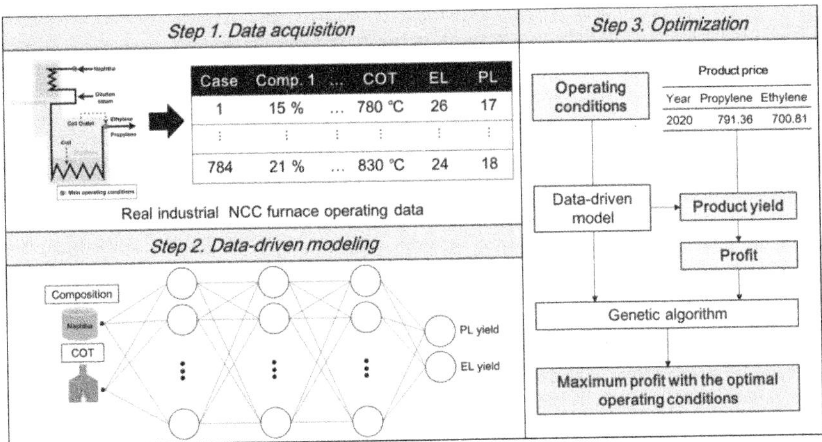

Figure 3. Overview of study

2. Method

2.1. Deep neural networks (DNNs)

DNNs have been used widely for data-driven modeling. A DNN consists of layers, including nodes and edges, that contain mathematical relationships. During data training, these relationships are updated by backpropagation. After training, the updated relationships are used as the equations for predicting the output variables based on the input variables. Therefore, a significant advantage of DNNs is that they can express the relationships that exist in a system regardless of the nonlinearity and complexity of the system.

Figure 4 shows the structure of the DNN-based model used in this study. The model consists of one input layer, three hidden layers, and one output layer, and these layers include 26 nodes, 100 nodes, and two nodes, respectively. This means that the 26 input variables, namely, the composition of the 25 naphtha components and the COT, were used to calculate the two output variables, namely, the EL and PL yields, while considering 300 functions.

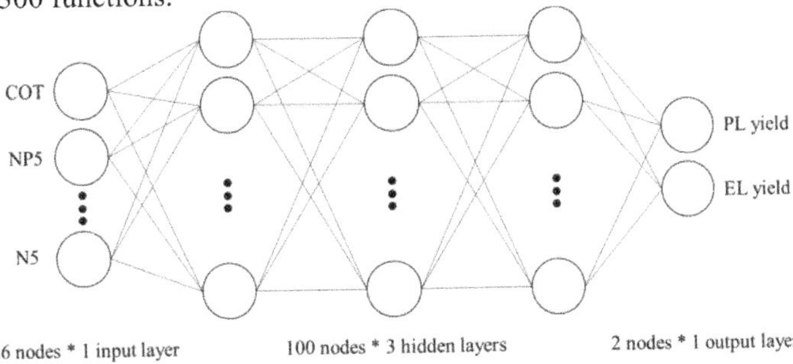

Figure 4. Structure of DNN-based model used in this study

2.2. Genetic algorithm

A genetic algorithm (GA) is a heuristic search algorithm inspired by the biological evolution process. GAs repeat three steps for optimization, as shown in Figure 5: selection, crossover, and mutation. In the selection step, a few data points, called parents, are selected randomly and contribute to the population of the next generation. Here, "Obj_n" is the result of the objective function of the nth data point. In the crossover step, two parents are combined to form the offspring for the next generation by switching some of the input variables. Lastly, the random changes in the input variables are applied to obtain different offspring. By repeating these steps, "Obj_n" is enhanced gradually, and the optimal data point is suggested eventually.

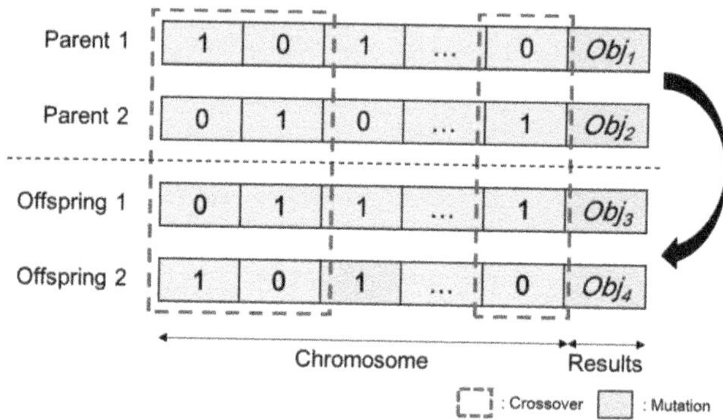

Figure 5. Main steps of GA: selection, crossover, and mutation

This study aimed to optimize the operating conditions of NCC furnaces for profit maximization. Thus, the objective function was defined as shown in Eq. (1), and a GA was employed to solve the optimization problem.

Objective function

$$\textit{maximize Profit} = max\ (flowrate \times \frac{yield_{EL} \times Price_{EL} + yield_{PL} \times Price_{PL}}{yield_{total}}) \qquad (1)$$

where $flowrate$ is assumed to be 29 ton/h in the steady state, and $yield_{EL}$, $yield_{PL}$, and $yield_{total}$ indicate the yields of EL, PL, and the total product, respectively. $Price_{EL}$ and $Price_{PL}$ are the product prices in 2020, as shown in Table 1.

Table 1. Prices of primary naphtha cracking products in 2020

Year	Propylene	Ethylene
2020	791.36 $/ton	700.81 $/ton

3. Results and discussion

3.1. Model development

To develop the data-driven model, the 784 industrial data points was randomly split into 60% training, 20% validation, and 20% test datasets. In other words, 470 data points were used for model training, 157 data points are used for model validation, and the rest (157 data points) were used for model testing.

After the model had been developed using the training and validation datasets, its performance was evaluated using the test dataset based on the coefficient of determination (R^2), which is a representative evaluation criterion for regression. R^2 was calculated using Eq. (2):

$$R^2 = 1 - \frac{\sum_{i=1}^{N}(S_{ia} - S_{ip})^2}{\sum_{i=1}^{N}(S_{ia} - \bar{S}_{ip})^2} \quad (2)$$

where N indicates the number of data points, S_{ia} indicates the ith actual value, and S_{ip} indicates the ith value predicted by the model.

The results of the model performance evaluation are shown in Figure 6. As can be seen, even though the data point within the red box shows a relatively large error, the DNN-based model predicted the EL and PL yields with high R^2 values (0.965 and 0.900, respectively). Moreover, the figure shows that the more EL the NCC furnace produces, the less the amount of PL produced. This is because it is likely that PL is decomposed into EL owing to the high COT or a lack of heavy components in naphtha. Therefore, the model showed good prediction performance and confirmed that the product yields could be changed by varying the cracking reaction conditions.

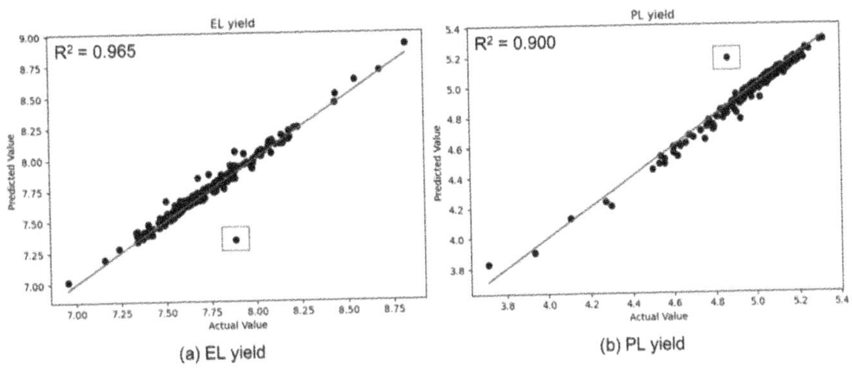

Figure 6. Parity plots and R^2 values for (a) EL yield and (b) PL yield

3.2. Profit optimization

Using the DNN-based model and the GA, the operating conditions, that is, the naphtha composition corresponding to 25 components and the COT, were optimized to maximize the profit of the NCC plant. In addition, the base case, that is, the actual operating data for 2020, were compared with the optimization results for validation.

Table 2 shows the operating conditions and the profits for the base and optimal cases; the 25 components are encoded as "Comp.N" for reasons of confidentiality. The optimal case corresponds to a profit of 12,202 \$/h, which is 30% higher than that for the base case, which is 9,382 \$/h. Moreover, for the optimal case, COT is 42 °C higher than that for the base case, and some of the naphtha components are very different from those for the base case. This means that a high COT will increase the profit because it accelerates the cracking reaction. Furthermore, the use of the appropriate naphtha composition also increases the profit. Thus, the blending process can be considered to increase the profit.

It should be noted that the optimization process only considered the product prices. However, other factors, such as the fuel price, energy consumption, and coke formation, should also be considered in the future to obtain a more practical model.

Table 2. Operating conditions and profits for base and optimal cases

Case	COT (°C)	Comp.1	Comp.2	Comp.3	Comp.4	Comp.5 (wt%)	Comp.6	Comp.7	Comp.8
Base	808	3.20	8.09	2.28	0.01	0.00	2.86	0.21	0.00
Optimal	850	2.79	6.70	0.76	2.02	0.55	1.28	0.03	0.00
	Comp.9	Comp.10	Comp.11	Comp.12	Comp.13 (wt%)	Comp.14	Comp.15	Comp.16	Comp.17
Base	0.00	0.00	0.13	5.12	22.20	11.97	1.06	0.00	0.00
Optimal	0.00	0.00	1.93	7.84	15.87	9.14	3.87	2.55	1.10
	Comp.18	Comp.19	Comp.20	Comp.21 (wt%)	Comp.22	Comp.23	Comp.24	Comp.25	**Profit** ($/h)
Base	0.00	0.51	22.34	16.86	3.13	0.01	0.01	0.00	9,382.01
Optimal	0.29	2.32	19.20	11.47	5.56	3.06	0.88	0.81	12,202.90

4. Conclusions

The profit of NCC plants depends on the operating conditions of the cracking furnace, composition of the 25 naphtha components, and COT. Thus, these conditions must be optimized to maximize the profit. However, it is challenging to optimize the conditions because developing a high-fidelity model for optimization is difficult owing to the nonlinearity and complexity of the cracking process. In this study, we used machine learning to optimize the operating conditions in a naphtha cracking furnace to maximize the profit. First, a DNN-based model was developed using 784 industrial operating data points. The model could predict the EL and PL yields with high R^2 values (0.965 and 0.900, respectively).Next, the model was integrated with a GA for optimization. The profit under the optimal operating conditions was 30% higher than that for the base case.

Despite the success in profit maximization, this study has a limitation in that it only considers the product prices for profit calculation. Thus, in the future, the model should be improved by considering the other factors that also affect the profit. The updated model should yield more practical solutions for maximizing the profit.

Acknowledgments

This study has been conducted with the support of the Korea Institute of Industrial Technology as “Development and application of carbon-neutral engineering platform based on carbon emission database and prediction model (kitech KM-22-0348)”.

References

E. Joo, K. Lee, M. Lee, S. Park, CRACKER — a PC based simulator for industrial cracking furnaces. Comput Chem Eng 2000;24:1523–8. https://doi.org/10.1016/S0098-1354(00)00558-5.

M. Masoumi, M. Shahrokhi, M. Sadrameli, J. Towfighi, Modeling and control of a naphtha thermal cracking pilot plant. Ind Eng Chem Res 2006;45:3574–82. https://doi.org/10.1021/ie050630f.Y. Brown, Year, Article or Chapter Title, etc.

Global ethylene demand & capacity 2015-2022 | Statista,(n.d.).https://www.statista.com/statistics/1246694/ethylene-demand-capacity-forecast-worldwide/ (accessed November 23, 2022).

Antonis Kokossis, Michael C. Georgiadis, Efstratios N. Pistikopoulos (Eds.)
PROCEEDINGS OF THE 33rd European Symposium on Computer Aided Process Engineering (ESCAPE33), June 18-21, 2023, Athens, Greece
 http://dx.doi.org/10.1016/B978-0-443-15274-0.50223-7

Methanol synthesis and hematite reduction using waste tires

Athi-enkosi Mavukwana,[a] Celestin Baraka Sempuga,[b]

[a]*Department of Civil and Chemical Engineering, College of Science, Engineering, and Technology, University of South Africa (UNISA), c/o Christiaan de Wet & Pioneer Avenue, Florida Campus 1710, Johannesburg, South Africa*

[b]*Institute for the Development of Energy for African Sustainability, College of Science, Engineering, and Technology, University of South Africa (UNISA), c/o Christiaan de Wet & Pioneer Avenue, Florida Campus 1710, Johannesburg, South Africa*

Abstract

Iron production and Steelmaking are very energy-intensive processes with an estimated global energy intensity of 16.5 GJ/ton and global CO_2 emissions of 1.8 tons per ton of steel produced. Key solutions are required to optimize this industry but maintain its output or production. This work proposes to supplant the coal used in steelmaking with waste tires to reduce operational costs and CO_2 emissions through the polygeneration of products. Process synthesis techniques are used to determine the fundamental thermodynamic limitations of such a process through process performance targeting. It is found that one ton of iron and 430 kg of methanol can be potentially produced from 1.42 tons of haematite and 479 kg of waste tires. However, 63% of the carbon resource will end up as carbon dioxide, equivalent to 985 kg/ton of iron produced. This is lower than the coal route which releases about 1.05 CO_2 tons/ ton of iron.

Keywords: Waste tires, Reduction, Methanol, Iron, Thermodynamics.

1. Introduction

Global energy needs are constantly increasing and as a result, the natural resources that sustain these energy demands are becoming scarcer and more expensive, whilst human-generated emissions and other solid waste are increasing to critical levels. All companies are now under pressure from societies to decarbonize or reduce emissions from their processes. One such process industry is iron ore reduction and steelmaking. Iron production and Steelmaking are very energy-intensive processes with an estimated global energy intensity of 16.5 GJ/ton [1]. The whole process is dependent on natural resources for both energy needs and products. This results in global CO_2 emissions of 1.8 tons per ton of steel produced [2]. Key solutions are required to optimize this industry but maintain its output or production. The major contributor to carbon emissions from the industry is the use of coal. This work proposes to reduce the coal used in steelmaking with waste tires. Waste tires are derived from fossil fuels and currently at their end-of-life tires have no fundamental use and end up being dumped in landfills creating an unnecessary environmental problem. For example, South Africa produces an estimated 250 000 tonnes of waste tires per annum adding to an existing stockpile excess of 900 000 tonnes spread across 26 national storage depots without a solid plan for reprocessing or recycling [3]. Tires possess a higher volatile content, higher heating value, and low ash content than coal [4], and can be used to replace or reduce coal in thermochemical conversion processes. This paper evaluates and determines theoretical performance targets for combined methanol synthesis and iron ore reduction using waste tires as a carbon source. The main constituents of tires are carbon and hydrogen and thus the reduction process simply involves reacting the carbon and hydrogen in waste tires with the oxygen in the iron ore to produce iron in its reduced form and carbon dioxide and water. However, in this work, we explore co-production opportunities to improve the efficiency of the process by reducing CO_2 and H_2O production and thereby improving the material and energy efficiency of the process.

2. Methodology

Two thermodynamic properties, i.e., the enthalpy (ΔH) and the Gibbs free energy (ΔG) are used for process targeting. Where, ΔH is the amount of energy requirement (input or output) across the process boundary, and ΔG is the quality of the energy expressed in terms of the equivalent amount of mechanical work required to effect the change from feed to the product [5]. We use a systematic approach to identify feasible process targets by first defining the attainable region of the process and systematically identifying targets of interest subject to material and energy balances and thermodynamic constraints. The feasibility of the process can be analyzed in terms of their heat and work requirements. The target mass balance for all processes discussed here is met by setting two limits; the process must be adiabatic (ΔH = 0), and the mass balance is based on one mole of a tire. The utilization of raw materials, waste generation, and the thermodynamic efficiency of the process will be assessed using the following three metrics [6].

$$Carbon\ Efficiency = \frac{moles\ of\ C\ in\ the\ desired\ product}{moles\ of\ C\ in\ the\ feed} \quad (1)$$

$$Atom\ Economy = \frac{mass\ of\ desired\ product}{Total\ mass\ of\ feed} \quad (2)$$

$$E-factor = \frac{Mass\ of\ waste\ produced}{Mass\ of\ desired\ product} \quad (3)$$

Where equation 2 quantifies how much of the carbon in the feed stream ends up in the desired products. Any carbon that is not contained within a product will be lost in undesired by-products. Equation 3 provides useful information on the utilization of all the feed and waste generation. Table 1 shows the ultimate and proximate analysis of waste tires and coal used in this study.

Table 1: Ultimate and proximate analysis [7].

Ultimate Analysis wt.%			Proximate Analysis wt.%			
	Tire	**Coal**		**Tire**	**Coal**	
C	85.05	67.69	Moisture	1.14	7.76	
H	6.79	4.59	Fixed Carbon	32.28	47.14	
N	0.5	1.13	Volatile Matter	62.24	34.05	
O	1.75	5.48	Ash	4.35	11.05	
S	1.53	2.3	LHV	34.9	30	MJ/kg

3. Results and discussion

3.1. Process targeting

Let us consider the case where waste tires, are used for the direct reduction of iron ore. The material balance is based on one mole of a tire. For simplicity of theoretical calculation, the S and N in the tire feed are assumed to be transferred only to H_2S and NH_3. Equation 4 shows the mass balance for iron ore reduction.

$$CH_{0.95}O_{0.02}N_{0.005}S_{0.007} + 0.814Fe_2O_3 = 1.628\,Fe + CO2 + 0.4605\,H2O + 0{,}007H_2S + 0{,}005NH_3 \quad (4)$$

$$\Delta H = 223.393\,kJ, \qquad \Delta G = 150.680\,kJ$$

Equation 4 shows that the overall process requires energy ($\Delta H > 0$ and $\Delta G > 0$) for it to be feasible. A positive ΔG across a process indicates the amount of energy equivalent to mechanical work that must be supplied for the process to be feasible, while a positive ΔH indicates the total amount of energy to be supplied in the form of heat or work. An additional feed of oxygen must be added to combust some of the tire feed for heat. The addition of oxygen must meet the target of $\Delta H = 0$ and one-mole tire. As shown in figure 1, one mole of the tire, can reduce 0.542 mols of iron ore and produce 1.084 mols of Fe. The process is feasible with excess work, but all the carbon available in the feed is transferred to CO_2. Thus, the carbon efficiency of the process is zero. This corresponds to about 730 kg of CO_2 released to the environment per ton of iron produced during this process. When compared to coal, tires marginally perform better in terms of iron production and carbon dioxide emissions. Figure 2 shows that for one mole of coal, the iron production decreases by 5% and the carbon dioxide emissions increase to 763 kg CO_2 per ton of iron produced. The Atom efficiency as well as the E-factor for the waste tire process is 54% and 0.87 compared to coal which is 53% and 0.89. This indicates the ability of waste tires to replace coal. However, this is not a solution for the industry if tires can only reduce carbon emissions by 5%. An alternative route is required where the carbon emissions per ton of iron are reduced as an incentive to transition to a waste tire feed process. One way is to consider a co-production of methanol and iron using waste tires. In this way, not all the carbon feed in the tire is used for reduction, but an indirect route is utilized where the tires are converted to syngas, and the syngas is used for the reduction and the excess syngas is used to produce methanol in a downstream process unit.

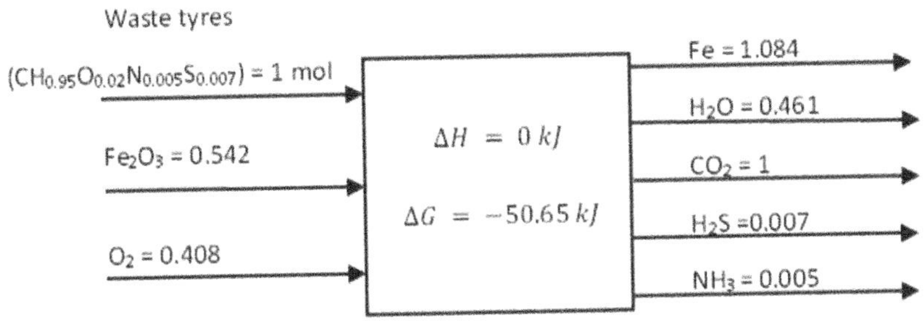

Figure 1: Reduction of iron ore with waste tires

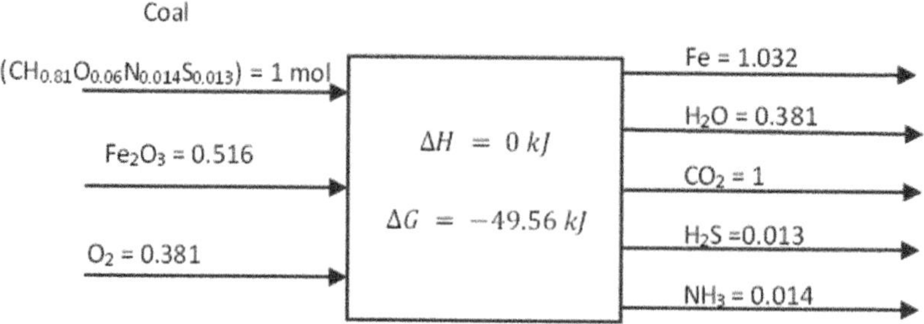

Figure 2: Reduction of iron ore with coal

For 100% raw material conversion efficiency and minimal impact on the environment, the law of conservation of mass implies that the elemental composition of the desired products must match that of the feed. Equation 5 shows the material balance for an integrated methanol synthesis and ironmaking process with 100% raw material conversion efficiency on tire feed. Tires contain a ratio of carbon to hydrogen close to one-to-one, therefore the eventual set of desired products (in combination), regardless of the process, must contain the same ratio of carbon to hydrogen, because any deviation would lead to the formation of unwanted products (i.e., CO_2), thus reducing the process material efficiency. The overall material balance for this process based on one mole of waste to produce methanol and iron ore from using waste tires for reduction is as follows:

$$CH_{0.95}O_{0.02}N_{0.005}S_{0.007} + 0.328Fe_2O_3 = 0.657\,Fe + 0.77CO + 0.23CH_4O + 0{,}007H_2S + 0{,}005NH_3 \quad (5)$$

$$\Delta H = 189.361\ kJ, \qquad \Delta G = 146.83\ kJ$$

Equation 5 shows that under ideal conditions one mole of tire feed can reduce 0.328 moles of iron ore to produce 0.657 moles of iron and 0.23 moles of methanol. However, in equation 5 we see that excess oxygen from iron ore leaves as CO and that the small ratio of hydrogen to carbon in the feed constrains the full conversion of Carbon to methanol. One can immediately see from the above equation that waste tires can never achieve the 2:1 ratio of hydrogen to carbon monoxide required for liquid fuel production. To achieve this ratio an additional amount of feed containing hydrogen often water is required, however often this additional feed comes in with extra oxygen which must leave the system as carbon dioxide. Equally, the system described in equation 5 requires energy input for it to be feasible as indicated by the positive ΔH and ΔG. Thus, additional oxygen can be fed to achieve a zero-energy requirement as shown in eq.6.

$$CH_{0.95}O_{0.02}N_{0.005}S_{0.007} + 0.328Fe_2O_3 + 0.335O_2 = 0.657\,Fe + 0.101CO + 0.669CO_2 + 0.23CH_4O + 0.007H_2S + 0.005NH_3 \quad (6)$$

$$\Delta H = 0\ kJ, \qquad \Delta G = -25.219\ kJ$$

Therefore, supplying the required energy needed by the process in eq.5 by feeding more oxygen generates 0.669 mols of carbon dioxide thus reducing the carbon efficiency to just 33% and the chemical potential efficiency decreases from 86% to 70%. In the process described by eq.6, further optimization can be done to convert the excess CO to methanol through the addition of water. The easiest route is to burn an additional number of tires and use water to increase the ratio of hydrogen to carbon. However, this approach also leads to increased CO_2 emissions. Thus, the overall material balance for a hypothetical fully integrated and adiabatic process to produce methanol and iron is shown in eq.7 and figure 3:

$$CH_{0.95}O_{0.02}N_{0.005}S_{0.007} + 0.252Fe_2O_3 + 0.285O_2 + 0.287H_2O = 0.504\,Fe + 0.627CO_2 + 0.373CH_4O + 0.007H_2S + 0.005NH_3 \tag{7}$$

$$\Delta H = 0\ kJ, \qquad \Delta G = -9.69\ kJ$$

$$Carbon\ Efficiency = 37.3\%$$
$$Atom\ Economy = 59\%$$
$$E - factor = 0.7$$

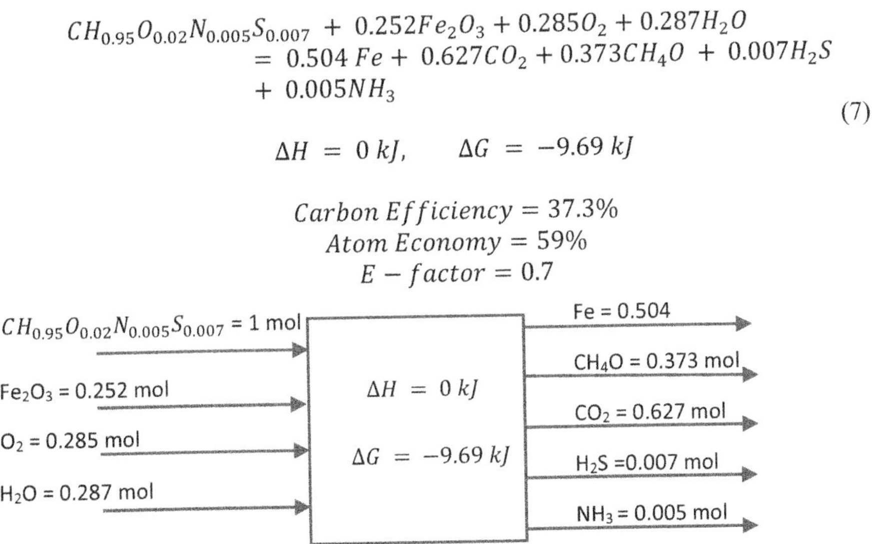

Figure 3: Process diagram for methanol and iron production from waste tires.

Eq. 7 shows that having an additional amount of water increases methanol production by 62%. The carbon efficiency is 37.3%. However, the amount of iron ore reduced decreases thus which decrease elemental iron production by 23%, and by 50% when compared to the direct routes. This, therefore, means that combined iron ore reduction with methanol synthesis with waste tires as carbon feed is achievable. The atom economy is 59% an improvement of about 10% compared to the no-methanol case discussed in figure 1. The E-factor also reduced by 25% suggesting that more feed is transferred to the desired product than left as waste. By equivalence, 1 ton of iron produced consumes 478 kg of waste tires, and 1.42 tons of iron ore and 430 kg of methanol are co-produced. The production of methanol reduces the molar carbon capacity to reduce iron ore as such, the quantity of iron reduces by almost 50% this, therefore, means that per ton of iron, the process produces 980 kg of CO_2. This is still lower than the coal route which is 1.05 tons per ton of iron.

3.2. Effect of temperature

The process described in figure 3 is an ideal process where all units in it operate at Carnot temperatures. The Carnot temperature is the single temperature at which heat can only be used to satisfy the work requirements of a process and such a process will not lose or take additional energy to the environment, in that it will reversible. However, in real processes reaction kinetics and materials of construction limits the use of Carnot temperature which necessitates a deviation from the ideal. The blast furnace temperature is > 1200 °C, whereas the water gas shift and methanol synthesis reactor operate at 450 °C and 220 °C respectively. Operating at a temperature other than the Carnot temperature results in the loss of some of the potential work and the energy released from other process units cannot be integrated directly into the blast furnace due to the temperature gradient. To offset the lost work, more tires are burned. However, this reduces iron production by 15% and increases CO_2 emissions by 6%.

3.3. Revenue potential

A high-level market-related revenue potential analysis is conducted using the production rates in figure 3. The revenue potential of the process is obtained by a difference in the price of selling products and the cost of feed materials. The analysis excludes the capital costs associated with each pathway. During the time of writing this manuscript, the price of pig iron was $ 1000 per ton, methanol was $ 370/ ton, and iron ore was $ 101.5 per ton. The revenue analysis of the different process pathways suggests the polygeneration of methanol and iron is the most profitable. A revenue of $1015 per ton of iron can be achieved for the polygeneration route compared to $855 per ton for the iron-only route. But more capital investment is required for the polygeneration due to the different units required to achieve the required products.

4. Conclusion

The fundamental thermodynamic analysis was used to investigate the performance limits of an idealized process of co-production of methanol and iron from a feed of waste tire and iron ore. The results show that co-production with methanol reduces CO_2 emissions by 37.3% (from 1 mole CO_2 to 0.627 CO_2 for the methanol case) and increases revenue by 15.6% (from 855 $ to 1015$). The purpose of this study is to present a method that takes energy, the environment, and economics into account, that can be used to select an acceptable process pathway. It is intended to give the process engineer a macroscopic perspective of the process and its possibilities rather than to illustrate the kinetics or the reaction path.

References

[1] Gupta, R. C., 2014, *Energy Resources, Its Role and Use in Metallurgical Industries*, Elsevier Ltd.

[2] Holappa, L., 2020, "A General Vision for Reduction of Energy Steel Industry," Metals (Basel)., **10**, p. 1117.

[3] Jenkin, N., 2022, "Industrial Usage Opportunities for Products Derived from Waste Tyres in South Africa," (April 2022), pp. 1–60.

[4] Portofino, S., Casu, S., Iovane, P., Russo, A., Martino, M., Donatelli, A., and Galvagno, S., 2011, "Optimizing H2 Production from Waste Tires via Combined Steam Gasification and Catalytic Reforming," Energy and Fuels, **25**(5), pp. 2232–2241.

[5] Mavukwana, A. E., Fox, J. A., and Sempuga, B. C., 2020, "Waste Tyre to Electricity: Thermodynamics Analysis," J. Environ. Chem. Eng., **8**(4).

[6] Mavukwana, A. E., Stacey, N., Fox, J. A., and Sempuga, B. C., 2021, "Thermodynamic Comparison of Pyrolysis and Gasification of Waste Tyres," J. Environ. Chem. Eng., **9**(2), p. 105163.

[7] Martínez, J. D., Puy, N., Murillo, R., García, T., Navarro, M. V., and Mastral, A. M., 2013, "Waste Tyre Pyrolysis - A Review," Renew. Sustain. Energy Rev., **23**, pp. 179–213.

Antonis Kokossis, Michael C. Georgiadis, Efstratios N. Pistikopoulos (Eds.)
PROCEEDINGS OF THE 33rd European Symposium on Computer Aided Process Engineering (ESCAPE33), June 18-21, 2023, Athens, Greece
 http://dx.doi.org/10.1016/B978-0-443-15274-0.50224-9

Unlocking the potentials of integrating direct air capture with HVAC system

Yasser M. ABDULLATIF *1, 2, Odi Fawwaz Alrebei2, Ahmed Sodiq2, Tareq AL-ANSARI1 , and Abdulkarem I. Amhamed*2
1 College of Science and Engineering, Hamad Bin Khalifa University, Qatar Foundation, Education City, Doha, Qatar.
2Qatar Environment and Energy Institute (QEERI), Doha, Qatar.
*Corresponding author's email: Yabdellatif@hbku.edu.qa and aamhamed@hbku.edu.qa

Abstract

The increasing concentration of carbon dioxide (CO_2) in the atmosphere has compelled researchers and policymakers to seek urgent solutions to address the current global climate change challenges. In order to keep the global mean temperature at approximately 1.5°C above the preindustrial era, the world needs increased deployment of negative emission technologies. Among all the negative emissions technologies reported, direct air capture (DAC) is positioned to deliver the needed CO_2 removal in the atmosphere. DAC technology is independent of the emissions origin, and the capture machine can be located close to the storage or utilization sites or in a location where renewable energy is abundant or where the price of energy is low. Notwithstanding these inherent qualities, DAC technology still has few drawbacks that need to be addressed before the technology can be widely deployed. The main findings point to undeniable facts that DAC overall system energy requirement is high, and it is the main bottleneck in DAC commercialization. The proposal of this study is to unlock the potential of integrating DAC with HVAC systems. DAC with HVAC system integration can be proposed to allow for reducing indoor CO_2 levels to improve indoor air quality (IAQ), HVAC energy reduction using higher air recirculation ratios, a more efficient DAC system by capturing from more elevated CO_2 concentrated streams (indoor air), applying cooler adsorption, and adsorption at different relative humidity points. The expected reduction in the net present value of tons of CO_2 after 50 years is estimated to be 3000\$/tonCO_2 compared to the highest in the literature of 1000 \$/tonCO_2 [1]. This reduction in NPV is estimated based on possible energy recovery by quantifying only one benefit which is HVAC energy reduction using higher air recirculation ratios. Therefore, coupling DAC with HVAC can be economically favorable in comparison to only DAC. In the present work, a comparison between efficiencies of the stand-alone DAC unit and the one integrated with Air Handling Unit are quantitively evaluated based on validated mathematical model. The positions of direct air capture unit within the HVAC are chosen to maximize the possible benefits. The DAC unit efficiencies throughout the hot months in Qatar are improved to be within the range of 21% to 22% compared to the range of 16% to 21% by only positioning the DAC at point 2 in the Air Handling Unit.

Keywords: DAC; green sorbents; HVAC; indoor air quality

1 Introduction

The high energy requirements of DAC lead to high capital and operating costs, which is a major challenge of the existing technologies [2]. As a possible solution, the DAC and HVAC integration was proposed as it bring benefits for both sub-systems. These possible benefits are; reducing indoor CO_2 levels to improve indoor air quality (IAQ), HVAC energy reduction using higher air recirculation ratio and finally, more efficient DAC system by capturing from more elevated CO_2 concentrated streams (indoor air) and applying cooler adsorption. The integration between direct air capture unit and HVAC system is shown in Figure 2.

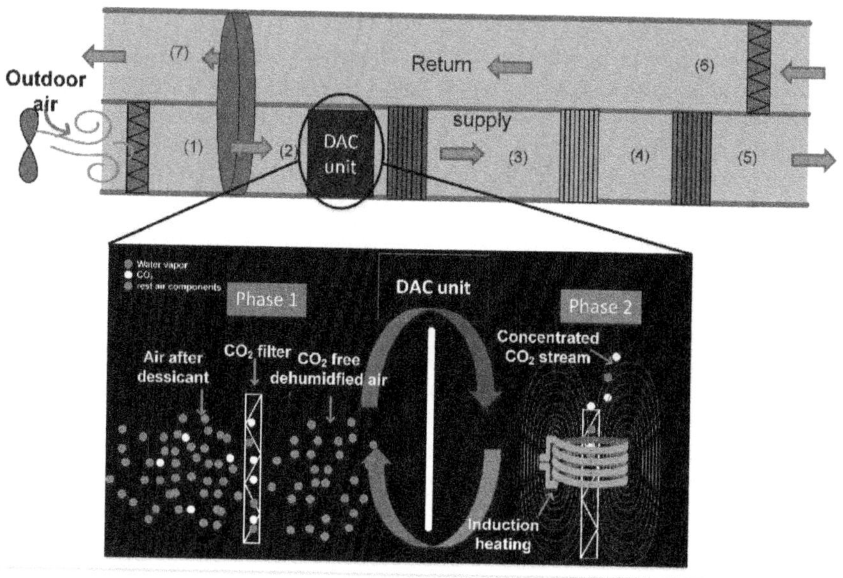

Figure 1: Air handling unit and Direct air capture Integration scheme

The current work includes two subsystems, which are a direct air capture unit and a building air handling unit. Both systems will be thermodynamically simulated for evaluating their performance and energy consumption. These two systems are expected to have better performance when combined, which will be assessed in the present work. The main AHU parameters and setpoints are presented in Table 1. A benchmark direct air capture cycle that uses temperature vacuum swing regeneration will be used in the present study. This cycle was chosen as its model accounts for the role of humidity, it uses commercial adsorbent material, and the model was experimentally validated [3].

Table 1: Air Handling Unit technical data during hot months obtained from Qatar foundation facilities

Parameter	Value
Humidity ratio	Same as point after energy recovery wheel
Relative humidity After precooler (%)	95
Dry Bulb temperature After Cooling coil (° C)	13.2
Relative humidity After Cooling coil (%)	95
Dry Bulb temperature After Reheater (°C)	21.66
Relative humidity After Reheater (%)	56
Air pressure drop through the Cooling coil (Pa)	245
Supply volume flow rate (m^3/h)	39906
Exhaust volume flow rate (m^3/h)	20500
Pressure drop through energy wheel (Pa)	239
Sensible energy wheel effectiveness ε_s	0.40
Latent energy wheel effectiveness ε_L	0.32

1.1 Methodology/Analysis

The main components of air handling unit such as energy recovery wheel, precooler, cooler, heater and fan was simulated based on the preset points in Table 1 using mass balance equation, first and second laws of thermodynamics, properties such as enthalpy, entropy, and specific exergy are determined for each component. The general energy balance equation for each component is described in [4].

$$\dot{m}_i \times h_i + \dot{Q}_i + \dot{W}_i = \dot{m}_o \times h_o + \dot{Q}_o + \dot{W}_o \quad (1)$$

The sensible and latent energy recovery wheel effectiveness ε_s , ε_l were used to calculate temperature and humidity after the energy wheel [5].

$$\varepsilon_s = \frac{\dot{m}Cp}{(\dot{m}Cp)_{max}} \frac{T_{s,o} - T_{s,i}}{T_{s,i} - T_{e,i}} \quad (2)$$

$$\varepsilon_t = \frac{\dot{m}Cp}{(\dot{m}Cp)_{max}} \frac{\omega_{s,o} - \omega_{s,i}}{\omega_{s,i} - \omega_{e,i}} \quad (3)$$

Where $\dot{m}$ is air mass flow rate in (kg/s), h is the enthalpy in (J/(kg K), $\dot{Q}$ is heat in kW, $\dot{W}$ is the power in kW, Cp is the air specific heat in (J/(kg K), T is the temperature in (K) and ω is the humidity ratio in (g/kg). The subscripts s, e, i and o stand for supply, exhaust, inlet, and outlet respectively.

The adsorbent isotherm was simulated based on mechanistic co-adsorption model, which accounts for the humidity effect on CO_2 adsorption and was validated by experimental work [3]. The energy required for regeneration was calculated from work in reference [6]. The performance of DAC unit in different position within AHU was evaluated. The proposed positions are stand-alone DAC and at state points 2, 4, 5, and 6 which were chosen to minimize adsorption temperature and humidity while maximize achieving other benefits such as better IAQ and less HVAC cooling load. To assess the performance of DAC, the second-law efficiency of TVSA cycle-based DAC system was calculated as follow [7]:

$$\eta_{TVSA} = \frac{Wmin}{W_{pump}+Q_H\left(1-\frac{T_0}{T_H}\right)-Q_L\left(1-\frac{T_0}{T_L}\right)} \quad (4)$$

η_{TVSA} is the efficiency of DAC unit uses temperature vacuum swing adsorption, Q_H is the heat input per cycle, Q_L is the cooling load required for reaching the cycle adsorption temperature and W_{pump} is the work required for reaching the vacuum pressure, T_0 is the ambient temperature, T_H is the desorption temperature and T_L is the adsorption temperature.

2 Results and discussion

The efficiencies of the direct air capture unit based on different relative humidity points and temperatures were shown in Figure 1. The range of temperatures and relative humidity values were chosen to cover the weather data for Qatar and the points within the air handling unit. The efficiencies in Figure 2 below were calculated for temperature vacuum swing adsorption system and a chemosorption material.

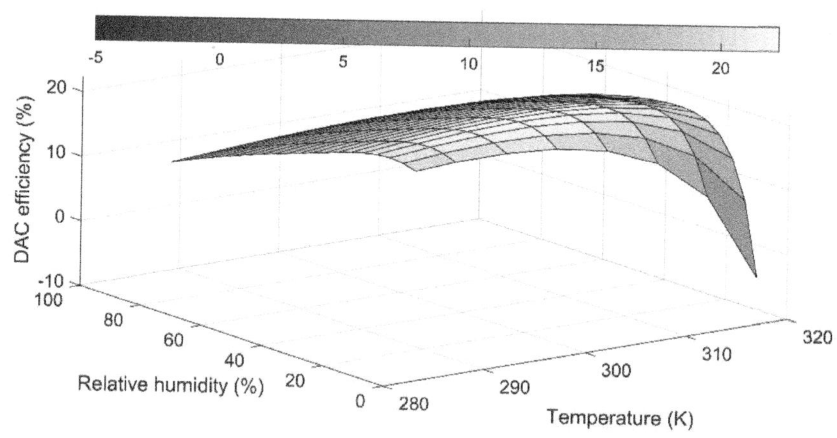

Figure 2: Direct Air Capture efficiency for different temperatures and relative humidity values

Figure 1 shows that the DAC efficiency increases as the adsorption temperature and relative humidity decrease. Based on the mentioned trend, three position of DAC in AHU were chosen and assessed to get a more efficient DAC system.

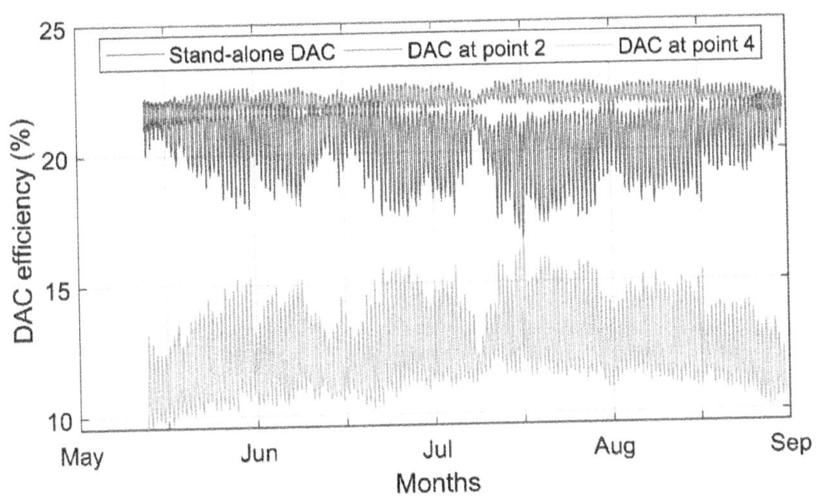

Figure 3: DAC efficiency for different positions in AHU

The efficiencies of three direct air capture units and their variation through the hot months in Qatar were presented in Figure 3. The three positions equally achieve lower indoor CO_2 levels and less HVAC cooling load by allowing air recirculation however, the adsorption conditions such as temperature and relative humidity is different for each position leading to different efficiencies. The calculated efficiency in Figure 3 considers the effect of these adsorption conditions only without considering other benefits. The figure shows that the best position between the three is at point 2. Although the temperature is lowest at point 4, the DAC unit efficiency was low because of the high relative humidity reaching 95%. The efficiency in position 2 was better than the stand-alone DAC unit because in position 2, the temperature and relative humidity is always lower than the ambient temperature during the hot months in Qatar. The analysis shows that the range of efficiencies can be raised from 16%-21% to 21% -22% by only positioning the DAC unit in point 2 in the Air Handling Unit.

3 Conclusion

In conclusion, buildings consume almost 40% of the world's energy demands, and about 50 to 82% of buildings' energy is consumed in the HVAC systems. The necessity to provide fresh indoor air, especially during and after COVID-19, and the rising energy demands, challenge researchers to reduce energy losses in the HVAC systems. The DAC and HVAC integration provides a solution for achieving better indoor air quality, lower HVAC energy consumption and a more efficient DAC system. The present work investigated the effect of positioning the DAC unit within HVAC system and provided a higher efficiency prediction, however, the other benefits need to be quantified and a technoeconomic analysis for the integration should be conducted to compare the integration to the normal DAC system. Moreover, improving the adsorbent material CO_2 uptake and affinity to water could further enhance the performance of the integration.

References

[1] L. Baus, and S. Nehr, "Potentials and limitations of direct air capturing in the built environment," *Building and Environment,* vol. 208, pp. 108629, 2022.

[2] O. S. Board, E. National Academies of Sciences, and Medicine, *Negative emissions technologies and reliable sequestration: A research agenda*: National Academies Press, 2019.

[3] J. Young, E. García-Díez, S. Garcia, and M. van der Spek, "The impact of binary water–CO2 isotherm models on the optimal performance of sorbent-based direct air capture processes," *Energy & Environmental Science,* vol. 14, no. 10, pp. 5377-5394, 2021.

[4] Y. M. Abdullatif, E. C. Okonkwo, Y. Bicer, and T. Al-Ansari, "Thermodynamic analysis of gravity assisted solar-powered reverse osmosis unit for greenhouses situated in a depleted zone," *Case Studies in Thermal Engineering,* vol. 25, pp. 100990, 2021/06/01/, 2021.

[5] A. S. Al-Ghamdi, "ANALYSIS OF AIR-TO-AIR ROTARY ENERGY WHEELS."

[6] J. A. Wurzbacher, C. Gebald, N. Piatkowski, and A. Steinfeld, "Concurrent Separation of CO2 and H2O from Air by a Temperature-Vacuum Swing Adsorption/Desorption Cycle," *Environmental Science & Technology,* vol. 46, no. 16, pp. 9191-9198, 2012/08/21, 2012.

[7] R. Zhao, L. Liu, L. Zhao, S. Deng, S. Li, Y. Zhang, and H. Li, "Thermodynamic exploration of temperature vacuum swing adsorption for direct air capture of carbon dioxide in buildings," *Energy Conversion and Management,* vol. 183, pp. 418-426, 2019.

Antonis Kokossis, Michael C. Georgiadis, Efstratios N. Pistikopoulos (Eds.)
PROCEEDINGS OF THE 33rd European Symposium on Computer Aided Process Engineering (ESCAPE33), June 18-21, 2023, Athens, Greece
 http://dx.doi.org/10.1016/B978-0-443-15274-0.50225-0

Model development of amine scrubbing and CO_2 liquefaction process using waste LNG cold energy

Jonghun Lim[a,b], Yurim Kim[a,b], Yup Yoo[a,b], Hyungtae Cho[a], Jaewon Lee[a], Jinwoo Park[a,c] Junghwan Kim[a,b,*]

[a]*Green Materials and Processes R&D Group, Korea Institute of Industrial Technology, 55, Jonga-ro, Ulsan, 44413, Republic of Korea*

[b]*Department of Chemical and Biomolecular Engineering, Yonsei University, 50, Yensei-ro, Seoul, 03722, Republic of Korea*

[c]*Department of Chemical Engineering, Kongju National University, 1223-24, Cheonan-daero, Seobuk-gu, Cheonan-si, Chungcheongnam-do, 31080, Republic of Korea*

Abstract

Amine scrubbing and CO_2 liquefaction processes are generally employed in most power plants for CO_2 capture and storage (CCS) because they are suitable for large-scale plants. However, a significant amount of cold energy is required to cool the lean amine solution in the amine scrubbing system and liquefy compressed CO_2. Hence, liquefied natural gas (LNG) cold energy, which is generally wasted in seawater during LNG regasification, should be recovered. In this study, we developed an amine scrubbing and CO_2 liquefaction process using waste LNG cold energy. The proposed process model comprises the following three steps: (1) LNG regasification, (2) CO_2 capture and regeneration, and (3) CO_2 liquefaction. First, in LNG regasification, the boil-off gas (BOG) is compressed, and the LNG with compressed BOG is regasified in an LNG vaporizer. To regasify the LNG, seawater is used as a heat source, and the waste cold energy in the chilled seawater is recovered in steps 2 and 3. Second, the CO_2 capture and regeneration process involves the use of an absorber and a regenerator, and the waste cold energy is recovered at the lean amine cooler to reduce the temperature of the lean amine solution. Finally, to liquefy CO_2, cold energy derived from LNG regasification is recovered to a heat exchanger to cool the compressed CO_2. Consequently, the total cold energy consumption can be decreased by 31% by recovering waste LNG cold energy. This increases the overall energy efficiency of the amine scrubbing and CO_2 liquefaction processes for CCS using waste LNG cold energy, thus resulting in less fuel use. We believe that the proposed approach will afford significant economic improvements and environmental protection effects.

Keywords: Amine scrubbing, CO_2 liquefaction, Liquefied natural gas, CO_2 capture and storage

1. Introduction

In many countries, natural gas (NG) emits less air pollutants than other fossil fuels and is considered the energy resource that is developing the most rapidly. NG resources are generally located in specific areas; thus, to ensure the safe long-distance transportation of

NG, it is liquefied by cooling to reduce its volume. However, a significant amount of cold energy is wasted during liquefied natural gas (LNG) regasification; thus, waste LNG cold energy recovery is crucial for improving energy efficiency. Waste LNG cold energy derived from LNG regasification can be used for CO_2 capture and storage (CCS). In general, amine scrubbing and CO_2 liquefaction processes are employed for CCS in LNG power plants. The amine scrubbing system is primarily composed of an absorber and a regenerator, and CO_2 capture depends on the variation in equilibrium constants based on the pKa difference between the amine and CO_2 via temperature control. The CO_2 captured during amine scrubbing is liquefied, which requires a significant amount of cold energy. To control the temperature of the absorber during amine scrubbing and liquefy CO_2, waste LNG cold energy should be recovered. However, most previous studies have only focused on waste hot energy recovery, which is generated from LNG power plants, to reduce the steam energy consumed by the regenerator in the amine scrubbing system. Dual et al. conducted an optimization study to reduce steam energy consumption by a regenerator reboiler. By recovering waste hot energy from an LNG power plant, the energy consumption of the regenerator reboiler was reduced by 56%. Garlapalli et al. proposed a heat recovery unit that directly recovers waste hot energy in flue gas for CCS. Using this unit, the steam energy consumption of the regenerator reboiler can be reduced by 15% to 31%. Despite the significant amount of relevant studies performed hitherto, the following challenges are yet to be addressed. Although waste LNG cold energy can be recovered during amine scrubbing and CO_2 liquefaction, it was not recovered in previous studies. Cold energy constitutes approximately 50% of the total energy required. To increase the overall energy efficiency, waste LNG cold energy must be recovered. Hence, a process model for amine scrubbing and CO_2 liquefaction using waste LNG cold energy is developed in this study. The objective of this study is to overcome the challenge of conventional studies to increase the overall energy efficiency of LNG power plants for CCS by efficiently recovering waste LNG cold energy, which will ultimately result in less fuel use.

2. Methods

This section describes the process model developed for amine scrubbing and CO_2 liquefaction using waste LNG cold energy. Section 2.1 presents an overview of the proposed process, and Section 2.2 presents a detailed description of the process model.

2.1. Overview of proposed process

Figure 1 shows a simplified diagram of the developed amine scrubbing and CO_2 liquefaction process model based on waste LNG cold energy. The proposed process comprises three steps: (1) LNG regasification, (2) amine scrubbing, and (3) CO_2 liquefaction. First, in the LNG regasification process, the transported LNG is assumed to be burned with boil-off gas (BOG) in an LNG power plant. Second, the CO_2 in the flue gas emitted at the combustor of an LNG power plant is captured by the amine scrubbing system. In the amine scrubbing system, waste LNG cold energy is recovered at the lean amine cooler. Finally, the captured CO_2 is liquefied during the CO_2 liquefaction process, and waste LNG cold energy is used to reduce the temperature of the captured CO_2.

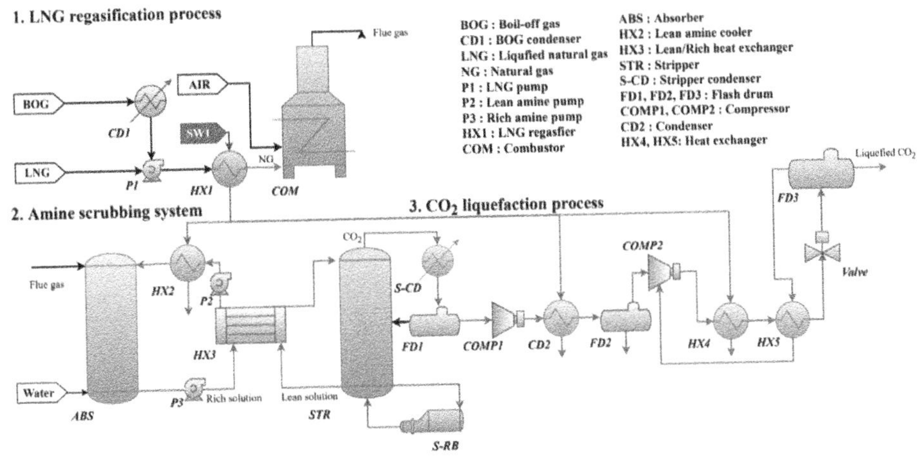

Figure 1. Simplified diagram of proposed process

2.2. Process model

Figure 2 shows the proposed process model for amine scrubbing and CO_2 liquefaction using waste LNG cold energy. To design the proposed model, Aspen Plus V10.0, and ELECNRTL thermodynamic models were employed to consider the liquid–liquid equilibria of amine solution with captured CO_2 and vapor–liquid equilibria of amine solution with flue gas. Subsequently, the following assumptions were considered for model development: (1) The developed process was in a steady state; (2) the temperature of the seawater for the LNG regasifier was 15 °C; (3) the LNG composition was CH_4 (97 wt%), C_2H_6 (1 wt%), C_3H_8 (1 wt%), and N_2 (1 wt%).

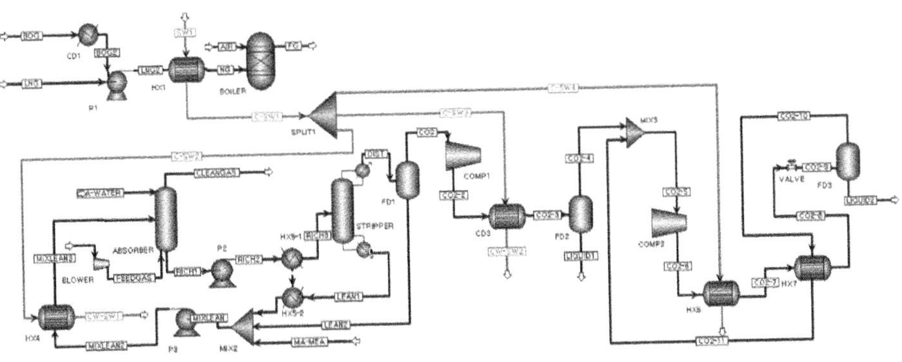
Figure 2. Developed process model of amine scrubbing and CO_2 liquefaction using waste LNG cold energy

2.2.1. LNG regasification

First, to obtain the driving force that can be supplied to an LNG power plant, the transported LNG is input to a high-pressure LNG pump with a BOG. In this pump, the pressure of the LNG is increased from 1 to 60 bar. Subsequently, the compressed LNG is input to an LNG vaporizer, and the temperature of the LNG is increased from -159 °C to

20 °C using seawater as the heat source. During LNG vaporization, a significant amount of LNG cold energy is wasted via dissipation into the sea. Thus, the waste LNG cold energy is recycled to the lean solution cooler (HX2) and each heat exchanger (CD2 and HX4) during LNG liquefaction.

2.2.2. Amine scrubbing system

The CO_2 from the LNG power plant is captured by the amine scrubbing system, which uses monoethanolamine (MEA) as a CO_2 absorbent. An amine scrubbing system is generally composed of an absorber and a regenerator. In the absorber, CO_2 is captured using a lean amine solution via an exothermic reaction (Equations 1–4).

$$2H_2O \leftrightarrow H_3O^+ + OH^- \tag{1}$$
$$HCO_3^- + H_2O \leftrightarrow CO_3^{-2} + H_3O^+ \tag{2}$$
$$MEAH^+ + H_2O \leftrightarrow MEA + H_3O^+ \tag{3}$$
$$CO_2 + OH^- \leftrightarrow HCO_3^- \tag{4}$$
$$MEA + CO_2 + H_2O \leftrightarrow MEACOO^- + H_3O^+ \tag{5}$$

The rich amine solution that captures CO_2 is discharged into the lower section of the absorber and enters the regenerator. Subsequently, it is preheated via the lean/rich heat exchanger (HX3) before entering the regenerator, which reduces the heat load on the regenerator reboiler. Finally, the rich amine solution is regenerated at the regenerator by supplying heat energy from the reboiler.

2.2.3. LNG liquefaction

Finally, the absorbed CO_2 is liquefied, and the liquefied CO_2 requires a significant amount of cold energy for storage. In this study, a Linde dual pressure system that utilizes a cold gas stream, which is generated by controlling a Joule–Thompson valve to cool CO_2, is employed for CCS. First, the captured CO_2 is compressed at a low-pressure compressor (COMP1); subsequently, a high-pressure compressor (COMP2) is used to increase the pressure of the CO_2, which contains CO_2 recycled from the flash drum (FD1). Subsequently, the temperature of the compressed CO_2 is decreased at each heat exchanger (HX4 and HX5) using waste LNG cold energy. Finally, the cooled compressed CO_2 expands through the Joule–Thomson valve (valve).

3. Results and discussion

This section presents the simulation results obtained using the proposed amine scrubbing and CO_2 liquefaction process model, which is based on waste LNG cold energy. The process for the base case involving the non-recovery of waste LNG cold energy from LNG regasification is compared with the proposed process. Table 1 shows the simulation results obtained using the proposed and base processes. Based on the results, the energy consumption of the cooler (HX2), condenser (CD2), and each heat exchanger (HX4 and HX5) was completely substituted by waste LNG cold energy. The total cold energy consumption decreased from 261 to 179 MW; thus, the cold energy saving rate was calculated to be 31% based on the recovery of the waste LNG cold energy. Finally, to conserve waste LNG cold energy, the total energy consumed by the proposed process should be 655 MW, which is 11% lower than that consumed by the base process.

Table 1. Simulation results of proposed and base processes

Classification		Base process	Proposed process	Units
LNG regasification process	CD1 (Condenser)	1.5	1.5	kW
	P1(Pump)	13.7	13.7	MW
CO_2 capture process	HX2 (Lean amine cooler)	13.9	0	MW
	P2 (Pump)	0.5	0.5	MW
	P3 (Pump)	3.8	3.8	kW
	Regenerator reboiler	420	420	MW
	Regenerator condenser	179	179	MW
CO_2 liquefaction process	COMP1 (Compressor)	30	30	MW
	COMP2 (Compressor)	12	12	MW
	CD2 (Condenser)	35	0	MW
	HX4,5(Cooler)	33	0	MW
Total energy consumption		741	659	MW

Figure 3 shows the total energy consumed by the proposed and base processes. By operating the condenser (CD1) and pump (P1), the energy consumed by the base and proposed processes was 14 MW each. In the amine scrubbing system, the energy consumed by the base process was 613 MW. Meanwhile, the energy consumed by the proposed process was calculated by 600 MW. Thus, based on the recovery of waste LNG cold energy to the lean amine cooler (HX2), the energy consumed by the proposed process is 2% lower than that consumed by the base process. The energy consumed during CO_2 liquefaction based on the base and proposed processes was 110 and 42 MW, respectively. Based on the recovery of the waste LNG cold energy to the condenser (CD2) and each heat exchanger (HX4 and HX5), the energy consumed during CO_2 liquefaction based on the proposed process was 61% lower than that consumed by the base process. Finally, the total energy consumed by the proposed process was 11% lower than that consumed by the base process, which indicates a high energy efficiency. Therefore, we believe that this study can provide valuable insights into the efficient recovery of waste LNG cold energy for CCS.

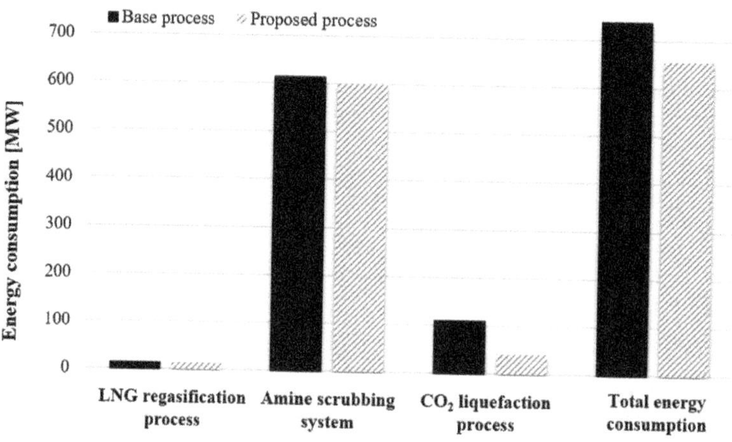

Figure 3. Total energy consumed by proposed and base processes

4. Conclusion

In this study, novel amine scrubbing and CO_2 liquefaction processes using waste LNG cold energy was developed. The main contributions of this study are twofold. First, waste LNG cold energy wasted during LNG regasification was efficiently reused for amine scrubbing and CO_2 liquefaction; thus, a higher overall energy efficiency was obtained. Second, the results showed a higher energy efficiency compared with that afforded by the base process, wherein waste LNG cold energy was not recovered; thus, the proposed processes can result in significant energy and economic improvements. Additionally, the results showed that the energy consumed during CO_2 amine scrubbing and CO_2 liquefaction based on the proposed process was lower than that based on the base process by 2% and 61%, respectively; thus, the total energy consumed by the proposed process was 11% lower than that consumed by the base process. Therefore, we believe that this study can provide valuable insights into the efficient recovery of waste LNG cold energy for CCS.

Acknowledgement

This research was supported by the Korean Institute of Industrial Technology within the framework of the following projects: "Development and application of carbon-neutral engineering platform based on carbon emission database and prediction model [grant number KM-22-0348] and Development of complex parameter smart analysis modules for color customering [grant number EH-22-0011]".

References

J Lim, J Lee , I Moon, H Cho, J Kim, Techno-economic comparison of amine regeneration process with heat-stable amine salt reclaiming units, Energy Science & Engineering, 9, 12, 2529-2543.
J Lim, J Lee, H Cho, J Kim, Model development of amine regeneration process with electrodialysis reclamation unit, Computer Aided Chemical Engineering, 50, 579-584.
Y Kim, J Lim, JY Shim, H Lee, H Cho, J Kim, Optimizing wastewater heat recovery systems in textile dyeing processes using pinch analysis, Applied Thermal Engineering, 214, 118880.

Antonis Kokossis, Michael C. Georgiadis, Efstratios N. Pistikopoulos (Eds.)
PROCEEDINGS OF THE 33rd European Symposium on Computer Aided Process Engineering (ESCAPE33), June 18-21, 2023, Athens, Greece
 http://dx.doi.org/10.1016/B978-0-443-15274-0.50226-2

A parametric analysis of the co-gasification of biomass and plastic waste using Aspen Plus

Nomadlozi L. Khumalo[a], Bilal Patel[a,]

Department of Chemical Engineering, University of South Africa, Corner of Christiaan de Wet Road and Pioneer Avenue, Florida 1709, South Africa

Abstract

The syngas produced from biomass (BM) gasification is typically of low quality and energy content. This inhibits it from being utilized for downstream application for manufacturing of high value products. Plastic waste (PW) is a suitable co-feed since it possesses useful properties, such as high hydrogen content, low oxygen, and low moisture content.

In this study, the effects of blend ratios of (25% PE - 50% PE), and operating condition such as Equivalence Ratio (ER)) on the product gas composition, H_2/CO ratio and the LHV, were evaluated using an oxygen-steam mixture as a gasifying agent. A non-stoichiometric equilibrium model for the co-gasification of BM and PW was developed in Aspen Plus. The co-gasification of BM and PW has led to high Lower Heating Value (LHV) of 9.6 MJ/Nm^3 and H_2/CO ratio of 2 was achieved at low ER values.

Keywords: co- gasification, blend ratio, equivalence ratio; synergistic effect.

1. Introduction

Biomass is an attractive renewable resource, particularly since it is the only renewable source of carbon, that can be used to generate carbon- based fuels and chemicals. Thus, biomass can be utilized to produce a wide variety of products including biofuels, chemicals, bioproducts and energy. There are various technologies to convert biomass, however, among these technologies, gasification is a versatile technology that allows the production of a wide range of products. Gasification process is a thermochemical process that converts carbonaceous material into syngas in the presence of a gasifying agent (s) at high temperatures. The syngas needs to be of good quality and high energy content, so that it can be suitable for downstream applications. However, the syngas produced from biomass gasification has low energy density and low quality, for instance its H_2/CO ratio needs to be adjusted by adding a feedstock with high hydrogen content to increase in quality.

Plastic waste has been identified as a suitable gasification feedstock It is estimated that global plastic usage will triple by 2060, with almost half ending up in landfills and less than a fifth recycled. (OECD, 2022). Therefore, valorization of waste plastic, via gasification, is seen as a viable option for large-scale implementation (Lopez et al. 2018). Plastic waste can compensate for qualities lacking in biomass such as high oxygen content, low energy density. Plastic waste has high volatile matter and is rich in hydrogen and does not contain oxygen (Al Amoodi et al. 2013). When biomass and plastic waste are combined the syngas produced is of improved quality and has high energy density, as compared to that of pure biomass (Singh et al. 2022). The blended feedstocks of biomass and plastic waste may enhance hydrogen content in the syngas, resulting in an increased energy density of the product gas (Huber et al. 2006). During the co-gasification technology tar is reduced in the syngas as the plastic percentage increases in the feedstock mixture (Lopez et al. 2015). The quantity of syngas and the composition (e.g. H_2/CO) ratio improves (Block et al. 2019). Co-gasification of biomass and plastic waste also assists in overcoming difficulties of seasonal biomass availability. Moreover, co-gasification eliminates operational challenges that are encountered during plastic waste gasification.

This study addresses a gap in literature in terms of determining the effect of using oxygen - steam mixture as a gasifying agent instead of single gasifying agent during the co-gasification of biomass (pine sawdust) and plastic waste (Low Density Polyethylene). Previous studies such as Pinto et al. (2002), Burra et al. (2018) and Ahmed et al. (2011), have considered the co-gasification of biomass and plastic waste, typically, using one gasifying agent such as air, steam, or oxygen, but not mixtures of gasifying agents. The effect of feedstock composition (biomass and plastic waste and their mixtures (25% PE – 50% Biomass)) in the presence of oxygen – steam mixture as gasifying agent on the product gas composition (H_2, CO), on the H_2/CO ratio of the syngas, on the Lower Heating Value (LHV) of the syngas, gas yield (GY) and on the cold gas efficiency (CGE). The synergistic interaction between the biomass and plastic waste will be evaluated. This study also aims to determine the effect of the equivalence ratio at a fixed SFR of 0.6 At value of SFR equal to 0.6 is considered optimal for the steam flowrate to enhance the steam reforming reaction and water gas shift reaction and the production of H_2 and CO in the product gas (Arjoloo et al. 2022). Increasing SFR above 0.6, leads to a decrease in process temperature and that affects the syngas quality as well as it reduces the process efficiency.

2. Modelling framework

Aspen Plus was used to develop a non- stoichiometric equilibrium model to determine the effect of feedstock composition and operating conditions such as the equivalence ratio on the product gas composition, H_2/CO ratio of the syngas and the Lower Heating value (LHV) of the syngas. The synergistic interaction between biomass and plastic waste is also evaluated. The ultimate analysis of biomass (sawdust) was determined using an elemental analyser (C; 45.5%, H; 5%, O; 47.1%, N;0.05 % and 2.35% Ash) and the proximate analyses using a thermogravimetric method (18.45%, VM, 79.2% and Ash 2.35%). Data for plastic waste (low-density polyethylene) was sourced from the Al Amoodi et al. (2013).

Various components such as H_2, H_2O, S, N, O_2, CH_4, where specified as conventional, whereas, biomass, ash and plastics were specified as non- conventional species. HCOALGEN and DCOALIGT algorithm were used as enthalpy and density functions for biomass and plastic waste. The Peng-Robinson – Boston – Mathias (PR-BM) thermodynamic model was employed. This property method was selected based on its ability to correlate the low operating pressure with the higher operating temperatures >700°C that are typically used in the co- gasification process (Ramzan et al. 2011). In addition, the process was assumed to operate at steady state, the gasifier is operated at atmospheric pressure, tar and carbon formation is not considered. The model was developed based on three-unit blocks; the first unit is the RYield reactor which represents the decomposition of the non-conventional components into conventional constituents. A FORTRAN statement is utilised within Aspen Plus to accomplish this. The ultimate analyses of the feedstocks is used to determine the yield distribution. The product stream from the RYield reactor enters the second unit, modelled by the RGibbs reactor, uses the minimization of Gibbs energy to predict the product distribution from the gasifier. Gasifying agents such as steam and oxygen were introduced to the gasifier (RGibbs). The equivalence ratio (ER) and steam – to – fuel ratio (SFR) are the operating parameters that controls the amount of oxygen and steam that enter the gasifier. The gasification products from RGibbs reactor (mainly syngas) are sent to the third unit (Cyclone) which separates solids from gaseous product. The validity of the model was assessed using the relative error (ε_r) method. The output values of the product gas obtained from the experimental work (Yu et al. 2014), were compared with the values obtained from Aspen Plus for the co-gasification of biomass and plastic waste. Based on the relative error, the model and experimental values for the syngas composition are in good agreement and are considered acceptable i.e. (the relative errors for each of the variable were within +/- 15%) (De Andres et al. 2019).

3. Results and Discussion

3.1 Effect of the equivalence ratio on H_2 and CO composition in the product gas for various blend ratios

Figure 3.1 (a) shows that an increase in the equivalence ratio (ER) at fixed SFR equal to 0.6 causes the H_2 to increase and reach maximum values and thereafter decrease as ER increases further, for feedstocks that consists of higher plastic percentage such as 50% PE and 100% PE. In Figure 3.1(b) the same pattern is observed for the CO composition for all the blended feedstocks except for biomass feedstock. It is observed that high H_2 and CO composition in the product gas are favoured at low ER values below 0.4. This is because at low ER values the chemical reactions such as the steam reaction, partial oxidation reaction, water – gas shift reaction, and steam reforming reaction and methanation reaction, which are responsible for the production of H_2 and CO are favoured.

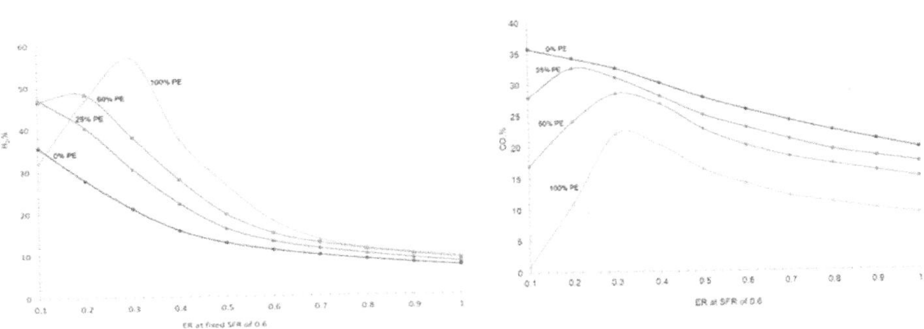

Figure 3.1 Effect of the equivalence ratio (ER) on the (a) H_2 and (b) CO composition in the product gas.

However, at high ER values above 0.4, the combustion reactions are favoured, which are responsible for the production of H_2O and CO_2.An increase in the percentage of plastic waste in the feedstock mixture increases the H_2 composition in the product gas and reduces the CO composition in the product gas. The reason for this is that plastic waste consists of high volatile matter, which enhances the H_2 composition in the syngas but does not consist of oxygen content. Figure 3.1 (a) shows that a H_2 composition of 49% is obtained at ER equal to 0.18 and SFR equal to 0.6, from the blend ratio of (50% PE +50 % Biomass). From the blend ratio of (25% PE + 75% Biomass) a H_2 composition of 46% is attained at ER equal to 0.1 and SFR equal to 0.6. The attainment of a higher H_2 composition in the product gas from the gasification of blended feedstocks indicates that there is a synergistic interaction between the biomass and plastic waste feedstocks. The highest CO composition is attained from the single feedstock (100% Biomass), which shows that in terms of the CO composition in the product gas, the co- gasification of biomass and plastic waste does not have the synergistic effect.

3.2 Effect of the equivalence ratio on the H_2/CO ratio and the Lower Heating Value (LHV) of the syngas.

Figure 3.2 (a) and (b) exhibits that an increase in the equivalence ratio (ER) from 0.1 – 1 at fixed SFR equal to 0.6 causes the H_2/CO ratio and the Lower Heating Value (LHV) of the syngas to decrease, for all the feedstock compositions.

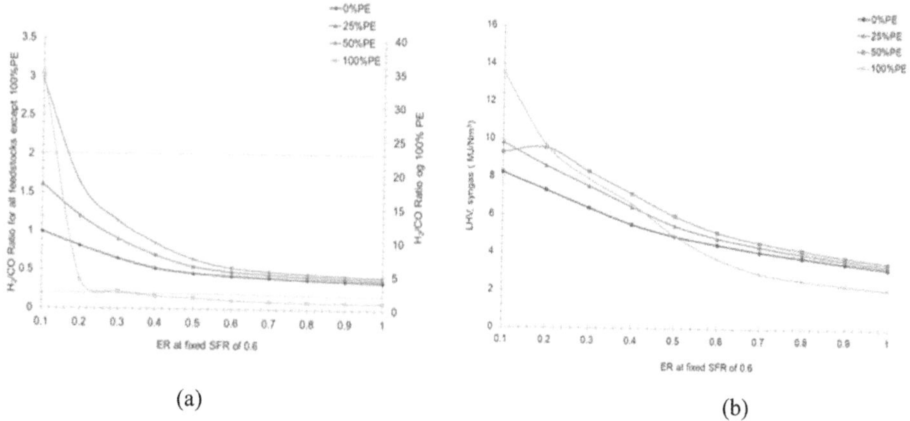

Figure 3.2 Effect of the equivalence ratio (ER) on the (a) H_2/CO ratio of the syngas and (b) Lower Heating Value (LHV) of the syngas.

This is due to the oxidation reactions that are enhanced by the increase in the amount of oxygen that enters the reactor. In Figure 3.2 (a) an increase in the H_2/CO ratio at low ER values and a fixed SFR equal to 0.6, can be attributed to the enhancement of the partial oxidation reaction, and the addition of steam which favours the endothermic reactions such as the steam-methane reforming reactions and the water gas shift reaction, which results in more H_2 being produced, and this increases the H_2/CO ratio of the syngas Moreover, the enhanced H_2/CO ratio of the syngas, is also caused by the addition of plastic waste, which contains high volatile matter and is rich in hydrogen. and the promotion of the water gas shift reaction, are the main factors that causes H_2 to increase while CO decreases, thus leading to an enhanced H_2/CO ratio of the syngas. Figure 3.2 (b) shows that as ER increases the Lower Heating Value (LHV) of the syngas decreases. The high Lower Heating Value (LHV) of the syngas is favoured at low ER values below

0.4. This patten can be attributed to the fact that at low ER values, the production of combustible gases is high, thus increasing the LHV of the syngas. In addition, an increase in plastic waste in the feedstock also contributes towards the increase in the LHV of the syngas, since increasing plastic waste results in a high hydrogen and methane composition in the syngas. Figure 3.2 (a) and (b) shows that an LHV of 9.6 MJ/Nm^3 was obtained from a feedstock with the blend ratio of (50% PE + 50% Biomass). The recommended value of 2 for the H_2/CO ratio of the syngas and the high Lower Heating Value (LHV) of the syngas attained from the blended feedstock in the blend ratio of (50% PE + 50% Biomass) when compared to pure feedstocks indicates that there is a synergistic effect between biomass and plastic waste.

3.3 Effect of the Equivalence ratio on the GY and CGE

Figure 3.3 (a) shows that an increase in the equivalence ratio (ER) at fixed SFR of 0.6, causes the gas yield to increase. This is attributed to the linear relationship between the equivalence ratio and the gasifier temperature. An increase in ER, increases the gasifier temperature.

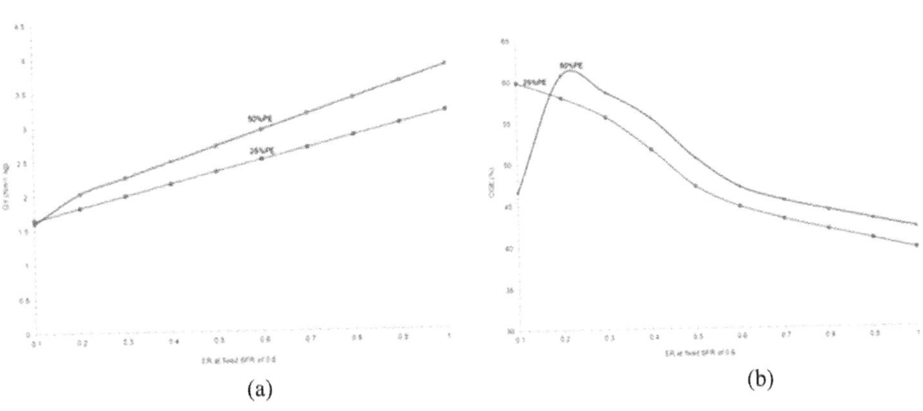

Figure 3.3 Effect of the equivalence ratio (ER) on the (a) gas yield (GY) and (b) cold gas efficiency.

The increase in the gasifier temperature increases the rate of devolatilization reactions, thus enhancing the gas production in the initial pyrolysis stage. Furthermore, the cracking and endothermic char gasification reactions are favoured at high temperatures, these reactions contribute toward the increase of the syngas yield. Figure 3.3 (b) exhibits that at low ER values, the cold gas efficiency is high, however as ER increases above 0.3, the cold gas efficiency decreases. This is because at low ER values the combustible gases such as H_2, CO and CH_4 are high in the product gas, and that increases the LHV of the syngas, which consequently increases the cold gas efficiency. Whereas, at high ER values, there is a decrease in the combustible gases in the product gas, as these are oxidized to form CO_2 and H_2O and thus resulting in a lower LHV of the syngas, which leads to a low cold gas efficiency. The LHV and cold gas efficiency have a linear relationship. Figure 3.3 (a) and (b) shows that increasing plastic waste fraction in the feedstock mixture increases both the gas yield and cold gas efficiency. As a result, high gas yield and high cold gas efficiency were obtained from the blend ratio of (50% PE + 50% Biomass) instead of from the blend ratio of (25% PE + 75% Biomass). This is expected due to the nature of the plastic waste by nature which they contain high volatile matter and no oxygen content. A high cold gas efficiency of 61% is attained at ER equal to 0.24 and SFR equal to 0.6 from the blend ratio of (50% PE + 50% Biomass) (Figure 3.3 (b).

4. Conclusion

The co-gasification of biomass and plastic waste in the presence of an oxygen – steam mixture as gasifying agent has shown that the increase in polyethylene percentage in the blends, increases the H_2, the H_2/CO ratio of the syngas, the Lower Heating Value (LHV) (9.6 MJ/Nm^3) of the syngas, the gas yield (GY) and the cold gas efficiency (GCE) (61%) of the system. This can be attributed to the low fixed carbon of polyethylene materials and that it has no oxygen. The improved outputs are determined from the blend ratio of (50% PE + 50% Biomass) instead of (25% PE + 75% Biomass) or from the biomass or plastic waste. The high yield of syngas composition, quality and energy content are favored at low ER values below 0.4, at fixed SFR equals to 0.6.

References

I.I Ahmed, N. Nipattummakul, A.K. Gupta, (2011), Characteristics of Syngas from Co-gasification of Polyethylene and Wood – Chips. *Applied Energy*, 88,165 – 174.

M. Arjorloo, M. Ghordrat, J. Scott, V. Strezov, (2022), Modelling and Statistical Analysis of Plastic Biomass Mixture Co-gasification, Energy, 256,124638.

N. Al Amoodi, P. Kannan, A. Al Shoaibi, C. Shrinivasakannan, (2013), Aspen Plus Simulation of Polyethylene Gasification under Equilibrium Conditions, Chemical Engineering Communication, 200, 7 – 9.C. Block, A. Ephraim, E. Weiss-Hortala, D. Pham Minh, A. Nzihou, C. Vandecasteele., (2019), Co- pyrogasification of Plastic and Biomass, a Review, Waste and Biomass Valorization, 10, 483 – 509.

K.G. Burra, A.K..Gupta, (2018), Synergistic Effects in Steam Gasification of Combined Biomass and Plastic Waste Mixtures, Applied Energy, 211, 230 – 236.

J. M. De Andres, M. Vedrenne, M. Brambilla, E. Rodriguez, (2019), Modeling and Model Performance in Fluidized Bed Gasifier using Aspen Plus, Journal of Air and Waste Management Association, 6901,23 – 33.

G. Lopez, A. Erkiaga, M. Amutio, J. Bilbao, M. Olazar, (2015), Effect of Polyethylene Co-feeding in the Steam Gasification of Biomass in a Conical Spouted Bed Reactor, Fuel, 393 – 401.

F. Pinto, C. Franco, R.N. Andre, M. Miranda, I. Gulyurtlu, I. Cabrita, (2002), Co-gasification Study of Biomass mixed with Plastic Wastes. *Fuel*, 81,291 – 297.

N. Ramzan, A. Ashraf, S. Naveed, A. Malik, (2011), Simulation of Hybrid Biomass Gasification Using Aspen Plus: A Comparative Performance Analysis for Food, Municipal Solid and Poultry Waste, Biomass and Bioenergy ,35, 3962 – 3969.

M. Singh, S.A. Salaudeen, B.H. Gilroyed, A. Dutta, (2022), Simulation of Biomass – Plastic Co-gasification in a Fluidized Bed Reactor using Aspen Plus, Fuel, 319,123708.

The Organization for Economic Co-operation and Development (OECD), Plastic waste, (2022), https://www.oecd.org/newsroom/global-plastic-waste-set-to-almost-triple-by-2060.htm.

G. W. Huber, S. Iborra, A. Corma, (2006), Synthesis of Transportation Fuels from Biomass: Chemistry, Catalysts and Engineering, Chemical Reviews, 106,4044 -98.

H. Yu., X. Yang., L. Jiang, Chen. D, (2014), Experimental Study on Co-gasification Characteristics of Biomass and Plastic Wastes, Bioresources, 9, 5615 – 5626.

Antonis Kokossis, Michael C. Georgiadis, Efstratios N. Pistikopoulos (Eds.)
PROCEEDINGS OF THE 33rd European Symposium on Computer Aided Process Engineering (ESCAPE33), June 18-21, 2023, Athens, Greece
 http://dx.doi.org/10.1016/B978-0-443-15274-0.50227-4

Food security in an oligopolistic EWF nexus system: a cooperative vs a non-cooperative case

Sarah Namany[a], Maryam Haji[a], Mohammad Alherbawi[a], Tareq Al-Ansari[a]

[a]*College of Science and Engineering, Hamad Bin Khalifa University, Qatar Foundation, Doha, Qatar*

**talansari@hbku.edu.qa*

Abstract

The ever-increasing population growth along with the diverse socio-economic pressures it creates, have led to rising demands for energy, water and food (EWF) resources. This pressure to satisfy the needs for such products has paved the way to competitive behaviors amongst the sectors' main players which might influence the efficient achievement of their objectives. Food security is one critical target that is heavily affected by the dynamics within the EWF nexus sectors. In fact, food production and supply depend primarily on the performance of water and energy systems, which involve diverse stakeholders characterised with both diverging and converging goals. Therefore, there is a need to understand and capture the competitive and collusive behaviors of each nexus sub-system as means to avoid any shortages and enhance the EWF nexus overall efficiency. The purpose of this paper is to optimise the economic and environmental performance of an EWF nexus system supplying a local food basket considering a set of renewable and non-renewable energy technologies and a range of water systems. The model is formulated as a multi-objective integer linear program that minimizes the economic costs and emissions. The model is run under two different scenarios investigating the impact of competition and cooperation on the optimal solution. Competition is modeled using Cournot game and cooperation is formulated as a Cartel, and results of the games are used as constraints restricting the optimisation framework. Results of the study assert that under competition and cooperation the water mix is mostly supplied by GW and TSE with more than 80%, while the energy production is evenly distributed amongst the three technologies. These results are highly influenced by the affordability and environmental emissions and less impacted by the behavior of the market. Findings of this study can be used to quantify the impact stakeholders' behaviors on the achievement of food security targets.

Keywords: EWF nexus, food security, game theory, Cournot, Cartel, optimisation

1. Introduction

Food security is one of the most challenging targets that many nations are striving to achieve and maintain. Food systems are governed by multiple pressures inflicted by the surrounding environment, counting climate change, political instability and market dynamics (Namany & Al-Ansari, 2021). Heavily relying on the energy and water resources to operate, the food sector's efficient functioning is also subject to the challenges faced by the water and energy sectors. Competition between water and energy industries is one of the critical issues that is often neglected when tackling food security. Competition amongst power generators and water producers can lead to fluctuations in the quantities produced and supplied to the food system, which can disturb its operations and productivity, hence threatening the food security target. A proper modeling of competitive and cooperative behaviors is essential to reduce the impact of such volatility on the water and energy supplies to the food sector. In this regard, game theoretic techniques have been widely adopted in literature as a tool to capture competition. These models have demonstrated their efficacy in simulating competitive and collusive behaviors and in identifying the best course of action that results in mutually beneficial outcomes for the contending parties in many domains, including resources sectors (Ho et al., 2021). Considering the food sector, Saghaian and Asgari (2013) examined the non-cooperative import system of pistachios to Japan in the context of a duopolistic market made up of two significant exporting nations. The technique examined quantity competition between American and Iranian pistachio suppliers using Cournot and Stackelberg games. Addressing cooperation within the energy system, oil and gas more specifically, Gausmann (2012) proposed a model to analyze the strategic motivations for cooperation among the three major players in the European gas market, namely Qatar, Algeria, and Russia. A game-theoretic model was developed to determine each agent's interest in gas policy. Since each of these nations has distinct geopolitical goals, they will each look for a certain kind of alliance. The goal of this work is to investigate the impact of competition and cooperation in oligopolistic energy and water markets on the performance of a food system in charge of supplying a diversified food basket. Competition and cooperation between the water and energy technologies are modeled using Cournot and Cartel games and the resulting payoffs are used as constraints for an optimisation framework. The latter consists of a multi-objective integer linear program which seeks to minimize both the economic costs and environmental emissions of a configuration of renewable and non-renewable technologies. This paper is constructed such that the following section 2 presents the scenarios formulation and the mathematical representation of the optimisation model. Then, section 3 describes the main findings of the study.

2. Data and Methods

The methodology developed in this work suggests a framework that assesses the influence of resource sectors' behaviors on the economic and environmental performances of an EWF nexus system delivering a food basket for the country of Qatar. Food groups considered in this study consist of perishable products, counting fruits, vegetables, poultry and meat products (Namany et al., 2019). The energy and water sectors in charge of supplying irrigation water, electricity for cultivation and fertilisers production are represented by a set of renewable and non-renewable technologies that are driven with the need of satisfying the food sector's demand for water and energy while maximising their individual profits. Two different scenarios are investigated wherein the set of energy and water technologies are either competing or cooperating to meet the food demands. Competition is modeled using Cournot game while cooperation is described by a Cartel game. Results of the games are incorporated as constraints in a multi-objective optimisation model that determines the optimal water and energy mix that would supply 40% self-sufficiency while minimising the cost and environmental impacts. The following sections describes the two different scenarios along with the optimisation model formulation.

2.1. Scenarios formulation

2.1.1. Cournot competition

Industries producing the same goods tend to compete as means to maximise their individual economic benefits, either through rising prices or increasing their outputs. In this first scenario investigated, technologies within the water and energy sectors compete between each other in order to maximise the amounts of water or power generated. The competition and the payoff of each competing technology are modeled using the Cournot game. The latter is founded on a set of assumptions entailing that players decide simultaneously and independently on the quantities to be produced and the market price of the goods is set once all players disclose their optimal quantities. In addition, Cournot is only applied when rivals are producing identical products. In this work, the water technologies counting reverse osmosis (RO), multi-stage flash (MSF), treated sewage effluent (TSE) and ground water pumping (GW) compete to produce water, and the power producers including combined-cycle gas turbine (CCGT), photovoltaics (PV) and biomass integrated gasification combined cycle (BIGCC) compete to generate power. The best response function for each water and energy technology is depicted by the following equation (1) where q is the individual Nash equilibrium quantity produced by each player *i*. *c* is the marginal cost of production of either water or energy, as for *a* and *b* are the parameters representing the linear inverse demand function described by equation (2), where *P* is the price and *Q* is the total quantity demanded.

$$q_i^* = \frac{a-c_i}{2b} - \frac{\sum_{i\neq j} q_j}{2} \tag{1}$$

$$P = a - b \times Q \tag{2}$$

2.1.2. Cartel cooperation

In the second scenario, each sector's set of technologies decide to collaborate and form cartels. Cartels occur when competitors decide to collude in order to rise prices and maintain their stability in the market. Players act in a monopoly where the optimal payoffs are distributed based on agreed-upon percentages. In this scenario, the water sector technologies and the power plants form two different cartels to maximise profits and restrict quantities. The optimal quantities produced under each scenario can be described by the following equations (3) and (4), wherein p_i is the pre-determined distribution percentage for each technology, which in this case is based on the current share of the technology in the market. As for j, it represents the index of the remaining technologies.

$$q_i = p_i \times Q \tag{3}$$

$$Q = \frac{\frac{a-c_i}{2b}}{p_i - \frac{\sum p_j}{b}} \tag{4}$$

2.2. Optimisation model formulation

Results of the games, which are mainly payoffs of each participating technology are used as constraints for the optimisation model. The contribution of each technology to the mix is bounded by the level of the expected profit that each techonology would make. If the quantity produced while participating in the mix is less than the Nash equilibrium quantity q_i^*, the technology has no incentive to be part of the mix.

The optimisation model is formulated as a multi-objective linear program aiming to minimise greenhouse gases (GHG) and the total economic costs of a configuration of water and energy technologies supplying a 40% of Qatar's food sector. The optimisation model can be described as follows:

Objective functions:

Minimise: $Env = Q_w \sum_{i=1}^{m} x_i^w e_i^w + Q_e \sum_{i=1}^{n} x_i^e e_i^e$ (5)

Minimise: $Eco = Q_w \sum_{i=1}^{m} x_i^w c_i^w + Q_e \sum_{i=1}^{n} x_i^e c_i^e$ (6)

Constraints:

$Q_w x_i^w \geq q_i^{*w}$ (7)

$Q_e x_i^e \geq q_i^{*e}$ (8)

$\sum_{i=1}^{m} x_i^w = 100\%$ (9)

$\sum_{i=1}^{m} x_i^e = 100\%$ (10)

$x_i^w;\ x_i^e > 0$

Such that:
Env is the total environmental emissions in kg of CO_{2eq};
Eco is the total economic cost in \$;
x_i^w and x_i^{CC} are the decision variables representing the percentage contribution of each water and energy technology, respectively;
Q_w and Q_e are the total quantities of water and energy required by the food sector in m^3 and kWh, respectively;
e_i^w and e_i^e represent the unit carbon footprint associated the water production and energy generation in in kg of CO_{2eq}/m^3 and kg of CO_{2eq}/kWh, respectively (table 1);
c_i^w and c_i^e are the unit costs associated with water production and energy generation in kg of \$/m^3 and kg of \$/kWh , respectively (table 1);
m is the number of water technologies, counting RO, MSF, TSE and GW;
n is the number of energy technologies, counting CCGT, PV and BIGCC;

Table 1. The optimisation model input data

Objective function	CCGT	PV	BIGCC	RO	MSF	GW	TSE
Cost(\$/kWh-\$/m³)	0.093	0.215	0.11	0.80	2.43	0.07	0.11
GWP(kg of CO_{2eq})	9.94E+08	9.83E+04	1.31E+08	1.71E+09	2.02E+09	1.90E+08	4.99E+08

3. Results and discussion

3.1. Results of the games

Under Cournot competition, TSE and GW are allowed to produce larger quantities compared to the desalination technologies, due to their relatively small marginal costs. As for Cartel, the largest quantities are granted to GW, RO and MSF, thanks to their important market shares. Considering the energy technologies, the same pattern is noticed for both scenarios. This means that low marginal costs and high market shares are the factors influencing the quantities produced and hence CCGT is generating the largest quantity of energy and GW is producing the highest amount of water in both scenarios (table 2).

Table 2. Cournot and Cartel quantities results

		Cournot	Cartel
Water Technologies (Mm^3)	RO	42.93	50.01
	MSF	10.68	50.01
	GW	160.68	125.04
	TSE	154.22	25.00
Energy Technologies (MW)	CCGT	2810.00	1751.29
	PV	1450.00	350.25
	BIGCC	2610.00	1401.03

3.2. Results of the optimisation model

Results of the multi-objective optimisation model suggest 5 optimal solutions for both Cournot and Cartel cases. Considering Cournot, the average optimal mix costs around $210M and generates 715,6M kg of CO_{2eq}. The optimal water configuration under this scenario is mainly consisting of GW and TSE with a slight contribution of desalination. This can be explained by the affordability of these technologies and their relatively low environmental impact. As for the energy mix, the three power generators are contributing with almost the same share, with CCGT slightly leading thanks to its low price (Figure 1). Considering the Cartel cooperation scenario, the average optimal solution has an economic cost of $203M and emits around 792M kg of CO_{2eq}. The slight increase in environmental impacts and decrease of the economic costs is due to the slightly higher contribution of CCGT which is affordable, yet environmentally damaging (Figure 2). Under both scenarios the participation of the water and energy technologies is almost the same, showing that the competition has almost no impact on the technology selection in this case study. This can also be explained by the fact that the equilibrium quantities generated when playing the games are profitable for all the studied technologies under both scenarios.

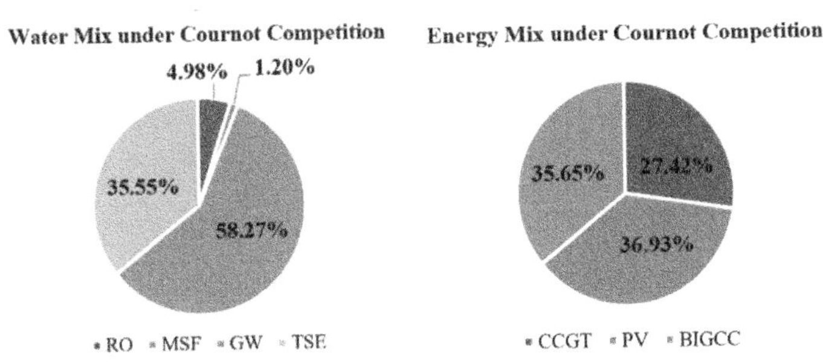

Figure 1. The water and energy configuration under Cournot competition.

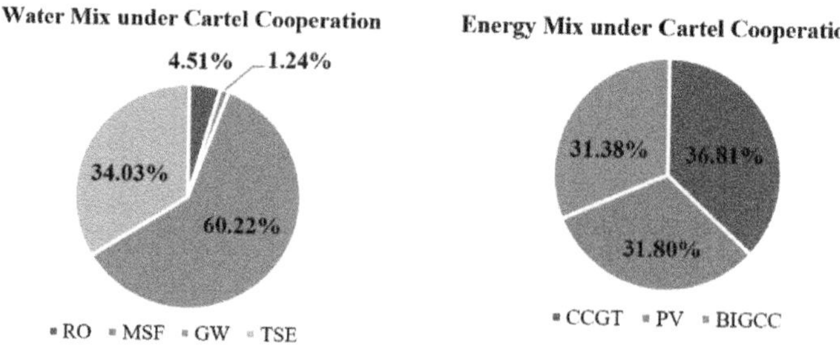

Figure 2. The water and energy configuration under Cartel competition.

4. Conclusion

Food security is a multi-faceted target involving the contribution of multiple sectors and entailing numerous risks. Competition between water technologies and energy technologies is one of the risks that can influence the productivity of the food sector due to the uncertain supplies of resources. In this paper, competitive and cooperative behaviors of these two sectors are modeled by means of Cournot non-cooperative game and Cartel cooperative game and the generated payoffs were used to restrict the contribution of some water and energy technologies in an EWF mix as part of an optimisation framework. Results of the study assert that under competition and cooperation the water mix is mostly supplied by GW and TSE with more than 80%, while the energy production is evenly distributed amongst the three technologies. These results are highly influenced by the affordability and environmental emissions and less impacted by the behavior of the market.

5. Acknowledgment

This research is supported by Hamad Bin Khalifa University (HBKU) and the Qatar National Research Fund (QNRF), members of Qatar Foundation, through the National Priorities Research Program (NPRP) grant No NPRP11S-0107–180216. The findings herein reflect the work, and are solely the responsibility, of the authors.

6. References

Gausmann, J. P. (2012). *A Game-Theoretic Approach to Cooperation in the European Gas Market. Why a Gas Exporters' Cartel Has Not Yet Materialized.*

Ho, E., Rajagopalan, A., Skvortsov, A., Arulampalam, S., & Piraveenan, M. (2021). *Game Theory in defence applications: a review.* https://doi.org/10.3390/S22031032

Namany, S., & Al-Ansari, T. (2021). Energy, water, food nexus decision-making for sustainable food security. In *Environmental Footprints and Eco-Design of Products and Processes* (pp. 191–216). https://doi.org/10.1007/978-981-16-0239-9_7

Namany, S., Al-Ansari, T., & Govindan, R. (2019). Optimisation of the energy, water, and food nexus for food security scenarios. *Computers & Chemical Engineering*, *129*, 106513. https://doi.org/10.1016/J.COMPCHEMENG.2019.106513

Saghaian, S. H., & Asgari, M. (2013). Oligopolistic market structure in the Japanese pistachio import market. *Journal of Agricultural and Food Industrial Organization*, *11*(1), 87–99. https://doi.org/10.1515/jafio-2013-0005

Antonis Kokossis, Michael C. Georgiadis, Efstratios N. Pistikopoulos (Eds.)
PROCEEDINGS OF THE 33rd European Symposium on Computer Aided Process Engineering (ESCAPE33), June 18-21, 2023, Athens, Greece
 http://dx.doi.org/10.1016/B978-0-443-15274-0.50228-6

Multiple-output Gaussian Process based Global Sensitivity Analysis for the cost-performance of electric vehicle in the United Kingdom

Min Tao[a], Jude O. Ejeh[a], Robert A. Milton[a], Joan Cordiner[a], Solomon F. Brown[a*]

[a]*Department of Chemical & Biological Engineering, The University of Sheffield, S102TN, United Kingdom*

Abstract

Transportation is one of the major sources of greenhouse gas (GHG) in the United Kingdom (UK), which motivates the government's policies to perform electrification of transportation and stimulate the use of electric vehicle (EV). However, many practical factors including battery costs and charging infrastructure are significantly affecting the wide utilisation of EVs. In this work, a multiple-output Gaussian processes (GPs) based Global Sensitivity Analysis (GSA) method is employed to investigate the relationship between these EV properties and the cost-performance. The variance-based GSA computations provide the detailed understandings of these key vehicle parameters for EV end-users and manufacturers. It is initially found that the total travel distances and its relevant variables available ID and fraction, and battery efficiency are the dominant factors for the cost of performance within all optimal solutions.

Keywords: Gaussian Process, greenhouse gas emissions, electric vehicle, global sensitivity analysis

1. Introduction

The impact of greenhouse gas (GHG) emissions has led to the growing commitments to net-zero emission by 2050. According to the BEIS (BEIS, 2020), Dominguez, et al, 2020), transportation is a significant source of GHG emission. The UK has therefore chosen to extensively electrify transportation and stimulate the use of electric vehicle (EV) to alleviate the GHG emissions. However, many practical factors, such as battery and ownership costs, and charging infrastructure, still inhibit the desired adoption of EVs. Thus, it is of importance to understand the relationships between the EV properties and their cost and performance. Such relationship can be obtained via Global Sensitivity Analysis (GSA). GSA calculates how much the uncertainty in outputs is attributed to the uncertainty in its individual inputs, or combinations thereof (Saltelli et al, 2008), and is a powerful tool to identify the key input variables to system outputs.

In this work, we adopt a variance-based GSA method using Gaussian Processes (GPs) (Williams & Rasmussen, 2006) to obtain the order of importance of vehicle properties to the cost and performance of EVs. The multi-output GP model is generated through the open-source ROM-COMMA software library (ROM-COMMA, 2022). GPs with automatic relevance determination (ARD) kernels allow for the semi-analytic calculation of Sobol' indices which measures the proportion of output variance attributable to each (combination of) input parameter Sobol' indices (Yeardley et al., 2021). The multiple-output GP-based GSA computations can thus provide an accurate and efficient understanding of key vehicle parameters for EV end-users and manufacturers.

2. Gaussian process based global sensitivity analysis

2.1. Gaussian process regression

In this work, we utilised GP surrogate models to build explicit relationship between the vehicle parameters $\boldsymbol{x}$ and the cost-performance $\boldsymbol{y} = \boldsymbol{f}(\boldsymbol{x}) + \boldsymbol{\sigma}^2{}_e\, \boldsymbol{I}_f$, along with the observe data noises $\boldsymbol{\sigma}^2{}_e$ and the unit matrix $\boldsymbol{I}_f$. GP regression is an efficient non-parametric model technique, which also work as one statistic machine learning approach. One key assumption of GP regression is that the random output variables $\boldsymbol{f}(\boldsymbol{x}_i)$ at any location point $\boldsymbol{x}_i$ follows the normal distribution with mean value $\mu(\boldsymbol{x}_i)$ and variance $\boldsymbol{\sigma}^2(\boldsymbol{x}_i)$, and then prior joint Gaussian distribution $\boldsymbol{f}(\boldsymbol{X})$ for the finite random variables $\boldsymbol{X}$:

$$\boldsymbol{f}(\boldsymbol{X}) \sim N(\mu(\boldsymbol{X}), \boldsymbol{\sigma}^2(\boldsymbol{X}))$$
$$\boldsymbol{y}|\boldsymbol{f} \sim N(\boldsymbol{f}, \boldsymbol{\sigma}^2{}_e) \tag{1}$$

where $\boldsymbol{f}(\boldsymbol{X}) = (\boldsymbol{f}(\boldsymbol{x}_1), \boldsymbol{f}(\boldsymbol{x}_2) \dots \boldsymbol{f}(\boldsymbol{x}_n))$, $\boldsymbol{\mu}(\boldsymbol{X}) = (\boldsymbol{\mu}(\boldsymbol{x}_1), \boldsymbol{\mu}(\boldsymbol{x}_2) \dots \boldsymbol{\mu}(\boldsymbol{x}_n))$ and $\boldsymbol{n}$ is the number of samples. which are typically replaced by the zero mean function, and the covariance matrix $\boldsymbol{\sigma}^2(\boldsymbol{X})$ are expressed as the kernel matrix $\boldsymbol{K}(\boldsymbol{X}, \boldsymbol{X})$ along with each element $\boldsymbol{K}_{i,j} = \boldsymbol{k}(\boldsymbol{x}_i, \boldsymbol{x}_j)$.

Here, the common radial basis function kernel was used since the automatic relevance determination kernel could provide semi-analytical computations for further variance-based GSA. The GP kernel function formulation indicates another assumption that when the inputs $\boldsymbol{x}_i, \boldsymbol{x}_j$ are close, the corresponding output functions $\boldsymbol{f}(\boldsymbol{x}_i)$ and $\boldsymbol{f}(\boldsymbol{x}_j)$ are close:

$$\boldsymbol{k}(\boldsymbol{x}_i, \boldsymbol{x}_j) = \boldsymbol{\sigma}^2{}_f \boldsymbol{exp}(-\frac{\mathbf{1}}{\mathbf{2}\boldsymbol{l}^2}(\boldsymbol{x}_i - \boldsymbol{x}_j)^T(\boldsymbol{x}_i - \boldsymbol{x}_j)) \tag{2}$$

where the kernel hyperparameter $\boldsymbol{\sigma}^2{}_f$ is global scale parameter while $\boldsymbol{l} = (\boldsymbol{l}_1, \boldsymbol{l}_2 \dots \boldsymbol{l}_d)$ represents the bandwidth parameters, $\boldsymbol{d}$ is the dimension of inputs $\boldsymbol{x}$.

If the prior samples $(\boldsymbol{X}, \mathbf{y})$ are collected, the predictive variable $\mathbf{f}(\mathbf{X}')$ at the new location $\mathbf{X}'$ and $\mathbf{y}$ would satisfy a joint Gaussian distribution:

$$\begin{bmatrix} \boldsymbol{f}(\boldsymbol{X}') \\ \boldsymbol{f}(\boldsymbol{X}) \end{bmatrix} \sim \mathrm{N}(\begin{bmatrix} \mathbf{0} \\ \mathbf{0} \end{bmatrix}, \begin{bmatrix} \boldsymbol{K}(\boldsymbol{X}', \boldsymbol{X}') & \boldsymbol{K}(\boldsymbol{X}', \boldsymbol{X}) \\ \boldsymbol{K}(\boldsymbol{X}, \boldsymbol{X}') & \boldsymbol{K}(\boldsymbol{X}, \boldsymbol{X}) + \boldsymbol{\sigma}^2{}_e\, \boldsymbol{I}_f \end{bmatrix}) \tag{3}$$

Then posterior predictive distribution $\boldsymbol{f}(\boldsymbol{X}')|(\boldsymbol{X}', \boldsymbol{X}, \boldsymbol{y})$:

$$\boldsymbol{f}(\boldsymbol{X}')|(\boldsymbol{X}', \boldsymbol{X}, \boldsymbol{y}) \sim N(\boldsymbol{K}(\boldsymbol{X}', \boldsymbol{X})(\boldsymbol{K}(\boldsymbol{X}, \boldsymbol{X}) + \boldsymbol{\sigma}^2{}_e\, \boldsymbol{I}_f)^{-1}\boldsymbol{y},\ \boldsymbol{K}(\boldsymbol{X}', \boldsymbol{X}') - \boldsymbol{K}(\boldsymbol{X}', \boldsymbol{X}')(\boldsymbol{K}(\boldsymbol{X}, \boldsymbol{X}) + \boldsymbol{\sigma}^2{}_e\, \boldsymbol{I}_f)^{-1}\, \boldsymbol{K}(\boldsymbol{X}, \boldsymbol{X}')) \tag{4}$$

The hyperparameters $\boldsymbol{\sigma}^2{}_f$, $\boldsymbol{l}$ and $\boldsymbol{\sigma}^2{}_e$ can be obtained through maximizing the marginal likelihood on the collected data $(\boldsymbol{X}, \mathbf{y})$.

2.2. Variance-based global sensitivity analysis

With the built explicit GP surrogate model, the variance-based GSA method, also referred to as the Sobol' indices, is then performed on the mean predictive function $\boldsymbol{\mu}(\boldsymbol{X}') = \boldsymbol{K}(\boldsymbol{X}',\boldsymbol{X})(\boldsymbol{K}(\boldsymbol{X},\boldsymbol{X}) + \sigma^2{}_e\, \boldsymbol{I}_f)^{-1}\boldsymbol{y}$. From the probabilistic perspective, the inputs-outputs $(\boldsymbol{X}', \boldsymbol{\mu}(\boldsymbol{X}')$ of GP mean function are considered as the random variables. The key idea of Sobol indices method is to decompose the variance of the random outputs $\boldsymbol{\mu}(\boldsymbol{X}')$:

$$\boldsymbol{Var}(\boldsymbol{\mu}) = \sum_{i=1}^{d} \boldsymbol{V}_i + \sum_{i<j}^{d} \boldsymbol{V}_{ij} + \cdots + \boldsymbol{V}_{12\ldots d} \tag{5}$$

where $\boldsymbol{V}_i = \boldsymbol{Var}_{\boldsymbol{X}'_i}(\boldsymbol{E}(\boldsymbol{Y}|\boldsymbol{X}'_i), \boldsymbol{V}_{i,j} = \boldsymbol{Var}_{\boldsymbol{X}'_{i,j}}(\boldsymbol{E}(\boldsymbol{Y}|\boldsymbol{X}'_i, \boldsymbol{X}'_j) - \boldsymbol{V}_i - \boldsymbol{V}_j$ and so on.

Then the important first-order index $\boldsymbol{S}_i$, quantifying the effect of the single input $\boldsymbol{X}'_i$, can be computed as follows:

$$\boldsymbol{S}_i = \frac{\boldsymbol{V}_i}{\boldsymbol{Var}(\boldsymbol{\mu})} \tag{6}$$

Then the cumulative first-order index $\boldsymbol{S}^j$ quantifying the sum effects of the single inputs $(\boldsymbol{X}'_1, \boldsymbol{X}'_2, \ldots, \boldsymbol{X}'_j)$ on the total variance, can be calculated

$$\boldsymbol{S}^j = \sum_{i=1}^{j} \boldsymbol{S}_i \tag{7}$$

where $\boldsymbol{j}\ (\boldsymbol{j} = \boldsymbol{1}, \boldsymbol{2} \ldots, \boldsymbol{d})$ is the ranking position in the input index order $\alpha = \{\boldsymbol{1}, \boldsymbol{2}, \boldsymbol{3}, \ldots, \boldsymbol{d}\}$.

Meanwhile, we computed the closed index $\boldsymbol{S}^{\boldsymbol{C}}_{\ell_j}$ to represent the single and interaction effects of the inputs $(\boldsymbol{X}'_1, \boldsymbol{X}'_2, \ldots, \boldsymbol{X}'_j)$, following the equation:

$$\boldsymbol{S}^{\boldsymbol{C}}_{\ell_j} = \sum_{\ell_i} \boldsymbol{S}_{\ell_i}, \boldsymbol{\ell}_i \subseteq \boldsymbol{\ell}_j \,\&\, \boldsymbol{\ell}_i \nsubseteq \emptyset \tag{8}$$

where the $\boldsymbol{\ell}_j = \{\boldsymbol{1}, \boldsymbol{2}, \ldots, \boldsymbol{j}\} \subseteq \alpha$ is the index set of inputs $(\boldsymbol{X}'_1, \boldsymbol{X}'_2, \ldots, \boldsymbol{X}'_j)$.

Thus the interaction effects $\boldsymbol{S}_{\sim j}$ among the first $\boldsymbol{j}$ inputs $(\boldsymbol{X}'_1, \boldsymbol{X}'_2, \ldots, \boldsymbol{X}'_j)$ could be obtained as

$$\boldsymbol{S}_{\sim j} = \boldsymbol{S}^{\boldsymbol{C}}_{\ell_j} - \boldsymbol{S}^j \tag{8}$$

In this paper, we focus on the effects of the key parameters on the outputs. Through comparing the cumulative first-order indices $\boldsymbol{S}^j$ and the closed indices $\boldsymbol{S}^{\boldsymbol{C}}_{\ell_j}$ within the order α by decreasing the first-order index in a GSA test computation, the important input group is identified.

Here, all the Gaussian Process based Global Sensitivity Analysis computations were implemented in our open-sources ROM-COMMA software library (ROM-COMMA, 2022).

3. Computational results

3.1. Process introduction

The objective of this work is to seek the key EV parameters for the cost and CO2 emission. Here, firstly the important EV properties are considered including EV pool, 2WP (two-way power), acceleration, battery capacity, battery efficiency, battery power, range of photovoltaic (PV) output profiles, vehicle capital cost, vehicle maintenance cost, total travel distance, availability faction, peak PV and PV cost. While the major attributions for the cost and CO2 emission contain CO2 emission, electricity import cost, electricity export revenue, total power imported, total power exported, total power of battery charging, total cost and total power of battery discharging. The multiple inputs-outputs relationship will be built using GP surrogate and further be analysed through the models.

3.2. Data preprocessing

Through a bio-objective optimisation procedure for the cost-performance under multiple scenarios, around 2000 data pairs are generated for the multiple cost-performance under the different EV properties. By filtering the dependent inputs-outputs, the inputs of acceleration, battery capacity, battery efficiency, battery power, range of PV output profiles, vehicle capital cost, vehicle maintenance cost, total travel distance) and the outputs of CO2 emission, electricity import cost, total power imported, total power of battery charging, total power of battery charging, total cost under two different optimisation strategies are chosen to construct the GP models.

3.3. Computational results: GP models

For the GP model training process, a two-fold cross-validation was implemented to ensure the generalization ability of the obtained GP models. Table 1 displays the results of prediction validation. all validation criteria (the roots of mean square errors, R^2 value and outlier) shows that the errors of the surrogate of the selected CO2 emission, total cost, the total power of battery charging and the total power imported are small enough. All the R^2 values are larger than 0.8, the RMSE errors are lower while the percentage of outliers are less 10%. The computational results verify the accuracy of the generated GP models. Meanwhile, Figure 1 compared the prediction of GP models and the actual observes. In generally, almost all the prediction points from the selected four parameters are close to the original observes, which indicates the efficiency of the generated GP models. Furthermore, small deviations are found for the predicted CO2 emission and total imported power. That may be caused by the noises inside of the observed data.

Table 1: Prediction validation of the generated GP models for the selected output attributes

Output attributes	RMSE	R^2	Outlier (%)
CO2 emission (g)	0.41	0.830	0.564
Total cost (GBP)	0.037	0.999	0.410
Total Power, Battery charging (kW)	0.061	0.996	6.410
Total Power imported (kW)	0.36	0.873	0.615

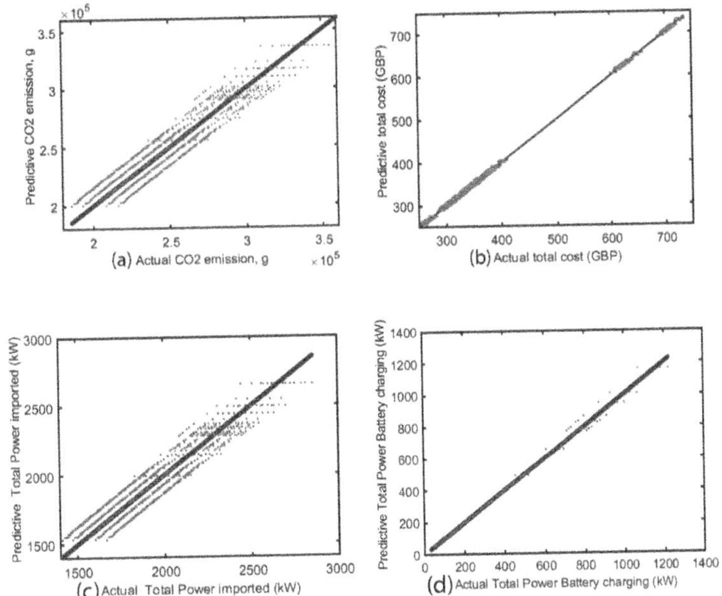

Figure 1: Tests of the GPs prediction (a) CO2 emission, (b) The total cost (c) The total power imported and (d) The total power battery charging

3.4. Computational results: GSA computations

With the validated GP models, GSA computations were then performed. Explicit ARD kernels contributes to the semi-analytical computations of GSA calculations, reducing the computational costs. Figure 2 summarises the results of cumulative first-order and closed indices for the four selected parameters (CO2 emission, total cost, total power imported and total power battery charging) on the eight inputs (acceleration, battery capacity, battery efficiency, battery power, range, vehicle capital cost, vehicle maintenance cost and total travel distance). It can be found, the cumulative closed indices of the first two terms all are larger than 94%, very close to 1 for all the selected four outputs. That indicates the first two input parameters total travel distance and battery efficiency are the dominant factors for the outputs. Furthermore, the small difference between the cumulative first-order and closed indices implies weak interactions effects among the input parameters. Thus, the total travel distance and its dependent variables available ID and fraction, together with another input battery efficiency are the key parameters for the cost and CO2 emission with the optimal solutions.

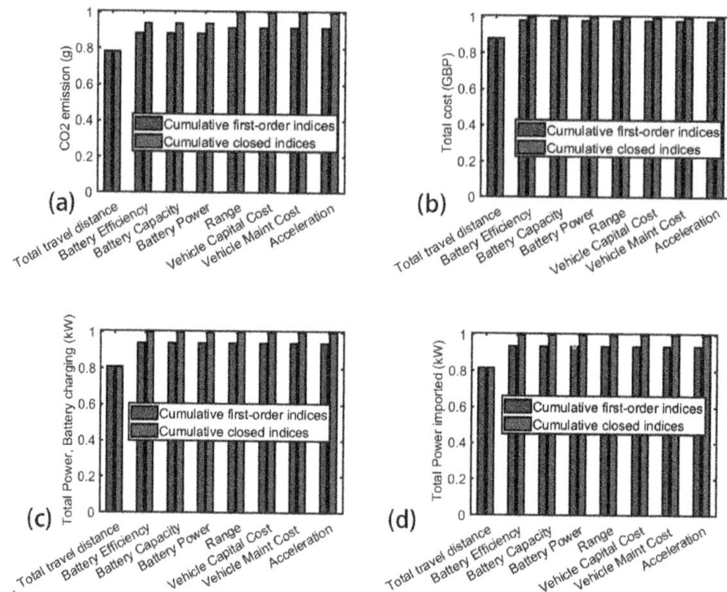

Figure 2: Computational results of the Sobol' indices (a) CO2 emission, (b) The total cost (c) The total power imported and (d) The total power battery charging

4. Conclusions

In this work, the multiple-output Gaussian Process based global sensitivity analysis computations are implemented between the EV properties and the cost-performance using different optimal solutions. The prediction validations indicate the high accuracy of the generated GP models while the GSA computations identify the dominant factors. It was initially found that the total travel distance and its highly relevant available ID and fraction, and battery efficiency are the key parameters for the cost and CO2 emission within the optimal solutions. Further conclusions will be reached through more and general data for detailed analysis.

References

A. S. Yeardley, S. Bellinghausen, R. A. Milton, J. D. Litster, S. F. Brown, 2021. Efficient global sensitivity-based model calibration of a high-shear wet granulation process. Chemical Engineering Science 238, 116569.

A. Saltelli, M. Ratto, T. Andres, F. Campolongo, J. Cariboni, D. Gatelli, ... & S. Tarantola, 2008. Global sensitivity analysis: the primer. John Wiley & Sons.

BEIS, 2020, Electricity Generation Costs.

C. K. Williams, C. E. Rasmussen, 2006, Gaussian processes for machine learning.

D. A. Dominguez, A. Dunbar, S. Brown, 2020, The electricity demand of an EV providing power via vehicle-to-home and its potential impact on the grid with different electricity price tariffs, Energy reports, 6, pp.132-141

S. Razavi, A. Jakeman, A. Saltelli, C. Prieur, B. Iooss, E. Borgonovo, E. Plischke, S.L. Piano, T.Iwanaga, W. Becker, S. Tarantola, 2021, The future of sensitivity analysis: an essential discipline for systems modeling and policy support. Environmental Modelling & Software, 137, p.104954

ROM-COMMA, 2022. ROM-COMMA. URL https://github.com/C-O-M-M-A/rom-comma

Antonis Kokossis, Michael C. Georgiadis, Efstratios N. Pistikopoulos (Eds.)
PROCEEDINGS OF THE 33rd European Symposium on Computer Aided Process Engineering (ESCAPE33), June 18-21, 2023, Athens, Greece
 http://dx.doi.org/10.1016/B978-0-443-15274-0.50229-8

Extended Multiple-Curve Resolution framework for the calibration of first-principles models

Daniel Casas-Orozco[a], Jaron Mackey[a], Ilke Akturk[a], Gintaras V. Reklaitis[a], Zoltan K. Nagy[a]*

[a]*Davidson School of Chemical Engineering, Purdue University, West Lafayette, IN, 47907,USA*

* *Corresponding author: zknagy@purdue.edu*

Abstract

In this work, a parameter estimation framework is presented for applications where both spectral and non-spectral information are used for model parameter estimation purposes. We illustrate our methodology by studying a crystallization system, where the liquid phase is monitored by spectroscopic means, while the solid phase is characterized by particle sizing techniques. The obtained results show that our framework is capable of handling experimental data of different types, and that crystallization model parameters and associated uncertainty regions can be successfully estimated. We also found a strong influence of data uncertainty on the accuracy of the estimated parameters for secondary nucleation.

Keywords: Multiple-Curve Resolution, parameter estimation, crystallization kinetics, PharmaPy

1. Introduction

Construction of digital twins in the pharmaceutical industry is a current trend motivated by the flexibility offered by in-silico tools, allowing exploration of process conditions and configurations early in process design/planning, and by the decision support digital twins can provide for process operation in the form of process control and monitoring. In both cases, mathematical models describing the unit operations involved typically include parameters which must be estimated by comparing model predictions to experimental data collected at carefully selected process conditions.

Some of the experimental data arise in the form of spectral information (UV-Vis, IR, Raman) gathered in real time for monitoring or parameter estimation purposes. In particular, spectra from the manufacturing of active pharmaceutical ingredients (APIs) is abundant in operations such as reaction and crystallization, and also in solid processing and product formulation. The availability of such data calls for frameworks that fully utilize the rich information provided by spectroscopic data in the parameterization of first-principle models. A layer of complexity in systems characterized by spectral data lies in need for the joint treatment of spectral and non-spectral data. For instance, building first-principles crystallization models usually requires information regarding both liquid and solid phases, the former often measured by spectroscopic techniques, and the latter by particle sizing instruments. In these cases, state and sensitivity information has to be utilized in such a way that general-purpose parameter estimation frameworks are still usable. Although selecting fixed wavelengths as traditionaly practiced is also an option, fully utilizing spectroscopic information offers a more robust and direct way of calibrating the models of interest [1], and also facilitates the analysis of multicomponent systems, of special interest in crystallization for impurity analysis.

In this work, we present a framework that uses spectral and non-spectral information from a crystallization system, to estimate kinetic parameter and their uncertainty regions. For this purpose, we expand a multiple-curve resolution/alternating least-squares (MCR-ALS) methodology [1] to allow the use of both liquid-phase and solid-phase batch cooling crystallization data within a parameter estimation framework built into the PharmaPy numerical library [2]. A case study is presented demonstrating the potential of the framework for handling different data sources and types in modeling of pharmaceutical systems. Moreover, the model parametric uncertainty estimated with this framework can be further utilized in probabilistic design space determination and robust optimization.

2. Mathematical framework

2.1 Crystallization modeling

The crystallization system is modeled by using the standard method of moments:

$$\frac{\mathrm{d}\mu_0}{\mathrm{d}t} = B \cdot V, \qquad \frac{\mathrm{d}\mu_n}{\mathrm{d}t} = n \cdot \mu_{n-1} G + B L_0^n, \quad n \in \{1,2,3\}, \tag{1}$$

with the liquid phase described by component and global material balances:

$$\begin{aligned} \frac{\mathrm{d}C_j}{\mathrm{d}t} &= -\frac{tr}{V_L}\left(\delta_{j,API} - \frac{C_j}{\rho_L}\right), \qquad j \in \{1,\ldots,n_{comp}\}, \\ \frac{\mathrm{d}V_L}{\mathrm{d}t} &= -\frac{tr}{\rho_L}. \end{aligned} \tag{2}$$

Liquid-to-solid mass transfer is given by $tr = 3k_v \rho_s G \mu_2$, and $\delta_{j,API} = 1$ for $j = API$ and 0 otherwise. Cooling profile $T(t)$ was modeled as a linear trajectory at constant cooling rate. Primary and secondary nucleation ($B = B_p + B_s$) and growth (G) crystallization mechanisms are modeled by power law expressions of the form:

$$B_p = k_p S^p, \qquad B_s = k_s S^s (k_v \mu_3), \qquad G = k_g S^g, \tag{3}$$

where the crystallization driving force is relative supersaturation $S = (C_{API} - C_{sol}) / C_{sol}$, with solubility $C_{sol}(T, x_{solvent})$. With the included mechanisms, the model parameter set is $\boldsymbol{\theta} = \{k_p, p, k_s, s, k_g, g\}$. An enlarged ODE system was constructed by including the sensitivity system $\mathbf{S} \in \mathbb{R}^{n_y \times n_\theta}$ of the crystallization model:

$$\frac{\mathrm{d}\mathbf{S}}{\mathrm{d}t} = \frac{\partial \mathbf{f}}{\partial \mathbf{y}}\mathbf{S} + \frac{\partial \mathbf{f}}{\partial \boldsymbol{\theta}}, \qquad \mathbf{S} \equiv \frac{\partial \mathbf{y}}{\partial \boldsymbol{\theta}}, \qquad \mathbf{y} = \{\mu_1, \ldots, \mu_n, C_1, \ldots, C_n, V_L\}, \tag{4}$$

where the vector function $\mathbf{f}$ in Eq. (4) represents the right hand sides of Eqs. (1) and (2).

Analytical forms for matrices $\partial \mathbf{f} / \partial \mathbf{y} \in \mathbb{R}^{n_y \times n_y}$ and $\partial \mathbf{f} / \partial \mathbf{y} \in \mathbb{R}^{n_y \times n_\theta}$ were adapted from [3] and incorporated into PharmaPy.

2.2 Parameter estimation framework

The parameter estimation problem was formulated as

$$\min_{\boldsymbol{\theta}} \frac{1}{2} \sum_{e \in E} \sum_{\ell \in L_e} (Y - Y_{data})_\ell^T W^{-1} (Y - Y_{data})_\ell, \tag{5}$$

$$\text{subject to} \qquad Y = \mathbf{h}(\mathbf{y}). \tag{6}$$

with $\mathbf{y}$ are the states of the dynamic mode (see Eq. (4)). Summations in Eq. (5) are indexed for a set of experiments $e \in E$, each experiment measured at L_e sampling times. $Y \in \mathbb{R}^{L_e \times n_y}$ collects the monitored states computed from the dynamic model, $Y_{data} \in \mathbb{R}^{L_e \times n_y}$ collects the corresponding experimental data, and $W \in \mathbb{R}^{n_y \times n_y}$ is a diagonal matrix of weighting factors. Since our methodology includes spectral (liquid phase) and non-spectral (solid phase) information, states are arranged as $Y = [Y_{sp}, Y_{nsp}]$. Following Eq. (6), simulated spectra [1] and (number-based) mean crystal size are given by:

$$Y_{sp} = \mathbf{C}(\mathbf{C}^{+} A), \qquad Y_{nsp} = \boldsymbol{\mu}_1 / \boldsymbol{\mu}_0, \tag{7}$$

where $A \in \mathbb{R}^{L_e \times n_{sp}}$ is a matrix of liquid phase absorbance measured at n_{sp} wavelength values, and $\mathbf{C}^{+}$ is the Moore-Penrose inverse of the concentration matrix $\mathbf{C} = [\mathbf{C}_1, \ldots, \mathbf{C}_{n_{comp}}]$, $\mathbf{C} \in \mathbb{R}^{L_e \times n_{comp}}$, $\boldsymbol{\mu}_0, \boldsymbol{\mu}_1 \in \mathbb{R}^{L_e}$.

Derivative information required to solve Eq. (5) was also decomposed into spectral and non-spectral contributions (Eq. (9)), with related dynamic sensitivities $\partial \mathbf{C} / \partial \theta_i (t)$ and $\mathbf{S}_{\mu_0}(t) = \partial \mu_0 / \partial \boldsymbol{\theta}$, $\mathbf{S}_{\mu_1}(t) = \partial \mu_1 / \partial \boldsymbol{\theta}$ retrieved after solving Eq. (4).

$$J = \begin{bmatrix} J_{sp} \\ J_{nsp} \end{bmatrix}, \qquad J_{sp} = [J_{sp,\theta_1}, J_{sp,\theta_2}, \ldots, J_{sp,\theta_{n\theta}}]. \tag{8}$$

Each column vector $J_{sp,\theta_i} \in \mathbb{R}^{L_e \cdot n_{sp}}$ in J_{sp} is obtained by vertically stacking the columns of the corresponding matrix given by the variable projection shown in Eq. (10) [4]:

$$J_{sp,\theta_i} = \left[\left(P_C^{\perp} \frac{\partial \mathbf{C}}{\partial \theta_i} \mathbf{C}^{+} \right) + \left(P_C^{\perp} \frac{\partial \mathbf{C}}{\partial \theta_i} \mathbf{C}^{+} \right)^T \right] \cdot A, \quad \forall \theta_i \in \boldsymbol{\theta}. \tag{9}$$

where $P_C^{\perp} = (\mathbf{I} - \mathbf{C} \cdot \mathbf{C}^{+})$ is an orthogonal projector of the column space of $\mathbf{C}$ [4]. On the other hand, the non-spectral contribution $J_{nsp} \in \mathbb{R}^{L_e \times n_\theta}$ is given by:

$$J_{nsp} = \frac{\mathbf{S}_{\mu_1} \circ \boldsymbol{\mu}_0 - \mathbf{S}_{\mu_0} \circ \boldsymbol{\mu}_1}{\boldsymbol{\mu}_0^2} \tag{10}$$

where the $\circ$ operator represents element-wise column multiplication.

2.3 Materials and methods

Table 1. Experimental design

Exp	%seed	Cooling rate (K/s)
1	0	-0.1
2	0	-0.3
3	5	-0.1
4	5	-0.3

Four in-silico API concentration and mean size (μ_1/μ_0) datasets were generated by simulating the system described by Eqs. (1) and (2) at the experimental conditions shown in Table 1, and using the nominal parameter values shown in Table 2. Time was sampled every 5 seconds until a final temperature T_f = 278 K was reached for all the simulations. Normal noise $\mathcal{N}(0,3)$ was added to the resulting concentration and mean size time profiles. Using the simulated concentration **C**, in-silico spectra was calculated by the Beer-Lambert law as $Y_{sp}^{data} = \mathbf{C} \cdot \mathbf{a}$, where a is a vector of pure component molar absorptivity, modeled as the summation of gaussian peaks centered at the wavenumber where maximum absorbance for the three-component system (API + THF + heptane) occurs. Only initial and final mean size data were passed to the parameter estimation framework, as in practice these are the typical measurement gathered in crystallization experiments.

Initial temperature was set as the saturation temperature at a given solvent composition. Values of initial API concentration, and molar fraction of the THF + heptane solvent mixture were fixed at $C_{API,init}$ = 167 kg m^{-3} (0.7 mol L^{-1}) and x_{THF} = 0.4 (30 %v/v), respectively according to previous works [5]. Furthermore, initial seed mass was calculated from the theoretical crystal yield using *%seed* in Table 1 as $m_{seed} = (C_{API} - C_{solub}(T_f))V(\%seed/100)$.

For experiments 3 and 4 (seeded experiments), initial crystal size distribution was assumed lognormal, $f_{init} \sim \ln\mathcal{N}(\mu = 35, \sigma^2 = 150)$, resulting in $\mu_1/\mu_0 = 25\,\mu\text{m}$ $(t = 0)$, values in the reported ranges for continuous lomustine crystallization [5].

The parameter estimation problem, Eq. (5) was solved using an in-house implementation of the Levenberg-Marquardt algorithm in PharmaPy. A weighing matrix $W = \text{diag}([W_{sp}^T, 1])$ where $W_{sp} \in \mathbb{R}^{n_{sp}=113}$ is a vector of 0.1's was used to account for the difference in the scales of the spectra versus crystal size, and was tuned by testing the optimizer performance at different candidate values.

2. Results

In order to decrease the chance of terminating at local minima, Eq. (5) was evaluated at 2^6 parameter realizations according to a Sobol sampling scheme, within the parameter bounds shown in Table 2. Then, the parameter realization that lead to the lowest objective function value was used as an initial estimate to perform optimization. After solving the optimization problem, the converged parameter estimates and their corresponding 95% asymptotic confidence intervals were obtained, as shown in Table 2.

The simulated dynamic profiles using the converged parameters are shown in Figure 1, where a close agreement between data and model prediction is observed. It must be emphasized that our optimization framework uses spectra as the simulated state (Y_{sp} in Eq. (7)), thus API concentration is used in Figure 1 only for illustration purposes. It is also noteworthy that primary nucleation and growth parameter estimates are very close to their corresponding nominal values, with some of their nominals being included in their corresponding confidence intervals. This shows how the methodology is effective in recovering the original parameters used to generate the in-silico data.

Table 2. Parameter information and optimization results

		Lower bound	Upper bound	Nominal	Initial	Converged
Primary nucleation	$\ln(k_p)$	11.513	34.539	19.519	21.543	19.39 ± 0.18
	p	0.500	5.000	3.000	1.270	2.78 ± 0.16
Secondary nucleation	$\ln(k_s)$	11.513	34.539	17.613	13.726	22.02 ± 0.32
	s_1	0.500	5.000	2.000	2.707	1.39 ± 0.25
Growth	$\ln(k_g)$	-36.148	2.303	-0.693	-0.665	-0.65 ± 0.024
	g	0.000	2.000	1.000	1.054	1.05 ± 0.026
WSSE		--	--	--	12041.82	367.46

On the other hand, it can be seen that secondary nucleation parameters significantly deviate from their nominal values, even though the model closely follows the provided data using the converged parameter estimates. As seen in Figure 1 (bottom), the level of noise used for the mean size profile significantly distorts the general trend of the in-silico data, which can have a pronounced effect on the final estimates, especially when only initial and final mean size information is being used. As a way to check the effect of noise, new in-silico datasets were generated, this time setting noise to zero for mean size. After parameter estimation (started at the initial parameter values shown in Table 2) the converged secondary nucleation parameters were closer to their corresponding nominal values ($\ln(k_s) = 19.84 \pm 1.69$, $s = 1.62 \pm 1.35$).

These results indicate how challenging it can be to determine true secondary nucleation parameters, and also call attention to the feasibility of decoupling this mechanism from primary nucleation. In this regard, the order of magnitude of primary and secondary nucleation can lead to one of them being underrepresented for a particular application. For illustration, the smallest ratio between the nucleation mechanisms at the highest supersaturation for the in-silico data occurs in experiment 3 ($B_p / B_s = 95$). This suggests that the effect of secondary nucleation on the system dynamics will be overshadowed by primary nucleation and growth, making sensitivity with respect to secondary nucleation parameters particularly small. This has been proven to be detrimental in the context of optimization, making the numerical system ill-conditioned, which in turn makes optimization problems harder to solve. This clearly indicates the need for more experimental data that allows one to better characterize secondary nucleation, for which different model-based designs of experiment frameworks exist in the literature [6].

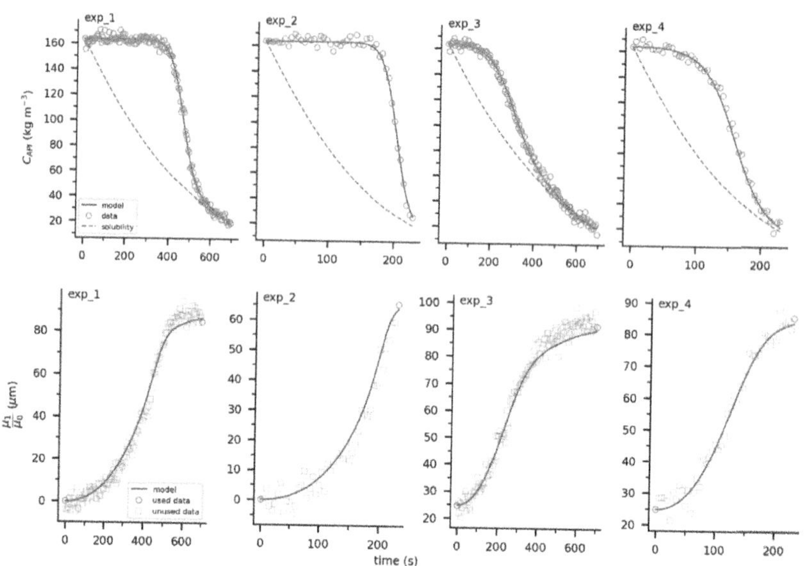

Figure 1. Predicted dynamic concentration and mean size profiles vs. corresponding in-silico data

3. Conclusions

In this work, we described a framework that allows the use of spectral and non-spectral information in the context of parameter estimation, and use it for the first time (to the best of our knowledge) in the context of crystallization systems. Two key elements contributed to the successful application of the framework, namely i) scaling the spectral and non-spectral contributions to the objective function and ii) properly arranging the sensitivity sources of the numerical system. This methodology can be applied to other studies in which both spectral and non-spectral information are gathered.

4. Acknowledgements

Funding for this work was made possible, in part, by the Food and Drug Administration through grant (U01FD006738). Views expressed herein do not necessarily reflect the official policies of the Department of Health and Human Services; nor does any mention of trade names, commercial practices, or organization imply endorsement by the United States Government.

References

[1] G. Puxty, M. Maeder, and K. Hungerbühler, "Tutorial on the fitting of kinetics models to multivariate spectroscopic measurements with non-linear least-squares regression," *Chemom. Intell. Lab. Syst.*, vol. 81, no. 2, pp. 149–164, 2006.

[2] D. Casas-Orozco *et al.*, "PharmaPy: An object-oriented tool for the development of hybrid pharmaceutical flowsheets," *Comput. Chem. Eng.*, vol. 153, p. 107408, Oct. 2021.

[3] Z. K. Nagy and R. D. Braatz, "Distributional uncertainty analysis using power series and polynomial chaos expansions," *J. Process Control*, vol. 17, no. 3, pp. 229–240, 2007.

[4] G. Golub and V. Pereyra, "Separable nonlinear least squares: the variable projection method and its applications," *Inverse Probl.*, vol. 19, no. 2, pp. R1–R26, Apr. 2003.

[5] J. Mackey *et al.*, "End-to-end reconfigurable process development for the cancer drug Lomustine," 2021. In: AIChE Meeting, Boston MA.

[6] S. P. Asprey and S. Macchietto, "Statistical tools for optimal dynamic model building," *Comput. Chem. Eng.*, vol. 24, no. 2–7, pp. 1261–1267, 2000.

Antonis Kokossis, Michael C. Georgiadis, Efstratios N. Pistikopoulos (Eds.)
PROCEEDINGS OF THE 33rd European Symposium on Computer Aided Process Engineering (ESCAPE33), June 18-21, 2023, Athens, Greece
 http://dx.doi.org/10.1016/B978-0-443-15274-0.50230-4

Energy analysis of CO_2 capture by flash and distillation

Giorgia De Guido[a]

[a]*GASP – Group on Advanced Separation Processes & GAS Processing, Dipartimento di Chimica, Materiali e Ingegneria Chimica "G. Natta", Politecnico di Milano, Piazza Leonardo da Vinci, I-20133, Milano, Italy*
giorgia.deguido@polimi.it

Abstract

The aim of this work is to compare the performances of CO_2 cryogenic separation from a flue gas stream by means of flash and of distillation. Simulations have been carried out with the process simulator Aspen HYSYS® V11 (AspenTech, 2019), selecting the Peng-Robinson Equation of State and a sensitivity analysis has been performed on the CO_2 concentration in the feed gas and on the operating pressure (in the range 1.5-4 MPa) of the unit. Given the better performances of cryogenic distillation, which allows obtaining a CO_2 product at high purity at lower energy expenses, distillation has been considered for CO_2 capture from a flue gas from an oxy-fuel combustion power plant. A process scheme has been developed, properly selecting the operating conditions so to avoid the formation of solid CO_2, demonstrating the better energy performances (ca. 10 % lower "Net Equivalent Methane" consumption) compared with those of a flash-based process.

Keywords: CCUS, cryogenic CO_2 capture, distillation, oxy-fuel combustion

1. Introduction

Carbon Capture Utilization and Storage (CCUS) is a key strategy to reach the goal of the Paris Agreement for the path to net-zero or even negative emissions in some scenarios. CCUS involves capturing CO_2 from a gas source and utilizing or storing it, with transportation being an essential step (Mazzoccoli et al., 2013). Among the known carbon capture technologies, low-temperature/cryogenic separation processes - recently also considered for a profitable exploitation of low-quality natural gas reserves (Pellegrini et al., 2019) - are particularly suitable for gases with high CO_2 concentrations. Thus, it is one of the best options for treating flue gases from oxy-combustion processes or as a finishing treatment in hybrid systems (De Guido et al., 2021). Two aspects, namely high CO_2 recovery rates and high purity have contributed to the considerable attention cryogenic CO_2 capture technologies are gaining (Font-Palma et al., 2021). This is the motivation for this work, which focuses on distillation-based CO_2 capture not involving dry ice formation by a careful selection of the proper operating conditions. To the author's knowledge, only few literature works on this topic deal with distillation techniques. After explaining the methods for the analysis (section 2), the advantage of distillation over flash has been demonstrated for further purification of a flue gas downstream of a bulk removal step. Then, a distillation-based process has been considered for CO_2 capture from a flue gas from an oxy-fuel combustion power plant for different CO_2 contents (66.8-85.5 mol%) in it and different recovery rates.

2. Methods

Process simulations have been carried out with Aspen HYSYS® V11 (AspenTech, 2019), selecting the Peng-Robinson Equation of State to properly represent the thermodynamic behavior of the CO_2-N_2 system, which is representative of a flue gas mixture. In order to identify and avoid the operating conditions causing solid CO_2 formation, the "CO_2 Freeze Out" utility (De Guido et al., 2019; Spatolisano and Pellegrini, 2021a, b) was used.
The comparison among the different process schemes analyzed in this work has been carried out on the basis of an energy analysis. Since two types of energy consumption are involved in a typical cryogenic CO_2 capture process, both thermal and electrical, to determine the overall energy consumption the different contributions have been expressed on the same basis adopting the "Net Equivalent Methane" (NEM) method (De Guido et al., 2018). In selecting the most suitable flue gas for cryogenic CO_2 separation, previous findings have been taken into account. For example, van Benthum et al. (van Benthum et al., 2012) reported that a minimum concentration of 40% of CO_2 in the feed gas is necessary to achieve recovery higher than 70%. Therefore, according to typical flue gas compositions (Berstad et al., 2013), suitable flue gases for CO_2 capture at low temperature are those from oxy-fuel combustion.

3. Process Description and Analysis

3.1. Cryogenic CO_2 capture by flash or distillation

Cryogenic separation of CO_2 from a flue gas mixture by means of flash or distillation downstream of a bulk removal step is illustrated in Figure 1.

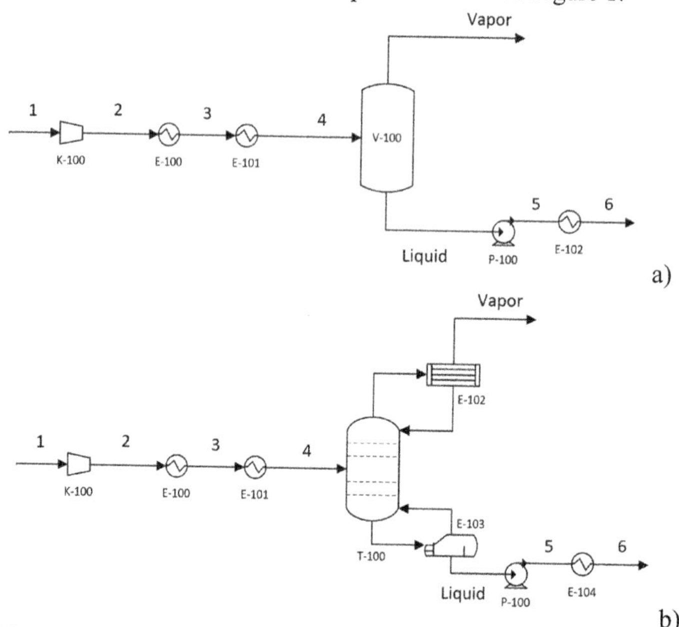

Figure 1. CO_2 separation from a CO_2-N_2 mixture by: a) flash expansion; b) distillation.

To this purpose, a flue gas mixture consisting of CO_2 and nitrogen with a CO_2 content in the range 40-60 mol%, available at 0.15 MPa, has been compressed to different pressures ranging from 1.5 to 4.0 MPa. In the case of separation carried out by flash expansion, such stream has been cooled down to a suitable operating temperature selected as the lowest allowing the liquid-vapor separation without formation of solid CO_2, which maximizes CO_2 recovery. Such temperature was determined using the "CO_2 Freeze Out" utility considering a 2 K margin from the freezing temperature. Two heat exchangers are shown in Figure 1 for cooling with cooling water (CW) first (in E-100) – not accounted for in the energy analysis – and, then, in E-101 with a suitable refrigeration cycle (De Guido et al., 2015). In the case of separation by distillation, the flue gas stream has been compressed to the operating pressure of the tower (in the range 1.5-4.0 MPa), and cooled down to the corresponding dew-point temperature. The saturated vapor stream enters the distillation column, equipped with a full reflux condenser, for which the number of trays, the liquid CO_2 purity (99.99 mol%) and the liquid CO_2 recovery (determined as the highest not forming solid CO_2 in any tray) have been specified. The number of trays has been determined using the short-cut method implemented in the process simulator Aspen HYSYS® V11 (AspenTech, 2019), knowing CO_2 product purity and recovery and assigning an actual reflux ratio 30 % higher than the minimum one. In both cases, the liquid CO_2-rich product is pumped to 6 MPa and heated to 293.15 K (Hasse et al., 2013). As shown in Figure 2a, increasing the operating pressure of the flash drum, the CO_2 product purity is lower and it is not significantly affected by feed gas composition, since the flash operating temperature is about the same (215.86 K at 1.5 MPa and 214.26 K at 4.0 MPa) for all considered pressures. On the contrary, when separation is performed by distillation, the CO_2 product purity remains constant and equal to the specified value. Figure 2b shows that the CO_2 product recovery is higher as the operating pressure increases at constant feed gas composition, and it is lower for lower CO_2 content in the feed gas at constant pressure, with a little difference (within 1.7 %) between separation by flash and by distillation.

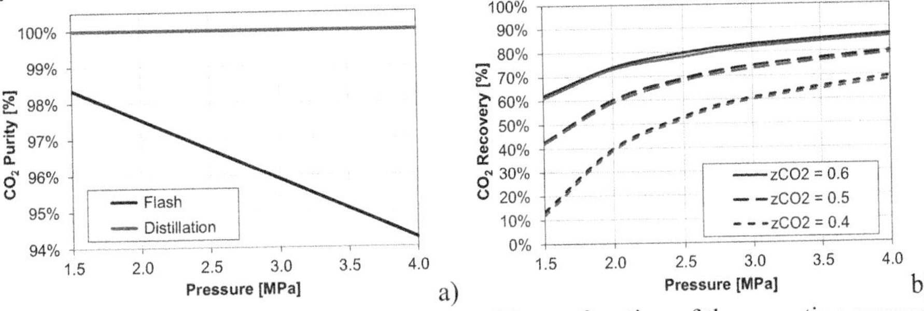

Figure 2. Liquid CO_2 purity (a) and recovery (b) as a function of the operating pressure of the flash unit / distillation column (1.5-4.0 MPa) and for different feed gas compositions. Comparison between flash (black) and distillation (blue).

Separation by distillation allows obtaining a high-purity CO_2 product with no penalty on energy consumptions, as shown in Figure 3. Indeed, the energy requirement in terms of NEM of the distillation column (related to flue gas compression, refrigeration both at the inlet and at the condenser of the column, and pumping of the CO_2 product) is lower in all the simulations, and the difference with the energy requirement of the flash (related to flue gas compression, refrigeration at the inlet of the drum, and pumping of the CO_2 product) is higher for higher operating pressures, which correspond to higher CO_2 recoveries (Figure 2b). The lower energy requirement of the distillation process could be explained by taking into account the distribution of the cooling duty in the two cases. In flash separation, the whole cooling duty is due to cooling down the feed to the flash operating temperature (ca. 215 K), while in the distillation process the cooling duty is partly due to cooling down the feed to its dew-point temperature, that is higher than flash temperature, and only a portion of the cooling duty is due to refrigeration in the column condenser, resulting in lower equivalent methane consumption.

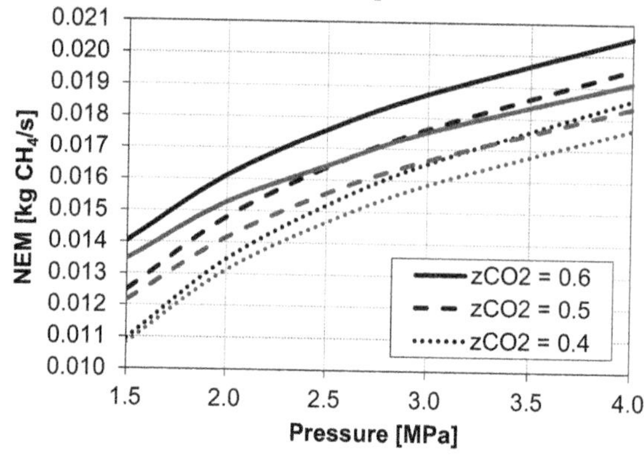

Figure 3. Net Equivalent Methane (NEM) comparison between flash (black) and distillation (blue).

3.2. Distillation-based CO_2 capture process

Following the results reported in the previous section, a distillation-based CO_2 capture process is proposed (both in a base and in an improved configuration) selecting an oxy-fuel combustion flue gas. The specific composition considered has been taken from Amann et al. (Amann et al., 2009) and is reported in Table 1.

Table 1. Composition (mol%) of the flue gas (Amann et al., 2009) entering the distillation-based CO_2 capture process.

CO_2	H_2O	N_2	O_2	Ar
66.8	3.1	18.9	5.5	5.5
74.1	3.0	11.0	5.7	6.0
85.5	3.0	0.4	5.7	5.4

The process scheme is illustrated in Figure 4 and differs from the one proposed by Amann et al. (Amann et al., 2009) for the replacement of the flash units by a distillation column for the reasons reported in the previous section. The feed gas is compressed in a three-stage process to 3.5 MPa, with intermediate cooling to 303.15 K using CW. The pressure of each stage has been determined as explained in a previous work (Pellegrini et al., 2015). After cooling in the first two stages, water is separated as liquid in two drums (V-100 and V-101), whereas the vapor stream is fed to the next compression stage. The stream exiting the compression train enters a dehydration unit (not accounted for in the energy analysis that follows, for comparison purposes with the flash-based process (Amann et al., 2009)), where the reduction of the water content to 20 ppm is accomplished. The dehydrated stream is, then, cooled to its dew-point temperature and is fed to the distillation column, equipped with a full reflux condenser, where an external refrigerant is used. The reboiler employs CW as heating medium because the heating is below ambient temperature. For all the investigated case studies (i.e., different CO_2 contents in the flue gas and different recoveries), the number of theoretical trays turned out to be 4 from a short-cut calculation and has been specified together with the CO_2 purity (99 mol%) and recovery set to the same values considered in the flash-based process (Amann et al., 2009) for comparison purposes. The distillate stream, exiting the top of the distillation column, is heated (E-105) to 293.15 K using CW, then it is heated

(E-106) to 423.15 K with steam and is expanded to atmospheric pressure in a turbine. The stream exiting the bottom of the column is pumped (P-100) to 15 MPa, heated (E-108) to 293.15 K with CW and finally heated (E-109) to 293.15 K.

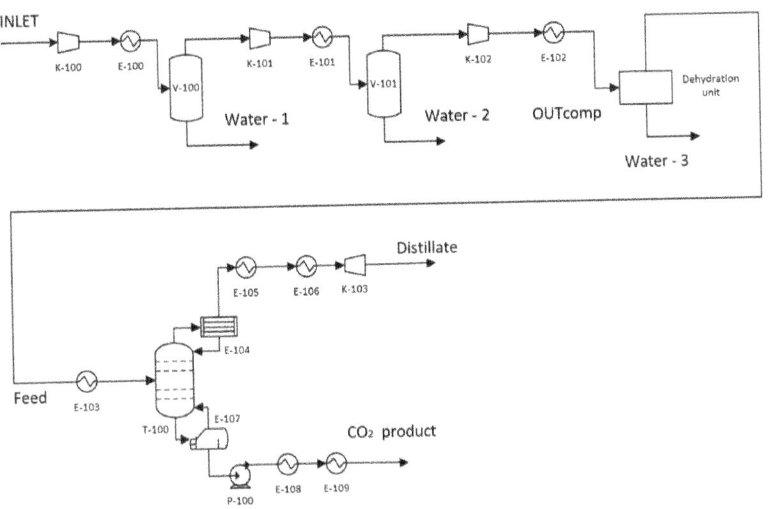

Figure 4. Scheme of the distillation-based CO_2 capture process without heat integration.

The process scheme in Figure 4 (in the following, referred to as "distillation-based (A)") can be improved from an energy point of view by using the distillate product for pre-cooling of the feed stream to the distillation column, so reducing the cooling duty supplied by an external refrigerant for bringing the feed to its dew-point temperature. Such an improved configuration is referred to as "distillation-based (B)" in the following. The results of the energy analysis are summarized in Table 2, where the relative difference has been computed with respect to the flash-based process presented in the literature (Amann et al., 2009).

Table 2. Results of the energy analysis ("Net Equivalent Methane" method).

		NEM [kgCH$_4$/s]			Relative difference [%]	
CO_2 content [mol%]	**CO_2 recovery [%]**	**flash-based (Amann et al., 2009)**	**distillation-based (A)**	**distillation-based (B)**	**distillation-based (A)**	**distillation-based (B)**
66.8	75	0.6945	0.7257	0.6366	+ 4	- 8
74.1	85	0.6873	0.7138	0.6603	+ 4	- 4
85.5	75	0.5527	0.5416	0.4930	- 2	- 11
85.5	90	0.6364	0.6274	0.6025	- 1	- 5

The "distillation-based (A)" process is more energy demanding than the flash-based separation for low CO_2 contents in the flue gas, whereas the configuration with heat integration ("distillation-based (B)") is less energy demanding in all cases, with the highest difference at low carbon dioxide recovery due to a decrease of column condenser and cooling (E-103) duties. An important advantage of the proposed process, compared with the flash-based one, is that a liquid CO_2 product ready for pumping is obtained, which is suitable for further utilization (e.g., for Enhanced Oil Recovery). Moreover, it allows specifying even a stricter specification (by carefully selecting the proper operating conditions) on the CO_2 product to meet the quality requirements, in particular with respect to the residual O_2 content, for food-grade CO_2, which will be the subject of a future work together with a techno-economic analysis.

4. Conclusions

The comparison between cryogenic CO_2 capture by flash and distillation, not addressed in previous literature works, has shown that a CO_2 product at higher purity can be obtained in the latter case with lower energy consumptions, especially at higher operating pressures, although with a slightly lower (within 1.7 %) recovery. Therefore, a distillation-based process has been proposed for separating CO_2 out of flue gas streams typical of oxy-fuel combustion power plants. An energy comparison with the flash-based analogous process has demonstrated the better performances of the distillation-based process for several CO_2 concentrations (66.8-85.5 mol%) in the flue gas to be treated and different recovery rates. An important advantage of the proposed process, compared with the flash-based one, is that a high purity CO_2 product is obtained, which is suitable for further utilization, including in the food and beverage industry if a stricter specification, in particular with respect to the residual O_2 content, is assigned.

References

J.-M. Amann, M. Kanniche, C. Bouallou, 2009, Natural gas combined cycle power plant modified into an O_2/CO_2 cycle for CO_2 capture, Energy Conversion and Management, 50, 510-521.

AspenTech, 2019, Aspen HYSYS®, Burlington (MA), United States.

D. Berstad, R. Anantharaman, P. Nekså, 2013, Low-temperature CO_2 capture technologies–applications and potential, International Journal of Refrigeration, 36, 1403-1416.

G. De Guido, M.R. Fogli, L.A. Pellegrini, 2018, Effect of Heavy Hydrocarbons on CO_2 Removal from Natural Gas by Low-Temperature Distillation, Industrial & Engineering Chemistry Research, 57, 7245-7256.

G. De Guido, M. Gilardi, L.A. Pellegrini, 2021, Novel technologies for low-quality natural gas purification, Computer Aided Chemical Engineering, 50, 241-246.

G. De Guido, S. Langè, L.A. Pellegrini, 2015, Refrigeration cycles in low-temperature distillation processes for the purification of natural gas, Journal of Natural Gas Science and Engineering, 27, 887-900.

G. De Guido, F. Messinetti, E. Spatolisano, 2019, Cryogenic Nitrogen Rejection Schemes: Analysis of Their Tolerance to CO_2, Industrial & Engineering Chemistry Research, 58, 17475-17488.

D. Hasse, S. Kulkarni, E. Sanders, E. Corson, J.-P. Tranier, 2013, CO_2 capture by sub-ambient membrane operation, Energy Procedia, 37, 993-1003.

M. Mazzoccoli, G. De Guido, B. Bosio, E. Arato, L.A. Pellegrini, 2013, CO_2-mixture properties for pipeline transportation in the CCS process, Chemical Engineering Transactions, 32, 1861-1866.

L.A. Pellegrini, G. De Guido, V. Valentina, 2019, Energy and exergy analysis of acid gas removal processes in the LNG production chain, Journal of Natural Gas Science and Engineering, 61, 303-319.

L.A. Pellegrini, S. Langè, M. Baccanelli, G. De Guido, 2015, Techno-Economic Analysis of LNG Production Using Cryogenic vs Conventional Techniques for Natural Gas Purification, Offshore Mediterranean Conference and Exhibition, Ravenna, Italy, 25-27 March 2015.

E. Spatolisano, L.A. Pellegrini, 2021a, CO_2-Tolerant Cryogenic Nitrogen Rejection Schemes: Analysis of Their Performances, Industrial & Engineering Chemistry Research, 60, 4420-4429.

E. Spatolisano, L.A. Pellegrini, 2021b, Solid–Liquid–Vapor Equilibrium Prediction for Typical Helium-Bearing Natural Gas Mixtures, Journal of Chemical & Engineering Data, 66, 4122-4131.

R. van Benthum, H. van Kemenade, J. Brouwers, M. Golombok, 2012, Condensed rotational separation of CO_2, Applied Energy, 93, 457-465.

Antonis Kokossis, Michael C. Georgiadis, Efstratios N. Pistikopoulos (Eds.)
PROCEEDINGS OF THE 33rd European Symposium on Computer Aided Process Engineering (ESCAPE33), June 18-21, 2023, Athens, Greece
 http://dx.doi.org/10.1016/B978-0-443-15274-0.50231-6

Feature Embedding of Molecular Dynamics-Based Descriptors for Modeling Electrochemical Separation Processes

H. K. Gallage Dona,[a, 1] T. Olayiwola,[b, 1] L. A. Briceno-Mena,[b] C. G. Arges,[c] R. Kumar,[a] J. A. Romagnoli,[b, *]

[a]*Department of Chemistry, Louisiana State University, Baton Rouge, Louisiana 70803, United States.*

[b]*Cain Department of Chemical Engineering, Louisiana State University, Baton Rouge, Louisiana 70803, United States.*

[c]*Department of Chemical Engineering, The Pennsylvania State University, University Park, PA 160802, United States.*

**corresponding author: jose@lsu.edu*

[1]*Equal contribution*

Abstract

Machine learning is increasingly being used as a modeling technique for the analysis of complex systems even when limited knowledge is available. Here, we demonstrate the application of machine learning models to predict macroscopic properties of electrochemical polymers from molecular attributes. The computational framework proposes a novel approach to estimate molecular dynamic simulations (MD) attributes and experimental activity coefficients. Results showed that the data augmentation and embedding strategy effectively produce unique representations of each polymer. The findings from this study allow for the estimation of activity coefficients of novel polymers without the need for new time-consuming MD simulation runs. This data processing technique could then be used map to material properties that influence for electrochemical ionic separation units.

Keywords: Feature embedding, machine learning modeling, molecular dynamics simulations, electrochemical separations.

1. Introduction

It is estimated that the world's population will clock 9 billion by 2050; these fast-growing populations lead to an increase in demand for energy and water supply. For example, over 4 billion people lives in areas with severe scarcity of water (Mekonnen and Hoekstra, 2016). Also, in the energy industry, there is strong demand for greater energy supply and to reduce the carbon footprints of these processes. Both energy and water industries require the selective separation of millions of liquid components. These industries have a high demand for effective separation procedures to obtain clean water, gases, or solvents, which today make up a sizable share (10–15%) of global energy consumption (Sholl and Lively, 2016). Thus, there is a strong interest on the design and optimization of high efficiency separation processes. In ionic separation using electrodialysis and membrance capactive deionization, the selection of the ion-exchange membrane (IEM) is very important as it can affect the energy efficiency during deionization (Palakkal et al., 2018).

Computational models that can bridge molecular descriptors to material properties and performance in electrochemical separation platforms are vital to advancing the field (Briceno-Mena et al., 2021). To understand the ionic transport between the solution and polymer, there exists a need to model the relationship defining this transport mechanism. Our primary macroscopic property of interest are the ion activity coefficients in polyelectrolytes. Activity coefficients influence the partitioning coefficient of an ion between the polymeric ion-exchange membrane and the liquid solution, and it can also affect ionic conductivity. However, accurate ion activity coefficient models based upon the physics and chemistry of the ion-exchange membranes is lacking. In fact, most activity coefficient models for ions in liquids rely upon empirical and semi-empirical approaches (e.g., Debye-Hückel or Pitzer models). These empirical and semi-empirical models often need large data sets for estimating the empirical constants in these models. Because there is a lack of data for ion activity in various ion-exchange membrane chemistries and structures, attaining effective and accurate activity coefficient models for these soft materials are elusive. Machine learning represents a unique tool to streamline model development for attaining accurate activity coefficients of polymeric ion-exchange membranes.

The Proposed ML network will bridge molecular scale attributes from molecular dynamics (MD) simulations to the ionic activity coefficients in polymeric ion-excahnge membrane materials. Machine learning methods such as Support Vector Regression (SVR) and Random Forest (RFR) were used to connect the polymer fingerprint (based on chemical structure) and molecular level attributes (based on MD descriptors) to the macroscopic attribute of polymer electrolytes (i.e., activity coefficient). This work developed a ML framework that involves predicting the MD descriptors of ionomers. Ionomer molecular fingerprints, salt concentration and one-hot encoded categorical features of the salt ions were used. The main goal of this work is to predict ion activity coefficient in polymer electrolytes from the predicted MD descriptors, thereby serving as a tool for future users when there exists no MD data and experimental activity coefficient for a chosen polymer and salt ions.

2. Methods

More than ten original experimental reports contanining ion activity coefficient values in ion-exchange membranes and thin films were collected from literature. We restricted our search to publications with detailed information about the structure of the copolymer, ion exchange capacity, water uptake and experimental ion activity coefficient. Polymer structures and salt ions represented by their SMILES representations and categorical variables, respectively were converted into numeric arrays using *RDKit* and the OneHotEncoder method in *Sckit-Learn*, respectively as shown in Figure 1.

With explicit counterions and salt ions, we simulated polyelectrolyte systems using molecular dynamics simulations. To represent polyelectrolytes, a system with 17 polymer chains, water and salt ions was prepared with PACKMOL package in conformance with experimental conditions. We adopted the OPLSAA or GAFF2 forcefield parameters and

TIP3P model for polymer and water respectively. The MD simulations were implemented in LAMMPS package. The simulations were then analyzed to compute the solvation properties of these polymer electrolyte systems including radial distribution function, coordination numbers and diffusion coefficients using final production run of 40ns long in the isothermal (fixed temperature at 300K using a Nose Hoover thermostat, constant number of particles and volume) ensemble.

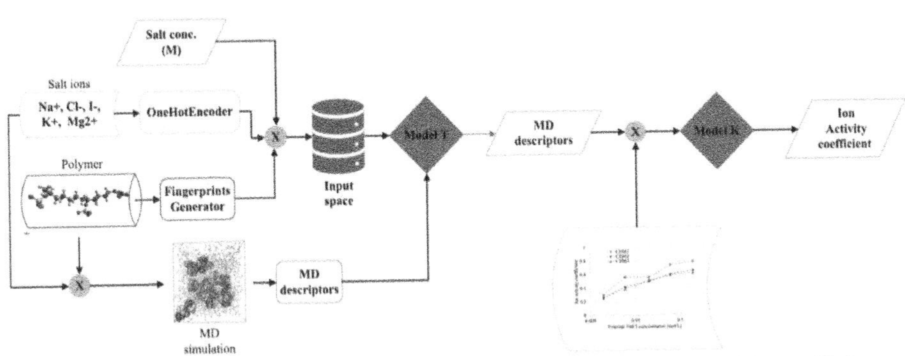

Figure 1: Workflow for modeling the prediction of MD descriptors and ion activity coefficient.

In this contribution, two regression models (Random Forest regression, RFR and Support Vector Regression, SVR) are compared to estimate the MD descriptors and ion activity coefficient. Figure 1 shows the framework for the model development. First, an ML model (*Model T*) was created to predict MD descriptors based on an input vector consisting of arrays of polymer fingerprints, one-hot encoded salt ions, salt concentration and number of water molecules per tethered ions. Secondly, another regression model (*Model K*) was created as shown in Figure 1 to predict the experimental ion activity coefficient. The input vectors consist of the MD descriptors (predictable from *Model T*) and the input matrix of model T. To train these models, the available dataset was split into training and test data. The model hyperparameters are tuned using the NSGA-II algorithm as implemented by *Pymoo*. The final model performance was quantified using coefficient of determination (R^2), mean square error (MSE) and mean absolute error (MAE).

3. Results and Discussions

3.1. Polymer space

Our studied dataset consists of 8 block copolymers and 3 random copolymers linked to experimental ion activity coefficients. To assess the similarity between the available copolymer, we computed the Tanimoto similarity (Tc) which is the most popular and widely accepted similarity metrics for polymer chemistry. The results of the analysis are depicted in Figure 2. Two polymers are considered identical when the Tc equals 1 and completely different when the Tc equals 0. The similarity matrix showed that our selected polymers are very different from one another. Only the RCE and BCE forms of a polymer have high similarity values (>0.7). This is expected because these polymer forms have the same chemistry except with different arrangement of repeat units. Thus, a model generated from these polymers and will have a good generalization to other polymers with similar constituents because the sampled polymers exhibit great variation.

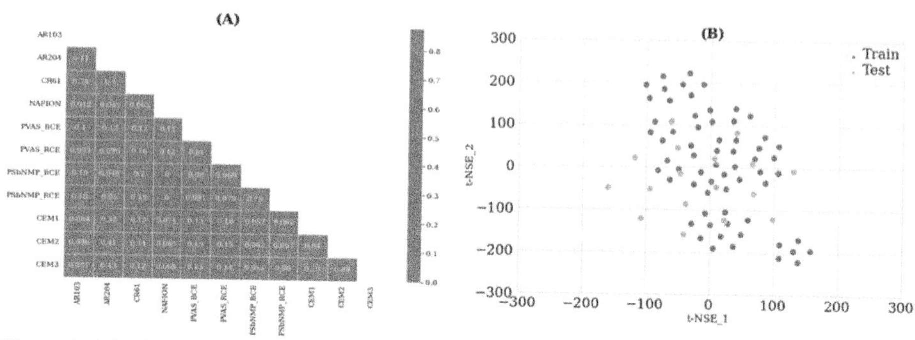

Figure 2: (a) Visualization of the similarity among the studied polymers and (b) sample space of the training and test input data used in machine learning.

During ML model development, greater attention must be given to the sample data. Specifically, by using similar train and test data, the impact of data discrepancies can be ameliorated and thus, gain a better understanding of the model's properties. To assess the conformational space of the training and test data, we computed the t-SNE plot, and the result is shown in Figure 2b. The analysis showed that the proportion of the sample space of the test data aligns strongly with that of the training set. Also, the t-SNE profile showed that the input matrix is well scattered and thus, an accurate ML model based on this input matrix will exhibit the better generalization to unseen data.

3.2. Prediction of MD descriptors

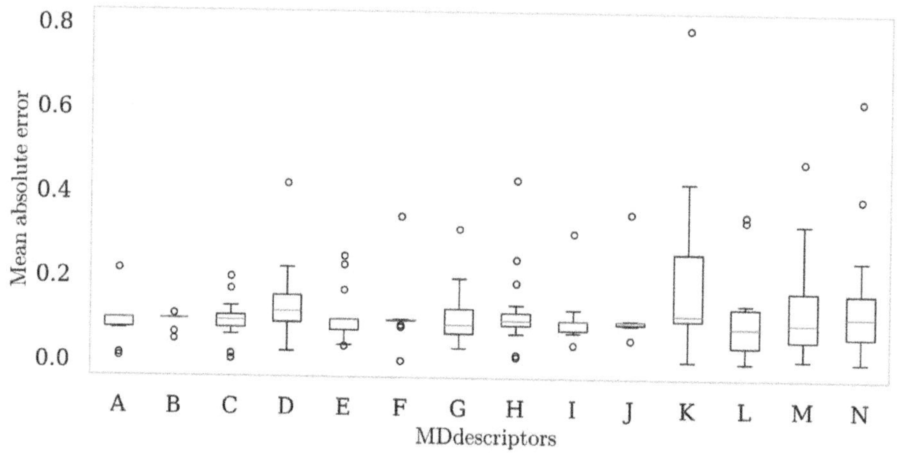

Figure 3. Boxplot showing the performance of the SVR model. Table 1 shows the description of the legends (A-N).

The SVR model was developed for a multi-output system. Here, the input features were transformed using principal component analysis to correct the effect of dimensionality reduction. Also, the target parameters (i.e., MD descriptors) were converted into a range (-1, 1). Afterwards, the transformed input and target data were imported to a SVR model and then trained based on the training set only. Figure 3 shows the statistical measure of the best model. From the perspective of the ML algorithm, the predicted MD descriptors exhibit close agreement with the target MD descriptors. Furthermore, the ML model based on the MF fingerprints shows a better predictive performance than the RDKit descriptor (not shown here); thus, a better representation of the studied polymer systems.

Table 1. The descriptions of the MD descriptions shown by the boxplot in Figure 3. (RDF refers to radial distribution function).

A	*first minima of RDF between counterion and oxygen of water.*	H	*coordination number (at first minima) from RDF between charge bearing group (in polymer) and oxygen (in water).*
B	*peak position of RDF between counterion and oxygen of water*	I	*first minima of RDF between charge bearing group (in polymer) and counterion (in salt).*
C	*peak height of RDF between counterion and oxygen of water*	J	*peak position of RDF between charge bearing group (in polymer) and counterion (in salt).*
D	*coordination number (at first minima) of RDF between counterion and oxygen of water.*	K	*peak height of RDF between charge bearing group (in polymer) and counterion (in salt)*
E	*first minima of RDF between charge bearing group (in polymer) and oxygen of water.*	L	*coordination number (at first minima) from RDF between charge bearing group (in polymer) and counterion (in salt).*
F	*peak position of RDF between charge bearing group (in polymer) and oxygen of water*	M	*diffusion coefficient of water in polymer electrolyte per diffusion coefficient of pure water.*
G	*peak height of RDF between charge bearing group (in polymer) and oxygen of water*	N	*diffusion coefficient of counterion in polymer electrolyte per diffusion coefficient of pure water*

3.3. Activity Coefficients Prediction

The model K was validated against existing datasets from the literature, giving confidence for generating additional ML models with both RFR and SVR. The test set is used to assess the model's performance on previously unexplored data. Prediction accuracy of Both RFR and SVR regression models were compared for predicting activity coefficients in figure 4a. Figure 4b illustrates the correlation between the model K predictions of the activity coefficients using RFR. An RFR architecture with max depth (D = 82.95) with minimum sample split (W = 2) was found to have the best hyperparameters for the model K implementation. With the optimized neural network, MSE values were 0.002 for fivefold cross-validation model training with RFR. This model might further be improved by expanding the scope of data set.

(a)

Data	Metric	RFR	SVR
Train	MSE	0.002	0.001
	R^2	0.96	0.98
Test	MSE	0.002	0.008
	R^2	0.94	0.80

(b)

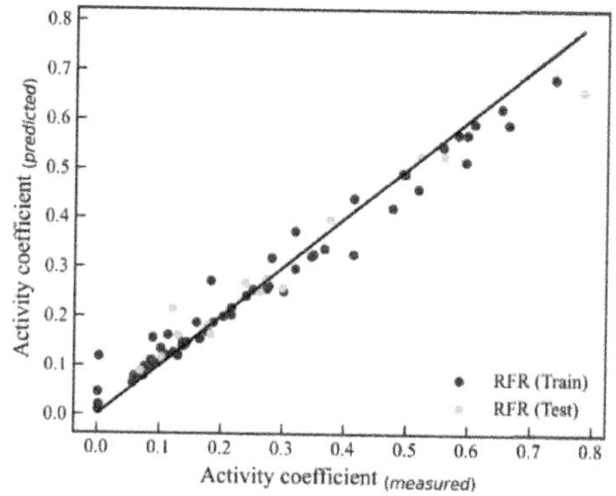

Figure 4. (a) Comparison of different model K for predictions of activity coefficients using RFR and SVR (b) Parity plot of predicted and experimental activity coefficients using the RFR.

4. Conclusions

Machine learning identified physically meaningful patterns between molecular structure, molecular dynamics data and experimentally determined ionic activity coefficients in polymeric ion-exchange membranes and thin films. The outcomes of this framework can inform the rationale design of the materials that enable energy efficien and potentially selective ionic separations.

Acknowledgments

This material is based upon work supported by the U.S. Department of Energy, Office of Science, under the Office of Basic Energy Sciences Separation Science program under Award No. DE- SC0022304. We gratefully acknowledge the computer time allotted by the high-performance computing center at LSU and the Louisiana Optical Network Initiative.

References

Briceno-Mena, L.A., Venugopalan, G., Romagnoli, J.A., Arges, C.G., 2021. Machine learning for guiding high-temperature PEM fuel cells with greater power density. Patterns 2, 100187.

Mekonnen, M.M., Hoekstra, A.Y., 2016. Four billion people facing severe water scarcity. Science Advances 2, e1500323. https://doi.org/10.1126/sciadv.1500323

Palakkal, V.M., Rubio, J.E., Lin, Y.J., Arges, C.G., 2018. Low-Resistant Ion-Exchange Membranes for Energy Efficient Membrane Capacitive Deionization. ACS Sustainable Chem. Eng. 6, 13778–13786. https://doi.org/10.1021/acssuschemeng.8b01797

Sholl, D.S., Lively, R.P., 2016. Seven chemical separations to change the world. Nature 532, 435–437. https://doi.org/10.1038/532435a

Antonis Kokossis, Michael C. Georgiadis, Efstratios N. Pistikopoulos (Eds.)
PROCEEDINGS OF THE 33rd European Symposium on Computer Aided Process Engineering
(ESCAPE33), June 18-21, 2023, Athens, Greece
 http://dx.doi.org/10.1016/B978-0-443-15274-0.50232-8

An NLP-based framework for extracting the catalysts involved in Hydrogen production from scientific literature

Avan Kumar,[a] Hariprasad Kodamana,[a,b,*]

[a]Department of Chemical Engineering, Indian Institute of Technology Delhi, New Delhi-110016, India

[b]Yardi School of Artificial Intelligence, Indian Institute of Technology Delhi, New Delhi-110016, India

*corresponding author: - *kodamana@chemical.iitd.ac.in*

Abstract

Hydrogen production as a renewable fuel source is an active area for the industry and researchers. NLP has been growing for the last few years as the need for extracting non-structural text data is accelerating with time. Here, in this work, we focused on the peer review articles on "Hydrogen production from alcohol" as a keyword for search in an Elsevier database. The abstracts are retrieved with the help of the "Elsevier API" key. We mainly focus on the article's abstract because it contains crucial points. The collated text data set combines two broad classes, i.e., relevant abstracts and non-relevant abstracts. The relevant abstract class stands for hydrogen production from alcohol-based abstracts. Other left-over abstracts will belong to the non-relevant class. The relevant abstracts are followed by the task of "Question and Answering" (Q&A), where we have selected a question, e.g. (a) What is the catalyst material used? Transfer learning implemented on the emerging model "BERT" will be utilized for these tasks. The finetuned BERT model is renamed H_2-BERT. The H_2-BERT performs well, with an accuracy of 0.980 for the train set and 0.974 for the test dataset for the classification task. For the task of Q&A, the data is prepared, fed to the "BERT" model and retrained the last layer weights. This downstream task gave outstanding results with 1.000 accuracy for the train set and an accuracy of 0.823 for the test dataset for accurately predicting the answer to their question based on the given context. This model will help readers extract the utilized catalyst name from the given text and make the process less time-consuming.

Keywords: Hydrogen production, NLP, H_2-BERT, Classification, Q&A

1. Introduction

Nowadays, the requirement for energy is increasing with time, and the depletion of non-renewable energy resources is also happening [1]. Hydrogen is an excellent energy carrier and an environmental-friendly energy source for the future. Recently, many researchers have focused on the production methodology search of H_2 to fulfil the energy demand [2]. A quick search on google scholar will give us millions of patents and research articles. Accessing all of those sources of knowledge and digging deep will be challenging and tedious. It requires expertise in this field and a considerable time to get essential information, such as the catalyst used. All articles hide some crucial parameters that may help the industries to grow. A quick recommendation system can be built for the above reference to extract all the crucial information from available articles to ease access for academicians and industry practitioners. Machine learning and Natural language process (NLP) fields are gaining high popularity to help extract non-structural text data into structural information.

The BERT is a standard model in the NLP field, which stands for "Bi-directional encoder representation of transformers developed by google in 2018. The BERT model is well-trained on Wikipedia, Books, Journals corpus, etc. The downstream tasks performed by the BERT model are Classification, Name entity recognition (NER), and Q&A. It has two classes based on the size of layers of the encoder (i) $BERT_{large}$ (total parameters= 340 M, encoder layers =24) and (ii) $BERT_{base}$ (total parameters = 110 M, encoder layers =12). The performance of BERT encourages many researchers, so many versions of BERT have been launched. The most famous version of BERT is "Sci-BERT". It adopts the architecture of BERT and is trained from scratch with the vocab of computer science (85%) and biomedical (15%) literature. Other finetuned "BERT" are BioBERT, clinical-BERT, m-BERT, and FinBERT [3].

Implementing transfer learning on the BERT model has helped learn the hidden information from a new corpus of data. In this context, Kumar et al. have reported Classification followed by the with an accuracy of 0.997(test dataset) for Classification, and NER also has a good 0.890 (test dataset) accuracy for a finetuned BERT model called "Ex-SciBERT" [4]. Huang et al. created six battery-related BERT models with downstream tasks, Classification and Q&A [5]. Qu et al. have built a ConQA model based on the BERT model; it helps to get insights into historical data [6]. Some recent projects introduced an attention-based deep learning approach for NER tasks for material science literature and disclosed a synthesis procedure for all-solid-state batteries using a 243-article literature corpus.

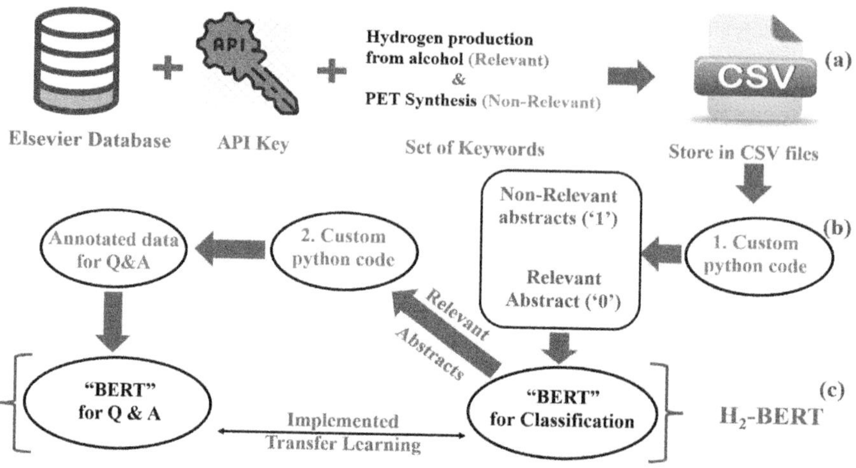

Figure: - (1) A Schematic diagram of the steps followed in this study. Where (a) signifies the steps for text extract, (b) stands for data annotation, and (c) shows transfer learning implementation on the BERT model.

In this study, we propose to develop a finetuned BERT mode for Q&A tasks related to the text that contains information on "Hydrogen production from alcohol" ". First, we will classify relevant class abstracts (means related to hydrogen production information) from the group of abstracts. The relevant abstracts will be processed, and a specific question will be asked. For clear visualization, see Figure 1, where each step of the proposed framework is showcased. The structure of the proposed study is as follows: Section 2 for data extraction and annotation, discussed the proposed model in Section 3, Section 4 for Result & Discussion and at last, a conclusive remark on the proposed study.

2. Data collection and annotation

We have utilized the application programming interface (API) API key to access the complete database of Elsevier. The keywords for the search are "Hydrogen production from alcohol" and a random keyword like "PET synthesis". The total number of accessible abstracts is 11397 using mentioned keywords.

The preliminary steps for cleaning all available abstracts are as follows; delete duplicate entries, some unwanted symbols, etc. These steps are implemented with the help of the "NLTK" library. For the task of Classification, relevant abstracts are marked as '0', and the rest of the other abstracts are labelled as '1'. Afterwards, only relevant abstracts are further processed for configuring the data for the task of Q&A. For both annotations of the dataset, two highly dedicated python scripts are written. The Q&A dataset has a specific question: (i) What is the catalyst material used?

The 'Chemdataextractor' library extracts the chemical entities from given relevant class abstracts, followed by a catalyst filter. It means a set of most probable catalyst elements. It helped to dig out only feasible chemical entities as catalyst material. In Figure 2, all reliable catalyst entities are shown with the help of a word cloud plot. After completion of data annotations, the datasets are fed to the BERT model for respective tasks.

Figure: - (2) All extracted catalyst entities are shown in the word cloud.

3. The proposed model: H_2-BERT

The Classification and Q&A tasks are performed using the BERT architecture and only updating the last layer weight. The BERT architecture that we have chosen has 12 encoder layers, with each encoder's input layer could have a maximum of 512 tokens. The attention layer is the significant difference that makes the "BERT" model more robust for text mining than other sequence-to-sequence models. The BERT model can quickly produce sentiment and contextual outputs.

3.1. Transfer Learning Framework

Transfer learning is a method to reuse the pre-trained weights of the model by skipping the high computational power and time. In our work, we utilize transfer learning to finetune the pre-trained BERT model's last layer weights on our annotated datasets. The training is carried out on a compute cluster with a 32GB GPU in the Python environment using the "PyTorch" framework. The hyperparameters during finetuning are learning rate, number of epochs, and sequence length. We optimized the following hyperparameters for the best performance: epoch= 3, maximum sequence length = 256, learning rate = 0.000003, hidden activation =" gelu", hidden dropout = 0.1, hidden size = 768, maximum position embedding 512, number attention heads = 12, vocabulary size = 28996, and loss function =" Cross-entropy loss". Similarly, for the Q&A task, we have optimized hyperparameters: epochs= 10, learning rate= 0.00004, maximum sequence length = 512, optimizer= AdamW, number of best sizes =3, and loss function is cross entropy. We have implemented the early stopping in both cases to avoid overfitting the model.

4. Results and Discussion

4.1. Statistical analysis of available relevant abstracts

We have obtained a total number of relevant abstracts equal to 5864 related to "Hydrogen production from alcohol" from an Elsevier database search. We have done some statistical analysis of the abstracts represented in the box plot. The box plot presents the dataset maximum, third quartile, median, first quartile, and minimum. The upper and lower bound of the median have covered the 50%-50% of the dataset, as indicated in Figure 3. The number of sentences in abstracts and the average length of each word in the abstract is shown in Figure 3(a); approximate mean values are 7 and 5, respectively. Similarly, the number of words and a number of stop words are represented in Figure 3(b), with mean values around 200 and 143, respectively.

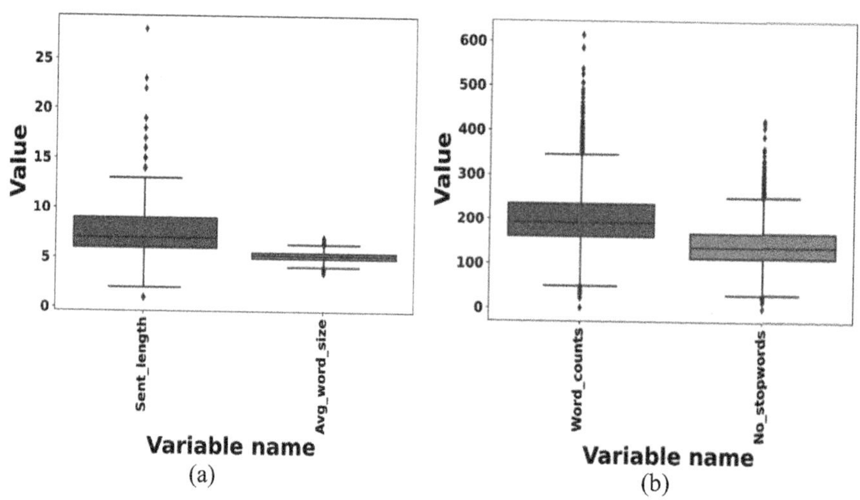

Figure: - (3) The box plots with statistical analysis of some key parameters are shown, (a) Number of sentences and Average length of words present, (b) Number of words and Count of stop words present in each relevant abstract.

4.2. Classification and Q&A using H_2-BERT

An annotated dataset for Classification has two classes: relevant and non-relevant. The dataset is well balanced, approximating 50%-50% of both classes. The data set is divided into the training dataset of 70% test dataset of 30%. While training, cross-validation is 30% of the training dataset to make the model robust. The learning curve with train loss and validation loss with values 0.0694 and 0.0785, respectively. The loss function is cross-entropy.

Moreover, the confusion matrix for the test dataset is shown in Figure 4. The final accuracy, precision and recall, and F1-score value of the model for the Train and test dataset are tabulated in Table 1. The accuracy of the Train set is 0.980, and the test has an accuracy value of 0.974.

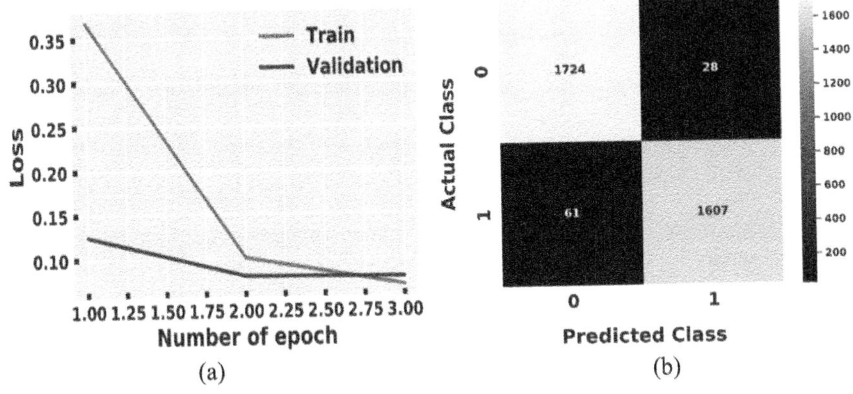

Figure: - (4) The performance evaluation of the classification task by the H_2-BERT model (a) The variation of loss per epoch (b) Confusion matrix for the test dataset.

Table: - (1) The performance metric values for the task of Classification of the H_2-BERT.

Type of data	Name of label	Precision	Recall	F1-score	Counts
Train	Relevant (0)	0.98	0.99	0.98	4112
	Non-relevant (1)	0.99	0.97	0.98	3865
Accuracy	0.980				
Test	Relevant (0)	0.97	0.98	0.97	1752
	Non-relevant (1)	0.98	0.96	0.97	1668
Accuracy	0.974				

Similarly, the Q&A task is performed on the architecture of "BERT" and finetuned weights of the last layer only. We have tabulated the performance of this model in Table 2, where Correct, Similar, and Incorrect predictions and total data points are mentioned. The correct answer means that the prediction and true answer are an exact match, and a similar answer means some portion of the prediction and true answer matched; the rest predictions are entirely wrong. If we consider [correct + similar] answers as right out of total data points, the calculated accuracy for Train and test are 1.000 and 0.823, respectively. At last, we took some random abstracts and checked the class of abstract, whether relevant or non-relevant abstract. Subsequently, we took only relevant classes for further processing. A question (What is the catalyst material used?) was asked, the given abstract as context. The predictions are tabulated in Table 3. These two tasks are performed sequentially.

Table: - (2) Performance of the H_2-BERT for the task of Q&A.

Type of data	Correct	Similar	Incorrect	Total
Train	1734	5	0	1739
Accuracy	1.000			
Test	243	64	66	373
Accuracy	0.823			

Table: - (3) The Q&A prediction on some random abstracts, **bold**, *italic*, and <u>underlined</u>, used for highlighting catalyst entities are present in the abstract, which are predicted correctly, similarly and incorrectly, respectively.

Abstract	Predicted-Catalyst
.... over a series of **y-Al2O3** supported metal NPs catalyst....	y-Al2O3

.... For bimetallic **nickel-indium** alloys, catalysts supported....	nickel-indium
Ni catalysts were prepared by wet impregnation of three....	Ni
.... Recoverable thiroporphyrazine catalyst *(CoPz(S-Bu)8/SiO2@Fe3O4)* was prepared...	CoPz(S-Bu)8/SiO2
...Species on the surface of the CuxNi3-xAlOy catalyst can be...	hydrotalcite

5. Conclusion

In this proposed work, we have implemented NLP tools for information extraction in the form of catalyst entities from scientific literature data. The text data are accessed with the help of the "Elsevier API key" from the complete database of Elsevier. The extracted abstracts are a mixture of relevant and non-relevant classes. The abstracts are annotated with python scripts for the classification task and Q&A task. These annotated datasets are utilized to retrain the weights of the last layer for task classification and Q&A of "BERT" architecture. The model's performance for Classification and the Train and test accuracy are 0.980 and 0.974, respectively.

Similarly, the accuracy of the finetuned model for Q&A for the Train is 1.000, and the test has 0.823, precisely answering the question. As a result, human effort is expected to be reduced in extracting the names of the catalyst entities for identifying catalysts from scientific text and making the process less time-consuming.

6. Acknowledgement

The authors acknowledge the Indian Institute of Technology, Delhi (IIT Delhi) for providing computational resources and a place to carry out this work.

7. References

1. Miller, H. A., Bellini, M., Vizza, F., Hasenöhrl, C., & Tilley, R. D. (2016). Carbon supported Au–Pd core–shell nanoparticles for hydrogen production by alcohol electroreforming. *Catalysis Science & Technology*, *6*(18), 6870-6878.
2. Kumar, A., Pant, K. K., Upadhyayula, S., & Kodamana, H. (2022). Multiobjective Bayesian Optimization Framework for the Synthesis of Methanol from Syngas Using Interpretable Gaussian Process Models. *ACS omega*.
3. Kuniyoshi, F., Makino, K., Ozawa, J., & Miwa, M. (2020). Annotating and extracting synthesis process of all-solid-state batteries from scientific literature. *arXiv preprint arXiv:2002.07339*.
4. Kumar, A., Ganesh, S., Gupta, D., & Kodamana, H. (2022). A text mining framework for screening catalysts and critical process parameters from scientific literature-a study on Hydrogen production from alcohol. *Chemical Engineering Research and Design*.
5. Huang, S., & Cole, J. M. (2022). BatteryBERT: A Pretrained Language Model for Battery Database Enhancement. *Journal of Chemical Information and Modeling*.
6. Qu, C., Yang, L., Qiu, M., Croft, W. B., Zhang, Y., & Iyyer, M. (2019, July). BERT with history answer embedding for conversational question answering.

Antonis Kokossis, Michael C. Georgiadis, Efstratios N. Pistikopoulos (Eds.)
PROCEEDINGS OF THE 33rd European Symposium on Computer Aided Process Engineering (ESCAPE33), June 18-21, 2023, Athens, Greece
 http://dx.doi.org/10.1016/B978-0-443-15274-0.50233-X

Simulation of low-temperature district heat networks from mine water energy

Thomas Cowley,[a] Timothy Hutty,[a] Solomon Brown[a]

[a] *Department of Chemical and Biological Engineering, University of Sheffield, Sheffield, S1 3JD, United Kingdom,*
s.f.brown@sheffield.ac.uk

Abstract

Mine water sourced district heat networks (DHN) are a pathway for reducing emissions of the heating sector. Here, an agent-based model for network dynamics is developed and showcased through a case study in the South Yorkshire Mayoral Combined Authority (SYMCA), consisting of 3437 domestic buildings and 8 non-domestic buildings used as anchor loads. The results show that the mine water DHN achieved a levelised heat (LCOH) cost of 12.6p/kWh compared to the boiler-only scenario of 6.8p/kWh. Emissions in the region could be reduced by 83.6% and can compete with boilers with a natural gas price of 10.8 p/kWh, highlighting the potential of mine water energy for decarbonising the heating sector.

Keywords: energy, heat decarbonisation, district heat networks, agent-based modelling

1. Introduction

To meet Net-Zero by 2050, the heating sector in the UK needs innovation and development. Currently, 85% of residential buildings utilise a natural-gas grid connection composed of standalone boilers (BEIS, 2022). The UK's Sixth Carbon Budget restricts new gas grid connections by 2025 and prescribes the phase-out of boilers by 2035 (Committee on Climate Change, 2020). This allows district heat networks (DHN) to penetrate the heating sector. Approximately 25% of UK homes are built above abandoned mine workings which are flooded with water between 10-20°C (The Coal Authority, 2021). With this source of low-grade heat, large-scale heat pumps can achieve a higher coefficient of performance (COP) due to the warmer energy source and better thermal efficiency due to the proximity of the source to heat demand.

Mine water networks are susceptible to location-specific parameters, and models should include features such as GIS to improve accuracy and spatial interactions between agents. A review (J. Keirstead et al, 2012) of 219 publications found that only 44% of models incorporated spatial features, and 58% used a yearly temporal scale or greater. Additionally, only a few studies incorporate fully dynamic models, most focusing on techno-economic studies, singular node simulation, and steady-state models (J. Wang et al, 2016). The work of E. Guelpa et al (E. Guelpa, 2020) analyses the relevance of implementing network dynamics such as thermal losses, time delay propagation and thermal transience. The results found that these phenomena had a significant impact when changes to the operating mode (i.e., during the night, start-up, shut-down), demand peaks and large variations in volumetric flow rate. Similarly, work by J. Allegreni (J. Allegrini, 2015) confirms that network dynamics contributed to a 20% discrepancy between the simulated thermal peak request and the measured data. Capturing these dynamics will vastly improve the accuracy of an extensive network simulation with variable flow at a short time resolution.

It will be of interest to project planners to gauge if a heat network project utilising mine water can reduce emissions and provide competitive heat prices to consumers (BEIS, 2018). Therefore, this work aims to model the network dynamics using an agent-based approach to decrease computational intensity, allow flexibility of location-based parameters, and incorporate network dynamics in an extensive network. Undertaking techno-economic analysis while capturing these features will help understand the trade-off between the cost and emissions of a region with mainly standalone gas boilers. Hence, the objectives are to:

- Develop an agent-based model to capture the network dynamics of DHNs
- Showcase the model through a techno-economic case study and determine at what price of natural gas the proposed heat network becomes a viable investment.

2. Methodology

This work aims to simulate the extraction of mine water energy from unused coal mine workings to supply domestic and non-domestic heating demand. The mine water heat is upgraded via large-scale heat pumps at a geothermal plant located above the mine workings, intersected by a borehole, and distributed around a network of pipes. A visual representation of the model is shown in Figure 1. Simulation of a DHN is an inherently complex system; therefore, to reduce the computational intensity of the model, only the phenomena that have a significant impact on the accuracy of the network dynamics are included. Thus, the following assumptions have been made:

- Water properties are constant parameters.
- Water is considered an incompressible fluid.
- Flow within pipeline agents is one-dimensional.
- Heat loss to the surroundings is one-dimensional.
- There is no degradation of piping material or insulation over the simulation time
- Axial conduction in each pipe is considered insignificant.
- The pressure drops in the pipeline agents have negligible viscous heating effects.
- Perfect mixing and adiabatic operation when flow combines at junctions.
- Non-constant terms in the momentum equation are neglected
- Pressure loss and efficiency losses within the HIU are considered insignificant

The model uses node agents, which consist of geothermal plant centres (GCs), demand centres (DCs), junctions (J) and pipes with start and end points at nodes and are classed as pipeline agents (P). An arbitrary number of pipeline agents can connect to a single node. Each node agent has several parameters, which are pre-determined and describe its location and whether there is a physical connection to the network. Each pipeline agent is tied to the corresponding start and end node agents. The length of each pipeline agent is calculated by geolocating the distance between the start and end node agents. All primary agents (GC, DC, J, P) interact with each other within a GIS space. Water packet agents are classed as secondary agents and are instantaneously generated in the case of volumetric flow or temperature changes in the network. The water packet agents allow heat propagation from a source at a finite speed, so that time delays over large distances are captured. The pipeline agents form the distribution network that follows the network's topology, while junction agents allow the splitting of flow and integration of loops. Generation centres and junctions have pumping stations that maintain the network's pressure, and the pressure block calculates its pumping requirements. The demand model block uses a linear regression model at multiple time intervals and assigns buildings to five archetypes: non-domestic (housing), leisure, education, commercial and residential. The thermal sub-model calculates the radial heat loss from the pipe to the soil and accounts for transient heat loss when the water in the pipes is stationary, and there is no flow. The mass flow model uses linear programming to maximise the flow through the distribution network.

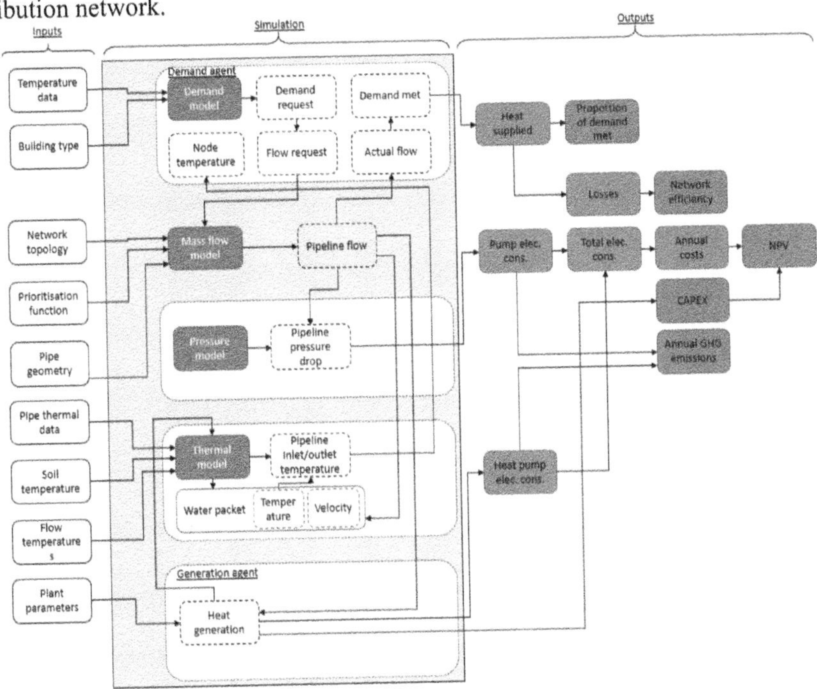

Figure 1 - Flowchart of heat network dynamic model

2.1. Case study

The case study location has many flooded mine workings; this provides an ambient and stable body of renewable low-grade heat. A borehole was chosen from the Coal Authority BGS database (British Geological Survey, 2022) nearest the heat demand and intersected the flooded mine workings at a temperature of 18 °C and a depth of 332 m. 3445 buildings were allocated to the five building archetypes (Commercial: 2, Education: 4, Leisure: 1, Residential: 1), with 3437 domestic buildings aggregated into 13 different demand agents. Eight non-domestic buildings are used as anchor loads to ensure consistent demand on the network. Analysis of the costs incurred by negating emissions from standalone boilers will inform the region of the use of mine water energy to decarbonise its heating sector; therefore, the case study is run with the same demand profiles generated from the demand model block for both the mine water heat network and boilers-only scenarios, while assuming an 80% efficiency for each individual boiler. Table 1 shows the parameters for the case study. A COP of 4.9 (A. David et al, 2017) has been selected as it is aligned with values from the literature that aggregate the supply and return temperatures of 58 different European heat networks. Table 2 shows the cost parameters used for the techno-economic analysis. Temperature data from 2021 (Met Office, 2021) is used as input into the demand sub-model. The network length is generated by the model and is ca. 12 km. Pipe dimensions were calculated pre-simulation using insulation class 1 pipes (Fintherm, 2019). The maximum pipe velocity was 5 m/s to prevent rapid degradation of piping material and to adhere to noise, safety and vibration regulation.

Table 1 – Network parameters

	Value
Pump efficiency (%)	80
Motor efficiency (%)	80
Network pressure (kPa)	1,500
Supply temperature (°C)	68
Return temperature (°C)	18
Soil temperature (°C)	10
Water conductivity (W/mK)	0.598
Water diffusivity (mm^2/s)	0.16
Lifetime (years)	30
Carbon factor natural gas (t/MWh)	0.210
Average carbon factor electricity (t/MWh)	0.136

Table 2 – Cost parameters

	Value
Electricity price (p/kWh)	13
Plant cost (£k/MW)	666.8
Gas price (p/kWh)	4.94
Operation costs (£/kWh)	35.9
Demand Centre costs (£/kWh)	919
Substation cost (£/kWh)	16
Ancillary cost (£/kWh)	68
Pipework cost (£/kWh)	516
Boiler CAPEX and OPEX cost (£/kWh)	2.28
Inflation rate (%)	2.7
Discount rate (%)	3.5

3. Results & Discussion

The simulation was run for one year (8760 hours) for the mine water scenario and the boiler-only scenario using temperature data from 2021. Tables 3 and 4 show the KPIs (Key performance indicators) and costs for both scenarios. The mine water scenario showed a high efficiency of 96.3% by the heat network and met 99.98% of the region's demand. This highlights the model's success in distributing the heat effectively to consumers while minimising the heat lost to the surrounding soil compared to the generated heat. 0.02% of the demand was not supplied to the consumers due to the water velocity constraints. These consumers tended to be at the end of the network.

Figure 2 shows the total demand on the network across the year and the variations over each day. As the network consists of mainly non-domestic buildings, this has a more significant effect on the demand profile of the network. The demand on the network during winter is greater than during summer, so the buildings are maintained at a comfortable temperature. Demand peaks between 6:30 and 10 am as more hot water is needed for showers before people go to work, and building heating is turned on. There are also peaks between 16:00 and 23:00, as the heating is turned back on as people get back from work. This is less prominent in the summer months, but there is still, on average, some smaller peaks in the morning. During the night, the network cools, and this is due to heat being lost when the water in the network is stagnant; this is the effect of transience in the network. Positive demand is seen on the network throughout the year due to the anchor loads provided by the non-domestic building types.

BEIS reported that the average household energy consumption of a typical domestic dwelling was 12.3 MWh/year (A. O'Mahoney, 2020); this corroborated the demand model results with an average of 12.2 MWh/year over the 3437 domestic buildings. The mine water network scenario emitted 1807 tCO_2e, whereas the boiler-only scenario 11013 tCO_2e, thus achieving achieved 83.59% reduction in emissions compared to boiler-only scenario. Over a 30-year lifetime, 276 million tCO_2e could be saved by 2050. Mine water LCOH (Levelised cost of heat) obtained was 12.4 p/kWh, which is slightly more economical than work by O. Gudmundsson et al (O. Gudmundsson et al, 2013), which found that geothermal DHN achieved 14.0 p/kWh for outer city projects. The boiler-only scenario obtained an LCOH of 6.8 p/kWh, significantly lower than the literature value of 11p/kWh. The disparity may have been due to lower efficiencies in actual gas boiler installation or underestimated cost parameters. The NPV (Net present value) of the boiler-only scenario is -£90.5 million, whereas the heat network NPV requires 46.3% more investment totalling -£132.4 million. The disparity in NPV is primarily due to the already extensive gas infrastructure in the UK and the price of natural gas. The mine water network scenario can abate emissions at an average cost of 151.7£/tCO_2e over its lifetime.

Table 3 – KPIs from the case study

	Mine water	Boiler-only
Efficiency (%)	96.3	80
Heat demand (MWh/year)	41,964	-
Heat supplied (MWh/year)	41,955	-
Heat generation (MWh/year)	43,594	52444
Demand supplied (%)	99.98	100
Pumping power (MWh)	529.55	-
Heat pump output (MW)	18.75	-
Emissions (tCO_2)	1,807	11,013
Emission savings (%)	83.59	0

Table 4 – Costs over a 30-year lifetime

	Mine water	Boiler-only
Total capital cost (£)	60,880,000	24,421,000*
Total operating cost (£)	43,077,000	-
Total energy cost (£)	28,418,000	66,054,000
Total cost over lifetime (£)	132,379,000	90,475,000
LCOH (p/KWh)	12.38	6.77

*Both capital and operational cost

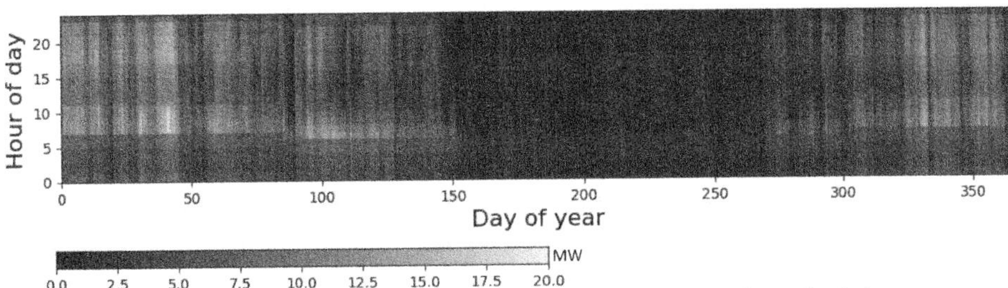

Figure 2 – Average heat demand of nodes in the heat network over the year (and over the day)

The wholesale gas price has seen a significant increase in the final quarter of 2021 due to the high demand for natural gas, and even more so in 2022 with the start of the war in Ukraine. 2020 saw the wholesale gas price at a minimum of 4.39 p/kWh, and a maximum of 27.30 p/kWh (Trading Economics, 2021); therefore, DHN would provide a more stable price of heat and improved energy security for customers by providing protection from global fluctuating gas prices. If the gas price trends upward, the mine water scenario becomes more economically attractive. The sensitivity analysis in Figure 3 shows that mine water can provide competitive prices at a natural gas price of 10.8p/kWh.

Figure 4 shows the percentage change in the LCOH when the total heat demand, capital cost and operational costs are modified by plus or minus 50%. The network is more sensitive to demand changes than operational costs, as it increases by roughly 18%

when the demand on the network is reduced by half. In contrast, it decreases by roughly 5% when demand is increased by half. Moreover, decreasing the capital and operational costs by 50% leads to a decrease in the LCOH of 16% and 24%, respectively. The energy costs of the Levelized cost method are assumed to be constant over the technologies' lifetimes. They are not the prices to sell energy. Hence, it does not reflect energy prices' short-term or long-term volatilities.

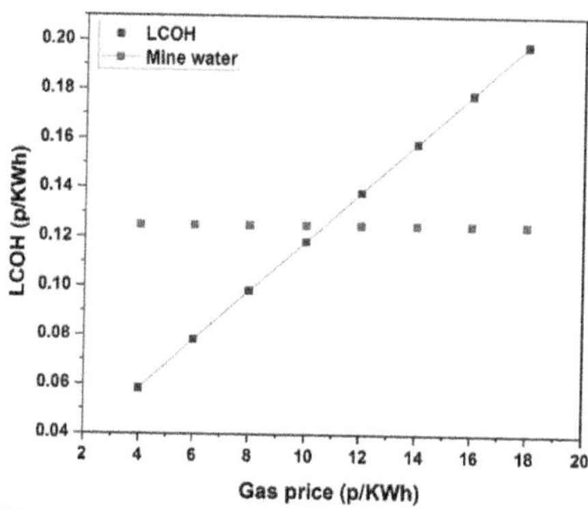

Figure 3 – LCOH as a function of gas price

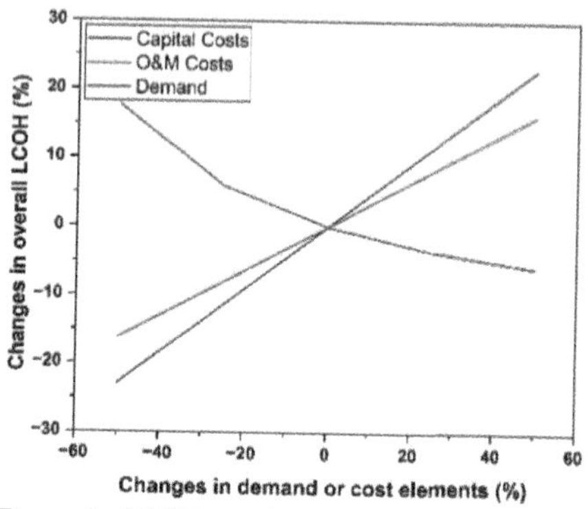

Figure 4 – LCOH as a function of changes to costs and demand

4. Conclusion

Utilising renewable energy sources with DHN is a possible way of decarbonising the UK's heating sector. Mine water energy paired with large-scale heat pumps is proven throughout European projects. Current research lacks accurate modelling of network dynamics. The model successfully captured the network dynamics and expected network demand profiles and obtained an efficiency of 96.3%. The case study was found to reduce the region's carbon emissions by 83.59% and provide an average LCOH of 12.38 p/kWh over the project's lifetime. Sensitivity analysis shows that the project became economically feasible at a natural gas price of 10.8p/kWh, highlighting a promising pathway to reduce emissions.

5. References

A. David, et al, 2017, "Heat Roadmap Europe: Large-Scale Electric Heat Pumps in District Heating Systems", Energies, 10, 4, 578.

A. O'Mahoney, 2020, "Review of Typical Domestic Consumption Values 2021," The Office of Gas and Electricity Markets.

BEIS, 2018, "Heat Networks: Ensuring sustained investment and protecting consumers," GOV.UK.

BEIS, 2022, "Decarbonising heat in homes," GOV.UK.

British Geological Survey, "Coal Authority Borehole Log Data," GOV.UK.

Committee on Climate Change, 2020, "The Sixth Carbon Budget,".

E. Guelpa, 2020, "impact of network modelling in the analysis of district heating systems," energy, vol. 213, p. 118292.

FINTHERM, 2019, "Pre-insulated pipes and accessories catalogue,".

J. Allegrini, K. Orehounig, G. Mavromatidis, F. Ruesch, V. Dorer, a. R and Evins, 2015, "A review of modelling approaches and tools for the simulation of district-scale energy systems," Renewable and Sustainable Energy Reviews, pp. 1391-1404.

J. Keirstead, M. Jennings and A. Sivakumar, 2012, "A review of urban energy system models: Approaches, challenges and opportunities," Renewable and Sustainable Energy Reviews, vol. 16, no. 6, pp. 3847-3866.

J. Wang, Z. Zhou and J. Zhao, 2016, "A method for the steady-state thermal simulation of district heating systems and model parameters calibration," Energy Conversion and Management, vol. 120, pp. 294-305.

Met Office, 2021, "MIDAS Open: UK daily temperature data, v202107", NERC EDS Centre for Environmental Data Analysis.

O. Gudmundsson et al, 2013, "Cost analysis of district heating compared to competing technologies", WIT Transactions on Ecology and the Environment, 176 107-118.

The Coal Authority, 2021, "Mine Water Heat," GOV.UK.

Trading Economics, 2021, "UK Natural Gas,".

Antonis Kokossis, Michael C. Georgiadis, Efstratios N. Pistikopoulos (Eds.)
PROCEEDINGS OF THE 33rd European Symposium on Computer Aided Process Engineering (ESCAPE33), June 18-21, 2023, Athens, Greece
 http://dx.doi.org/10.1016/B978-0-443-15274-0.50234-1

In-silico investigation of microalgae culture performance in lab and pilot scale: The trade-off between reactor size and productivity.

Christos Chatzidoukas*, Vasileios Parisis
Department of Chemical Engineering, Aristotle University of Thessaloniki (AUTH), P.O. Box: 472, 54124, Thessaloniki, Greece
chatzido@auth.gr

Abstract

The biomass productivity of photo-autotrophic microalgae is known to be influenced by a plethora of environmental factors, with light being the most notable one. Given the critical role of light in the complex metabolic network of microalgae, conditionally directing the carbon flow either towards biomass growth or lipid accumulation, the empirical operational design of a photobioreactor system can only be suboptimal. This task becomes even more mazy in larger scales, where the culture growth results in more intensive light attenuation. The present work analyses the importance of a structured, distributed mathematical model for the design of optimal lighting policies of a photobioreactor, and investigates the successful shift of microalgae cultures from a lab to a pilot scale unit. The key contribution of this study is to help realize the performance bounds of a photobioreactor depending on its geometry and size and to showcase how to deal with the lighting policy when scaling-up of a microalgae bioreactor is desired.

Keywords: Lab/pilot scale photobioreactor; Structured macroscopic/kinetic model; Dynamic optimization; Light path; Photobioreactor design.

1. Introduction

Despite the ever increasing academic and industrial attention and research on microalgae based biorefineries over the last decades, the commercialization of microalgae products is hindered by the low productivity of photobioreactors and open culture systems. Particularly in autotrophic cultures, light is the most notable limiting growth factor. Its availability inside the culture volume is significantly limited by the high optical density of microalgae suspensions. Since the measurement of light intensity inside the culture often requires expensive equipment, the empirical deduction of lighting polices that maximize biomass production is unfeasible. Model-based optimization has the potential to facilitate the above task, supporting the sustainable use of microalgae in nutritional and biofuel applications, and fill in the gaps in our understanding of the impact of the lighting on microalgae growth.

In this study, a multi-physics mathematical model is developed and validated, with the view to systematically obtain the optimal lighting policies that maximize biomass growth inside stirred photobioreactors. Three scenarios are devised that reflect the impact of scale up on process performance, which is assessed by proper process KPIs.

2. Model development

In a photoautotrophic culture, microalgae fix CO_2 into organic carbon photosynthetically, and subsequently mobilize it, in the presence of nitrogen and phosphorus, to synthesize the cellular machinery needed for cell division and maintenance. Furthermore, microalgae store their organic carbon in the form of carbohydrates and/or lipids, especially when their environment becomes limiting, i.e., nutrient exhaustion. A distributed/structured dynamical model is developed to describe the dynamic operation of typical stirred photobioreactors (PBRs) employed for microalgae cultures. The model has been built up upon the well-established Droop model (Droop, 1983), incorporating biological phenomena like photoinhibition and photo acclimation, via chlorophyll-a dynamics. Carbon allocation and product synthesis (i.e., carbohydrates, proteins, and lipids) are dependent on the internal state of the cell as well as the nutrient availability in the culture medium. The one-dimensional Cornet model (Cornet et al., 1992a, 1992b) is adopted and modified in order to describe the light gradients that develop inside the photobioreactor along with the evolution of microalgal biomass and pigment content. Tables 1 and 2 summarize the dynamic mass balances and the specific rates of the model.

Table 1: Dynamic mass balances of the model.

$$\frac{d(V_l[O_{2,l}])}{dt} = Q_{l,fd}[O_{2,l,fd}] - Q_{l,out}[O_{2,l}] + V_l OTR + V_l Y_{CO_2}\langle\mu\rangle X - V_l Y_{CO_2} R X$$

$$\frac{d(V_l[CO_{2,l}])}{dt} = Q_{l,fd}[CO_{2,lfd}] - Q_{l,out}[CO_{2,l}] + V_l CTR + V_l Y_{CO_2} R X - V_l Y_{CO_2}\langle\mu\rangle X$$

$$\frac{d(V_l[O_{2,g}])}{dt} = Q_{g,fd}[O_{2,g,fd}] - Q_g[O_{2,g}] - V_l OTR$$

$$\frac{d(V_g[CO_{2,g}])}{dt} = Q_{g,fd}[CO_{2,g,fd}] - Q_g[CO_{2,g}] - V_l CTR$$

$$\frac{d(V_l)}{dt} = -\frac{d(V_g)}{dt} = Q_{l,fd} - Q_{l,out}$$

$$\frac{d(V_l[N_l])}{dt} = Q_{l,fd}[N_{l,fd}] - Q_{l,out}[N_l] - V_l\langle\rho_N\rangle X$$

$$\frac{d(V_l[P_l])}{dt} = Q_{l,fd}[P_{l,fd}] - Q_{l,out}[P_l] - V_l\rho_P X$$

$$\frac{d(V_l X q_N)}{dt} = \langle\rho_N\rangle$$

$$\frac{d(V_l X q_p)}{dt} = \rho_P$$

$$\frac{d(V_l X)}{dt} = Q_{l,fd}X_{in} - Q_{l,out}X + V_l(\langle\mu\rangle - R)X$$

$$\frac{d(V_l\theta X)}{dt} = V_l\frac{K_{accl\theta}}{\theta_{ad}}(\theta_{ad} - \theta), \frac{\theta_{ad}(I)}{q_N} = \varphi_{max}\frac{k_{I^*}}{k_{I^*} + \langle I\rangle}$$

$$\frac{d(V_l C)}{dt} = Q_{l,fd}C_{fd} - Q_{l,out}C + V_l\langle\mu\rangle X - V_l r_{C,Pr}X - V_l r_{C,L}C - V_l R X$$

$$\frac{d(V_l L)}{dt} = Q_{l,fd}L_{fd} - Q_{l,out}L + V_l r_{C,L}C + V_l D_p Pr$$

$$\frac{d(V_l Pr)}{dt} = Q_{l,fd}Pr_{fd} - Q_{l,out}Pr + V_l r_{C,Pr}X - V_l D_p Pr$$

where: $[O_{2,i}]$ $[CO_{2,i}]$ oxygen and carbon dioxide cocnentrations (gas & liquid phase); $[N_l]$, $[P_l]$ nitrogen and phosphorus consentrations; q_N, q_P, θ nitrogen and phosphorus and chlorophyll-a cellular quotas; X: biomass; C: carbohydrates; L: lipids; Pr: proteins; R: respiration rate; OTR, CTR: O_2 and CO_2 transfer rates; Y_{CO_2}: CO_2 to biomass yield coefficient;

3. Model validation and biomass optimization.

Experimental data from the work of Psachoulia and Chatzidoukas, (2021) were utilised in the scope of parameter tuning. The datasets contain dynamic information about the main macroscopic process variables (nutrients, biomass, intracellular products) in *Stichococcus sp.* photoautotrophic cultures in a lab stirred photobioreactor. The modified Cornet sub-model was validated offline using the experimental data from Ma et al., (2022).

Table 2: Specific growth/uptake/production/consumption rates of the model.

$$\rho_P = \rho_{P,max}\left(\frac{[P_l]}{K_{S,P}+[P_l]}\right)\left(\frac{q_{P,max}-q_P}{q_{P,max}-q_{P,min}}\right)$$

$$\rho_N(r) = \rho_{N,max}\left(\frac{[N_l]}{K_{S,N}+[N_l]}\right)\left(\frac{q_{N,max}-q_N}{q_{N,max}-q_{N,min}}\right)\left(\frac{q_P-q_{P,min}}{q_{P,max}-q_{P,min}}\right)\left(\eta+(1-\eta)\frac{I}{I+\varepsilon_I}\right)$$

$$\mu(r) = \mu_{max}\frac{I}{I+K_{s,I}+\frac{I^2}{K_{iI}}}\frac{S_{CO_2,l}}{K_{S,C}+S_{CO_2,l}}\left(1-\frac{[O_{2,l}]}{[O_{2,l,max}]}\right)\left(1-\frac{q_{N,min}}{q_N}\right)\left(1-\frac{q_{P,min}}{q_P}\right)$$

$$r_{C,Pr} = \alpha\langle\rho_N\rangle$$

$$r_{C,L} = \mu_{c,s}+\mu_{c,m} = \mu_{c,m,max}\frac{q_N}{q_{N,max}}+\mu_{c,s,max}\left(1-\frac{q_N}{q_{N,max}}\right)$$

$$D_p = \gamma\mu_{c,s}$$

$$I(r) = I_0\frac{4\alpha_1}{(1+\alpha_1)^2e^{\alpha_2 r}-(1-\alpha_1)^2e^{-\alpha_2 r}};\ \alpha_1 = \sqrt{\left(\frac{E_a' k_a\theta}{E_a' k_a\theta+E_s}\right)};\ \alpha_2 = (E_a' k_a\theta+E_s)\alpha_1 X$$

$$\langle z_i\rangle = \frac{1}{V_l}\oiiint_{V_l} z_i(\boldsymbol{x})\,d\boldsymbol{x},\quad where\ \ z_i: \rho_N(r), or\ \mu(r), or\ I(r)$$

where: ρ_P: specific phosphorus uptake rate; $\rho_N(r), \mu(r)$: specific nitrogen uptake rate & specific growth rate at depth r; $r_{C,Pr}, r_{C,L}$ specific carbon allocation rate to proteins & lipids; D_p specific protein degradation rate to lipids; $I_0, I(r)$ Incident light intensity & light intensity at depth r.

Subsequently, the validated model was used to guide the optimization of a batch photobioreactor, in the context of biomass production, on three different spatial scales. The incident light intensity on the photobioreactor lateral surface, I_0, is the sole control variable of the problem, and is modelled as a piecewise-constant function using ten 1-day time intervals. The three different scenarios are presented below, whereas the characteristic dimensions of the three photobioreactors are detailed in Table 3:

Table 3: Characteristic dimensions and optimization strategies of the three PBRs.

PBR ID	Diameter, T (m)	Height, H (m)	Working volume, V_l (L)
1	0.16	0.16	3.2
2	0.32	0.32	25
3	0.16	9.9	200

<u>Scenario 1</u>: Maximization of the final biomass concentration in the lab PBR.

Task	Objective	Operational constraint
Task: $\max_{I_0} \boldsymbol{Obj_1}$	$Obj_1 = X_{PBR1}$	$50 \le I_0 \le 2{,}000$ $\mu mol\ m^{-2}\ s^{-1}$

w.r.t model equations of **Tables 1** and **2**

<u>Scenario 2</u>: Optimal tracking of the biomass profile achieved in the lab scale by the semi-pilot PBR.

Task	Objective	Operational constraint
Task: $\min_{I_0} \boldsymbol{Obj_2}$	$Obj_2 = \int_0^{t_f}\left(\frac{X_{PBR1}-X_{PBR2}}{X_{PBR1}}\right)^2 dt$	$50 \le I_0 \le 3{,}000$ $\mu mol\ m^{-2}\ s^{-1}$

w.r.t model equations of **Tables 1** and **2**

<u>Scenario 3</u>: Optimal tracking of the biomass profile achieved in the lab scale by the pilot PBR.

Task	Objective	Operational constraint
Task: $\min_{I_0} \boldsymbol{Obj_3}$	$Obj_3 = \int_0^{t_f}\left(\frac{X_{PBR1}-X_{PBR3}}{X_{PBR1}}\right)^2 dt$	$50 \le I_0 \le 3{,}000$ $\mu mol\ m^{-2}\ s^{-1}$

w.r.t model equations of **Tables 1** and **2**

An energy efficiency factor, adopted to compare the performance of the three PBRs, is defined as the produced biomass per unit of lighting power input: $\boldsymbol{e} = \frac{X}{a \cdot I_0 \cdot A_{lat}}, \left(\frac{gL^{-1}}{W}\right)$, where $\boldsymbol{A_{lat}}$ is the illuminated surface area of the PBR and $\boldsymbol{a} = \mathbf{0.219}$ is a conversion factor between $\boldsymbol{\mu mol\ m^{-2}\ s^{-1}}$ and $\boldsymbol{W\ m^{-2}}$.

4. Results and discussion.

The model predictions are plotted against the experimental data in Figure 1. The dynamic trajectory evolution of the main state variables is represented by the model, thereby establishing a sound basis for further exploration of the optimal lighting policies that maximize production on different scales.

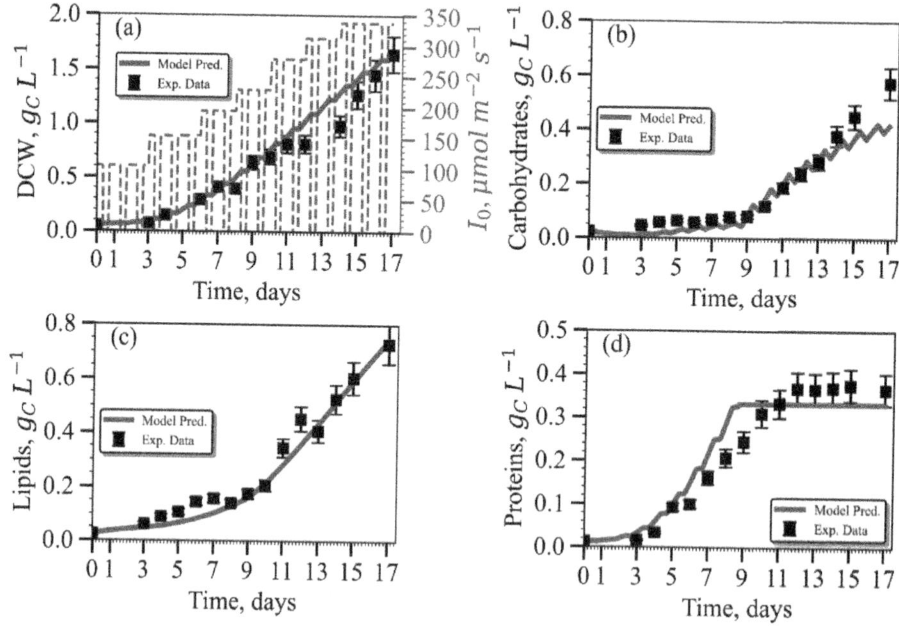

Figure 1: Comparison of the model predicted time trajectories to the experimental data: a) biomass dry cell weight concentration; b) carbohydrates; c) lipids and d) proteins.

Figure 2 depicts the optimal lighting policy for a batch mode cultivation along with the corresponding biomass evolution trajectory, on the three spatial scales. The increasing culture density promotes progressively steeper light gradients inside the culture volume. This results in constantly escalating lighting demand until the point of saturation. In Scenario 1, the overall increased energy input to the system, compared to the validation experiment, results in approximately a 35% increase in volumetric productivity, demonstrating the value of model-based optimization. In Scenario 2, the lighting profile that will steer the semi-pilot PBR on the respective biomass growth trajectory attained in the lab PBR is sought. Despite the increase in light provision in this setup serious loss in biomass productivity is observed. This is entirely attributed to the large proportion of "dark volume" in the semi-pilot PBR as one can deduce inspecting the spatial profile of light distribution reducing sharply over time (Figure 2 d-f). On the contrary, in Scenario 3,

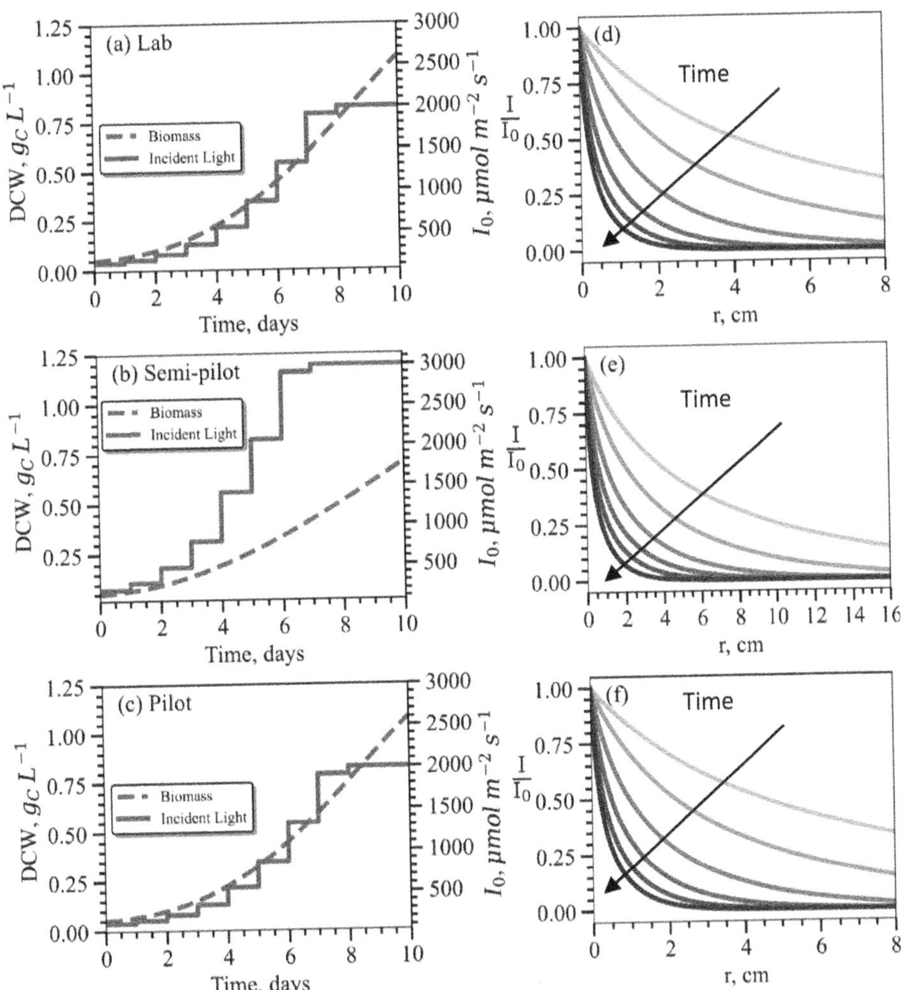

Figure 2: Light optimization in three spatial scales. (a-c): biomass trajectory and incident light profile; (d-f): Light gradients inside the PBRs during a batch cultivation.

the biomass evolution in the PBR with the largest height/diameter ratio is identical to its lab scale counterpart. Though it may seem that tall and skinny geometries can pave the way to a successful scale-up of microalgae cultures, one should address this matter from a techno-economic viewpoint as well. Indeed, Figure 3 demonstrates that the photobioreactor size is practically correlated with a reduced lighting efficiency, and this is consistent across the culture time horizon. This point is in agreement with the findings of Nikolaou et al., (2016). The value of 'e' decreases by about one order of magnitude during scale up from the lab scale to the semi-pilot scale (increase in reactor radius from 8 cm to 16 cm), and furthermore decreases by about two orders of magnitude on the pilot scale, compared to the lab scale. This comparison indicates that PBRs with the same specific illuminated area per unit volume can attain the same productivity with identical light flux, regardless of their volume. However, in our viewpoint, the absolute light power (W) and not the luminous flux (W m^{-2}) spent is a more appropriate criterion to gauge PBR efficiency on a techno-economic basis.

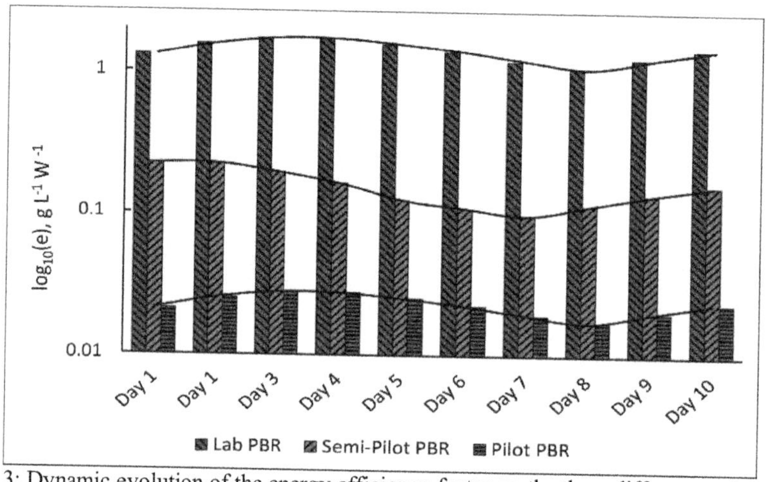

Figure 3: Dynamic evolution of the energy efficiency factor on the three different scales.

5. Conclusions.

In this work the notion of light limitation on photobioreactor is confirmed and quantified, and an attempt to explore strategies that maximize biomass production in photobioreactors on three different scales has been made. Results suggest that the industrial cultivation of microalgae on "traditional" fermenter like photobioreactors seems unrealistic. Furthermore, it is considered that is of utmost importance that photobioreactor design, process optimization and scale-up should be coupled activities, due to the dominant effect of the characteristic length of light distribution on the performance of these systems. We are confident that the present modeling framework can serve at the epicenter of a multi-physics platform that can provide valuable insight and guide the successful commercialization of microalgae based biorefineries.

References

Cornet, J.F., Dussap, C.G., Cluzel, P., Dubertret, G., 1992a. A structured model for simulation of cultures of the cyanobacteriumSpirulina platensis in photobioreactors: II. Identification of kinetic parameters under light and mineral limitations. Biotechnol. Bioeng. 40, 826–834.

Cornet, J.F., Dussap, C.G., Dubertret, G., 1992b. A structured model for simulation of cultures of the cyanobacteriumSpirulina platensis in photobioreactors: I. Coupling between light transfer and growth kinetics. Biotechnol. Bioeng. 40, 817–825.

Droop, M.R., 1983. 25 Years of Algal Growth Kinetics A Personal View. Bot. Mar. 26.

Ma, S., Zeng, W., Huang, Y., Zhu, Xianqing, Xia, A., Zhu, Xun, Liao, Q., 2022. Revealing the synergistic effects of cells, pigments, and light spectra on light transfer during microalgae growth: A comprehensive light attenuation model. Bioresour. Technol. 348, 126777.

Nikolaou, A., Booth, P., Gordon, F., Yang, J., Matar, O., Chachuat, B., 2016. Multi-Physics Modeling of Light-Limited Microalgae Growth in Raceway Ponds. IFAC-Pap. 49, 324–329.

Psachoulia, P., Chatzidoukas, C., 2021. Illumination Policies for Stichococcus sp. Cultures in an Optimally Operating Lab-Scale PBR toward the Directed Photosynthetic Production of Desired Products. *Sustainability* 13, 2489.

Antonis Kokossis, Michael C. Georgiadis, Efstratios N. Pistikopoulos (Eds.)
PROCEEDINGS OF THE 33rd European Symposium on Computer Aided Process Engineering (ESCAPE33), June 18-21, 2023, Athens, Greece
 http://dx.doi.org/10.1016/B978-0-443-15274-0.50235-3

Process modelling of the hydrothermal liquefaction of oil-palm waste for biocrude and hydrochar production

Muhammad Shahbaz[a*], Mohammad Alherbawi[a] Prakash Parthasarathy[a], Gordon McKay[a], Tareq Al-Ansari[a*]

[a] *College of Science and Engineering, Hamad Bin Khalifa University, Qatar Foundation, Doha, Qatar*

[b]*Centre of Bio-Fuel and Biochemical Research, Department of Chemical Engineering, Universiti Teknologi PETRONAS, 32610 Bandar Seri Iskandar, Perak, Malaysia.*

Email: talansri@hbku.edu.qa, mshahbaz@hbku.edu.qa

Abstract

Energy production from biomass has promising prospects in the domain of clean and sustainable sources. Hydrothermal Liquefaction (HTL) is an emerging process to convert biomass into energy products and is less energy intensive than other thermal conversion processes including gasification and pyrolysis, as HTL does not require pre-drying of biomass. In this study, a process simulation is developed using Aspen Plus® V10 to quantify the yield of liquefaction products including bio-crude, hydrochar and gas from various palm oil wastes; such as palm kernel shell (PKS), empty fruit bunches (EFB), palm oil fronds (POF) and their blends. The use of PKS as feedstock for HTL resulted in the highest biocrude yield of ~441 kg/tonne, whereas the EFB was linked to the lowest biocrude yield of ~293 kg/tonne. Meanwhile, the highest yield of hydrochar was achieved at ~193 kg/tonne for EFB. The gas production was also higher in the case of EFB, at about 317 kg/tonne. The blend of all three wastes yielded ~356 kg/tonne of biocrude, 102 kg/tonne of hydrochar and 387 kg/tonne of gas. In addition, the produced gas composition is dominated by CO_2 followed by CH_4. The blended feed yielded the lowest CO_2 composition, which indicates the importance of feedstock blending in CO_2 mitigation. POF is found to be a more economical feedstock which has shown the lowest cost of 0.331USD/kg and 0.045 USD/kg. The generated biocrude replicates several petroleum properties, granting it a high potential for conventional refining into clean transportation fuels. This study provides insights and technical knowledge on HTL technology and its possible upgradation from lab and pilot into commercial scales.

Keywords: Liquefaction; Oil-palm waste; Aspen Plus; Biocrude; Hydrochar.

1. Introduction

The existence of modern life standard is in danger due to the increasing energy demand and its associated environmental impact due to the high dependency on fossil fuels. The domino effect of these issues is not only extended to energy deficiency, and the world existence inquest is due to climate change, which caught the world's attention at the research and policy level to explore new and alternative renewable and sustainable energy sources (Ali et al., 2022). In all alternative energy resources, biomass emerges as an attractive and viable source due to its wide availability and possible conversion into all forms of energy, materials, and chemicals that are currently obtained from fossil fuels (AlNouss et al., 2022).

Biochemical and thermochemical routes can convert biomass into energy derivatives and chemicals. Biomass can be converted into gaseous, liquid, and solid fuels and value-added chemicals through thermochemical and biochemical conversion processes (Chan et al., 2019; Chan et al., 2015). The thermochemical conversion processes including gasification and incineration are used for gaseous fuel, heat, and electricity production. In contrast, pyrolysis and liquefaction are utilised for liquid fuel and char generation. All the above processes are relatively energy-extensive, with liquefaction being the least-energy consuming. Liquid fuel is vital in energy products due to its high energy density, ease of storage and transportation (Chan et al., 2015). Liquefaction of biomass is a relatively new process compared to the other process. Still, it appears to be very effective for liquid fuel production due to its low energy requirement, no drying and pre-treatment process (Chan et al., 2015). In the liquefaction process biomass is treated at a lower temperature than 400 °C and high pressure of 4 - 20 MPa using an appropriate solvent such as; water, organic solvent, and their mixture (Chan et al., 2018).

Palm oil waste (POW) is Southeast Asia's major crop, supplying 79% of the world's palm oil demand (Umar et al., 2021). POW is primarily used in the industry for electricity and heating purposes. Chan et al. (2019) reviewed the utilization of waste in Malaysia for energy production via thermochemical conversion pathway and noticed that most of POW were investigated for gasification and pyrolysis processes for gaseous fuel and bio-oil production. Umar et al. (2021) recently reviewed the POW utilisation for energy production and found that the gasification and pyrolysis processes are well investigated at pilot and commercial scales for gaseous fuel generation. On the other hand, liquefaction was mainly investigated at the lab and pilot scale levels (Chan et al., 2019). The first objective of this study is to develop a flowsheet model of liquefaction process using Aspen Plus® to investigate the process parameters on the yield of solid crude, hydro crude, and gas for each type of POW and their blend. HTL is used to procude for liquid production and gas is obtained as a by product or waste.While the second objective is to conduct a technoeconomic analysis using Aspen Plus Economic Analyzer to evaluate the production cost of each product for all types of selected feedstocks. This study may provide insights for researchers, as well as policymakers to make decisions on the selection of appropriate wastes for suitable products generation.

2. Methodology

2.1 Description of process modelled

The high tech, advanced and reliable simulation modelling software are in research and commercial use for many polymer, chemicals, oil and gas, power, and biofuel sectors. (Shahbaz et al., 2021). Aspen plus is one of the most important simulation modelling software that can perform high-level simulations for solid, fluids, and gases by providing suitable fluid packages and properties. Hydrothermal liquefaction process is developed using Aspen Plus® V10 to yield the biocrude and hydrochar as illustrated in the process flow diagram presented in Figure 1. To perform a steady state smooth simulation process, some important assumption were considered, including no temperature and pressure gradient and carbon is graphite carbon (Alherbawi et al., 2022). The feedstock including

EFB, PKS, POF, and their blend are introduced into the reactor as biomass slurry where non-conventional components are converted into conventional elements using Fortran routine based on the composition presented in Table. 1 as presented in previous study (Parthasarathy et al., 2022). Fortran routine is also employed to calculate the quantity of water to make a 20% solid concentrated slurry. The pressure of the slurry is raised up to 100 bar and hated up to 350 °C before being fed into the HTL reactor for the hydrothermal liquefaction process. The maximum biocrude and minimum hydrochar yields are restricted using a Fortran code calculation based on previous prediction models (Alherbawi et al., 2021).

Table 1. Composition of palm waste (PKS, EFB, POF and blends).

	PKS	EFB	POF	Blend
		Proximate analysis		
Moisture	9.70	5.30	5.30	6.77
FC	14.42	20.61	17.74	17.59
VM	80.57	76.17	74.97	77.24
Ash	5.00	3.22	7.29	5.17
		Ultimate analysis		
C	46.44	43.70	45.03	45.06
H	5.43	4.74	4.92	5.03
N	0.96	0.76	0.59	0.77
Cl	-	-	-	-
S	-	-	-	-
O	42.17	47.59	42.17	43.98

The HTL reactor is set to operate at Gibbs free energy minimization approach in which all possible products and biocrude components have been defined. Hydrochar is separated from the product stream in a hydro cyclone unit as depicted in flow diagram. The solid free product stream is directed to the three-stage flash where biocrude (liquid) and gases are separated along with aqueous phase which contain other dissolved organics.

2.2. Cost analysis

A techno economic analysis is performed using the Aspen Plus Economic Analyzer and based on the correlation and values presented in previous study (Parthasarathy et al., 2022). The levelized cost of biocrude and hydrochar per kg feed of PKS, EFB, POF, and their blend is calculated based on the levelized cost of energy concept.

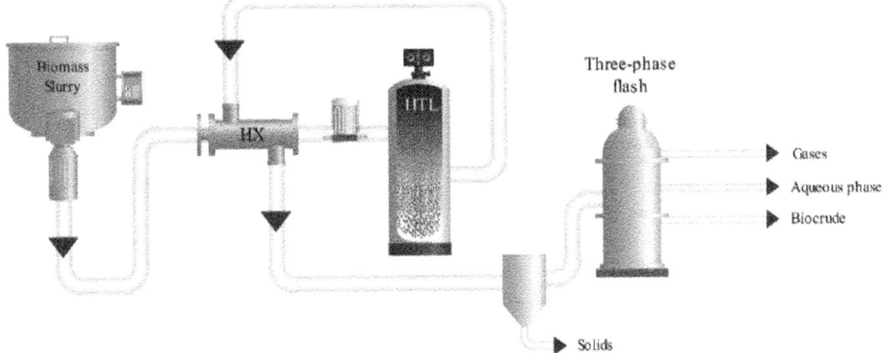

Fig.1 Process flow diagram of hydrothermal liquefaction of (PKS, EFB.POF. and Blend).

3. Results and discussion

3.1 Product yield distribution and gas composition

The product yield of biocrude, hydrochar and gas per tonne of all feedstocks (PKS. EFB, POF and Blend) are presented in Figure 2. It can be noticed that biocrude is the dominant product for all cases except EFB. The higher yield of biocrude is due to the higher degradation rate for the cracking of the matrix structure of lignocellulosic biomass at higher temperature of 500 °C (Chan et al., 2015). PKS has shown the highest biocrude yield of 441.56 kg/tonne. POF and the blend yielded biocrude at 359.69 and 356.02 kg/tonne, respectively. Chen et al. (2018) have also observed the increase of bio oil yield with the increase in temperature for the liquefaction of PKS. A similar trend is also reported for bio-oil production from the liquefaction of giant fennel with the increase in temperature (Aysu & Küçük, 2013). Hydrochar production is found to be the highest in the case of EFB at about 192.95 kg/tonne. The higher hydrochar production for EFB is due to higher fixed carbon content of 20% as presented in Table 1. POF was associated to hydrochar production of 159.62 kg/tonne, which is the second highest amongst all feedstocks. PKS and the blend of all palm waste have a significant lower yield of 100.73 and 101.25 kg/tonne, respectively. Although liquefaction is not known for hydrochar production, it is obtained as a by-product. In this study, the yield of hydrochar is still significant, contributing to the process's commercial viability. Gas production is an important by-product that can be obtained significantly, but its usage is limited due to the domination of CO_2 in gas yield. The gas production for EFB, PKS, and POF are 316.48, 236.14, and 259.85 kg/tonne, respectively. The blend of feedstocks generated highest quantity of gas at 387.48 kg/tonne, which is the highest within the products array.

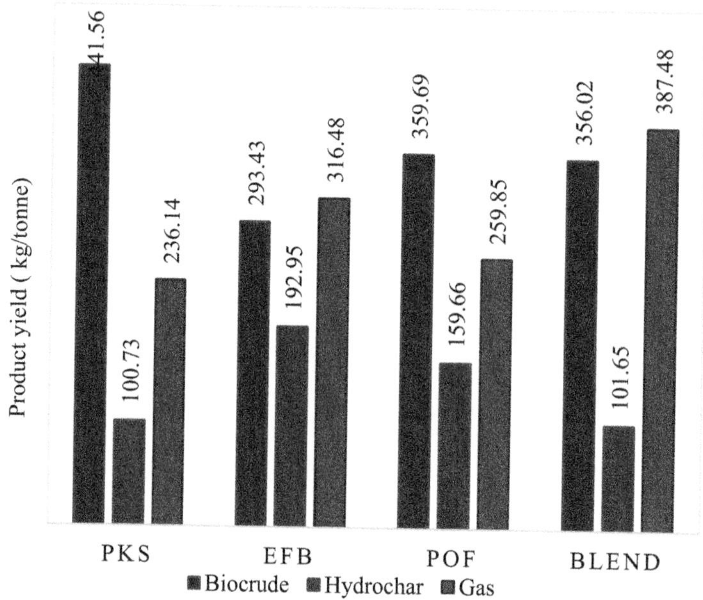

Figure 2. Product yield at temperature of 500 °C for PKS, EFB, POF and Blend.

Figure 3 demonstrates the composition of the syngas attained from liquefaction of PKS, EFB, POF and their blend. It is very clear from Figure 3 that CO_2 is the dominant gas in the array of gasses present in syngas. The CO_2 content is more than 90% in syngas for all feedstocks, being the highest in PKS case at about 98.14 %. However, the palm waste blend has yielded the lowest CO_2 composition at 90.14%. CH_4 content is the second highest component which is in the range of 1 to 10% in which the blend has also shown the highest syngas composition of CH_4 at 9.73%. The content of H_2 and CO is very low (less than 0.1%) which makes it insignificant. Due to this reason, the composition of syngas and their yield for liquefaction is very rarely reported in the literature, as such gas composition is incomparable to gasification but may be compared with hydro-pyrolysis.

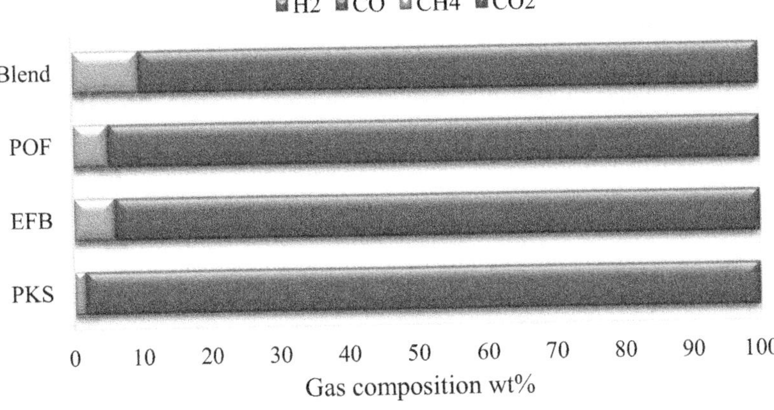

Figure 3. Gas composition at a temperature of 500 °C for PKS, EFB, POF and Blend.

3.2 Cost analysis

The cost of biocrude and hydrochar production per kg of PKS, EFB, POF and their blend is presented in Figure 4. The highest cost of biocrude and hydrochar are found for EFB which are at 0.402 USD/kg and 0.203 USD/kg, repeatedly. Whereas the second highest cost is noticed for palm oil waste blend. For POF, the cost of biocrude and hydrochar production is estimated at 0.391 USD/kg and 0.045 USD/kg, respectively. In the case of PKS, the cost of production for biocrude is achieved at 0.345 USD/kg, whereas hydrochar is obtained as a complementary product. It can be concluded that the POF can be a good source in terms of cost for biocrude and hydrochar production, which is linked to lower cost of feedstock.

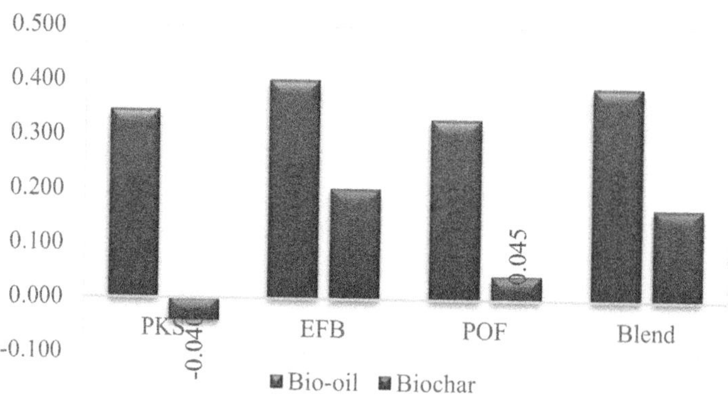

Figure 4. Cost of Biocrude and Hydrochar at 500 °C using PKS, EFB, POF and Blend.

4. Conclusion:

A process simulation model of Hydrothermal Liquefaction (HTL) of PKS, EFB, POF and their blend is instituted for biocrude, hydrochar, and gas production using Aspen Plus® V10. The highest production yield of biocrude is achieved at 441.56kg/tonne using PKS, with hydrochar and gas yields at 100.73 kg/tonne and 236.14 kg/tonne, respectively. Whereas the lowest yield of biocrude is attained at 293.43 kg/tonne using EFB. The blend of all three palm wastes is found to be a good option for enhanced composition and yield of gas. Where the blended feedstock yielded 387.48 kg/tonne of gas with the lowest CO_2 composition. While it yielded biocrude and hydrochar at 356 kg/tonne and 101.65 kg/tonne, respectively. The gas crude composition is dominated by the CO_2 Content which is within the range of 90 to 98% for all feedstocks. The economic analysis indicated that the lowest biocrude cost is attained using POF at 0.331 USD/kg, while PKS was associated to the highest biocrude cost at 0.345 USD/kg.

Acknowledgment: The authors thank Hamad Bin Khalifa University (HBKU) for providing financial and technical support.

References

M. Alherbawi, T. Al-Ansari, H.R. Mackey,G. McKay, 2021, A technoeconomic assessment of an on-site biocrude production from sewage sludge in Qatar's wastewater treatment plants. Computer Aided Chemical Engineering, 50, 1929-1935.

M. Alherbawi, P. Parthasarathy, G. McKay, H.R. Mackey, T. Al-Ansari, 2022, Investigation of optimal blending of livestock manures to produce biocrude via hydrothermal liquefaction. In Computer Aided Chemical Engineering (Vol. 51, pp. 1243-1248): Elsevier.

A.M. Ali, M. Inayat, A.A. Zahrani, K. Shahzad, M. Shahbaz, S.A. Sulaiman, H. Sadig, 2022, Process optimization and economic evaluation of air gasification of Saudi Arabian date palm fronds for H2-rich syngas using response surface methodology. Fuel, 316, 123359.

A. AlNouss, M. Shahbaz, G. McKay, T. Al-Ansari, 2022, Bio-methanol production from palm wastes steam gasification with application of CaO for CO2 capture: techno-economic-environmental analysis. Journal of Cleaner Production, 341, 130849.

T. Aysu, M.M. Küçük, 2013, Liquefaction of giant fennel (Ferula orientalis L.) in supercritical organic solvents: Effects of liquefaction parameters on product yields and character. The Journal of Supercritical Fluids, 83, 104-123.

Y. Chan, K.W. Cheah, B. How, A. Loy, M. Shahbaz, H. Singh, A. Shuhaili, S. Yusup, W. Ghani, J. Rambli, 2019,. An overview of biomass thermochemical conversion technologies in Malaysia. Science of The Total Environment, 680, 105-123.

Y. Chan, A. Quitain, S. Yusup, Y. Uemura, M. Sasaki, T. Kida, 2018, Optimization of hydrothermal liquefaction of palm kernel shell and consideration of supercritical carbon dioxide mediation effect. The Journal of Supercritical Fluids, 133, 640-646.

Y. Chan, S. Yusup, A. Quitain, R. Tan, M. Sasaki, H. Lam, Y. Uemura, 2015, Effect of process parameters on hydrothermal liquefaction of oil palm biomass for bio-oil production and its life cycle assessment. Energy Conversion and Management, 104, 180-188.

P. Parthasarathy, M. Alherbawi, M. Shahbaz, H.R. Mackey, G. McKay, T. Al-Ansari, 2022, Conversion of oil palm waste into value-added products through pyrolysis: a sensitivity and techno-economic investigation. Biomass Conversion and Biorefinery.

M. Shahbaz, A. AlNouss, S. Yusup, G. McKay, T. Ansari, 2021, Techno-economic evaluation of sorption enhanced steam gasification of PKS system for syngas using CaO for CO_2 capture. In M. Türkay & R. Gani (Eds.), Computer Aided Chemical Engineering (Vol. 50, pp. 129-134): Elsevier.

H. Umar, S. Sulaiman, M. Meor Said, A. Gungor, M. Shahbaz, M. Inayat, R. Ahmad, 2021, Assessing the implementation levels of oil palm waste conversion methods in Malaysia and the challenges of commercialisation: Towards sustainable energy production. Biomass and Bioenergy, 151, 106179.

Antonis Kokossis, Michael C. Georgiadis, Efstratios N. Pistikopoulos (Eds.)
PROCEEDINGS OF THE 33rd European Symposium on Computer Aided Process Engineering (ESCAPE33), June 18-21, 2023, Athens, Greece
 http://dx.doi.org/10.1016/B978-0-443-15274-0.50236-5

Heuristic approaches to solving multistage stochastic programs with type II endogenous uncertainty

Yasuhiro Shoji,[a] Selen Cremaschi*,[a]
[a]*Department of Chem. Eng., Auburn University, Auburn, AL 36849, United States*
szc0113@auburn.edu

Abstract

Many optimization problems in the process industry, such as production planning and scheduling, contain endogenous (decision-dependent) uncertainties. The endogenous uncertainties can be represented as scenarios based on the realized outcomes; hence, by nature, multistage stochastic programming (MSSP) is one approach to modeling and solving these problems. MSSP problems quickly grow and become computationally intractable as the problem size increases. We solve four MSSP problems with Type II endogenous uncertainty using absolute expected value solution (AEEV) and generalized Knapsack-problem-based decomposition algorithm (GKDA) to address the computational tractability issues. We bound the solutions using a modified Lagrangian relaxation (mLR). The results reveal that AEEV provides the smallest optimality gap.

Keywords: multistage stochastic program, endogenous uncertainty, heuristic approaches

1. Introduction

MSSP is a mathematical decision-making approach used in the process industry, such as for process synthesis (Tarhan and Grossmann, 2008), to solve optimization problems with the sequential realization of uncertainties. Uncertainty in optimization problems is classified as exogenous or endogenous, which can be further classified into Type I: decisions affect the probability distribution, and Type II: decisions affect the realization time (Goel and Grossmann, 2006). For MSSPs with type II endogenous uncertainty, implicitly enforcing non-anticipativity (NA) in the model formulation is not possible. Hence, NA constraints (NACs) are added to the model to prevent decisions from anticipating unrealized future outcomes. However, the number of NACs grows exponentially as the number of scenarios increases, resulting in computational intractability (Apap and Grossmann, 2017). Various heuristic, approximation, and decomposition methods exist for solving MSSPs with endogenous uncertainty, e.g., sample average approximation algorithm (Solak et al., 2010), improved Lagrangian decomposition framework (Gupta and Grossmann, 2014), and rolling-horizon heuristic approach (Colvin and Maravelias, 2009).

This paper applies two heuristic primal bounding approaches, AEEV (Zeng and Cremaschi, 2019) and GKDA (Zeng et al., 2018), to solve MSSP problems with Type II endogenous uncertainty and bound the solutions using mLR (Zeng and Cremaschi, 2020). It has been suggested that AEEV and GKDA yield tight primal bounds quickly, and the goal is to investigate their performance for a wider range of MSSP problems.

2. A general MSSP formulation under endogenous uncertainties

A general MSSP formulation under endogenous uncertainties is given in Eqs. 1-6.

$$\mathrm{OV} = \min \sum_{s} p_s \sum_{i,t} G_{i,t,s}\left(V_{i,t}, \theta_i^s, b_{i,t}^s, \gamma_t^s\right) \tag{1}$$

$$g_{i,t,s}\left(b_{i,t}^s, \gamma_t^s, \theta_i^s\right) \leq 0 \quad \forall i \in I, t \in T, s \in S \tag{2}$$

$$h_{i,t,s}\left(b_{i,t}^s, \gamma_t^s, \theta_i^s\right) = 0 \quad \forall i \in I, t \in T, s \in S \tag{3}$$

$$b_{i,1}^s = b_{i,1}^{s'} \quad \forall i \in I, \forall (s, s') \in S \tag{4}$$

$$\begin{bmatrix} Z_{i,t}^{s,s'} \\ b_{i,t}^s = b_{i,t}^{s'} \end{bmatrix} \vee \left[\neg Z_{i,t}^{s,s'}\right] \quad \forall (i, s, s') \in S_E, \forall t \in T, t > 1 \tag{5}$$

$$\gamma_t^s \in \mathbb{R},\ b_{i,t}^s, Z_{i,t}^{s,s'} \in \{0,1\} \quad \forall (i, s, s') \in S_E, \forall t \in T, \forall i \in I \tag{6}$$

The objective function (Eq. (1)) calculates the optimum OV. The model has two decision variables, $b_{i,t}^s$ and γ_t^s, with uncertain resources $i \in I = \{1,2, \dots, I\}$, planning horizon $t \in T = \{1,2, \dots, T\}$, and scenarios $s \in S = \{1,2, \dots, S\}$. The binary decision variable $b_{i,t}^s$ is enforced to be identical by either initial (Eq. (4)) or conditional NACs (Eq. (5)) until scenario pair (s, s') becomes distinguishable. The decision variable γ_t^s is determined by scenario-specific constraints (Eqs. (2)-(3)). $V_{i,t}$ and θ_i^s represent deterministic and endogenous uncertain parameters. The function $G_{i,t,s}\left(V_{i,t}, \theta_i^s, b_{i,t}^s, \gamma_t^s\right)$ calculates the contribution of a specific scenario s with probability p_s to the objective. The conditional NACs (Eq. (5)) are enforced depending on the values of the Boolean variable $Z_{i,t}^{s,s'}$, which is equal to one if scenario pair (s, s') is indistinguishable, and zero otherwise.

3. Solution approaches

3.1. Absolute Expected Value Solution Approach (AEEV)

AEEV (Zeng and Cremaschi, 2019) is a primal bounding approach that converts an MSSP problem into two sub-problems, deterministic expected value sub-problems (*DEVSPs*) and recourse deterministic expected value sub-problems (*DEVSPs*^{recourse}), by removing all NACs (Eq. (4), (5)) and indices s from the MSSP formulation. *DEVSPs* and *DEVSPs*^{recourse} determine here-and-now decisions and corresponding recourse actions. AEEV starts by constructing a *DEVSP* at the initial time, as shown in Figure 1.

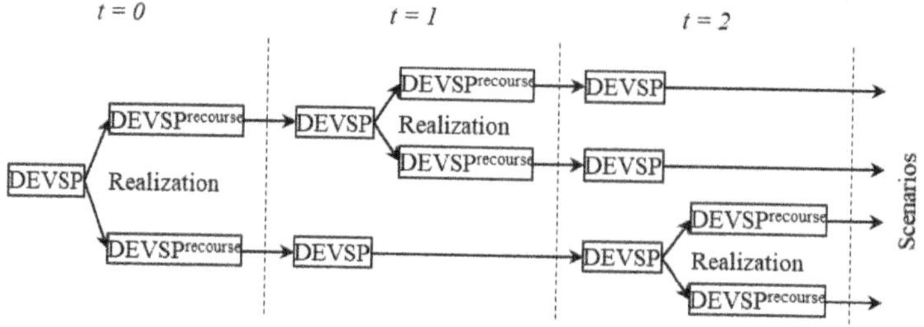

Figure 1 The schematic diagram of AEEV

The values of endogenous uncertain parameters are set to their expected values over all scenarios because there are no realizations before solving the initial *DEVSP*. AEEV stores the here-and-now decisions obtained from the *DEVSP* solutions and determines which scenarios can be differentiated based on the realized outcomes. If realizations occur and the MSSP problem has decision variables associated with recourse actions, *DEVSPs*^{recouse} is generated and solved to obtain these decision variables. If there is no realization, AEEV skips constructing *DEVSPs*^{recouse}. AEEV continues to construct and solve *DEVSPs* and *DEVSPs*^{recourse} for the subsequent time periods with stored decisions set to solutions of prior *DEVSPs* and *DEVSPs*^{recourse} until the end of the time horizon. The values of endogenous uncertain parameters are expected-valued over possible scenarios at each *DEVSP* and *DEVSP*^{recourse}. At termination, AEEV provides a feasible solution and an upper bound for the MSSP model.

3.2. Generalized Knapsack-Problem based Decomposition Algorithm (GKDA)

GKDA (Zeng et al., 2018) is a similar primal bounding approach. It translates the MSSP problem into two types of knapsack sub-problems (KSPs) by decomposing scenario and time indices, i.e., it removes all NACs (Eq. (4), (5)) and indices s and t from the MSSP formulation. The remaining index i is translated into an item set of KSPs by enumerating all eligible decisions. Scenario-specific constraints (Eq. (2), (3)), except sequencing constraints, are transformed into KSP weight constraints. Sequencing constraints and bounds are used to determine the allowable items in a KSP from the eligible items. Item values are generally objective function coefficients, i.e., $V_{i,t}$ in Eq (1) translates into V_e as item values in KSPs, where e notates item $e \in E$. Under certain conditions, this translation does not accurately approximate the expected contribution of the associated decision to the MSSP objective function. Zeng et al. (2018) defined two conditions with approaches to better approximate item values. Condition 1 applies to problems in which gains/revenues are only realized at certain stages or are only associated with certain decision variables, and the item values are generated using the maximum potential gain approach (MPGA). Condition 2 applies to problems where an item represents a capital investment decision variable in MSSP, and the item value is generated using annual capital charges. Similar to AEEV, GKDA continues to construct and solve KSPs along the planning horizon with stored decisions of prior KSPs. At termination, GKDA provides a feasible solution and an upper bound for the MSSP problem.

3.3. Modified Lagrangian Relaxation (mLR)

Lagrangian relaxation (LR) is a common approach to removing complicating constraints and adding them to the objective function as penalty terms (Lin et al., 2011). For MSSP problems, NACs (Eq. (4), (5)) are complicating constraints. A standard LR dualizes these NACs to the MSSP objective function, as shown in Eq. (7).

$$
\begin{aligned}
\mathrm{OV}_{LR} = \min \sum_{s} p_s \sum_{i,t} G_{i,t,s}\left(V_{i,t}, \theta_i^s, b_{i,t}^s, \gamma_t^s\right) + \sum_{i,s,s'} \lambda 1_{i,s,s'} \left(b_{i,1}^{s} - b_{i,1}^{s'}\right) \\
+ \sum_{i,t,s,s'} \lambda 2_{i,t,s,s'} \left(1 - Z_{i,t}^{s,s'} - b_{i,t}^{s} + b_{i,t}^{s'}\right) \\
+ \sum_{i,t,s,s'} \lambda 3_{i,t,s,s'} \left(1 - Z_{i,t}^{s,s'} + b_{i,t}^{s} - b_{i,t}^{s'}\right)
\end{aligned}
\tag{7}
$$

In Eq. (7), $\sum_{i,s,s'} \lambda 1_{i,s,s'} \left(b_{i,1}^{s} - b_{i,1}^{s'}\right)$ corresponds to initial NACs (Eq. (4)), and $\sum_{i,t,s,s'} \lambda 2_{i,t,s,s'} \left(1 - Z_{i,t}^{s,s'} - b_{i,t}^{s} + b_{i,t}^{s'}\right)$ and $\sum_{i,t,s,s'} \lambda 3_{i,t,s,s'} \left(1 - Z_{i,t}^{s,s'} + b_{i,t}^{s} - b_{i,t}^{s'}\right)$ correspond to conditional NACs (Eq. (5)). $\lambda 1_{i,s,s'}$, $\lambda 2_{i,t,s,s'}$, and $\lambda 3_{i,t,s,s'}$, called multipliers, are updated over the iterations to obtain tighter bounds. LR provides a lower bound. For $\sum_{i,t,s,s'} \lambda 2_{i,t,s,s'} \left(1 - Z_{i,t}^{s,s'} - b_{i,t}^{s} + b_{i,t}^{s'}\right)$ and $\sum_{i,t,s,s'} \lambda 3_{i,t,s,s'} (1 -$

$Z_{i,t}^{s,s'} + b_{i,t}^{s} - b_{i,t}^{s'}$) in Eq. (7), NACs are always satisfied when scenario pairs (s,s') are distinguishable, i.e., $Z_{i,t}^{s,s'} = 0$, thereby leading to looser bounds. To tighten it, Zeng and Cremaschi (2020) proposed mLR by transforming Eq. (7) into Eq. (8) where the conditional NACs are dualized only when $Z_{i,t}^{s,s'} = 1$.

$$OV_{LR} = \min \sum_{s} p_s \sum_{i,t} G_{i,t,s}(V_{i,t}, \theta_i^s, b_{i,t}^s, \gamma_t^s) + \sum_{i,s,s'} \lambda 1_{i,s,s'} \left(b_{i,1}^s - b_{i,1}^{s'}\right) + \sum_{i,t,s,s'} \lambda_{i,t,s,s'} Z_{i,t}^{s,s'} (b_{i,t}^s - b_{i,t}^{s'}) \quad (8)$$

Multiplier updating schemes are critical for obtaining tighter bounds. The sub-gradient optimization method (Eqs. (9) and (10)) is a commonly used technique (Fisher, 1985).

$$\lambda_n^{k+1} = \max\{0, \lambda_n^k - \gamma^k (Ax_n^k - b)\} \quad (9)$$

$$\gamma^k = scale \frac{Z(\lambda_n^k) - Z^*}{\sum_n (Ax_n^k - b)^2} \quad (10)$$

The total $|N|$ number of multipliers at $k + 1^{th}$ iteration, λ_n^{k+1} ($n \in N$), is calculated with Eq. (9) (Fisher, 1981), where $Ax_n^k - b$ is the dualized constraints transferred to the objective function, where x_n^k is solved variables, and A and b are parameters of dualized constraints. The step size at iteration k, γ^k, is calculated by Eq. (10) based on the gap between the best-known primal Z^* and dual bound $Z(\lambda_n^k)$. $scale$ in Eq. (10) is an adjustable parameter in $0 < scale < 2$, which is often set via trial and error. To achieve tighter bounds efficiently, Zeng and Cremaschi (2020) proposed updating $scale$ at the kth iteration using Eq. (11) if the ratio of dualized constraints value D to the dualized objective function value OV_{LR} is greater than or equal to 0.1. The multipliers are updated using Eq. (12), where N^v is the total number of violated dualized constraints.

$$\text{If } \frac{D}{OV_{LR}} \geq 0.1 \text{ Then } scale^k = \frac{scale^{k-1}}{1.618} \quad (11)$$

$$\gamma^k = scale^k \frac{Z(\lambda_n^k) - Z^*}{N^v} \quad (12)$$

4. Case study descriptions, results, and discussion

We applied the solution approaches to four MSSP problems with Type 2 endogenous uncertainty. All problems are Mixed Integer Linear Programming (MILP) problems. Clinical Trial Planning (CTP) (Colvin and Maravelias, 2009) problem determines the timing of the clinical trials, maximizing expected profit. Artificial Lift Infrastructure Planning (ALIP) (Zeng and Cremaschi 2017) problem determines installed equipment and installation timing for exploiting shale gas, maximizing expected profit. Vehicle Routing (VR) (Hooshmand and MirHassani, 2016) problem determines customer visit order, minimizing expected cost. Open Pit Mining (OPM) (Boland et al., 2008) problem determines aggregates to be excavated and processed and their timing, maximizing expected profit. We modeled all instances in Pyomo and solved them with CPLEX 20.10 using 48 processors on a node of Auburn University Easley Cluster.

Figure 2 represents the result of mLR for an instance of ALIP with a 12-month planning horizon. mLR required significantly longer times for other instances, yielding a few to 40 iterations within 48 hours. Figure 2 A shows that mLR yields a tighter bound than LR and

LR with the new scale updating scheme only (LR-SU). However, mLR takes longer than LR and LR-SU, as can be seen in Figure 2 B. The objective function of mLR is nonlinear and cannot be decomposed if linearized due to scenario pairs (s, s') in term $\sum_{i,t,s,s'} \lambda_{i,t,s,s'} Z_{i,t}^{s,s'} (b_{i,t}^{s} - b_{i,t}^{s'})$ of Eq. (8). We only applied LR-SU to other instances.

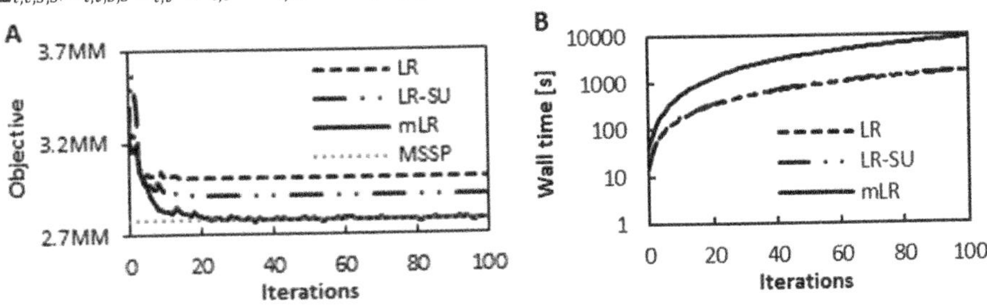

Figure 2 A) Objective value of LR, LR-SU, mLR, and MSSP and B) Computational time of LR, LR-SU, and mLR for ALIP with a planning horizon of 12 months.

Figures 3 and 4 represent the optimality gap and computational times. LR and LR-SU results are for iteration 50. The optimality gap is defined from MSSP optimum solution. AEEV provided the best feasible solutions, and LR-SU provided tighter bounds than LR. GKDA yielded better feasible solutions for CTP and ALIP than VR and OPM because it employed Conditions 1 and 2 for item value estimation for CTP and ALIP, respectively. These results indicate the importance of item value estimation for GKDA, which is the fastest (Figure 4). LR-SU takes almost the same time as LR but yields tighter bounds. Computational time increased milder for all approaches than for MSSP.

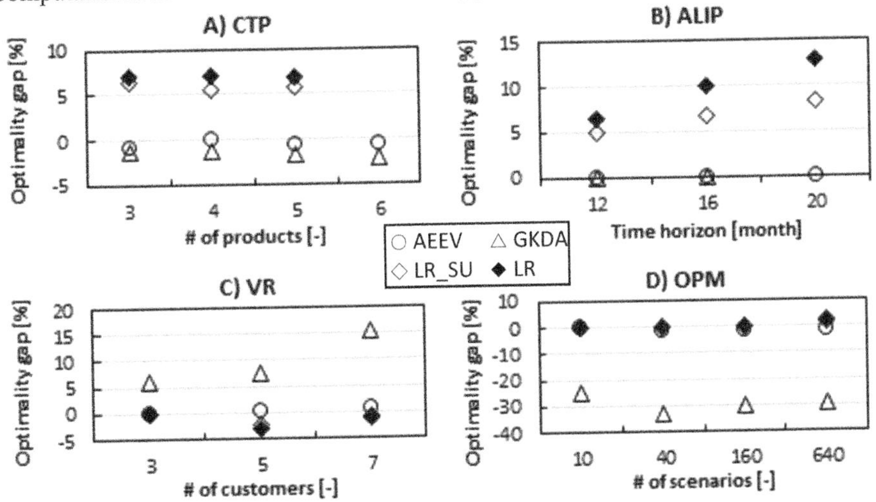

Figure 3 Optimality gap of solutions yielded by approaches from the MSSP solution

5. Conclusion and Future Directions

We applied two heuristics, AEEV, GKDA, and two bounding approaches, mLR and LR-SU, to four MSSP problems with Type II endogenous uncertainty. The results revealed AEEV provides the tightest feasible solutions. The bound yielded with mLR was tighter than LR; however, mLR required a long computational time. Computational time increased milder for all approaches than for MSSP. We plan to apply the approaches to other MSSP problems and study avenues for improving them.

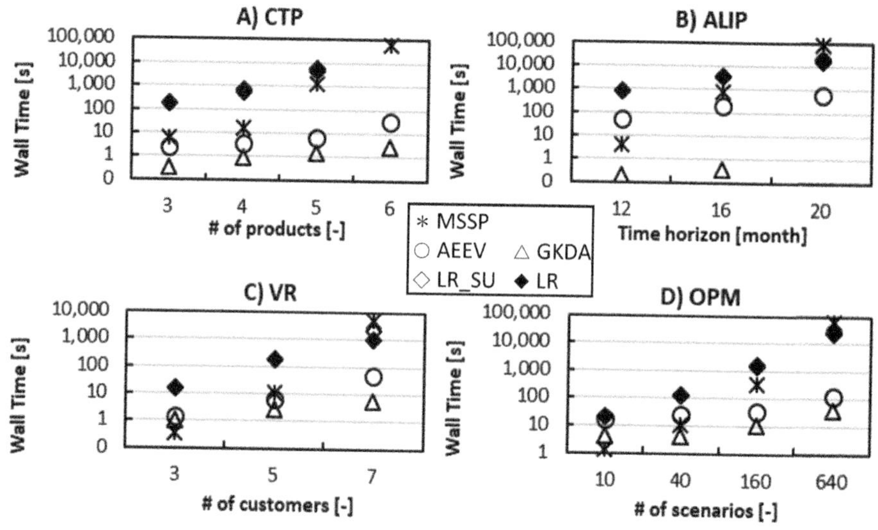

Figure 4 Computational time of MSSP and solution approaches

References

Apap, R.M. and Grossmann, I.E. (2017), Models and computational strategies for multistage stochastic programming under endogenous and exogenous uncertainties. Comput. Chem.Eng., 103; pp.233-274

Boland, N., Dumitrescu, I., & Froyland, G. (2008), A multistage stochastic programming approach to open pit mine production scheduling with uncertain geology. Optimization online, 1-33.

Colvin, M., & Maravelias, C. T. (2009), Scheduling of testing tasks and resource planning in new product development using stochastic programming. Comput. & Chem. Eng., 33(5), 964-976.

Fisher M. L. (1981), The Lagrangian Relaxation Method for Solving Integer Programming Problems. Management Science., 27(1)

Fisher M. L. (1985), An applications oriented guide to Lagrangian relaxation. Interfaces 15, 10-21Hooshmand Khaligh, F., & MirHassani, S. A. (2016), A mathematical model for vehicle routing problem under endogenous uncertainty. Int. J. Prod. Res., 54(2), 579-590.

Goel, V., & Grossmann, I. E. (2006), A class of stochastic programs with decision dependent. uncertainty Math. Program., 108(2-3), 355-394.

Gupta, V., & Grossmann, I. E. (2014), Multistage stochastic programming approach for offshore oilfield infrastructure planning under production sharing agreements and endogenous uncertainties. Journal of Petroleum Science and Engineering, 124, 180-197.

Lin, Z., Liu, R., & Su, Z. (2011), Linearized Alternating Direction Method with Adaptive Penalty for Low-Rank Representation. In Advances in neural information processing systems (pp. 612-620).

Solak, S., Clarke, J. P. B., Johnson, E. L., & Barnes, E. R. (2010), Optimization of R&D project portfolios under endogenous uncertainty. European Journal of Operational Research, 207(1), 420-433.

Tarhan, B. and I.E. Grossmann. (2008), A multistage stochastic programming approach with strategies for uncertainty reduction in the synthesis of process networks with uncertain yields. Comput. & Chem. Eng., 32(4-5), 766-788.

Zeng, Z., Christian, B., & Cremaschi, S. (2018), A generalized knapsack-problem based decomposition heuristic for solving multistage stochastic programs with endogenous and/or exogenous uncertainties. Ind. Eng. Chem. Res., 57(28), 9185-9199.

Zeng, Z., & Cremaschi, S. (2017). Artificial lift infrastructure planning for shale gas producing horizontal wells. Proceedings of the FOCAPO/CPC, Tuscan, AZ, USA, 8-12.

Zeng, Z., & Cremaschi, S. (2019), A general primal bounding framework for large-scale multistage stochastic programs under endogenous uncertainties. Chem. Eng. Res. Des.: 141, 464-480.

Zeng, Z., & Cremaschi, S. (2020), A new lagrangean relaxation approach for multistage stochastic programs under endogenous uncertainties. 30th ESCAPE, 47.

Antonis Kokossis, Michael C. Georgiadis, Efstratios N. Pistikopoulos (Eds.)
PROCEEDINGS OF THE 33rd European Symposium on Computer Aided Process Engineering (ESCAPE33), June 18-21, 2023, Athens, Greece
 http://dx.doi.org/10.1016/B978-0-443-15274-0.50237-7

Hybrid modeling of the catalytic CO_2 methanation using process data and process knowledge

Luisa Peterson[a] , Jens Bremer[b] , Kai Sundmacher[a,c]

[a] *Max Planck Institute for Dynamics of Complex Technical Systems, Process Systems Engineering, Sandorstraße 1, Magdeburg, 39106, Germany*

[b] *Clausthal University of Technology, Institute of Chemical and Electrochemical Process Engineering, Leibnizstraße 17, Clausthal-Zellerfeld, 38678, Germany*

[c] *Otto von Guericke University Magdeburg, Chair for Process Systems Engineering, Universit ̈atsplatz 2, Magdeburg, 39106, Germany*

Abstract

Heterogeneously catalyzed CO_2 methanation is an essential process in the power-to-methane concept. To overcome the limitations of both mechanistic and data-driven models, this study presents three modeling approaches for predicting the CO_2 conversion using different techniques for integrating pseudo-experimental process data into process models. One approach is a data-driven model that uses multilayer perceptron regression (MLPR), a type of feedforward artificial neural network, to represent the data. Another approach is a hybrid model that integrates MLPR to estimates the reaction rate within a mechanistic model. The third approach is another hybrid model that uses MLPR to correct for a simplified mechanistic model that does not fully capture the physical behavior of the system. All models are evaluated in terms of accuracy, complexity, and reliability, and the results show that they all predict the conversion of CO_2 with only minor deviations. Hybrid models have the advantage of combining the strengths of mechanistic and data-driven models and, with basic process knowledge, can make accurate predictions beyond the training data. However, they also require both process knowledge and large amounts of process data.

Keywords: Power-to-X, Reactor Modeling, Hybrid Modeling, Machine Learning

1. Introduction

The intermittent nature of solar and wind power underscores the need for effective energy storage technologies. Power-to-methane (PtM) is a process that uses electrolysis to convert energy into hydrogen, which is then chemically bonded with carbon to form synthetic methane. This CO_2-neutral fuel can be used in households and industries through existing distribution infrastructure (Götz, et al., 2016). However, the low efficiency, complexity, and high cost of the PtM process pose challenges to its widespread adoption (Er-rbib & Bouallou, 2013). Optimization of electrolyzer, reactor, and downstream process operation and design can address these challenges. Accurate process models are required to improve the performance of PtM systems. Traditionally, process models are established based on physical and chemical principles, but building first-principle models requires extensive data from specific laboratory experiments and is time-consuming (Zimmermann, et al., 2020). In addition, quantifying the kinetics that depend on elementary reactions and intermediates is challenging using traditional methods (Zahedi, et al., 2005). Data-driven regression models provide an alternative to mechanistic models by identifying relationships between the dependent variable and one or more independent inputs (Kahrs & Marquardt, 2007). They can be generated quickly without prior process knowledge, but require large amounts of data and may have limited physical interpretation. Hybrid models, which combine mechanistic models with data-driven models, can describe unknown phenomena by integrating data-driven approaches into mass and energy balances or by compensating for discrepancies between the model and the plant (Von Stosch, et al., 2014) Multilevel perceptron regression (MLPR), a type of feedforward artificial neural network, represents all data-driven relationships. In MLPR, data flows from the input layer through the hidden layers to the output layer, where the hidden layers weights the inputs and pass them through a nonlinear activation function to the output. The MLPR is trained using backpropagation by describing the target with input representations. With the advancement of data-driven models, the number of parameters, runtime, and computational resources have increased significantly. Thus, it is imperative to consider metrics such as accuracy, complexity, and reliability in addition to the accuracy of the model (Menghani, 2021). In this study, the models are evaluated based on these metrics.

2. Methodology

The modeling approach involves generating pseudo-experimental process data using a mechanistic CSTR model for the CO_2methanation and predicting this data using three models: one data-driven model and two hybrid models.

2.1. Data Generation via CSTR Model

The mass and energy balances for modeling a CSTR are based on the work of Bremer and Sundmacher (Bremer & Sundmacher, 2021). The balances were constructed under the assumptions that (1) the CO_2 methanation is a single reaction, that (2) it occurs in a single phase, and that (3) the pressure p, the gas heat capacity c_p, the heat of reaction $\Delta_R \widetilde{H}$ and the effectiveness η for limiting catalyst mass transport are constant. Using the conversion of CO_2 (X_{CO_2}), the residence time τ, the void fraction ε, the inlet concentration of CO_2 c_{CO_2}, and the mass flow fraction of the each component ω_i, the steady state mass balance of the CSTR is described as

$$0 = X_{CO_2} - \frac{\tau}{\epsilon} \frac{R_{meth}(T, p, \omega_\alpha)}{c_{CO_2,in}}. \tag{1}$$

The intrinsic reaction rate, R_{meth}, is a LHHW-type equation proposed by Koschany et al. (Koschany, et al., 2016). R_{meth} depends on the reaction temperature T and the partial pressures p_i of the components.

$$R_{meth}^{LHHW} = k p_{CO_2}^{0.5} p_{H_2}^{0.5} \left(1 - \frac{p_{CH_4} p_{H_2O}^2}{K_{eq} p_{CO_2} p_{H_2}^4}\right) / \left(1 + K_{OH} + \frac{p_{H_2O}}{p_{H_2}^{0.5}} + K_{H_2} p_{H_2}^{0.5} + K_{mix} p_{CO_2}^{0.5}\right). \tag{2}$$

The state variables for the energy balance of the CSTR are T and X_{CO_2}, along with the Stanton number St, the adiabatic temperature rise ΔT_{ad}, and the operation temperature T_{op}. The energy balance is solved in steady state.

$$0 = X_{CO_2} - \frac{(1 + St)}{\Delta T_{ad}} (T - T_{op}). \tag{3}$$

The steady-state mass and energy balances are solved simultaneously for X_{CO_2} and T_{op}. To obtain the training data, 1000 different operating conditions are selecting using statistical Latin Hypercube Sampling. The sample space includes pressure (1 bar $< p <$ 10 bar), inlet temperature (400 K $< T_{in} <$ 800 K), residence time (0.04 s $< \tau <$ 0.4 s), and the ratio of products in the feed ($0 < x_i < 0.5$). The test set is created by selecting 100 evenly distributed data points across the operating temperature range.

2.2. Training Strategy

The pipeline used in the study combines data processing and model training. The process data are first scaled using a min-max scaler, ensuring that all representations and target variable of the MLPR were between 0 and 1. In the next step, the NAdam optimization algorithm is used to minimize the mean squared error (MSE) of the CO_2 predictions by training a multi-layer perceptron regression (MLPR) model using the PyTorch library. The SELU activation function is used to compute the output of the hidden layer. The MLPR hyperparameters are determined using a Bayesian optimizer that used Gaussian processes, including the number of hidden layers (1 or 2), the number of neurons per hidden layer (between 5 and 30), the batch size for the optimizer (8, 16, 32, or 64), and the learning rate (1e-2 or 1e-3). The set of hyperparameters that produced the lowest average MSE on the test set when trained with five different sets of initial weights for the MLPR parameters is selected.

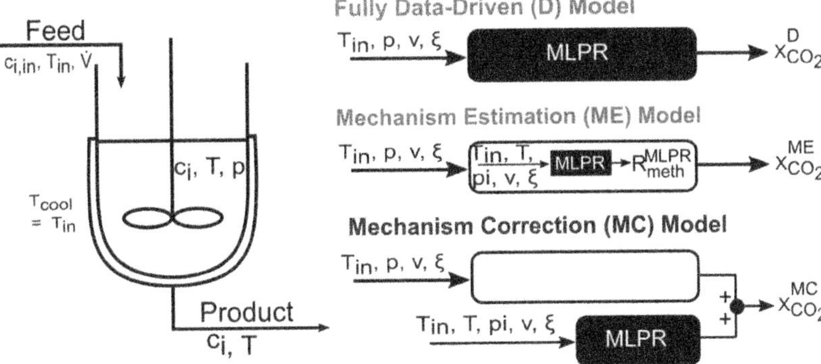

Figure 1: Illustrated schematic depicts CSTR (left) and hybrid modeling approach (right), with white boxes representing first-principle models and black boxes representing MLPR models.

In total, three different models with a data-driven component are set up. Figure 1 shows the models schematically. The fully data-driven (D) model represents mass and energy balances completely with a MLPR. $X_{CO_2}^{D}$ is a function of the input representations (*repr*) namely, p, T_{in}, τ, ξ as well as the MLPR weights (φ) optimized during MLPR training, and can be expressed as:

$$X_{CO_2}^{D} = MLPR(repr, \phi). \tag{4}$$

In the Mechanism Estimation (ME) model, the mechanistic part employs the mass and energy balances, while R_{meth} is represented by an MLPR. The input representations are p_i, T_{in}, T, τ, and x_i. For the model, $X_{CO_2}^{ME}$ is calculated from the mechanistic arguments (*args*), mechanistic parameters (θ), and $R_{meth}^{MLPR} = MLPR(repr, \phi)$

$$X_{CO_2}^{ME} = f\big(args, \theta, MLPR(repr, \phi)\big). \tag{5}$$

Minimizing the MSE of the X_{CO_2} predictions determines R_{meth}. The third model, the Mechanism Correction (MC) model, utilizes a correction term to compensate for a mechanistic model that does not fully capture system behavior. The simplified mechanistic model calculates $X_{CO_2}^{PL}$ using a mechanistic model with a fitted power law R_{meth}^{PL} as the reaction rate. MLPR serves as the correction term $\kappa = MLPR(repr, \varphi)$, learning the difference between $X_{CO_2}^{PL}$ and X_{CO_2} using the representations p_i, T_{in}, T, τ, and x_i. $X_{CO_2}^{MC}$ is calculated by summing $X_{CO_2}^{PL}$ and κ:

$$X_{CO_2}^{MC} = f(args, \theta, R_{meth^{PL}}) + MLPR(repr, \phi) \text{ with } R_{meth}^{PL} = k p_{H_2}^{n_{H_2}} p_{CO_2}^{n_{CO_2}} \left(1 - \frac{p_{CH_4} p_{H_2O}^2}{K_{eq} p_{CO_2} p_{H_2}^4}\right). \quad (6)$$

2.3. Model Evaluation

The accuracy of the regression is determined by the MSE. The MSE depends on the random initialization of the MLPR parameter weights, so the model is trained 101 times with different initial weights to determine the distribution of the MSE. To ensure reproducibility, a seed is assigned to each set of initial weights, and the model with the mean MSE value among the test models is being further evaluated. Model complexity is defined by the number of parameters in the MLPR and the average CPU time required to fit the training set and predict the test set in 101 runs on a single node of a computing cluster. In addition, the performance of the model under varying amounts of test data is being examined. Fitted models are used to predict X_{CO_2} from unknown inlet temperature distributions (800 K < T < 900 K) to assess extrapolation capability. Both training and test data are subjected to 5 % noise to test robustness, and the effect of noise on the predicted X_{CO_2} data is analyzed.

3. Results and Discussion

Three models are trained on pseudo-experimental process data as described in Section 2.2. Table 1 shows the hyperparameters, average MSEs, and average fitting times for the test and training set of all models. The left part of Figure 2 exhibits the distribution of the MSE over 101 model runs with different MLPR initial values. The results indicate that the MC model achieves the lowest average MSE on the test set, followed by the D model and the ME model. The trained models are applied to pseudo-experimental process data collected at various inlet temperatures with all other variables held constant. Figure 3 demonstrates the prediction of X_{CO_2} versus the operating temperature for each model. Despite its second-best performance in the training and test sets, the D model fails to correctly predict the pseudo-experimental temperature curves at small temperatures and residence times, due to underrepresentation of certain operating temperatures in the training set. The D model appears to overfit rather than interpolate the underrepresented data. In contrast, the ME model and MC model accurately describe the temperature curves.

The complexity of the models is determined by the MLPR architecture and the average training and prediction time (refer to Table 1). The D model trains faster. The time required to fit the data correlates with the architecture of the MLPs. Faster fitting times can be achieved by using fewer nodes per layer, larger batches, and lower learning rates. Although the ME model has the same architecture as the other models, it requires more time to train because the reaction rate is calculated using MLPR on X_{CO_2} as the target, resulting in more computational steps. The optimization of the models in this work is focused solely on accuracy. Reducing complexity by finding a Pareto front between accuracy and complexity is possible, but not the focus of this study.

Table 1: Selected hyperparameters, average MSEs, and average run times of training and test data for each model.

Model	Hyperparameters			Training		Test			
	Nodes per layer	Learn rate	Batch size	$\overline{MSE}$	Time in s	$\overline{MSE}$	Time in s	MSE extrap.	MSE 5% noise
D	4-25-25-1	1e-3	16	5.32e-5	61.8	4.62e-5	0.0051	4.39e-3	6.30e-4
ME	8-30-30-1	1e-3	8	6.83e-5	254.1	5.08e-5	0.0066	6.87e-3	5.43e-4
MC	8-25-25-1	1e-3	8	4.13e-5	117.7	2.86e-5	0.0071	8.39e-3	0.186

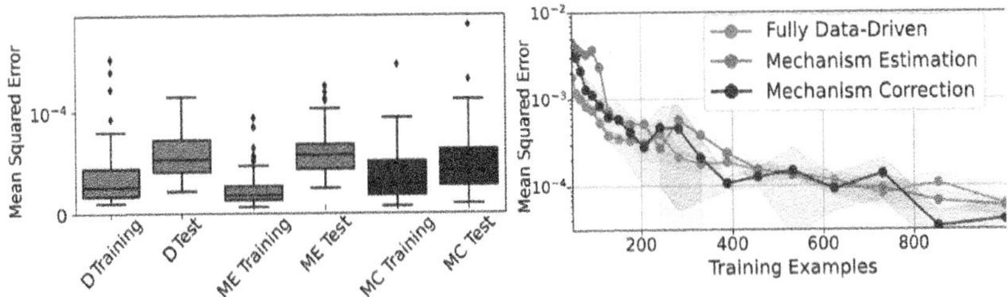

Figure 2: Left: Distribution of MSEs for models with fixed hyperparameters. Right: Model's loss function on the test data against different amounts of training data.

When extrapolating to a temperature range not covered by the training data (see Table 1), the D model performs significantly worse than when predicting unknown data within the trained temperature range. Fully data-driven models perform well only within the trained parameter range. The ME model and the MC models perform reasonably well in predicting the extrapolated data. The mechanistic framework (ME model) or the mechanistic basis (MC model) seems to improve not only the interpretability of the models but also their extrapolation ability. The MLPR part of the ME model can also be used for dynamic simulations. Since R is independent of time, $R_{\mathrm{meth}}^{\mathrm{MLPR}}$ can be solved in a steady state. The predictive ability of the D model and the ME model decreases only slightly when noise is added to the training and test data. The MC model is unable to handle noisy data. In the κ-term of the MC model, the structural error due to the simplified model and the error due to the random noise are superimposed, resulting in an incorrect prediction of the MC model.

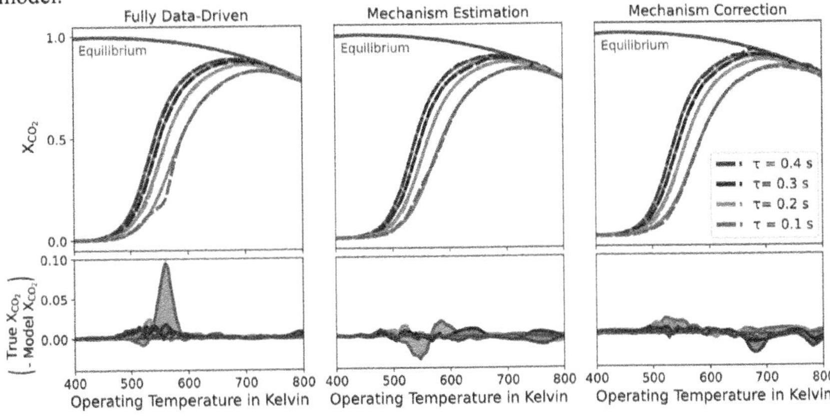

Figure 3: Model prediction of XCO_2 over the operating temperature. Model predictions are represented by dashed lines.

4. Conclusion

Hybrid models are a promising approach to integrate data-driven and mechanistic models. This study proposes three modeling architectures for CO_2-methanation in a CSTR. The models are trained and tested using pseudo-experimental process data and predict X_{CO_2}

with relatively small deviations. However, obtaining sufficient and high-quality process data is a challenge for all three models. The first model (D model) is straightforward to implement, but its performance is limited to the trained parameter range. The second model (ME model) uses a mechanism estimation approach and integrates a multilevel perceptron regression (MLPR) into the mass and energy balances. This approach provides a better representation of the reaction rate, including extrapolation data and noisy data, without requiring kinetic experiments. The third model (MC model) employs a mechanism correction approach, where a data-driven correction term is added to a mechanistic model. The MC model demonstrates good performance in both the trained and untrained parameter space but struggles with handling noisy data. In conclusion, the proposed hybrid models offer a way to improve the accuracy and reliability of process models for CO_2-methanation

Acknowledgment

This contribution is funded by the Bundesministerium für Bildung und Forschung (BMBF) and Project Management Jülich (PtJ) under grant 03HY302R.

References

Bremer, J. & Sundmacher, K., 2021. Novel Multiplicity and Stability Criteria for Non-Isothermal Fixed-Bed Reactors. *Frontiers in Energy Research,* Volume 8.

Er-rbib, H. & Bouallou, C., 2013. Modelling and simulation of methanation catalytic reactor for renewable electricity. *Chemical Engineering Transactions,* pp. 541-546.

Götz, M. et al., 2016. Renewable Power-to-Gas: A technological and economic review. *Renewable energy,* Volume 85, pp. 1371-1390.

Kahrs, O. & Marquardt, W., 2007. The validity domain of hybrid models and its application in process optimization. *Chemical Engineering and Processing: Process Intensification,* Volume 46(11), pp. 1054-1066.

Koschany, F., Schlereth, D. & Hinrichsen, O., 2016. On the kinetics of the methanation of carbon dioxide on coprecipitated NiAl (o) x.. *Applied Catalysis B: Environmental,* Volume 181, pp. 504-516.

Menghani, G., 2021. Efficient deep learning: A survey on making deep learning models smaller, faster, and better. *arXiv preprint, arXiv:2106.08962.*

Von Stosch, M., Oliveira, R., Peres, J. & de Azevedo, S. F., 2014. Hybrid semi-parametric modeling in process systems engineering: Past, present and future. *Computers & Chemical Engineering,* Volume 60, pp. 86-101.

Zahedi, G. et al., 2005. Hybrid artificial neural network-first principle model formulation for the unsteady state simulation and analysis of a packed bed reactor for CO2 hydrogenation to methanol. *Chemical Engineering Journal,* Volume 115, pp. 113-120.

Zimmermann, R. T., Bremer, J. & Sundmacher, K., 2020. Optimal catalyst particle design for flexible fixed-bed CO2 methanation reactors. *Chemical Engineering Journal,* Volume 387.

Antonis Kokossis, Michael C. Georgiadis, Efstratios N. Pistikopoulos (Eds.)
PROCEEDINGS OF THE 33rd European Symposium on Computer Aided Process Engineering (ESCAPE33), June 18-21, 2023, Athens, Greece
 http://dx.doi.org/10.1016/B978-0-443-15274-0.50238-9

Dynamic Optimization of Active Pharmaceutical Ingredient (Semi-)Batch Crystallization using Population Balance Modelling

Gustavo L. Quilló[a,b], Jan F.M. Van Impe[a*], Alain Collas[b], Christos Xiouras[b*], Satyajeet S. Bhonsale[a]

[a]*BioTeC+, Chemical and Biochemical Process Technology and Control, KU Leuven Campus Ghent, Gebroeders De Smetstraat 1, 9000 Ghent, Belgium*
[b]*Chemical & Pharmaceutical Development and Supply, Janssen R&D, Turnhoutseweg 30, 2340 Beerse, Belgium*
cxiouras@its.jnj.com, jan.vanimpe@kuleuven.be

Abstract

Current industrial (semi-)batch crystallization processes follow recipe-based temperature and antisolvent dosing trajectories obtained by extensive exploratory experimentation. This work optimizes the trajectories based on a 1-D Population Balance Equation coupled with the mass balance, (secondary) nucleation and crystal growth kinetics. The system of differential equations is solved by a High-Resolution Finite Volume Method. The method is demonstrated in a combined cooling-antisolvent crystallization that maximizes the number-average crystal size while satisfying process constraints (e.g., minimum yield). The dynamic optimization problem is solved via single shooting, where the control trajectory is discretized as piecewise linear, and the global solution is determined by multi-start single-objective optimization.

Keywords: crystallization, dynamic optimization, thermodynamics, population balance modelling, high-resolution finite volume method.

1. Introduction

The Population Balance Equation (PBE) presents a viable mathematical framework for the description of many chemical processes, such as crystallization, deposition, granulation, flocculation, milling, drying, mixing, and reactive multiphase systems such as polymerization and bioreactors (Randolph and Larson, 1971). In all these processes, it is vital to track the time evolution of a distributed property, e.g., the particle size distribution (PSD) in crystallization applications. The formulation for batch processes in the absence of breakage and agglomeration results in a hyperbolic partial differential equation for the crystal number density f [#/(kg-neat solvent.m)] in the form of eq. (1),

$$\frac{\partial f(L,t)}{\partial t} + \frac{\partial [G(L,t) f(L,t)]}{\partial L} + \frac{f(L,t)}{M}\frac{\partial M}{\partial t} = 0 \quad (1)$$

with boundary conditions $f(L_0,t) = B(L_0,t)/G(L_0,t)$, $f(\infty,t) = 0$ and initial condition $f(L,0) = f_{\text{seed}}(L)$. This semi-mechanistic approach is then used to engineer processes to satisfy a set of critical quality attributes by numerically solving a system of equations.

2. Problem formulation and computational methods

The PBE is supplied with several auxiliary functions. First, the mole-fraction, activity dependent supersaturation (driving force) for crystal nucleation and growth is given by eq. (2), where the denominator refers to the saturated state and γ and x are the activity coefficient and molar fraction of the API, respectively. The procedure to estimate the activity at supersaturated state γ_{API} is available elsewhere (Quilló et al., 2021).

$$\ln(\sigma_{API,MFAD}) = \ln\left(\frac{\gamma_{API}x_{API}}{\gamma_{API}^{sat}x_{API}^{sat}}\right) \tag{2}$$

The kinetic expressions for the overall linear growth rate G and secondary nucleation are taken as empirical power law functions in this work as in eq. (3).

$$\bar{G} = k_g\,[\ln(\sigma_{API})]^g \qquad B = k_b\,[\ln(\sigma_{API})]^b \qquad \sigma_{API} \geq 1 \tag{3}$$

2.1. High-resolution Finite Volume Scheme (HR-FVM)

The HR-FVM approach allows tracking the PSD, although at additional computational effort (LeVeque, 2002). The PBE is discretized in a static, nonuniform, non-smooth, structured spatial grid which results in eq. (4), where N is the number of cells (see **Figure 1**). The growth rate G_n can be size-dependent and is evaluated at the upper boundary of cell n, where it can assume positive (growth, G_n^+) or negative (dissolution, G_n^-) values. The formulation allows thermal cycling if the dissolution kinetics is known by applying absorbing boundary conditions on both sides of the domain if $\sigma_{API} < 1$, which imply in complete dissolution of any clusters of the lowest size range.

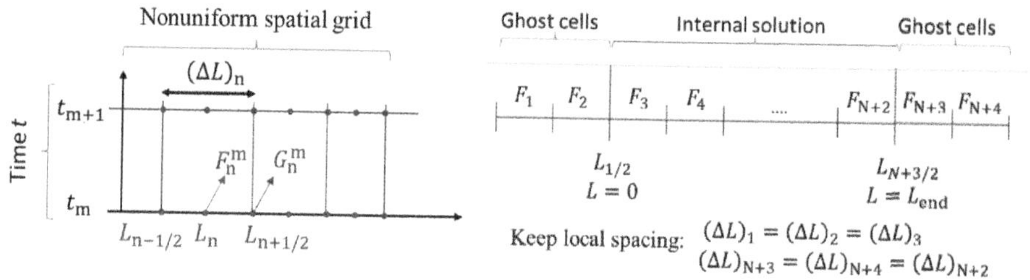

Figure 1. Nonuniform grid in the characteristic size coordinate extended by ghost cells for the implementation of boundary conditions, along with indexation. The internal solution is extended by two ghost cells on each side because the flux limiter implies in a stencil of five cells.

$$\begin{aligned}
\hat{F}_n^{m+1} = F_n^m &- \frac{\Delta t^m}{\kappa_{n-1/2}\Delta\xi}\left(G_n^+F_n^m - G_{n-1}^+F_{n-1}^m + G_{n+1}^-F_{n+1}^m - G_n^-F_n^m\right)\\
&- \left[\frac{\Delta t^m G_n^+}{2\kappa_{n-1/2}\Delta\xi}\left(\frac{\kappa_{n-1}}{\kappa_{n-1/2}} - \frac{\Delta t^m G_n^+}{\kappa_{n-1/2}\Delta\xi}\right)(F_{n+1}^m - F_n^m)\phi_n\right.\\
&\left. - \frac{\Delta t^m G_{n-1}^+}{2\kappa_{n-1/2}\Delta\xi}\left(\frac{\kappa_{n-2}}{\kappa_{n-3/2}} - \frac{\Delta t^m G_{n-1}^+}{\kappa_{n-1/2}\Delta\xi}\right)(F_n^m - F_{n-1}^m)\phi_{n-1}\right]\\
&+ \left[\frac{\Delta t^m G_n^-}{2\kappa_{n-1/2}\Delta\xi}\left(\frac{\kappa_{n-1}}{\kappa_{n-1/2}} + \frac{\Delta t^m G_n^-}{\kappa_{n-1/2}\Delta\xi}\right)(F_{n+1}^m - F_n^m)\phi_n\right.\\
&\left. - \frac{\Delta t^m G_{n-1}^-}{2\kappa_{n-1/2}\Delta\xi}\left(\frac{\kappa_{n-2}}{\kappa_{n-3/2}} + \frac{\Delta t^m G_{n-1}^-}{\kappa_{n-1/2}\Delta\xi}\right)(F_n^m - F_{n-1}^m)\phi_{n-1}\right]
\end{aligned} \tag{4}$$

The intermediate solution $\hat{F}_n^{m+1}$ is corrected by the fractional step method in eq. (5), which allows sequential solution of the homogenous PBE and source terms (Bosetti and Mazzotti, 2020; LeVeque, 2002). The coupling with the mass balance of the solute in eq. (6) and the solvent in eq. (7) occurs by the supersaturation used to compute nucleation and growth. Moreover, the solvent mass balance leads to dilution and changes the API solubility by altering the solution composition through the dosing stream Q_{in} [kg-neat solvent] with mass composition w_{in}. Further, the mixture volume V is obtained by eq. (8).

$$F_{\text{n}}^{\text{m}+1} = \hat{F}_{\text{n}}^{\text{m}+1}(M^{\text{m}}/M^{\text{m}+1}) \tag{5}$$

$$c_{\text{API}}^{\text{m}+1} = c_{\text{API}}^{\text{m}} - \rho_{\text{c}} k_{\text{v}} \left(3 \sum_{\text{n}=1}^{\text{N}} [G_{\text{n}}^{\text{m}} F_{\text{n}}^{\text{m}} L_{\text{n}}^{2} (\Delta L)_{\text{n}}] + B^{\text{m}} L_{0}^{3} \right) \Delta t^{\text{m}} - \frac{c_{\text{API}}^{\text{m}}}{M^{\text{m}}} (M^{\text{m}+1} - M^{\text{m}}) \tag{6}$$

$$M_{i}^{\text{m}+1} = M_{i}^{\text{m}} + w_{\text{in,i}} Q_{\text{in}} \Delta t^{\text{m}} \tag{7}$$

$$V^{\text{m}} = \frac{\sum M_{\text{i}}^{\text{m}}}{\rho_{\text{mix}}} = \frac{M^{\text{m}}}{1/\sum(w_{\text{i}}/\rho_{\text{i}})} \tag{8}$$

The left- and right-going propagating waves are retrieved separately by eq. (9).

$$G_{\text{n}}^{-} = \min(G_{\text{n}}, 0), \qquad G_{\text{n}}^{+} = \max(G_{\text{n}}, 0) \tag{9}$$

The nonuniformity in the spatial grid is tackled by the cell capacity κ, which is the ratio of the true cell size $(\Delta L)_{\text{n}}$ and a computational step $\Delta\xi$ as in eq. (10). The non-smoothness is addressed by the mean of adjacent cell capacities as in eq. (11), where L_{end} and L_0 are the right and left boundaries of the discretized spatial domain, respectively.

$$\kappa_{\text{n}} = (\Delta L)_{\text{n}}/\Delta\xi, \text{ where } \Delta\xi = (L_{\text{end}} - L_0)/\text{N} \tag{10}$$

$$\kappa_{\text{n}-1/2} = (1/2)(\kappa_{\text{n}-1} + \kappa_{\text{n}}) \tag{11}$$

The flux limiter function ϕ in eq. (12) depends on the smoothness θ defined in eq. (13), where ε is a small number. Many possibilities of high-resolution flux limiters are available in the literature (LeVeque, 2002) (e.g. minmod, superbee, MC, Koren, van Leer), all of which are algebraic total variation diminishing (TDV) but imply in mildly different capabilities to mitigate numerical problems (e.g., diffusion, dispersion) in the obtained solution. This work uses the superbee flux limiter, as it displays minimal solution smearing while retaining near 2nd-order accuracy.

$$\phi_{\text{n,Superbee}}(\theta_{\text{n}}) = \max(0, \min((1 + \theta_{\text{n}})/2, 2, 2\theta_{\text{n}})) \tag{12}$$

$$\theta_{\text{n}} = \frac{F_{\text{n}} - F_{\text{n}-1} + \varepsilon}{F_{\text{n}+1} - F_{\text{n}} + \varepsilon} \tag{13}$$

The Dirichlet boundary conditions are applied by ghost cells (see **Figure 1**). The left side of the domain is limited by nuclei size L_0, and the values assigned to the ghost cells are calculated by linear extrapolation as in eq. (14) to eq. (16), i.e., by the upwind method.

$$F_{L=L_0} = [2/(\Delta L)_1]\big(F_3 - B(t)/G(L_0, t)\big) \tag{14}$$

$$F_1 = -2(\Delta L)_1 F_{L=L_0} + F_3 \tag{15}$$

$$F_2 = -(\Delta L)_1 F_{L=L_0} + F_3 \tag{16}$$

The ghost cells on the right-side of the domain embody an absorbing boundary condition, where the L coordinate is artificially truncated at a sufficiently high value L_{end} to avoid particle loss to the boundary over the process time frame. The boundary condition is implemented via zero-order extrapolation, as in eq. (17).

$$F_{N+3} = F_{N+4} = F_{N+2} \tag{17}$$

The time coordinate is discretized by a forward Euler scheme, where the time step Δt^{m} is selected based on the most stringent criterion among the Courant condition for numerical stability $\min((\Delta L)_n/G_n) \leq 1$, the acceleration limit based on the previous time step, the minimum and maximum user-specified integration steps and the query time points. After the application of the vectorized HR-FVM scheme, the mass balance for the solute API and the dilution effect are updated using the fractional-step method. Next, the applied time step is approved or rejected based on relative tolerance checks in reference to the previous step supersaturation, solvent mass, and concentration and negative F due to small oscillations. The rejection cycle is applied up to three times before the time step is forcefully accepted, and each cycle reduces the applied time step (α_t = 1/3) prior to reapplication of HR-FVM. The summary of the algorithm is provided in **Figure 2**.

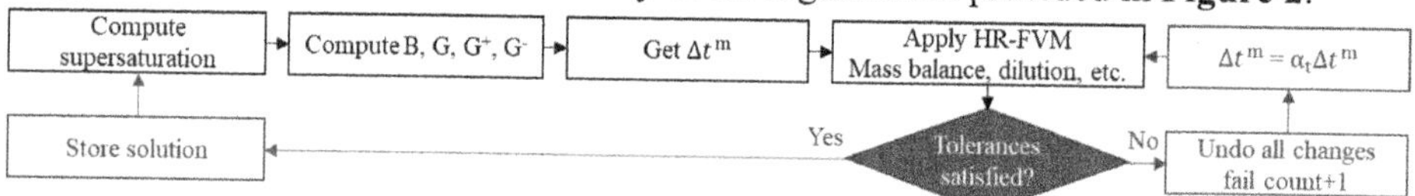

Figure 2. Implementation of the adaptative time scheme for the HR-FVM with error control for the custom-built partial differential equation integrator.

3. Dynamic optimization of cooling-antisolvent seeded (semi-)batch crystallization

The model is used to find optimal temperature and/or antisolvent trajectories that minimize (or maximize) a given goal under constraints that reflect equipment operation limits. In this case, the objective is to maximize the number average crystal size, defined by the ratio of the first to the zeroth moment of the PSD at final batch time t_f given the system parameters $\boldsymbol{\theta}$. The nonlinear programming problem (NLP) is stated in eq. (18).

$$\max_{T(k),\, M_{in}(k)} \quad f(t_f, \boldsymbol{\theta})$$

subject to:

$$\begin{gathered} T_{min}(k) \leq T(k) \leq T_{max}(k) \\ R_{min}(k) \leq \mathrm{d}T/\mathrm{d}t(k) \leq R_{max}(k) \\ Q_{min}(k) \leq \mathrm{d}M_{in}/\mathrm{d}t(k) \leq Q_{max}(k) \\ 0.9\max(\text{Yield}_{\text{theoretical}}) \leq \text{Yield}_{\text{process}}(t_f) \\ V(t_f) \leq V_{max} \end{gathered} \tag{18}$$

The trajectories are obtained by optimizing the piecewise linear control variables at specified nodes $T(k), M_{in}(k)$ located in regular time intervals and solving the system by HR-FVM. In this case study, the control variables were discretized in 16 nodes after preliminary numerical analysis. The optimal trajectory is determined by derivative-free, trust-region-based, multi-start optimization using the COBYLA algorithm (Powell, 1994). The method is implemented in MATLAB® 2022b, although the optimization and objective function evaluations are compiled in FORTRAN and in C++ by MEX files, respectively, which speed up calculations by at least threefold. The data for the case study with fixed batch time is given in **Table 1**.

Table 1. Summary of parameters for the case study.

$k_b = 7.92 \times 10^{14}\ \#/(\text{kg.s})$	$b = 6.22$	$k_g = 2.38 \times 10^{-5}$ m/s	$g = 2.37$
$\rho_c = 1300$ kg/m^3	$k_v = 0.1$	$c_{API,t=0} = 0.33$ kg/kg	$M_{t=0} = 0.288$ kg
$V_{max} = 400$ mL	$t_{batch} = 16$ h	$T_{min} = 33$ °C	$T_{max} = 60$ °C
$R_{min} = -1$ °C/min	$R_{max} = 0$ °C/min	$L_{end} = 1000$ μm	$N = 4000$

The API solubility is calculated by the Van't Hoff Jouyban-Acree model as in eq. (19).

$$\ln(x_{\text{API}}^{\text{sat}}) = \sum_{p\in\{1,2,3\}} \left(\psi_{0,p} w_p + \psi_{1,p}\frac{w_p}{T}\right) + \sum_{p,q\in\{1,2,3\};\ p\neq q} \frac{w_p w_q}{T}\left[\psi_{2,p,q} + \psi_{3,p,q}(w_p - w_q) + \psi_{4,p,q}(w_p - w_q)^2\right] \quad (19)$$

Table 2 Van't Hoff Jouyban Acree solubility parameters. Temperature is in degrees Kelvin.

$\psi_{0,1} = -8.77$	$\psi_{1,1} = 9.43 \times 10^6$	$\psi_{2,1,2} = 7.80 \times 10^6$	$\psi_{3,1,2} = -3.40 \times 10^7$	$\psi_{4,1,2} = -3.47 \times 10^7$
$\psi_{0,2} = 17.2$	$\psi_{1,2} = -1.39 \times 10^7$	$\psi_{2,1,3} = -1.55 \times 10^7$	$\psi_{3,1,3} = -6.93 \times 10^6$	$\psi_{4,1,3} = -8.03 \times 10^5$
$\psi_{0,3} = 3.36$	$\psi_{1,3} = -3.06 \times 10^3$	$\psi_{2,2,3} = 2.07 \times 10^7$	$\psi_{3,2,3} = 7.94 \times 10^6$	$\psi_{4,2,3} = 1.15 \times 10^6$

The initial ternary solvent mass composition is $w_1 = 0.503$, $w_2 = 0.313$ and $w_3 = 0.184$ and pure antisolvent is dosed ($w_{\text{in},2} = 1$). The theoretical maximum yield is $\approx$100% due to the strong antisolvent effect. The optimization results are summarized in **Figure 3**.

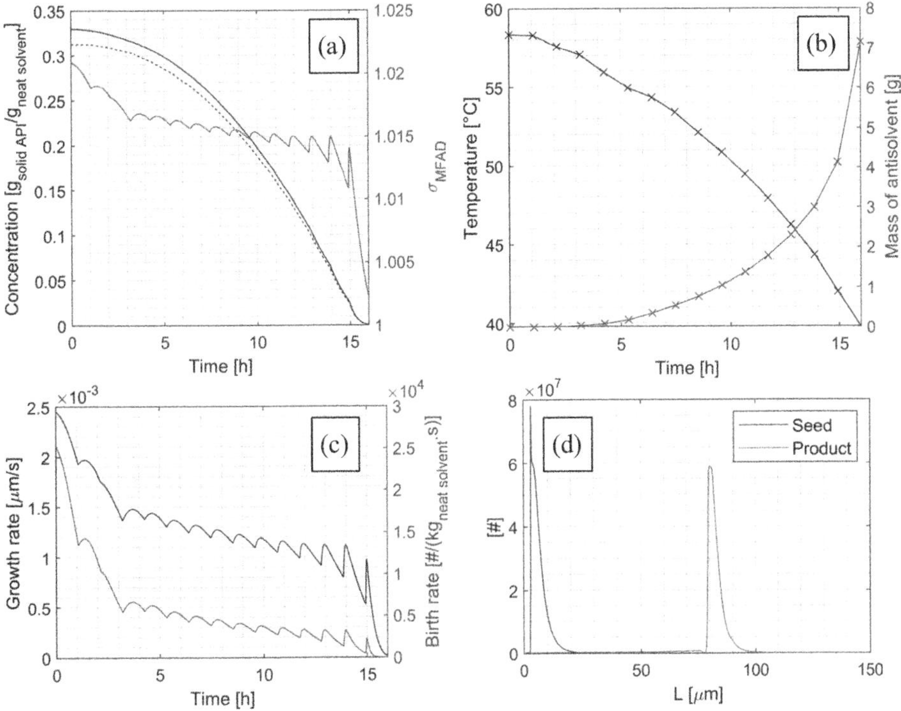

Figure 3. (a) Solution for the mass balance for HR-FVM. Solid line and dashed black line represent the crystallizer concentration and the solubility, respectively, while the red line depicts the supersaturation. (b) Optimal piecewise linear temperature trajectory. (c) Time evolution of the growth and birth rates. (d) Seed and product number density.

The HR-FVM method using the custom-built integrator was validated by comparison with standard ODE integrators in MATLAB (i.e., ode45, ode23s) solving the equivalent NLP by the Standard Method of Moments (SMOM) with adequate solution juxtaposition.

The problem identifies that the initial supersaturation is too high and favors nucleation excessively, so the optimal trajectory starts with an isothermal hold and no antisolvent addition at the beginning. Then, cooling starts before dosing since it is a gentler form of managing the supersaturation, followed by the combined effect of both controls, which are required to satisfy the imposed yield constraint. Furthermore, the optimal profile is consisted of increasingly larger cooling and dosing rates which cause supersaturation to be nearly constant at a level that suppresses nucleation for most of the simulation while growth is consuming the supersaturation. The long but small tail in the product PSD is caused by nucleation, which constantly generates fines, whereas the seeds PSD simply propagate in the size coordinate at the speed determined by the growth rate due to the absence of dissolution and particle death events (breakage, agglomeration). The peak artifact present in the seeds PSD is smeared during the solution, but the shape of the initial PSD is mostly well-retained. The piecewise linear discretization of the controls provides significantly better results than the piecewise constant counterparts (not shown), which implies in supersaturation spikes close to the nodes due to the sudden change in solubility.

4. Conclusions

Piecewise linear optimal temperature and antisolvent addition trajectories were successfully obtained from a custom-built integrator using a versatile and easily extensible implementation of the HR-FVM method to solve a PBE for a (semi-)batch crystallizer. The method at current allows grid tuning, size-dependent growth, and thermal cycles, and can be expanded to cope with more complicated source terms from breakage and agglomeration events by the fractional-step method.

The custom-built integrator for HR-FVM presented comparable accuracy to standard MATLAB integrators using SMOM while delivering more information, however, it needed 2-100 times more optimization time depending on the selected spatial grid resolution. Therefore, there are opportunities to tune the discretization (e.g., via Runge-Kutta, Crank-Nicholson for the time, adaptative spatial grid) to improve the calculations speed. The obtained control trajectory profiles led to crystals that are on average $\approx$12.4 times larger than the seeds and low quantity of fines, as nucleation was suppressed.

5. Acknowledgements

This work was supported by VLAIO (Agentschap Innoveren & Ondernemen) and Janssen Pharmaceutica [Baekeland grant number HBC.2020.2214].

References

L. Bosetti., M. Mazzotti, 2020. Population Balance Modeling of Growth and Secondary Nucleation by Attrition and Ripening. Cryst. Growth Des. 20, 307–319.

R.J. LeVeque, 2002. Finite Volume Methods for Hyperbolic Problems, 1st ed. Cambridge University Press, Cambridge, United Kingdom.

M.J.D Powell., 1994. A Direct Search Optimization Method That Models the Objective and Constraint Functions by Linear Interpolation, in: Advances in Optimization and Numerical Analysis. Springer Netherlands, Dordrecht, 51–67.

G.L. Quilló, S. Bhonsale, B. Gielen, J.F. Van Impe., A. Collas, C. Xiouras, 2021. Crystal Growth Kinetics of an Industrial Active Pharmaceutical Ingredient: Implications of Different Representations of Supersaturation and Simultaneous Growth Mechanisms. Cryst. Growth Des. 21, 5403–5420.

A.D. Randolph, M.A. Larson, 1971. Theory of Particulate Processes: Analysis and Techniques of Continuous Crystallization, 1st ed. Academic Press, Inc., New York, NY.

Antonis Kokossis, Michael C. Georgiadis, Efstratios N. Pistikopoulos (Eds.)
PROCEEDINGS OF THE 33rd European Symposium on Computer Aided Process Engineering (ESCAPE33), June 18-21, 2023, Athens, Greece
 http://dx.doi.org/10.1016/B978-0-443-15274-0.50239-0

A Systematic Framework for Iterative Model-Based Experimental Design of Batch and Continuous Crystallization Systems

Hemalatha Kilari, Yash Barhate, Yung-Shun Kang, Zoltan K. Nagy*

*Davidson School of Chemical Engineering, Purdue University, West Lafayette, Indiana 47907, United States. *znagy@purdue.edu*

Abstract

A generalized and systematic framework is proposed for rapid model development of batch and continuous crystallization processes. The proposed approach uses iterative model-based experimental design (IMED or MBDoE), whereby the experiments are guided by an imprecise-model to produce information rich data that will enhance its prediction performance for further improvement of model structure, parameters, robustness and accuracy. This enables to generate a precise model that can be used for *in silico* design of experiments in broader range of process parameters as well as for efficient digital design. This paper elucidates various steps for model discrimination, selection as well as precise parameter estimation with a combination of statistical tests and optimal experimental design approaches to arrive at an accurate model. The application benefits of the proposed strategy are demonstrated through case studies for batch and continuous crystallization systems of commercial active pharmaceutical ingredients.

Keywords: Optimal experimental design; Model-based experimental design, Design of Experiments; Batch crystallization; Continuous crystallization.

1. Introduction

Crystallization is a complex solid-liquid-separation technique that is widely applied in the manufacturing of solid products. High fidelity models are necessary for process improvements to achieve customized product specifications and properties. The usual industrial practice for modelling a crystallization process involves a traditional Quality-by-design (QbD) approach where a predetermined set of crystallization experiments are designed either by prior-knowledge/intuition or traditional statistical design of experiments (DoE) that considers the effect of critical process parameters (CPPs) such as seed, supersaturation, temperature, on the critical quality attributes (CQAs) such as mean crystal size, crystal size distribution (CSD), shape and purity. When large number of CPPs are considered, this approach results in a large set of experiments that is resource intensive. This calls for a more systematic procedure for rapid model development.

Application of model-based experimental design approach for rapid model development has been demonstrated to crystallization processes in earlier studies (Chung et al., 2000; Chen et al. 2004; Pal et al., 2020). It is evident from their studies that such approach is capable of predicting information rich experiments and reduce number of experiments to obtain accurate model with precise estimates of parameters. However, the focus in those studies was on either a particular case study involving batch or a particular type of crystallization process concentrating on either model discrimination or improving parameter precision only and are sometimes case specific. Also, with increased adoption of continuous manufacturing in the industries, it is necessary to extend this procedure to continuous crystallization systems.

The objective of this work is to propose a generalized framework that gives guidelines to follow a model building procedure for crystallization processes starting with model screening from a model structure database and conduct an iterative sequence of statistical approaches to discriminate between the models along with optimal experimental design criterion to obtain a best model with precise parameters. The experimental design in crystallization processes is posed as an optimal control problem with in a framework that enables to choose cooling recipe, seed recipe, antisolvent flow profile, experiment time, residence time, number of crystallizers, number of sampling points etc. with ease of allowing constraints on inputs and outputs. In the proposed framework, methods are presented accounting to parameter uncertainty to improve model robustness. This work presents the first generalized comprehensive study that is applied to both batch and continuous crystallization processes.

2. Framework Development Methodology

2.1. Crystallization Model

Population balance equation (PBE) governing the CSD assuming nucleation with constant nucleation size, growth and agglomeration mechanisms can be represented as follows (Szilagyi et al., 2020; Szilagyi et al., 2022).

$$\frac{\partial[(Vn(L,t))]}{\partial t}+\frac{\partial(Gn(L,t)V)}{\partial L}=V[B\delta(L-L_n)+B_{agg}-D_{agg}+B_{bre}-D_{bre}]+F_f n_f(L,t)-Fn(L,t) \tag{1}$$

$$\frac{d(Vc)}{dt}=-3Vk_v\rho_c\left[(B_p+B_s)L_n^3+\int_0^\infty GL^2n(L)dL\right]+F_f c_f-Fc \tag{2}$$

$$B_p=k_p(1+\gamma_p A)\sigma^p\exp\left(-\frac{k_e}{\ln^2(\sigma+1)}\right)\exp\left(\frac{-E_p}{RT}\right);B_s=k_s(1+\gamma_s A)\sigma^s\exp\left(-\frac{k_e}{\ln^2(\sigma+1)}\right)\exp\left(\frac{-E_s}{RT}\right);G=k_g(1+\gamma_g A)\sigma^g\left(\alpha_g+\beta_g L^{\gamma_g}\right)\exp\left(\frac{-E_g}{RT}\right) \tag{3}$$

From Eq (1), B_{agg} *and* B_{bre} are birth functions that describe production of new crystals of size L by agglomeration and breakage. D_{agg} and D_{bre} represent Death functions for the rate of consumption of crystals of size L by agglomeration and breakage. Eq (2) represents mass balance equation for crystallizing compound that accounts for solid-liquid mass transfer generated by nucleation and growth with feed and outflow streams in a crystallizer. Generalized expressions for primary & secondary nucleation and growth are shown in Eq (3). As a first step to modeling crystallization processes, experimental data is obtained for a preliminary set of crystallization experiments by monitoring the process variables throughout the experiment using in situ or inline process analytical technology (PAT) tools and offline characterization methods. An optimization problem for parameter estimation is formulated (Eq (4)) with the objective of minimizing the weighted sum of squared error (*WMSE*) between experimental and model predicted values of the process parameters.

$$\min_{\theta} WMSE=\frac{1}{N_E N_S}\sum_{h=1}^{N_v}\sum_{i=1}^{N_E}\sum_{j=1}^{N_S}w_h(y_{h,i,j}^{\exp}-y_{h,i,j}^{\mathrm{mod}})^2 \tag{4}$$

Where N_v, N_E and N_S are number of process variables, experiments and samples or data points respectively; w_h are the weights chosen to normalize the value of the objectives, y represents the process variables such as concentration, online crystal counts /sec during the operation time, final time mean size or D_{10}, D_{50} and D_{90} values, or the entire CSD, depending on measurement and modelling techniques used.

2.2. Optimal Experimental design for Model Identification, Discrimination and Parameter precision

Selecting a best model from a crystallization model library consisting of combinations of various expressions for different mechanisms is a challenging task. Initial kinetic parameter estimation for some of the chosen models may give an incorrect model that still can closely represent the experimental data. It is of primary importance to statistically assess the reliability and adequacy of the model and if necessary, of the new parameter estimates. This can be done using standard checks (i.e. on residuals distributions) and statistical tests. Statistical goodness-of-fit tests such as chi-square test, F-test, student-t test, etc. can be used for assessing model identifiability and parameter estimability. The problem of over fitting and underfitting a model can be tackled well by using Akaike information criterion (*AIC*) which estimates prediction error and thereby provides the relative quality of statistical models for a given set of data. When all these statistical tests fail to screen one model as the best model, optimal experimental design criterion for model discrimination (MD) can be used, which determines experimental conditions for additional experiments to be conducted in order to maximize the divergence between models. There have been many such approaches reported so far and a modified Hunter-Reiner criterion for many rival models similar to the one reported by Buzzi-Ferraris-Forzatti (1992) has been used in our work. When a best model is selected, it is subjected to parameter refinement (PR) to reduce the parameter uncertainties and improve the predictive capability of the model. Various steps involved in the IMED procedure are illustrated in the Fig.1.

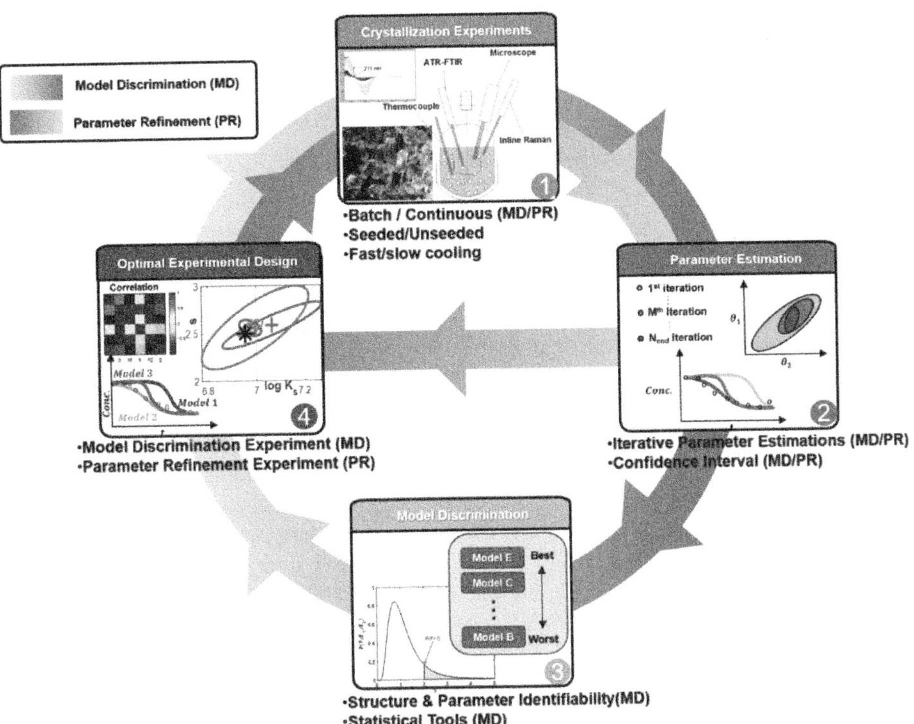

Fig.1. Iterative model-based experimental design procedure for crystallization processes

D-optimal design is a popular design procedure widely used for improving parameter precision that uses Fisher Information Matrix (FIM) as a metric to assess experimental information of the present experiments and prescribe a future experiment that provides with more information rich data. FIM is constructed from sensitivity matrix of process states and model parameters (Yang et al., 2022). The new experiment is obtained by maximizing the determinant of the *FIM* (which is an indirect approach of minimizing confidence region of parameter estimates). The generic mathematical representation of this *D*-optimal approach is shown below for a batch cooling crystallization and for a continuous MSMPR crystallization system.

Batch Cooling Crystallization	*Continuous MSMPR Crystallization*
$\max\limits_{N_E, CR, Ms} \left[\det(FIM)\right]$	$\max\limits_{N_c, \tau_i, T_i} \left[\det(FIM)\right]$
Subject to:	Subject to:
$c_1 : Ms_{\min} \leq Ms \leq Ms_{\max}$	$c_1 : \tau_{\min} \leq \tau_i \leq \tau_{\max}$
$c_2 : CR_{\min} \leq CR \leq CR_{\max}$	$c_2 : T_{N_c} \leq T_i \leq T_f$
$c_3 : CR \leq 0$	$c_3 : T_i < T_{i-1} \ (i=1 \ldots N_c)$

(5)

Where *CR* is the cooling rate for a linear temperature profile, *Ms* is the seed loading, N_E is the number of experiments necessary for another iteration for the batch case and N_c is number of crystallizers in a cascade of MSMPRs for the continuous case with τ_i being the residence time and T_i, the temperature of ith Crystallizer. *FIM* is the fisher information matrix and is constructed from the sensitivity matrix with sensitivity coefficients of process variables at various sampling instants with respect to each parameter. This step yields new experimental conditions to be implemented for data collection and for parameter refinement. This procedure is iterated until a predefined model accuracy is obtained with precise parameter estimates.

3. Application of the Proposed Framework: Results

The benefits of the framework are tested first by application to a batch cooling crystallization case study with a known model. *In silico* experiments are generated for a combination of high/low seed loading and fast/slow linear cooling experiments. This set of four seeded experiments are used to obtain initial parameter estimates for six plausible model candidates (Table 1) that are subjected to model discrimination. It is observed that all these models closely fit the concentration and final mean size data. The AIC criteria correctly chooses model 6 as the best model. However, WMSE values are very close for other models too. Hence, it is decided to allow these models for an iteration of discriminatory experimental design (MD-OED). F-test (Sobek and Werle, 2020) is efficient in screening 3 models out of 6 for MD-OED procedure. With one iteration of the MD-OED, an unseeded experiment is obtained and this experiment is carried out *in silico* to collect experimental data for the next iteration of parameter estimation procedure. This experiment is added to the initial parameter set and a new parameter estimation is carried out for all 3 models. The unseeded experiment seems to create a better divergence between the 3 models and helps to choose the best model amongst them. Model 6 is chosen as the best model (true model) according to AIC and also F-test rejects other models choosing model 6 as the best one. Model 6 is now chosen for iterative optimal experimental design procedure for improving precision of parameters.

Table 1: Model discrimination and Selection

Model	Models	N_θ	WMSE $*10^3$	AIC	F-value ($F_{Critical}$=2.21)	Decision
Initial Parameter Estimation with a preliminary set of 4 seeded experiments						
1	SN+SIG	3	1.11	-15.07	7.8	Reject
2	SN+SIG^T	4	1.68	-14.65	11.8	Reject
3	SN+SDG	4	0.22	-16.69	1.5	Accept
4	PN+SN+SIG	5	0.27	-16.48	1.9	Accept
5	PN+SN+ SIG^T	6	0.76	-15.44	5.3	Reject
6	PN+SN+SDG *True model	6	0.14	-17.12	1.0	Accept
Re-Parameter estimation after including MD-OED experiment						
3	SN+SDG	4	3420.0	-1.88	1.89E+07	Reject
4	PN+SN+SIG	5	20410.0	-1.10	1.12E+08	Reject
6	PN+SN+SDG *True model	6	1.82E-04	-9.15	1.0	Accept

#PN: Primary nucleation; SN: Secondary Nucleation; SIG: Size Independent Growth, SIG^T: SIG with Temperature term; SDG: Size dependent growth

The IMED approach for parameter precision or parameter refinement can be carried out using either a sequential or a simultaneous approach. In the simultaneous approach a complete set of experiments with either a fixed number (e.g. in our case $n_{exp} = 4$) or using the number of experiments also as a decision variable can be used to obtain the new set of experimental conditions for all experiments within a single optimization that could be implemented to collect new experimental data for parameter refinement with new parameter estimation. In the sequential approach only one new experiment is obtained at each iteration that can be implemented and the results are added to previous experimental data to carry out the new parameter estimation.

Sample results are shown for the sequential approach where one experiment is obtained with each iteration and new parameters are estimated and analyzed for parameter uncertainty. The confidence region (Fig. 2) plotted for a pair of initial parameter estimates ($\theta_1 = logK_p$ & $\theta_2 = p$; primary nucleation parameters shown in Eq(3)) is very broad. With each iteration of the optimal experimental design, the parameters are improved as shown by shrinkage of the confidence ellipse and moving closer to the true parameter values after each iteration. Fig 2(a) shows comparison of confidence ellipse with initial parameter estimation and a model discriminatory experiment. Adding the discriminatory experiment to the parameter estimation set improved the model as it can be clearly observed that the parameter values shifted closer to the true parameter region.

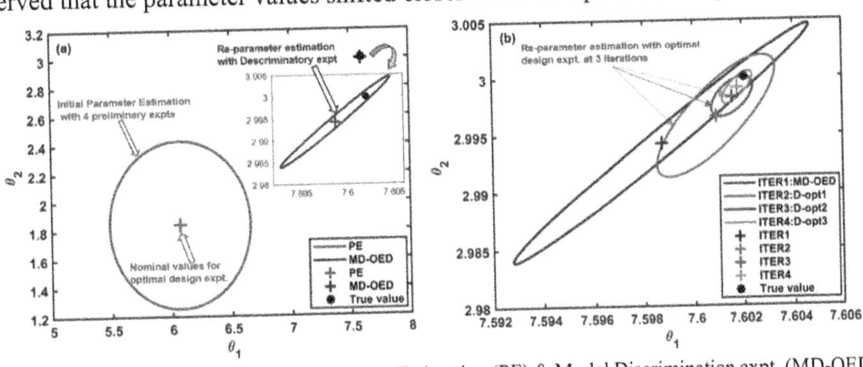

Fig.2. Confidence Ellipse plots (a) Initial Parameter Estimation (PE) & Model Discrimination expt. (MD-OED) (b) Optimally designed experiments using D-design Criterion (D-opt1, D-opt2, D-opt3) at each iteration (ITER)

Fig 2(b) shows the improvement of confidence region and corresponding parameters with each iteration of re-parameter estimation including respective D-optimal experiment. The IMED procedure is continued until no further improvement in the confidence region is obtained. In this case, 3 iterations of IMED brought the parameters closer to the true value of the parameters. Precise parameters with very narrow confidence intervals are obtained at the end of 3 iterations and found to be very close to the true parameters of the model considered. Thus, this batch case study demonstrates powerful predictive capabilities of the proposed framework. Similarly, another case study for continuous crystallization of Diphenhydramine hydrochloride is carried out and it is observed to be effective in arriving at an accurate model with precise parameters and very small confidence intervals.

4. Conclusions

A generalized systematic framework is proposed for batch and continuous crystallization model development using an iterative model-based experimental design framework. The framework combines model discrimination and parameter refinement steps and is able to identify the best model structure that constitutes relevant crystallization mechanisms and drive the model with imprecise initial parameter estimates towards an accurate model with precise parameters. Future work is focused on expanding this knowledge to application scenarios that include other types of crystallization processes and multi-dimensional models to cater to a wider range of applications of crystallization modeling.

Acknowledgements

Funding for this work was made possible, in part, by the US Food and Drug Administration (FDA) through grant number 75F40121C00106. Views expressed herein do not necessarily reflect the official policies of the Department of Health and Human Services; nor does any mention of trade names, commercial practices, or organization imply endorsement by the United States Government.

References

S. H. Chung, D. L., Ma., R. D., Braatz, 2000, Optimal model-based experimental design in batch crystallization. *Chemometrics and Intelligent Laboratory Systems*, *50*(1), 83-90.

B.H. Chen, S., Bermingham, A.H., Neumann, H.J. Kramer, S.P., Asprey, 2004, On the design of optimally informative experiments for dynamic crystallization process modeling, *Industrial & engineering chemistry research*, *43*(16), 4889-4902.

K. Pal, B. Szilagyi, Burcham, C.L., D.J. Jarmer, Z.K. Nagy, 2021, Iterative model-based experimental design for spherical agglomeration processes. *AIChE Journal*, *67*(5), e17178.

B. Szilagyi, A. Majumder, Z. K. Nagy, 2020. Fundamentals of Population Balance Based Crystallization Process Modeling, In *The Handbook of Continuous Crystallization* (pp. 51-101).

B. Szilagyi, ,W. L. Wu, A. Eren, J. Mackey, S. Kshirsagar, E. Szilagyi, Z. K. Nagy, 2022, Cross-Pharma Collaboration for the Development of a Simulation Tool for the Model-Based Digital Design of Pharmaceutical Crystallization Processes (CrySiV). *Crystal Growth & Design*, *21*(11), 6448-6464.

G. Buzzi-Ferraris, P. Forzatti, 1990, An improved version of a sequential design criterion for discriminating among rival response models. *Chem. Eng. Sci. 45* (2), 477–481.

L. Yang, Y. Zhang, P. Liu, C. Wang, Y. Qu, J. Cheng, C. Yang, 2022, Kinetics and population balance modeling of antisolvent crystallization of polymorphic indomethacin. *Chemical Engineering Journal*, *428*, 132591.

S. Sobek, S. Werle, 2020, Kinetic modelling of waste wood devolatilization during pyrolysis based on thermogravimetric data and solar pyrolysis reactor performance. *Fuel*, *261*, 116459.

Antonis Kokossis, Michael C. Georgiadis, Efstratios N. Pistikopoulos (Eds.)
PROCEEDINGS OF THE 33rd European Symposium on Computer Aided Process Engineering (ESCAPE33), June 18-21, 2023, Athens, Greece
 http://dx.doi.org/10.1016/B978-0-443-15274-0.50240-7

Evaluation of Factors Affecting Novel Technology Development using Mathematical Programming

Pooja Zen Santhamoorthy[a], Selen Cremaschi[a,*]

[a]*Department of Chemical Engineering, Auburn University, AL 36849, USA*
scz0113@auburn.edu

Abstract

The successful development and timely deployment of novel technologies to compete with existing ones depend on the availability of necessary resources and associated costs. This paper introduces a multiperiod optimization model that integrates processing plants, technologies, and resources under a single network to study the impact of novel technology development on the process network design to meet the annual product demands over a planning horizon. A case study of developing post-combustion adsorption for carbon capture (CC) while meeting the annual capture targets reveals that adsorption should become commercially available within 5 years to compete with the commercially available absorption technology. The total CC cost if only absorption were available is \$13.9 billion, 6% of which could be used towards the research and development (R&D) of the adsorption technology while still making it an economically viable option.

Keywords: technology development; multiperiod optimization; CO_2 capture planning.

1. Introduction

Technology development requires resources for R&D, pilot plants, and demonstration projects. The availability of such resources plays a major role in the successful development and timely implementation of new technologies to compete with existing and other novel technologies (Anderson Jr and Joglekar, 2005). For example, a high cost for a necessary resource can make a novel technology economically unviable, or prolonged development time can make it uncompetitive against technologies developing at a faster pace (Davila and Wouters, 2004). Hence, it is important to study the impacts of resource availability, the associated costs, and the development time on novel technology development along with the operation of the existing technologies. This work develops a multiperiod optimization model for production planning and resource allocation for developing novel technologies and operating implemented technologies to meet the annual product demands over time.

2. Problem Statement

The goal is to develop an optimization-based framework to study the development of novel technologies and deployment and operation of commercial technologies by determining strategic and tactical decisions: (i) allocation of resources for technology development, (ii) selection of processes and commercial technologies to produce products, (iii) determination of production rate based on the annual product demands, and (iv) addition of expansions to the process facilities, if necessary, over time. For given yearly product demands, the objective is to minimize the total cost of technology development, implementation, and operation.

The superstructure in Figure 1 illustrates the overarching network. It consists of a set of processes ($i \in I$), technologies ($j \in J$), resources ($r \in R$), development stages called Technology Stages ($s \in S$), expansions ($n \in N$), and time periods ($t \in T$). The technologies T_j are in different Technology Stages ($TS_{j1}, TS_{j2},.., TS_{js}$). They mature to subsequent stages if the necessary resources R_{rt} are allocated during development. The technically developed and commercially available technologies at TS_{js} are used by processes P_i in the facilities C_{ij} to produce products X_{it} over the planning period t. Expansion facility E_{nij} using technology j is added for process i when demand for product X_{it} increases at time t.

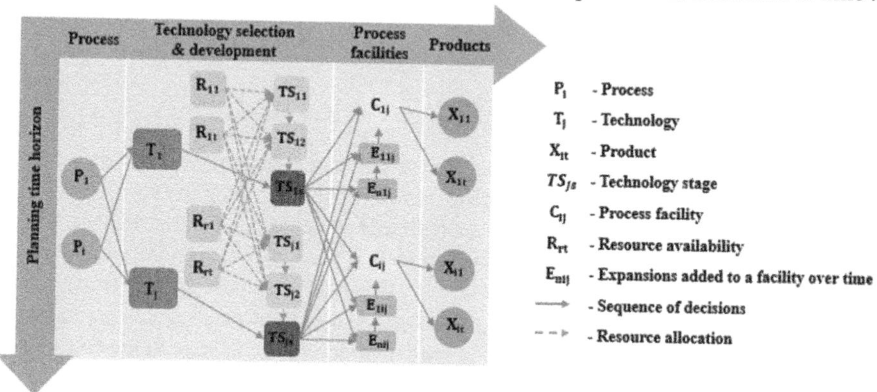

Figure 1. Superstructure of technology development and deployment to produce products.

3. Mathematical Model Formulation

3.1. Objective

The objective is to minimize the total cost of producing a set of products over the planning horizon while meeting annual demands. The cost consists of annualized investment and annual operating costs, and R&D and demonstration project costs as given in Eq. (1).

$$Min\,TC = \sum_{i,j,t,n}\left(IC_{i,j,t,n} + OC_{i,j,t,n}\right) + \sum_{i,j,t} DC_{i,j,t} \quad \forall i \in I, j \in J, t \in T, n \in N \qquad (1)$$

The sets $i \in I$, $j \in J$, $t \in T$, and $n \in N$ represent the processes, technologies, planning periods, and capacity expansions. For process i, $n=1$ represents the initial facility capacity, and $n>1$ represents the expansions. The variable TC is the total cost of investment and operation of implemented technologies and the cost of developing novel technologies. The variables $IC_{i,j,t,n}$ and $OC_{i,j,t,n}$ are the investment and operating costs of facility/expansion n of process i using technology j at time t. The variable $DC_{i,j,t}$ is the technical development cost of technology j for process i at time t.

3.2. Binary Variables

A binary variable $yc_{i,j}$ is defined in Eq. (2) to represent the compatibility of technology j with process i. Technology Stages are represented by set $s \in S$, and a binary variable $x_{i,j,t,s}$ is introduced in Eq. (3) to represent if technology j has been developed to at least Technology Stage s by time t for process i. If a technology is not compatible with process i, it remains in the initial stage of development ($x_{i,j,t,1}=1\ \forall t \in T$ and $x_{i,j,t,s}=0\ \forall t \in T,\ s \in \{2,3,..,S\}$). Another binary variable $x'_{i,j,t,s}$ is introduced in Eq. (4) to denote if the resources necessary for development to stage s are available at time t. Once the technology j is commercially developed for process i ($x_{i,j,t,s}=1$), a binary variable, $y_{i,j,t,n}$, in Eq. (5), represents the technology implementation for facility/expansion n of process i at time t.

$$yc_{i,j} = \begin{cases} 1, \text{if tech } j \text{ is compatible with process } i \\ 0, \text{otherwise} \end{cases} \quad (2)$$

$$x_{i,j,t,s} = \begin{cases} 1, \text{if tech } j \text{ is developed to at least stage } s \text{ for process } i \text{ by time } t \\ 0, \text{otherwise} \end{cases} \quad (3)$$

$$x_{i,j,t,s} = \begin{cases} 1, \text{if resources necessary for tech } j \text{ to develop to stage } s+1 \\ \quad \text{for process } i \text{ is available by time } t \\ 0, \text{otherwise} \end{cases} \quad (4)$$

$$y_{i,j,t,n} = \begin{cases} 1, \text{if technology } j \text{ is used in facility } n \text{ of process } i \text{ during time } t \\ 0, \text{otherwise} \end{cases} \quad (5)$$

3.3. Parameters

The initial Technology Stage of technology j for process i is denoted by the parameter $ITS_{i,j}$. The resources required for developing technologies are represented by set $r \in R$. The amount of resource r necessary for technology j to be developed to Technology Stage s for process i is denoted by parameter $Q_{r,i,j,s}$. These resources should be available for $R_{i,j,s}$ periods for successful progression to stage s. The amount of resource r available at time t is denoted by parameter $QT_{r,t}$. The cost of resources at each period t is denoted $QC_{r,t}$.

3.4. Constraints and Other Variables

The Technology Stage of technology j for process i during time t is represented by the variable $TS_{i,j,t}$. It is computed from the binary variable $x_{i,j,t,s}$. Another variable called Technology Integration Level ($TIL_{i,j,t}$) is introduced to incorporate a commercially developed technology j into the network for process i. It is a function of the technology's compatibility with the process ($yc_{i,j}$) and its development stage for the process ($TS_{i,j,t}$). If the final development stage is S, variable $TS_{i,j,t}$ takes a value from 1 to S. Variable $TIL_{i,j,t}$ is expressed in terms of $TS_{i,j,t}$ and takes a value from 1 to S if the technology is compatible with the process ($yc_{i,j}=1$), else $TIL_{i,j,t}=0$ when $yc_{i,j}=0$. A disjunctive model is used to implement these conditions in Eq. (6). For technology j that is commercially developed for process i by time t, $TS_{i,j,t}=S$ and $TIL_{i,j,t}=S$, while for an incompatible technology for process i, $TS_{i,j,t}=1$ and $TIL_{i,j,t}=0$. Equation (7) ensures that only the technologies commercially developed for the process are implemented in process facility/expansion. The amount of each resource allocated for developing technology j for source i to stage $s+1$ during time t is given by variable $q_{r,i,j,t,s}$. The cost parameter for the resources is $QC_{r,t}$ The development cost for technology j for process i at time t is computed by Eq. (8).

$$\begin{bmatrix} yc_{i,j} = 1 \\ TIL_{i,j,t} = TS_{i,j,t} = \sum_{s \in S} x_{i,j,t,s} \end{bmatrix} V \begin{bmatrix} yc_{i,j} = 0 \\ TIL_{i,j,t} = 0 \end{bmatrix} \quad \forall i \in I, j \in J, t \in T \quad (6)$$

$$TIL_{i,j,t} / S \geq y_{i,j,t,n} \quad \forall\, i \in I, j \in J, t \in T, n \in N \quad (7)$$

$$DC_{i,j,t} = \sum_{r \in R, s \in S} q_{r,i,j,t,s} QC_{r,t} \quad \forall\, i \in I, j \in J, t \in T, r \in R, s \in S \quad (8)$$

Equation (9) ensures that developing technology j is at least at stage $s-1$ before it progresses to the subsequent stage s. Equation (10) computes the Technology Stage of technology j for process i at the first planning period using the initial Technology Stage parameter. The resource allocation for Technology Stage progression is represented with the disjunctive model in Eq. (11). The binary variable $x'_{i,j,t,s}$ takes a value of 1 when all necessary resources for Technology Stage progression of technology j to stage s for process i are available at time t. Equation (12) enables the progression of a technology to stage s at time t if the necessary resources are available throughout $R_{i,j,s}$ by time t-1. A technology can progress to only one subsequent development stage for a process per period. Equation (13) ensures this condition by allowing resource allocation for only one Technology Stage progression of technology j for process i at time t. Resource allocation at every period ($q_{r,i,j,t,s}$) is limited by resource availability at that period ($QT_{r,t}$) in Eq. (14).

$$x_{i,j,t,s} \leq x_{i,j,t,s-1} \quad \forall\ i \in I, j \in J, t \in T, s \in S \tag{9}$$

$$TS_{i,j,1} = \sum_{s \in S} x_{i,j,1,s} = ITS_{i,j} \qquad \forall\ i \in I, j \in J \tag{10}$$

$$\begin{bmatrix} x'_{i,j,t,s} = 1 \\ q_{r,i,j,t,s} = Q_{r,i,j,s} \end{bmatrix} V \begin{bmatrix} x'_{i,j,t,s} = 0 \\ q_{r,i,j,t,s} = 0 \end{bmatrix} \quad \forall r \in R, i \in I, j \in J, t \in T, s \in S \tag{11}$$

$$\sum_0^{t-1} x'_{i,j,t,s} / R_{i,j,s} \geq x_{i,j,t,s} \quad \forall i \in I,\ j \in J, t \in T, s \in S \tag{12}$$

$$\sum_{s \in S} x'_{i,j,t,s} \leq 1 \quad \forall\ i \in I, j \in J, t \in T \tag{13}$$

$$\sum_{i \in I, j \in J,\ s \in S} q_{r,i,j,t,s} \leq QT_{r,t} \qquad \forall\ r \in R, t \in T \tag{14}$$

Technology j commercially developed for process i by time t ($TIL_{i,j,t}=S$) can be implemented in its facility at time t. If it is implemented in facility n, $y_{i,j,t,n}=1$. The facility capacity is denoted by variable $c_{i,j,t,n}$, and feed flow rate is denoted by variable $f_{i,j,t,n}$. The investment cost ($CIC_{i,j,t,n}$) is a function of $c_{i,j,t,n}$, while the operating cost ($COC_{i,j,t,n}$) is a function of $f_{i,j,t,n}$. If the technology is not implemented in the facility, $y_{i,j,t,n}=0$, and variables $c_{i,j,t,n}$ and $f_{i,j,t,n}$. become zero. These conditions are implemented with disjunctive model Eq. (15). Equations (16) and (17) ensure process i uses at most one technology in a processing facility, and the same technology is used in the subsequent periods.

$$\begin{bmatrix} y_{i,j,t,n} = 1 \\ CIC_{i,j,t,n} = f\left(X_i, c_{i,j,t,n}\right) \\ COC_{i,j,t,n} = f\left(X_i, f_{i,j,t,n}\right) \end{bmatrix} V \begin{bmatrix} y_{i,j,t,n} = 0 \\ CIC_{i,j,t,n} = 0 \\ COC_{i,j,t,n} = 0 \end{bmatrix} \quad \forall i \in I, j \in J, t \in T, n \in N \tag{15}$$

$$\sum_{j \in J} y_{i,j,t,n} \leq 1 \qquad \forall i \in I, t \in T, n \in N \tag{16}$$

$$y_{i,j,t,n} \geq y_{i,j,t-1,n} \quad \forall i \in I, j \in J, t \in T, n \in N \tag{17}$$

4. Results and Discussion

A case study of planning CC from 10 different sources over a planning period of 10 years (discretized into 10 equal time periods) to meet annual capture targets that increase biannually by 10% (from 30% to 70%) is considered. Post-combustion capture using absorption using aqueous monoethanolamine (ABS-MEA) is assumed to be commercially available, and pressure swing adsorption using methyl viologen exchanged zeolite Y (PSA-MVY) is under development. Previous studies have shown that PSA-MVY is more cost-effective than ABS-MEA for sources with CO_2 composition over 5% (Hasan et al. 2014, Santhamoorthy et al. 2023). The total CC cost should decrease if PSA-MVY is commercially developed during the planning period. The degree of the decrease would depend on when PSA-MVY becomes available and the emission sources using it. We demonstrate the model's capabilities by studying these factors.

Three Technology Stages, lab-scale R&D, pilot-scale testing, and commercialization, are considered. The pilot-scale plants are assumed to capture 0.1 million tons of CO_2 per annum. The research funds and time for successful lab-scale R&D and pilot-scale testing are not specified as parameters in the model. Instead, it is assumed that each Technology Stage can take from one to ten years for successful testing. The total development time is the total time taken for testing these stages. The maximum research fund available is computed as the difference between total costs with and without the technology.

The emission sources data and the cost models and carbon composition limitations for the capture technologies from Hasan et al. (2014) are used to construct model parameters. The costs of pilot-scale plants are estimated by escalating the cost by a factor of 1.3 to account for the overhead costs associated with testing new technology. The nonlinear cost models for CC yielded a mixed-integer nonlinear programming model (MINLP). The solution approach developed by Santhamoorthy et al. (2023) is used to solve the problem to an optimality gap of 5%. It is a two-step approach that solves a relaxed MILP model to obtain a lower bound and an initialized MINLP model to obtain an upper bound. The steps are employed iteratively with an increasing number of segments in the relaxation step to generate tight bounds by updating the lower and upper bounds.

The optimization models are formulated in Python V3.8.6 using PYOMO V6.4.1. The MILP models are solved using CPLEX V20.10, and the initialized MINLP models are solved using DICOPT V2 through GAMS V24.8.5, all on an Intel Xeon Gold 6248R 3 GHz processor with 48 cores and utilizing a maximum of 50 GB RAM.

Different development times are assumed for PSA-MVY, and the model is solved to generate the results summarized in Figure 2. In Figure 2(a), the marker star shows the cases where there is a commercial development of PSA-MVY during the planning period. There is successful development for all cases with at most 8 years of total development time. The total CC planning cost increases as the total development time increases. Figures 2(b) and (c) show the number of sources using only PSA-MVY and t both ABS-MEA and PSA-MVY for capture. More sources use only PSA-MVY for shorter total development times, leading to a lower total CC cost. As the development time increases, ABS-MEA is implemented for capture to meet the annual capture targets, while PSA-MVY is developed and implemented in an expansion facility over time. The number of sources using PSA-MVY is high when the total development time is at most 6 years. The number of cases in which different emission sources use either only PSA-MVY or PSA-MVY in the expansion capacity is shown in Figure 2(d). When the development time is longer for PSA-MVY, sources 1-4, having high emission rates, use it in the expansion capacity, while sources 5-7 still use only PSA-MVY for capture. Sources 8-10 do not use PSA-MVY since they have a very low CO_2 composition. The total cost of CC planning for 10 years using only ABS-MEA technology is \$13.9 billion. The total cost when PSA-MVY is developed in 2, 5, or 9 years is \$12.0, \$13.1, and \$13.7 billion, respectively. Thus, the funds that can be used towards initial R&D and lab-testing of PSA-MVY decrease as its development time prolongs.

5. Conclusion and Future Direction

This paper developed a multi-period optimization model that assists the planning decisions related to technology development, implementation, and operation. A case study of CC demonstrated the capability of the model to minimize the total cost of capture planning while developing a novel technology for implementation in the network. The funds that can be allocated towards R&D of the new technology were determined based on its anticipated development time. The results revealed that some sources used only the new technology for capture despite its long development time, while others used it only in the expansions. This work will be extended to study the effect of uncertainty in the successful progression of the technologies to subsequent development stages.

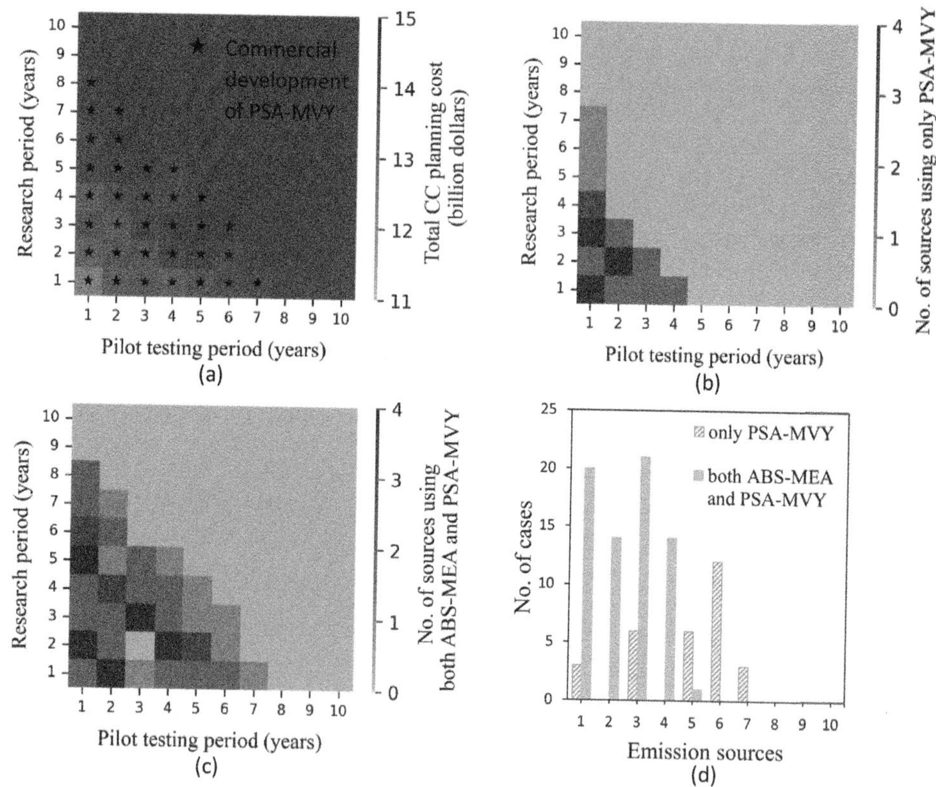

Figure 2. (a) Total cost of CC, (b) No. of sources using only PSA-MVY, and (c) No. of sources using both ABS-MEA and PSA-MVY for different cases of different development periods. (d) No. of cases in which sources use either only PSA-MVY or both ABS-MEA and PSA-MVY.

Acknowledgment

The authors acknowledge the financial support from Chevron Technical Center.

References

Anderson Jr, E. G., & Joglekar, N. R. (2005). A hierarchical product development planning framework. Production and Operations Management, 14(3), 344-361.

Davila, A., & Wouters, M. (2004). Designing cost-competitive technology products through cost management. Accounting horizons, 18(1), 13-26.

Fahmi, I., Nuchitprasittichai, A., & Cremaschi, S. (2014). A new representation for modeling biomass to commodity chemicals development for chemical process industry. Computers & chemical engineering, 61, 77-89.

Hasan, M. F., Boukouvala, F., First, E. L., & Floudas, C. A. (2014). Nationwide, regional, and statewide CO_2 capture, utilization, and sequestration supply chain network optimization. Industrial & Engineering Chemistry Research, 53(18), 7489-7506.

Santhamoorthy, P., Subramani, H. J., & Cremaschi, S. (2023). A multiperiod model for portfolio optimization of carbon capture and utilization. FOCAPO/CPC.

Antonis Kokossis, Michael C. Georgiadis, Efstratios N. Pistikopoulos (Eds.)
PROCEEDINGS OF THE 33rd European Symposium on Computer Aided Process Engineering (ESCAPE33), June 18-21, 2023, Athens, Greece
 http://dx.doi.org/10.1016/B978-0-443-15274-0.50241-9

An Encrypted MPC Framework for Security to Cyber-Attacks

Atharva V. Suryavanshi[a], Aisha Alnajdi[b], Mohammed S. Alhajeri[a], Fahim Abdullah[a], Panagiotis D. Christofides[a,b,*]

[a] *Department of Chemical and Biomolecular Engineering, University of California, Los Angeles, CA, 90095-1592, USA*

[b] *Department of Electrical and Computer Engineering, University of California, Los Angeles, CA 90095-1592, USA*

pdc@seas.ucla.edu

Abstract

This work focuses on the design and development of a secure and private communication network using semi-homomorphic encryption to ensure cyber-security of model predictive control (MPC) systems. Specifically, Paillier encryption algorithm is used for encrypting-decrypting signals in the communication links. This implementation involves quantization of signals, which may result in significant errors. Thus, the MPC system is designed to ensure a certain degree of robustness with respect to quantization errors. The Encrypted Lyapunov-based MPC scheme is tuned to ensure that the calculations can be done with the available computational resources. Finally, the proposed design is implemented on a chemical process example.

Keywords: Model predictive control (MPC), cyber-security, encrypted control, semi-homomorphic encryption, quantization

1. Introduction

Cybersecurity of networked and distributed control systems has become crucial in recent years due to their increased vulnerability to targeted cyber-attacks. The integration of cloud-computing technologies with physical control systems has significantly increased process performance and decreased computational time, thus allowing real-time optimization. However, these advantages do not come without the increased threats of targeted cyber-attacks. Unsecure communication channels are vulnerable to data-interception, eavesdropping on sensitive information, and data-manipulation. These cyber-attacks compromise the integrity, stability, and profitability of control systems. 77 cyber-security related incidents, on critical infrastructure, were reported from 2000 to 2019. Out of these reported incidents, a vast number of attacks were carried out on oil and energy sector industries (Parker et al., 2023). Clearly, cyber-attacks on these systems are extremely dangerous and hence it is of crucial importance to design and develop cyber-secure architectures for networked control systems.

In this work, we design and develop an Encrypted Lyapunov-based Model Predictive Control (LMPC) framework for nonlinear systems in which secure communication channels are established between sensor-controller and controller-actuator links. Specifically, Paillier cryptosystem is used to encrypt-decrypt sensitive state and input data in the network links. The designed controller is robust to the quantization errors arising as a consequence of encryption. The computational cost associated with different quantization parameters is analyzed. Finally, the proposed framework is implemented on a multi-input, multi-output chemical process example.

2. Preliminaries

2.1. Class of Systems

In this work, we consider continuous-time nonlinear systems, described by the following nonlinear ordinary differential equation,

$$\dot{x}(t) = F(x(t), u(t)) = f(x) + g(x)u \tag{1}$$

where $u^T = [u_1, u_2, u_3, ..., u_m] \in \mathbb{R}^m$ is the manipulated input vector denoting m inputs and, $x^T = [x_1, x_2, x_3, ..., x_n] \in \mathbb{R}^n$ is the n-dimensional state vector. The manipulated inputs are constrained such that $u \in U$, where $U := \{u \in U \mid u_{\min,i} \leq u_i \leq u_{\max,i}, \ \forall i = 1, 2, ..., m\}$, with $u_{\min,i}$ and $u_{\max,i}$ being the physical constraints on the ith manipulated input. $f(\cdot)$ and $g(\cdot)$ are sufficiently smooth vector and matrix functions, respectively. The origin is a steady state of the nonlinear system and, without loss of generality, it is assumed that $f(0) = 0$.

2.2. Stabilizability Assumptions

We assume that there exists a feedback control law $u = \Phi(x)$ that can render the steady-state of the nonlinear system of Eq. (1) exponentially stable in the sense that there exists a continuously differentiable control Lyapunov function $V(x)$ such that the following inequalities hold for all $x \in D$, where D is an open neighborhood around the origin (Wu et al., 2019a):

$$c_1 |x|^2 \leq V(x) \leq c_2 |x|^2, \quad \frac{\partial V(x)}{\partial x} F(x, \Phi(x)) \leq -c_3 |x|^2, \quad \left| \frac{\partial V(x)}{\partial x} \right| \leq c_4 |x| \tag{2}$$

where $c_1, c_2, c_3, c_4 > 0$. The stabilizing controller can be constructed using the universal Sontag control law (Lin & Sontag, 1991). The closed-loop stability region for the nonlinear system of Eq. (1) is characterized as a level set of the Lyapunov function $V(x)$, and this closed-loop stability region is denoted as Ω_ρ, where $\Omega_\rho := \{x \in D \mid V(x) \leq \rho\}$, $\rho > 0$.

2.3. Paillier Cryptosystem and Quantization

In this paper, Paillier cryptosystem (Paillier, 1999) is used for encryption and decryption operations of state and input signals. Paillier cryptosystem is a semi-homomorphic encryption scheme in the sense that addition operations can be carried out in the encrypted space. The first step in the Paillier cryptosystem is to generate a public key (M, g) which is used for encrypting plaintext messages. The next step is to generate a private key (λ, u) that is used to decrypt ciphertexts. The detailed key generation steps can be found in Paillier (1999). The encryption-decryption operations in the Paillier cryptosystem are carried out on a subset of $\mathbb{Z}_+$. This subset is denoted as $\mathbb{Z}_M$ and contains positive integers less than M. Thus, it is important to map the real-valued state and input data to the set $\mathbb{Z}_M$ (Darup et al., 2017). Here, we consider signed-fixed point numbers in base 2, where l_1 is the number of total bits and d is the number of fractional bits. In the first step, the function $g_{l_1,d}(\cdot): \mathbb{R} \to \mathbb{Q}_{l_1,d}$, $g_{l_1,d}(a) := \arg\min_{q \in \mathbb{Q}_{l_1,d}} |a - q|$ is used to map the real-valued data to the set $\mathbb{Q}_{l_1,d}$, where the elements of the set $\mathbb{Q}_{l_1,d}$ are rational numbers from

-2^{l_1-d-1} to $2^{l_1-d-1}-2^{-d}$ and are separated from each other by a resolution of 2^{-d}. Next, we map the quantized signals to a subset of $\mathbb{Z}_M$ using the bijective mapping $f_{l_2,d}(\cdot):\mathbb{Q}_{l_1,d}\rightarrow\mathbb{Z}_{2^{l_2}}$, $f_{l_2,d}(q):=2^d q \bmod 2^{l_2}$. Once the quantized signals have been mapped to the set $\mathbb{Z}_{2^{l_2}}$, our data is ready for encryption. It is important to note that the encrypted data will be decrypted at the controller and at the actuator and, in that case, the resulting data will be in the set $\mathbb{Z}_{2^{l_2}}$. Hence, it is important to map the decrypted integer messages to the set of quantized signals, $\mathbb{Q}_{l_1,d}$. The inverse mapping $f_{l_2,d}^{-1}$, is defined as,

$$f_{l_2,d}^{-1}(\cdot):\mathbb{Z}_{2^{l_2}}\rightarrow\mathbb{Q}_{l_1,d},\; f_{l_2,d}^{-1}(m):=\frac{1}{2^d}\begin{cases} m-2^{l_2} & \text{if } m\geq 2^{l_2-1} \\ m & \text{otherwise}\end{cases}.$$

3. Encrypted MPC Design

In this section, we propose an encrypted model predictive controller (MPC) scheme. In the closed-loop encrypted MPC design of Figure 1, state measurements $x(t)$ are encrypted at the sensor, and the encrypted data is then sent to the MPC. At the controller, before performing the nonlinear optimization calculations, the encrypted state data is decrypted to obtain the quantized states $\hat{x}(t)$. These quantized states are provided as measurement-feedback to the MPC optimization problem. The optimal control inputs $u(t)$ are calculated by solving an optimal control problem and are encrypted before being transmitted to the actuator. At the actuator, the encrypted control inputs data is decrypted to obtain the quantized control actions, $\hat{u}(t)$. These quantized control actions are applied on the nonlinear system.

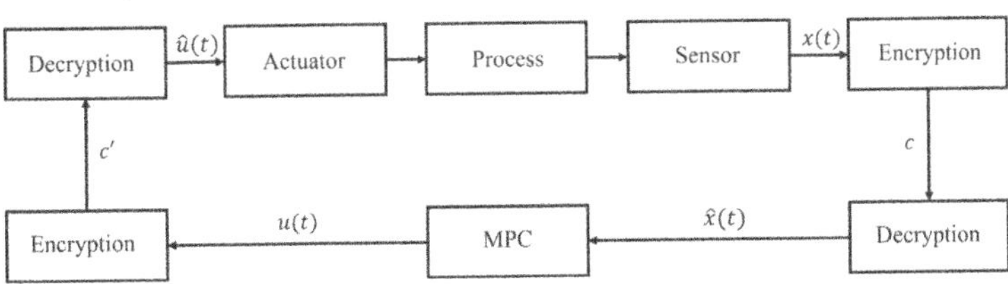

Figure 1: Schematic of nonlinear system under closed-loop encrypted MPC

In the closed-loop design of Figure 1, secure communication is established in the sensor-controller and controller-actuator links. However, two types of quantization errors arise. There is a loss of information in the sensor-controller link due to quantization of states, and in the controller-actuator link due to quantization of inputs. Based on the quantization mappings and the structure of the set $\mathbb{Q}_{l_1,d}$, these quantization errors are bounded such that:

$$|x(t)-\hat{x}(t)|\leq\eta_1 2^{-d} \quad\text{and}\quad |u(t)-\hat{u}(t)|\leq\eta_2 2^{-d} \tag{3}$$

where $\eta_1,\eta_2>0$ and d is the number of fractional bits (quantization parameter). Thus, the larger the magnitude of d, the smaller are the quantization errors in the design.

4. Encrypted Lyapunov-based Model Predictive Control

In this section, we present the mathematical formulation of an encrypted LMPC for the nonlinear system of Eq. (1), with secure controller-actuator and sensor-controller communication channels. The proposed formulation is as follows:

$$J = \min_{u \in S(\Delta)} \int_{t_k}^{t_k+N} L(\tilde{x}(t), u(t))\, dt \tag{4a}$$

$$\text{s.t.} \quad \dot{\tilde{x}}(t) = F(\tilde{x}(t), u(t)) = f(\tilde{x}) + g(\tilde{x})u \tag{4b}$$

$$u(t) \in U, \forall t \in [t_k, t_{k+N}) \tag{4c}$$

$$\tilde{x}(t_k) = \hat{x}(t_k) \tag{4d}$$

$$\dot{V}(\hat{x}(t_k), u) \leq \dot{V}(\hat{x}(t_k), \phi(\hat{x}(t_k)))\,,\ \text{if}\ \hat{x}(t_k) \in \Omega_{\hat{\rho}} \setminus \Omega_{\rho_{\min}} \tag{4e}$$

$$V(\tilde{x}(t)) \leq \rho_{\min},\ \forall t \in [t_k, t_{k+N}),\ \text{if}\ \tilde{x}(t_k) \in \Omega_{\rho_{\min}} \tag{4f}$$

where $\tilde{x}(t)$ is the state trajectory predicted by the process model, U is the set of physical constraints on the control inputs, and $\Omega_{\rho_{\min}}$ is the level-set of Lyapunov function V in which the system states are ultimately bounded. The sampling period is denoted by Δ, and $S(\Delta)$ is the set of piecewise constant functions with a period Δ. The optimal control input sequence, $u^*(t \mid t_k)$, is calculated by solving the MPC optimal control problem, and the first control action from the sequence is applied on the process. In the encrypted LMPC formulation, the constraint of Eq. (4b) defines the process model, Eq. (4c) defines the physical constraints on the control inputs, and Eq. (4d) is the state measurement feedback. The constraint of Eq. (4e) drives the closed-loop system state to a small neighborhood around the origin, and the constraint of Eq. (4f) ensures that the system states remain ultimately bounded in a small neighborhood, $\Omega_{\rho_{\min}}$ around the origin. Thus, starting with an initial state $x_0 \in \Omega_{\hat{\rho}}$, the proposed Encrypted LMPC design based on the stabilizing control law $u = \Phi(x)$ ensures that the states of the nonlinear system of Eq. (1) remain in the stability region $\Omega_{\hat{\rho}}$ and are ultimately driven to a small neighborhood around the origin $\Omega_{\rho_{\min}}$.

5. Chemical Process Example

In this section, we implement the proposed Encrypted LMPC design to a chemical process, specifically the system from Wu et al. (2019b). In this process, a second-order, irreversible, exothermic reaction occurs in a non-isothermal CSTR. The reaction converts a reactant, *A*, to the product, *B*, via the reaction $A \rightarrow B$. A heating jacket is provided to the CSTR that supplies/removes heat at a rate *Q*. Using mass and energy balances, the system of ODEs that describe the nonlinear process can be written as

$$\frac{dC_A}{dt} = \frac{F}{V}(C_{A0} - C_A) - k_0 e^{\frac{-E}{RT}} C_A^2, \quad \frac{dT}{dt} = \frac{F}{V}(T_0 - T) + \frac{(-\Delta H)}{\rho_L C_p} k_0 e^{\frac{-E}{RT}} C_A^2 - \frac{Q}{\rho_L C_p V} \tag{5}$$

where C_A is the reactant concentration, T is the temperature inside the reactor, C_{A0} is the inlet feed concentration, and Q is the heat added/removed from the reactor. The control objective is to maintain the CSTR at the unstable steady state $(C_{As}, T_s) = (1.95\,\text{kmol/m}^3, 402\,\text{K})$. This is achieved by manipulating the control inputs

ΔC_{A0} and ΔQ. The state and input vectors of the closed-loop system are defined in terms of their deviation from steady state values: $x^T = [C_A - C_{As}, T - T_s]$ and $u^T = [C_{A0} - C_{A0s}, Q - Q_s]$. Based on the smallest and largest values of the permissible inputs and possible state values (in the operating region), we get the optimal value of $l_1 - d$ to be 18. For closed-loop simulations under encrypted LMPC, we vary d from 1 to 7 in increments of 2, and l_1 changes accordingly. For l_2, it is important to have $l_2 > l_1$, which makes it possible to map quantized numbers from the set $\mathbb{Q}_{l_1,d}$ to the set $\mathbb{Z}_M$. Hence, we select the value $l_2 = 29$. The prediction horizon, *N*, for MPC is chosen to be 2. The implementation of the encrypted MPC is done in a sample-and-hold fashion with a receding horizon. A first-principles model for the nonlinear CSTR system of Eq. (5) is used in the MPC and the nonlinear optimization package IPOPT (Wächter & Biegler, 2006) is used to solve the MPC optimization problem. The Explicit Euler method is used to simulate the nonlinear system of Eq. (1) with an integration time step of $h_c = 10^{-4}$ hr. The sampling period is $\Delta = 10^{-2}$ hr. The closed-loop encrypted MPC simulation results are presented in Figures 2 to 4.

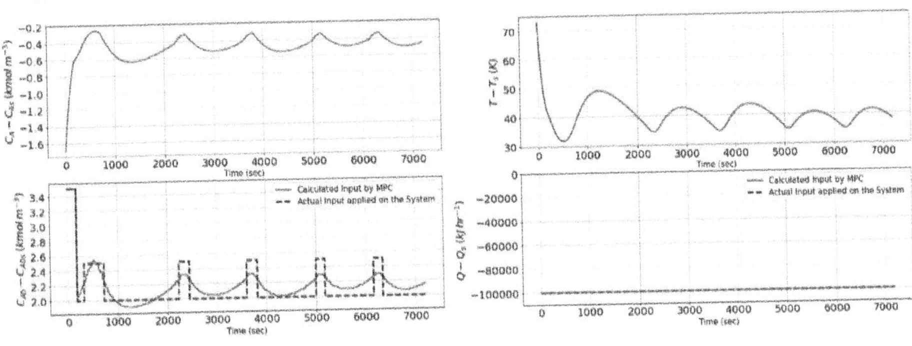

Figure 2: State and input profiles of closed-loop simulations under encrypted LMPC, with $d = 1$.

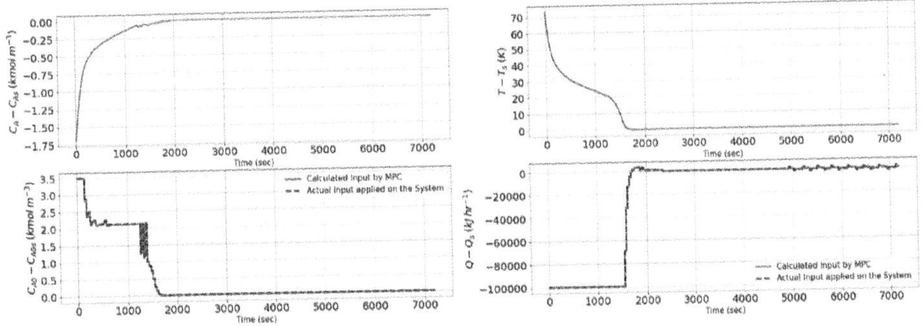

Figure 3: State and input profiles of closed-loop simulations under encrypted LMPC, with $d = 3$.

From the results of closed-loop encrypted MPC simulations, it can be observed that, for some small value of the quantization parameter d , the encrypted MPC is not able to stabilize the nonlinear system around a small neighborhood around the origin. Instead, we observe the oscillations of states around a point other than the unstable steady state.

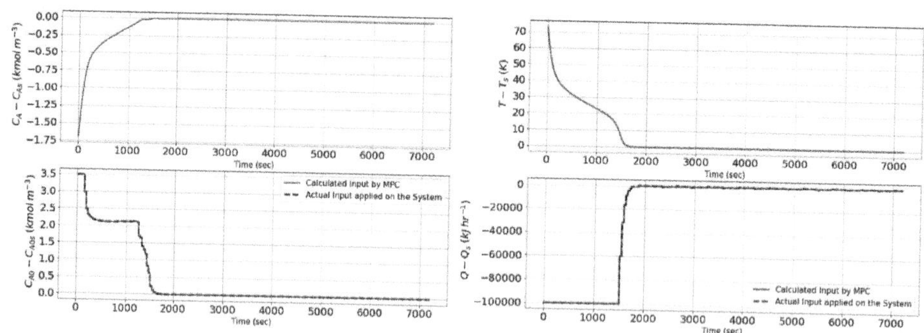

Figure 4: State and input profiles of closed-loop simulations under encrypted LMPC, with $d = 7$.

Thus, for systems being operated at an unstable equilibrium, it is possible that the encrypted MPC cannot practically stabilize the system for $d \leq d_{\text{critical}}$, and it is important to identify this critical value of the quantization parameter. For the nonlinear system of Eq. (1), we have $d_{\text{critical}} = 1$. Additionally, for $d > d_{\text{critical}}$ as the value of the quantization parameter d increases, we see improvement in MPC performance in the sense that we achieve faster convergence of states to a small neighborhood, $\Omega_{\rho_{\min}}$, around the origin. The MPC performance improved because the quantization error significantly decreases with the increase in d. However, this improvement in performance comes at a computational cost. The computational cost increases significantly with the increase in d primarily for two reasons. Firstly, with an increase in d, the number of elements in the set $\mathbb{Q}_{l_1,d}$ increase and, hence, the computational cost required to construct such a set increases significantly. Secondly, with a large number of elements in the set $\mathbb{Q}_{l_1,d}$, the number of search operations required to map real valued signals from $\mathbb{R}$ to $\mathbb{Q}_{l_1,d}$ also increase significantly. Thus, if the computational resources are limited, the MPC performance that can be achieved is also limited but one must ensure that the chosen value of quantization parameter d is larger than the critical value, d_{critical}.

References

S. Parker, Z. Wu, and P. D. Christofides, 2023, "Cybersecurity in Process Control, Operations, and Supply Chain," *Proceedings of Foundations of Computer Aided Process Operations/Chemical Process Control*, San Antonio, Texas.

Z. Wu, A. Tran, D. Rincon, and P. D. Christofides, 2019a, "Machine learning-based predictive control of nonlinear processes. part I: Theory," AIChE Journal, vol. 65, e16729.

Z. Wu, A. Tran, D. Rincon, and P. D. Christofides, 2019b, "Machine learning-based predictive control of nonlinear processes. part II: Computational implementation," AIChE Journal, vol. 65, e16734.

Y. Lin and E. D. Sontag, 1991, "A universal formula for stabilization with bounded controls," Systems & Control Letters, vol. 16, 6, pp. 393–397.

P. Paillier, 1999, "Public-key cryptosystems based on composite degree residuosity classes," in Proceedings of the International conference on the theory and applications of cryptographic techniques. Berlin, Heidelberg: Springer, pp. 223–238.

M. S. Darup, A. Redder, I. Shames, F. Farokhi, and D. Quevedo, 2017, "Towards encrypted MPC for linear constrained systems," IEEE Control Systems Letters, vol. 2, 2, pp. 195–200.

A. Wächter, L. T. Biegler, 2006, "On the implementation of an interior-point filter line-search algorithm for large-scale nonlinear programming," Mathematical Programming, vol. 106, pp. 25-57.

Antonis Kokossis, Michael C. Georgiadis, Efstratios N. Pistikopoulos (Eds.)
PROCEEDINGS OF THE 33rd European Symposium on Computer Aided Process Engineering (ESCAPE33), June 18-21, 2023, Athens, Greece
 http://dx.doi.org/10.1016/B978-0-443-15274-0.50242-0

Machine learning-based product concentration estimation, real-time optimization, and multivariable control of an experimental electrochemical reactor

Berkay Çıtmacı[a], Junwei Luo[a], Joon Baek Jang[a], Carlos Morales-Guio[a], and Panagiotis D. Christofides[a,b]

[a]*Dept. of Chemical and Biomolecular Engineering, University of California, Los Angeles, Los Angeles 90095, USA.*

[b]*Dept. of Electrical and Computer Engineering, University of California, Los Angeles, Los Angeles 90095, USA; pdc@seas.ucla.edu.*

Abstract

Electrochemical reduction of CO_2 gas is a novel carbon capture technology that can potentially not only relieve the global climate crisis caused by increasing CO_2 gas emissions but can also restore electricity generated from clean energy sources in the form of carbon-based fuels. However, due to the complexity of the electrochemical reactions, there has been a limited effort to gain process control intuition for electrochemical reactors and the process development has been restricted to lab scale. This work aims to develop a multivariable control scheme for an experimental electrochemical reactor at UCLA by integrating artificial neural network modeling, real-time optimization, and process controller design. UCLA's experimental reactor is utilized to carry out the experiments required to construct a database to address the modeling, optimization and control tasks.

Keywords: Electrochemical CO_2 reduction; Multivariable Control; Experimental data modeling; Real-time optimization; Neural Networks.

1. Introduction and Reactor Description

The percentage of renewable energy in the electricity grid has been increasing as a part of decarbonization efforts against the detrimental effects of global warming. With the current advancing awareness of sustainability, electrochemical methods have emerged as an attractive alternative to transform CO_2 gas into organic chemicals and synthetic fuels (De Luna et al. (2019)). However, scaling up the process to the industrial scale is limited by the lack of an effective control system and a solid understanding of the underlying physico-chemical phenomena of the process. To this end, a machine learning method is adopted to construct data-driven models mapping the critical input-output variables of the reactor and providing predictions to establish a feedback control scheme. As a continuation of the results from the previous study (Çıtmacı et al. (2022)) that focused on single-input single-output control, this work aims to develop an operational model of an electrochemical reactor with a nanoporous copper electrode using experimental data and perform multivariate control.

A gastight rotating cylinder electrode (RCE) cell reactor was used to carry out electrocatalytic experiments in this work. This electrochemical reactor consists of six major components: two electrode (working and counter electrode) chambers separated by an anion-exchange membrane, a potentiostat, a mass flow controller (MFC), a modulated speed rotator (MSR), and a temperature control block. During the operation, gas is bubbled through 0.2 M $KHCO_3$ electrolyte in both electrode chambers at a fixed flowrate of 20 sccm controlled by MFC. Nanoporous catalyst structure was synthesized directly on the copper cylinder electrode through the redox cycling in 4 mM KCl, and this roughened copper electrode was used as the cathode (working electrode) to electrochemically reduce CO_2 into 17 different products. There are 5 gas-phase products (H_2, CO, CH_4, C_2H_4 and C_2H_6) including hydrogen from the competing hydrogen evolution reaction and 12 liquid-phase products, which are produced on the electrode surface through multiple proton-electron transfer processes. Hydrodynamics and convective mass transport can be controlled systematically with the MSR, to and from the electrode for all species including reactant, intermediate, and final products. Under well-defined hydrodynamics, now the potential applied to the working electrode can be adjusted by the potentiostat, and this potential against the reference electrode and the electrical current are continuously measured and recorded. Therefore, by controlling the electrode rotation speed and the applied potential, the reaction kinetics and mass transfer effects of this process can be decoupled and studied, respectively. Lastly, the reactor is gastight such that online quantification of gas products can be performed using a gas chromatograph (GC). Liquid products, on the other hand, accumulate in the electrolyte solution and are analyzed using nuclear magnetic resonance (NMR) spectroscopy after each experiment.

2. Weighted-FNN Steady-State Modeling and Real-Time Optimization for Feasible Setpoints

2.1. ML Model Trained with Statistical Weight

With the development of data and computation technologies, artificial neural networks (ANN), that have the ability to perform universal approximation for any linear/nonlinear relations (Csáji et al., (2001)), have become a popular method to address chemical and process control challenges. A feed-forward neural network (FNN) is trained with a database generated from open-loop and input step-change experiments to predict steady state concentrations of 17 products based on the applied potential and electrode rotation speed. The FNN model has one hidden layer that contains 64 neurons and a ReLu activation function. The Softplus function is selected as the activation function of the output layer to ensure non-negative prediction. Additionally, the Softplus function predicts the output with a smoother curve which aligns the physical expectation better than other activation functions with similar property such as ReLu. The mean squared error (MSE) function (Eq. 1) is utilized as the cost function to train the FNN model as:

$$L(y,\hat{y}) = \frac{1}{n}\sum_{i=1}^{n}(y_i - \hat{y}_i)^2 \quad (1), \qquad L(y,\hat{y}) = \frac{1}{n}\sum_{i=1}^{n}\frac{1}{v_i^2}(y_i - \hat{y}_i)^2 \quad (2)$$

Although the trained FNN model can give accurate predictions (i.e., low MSE), it treats every data point in the dataset equally which increases the probability of overfitting high variance data points due to the stochastic nature of the electrochemical reaction and experimental uncertainties. To account for data variance, a weighted MSE function (Eq. 2) is adopted during the training process, where v_i is the coefficient of variance for the ith data point. Specifically, data points are grouped by similar operating conditions such that averages and standard deviations under certain input conditions can be calculated and included in the weighted loss function. Therefore, the weighted-FNN is granted more tolerance of prediction error if the corresponding data point has a higher standard deviation. The weighted-FNN predictions for four gas products are represented in Fig. 1, which displays the ability of FNN to not only provide accurate prediction based on given input conditions, but to also capture the relationship between inputs and outputs.

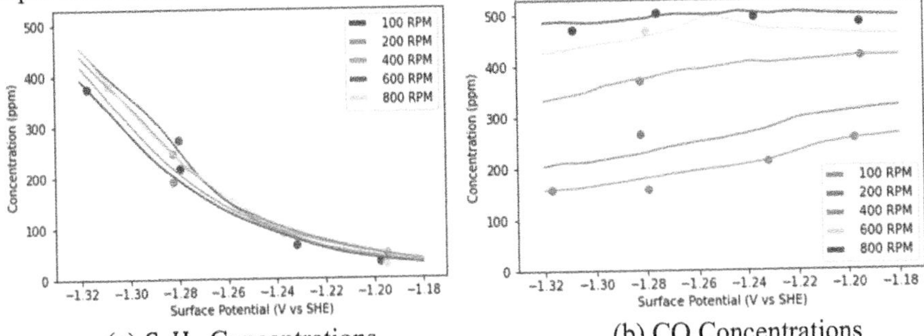

(a) C_2H_4 Concentrations (b) CO Concentrations

Figure 1: Weighted-FNN prediction of (a) C_2H_4 and (b) CO concentrations under different electrode surface potential and rotation speed conditions.

2.2. Real-Time Optimization

An optimization problem can be established based on the neural network model prediction to locate the optimum operational setpoint. Specifically, by assuming that electricity consumption is the only operating cost of this process, the optimum setpoint is where the economic profit of the operation is maximized. In this work, the concentrations of C_2H_4 and CO are chosen to be the controlled variables (outputs), due to their strong correlations with surface potential and electrode rotation speed (inputs), respectively. A third-party software, IPOPT is utilized to continuously solve for the optimum setpoint based on the neural network prediction and varying electricity price.

3. Recurrent Neural Network Dynamic Modeling of C_2H_4 and CO Concentrations

3.1. Data Fitting for Experimental Dynamics Approximation

The experimental concentration data from the GC is discrete and deficient to train a neural network model due to the limitation of separation time of chromatography. To tackle this, polynomial and sigmoid best-fits were used to enhance the training data by approximating a probable experimental trajectory between GC results. This significantly increases the number of data points and enables the training of a neural network model. The polynomial and sigmoid best-fits interpolation is demonstarted in Fig. 2, where the original dataset contains only 16 data points collected from the GC readings. After data interpolation, the available concentration estimations increase to 18000.

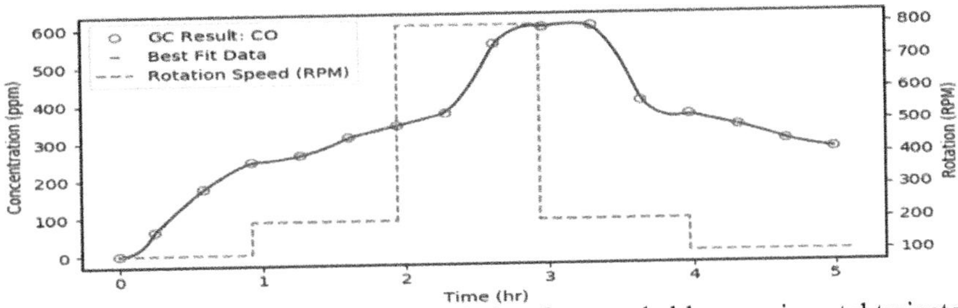

Figure 2: Example of GC data interpolation for a probable experimental trajectory from an input (rotation speed) step change experiment under constant applied potential.

3.2. Machine Learning Modeling of Process Dynamics

Recurrent neural network (RNN) is a popular structure to model the dynamic behaviors of chemical processes for its proven ability to fit time-series data effectively. Long-Short-Term-Memory (LSTM) models, designed to restore the exploding/vanishing gradients in basic RNN models, have a better performance in predicting time-series data and thus are used in our work as the alternative of explicit operation model in a feedback control loop (Wu et al. (2019)). Specifically, two LSTM models are trained using the enhanced dataset to predict the production rates of C_2H_4 and CO respectively with three input features (i.e., surface potential, rotation speed, and current). The input data sequence contains 36 timestamps to represent a window of one-hour historical data. The time step size is chosen to be 100 seconds to reduce the computational cost. In addition, the model is trained to predict the next 800 seconds of the production rates, with 100 seconds time step. In this way, LSTM models can learn the delay response and process dead time as it is trained with a sequence of future data.

The LSTM model contains one hidden recurrent layer with 200 neurons. Simple grid search is performed to tune the hyper-parameters which demonstrates that the prediction accuracy decreases with decreasing the number of neurons and starts overfitting the data with more neurons or hidden layers. Regularization methods including recurrent dropout and L_2 kernel regularization were performed to prevent overfitting further. A sigmoid function is used as the activation function of the output layer to ensure the model prediction is constrained by the maximum and minimum values in the training set. Finally, one experiment from constant input parameters and two step change experiments were used to test the model. The testing performance is shown in Fig. 3 where the predictions demonstrate a good overlap with the probable experimental trajectories.

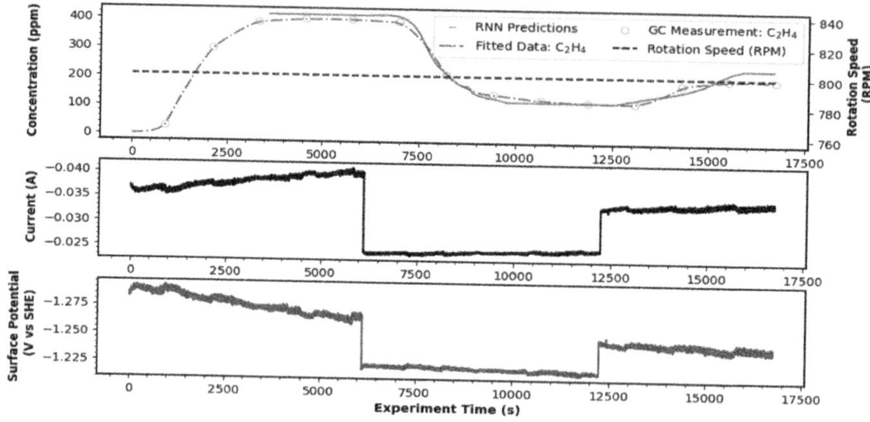

Figure 3: LSTM predictions for C_2H_4 production rates based on plotted input parameters, where RNN predictions are represented in continuous blue and probable experimental trajectory in dashed gray line.

4. Controller Tuning and Multivariable Control Architecture

4.1. Process Model Parameter Extraction Using Experimental Data

Two first-order plus time delay (FOPTD) models are used to describe the reaction processes, and the corresponding process parameters are extracted from experimental data. Specifically, various open-loop experiments were conducted by applying single-variable step changes to either applied potential or rotation speed. Subsequently, the response in C_2H_4 and CO concentrations, captured by the GC, are enhanced with a sigmoid function to configure a continuous trajectory of the real-time response as shown in Fig. 4. The responses are normalized for an easy comparison between various step changes. With step change responses, the dead time (θ) is extracted by drawing a tangent line through the inflection point, and time required to reach 63% of the final steady state is defined to be the time constant (τ). The mathematical expression of FOPTD model is demonstrated in Eq. 3, where *K* denotes the steady state gain.

$$G(s) = \frac{Ke^{-\theta}}{\tau s + 1} \quad (3)$$

However, a single FOPTD model is not enough to model the entire operation range due to the nonlinearity of the electrochemical CO_2 reduction. To account for that, various FOPTD models are developed to describe different segment of the overall operational range (Fig. 5). We should check general correlations for the dead time and τ for different step changes of rotation and potential. If there is a general correlation for the time parameters, it is possible to extract steady state gains from the weighted-FNN model. Fig. 4 is used to observe dynamic responses of C_2H_4 under various rotation speed and potential changes.

The first plot clearly shows that the concentration change has similar dynamic behavior under constant rotation speed. Therefore, we can assume the same dead time and τ for the tuning process. The second plot shows that similar dead times and τ values can be attained for the rotation changes less than 200 RPM. These values are used in FOPTD model. A similar procedure is applied to develop the transfer function of CO concentration.

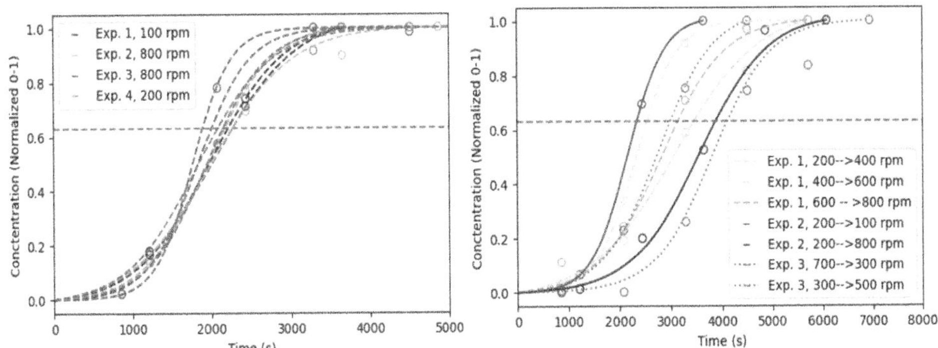

(a) Potential change under constant rotation speed (b) Rotation speed change under constant potential

Figure 4. Sigmoid functions fitted to dynamic C_2H_4 concentration change data, normalized between 0 and 1, to observe dead time and τ values for FOPTD models

4.2. Relative Gain Array

There are two options to implement the two-input-two-output control for this reactor, which are using 2 PI controllers or a model predictive controller (MPC). The control strategy is selected based on the individual manipulated variable-controlled variable correlations. Specifically, if there is no significant cross-coupling between the input-output pairs, it is feasible to implement the Multi-Input-Multi-Output (MIMO) control with 2 PI controllers. In the opposite case, an MPC should be considered to drive the system to the specific combination of parameters for the targeted set-points. A relative gain array (RGA) is useful to find if there are direct correlations between a specific input and output. The RGA only needs the steady state gains to be used. However, due to the nonlinearities, we divide the operation range into 9 regions with respect to the input ranges as follows:

- Potential ranges = (-1.19, -1.26 V), (-1.26, -1.30 V), (-1.30, -1.32 V);
- Rotation ranges = (100-200 RPM), (200-400 RPM), (400-800 RPM).

These ranges are anticipated to satisfactorily linearize the process. The steady state gains are calculated from the weighted-FNN model. As a result, we found that rotation speed has a high correlation to CO concentration, whereas the surface potential is strongly correlated to C_2H_4 concentration, which is shown in Fig. 5 that the diagonals of RGA are close to 1.

	100-200 RPM		200-400 RPM		400-800 RPM	
-1.19 - -1.26 V vs SHE	0.9704	0.0296	0.9706	0.0294	1.0181	-0.018
	0.0296	0.9704	0.0294	0.9706	-0.018	1.018
-1.26 - -1.30 V vs SHE	0.9444	0.0556	0.9610	0.0390	0.9977	0.0023
	0.0556	0.9444	0.0390	0.9610	0.0023	0.9977
-1.30 - -1.32 V vs SHE	0.9693	0.0307	0.9669	0.0331	0.9908	0.0092
	0.0307	0.9693	0.0331	0.9669	0.0092	0.9908

Figure 5. Relative gain array analysis for 100-200 RPM & -1.19 to -1.26 V vs SHE.

$$\begin{bmatrix} C_{C2H4}(s) \\ C_{CO}(s) \end{bmatrix} = \begin{bmatrix} \dfrac{-1427e^{-1113s}}{952s+1} & \dfrac{-0.03e^{-1557s}}{1395s+1} \\ \dfrac{919.4e^{-671s}}{1409s+1} & \dfrac{0.64e^{-1234s}}{540s+1} \end{bmatrix} \begin{bmatrix} E(s) \\ R(s) \end{bmatrix}$$

(4) Multivariable control array example of FOPTD models

Therefore, two PI controllers are sufficient to control this process, in which one of the controllers will manipulate the applied potential to control C_2H_4 production while the other controller will manipulate the rotation speed to control CO production rate. Furthermore, the arrays of FOPTD transfer functions will be used to determine the initial values of the PI tuning parameters using the Cohen-Coon tuning method.

5. Conclusion

This study carried out data-based modeling, set-point optimization and preliminary feedback control design using the open-loop experiments conducted using UCLA's experimental electrochemical reactor. Various data fitting methodologies were used to increase the size of the dataset for dynamic ML model development and extract feedback control parameters. Using the calculations presented in this paper, the closed-loop experiments will soon be conducted to drive the process between optimized set-points for C_2H_4 and CO production rates.

References

[1] De Luna, P., Hahn, C., Higgins, D., Jaffer, S.A., Jaramillo, T.F., Sargent, E.H., 2019. What would it take for renewably powered electrosynthesis to displace petrochemical processes? Science 364, 3506.

[2] Jeng, E., Qi, Z., Kashi, A.R., Hunegnaw, S., Huo, Z., Miller, J.S., BayuAji, L.B., Ko, B.H., Shin, H., Ma, S., Kuhl, K.P., Jiao, F., Biener, J., 2022. Scalable gas diffusion electrode fabrication for electrochemical CO_2 reduction using physical vapor deposition methods. ACS Appl. Mater. Interfaces 14, 7731–7740

[3] Çıtmacı, B., J. Luo, J. B. Jang, V. Canuso, D. Richard,Y. M. Ren, C. G. Morales-Guio, and P. D. Christofides(2022). Machine learning-based ethylene concentration estimation, real-time optimization and feedback control of an experimental electrochemical reactor. Chemical EngineeringResearch and Design 185, 87–107.

[4] Csáji, Balázs Csanád. Approximation with artificial neural networks. Faculty of Sciences, Etvs Lornd University, Hungary 24, no. 48 (2001): 7.

[5] Wu, Z., Tran, A., Rincon, D., Christofides, P.D. (2019). Machine learning-based predictive control of nonlinear processes Part I: Theory. AIChE Journal 65, e1672

Antonis Kokossis, Michael C. Georgiadis, Efstratios N. Pistikopoulos (Eds.)
PROCEEDINGS OF THE 33rd European Symposium on Computer Aided Process Engineering (ESCAPE33), June 18-21, 2023, Athens, Greece
 http://dx.doi.org/10.1016/B978-0-443-15274-0.50243-2

Symbolic regression based interpretable data-driven soft-sensor for process quality control

Harry Kay [a], Sam Kay[a] ,Max Mowbray[a], Amanda Lane[b], Cesar Mendoza[b], Philip Martin[a], Dongda Zhang[a,*]

[a]*Department of Chemical Engineering, University of Manchester, Oxford Road, Manchester, M1 3AL, UK.*
[b]*Unilever Research Port Sunlight, Quarry Rd East, Bebington, CH63 3JW, UK.*
*: *dongda.zhang@manchester.ac.uk;*

Abstract

Accurately predicting key performance indicators for product quality control is a research priority within the process industry. Physical models capable of predicting real-time process indicators of complex formulations are traditionally difficult to construct or retrofit, especially when the process is operated under dynamic process conditions and incorporates complex mixing phenomena. A solution to this employs the use of machine learning models to develop a soft-sensor using the information rich process data available, however this method lacks interpretability as it is difficult to extract physical information from the machine learning models, and the models are less user-friendly to implement within an existing factory advance process control system. To fill this gap, in this study we propose the use of symbolic regression, an interpretable machine learning technique, which aims to search the space over a subset of pre-defined mathematical functions to make accurate predictions of product quality. The high dimensional historical industrial data is first screened, and partial least squares is used to extract critical information regarding important operating regions and process parameters. Statistical features of critical process parameters are then employed within different symbolic regression based frameworks to obtain suitable models for batch quality prediction. Moreover, Monte Carlo sampling is adopted to estimate model uncertainty and evaluate reliability of the constructed interpretable models. To evaluate the model's performance, predictions were made for batches operated over different seasons and formulations. In addition, these models are benchmarked against a recently developed heteroscedastic noise neural network based data-driven soft-sensor. It is found that symbolic regression provided expressions with high accuracy, robustness, and responsiveness to changing process data suggesting that information about the true underlying correlation between the process data and a key performance indicator has been extracted. This research highlights the great potential of interpretable machine learning techniques for future industrial application.

Keywords: Machine learning, data analytics, quality control, symbolic regression, uncertainty estimation.

1. Introduction

Historically, mathematical models have been formulated to express relationships describing physical phenomena in order to evaluate functions at both current and future states. The predictive models were derived from first principles using conservation laws, experimental observations, and physical knowledge. However, many industrial processes contain too complex of interactions and phenomena to be described effectively via such methods. Within Industry 4.0, access to information rich data has become more common due to the large quantity of sensor data being recorded from the industrial processes. The paradigm of machine learning can be exploited with the intention to analyse complex process data and formulate equations to characterise nonlinear and dynamic systems.

Recent works have explored the use of parametric and non-parametric data-driven models as a soft-sensor for process predictions (Kay et al., 2022; Mowbray et al., 2022). These models have proven to mitigate false confidence, have high accuracy, make uncertainty estimations representative of the data, and generalise well. However, a problem associated with such black box machine learning models is the lack of interpretability. To resolve the issue, symbolic regression can be integrated with real process data to generate statistical models representative of the industrial process. In essence, symbolic regression attempts to infer both a model and its parameters such that, for a given set of input and output data $\{(x_i, y_i)\}_{i=1}^{n}$ for $\{(\boldsymbol{x}, \boldsymbol{y}) \in \mathbb{R}\}$, a function f with parameters θ can be obtained where $\boldsymbol{y} \approx f(\boldsymbol{x}|\theta)$. A problem associated with this method is the that the search space increases exponentially with an increasing domain of operators. Knowledge of the system can be integrated to reduce this search space and reduce the dimensionality of the problem. A further complex challenge is the compromise between complexity and accuracy. Parsimonious expressions are traditionally preferred as they hold more relevance to providing understanding of the system at hand. In addition to this, more complex expressions are less robust to process changes and perform worse at extrapolation as they tend to overfit to the training data. Other works have explored the development of symbolic regression models in conjunction with neural networks to overcome the limitations of the combinatorial nature of genetic (Kim et al., 2020; Zhang et al., 2022). However, little attempts have been made to explore the use of symbolic regression in the context of soft sensing and monitoring for batch process.

2. Problem statement

This study focuses on the development of a statistical model based soft sensor and attempts to explore the feasibility of using symbolic regression to extract symbolic equations from a non-Newtonian fluid in a batch mixing process. Three datasets are available with the following labels: Alpha, Beta, and Gamma. Alpha and Beta were recorded from the same process line at different seasons of the year, and the Gamma data corresponds to a different formulation obtained from a similar process. We can assume process data of the form $\boldsymbol{D} = \{\boldsymbol{X}, \boldsymbol{y}\}$ with the predictor matrix, $\boldsymbol{X}$ being configured with $\boldsymbol{N}$ batches, $\boldsymbol{T}$ timesteps, and $\boldsymbol{J}$ sensors for $\boldsymbol{X} \in \mathbb{R}^{N \times T \times J}$. The subsequent response matrix contains a single process measurement per batch, expressed as $\boldsymbol{y} \in \mathbb{R}^{N \times n_p}$ where $\boldsymbol{n_p}$ represents the number of process qualities measured. To ensure generalisability within the models, the data must show uniformity, thus all sensors must be shared, so $\boldsymbol{J} = \mathbf{28}$ for all datasets. To further ensure uniformity, an indicator depicting the beginning of a new phase of operation is used for alignment of batches. Dataset Alpha is characterised by $\boldsymbol{N} = \mathbf{25}, \boldsymbol{J} = \mathbf{28}$ process sensors, and $\boldsymbol{T} = \mathbf{8000}$ timesteps. Dataset Beta uses the same formulation and process equipment as Alpha in a different season with dimensionality of $\boldsymbol{N} = \mathbf{14}$ batches, $\boldsymbol{J} = \mathbf{28}$ sensors, and $\boldsymbol{T} = \mathbf{3600}$ timesteps. The Gamma data is comprised of $\boldsymbol{N} = \mathbf{9}$ batches, $\boldsymbol{J} = \mathbf{28}$ sensors, and $\boldsymbol{T} = \mathbf{3700}$ timesteps. Although, two different product formulation are tested, the target product quality of both is identical.

Typical problems associated with high dimensional datasets (Min, 2005) can be solved using dimensionality reduction, in which the data is reduced to remove redundant features and multi-collinearity, whilst still retaining sufficient information for quality prediction. We use the same approach as previous work (Hicks et al., 2021) where partial least squares (PLS) is employed, identifying 8 sensors, and 300 timesteps for accurately inferring batch quality predictions. Due to the poor capability of genetic algorithms for handling large datasets, the data is further reduced using timewise unfolding as mentioned in (Hicks et al., 2021) and taking either the mean and standard deviations or the principal components (PCA) of each individual sensor within the critical time region. This results in a 2-rank matrix of dimensions $\boldsymbol{X} \in \mathbb{R}^{N\times(16\times28)}$. This approach is adopted to preserve the ability of the symbolic regression models to generalise between the different datasets.

3. Methodology

First, we will introduce the concept of genetic algorithms and their function in symbolic regression. The principle of genetic algorithms follows Darwin's theory of survival of the fittest and through this, populations of symbolic equations can be bred (Koza, 1994) with the objective of fitting a predictor matrix to the response matrix. Within this case study, the individuals of the population are combinations of operators, functions, and numerical constants where the operator domain is restricted to $(+,-,*,\div,\sqrt{\ },\exp,\ln)$. An example of an individual is shown in tree diagram form in figure 1.

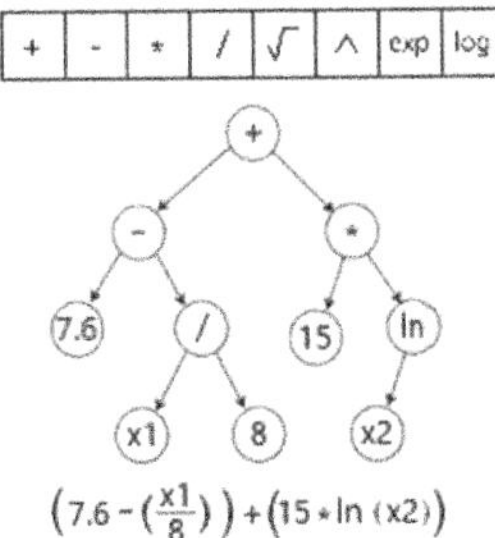

Figure 1: Tree diagram representation of an individual and its operator domain

Different metrics can be used to assess the generated model's performance including its accuracy and complexity. In this study, the accuracy is defined as the average percentage error (1) and the loss function as the mean squared error (2) such that, given a functional representation, f, with input data $\boldsymbol{x}$, parameters θ, and predictions $\boldsymbol{y}$:

$$MAPE = \frac{1}{N}\sum_{i=1}^{N}\frac{(y_i - f(x_i|\theta))}{y_i}\cdot 100 \; and \; MSE = \frac{1}{N}\sum_{i=1}^{N}(y_i - f(x_i|\theta))^2 . \qquad 1$$

The complexity is defined as having a linear dependency to the sum of operators, constants and process variables used within the model structure. This dependency can be altered by assigning unique weightings to specific operators or numerical constants, however, for this study, a value of 1 was assigned to all operators and numerical constants. Let us now discuss the procedure to assess the uncertainty of the statistical models generated via symbolic regression. In essence, we will exploit the fact that, under certain conditions, the covariance matrix is equal to the inverse of the information matrix. The negative log likelihood is employed as the objective function for the optimisation of model parameters and if we consider a gaussian random vector, $\boldsymbol{x}$ with mean $\boldsymbol{x}^*$ and covariance matrix E_θ, the objective function is (Hirschfeld et al., 2020):

$$obj(x) = argmin\left(\frac{N}{2}(log\, 2\pi\sigma^2) + \frac{1}{2}\frac{(\boldsymbol{x}-\boldsymbol{x}^*)^T(\boldsymbol{x}-\boldsymbol{x}^*)}{\sigma^2}\right) \quad 2$$

Given that the second derivative of the negative log likelihood objective function is independent of $\boldsymbol{x}$, the ith and jth components of the hessian matrix H can be obtained without knowing $\boldsymbol{x}^*$. The parameter variance can be extracted from the diagonal entries of the covariance matrix and used to estimate the model uncertainty. A gaussian distribution of parameters with mean and standard deviation $N(\theta, \sigma_\theta^2)$ can be sampled from via Monte Carlo sampling, and the statistical models queried to produce a normal distribution of batch quality predictions, $N(f(x_i|\theta), \sigma_f^2(\boldsymbol{x}))$. The requirements of the industrial collaborator specify that the $MAPE < 15\%$ for soft sensor prediction and the measurement uncertainty of the equipment was quantified as being approximately 10% per standard deviation. All statistical models were optimsied through using gradient based optimisation techniques in Pyomo, namely IPOPT, allowing for efficient exploration and exploitation of the parameter space such that high-quality solutions were ensured.

4. Results and Discussion

The data is split into two sets, training and testing, where the training set comprises of alpha and beta, and the model's prediction performance is tested using the gamma data. To reduce the dimensionality of the feature space and thus improve the quality of the solutions, gradient boosting was applied. Using this, feature importance charts were generated for selection of process variables to use for training within the symbolic regression models as shown in Figure 2.

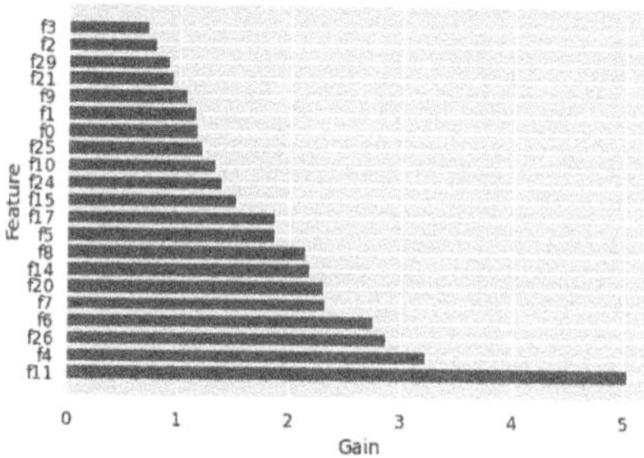

Figure 2: Feature importance chart generated using gradient boosting methods

The extracted symbolic equations that exhibit low complexity and high accuracy are presented in Table 1.

Table 1: MAPE and complexity values of generated symbolic expressions. Models 1-3 are constructed using the data's mean (X_{10}) and std (X_{11}). Model 4 is constructed using principal components (X_{10})

Model	Symbolic expression	MAPE (Training) / %	MAPE (Testing) / %	Complexity
1	$A\ln(\lvert BX_{10} + C\rvert)$	9.23	13.89	6
2	$A\ln(\lvert\frac{BX_{10} + C}{DX_{11}^2 + E}\rvert)$	8.53	12.56	12
3	$Ae^{-BX_{11}}\ln(\lvert\frac{CX_{10} + D}{EX_{11}^2 + F}\rvert)$	8.44	13.03	15

4	$Ax_{10} + B$	10.87	10.89	4

Models 2 and 4 were selected for further analysis due to enhanced performance within their respective groups. The predictions from models 2 and 4 on the training and validation sets are shown below in Figures 2a and 2b after being normalised.

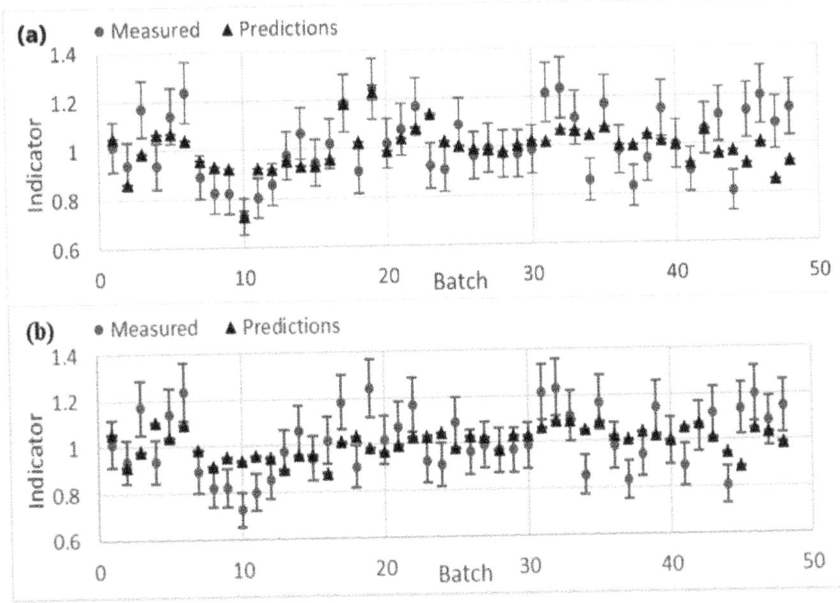

Figure 3: Soft sensor predictions against measured data for the alpha, beta and gamma for models 2 (a) and 4 (b). The error bars represent one standard deviation.

In Figure 2, the alpha, beta and gamma data are shown as batches 1-25, 26-39, and 40-49 respectively. The soft-sensors for both models 2 and 4 are shown to be capable of predicting the training and validation batches with high accuracy with training MAPE's of 8.53% and 10.87%, and validation MAPE's of 12.56% and 10.89% respectively. Notably, the uncertainty estimations of the generated models are far lower than the measured uncertainty so does not provide an accurate representation of the data uncertainty. This is predominantly due to the simple structure used to approximate the required performance indicator. All the statistical model predictions are within three standard deviations of the measured indicator so lie within an acceptable range of values. Model 2 provides a higher training accuracy than its validation error, indicating that some overfitting to the training data has occurred, however the model seems to be more responsive to process variation within the data as the predictions cover a larger domain. Model 4 provides almost identical errors for training and validation, indicating that enough information has been captured from the training set to prevent underfitting and make informed inferences of product quality without overfitting to batch specific trends.

Overall, both models have shown good capacity for predicting on data derived from different process lines, showing that the symbolic regression technique has successfully learned correlations between the process data and batch quality.

A further conclusion to draw is that it is possible to represent the process accurately using minimal features and process sensors. Both models required the use of one shared key process variable, with model 4 only using one principal component (X_{10}) from the sensor and model 2 using the mean (X_{10}) and standard deviation (X_{11}). This is indicative of the sensors individual importance to inferring batch quality and for process monitoring. This conclusion corroborates the results obtained from the generation of importance charts using gradient boosting where the same feature was identified as being highly influential.

4. Conclusion

In conclusion, symbolic regression proved to be a useful tool for developing accurate, interpretable statistical models for predicting process indicators. In addition, the expressions generated were simple and could easily be integrated into an industrial process control system to assist operators with process monitoring and control presenting a substantial benefit in comparison to traditional black box machine learning techniques which are difficult to extract knowledge from. Through the use of information rich statistical features, derived from industrial data, and data driven models, it was shown possible to identify high quality solutions at low computational cost. Upon validation of the optimised statistical expressions, using the training and testing data, the soft-sensor was shown to exhibit good predictive performance and generalisation capabilities. It was also seen to be able to capture relatively large process variation within the industrial data and filter out noise. This study has presented the potential of interpretable machine learning models for use in batch quality prediction in industrial processes.

References:

Hicks, A., Johnston, M., Mowbray, M., Barton, M., Lane, A., Mendoza, C., Martin, P., & Zhang, D. (2021). A two-step multivariate statistical learning approach for batch process soft sensing. *Digital Chemical Engineering*, *1*, 100003.

Hirschfeld, L., Swanson, K., Yang, K., Barzilay, R., & Coley, C. W. (2020). Uncertainty Quantification Using Neural Networks for Molecular Property Prediction. *J. Chem. Inf. Model*, *2020*, 3770–3780.

Kay, S., Kay, H., Mowbray, M., Lane, A., Mendoza, C., Martin, P., & Zhang, D. (2022). Integrating Autoencoder and Heteroscedastic Noise Neural Networks for the Batch Process Soft-Sensor Design. *Industrial and Engineering Chemistry Research*, *61*(36), 13559–13569.

Kim, S., Lu, P., Mukherjee, S., Gilbert, M., Jing, L., Ceperi´c, V. C., & Soljači´c, M. S. (2020). *Integration of Neural Network-Based Symbolic Regression in Deep Learning for Scientific Discovery*.

Koza, J. R. (1994). Genetic programming as a means for programming computers by natural selection. *Statistics and Computing*, *4*, 87–112.

Min, R. (2005). *A Non-linear Dimensionality Reduction Method for Improving Nearest Neighbour Classification*.

Mowbray, M., Kay, H., Kay, S., Caetano, P. C., Hicks, A., Mendoza, C., Lane, A., Martin, P., & Zhang, D. (2022). Probabilistic machine learning based soft-sensors for product quality prediction in batch processes. *Chemometrics and Intelligent Laboratory Systems*, *228*.

Zhang, M., Kim, S., Lu, P. Y., & Soljači´c, M. S. (2022). *Deep Learning and Symbolic Regression for Discovering Parametric Equations*.

Antonis Kokossis, Michael C. Georgiadis, Efstratios N. Pistikopoulos (Eds.)
PROCEEDINGS OF THE 33rd European Symposium on Computer Aided Process Engineering (ESCAPE33), June 18-21, 2023, Athens, Greece
 http://dx.doi.org/10.1016/B978-0-443-15274-0.50244-4

Oligomerization Using Reactive Distillation, Design and Model Predictive Control

Gabriel Contreras-Zarazúa1,[a,b] Juan José Quiroz Ramírez,[a] Eduardo Sánchez-Ramirez,[b] Esteban Abelardo Hernández-Vargas,[c] Juan Gabriel Segovia-Hernández[b]

[a]*Country CONACyT- CIATEC, Center for Applied Innovation in Competitive Technologies, Leon Guanajuato*
[b]*Department of Chemical Engineering University of Guanajuato, Noria Alta S/N,36000, Guanajuato,Gto., Mexico.*
[c]*Department of Mathematics and Statistical Science University of Idaho, 875 Perimeter Drive, MS 1103 Moscow*

Abstract

Biojet-fuel is a promising alternative to reduce emissions in the aviation industry. The alcohol to jet (ATJ) is considered one of the most attractive alternatives. However, the economic feasibility is an important drawback of this process. With this in mind, the process intensification offers an interesting alternative to improve the ATJ process. Nevertheless, in many cases, the intensification could lead to complex equipment that are difficult to operate. This work proposes a novel reactive catalytic distillation column to improve the oligomerization stage of ATJ process. This intensified process was designed using a sensitivity analysis. In order to determine the feasibility of operating this intensified column, a control study using PI controller and model predictive control was performed. The results showed that a reactive distillation column to intensify the oligomerization zone is feasible. This equipment only consumes 200kW. Finally, the control studies indicated that it is possible to operate the column using a traditional feedback control and model predictive control techniques with good performance.

Keywords: Biojet-fuel, reactive distillation, model predictive control, process intensification.

1. Introduction

Energy has an utmost important role in our society. Energy is used to produce almost all the necessary commodities such as food, polymers, fertilizers. This energy employed to produce these commodities has several forms such as electricity, natural gas, liquid fuels, among others. In this sense, one sector that is essential for our society and which is characterized by an intensive use of energy is the transport sector. This sector consumes around 21% of the total energy demand annually, and its energy consumption is expected to increase up to 130% by 2050 which also represents an increase in CO_2 emissions. Among all the transport means, the aviation sector is expected to grow the fastest, doubling its energy consumptions and CO2 emission in the next 20 years. For this reason, the aviation industry has proposed to reduce its CO_2 emissions in order to achieve a more sustainable aviation industry. The use of alternative biofuels for replacing oil-derived fuels is considered by the International Air Transport Association (AITA) as the most interesting alternative to reduce emissions (Rivas-Interian et al., 2022). These biofuels are commonly called biojet-fuels and can be produced from biomass. Nowadays, there are several routes to produce biojet-fuel such as Fischer-Tropsch (FT-SPK), Hydroprocessed Esters and Fatty Acids (HEFA-SKP), Fisher-Tropsch with Aromatics (FT-SPK/A) and Alcohol to Jet (ATJ-SPK). In this sense, the ATJ process is considered one of the most attractive, since the alcohols are produced in abundance, offering logistical flexibility. The ATJ process consist of three reaction steps: alcohol dehydration, olefin oligomerization and hydrogenation. In this regard, oligomerization is a common stage in many jet fuel production routes, such as ATJ or aqueous phase processing. This stage is utmost importance, because it allows to achieve hydrocarbons with the exact number of carbons required by the aviation industry. This stage has two steps: a reaction stage in which ethylene is converted into larger hydrocarbons and a separation stage, in which the heavy hydrocarbons are separated from the lighter ones. This process stage is characterized by an intense use of energy, both in the reaction stage and in the separation. Therefore, one way to reduce energy consumption and increase the profitability of the ATJ process is through process intensification using reactive distillation.

Nowadays, the reactive distillation has not been explored as alternative to intensify the oligomerization zone. With this mind, this work proposes the design of a novel reactive distillation column for the oligomerization stage of ATJ process. The design parameters and operating conditions of the column were selected to increase the hydrocarbon production required for aviation industry; these hydrocarbons are in the range of C8 to C14. Since this equipment integrates the reaction and separation in a single unit and many reactions are performed inside it, a control study was carried out to determine the operation and control feasibility of this column. The control study consists of two parts: a study using PI controllers, in order to analyze the column control with common controllers in the industry. The second part is a model predictive control study in order to compare its performance with traditional feedback control using PI controllers. The integral absolute error (IAE) was selected as metric to quantify and compare the closed-loop performance for both PI and MPC controllers.

2. Design of reactive distillation equipment

The ATJ process considered for this study uses ethanol as raw material. This ethanol is dehydrated and converted to ethylene during the dehydration stage. In this work, it is only considered the oligomerization stage thus the ethylene is the raw material considered to this stage. Additionally, it is assumed that only linear olefins are produced during the oligomerization, this assumption is realistic because the catalysts reported in previous literature produce linear olefins, also it is a common practice to incorporate an isomerization stage after oligomerization stage to generate branched hydrocarbons. On the other hand, in this work, a catalytic distillation column is used to intensify the oligomerization stage, it consists of a distillation column with an intermediate reactive zone where the catalyst is placed and the whole set of reactions take place. The separation stages are located below and above of this reaction zone. Figure 1 shows a scheme of the reactive distillation process. The reactions that occur in the reaction zone and their respective kinetic parameters were taken from Sánchez-Ramírez et al., 2022. It is important to highlight, that the kinetic model corresponds to a set of elementary reactions of second order. A ethylene feed flowrate of 2100 kg/hr was chosen according to the information provided by Rivas-Interian et al., 2022.

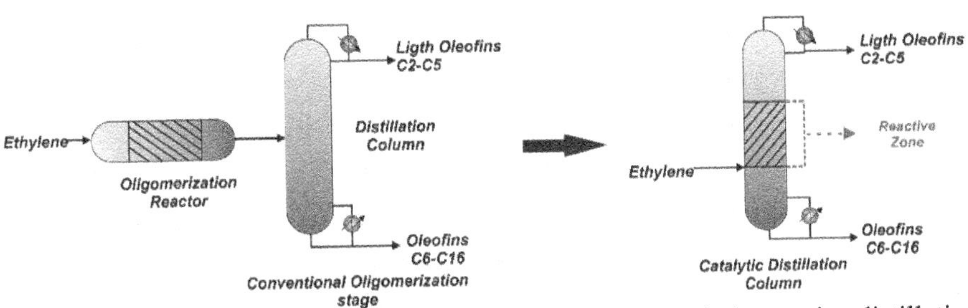

Figure 1. Intensification of oligomerization stage by catalytic reactive distillation.

The distillation column was designed and simulated in Aspen Plus using the RADFRAC module. The selection of thermodynamic model for the simulation was performed according to Carlson´s algorithm (Carlson, 1996). Consequently, the Soave-Redlich-Kwong (SRK) was used as thermodynamic model to simulate the system, owing to the mixture contains linear hydrocarbons and the pressure required for the oligomerization is around 10-40bar. The design of the reactive distillation was performed using a sensitivity analysis in order to achieve a suitable hydrocarbon distribution and maintain low energy consumption. It is important to highlight, that according to Li et al., (2016), a suitable hydrocarbon distribution corresponds to a hydrocarbon mixture in the range of C10 to C16, being the C12 the most abundant compounds in the blend with the aim of ensuring the suitable physiochemical properties. Finally, in order to determine the improvements of intensification, the total annual costs of both the conventional and intensified oligomerization process were calculated using the Guthrie´s method and data reported by Rivas-Interian et al., 2022.

3. Control of the reactive distillation column.

Reactive distillation columns are significantly more complex process due to integration of reaction and separation stages. The complexity of performing the separation and reaction processes in a single equipment could result in a process that is difficult to control and operate. Therefore, in order to evaluate the feasibility of operating this intensified process, a control study was performed. This control study considers the implementation of model predictive control (MPC) and compares this control strategy with traditional feedback control strategies using PI controllers. The control schemes using both MPC and PI controllers are shown in the Figure 2.

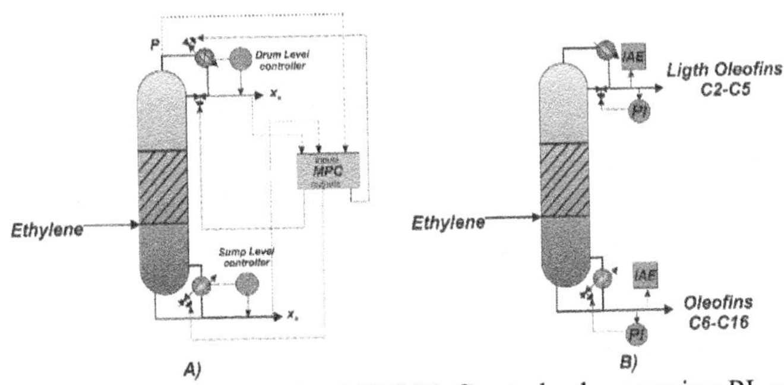

Figure 2. A) Control scheme using MPC B) Control scheme using PI controllers.

The integral absolute error (IAE) was chosen as metric to quantify and compare the closed-loop performance of the PI and MPC controllers. The tunning of controllers were carried out using the minimization of IAE method, to achieve this, disturbances of -1% were performed on the manipulate variables of distillation column. Mathematically, the IAE can be defined as follows:

$$\text{Integral of Absulote Error (IAE)} = \int_0^{\infty} \left| y(t) - y_{sp} \right| dt \quad (1)$$

Where y_{sp} is the value of the set point for *the* control variables, whereas *y(t)* is the value of control variable at time *t*. It is important to mention, that the MPC controller uses a state space (SS) model as predictor model, this state space model was obtained by subspace identification. The subspace identification is a common method used to obtain a dynamic model when the process is very complex (Li et al., 2021). The rigorous simulation of Aspen Dynamics was used as plant model. The MPC control was implemented in a hybrid platform, which links Aspen Dynamics and the Simulink software. On the order hand, the feedback control using PI controllers was fully implemented in Aspen Dynamics. Disturbances of -1% on manipulate variables were done in order to compare the performance of control systems and to tune the controllers.

4. Results

This section shows the design and control results of the catalytic column. As aforementioned, the design of reactive distillation column was performed by a sensitivity analysis in order to obtain a suitable olefin distribution. In this case, designs for 20, 30, and 40 stages were studied for the column. The results of the sensitivity analysis for the reflux ratio and reboiler duty as representative cases are shown in Figure 3. Please note, in Figure 3 as large values of reflux ratio have a positive effect on the production of C10 and C12, while more reflux ratio has a negative effect on C14. The C12 reaches a maximum concentration in reflux ratios close to 40, while for C10 and C14 the maximum concentrations are reached at higher and lower reflux ratios, respectively. Based on the results, it is concluded that a reflux ratio of 43 is the most adequate, because for this value, there is a suitable distribution of hydrocarbons, being C12 the most abundant and C10 and C14 olefins has a similar production. In addition, it was determined that a reboiler duty of 200kW is suitable since at this point the hydrocarbon distribution is very similar to that reported by Li et al., (2016). The sensitivity analysis was repeated for 30 and 40 stages finding that a greater number of stages does not benefit the production of more C10, C12 and C14. The Hydrocarbon distribution found with the best design is reported in Figure 3., also, some important design parameters are reported in Table 1. In the case of total annual costs (TAC), the results indicated that a conventional process has TAC $582,204.90 $/y. On the other hand, the intensified has a TAC of $511,128.62$/yr representing savings on costs around 13% in contrast to conventional process.

Table 1. Representative design parameters for the reactive distillation column.

Design Parameter	Value
Number of stages	20
Reboiler duty (kW)	200
Reflux ratio	43
Feed stage	7

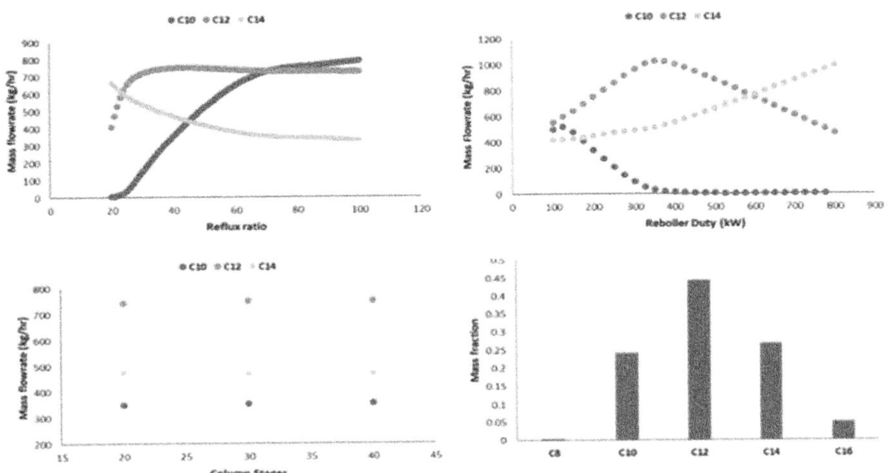

Figure 3. Design results of reactive distillation.

With respect to control results, the composition of the most abundant compound in domes and bottoms were used as control variables. In this case, the butane (C4) and dodecane (C12) are de most abundant compounds in the top and bottom of the column, respectively. The composition of C4 was controlled with reflux ratio, while the composition of C12 is controlled using the reboiler duty of bottoms. Once identified the control variables, the tunning of controllers was performed. As aforementioned, the minimization of IAE method was used for tuning the controllers, in this case the parameters that minimize the IAE for the PI controllers at the top of column are 40 min for integral time and 20 for gain whereas the parameters for the PI controller in the bottom of column are 11.443 for gain and 70.99min integral time. In the case of MPC control, it is necessary only one controller to control the composition at the top and bottoms of the column. In this case, the parameters that minimize the IAE are a sample time of 3 min, control horizon of 10 intervals and prediction horizon of 50 intervals. Once the suitable parameters were determined, the control performance for both PI and MPC controllers are compared using their respective IAE values. The results of disturbing the reboiler duty as representative case are shown in Figure 4 whereas the IAE value is shown in Table 2

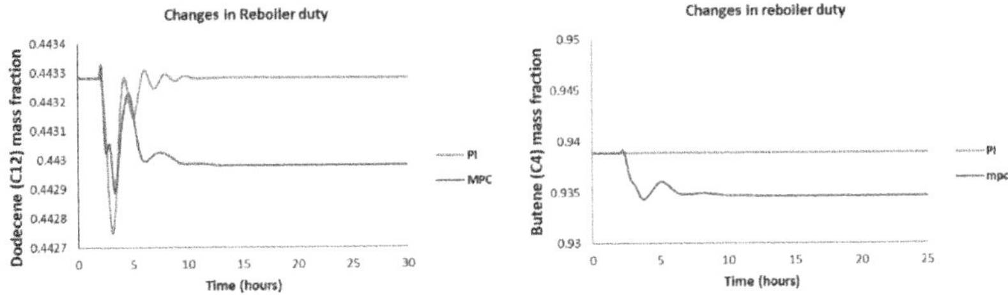

Figure 4. Disturbances of -1% in reboiler duty.

Table 2. IAE values for disturbances on reboiler duty

Control loop	C4 control loop (PI)	C4 control loop (MPC)	C12 control loop (PI)	C12 control loop (MPC)
Reboiler duty disturbance	2.86E-05	0.1091	0.0010875	0.00794

Based on the results, there is an appreciable difference between the two types of controllers. In this case, the butene composition can be controller better using a PI controller, having the PI controller up to 30% better performance than the MPC controller. However, for the dodecene control loop, the performance is very similar between both controllers. Note that the MCP control has the capability of controlling the process, however it do not fully fits to the setpoint, being the difference small due to the tolerance with which the objective function is solved in predictive control. However, it is concluded that this difference is not significant and both controllers are adequate to control the process.

5. Conclusions

In this work, a reactive distillation equipment was designed to intensify the oligomerization stage in the ATJ process. The intensified reactive distillation column has savings on total annuals costs of 13% in contrast to the conventional process. On the other hand, the control studies indicated that it is possible to operate this catalytic column using traditional control strategies, such as feedback control and other advanced control techniques, such as model predictive control. Finally, it is concluded that both controllers have similar performance, despite of this please note, that the PI controllers have a slight better performance due to they returned to exactly set-point value.

References

Rivas-Interian, Eduardo Sanchez-Ramirez, Quiroz-Ramírez, J.J., Segovia-Hernandez., J.G., 2022. Feedstock Planning and Optimization of a Sustainable Distributed Configuration Biorefinery for Biojet Fuel Production via ATJ Process. Biofuels, Bioprod. Biorefining.

Sánchez-Ramírez, E., Huerta-Rosas, B., Quiroz-Ramírez, J.J., Suárez-Toriello, V.A., Contreras-Zarazua, G., Segovia-Hernández, J.G., 2022. Optimization-based framework for modeling and kinetic parameter estimation. Chem. Eng. Res. Des. 186, 647–660.

Carlson, E.C., 1996. Succedding at Simulation. Chem. Eng. Prog. 35–46.

Li, J., Yang, G., Yoneyama, Y., Vitidsant, T., Tsubaki, N., 2016. Jet fuel synthesis via Fischer-Tropsch synthesis with varied 1-olefins as additives using Co/ZrO2-SiO2 bimodal catalyst. Fuel 171, 159–166. https://doi.org/10.1016/j.fuel.2015.12.062

Antonis Kokossis, Michael C. Georgiadis, Efstratios N. Pistikopoulos (Eds.)
PROCEEDINGS OF THE 33rd European Symposium on Computer Aided Process Engineering (ESCAPE33), June 18-21, 2023, Athens, Greece
 http://dx.doi.org/10.1016/B978-0-443-15274-0.50245-6

Fault Diagnosis with GAT and PageRank: From Local Attention to Global Attention

Deyang Wu,[a] Jinsong Zhao,[a,b*]

[a]State Key Laboratory of Chemical Engineering, Department of Chemical Engineering, Tsinghua University, Beijing 100084, China
[b]Beijing Key Laboratory of Industrial Big Data System and Application, Tsinghua University, Beijing 100084, China
jinsongzhao@tsinghua.edu.cn

Abstract

Fault detection and diagnosis (FDD) is important to ensure process safety in chemical plants. In the past decades, different FDD methods have been proposed. And data-driven methods stood out due to their high diagnosis accuracy rate. When deep learning was introduced to this field, the diagnosis accuracy became higher and higher, but the models became more and more difficult to understand. The lack of explainability and interpretability limited practical application of deep learning-based FDD methods. In this paper, we applied graph attention neural network (GAT) for FDD and achieved high diagnosis accuracy. Attention mechanism in GAT can help understand the inference process of the network. Combining the attention mechanism with the PageRank algorithm, a visualization graph can be generated to highlight the variables that are most relevant to a certain type of fault. This can be used to explain the fault diagnosis results and help us understand deep learning-based fault diagnosis better.

Keywords: chemical fault diagnosis, explainability, graph attention network, page rank.

1. Introduction

Fault detection and diagnosis (FDD) is quite critical to safe operations of chemical processes to identity abnormal events that can hardly be controlled by distributed control system. In decades, many researchers have been proposed different models of real-time FDD. These models can be categories into quantitative model-based, qualitative model-based and process history-based models (Venkatasubramanian et al., 2003). Process history-based models are further classified into qualitative and quantitative models. The latter is also commonly termed as data-driven models. Data-driven models based on deep learning have drawn much attention of researchers these years, such as DBN-based models (Zhang and Zhao, 2017), CNN-based models (Wu and Zhao, 2018), RNN-based models (Zhang et al., 2020), autoencoder-based models (Bi and Zhao, 2021) and GCN-based models (Wu and Zhao, 2021). These models achieved high diagnosis accuracy, but the lack of explainablity made them hard to understand by humans thus limited their practical application.

In this paper, graph attention neural network (GAT) was applied to FDD to achieved high diagnosis accuracy. Attention mechanism built in GAT gave out local attention weights to help understand the model itself. Then a classic algorithm PageRank was applied to extend the local attention to global attention. The global attention can be used to generate a visualization graph to highlight the variables that are most relevant to a certain type of fault. This was a great improvement of explainability of deep learning-based FDD model.

2. Graph Attention Network and PageRank

2.1. Graph Attention Network

Graph neural networks (GNNs) were proposed for processing graph-structured data, which are common in the real world. GNNs can be classified into recursive graph neural networks (RecGNNs) and convolutional graph neural networks (ConvGNNs), and the latter includes spectral-based and spatial-based ones. Because of many similarities with CNNs, spatial-based ConvGNNs have gained much attention from researchers, and the representative network structure is graph convolutional network (GCN) (Kipf and Welling, 2017). Spatial-based ConvGNNs define convolutional layers for information passing and aggregation operations. They propagate information through the connections, and aggregate the information of neighbor nodes and corresponding edges to a certain node to update its hidden embedding. Then graph readout technologies are used to obtain a node-level or graph-level representation of the information in the graph.

Graph attention network (GAT) (Veličković et al., 2018) introduced masked self-attentional layers to address the shortcomings of GCN and its variants. In convolutional operations, GAT allows for assigning different weights to different nodes within a neighborhood with efficient computation, as shown in Figure 1. The attention mechanism greatly improved the performance of the model, while bringing some model interpretability. The same advantages can be obtained by applying GAT to chemical process diagnosis.

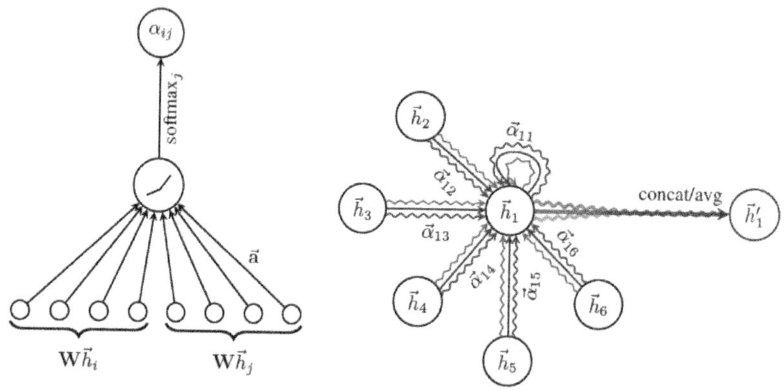

Figure 1: Left: The attention mechanism in GAT. Right: Multi-head attention by node 1 on its neighborhood. (Veličković et al., 2018)

2.2. PageRank

PageRank is the founding algorithm of Google's search engine and is one of the most famous algorithms used to rank web pages in order of importance. It's named after Larry Page, one of the founders of Google Inc. Essentially, the algorithm analyzes the importance of a web page by the number and quality of its hyperlinks.

In terms of quantity, the more hyperlinks from other pages point to a web page, the more important the page is. In terms of quality, the more important hyperlinks point to a web page, the more important the page is. This is very similar to the concept of citation networks.

In addition to web ranking, PageRank has a wide range of other application scenarios. When combing with the attention mechanism in GAT, it will be possible to analyze the importance of the nodes in a graph.

3. Fault Diagnosis with GAT

3.1. Fault Diagnosis of Chemical Processes

When applying deep learning-based models, fault diagnosis of a chemical process is generally regarded as a classification problem. Given an observation $\boldsymbol{X}_t \in \mathbb{R}^{v \times w}$ at time t, the current operating state $y \in \mathbb{N}$ of the process should be identified out of a set of operating states consisting of normal state and different types of faults. The observation $\boldsymbol{X}_t$ is a matrix with the time window and process variable dimensions. The operating state or label y is an integer between 0 and N_c (the number of different fault states). $y = 0$ means the process is under normal operating state, and $y = c$ $(1 \leq c \leq N_c)$ means the fault c has occurred. With enough pairs of observations and labels, a dataset $\left\{\boldsymbol{X}_t^{(i)}, y^{(i)}\right\}_{i=1}^{N}$ is obtained for training and testing a deep learning-based fault diagnosis model.

3.2. Integrated with Process Topology

The experiments were conducted on the Tennessee Eastman (TE) process (Downs and Vogel 1993). It's a simulation process model modified based on an actual industrial process of Eastman Chemical Company in Tennessee, USA. TE process mainly consists of five unit operations and defines 20 different process disturbances. In the simulation program of TE process, 52 variables can be observed, and 20 different disturbances can be inserted at any time to make the process operate in a fault state.

GAT performs convolutional operations on a graph. Naturally, the topology between process variables can be considered as a graph in a chemical process. The graph structure used in this paper was consistent with PTCN in (Wu and Zhao, 2021). When training the model with process data, GAT performs graph convolutional operations based on the adjacency matrix corresponding to the graph.

3.3. Data Preparation and GAT Model Training

To obtain enough data for training the GAT model, the simulation program of TE process ran for 3000 h in normal state. Under every different fault, the simulation program ran for 20 h after the fault inserted, and 10 sets of parallel simulations were carried out. All the data have been normalized with the mean and standard deviation calculated from data in the normal state. The normalized data were then cut to slices with the time widow of 1 h, and the data are sampled every 3 min. Data from 8 sets of parallel simulations were used for training and the rest 2 sets were used for testing.

The network structure is *GATConv(20)-GATConv(20)-GATConv(20)-FC(300, dropout=0.5)-FC(21)*. It means that the network includes 3 GAT convolutional layers with node embedding of 20 dimensions. This is followed by a fully connected layer with 300 neurons (dropout technique is used with a 0.5 ratio) and a fully connected layer with 21 neurons. It was trained for 50 epochs with Adam optimizer, the mini-batch size was set to 128 and the learning rate was set to 0.001. The trained GAT model finally got an average classification rate of 0.9399 over all the 21 classes.

The confusion matrix of the diagnosis results is shown in Figure 2. Table 1 gives FDR (fault diagnosis rate) and ACR (accurate classification rate) comparison between different types of deep learning models. It shows that the GAT model has the same excellent diagnosis performance as PTCN. If fault 9 and 15 are not considered, the ACR is as high as 0.9731. Although the GAT model has only a very weak advantage in diagnosis performance compared to PTCN, the attention weights in the network give us an entry for an interpretability study.

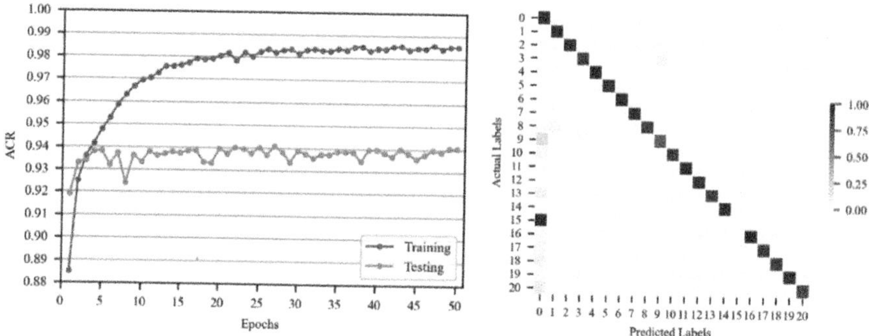

Figure 2: Left: Changes of ACR when training. Right: Confusion matrix of diagnosis results.

Table 1: FDR and ACR comparison of fault diagnosis experiments results on testing dataset

Faults	**DBN** (Zhang and Zhao, 2017)	**DCNN** (Wu and Zhao, 2018)	**BiGRU** (Zhang et al., 2020)	**PTCN** (Wu and Zhao, 2021)	**GAT** (ours)
Normal	-	0.978	0.969	0.9924	0.9927
Fault 1	1	0.986	0.986	0.9931	0.9944
Fault 2	0.99	0.985	0.972	0.9819	0.9817
Fault 3	0.95	0.917	0.935	0.8804	0.8872
Fault 4	0.98	0.976	0.974	0.9956	0.9925
Fault 5	0.86	0.915	0.998	0.9786	0.9785
Fault 6	1	0.975	1	1	0.9984
Fault 7	1	0.999	1	1	1
Fault 8	0.78	0.922	0.753	0.9160	0.9136
Fault 9	0.57	0.584	0.807	0.6601	0.6814
Fault 10	0.98	0.964	1	0.9276	0.9288
Fault 11	0.87	0.984	0.965	0.9798	0.9807
Fault 12	0.85	0.956	0.961	0.9704	0.9664

Fault 13	0.88	0.957	0.953	0.8969	0.8971
Fault 14	0.87	0.987	0.996	0.9964	0.9947
Fault 15	0	0.28	0.541	0.0035	0.0029
Fault 16	0	0.442	0.788	0.9685	0.9682
Fault 17	1	0.945	0.97	0.9254	0.9261
Fault 18	0.98	0.939	0.923	0.9049	0.9049
Fault 19	0.93	0.986	0.926	0.9650	0.9682
Fault 20	0.93	0.933	0.981	0.8825	0.8828
ACR	0.821	0.882	0.927	0.9392	0.9399
ACR w/o Fault 9 & 15	0.889	0.934	0.952	0.9729	0.9731

4. Extend from Local to Global Attention with PageRank

After the GAT model was trained well, the attention weights in it were also fixed. These weights can be used to investigate exactly which nodes or variables the network is focusing on when making inferences. Visualization of attention weights in GAT is shown in Figure 3. Every node in the attention weight graph represents a variable in TE process. The nodes are connected by edges to represent the physical connection between variables, i.e. the topology of the process. The magnitude of the attention weight is expressed with the thickness of the edges. Note that this attention mechanism only makes sense for arbitrary nodes and their neighbor nodes. For each node, the attention weights of all its incoming edges are normalized. This means that it's a local attention mechanism in GAT.

The attention weight graph also follows the same quantitative and qualitative assumptions as PageRank. When applying PageRank to the graph, it can be iterated until convergence and finally give the importance ranking of the nodes in the whole graph, as shown in Figure 3. The size of the nodes is used to indicate the importance of the variables. This importance ranking is for the whole graph and is therefore a kind of global attention. This helps us to analyze which variable is more important for fault diagnosis. In Table 2, the most important process variables for diagnosing each fault are listed and compared with the results of the manual process analysis. For some faults, affected variables can be located with the importance ranking. For those faults that are hard to identity (such as fault 9 and 15), all variables only change slightly after the disturbance is inserted. Thus, it can hardly localize some certain variables to determine the type of fault. This result indicates that the GAT model, when performing fault diagnosis, does not necessarily focus directly on the root cause of the fault, but rather identifies those sensitive variables that are strongly affected by the root cause.

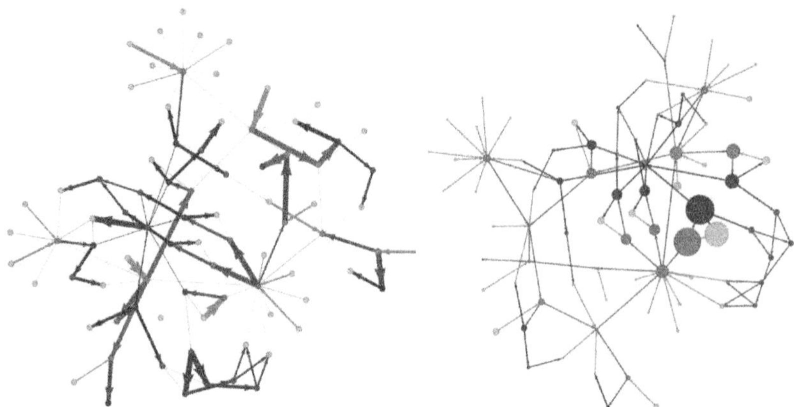

Figure 3: Left: Visualization of attention weights in GAT. Right: Visualization of global attention after applying PageRank to attention weight graph.

Table 2: Comparison between variables affected by disturbances firstly and variables the GAT model focused on when diagnosing faults. T, P, L, Q, x, u mean temperature, pressure, level, flow rate, composition, control signal respectively. Subscript means unit operations or streams. Superscript means characteristics or components.

Faults	Variables affected firstly	Variables GAT focused on
Fault 1	$x_{feed}^{A}, x_{feed}^{C}, \boldsymbol{Q_{s1}}, \boldsymbol{Q_{s4}}, \boldsymbol{u_{s1}^{Q}}, \boldsymbol{u_{s4}^{Q}}$	$\boldsymbol{Q_{s1}}, \boldsymbol{Q_{s4}}, \boldsymbol{u_{s1}^{Q}}, \boldsymbol{u_{s4}^{Q}}$
Fault 2	$\boldsymbol{x_{feed}^{A}}, x_{feed}^{B}, \boldsymbol{x_{feed}^{C}}, \boldsymbol{Q_{s1}}, \boldsymbol{Q_{s4}}, \boldsymbol{u_{s1}^{Q}}, \boldsymbol{u_{s4}^{Q}}$	$\boldsymbol{x_{feed}^{A}}, \boldsymbol{x_{feed}^{C}}, \boldsymbol{Q_{s1}}, \boldsymbol{Q_{s4}}, \boldsymbol{u_{s1}^{Q}}, \boldsymbol{u_{s4}^{Q}}$
Fault 3	$T_{reac}, P_{reac}, L_{reac}, T_{reac}^{cwr}, u_{reac}^{cwr}$	$x_{feed}^{A}, x_{feed}^{C}, Q_{s1}, Q_{s4}, u_{s1}^{Q}, u_{s4}^{Q}$
Fault 4	$\boldsymbol{T_{reac}^{cwr}}, \boldsymbol{u_{reac}^{cwr}}, \boldsymbol{T_{reac}}, P_{reac}, L_{reac}$	$\boldsymbol{T_{reac}^{cwr}}, \boldsymbol{u_{reac}^{cwr}}, \boldsymbol{T_{reac}}, x_{feed}^{A}, Q_{s1}, u_{s1}^{Q}$
Fault 5	$T_{cond}^{cwr}, \boldsymbol{u_{cond}^{cwr}}, T_{sep}, P_{sep}, L_{sep}$	$\boldsymbol{u_{cond}^{cwr}}, Q_{s4}, u_{s4}^{Q}$
Fault 6	$Q_{s1}, \boldsymbol{u_{s1}^{Q}}, Q_{feed}, x_{feed}^{A}$	$T_{reac}^{cwr}, \boldsymbol{u_{s1}^{Q}}$
Fault 7	$\boldsymbol{Q_{s4}}, \boldsymbol{u_{s4}^{Q}}, T_{strip}, P_{strip}, L_{strip}$	$\boldsymbol{Q_{s4}}, \boldsymbol{u_{s4}^{Q}}, T_{reac}^{cwr}$
Fault 8	$x_{feed}^{A}, x_{feed}^{B}, x_{feed}^{C}, \boldsymbol{Q_{s1}}, \boldsymbol{Q_{s4}}, \boldsymbol{u_{s1}^{Q}}, \boldsymbol{u_{s4}^{Q}}$	$\boldsymbol{Q_{s1}}, \boldsymbol{Q_{s4}}, \boldsymbol{u_{s1}^{Q}}, \boldsymbol{u_{s4}^{Q}}$
Fault 9	$T_{reac}, P_{reac}, L_{reac}, T_{reac}^{cwr}, u_{reac}^{cwr}$	$Q_{s1}, Q_{s4}, u_{s1}^{Q}, u_{s4}^{Q}$
Fault 10	$\boldsymbol{T_{strip}}, P_{strip}, L_{strip}, u_{strip}^{steam}, Q_{strip}^{steam}$	$\boldsymbol{T_{strip}}, Q_{s1}, Q_{s4}$
Fault 11	$T_{reac}^{cwr}, u_{reac}^{cwr}, T_{reac}, P_{reac}, L_{reac}$	Q_{s1}, u_{s1}^{Q}
Fault 12	$T_{cond}^{cwr}, u_{cond}^{cwr}, \boldsymbol{T_{sep}}, P_{sep}, L_{sep}$	$T_{reac}, \boldsymbol{T_{sep}}$

Fault 13	$\boldsymbol{T_{reac}}, P_{reac}, \boldsymbol{L_{reac}}, T_{reac}^{cwr}, u_{reac}^{cwr}$	$\boldsymbol{T_{reac}^{cwr}}, \boldsymbol{L_{reac}}$
Fault 14	$\boldsymbol{T_{reac}^{cwr}}, u_{reac}^{cwr}, T_{reac}, P_{reac}, L_{reac}$	$\boldsymbol{T_{reac}^{cwr}}, Q_{s1}, u_{s1}^{Q}$
Fault 15	$T_{cond}^{cwr}, u_{cond}^{cwr}, T_{sep}, P_{sep}, L_{sep}$	$Q_{s1}, Q_{s4}, u_{s1}^{Q}, u_{s4}^{Q}$

5. Conclusion

In this paper, GAT was applied to fault diagnosis and achieved high diagnosis accuracy. After the model was trained well, PageRank was used to extend the local attention of GAT to global attention and generated the importance ranking of variables for each type of fault. This ranking helps us to understand the diagnosis process of the GAT model more deeply.

The lack of interpretability and explainability of the deep learning models limits their practical application. Future research on fault diagnosis should pay more attention to inventing more interpretable models, or better integrating first principles into data-driven models. More explainability research on deep learning models should also be conducted.

References

Bi, X., Zhao, J., 2021. A novel orthogonal self-attentive variational autoencoder method for interpretable chemical process fault detection and identification. Process Safety and Environmental Protection.

Kipf, T.N., Welling, M., 2017. Semi-Supervised Classification with Graph Convolutional Networks. arXiv:1609.02907 [cs, stat].

Veličković, P., Cucurull, G., Casanova, A., Romero, A., Liò, P., Bengio, Y., 2018. Graph Attention Networks. arXiv:1710.10903 [cs, stat].

Venkatasubramanian, V., Rengaswamy, R., Yin, K., Ka, S.N., 2003. A review of process fault detection and diagnosis Part I: Quantitative model-based methods. Computers and Chemical Engineering 19.

Wu, D., Zhao, J., 2021. Process topology convolutional network model for chemical process fault diagnosis. Process Safety and Environmental Protection 150, 93–109.

Wu, H., Zhao, J., 2018. Deep convolutional neural network model based chemical process fault diagnosis. Computers & Chemical Engineering 115, 185–197.

Zhang, S., Bi, K., Qiu, T., 2020. Bidirectional Recurrent Neural Network-Based Chemical Process Fault Diagnosis. Industrial & Engineering Chemistry Research 59, 824–834.

Zhang, Z., Zhao, J., 2017. A deep belief network based fault diagnosis model for complex chemical processes. Computers & Chemical Engineering 107, 395–407.

Antonis Kokossis, Michael C. Georgiadis, Efstratios N. Pistikopoulos (Eds.)
PROCEEDINGS OF THE 33rd European Symposium on Computer Aided Process Engineering (ESCAPE33), June 18-21, 2023, Athens, Greece
 http://dx.doi.org/10.1016/B978-0-443-15274-0.50246-8

Model predictive control of compact combined cycles in offshore power plants integrating a wind farm

Kang Qiu [a], Leif E. Andersson [a*], Cristina Zotică [a], Adriana Reyes-Lúa [a], Rubén M. Montañés[a], Adriaen Verheyleweghen[b], Valentin Chabaud [a], Til Kristian Vrana [a]

[a] *SINTEF Energy Research, Sem Sælands vei 11, 7034 Trondheim, Norway*
[b] *Cybernetica AS, Leirfossvegen 27, 7038 Trondheim, Norway*
* *Corresponding author: leif.andersson@sintef.no*

Abstract

Combined cycle gas turbine plants (CCGTs) fulfill an important role in emission reduction of offshore power systems as the bottoming cycle (BC) produces additional power from exhaust heat of the gas turbines (GTs). With increasing integration of wind turbines, CCGTs offshore must be flexible and provide variation management to the offshore energy system across multiple time scales. This work proposes a model predictive controller (MPC) sending setpoints to the CCGT to satisfy demand in the offshore power system under fluctuating wind power. A high-speed surrogate model suitable for optimizing in an MPC is identified. A linear MPC using a quadratic cost function with process constraints is formulated. The model-based control structure is then validated in simulation for satisfying a constant power demand under disturbances introduced by fluctuating wind power.

Keywords: Model predictive control; steam bottoming cycle; surrogate modelling; feedforward disturbance rejection; PI-control; wind power

1. Introduction

Currently, natural gas fueled gas turbines (GTs) are the main source of power in offshore oil and gas installations. National and international CO_2 reduction targets are incentivizing offshore operators to develop emission reduction strategies such as increasing energy efficiency and low emission power generation solutions like offshore wind power. It is likely that more than one technology will be implemented (Voldsund et al., 2023), resulting in a hybrid integrated offshore power generation system (Riboldi et al., 2020). In this work, we study a system with a combined cycle (CCGT) and power from wind. In CCGTs, the bottoming cycle (BC) uses the exhaust heat from GTs to produce additional power in a steam turbine (ST) and increase thermal energy efficiency (Nord and Bolland, 2013). Due to weight and space limitations, BCs on offshore installations are designed to be low weight and compact, which affects their dynamic response (Montañés et al., 2021). As simple cycle GTs, CCGTs offshore must provide variation management to the offshore energy system across multiple time scales, stabilizing the power generation system. Decentralized control strategies for compact BCs and CCGT based on PI- and feedforward controllers were studied by Nord and Montañés (2018), and model-based nonlinear feedforward in combination with PI-controllers was developed by Zotică et al. (2022).

The requirements for disturbance rejection of CCGTs are stringent when integrating non-dispatchable wind turbines to the power generation system offshore, due to the increased variability of the net load to be covered by the CCGT. A model predictive controller (MPC) is proposed to control the setpoints to the GTs and provide setpoints to the lower PID control layer of the BC by exploiting information, e.g., of the required system demand, wind profile forecasts, and a model of the disturbance. This enables the CCGT power output of the integrated system to minimize generation and consumption mismatches, thereby stabilizing the electrical frequency of the power system. Furthermore, the MPC framework allows for additional control objectives, such as minimizing CO_2 emissions and inherently handle steam temperature and pressure constraints.

2. System description and model

Figure 1 depicts the system considered in this work, consisting of an integrated wind-thermal electricity generation system. We assume that the thermal system should compensate for variations in wind power generation. The thermal system, a CCGT, consists of two GTs and a BC with two once-through steam generators (OTSG) connected to a common steam turbine (ST) and condenser system. A wind farm produces non-dispatchable power and the CCGT stabilizes the integrated power output. We simulate the power from the wind farm and CCGT using high-fidelity models and develop a surrogate model of the CCGT for the MPC.

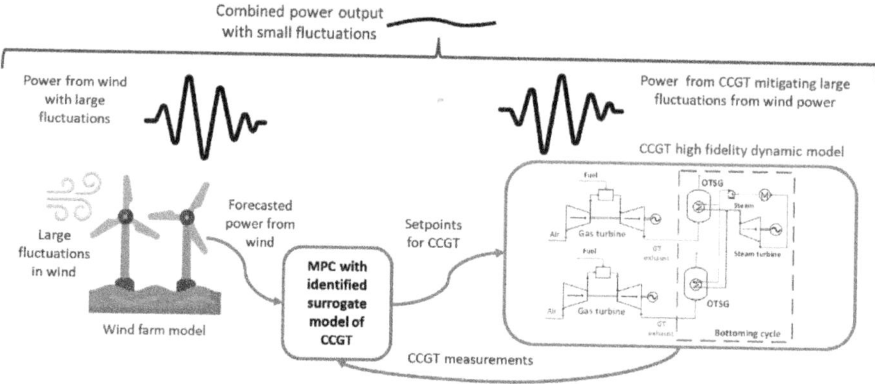

Figure 1. Conceptual description of the system.

2.1. Power from wind farm

The considered wind power plant consists of three wind turbines with 8 MW power rating each. The specifications are inspired by the Siemens Gamesa SG 8.0-167 DD wind turbine. The control of the wind turbines is taken as state-of-the-art grid-following maximum-power-point-tracking control, where the turbine actions are almost purely driven by the incoming wind, from which as much energy as possible is harvested. The wind turbines do not actively attempt to stabilize the electric power system. The task of stable control of the electric system is purely left to the gas turbines. This principle reflects the control of commercially available wind turbines, and the typical variation management strategy in wind-thermal systems (Burton, 2011).

The turbine layout is taken as a straight line with a spacing of ten diameters between turbines (standard, conservative spacing). A wind speed of 10 m/s with direction perpendicular to the layout (no wake) and with standard turbulence level is chosen. The incoming wind is then obtained from a farm-scale synthetic turbulence generator providing the rotor-averaged (power-equivalent) wind speed with consistent low-frequency fluctuations and correlation between turbines (Chabaud, 2023), for a consistent representation of wind power variability.

2.2. High fidelity dynamic model of the combined cycle

SINTEF Energy Research developed a high-fidelity dynamic Modelica model of the CCGT (Montañés et al., 2021; Zotică et al., 2022), using the commercial modeling and simulation environment Dymola. We utilize Modelica models from the Thermal Power Library 1.21 and adapt them to build up the combined cycle power plant models. In this work we model the thermo-hydraulics of the steam cycle of the CCGT with high-fidelity principles and apply a simplified quasi-static model for the GT based on data provided by Siemens Energy through the LowEmission Consortium. The bottoming cycle is based on the design presented in Zotică et al. (2022). For further details on the underlying models for turbine islands and steam generators, refer the work by Montañés et al. (2017) and Zotică et al. (2022). OTSG models are explained in the work by Montañés et al. (2021).

2.3. MPC surrogate model of the combined cycle

In this work, the surrogate model used for the predictive controller is a linear, autoregressive model of the CCGT predicting the steam turbine power output P_s, and CO_2 emissions from the operation of the GTs. The parameters of the surrogate model are determined with a stability-constrained linear regression performed on data gathered from simulation of the high-fidelity dynamic model of the CCGT.

The GT power x_{GT} and GT CO_2 emissions x_{CO_2} are modelled dependent on the current GT load, u_{GT},

$$x_{GT} = f(u_{GT}) \tag{1}$$

$$x_{CO_2} = f(u_{GT}). \tag{2}$$

The dynamics of the BC are highly nonlinear. However, the lower PID control layer results in more linearized dynamics (see section 3.1) and enables the use of a linear prediction model for the MPC. In the BC, the OTSG consists of relatively large volumes and mass of metal walls that accumulate heat over time. Hence, the BC exhibits high thermal inertia at the interface of the gas turbine exhaust and steam generation in the OTSG (Montañés et al., 2021). Therefore, the surrogate model necessitates a higher order, n, for modelling the predicted ST power in the next timestep $k + 1$. The ST power is predicted from the temperature setpoints u_t and pressure setpoint u_p of the BC

$$x_{ST,k+1} = f\left(x_{BC,k-n}, u_{GT,k-n}, u_{t,k-n}, u_{p,k-n}\right),\ n = 0,..,5 \tag{3}$$

3. Control structure

The control structure proposed in this paper consists of a lower PID control layer of the BC and GT and an upper MPC control layer to provide setpoints to the lower layer.

3.1. PID control layer

The lower PID control layer keeps the superheated steam temperature at the outlet of the two OTSGs and superheated steam pressure at a setpoint given by the upper supervisory control layer (MPC). The power output of the ST is highly dependent on the enthalpy at the inlet of the steam turbine. The BC heat input is mostly given by the gas turbine operation specified as GT load. The live steam pressure is controlled by manipulating the valve upstream the ST using a pure I-controller. The temperature is controlled by manipulating the feedwater flowrate, using a combination of model-based nonlinear feedforward for disturbance rejection and a feedback PI-controller to reject unmeasured disturbances and account for plant model mismatch. The nonlinear feedforward is derived from a simple steady-state energy balance over the OTSG. It receives measurements from the hot gas side flowrate and temperatures which are the main disturbances for the BC. The I and PI controllers are tuned from step responses using the SIMC tuning rules (Skogestad, 2003).

3.2. MPC control layer

The MPC provides GT setpoints u_{GT}, superheated steam temperature setpoints u_t, and the superheated steam pressure setpoint u_p to the lower PID control layer. A linear MPC using a quadratic cost function with input constraints, and the surrogate model described in 2.3 is formulated

$$\begin{aligned} &\min_{u} \ \omega_d J_d + \omega_{CO_2} J_{CO_2} + \omega_a J_a \\ s.t.\ & x_{k+1} = f(x_{k-n}, u_{k-n}),\ n = 0,..5,\ k = 0, \dots, N-1 \\ & g(u_k) \leq 0,\ \ k = 0,\ \dots, N-1 \\ & x_0 = \hat{x}_0. \end{aligned} \quad (4)$$

The objective of the MPC consists of, in order priority, a power tracking cost J_d to satisfy a power demand, a CO_2 emission penalty J_{CO_2}, and an actuation cost J_a. The power tracking cost is calculated from the difference between the total predicted power output, consisting of the power output from the GTs x_{GT} and ST x_{ST}, forecasted wind power p_w, and the power demand p_d

$$J_d = \sum_{k=0}^{N} (p_{d,k} - p_{w,k} - x_{GT,k} - x_{ST,k})^2. \quad (5)$$

The CO_2 emission penalty is a quadratic cost on the predicted CO_2 emissions

$$J_{CO_2} = \sum_{k=0}^{N} x_{CO_2}{}^2. \quad (6)$$

Lastly, the actuation cost penalizes the change in the GT setpoints u_{GT}, BC temperature setpoints u_t, and BC pressure setpoint u_p with a quadratic cost

$$J_a = \sum_{k=0}^{N-1} (u_{GT,k+1} - u_{GT,k})^2 + (u_{t,k+1} - u_{t,k})^2 + (u_{p,k+1} - u_{p,k})^2. \quad (7)$$

To simplify the tuning of the weights ω, both the state and input variables are normalized to ensure comparability of the subobjectives J_{CO_2} and J_a. As J_d is the most important subobjective, the ω_d is set first. Afterwards ω_{CO_2} and ω_a are tuned by trial and error.

4. Results and discussion

The proposed MPC is tested in a simulation environment introduced in 4.1. The results (see 4.2) illustrate the disturbance rejection the MPC provides given fluctuating wind power and ST power output.

4.1. Simulation environment

The MPC is simulated in loop with a Gaussian process model of the CCGT regressed from data gathered from the high-fidelity model. The sampling time of 5 s and prediction horizon of 5 min were chosen to account for the vastly different time dynamics of the GTs and the BC, which exhibited a closed loop settling time of 10^0-10^1 s and 10^2-10^3 s, respectively (Montañés et al., 2021). The simulation environment is assumed to provide full-state feedback of the ST power, GT power, and CO_2 emissions of the GTs. Realistic wind profiles as described in Section 2.1, assumed to be forecasted exactly, are utilized.

The simulations are formulated with CasADi (Andersson et al., 2019) and solved with IPOPT (Wächter and Biegler, 2006), where each MPC iteration is solved within 0.05 s.

4.2. Results

Figure 2(a) shows the cumulative power output of the wind turbines and the CCGT. The proposed MPC can control the CCGT to provide variation management under fluctuating wind power to satisfy a constant demand of 70 MW. Figure 2(b), showing the relative power of the wind turbines, GTs, and ST scaled to the total power demand indicates that fluctuations in wind power are mainly compensated by the power output of the GTs due to their fast dynamics.

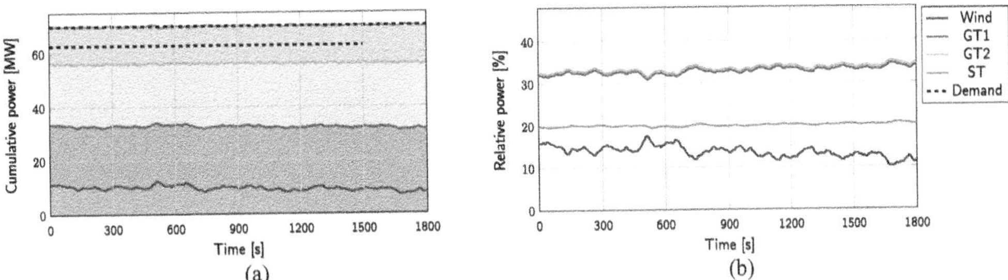

Figure 2. (a) Cumulative power output of the combined wind-thermal system. (b) Relative power output of CCGT and wind scaled to total demand.

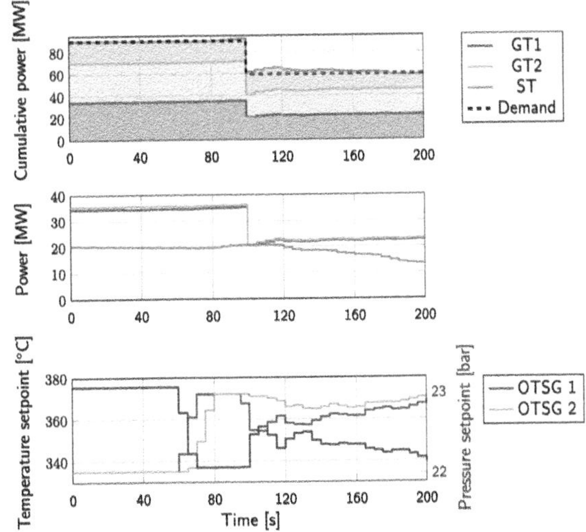

Figure 3. MPC step response showing cumulative power, GT load inputs, and steam temperature and pressure setpoints chosen by the MPC.

Figure 3 illustrates the working principle of the MPC when the power demand of the CCGT changes, e.g., due to a change in wind power or overall wind-thermal system demand. Overall, the CCGT responds sufficiently fast to the step change in power demand. The GT loads are adjusted at time of demand change, as GTs have a fast dynamic response. Furthermore, the MPC accounts for the slower dynamic response of the BC by regulating the gas turbine loads at $t = 100\ s$ lower than the steady state gas turbine load at $t \geq 200\ s$.The dynamic response of the BC is improved by MPC as it regulates the setpoints to the lower PID control layer at $t = 60\ s$ before the demand step change at $t = 100\ s$ to account for the slower BC dynamics

5. Conclusion

This work proposes an MPC for CCGT control with variation management, and CO_2 emission reduction objectives. To formulate the MPC, a linear surrogate model is identified. As the CCGT process and the high-fidelity Modelica model are highly nonlinear, it is challenging to represent the dynamics correctly in a linear model.
Simulations show that the MPC helps to achieve variation management of the integrated wind-thermal system by controlling the CCGT under fluctuating, realistic wind profiles. Further work should address uncertain wind profile forecasts and investigate different objectives to the MPC.

Acknowledgements

This publication has been produced with support from the LowEmission Research Centre (www.lowemission.no), performed under the Norwegian research program PETROSENTER. The authors acknowledge the industry partners in LowEmission for their contributions and the Research Council of Norway (296207). We acknowledge SIEMENS Energy AG for providing the gas turbine model used as reference.

References

Andersson, J.A.E., Gillis, J., Horn, G., Rawlings, J.B., Diehl, M., 2019. CasADi: a software framework for nonlinear optimization and optimal control. Math. Prog. Comp. 11, 1–36.

Burton, T. (Ed.), 2011. Wind energy handbook, 2nd ed. ed. Wiley, Chichester, West Sussex.

Chabaud, V., 2023. Synthetic turbulence modelling for offshore wind farm engineering models using coherence aggregation (preprint). Preprints.

Montañés, R.M., Skaugen, G., Hagen, B., Rohde, D., 2021. Compact Steam Bottoming Cycles: Minimum Weight Design Optimization and Transient Response of Once-Through Steam Generators. Front. Energy Res. 9, 687248.

Nord, L.O., Bolland, O., 2013. Design and off-design simulations of combined cycles for offshore oil and gas installations. Applied Thermal Engineering 54, 85–91.

Riboldi, L., Alves, E.F., Pilarczyk, M., Tedeschi, E., Nord, L.O., 2020. Optimal Design of a Hybrid Energy System for the Supply of Clean and Stable Energy to Offshore Installations. Front. Energy Res. 8, 607284.

Skogestad, S., 2003. Simple analytic rules for model reduction and PID controller tuning. Journal of Process Control 13, 291–309.

Voldsund, M., Reyes-Lúa, A., Fu, C., Ditaranto, M., Nekså, P., Mazzetti, M.J., Brekke, O., Bindingsbø, A.U., Grainger, D., Pettersen, J., 2023. Low carbon power generation for offshore oil and gas production. Energy Conversion and Management: X 17, 100347.

Wächter, A., Biegler, L.T., 2006. On the implementation of an interior-point filter line-search algorithm for large-scale nonlinear programming. Math. Program. 106, 25–57.

Zotică, C., Montañés, R.M., Reyes-Lúa, A., Skogestad, S., 2022. Control of steam bottoming cycles using nonlinear input and output transformations for feedforward disturbance rejection. IFAC-PapersOnLine 55, 969–974.

Antonis Kokossis, Michael C. Georgiadis, Efstratios N. Pistikopoulos (Eds.)
PROCEEDINGS OF THE 33rd European Symposium on Computer Aided Process Engineering (ESCAPE33), June 18-21, 2023, Athens, Greece
 http://dx.doi.org/10.1016/B978-0-443-15274-0.50247-X

Application of Bayesian Optimization in HME Batch Concentration Process

Chong Liu[a], Cheng Ji[a], Chengyu Han[a], Chenxi Gu[b], Jindong Dai[a], Wei Sun[a,*], Jingde Wang[a,*]

[a]*College of Chemical Engineering, Beijing University of Chemical Technology, North Third Ring Road 15, Chaoyang District, Beijing, 100029, China*

[b]*Shijiazhuang Lonzeal Pharmaceutical Co., Ltd., No. 16 West Ring Road, Shenze County Industrial Park, Shijiazhuang, Hebei Provice, 052560, China*

**Corresponding Author's E-mail: jingdewang@mail.buct.edu.cn*

**Corresponding Author's E-mail: sunwei@mail.buct.edu.cn*

Abstract

Pharmaceutical intermediates are chemical products that play an important role in the synthesis of drugs. Different from the general chemical production process, the production process of pharmaceutical intermediates is characterized by flexible small scale, batch mode and multifunction. Therefore, optimal operation strategies are required at different stages, usually implemented by adjusting operation parameters, such as temperature and pressure along the batch for particular units and instruments. However, it is challenging to obtain the best combination of multiple parameter trajectories. In this paper, the concentration of HME, a pharmaceutical intermediate, is studied as an industrial case. A dynamic model is established to simulate the whole concentration process of HME. To promptly achieve optimal temperature and pressure in HME batch concentration, a Bayesian optimization algorithm based on Gaussian process regression model is used with the goal of minimizing the total cost. It can be seen from the results that Bayesian optimization is very competitive in dynamic simulation.

Keywords: batch concentration, Bayesian optimization, pharmaceutical intermediates, dynamic process simulation

1. Introduction

Lamivudine is an antiviral drug of nucleoside reverse transcriptase inhibitor, which can selectively inhibit the replication of hepatitis B virus. At present, the chemical synthesis method using L-menthol as the starting material is often adopted in industry (Xue et al., 2010). (2S,5R)-5-Hydroxy- [1,3] -oxathiolane-2-carboxylic acid methyl ester (HME) is an important chemical synthesis intermediate of Lamivudine, and the concentrate process of HME has a significant impact on the synthesis of Lamivudine. The synthesis route is shown in Figure 1 (where I is Lamivudine and III is HME.). Based on industrial projects, this paper focuses on the concentration of HME from its solvent mixture system and optimization of operating parameters, in order to minimize the total production cost.

The dynamic optimization of batch processes has been concerned by many researchers, and many solutions have been proposed. Yang et al. applied iterative particle swarm optimization (PSO) algorithm to the comprehensive optimization problem of batch processes, in which objective function was set as the sum of product concentration, reciprocal of reaction time and reciprocal of energy loss, and the optimal solution was obtained by searching for optimal trajectory(Yang et al., 2017). Amini et al. proposed a new method to optimize the control of batch cooling crystallizers by genetic algorithm, and obtained the minimum and maximum values of the objective function (Amini et al., 2016). Mujtaba et al. solved the dynamic optimization of the batch reactor with a simple model through neural network, and obtained the optimal operation strategy in terms of reactor temperature (Mujtaba et al., 2006). However, the above algorithms are easy to fall into local optimum or less interpretability. Compared with the traditional optimization methods, the surrogate model of Bayesian optimization has stronger fitting ability and better interpretability, and therefore it is applied for the dynamic optimization of an HME batch concentration process in this work (Shahriari et al., 2016).

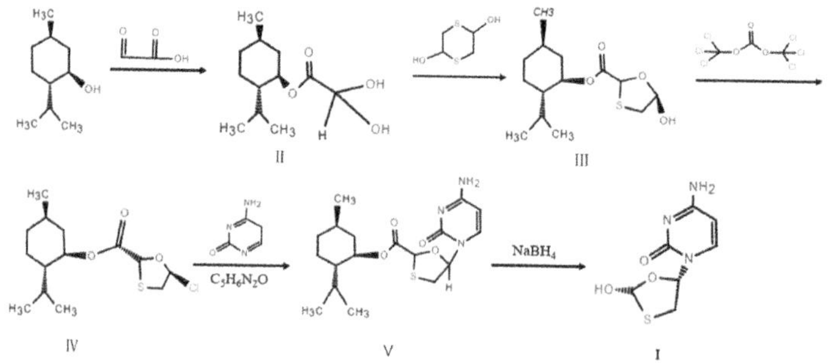

Figure 1 Synthesis route of Lamivudine

2. Process Modeling

HME concentration system is mainly composed of toluene, acetic acid, HME and water. The mass and mass fraction of each component are shown in Table 1.

Table 1 Mass and mass fraction of components

Component	Component quality/kg	Mass fraction
Toluene	2480	0.6727
Acetic acid	26.7	0.0072
HME	1080	0.2930
Water	100	0.0271

The dynamic concentration model is established as shown in the Figure 2. The whole concentration system consists of a jacketed distillation tank, two coolers, and two collection tanks. Instruments and controllers are equipped in necessary pipelines and equipment. Details on the control strategies are shown in Table 2.

Table 2 Details on the control strategies

Letter	Name	Control variable	Manipulate variable
A	TICT01	Temperature of T01	Heat medium flow
B	PICT01	Pressure of T01	Gas outlet flow of T01
C	TICE01	Outlet temperature of E01	Refrigerant 1 flow
D	TICE02	Outlet temperature of E02	Refrigerant 2 flow

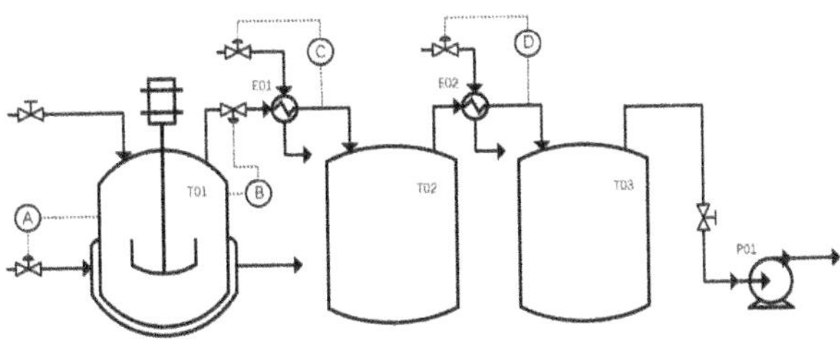

Figure 2 Concentration model

3. Bayesian optimization

3.1. Parameter space and Objective function

3.1.1. Manipulated variable selection

In the dynamic batch simulation process, the final production cost of the concentration process is highly correlated with the temperature and pressure control. First of all, pressure has a significant impact on the bubble dew point of the system. The lower the pressure, the lower the bubble point temperature, and the easier it is for concentration. However, to ensure the lower pressure, it is inevitable to increase the load of the vacuum pump. Therefore, it is necessary to balance the pressure in the distillation tank. Secondly, due to the intermittent feature of the concentration process, different temperature control schemes are selected in industry as time changes. Therefore, difference between the set value of temperature controller and bubble point of the system in the concentration process ΔT is selected as one of optimization variables, so that temperature control value changes dynamically with the concentration time, making temperature control scheme more suitable for industrial practice. Similarly, difference between dew point at inlet of primary cooler and the set value of temperature controller Δt_1, and difference between dew point at inlet of secondary cooler and the set value of temperature controller Δt_2, are selected as optimization variables.

3.1.2. Objective function

The total production cost is defined as objective function, and it is mainly divided into two parts, one is the real operation cost A, and the other is cost B of lost solvent toluene. A mainly includes labor cost, vacuum pump cost and utility cost. B is the cost of lost solvent toluene pumped by vacuum pump, including the cost of solvent toluene and the cost of handling the extracted toluene in order to meet the requirements of environmental protection. The calculation of the cost is listed below.

$$c_1 = a \cdot t \cdot n \tag{1}$$

$$c_2 = \left(\frac{d}{10 \cdot 8000} + \frac{f \cdot (101.325 - P)/101.325}{\eta} \cdot e\right) \cdot t \tag{2}$$

$$c_3 = F_1 \cdot U_1 \tag{3}$$

$$c_4 = F_2 \cdot U_2 \tag{4}$$

$$c_5 = F_3 \cdot U_3 \tag{5}$$

$$A = c_1 + c_2 + c_3 + c_4 + c_5 \tag{6}$$

$$B = F_4 \cdot (U_4 + U_5) \tag{7}$$

$$Objective(P, \Delta T, \Delta t_1, \Delta t_2) = A + B \tag{8}$$

where, c_1, c_2, c_3, c_4, c_5: Cost of labor, vacuum pump, refrigerant 1, refrigerant 2, heat medium, ¥, a: Hourly salary, ¥, t: Concentration time, h, n: Number of workers, d: Fixed cost of vacuum pump, ¥, f: Rated power of vacuum pump, kW, η: Efficiency of vacuum pump, e: Electricity price, ¥/kw·h, F_1, F_2, F_3, F_4: Total flow of refrigerant 1, refrigerant 2, heat medium, solvent toluene, kg, U_1, U_2, U_3, U_4, U_5: Unit price of refrigerant 1, refrigerant 2, heat medium, solvent toluene, solvent toluene treatment, ¥/kg.

3.2. Algorithm introduction

In essence, the Bayesian optimization framework uses the probabilistic surrogate model to fit the real objective function, and actively selects the most 'potential' evaluation points for evaluation according to the fitting results to avoid unnecessary sampling. Therefore, Bayesian optimization is also called active optimization (Jones, 1998). Overall, Bayesian optimization framework can effectively use process dynamic information to improve the search efficiency. And it mainly includes two parts: probabilistic surrogate model and acquisition function.

3.2.1. Gaussian process regression

Gaussian process regression model aims to quantitatively estimate the probability distribution of the predicted value and obtain the confidence interval by establishing a nonparametric probability model based on the kernel function. It is considered as an extension of multivariate Gaussian function on infinite dimensional set variables. For the existing dataset, $D = \{(\boldsymbol{x}_i, y_i)\}_{i=1}^{n}$, $\boldsymbol{x}_\mathrm{i} \in R^d$ is the input variable, d is the dimension of the input variable, $y_\mathrm{i} \in R$ is the target variable. When there are n experimental points in the data set, the prior knowledge of the Gaussian process regression algorithm assumes that these n points obey the multivariate Gaussian distribution N in space shown in Equation (9).

$$\begin{bmatrix} y_1 \\ \vdots \\ y_n \end{bmatrix} \sim N\left(\begin{bmatrix} \mu(\boldsymbol{x}_1) \\ \vdots \\ \mu(\boldsymbol{x}_n) \end{bmatrix} \begin{bmatrix} k(\boldsymbol{x}_1, \boldsymbol{x}_1) & \cdots & k(\boldsymbol{x}_1, \boldsymbol{x}_n) \\ \vdots & \ddots & \vdots \\ k(\boldsymbol{x}_n, \boldsymbol{x}_1) & \cdots & k(\boldsymbol{x}_n, \boldsymbol{x}_n) \end{bmatrix}\right) \tag{9}$$

where μ is the mean value; $k(\cdot,\cdot)$ is a kernel function used to measure the correlation between two input variables. The kernel function selected in this paper is Matern, which is a stationary kernel and a generalization of the radial basis function (RBF). There is an additional parameter v, which controls the smoothness of the resulting function. In this work, $v = 5/2$, and specific expression is as follows.

$$k(x_i, x_j) = \left(1 + \frac{\sqrt{5}}{l} d(x_i, x_j) + \frac{5}{3l} d(x_i, x_j)^2\right) exp\left(-\frac{\sqrt{5}}{l} d(x_i, x_j)\right) \tag{10}$$

where $\mathrm{d}(\cdot,\cdot)$ is the Euclidean distance.

3.2.2. Confidence Boundary Strategy

Srinivas et al. proposed a confidence boundary strategy for Gaussian processes, GP-UCB (Srinivas et al., 2010). When solving the minimum value of the objective function, the acquisition function of the UCB strategy is:

$$\alpha_t(x; D_{1:t}) = \mu_t(x) + \sqrt{\beta_t}\sigma_t(x) \tag{11}$$

where, parameter β_t balances expectation and variance.

Then Bayesian optimization algorithm is embedded into the batch concentration dynamic model, logic diagram is shown in Figure 3. The left side of the block diagram shows the process of Bayesian optimization algorithm using simulation data to recommend the next point to be evaluated. The batch concentration dynamic model on the right side is responsible for running under the working conditions of the parameter combination recommended by Bayesian optimization algorithm to obtain new cost simulation data, and then adding the simulation data to the dynamic database to iterate until convergence. In order to achieve data interaction between Python and C++, a set of communication master control for data transmission between both parties has been developed, which can greatly improve the speed and reliability of data transmission, and remove the barriers between the algorithm and model operations.

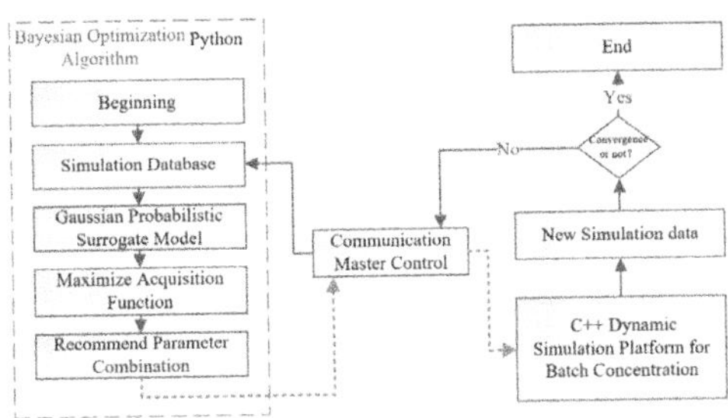

Figure 3 Logic diagram of Bayesian optimization for batch concentration

4. Simulation results and discussion

4.1. Raw simulation data set

Table 3 Raw Simulation Data

	P	ΔT	Δt_1	Δt_2	Total cost		P	ΔT	Δt_1	Δt_2	Total cost
1	5	0	10	20	99999	6	15	20	10	20	5736.6
2	5	10	10	20	7004.58	7	25	0	10	20	99999
3	5	20	10	20	7232.51	8	25	10	10	20	12764.32
4	15	0	10	20	99999	9	25	20	10	20	12764.57
5	15	10	10	20	4531.47						

Table 4 Iterative data

Iterations	P	ΔT	Δt1	Δt2	Total cost	Iterations	P	ΔT	Δt1	Δt2	Total cost
44	17.38	8.22	11.98	3.96	4443.72	285	5.04	8.39	12.68	4.25	4483.76
71	18.99	9.19	9.59	2.40	4409	320	16.55	2.02	6.65	5.97	4457.54
144	15.38	14.79	9.35	5.62	4407.43	350	24.77	6.19	8.04	7.64	4439.91
162	17.83	3.63	11.21	1.33	4442.43	428	19.05	18.22	11.34	2.16	4462.79
234	13.84	16.88	13.64	6.09	4433.9	467	15.88	14.20	10.14	5.24	4413.62

In order to obtain the initial data set of Bayesian optimization, 9 groups of different parameter combinations are selected, and target values of different combinations are calculated. As shown in Table 3, the selection of initial parameter combinations is kept in discrete state as far as possible, which is conducive to the rapid convergence of subsequent Bayesian optimization iterations.

4.2. Results and discussion

Based on the initial data set, 500 iterations of Bayesian optimization calculations have been carried out for HME batch concentration process. Table 4 shows the optimal target values for every 50 iterations. It can be observed from the Table 4 that after 500 iterations, a lower target value is found, and the total cost of batch concentration is reduced to 4407.43 from 4531.47, the lowest in the initial dataset.
In terms of parameter selection, it can be seen from the iteration selection histogram of four parameters in Figure 4 that, with the support of simulation database and comprehensive consideration of the uncertainty of prediction in the Bayesian optimization iteration process, once a lower value is found, the search can be frequently continued near the point, which greatly improves the search efficiency. In terms of iteration results, after 144 iterations, it can be found that the best parameter combination is P=15.38kpa, ΔT=14.79 ℃, Δt_1=9.35 ℃, Δt_2=5.62 ℃, the minimum value of the objective function is 4407.43. As shown in Figure 5, the target value close to the lowest value has been obtained after 71 iterations. However, in the subsequent iteration process, there is no lower target value. The reason could be that as the times of iterations increases, the acquisition of the known value of the function increases, and the surrogate model is prone to over fitting. Therefore, the number of iterations can be appropriately reduced to avoid over fitting.

5. Conclusion

In this paper, Bayesian optimization algorithm based on Gaussian regression process and UCB is applied to optimize the batch concentration process of HME and obtain the optimal combination of control parameters, so as to minimize the total cost of batch concentration. After 144 iterations, the best control scheme and the lowest total cost value 4407.43 are obtained. It can be seen that Bayesian optimization algorithm has strong competitiveness in dynamic optimization, especially in the screening of control parameters. The surrogate model of Bayesian optimization also has strong fitting ability and good interpretability with the support of simulation database and comprehensive consideration of the uncertainty of prediction in iterations. In the subsequent work, the impact of different probabilistic surrogate models and acquisition functions on the optimization results can be compared.

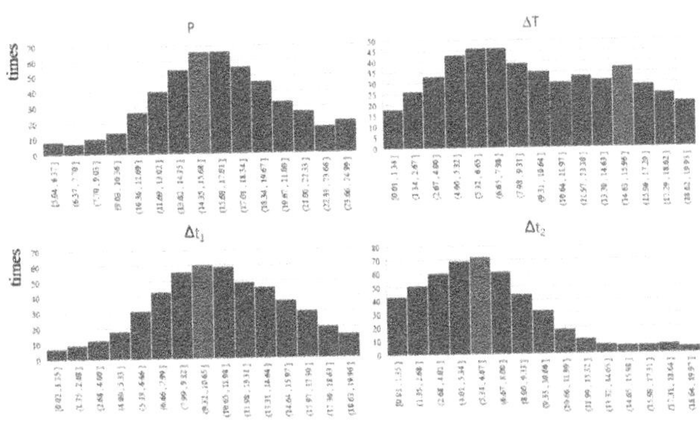

Figure 4 Iterations of each parameter

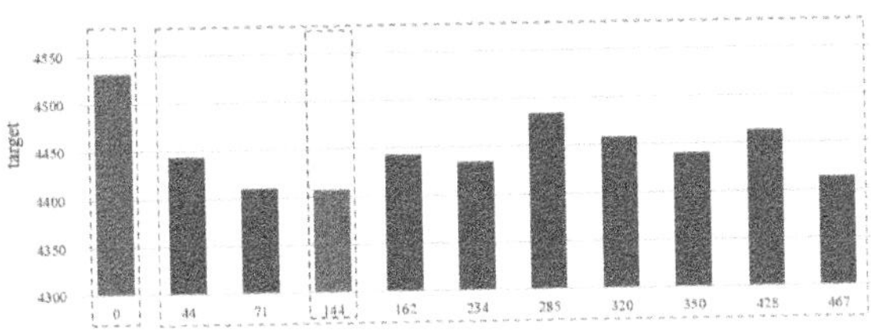

Figure 5 Optimal target values for every 50 iterations

Acknowledgments

This work was supported by the National Natural Science Foundation of China (grant numbers 2187081029).

References

F. J. Xue, Y. Z. Yang, X. M. Liu, 2010, Synthesis of Lamivudine, Fine Chemicals,27, 6, 589-592.

L. Yang, H. P. Pan, E. B. Zhang, 2017, Comprehensive Optimization of Batch Process based on Particle Swarm Optimization Algorithm, 29th Chinese Control And Decision Conference (CCDC), pp. 4504-4508.

Y. S. Amini, M. B. Gerdroodbary, 2016, Optimal Control of Batch Cooling Crystallizers by using Genetic Algorithm, Case Studies in Thermal Engineering, 8, 8, 300-310.

I. M. Mujtaba, N. Aziz, M. A. Hussain, 2006, Neural Network Based Modelling and Control in Batch Reactor, Chemical Engineering Research and Design, 84, 8, 635-644.

B. Shahriari, K. Swersky, Z. Wang, 2016, Taking the human out of the loop: A review of Bayesiaoptimization, Proc. of the IEEE, 104, 1, 148–175.

D. R. Jones, M. Chonlau, W. J. Welch, 1998, Efficient global optimization of expensive black-box functions, Journal of Global Optimization, 13, 4, 455–492.

N. Srinivas, A. Krause, S. M. Kakade, 2010, Gaussian process optimization in the bandit setting: No regret and experimental design, International Conference on Machine Learning.

Antonis Kokossis, Michael C. Georgiadis, Efstratios N. Pistikopoulos (Eds.)
PROCEEDINGS OF THE 33rd European Symposium on Computer Aided Process Engineering (ESCAPE33), June 18-21, 2023, Athens, Greece
 http://dx.doi.org/10.1016/B978-0-443-15274-0.50248-1

Symbolic regression-based method for developing a physics-informed surrogate model for a manufacturing process

Utsav Awasthi,[a] and George M. Bollas,[a]

[a]*Department of Chemical and Biomolecular Engineering, Pratt and Whitney Institute of Advanced Systems Engineering, University of Connecticut, 159 Discovery Dr, Storrs, CT, 06269, USA.*

Abstract

Hybrid models are used increasingly for the simulation of complex physical processes, because they combine the knowledge of the physics of a process and the data obtained from process measurements. This fusion of domain expert knowledge with data leads to hybrid models that are more accurate, can extrapolate, and obey laws of physics. Surrogate modeling can be cast as a hybrid modeling approach that builds reduced-order models for complex physical phenomena with simple model structure and low computational complexity. This study presents the application of hybrid modeling developed through a simple and accurate recursive symbolic regression algorithm for the complex and ill-understood phenomenon of tool wear in precision machining.

Keywords: Surrogate model, Symbolic Regression, Physics-explained AI.

1. Introduction

With the advent of Industry 4.0, manufacturing industries have undergone the digitalization of manufacturing processes. This digitalization has enabled the collection of better and large sets of sensor data to build data-driven models. Some popular methodologies for building data-driven models are neural networks and support vector machines (Han et al. 2021, 2022) . Though these models are widely used for process health monitoring in the manufacturing industries, they tend to be limited due to data idiosyncrasies. Hybrid models overcome this shortcoming. Hybrid models infuse domain knowledge of the physics of the process in the data-driven methodologies (Yang et al. 2022). Sansana et al. (2021) classified hybrid models based on the structure of the models broadly into three categories, serial structure, parallel structure, and surrogate models. A serial structure combines a hybrid model where the mechanistic part of the model either takes care of the known process behavior, and the data-driven part is used to capture the behavior, a feature of the process that cannot be directly estimated from the mechanistic model. A surrogate model is an approximate process model, and it is a simple representation of a complex phenomenon. These attributes of the surrogate models are helpful in building models for processes/phenomena that are difficult to model directly using data from the sensors or in cases where the physics of the process/phenomenon is too complex to model. In this study, a surrogate model is built to capture the progression of a fault in a manufacturing process. As is often the case in physical processes, the progression of a process fault is dynamic and depends on its state at any given time. Therefore, incorporating the fault state to predict its future values is important in modeling the process and fault accurately. This is accomplished by building dynamic or recursive models. This work illustrates the development of a recursive symbolic regression model.

2. Surrogate modeling approaches

Surrogate models are approximate models of complex physical phenomena that can be used to represent an actual physical process with low complexity. These models require less computational effort and are used in process modeling and optimization. Over the years, several mathematical software have been developed to build surrogate models. Some of the commercially available software are Eureqa (Schmidt and Lipson 2009) (now DataRobot), AI Feynman: A physics-inspired method for symbolic regression (Udrescu and Tegmark 2020), Automated Learning of Algebraic Models for optimization ALAMO (Cozad, Sahinidis, and Miller 2014, 2015; Wilson and Sahinidis 2017), and the Genetic Programming Toolbox for Identification of Physical Systems (GPTIPS) (Searson, Leahy, and Willis 2010; Searson 2015). These software toolboxes exploit symbolic regression to develop surrogate models of response variables for selected input variables. Symbolic regression gives an algebraic expression of input variables. This approach is different from traditional regression techniques because the model structure and the regression coefficients are both evolved in symbolic regression. This helps in finding unique mathematical expressions that would best represent the data. Here, we choose GPTIPS to illustrate a methodology for recursive symbolic regression.

2.1. GPTIPS (Genetic programming toolbox for identification of physical systems)

GPTIPS is a MATLAB toolbox based on genetic programming (Searson 2015). Genetic programming is a biologically inspired machine learning method that uses the principles of evolution to perform a task or operation. To perform a task/operation, GP generates a random population of tree structures and then performs mutation, selection, and crossover over the best-performing tree to build a new population. This process of GP is repeated until a tree structure is obtained that fulfills the objective or, in the case of symbolic regression, gives a model of desired complexity and accuracy. GPTIPS utilizes the aforementioned features to build empirical mathematical models using data. GPTIPS generates models using multigene genetic programming with a linear combination of nonlinear transformation of input variables. GPTIPS has a variety of tournament selection methods, such as regular tournament selection, Pareto tournament selection, and lexicographic tournament selection. To build models for symbolic regression, GPTIPS has a large set of basis functions, +, -, *, \, mult3(product of three terms), add3(sum of three terms), tanh, cos, sin, exp, ln (natural logarithm), x2, abs, x3, Ö, exp(-x), if-then-else, >, <, exp(x2), threshold and step functions, for model building. GPTIPS provides options to select a user-defined fitness criterion for model building (the default fitness metric is the root mean squared error, RMSE). GPTIPS provides an extensive list of hyperparameters, such as the number of generations, number of populations, maximum tree depth, maximum mutation depth, tournament size, crossover, mutation, elitism, and direct reproduction, that can be tuned to improve the model fitness. These hyperparameters decide on the complexity, and the model fitness is adjusted to get a model that does not overfit the data.

3. Application of surrogate modeling in manufacturing

3.1. Precision machining and tool wear

Precision machining is performed using Computer Numerical Control (CNC) machines (Awasthi and Bollas 2020). CNC machines use a circular tool with high-grade metal inserts, rotating at high speed and removing material from the workpiece as metal chips. This process results in high shear stress, friction, and high temperature at the tool interface and workpiece interface (Awasthi et al. 2022). These aggressive conditions lead to the wear of the insert surface. This gradual wear of the insert surface is called tool wear. Analysis of data and understanding of the physics of the tool wear mechanism have identified that tool wear is a function of the operating conditions and the state of the tool. Using this insight, a recursive surrogate model of the tool wear as a function of the tool wear state was developed using GPTIPS. In addition to the tool wear state, the cutting conditions impact tool wear and were considered for model building. Hence, in this study, we utilize the recursive symbolic regression methodology to build a model for tool wear.

Table 1. Machine setting for experiments performed on the HAAS machine

Machine	# of parts	Diameter (mm)	Teeth	Width of cut (mm)	Depth of cut (mm)	Feed rate mm/min	Spindle speed (RPM)
HAAS	#1-#4	20	2	10.16	2.54	710.184	2330
HAAS	#5-#7	20	2	10.16	2.54	970.483	3184

3.2. Machining data collection

Machining data were collected by performing experiments in a HAAS Mini Mill machining center. The HAAS mini mill is a 3-axis compact machining center. To generate data, milling was performed on an AISI 4340 cylindrical steel block of 177.8 mm diameter, and 20 HRC hardness using a tool with two Kennametal inserts of grade KC725M and a lead angle of 90^0. Each of these steel blocks was called a "Part." A spiral tool path was selected to generate tool wear data and to make a cylindrical boss height of 10.16 mm and a diameter of 76.2 mm. One complete spiral tool path was called a "Run," and while following the path, the tool went around the workpiece five times to reduce the diameter from 177.8 mm to 76.2 mm. The width of cut for this spiral path was 10.16 mm, and the depth of cut was 2.54 mm. At the end of the spiral tool path, when a boss of 76.2 mm diameter was obtained, the machining was stopped, and the flank wear measurement was taken. This whole process was repeated four times on a part to form a boss height of 10.16 mm and a diameter of 76.2 mm. This process was performed on the HAAS machine, and tool wear data were collected. A total of seven parts were machined, parts #1-#4 were machined for the machine settings of feed rate 710.184 mm/min and spindle speed 2330 RPM, and parts #5-#7 were machined for a feed rate of 970.483 mm/min and a spindle speed of 3184 RPM. Table 1 shows the machine settings for the experiments. In the case of the HAAS machine, the flank wear measurements were estimated using the values of the tool radius change. The tool diameter was measured after each run, and for the last run of part#7, the flank wear measurements were taken using a Keyence VHX-500, with a VH-Z100 lens and OP-72402 light ring microscope. To determine the flank wear, the tool radius change was calculated from the tool diameter measurements. In milling, the tool radius change follows the same trend as that of the flank wear. Therefore, by using the tool radius change values and the flank wear value from the last run, the flank wear values were estimated because both were assumed to have the same profile. Table 1 shows the values of the machine settings of spindle speed, width of cut, depth of cut, and feed rate, which are used to build a surrogate model for tool wear using GPTIPS.

3.3 Methodology for recursive symbolic regression

As shown in Fig. 1, the data of inputs and response variable were collected by performing experiments. Then GPTIPS was used to develop a recursive model of the input response variable, which contains the machine settings and an initial guess of the state of wear. This guess was updated iteratively, and a new recursive model was generated for the updated value of the state. To update the response variable in the input, the symbolic regression model obtained from GPTIPS was used to calculate the actual response variable values using the machine settings and the predictions of the response variables recursively for all the time steps. Fig. 1 shows the methodology for building a recursive symbolic regression model for the case of machining. Machine settings were the inputs, and the tool wear was the response variable. This method was applied to build a recursive model for tool wear for a milling operation. In the figure, "i" and "j" are the counters, where the maximum value of j is j_{max} and the maximum value of i is N (number of datapoints). $\widehat{W}$ is the predicted and W the experimental tool wear, respectively.

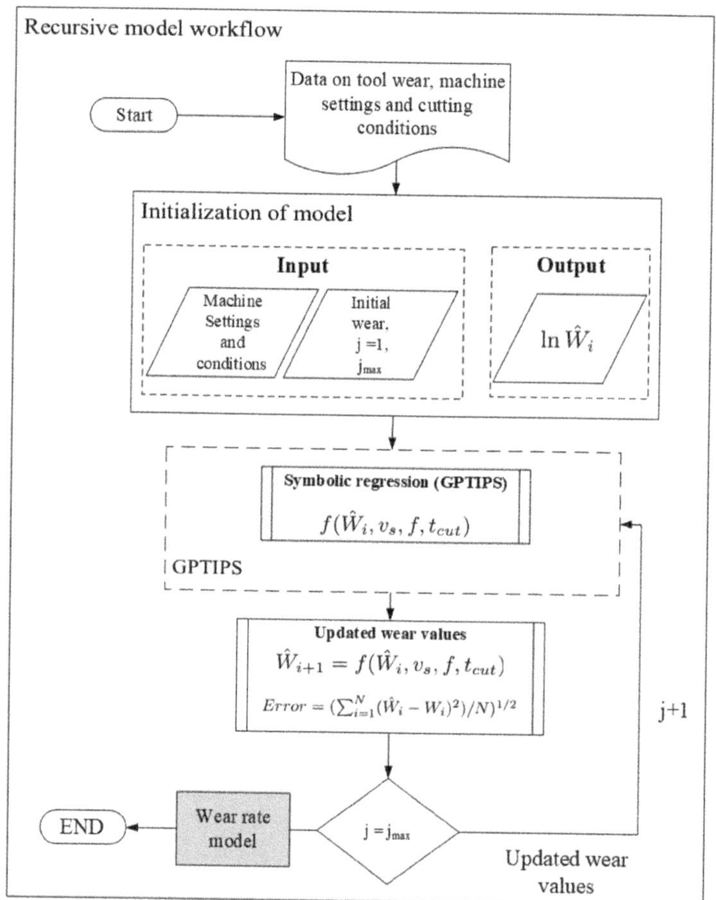

Figure 1: Workflow of the recursive model building methodology.

3.4 Surrogate model for tool wear

The recursive surrogate model for tool wear was developed using the methodology shown in Fig. 1. The inputs for model building were the spindle speed (v_s), feed rate (f), cutting time (t_{cut}) and the natural logarithm of the tool wear ($\ln \widehat{W}_i$) at the current time step. The output of the model was the natural logarithm of the current wear ($\ln \widehat{W}_{i+1}$). The logarithm of the output was taken to reduce the variability of the data and assist in model fitting. Spindle speed and feed rate were the two input settings that changed during machining and hence are included in model building. The complete experimental dataset was used for model building. Table 2 shows the basis function. exp, ln, x^3, +, *, \, used for model building. Default values were selected for crossover, elitism, mutation, and direct reproduction per-centages. The population size, number of generations, max. tree depth, and max. mutation depth were the parameters that were tuned to obtain good model fit. The number of iterations, j_{max}, were 10. After tuning the hyperparameters the best model obtained are shown in Table 2. The number of generations was 100, population size was 100, maximum tree depth was 3 and the maximum mutation depth was 2.

Table 2: GPTIPS toolbox settings for model building.

Basis functions: exp, ln, x^3, +, *, \
Population size : 100
Number of generations : 150
Tournament size : 10
Max. tree depth : 3
Max. mutate depth : 2
Crossover (%) : 84 (default)
Mutation (%) : 14 (default)
Elitism (%) : 15 (default)
Direct reproduction (%) : 2 (default)

$$\ln \widehat{W}_{i+1} = 108.438 \ln(t_{cut} + 3.0142) - 10.051\, t_{cut} - 10.051 \ln t_{cut} - 1.768 \ln t_{cut}{}^2 - 4.856 t_{cut}{}^3 \ln \widehat{W}_i - 140.614 \quad \ldots(1)$$

Eq. (1) shows the expression of the recursive tool wear model obtained. Fig. 2 shows the plot of the actual tool wear values obtained from the experiment performed on the HAAS machine and the predictions obtained from the tool wear model. The model had an R^2 value of 0.989, and as seen in the figure, the recursive tool wear model was able to capture the progression of tool wear in the machining process. The model shown in Eq. (1) is a function of the tool wear at the previous time step and cutting time. The tool wear increases with the increase in the cutting time and the term $t_{cut}{}^3 \ln \widehat{W}_i$, adds nonlinearity to the cutting profile to capture the effect of the state of the cutting tool and the cutting conditions. GPTIPS added terms t_{cut}, $\ln t_{cut}$, and $\ln t_{cut}{}^2$ to the term to fit the data. Overall, the model gives excellent agreement between predicted and measured tool wear for this process.

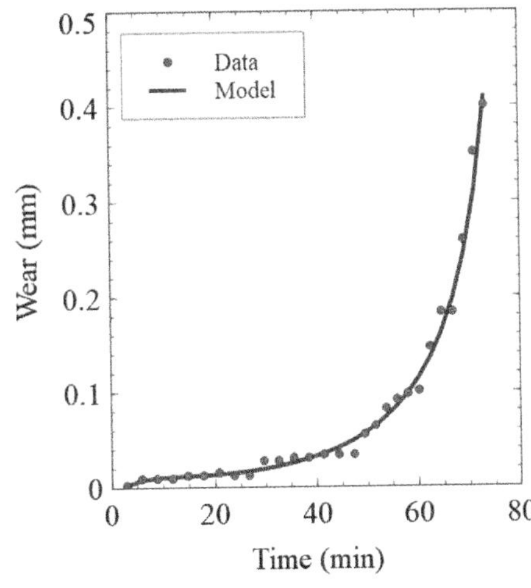

Figure 2: Tool wear predictions of the surrogate model for the wear data of the experiments performed on HAAS.

4. Conclusion

This study demonstrated the use of symbolic regression to build a recursive surrogate model using an open-source genetic programming-based toolbox GPTIPS to perform symbolic regression. This approach helped to build models in the form of simple, transparent expressions, which are interpretable due to their algebraic form. This approach was applied to a manufacturing process to build a model for tool wear. A recursive tool wear model was developed, as a function of the tool wear state at any point in the manufacturing process. The model gave excellent predictions of the tool wear with a very simple and easy to interpret function.

Acknowledgement

This material is based upon work supported by the U.S. Department of Energy's Office of Energy Efficiency and Renewable Energy (EERE) under the Advanced Manufacturing Office Award Number DE-EE0007613.We also gratefully acknowledge the Air Force Research Laboratory, Materials and Manufacturing Directorate (AFRL/RXMS) for support via Contract No. FA8650-20-C-5206Disclaimer: This report was prepared as an account of work sponsored by an agency of the United States Government. Neither the United States Government nor any agency thereof, nor any of their employees, makes any warranty, express or implied, or assumes any legal liability or responsibility for the accuracy, completeness, or usefulness of any information, apparatus, product, or process disclosed, or represents that its use would not infringe privately owned rights. Reference herein to any specific commercial product, process, or service by trade name, trademark, manufacturer, or otherwise does not necessarily constitute or imply its endorsement, recommendation, or favoring by the United States Government or any agency thereof. The views and opinions of authors expressed herein do not necessarily state or reflect those of the United States Government or any agency thereof.

References

1. Searson, D.P., Leahy, D.E. & Willis, M.J., GPTIPS: an open source genetic programming toolbox for multigene symbolic regression, Proceedings of the International MultiConference of Engineers and Computer Scientists 2010 (IMECS 2010), Hong Kong, 17-19 March, 2010.
2. Searson, D.P., GPTIPS 2: an open-source software platform for symbolic data mining, Chapter 22 in Handbook of Genetic Programming Applications, A.H. Gandomi et al., (Eds.), Springer, New York, NY, 2015.
3. Awasthi, U., Wang, Z., Mannan, N., Pattipati, K. R., & Bollas, G. M. (2022). Physics-based modeling and information-theoretic sensor and settings selection for tool wear detection in precision machining. Journal of Manufacturing Processes, 81(December 2021), 127–140.
4. Awasthi, U., and Bollas, G. M. (2020). "Sensor Network Design for Smart Manufacturing – Application on Precision Machining." *IFAC PapersOnLine* 53 (2): 11440–45.
5. Han, S., Mannan, N., Stein, D. C., Pattipati, K. R., and Bollas, G. M. (2021). "Classification and Regression Models of Audio and Vibration Signals for Machine State Monitoring in Precision Machining Systems." *Journal of Manufacturing Systems*: 45–53.
6. Han, S., Yang, Q., Pattipati, K. R., and Bollas, G. M. (2022). "Sensor Selection and Tool Wear Prediction with Data-Driven Models for Precision Machining." *Journal of Advanced Manufacturing and Processing* 4 (4): e10143.
7. Yang, Q., Pattipati, K. R., Awasthi, U., and Bollas, G. M. (2022). "Hybrid Data-Driven and Model-Informed Online Tool Wear Detection in Milling Machines." *Journal of Manufacturing Systems* 63 (December 2021): 329–43.
8. Cozad, A., Sahinidis, N. V., and Miller, D. C. (2015). "A Combined First-Principles and Data-Driven Approach to Model Building." *Computers and Chemical Engineering* 73: 116–27. https://doi.org/10.1016/j.compchemeng.2014.11.010.
9. Cozad, A., Sahinidis, N. V., and Miller, D. C. (2014). "Learning Surrogate Models for Simulation-Based Optimization." *AIChE Journal* 60 (6): 2211–27.
10. Sansana, J., Joswiak, M. N., Castillo, I., Wang, Z., Rendall, R., Chiang, L. H., and Reis, M. S. (2021). "Recent Trends on Hybrid Modeling for Industry 4.0." *Computers and Chemical Engineering* 151: 107365.
11. Schmidt, M., and Lipson., H. (2009). "Distilling Free-Form Natural Laws from Experimental Data." *Science* 324 (5923): 81–85. https://doi.org/10.1126/science.1165893.
12. Udrescu, S. M., and Tegmark, M. (2020). "AI Feynman: A Physics-Inspired Method for Symbolic Regression." *Science Advances* 6 (16). https://doi.org/10.1126/sciadv.aay2631.
13. Wilson, Z., and Sahinidis, N. (2017). "ALAMO: Machine Learning from Data and First Principles." *Process Development Symposium 2018: Applying New Technologies in Process Development*, 38–48.

Antonis Kokossis, Michael C. Georgiadis, Efstratios N. Pistikopoulos (Eds.)
PROCEEDINGS OF THE 33rd European Symposium on Computer Aided Process Engineering (ESCAPE33), June 18-21, 2023, Athens, Greece
 http://dx.doi.org/10.1016/B978-0-443-15274-0.50249-3

Demand Response in Microgrids with Attention-Based Deep Reinforcement Learning

Jiahan Xie, Akshay Ajagekar, Fengqi You

Cornell University, Ithaca, NY 14853, USA

Abstract

In this work, we propose a novel multi-agent deep reinforcement learning (MA-DRL) based approach to facilitate efficient load shaping through automated demand response in a microgrid. Achieving real-time autonomous demand response with energy management systems in buildings can be challenging due to factors like uncertain system parameters, the dynamic market price, and the complex coupled operational constraints. To develop a scalable approach for automated demand response in a microgrid, it is necessary to allow for coordination between buildings in the grid to smoothen the overall demand curve. We present a MA-DRL agent that utilizes an actor-critic algorithm incorporating a shared attention mechanism to enable an effective and scalable real-time cooperation to prevent the non-coordinated peak shifting actions in the complicated building system. A computational case study conducted with nine residential buildings revealed the attention-based MA-DRL agent's ability to execute decentralized cooperative policies without knowledge of building systems models or electricity price dynamics. Viability of the proposed control approach was also demonstrated by a reduction in net electricity consumption of over 6% accompanied by reduction in carbon emissions as compared to both conventional and state-of-the-art reinforcement learning approaches for automated demand response.

Keywords: Deep reinforcement learning, multi-agent, demand response, microgrids.

1. Introduction

As the integration of renewable energy sources into microgrids increases, demand response presents an opportunity to reduce the strain on the electric grid by reducing electricity usage in turn lowering greenhouse gas emissions (Pistikopoulos et al., 2021). By promoting the responsiveness of the buildings, demand response can lower the peak demand and the electricity consumption and thus improve the grid's stability in the long term (Yang et al., 2022). Factors like the variability of renewable energy sources, uncertainty associated with weather and load demand, as well as fluctuation of electricity prices further complicate the demand response problem (Silva et al., 2022). Demand response can be framed as an optimization problem to optimize electricity consumption or associated costs subject to coupled spatial and temporal constraints (Sun et al., 2021). Several math programming-based approaches have been previously applied in demand response for optimal scheduling of resources in a microgrid (Jordehi, 2019; Qiu, 2021). However, as the size of these problems increases with the number of buildings and the energy storage devices (Chen, 2022), metaheuristic solution techniques have also been proposed to obtain approximate solutions at the expense of reduced computational times (Shewale et al., 2020). Optimal control techniques like model predictive control have also been successfully incorporated for demand response (Qureshi et al., 2014). Such model-based approaches rely on first-principle models that are often simplified and may fail in a practical setting owing to the non-ideal behavior of individual energy components (Chu and You, 2015). As a result, it is necessary to develop a comprehensive solution approach that is capable of handling various sources of uncertainties present in a microgrid while overcoming the need for an accurate first-principle model.

Reinforcement learning (RL) is a promising candidate for demand response problems owing to its model-free, and adaptive learning ability (Vázquez-Canteli and Nagy, 2019). Deep reinforcement learning (DRL) utilizes deep neural networks as nonlinear function approximators to overcome computational intractability caused by the increase in problem size. Many DRL-based techniques have been proposed for demand response and energy management in smart microgrids. DRL techniques assume no knowledge of the system dynamics and have been shown to optimize peak load in residential buildings (Mocanu et al., 2018) and minimize energy costs (Yu et al., 2020) by inducing coordination between the HVAC system and energy storage devices. Multi-agent variants of DRL techniques have also been explored to reduce operating costs and peak load through distributed scheduling of energy storage systems (Yang et al., 2019). Multi-agent deep reinforcement learning (MA-DRL) techniques can be directly applied to demand response problems (Vazquez-Canteli et al., 2020), however, they may add dynamicity to the microgrid operation and yield sub-optimal policies. Another challenge lies in ensuring coordination and information exchange between the agents in a multi-agent setting to obtain joint optimal policies. To overcome these challenges, we propose a novel multi-agent approach for demand response in grid-responsive buildings that employ an actor-critic based deep reinforcement learning technique to produce efficient energy management policies resulting in reduced overall electrical demand. Utilizing the attention mechanism in DRL (Iqbal and Sha, 2019) allows for centralized training with a shared critic network while ensuring decentralized execution governed by the multi-agent setting. We also perform several computational experiments to demonstrate that buildings in a microgrid can better learn how to coordinate to decrease the net electricity consumption accompanied by a reduction in carbon emissions as compared to both conventional and state-of-the-art DRL approaches for automated demand response.

2. Attention-based MA-DRL for demand response

2.1. Problem formulation

Multi-agent Markov decision processes (MMDPs), the generalization of the Markov decision process to the multi-agent setting, is applied to formulate the demand response problem in a grid of interconnected buildings. An MMDP is a tuple $<S, A_1, \dots, A_N, T, R_1, \dots, R_N>$, where S is a set of states, $A_1, \dots, A_N$ is a collection of action sets for N agents, $T: S \times A_1 \times \dots \times A_N \rightarrow P(S)$ is the state transition function, which defines the probability distribution over possible next states given the current state and actions for each agent, and $R_i: S \times A_1 \times \dots \times A_N \rightarrow \mathbb{R}$ is the reward function for each agent that is a function of global state and actions of all agents. As real-world settings often feature incomplete observations, each agent i receives its private observation $o_i \in O_i$, which contains partial information from the global state, $s \in S$.

For each building i, the agent receives observation $o_{i,t}$ at time t. The observations contain the time variable accurate to an hour, district's temperature and humidity along with their associated predictions, district's solar radiation along with its prediction, buildings' indoor temperature and humidity, buildings' non-shiftable load, buildings' storage device state of charge (SoC). Each agent has up to three continuous actions:

cooling device storage, DHW device storage, and electrical device storage. Since our goal is to minimize the net electricity consumption and the carbon emissions in the district, the reward function is related to the carbon intensity in the district at time t, CI_t, net electricity consumption of building i at time t, $E_{i,t}$, and the total net electricity consumption in the whole district, $E_t = \sum_{i=1}^{N} E_{i,t}$. Empirically, we found the following reward function is helpful to reduce the electricity consumption and carbon emissions and to perform more aggressive load flattening: $500 \min(0, E_t) \times CI_t - 0.01 sign(E_{i,t}) \times |E_{i,t}| \times E_t^3$.

2.2. Multi-agent attention-based control (MAAC)

In reinforcement learning, methods that learn approximations to both policy and value functions are called actor-critic methods, where 'actor' refers to the learned policy, and 'critic' refers to the learned value function (Li, 2017). The bias introduced through bootstrapping the expected returns with temporal-difference (TD) methods is often beneficial because it reduces variance introduced by the policy gradient estimator. Specifically, the policy gradient method aims to learn a parameterized policy that can select actions without consulting a value function. The gradient estimate is expressed as follows, where θ parameterizes the policy network π.

$$\nabla_\theta J(\pi_\theta) = \nabla_\theta log(\pi_\theta(a_t \mid s_t)) \sum_{t'=t}^{\infty} \gamma^{t'-t} r_{t'}(s_{t'}, a_{t'}) \tag{1}$$

For the learned value function, the action-value function is typically used, $Q_\psi(s_t, a_t) = \mathbb{E}[\sum_{t'=t}^{\infty} \gamma^{t'-t} r_{t'}(s_{t'} - a_{t'})]$. $Q_\psi(s_t, a_t)$ is learned through TD learning by minimizing the mean squared Bellman error based on the past system transitions sampled from the experience replay buffer $\mathcal{D}$:

$$\begin{gathered}\mathcal{L}_Q(\psi) = \mathbb{E}_{(s,a,r,s')\sim\mathcal{D}}[(Q_\psi(s,a) - y)^2] \\ \text{where } y = r(s,a) + \gamma \mathbb{E}_{a'\sim\pi(s')}[Q_{\bar{\psi}}(s',a')]\end{gathered} \tag{2}$$

To encourage exploration and avoid converging to non-optimal policies, the maximum entropy framework is adopted in the soft-actor-critic algorithm [11]:

$$\begin{aligned}\nabla_\theta J(\pi_\theta) = \mathbb{E}_{s\sim\mathcal{D},a\sim\pi}[&\nabla_\theta log(\pi_\theta(a \mid s))(-\alpha log(\pi_\theta(a \mid s)) \\ &+ Q_\psi(s,a) - b(s))]\end{aligned} \tag{3}$$

where $b(s)$ is a state-dependent baseline. The loss function for TD learning of the value function is also revised accordingly:

$$y = r(s,a) + \gamma \mathbb{E}_{a'\sim\pi(s')}[Q_{\bar{\psi}}(s',a') - \alpha log(\pi_{\bar{\theta}}(a' \mid s'))] \tag{4}$$

The attention mechanism is used for each agent to query other agents for information about their observations and actions and incorporates that information into calculating its action-value function $Q_i^\psi(o,a)$ for agent i. Precisely, the action-value for agent i is calculated as follows:

$$Q_i^\psi = f_i(g_i(o_i, a_i), x_i) \tag{5}$$

where f_i is a two-layer multi-layer perceptron (MLP), g_i is a one-layer MLP embedding function, x_i denotes the total contribution from other agents, h is an MLP with V as the value matrix.

$$x_i = \sum_{j \neq i} \alpha_j v_j = \sum_{j \neq i} \alpha h(V g_j(o_j, a_j)) \tag{6}$$

The attention weight α_j compares the embedding e_j with $e_i = g_i(o_i, a_i)$, using the query-key system (Vaswani et al., 2017) and apply softmax over the similarity value between these two embeddings,

$$\alpha_j = \frac{exp\left((W_k e_j)^T W_q e_i\right)}{\sum_{j=1}^{N} exp\left((W_k e_j)^T W_q e_i\right)} \tag{7}$$

where W_q transforms e_i into a "query" and W_k transforms e_j into a "key". Because of the sharing of the set of parameters W_q, W_k, and V, all critics are updated to minimize a joint regression loss function:

$$\begin{gathered} \mathcal{L}_Q(\psi) = \sum_{i=1}^{N} \mathbb{E}_{(o,a,r,o') \sim \mathcal{D}}[(Q_i^{\psi}(o,a) - y_i)^2] \\ \text{where } y_i = r_i + \gamma \mathbb{E}_{a' \sim \pi_{\bar{\theta}}(o')}[Q_i^{\bar{\psi}}(o',a') - \alpha log(\pi_{\bar{\theta}_i}(a_{i'} \mid o_{i'}))] \end{gathered} \tag{8}$$

Likewise, the individual policies are updated by gradient ascent on the following gradient:

$$\begin{aligned} \nabla_{\theta_i} J(\pi_\theta) = \mathbb{E}_{o \sim \mathcal{D}, a \sim \pi}[\nabla_{\theta_i} log(\pi_{\theta_i}(a_i \mid o_i))(-\alpha log(\pi_{\theta_i}(a_i \mid o_i)) \\ + Q_i^{\psi}(o,a) - b(o, a_{\setminus i}))] \end{aligned}$$

where $\setminus i$ is denoted as the set of all agents except i, and $b(o, a_{\setminus i})$ is the multi-agent baseline. As a controller, each building will store transition tuples $(o_t, a_t, o_{t+1}, r_{t+1})$ into its corresponding replay buffer $\mathcal{D}$. When the replay buffer stores enough experience and the system reaches the update time T_{update}, we sample a mini-batch $\mathcal{B}$ and use the batch to update the critics and actors networks. The critics will take in all policy networks and perform enhanced updates through the attention mechanism. During the execution phase, each agent will download the latest updated policy networks and take actions only according to their local observations. This decentralized execution has manifested better computational efficiency.

3. Case study: Computational results

We compare our results with three state-of-the-art RL controllers and one Rule-based controller (RBC). The three RL controllers include Deep Deterministic Policy (DDPG) controllers, Soft-Actor-Critic (SAC) controllers, and the Multi-Agent RL with Iterative Sequential Action Selection (MARLISA) controllers [10]. The DDPG controllers and the SAC controllers do not have explicit information sharing and coordination, while the MARLISA controllers have information sharing of buildings' own future electricity consumption following a leader-follower schema. The RBCs take uniform and greedy actions, they charge 9.1% of their storage devices' maximum capacities every hour from 10pm to 8am, and release 8% of their maximum capacities every hour from 9am to 9pm. We conduct simulations with the energy demand for each building in New Orleans using a pre-recorded dataset of nine buildings. Each building consists of an air-to-water heat pump, as well as an electric heater that supplies them with DHW. The primary grid provides electricity for these devices, other electrical equipment, and appliances. Meanwhile, buildings can generate electricity using photovoltaic generation to offset some of their electricity consumption. We run simulations on the same weather data for 10 epochs to test the performance and the convergence of different RL controllers. All the RL controllers perform random action exploration for the first 250 days during the first epoch. For the SAC, MARLISA, and MAAC controllers, they perform stochastic policies for 300 more days by sampling from the policies distribution with entropy regularization. For the DDPG controllers, they have uncorrelated, mean-zero Gaussian noise for 300 more days of exploration. After 550 days into the simulation, the SAC, MARLISA, and MAAC controllers remove stochasticity and exploit their policies by taking the mean action. The DDPG controllers' actions noises scale to 0 for taking deterministic actions. The Adam optimizer is employed to update all neural networks.

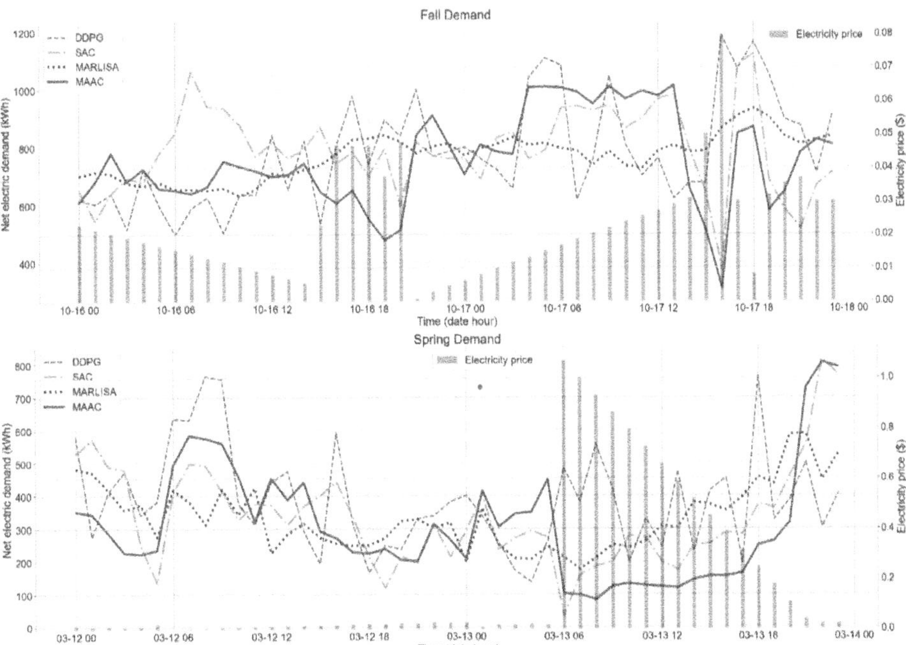

Figure 1. Net electricity consumption of the buildings in the fall and spring seasons over a period of 48 hours along with the electricity pricing.

In Figure 1, we visualize the electricity demand under different RL control algorithms for a fixed period of 48 hours during the fall and spring seasons. With the proposed controller, the electricity demand starts decreasing concurrently with the increase in price, and the demand rises back when the price is lower, during which the agent can store more energy for the preparation of the next peak price period. As the simulation results have demonstrated, all the RL controllers react appropriately to price fluctuations and reduce electric demand except the DDPG method. Since DDPG relies on all kinds of tuning and tends to overestimate the Q-values, it is common to see that the controllers exploit problematic Q-functions and have learned poor policy. Furthermore, during this high-demand period, the MAAC controllers demonstrate their strong ability compared to two other RL algorithms that are also using the soft-actor-critic framework. As the demand curves have shown, the attention-based MA-DRL controllers ensure stability and less fluctuation of the demand curve. We can conclude that the attention-based MA-DRL controllers perform better in minimizing electricity consumption while performing load shaping in the meantime. Their superior abilities to reduce energy consumption and flatten the energy demand curve are essential for the reliability of the microgrids.

4. Conclusion

In this paper, we introduced an attention-based MA-DRL demand response algorithm that can achieve better cooperation between the interconnected buildings to reduce electricity consumption and provide effective demand flattening in a decentralized way in the absence of any prior knowledge of the system model. Simulation results demonstrated the convergence and the efficacy of our proposed attention-based MA-DRL controllers. They achieve a significant reduction in net electricity consumption accompanied by a reduction in carbon emissions compared to both conventional and state-of-the-art RL approaches for automated demand response.

References

W. Chen, F. You, 2022, Sustainable building climate control with renewable energy sources using nonlinear model predictive control. Renewable & Sustainable Energy Reviews, 168, 112830.

Y. Chu, F. You, 2015, Model-based integration of control and operations: Overview, challenges, advances, and opportunities. Computers & Chemical Engineering, 83, 2-20.

J. Gong, F. You, 2015, Sustainable design and synthesis of energy systems. Current Opinion in Chemical Engineering, 10, 77-86.

S. Iqbal and F. Sha, 2019. Actor-attention-critic for multi-agent reinforcement learning. International conference on machine learning, PMLR.

A. R. Jordehi, 2019, Optimisation of demand response in electric power systems, a review, Renewable and Sustainable Energy Reviews 103: 308-319.

Y. Li, 2017, Deep reinforcement learning: An overview, arXiv preprint arXiv:1701.07274.

E. Mocanu, D. C. Mocanu, P. H. Nguyen, A. Liotta, M. E. Webber, M. Gibescu and J. G. Slootweg, 2018, On-line building energy optimization using deep reinforcement learning, IEEE Transactions on Smart Grid 10(4): 3698-3708.

E.N. Pistikopoulos, A. Barbosa-Povoa, J.H. Lee, et al., 2021, Process systems engineering – The generation next? Computers & Chemical Engineering, 147, 107252.

H. Qiu, W. Gu, F. You, 2021, Bilayer Distributed Optimization for Robust Microgrid Dispatch With Coupled Individual-Collective Profits. IEEE Transactions on Sustainable Energy, 12, 1525-1538.

H. Qiu, F. You, 2020, Decentralized-distributed robust electric power scheduling for multi-microgrid systems. Applied Energy, 269, 115146.

F. A. Qureshi, T. T. Gorecki and C. N. Jones, 2014, Model Predictive Control for Market-Based Demand Response Participation, IFAC Proceedings Volumes 47(3): 11153-11158.

C. Shang, F. You, 2019, Data Analytics and Machine Learning for Smart Process Manufacturing: Recent Advances and Perspectives in the Big Data Era. Engineering, 5, 1010-1016.

A. Shewale, A. Mokhade, N. Funde and N. D. Bokde (2020) An Overview of Demand Response in Smart Grid and Optimization Techniques for Efficient Residential Appliance Scheduling Problem. Energies 13, DOI: 10.3390/en13164266

C. Silva, P. Faria, Z. Vale and J. M. Corchado, 2022, Demand response performance and uncertainty: A systematic literature review, Energy Strategy Reviews 41: 100857.

L. Sun, F. You, 2021, Machine Learning and Data-Driven Techniques for the Control of Smart Power Generation Systems: An Uncertainty Handling Perspective. Engineering, 7, 1239-1247.

A. Vaswani, N. Shazeer, N. Parmar, J. Uszkoreit, L. Jones, Ł. Kaiser and I. Polosukhin, 2017, Attention is all you need, Advances in Neural Information Processing Systems 30.

J. R. Vazquez-Canteli, G. Henze and Z. Nagy, 2020. MARLISA: Multi-agent reinforcement learning with iterative sequential action selection for load shaping of grid-interactive connected buildings. Proceedings of the 7th ACM international conference.

J. R. Vázquez-Canteli and Z. Nagy, 2019, Reinforcement learning for demand response: A review of algorithms and modeling techniques, Applied Energy 235: 1072-1089.

S. Yang, H. Oliver Gao, F. You, 2022, Model predictive control for Demand- and Market-Responsive building energy management by leveraging active latent heat storage. Applied Energy, 327, 120054.

Y. Yang, J. Hao, Y. Zheng and C. Yu, 2019. Large-Scale Home Energy Management Using Entropy-Based Collective Multiagent Deep Reinforcement Learning Framework. IJCAI.

L. Yu, W. Xie, D. Xie, Y. Zou, Z. Sun, Y. Zhang and T. Jiang, 2020, Deep Reinforcement Learning for Smart Home Energy Management, IEEE Internet of Things Journal 7(4): 2751-2762.

Antonis Kokossis, Michael C. Georgiadis, Efstratios N. Pistikopoulos (Eds.)
PROCEEDINGS OF THE 33rd European Symposium on Computer Aided Process Engineering (ESCAPE33), June 18-21, 2023, Athens, Greece
 http://dx.doi.org/10.1016/B978-0-443-15274-0.50250-X

Multi-Kernel Canonical Variate Analysis with Bayesian Optimized Kernel Designs for Nonlinear System Identification

Jan Vincent Madayag,[a] Karl Ezra Pilario,[a,*]

[a]*Process Systems Engineering Laboratory, University of the Philippines, Diliman, Quezon City, 1101, Philippines*

**Corresponding author, email: kspilario@up.edu.ph*

Abstract

Kernel Canonical Variate Analysis (KCVA) is a widely used method for nonlinear system identification. Recently, it was shown that combining the Gaussian kernel with a linear kernel can improve the generalization ability of KCVA models. However, there are no known results for multi-kernel CVA as of yet, that is, when more than two kernels are combined to capture the nonlinearities from the process data. In this work, a multi-kernel designs involving two to four kernel functions are investigated in KCVA. Bayesian optimization (BayesOpt) is also proposed to tune kernel parameters by fitting a surrogate Gaussian process on the loss surface, and then searching the next best kernel mixture efficiently via the Expected Improvement policy. We demonstrate that BayesOpt is more data-efficient than other stochastic search methods in finding kernel designs. Results were verified using a series of empirical trials in an benchmark evaporator case study.

Keywords: system identification, multiple kernel learning, canonical variate analysis, Newll-Lee evaporator, Gaussian process regression.

1. Introduction

Most aspects of process control and operations inherently require a model of the physical process that accurately describe the process dynamics. In order to build these models, two main approaches exist, namely, first-principles modelling or system identification. Due to the highly complex nature of physical systems, first-principles modelling can become too tedious to perform. Hence, system identification (SysID) techniques are more preferred, since they only require an assumed general model structure to be fitted to the real process input-output data.

Many SysID methods already exist, but among these, the subspace methods are favorable since they only require matrix algebra to estimate the model from data, making them fast and non-iterative, whereas prediction error methods require an iterative minimization of a cost function to find an accurate model (Qin, 2006). For processes operating at widely varying operating conditions, nonlinear model structures must be assumed, since nonlinear dynamic behaviors are expected to appear in the data statistics. One of the recently developed nonlinear subspace methods is the feature-relevant mixed kernel canonical variate analysis (FR-MKCVA) (Pilario *et al.*, 2020), which proposed a new kernel design for kernel canonical variate analysis (KCVA) based subspace identification. The proposed kernel is a mixture of a global kernel, given by the linear kernel, and a local kernel, given by the Gaussian radial basis function (RBF) with automatic relevance determination. Since the kernel is a convex combination of of global and local types, it has been demonstrated to improve the generalization ability of the fitted model for the Newell-Lee evaporator case study (Al-Seyab & Cao, 2008). Even with a 25% increase in the input amplitude, the FR-MKCVA model provided superior fitness than linear CVA and recurrent neural nets. The KCVA framework has also been applied to fault detection (Samuel & Cao, 2015) due to its ability to handle nonlinear dynamic processes.

In this work, the FR-MKCVA model is extended by analyzing the impact of other kernel designs such as when the sigmoid and Laplacian kernels are included. We investigate kernel mixtures consisting of two to four kernel functions being combined in a convex manner. Hence, we propose a novel multi-kernel CVA algorithm. Note, however, that the search for kernel parameters become increasingly difficult with larger kernel mixtures. Hence, we also propose in this work a more robust and efficient way to tune kernel parameters via the use of Bayesian Optimization (BayesOpt). We compared the computational efficiency and accuracy of the subspace models when tuned by BayesOpt to those tuned by alternative stochastic algorithms commonly used in literature, such as particle swarm optimization (PSO), genetic algorithm (GA), simulated annealing (SA), and random search (RS). Using the Newell-Lee evaporator case study, the results show that BayesOpt is more efficient, hence, more recommended than other search methods, and that the addition of new kernel mixtures can help improve the accuracy of the subspace models as well.

This paper is organized as follows. Section 2 covers preliminary background on FR-MKCVA. Section 3 presents the BayesOpt method used in this work, as well as the kernel designs being investigated. Section 4 gives the results and discussion for each case study. Finally, Section 5 contains a summary of this work.

2. Feature-Relevant Mixed Kernel Canonical Variate Analysis

The FR-MKCVA method is an improved subspace identification method from the original canonical variate analysis (CVA) proposed by Larimore (1990). The idea in FR-MKCVA is to identify a nonlinear dynamic model with the state-space structure:

$$\boldsymbol{x}_{k+1} = \boldsymbol{A}\boldsymbol{x}_k + \boldsymbol{B}\boldsymbol{f}(\boldsymbol{u}_k, \boldsymbol{y}_{k-1}) + \boldsymbol{w}_k \tag{1}$$

$$\boldsymbol{y}_k = \boldsymbol{C}\boldsymbol{x}_k + \boldsymbol{D}\boldsymbol{f}(\boldsymbol{u}_k, \boldsymbol{y}_{k-1}) + \boldsymbol{v}_k \tag{2}$$

where $\boldsymbol{x}_k, \boldsymbol{y}_k, \boldsymbol{u}_k$ are the states, outputs, and inputs, respectively, $\boldsymbol{w}_k, \boldsymbol{v}_k$ are the process noise and measurement noise vectors, respectively, $\boldsymbol{A}, \boldsymbol{B}, \boldsymbol{C}, \boldsymbol{D}$ are the state-space matrices, and $\boldsymbol{f}(\boldsymbol{u}_k, \boldsymbol{y}_{k-1})$ is some nonlinear mapping of the input-output data where the mapping is defined by a set of kernel functions (Pilario *et al.*, 2020). To identify the model, the input-output data are first projected onto the kernel subspace using kernel principal components analysis (KPCA). KPCA employs a kernel function $K(\boldsymbol{x}, \boldsymbol{x}')$ to act as a dot product in some nonlinear space. Through KPCA, a nonlinear set of features can be extracted from the data as follows:

$$\boldsymbol{f}(\boldsymbol{u}_k, \boldsymbol{y}_{k-1}) = [\boldsymbol{t}_k] = \left[\sum_{j=1}^{N} \boldsymbol{\alpha}_i^T K(\boldsymbol{z}_k, \boldsymbol{z}_j')\right]_{i=1,\dots,l} \tag{3}$$

where $\boldsymbol{z}_k = [\boldsymbol{u}_k \ \boldsymbol{y}_k]^T$ is the input-output data set at the kth sampling instant, N is the number of training samples, $\boldsymbol{\alpha}$ are the KPCA loadings, and l is the number of principal components. The l is chosen according to the 99% cumulative percent variance rule. The kernel design in Eq. (3) that was proposed in Pilario *et al.* (2020) is given in Table 1. After performing KPCA, the CVA algorithm is now employed. First, the features from Eq. (3) are organized into past and future Hankel matrices, $\boldsymbol{Y}_p$ and $\boldsymbol{Y}_f$. Their covariances are given by $\boldsymbol{\Sigma}_{pp} = \text{cov}(\boldsymbol{Y}_p, \boldsymbol{Y}_p)$, $\boldsymbol{\Sigma}_{ff} = \text{cov}(\boldsymbol{Y}_f, \boldsymbol{Y}_f)$, and $\boldsymbol{\Sigma}_{fp} = \text{cov}(\boldsymbol{Y}_f, \boldsymbol{Y}_p)$. CVA proceeds by taking the singular value decomposition (SVD) of the Hankel matrix, that is $\boldsymbol{\Sigma}_{ff}^{-1/2}\boldsymbol{\Sigma}_{fp}\boldsymbol{\Sigma}_{pp}^{-1/2} = \boldsymbol{\mathcal{U}}\boldsymbol{\Sigma}\boldsymbol{V}^T$. The CVA states are now estimated from the data as $\boldsymbol{X} = [\boldsymbol{x}_k] = \boldsymbol{V}_n^T\boldsymbol{\Sigma}_{pp}^{-1/2}\boldsymbol{Y}_p$ where $\boldsymbol{V}_n$ represents the first n singular vectors from the matrix $\boldsymbol{V}$, signifying that only the subspace of the first n states are taken as the principal subspace. Finally the state-space matrices are computed from $\boldsymbol{X} = [\boldsymbol{x}_k]$, $\boldsymbol{U} = [\boldsymbol{u}_k]$, and $\boldsymbol{Y} = [\boldsymbol{y}_k]$, as follows:

$$\begin{bmatrix} \boldsymbol{A} & \boldsymbol{B} \\ \boldsymbol{C} & \boldsymbol{D} \end{bmatrix} = \text{cov}\left[\begin{pmatrix} \boldsymbol{x}_{k+1} \\ \boldsymbol{y}_k \end{pmatrix}, \begin{pmatrix} \boldsymbol{x}_k \\ \boldsymbol{t}_k \end{pmatrix}\right] \cdot \left\{\text{cov}\left[\begin{pmatrix} \boldsymbol{x}_k \\ \boldsymbol{t}_k \end{pmatrix}, \begin{pmatrix} \boldsymbol{x}_k \\ \boldsymbol{t}_k \end{pmatrix}\right]\right\}^{-1} \tag{4}$$

where $\boldsymbol{t}_k$ are the KPCA-extracted nonlinear features from Eq. (3).

3. Investigating Multi-Kernel Canonical Variate Analysis

3.1. Kernel Designs

The design of the kernel function in Eq. (3) has a large impact in the performance of the identified model. In this work, various kernel designs are tested, shown in Table 1. The first kernel, namely Linear + Isotropic Gaussian RBF, only has 2 kernel parameters to tune, which are the mixture weight ω and the kernel width γ. Since the hyper-parameter search space is only 2-D, this mixed kernel was used to compare the performance of BayesOpt against other search methods: Particle swarm optimization, genetic algorithm, simulated annealing, and random search.

The main metric for comparison is the R^2 fitness of the model to the validation data:

$$R^2 = 100\% \times \left(1 - \frac{\|\boldsymbol{y}_m - \boldsymbol{y}\|^2}{\|\boldsymbol{y}_m - \text{mean}(\boldsymbol{y}_m)\|^2}\right) \tag{5}$$

where $\boldsymbol{y}$ are the predicted outputs from the model and $\boldsymbol{y}_m$ are the reference measured outputs. The closer the R^2 value to 100%, the more accurate the model. Aside from the R^2, the optimizers are also evaluated by their computational time given that all of them are provided the same budget of 100 maximum objective function evaluations.

3.2. Bayesian Optimization

Bayesian optimization is a fast and efficient optimizer for black-box functions that are computationally expensive to sample. It has found many current applications such as in recommender systems, reinforcement learning, and hyper-parameter tuning in machine learning models (Shahriari *et al.*, 2016). For these reasons, BayesOpt is proposed in this work for tuning kernel parameters in multi-kernel CVA.

In BayesOpt, the adopted objective function is the fitness of the multi-kernel CVA model to the validation data. The goal is to find the kernel parameters $\boldsymbol{\theta}$ that maximize the R^2

value on a held-out validation data set. A surrogate Gaussian process (GP) model is chosen to estimate the mean $m(\boldsymbol{\theta})$ and variance $var(\boldsymbol{\theta})$ of the validation R^2 as follows:

$$m(\boldsymbol{\theta}) = \boldsymbol{K}_*^T[\boldsymbol{K} + \sigma^2\boldsymbol{I}]^{-1}\boldsymbol{y} \tag{6}$$
$$var(\boldsymbol{\theta}) = k(\boldsymbol{\theta}_*, \boldsymbol{\theta}_*) - \boldsymbol{K}_*^T[\boldsymbol{K} + \sigma^2\boldsymbol{I}]^{-1}\boldsymbol{K}_* \tag{7}$$

where $\boldsymbol{K}_* = k(\boldsymbol{\theta}_*, \boldsymbol{\theta})$ is the kernel matrix, $k(\boldsymbol{\theta}, \boldsymbol{\theta}')$ denotes the kernel function in the GP, σ^2 is the assumed noise variance and $\boldsymbol{y}$ contains the validation R^2 values. In this work, the chosen GP kernel is the *squared exponential* function, which is equivalent to the Gaussian RBF. Next, the Expected Improvement (EI) acquisition function is used to find the next best set of kernel parameter values $\boldsymbol{\theta}_{next}$ to sample:

$$\boldsymbol{\theta}_{next} = \text{argmax } EI(\boldsymbol{\theta}) \tag{8}$$
$$EI(\boldsymbol{\theta}) = \Delta(\boldsymbol{\theta})\Phi\left(\frac{\Delta(\boldsymbol{\theta})}{\sqrt{var(\boldsymbol{\theta})}}\right) + \sqrt{var(\boldsymbol{\theta})}\phi\left(\frac{\Delta(\boldsymbol{\theta})}{\sqrt{var(\boldsymbol{\theta})}}\right) \tag{9}$$

where $\Delta(\boldsymbol{\theta}) = m(\boldsymbol{\theta}) - y_i^* - \xi$, the ξ is an exploration/exploitation parameter, the y_i^* is the maximum R^2 observed so far at the ith iteration, and the Φ and ϕ are the cumulative density function and probability density function of the standard normal distribution, respectively. In this work, the max number of function evaluations is set to 100 and the ξ is made to decrease linearly from 20 to 0. The EI policy is to first explore the search space by favoring regions with high uncertainty, and then later on exploit on the regions where previous maxima were encountered. This balance between exploration and exploitation makes BayesOpt a data-efficient search algorithm (Shahriari *et al.*, 2016).

Table 1: Kernel functions investigated in this work.

Name and Function	Optimizer
Linear + Isotropic Gaussian RBF $K(\mathbf{z}, \mathbf{z}') = \omega(\mathbf{z}^T\mathbf{z}' + 1) + (1 - \omega)\exp\left(\frac{-\lVert\mathbf{z} - \mathbf{z}'\rVert^2}{\gamma}\right)$	RS, GA, SA, PSO, BayesOpt
Anisotropic Gaussian RBF only $K(\mathbf{z}, \mathbf{z}') = \exp\left(-(\mathbf{z} - \mathbf{z}')^T\boldsymbol{\Gamma}(\mathbf{z} - \mathbf{z}')\right)$ $\boldsymbol{\Gamma} = \text{diag}(1/\gamma_1, 1/\gamma_2, \dots)$	BayesOpt
Linear + Anisotropic Gaussian RBF (Pilario *et al.*, 2020) $K(\mathbf{z}, \mathbf{z}') = \omega(\mathbf{z}^T\mathbf{z}' + 1) + (1 - \omega)\exp\left(-(\mathbf{z} - \mathbf{z}')^T\boldsymbol{\Gamma}(\mathbf{z} - \mathbf{z}')\right)$ $\boldsymbol{\Gamma} = \text{diag}(1/\gamma_1, 1/\gamma_2, \dots)$	BayesOpt
Linear + Isotropic Gaussian RBF + Sigmoid + Laplacian $K(\mathbf{z}, \mathbf{z}') = \omega_1(\mathbf{z}^T\mathbf{z}' + 1) + \omega_2\exp\left(\frac{-\lVert\mathbf{z} - \mathbf{z}'\rVert^2}{\gamma_1}\right)$ $+ \omega_3\tanh(a\mathbf{z}^T\mathbf{z}' + b) + \omega_4\exp\left(\frac{-\lVert\mathbf{z} - \mathbf{z}'\rVert}{\gamma_2}\right)$ $\omega_1 + \omega_2 + \omega_3 + \omega_4 = 1$	BayesOpt

4. Case Study

The full description of the Newell-Lee evaporator system can be found in Pilario *et al.* (2020). It consists of multiple subsystems having 3 inputs and 3 outputs overall. The system was simualted in MATLAB Simulink and it is available online through the following link: https://www.mathworks.com/matlabcentral/fileexchange/68641-newell-lee-evaporator-system-for-system-identification.

Since the process has six variables in total, the corresponding search space dimensions for kernel parameter tuning are 2-D, 6-D, 7-D, and 8-D for each kernel mixture proposed in Table 1, respectively. The number of states for this case study is fixed at $n = 4$, based on results in Pilario *et al.* (2020).

4.1. Results on Bayesian Optimization

Fig. 1(a) compares various hyper-parameter optimizers in terms of computation time and maximum found, given the same budget of 100 max objective function evaluations. The box plots in this figure summarize the result of 10 trials in each optimizer. BayesOpt-2D is clearly more efficient than the other 2-D optimizers due to having one of the lowest computation times, yet highest maximum found. PSO was found to be the slowest, but it also found consistently high max values. Random Search, although fast, struggles to find the maximum. SA has a better performance than GA, but both are still less efficient than BayesOpt. To further illustrate the efficiency of BayesOpt-2D, Fig. 1(b) and (c) show, respectively, the EI surface and the surrogate GP surface laden with the sampled points. As shown, BayesOpt-2D concentrated more on the promising area near the maximum rather than needlessly exploring across the entire surface. Fig. 1(d) shows the *exact* objective function surface, which, upon comparing with Fig. 1(c), indicates that the surrogate GP can closely estimate the 2-D function given only the sampled points. The consistency of the result of BayesOpt-2D across 10 trials can also be verified in Fig. 1(e), since its converged solutions are more consistent than those from other optimizers.

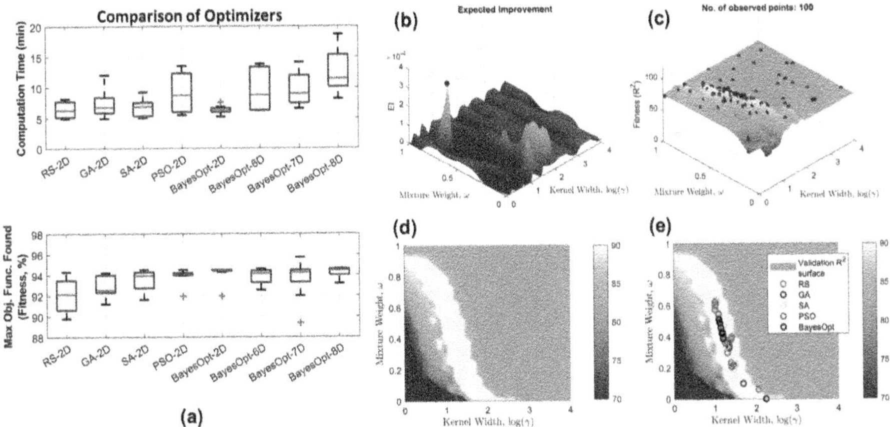

Fig. 1. Optimization results for the evaporator case study: (a) Comparison of computation time and max objective function values obtained from the 2-D, 6-D, 7-D, and 8-D hyper-parameter search repeated for 10 trials each; (b) Sample EI surface at the last iteration of a BayesOpt 2-D run; (c) A surrogate validation R^2 surface showing the 100 points where the algorithm sampled the objective function; (d) The exact validation R^2 surface, top view; (e) Same as (d) but showing the final converged solutions of the various optimizers.

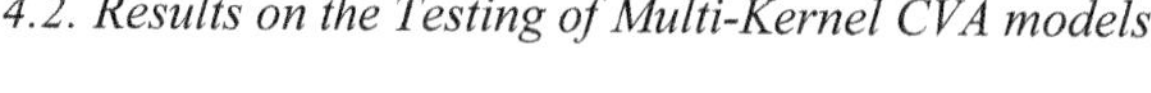

4.2. Results on the Testing of Multi-Kernel CVA models

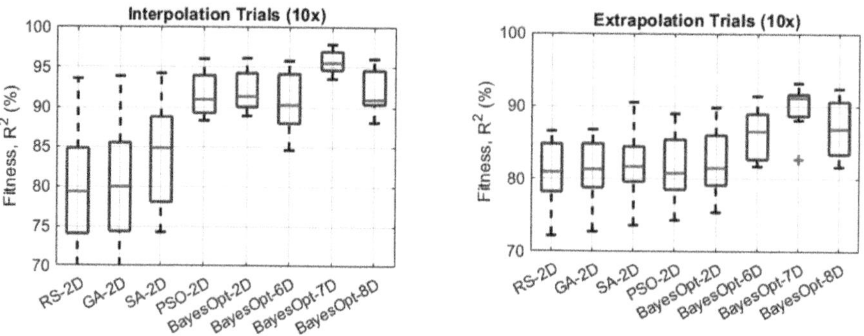

Fig. 2. Testing phase results in the evaporator case study. Each box plot summarizes 10 trials.

In Fig. 2, the validated multi-kernel CVA models are subjected to 10 interpolation data sets and 10 extrapolation data sets (with 25% increase in input amplitude) for testing. These data sets differ in the random input sequences and the random seed for noise generation. Results show that the best model is still the FR-MKCVA, but now Bayesian optimized (BayesOpt-7D) rather than subjected to random search as in Pilario *et al.* (2020). Also, the multi-kernel CVA with 4 kernels (BayesOpt-8D) was seen to be overfitting, since its performance is worse than FR-MKCVA. This indicates that 4 kernel functions may be too complex to validate with such a limited validation data set.

5. Conclusion

In this work, various multi-kernel designs were investigated for KCVA based nonlinear system identification. In addition, Bayesian optimization was demonstrated to be an efficient hyper-parameter search method for tuning these kernel designs. Using the benchmark Newell-Lee evaporator case study, the Bayesian optimized multi-kernel consisting of the linear and anisotropic Gaussian RBF was found to provide the best accuracy, as proven by both interpolation and extrapolation test data sets. However, our results also show that more complex multi-kernel designs are more difficult to tune as they become more prone to overfitting. In the future, this issue can be addressed by improving the search policy within BayesOpt for tuning the kernel parameters. Future work also includes the application of multi-kernel CVA to real-world case studies.

References

R. Al-Seyab and Y. Cao, 2008, Differential recurrent neural network based predictive control, Comput. Chem. Eng., vol. 32, no. 7, pp. 1533-1545.

W. E. Larimore, 1990. Canonical variate analysis in identification, filtering, and adaptive control. Proceedings of the IEEE Conference on Decision and Control 2, 596–604.

K.E.S. Pilario, Y. Cao, and M. Shafiee, 2020, A Kernel Design Approach to Improve Kernel Subspace Identification, IEEE Trans. On Industrial Electronics, vol. 68, no. 7, pp. 6171-6180.

S.J. Qin, 2006, An Overview of Subspace Identification, Computers and Chemical Engineering, vol. 30 (2006), pp. 1502-1513.

R. T. Samuel and Y. Cao, 2015, Kernel canonical variate analysis for nonlinear dynamic process monitoring, IFAC-PapersOnLine, vol. 28, no. 8, pp. 605-610.

B. Shahriari, K. Swersky, Z. Wang *et al.*, 2016, Taking the Human Out of the Loop: A Review of Bayesian Optimization, Proceedings of the IEEE, vol. 104, no. 1, pp. 148-175.

Antonis Kokossis, Michael C. Georgiadis, Efstratios N. Pistikopoulos (Eds.)
PROCEEDINGS OF THE 33rd European Symposium on Computer Aided Process Engineering (ESCAPE33), June 18-21, 2023, Athens, Greece
 http://dx.doi.org/10.1016/B978-0-443-15274-0.50251-1

Modular Development of Condition Monitoring Systems for the Tennessee Eastman Process

Rexonni B. Lagare,[a] Marcial Gonzalez,[b] Zoltan K. Nagy,[a] Gintaras V. Reklaitis[a]

[a]*Davidson School of Chemical Engineering, Purdue University, West Lafayette, IN 47907, USA*

[b]*School of Mechanical Engineering, Purdue University, West Lafayette, IN 47907, USA*

Abstract

This paper presents a condition monitoring system development framework that incorporates process knowledge to enhance the performance of machine learning models. Essentially, the framework uses information about the process to create a representation of the process condition, which can be broken down into modules using concepts borrowed from probabilistic graphical modeling. These modules represent simpler problems for fault detection and diagnosis, which allows traditional machine learning (ML) models to perform better without the need for a larger set of training data. Using the Tennessee Eastman Process (TEP) as a case study, the framework was shown to improve detection and diagnosis of all fault types in the TEP fault library under relevant metrics for evaluating condition monitoring systems.

Keywords: Tennessee Eastman Process, Machine Learning, Condition Monitoring, Fault Detection and Diagnosis, Condition-based Maintenance

1. Introduction

Continuous manufacturing processes often require a condition-based maintenance approach in order to fully realize its benefits. This requires the effective implementation of a condition monitoring (CM) system that can holistically oversee the condition of a process.(Schenkendorf, 2016) It is often challenging to obtain a mechanistic model of the condition of a process, so most of the work in literature focused on data-driven methods such as machine learning (ML).(Yin et al., 2012) The workflow for this development can be visualized in Figure 1, where data from the equipment and sensors of a continuous tableting line are used to train a ML model that can detect and diagnose faults. Given the right dataset, this workflow can be very effective in creating high performing fault detection and diagnosis algorithms. The problem is real scenarios rarely have the right dataset available. Hence, it is often the case that traditional ML models developed in this manner underperform, and the course of action is to acquire a better dataset, which might be prohibitively expensive, and/or to use a better ML model.

Another course of action that one could take is to build on the existing knowledge about the process to develop a mechanistic model. However, level of process knowledge that is required to do this could be even more expensive than acquiring a better dataset for training a machine learning model. Hence, any available process knowledge is often neglected because it is not enough for mechanistic modeling, and modeling efforts would be directly to a purely data-driven approach. An innovative solution would be to find a way to apply this knowledge in enhancing the performance of the ML model development workflow in Figure 1. Such a framework was developed and found to be effective for a continuous pharmaceutical tableting pilot plant. (Lagare et al., 2022a)

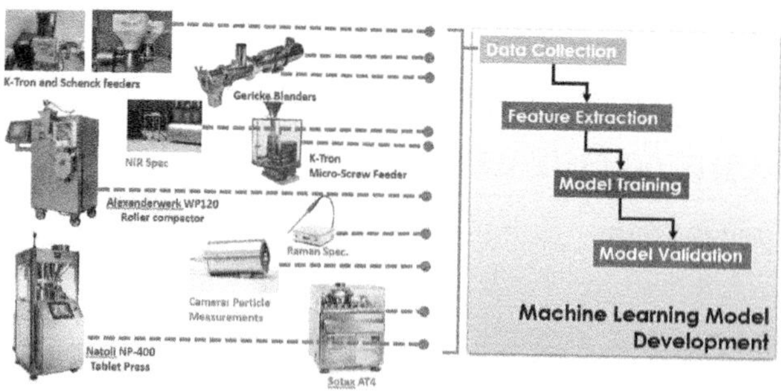

Figure 1. Machine Learning Model Development Workflow Applied to the Continuous Tableting Pilot Plant at Purdue University

It is now interesting to question the limitations of the framework, especially in its applicability towards larger processes with a larger fault library. For this purpose, the Tennessee Eastman Process (TEP) will be considered,(Downs and Vogel, 1993) which is a very different process than a continuous tableting line. First, the material streams are no longer solids, but either liquids or gases. The TEP also has more chemical components and chemical transformations are involved. There are more unit operations in the TEP and to add to its complexity, includes a recycle stream, which the pharmaceutical solids processing system does not have. Finally, the fault library of the TEP is much larger, with 22 conditions that need to be determined from 53 input variables taken from sensors and equipment across the process. (Chiang et al., 2000)

As expected, the condition monitoring system for the TEP, developed using workflow in Figure 1, showed poor performance using metrics relevant to condition monitoring. This study shows that the condition monitoring system development framework, which will be discussed in Section 2.1, was effective in improving the performance of the machine learning models in monitoring the condition of the process.

2. Methods

2.1. Condition Monitoring System Development Framework

The proposed framework adds several steps to the ML model development workflow in Figure 1, which is now depicted as a node (white) in Figure 2. The two additional steps, i.e., representation and modularization, is mainly responsible for improving the ensuing ML model development step and will be the main subject of this paper.

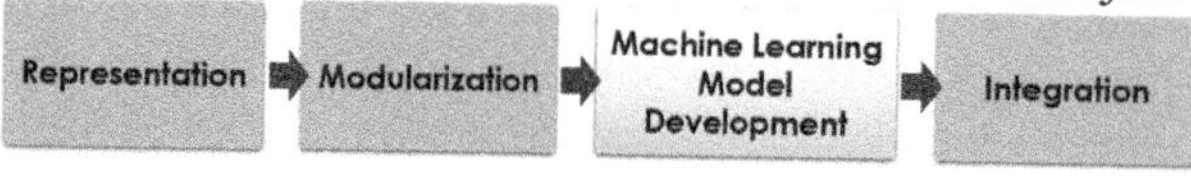

Figure 2. Proposed Condition Monitoring System Development Workflow

The final step of the proposed framework—integration—is critical in holistically interpreting the predictions of the modules. The result is a more robust condition monitoring system that can still function amid sensor maintenance repairs, reducing the need for product diversions and/or process shutdowns.(Lagare et al., 2022b)

2.1.1. Process Representation

Process representation is the first step in the CM system development framework. This step is responsible for incorporating available process knowledge into the workflow, and its key components are shown in Figure 3.

Based on these components, the minimum process knowledge requirement to be able to perform process representation is a process flow diagram (PFD). The PFD shows the material transformations that taking place, the unit operations responsible for the material transformations, the locations of the measured and manipulated variables, and the locations of the faults in the fault library. Based on the PFD of the TEP (Chiang et al., 2000), the process condition may be represented as in Figure 4.

Figure 3. Process Representation Workflow

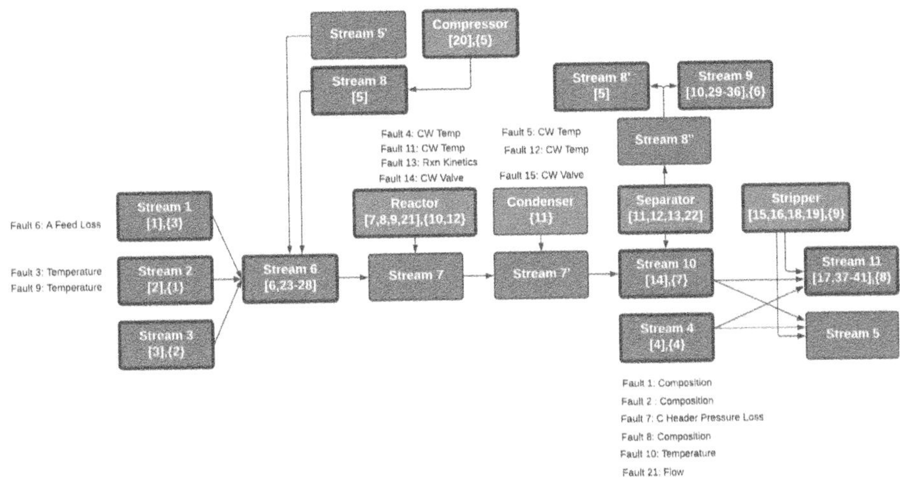

Figure 4. Process Condition Representation of the TEP depicting material and unit operation condition as blue and green nodes respectively. (The numbers in straight brackets are the designated numbers of the measured variables, and the numbers in curly brackets are the designated numbers of the manipulated variables)

2.1.2. Process Modularization

Once the process representation is available, it is now possible to modularize the process—i.e., break it down into smaller sections. Although there would be countless ways to do this, there are logical limitations to this combinatorics problem. First, a module must contain at least one fault. Without a fault, there is no condition monitoring job. It is thus interesting to see in Figure 4 that the faults are not dispersed throughout the process but are concentrated on certain nodes. With this first constraint, one can see that the condition representation of the TEP can be broken down into five modules, where the nodes that contain the fault can be considered the central node of the module.

Nodes directly adjacent to the central node may be included in the module, especially if the central node does not have input variables for ML model development. If these secondary nodes, do not have any measured variables, then other nodes that are directly adjacent to the secondary nodes may be added to the module. This process may be repeated until the added node has a measured variable or if there are no more nodes to add. This methodology is represented in Figure 5, and is consistent with the concept of d-separation in probabilistic graphical modeling, where observations in one probabilistic variable/node removes the probabilistic relationship between the parent and the child of the observed node.(Bishop and Nasrabadi, 2006) Hence, if a measured variable is involved in an added node, adding a node that is conditionally dependent on that measured variable becomes unnecessary since it gives you no further information about the central node that involves the faults for the module.

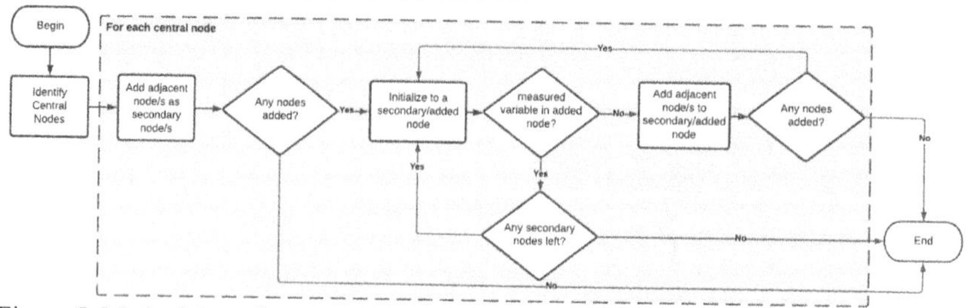

Figure 5. Methodology for modularization of process condition representation

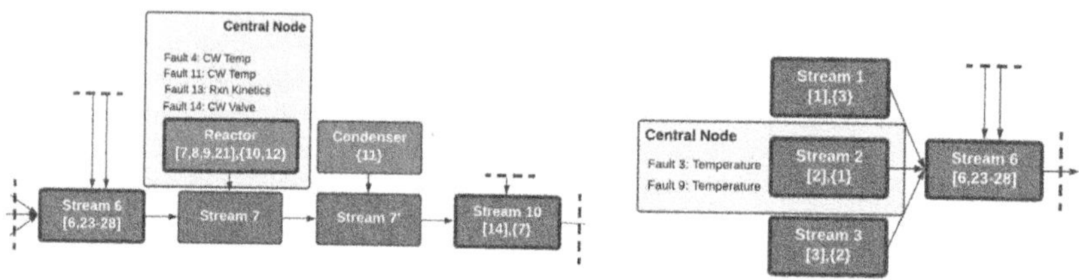

Figure 6. Reactor Condition Module

Figure 7. Stream 2 Condition Module

Another noteworthy implication of the workflow in Figure 5 is that only the central nodes are mutually exclusive among the modules, and the measured/manipulated variables in the secondary nodes may be shared. An example of a module based on the reactor condition as the central node is shown in Figure 6 and a module based on the feed stream condition is shown in Figure 7. Notice that both modules have "Stream 6" condition node as a secondary node. The dashed lines indicate the conditional independence of the central node to the rest of the condition nodes in the process condition representation.

2.1.3. Machine Learning Model Development Workflow

After the modularization step, ML classifiers are developed using the Model Builder feature by the ML.NET application. Model Builder can take the labelled input variables from each condition module to explore different traditional machine algorithms (i.e., not neural networks) and recommend the best one based appropriate classification metrics.(Microsoft, 2022)

2.2. Performance Metrics for Condition Monitoring

In this study, condition monitoring performance will be determined using five metrics: fault detection rate, false alarm rate, accuracy, normal condition certainty index (NCCI), and the overall prediction certainty index (OPCI). Fault detection rate, false alarm rate, and accuracy are standard metrics used for evaluating classifiers(Fawcett, 2006) and fault detection algorithms for the TEP.(Yin et al., 2012)

However, the two certainty indices are new metrics that is a unique contribution of this paper. These indices assume a threshold and work with classifiers (machine learning classification algorithms) that produces a probability for each possible condition in a module. For such classifiers, the condition with the highest assigned probability is considered the prediction condition and its assigned probability can be considered its prediction certainty. If the OPCI for a classifier is 0.95 for a threshold of 0.90, this means

95 out of 100 predictions made had a prediction probability/certainty higher than 0.90. The NCCI works similarly, but only considers the normal condition predictions. These certainty indices reflect the confidence of the predictions by the machine learning algorithm, which the operator can use to evaluate classifier performance. Interestingly, the NCCI has implications for novel fault detection capabilities. If the NCCI is close to 1.0, then future classification predictions that are lower than the threshold could be labelled as novel faults that would require further action by a human operator. Formulas for the OPCI and NCCI are shown below.

$$OPCI = \frac{\sum Predictions|_{Prediction\ Probability>Threshold}}{\sum Predictions} \quad \text{Equation 1}$$

$$NCCI = \frac{\sum Normal\ Condition\ Predictions|_{Prediction\ Probability>Threshold}}{\sum Normal\ Condition\ Predictions} \quad \text{Equation 2}$$

3. Results and Discussion

To evaluate the effectiveness of the framework, a base case is considered where a machine learning model is developed to use all 53 of the input variables in the TEP for predicting all 22 possible conditions (i.e., normal condition plus 21 fault types). With the framework applied, the ML development task is much simpler after modularization; for the case of the reactor module in Figure 6, the classifier only needs to classify 5 conditions (i.e., normal condition plus 4 fault types) from 16 input variables.

To determine the impact of the d-separation concept in the modularization workflow, a modified case of the framework was considered so that for each module, all the TEP variables would be used as input to determine the faults that are local to the central node of the module. For the reactor module in Figure 6, this would be using all 53 input variables to classify the 5 conditions considered in the module. This results in three cases that can be considered when evaluating the CM performance of each module. The details of these three cases when evaluating the reactor module is summarized in Table 1.

Table 1. Reactor Module Comparison Cases

	Base Case	Modular Case	Modified Modular Case
No. of Faults	21	4	4
No. of Input Variables	53	4	53

Table 2. Performance Summary for Reactor Module Faults

	Base Case	Modular Case	Modified Modular Case
Fault Detection Rate (%)	99-100	88-99	80-100
False Alarm Rate (%)	19-67	0-1	0-2
Accuracy (%)	33-99	89-100	94-99
NCCI	0.57-0.70	0.97-1.00	0.98-1.00
OPCI	0.33-0.89	0.95-1.00	0.92-1.00

For the faults local to the reactor module, the base case performance can be summarized in Table 2. While the fault detection rate is perfect, the other metrics, particularly the high rates of false alarms, make it unusable for condition monitoring. On the contrary, the modular cases show better performance across all metrics, with negligible false alarm rates and high certainty indices which indicate novel fault detection capabilities. For brevity, the performance summaries of the other modules would not be explicitly shown in this paper, albeit the aforementioned trends still hold for other modules. Overall, there seem to be no significant difference between the two modular cases, except for some faults like fault 5 which saw a lower false alarm rate by 14% compared to the modified modular case.

4. Conclusions

The condition monitoring system development framework that was originally developed for a continuous pharmaceutical tableting line proved to be effective for the Tennessee Eastman Process. The process representation workflow allowed the incorporation of process knowledge into machine learning model development. The ensuing modularization of the process representation simplified the machine learning model development task, which resulted in high performance classifiers that have novel fault detection potential.

For the impact of applying d-separation in the modularization workflow, it was not significant across all faults, although it yielded major improvements for some. Ultimately, this case study validates the applicability of the development framework across two different kinds of continuous manufacturing processes and suggests that it could be effective for other kinds of continuous processes.

5. Acknowledgements

The authors acknowledge Dr. Ziyan Sheriff for sharing valuable insights and knowledge about the Tennessee Eastman Process, which proved valuable in shaping the direction of this research.

This work was supported by the NSF under grant #2140452.

References

Bishop, C.M., Nasrabadi, N.M., 2006. Pattern recognition and machine learning. Springer.

Chiang, L.H., Russell, E.L., Braatz, R.D., 2000. Fault detection and diagnosis in industrial systems. Springer Science & Business Media.

Downs, J.J., Vogel, E.F., 1993. A plant-wide industrial process control problem. Comput Chem Eng 17, 245–255.

Fawcett, T., 2006. An introduction to ROC analysis. Pattern Recognit Lett 27, 861–874.

Lagare, R.B., Nagy, Z., Reklaitis, G. v., 2022a. Applying process knowledge for more powerful data-driven condition monitoring systems. Under preparation.

Lagare, R.B., Sheriff, M.Z., Gonzalez, M., Nagy, Z., Reklaitis, G. v., 2022b. A Comprehensive Framework for the Modular Development of Condition Monitoring Systems for a Continuous Dry Granulation Line. Computer Aided Chemical Engineering 49, 1543–1548.

Microsoft, 2022. What is ML.NET and how does it work? [WWW Document]. URL https://learn.microsoft.com/en-us/dotnet/machine-learning/how-does-mldotnet-work?WT.mc_id=dotnet-35129-website (accessed 12.1.22).

Schenkendorf, R., 2016. Supporting the shift towards continuous pharmaceutical manufacturing by condition monitoring. Conference on Control and Fault-Tolerant Systems, SysTol 2016-November, 593–598.

Yin, S., Ding, S.X., Haghani, A., Hao, H., Zhang, P., 2012. A comparison study of basic data-driven fault diagnosis and process monitoring methods on the benchmark Tennessee Eastman process. J Process Control 22, 1567–1581.

Antonis Kokossis, Michael C. Georgiadis, Efstratios N. Pistikopoulos (Eds.)
PROCEEDINGS OF THE 33rd European Symposium on Computer Aided Process Engineering (ESCAPE33), June 18-21, 2023, Athens, Greece
 http://dx.doi.org/10.1016/B978-0-443-15274-0.50252-3

Data-driven predictive model for irrigation management in greenhouses under CO_2 enrichment and high solar radiation

Ikhlas Ghiat, Rajesh Govindan, Tareq Al-Ansari*

College of Science and Engineering, Hamad Bin Khalifa University, Qatar Foundation, Doha, Qatar

**Corresponding author: talansari@hbku.edu.qa*

Abstract

Machine learning models have emerged as a viable method to predict plant water requirements with the ability to form non-linear correlations between plant response variations and microclimate conditions. An artificial neural network is used in this work to predict irrigation water requirements in a greenhouse located in a hyper-arid region with high solar radiation and encompassing an HVAC cooling system and CO_2 enrichment. The prediction model is developed from direct gas exchange measurements of transpiration and takes as input parameters microclimate data including greenhouse solar radiation, temperature, humidity and CO_2 concentration along with hyperspectral imaging-based vegetation indices. Results demonstrate the high performance of the data-driven artificial neural network (ANN) model with microclimate and vegetation index (VI) features. The ANN model resulted in a comparatively higher performance than the FAO56 Penman Monteith empirical model and linear regression with an R^2 of 91.2%, RMSE and MAE of 0.0648 mm/h 0.0528 mm/h respectively. The proposed data-driven predictive model will enable the determination of irrigation water supply in greenhouses under CO_2 enrichment and with varying microclimatic conditions, and support the implementation of precision irrigation.

Keywords: Artificial neural network, Data driven model, Vegetation index, Precision irrigation, CO_2 enrichment.

1. Introduction

Agriculture in hot and arid regions faces many challenges, primarily due to unfavorable weather conditions, inarable lands, and water scarcity. The continuous development of the agricultural system and the expansion of food production plays an important role in achieving food security (Al-Ansari et al., 2018; Karanisa et al., 2021). Hence, agricultural greenhouse systems have emerged as a viable solution to help secure food by controlling the growing medium microclimate, reducing uncertainties related to the external environment and optimising resource utilisation (Ghiat et al., 2021). Optimum irrigation scheduling is particularly necessary in regions where freshwater resources are limited and require large amounts of energy for desalination and treatment. Agricultural greenhouses in hot and arid climates still face low water use efficiencies due to the inapplicable irrigation schedules applied to these systems, and the various and complex physiological and environmental interactions, which have not been rigorously studied. The current state of irrigation scheduling in hot and arid regions is based on historical climatic data and expert observations which still poses limitations (Ali, 2010). Intelligent irrigation management such as precision irrigation has emerged as a promising method to improve the operations of agricultural greenhouses and achieve optimum water use efficiencies by ensuring the accurate amount of water inputs at the right time and space (Abioye et al., 2022). Current trends in agriculture are focused on the shift towards data driven models rather than heuristic physical models for system control and management, as they offer better predictions (Mahmood et al., 2021).

The most common methods of irrigation management are based on empirical or mechanistic evapotranspiration models that calculate crop water requirements, with the mostly used one as the Penman Monteith combination model which includes different climatic inputs in the prediction of evapotranspiration mainly solar radiation, relative humidity, temperature, and wind speed (Allen et al., 1998). Crop transpiration is a physiological measure that depends on different microclimatic factors in the greenhouse which interact nonlinearly and dynamically with each other (Stanghellini, 1987).

Machine learning methods have been adopted to estimate evapotranspiration of different vegetation crops and in different growing conditions. Models such as single layer artificial neural networks (ANN) and support vector machines (SVM) have been used because they can provide non-linear correlations between variables and eliminate the parametrisation complexities of internal factors in physical models (Fan et al., 2021). Many studies compared between different machine learning algorithms to predict plant responses. For example, Liu et al. (2009) compared artificial neural networks (ANN) and multiple linear regression (MLR) for the prediction of sap flow of pear trees based on meteorological data and soil water content. Findings from this study indicate the that the ANN model had a better performance in predicting sap flow than MLR due to its superiority in predicting non-linear relationships.

This work entails the adoption of artificial neural networks (ANN) for the prediction of transpiration rates in greenhouses under CO_2 enrichment using three different datasets: microclimate data, gas exchange data, and hyperspectral imaging data. The ANN model is compared with the FAO 56 Penman Monteith model and linear regression using statistical performance indicators.

2. Greenhouse data

The study was conducted in a semi-closed, Venlo shaped, and glass-based greenhouse located in the State of Qatar. The microclimate of the greenhouse encompasses CO_2 enrichment (600-800 ppm) and temperature and humidity are controlled by a heating, ventilation, and air conditioning (HVAC) system. The greenhouse is characterised by a hydroponic culture with cucumber as crops. To describe the fast-changing dynamics of plant responses, three different types of data were collected; 1) microclimate data, 2) transpiration rates; 3) hyperspectral-based vegetation indices. Microclimate data including greenhouse temperature, relative humidity and CO_2 concentration were collected using an aspirator box which consists of different sensors, and solar radiation using a pyranometer sensor outside the greenhouse. A gas exchange measurement system (CIRAS-3, PP systems) is used to collect transpiration measurements at the leaf level in the greenhouse. The gas exchange system is based on infrared gas analysers (IRGA) that trace gases through their absorption in the infrared region. Hyperspectral images were collected using a HSC-2 SENOP camera that takes snapshots in the visible and near-infrared region (400-1000 nm).

3. Methodology

A data-based model for transpiration is developed based on microclimate parameters including solar radiation, temperature, relative humidity, and CO_2 concentration along with vegetation indices inside the greenhouse. Several vegetation indices (VIs) belonging to different vegetation categories including broadband greenness, narrowband greenness, light use efficiency, dry or senescent carbon, and canopy water content, were calculated for greenhouse HIS images.

The predictive model is constructed using an artificial neural network (ANN) model which accounts for nonlinearity between parameters. The ANN model is a multi-layer connected network. The input layer in the ANN model receives the input signals and passes them through the hidden layer neurons which in turn transfer them to the output layer to generate the predicted output (Figure 1). Weights are initially assigned between the input layer and the first hidden layer for each input and are passed through the activation function along with the added bias in each neuron. Weights are then updated after acquiring the first predicted output to match the actual value by minimising the loss function through an optimizer.

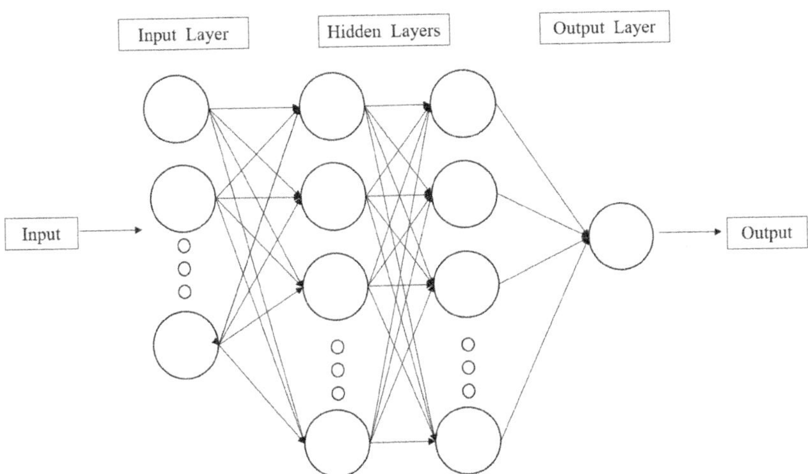

Figure 1: ANN model architecture.

The ANN model is implemented for the prediction of transpiration in the greenhouse. Two activation functions are adopted; the Rectified Linear Unit (Relu) for the input layer (Equation 1) and the hyperbolic tangent (tanh) for the hidden layers (Equation 2). Using different activation functions in the different layers can bring different non-linearities to the problem which can help better solve it. A loss function is used to measure how close the predicted transpiration value is to the actual value. The loss function used is the mean squared error. The ANN model updates the weights through the use of the Adam optimiser.

$$f(x) = \frac{2}{1+e^{-2x}} - 1 \quad [1]$$

$$f(x) = \frac{e^{x}-e^{-x}}{e^{x}+e^{-x}} \quad [2]$$

The optimal number of hidden layers and nodes were experimentally investigated through trial and error. The testing and training sets were chosen as 30% and 70% of the total dataset respectively. Moreover, a hyperparameter optimisation was conducted to select the optimal combination of number of batch size (5, 10, 15, and 20) and Epochs number (5, 10, 50, and 100). Python 3.7 was used in the implementation of the ANN model.

The ANN model is compared against the FAO56 Penman Monteith (Equation 3) empirical model and a linear regression model (Allen et al., 1998).

$$ET_c = k_c \frac{\left(0.408(R_n - G) + \gamma \frac{900}{T_{mean}+273} uVPD\right)}{\Delta + \gamma(1+0.34u)} \quad [3]$$

The comparison between the models is assessed using four statistical indicators: the coefficient of determination (R^2), the root mean square error (RMSE), and the mean absolute error (MAE) (Equations 4-6). The higher the values of R^2 (closer to 1) and the lower the values of RMSE and MAE indicate good prediction results.

$$MSE = \frac{\sum_{i=1}^{n}(y_i - \hat{y}_i)^2}{n} \quad [4]$$

$$RMSE = \sqrt{\sum_{i=1}^{n} \frac{(y_i - \hat{y}_i)^2}{n}} \quad [5]$$

$$MAE = \frac{\sum_{i=1}^{n}|y_i - \hat{y}_i|}{n} \quad [6]$$

4. Results and Discussion

In this work, an artificial neural network (ANN) model was implemented for the prediction of transpiration. The number of hidden layers and neurons were decided based on an experimental trial and error method. Two hidden layers with 3 neurons each were decided having only microclimate as input features. However, with the inclusion of vegetation indices as inputs, the dimensionality of the input layer increased, and the optimal number of hidden layers was found as 2 with 6 neurons. The next step entailed the optimisation of the hyperparameters batch size and Epoch number. The hyperparameter optimisation on the first dataset led to a batch size of 5 and an Epochs of 100 as the optimal parameters having the lowest RMSE value.

The ANN model reported good forecasting accuracy as compared to both the FAO56 Penman Monteith (PM) empirical model and the linear regression as shown in Figure 2. The linear regression model reported an R^2 of 0.680, and an RMSE and MAE values of 0.114 mm/h and 0.0893 mm/h respectively (Table 1). The data-driven model, although it is a simple linear regression demonstrated a better performance than the empirical PM model. With the inclusion of vegetation indices, the performance of the linear regression model improved to an R^2 of 0.723, an RMSE of 0.106 mm/h and MAE of 0.082 mm/h.

The ANN model with microclimate data reported values of R^2 of 0.747, RMSE of 0.109 mm/h and MAE of 0.0868 mm/h. In the case where vegetation indices were included as input features, the ANN model was able to fit 91.2% of the data with low tabulated errors of 0.0648 mm/h for RMSE and 0.0528 for MAE (Table 1). The inclusion of vegetation indices increased the ability of the ANN model to predict transpiration variability by 22% (Figure 3). Moreover, the comparative performance of the ANN model including VIs with the FAO56 Penman-Monteith model reveals a higher R^2 by 44%, and lower RMSE and MAE by 68% and 66% respectively.

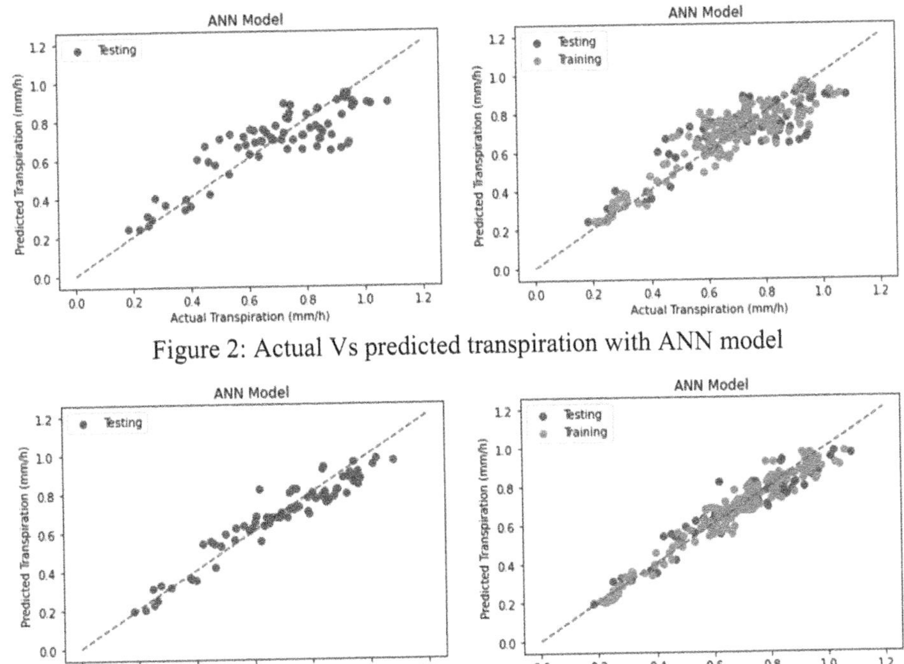

Figure 2: Actual Vs predicted transpiration with ANN model

Figure 3: Actual Vs predicted transpiration with ANN model with VIs

Table 1: Model performance on testing set.

Model	RMSE (mm/h)	MAE (mm/h)	R^2
Linear regression model	0.114	0.0893	0.680
Linear regression with VIs	0.106	0.0820	0.723
ANN model	0.109	0.0868	0.747
ANN model with VIs	0.0648	0.0528	0.912

Conclusion

Food insecurity is one of the most pressing global challenges. Hot and arid regions are faced to additional environmental limitations for local food production. These regions suffer from climate dependent food production systems that entail large utilisation of natural resources. Irrigation water requirements is one of the most critical inputs in agricultural food systems and needs to be closely assessed, especially for water scarce countries. This work entails the prediction of irrigation water requirements in terms of plant transpiration for a greenhouse with CO_2 enrichment located in a hot and arid climate. A data-driven model is developed from microclimate, physiological, and hyperspectral data using artificial neural networks to predict transpiration in the greenhouse. Results demonstrate the low performance of the FAO56 Penman Monteith empirical model in predicting transpiration for agricultural greenhouses with CO_2 enrichment under high solar radiations. The data-driven artificial neural network (ANN) model with microclimate and vegetation index (VI) features presented a comparatively higher performance than the empirical ET model by increasing the model's ability to explain data variability by 44%, and reducing errors corresponding to RMSE and MAE by 68% and 66% respectively.

References

E.A. Abioye, O. Hensel, T.J. Esau, O. Elijah, M.S.Z. Abidin, A.S. Ayobami, O. Yerima, & A. Nasirahmadi, 2022, Precision Irrigation Management Using Machine Learning and Digital Farming Solutions, AgriEngineering, 4, 1, 70–103. https://doi.org/10.3390/agriengineering4010006

T. Al-Ansari, R. Govindan, A. Korre, Z. Nie, & N. Shah, 2018, An energy, water and food nexus approach aiming to enhance food production systems through CO2 fertilization. Computer Aided Chemical Engineering, 1487–1492. https://doi.org/10.1016/B978-0-444-64235-6.50259-X

M.H. Ali, 2010, Crop Water Requirement and Irrigation Scheduling. Fundamentals of Irrigation and On-farm Water Management, *1*, 399–452, Springer New York. https://doi.org/10.1007/978-1-4419-6335-2_9

R. Allen, L. S. Peirera, D. Raes, & M. Smith, 1998, Crop evapotranspiration - Guidelines for computing crop water requirements, FAO - Food and Agriculture Organization of the United Nations.

J. Fan, J. Zheng, L. Wu, & F. Zhang, 2021, Estimation of daily maize transpiration using support vector machines, extreme gradient boosting, artificial and deep neural networks models, Agricultural Water Management, 245, 106547. https://doi.org/10.1016/j.agwat.2020.106547

I. Ghiat, H. R. Mackey, & T. Al-Ansari, 2021, A Review of Evapotranspiration Measurement Models, Techniques and Methods for Open and Closed Agricultural Field Applications, *Water*, *13*, 18, 2523. https://doi.org/10.3390/w13182523

T. Karanisa, A. Amato, R. Richer, S. Abdul Majid, C. Skelhorn, & S. Sayadi, S, 2021, Agricultural Production in Qatar's Hot Arid Climate, *Sustainability*, *13*, 7, 4059. https://doi.org/10.3390/su13074059

X. Liu, S. Kang, & F. Li, 2009, Simulation of artificial neural network model for trunk sap flow of Pyrus pyrifolia and its comparison with multiple-linear regression, Agricultural Water Management, 96, 6, 939–945. https://doi.org/10.1016/j.agwat.2009.01.003

F. Mahmood, R. Govindan, & T. Al-Ansari, 2021, Predicting Microclimate of a Closed Greenhouse Using Support Vector Machine Regression, Computer Aided Chemical Engineering, 1229–1234. https://doi.org/10.1016/B978-0-323-88506-5.50189-3

C. Stanghellini, 1987, Transpiration of Greenhouse Crops: an aid to climate management. Wageningen University.

Antonis Kokossis, Michael C. Georgiadis, Efstratios N. Pistikopoulos (Eds.)
PROCEEDINGS OF THE 33rd European Symposium on Computer Aided Process Engineering (ESCAPE33), June 18-21, 2023, Athens, Greece
 http://dx.doi.org/10.1016/B978-0-443-15274-0.50253-5

Greenhouse temperature regulation in the presence of uncertainties using data-driven robust model predictive control

Farhat Mahmood, Rajesh Govindan, Amine Bermak, David Yang, *Tareq Al-Ansari

College of Science and Engineering, Hamad Bin Khalifa University, Qatar Foundation, Doha, Qatar
**Corresponding author: talansari@hbku.edu.qa*

Abstract

Closed environment agriculture is gaining popularity due to the exponentially increasing demand for food, a growing shortage of water, urbanization, reduction in arable land, and degrading soil quality. Greenhouses are closed environment agriculture that provides a viable solution to these problems by maintaining favorable growing conditions. However, maintaining optimum conditions inside a greenhouse is a resource-intensive process, especially in hot and arid climates. Microclimate management, especially the temperature, requires a model-based systematic approach for consistent performance. Therefore, model predictive control is an effective method of managing greenhouse temperature; however, it requires a detailed system model. Moreover, the model predictive control strategy considers the perfect knowledge of the system while not accounting for uncertainties and disturbances. This leads to sub-optimal temperatures inside the greenhouse in the presence of uncertainties. Therefore, this study proposes a robust model predictive control framework to regulate the greenhouse temperature in the presence of external disturbances/uncertainties. An artificial neural network represents the nonlinear and dynamic greenhouse system. The model inputs are solar irradiance, ambient temperature, fan speed, and HVAC control, while the greenhouse temperature is the output. Results illustrate that the robust model predictive control algorithm has superior temperature control compared to a basic model predictive control and the existing greenhouse climate management system with a root mean squared error of 0.25 °C.

Keywords: Robust model predictive control, Energy Assessment, Greenhouse control, Robust optimization, Artificial neural network.

1 Introduction

Rapidly increasing global population and climate change are adversely affecting the energy, water, and food security of different regions worldwide. Agriculture, which consumes 70% of the total water, is one of the most significantly affected sectors due to irregular weather patterns, high temperatures, droughts, etc. Furthermore, producing crops through open-field agriculture in arid regions is difficult due to the scarcity of freshwater and the harsh external environment. Therefore, efficient and protected crop production environments are required to increase crop yield and minimize resource consumption.

In this context, greenhouses provide a favorable option by protecting the crop from the external environment. Greenhouses regulate the internal climate leading to increased production and reduction of utilized resources by 10-20% and 25-35%, respectively [1].

Greenhouses also provide the benefit of year-round crop production as the microclimate is maintained in the favorable range irrespective of the external environment. The psychrometric conditions of the greenhouse's air significantly affect the cultivated crop's development and yield. Therefore, maintaining optimum conditions, especially the temperature inside the greenhouse, is critical for crop production.

Greenhouses in arid regions require cooling for almost 9-10 months of the year due to high solar irradiance and ambient temperature. In summer, evaporative cooling becomes ineffective; therefore, some greenhouses use active mechanical heat, ventilation, and air conditioning (HVAC) technologies. HVAC technologies perform better than conventional approaches such as evaporative cooling, fogging, etc. however they have a higher capital cost and are more complex to operate.

Different automatic control approaches have been adopted to operate HVAC technologies to improve their performance. One of the most effective control approaches is model predictive control (MPC) which predicts and determines the optimum control. MPC has been successfully implemented in many fields, such as agriculture, aviation, chemical plants, etc.

Mahmood et al. [2] adopted an MPC approach based on a multi-layer perceptron model for a semi-closed greenhouse. Results demonstrated that the MPC performed better than the existing greenhouse climate management system for winter and summer. Similarly, Jung et al. [3] developed an MPC strategy based on an output feedback artificial neural network (ANN) to manage the greenhouse temperature. Results illustrated that the MPC had a root mean squared error (RMSE) value of 2.45 °C while the conventional control strategy had a value of 3.01 °C. However, the performance of MPC drops in the presence of disturbances and uncertainties in the system, leading to the sub-optimal temperature inside the greenhouse. Robust MPC (RMPC) ensures consistent and effective performance in system disturbances. Chen et al. [4] developed a particle swarm optimization-based RMPC strategy for greenhouse temperature control. A nonlinear affine model represents the greenhouse temperature. Results illustrate that the proposed strategy has a superior temperature control performance in the presence of uncertainties. Similarly, González et al. [5] proposed an MPC strategy using a state space model to control the greenhouse temperature. The proposed control strategy reduced the operational cost of the greenhouse; however, it was only applicable to a linear system. Few studies have utilized a data-driven method for greenhouse temperature control; therefore, this study proposes an RMPC framework using an artificial neural network as the system model.

2 Methodology

The following methodology is adopted in the present study.

2.1 Case study: semi-closed greenhouse

The research was conducted in a semi-closed greenhouse with an active mechanical HVAC system to control the greenhouse microclimate. The greenhouse temperature and relative humidity are measured by a sensor box installed inside the greenhouse. The external climate, such as ambient temperature, solar irradiance, and external relative humidity, is measured by different sensors installed on the outside. The measured data is stored in data loggers and, based on the external and internal climate signals, is sent to the greenhouse actuators. The speed of the ventilation fan and the temperature of the chiller are used to control the microclimate of the greenhouse. The fan speed and chiller temperature are control variables, while the solar irradiance and ambient temperature are the disturbances. The values of the measured variables are given in Table 1.

Table 1: Model parameters measured at the greenhouse site.

Parameter	**Value**	**Unit**
Solar irradiance	0 – 1051	Wm^{-2}
Ambient temperature	8.1 – 46.5	°C
Fan speed	25 – 100	%
Chiller temperature	17 – 22	°C
Greenhouse temperature	12.1 – 26.8	°C

The greenhouse operates in a closed cycle system where moisture is extracted from the air through a humidifier and then heated to a suitable temperature using an electric heater before returning it to the plantation area. The greenhouse schematic is shown in Figure 1.

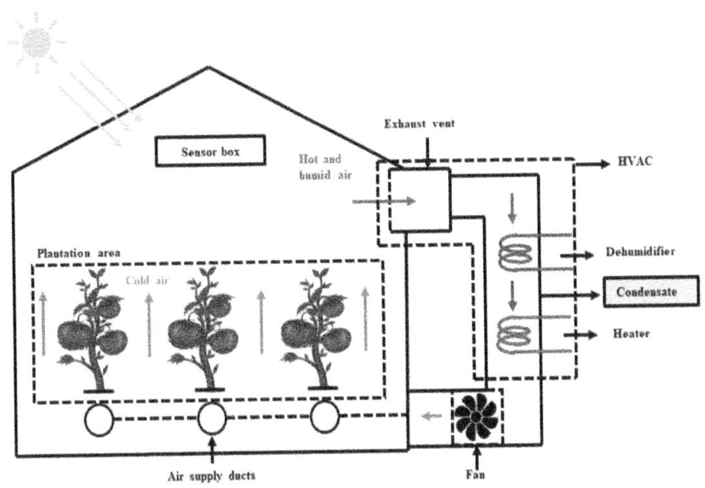

Figure 1: Greenhouse operational schematic. The hot and humid air from the greenhouse is processed through a dehumidification system to remove the moisture. The treated air is then subjected to heating to reach a suitable temperature before being circulated back into the plantation area.

2.2 Greenhouse model

An ANN is utilized to predict the dynamic greenhouse temperature. The ANN has solar fan speed, irradiance, ambient temperature, and chiller temperature as the inputs, while the greenhouse temperature is the output. The greenhouse temperature depends on the previous and current values of the control variables (u) and system disturbances (d). The output is given as follows:

$$\hat{y}(t+1) = f(u(t), u(t-1) \dots u(t-n_t), d(t-1), \dots d(t-n_d)) \qquad (1)$$

The structure of the ANN is illustrated in Figure 2.

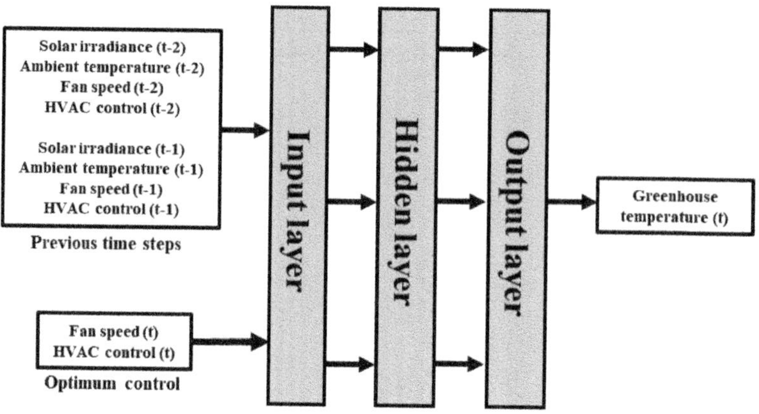

Figure 2: ANN structure

2.3 Robust model predictive control

MPC control strategy does not account for the possible uncertainties existing in the system, which may lead to sub-optimal temperatures inside the greenhouse. MPC strategy can be converted into an RMPC by considering the uncertainties in the system. In the present study, a minimax objective function is utilized, which minimizes the worst-case cost and is given by:

$$\underset{u}{minimize} \quad \underset{k=1,2,...K}{maximize} \left\{ \sum_{i=1}^{P} ||r(t+i) - \hat{y}(t+1)||^2 \right. \tag{2}$$

Where r is the reference trajectory, $\hat{y}$ is the system output, and k= 1, 2..., K are the different realizations of the system in the presence of uncertainty to which robust optimization is applied to find the control setting.

$$u_{1min} \leq u_1(t+i) \leq u_{1max} \qquad p = 0, \ldots, N_u = 1, \tag{3}$$

$$u_{2min} \leq u_2(t+i) \leq u_{2max} \qquad p = 0, \ldots, N_u = 1, \tag{4}$$

The ventilation fan (u_1) operates between 25-100% while the chiller temperature (u_2) operates between 17-22 °C. The particle swarm optimization (PSO) algorithm determines the optimum control setting at each time step.

3 Results and discussions

The proposed control strategy was implemented in Python 3.7. The proposed RMPC strategy is compared with the greenhouse climate management system (GCMS) and a basic MPC strategy.

3.1 ANN prediction performance

The accuracy of the system model is critical in the performance of the control strategy. The model is evaluated for 17-18th June using 21312 data samples, having ten inputs and the greenhouse temperature as the output. The data set is split into 70% training and 30% testing data sets. The performance of ANN for two days is illustrated in Figure 3. The ANN slightly underestimates the greenhouse temperature during the peak hours while showing very high accuracy at night. The ANN has an RMSE value of 0.39 °C and R^2 value of 0.98.

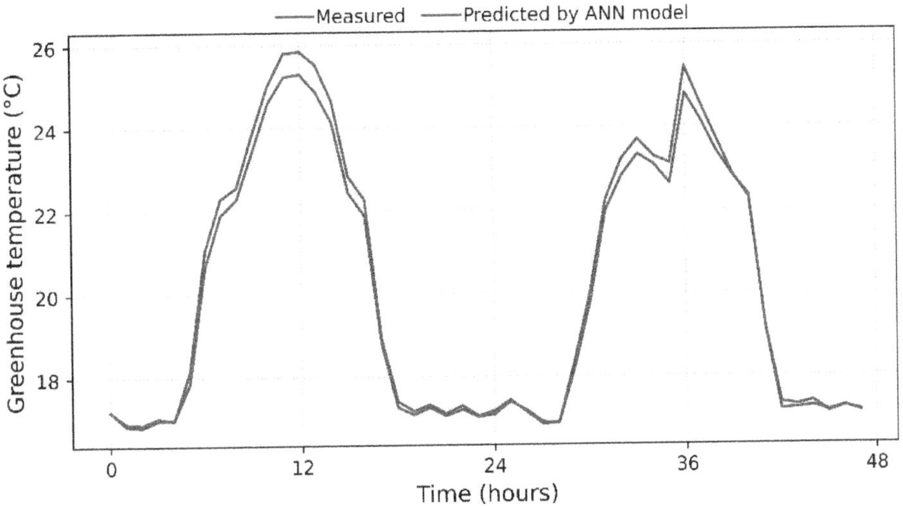

Figure 3: ANN temperature prediction comparison

3.2 Robust model predictive control

The greenhouse is in a desert area, and due to higher temperatures, sandstorms are frequent. The solar irradiance reaching the greenhouse fluctuates during the sandstorm due to the dust particles present in the air. The varying solar irradiance affects the greenhouse temperature as the existing climate management does not account for the changing cooling load. The fluctuation leads to sub-optimal temperature during the day, and inefficient use of resources as the amount of cooling required varies. The GCMS also fails to maintain the desired temperature at night as it operates on predetermined control settings. Therefore, an RMPC strategy is proposed to improve the greenhouse's performance in the presence of uncertainty and disturbances. To analyse the performance of the proposed control strategy, additive noise is added to the SI to replicate the sandstorm fluctuations. Figure 4 illustrates the temperature control comparison between the existing greenhouse climate management system, MPC, and RMPC strategy for a two-period (30-31st March). RMPC leads to better temperature control performance as compared to GCMS and MPC. RMPC, MPC, and GCSM have RMSE values of 0.25, 0.285, and 2.32 °C.

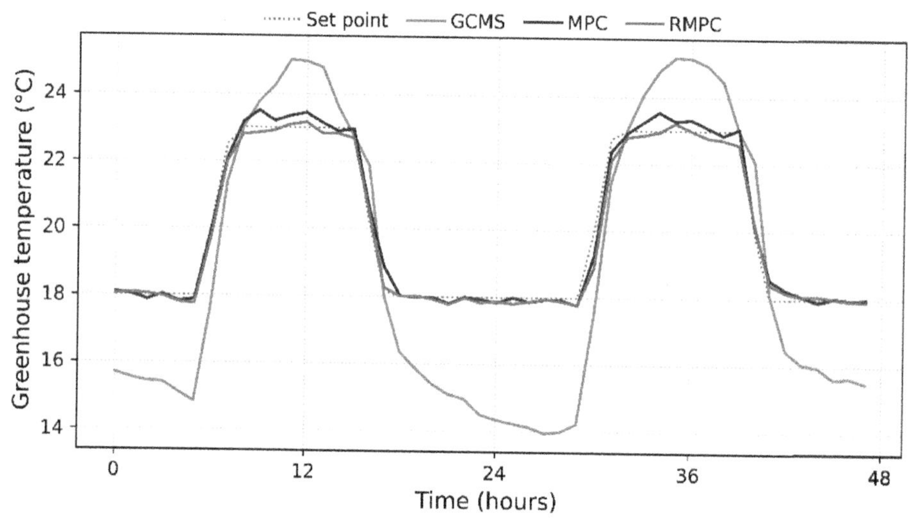

Figure 4: Temperature control performance comparison

4 Conclusion

This study proposes a data-driven robust model predictive control strategy for controlling the greenhouse temperature in the presence of uncertainties. An artificial neural network is used to predict the nonlinear greenhouse temperature. Results illustrate that the minimax-based robust model predictive control has superior performance compared to the greenhouse climate management system and basic model predictive control strategy in the presence of uncertainties due to the external environment. In future studies, the performance of the control strategies in terms of energy consumption will be analysed along with temperature control.

Acknowledgment

The authors acknowledge the support provided by Hamad bin Khalifa University, Education, City Doha, Qatar, a member of the Qatar foundation. The research is funded by Qatar National Research Fund (MME01-0922-190049).

References

[1] Mahmood F, Al-Ansari T. Design and analysis of a renewable energy driven greenhouse integrated with a solar still for arid climates. Energy Convers Manag 2022;258:115512. doi:10.1016/j.enconman.2022.115512.

[2] Mahmood F, Govindan R, Bermak A, Yang D, Khadra C, Al-Ansari T. Energy utilization assessment of a semi-closed greenhouse using data-driven model predictive control. J Clean Prod 2021;324. doi:10.1016/j.jclepro.2021.129172.

[3] Jung DH, Kim HJ, Kim JY, Lee TS, Park SH. Model predictive control via output feedback neural network for improved multi-window greenhouse ventilation control. Sensors (Switzerland) 2020;20. doi:10.3390/s20061756.

[4] Chen L, Du S, He Y, Liang M, Xu D. Robust model predictive control for greenhouse temperature based on particle swarm optimization. Inf Process Agric 2018;5:329–38. doi:10.1016/j.inpa.2018.04.003.

[5] González R, Rodríguez F, Guzmán JL, Berenguel M. Robust constrained economic receding horizon control applied to the two time-scale dynamics problem of a greenhouse. Optim Control Appl Methods 2014;35:435–53. doi:10.1002/oca.2080.

Antonis Kokossis, Michael C. Georgiadis, Efstratios N. Pistikopoulos (Eds.)
PROCEEDINGS OF THE 33rd European Symposium on Computer Aided Process Engineering (ESCAPE33), June 18-21, 2023, Athens, Greece
 http://dx.doi.org/10.1016/B978-0-443-15274-0.50254-7

A model predictive control framework for the production of functional, safe and sustainable nanomaterials

Argyri Kardamaki[a], Athanassios Nikolakopoulos[a], Mihalis Kavousanakis[a], Philip Doganis[a], Haralambos Sarimveis[a]

[a]*9 Heroon Polytechniou Street, School of Chemical Engineering, National Technical University, Athens 15780, Greece*

hsarimv@central.ntua.gr

Abstract

In recent years, the production of nanomaterials (NMs) has gained a considerable amount of attention and it has emerged as a flourishing field of research. The increasing interest in nanotechnology and the wide range of innovative industrial applications that NMs offer have highlighted the need to further study NM formation and develop advanced control schemes that can address multiple industrial objectives. This paper presents a new control framework for the production of functional, safe and sustainable silver NMs (Ag NMs) which combines dynamic first-principles modeling, the Model Predictive Control (MPC) architecture and the ability to integrate sensors that provide real-time toxicity data regarding the safety of the produced NMs.

Keywords: Nanomaterials, safety, control

1. Introduction

The NM case study presented in this paper focuses on the production of Ag NMs through a series of five plug flow (PFR) reactors. The inlet streams of the 1st reactor are solutions of the reducing and stabilizing agents, sodium citrate and tannic acid, and the precursor, silver nitrate. All subsequent PFRs are only fed with the precursor solution. The six flows of the above inlet streams are used as the manipulated variables of the process whereas the corresponding concentrations are regarded as measured disturbances that can occur during production. The main objective of this study is to produce Ag NMs of specific diameters that comply with the production and safety specifications. To that end, the mean diameter of the Ag NMs in the product stream of the last PFR is set to be the controlled variable.

2. Methods

2.1. First principles modeling

The development of a dynamic process model is essential for the design of the control strategy. A detailed dynamic model of the process was developed based on the discretisation of the PFRs corresponding to the nucleation and growth stages of the Ag NMs production system and the formulation of the respective mass balances. The kinetic expressions were estimated using the available steady state experimental data. It was assumed that the formation of Ag NMs follows the Finke-Watzky two Step Mechanism (Thanh et al, 2014), which is a process of nucleation and growth where both steps occur simultaneously. The first is a slow continuous nucleation (Eq. 1), and the second is the fast autocatalytic surface growth (Eq. 2) which is not diffusion controlled (Sandoe et al, 2019). The first-principles model describing the system consists of the mass balances of the nuclei (*A*) and the NMs (*B*), while the ordinary differential equation describing the diameter growth depends on the NMs concentration and the autocatalytic surface growth rate.

$$A \overset{k_1}{\rightarrow} B \ (1), \qquad A + B \overset{k_2}{\rightarrow} B \ (2)$$

The above system results in sigmoidal kinetic curves with the FW two-step kinetic model of slow nucleation and fast autocatalytic growth (Watzky and Finke, 2016), with respective, pseudo-elementary nucleation average rate constant, k_1 and surface growth average rate constant, k_2. Experiments show that with increasing concentration of sodium citrate and/or tannic acid a decrease is observed in both the particle size and UV-vis peak position. Also, the higher concentration of metal atoms within the growing nanoparticle an increase in the particle size was seen (growth favored over nucleation). For the PFR of volume *V*, the mass balances for *A* and *B* and the diameter *D* changes are described by the Partial Differential Equations (PDEs) (3), (4) and (5), where vo is the volumetric flow rate of the sum of the solutions flowing in the reactor:

$$\frac{\partial C_A(t)}{\partial t} = -vo\,\frac{\partial C_A(t)}{\partial V} + - k_1 C_A(t) - k_2 C_A(t)\, C_B(t) C_{TA}(t)^{n_{TA}} C_{SC}(t)^{n_{SC}} \quad (3)$$

$$\frac{\partial C_B(t)}{\partial t} = -vo\,\frac{\partial C_b(t)}{\partial V} + k_1 C_A(t) + k_2 C_A(t)\, C_B(t) C_{TA}(t)^{n_{TA}} C_{SC}(t)^{n_{SC}} \quad (4)$$

$$\frac{\partial D(t)}{\partial t} = -vo\,\frac{\partial D(t)}{\partial V} + k_3 C_A(t) C_B(t) C_{TA}(t)^{n_{TA}} C_{SC}(t)^{n_{SC}} \quad (5)$$

where $C_A(t), C_B(t), C_{TA}(t), C_{SC}(t)$ are the concentrations of nuclei, Ag NM, tannic acid and sodium citrate respectively. The constants k_1, k_2 and k_3 are given by the Arrhenius equation, where the parameters k_{o_1}, k_{o_2} and k_{o_3} are the pre-exponential factors and E_1, E_2 and E_3 are the activation energies. The values of the parameters are estimated using steady state experimental data from the process and the Thompson Sampling Efficient Multi-Objective (TSEMO) optimization procedure (Bradford et al, 2018). Two objectives, namely the error between the predicted NM diameter values and the experimental NM values f_1 and the standard deviation of the wavelength of the peak UV-Vis absorbance f_2, are placed on the pareto front of the bi-objective optimization problem. From the pareto front, the point that is used for defining the parameter values is $x_2 = \min_{x_1}(f_2)$ $, where\ x_1 = \min_{x}(f_1)$.

The system of Eqs. (3), (4) and (5) is solved using finite differences and the Method of Lines. Each PFR is discretized along the horizontal axis into *N* intervals; elementary Continuous Stirred Tank Reactors, where the solutions are considered homogeneous. Thus, the initial system of PDEs (3 – 5) where vo is the volumetric flow rate of the solutions in the reactors is reformulated as a system of Ordinary Differential Equations (ODEs):

$$\frac{dC_{A,i}(t)}{dt} = -vo\,\frac{C_{A,i}(t) - C_{A,i-1}(t)}{\Delta V} - k_1 C_{A,i}(t) - k_2 C_{A,i}(t)\, C_{B,i}(t) C_{TA,i}(t)^{n_{TA}} C_{SC,i}(t)^{n_{SC}} \quad (6)$$

$$\frac{dC_{B,i}(t)}{dt} = -vo\,\frac{C_{B,i}(t) - C_{B,i-1}(t)}{\Delta V} + k_1 C_{A,i}(t) + k_2 C_{A,i}(t)\, C_{B,i}(t) C_{TA,i}(t)^{n_{TA}} C_{SC,i}(t)^{n_{SC}} \quad (7)$$

$$\frac{dD(t)}{dt} = -vo\,\frac{D_i(t) - D_{i-1}(t)}{\Delta V} + k_3 C_{A,i}(t) C_{B,i}(t) C_{TA,i}(t)^{n_{TA}} C_{SC,i}(t)^{n_{SC}} \quad (8)$$

where $C_{A,i}$, $C_{B,i}$, $C_{TA,i}$, $C_{SC,i}$ and D_i are the concentrations and the particle diameter at the ith interval.

2.2. Model Predictive Control Framework

Model Predictive Control (MPC) is a control strategy that solves an online optimization algorithm at each sampling instant, to compute the optimal control action that drives the predicted output to the reference. MPC stands out among other control strategies because it can handle with ease multi-input multi-output systems and can incorporate explicitly multiple constraints on the state and input variables. The MPC configuration requires a dynamic model of the plant to make predictions about the system's future behavior over a finite time horizon (Camacho et al., 2007). The rolling-horizon concept of MPC is presented graphically in Fig. 1.

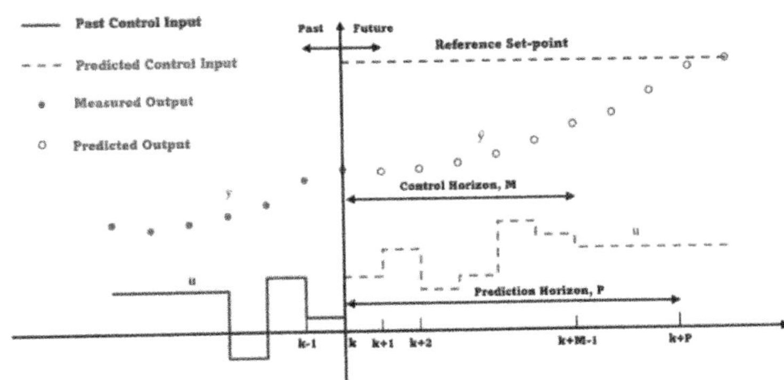

Figure 1: The rolling horizon concept of MPC

2.2.1. Dynamic Matrix Control

This study employs Dynamic Matrix Control (DMC) as the control algorithm, which is a popular MPC variant. The values of the optimized future manipulated variables are derived by minimizing the quadratic objective function featuring the output prediction errors and the incremental changes of the manipulated inputs. The optimization problem is solved and reformulated at each time instant while the controller only implements the first control action, that has been computed.

In DMC, discrete time step-response models of the plant are integrated in the control algorithm to provide the predictions for the system's future behavior. The discrete-time step response of the system that describes the model output, *y*(t), is given by Eq. (9) where g_i is the step response coefficient at each time instant *i* and Δu is the change of the manipulated variable u between two consecutive discrete time instances (Tatjewski, 2007; Camacho et al., 2007; Mulholland, 2016):

$$y(t) = \sum_{i=1}^{\infty} g_i \Delta u(t-i) \quad (9)$$

In systems that reach a steady state within a finite horizon, l_{pin}, the prediction of the output at each time instant k can be described by Eq. (10):

$$\hat{y}(t+k|t) = \sum_{i=1}^{k} g_i \Delta u(t) + \sum_{i=1}^{k} g_i{}' \Delta d(t) + \sum_{i=1}^{l_{pin}} (g_{k+i} - g_i) \Delta u(t-i) + \sum_{i=1}^{l_{pin}} (g_{k+i}{}' - g_i{}') \Delta d(t-i) + y_m(t) - y_{ss} \quad (10)$$

where $y_m(t)$ is the current state measurement, y_{ss} is the steady state, $g_i{}'$ is the step response coefficient for the measured disturbance prediction model and Δd is the change of the measured disturbance d between two consecutive discrete time instances. The measured disturbances are incorporated into the prediction model as additional input variables that cannot be manipulated.

DMC optimizes the future manipulated variables by minimizing the quadratic optimizing function presented in Eq. (11):

$$\min_{\Delta u(k),\ \Delta u(k+1),\dots,\Delta u(k+N_c)} \left\{ \sum_{k=1}^{N_p} \|\hat{y}(t+k|t) - y_{sp}\|_Q^2 + \sum_{k=0}^{N_c} \|\Delta u(t+k|t)\|_R^2 \right\} \quad (11)$$

where $\hat{y}(t)$ is the model prediction described by Eq. (10), y_{sp} is the set-point value, N_p and N_c are lengths of the prediction and control horizons and Q, R are positive definite matrices weighting the output deviations from the set-points and the incremental changes of the manipulated variables, which are defined as follows:

$$\Delta u(t+k|t) = u(t+k) - u(t+k-1) \quad (12)$$

The manipulated variables are constrained within upper and lower bounds with the following inequalities:

$$u_{min} \leq u(t+k) \leq u_{max} \quad (13)$$

The DMC algorithm was developed in MATLAB/Simulink using the external YALMIP library (Löfberg, 2004) for modelling the optimization problem. The GUROBI optimizer (Gurobi Optimization, LLC., 2022) was used to solve the problem. The step response coefficients, g_i and $g_i{}'$, were generated by introducing a step change of +10% on the initial steady state value of each input variable and by descritising the dynamic responses of the controlled variables.

2.3. Configuration of control simulation platform

The first-principles model described in 2.1 plays the role of the process plant in the integrated MATLAB/Simulink platform. Fig. 2 presents the control loop including the NM production process plant and the DMC controller. The platform can be used both as an open loop simulation to observe the dynamic behavior of the system and as a closed loop simulator, after connecting the DMC controller.

3. Results

The tuning parameters of the DMC controller were selected by implementing random sets of disturbances and different setpoint changes on the plant and observing and evaluating the responses of the controlled outputs. Fig. 3 presents an indicative disturbance rejection case study that demonstrates the successful design of the controller, if the system starts from a steady state corresponding to the production of 50 nm Ag NMs. Fig.3 presents the disturbances on the solution concentrations (bottom right), the optimal sequences of manipulated variables (bottom left) and the dynamic response of the mean diameter (top).

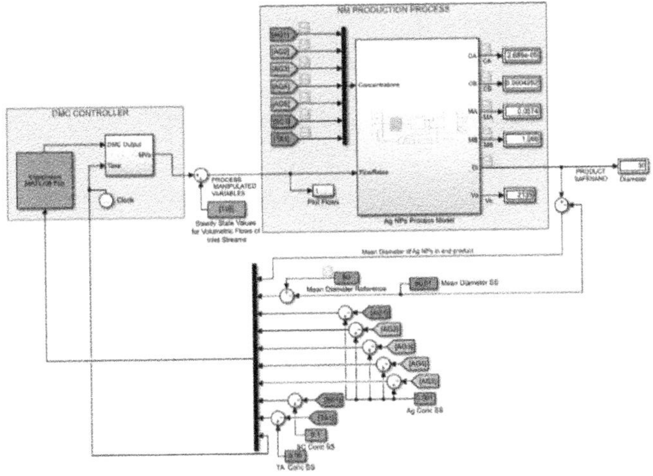

Figure 2: Integrated components of closed loop control platform

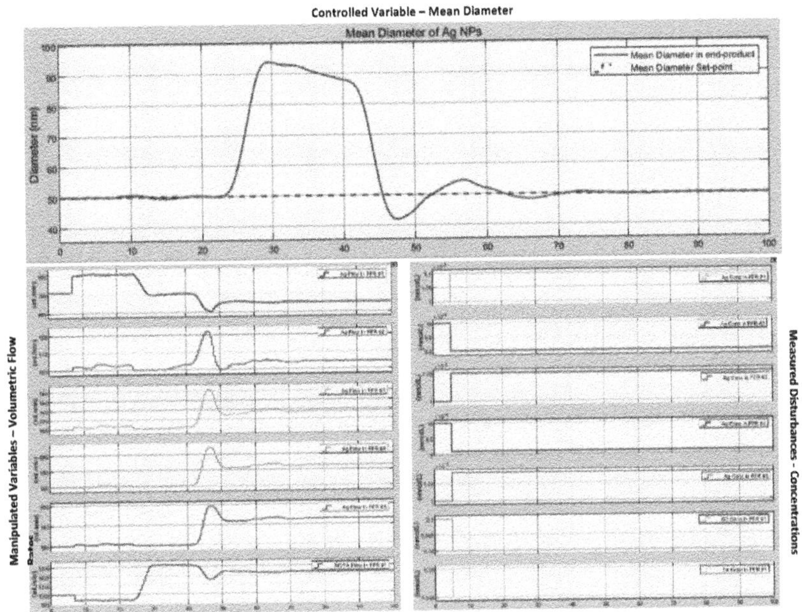

Figure 3: Dynamic response of the DMC control framework at random step changes

After five minutes of simulation time, the system is subject to a series of random step changes in all seven concentrations of the inlet streams (+10%, -4%, +5%, -9%, +8%, -10%, -10% accordingly). These disturbances tend to shift the controlled variable away from its initial state but, as Fig.3 clearly illustrates, the controller succeeds in driving it back to the desired set-point with zero offset by properly manipulating the flow rates of the inlet streams.

4. Discussion

The results demonstrate the effectiveness of the DMC controller in driving the system towards desired set-points while effectively rejecting disturbances that affect closed-loop performance. The proposed approach not only offers the standard benefits of MPC formulations, such as considering constraints on the system variables, but also allows for the incorporation of additional controlled variables relevant to safety specifications for produced NMs. The method is now extended to include feedback from a UV-Vis sensor for real-time information on NM size distribution and a biomembrane sensor for efficient toxicity screening.

Acknowledgements: This work has been supported by the SABYDOMA project which has received funding from the European Union's HORIZON 2020 Research & Innovation Programme under grant agreement no. 862296.

References

E. Bradford, A.M. Schweidtmann, A. Lapkin, 2018, Efficient multiobjective optimization employing Gaussian processes, spectral sampling and a genetic algorithm. Journal of Global Optimization, 71, 2 , p. 407–438.

E.F. Camacho, C. Bordons, 2007, Model Predictive Control. Springer London

R.J. Field, R.M. Noyes, 1977, Oscillations in chemical systems. 18. Mechanisms of chemical oscillators: Conceptual bases. Accounts of Chemical Research, 10,6, p. 214–221.

Gurobi Optimization, LLC, 2022, Gurobi Optimizer Reference Manual.

J. Löfberg, 2004, YALMIP: A Toolbox for Modeling and Optimization in MATLAB. In Proceedings of the CACSD Conference.

M. Mulholland, 2016, Applied Control: Essential Methods. John Wiley & Sons

R.M. Noyes, R.J. Field, 1977, Oscillations in chemical systems. 19. Mechanisms of chemical oscillators: Experimental examples. Accounts of Chemical Research, 10, 8, p. 273–280.

H.E. Sandoe,M.A.Watzky, M. A., S.A. Diaz, 2019, Experimental probes of silver metal nanoparticle formation kinetics: Comparing indirect versus more direct methods. International Journal of Chemical Kinetics, 51,11, p. 861–871.

P. Tatjewski, 2007, Advanced control of industrial processes. Springer London.

N.T.K. Thanh, N. Maclean, S. Mahiddine, 2014, Mechanisms of nucleation and growth of nanoparticles in solution. Chemical Reviews, 114,15, p. 7610–7630.

M.A. Watzky, R.G. Finke, 1997, Nanocluster size-control and "magic number" investigations. Experimental tests of the "living-metal polymer" concept and of mechanism-based size-control predictions leading to the syntheses of iridium(0) nanoclusters centering about four sequential magic numbers. Chemistry of Materials, 9,12, p.3083–3095.

Antonis Kokossis, Michael C. Georgiadis, Efstratios N. Pistikopoulos (Eds.)
PROCEEDINGS OF THE 33rd European Symposium on Computer Aided Process Engineering (ESCAPE33), June 18-21, 2023, Athens, Greece
 http://dx.doi.org/10.1016/B978-0-443-15274-0.50255-9

Multi-period optimisation of oligopolies with contracts: A cooperative approach to customer fairness

Asimina Marousi[a], Jose M. Pinto[b], Lazaros G. Papageorgiou[a], Vassilis M. Charitopoulos[a]

[a]*Department of Chemical Engineering, Sargent Centre for Process Systems Engineering, University College London, Torrington Place, London WC1E 7JE, UK*
[b]*Linde Digital Americas, 10 Riverview Drive, Danbury CT 06810, United States*
v.charitopoulos@ucl.ac.uk

Abstract

In an ever-increasing economic instability, cooperation among members of the same market can be proven an efficient way to build resilience. A growing interest has been observed among the process systems engineering community for integrating process design with game theory towards decentralised decision making. In this work, we aim to find the optimal supply chain design in oligopoly case studies by formulating a multi-period cooperative game among manufacturers and customers. Different fairness approaches are examined to maximise the firms' profits and customers' savings by determining the optimal customer allocation and sequencing of different contracts. Since the Nash fairness scheme results in a nonlinear objective, SOS2 approximation is employed resulting in an MILP class. The examined case studies are derived from a duopoly and an oligopoly in the liquid gas market. Computation results suggest that a cooperation can be beneficial for the payoffs of the firms and customers.

Keywords: Game theory, Fairness schemes, Supply chain optimisation, Customer allocation

1. Introduction

Fairness considerations within supply chain optimisation problems have increasingly attracted the researcher's attention from the PSE community. The interest in this direction stems from the fact that conventional centralised optimisation approaches fail to explicitly consider individual and potentially conflicting objectives of the different parties involved. Game theory is one approach that has been employed by researchers in the field of process design to evaluate both centralised and decentralised decision making. The case of decentralised decision making can be modelled as a cooperative game where all stakeholders' payoffs are considered in the objective function. Different fairness principles can be applied to allocate the payoff to the players of the game, the two main categories are egalitarian and utilitarian schemes. Egalitarian fairness schemes such as Rawslian Welfare and Schmeidler's nucleolus, aim to equally split the payoff among the players of the game. In contrast, in utilitarian schemes i.e., Nash bargaining and Shapley value, the payoff of a player depends on the contribution of the player to the coalition's payoff.

An early sample of fairness consideration in PSE can be traced in the work of Gjerdrum et al. (2001), where a Nash bargaining approach was employed to solve the problem of fair profit allocation in a multi-echelon supply chain. A common implementation of game

theory can be found in supply chain applications were stakeholders have conflicting objectives (Barbosa-Povoa and Pinto (2020)). To this end, Liu and Papageorgiou (2018) employed a lexicographic min-max approach for the fair profit distribution within a multi-echelon supply chain with firms holding different bargaining powers in the market. Finding the optimum supply chain design of chemical industries under a carbon trading policy framework was examined by Salcedo-Diaz et al. (2021). Different coalition formations were evaluated among the manufacturing firms aiming to maximise their profit while at the same time comply with CO_2 emission caps. Salcedo-Diaz and co-workers suggested that a cooperative framework is beneficial both from an economic and environmental perspective in the examined application.
The impact of the choice of different farness criteria for the profit allocation of a three-echelon closed loop supply chain was evaluated by Zheng et al. (2019) using the Shapley value, Schmeidler's nucleolus and equal satisfaction mechanism. Charitopoulos et al. (2020) considered a cooperative game to evaluate the fair profit allocation between manufacturing firms in oligopolies using Nash bargaining and a Naïve fairness scheme. The difference between an egalitarian and utilitarian approaches in the formation of an eco-industrial park for fair cost allocation was evaluated by Cruz-Avilés et al. (2021); the examined schemes where Nash bargaining, Social Welfare and Rawslian Welfare.

2. Problem Statement

The present work aims to build on the work of Charitopoulos et al. (2020) by considering a multi-period framework in which customers can be serviced by the different firms and different contracts (Calfa and Grossmann (2015)). In addition, while in Charitopoulos et al. (2020), only the firms were considered as players of the game, here we introduce the customers as additional players aiming to maximise their savings. The proposed methodology can be implemented in different oligopolistic markets of chemical industries, for that reason the model formulation introduced in the following section will be of a generic form.

2.1. Model formulation outline

The key elements of the proposed model follow the outline of the model proposed by Charitopoulos et al. (2020) and entail: a) contract modeling, b) material & inventory balances, c) power consumption constraints, d) inter-firm swapping agreements, e) plant production short-cut model, f) service cost. The modeler needs to decide on the sets: i) of the planning horizon of P time periods, ii) of firms F, producing I products at known plant capacities, iii) of existing and new customers C with deterministic demand and delivery costs, iv) of distinct type of contracts K, along with the deterministic electricity costs and power consumption limits. The aim of the study is to find the optimal customer allocation and sequencing of contracts, production planning, inventory, swaps, outsourcing and energy consumption levels in order to fairly maximise the firms' profit and customers' savings.

2.2. Fair objective function

In the present study two fairness schemes are evaluated: the Naïve and Nash bargaining. Let the parameter π_f^{sq} be the profit of firm f at the status quo market, i.e., by fixing the customer allocation to a firm and assume no new customer acquisition, and π_f the profit variable of firm f in a cooperative framework. Both π_f^{sq} and π_f are calculated for the entire duration of the examined time horizon. For the customers' savings consideration, let the parameter CC_c^{max} be the maximum cost from purchases of a customer c, and the variable CC_c be the purchase cost of a customer, both are calculated for the duration of the examined time horizon.

2.2.1. Naïve scheme

The objective function for the fair customer allocation with the Naïve fairness scheme is formulated in Eq. (1). It can be observed that in this formulation no consideration of the status quo of the market are considered nor of the maximum cost to be inflicted on the customers.

$$\Omega = \sum_f \pi_f - \sum_c CC_c \tag{1}$$

2.2.2. Nash bargaining scheme

The Nash bargaining scheme can be perceived as maximising the geometric mean of the selected payoffs and thus results in a nonlinear objective function (Eq. (4)).

$$\Phi = \prod_f (\pi_f - \pi_f^{sq}) * \prod_c (CC_c^{max} - CC_c) \tag{2}$$

A common reformulation to the Nash fairness objective is separable programming and logarithmic transformation (Liu and Papageorgiou (2018); Charitopoulos et al. (2020); Cruz-Avilés et al. (2021)). Accordingly, it can be observed that Eq. (2) is separable in f and c, thus a logarithmic approximation can be applied to the nonlinear terms respectively.

$$\Psi_1 = \ln \Phi_1 = \sum_f \ln(\pi_f - \pi_f^{sq}) \tag{3}$$

$$\Psi_2 = \ln \Phi_2 = \sum_c \ln(CC_c^{max} - CC_c) \tag{4}$$

The reformulated objective function can be derived as the summation of Eq. (3), (4) as follows:

$$\Psi = \Psi_1 + \Psi_2 = \sum_f \ln(\pi_f - \pi_f^{sq}) + \sum_c \ln(CC_c^{max} - CC_c) \tag{5}$$

The given optimisation problem with Eq. (5) as an objective function is of MINLP class, where the nonlinear terms stem from the objective only. In order to transform the problem in a tractable MILP class, a piecewise linear approximation is performed respectively in Eq. (3), (4) using SOS2 variables (Liu and Papageorgiou (2018); Charitopoulos et al. (2020)). It is noteworthy that SOS2 approximations where initially proposed to retrieve global optimum solutions of problems with a single nonlinear function in an otherwise linear programming problem (Beale and Forrest (1976)), which makes it suitable for the examined model. Let $\widetilde{\Psi}_1$ be the linear approximation of Ψ_1, over a number of grid points n, parameter $\tilde{\pi}_{fn}$ the profit of firm f at grid point n and λ_{fn} the auxiliary SOS2 variable. Hence, the approximation of Eq. (3) is derived in Eq. (6) with the complementary constraints for the SOS2 variable in Eq. (7), (8).

$$\widetilde{\Psi}_1 = \sum_f \sum_n \ln(\tilde{\pi}_{fn} - \pi_f^{sq}) \lambda_{fn} \tag{6}$$

$$\sum_f \lambda_{fn} = 1 \qquad \forall f \tag{7}$$

$$\pi_f = \sum_n \tilde{\pi}_{fn} \lambda_{fn} \quad \forall f \tag{8}$$

Analogously the approximation of the customer cost can be reformulated in Eq. (4) with based auxiliary SOS2 variable μ_{cn} and the corresponding complementary constraints over the same number of grid points n. The final linearised objective for the fair customer allocation is formulated in Eq. (9).

$$\widetilde{\Psi} = \widetilde{\Psi}_1 + \widetilde{\Psi}_2 = \sum_f \sum_n \ln(\tilde{\pi}_{fn} - \pi_f^{sq}) \lambda_{fn} + \sum_c \sum_n \ln(CC_c^{max} - \widetilde{CC}_{cn}) \mu_{cn} \tag{9}$$

3. Case studies

For the purpose of this paper two case studies from an industrial liquid market will be examined. The first case study is a duopoly with 2 firms and 97 customers. At the status quo, Firm A serves 44 customers and Firm B 37, while there are 16 free customers that allow a market share growth. In the second case study, an oligopoly with 3 firms and 81 customers, of whom 13 are new. At the status quo, Firm A holds 21 customers, Firm B 17 and Firm C 30. The selected time horizon is one year discretised into 12 monthly time intervals. Three different contract formulations are evaluated i.e., Open, Formula and Firm, with different duration and pricing as proposed by Marousi et al. (2023). The computational experiments were carried in an Intel®Core™i9-10900K CPU @ 3.70GHZ machine using GAMS-IDE with CPLEX v.38.2 with 12 threads. The optimality gap was set to 1%. The number of grid points for both case studies was selected as *n=1000*. The original MINLP problem was addressed with the use of global solver BARON, however no convergence was achieved in reasonable computational time. The aspects that are of interest in this study is the impact of the different fairness schemes on the profit allocation between the firms and the savings of the customers and at the same time how does contract selection facilitate fairness allocation.

3.*1*. *Duopoly*

Initially it can be observed from Table 1, for the Naïve objective formulation (Eq. (1)), that Firm B has a significant profit increase while Firm A has an only a marginal improvement. This results in the reversal of the market share structure compared to the status quo. In contrast, using the Nash bargaining scheme (Eq. (9)), results in a notable profit increase for both firms, 17% for Firm A and 36% for Firm B. Since Nash scheme follows a utilitarian approach the profit allocation is distributed based on the player's contribution to the total payoff; thus, the resulting market structure closer approximates that of status quo.

Table 1 Profit increase compared to status and market share for different fairness schemes in duopoly market

	%Profit change A	%Profit change B	Market share A	Market share B
SQ	-	-	0.63	0.37
Naïve	+1.8	+80.9	0.49	0.51
Nash	+16.9	+36.4	0.59	0.41

To have a broader picture of the impact of the fairness schemes in the proposed model, it is necessary to evaluate the customer savings. Figure 1 displays the customer savings of the customers with the highest purchasing costs in the status quo, for different schemes. Results in Figure 1 suggest that Nash bargaining scheme leads to increased cost savings ranging from 5 to 8 % for the 5 top customers, in terms of purchasing cost, while the Naïve scheme results in lower customer savings compared to the Nash scheme.

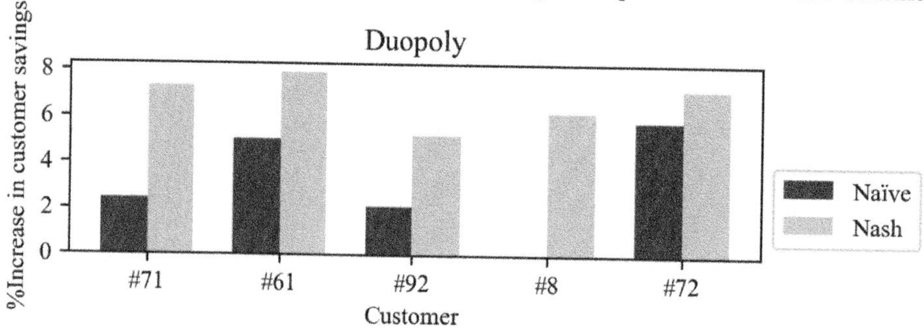

Figure 1 Customer savings compared to status quo for customers with higher cost in duopoly for different fairness schemes

From the Gantt charts of customers #71 and #61 in Figure 2, the savings of the customers can be related to the contract choice. For both customers, status quo allocation results in the use of Formula contract, which inflicts high purchasing cost, in contrast in the Nash fairness scheme the Firm contract prevails in both customers. In the Naïve scheme a combination of all three contracts is used for customer #71 while only Formula and Firm for #61. Consequently, the lower customer savings in the Naïve scheme, compared to the Nash, can be attributed to the fact that Formula contract is selected.

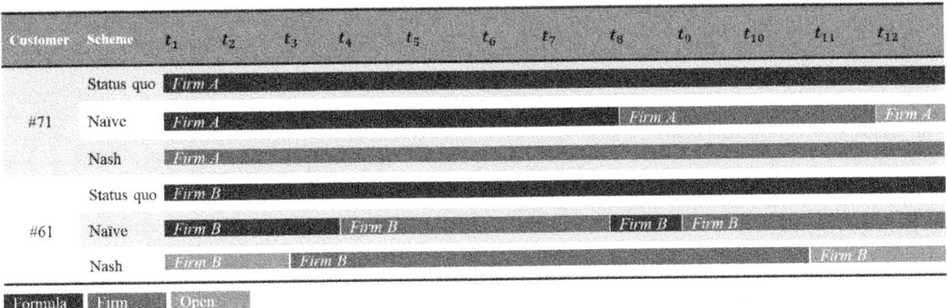

Figure 2 Duopoly Gantt chart for customer allocation in firms and contracts under different fairness schemes

3.2. *Oligopoly*

In the oligopoly case study, an interesting aspect is that there is an imbalance in the market share of the status quo, as suggested by Table 3. Firm C holds more than half of the market share in terms of total profit, while Firm A has a slight advantage over Firm B. The Naïve scheme results in a more balanced market allocation, which is achieved by increasing by 100% the profit of Firm B while maintaining the profit of Firm C at the status quo levels. Although the Nash scheme allows for only a marginal increase for the oligopoly firms, the market share remains stable compared to the status quo.

Table 2 Profit increase compared to status and market share for different fairness schemes in oligopoly market

	%Profit change A	%Profit change B	%Profit change C	Market share A	Market share B	Market share C
SQ	-	-	-	0.25	0.20	0.55
Naïve	+25.0	+104.9	+0.2	0.24	0.32	0.44
Nash	+4.8	+2.8	+6.3	0.25	0.21	0.54

Concerning the customer savings displayed in Figure 3, Nash scheme outperforms the Naïve scheme for up to 10 times for examined customers. On average the Nash scheme results in savings of 22%, while Naïve only to 4%, suggesting that when there is a greater disproportion in the market share of the firms the Nash scheme can result in higher payoffs for the customers.

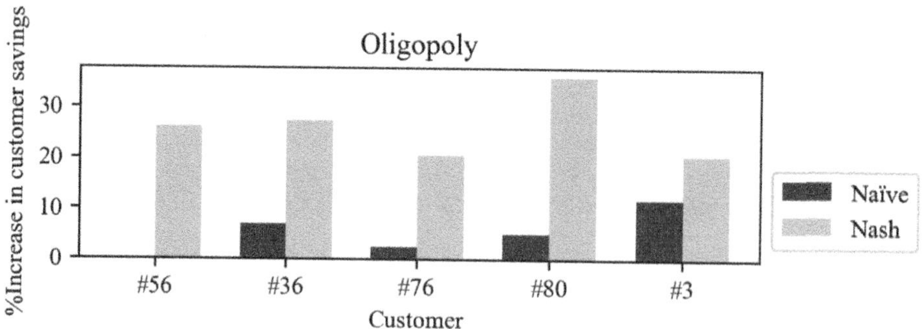

Figure 3 Customer savings compared to status quo for customers with higher cost in oligopoly for different fairness schemes

4. Conclusions

Game theory has been widely used to facilitate supply chain optimisation problems under cooperation. However, the allocation of the coalition's payoff is not always straightforward and different fairness schemes can be evaluated. This paper examines the effect of different fairness schemes on the maximisation of the payoff of manufacturers and customers in cooperative oligopolies, by deciding the optimum contract and customer allocation. For the examined case studies of liquid gas market, Nash fairness resulted in higher customer savings while maintaining the market structure of the status quo.

5. Acknowledgments

The authors gratefully acknowledge financial support from EPSRC grants EP/V051008/1, EP/T022930/1, EP/W003317/1, EP/V050168/1.

References

A. P. Barbosa-Povoa, J. M. Pinto, 2020. Process supply chains: Perspectives from academia and industry. Comp. Chem. Eng. 132, 106606.

E. M. L. Beale, and J. J. H. Forrest, 1976. Global optimization using special ordered sets. Math. Program. 10, 52–69.

B. A. Calfa, I. E. Grossmann, 2015. Optimal procurement contract selection with price ptimization under uncertainty for process networks. Comp. Chem. Eng. 82, 330–343.

V. M. Charitopoulos, V. M. Dua, J. M Pinto, and L. G. Papageorgiou, 2020. A game-theoretic optimisation approach to fair customer allocation in oligopolies. Opt. Eng. 21, 1459–1486.

D. J. Cruz-Avilés, A. del Carmen Munguía-López, J. M. Ponce-Ortega, 2021. Optimal design of water networks in eco-industrial parks incorporating a fairness approach. Ind. Eng. Chem. Res. 275, 8844–8860.

J. Gjerdrum, N. Shah, L. G. Papageorgiou, 2001 Transfer prices for multienterprise supply chain optimization. Ind Eng Chem Res 40:1650–1660

S. Liu, L. G. Papageorgiou, 2018. Fair profit distribution in multi-echelon supply chains via transfer prices. Omega 80, 77–94.

A. Marousi , J. M. Pinto, L. G. Papageorgiou, V. M. Charitopoulos, 2023. Multi-period fair customer allocation in oligopolies. FOCAPO/CPC 2023.

R. Salcedo-Diaz, J. R. Ruiz-Femenia, A. Amat-Bernabeu, J. A. Caballero, 2021. A cooperative game strategy for designing sustainable supply chains under the emissions trading system. J. Clean. Prod. 285, 124845.

X. X. Zheng, Z. Liu, K. W. Li, J. Huang, J. Chen, 2019. Cooperative game approaches to coordinating a three-echelon closed-loop supply chain with fairness concerns. Int. J. Prod. Econ. 212, 92–110.

Antonis Kokossis, Michael C. Georgiadis, Efstratios N. Pistikopoulos (Eds.)
PROCEEDINGS OF THE 33rd European Symposium on Computer Aided Process Engineering (ESCAPE33), June 18-21, 2023, Athens, Greece
 http://dx.doi.org/10.1016/B978-0-443-15274-0.50256-0

Artificial Neural Network-Based Real-Time PID Controller Tuning

Tate Bestwick,[a] Kyle V. Camarda,[a]

[a]*University of Kansas, Department of Chemical and Petroleum Engineering, 1530 West 15th Street, Lawrence, Kansas 66045, United States*

Abstract

Proportional-integral-derivative (PID) control is well established and is still highly prevalent in chemical industries. Many methods for tuning PID controllers to keep a process running at steady-state are also well established; however, they must be retuned if the setpoint must be altered or the system dynamics change at any point in time. The purpose of this work is to use an artificial neural network (ANN) to tune a PID controller on a third order chemical system. It was found that the ANN generated parameters that resulted in smaller overshoot and less settling time than either the Ziegler-Nichols tuning method or the Tyreus-Luyben tuning method in one hundred percent of randomly generated systems.

Keywords: Machine Learning, Process Control, Renewable Energy, Neural Network

1. Introduction

The focus of this paper is the use of artificial neural networks (ANNs) to predict proportional-integral-derivative (PID) controller tuning parameters for a second-order chemical process which has a third-order closed-loop control scheme. The performance of the PID parameters determined by the ANN are compared with the performance of those obtained from the Ziegler-Nichols tuning method as well as those obtained from the Tyreus-Luyben tuning method.

PID control has been utilized and developed for at least 100 years [1]. In the last century, numerous tuning methods for PID controllers have been developed. The most famous of these methods is likely the Ziegler-Nichols method [2]. However, since 1942, numerous other methods have been introduced, including the integral of the time-weighted absolute error (ITAE) method, Fruehauf's and Chien's internal model control (IMC) method, and the Tyreus-Luyben method [3-5]. All of these methods use system dynamics to obtain heuristical controller parameters which can either be used directly or used as a baseline for finding finer-tuned parameters.

All of the aforementioned tuning methods are primarily for determining values for maintaining steady-state for a single setpoint; in order to transition to a different setpoint, the process must be repeated to obtain new values. Other control methods have been examined in more recent years which allow for more complex and wholistic process control or for automation of PID control tuning. Model predictive control (MPC), fuzzy logic PID control, and active disturbance rejection control (ADRC) are all topics of recent research. [6-8]. However, conservative industries still prefer traditional PID control, particularly if the process is of a more hazardous nature [9]. This has led to the development of using artificial intelligence (AI) to tune PID control parameters. AI-tuned controllers have been demonstrated to be an effective means of process control for a chemical system [10].

Artificial Neural Networks are a branch of machine learning with considerable predictive power. The input nodes to the network send a signal to the first hidden layer of nodes. This signal is propagated through the first layer to a second layer, and so on until reaching the output layer. Each connection between two nodes is weighted by some function. Finally, each node has a threshold of input signals required to elicit an output greater than zero; this threshold is known as the bias. The number of hidden layers, the number of nodes in each hidden layer, and the weights and biases of each connection are tuned to best apply to the problem of interest. [11-13]

2. Physical System and Mathematical Model

2.1. Physical System Description

The system examined is two stirred tanks in series. The first tank has two inlets; an initial stream at some concentration and flowrate, and a correction stream which is

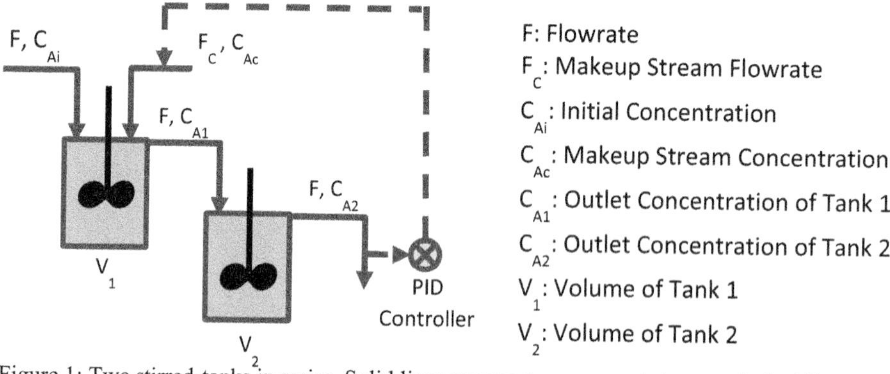

Figure 1: Two stirred-tanks in series. Solid lines represent component streams, dashed lines represent measurements or signals.

highly concentrated and at a much lower flowrate than the previously mentioned initial stream. The two are mixed together in the first tank. The outlet of this first tank is fed into the second stirred tank. The outlet of the second tank is measured, and this signal is sent to a PID controller which is used to adjust the flowrate of the correction stream. The system is depicted in Figure 1.

The following assumptions apply:

1) Both tanks are well-mixed at all times.
2) The flowrate of the correction stream is orders of magnitude smaller than the flowrate of the initial stream. Thus, the following equation applies:

$$F + F_c \approx F \qquad 1$$

3) The valve on the correction stream has first order dynamics. This causes the overall control scheme to have third order dynamics.

2.2. *Mathematical Model*

The mass balance over each tank is written in equations 2 and 3 with the second assumption applied.

$$\frac{V_1}{F}\frac{d}{dt}C_{A1} + C_{A1} = \tau_1\frac{d}{dt}C_{A1} + C_{A1} = C_{Ai} + \frac{C_{Ac}}{F}F_c \qquad 2$$

$$\frac{V_2}{F}\frac{d}{dt}C_{A2} + C_{A2} = \tau_2\frac{d}{dt}C_{A2} + C_{A2} = C_{A1} \qquad 3$$

Combining these two equations yields a second order equation in which the outlet of tank 2 (measured variable) is obtained from the inlet (disturbance variable) and correction stream (manipulated variable) specifications. Solving for the concentration of the outlet of tank 2 in terms of deviation variables yields equation 4, which is written in the Laplace domain. Deviation variables are denoted as prime variables. Equation 4b relates provides a simplified notation for the inlet variable dynamics and the disturbance variable dynamics.

$$C'_{A2}(s) = \frac{1}{(\tau_1 s + 1)(\tau_2 s + 1)} C'_{Ai}(s) + \frac{C_{Ac}/F}{(\tau_1 s + 1)(\tau_2 s + 1)} F'_c(s) \quad 4a$$

$$C'_{A2}(s) = G_d(s) C'_{Ai}(s) + G_m(s) F'_c(s) \quad 4b$$

The valve on the correction stream was assumed to have dynamics which are modeled in the Laplace domain in equation 5.

$$\frac{0.01}{0.1s + 1} = G_v(s) \quad 5$$

The sensor which measures the concentration of the outlet of the second tank and converting the setpoint from engineering units to electrical signal was assumed to have negligible dynamics (equation 6). Finally, equation 7 models the PID controller.

$$\frac{1}{1} = G_s(s) = G_{sp}(s) \quad 6$$

$$\frac{K_D s^2 + K_p s + K_I}{s} = G_c(s) \quad 7$$

Equations 4-7 allow us to obtain transfer functions which describe the outlet variable response to a change in the setpoint (equation 8), the outlet variable response to the change in the disturbance variable (equation 9), the manipulated variable response to a change in the setpoint (equation 10), and the manipulated variable response to a change in the disturbance variable (equation 11).

$$\frac{G_m(s) G_v(s) G_c(s) G_{sp}(s)}{1 + G_m(s) G_v(s) G_c(s) G_s(s)} = H_{os} \quad 8$$

$$\frac{G_d}{1 + G_m(s) G_v(s) G_c(s) G_s(s)} = H_{od} \quad 9$$

$$\frac{G_v(s) G_c(s) G_{sp}(s)}{1 + G_v(s) G_c(s) G_s(s) G_m(s)} = H_{ms} \quad 10$$

$$\frac{-G_v(s) G_c(s) G_s(s) G_d(s)}{1 + G_v(s) G_c(s) G_s(s) G_m(s)} = H_{md} \quad 11$$

3. Methodology

The system described above was modeled in Python using the Python Controls Module [14]. Once modeled, 100,000 variations of the system above were generated. The space-time for tank 1 and the space-time for tank 2 were varied from 1.5 to 11.5 minutes at 0.2-minute increments, and the ratio of the concentration of the correction stream over the inlet flowrate (C_{Ac}/F) varied from 20 to 1400 kg/min at an 35 kg/min increments. The Tyreus-Luyben method was used to generate baseline PID tuning parameters which could aptly respond to a setpoint step change for each system. These parameters were then further adjusted to ensure the overshoot for a setpoint step change did not exceed ten percent and the settling time for the system did not exceed the sum of the space-time for tank 1 and the space-time for tank 2. The dataset containing the further adjusted tuning parameters was randomly split such that eighty percent of it served as the training dataset and twenty percent of it served as the testing data for the ANN which was used to predict the PID controller parameters.

The ANN was generated and trained using Keras's Python interface [15]. The neural network was structured to have three inputs, which can be easily calculated from variables which are directly measurable; τ_1, τ_2, and $\frac{C_{Ac}}{F}$. The ANN predicts three outputs; K_p, τ_I, and τ_D.

$$\frac{K_p}{\tau_I} = K_i \tag{12}$$

$$K_p \tau_D = K_D \tag{13}$$

The architecture and the hyperparameters related to training the ANN were varied to find a network which was both accurate and reproducible despite the randomness associated with the training method.

Once the ANN was trained, one hundred variations of the system described in section 2 were randomly generated with the constraint that τ_1, τ_2, and $\frac{C_{Ac}}{F}$ were always within the range set for the generation of the training dataset. The trained ANN was used to predict K_p, τ_I, and τ_D for each system. The response to a step change in the setpoint was measured for each system using the PID controller parameters obtained by the ANN. The response from the system using PID controller parameters generated by the ANN were compared to the responses from the system using PID controller parameters obtained by the Ziegler-Nichols tuning method, and also compared with those obtained by the Tyreus-Luyben tuning method.

4. Results

The ANN generated was able to be trained within a few minutes to predict the tuning parameters of the testing set with an overall R^2 score of 0.919. See Table 1 (below) for ANN performance predicting each controller parameter. The PID parameters predicted by the ANN always resulted in smaller percent overshoot and a shorter settling time than parameters obtained by the Ziegler-Nichols method or those obtained by the Tyreus-Luyben method. The percent overshoot from the system using the Ziegler-Nichols

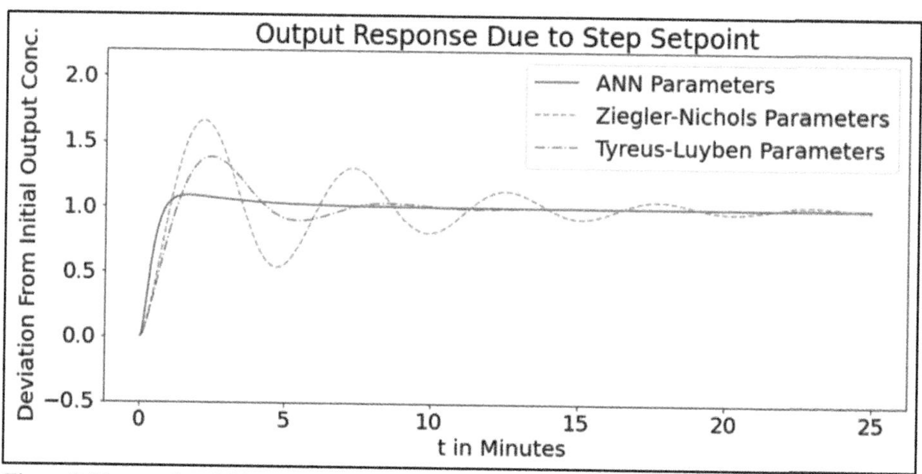

Figure 2:Measured variable response to a setpoint step-change using PID parameters predicted by the ANN, the Ziegler-Nichols method, and the Tyreus-Luyben method.

parameters was always at least 5 times larger than the percent overshoot of the system using the ANN predicted parameters; the percent overshoot from the system using the Tyreus-Luyben parameters was always at least 2.8 times larger than the percent overshoot of the system using the ANN predicted parameters; the settling time of the systems using the Ziegler-Nichols method was always at least 3.8 times larger than the settling time of the system using parameters predicted by the ANN; the settling time of the systems using the Tyreus-Luyben method was always at least 1.6 times larger than the settling time of the system using parameters predicted by the ANN. Figure 2 depicts a comparison of the response of the system using each of the three methods to obtain PID parameters. The trends which are seen in Figure 2 were consistent in all one hundred trials.

Parameter	K_p	τ_I	τ_D	Overall
R^2 Score	1.00	0.876	0.881	0.919

Table 1: R^2 score for all ANN performance on all parameters in testing set, as well as overall R^2 score.

5. Discussion

The ANN was trained using a dataset of PID controller parameters which outperformed parameters obtained from the Ziegler-Nichols method as well as those obtained from the Tyreus-Luyben method. The results verify that the ANN preserves this quality for PID controller parameters which it predicts for any system which is of the same structure as the one used to generate the training data so long as it has system parameters (space-time, inlet flowrates, and inlet concentrations) which fall within the boundaries specified by the training set. It is quite feasible that this implication extends to any system which is modelled and used to generate training data; the procedure implemented for the system detailed in this paper could be repeated for other systems, including for more complex systems. Furthermore, the procedure for generating training and testing data could be altered to meet standards for specific processes; for example, if a system requires never exceeding 1% overshoot but can tolerate a longer rise time, the ANN could be trained using parameters which meet this criteria.

It is important to note that the ANN takes inputs which capture all of the system variables which could vary across time as they appear in the transfer functions. For example, all three ANN inputs change if the disturbance variable flowrate changes. If the disturbance variable fluctuates over time, the ANN is able to predict PID controller parameters which account for the change in system dynamics. Additionally, the ANN architecture was simplified by using ratios of system variables rather than all of the system variables directly. This was possible because the system variables always appear in the transfer functions in those ratios. Specifically, the volume of tank 1 never appears in the transfer function except for when it appears in the space-time τ_1.

Plants which run on renewable energy [16] are excellent candidates for the application of using an ANN to obtain real-time PID controller parameters. Solar and wind energy are intermittent energy sources; the amount of energy is dependent on how much sunlight or wind there is any given day. Thus, plants which run, or which will run, entirely on these energy sources do not have a consistent energy supply. While batteries may be used to store energy, the production and decommissioning of batteries is currently damaging to the environment. Having ANNs which could update tuning parameters for process setpoints would allow for adjusting process output real-time in response to energy availability. The number of batteries necessary would be considerably lowered because the process would be far more responsive to the variability of the energy sources.

The most significant implication of this work is with regard to dynamic operation or simulation. There are many scenarios in which it is preferable to account for a dynamic setpoint or time-varying inputs. Whether the inputs are inconsistent or product demand varies with time, the ability to obtain real-time PID controller parameters which meet the process performance standards could allow for a more efficient process than a static setpoint would allow. Since many processes currently use PID controllers, this implementation would not require drastic changes to the process control scheme. The computation time of ANNs is small enough that PID controller parameters could be updated every few seconds.

6. Conclusion

In this work, one hundred thousand variations of a third order chemical system were modelled and the parameters of the PID controller in each one were tuned to obtain previously determined acceptable output responses. These values were then used to train an ANN to predict PID controller parameters for any variation of the modelled system, and was then tested on one hundred randomly generated variations of the system. The ANN was found to predict PID controller parameters which resulted in better system performance than either the Ziegler-Nichols method or the Tyreus-Luyben method in all one hundred cases. This implies that ANNs could be used to tune PID controllers in more complex systems. Furthermore, it implies that ANNs could be used to provide real-time tuning for PID controllers in processes which could have time-varying system dynamics or setpoints.

References

[1] Minorsky, N. 1922. Directional Stability of Automatically Steered Bodies, Nav. Eng. J. 34 (2): 280-309

[2] Ziegler, J. B, Nichols, N. B., 1942, Optimum settings for automatic controllers, ASME Transactions, 64 , pp. 759-768.

[3] Maiti, D., Acharya, A., Chakraborty, M., Konar, A., Janarthanan, R., 2008, Tuning Pid and $PI^{\lambda}D^{\delta}$ Controllers Using the Integral Time Absolute Error Criterion. In Proceedings of the 2008 4th International Conference on Information and Automation for Sustainability, Colombo, Sri Lanka, 12–14 December 2008; pp. 457–462.

[4] Fruehauf, P. S, Chien, I. L., 1994, Simplified IMC-PID Tuning Rules, 33 (1), 43-59.

[5] Tyreus, B. D., Luyben, W. L., 1992, Tuning PI Controllers for Integrator/ Deadtime Processes. Ind. Eng. Chem. Res., 31, 2625-2628.

[6] Oyama, H., Durand, H., 2020, Interactions between control and process design under economic model predictive control, 92, 1-18.

[7] Turman, E. M., Strasser, W., 2022, Leveraging fuzzy Logic PID controllers for accelerating chemical reactor CFD. Chem. Eng. Sci., 262.

[8] Martinez, B. V, Sanchis, J., Garcia-Nieto, S., Martinez, M., 2021, Active disturbance rejection control; a guide for design and application. Rev Iberoam Autom In., 18 (3), 201-217.

[9] Lee, Y. S., Jang, D. W., 2021, Optimization of Neural Network-Based Self-Tuning PID Controllers for Second Order Mechanical Systems. Appl Sci-Basel, 11 (17).

[10] Sousa, B. S., Silva, F. V., Fileti, A. M. F., 2020, Level Control of Coupled Tank System Based on Neural Network Techniques. Chem Prod Process Model., 15 (3).

[11] Hinton, G. E., 1989, Connectionist Learning Procedures, Artif Intell, 40 (1-3), 185-234.

[12]https://scikitlearn.org/stable/modules/generated/sklearn.neural_network.MLPRegressor.html?highlight=neural+network (Accessed 2022)

[13] Rajapakse, D., Meckstroth, J., Jantz, D. T., Camarda, K. V., Yao, Z., Leonard, K. C., 2022, Deconvoluting Kinetic Rate Constants of Catalytic Substrates from Scanning Electrochemical Approach Curves with Artificial Neural Networks, ACS meas. sci. au.

[14] https://python-control.readthedocs.io/en/0.9.2/ (Accessed 2022)

[15] https://keras.io/ (Accessed 2022)

[16] Ojo, G.; Camarda, KV; Sustainable Ammonia Production via Electrolysis and Haber-Bosch Process. ESCAPE Proc. of 2022, vol 32, pp. 229-234

Antonis Kokossis, Michael C. Georgiadis, Efstratios N. Pistikopoulos (Eds.)
PROCEEDINGS OF THE 33rd European Symposium on Computer Aided Process Engineering (ESCAPE33), June 18-21, 2023, Athens, Greece
 http://dx.doi.org/10.1016/B978-0-443-15274-0.50257-2

State of the Art Flexibility Analysis for Natural Gas Monetization Production Processes Under Uncertainties

Noor Yusuf, Ahmed AlNouss and Tareq Al-Ansari

College of Science and Engineering, Hamad Bin Khalifa University, Qatar Foundation, Education City P.O. Box 5825, Doha, Qatar

Abstract

The increased demand for cleaner energy resources places a significant emphasis on monetizing natural gas into value-added products such as liquified natural gas (LNG), ammonia and power to target different market needs. Despite the estimated growth in the demand for the different cleaner energy resources, each monetization process is subject to market uncertainties induced by global events such as pandemics, energy crises and supply chain disruptions. In this context, this study considers a promising strategic approach to investigate the economic profitability of embedding production flexibility in three natural gas monetized products (i.e., LNG, ammonia and power) subject to forecasted demand and price, and technical flexibility of each process (upper and lower production limits). The results provide decision-makers with a systematic approach to assess the economic and technical aspects of a flexible integrated LNG, ammonia, and power plant design in today's turbulent markets.

Keywords: Flexibility, natural gas monetization, uncertainty, LNG, ammonia.

1. Introduction

Natural gas (NG) is a crucial input to different processes and sectors such as household, transportation, petrochemical, agricultural fertilizers, and plastics industries. Within local markets, NG is mainly used to generate electricity using the gas-to-power (GTP) process. Furthermore, different chemical and physical monetization routes have been deployed to monetize NG into economically viable products. The monetization routes are classified based on the required final product: service-based monetization route (i.e., GTP), physical monetization route (i.e., liquefied natural gas - LNG and compressed natural gas - CNG), and the chemical monetization route (i.e., chemical conversion of natural gas into liquid products such as ammonia and methanol). The selection of a monetization route depends on factors such as international market requirements, the geographical location of gas reserves, climate in the region, gas field characteristics and gas quality.

Although renewables are expected to continue being the fastest-growing class of energy resources in the global market share, NG is expected to be the fastest-growing fossil-based energy resource due to its resilient characteristics (BP, 2014). On the other hand, energy production management has been essential to guarantee global energy sustainability, environmental sustainability, effective resources management, and economic profitability of projects. In terms of economic sustainability, producing companies must strategically decide on the final product portfolio, targeted markets, pricing, and shipping routes to optimize the project's profitability and hedge against risks. Following the various events that disrupted the global energy trade and supply chains, such as the financial crisis of 2008, the oil markets crush of 2015, the COVID-19 pandemic in 2020, and the most recent Russian-Ukrainian crisis, the concept of production flexibility has become increasingly vital to increase the responsiveness of projects to exogenous uncertainties (Heffron et al., 2021; Sarkar et al., 2021; Sarkar & Seo, 2021). In a flexible production system, the decision-maker has the right, but not the obligation, to change the production capacity based on market performance to sustain profitability. In the literature, most studies focused on optimizing the production and planning of a single-product NG project subject to production flexibility based on varying demand (Bhosekar & Ierapetritou, 2020; Verleysen et al., 2021; Yusuf et al., 2022). There is a lack of studies that investigate the optimal production and management of integrated NG production systems. In this study, global historical and forecasted demand and price data for three NG monetization routes: LNG, power, and ammonia, are used as input to investigate the yearly optimal NG capacity allocation to the three production processes for maximum overall project profitability. Moreover, constraints in production flexibility (upper and lower limits) are considered. Fundamentally, capacity flexibility allows decision-makers to change the production level throughout the project lifetime based on market demand as a proactive response to uncontrolled exogenous uncertainties.

2. Methodology and Data

The methodology developed as part of this study involves performing sustainable planning within the downstream energy sector in Qatar. The designed framework serves as a decision-making tool that enables the prediction of flexible decisions for the downstream production of power, LNG, and ammonia to satisfy the local power demand and participate in the international demand for LNG and ammonia. The benefits of deploying a sustainable planning system as an allocation tool to enhance the overall economic profit are also investigated through an economic analysis of capital investment cost, operating cost, and sales revenue. This section describes the model development of the different involved processes, the tool used to execute the simulation, and the historical and forecasted data sets available to run the model.

The commercial software Aspen HYSYS is used for process modelling and simulation. The NG liquefaction process scheme evolved from condensation at constant pressure, using the Joule-Thompson method, to the energy-efficient heat exchange process using refrigeration systems. The latter is widely used in different projects worldwide. Licensors developed different technologies in the market to design cost-effective and energy-efficient processes for large-scale commercial deployment. The technologies differ based on the production capacity, refrigerants, liquefaction scheme, and equipment capacity (Al-Mutaz et al., 2016; Castillo et al., 2010; Danilov et al., 2019). Fundamentally, the AP-X technology, an extension of the C3MR process developed by Air Products, is considered for LNG production in Qatar. While for ammonia production, NG is first reformed into syngas through steam and oxygen reforming reactions and the effluent syngas is assured of reaching the necessary 1:3 nitrogen to hydrogen ratio. After which, the syngas is cleaned from CO_2 and water and introduced in the ammonia synthesis loop, where ammonia is generated from the reaction of hydrogen and nitrogen at high pressures ranging from 150 to 350 bar and temperatures ranging from 430 to 480 °C. After which, the ammonia is chilled to be removed from the loop. Whereas, for power generation, the syngas generated from NG is initially conditioned before being delivered to the combustion reactor, where it burns with air. To create electricity, the effluent stream is sent into a gas turbine. The expanded gas is used to produce various forms of steam through a series of heat exchangers, which are then utilised to create electricity utilising a steam turbine. The simulated LNG, power and ammonia models using Aspen HYSYS are assessed using a built-in economic evaluation tool. The results of the economic evaluation tool are illustrated in Table 1.

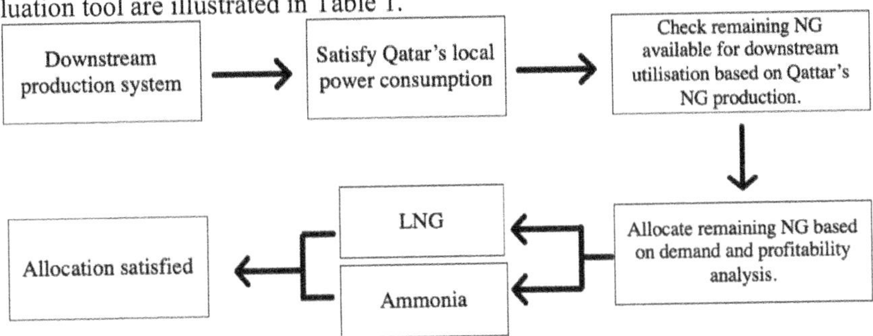

Figure 1: High-level representation of planning methodology.

Table 1. An economic evaluation of LNG, power and Ammonia using Aspen HYSYS.

Process	Power	Ammonia	LNG
NG Flow (kg/h)	5.40E+03	6.41E+05	1.38E+07
Product Amounts	Power (kWh)	Ammonia (kg/h)	LNG (kg/h)
	4.85E+08	7.26E+05	9.98E+06
	Steam (tonne/y)		
	5.21E+05		
Product Prices	Power ($/kWh)	Ammonia ($/T)	LNG ($/T)
	0.05724	400	1785
	Steam ($/tonne)		
	7.5		
Total Capital Cost ($)	1.01E+08	1.59E+08	3.16E+08
Total Operating Cost ($/y)	7.54E+06	1.41E+08	1.15E+08
Total Raw Materials Cost ($/y)	5.78E+05	1.18E+09	2.54E+10
Equipment Cost ($)	5.66E+07	6.06E+08	1.13E+08
Total Installed Cost ($)	6.48E+07	1.03E+09	1.92E+08
Total Annualized Cost ($/y)	2.89E+07	1.35E+09	2.56E+10
Revenue ($/y)	3.16E+07	2.54E+09	1.56E+11

The simulation results and historical and forecasted market data are used as data sets for the planning and allocation formulated in Python using the MESA library (Mesa, 2016). The historical and forecasted demand and production data for NG, LNG, power and ammonia are illustrated in Figure 2. Whereas the trends of global prices of ammonia and LNG are illustrated in Figure 3.

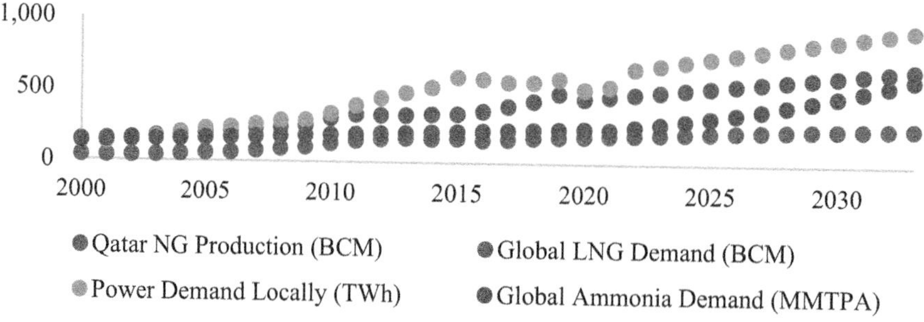

Figure 2: Historical and forecasted demand and production data.

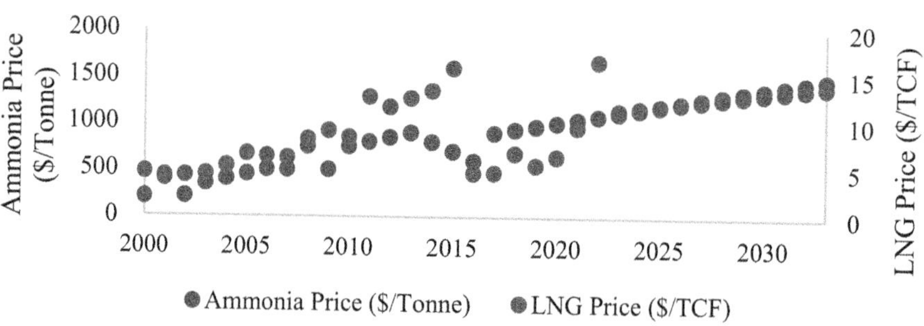

Figure 3: Trends of global prices of Ammonia and LNG.

The ammonia and LNG production must satisfy a minimum operating limit per train capacity and operational flexibility lower limits, as indicated in Table 2.

Table 2. Minimum operating limits (Cheema & Krewer, 2018; Yusuf et al., 2022).

Process	**LNG (MTPA)**	**Ammonia (kg/h)**
Lower limit Production capacity	5	708

3. Results and Discussion

With the influence of global events and supply chain disruptions on the planning and operation of different NG projects, the integration of various tools in the early design stages of a project is crucial to design a responsive integrated NG production system responsive to market changes. In this study, the results of simulated NG to ammonia, LNG, and power processes, along with historical and forecasted local power requirements in Qatar and global LNG and ammonia demand, are considered to investigate the optimal annual NG production capacity allocation to the three production processes using the python MESA library based on the optimal final product portfolio for maximum overall profitability. The optimal allocation is subject to primarily satisfying the local power demand and then distributing the remaining NG production capacity to the LNG and ammonia production processes to meet international demand. Moreover, lower production capacity limits for LNG and ammonia production processes must be met, wherein a production train cannot operate below the lower flexibility threshold due to technical infeasibility. As illustrated in Figure 4, the optimal NG distribution to LNG, ammonia and power processes from 2000 to 2033 is changing annually, subject to NG production and demand for each product. The simulation indicates the common trend was the dominance of NG allocation to power and LNG.

In contrast, in the years 2017, 2019, and 2020, a shift to NG allocation for ammonia production is identified due to factors such as the decrease in local power requirements, the decline in global LNG demand in 2020 influenced by the COVID-19 pandemic, and the growth in NG production. Figure 5 illustrates the ammonia and LNG capacities produced in MTPA and Power generation in TWh. The associated annual profits and annualized costs for the optimal NG distribution to LNG, ammonia, and power are illustrated in Figures 6 and 7, respectively, which demonstrate that a significant share of the total profits is associated with international LNG and local power sales. Despite the minimal allocation for NG to ammonia production compared to the other monetization routes, the high selling price of ammonia in international markets reflected positively on the project's profitability. However, ammonia production is associated with high annualized costs due to the high costs of raw materials, equipment, and installation. Hence, annualized costs are an essential factor influencing the economic sustainability of a monetization route under turbulent markets.

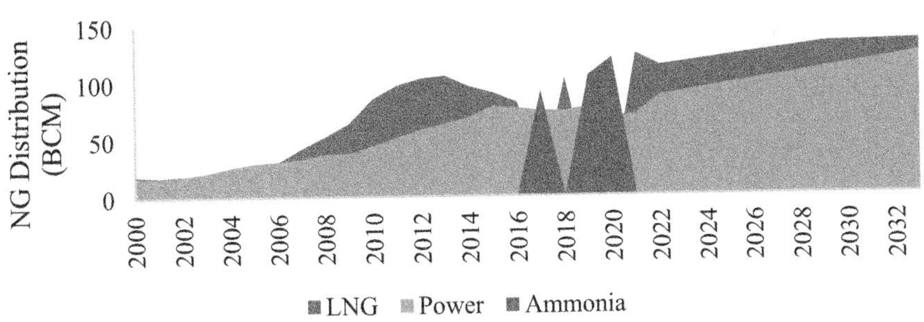

Figure 4: Annual NG distribution (BCM) to LNG, Ammonia, and Power between 2000 and 2033.

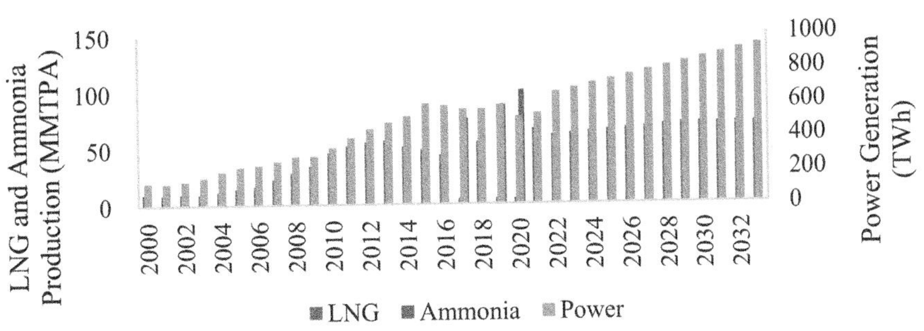

Figure 5: Annual production capacity of LNG, Ammonia, and Power between 2000 and 2033.

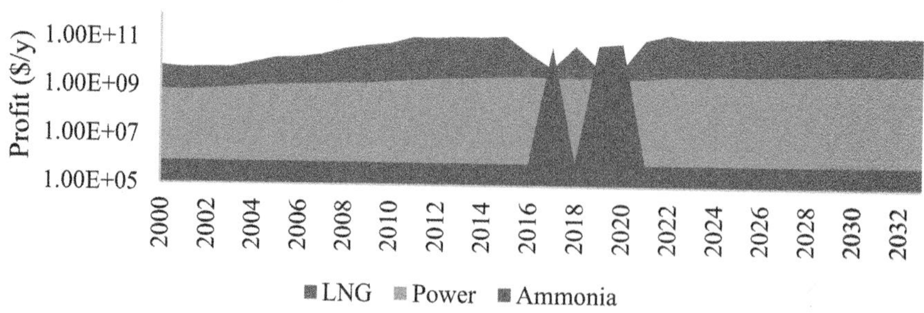

Figure 6: Annual profits for LNG, Ammonia and Power sales based on optimal NG distribution.

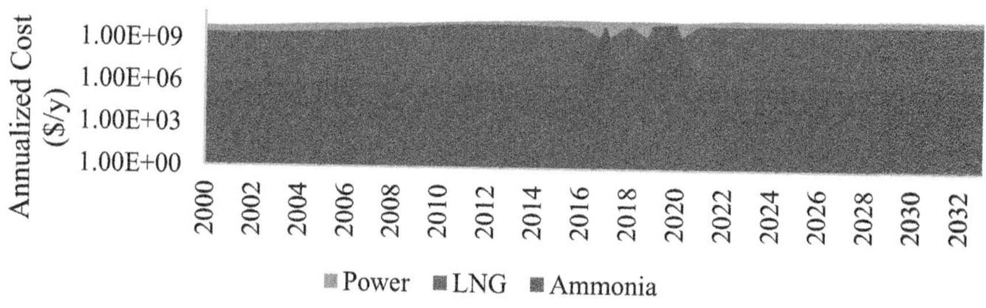

Figure 7: Annualized costs for LNG, Ammonia and Power processes.

4. Conclusion

Integrating different tools for optimal NG allocation to products is crucial in designing integrated NG production systems in turbulent markets. Despite the trend for shifting to ammonia production, this study demonstrates the importance of prioritizing power and LNG production and varied production flexibility in the studied period to achieve maximum profitability and hedge against risks. The technical limitations of production were considered for optimal NG distribution to open the door for future comprehensive studies in assessing the value of flexibility associated with each monetization route.

References

Al-Mutaz, I. S., Liu, X., & Mazza, G. (2016). Natural Gas Liquefaction Technologies-An Overview. *OIL GAS-EUROPEAN MAGAZINE*, *42*(4), 213–218.

Bhosekar, A., & Ierapetritou, M. (2020). Modular design optimization using machine learning-based flexibility analysis. *Journal of Process Control*, *90*, 18–34.

BP. (2014). *BP Energy Outlook 2035*.

Castillo, L., Nadales, R., González, C., Dorao, C. A., & Viloria, A. (2010). Technology Selection for Liquefied Natural Gas (Lng) on Baseload Plants. *XIX International Gas Convention AVPG 2010, May 24th - 26th Caracas, Venezuela*, 1–15.

Cheema, I. I., & Krewer, U. (2018). Operating envelope of Haber–Bosch process design for power-to-ammonia. *RSC Advances*, *8*(61), 34926–34936.

Danilov, R., Arabyan, M., & Usov, D. (2019). Influence of technologies on LNG market development. *IOP Conference Series: Materials Science and Engineering, 537*(4). https://doi.org/10.1088/1757-899X/537/4/042030

Heffron, R.., Körner, M., Schöpf, M., Wagner, J., & Weibelzahl, M. (2021). The role of flexibility in the light of the COVID-19 pandemic and beyond: Contributing to a sustainable and resilient energy future in Europe. *Renewable and Sustainable Energy Reviews, 140*, 110743.

Mesa, P. (2016). *Mesa: agent-based modeling in python 3+.*

Sarkar, B., Mridha, B., Pareek, S., Sarkar, M., & Thangavelu, L. (2021). A flexible biofuel and bioenergy production system with transportation disruption under a sustainable supply chain network. *Journal of Cleaner Production, 317*, 128079.

Sarkar, M., & Seo, Y. W. (2021). Renewable energy supply chain management with flexibility and automation in a production system. *Journal of Cleaner Production, 324*, 129149.

Verleysen, K., Parente, A., & Contino, F. (2021). How sensitive is a dynamic ammonia synthesis process? Global sensitivity analysis of a dynamic Haber-Bosch process (for flexible seasonal energy storage). *Energy, 232*, 121016.

Yusuf, N., Govindan, R., & Al-Ansari, T. (2022). Techno-economic analysis of flexible AP-X LNG production process under risks and uncertainties. In *Computer Aided Chemical Engineering* (Vol. 51, pp. 367–372). Elsevier.

Antonis Kokossis, Michael C. Georgiadis, Efstratios N. Pistikopoulos (Eds.)
PROCEEDINGS OF THE 33rd European Symposium on Computer Aided Process Engineering (ESCAPE33), June 18-21, 2023, Athens, Greece
 http://dx.doi.org/10.1016/B978-0-443-15274-0.50258-4

Online Monitoring and Robust, Reliable Fault Detection of Chemical Process Systems

Zheyu Jiang[a]*

[a]*Oklahoma State University, 420 Engineering North, Stillwater, Oklahoma, USA 74078*
**Corresponding author: zheyu.jiang@okstate.edu*

Abstract

Nowadays, large amounts of data are continuously collected by sensors and monitored in chemical plants. Despite having access to large volumes of historical and online sensor data, industrial practitioners still face several challenges in effectively utilizing them to perform process monitoring and fault detection, because: 1) fault scenarios in chemical processes are naturally complex and cannot be exhaustively enumerated or predicted, 2) sensor measurements continuously produce massive arrays of high-dimensional big data streams that are often nonparametric and heterogeneous, and 3) the strict environmental, health, and safety requirements established in the facilities demand uncompromisingly high reliability and accuracy of any process monitoring and fault detection tool. To address these challenges, in this paper, we introduce a robust and reliable chemical process monitoring framework based on statistical process control (SPC) that can monitor nonparametric and heterogeneous high-dimensional data streams and detect process anomalies as early as possible while maintaining a pre-specified in-control average run length. Through an illustrative case study of the classical Tennessee Eastman Process, we demonstrate the effectiveness of this novel chemical process monitoring framework.

Keywords: Process monitoring, fault detection, statistical process control, CUSUM, big data streams

1. Introduction

Digitalization is transforming chemical and process industries. Modern chemical plants are equipped with sophisticated digital tools and infrastructures, including numerous sensors and advanced distributed control systems (DCSs), which continuously monitor the plants' equipment performance, manufacturing processes, and mass, energy, and information flows. Together, these sensors generate massive arrays of online data streams that are often *nonparametric* (i.e., data streams do not necessarily follow any specific distribution) and *heterogeneous* (i.e., data streams do not necessarily follow the same distribution). Over the past decades, a number of algorithmic approaches have been developed to effectively utilize the large volumes of historical and online sensor data for reliable online process monitoring and early fault detection. Among them, dimensionality reduction techniques, such as principal component analysis (Jackson and Mudholkar, 1979; Fezai et al., 2018) and partial least squares regression (Geladi and Kowalski, 1986), are the most popular ones in the literature (Russell et al., 2000). These dimensionality reduction-based approaches assume that the statistics characterizing the in-control profiles must also span the subspace defining the out-of-control states or faults (Woodall et al., 2004). In other words, to use the features (e.g., principal components) obtained from historical in-control process data (known as Phase I) for online monitoring (known as Phase II), one must ensure there is no profile shift during online monitoring in some otherwise undetectable direction. However, this assumption is not guaranteed as the chemical process dynamics typically are quite complex and out-of-control states cannot be fully enumerated or anticipated a priori. Another shortcoming of dimensionality reduction-based methods is that operators and process engineers often have a hard time interpreting the results obtained because the features are in the reduced space, which does not have a one-to-one mapping to the original big data streams. Furthermore, since the number of possible distillation fault scenarios can be quite large, monitoring only the most significant subset of features can cause significant error, as the fault may not be noticeable in the selected features.

More recently, advancements in machine learning, such as support vector machine (Onel et al., 2018) and artificial neural network (Heo and Lee, 2018), offer new pathways toward chemical process monitoring and fault detection. Nevertheless, state-of-the-art machine learning-based approaches still face problems such as overfitting and poor predictive accuracy. For example, while most of the published machine learning methods perform well on training and validation sets, their fault detection accuracies rarely exceed 95% in test sets. Considering the strict EHS requirements on plant site and the severity of consequences in case of fault detection failure, such predictive accuracies are unacceptable and can be catastrophic. Furthermore, machine learning methods do not scale well for new fault scenarios that have not been encountered before. In summary, existing distillation process monitoring and fault detection frameworks are inadequate and unsuitable to address practical, sophisticated data stream characteristics and fault scenarios encountered in chemical process industries.

In this work, we introduce a generic chemical process monitoring and fault detection framework featuring nonparametric and heterogeneous big data streams. This framework can detect process mean shifts or anomalies as early as possible while maintaining a user-specified false alarm rate (or in-control average run length, IC-ARL). Specifically, we adopt and simplify the quantile-based statistical process control (SPC) framework that generalizes the proven and reliable multivariate cumulative sum (CUSUM) control charts for nonparametric, heterogeneous big data systems (Ye and Liu, 2022). To demonstrate its effectiveness in chemical process monitoring, we apply this new framework to the classic problem of Tennessee Eastman Process (Downs and Vogel, 1993) and compare the fault detection performance with PCA and SVM-based approaches for the first time.

2. Recent Advancements in Multivariate SPC

We use $X(t) = (X_1(t), \cdots, X_p(t))$ to denote the measurement of p data streams over the observation time $t = 1,2,\cdots$. We assume that the local statistic $X_j(t)$ is i.i.d. across time t for every j. Note that the i.i.d. assumption of the data streams is often satisfied when $X_j(t)$ measures the residual value (Zou et al., 2015). Also, we emphasize that the independency across different data streams is not required. To combine these individual local statistics into a single global monitoring statistic for fault detection, Tartakovsky et al. (2006) and Mei (2010) proposed using the maximum and the sum of local statistics to form the global monitoring statistic, respectively. Mei (2011) further proposed another global monitoring scheme known as the "top-r approach" based on the sum of the largest r local statistics. Unfortunately, all these methods were developed under the assumption that all data streams follow a normal distribution, which is rarely encountered in chemical and process industries. To address this issue, various nonparametric multivariate CUSUM procedures (e.g., Liu et al., 2015) have been developed following the pioneering work of Qiu and Hawkins (2001, 2003). Although these multivariate CUSUM methods relax the normality assumption, they still assume that all data streams follow the same distribution, thereby limiting its applicability to monitoring homogeneous data streams. Earlier this year, Ye and Liu (2022) proposed a quantile-based nonparametric SPC algorithm to construct the local statistic for each data stream. The idea is to incorporates the ordering information of in-control data measurements by categorizing them into a number of quantiles (e.g., 10 or 15) for each data stream in Phase II (Qiu and Li, 2011). Next, in Phase II, online measurements are categorized into the learnt quantiles to generate a quantile-based distribution, which is compared with the quantile-based distribution of in-control data for the detection of any mean shift. Thus, this new quantile-based SPC approach can effectively monitor heterogeneous data streams, and it does not require any prior knowledge about the fault scenarios. In the next section, we present a simplified formulation of the original quantile-based SPC framework of Ye and Liu (2022).

3. Formulation and Methodology

Due to space limitations, we only highlight the key results in this quantile-based SPC framework. First, in Phase I, for each data stream $j = 1, \cdots, p$, in-control measurements are ordered and partitioned into d quantiles: $I_{j,1} = (-\infty, q_{j,1}]$, $I_{j,2} = (q_{j,1}, q_{j,2}]$, …, $I_{j,d} = (q_{j,d-1}, +\infty)$, such that each quantile contains exactly $\frac{1}{d}$ of the in-control measurements. Therefore, one can define cumulative intervals as $CI_{j,i}^{+} = [q_{j,i}, +\infty)$ and $CI_{j,i}^{-} = (-\infty, q_{j,i}]$ for $i = 1, \cdots, d-1$. This information is then used in Phase II monitoring. Specifically, in Phase II, a vector $\boldsymbol{Y}_j(t) = (Y_{j,1}(t), \cdots, Y_{j,d}(t))$ is defined for each data stream j, where $Y_{j,q}(t) = \mathbb{I}\{X_j(t) \in I_{j,q}\}$ and $q = 1, \cdots, d$. Here, $\mathbb{I}\{X_j(t) \in I_{j,q}\}$ is the indicator function that equals 1 when $X_j(t) \in I_{j,q}$ and 0 otherwise. Correspondingly, one can further define two vectors, $\boldsymbol{A}_j^{+}(t) = (A_{j,1}^{+}(t), \cdots, A_{j,d-1}^{+}(t))$ and $\boldsymbol{A}_j^{-}(t) = (A_{j,1}^{-}(t), \cdots, A_{j,d-1}^{-}(t))$, for each data stream j, such that $A_{j,i}^{+}(t) = \mathbb{I}\{X_j(t) \in CI_{j,i}^{+}\}$ and $A_{j,i}^{-}(t) = \mathbb{I}\{X_j(t) \in CI_{j,i}^{-}\}$. One can show that $A_{j,i}^{+}(t) = 1 - \sum_{k=1}^{i} Y_{j,k}(t)$ and $A_{j,i}^{-}(t) = \sum_{k=1}^{i} Y_{j,k}(t)$. And $\mathbb{E}(A_{j,i}^{+}(t)) = 1 - i/d$ and $\mathbb{E}(A_{j,i}^{-}(t)) = i/d$ for $j = 1, \cdots, p$ and $i = 1, \cdots, d$. Therefore, detecting the mean shifts in the distribution of $X_j(t)$ is equivalent to detecting shifts in the distribution of $A_{j,i}^{+}(t)$ and $A_{j,i}^{-}(t)$ with respect to their expected values. Specifically, $A_{j,i}^{+}(t)$ (resp. $A_{j,i}^{-}(t)$) is more sensitive to upward (resp. downward) mean shift (Ye and Liu, 2022). Thus, to detect upward (+) and downward (−) mean shifts, the multivariate CUMSUM procedure developed of Qiu and Hawkins (2001, 2003) was adopted by defining $C_j^{\pm}(t)$ as:

$$
\begin{aligned}
C_j^{\pm}(t) = &\left[\left(\boldsymbol{S}_j^{\pm,\mathrm{obs}}(t-1) + \boldsymbol{A}_j^{\pm}(t)\right) - \left(\boldsymbol{S}_j^{\pm,\mathrm{exp}}(t-1) + \mathbb{E}\left(\boldsymbol{A}_j^{\pm}\right)\right)\right]^T \\
&\cdot \operatorname{diag}\left(\boldsymbol{S}_j^{\pm,\mathrm{exp}}(t-1) + \mathbb{E}\left(\boldsymbol{A}_j^{\pm}\right)\right)^{-1} \\
&\cdot \left[\left(\boldsymbol{S}_j^{\pm,\mathrm{obs}}(t-1) + \boldsymbol{A}_j^{\pm}(t)\right) - \left(\boldsymbol{S}_j^{\pm,\mathrm{exp}}(t-1) + \mathbb{E}\left(\boldsymbol{A}_j^{\pm}\right)\right)\right],
\end{aligned} \tag{1}
$$

where $\boldsymbol{S}_j^{\pm,\mathrm{obs}}(t)$ and $\boldsymbol{S}_j^{\pm,\mathrm{exp}}(t)$ are two $(d-1)$-dimensional vectors with $\boldsymbol{S}_j^{\pm,\mathrm{obs}}(t) = \boldsymbol{S}_j^{\pm,\mathrm{exp}}(t) = \boldsymbol{0}$ if $C_j^{\pm}(t) \le k$, whereas $\boldsymbol{S}_j^{\pm,\mathrm{obs}}(t) = \frac{C_j^{\pm}(t)-k}{C_j^{\pm}(t)}\left(\boldsymbol{S}_j^{\pm,\mathrm{obs}}(t-1) + \boldsymbol{A}_j^{\pm}(t)\right)$ and $\boldsymbol{S}_j^{\pm,\mathrm{exp}}(t) = \frac{C_j^{\pm}(t)-k}{C_j^{\pm}(t)}\left(\boldsymbol{S}_j^{\pm,\mathrm{exp}}(t-1) + \mathbb{E}\left(\boldsymbol{A}_j^{\pm}\right)\right)$ if $C_j^{\pm}(t) > k$. With this, the local statistic $W_j^{+}(t)$ (resp. $W_j^{-}(t)$) for detecting upward (resp. downward) mean shift is: $W_j^{\pm}(t) = \left(\boldsymbol{S}_j^{\pm,\mathrm{obs}}(t) - \boldsymbol{S}_j^{\pm,\mathrm{exp}}(t)\right)^T \cdot \operatorname{diag}\left(\boldsymbol{S}_j^{\pm,\mathrm{exp}}(t)\right)^{-1} \cdot \left(\boldsymbol{S}_j^{\pm,\mathrm{obs}}(t) - \boldsymbol{S}_j^{\pm,\mathrm{exp}}(t)\right)$, which is shown to be equivalent to $W_j^{\pm}(t) = \max\{0, C_j^{\pm}(t) - k\}$ (Qiu and Hawkins, 2001). Here, k is a pre-computed allowance parameter that restarts the CUSUM procedure by resetting the local statistic back to 0 if there is no evidence of upward or downward mean shift after a while (Xian et al., 2021). To detect both upward and downward mean shifts in a data stream j, we simply define a two-sided local statistic $W_j(t) = \max\{W_j^{+}(t), W_j^{-}(t)\}$ (Li, 2020). And the initial condition is $W_j(0) = W_j^{+}(0) = W_j^{-}(0) = 0$ for all $j = 1, \cdots, p$. Finally, we ranklist $W_j(t)$ for the current time t based on its magnitude: $W_{(1)}(t) \geq W_{(2)}(t) \geq \cdots \geq W_{(p)}(t)$, where $W_{(j)}(t)$ denotes the jth largest estimated local statistic. With this, we generalize the top-r approach (Mei, 2011) to determine the stopping time T for raising an alarm and declaring the system is out of control for monitoring heterogeneous data streams: $T = \inf\{t > 0: \sum_{(j)=1}^{r} W_{(j)}(t) \geq h\}$, where r is typically much less than q (Mei, 2011), and h is a constant threshold value related to false alarm rate (Liu and Shi, 2013). A commonly used h corresponds to the false alarm (Type-I error) rate of no more than 0.0027 (classic 3σ limit). Overall, this quantile-based SPC framework developed by Ye and Liu (2022) offers strong statistical justifications and great flexibility as process engineers can customize the choice of h based on the severity of potential failures.

4. Case Study: Tennessee Eastman Process

The Tennessee Eastman Process (TEP, see Figure 1) is an extensively used benchmark case for comparative assessment of process monitoring algorithms. It consists of a reactor, a product condenser, a separator, and a stripping column. The TEP takes four feed streams (streams 1-4) and partially converts them into desired products G and H and byproduct F. As illustrated in Figure 1, the model contains 11 manipulated and 41 measured variables, as well as 28 predefined fault scenarios to choose from. To generate in-control and out-of-control process data, we utilize the MATLAB/Simulink-based GUI developed by Anderson et al. (2022). In total, 50 hours of in-control process data were generated and collected in Phase I for quantile learning and threshold value h determination.

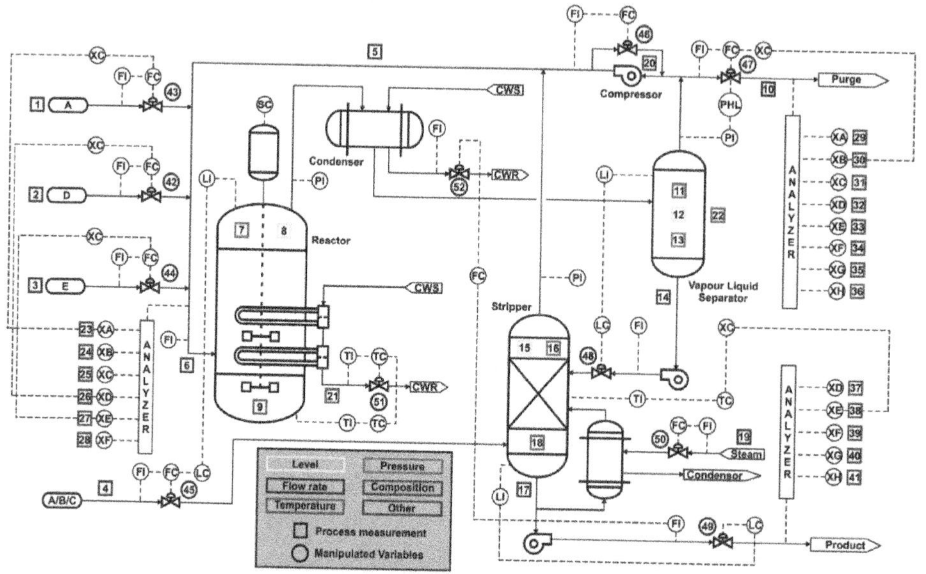

Figure 1. Schematic of Tennessee Eastman Process (source: Ma et al., 2020)

As an illustrative example, we perform process monitoring and fault detection of three representative faults, namely IDV 2, 3, and 13, as summarized in Table 1. In Phase II, we first run the process at normal operations and collect 2 hours of in-control data, followed by starting the fault and operate the process and collect 3 hours of out-of-control measurements. We compare our SPC approach with two other process monitoring algorithms commonly used in chemical industries, namely PCA and SVM. Specifically, we utilize the open source Pyphi package developed by García-Muñoz at Eli Lilly that performs multivariate PCA and Hotelling's T^2 analyses (López-Negrete et al., 2010). We select and keep all principal components whose eigenvalues are greater than 1, thereby leading to a total of 15 principal components four all four fault scenarios. For SVM, we adopt the method and comparison results of Onel et al. (2018).

Table 1. Summary of fault scenarios considered in this comparison study.

Fault No.	Description	Fault Type
IDV 2	Stream 4 composition of B (with constant A/C ratio)	Step
IDV 3	Temperature in stream 2	Step
IDV 13	Reaction kinetics	Slow drift

The comparison results of fault detection speed and the corresponding false alarm rate of all three monitoring frameworks, quantified by how many additional observations (after fault is introduced) are needed for each algorithm before it realizes the process's out-of-control status and raises an alarm, are tabulated in Table 2. As we can see, among the three monitoring frameworks, quantile-based SPC framework yields the fastest fault detection speed in all three fault scenarios, while maintaining the lowest false alarm rate. Furthermore, quantile-based SPC framework only stores and uses quantile information ($I_{j,i}$) for online monitoring and fault detection and is thus very computationally efficient. This result is exciting, given that a lower false alarm rate will lead to reduced fault detection speed due to more conservative monitoring behavior. While more thorough and extensive comparison studies are still ongoing, preliminary results presented in this work clearly demonstrate the effectiveness and attractiveness of this novel SPC framework in effective online monitoring of big data streams and robust, reliable fault detection.

Table 1. Summary of fault detection speed (characterized by out-of-control run length) and false alarm rate using three monitoring frameworks.

Fault No.	SPC	PCA-T^2	SVM
IDV 2	125 (0.27%)	216 (0.5%)	180 (0.8%)
IDV 3	95 (0.27%)	366 (0.5%)	16815 (83%)
IDV 13	128 (0.27%)	1131 (0.5%)	675 (12.7%)

5. Conclusion

In this work, we present a novel, powerful chemical process monitoring and fault detection framework for nonparametric, heterogeneous big data streams. In particular, the heterogeneity nature of this framework is enabled by recent advancements in SPC, such as the quantile-based multivariate CUSUM (Ye and Liu, 2022). We also compare the performance of this SPC framework with two benchmark process monitoring algorithms, PCA-T^2 and nonlinear SVM, in the classic example of Tennessee Eastman Process. Compared with existing dimensionality reduction or machine learning based approaches, the SPC-based framework possesses several advantages, including high reliability and accuracy with customizable, precisely controlled false alarm rate, guaranteed detection of process anomalies or mean shifts, significantly faster fault detection speed, unique capabilities to handle nonparametric and heterogeneous big data streams, low computational costs, etc.

References

E.B. Andersen, I.A. Udugama, K.V. Gernaey, A.R. Khan, C. Bayer, M. Kulahci, 2022, An easy to use GUI for simulating big data using Tennessee Eastman process, *Quality and Reliability Engineering International*, 38, 1, 264-282.

J.J. Downs, E.F. Vogel, 1993, A plant-wide industrial process control problem, *Computers and Chemical Engineering*, 17, 3, 245-255.

R. Fezai, M. Mansouri, O. Taouali, M.F. Harkat, N. Bouguila, 2018, Online reduced kernel principal component analysis for process monitoring. *Journal of Process Control*, 61, 1-11.

P. Geladi, B.R. Kowalski, 1986, Partial least-squares regression: a tutorial. *Analytica Chimica Acta.*, 185, 1-17.

S. Heo, J.H. Lee, 2018, Fault detection and classification using artificial neural networks, IFAC-PapersOnLine, 51, 18, 470-475.

J.E. Jackson, G.S. Mudholkar, 1979, Control Procedures for Residuals Associated with Principal Component Analysis. *Technometrics*, 21, 341-349.

K. Liu, J. Shi, 2013, Objective-oriented optimal sensor allocation strategy for process monitoring and diagnosis by multivariate analysis in a Bayesian network, *IIE Trans.*, 45, 6, 630-643.

K. Liu, Y. Mei, J. Shi, 2015, An Adaptive Sampling Strategy for Online High-Dimensional Process Monitoring, *Technometrics*, 57, 3, 305-319.

J. Li, 2020, Efficient global monitoring statistics for high-dimensional data, *Qual. Rel. Eng. Int.*, 36, 1, 18-32.

R. López-Negrete, S. García-Muñoz, L.T. Biegler, 2010, An efficient nonlinear programming strategy for PCA models with incomplete data sets, 24, 301-311.

L. Ma, J. Dong, K. Peng, 2020, A novel key performance indicator oriented hierarchical monitoring and propagation path identification framework for complex industrial processes, *ISA Trans.*, 96, 1-13.

Y. Mei, 2010, Efficient scalable schemes for monitoring a large number of data streams, *Biometrika*, 97, 2, 419-433.

Y. Mei, 2011, Quickest detection in censoring sensor networks, *Proc. IEEE Int. Symp. Inf. Theory Proc.*, 2148-2152.

M. Onel, C.A. Kieslich, E.N. Pistikopoulos, 2018, A nonlinear support vector machine-based feature selection approach for fault detection and diagnosis: Application to the Tennessee Eastman process, *AIChE Journal*, 65, 3, 992-1005.

P. Qiu, D. Hawkins, 2001, A Rank-Based Multivariate CUSUM Procedure, *Technometrics*, 43, 2, 120-132.

P. Qiu, D. Hawkins, 2003, A nonparametric multivariate cumulative sum procedure for detecting shifts in all directions, *Journal of the Royal Statistical Society: Series D*, 52, 2, 151-64.

P. Qiu, Z. Li, 2011, On nonparametric statistical process control of univariate processes, *Technometrics*, 53, 4, 390-405.

E. Russell, L.H. Chiang, R.D. Braatz, 2000, Data-driven methods for fault detection and diagnosis in chemical processes. Springer, 10-45.

A.G. Tartakovsky, B.L. Rozovskii, R.B. Blažek, H. Kim, 2006, Detection of intrusions in information systems by sequential change-point methods", *Statist. Methodol.*, 3, 3, 252-293.

W.H. Woodall, D.J. Spitzner, D.C. Montgomery, S. Gupta, 2004, Using Control Charts to Monitor Process and Product Quality Profiles, *Journal of Quality Technology*, 36, 3, 309-320.

X. Xian, C. Zhang, S. Bonk, K. Liu, 2021, Online monitoring of big data streams: A rank-based sampling algorithm by data augmentation, *Journal of Quality Technology*, 53, 2,135-153.

H. Ye, K. Liu, 2022, A Generic Online Nonparametric Monitoring and Sampling Strategy for High-Dimensional Heterogeneous Processes, *IEEE Trans. Auto. Sci. Eng.*, 19, 3, 1503-1516.

C. Zou, W. Jiang, Z. Wang, X. Zi, 2015, An efficient on-line monitoring method for high-dimensional data streams, *Technometrics*, 57, 3, 374-387.

Antonis Kokossis, Michael C. Georgiadis, Efstratios N. Pistikopoulos (Eds.)
PROCEEDINGS OF THE 33rd European Symposium on Computer Aided Process Engineering (ESCAPE33), June 18-21, 2023, Athens, Greece
 http://dx.doi.org/10.1016/B978-0-443-15274-0.50259-6

Perspectives on New Trends of Statistical Process Monitoring for Industrial Process Safety

Cheng Ji, Jingde Wang, Wei Sun*

[a] *College of Chemical Engineering, Beijing University of Chemical Technology, 100029, North Third Ring Road 15, Chaoyang District, Beijing, China*
**emailOfTheCorrespondingAuthor: sunwei@mail.buct.edu.cn*

Abstract

Smart manufacturing with the help of internet of things and artificial intelligence has attracted significant attention as the modern process industry is highly integrated with information technology. Among them, statistical process monitoring (SPM) is a powerful big data analytics tool for ensuring industrial process safety. In this work, developments in multivariate SPM and corresponding monitoring indicators are briefly introduced. Researches regarding new trends of SPM, the feature-based SPM methods are reviewed and discussed, including statistics pattern analysis and statistics Mahalanobis distance. The feature-based SPM has been considered as one of possible directions in next generation SPM, which shows better process monitoring performance in both sensor faults and process faults through two case studies. Meanwhile, certain limitations and challenges also can be revealed. We discuss the reasons for these challenges from the perspective of industrial process data, and suggestions are provided accordingly to make feature-based SPM better contribute the intelligent manufacturing.
Keywords: multivariate statistical analysis, feature selection and extraction, sensor and process fault detection, intelligent manufacturing, internet of things.

1. Introduction

The widespread application of distributed control systems facilitates the rapid development of data-driven process monitoring. Statistical process monitoring (SPM), as a class of highly interpretable data-enabled modeling methods, have become a powerful tool for industrial process safety. This is achieved mainly by establishing feature extraction models from historical data to support prompt decision-making regarding the operating status of a process. Statistical process control (SPC) chart, such as cumulative sum chart (CUSUM) chart and exponentially weighted moving average (EWMA) chart could be considered as the earliest SPM, in which univariate upper and lower control limits are employed to determine whether the process is remaining in a state of statistical control(MacGregor, 1994). However, with the extensive arrangement of instruments since 1980s, the limitations of univariate SPM have been revealed. The most important one is that the correlations among process variables are not considered(MacGregor, 1994), since no model is exploited in SPC. Meanwhile, multivariate SPM (MSPM) combining latent variable models and monitoring indicators appears and rapidly develops(Kresta, et al., 1991). In last thirty years, people have witnessed various successful application of MSPM, such as principal component analysis (PCA), partial least squares (PLS) and their variants(Ji and Sun, 2022, Qin, 2012). Although MSPM has become a standard for industrial applications, there are still several challenges to be addressed. The first one is that data are assumed to be multivariate normally distributed in most MSPM methods, while the assumption is usually not satisfied because the complex features contained in industrial data, such as nonlinearity, process dynamics, and non-Gaussianity, etc(He and Wang, 2018). Although many efforts have been made to deal with this issue through kernel tricks and dynamic latent variable models, the monitoring performance still needs to be further improved. Moreover, the monitoring indicators and corresponding boundary of control region in most methods are still determined under the assumption of normal distribution. High missing alarm rate and false alarm rate will be triggered despite several improvements in monitoring indicators have been developed. On the other hand, traditional MSPM methods focus on process faults, but the incipient sensor faults, which refers to faults with tiny magnitudes(Ji, 2021), are generally not considered. Therefore, more advanced methods need to be proposed to address the above challenges. Wang and He proposed a new concept named statistics pattern analysis (SPA), in which features of process variables rather than the process variables themselves are employed as the monitoring statistics(Wang and He, 2010). Motivated by SPA, more feature-based SPM methods have been further proposed(Ji, 2021, Shang, et al., 2017). Some researchers suggest that it can be regarded as a candidate for next generation SPM(He and Wang, 2018). However, it can be observed that current feature-based SPM methods are mostly designed for incipient sensor fault, and their superiority in monitoring process faults cannot be clearly demonstrated compared to state-or-the-art MSPM. The reason could be the lack of sufficient analysis concerning the characteristics of industrial process data. Regarding this issue, this work compares common MSPM methods and feature-based SPM methods through two cases, by which the challenges in existing feature-based SPM are analyzed. Combined with the case studies, promising suggestions for overcoming these challenges are discussed, and the potential of feature-based SPM in monitoring various types of faults in industrial processes are expected.

2. Related Methods

In this section, common MSPM and feature-based SPM methods used in this work for comparison are briefly introduced.

2.1. Multivariate Statistical Process Monitoring

MSPM methods mostly employ multivariate linear projection to obtain latent features that could explain most information of the original process variables, and then monitoring indicators are established in the feature space to perform fault detection. Considering MSPM methods have been widely reviewed in literature, only PCA is briefly introduced, and the focus will be on the summary of monitoring indicators.

2.1.1. Principal Component Analysis

PCA is the most commonly used dimensionality reduction method by extracting the cross-correlation among process variables, which has been widely applied in SPM. Given a set of historical data $X_{n\times m}$ with n samples and m variables, PCA aims to find a projection $P_{m\times k}$ that could transform $X_{n\times m}$ into an orthogonal space, where most information of original data can be explained,

$$X_{n\times m} = TP_{m\times k}{}^{T} + E \tag{1}$$

where $T = X_{n\times m}P_{m\times k}$ is the score matrix, E is the residue matrix, and k represents the number of principal components. To perform fault detection, monitoring indicators could be established on the principal component space and residual space respectively.

2.1.2. Monitoring Indicators

T^2 statistic is a widely applied monitoring indicator in MSPM to measure changes in the principal component space, which is calculated as follows,

$$T^2 = \sum_{i=1}^{k} \frac{(p_i{}^T x)^2}{\lambda_i} \tag{2}$$

where p_i is the projection of the i_{th} principal component direction. Comparatively, squared prediction error (SPE) is commonly used to measure the changes in the residue space, which can be calculated as follows,

$$SPE = \sum_{i=k+1}^{m} (p_i{}^T x)^2 \tag{3}$$

Moreover, the Mahalanobis distance (MD) is also widely used to measure the changes in all projection directions.

$$MD = \sqrt{\sum_{i=1}^{m} \frac{(p_i{}^T x)^2}{\lambda_i}} \tag{4}$$

Raich and Cinar further proposed a combined indicator that could consider changes in principal component space and residue space simultaneously(Raich and Cinar, 1996),

$$\varphi_1 = c\frac{T^2}{\chi^2} + (1-c)\frac{SPE}{\delta^2} \tag{5}$$

where χ^2 and $\delta^2$are the control limit of T^2 and SPE statistic, and c is a weight coefficient. However, a high missing alarm rate will be triggered because φ_1 could be less than 1 even if either T^2 statistic or SPE statistic exceeds its threshold(Yue and Qin, 2001). Aiming at this issue, Yue and Qin suggested using the following combined indicator(Yue and Qin, 2001),

$$\varphi_2 = \frac{T^2}{\chi^2} + \frac{SPE}{\delta^2} \tag{6}$$

and the control limit is also derived distribution of quadratic forms. Regardless of which monitoring indicator is adopted, the challenges in MSPM mentioned before still need to be faced.

2.2. Feature-based Statistical Process Monitoring

The feature-based SPM is a concept proposed by Wang and He(Wang and He, 2010), where the features of process variables rather than the process variables themselves are employed as the monitoring statistics. Several related methods will be introduced in this part.

2.2.1. Dissimilarity of Process Data

Actually, the concept of feature-based SPM could be reflected in early methods, such as the dissimilarity of process data (DISSIM)(Kano, et al., 2002). In this work, the distribution of process variables, especially the correlation among variables, which can be regarded as a kind of feature, are monitored. Therefore, this method works better than traditional MSPM on detecting changes in correlation structure. However, certain faults cannot be effectively detected since the features of variables considered by this method are still limited.

2.2.2. Statistics Pattern Analysis

The concept of feature-based SPM is first proposed with SPA. In SPA, process variables are divided into moving windows, and the features of the variables including means, variance, etc. in the windows are monitored. Then the PCA is established on the selected features for fault detection. In the case study, hundreds of features are considered by SPA, and therefore it is applicable to most kinds of faults.

2.2.3. Statistics Mahalanobis Distance

Motivated by SPA, statistics Mahalanobis distance (SMD) is newly proposed for incipient sensor fault detection(Ji, 2021). Unlike SPA, the MD is directly established on the features of process variable as the monitoring indicator, and the results show that the SMD performs perfectly in three common incipient faults by only considering mean and variance, which greatly reduces the computational costs.

3. Case Studies

In this section, a common numerical example and Tennessee Eastman (TE) process are used to compare the process monitoring performance of MSPM and feature-based SPM. Datasets and the implementation of all the methods are available in literature, and the codes can be obtained by request to the corresponding author.

3.1. A Numerical Example of Incipient Sensor Faults

A simple multivariate process given in literature is used as the first case study(Ji, 2021). In this case, a set of 100,000 samples are generated under normal conditions as the training data, and three sets containing 4000 samples each are used as test data. The three test data sets correspond to three common incipient sensor faults, which are constant bias, gain degradation, and precision degradation. The faults are introduced in the 1500th sample, and the fault detection rate (FDR) and false alarm rate(FAR) are adopted to evaluate the fault detection performance.

Table 1 A summary of process monitoring results for incipient sensor faults

FDR/FAR %	PCA T^2	PCA SPE	MD	φ_1	φ_2	**SMD**	DISSIM	SPA T^2	SPA SPE
Fault 1	1/1	1.7/1.3	4.8/0.9	2/1	0.8/0.5	**98.7/0.5**	0.3/0	61.5/1.3	97/2.7
Fault 2	0.8/1	0.8/0.8	1.6/1.1	0.7/1.1	0.4/0.5	**97.8/3.6**	7.2/0.3	1/9.7	9/1.1
Fault 3	2.6/1.2	4.4/0.6	21.9/0.9	6.2/1.1	1.9/0.6	**97.5/3.7**	99.1/0.9	76.9/4.4	98.3/0.5

The results are summarized in Table 1. It can be found that all MSPM methods fail to detect the three incipient faults because the magnitude of the fault is low. By contrast, feature-based SPM methods perform much better. Among them, DISSIM only works on fault 3. It is reasonable because only variable correlation is considered and it has been proposed for twenty years. SPA works well in fault 1 and fault 3, but the FDR of fault 2 is low because the window in which the gain degradation is most statistically evident is when the fault has just been introduced. When the window moves forward, the dissimilarity of faulty data and normal data is not that obvious, resulting in a low FDR. Moreover, the features used in SPA are too much, which may cause redundancy and decrease the sensitivity of the model to faults. Comparatively, SMD achieves a more than 95% FDR for all faults by considering only the mean and variance, which indicating its superiority to monitoring incipient sensor faults.

3.2. Tennessee Eastman Process

TE process is the most commonly used benchmark process in process monitoring domain. Two faults including a random variation fault and a sensor fault are taken as examples, which both are introduced at the 160th sample.

Table 2 A summary of process monitoring results for TE process

FDR/FAR %	PCA T^2	PCA SPE	MD	φ_1	φ_2	**SMD**	SPA T^2	SPA SPE
Fault 10	29.6/0	25.8/0.6	88.4/0	45.5/0.6	28.6/0	**100/100**	79.5/-	73.9/-
Fault 21	39.3/0	47.2/3.1	54.7/1.9	49.6/3.7	40.6/0	**100/100**	99.9/-	90.8/-

The results are summarized in Table 2. The SMD cannot be applied to monitor process faults because it is developed for sensor faults, and different from sensor variables, the process variables are highly influenced by random factors. As shown in Figure 1, MD performs the best among all indicators in MSPM for fault 10 because changes in all principal components are added up to obtain the MD statistic. This fault is hard to detect for its low magnitude and the response of control systems, while the SPA still performs well as long as there are features deviated from normal operating conditions. For fault 21, a valve is fixed at a constant value. As shown in Figure 2, the fault can be detected by all MSPM methods only after it has affected process variables. It can only be timely detected by SPA because the fault is reflected in the variance of the valve, which is a monitored feature of SPA. Overall, feature-based SPM represented by SPA shows obvious advantages in monitoring both sensor faults and process variables. However, there are also several challenges should be focused on, which will be discussed in next section.

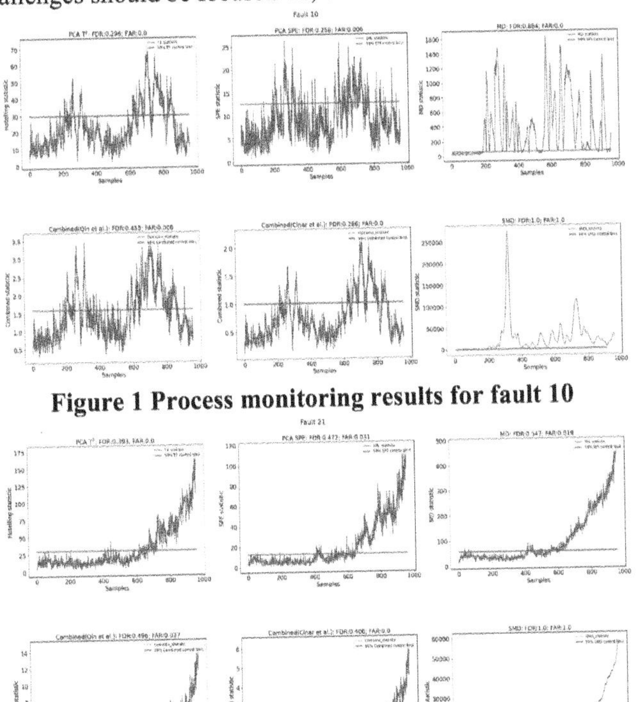

Figure 1 Process monitoring results for fault 10

Figure 2 Process monitoring results for fault 21

4. Challenges and Suggestions for Feature-Based Statistical Process Monitoring

4.1. Redundancy of Features

The feature selection plays the most important role in feature-based SPM. The deficiency or redundancy will affect the process monitoring performance. For example, there are 265 features used for TE process in SPA. The number could be even larger for industrial processes with more variables. It can not only lead to high computational costs but also affect the sensitivity of fault detection. In particular, variable correlation, although an important monitoring statistic, will contribute at least $m(m-1)/2$ features. It is suggested that features more than two orders could be excluded because they are regarding to the distribution of data, and the changes in the distribution could be reflected in other low-order statistics. Moreover, orthogonal transformation is recommended as a preprocessing to avoid the calculation of features of variable correlation.

4.2. Poor Detectability

Since feature-based SPM is a class of methods using moving window technique, the detectability is an unavoidable challenge. Generally, slowly changing faults cannot be early detected by common features of the window, such as the mean and variance. To address this issue, secondary features are suggested to be employed, especially for the maximum, minimum and slope of the window.

4.3. Parameter Selection

At last, the parameter selection is also a challenge that has not been effectively solved. The results of feature-based SPM are highly affected by the size of the moving window. Especially, it is not fixed, but determined according to target processes. It is suggested to be determined by a quantitative indicator obtained by information entropy or other mathematical methods in future research.

5. Conclusion

In this work, the performance of MSPM and feature-based SPM in monitoring both sensor faults and process faults is compared through two case studies. The results show that traditional MSPM is not applicable to monitor incipient sensor faults, while the FDR of the SMD reaches an average of 98%. Moreover, the SPA also achieves better process monitoring performance for process faults. It can be concluded that the feature-based SPM represented by SPA shows strong applicability to various types of faults in industrial processes and could be regarded as a candidate to the next generation of SPM. Existing challenges are also analyzed and corresponding solutions are discussed. Future work will focus on the development of more effective feature-based SPM method by addressing the challenges outlined in this paper.

Acknowledgments

This work was supported by the National Natural Science Foundation of China (grant numbers 22278018).

References

Q. P. He, J. Wang, 2018, Statistical process monitoring as a big data analytics tool for smart manufacturing, J. Process Control 67, 35-43.

C. Ji, W. Sun, 2022, A Review on Data-Driven Process Monitoring Methods: Characterization and Mining of Industrial Data, Processes 10(2), 335.

H. Ji, 2021, Statistics Mahalanobis distance for incipient sensor fault detection and diagnosis, Chem. Eng. Sci. 230.

M. Kano, S. Hasebe, I. Hashimoto, H. Ohno, 2002, Statistical process monitoring based on dissimilarity of process data, AIChE Journal 48(6), 1231-1240.

J. V. Kresta, J. F. MacGregor, T. E. Marlin, 1991, Multivariate statistical monitoring of process operating performance, The Canadian Journal of Chemical Engineering 69(1), 35-47.

J. F. MacGregor, 1994, Statistical Process Control of Multivariate Processes, IFAC Proceedings Volumes 27(2), 427-437.

S. J. Qin, 2012, Survey on data-driven industrial process monitoring and diagnosis, Annual Reviews in Control 36(2), 220-234.

A. Raich, A. Cinar, 1996, Statistical process monitoring and disturbance diagnosis in multivariable continuous processes, AIChE J. 42(4), 995-1009.

J. Shang, M. Chen, H. Ji, D. Zhou, 2017, Recursive transformed component statistical analysis for incipient fault detection, Automatica 80, 313-327.

J. Wang, Q. P. He, 2010, Multivariate statistical process monitoring based on statistics pattern analysis, Ind. Eng. Chem. Res. 49(17), 7858-7869.

H. H. Yue, S. J. Qin, 2001, Reconstruction-based fault identification using a combined index, Ind. Eng. Chem. Res. 40(20), 4403-4414.

Antonis Kokossis, Michael C. Georgiadis, Efstratios N. Pistikopoulos (Eds.)
PROCEEDINGS OF THE 33rd European Symposium on Computer Aided Process Engineering (ESCAPE33), June 18-21, 2023, Athens, Greece
 http://dx.doi.org/10.1016/B978-0-443-15274-0.50260-2

Aeration Optimization and Control for Wastewater Treatment Processes

Ioana Naşcu[a,b], Wenli Du[b], Ioan Naşcu,[a]

[a]*Department of Automation Technical University of Cluj Napoca, Romania*
[b]*Key Laboratory of Advanced Control and Optimization for Chemical Processes, East China University of Science and Technology, Shanghai 200237, China*
Ioana.Nascu@aut.utcluj.ro

Abstract

In this work we develop model based control strategies for the optimization and control of the aeration of a municipal wastewater treatment plant, intermittently operated by switching the aeration on and off. The proposed strategies aim to reduce energy consumption as well as to further improve effluent quality. A two-layer hierarchical control structure is implemented: (i) the lower layer using a modified version of the GPC to control the dissolved oxygen concentration and minimize the aeration energy and (ii) the higher level (optimization layer) where standard GPC algorithm is used to obtain the optimal values for the setpoint of the dissolved oxygen in the bioreactor. The developed control strategies show good performances and significant energy savings in comparison with the conventional control systems.

Keywords: wastewater, MPC, hierarchical control, energy efficiency

1. Introduction

The optimal treatment and administration of industrial and municipal wastewater is critical in maintaining a safe and clean environment as well as supporting the health of the community. When dealing with Wastewater Treatment Plants (WWTPs) it is important that the plant operation can satisfy the requirements for the effluent water quality. Unfortunately this leads to a large amount of energy consumption. Therefore, another important goal for the WWTP, which has become a hot research topic during the last years, is energy efficiency optimization (Daverey et al. 2019, Cardoso et al. 2021).
The main part of the WWTPs is the biological treatment where the most used technology is the Activated Sludge Process (ASP). This is due to its cost-effectiveness, its flexibility and high reliability as well as the capability to produce effluent of high quality. Usually, the activated sludge processes are highly complex, nonlinear and challenging to control. Various works have focused on developing accurate models used to design and tune controllers (Harja and Nascu 2019, Naşcu et al. 2022). There has also been some focus on the development of multiple model based controllers as well as predictive (Nascu and Nascu 2016, Bernardelli et al. 2020), adaptive (Vlad et al. 2010), fuzzy (Yelagandula and Ginuga 2022) or fractional order PID control (Harja et al. 2015).
To optimize the performances of WWTP, real time control is a generally accepted method but one must also consider model uncertainties as well as the complexity of the process. In this work a control system with hierarchical structure is developed: a lower layer with process level control loops as well as a higher level dedicated for optimization.
Model based control (MPC) strategies have a variety of advantages when trying to control a process and are widely used in WWTP. They are robust against variable and unknown

time-delay and have good disturbance rejection properties. Moreover, they are able to incorporate constraints explicitly and handle uncertainty in process model parameters and model mismatch (Harja et al. 2016, Eagalapati et al. 2022). In this work MPC will be used in both layers of the control system.

2. Plant description an simulation platform

2.1. Plant Structure

The WWTP presented in this work is a medium municipal WWTP (maximum flow - 4500 m3/day). For this plant there are three main stages: mechanical or pre-treatment, biological treatment, and sludge treatment. In this work we will focus on the biological treatment, the most complex process used in WWTPs. The typical procedure for this treatment, is the activated sludge technology.

The biological treatment part is composed of two circular ring tanks. The inner ring is given by the secondary settler and the outer cylindrical tank represents the biological treatment tank. By switching the aeration on and off, they are operated alternately for nitrification and denitrification. In small and medium WWTP, as the one presented in this work, the bioreactor tank operates for nitrification and denitrification, alternatively. This results in having both processes for removal of nitrogen not spatially separated but sequentially in time by having the bioreactor operate once aerobic and once anaerobic in turn. To keep the contents of the bioreactor completely mixed, the bioreactor has to be equipped with submersible mixers. The aeration is performed by using fine bubble porous diffusers that are situated at the bottom part of the tank. The aeration required by the bioreactor is given through blowers with a maximum airflow of Qmax=1400 m3/h as well as the capability to operate at variable airflows. After the bioreactor, in the secondary settler the cleaned water from the top of the settler will overflow to the pumping station of the effluent. The sludge that is settled is recirculated to the bioreactor. A small part of the sludge is evacuated to the sludge treatment stage. The aerated bioreactor has a volume of 6483m3 and the volume of the secondary clarifier is 2048m3.

2.2. Mathematical model

Two models are used for the modeling and simulation of the WWTP processes: (i) The Activated Sludge Model No 1 (ASM1) (Henze et al. 2000) to simulate the biological processes from the aerated bioreactor and (ii) the double-exponential settling velocity model to simulate the transfer between the multiple layers in the settler. It is assumed that in the settler no biological processes occur.

The bioreactor tanks succeeded by a secondary settler represents the WWTP layout in the Benchmark Simulation Model no. 1 (BSM1)(Alex et al. 2008) where the first biological reactor tanks are anoxic well mixed and the next biological reactor tanks are aerated. To model the settler ten layers are used. The plant is developed such that, using nitrification and denitrification processes, it removes the nitrogen in wastewater. The nitrification takes place in the aerated tanks while denitrification takes place in the anoxic tanks.

2.2.1. Modification of the Simulation Platform

The BSM1 simulation platform mentioned above has to be adjusted such that it accurately represents the real physical WWTP described. Hence, some bioreactors as well as the internal recirculation are excluded from the BSM1 model. An intermittent biological reactor tank having anoxic and aerated phases chronologically alternated, the secondary settler as well as the main recirculation remains in the model. The block diagram for the new Simulink model obtained from BSM1 is presented in Figure 1. The model initialization parameters are changed (according to the model calibration presented in (Harja and Naşcu 2020, Harja and Nascu 2019)) to match the real plant parameters.

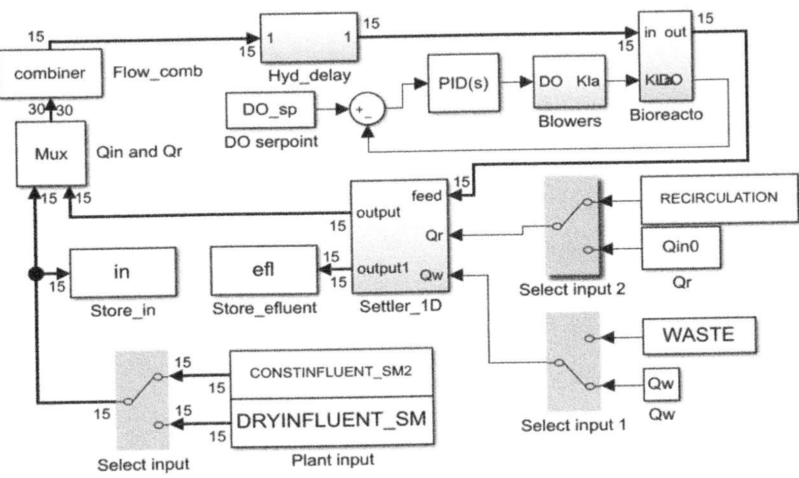

Figure 1 Modified Benchmark Simulation Model 1

To model the blowers and aeration system for the real process as accurately as possible, some modifications were performed. The variable from the BSM1 model that is used for the control of the dissolved oxygen concentration is the transfer coefficient KLa. Unfortunately, this variable cannot be used in a real plant. Therefore, to convert the aeration flow rate (further used as manipulated input for the bioreactor) into KLa, a new block is created. In technical literature, the relationship between these two variables is usually represented by an exponential model given by:

$$KLa(Q_{air}) = k1 \cdot (1 - exp^{-k2 \cdot Q_{air}})$$

Using the data collected from the real process for calibration to determine the values for k1 and k2 parameters, k1= 77 and k2=0.0015 were obtained.
To ensure the bioreactor with alternating nitrification and denitrification phases, the blowers are operating switching ON-OFF. To keep the DO concentration at the reference value determined by the optimization level, a control algorithm will be used.

3. Control Strategies

3.1. Hierarchical Control Structure

A simplified diagram of the hierarchical control structure used in this paper is presented in Figure 2. For the substrate concentration control in the bioreactor, the optimization control layer comprises of a MBPC strategy. The optimal values for the DO setpoint is given by the MBPC output.
The lower layer is composed of two main control loops: (i) control of the air pressure in the air piping system and (ii) the control of the DO in the aeration tank for which the optimization level generates the setpoint. The pressure control loop design is rather straightforward and is not presented in this work, unlike the control loop for the DO concentration which poses challenging issues. The ASPs are complex processes with significant nonlinearities, large disturbances as well as uncertainty in process model parameters. These challenges will be tackled by the MBPC algorithm which is used for the control loop of the DO concentration.
In this work, two versions for the Generalized Predictive Control (GPC) method are implemented. In the optimization layer, the standard GPC algorithm is used to obtain the optimal values for the setpoint of the DO in the aeration tank. In the lower layer, a

modified version of the GPC is used to control the DO concentration and to minimize the aeration energy.

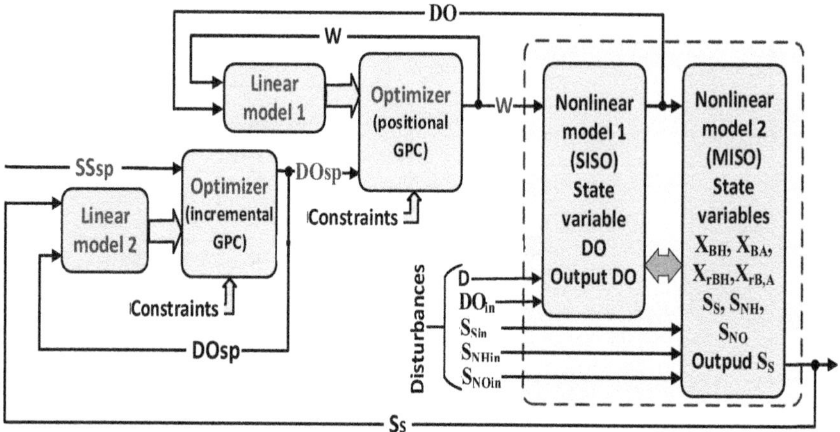

Figure 2. Hierarchical control system layers

3.2. GPC Algorithm

3.2.1. Standard GPC Algorithm

The GPC algorithm, developed by Clarke et al is a MBPC strategy that uses the process model to compute online the process output predictions and the optimal control. The optimal control is based on the minimization of the cost function (Clarke et al. 1987):

$$J = \sum_{j=N1}^{N2}[y(t+j) - y_r(t+j)]^2 + \sum_{j=0}^{Nu-1}\rho\,\Delta u(t+j)^2 \tag{1}$$

The design parameters represent: N_1 - the minimum costing horizon, N_2 - the maximum costing horizon, N_u - the control horizon, ρ - the control weight parameter. The process input and output are given by u and y respectively and y_r represents the future reference sequence.

3.2.2. Modified GPC Algorithm

An essential step to develop the control algorithm for DO concentration is to reparametrize the cost function (1) to include a measure of the energy that is consumed by aeration. This could lead to significant savings of energy by exploiting the fluctuation of the operating conditions. Because the aeration airflow W represents the manipulated process input given by the output, of the controller, the reparametrized cost function needs to include the controller output u, rather than the controller output increments Δu used in (1), to be able to minimize the airflow instead of its variations. This leads to a positional form of the cost function of the controller:

$$J = \sum_{j=N1}^{N2}[y(t+j) - y_r(t+j)]^2 + \sum_{j=0}^{Nu-1}\rho\, u(t+j)^2 \tag{2}$$

4. Results

The model as well as the control strategies presented in this paper are implanted in MATLAB Simulink. The model kinetic and stoichiometric parameters used in this paper are: YA =0.24, YH =0.7, fP =0.08 iXB =0.086, iXP =0.06, μH =4.59, KS =20.0, KO,H

=0.5, KNO =0.50, bH =0.4, ηg =0.6, ηh =0.8, kh =2.1, KX =0.025, μA =0.5, KNH =1.0, bA =0.05, KO,A =0.4, ka =0.05. In one complete aeration cycle with time period T, the bioreactor is operated 0.3T hours with aeration and 0.4T hours without. For the design of the standard and modified GPC controllers the following tuning parameters are used: (i) for the lower layer N1=1, N2=20, Nu=1, ro=0, and the sampling time is 0.01h (ii) for the higher layer N1=1, N2=10, Nu=1, ro=0 and the sampling time is of 1h.

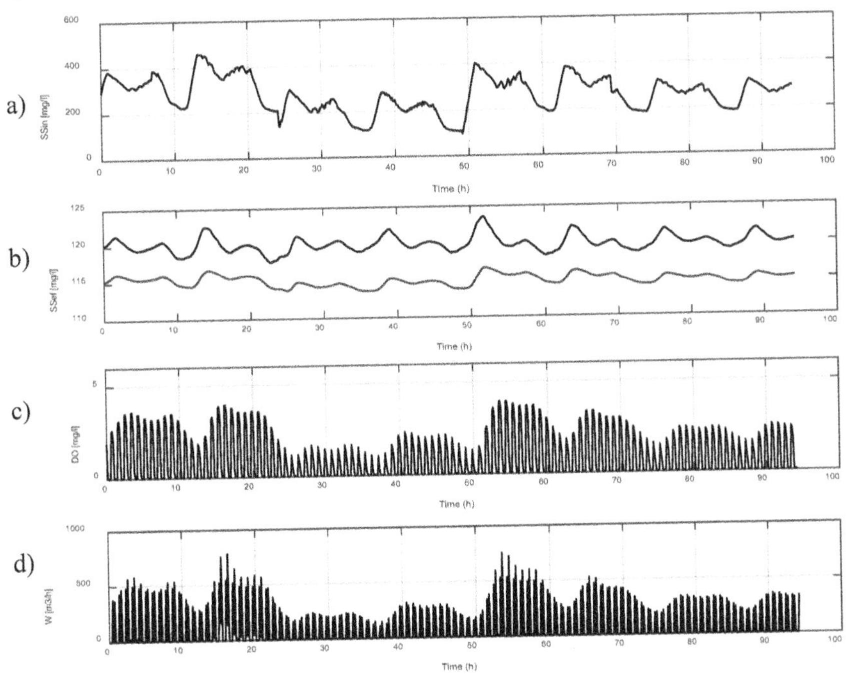

Figure 3. Simulation results. a) load disturbances, b) substrate concentration in the effluent, c) DO concentration in bioreactor, d) aeration flow

It can be observed that the developed control strategies are capable of reducing the concentration of organic substances in the effluent (SSe) in the presence of the variation of the loading of the influent with organic substance (SSin) and simultaneously aiming at a reduction of energy consumption. The variation of the loading of the influent, considered as disturbances are presented in Figure 3a. Figure 3d shows how the aeration flow changes depending on the aeration requirement. As a result, the concentration of dissolved oxygen in the bioreactor changes (Figure 3c) depending on the degree of loading of the influent so as to maintain the value of the concentration of organic substances in the effluent at the reference value (Figure 3b). Figure 3b shows the simulation results for a reference value of 120 mg/l (black line) and respectively 150 mg/l (blue line).

5. Conclusions

In this paper we develop model based control strategies to optimize the aeration of an intermittently operated municipal wastewater treatment plant to reduce the energy consumption as well as to further improve the quality of the effluent. A two-layer hierarchical control structure is implemented: (i) the lower layer using a modified version of the GPC to control the dissolved oxygen concentration and minimize the aeration energy and (ii) the higher level (optimization layer) where a predictive algorithm using the standard formulation of GPC is used to obtain the optimal setpoint values of the dissolved oxygen in the bioreactor. The developed strategies show good performances, the quality performances of the WWTP are maintained, disturbances from the influent are eliminated, and the content of organic substances is maintained within the prescribed limits while the energy consumption is significantly reduced.

Acknowledgments

Financial support from Ministry of Research, Innovation and Digitization, CCCDI - UEFISCDI, project number PN-III-P2-2.1-PED-2021-1147, within PNCDI III and with financial support from National Natural Science Fund for Distinguished Young Scholars (61725301), International (Regional) Cooperation and Exchange Project (61720106008).

References

Alex, J., Benedetti, L., Copp, J., Gernaey, K., Jeppsson, U., Nopens, I., Pons, M. N., Rieger, L., Rosen, C. and Steyer, J. P. (2008) 'Benchmark Simulation Model no. 1 (BSM1)', *Report by the IWA Taskgroup on Benchmarking of Control Strategies for WWTPs*.

Bernardelli, A., Marsili-Libelli, S., Manzini, A., Stancari, S., Tardini, G., Montanari, D., Anceschi, G., Gelli, P. and Venier, S. (2020) 'Real-time model predictive control of a wastewater treatment plant based on machine learning', *WATER SCIENCE AND TECHNOLOGY,* 81(11), 2391-2400.

Cardoso, B. J., Rodrigues, E., Gaspar, A. R. and Gomes, Á. (2021) 'Energy performance factors in wastewater treatment plants: A review', *Journal of Cleaner Production,* 322.

Clarke, D. W., Mohtadi, C. and Tuffs, P. S. (1987) 'Generalized predictive control-Part I. The basic algorithm', *Automatica,* 23(2), 137-148.

Daverey, A., Pandey, D., Verma, P., Verma, S., Shah, V., Dutta, K. and Arunachalam, K. (2019) 'Recent advances in energy efficient biological treatment of municipal wastewater', *Bioresource Technology Reports,* 7.

Eagalapati, S. S. T., Sheik, A. G. and Ambati, S. R. (2022) 'Fractional order-based hierarchical controller design and evaluation with Bürger-Diehl settler model in a total nitrogen removal wastewater treatment process', *Environmental Science and Pollution Research.*

Harja, G., Muresan, C., Nascu, I. and Vlad, G. (2015) *Fractional order PI control strategy on an activated sludge wastewater treatment process, 2015 19TH INTERNATIONAL CONFERENCE ON SYSTEM THEORY, CONTROL AND COMPUTING (ICSTCC).*

Harja, G. and Nascu, I. (2019) *Control of an activated sludge wastewater treatment process based on a Calibrated and modified BSM1 Model.*

Harja, G. and Naşcu, I. (2020) *Advanced control for nitrogen removal of an intermittently operated ASWWTP,* translated by 1-6.

Harja, G., Nascu, I., Muresan, C. and Nascu, I. (2016) 'Improvements in Dissolved Oxygen Control of an Activated Sludge Wastewater Treatment Process', *CIRCUITS SYSTEMS AND SIGNAL PROCESSING,* 35(6), 2259-2281.

Henze, M., Gujer, W., Mino, T. and van Loosdrecht, M. (2000) *Activated Sludge Models ASM1, ASM2, ASM2D, ASM3.*

Naşcu, I., Du, W. and Ioan, N. (2022) 'An Auto-tuning method for aeration control in activated sludge wastewater treatment processes', in *IEEE 2022 International Conference on Electrical, Computer, Communications and Mechatronics Engineering (ICECCME)*, Male, 16-18 nov.2022,

Nascu, I. and Nascu, I. (2016) 'Modelling and optimization of an activated sludge wastewater treatment process', *(ESCAPE26), PT A*, 1159-1164.

Vlad, G., Crisan, R., Muresan, B., Nascu, I. and Cosmin, D. (2010) *Development and application of a predictive adaptive controller to a wastewater treatment process.*

Yelagandula, S. and Ginuga, P. R. (2022) 'Control of a Waste Water Treatment Plant Using Fuzzy Logic Controller', *Journal of The Institution of Engineers (India): Series E,* 103(2), 167-177.

Antonis Kokossis, Michael C. Georgiadis, Efstratios N. Pistikopoulos (Eds.)
PROCEEDINGS OF THE 33rd European Symposium on Computer Aided Process Engineering (ESCAPE33), June 18-21, 2023, Athens, Greece
 http://dx.doi.org/10.1016/B978-0-443-15274-0.50261-4

Safe deployment of reinforcement learning using deterministic optimization over neural networks

Radu-Alexandru Burtea[a] and Calvin Tsay[a]

[a]*Department of Computing, Imperial College London, SW7 2AZ, United Kingdom*
[*]*c.tsay@imperial.ac.uk*

Abstract

Enabling reinforcement learning (RL) to explicitly consider constraints is important for safe deployment in real-world process systems. This work exploits recent developments in deep RL and optimization over trained neural networks to introduce algorithms for safe training and deployment of RL agents. We show how optimization over trained neural-network state-action value functions (i.e., a critic function) can explicitly incorporate constraints and describe two corresponding RL algorithms: the first uses constrained optimization of the critic to give optimal actions for training an actor, while the second guarantees constraint satisfaction by directly implementing actions from optimizing a trained critic model. The two algorithms are tested on a supply chain case study from OR-Gym and are compared against state-of-the-art algorithms TRPO, CPO, and RCPO.

Keywords: Constrained reinforcement learning, Optimization of neural network surrogates, Supply chain optimization, Optimization and machine learning toolkit

1. Introduction

Reinforcement learning (RL) has been central to many notable successes in machine learning, such as in self-driving cars, playing games, and operating data centers (Shin et al., 2019). Compared to traditional control strategies, e.g., model predictive control, RL does not typically consider state constraints explicitly. For many practical engineering applications, simply maximizing reward without considering the appropriateness of actions can lead to undesirable consequences. For instance, in supply chain applications, an RL algorithm may direct all goods to the cheapest warehouse without considering its maximum capacity, generating infeasible policies and/or creating massive backlogs.

Given the above, it is desirable to impose constraints on the range of behavior that can be explored in RL. This has inspired research into so-called safe RL (García and Fernández, 2015). Several methods for safe RL implicitly consider constraints using stage-wise reward or penalty functions. Alternatively, some safe RL techniques use external knowledge, e.g., imitation learning, and/or risk metrics during exploration.

This work takes advantage of two recent developments to enable explicit consideration of state constraints during deployment of RL: (i) incorporation of deep neural networks (NNs) into RL algorithms, known as deep RL, and (ii) techniques for deterministic optimization of trained NNs. We present two algorithms based on this combination. The first, called OMLT-DDPG, comprises an actor-critic method, where optimization over a critic NN gives optimal actions on which to train the actor. The second, called SAFE, is a strategy for deployment, wherein directly implementing actions from optimizing a trained critic NN guarantees constraint satisfaction.

The two proposed algorithms are applied to a multi-level supply chain case study from the `or-gym` library (Hubbs et al., 2020). Computational results demonstrate that OMLT-DDPG is significantly more sample-efficient compared to other RL methods, owing to the use of deterministic optimization. The results further show that SAFE explicitly satisfies constraints during RL deployment.

2. Safe Reinforcement Learning (RL) Background

Markov decision processes, or MDPs, are the foundation of RL problems. A given MDP is defined by a state space S, action space A, reward function $R: S \times A \times S \to \mathbb{R}$, and transition probability $P: S \times A \times S \to [0,1]$. The goal of RL is to learn a policy $\pi: S \to \mathcal{P}(A)$ that maximizes a performance metric $J(\pi)$, usually defined as the total expected reward R over an infinite time horizon, subject to a discount factor γ. Safe RL methods seek to additionally enforce constraints, typically by introducing some cost function(s), analogous to reward. The inclusion of cost functions results in constrained MDP, or CMDP. Recent methods for safe RL such as constrained policy optimization (Achiam et al., 2017) and interior-point policy optimization (Liu et al., 2020) can provide safety guarantees for CMDPs, but only as simple constraints on expected total discounted cost.

Trust-Region Policy Optimization (TRPO). TRPO (Schulman et al., 2015) serves as the basis for several safe RL algorithms: it is a policy iteration algorithm based on computing the advantage of one policy over another, i.e., the expected improvement in the performance metric between the iterations of a policy. Schulman et al. (2015) provide a method for approximating this. As these approximate updates resemble a first-order method when the policy is differentiable, the step size between policies should be constrained. This is typically done using a trust region method, such as constraining the KL divergence to the old policy when maximizing the advantage function.

Constrained Policy Optimization (CPO). Achiam et al. (2017) extend TRPO to CMDPs by adding constraints on the auxiliary cost functions to the above trust-region constraint on KL divergence, resulting in the following optimization problem to find policy π_{k+1}:

$$\begin{aligned} &\max_{\pi} J(\pi) \\ &\quad \text{s.t. } J_{C_i}(\pi) \leq d_i, \forall\, i = 1, \dots, m \\ &\quad\quad D_{KL}(\pi, \pi_k) \leq \delta \end{aligned} \tag{1}$$

where J_{C_i} is the total expected cost of the ith constraint over an infinite time horizon and δ limits the step size. The objective and constraints are replaced with surrogate functions, based on the algorithm hyperparameters. Again, this problem is computationally difficult and is solved using a primal-dual method after linearizing $J(\pi)$ and $J_{C_i}(\pi)$, and a second-order expansion for $D_{KL}(\pi, \pi_k)$. This approximation motivates a small step size δ.

Reward Constrained Policy Optimization (RCPO). Tessler et al. (2018) propose RCPO, which is similar to CPO but solves an unconstrained optimization problem with Lagrange multipliers. Updating the policy then becomes a bilevel optimization problem: $\min_{\lambda \geq 0} \max_{\pi} \big(J(\pi) - \lambda(J_c(\pi) - d)\big)$. This optimization problem can be viewed on two timescales: a faster one, where the policy is optimized, and a slower one, which involves increasing λ until the constraint is satisfied. This timescale decomposition is achieved by selection of different step sizes for the updates to the Lagrange multipliers and the policy.

Algorithm 1 OMLT-DDPG (one episode)

Require: $|R|, \sigma \geq 0$ ▷ Replay buffer, warm-up
Randomly initialize critic $Q(s, a|\theta^Q)$, actor $\mu(s|\theta^\mu)$, and target NNs $Q'(s, a|\theta'^Q)$, $\mu'(s|\theta'^\mu)$.
Observe: s_1
for $t = 1, \ldots, T$ **do**
    Take action: $a_t = \mu(s_t|\theta^\mu)$
    Observe: r_t, c_t, s_{t+1} and store: $R \leftarrow (s_t, a_t, r_t, c_t, s_{t+1})$
    Sample $\hat{R}$ from R, $|\hat{R}| = N$
    $y_i = r_i + \gamma Q'(s_{i+q}, \mu'(s|\theta'^\mu)|\theta'^Q), \forall i \in \hat{R}$
    Update $Q(s, a|\theta^Q)$ by minimizing MSE against $y_i, \forall i \in \hat{R}$
    if $t \geq \sigma$ **then**
        Optimize: $a_i^* = \text{argmax } Q(s_i, a_i|\theta^Q)$ s.t. constraints c_i ▷ OMLT Package
        Update $\mu(s|\theta^\mu)$ by minimizing MSE against $a_i^*, \forall i \in \hat{R}$
    end if
    $\theta'^Q \leftarrow \tau\theta^Q + (1 - \tau)\theta'^Q$
    $\theta'^\mu \leftarrow \tau\theta^\mu + (1 - \tau)\theta'^\mu$
end for

3. First Algorithm: OMLT-DDPG

Our approach presented in this section incorporates the Optimization and Machine Learning Toolkit (OMLT) into Deep Deterministic Policy Gradients (DDPG).

Deep Deterministic Policy Gradients (DDPG). DDPG (Lillicrap et al., 2016) was conceived as a continuous-space extension to the popular deep Q-learning framework (DQN). DQN cannot be directly applied to continuous action spaces, as selecting the action with maximum Q-value at a given state becomes inefficient for high-dimensional action spaces. DDPG is an off-policy, model-free algorithm that instead uses the Q-value function to estimate the policy gradient. Specifically, DDPG keeps a parameterized policy network, known as the actor, which deterministically maps states to actions. The other component is a critic network, which behaves as the Q-value function used in DQN. The actor is trained using gradients from the critic, and the critic is trained by minimizing the difference between the expected discounted rewards (following the greedy actor) and the current Q-value assigned by the critic to the state-action pair. Most implementations maintain a replay buffer to avoid "catastrophic forgetting" of previous transitions.

Optimization and Machine Learning Toolkit (OMlT). The challenge of selecting the action with maximum Q-value from a given neural network (NN) can be viewed as optimization over a trained neural network (the neural network parameters are fixed during an RL step). We propose to address this using the Optimization and Machine Learning Toolkit (OMLT), an open-source package for optimization over pre-trained machine learning models (Ceccon et al., 2022). OMLT enables users to easily translate learned machine learning models to optimization formulations. OMLT 1.0 supports GBTs through an ONNX (https://github.com/onnx/onnx) interface and NNs through both ONNX and Keras interfaces. OMLT transforms trained machine learning models into the Python-based algebraic modeling language Pyomo (https://github.com/pyomo/pyomo). The literature often presents different optimization formulations as competitors, but in OMLT, competing optimization formulations become alternative choices for users.

DDPG with Deterministic Optimization of NNs. Algorithm 1 gives pseudocode of our proposed algorithm. OMLT-DDPG preserves properties of the DDPG algorithm, but is extended for CMDPs and benefits from deterministic optimization of trained NNs.

The algorithm is initialized with actor and critic neural networks, as well as target actor and critic NNs. For each episode the agent observes the initial state and executes the

Algorithm 2 SAFE

Require: Trained critic $Q(s,a|\theta^Q)$
Observe: s_1, c_1
Optimize: $a_1^* = \text{argmax } Q(s_i, a_i|\theta^Q)$
Action: $a_1 = a_1^*$ s.t. constraints c_1 ▷ OMLT Package **for** $t = 2, \dots, T$ **do**
end
Optimize: $a_t^* = \text{argmax } Q(s_t, a_t|\theta^Q)$ s.t. constraints c_{t+1} ▷ OMLT Package
Select action: $a_t = a_t^*$

policy defined by the actor NN, storing transitions in the replay buffer. The algorithm then samples a batch of transitions of length N and updates the critic NN against the target. Up to this point, our algorithm closely resembles DDPG; however, DDPG uses the negative value of the critic for a given state-action pair as the loss for training the actor. DDPG-OMLT instead uses the optimal actions obtained by optimizing the critic NN, subject to the problem constraints. The algorithm preserves the theoretical properties of the policy gradients used in DDPG, as the gradient of the loss used for training the actor is obtained by applying a linear operation on the gradient of the policy network.

Note that this involves solving one constrained optimization problem for each sample from the replay buffer. Therefore, we use the predicted action from the actor as the initial guess to expedite optimization. We hypothesize that in the initial stages of training the actions chosen in this manner will be suboptimal, as the critic is not accurate enough to judge state-action pairs accurately. Nevertheless, this may prove beneficial for exploration. Given the above, we introduce a warm-up period σ, where only the actor is fixed and only the critic is updated. Following the actor and critic updates, the target networks are also updated using "soft" updates, i.e., only a certain fraction τ of the weights are updated. Lillicrap et al. (2016) found this to improve the stability of DDPG.

4. Second Algorithm: SAFE

The above DDPG-OMLT algorithm promotes safe exploration and exploitation by always incorporating environment constraints when optimizing over the critic network to select optimal actions. However, in deployment, the actor can still give an action that results in constraint violation. Here we describe a second algorithm, SAFE, that explicitly enforces constraints in deployment. In an actor-critic setting, the actor network usually gives the action to take (as implied by its name). OMLT (https://github.com/cog-imperial/OMLT) enables us to directly use the critic net- work to choose the optimal actions at each state. Specifically, we can optimize over the critic network to maximizing rewards, while enforcing environment constraints. If an actor NN is available, this can provide initial guesses for optimization. Algorithm 2 gives pseudocode for SAFE.

A key difference between OMLT-DDPG and SAFE is that OMLT-DDPG uses the constraint values in the previous timestep to evaluate the actor (after the action has been taken), while SAFE uses the constraint values in the current timestep to only take an action that is feasible. We note that SAFE is model-agnostic, meaning it can use any state-action value NN (supported by OMLT). While we only test SAFE with critic models from OMLT-DDPG, any Q-value NN that takes a state-action pair as input can be used.

5. Results and Discussion

We employ the multilevel supply chain case study from `or-gym` (Hubbs et al., 2020) to test OMLT-DDPG and SAFE, comparing against TRPO, CPO, and RCPO. The agent must place replenishment orders at nodes throughout levels of a supply chain, subject to lead times and uncertain customer demand at retail nodes. Inventories are subject to capacity constraints, with excess incurring a penalty cost. Likewise, penalties are incurred for unmet demand. Full details can be found in Hubbs et al. (2020). A three-level supply chain is selected, and algorithms are run for 150 episodes, except for RCPO, which required more episodes to converge. Each experiment is repeated with five random starts.

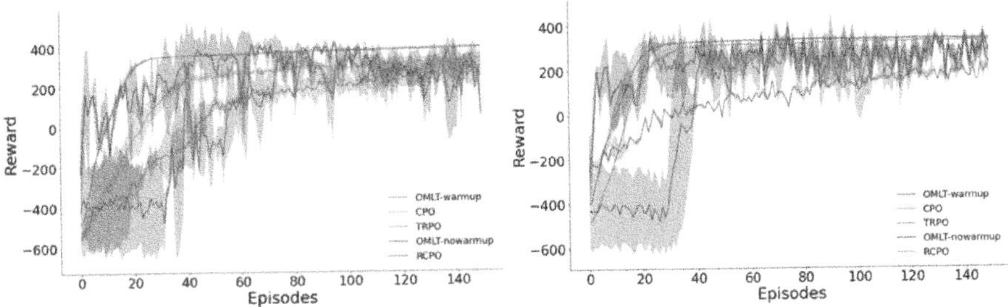

Figure 1: Rewards during training for nominal (left) and relaxed (right) supply chain.

DDPG is typically run with a lower learning rate for the actor than for the critic, allowing the two to converge on different "timescales." However, we found OMLT-DDPG to benefit from a faster learning rate for the actor, and we set the actor and critic learning rates to, respectively, 0.005 and 0.001. We used a batch size of 8, sampled from a replay buffer size of maximum 35000. We found a strong tradeoff between performance gain and computational time related to the batch size. We use a batch size of 40000 for CPO and TRPO. To investigate the effect of the feasible region size, we consider two settings for the supply chain case study: one with the nominal constraint, and a relaxed setting, where capacity constraints are increased by 40%.

Safe Reinforcement Learning. Figure 1 compares the performance of the various algorithms through 150 episodes. TRPO, CPO, and RCPO exhibit more stable behavior, while OMLT-DDPG seems to fluctuate. However, for the nominal case (left), OMLT-DDPG with warm-up achieves rewards only 6% and 28% lower than CPO and TRPO, respectively, which is noteworthy as OMLT- DDPG uses >1000x fewer samples. For the relaxed case (right), OMLT-DDPG performs inter- mediate to CPO and RCPO, while again using significantly fewer samples. This sample efficiency may be attributed to the information gain from the use of deterministic optimization. Indeed, OMLT-DDPG achieves a reward of 200 after only four episodes, and this steep learning curve is consistent in our experiments, suggesting behavior as a "few-shot" learner.

The incorporation of the warm-up period does not seem to help OMLT-DDPG in the nominal case (left), but greatly improves performance in the relaxed case (right). In general, the lack of warm-up period results in more aggressive behavior by the agent.

Safe Deployment. To simulate safe deployment in production, we deploy the models trained after the 150 episodes. Specifically, we deploy the trained models in a 30-episode simulation, varying the initial state randomly. The rewards and penalties incurred by using these trained models to operate the supply chain are shown in Figure 2. We find that OMLT-DDPG outperforms the other algorithms in deployment in terms of both maximizing rewards and avoiding penalties. Compared to CPO, OMLT-DDPG achieves 21% higher rewards and 20% less penalties in deployment in the nominal case. The SAFE algorithm results in more unstable rewards, but is the only algorithm to guarantee constraint satisfaction, avoiding all penalties during deployment. This demonstrates SAFE as the most appropriate option for guaranteed safe exploitation.

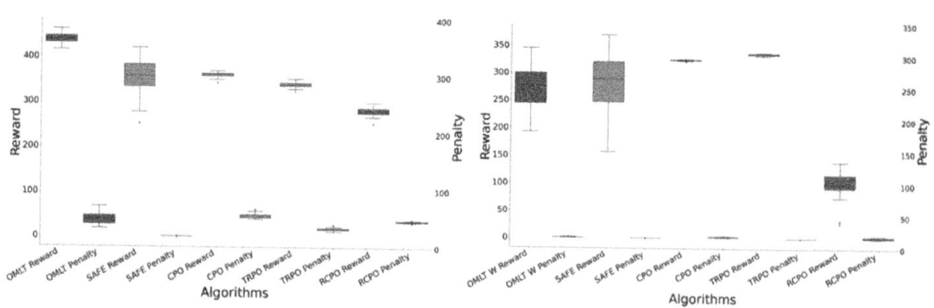

Figure 2: Deployment rewards/penalties for nominal (left) and relaxed (right) cases.

6. Conclusions

This paper introduces two algorithms, OMLT-DDPG and SAFE, for safe reinforcement learning and deployment, based on constrained optimization over trained critic networks. The first algorithm uses an actor-critic framework; constrained optimization over the critic network is used to provide targets on which the actor is trained. We show that this algorithm is very sample efficient, resembling behavior of a "few-shot" learner. The second algorithm uses constrained optimization over a pre-trained critic network to explicitly enforce process constraints during deployment.

7. Acknowledgments

The authors acknowledge support from the Engineering & Physical Sciences Research Council (EPSRC) grant EP/T001577/1 and an Imperial College Research Fellowship.

References

J. Achiam, D. Held, A. Tamar, P. Abbeel, 2017. Constrained policy optimization. In: International Conference on Machine Learning. PMLR, pp. 22–31.

F. Ceccon, J. Jalving, J. Haddad, A. Thebelt, C. Tsay, C. D. Laird, R. Misener, 2022. OMLT: Optimization Machine Learning Toolkit. Journal of Machine Learning Research 23, 1–8.

J. García, F. Fernández, 2015. A comprehensive survey on safe reinforcement learning. Journal of Machine Learning Research 16 (1), 1437–1480.

C. D. Hubbs, H. D. Perez, O. Sarwar, N. V. Sahinidis, I. E. Grossmann, J. M. Wassick, 2020. Or-gym: A reinforcement learning library for operations research problems. arXiv:2008.06319.

T. P. Lillicrap, J. J. Hunt, A. Pritzel, N. Heess, T. Erez, Y. Tassa, D. Silver, D. Wierstra, 2016. Continuous control with deep reinforcement learning. In: International Conference on Learning Representations.

Y. Liu, J. Ding, X. Liu, 2020. Ipo: Interior-point policy optimization under constraints. In: Proceedings of the AAAI Conference on Artificial Intelligence. Vol. 34. pp. 4940–4947.

J. Schulman, S. Levine, P. Abbeel, M. Jordan, P. Moritz, 2015. Trust region policy optimization. In: International Conference on Machine Learning. PMLR, pp. 1889–1897.

J. Shin, T. Badgwell, K. Liu, J.H. Lee, 2019. Reinforcement learning–overview of recent progress and implications for process control. Computers & Chemical Engineering 127, 282–294.

C. Tessler, D. J. Mankowitz, S. Mannor, 2018. Reward constrained policy optimization. In: International Conference on Learning Representations.

Antonis Kokossis, Michael C. Georgiadis, Efstratios N. Pistikopoulos (Eds.)
PROCEEDINGS OF THE 33rd European Symposium on Computer Aided Process Engineering (ESCAPE33), June 18-21, 2023, Athens, Greece
 http://dx.doi.org/10.1016/B978-0-443-15274-0.50262-6

Distributional Constrained Reinforcement Learning for Supply Chain Optimization

Jaime Sabal Bermúdez,[a] Antonio del Rio Chanona,[b] and Calvin Tsay[a,b*]

[a]*Department of Computing, Imperial College London, SW7 2AZ, United Kingdom*
[b]*Sargent Centre for Process Systems Engineering, Imperial College London, SW7 2AZ, United Kingdom*
**c.tsay@imperial.ac.uk*

Abstract

This work studies reinforcement learning (RL) in the context of multi-period supply chains subject to constraints, e.g., on inventory. We introduce Distributional Constrained Policy Optimization (DCPO), a novel approach for reliable constraint satisfaction in RL. Our approach is based on Constrained Policy Optimization (CPO), which is subject to approximation errors that in practice lead it to converge to infeasible policies. We address this issue by incorporating aspects of distributional RL. Using a supply chain case study, we show that DCPO improves the rate at which the RL policy converges and ensures reliable constraint satisfaction by the end of training. The proposed method also greatly reduces the variance of returns between runs; this result is significant in the context of policy gradient methods, which intrinsically introduce high variance during training.

Keywords: Safe reinforcement learning, Process operations, Inventory management

1. Introduction

Recent years have highlighted the importance of efficient supply chains in modern society. The field of inventory management deals with optimization of the ordering, storing, and selling of product inventory. There are many aspects of this problem that make it challenging, and a lot of benefits in doing so that make it worth investigating. Challenges include the presence of outdated "legacy" processes or the inefficient use of available physical capital. A variety of operations research methods have been used in tackling this family of problems, including dynamic programming, linear programming, and game theory. Given complex environments, however, modern inventory management also lends itself towards optimization through reinforcement learning (RL).

In the past few decades, machine learning has gained increasing popularity for its capacity to imitate the way humans learn. RL provides a framework for artificial agents to learn by interacting with an external environment and iteratively updating the way in which they act (i.e., their policy). Specifically, each interaction moves the agent to a new state and produces some quantifiable reward; the agent seeks to maximize cumulative reward (i.e., return) from its interactions. RL has proven very effective in areas such as gaming, demonstrating superhuman performance in incredibly complex games, or robotics and control, where problems can be too complex for analytical solution (Shin et al., 2019).

Real-world environments such as supply chains, may impose physical or safety limitations on the actions an agent can take. This work introduces a framework for RL in these circumstances, combining advantages of constrained and distributional RL.

Constrained RL algorithms extend the objective of an RL agent to include minimizing the expected costs below some threshold defined by the constraints in an environment, in addition to maximizing expected returns. In a related vein, *distributional* RL attempts to imitate the way in which organisms consider risk in decision-making, by estimating the distribution of values for all the states in the environment. The intrinsic stochasticity of many environments motivates using probability distributions to evaluate the possible future rewards associated with an action (Petsagkourakis et al., 2022). By estimating these distributions, distributional RL simplifies the parameterization of risk-adversity in RL, giving benefits in problems where constraints should be satisfied under uncertainty.

2. Reinforcement Learning (RL) Background

RL often employs a Markov Decision Process (MDP). An MDP is defined by the tuple $(\mathcal{S}, \mathcal{A}, \mathcal{R}, \mathcal{P}, \gamma)$, where $\mathcal{S}$ is the set of states in the environment, $\mathcal{A}$ is the set of possible actions, $\mathcal{R}: \mathcal{S} \times \mathcal{A} \times \mathcal{S} \rightarrow \mathbb{R}$ is a reward function, $\mathcal{P}: \mathcal{S} \times \mathcal{A} \times \mathcal{S} \rightarrow [0,1]$ is the transition probability function ($\mathcal{P}(s'|s,a)$ is the probability of transitioning to state s' by taking action a from state s), and γ is a discount rate for future rewards. A policy $\pi: \mathcal{S} \rightarrow \mathcal{P}(\mathcal{A})$ maps states to probability distributions over the possible actions ($\pi(a|s)$ is the probability of taking action a from state s). We denote the set of all possible stationary policies as Π. RL seeks a policy π that maximizes a performance measure, often taken as $J_R(\pi) = \mathbb{E}_{\tau\sim\pi}[\sum_t^\infty \gamma^t R(s_t, a_t, s_t + 1)]$, where τ is a trajectory sampled using π. Denoting $\hat{R}(\tau)$ as the discounted return of a trajectory, the on-policy state value function is denoted $V^\pi(s) = \mathbb{E}_{\tau\sim\pi}[\hat{R}(\tau)|s_0 = s]$, with optimal policy $\pi^* = \mathrm{argmax}_\pi V^\pi(s)$. Analogously, the state-action value function is $Q^\pi(s,a) = \mathbb{E}_{\tau\sim\pi}[\hat{R}(\tau)|s_0 = s, a_0 = a]$ and the advantage function of an action is $A^\pi(s,a) = Q^\pi(s,a) - V^\pi(s)$.

Trust Region Policy Optimization. Deep RL may parameterize the learned policy as a neural network (NN) with trainable parameters θ. Policy-gradient methods train the NN using the gradient $\nabla_\theta J(\theta) = \mathbb{E}_\pi[Q^\pi(s,a)\nabla_\theta \ln \pi_\theta(a|s)]$. This produces an unbiased update rule, but with large variance in the gradient estimates. Trust Region Policy optimization (TRPO) (Schulman et al., 2015a) deals with this noise by limiting the Kullback-Leibler (KL) divergence between consecutive policies to a threshold δ:

$$\max_\theta \mathbb{E}_{s\sim\rho^{\pi_k}, a\sim\pi_k}\left[\frac{\pi(a|s)}{\pi_k(a|s)} A^{\pi_k}(s,a)\right] \text{ s.t. } \mathbb{E}_{s\sim\rho^{\pi_k}}[\bar{D}_{\mathrm{KL}}(\pi_k(\cdot\,|s)||\pi(\cdot\,|s))] \leq \delta \tag{1}$$

where ρ^{π_k} is the state visitation probability density under π_k, and $\bar{D}_{\mathrm{KL}}$ the average KL divergence. Problem (1) is typically solved using Monte-Carlo estimates of $Q^{\pi_k}(s,a)$, linearization of the objective, and second-order approximation of the constraint.

Constrained Reinforcement Learning. Constrained RL algorithms often consider a constrained MDP (CMDP), which includes costs $\mathcal{C}: \mathcal{S} \times \mathcal{A} \times \mathcal{S} \rightarrow \mathbb{R}$ analogously to rewards, as well as a safety threshold d. The objective of the agent then becomes to maximize $J_R(\pi)$ such that $J_C(\pi) = \mathbb{E}_{\tau\sim\pi}[\sum_t^\infty \gamma^t C(s_t, a_t, s_{t+1})] \leq d$.

Constrained Policy Optimization (CPO) (Achiam et al., 2017) restricts policy updates similarly to TRPO, but to satisfy both the trust-region constraint from (1) and environment constraints. The proposed optimization problem can be solved efficiently using duality, where Lagrange multipliers for both the objective and environment constraints are estimated at each step. However, CPO is inherently subject to approximation errors in the estimate of the objective and constraints, which in practice lead it to converge to infeasible policies. Another way of enforcing constraints is to augment the state-space and reshape the objective of the problem. *Safety Augmented (SAUTE) MDPs* (Sootla et al., 2022) allow for a reformulation of the problem in terms of minimization of the reshaped objective, whereby constraints are incorporated through a safety budget. This approach only modifies the environment and is compatible with most off-the-shelf RL algorithms, but the state space increases with the number of constraints, leading to scalability issues and a loss in sampling efficiency as constraints are added.

3. Approach: Distributional Constrained Policy Optimization

In this work we apply distributional RL to the CPO setting to balance reliable constraint satisfaction and consistent sampling efficiency, independent of the number of constraints. Distributional RL (Bellemare et al., 2017) considers the probability distribution of returns over state-action pairs, which we extend to approximate distributions of costs to manage risk aversion. Algorithm 1 shows our distributional version of CPO, which we call DCPO.

Problem Formulation. Consider the CPO policy update with a single constraint:

$$\begin{aligned} \pi_{k+1} = &\underset{\pi \in \Pi_\theta}{\operatorname{argmax}}\, \mathbb{E}_{s\sim\rho^{\pi_k}, a\sim\pi_k}[A^{\pi_k}(s,a)] \\ &\text{s.t.} \quad J_C(\pi_k) + \frac{1}{1-\gamma}\mathbb{E}_{s\sim\rho^{\pi_k}, a\sim\pi_k}\left[A_C^{\pi_k}(s,a)\right] \le d \\ &\bar{D}_{\text{KL}}(\pi_k || \pi) \le \delta \end{aligned} \tag{2}$$

where $A_C^{\pi_k}$ denotes the estimated cost advantage under π_k for the cost function C. For small δ, we can linearize the objective and safety constraints around π_k. Denoting the gradients of the objective and constraint as g and b, respectively, the Hessian of the KL-divergence as H, and defining $c = J_C(\pi_k - d)$, (2) can be written in terms of the step direction $x = \theta - \theta_k$, giving: $x^* = \min_x g^T x$ s.t. $\{c + b^T x \le 0, x^T H x \le \delta\}$.

We note that g and b can be computed by back-propagation, and that H is always positive semi-definite, making (3) a convex optimization problem. Given a single cost function, the optimal point x^* (assuming one exists) is $x^* = \frac{-1}{\lambda^*}H^{-1}(g - v^* b)$, with dual variables:

$$v^* = \left(\frac{\lambda^* c - r}{s}\right);\ \lambda^* = \underset{\lambda \ge 0}{\operatorname{argmax}} \begin{cases} f_a(\lambda) := \frac{1}{2\lambda}\left(\frac{r^2}{s} - 1\right) + \frac{\lambda}{2}\left(\frac{c^2}{s} - \delta\right) - \frac{rc}{s}, & \text{if } \lambda c - r > 0 \\ f_b(\lambda) := \frac{-1}{2\lambda}\left(\frac{q}{\lambda} + \lambda\delta\right), & \text{otherwise} \end{cases}$$

where $q = g^T H^{-1} g, r = g^T H^{-1} b, s = b^T H^{-1} b$. In practice H^{-1} is expensive to compute so we approximately solve for $H^{-1}g, H^{-1}b$ using conjugate gradients. Since $\lambda \ge 0$, we must restrict the optimal λ from the two cases above through a projection to the sets $\Lambda_a := \{\lambda | \lambda c - r > 0, \lambda \ge 0\}$, and $\Lambda_b := \{\lambda | \lambda c - r \le 0, \lambda \ge 0\}$ respectively, such that:

$$\lambda^* \in \left\{\text{Proj}\left(\sqrt{\frac{q - r^2 s^{-1}}{\delta - c^2 s}}, \Lambda_a\right), \text{Proj}\left(\sqrt{\frac{q}{\delta}}, \Lambda_b\right)\right\} = \begin{cases} \lambda_a^*, & \text{if } f_a(\lambda_a^*) \ge f_b(\lambda_b^*) \\ \lambda_b^*, & \text{otherwise} \end{cases} \tag{3}$$

Note that H is computed using the quadratic constraint such that any step taken satisfies the KL-divergence constraint. In the infeasible case when $c^2 s - \delta > 0$ and $c > 0$, we apply a recovery method towards constraint satisfaction, $x^* = \sqrt{2\delta/(b^T H^{-1} b)}H^{-1}b$.

Distributional Value Function and Safety Baseline. Accurate estimates of the return and safety advantages A^{π_k} and $A_C^{\pi_k}$ in (2) are essential to reduce the variance associated with the policy-gradient update step. These advantage functions can be estimated, e.g.,

Algorithm 1: Distributional Constrained Policy Optimization (DCPO)

Input: Initial policy $\pi_0 \in \Pi_\theta$, safety margin ε, reshaping coefficient β
for $k = 0, 1, 2, \ldots$ **do**
  Sample a set of trajectories $\mathfrak{D} = \tau \sim \pi_k = \pi(\theta_k)$
  Form sample estimates $\hat{g}, \hat{b}, \hat{H}, \hat{c}$ using $\mathfrak{D}$, and reshaped $\tilde{J}_C(\pi_k), \tilde{d}$
  if *approximate CPO* (3) *is feasible* **then**
    Solve dual problem with reshaped constraint $\tilde{c} = \tilde{J}_C(\pi_k) - \tilde{d}$
    Compute new policy θ^* using $x^* = -\frac{1}{\lambda^*} H^{-1}(g - \nu^* b)$
  else
    Compute recovery policy θ^* using $x^* = -\sqrt{2\delta/(b^T H^{-1} b)} H^{-1} b$
  end
  Obtain θ_{k+1} through backtracking linesearch to enforce constraints in (2) over $\mathfrak{D}$
end

see Schulman et al. (2015b), and approximated with NNs. To model uncertainty, we choose to represent the return and cost value functions (used in estimating A^{π_k} and $A_C^{\pi_k}$) using NNs whose outputs are distributions parameterized by N equally spaced quantiles in a range $[V_{\min}, V_{\max}]$ (Bellemare et al., 2017). We employ a surrogate loss for these NNs comprising the average negative log-likelihood of obtaining the costs or returns sampled under policy π_k. To simplify computation of the log-likelihood, we approximate the quantile distribution as Gaussian. For a value function with parameters ϕ the loss is thus $\mathcal{L}(\phi) = -\log\big(p(J_R(\pi_k)|\phi)\big)$, which is trainable using stochastic gradient descent.

Cost Reshaping. To reliably satisfy constraints, we reshape the cost $J_C(\pi_k)$ and safety threshold d based on the agent's confidence that constraints are satisfied given a cost distribution with parameters ϕ_c. Specifically, we introduce a reshaping parameter ρ, whose value is set to $\beta\mathcal{P}(J_C^{\pi_k} > d; \phi_c)$ when $J_C^{\pi_k} > d$ (the constraints are violated), and to $-\beta\mathcal{P}(J_C^{\pi_k} < d; \phi_c)$ when the constraints are satisfied, with $\beta \geq 0$. In other words, ρ takes the value of β when the agent is fully confident that constraints are violated under the previous policy π_k, and $-\beta$ when it is fully confident the constraints are satisfied. We then reshape $J_C^{\pi_k}$ and d by multiplying each with $\big(1 + \text{clip}(\rho, -\epsilon, \infty)\big)$. The new parameter ε controls risk aversion. This reshaping only affects policy updates when constraints are satisfied, effectively avoiding large steps in directions that may increase costs. Intuitively, maximizing rewards in a constrained environment usually leads to the agent to the edges of the feasible region. We thus mitigate risk by adjusting the feasibility of the trust region depending on the confidence of constraint satisfaction.

4. Results and Discussion

We consider inventory management for a multi-echelon, multi-period supply chain: the goal is to maximize profits while satisfying constraints. We use *InvManagement-v0* in the `or-gym` library (Hubbs et al., 2020). At each period t, the environment is:

$$I_{t+1}^m = I_t^m + R_{t-L_m}^m - S_t^m, \qquad \forall\, m \in \mathcal{M}, t \in \mathcal{T} \tag{6a}$$
$$T_{t+1}^m = T_t^m - R_{t-L_m}^m + R_t^m, \qquad \forall\, m \in \mathcal{M}, t \in \mathcal{T} \tag{6b}$$
$$S_t^m = \begin{cases} R_t^{m-1}, \text{if } m > 0 \\ \min\big(I_t^0 + R_{t-L_m}^0, D_t + B_{t-1}^0\big), \text{otherwise} \end{cases} \qquad \forall\, m \in \mathcal{M}, t \in \mathcal{T} \tag{6c}$$
$$U_t^m = R_t^{m-1} - S_t^m, \qquad \forall\, m \in \mathcal{M}, t \in \mathcal{T} \tag{6d}$$
$$P_t^m = \alpha^n\big(p^m S_t^m - r^m R_t^m - k^m U_t^m - h^m I_{t+1}^m\big), \qquad \forall\, m \in \mathcal{M}, t \in \mathcal{T} \tag{6e}$$

Eqs. 6a—6b are mass balances for the on-hand (I) and pipeline (T) inventories for each stage m, where R and L denote replenishment orders and lead times, respectively.

Eq. 6c gives the sales (*S*) for the retailer ($m = 0$) and the rest of the pipeline ($m > 0$), given demand *D* and backlog *B*. Eq. 6d escribes the unfulfilled re-order quantities (*U*), and Eq. 6e calculates the profit *P* as the sales revenue minus procurement costs, unfulfilled demand penalties, and excess inventory penalties, all subject to a discount factor $\alpha = 0.97$. The unit sales price *p*, and penalties *k*, *h* are all known.

There are production capacity and inventory constraints on the re-order quantities R_t^m. To enforce these, we define a cost function that sums the number of constraint violations $C(s,a) = \sum_{m \in \mathcal{M}, t \in \mathcal{T}} C(I_t^{m+1}, R_t^m)$, where $C(I_t^{m+1}, R_t^m)$ is the number of constraints violated at node *m* at time *t*. We use this indicator cost function rather than a continuous one in order to limit the number of constraint violations, regardless of their degree.

Training DCPO. All of the algorithms use separate feed-forward NNs for the policy, return value distribution, and cost value distribution of size (64, 64) with tanh activations. Running DCPO (github.com/jaimesabalimperial/jaisalab) with N = 102 quantiles allowed for convergence of the learned distributions, as shown in Figure 1. Intuitively, at earlier periods the agent is more uncertain about the cost/value of a state (epistemic uncertainty). As samples are observed, the distribution converges, and remaining uncertainty likely stems from stochasticity of the MDP (aleatoric uncertainty).

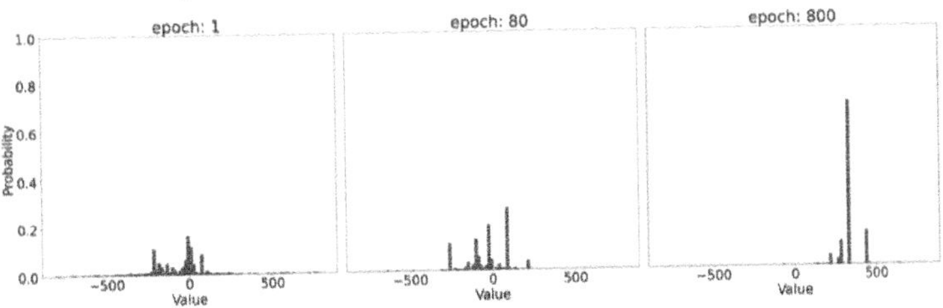

Figure 1: Learned quantiles (N=102) for the return distribution of the initial state s_0.

Importance of Cost Reshaping. Figure 2 shows the costs and returns incurred during training for CPO, DCPO, and 'ablation,' or DCPO without cost reshaping. We observe that using parametric return and cost value distributions improves the initial stages of training, with the 'ablation' agent learning more quickly (see episodes 0–100) to both minimize costs and maximize returns compared to CPO. Moreover, the standard deviation of the returns across replications by is reduced by 40.0% and 76.6% for 'ablation' and DCPO, respectively, compared to CPO at the end of training.

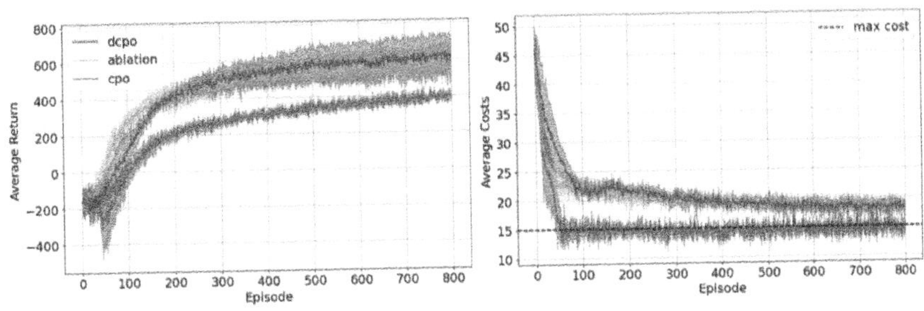

Figure 2: Returns (left) and costs (right) incurred by during training.

Comparison to TRPO and Saute. Figure 3 shows the costs and returns incurred during training for TRPO, Saute TRPO, CPO, and DCPO. TRPO is unconstrained and produces infeasible policies, more than doubling the permitted number of violations, but it attains the largest returns. On the other hand, Saute TRPO incurs costs well below the limit by the end of training, with a clear trade-off in the low returns. DCPO seems to balance the intrinsic trade-off between returns and costs, with mean costs converging to the defined limit. Moreover, DCPO showed a reduction of 70.8% in the standard deviation of the returns relative to TRPO by the end of training. Finally, CPO converged to intermediate values for both costs and returns, potentially owing to approximation errors.

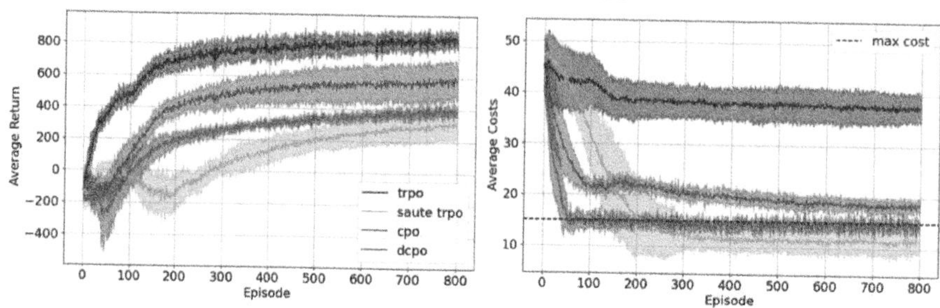

Figure 3: Returns (left) and costs (right) for algorithms over ten runs.

5. Conclusions

This paper introduces an approach for trust region optimization in CPO, by reshaping episodic discounted costs and the maximum allowed value using the agent's confidence in constraint satisfaction, in turn given by an approximated cost distribution. The developed algorithm, which we call DCPO, balances the trade-off between maximizing returns and minimizing costs in a supply chain case study, such that constraints are satisfied (within one standard deviation), while reducing run-to-run variance.

6. Acknowledgments

The authors acknowledge support from the Engineering & Physical Sciences Research Council (EPSRC) grant EP/T001577/1 and an Imperial College Research Fellowship.

References

J. Achiam, D. Held, A. Tamar, P. Abbeel, 2017. Constrained policy optimization. In: International Conference on Machine Learning. PMLR, pp. 22–31.

M. G. Bellemare, W. Dabney, R. Munos, 2017. A distributional perspective on reinforcement learning. In: International Conference on Machine Learning. PMLR, pp. 449–458.

C. D. Hubbs, H. D. Perez, O. Sarwar, N. V. Sahinidis, I. E. Grossmann, J. M. Wassick, 2020. Or-gym: A reinforcement learning library for operations research problems. arXiv:2008.06319.

P. Petsagkourakis, I. O. Sandoval, E. Bradford, F. Galvanin, D. Zhang, E. A. del Rio-Chanona, 2022. Chance constrained policy optimization for process control and optimization. Journal of Process Control 111, 35–45.

J. Schulman, S. Levine, P. Abbeel, M. Jordan, P. Moritz, 2015a. Trust region policy optimization. In: International Conference on Machine Learning. PMLR, pp. 1889–1897.

J. Schulman, P. Moritz, S. Levine, M. Jordan, P. Abbeel, 2015b. High-dimensional continuous control using generalized advantage estimation. arXiv:1506.02438.

J. Shin, T. Badgwell, K. Liu, J.H. Lee, 2019. Reinforcement learning–overview of recent progress and implications for process control. Computers & Chemical Engineering 127, 282–294.

A. Sootla, A. I. Cowen-Rivers, T. Jafferjee, Z. Wang, D. H. Mguni, J. Wang, H. Ammar, 2022. Sauté RL: Almost surely safe reinforcement learning using state augmentation. In: International Conference on Machine Learning. PMLR, pp. 20423–20443.

Antonis Kokossis, Michael C. Georgiadis, Efstratios N. Pistikopoulos (Eds.)
PROCEEDINGS OF THE 33rd European Symposium on Computer Aided Process Engineering (ESCAPE33), June 18-21, 2023, Athens, Greece
 http://dx.doi.org/10.1016/B978-0-443-15274-0.50263-8

Machine learning-based decomposition for complex supply chains

Niki Triantafyllou[a], Stavros Papaiakovou[b], Andrea Bernardi[a], Matthew Lakelin[c], Nilay Shah[a], Antonis Kokossis[b], Maria M. Papathanasiou[a,*]

[a]*The Sargent Centre for Process Systems Engineering, Imperial College London, London, United Kingdom, SW72AZ*

[b]*School of Chemical Engineering, National Technical University of Athens, 9, Iroon Polytechniou St, Athens, Greece, 15780*

[c]*TrakCel Limited, 10/11 Raleigh Walk, Cardiff, CF10 4LN UK*

**maria.papathanasiou11@imperial.ac.uk*

Abstract

Personalised medicine products represent a novel category of therapeutics often characterised by bespoke manufacturing lines and dedicated distribution nodes. An example of such products is Chimeric Antigen Receptor (CAR) T-cells, whose manufacturing poses challenges to volumetric scale-up, leading to increased production and supply chain costs. From a modelling perspective, such networks lead to complex large-scale supply chain models that grow exponentially as the demand increases and more therapies are tracked simultaneously throughout the supply chain. In this work, we present a hybrid model that utilizes the potential of machine learning for strategic planning by forecasting optimal supply chain structures and Mixed Integer Linear Programming (MILP) for detailed scheduling. The proposed model is robust to uncertain demand patterns and can reduce the number of linear constraints and binary variables in the original MILP by more than 64.7%.

Keywords: supply chain optimisation, MILP, artificial neural networks, hybrid models, personalised medicine

1. Introduction

Personalised medicine products are often characterised by 1:1 business models that consider parallel manufacturing lines and dedicated distribution routes. Such models are also adopted by personalised cancer therapies that use the patient's own cells as the starting material (Papathanasiou *et al.*, 2020). Chimeric Antigen Receptor (CAR) T-cells are an indicative example of patient-specific therapies that owing to their promising clinical results have pioneered the approval of cell and gene therapy products by the regulatory authorities (Young *et al.*, 2022). The commercially available CAR T-cell products are offered at high list prices ranging between \$300,000 and \$475,000. These prices can be attributed to the high manufacturing, distribution and therapy administration costs (Spink *et al.*, 2018). In addition, there is a pressing need to coordinate the manufacturing and distribution lifecycle with the patient schedule. Depending on the clinical condition and location of each patient, in-time delivery is of utmost importance for the patient's prognosis. Future scenarios indicate that cell and gene therapy products could be treating up to 60,000 patients per year by 2030 (Quinn *et al.*, 2019). Planning and scheduling the manufacturing and distribution of such therapeutics is therefore a challenging task, currently relying on white-glove logistics, which might not always ensure a responsive and resilient supply chain network. Process Systems Engineering tools can offer a systematic approach to assist decision-making.

Previous works on the optimization of CAR T-cell therapies via Mixed Integer Linear Programming (MILP) (Triantafyllou *et al.*, 2022a), heuristics-based decomposition algorithms (Triantafyllou *et al.*, 2022b) and metaheuristics-based algorithms (Karakostas *et al.*, 2020) highlight the patient-specific nature of these therapeutics, where the full-scale model may comprise complex sets of constraints and binary variables, translating into computationally expensive optimisation problems. In this space, methodologies that rely on Machine Learning (ML) can enhance supply chain planning in the area of Operations Research (Goettsch *et al.*, 2020; Abbasi *et al.*, 2020).

In this work, we present a hybrid model that describes the end-to-end supply chain of personalised therapeutics based on Mixed Integer Linear Programming (MILP) and Artificial Neural Networks (ANNs) (Figure 1). To reduce the computational complexity of the full-space MILP model, we harness the potential of ML for strategic planning in the supply chain by forecasting the optimal supply chain network structure.

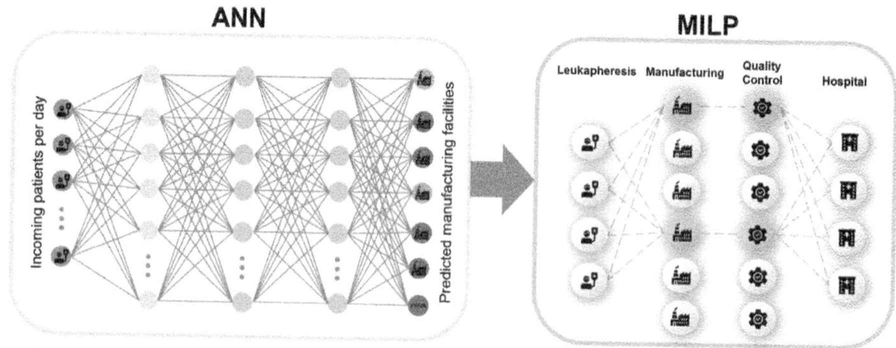

Figure 1. Proposed hybrid model for the CAR T-cell supply chain network with 4 nodes: leukapheresis centres, (b) manufacturing sites, (c) quality control sites, and (d) hospitals.

2. Materials & Methods

2.1. Mixed Integer Linear Programming (MILP) model

This study uses an in-house MILP model that describes the CAR T-cell therapy supply chain, and it is used for the identification of optimal supply chain network structures (Triantafyllou *et al.* 2022a). The supply chain superstructure consists of 4 nodes–leukapheresis centre, manufacturing site, Quality Control (QC), and hospital. CAR T-cell manufacturing starts at the leukapheresis centre, where T-cells are isolated from the patient's bloodstream and are cryopreserved before further processing. The leukapheresis sample is then shipped to the manufacturing facility, where it is genetically engineered to express the tumour-associated antigen receptor. Lastly, the therapy undergoes in-house Quality Control (QC) and once product quality and safety are ensured, it is shipped back to the hospital for administration to the patient. We consider 4 leukapheresis sites and 4 hospitals in the UK and 6 manufacturing sites located in the UK, US and Europe. The manufacturing facilities m have a capacity of 4 (m_1 and m_4), 10 (m_3 and m_6), or 31 (m_2 and m_5) parallel lines, and a forward-looking scenario of 8 days of manufacturing is considered. The model considers demand uncertainty, manufacturing capacity limitations, patient-specificity, time and location constraints, whilst the objective is to minimize the total supply chain cost. Finally, the total turnaround time is expressed as a non-monetary supply chain metric modelled as a constraint.

2.2. Hybrid model

The data-driven part of the model is responsible for strategic planning by forecasting the optimal supply chain network configuration based on the annualized demand. Specifically, the ANN model predicts the number and the location of the manufacturing facilities to optimally satisfy the demand. The mechanistic MILP model is then solved considering only the subset of manufacturing facilities chosen by the ANN, becoming a subproblem of the original planning and scheduling model. Therefore, the MILP accounts only for detailed scheduling in the supply chain, which entails optimal transport modes for the node-to-node connections, optimal allocation of patient samples in the manufacturing facilities and hospitals, and the optimal utilisation of the available parallel lines in the manufacturing sites with the scope of minimising the therapy cost and return time. The hybrid model can significantly reduce computational complexity by eliminating branches of the search space. The model's performance is assessed for different randomized annual demand scenarios (100, 200, 500, 1000, 1500 and 2000 patients).

2.2.1 ANN training

The ANN is trained based on sets of feasible and optimal solutions obtained from the MILP for 760 randomized demand profiles. The input features considered for the ANN are the total daily demands for a quarter of a year (90 days), assuming a recurrent demand profile per trimester. Based on three different probability distributions, namely uniform distribution, left triangular distribution, and right triangular distribution, 660 randomized demand profiles were generated for demands between 50 and 2000 patients per year (Figure 2). For each leukapheresis site *c*, we assume a daily capacity of 8 patients. Different probability distributions are used to account for the uncertainty in demand and lead to robust solutions as the demand distribution highly affects supply chain structures. To generate the training labels, the MILP model was solved for each demand scenario using CPLEX. All inputs were normalized.

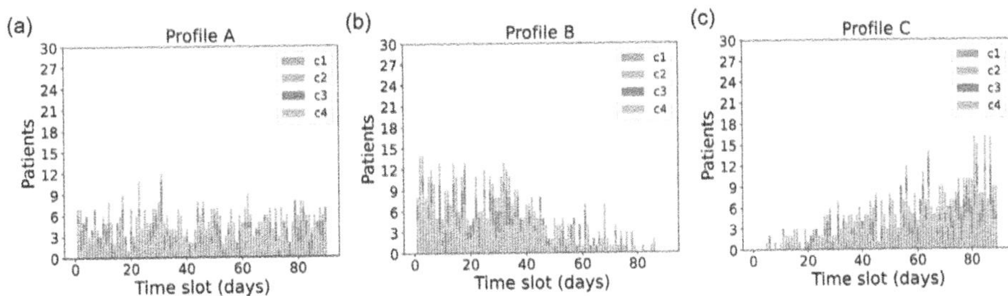

Figure 2. Randomized demand profiles for the 2000 patients' case for 3 different probability distributions: (a) uniform distribution (b) left triangular distribution (c) right triangular distribution.

A feed-forward ANN model is built in PyTorch for multi-label classification, where the labels are the 6 manufacturing facilities and the possibility of infeasible solutions due to limited facility capacity as seen in Figure 1. Multi-label classification is used so that more than one facility can be selected for each instance. However, because the original dataset was imbalanced, with classes m_6 and infeasible being underrepresented, extra 100 randomized demand scenarios were generated and equally distributed for the 350, 500 and 2000 patients per year cases to balance the data and capture more instances where the underrepresented classes are the solution. It should be noted that oversampling techniques for imbalanced classification datasets such as SMOTE do not apply to multi-label datasets. The data set is split into 80% for training and 20% for testing. The topology of the ANN was tuned using Grid Search and parameter tuning. To configure the feed-forward ANN, 90 input features, 3 hidden layers, 256 neurons at each hidden layer, and 7 output layers are used. Because multi-label classification is based on binary classification, the sigmoid activation function is used at each output layer node to predict a probability for each class in the output. The threshold that separates positive and negative classes is set at 0.5. Finally, the ReLU function is used in the hidden layers and the model is trained using the binary cross-entropy loss function and the Adam solver.

3. Results and discussion

The overall performance of the multi-label classification ANN model is measured by *accuracy* and additional extracted metrics from the *confusion matrix,* i.e. *precision, recall*, and *F-score* for each class and *micro*, *macro*, and *weighted* average of all classes (Table 2). Accuracy provides the number of correct predictions to the total number of labels (predicted and true). Precision and recall quantify the proportion of correctly predicted labels to the total number of true labels and the total number of predicted labels, respectively. On the other hand, the F1-score leads to a harmonic balance between precision and recall as it considers both. The ordinary 2-dimensional confusion matrix used in multi-class classification is undefined when it comes to multi-label classification, where each instance can be labelled with more than one class. To overcome this, the multi-label confusion matrix with one extra row (No True Label-NTL) and one extra column (No Predicted Label-NPL) is used (Heydarian *et al.*, 2022).

Table 1. Overall performance metrics for the ANN.

Class	Precision	Recall	F1-score	weight
m_1	0.84	0.76	0.80	50
m_2	0.92	0.88	0.90	67
m_3	0.81	0.81	0.81	47
m_4	0.85	0.73	0.79	15
m_5	0.88	0.88	0.88	25
m_6	0.78	0.90	0.84	20
infeasible	1	0.93	0.96	29
micro avg	0.84	0.84	0.84	253
macro avg	0.87	0.84	0.85	253
weighted avg	0.87	0.84	0.86	253

The resulting ANN has a train set overall accuracy of 98.5% and a test set overall accuracy of 95%. The rest of the performance metrics for each class are presented in Table 1. The ANN performs relatively well, as the average precision, recall and F1-score are above 84%. It should be noted that the ANN model is trained with three different demand distributions to better cope with demand uncertainty. This can lead to different supply chain structures for the same yearly demands. Hence, there is a possible trade-off between model accuracy and robustness to demand uncertainty.

It is observed that class m_4 has the lowest F1-score (0.79). The precision and recall for this class are 0.85 and 0.73, respectively, indicating more false negatives (*FN*) for this class than false positives (*FP*). By looking at the confusion matrix in Figure 3, it can be seen that the model confuses facility m_4 (4 parallel lines) with facility m_3 (10 parallel lines) and with the no-label prediction. Specifically, while the model predicts correctly in 73% of the instances that m_4 belongs to the optimal supply chain configuration, in 13% of the instances it confuses facilities m_4 and m_3 and in 13% of the instances, it fails to predict m_4 as an extra required facility. These mispredictions are sensible as in some cases the model overpredicts by choosing a slightly bigger facility to facilitate the demand and in other cases, it underpredicts by failing to add an extra facility. In the former case, the hybrid model leads to local optimum solutions, whereas in the latter case the model leads to infeasible solutions.

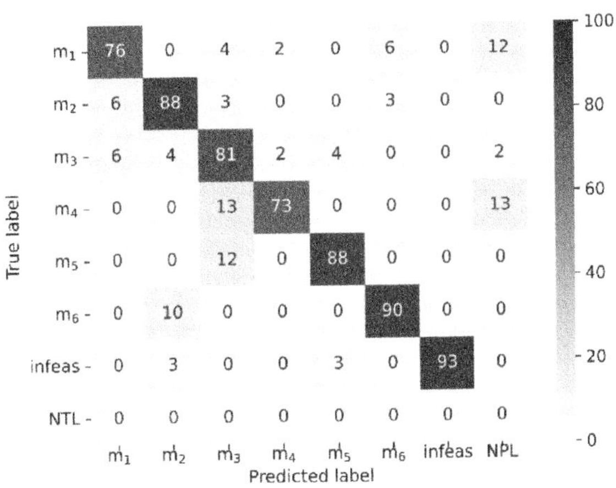

Figure 3. Normalised multi-label confusion matrix with one extra row for No True Labels (NTL) and one extra column for No Predicted Labels (NPL).

Similar behaviour is observed with facility m_1, where the recall is 0.76. Figure 3 shows that in 12% of the instances the model fails to predict that m_1 is a required extra facility to satisfy the demand, thus leading to infeasible solutions. It is important to note, that the classifier is very accurate in terms of infeasibilities. The precision for infeasible solutions is 100%, which means that the classifier never predicts feasible solutions as infeasible. In rare occasions (3% of the cases) the hybrid model assumes infeasible solutions as feasible. In those cases, the MILP part of the model will determine the infeasibility of the solutions.

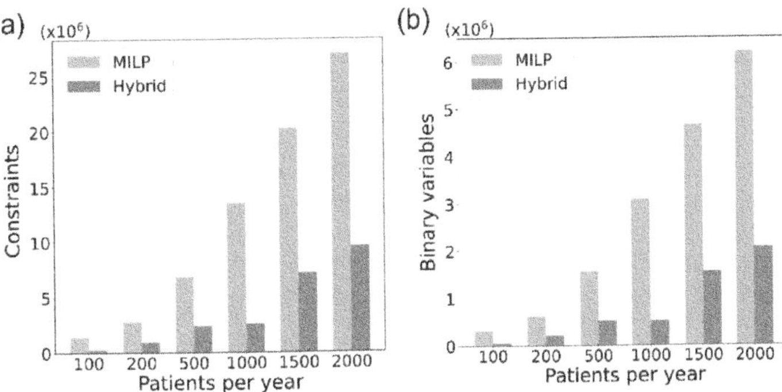

Figure 4. Comparison of the original MILP model with the hybrid model for an increasing number of annual demands in (a) the number of constraints, and (b) the number of binary variables.

In Figure 4, the performance of the hybrid model is showcased in terms of model complexity by evaluating the number of constraints and binary variables for both the original mechanistic model and the proposed hybrid model. The hybrid model can reliably reduce the computational burden by decreasing the total number of linear constraints by 65%–81% and the total number of binary variables by 67%–83% as shown in Figure 4. In cases of 500 and 1000 patients per year, the hybrid model is described by almost the same number of constraints and binary variables. This is attributed to the fact that in the 1000 patients per year case, the classifier predicted that only one large facility with 31 parallel lines (m_2) can satisfy the demand, and hence the feasible search space for the MILP model was significantly reduced.

4. Conclusions

In this study, a framework that integrates a multi-label classification feed-forward neural network with a MILP model in order to decompose large-scale supply chain problems is presented. The unique CAR T-cell therapy supply chain model is used as the case study. The data-driven part of the hybrid model accounts for strategic planning and the mechanistic part is responsible for detailed scheduling. The framework proposes a novel method to account for demand uncertainty using machine learning techniques, resulting in robust supply chain structures. The overall accuracy of the classifier is 95%, with just a few cases leading to suboptimal or infeasible solutions. The computational analysis exhibited significant improvements over the original MILP model. A key novelty of this work is the ability of the hybrid model to solve larger problem instances, for which the MILP problem did not manage to converge to a feasible solution. The efficiency of the hybrid model creates new possibilities to explore, such as the use of convolutional neural networks for improved classification accuracy.

Acknowledgements

Funding from the UK Engineering & Physical Sciences Research Council (EPSRC) for the Future Targeted Healthcare Manufacturing Hub hosted at University College London with UK university partners is gratefully acknowledged (Grant Reference: EP/P006485/1). Financial and in-kind support from the consortium of industrial users and sector organisations is also acknowledged.

References

M. M. Papathanasiou, C. Stamatis, M. Lakelin, S. Farid, N. Titchener-Hooker, N. Shah, 2020, Autologous CAR T-cell therapies supply chain: challenges and opportunities?, *Cancer Gene Therapy*, 27, 799–809.

C. M. Young, C. Quinn, M. R. Trusheim, 2022, Durable cell and gene therapy potential patient and financial impact: US projections of product approvals, patients treated, and product revenues, *Drug Discovery Today*, 27(1), 17–30.

K. Spink, A. Steinsapir, 2018, The long road to affordability: a cost of goods analysis for an autologous CAR-T process, *Cell & Gene Therapy Insights*, 4(11), 1105–1116.

C. Quinn, C. Young, J. Thomas, M. Trusheim, 2019, Estimating the Clinical Pipeline of Cell and Gene Therapies and Their Potential Economic Impact on the US Healthcare System, *Value in Health*, 22(6), 621–626.

N. Triantafyllou, A. Bernardi, M. Lakelin, N. Shah, M. M. Papathanasiou, 2022, A digital platform for the design of patient-centric supply chains, *Scientific Reports*, 12, 17365.

N. Triantafyllou, A. Bernardi, M. Lakelin, N. Shah, M. M. Papathanasiou, 2022, A bi-level decomposition approach for CAR-T cell therapies supply chain optimisation, *Computer Aided Chemical Engineering*, 49, 2197–2202.

P. Karakostas, N. Panoskaltsis, A. Mantalaris, and M. C. Georgiadis, 2020 Optimization of CAR T-cell therapies supply chains, *Computers & Chemical Engineering*, 139, 106913.

D. Goettsch, K. K. Castillo-Villar, M. Aranguren, 2020, Machine-Learning Methods to Select Potential Depot Locations for the Supply Chain of Biomass Co-Firing, *Energies*, 13, 6554.

B. Abbasi, T. Babaei, Z. Hosseinifard, K. Smith-Miles, M. Dehghani, 2020, Predicting solutions of large-scale optimization problems via machine learning: A case study in blood supply chain management, *Computers & Operations Research*, 119, 104941.

M. Heydarian, T. E. Doyle, R. Samavi, 2022, MLCM: Multi-Label Confusion Matrix, *IEEE Access*, 10, 19083–19095

Antonis Kokossis, Michael C. Georgiadis, Efstratios N. Pistikopoulos (Eds.)
PROCEEDINGS OF THE 33rd European Symposium on Computer Aided Process Engineering (ESCAPE33), June 18-21, 2023, Athens, Greece
 http://dx.doi.org/10.1016/B978-0-443-15274-0.50264-X

Optimal energy management in greenhouses using distributed hybrid DRL-MPC framework

Benjamin Decardi-Nelson,[a] Fengqi You[a]
[a]*Cornell University, Ithaca, New York, 14853, USA*

Abstract

In this work, we propose a framework for optimal energy management of greenhouses, which considers the presence of both discrete-valued and continuous-valued actuators in the controller design. To reduce the computational burden because of having to solve mixed-integer nonlinear optimal control problems online, we propose to decompose the system model based on the type of actuators. The underlying subproblems are solved using a hybrid deep reinforcement learning (DRL)-model predictive control (MPC) framework in a distributed manner. We use a greenhouse case study to demonstrate the effectiveness of the proposed control framework.

Keywords: MPC, greenhouse, energy management, deep reinforcement learning.

1. Introduction

Greenhouses are intensified crop production systems that have a potential to improve the resilience of the food system in a sustainable way (Chen et al., 2022). One of the main drawbacks of greenhouses is the enormous energy consumption of their microclimate control activities such as lighting, heating, ventilation, and air conditioning activities (Vadiee & Martin, 2012). Because the rule-based control methods traditionally used in greenhouses are unable to optimally manage the energy usage and ensure constraint satisfaction, optimization-based model predictive control (MPC) has been proposed as a better alternative (Hu & You, 2022). Unfortunately, the effective operation of greenhouses relies on equipment with discrete-valued actuation, like ventilation window opening and lights, which complicate the design and implementation of MPC (Chen et al., 2021). This is because the underlying centralized MPC optimization problem becomes a parametric mixed-integer nonlinear (possibly non-convex) optimization problem (MINLP), which may be challenging to solve repetitively online. To facilitate the adoption of MPC is greenhouse operation, efficient control methods are needed. The use of data-driven techniques to address robustness and computational issues in MPC is on the rise (Chen and You, 2022). In particular, the use of machine learning to assist in solving mixed-integer MPC (MIMPC) problems has received considerable attention (Löhr et al., 2020; Cauligi et al., 2021). However, the approaches focus on only linear problems. Moreover, the authors assume that solutions of the original MIMPC problem can be readily obtained. Often, these solutions for large scale MIMPC optimization problems are difficult to obtain. In this work, we propose a novel distributed hybrid deep reinforcement learning (DRL)-MPC framework to address the greenhouse optimal control problem. The proposed control framework exploits the attractive features of both DRL and MPC, that is, fast discrete-valued sequential decision making, and optimal control with constraint satisfaction (Ning et al., 2021), respectively. This is achieved by first decomposing the greenhouse optimal control problem into two sub-problems based on the type of actuation, namely discrete-valued or continuous-valued. Thereafter, we train the DRL to make the discrete-valued decisions while the MPC handles the continuous-valued decisions. Similar to the approaches used in distributed MPC (DMPC), the two controllers cooperatively determine the optimal input sequence to be sent to the greenhouse system at each time-step in a receding horizon manner. We demonstrate the effectiveness of the proposed framework using a greenhouse case study.

2. Greenhouse optimal control problem

In this section, we present the details of the greenhouse model and formulate the associated mixed-integer model predictive control problem.
The greenhouse model used in this work is a discrete-time version of the model used in the study by Van Henten (2003). The model is a nonlinear system of ordinary differential equations (ODEs) described by the equation

$$x_{k+1} = f(x_k, u_k, v_k) \tag{1}$$

where x_k denotes the states of the system at time-step k, u_k denotes the continuous-valued control actions, v_k represents the discrete-valued control actions and f represents the nonlinear ODEs. The states of the system are the crop dry weight, carbon dioxide concentration, air temperature and humidity in the greenhouse. The continuous-valued control actions are the carbon dioxide enrichment, heating and cooling systems, and the (de)humidification system. The discrete-valued control actions on the other hand are the supplemental lighting and vent opening, which can either be turned on/open or off/closed. Throughout this article, the states and the inputs are constrained to be in the sets

$$x_k \in X, u_k \in U, v_k \in V = \{0,1\} \tag{2}$$

The objective of the control system is to minimize the energy consumption while ensuring a higher crop growth rate. This control problem can be formulated as a mixed-integer MPC (MIMPC) described in Equation (3).

$$\min_{u_k, v_k} \; \textstyle\sum_{k=0}^{N-1} \mathrm{L}(x_k, u_k, v_k) \tag{3a}$$

$$s.t. \;\; x_{k+1} = f(x_k, u_k, v_k) \tag{3b}$$

$$x_0 = x(t) \tag{3c}$$

$$x_k \in X, u_k \in U, v_k \in V \tag{3d}$$

In (3), Equation (3a) is the objective function to be minimized, Equation (3b) is the nonlinear system model described in (1), Equation (3c) denotes the initial state constraint and Equation (3d) represents the state and input constraints. At each sampling time k, the MIMPC optimization problem in (3) is solved to obtain a sequence of optimal values of the continuous-valued and discrete-valued control actions. As consistent with an MPC implementation, only the first control actions (u_0^*, v_0^*) of the optimal sequence are applied to the system in a receding horizon manner (Mayne et al., 2000). The feasibility issue is guaranteed via soft constraints (Lu et al., 2020). As mentioned earlier, the presence of both continuous and discrete decisions in the MIMPC optimization problem (3) results in having to solve an MINLP online several times. This can be computationally difficult to solve (Shang et al., 2019). In this article, we proposed a computationally efficient hybrid distributed framework the takes advantages of the improvements in deep reinforcement learning (DRL) and standard model predictive control.

3. Optimal control using distributed hybrid algorithm

The presence of both discrete-valued and continuous-valued actions in the MPC formulation makes it computationally demanding to implement as an online control algorithm. In this section, we present a novel distributed hybrid algorithm that considers the strength of both model predictive control and deep reinforcement learning. The central idea we employ in this framework is to use a deep Q network (DQN) policy obtained from DRL to make the discrete-valued decisions. By fixing the discrete decisions in the original MIMPC optimization problem in (3), a simpler nonlinear problem (NLP) can easily be solved online. To achieve this, we reformulate the MIMPC optimization problem. We begin this section by first describing the reformulation. Thereafter, we present the details of the DRL. Finally, we describe and discuss the implementation details of the proposed framework. In general, it is difficult to solve mixed-integer optimization problems. This is because the complexity of the optimization problem increases exponentially in the discrete-valued decision variables. While there have been improvements in algorithms for solving such problems, using these algorithms in an online setting may be limiting. To address this, we propose to reformulate the optimization problem such that a deep reinforcement learning algorithm learns to make the discrete-valued decisions. These decisions are then fixed in the MPC optimization problem, thus resulting in having to solve online an NLP.

Let $\tilde{v}_k$ approximate the discrete-valued decision variables in the optimal control problem described in (3) at each time-step k. If the discrete-valued variables in (3) are fixed, then the resulting optimization problem becomes an NLP which can be solved efficiently using standard NLP solvers such as IPOPT. Notice that $N-1$ discrete decision variables need to be determined in the optimal control problem. In contrast to the approach proposed by Löhr et al. (2020) where only the first discrete variables are fixed with the approximated value, we take a distributed approach. The reformulated optimization problem is described in (4).

$$\min_{u_k, v_k} \sum_{k=0}^{N-1} \mathrm{L}(x_k, u_k, v_k) \tag{4a}$$

$$s.t.\ x_{k+1} = f(x_k, u_k, v_k) \tag{4b}$$

$$x_0 = x(t) \tag{4c}$$

$$u_k \in U, v_k \in V \tag{4d}$$

$$v_k = \tilde{v}_k \tag{4e}$$

The details of the equations in (4) remain the same as that of (3). Equation (4e) ensures that the discrete valued decisions remain fixed with the approximate values from the DQN policy. To ensure feasibility of the optimization problem, the state constraint is relaxed. The objective of the DRL neural network policy is to make one-step discrete-valued decisions $\tilde{v}_k$ given the state information x_k (Ajagekar et al., 2023). Let is denote the neural network policy as P, then the policy is described by the mapping $P: x_k \in R^{n_x} \rightarrow \tilde{v}_k \in R^{n_v}$ where n_x and n_v denote the dimension of the state and discrete-valued action spaces respectively. Because the state variables are continuous-valued and the actions are discrete-valued, we use DQN reinforcement learning algorithm (Minh et al., 2015).

The implementation detail of the control algorithm is summarized in Figure 1. At each time-step, an estimate of the discrete actions is used in the reformulated MPC optimization problem described in (4) to obtain the state predictions and the optimal continuous-valued actions. The state predictions are sent to the DQN policy to determine an improved estimate of the discrete-valued decisions. This procedure is repeated until the stopping criteria is achieved. Notice that the proposed framework is no different from the DMPC framework, where information is repeatedly exchanged between two or more MPC-NLPs (Christofides et al., 2013). The difference here is that the communication is between an MPC-NLP and an DQN policy. Again, like DMPC algorithms, an initial guess of the subsystem solutions is required at the start of the algorithm, and a termination criterion needs to be specified. The stopping criteria employed in this work are either the sum of the absolute difference between any two successive discrete actions from the DQN policy is less than 1 or the maximum number of iterations is reached.

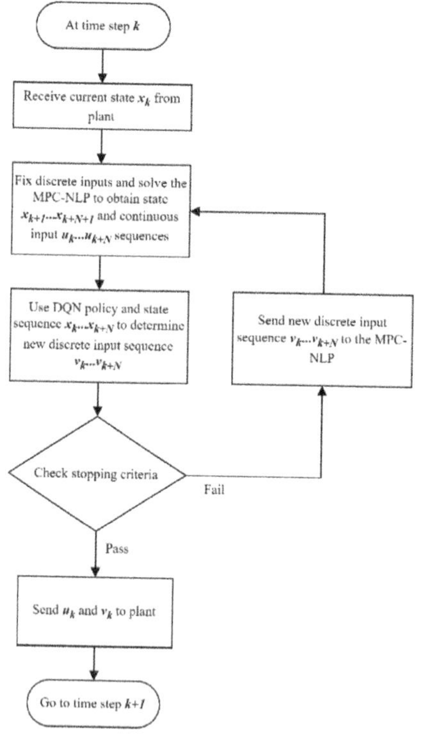

Figure 1. Implementation details of the distributed hybrid algorithm.

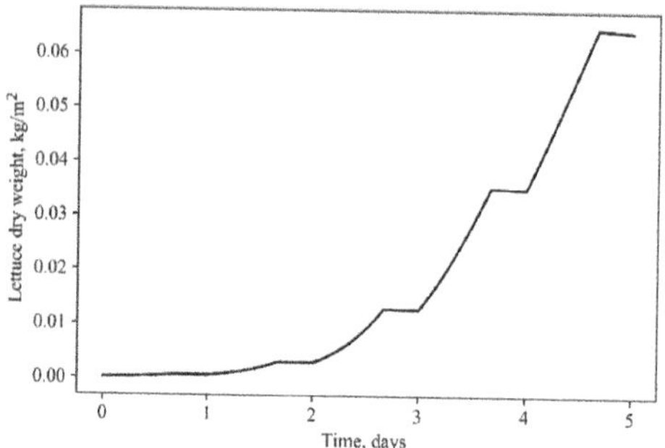

Figure 2. Crop growth profile after 5 days.

4. Simulation results

In this section, we demonstrate the efficacy of the proposed hybrid algorithm using a case study. We begin this section by first presenting the simulation settings. Thereafter, we present the results of the proposed algorithm. In this work, we assume that the external weather conditions are known and constant. The values for the external climate conditions include external temperature of 20 °C, external humidity of 400 ppm, the solar radiation at 100 W/m^2 and off at night (Yang et al., 2022). The control objective is to track the setpoints for the CO_2 concentration, relative humidity, and greenhouse air temperature which are determined as 1000 ppm, 70 % and 24 °C respectively. The photoperiod is set at 16 hours on period and 8 hours off period. The sampling time is fixed at 10 minutes and the MPC horizon is fixed at 12 which represents a two-hour prediction horizon. All computations were conducted on laptop with an Intel 2.2 GHz quadcore processor, 16 GB random access memory (RAM) and 2 GB of video RAM. The profiles of the greenhouse system under the action of the proposed control framework are presented in Figures 2 – 4. As can be observed in the profiles, the proposed control framework is able to ensure a steady crop growth rate. From Figure 2, the controller turns on the supplemental lighting in a similar fashion as the photoperiod to supplement the external light intensity entering the greenhouse. The controller optimally turns the supplemental lights off when it is not required, thus reducing energy consumption. Similarly, the controller uses the vent opening as an economizer to reduce the cooling requirements in the greenhouse. It is worth mentioning that the entire 5-day simulation took less than 5 minutes to complete.

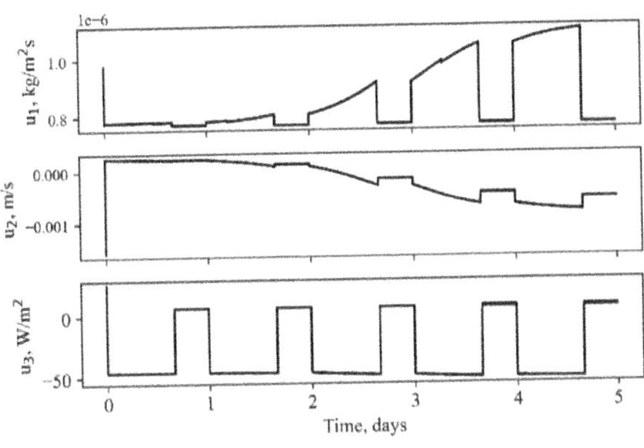

Figure 3. Control profiles for the continuous-valued actions. u_1, u_2 and u_3 denote the supplemental CO_2 supply rate, (de)humidification rate and the cooling/heating rates respectively.

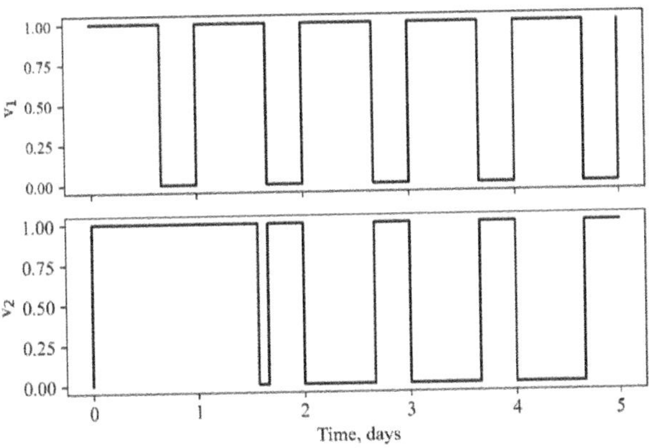

Figure 4. Control profiles for the discrete-valued actions. υ_1 and υ_2 denote the on/off state of the supplemental lighting and the open/close state of the vent opening respectively.

5. Concluding remarks

In this article, we have presented a novel MPC algorithm to optimally manage the energy resource in a greenhouse. A DRL neural network policy is trained to determine approximations of the discrete decisions, which is then fixed in the original MPC optimization problem. This reduces the online computational burden in the control system implementation. A simulation case study demonstrates the efficacy of the proposed approach to ensure steady crop growth in the greenhouse.

References

A. Ajagekar, N.S. Mattson, et al., 2023, Energy-efficient AI-based Control of Semi-closed Greenhouses Leveraging Robust Optimization in Deep Reinforcement Learning. Advances in Applied Energy, 9, 100119.

A. Ajagekar, F. You, 2023, Deep Reinforcement Learning Based Unit Commitment Scheduling under Load and Wind Power Uncertainty. IEEE Transactions on Sustainable Energy, DOI: 10.1109/TSTE.2022.3226106.

A. Cauligi, P. Culbertson, E. Schmerling, et al., 2021, CoCo: Online mixed-integer control via supervised learning, IEEE Robotics and Automation Letters, 7, 2, 1447-1454.

W.-H. Chen, N.S. Mattson, F. You, 2022, Intelligent control and energy optimization in controlled environment agriculture via nonlinear model predictive control of semi-closed greenhouse. Applied Energy, 320, 119334.

W.-H. Chen, F. You, 2022, Semiclosed Greenhouse Climate Control Under Uncertainty via Machine Learning and Data-Driven Robust Model Predictive Control. Ieee Transactions on Control Systems Technology, 30, 1186-1197.

W.-H. Chen, F. You, 2021, Smart greenhouse control under harsh climate conditions based on data-driven robust model predictive control with principal component analysis and kernel density estimation. Journal of Process Control, 107, 103-113.

W.-H. Chen, F. You, 2022, Sustainable building climate control with renewable energy sources using nonlinear model predictive control. Renewable & Sustainable Energy Reviews, 168, 112830.

P. D. Christofides, R. Scattolini, D. M. de la Pena, J. Liu, 2013, Distributed model predictive control: A tutorial review and future research directions, Computers & Chemical Engineering, 51, 21-41.

Y. Chu, F. You, 2015, Model-based integration of control and operations: Overview, challenges, advances, and opportunities. Computers & Chemical Engineering, 83, 2-20.

G. Hu and F. You, 2022, Renewable energy-powered semi-closed greenhouse for sustainable crop production using model predictive control and machine learning for Energy Management, Renewable and Sustainable Energy Reviews, 168, 112790.

Y. Löhr, M. Klaučo, M. Fikar, M. Mönnigmann, 2020, Machine learning assisted solutions of mixed integer MPC on embedded platforms, IFAC-PapersOnLine, 53, 2, 5195-5200.

S. Lu, J.H. Lee, et al., 2020, Soft-constrained model predictive control based on data-driven distributionally robust optimization. AIChE Journal, 66, e16546.

D. Q. Mayne, J. B. Rawlings, C. V. Rao, et al., 2000, Constrained model predictive control: Stability and optimality, Automatica, 36, 6, 789-814.

V. Mnih, K. Kavukcuoglu, D. Silver, et al., 2015, Human-level control through deep reinforcement learning, Nature, 518, 7540, 529-533.

C. Ning, F. You, 2021, Online learning based risk-averse stochastic MPC of constrained linear uncertain systems. Automatica, 125, 109402.

C. Shang, F. You, 2019, A data-driven robust optimization approach to scenario-based stochastic model predictive control. Journal of Process Control, 75, 24-39.

A. Vadiee, V. Martin, 2012, Energy management in horticultural applications through the closed greenhouse concept, Renewable and Sustainable Energy Reviews, 16, 7, 5087-5100

E. J. Van Henten, 2003, Sensitivity analysis of an optimal control problem in greenhouse climate management, Biosystems Engineering, 85, 3, 355-364

S. Yang, H. Gao, et al., 2022, Model predictive control in phase-change-material-wallboard-enhanced building energy management considering electricity price dynamics. Applied Energy, 326, 120023.

Antonis Kokossis, Michael C. Georgiadis, Efstratios N. Pistikopoulos (Eds.)
PROCEEDINGS OF THE 33rd European Symposium on Computer Aided Process Engineering (ESCAPE33), June 18-21, 2023, Athens, Greece
 http://dx.doi.org/10.1016/B978-0-443-15274-0.50265-1

Multi-Agent Reinforcement Learning and RL-Based Adaptive PID Control of Crystallization Processes

Qingbo Meng [a], Paul Danny Anandan [a], Chris D. Rielly [a], Brahim Benyahia [a*]

[a]*Department of Chemical Engineering, Loughborough University, Leicestershire, UK*
B.Benyahia@lboro.ac.uk

Abstract

In this work, two model-based reinforcement learning (RL) control strategies are investigated namely a multi-agent RL and RL-based adaptive PID control. An off-policy deep deterministic policy gradient (DDPG) was adopted in both cases to achieve optimal trajectory tracking control of crystallization processes. Two case studies were considered validate the new control strategies. The first is the cooling and antisolvent crystallization of aspirin in a mixture of ethanol and water, and the second is a 2-dimensional (2D) cooling crystallization of potassium dihydrogen phosphate in water. The optimal reference trajectories were identified using model-based dynamic optimization approaches which aim at maximizing the mean crystal size/minimizing the aspect ratio. Transfer Learning (TL) techniques and various reward-shaping strategies were also investigated to enhance the learning capabilities of the RL control. The results indicate that multi-agent RL saves massive training costs, compared to single agent, and RL-based adaptive PID exhibits excellent performance against state-of-the-art MPC.

Keywords: Multi-agent Reinforcement Learning; Transfer Learning; Adaptive PID; Semi-Batch Antisolvent and Cooling Crystallization; 2-Dimensional Crystallization Model.

1. Introduction and background

As one of the oldest purification technologies, crystallization has widespread applications in the pharmaceutical industry to separate and purify active pharmaceutical ingredients (API). One of the most critical features in pharmaceutical manufacturing is the stringent regulation associated with the product quality control where small deviations from the target quality profiles may significantly impact drug safety and efficacy and may lead to failures during clinical trials or even product recalls (Mascia et al., 2013). In recent years, thanks to the development of process analytical technologies (PAT), real-time measurement, monitoring and control of the crystal size and shape become achievable during the crystallization which opened new opportunities to implement advanced control strategies such as model predictive control and Artificial Intelligence based techniques such as Reinforcement Learning (Benyahia et al., 2021).

The most common control objective in crystallization processes is to maximize the mean crystal size due to its dramatic impact on the drug product's critical quality attributes, such as dissolution rate and efficacy, and on the downstream operations, such as filtration and drying (Fysikopoulos et al., 2019). Supersaturation is the fundamental driving force in crystallization, which is typically generated by either cooling, evaporation or antisolvent addition (Liu and Benyahia, 2022). Over the last two decades, model-free and model-based techniques have been extensively investigated to control and optimize crystallization processes (Lakerveld and Benyahia, 2020). However, there is still an increasing demand for more robust techniques for real time monitoring, optimization, and control to achieve precision and simultaneous control of several critical quality attributes such as purity, crystal size and shape distribution.

Over the last few years, machine learning (ML) applications have witnessed rapid growth in many research and industrial areas. Reinforcement learning (RL) has been successfully implemented in dynamic systems particularly in robotics and automotive industries because of its ability to solve a wide range of sequential decision-making problems (Canese et al., 2021). Most of the MARL applications were developed for unmanned aerial vehicles (UAV), optimal path planning (Qie et al., 2019); resource allocation and energy efficiency (Cui et al., 2020). These recent successful applications have motivated further exploration of RL capabilities as a new control strategy for crystallization (Benyahia et al., 2021).

In this study, two crystallization case studies are presented where different types of RL model-based control methods are implemented. The first case study presents an antisolvent and cooling crystallization of aspirin (acetylsalicylic acid, ASA) in a mixture of ethanol (solvent) and water (antisolvent) with a mathematical model based on population balance equations (PBE) is used. The second case study presents a 2-Dimensional (2D) cooling crystallization process of potassium dihydrogen-phosphate (KDP) in water modelled using a 2-dimensional (2D) PBE to enable crystal shape prediction and control. Firstly, a model-based optimization approach is formulated to determine the reference trajectories that maximize the critical quality attributes. A MARL and RL-PID are developed as control approaches for both case studies respectively. Both RL agents are designed based on an off-policy deep deterministic policy gradient (DDPG) then trained to track several reference trajectories, which include process temperature, supersaturation, and particle size or/and shape. The performance of the proposed RL-based control strategies is compared against a single agent RL and model predictive control (MPC).

2. Problem Formulation

2.1 Process Modelling

The ultimate objective of this work is to exploit multi-agent RL and RL-based techniques to build responsive and efficient control strategies for batch and semi-batch crystallization processes. The mathematical model associated with the first case study was obtained from the literature (Nagy et al., 2008). The growth and nucleation kinetic parameters obtained from Barik et al., 2020. The 2-dimensional mathematical model associated with the second case study was obtained from Fysikopoulos et al., 2019. Both mathematical models were experimentally validated in the original studies. A few assumptions were considered to build the models which includes: (i) the vessel is assumed to be well-mixed; (ii) crystal breakage and agglomeration of fine crystals are negligible; (iii) crystals growth is size-independent. (iv) crystal growth and dissolution do not affect the total volume. For the sake of brevity, the mathematical formulation and PBE for both models are not presented here.

2.2 Process optimization

The optimization strategy is required to ensure the best design options and optimal operational performance of the two crystallization simulation processes. It will also provide the reference trajectory profiles for control and operation purposes when applying the RL strategies. In this study, the model-based open-loop dynamic optimization method is implemented to maximize the mean crystal size/minimize the aspect ratio. The decision variables used in the first case study are the jacket temperature (T_j) and the antisolvent flowrates (F_A) whereas the process temperature is used in the second case study.

2.3 Reinforcement Learning

The development of effective two-agent RL and RL-PID are the primary objectives of this study. These objectives translate into tracking the reference trajectories that maximize crystal mean size, in case study one, and minimize the crystal aspect ratio in case study two. The reference profiles of the process temperature, supersaturation, particle size, and aspect ratio were computed using a model-based open-loop optimal control strategy. The RL agents in both case studies were developed using MATLAB Simulink and trained to reach the best control performance corresponding to the maximum rewards. Here, an off-policy actor-critic agent based on deep deterministic policy gradient (DDPG) was used to improve effective tracking control through computing the optimal control policies that maximize the total rewards (Benyahia et al., 2021).

2.3.1 Two-agent reinforcement learning

Figure 1 shows the structure of the two-agent RL. Jacket temperature and antisolvent flowrate were used separately as the manipulated variables for each RL agent to control the crystallization process.

The RL objective is to maximize the rewards/minimize penalties. RL agent 1 receives one observation (process temperature) and takes one action (cooling rate) whereas RL agent 2 observes two controlled outputs, which are supersaturation and particle size, and takes one action (antisolvent flowrate). The two actions work together to maximize the instantaneous and cumulative rewards of the agents to achieve effective training.

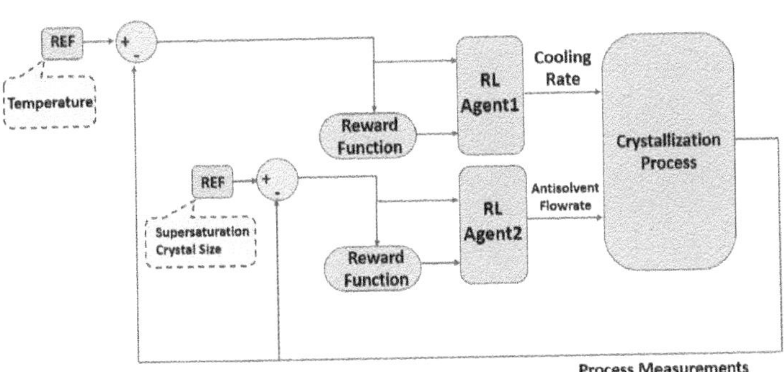

Figure 1. Block diagram of closed-loop two-agent RL.

2.3.2 RL-PID

The RL-PID is designed to combine an RL agent and a conventional PID control as a cascade structure to achieve effective control performance. Here, the cooling rate is used as the controlled variable and the tuning parameters of the PID, k_p, k_i and k_d, are the agent 'actions'. The reference tracking trajectories for the RL agent are the temperature, supersaturation, crystal size and aspect ratio.

2.4 Reward Shaping

The reward function used in both case studies is described in figure 2 below. It is defined based on penalties only. Therefore, zero is the maximum attainable theoretical value.

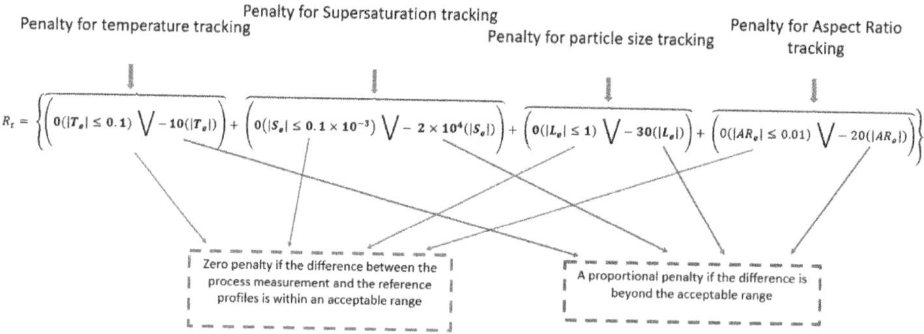

Figure 2. Simplified overview of the reward function.

2.5 Transfer Learning

Transfer learning (TL) is a technique which is commonly used in RL to allow the agent to learn from previous agent experiences. The main advantage of TL is to improve training performances by implementing a sequence of relatively simple and attainable training objectives to help achieve the final training target gradually (Benyahia et al., 2021). In this study, TL is utilized by training the agents with an increased number of reference trajectories sequentially. Furthermore, error tolerances were also tightened gradually to improve tracking accuracy.

3. Result and Discussion

The corresponding training results indicated that two-agent RL requires less training episodes compared to a single-agent RL. This key advantage may save massive time-cost in the design and implementation of advanced control systems. Figure 3 shows the validation results where the performance of the two-RL agent is compared against a single agent RL in case study one. The sum of the squared errors (SSE) is computed and shown in Figures 3d, e and f to provide more quantitative insights to compare the accuracy of both control strategies. Both control methods can achieve trajectory tracking objectives to maintain the error within the acceptable range (Figures 3a, b, and c). However, the SSE indicates that a single agent exhibits a more effective performance compared to the two-agent RL in terms of accuracy.

Figure 4 presents a comparison between RL-PID and MPC in the 2D PMB model in case study 2. Here, the MPC is also developed in MATLAB/SIMULINK using basic tuning parameters. Overall, the RL-PID shows better performance in accuracy compared to the MPC for the 2D crystallization process of KDP.

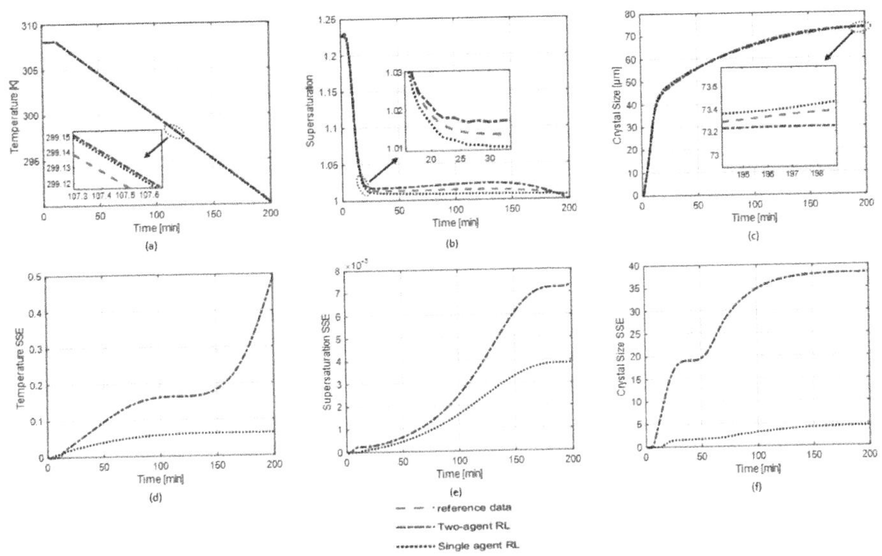

Figure 3. Validation results of the two RL agents vs single RL agent

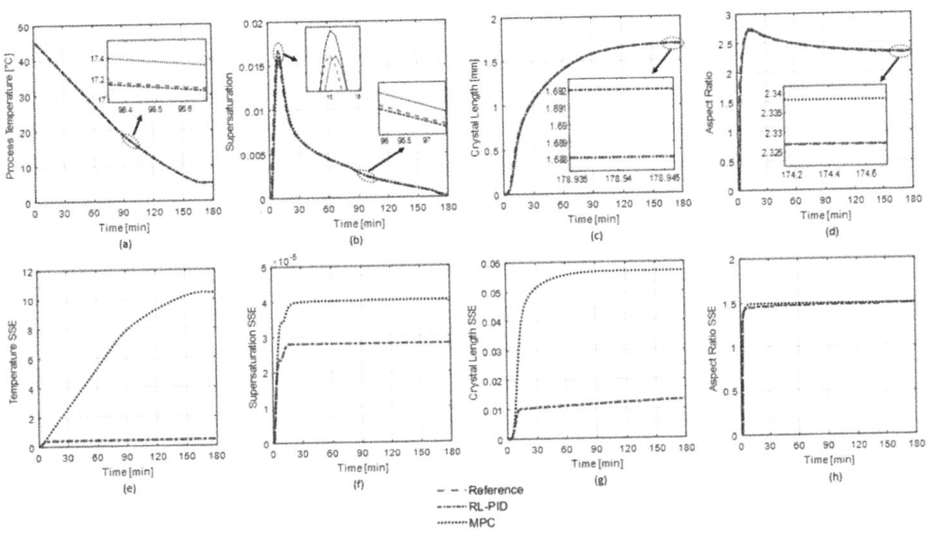

Figure 4. Overview of the comparisons between RL-PID and MPC

The simulation results depicted in Figures 4c and d show the tracking control performance of RL-PID with respect to one of the critical quality attributes, the crystal size and shape. Here. the RL-PID exhibits effectively and precise tracking performance. Figures 4e, f, g and h, provide more insights to further qualitatively investigate the performance of the RL-PID against MPC using the SSE. Overall, the RL-PID controller exhibits lower SSE compared to the MPC.

4. Conclusions

A model-based RL with multiple agents and an RL-PID strategy was designed to achieve trajectory tracking control of an antisolvent cooling semi-batch and a cooling-only batch crystallization process. The proposed RL agents were based on the DDPG algorithm and were effectively trained using penalties only rewarding and transfer learning. Overall, both RL approaches exhibited excellent tracking performance. The results associated with case study one demonstrated that the two RL agents require less training time compared to the TL support single agent which reduces the training costs as a potential benefit. The results associated with case study two revealed that RL-PID can achieve more effective and accurate crystal size and shape control by tracking the optimal trajectories compared to the model predictive control.

Acknowledgements: This work was funded by the EPSRC (EP/R032858/1) ARTICULAR project.

References

Barik, K., Prusti, P., Mohapatra, S. S. (2019). Single- and multi-objective optimisation for a combined cooling and antisolvent semi-batch crystallisation process with an ACADO toolkit, *Indian Chemical Engineer*, 62 (3), 287-300.

Benyahia, B., Anandan, P. D., Rielly, C. (2021). Robust Model-Based Reinforcement Learning Control of a Batch Crystallization Process. *9th International Conference on Systems and Control (ICSC)*, 89-94.

Canese, L., Cardarilli, G. C., Di Nunzio, L., Fazzolari, R., Giardino, D., Re, M., Spanò, S. (2021). Multi-Agent Reinforcement Learning: A Review of Challenges and Applications. *Appl. Sci.* 11, 4948.

Cui, J., Liu, Y., Nallanathan, A. (2020). Multi-Agent Reinforcement Learning-Based Resource Allocation for UAV Networks. *IEEE Trans on Wire Commun*, 19, 729–743.

Fysikopoulos, D., Benyahia, B., Borsos, A., Nagy, A. K., Rielly, C. D. (2019). A framework for model reliability and estimability analysis of crystallization processes with multi-impurity multi-dimensional population balance models, *Computers & Chemical Engineering*, 122, 275-292.

Lakerveld, R., Benyahia, B. (2020). *"CHAPTER 4: Process Control", The Handbook of Continuous Crystallization*. The Royal Society of Chemistry, 172-218.

Liu, J., Benyahia, B. (2022). Optimal start-up strategies of a combined cooling and antisolvent multistage continuous crystallization process, *Computers & Chemical Engineering*, 159, 107671.

Mascia, S., Heider, P. L., Zhang, H., Lakerveld, R., Benyahia, B., Barton, P., Braatz, R. D., Cooney, C. L., Evans, J. M. B., Jamison, T. F., Jensen, K. F., Myerson, A. S., Trout, B. L. (2013). End-to-End Continuous Manufacturing of Pharmaceuticals: Integrated Synthesis, Purification, and Final Dosage Formation. *Angewandte Chemie International Edition*, 52 (47), 12359-12363.

Nagy, Z. K., Fujiwara, M., Braatz, R. D. (2008). Modelling and control of combined cooling and antisolvent crystallization processes, *J. Process Contr*, 18(9), 856-864.

Qie, H., Shi, D., Shen, T., Xu, X., Li, Y., Wang, L. (2019). Joint Optimization of Multi-UAV Target Assignment and Path Planning Based on Multi-Agent Reinforcement Learning. *IEEE Access*, 7, 146264–1462.

Antonis Kokossis, Michael C. Georgiadis, Efstratios N. Pistikopoulos (Eds.)
PROCEEDINGS OF THE 33rd European Symposium on Computer Aided Process Engineering (ESCAPE33), June 18-21, 2023, Athens, Greece
 http://dx.doi.org/10.1016/B978-0-443-15274-0.50266-3

Globally Convergent Composite-Step Trust-Region Framework for Model-Based Real-Time Optimization

Duo Zhang[a], Xiang Li[b], Kexin Wang[a], Zhijiang Shao[a]

[a] *College of Control Science and Engineering, Zhejiang University, Hangzhou, 310027, China*

[b] *Department of Chemical Engineering, Queen's University, Kington, ON K7L 3N6, Canada*

Abstract

In model-based static real-time optimization (RTO), trust-region framework is globally convergent for unconstrained problems. However, it remains to be seen if the trust-region strategy can handle constraints directly with theoretical optimality guarantees. This paper addresses this issue and proposes a novel composite-step trust-region framework that guarantees global convergence for constrained RTO problems. In each iteration, the trial step is decomposed into a normal step that improves feasibility and a tangential step that reduces the cost function. Simulation results show that the proposed algorithm converges to the plant optimum in the presence of structural plant-model mismatches and wrong model curvatures. The proposed composite-step approach is more advantageous than the penalty approach because it needs a smaller penalty coefficient and does not suffer from ill-conditioned problems due to a large penalty coefficient.

Keywords: trust region, composite step, inequality constraints, global convergence, real-time optimization.

1. Introduction

Model-based optimization and control are important for achieving operational excellence. However, plant-model mismatches may influence practical performance and even cause convergence problems. Thus, handling mismatches by reliable optimization algorithms aroused great interest. In static model-based real-time optimization (RTO), algorithms based on gradient matching ideas can guarantee upon-convergence optimality in the presence of structural plant-model mismatch [1,2]. However, they are not globally convergent and may cause oscillation due to an inadequate model [3].

There are two ways to achieve global convergence in RTO. First, global convergence is a byproduct of feasible-side convergence. In this case, algorithms [4,5] rely on additional plant information such as Lipschitz constants or global upper bounds, which could be hard to get. Second, globalization strategies in nonlinear optimization are adopted. Among them, trust-region method [6] is the most natural for model-based RTO problems. Bunin [7] showed that the trust-region framework can be applied to unconstrained RTO problems. In each iteration, the next input is restricted to the trust region, where plant-model mismatches are small. However, constrained RTO problems have to be converted to unconstrained ones by the penalty function to be incorporated into this framework [8]. It remains unclear if the trust-region strategy can handle constrained RTO problems directly with theoretical optimality guarantees. There are several benefits to preserving inequality constraints. On the one hand, it is more intuitive and more compatible with existing RTO algorithms. On the other hand, it is also more reliable since the physical model restricts the trial step. This paper addresses this issue and proposes a novel composite-step trust-region framework for constrained RTO problems.

2. Preliminaries

2.1. Model-Based Static Real-Time Optimization

We consider the optimization of a continuous process operating at steady states. The plant optimization problem can be stated as problem (1). u is the input variable, ϕ is the cost function, and g is the constraints. The subscript p denotes the plant.

$$\begin{aligned} &\min_u \phi_p(u) \\ &\text{s.t. } g_p(u) \leq 0 \end{aligned} \tag{1}$$

The mathematical representation of the plant is unknown, so the model optimization problem (2) with subscripts m is solved in each RTO iteration to calculate optimal u.

$$\begin{aligned} &\min_u \phi_m(u) \\ &\text{s.t. } g_m(u) \leq 0 \end{aligned} \tag{2}$$

However, plant-model mismatches exist, i.e., $\phi_p \neq \phi_m$ and $g_p \neq g_m$, and may lead to suboptimal or even infeasible input. Therefore, the model needs to be adapted based on the local information around u_k at each iteration, and the model optimization problem (3) then uses this adapted model, where subscript k indicates the kth RTO iteration.

$$\begin{aligned} &\min_u \phi_{m,k}(u) \\ &\text{s.t. } g_{m,k}(u) \leq 0 \end{aligned} \tag{3}$$

To deal with structural plant-model mismatch, the functions $\phi_{m,k}$ and $g_{m,k}$ in problem (3) are supposed to satisfy condition (4) given the information of plant output ($\phi_p(u_k)$, $g_p(u_k)$) and plant derivatives ($\nabla\phi_p(u_k)$, $\nabla g_p(u_k)$). Condition (4) means gradient matching between the model and the plant and leads to KKT matching upon convergence. There are various ways to do this, e.g., [1,2].

$$\phi_{m,k}(u_k) = \phi_p(u_k) \tag{4a}$$
$$g_{m,k}(u_k) = g_p(u_k) \tag{4b}$$
$$\nabla\phi_{m,k}(u_k) = \nabla\phi_p(u_k) \tag{4c}$$
$$\nabla g_{m,k}(u_k) = \nabla g_p(u_k) \tag{4d}$$

2.2. Trust-Region Framework for Unconstrained Problems

Algorithms that solve problem (3) repeatedly based on local models only guarantee upon-convergence optimality. Inadequate models [3] and improper algorithm parameters [2] could lead to oscillation. To address this, Algorithm 1 based on the trust-region framework was proposed [7,8].

Algorithm 1, Trust-Region Algorithm for Unconstrained Problems

Step 0: Initialization. Choose an initial point u_0, an initial trust-region radius $\Delta_0 > 0$. Choose constants $0 < \eta_1 \leq \eta_2 < 1$, $0 < \gamma_1 < 1$. $k \leftarrow 0$.

Step 1: Model adaptation. Build a model satisfying Eqs. (4).

Step 2: Step calculation. Minimize the cost function of the model to obtain the step s_k.

$$\begin{aligned} s_k = &\arg\min_s \phi_{m,k}(u_k + s) \\ &\text{s.t. } \|s\| \leq \Delta_k \end{aligned} \tag{5}$$

Step 3: Acceptance of the trial point. Compute the ratio

$$\rho_k = \frac{\phi_p(u_k)-\phi_p(u_k+s_k)}{\phi_{m,k}(u_k)-\phi_{m,k}(u_k+s_k)} \tag{6}$$

If $\rho_k > \eta_1$, then define $u_{k+1} = u_k + s_k$; otherwise, define $u_{k+1} = u_k$.

Step 4: Trust-region radius update. Set

$$\Delta_{k+1} \in \begin{cases} \min(\Delta_k, \Delta_{\max}) & , \text{if } \rho_k > \eta_2, \\ \Delta_k & , \text{if } \rho_k \in [\eta_1, \eta_2), \\ \gamma_1 \Delta_k & , \text{if } \rho_k < \eta_1. \end{cases} \tag{7}$$

$k \leftarrow k + 1$. Go back to Step 1.

In Step 2, the trial step is determined by minimizing the adapted model cost within the trust-region constraint. After applying the trial step to the plant and getting new measurements, we compare the actual improvement to the value predicted by the model in Step 3. The trial point is accepted as the next iterate if sufficient improvement is made. Otherwise, the trust region shrinks, and hopefully, the model provides a better prediction in a smaller region.

2.3. Penalty Method for Constrained Problems

Constrained RTO problems are more complicated because there are two aims: minimizing the cost function and reducing infeasibility. The penalty trust-region algorithm [8] converts the constrained problem to an unconstrained one by the merit functions (8), where $c(u)$ is the infeasibility at the current point and $\sigma > 0$ is the penalty coefficient. $c(u)$ is calculated by Eqs. (9).

$$f_p(u,\sigma) = \phi_p(u) + \sigma c_p(u) \tag{8a}$$

$$f_{m,k}(u,\sigma) = \phi_{m,k}(u) + \sigma c_{m,k}(u) \tag{8b}$$

$$c_p(u) = \left\|\max\left(g_p(u), 0\right)\right\| \tag{9a}$$

$$c_{m,k}(u) = \left\|\max\left(g_{m,k}(u), 0\right)\right\| \tag{9b}$$

If the penalty coefficient σ is large enough, then Algorithm 1 with $\phi(u)$ replaced by $f(u,\sigma)$ is globally convergent [8]. However, determining σ is not easy because it depends on the specific problem, and a too large σ causes numerical difficulties.

3. Composite-Step Trust-Region Framework

In this section, we propose a novel trust-region algorithm for constrained RTO problems that deals with inequality constraints directly.

3.1. Algorithm Description

Algorithm 2, Composite-Step Trust-Region Algorithm

Step 0: Initialization. Choose an initial point u_0, maximum trust-region radius $\Delta_{max} > 0$, an initial trust-region radius $\Delta_0 \in (0, \Delta_{max})$, and a sufficiently large penalty parameter $\sigma > 0$. Choose constants $0 < \eta_1 \leq \eta_2 < 1$, $0 < \gamma_1 < 1$, $0 < \xi < 1$. $k \leftarrow 0$.

Step 1: Model adaptation. Build a model satisfying Eqs. (4).

Step 2: Normal step calculation. Move towards feasibility.

$$\begin{aligned} n_k = \arg\min_n \, & c_{m,k}(u_k + n) \\ \text{s.t.} \, & \|n\| \leq \xi \Delta_k \end{aligned} \tag{10}$$

Step 3: Tangential step calculation. Compute a step t_k by

$$\begin{aligned} t_k = \arg\min_t \, & \phi_{m,k}(u_k + n_k + t) \\ \text{s.t.} \, & c_{m,k}(u_k + n_k + t) \leq c_{m,k}(u_k + n_k) \\ & \|n_k + t\| \leq \Delta_k \end{aligned} \tag{11}$$

Step 4: Acceptance of the trial point. The composite step is $s_k = n_k + t_k$. Compute the ratio

$$\rho_k = \frac{f_p(u_k,\sigma) - f_p(u_k + s_k,\sigma)}{f_{m,k}(u_k,\sigma) - f_{m,k}(u_k + s_k,\sigma)} \tag{12}$$

If $\rho_k > \eta_1$, then define $u_{k+1} = u_k + s_k$; otherwise, define $u_{k+1} = u_k$.
Step 5: Trust-region radius update. Set Δ_{k+1} according to Eq. (7). Go back to Step 1.

The composite step idea comes from [6]. The overall step is decomposed into a normal step that improves feasibility and a tangential step that reduces the cost function. By adjusting ξ, we change the step size for cost reduction and infeasibility reduction, and the overall step size is bounded by Δ_k. Since the feasible set of problem (3) may not intersect with the trust region, small infeasibility has to be allowed during the iteration as long as there is a trend toward feasibility.
In Step 1, any model adaptation approach satisfying Eqs. (4) can be used, e.g., modifier adaption [1] and generalized parameter estimation [2]. In Step 2 and 3, the inequality-constrained subproblem is solved with the updated model. In Step 4, the progress towards a critical point is measured by the decrease of the merit function $f(u,\sigma)$.

3.2. Global Convergence

Global convergence is ensured because the trust-region framework enforces a monotonic decrease of the merit function. However, there are some prerequisites. First, all the subproblems are solved globally. Second, σ should be large enough, so the algorithm can avoid getting stuck at an infeasible point. Third, the model update approach should be consistent, i.e., it builds similar local models given similar measurements.

4. Case Studies

4.1. Simulation Settings

We illustrate the idea by a simple quadratic optimization problem (13) adapted from [9]. Input variables are bounded by $u_1 \in [-2,4]$ and $u_2 \in [-4,2]$. It is assumed that the measurements are noise-free and that the plant gradients are available.

$$\begin{aligned} &\min_u u_1^2 + u_2^2 + u_1 u_2 \\ &\text{s.t.}\, 1 - u_1 + u_2^2 + 2u_2 \le 0 \end{aligned} \tag{13}$$

Three models with different structural mismatches are considered. The first model optimization problem (14) has the right model curvature, while the other two problems (15) and (16) have the wrong curvature in the cost function and the constraint, respectively. Models with wrong curvatures are likely to be inadequate and cause oscillation.

$$\begin{aligned} &\min_u u_1^2 + u_2^2 \\ &\text{s.t.}\, 1 - u_1 + u_2^2 \le 0 \end{aligned} \tag{14}$$

$$\begin{aligned} &\min_u -u_1^2 + u_2^2 \\ &\text{s.t.}\, 1 - u_1 + u_2^2 \le 0 \end{aligned} \tag{15}$$

$$\begin{aligned} &\min_u u_1^2 + u_2^2 \\ &\text{s.t.}\, 1 - u_1 - 4u_2^2 \le 0 \end{aligned} \tag{16}$$

The proposed algorithm is compared with modifier adaptation [1] (no trust-region constraints, with filter gain $K = 0.1$) and penalty trust-region method. The simulation begins at point [2, -2]. The maximum step size for each iteration is 2. Parameters for Algorithm 1 are selected according to [6]. For the composite-step and the penalty methods, the penalty coefficient and the initial trust radius are 1. In Step 1 of each algorithm, the model is updated in the same way as in the modifier adaptation method. For the composite-step method, $\xi = 0.5$. The nonsmooth maximum operator is approximated by a smooth one [8]. Global convergence of the proposed algorithm and the influence of σ are studied.

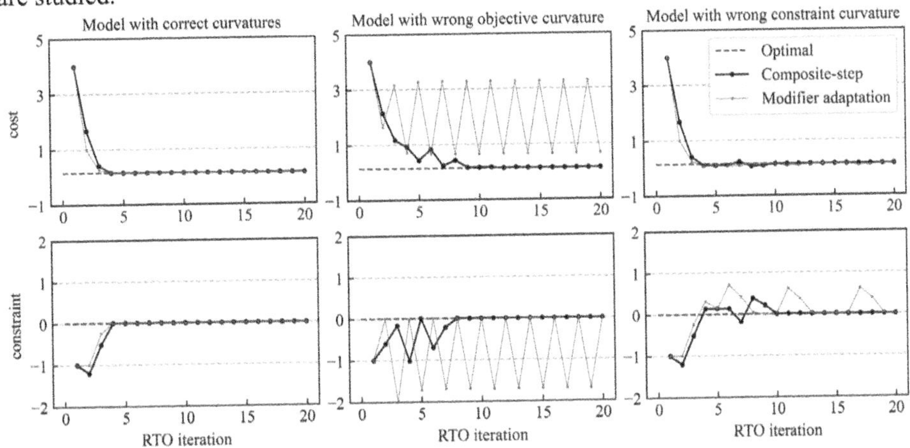

Figure 1 Evolution of plant cost and constraint using mismatched models during 20 RTO iterations. Left: simulation using model (14); middle: simulation using model (15); right: simulation using model (16). Upper: profiles of plant cost; lower: profiles of plant constraint. Black: Algorithm 2; red: modifier adaptation method [1].

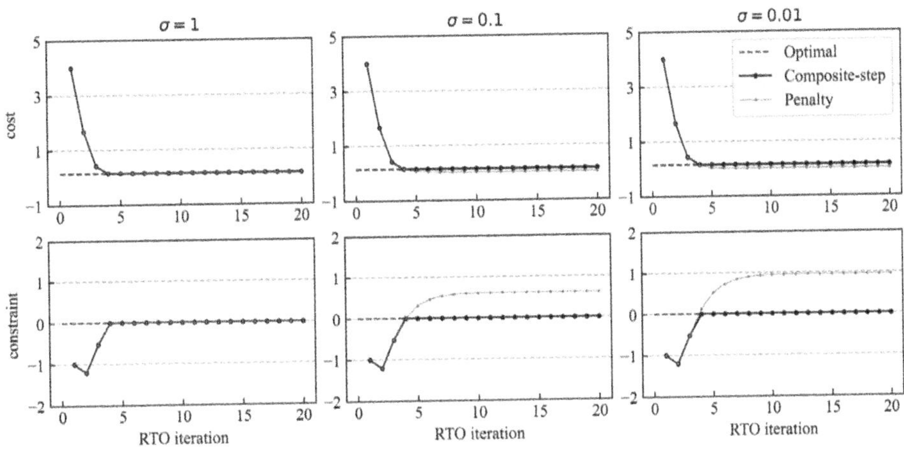

Figure 2 Evolution of plant cost and constraint with different penalty coefficients during 20 RTO iterations. Left: $\sigma = 1$; middle: $\sigma = 0.1$; right: $\sigma = 0.01$. Upper: profiles of plant cost; lower: profiles of plant constraint. Black: Algorithm 2; red: penalty trust-region method [8].

4.2. Global Convergence

In this subsection, we show that the proposed algorithm converges even when objective and constraints curvatures of the model are wrong. The three model optimization problems (14)-(16) are investigated. Model adaptation method (without trust-region constraints) serves as a comparison.

As Figure 1 shows, the proposed composite-step algorithm converges to the plant optimum in all three scenarios, while the modifier adaptation method fails to find the optimum with the wrong curvature model. For both methods, the model's KKT condition matches that of the plant. However, the second-order sufficient optimality condition may be violated for the modifier adaptation method due to the wrong curvature of the model. In other words, the trust-region framework lowers the requirement for model adequacy.

4.3. Penalty Coefficient

In this subsection, we investigate the effect of a small penalty coefficient on the two trust-region algorithms. RTO based on problem (14) using the composite-step and the penalty trust-region methods with different σ are illustrated in Figure 2.

Theoretically, both methods could be biased if σ is not large enough. However, our experience shows that the proposed method is less sensitive to small σ than the penalty method. This is because the model constraint is explicitly imposed in Step 3.

Moreover, if a very large σ is adopted to avoid the bias, the optimization subproblems in the penalty method may be ill-conditioned and hard to solve. The proposed composite-step approach does not suffer from this problem. The reason is that σ only appears when evaluating the trial point in Eq. (12), not in solving optimization subproblems (11) or (12).

5. Conclusions

This paper proposes a novel composite-step trust-region algorithm that directly handles inequality RTO constraints. The composite-step approach does not suffer from numerical problems caused by a large penalty coefficient in the merit function. Numerical simulations show that the proposed method is globally convergent and reduces model adequacy requirements. Future work will be focused on the rigorous proof and methods to find a proper penalty coefficient σ automatically.

Acknowledgements

This work was supported by the National Natural Science Foundation of China under grant number 62120106003, 62173301, 61873242, and China Scholarship Council under CSC NO. 202006320176.

References

[1] Marchetti, A., Chachuat, B. and Bonvin, D., 2009. Modifier-adaptation methodology for real-time optimization. Industrial & engineering chemistry research, 48(13), pp.6022-6033.

[2] Zhang, D., Wang, K., Xu, Z., Tula, A.K., Shao, Z., Zhang, Z. and Biegler, L.T., 2022. Generalized Parameter Estimation Method for Model-Based Real-Time Optimization. Chemical Engineering Science, 258, p.117754.

[3] Forbes, J.F. and Marlin, T.E., 1996. Design cost: A systematic approach to technology selection for model-based real-time optimization systems. Computers & chemical engineering, 20(6-7), pp.717-734.

[4] Bunin, G.A., François, G. and Bonvin, D., 2014. Implementation techniques for the SCFO experimental optimization framework. arXiv preprint arXiv:1406.3997.

[5] Marchetti, A.G., Faulwasser, T. and Bonvin, D., 2017. A feasible-side globally convergent modifier-adaptation scheme. Journal of Process Control, 54, pp.38-46.

[6] Conn, A.R., Gould, N.I. and Toint, P.L., 2000. Trust region methods. Society for Industrial and Applied Mathematics.

[7] Bunin, G.A., 2014. On the equivalence between the modifier-adaptation and trust-region frameworks. Computers & chemical engineering, 71, pp.154-157.

[8] Biegler, L.T., Lang, Y.D. and Lin, W., 2014. Multi-scale optimization for process systems engineering. Computers & Chemical Engineering, 60, pp.17-30.

[9] del Rio Chanona, E.A., Petsagkourakis, P., Bradford, E., Graciano, J.A. and Chachuat, B., 2021. Real-time optimization meets Bayesian optimization and derivative-free optimization: A tale of modifier adaptation. Computers & Chemical Engineering, 147, p.107249.

Antonis Kokossis, Michael C. Georgiadis, Efstratios N. Pistikopoulos (Eds.)
PROCEEDINGS OF THE 33rd European Symposium on Computer Aided Process Engineering (ESCAPE33), June 18-21, 2023, Athens, Greece
 http://dx.doi.org/10.1016/B978-0-443-15274-0.50267-5

Fault Detection and Diagnosis for Chemical Processes based on Deep Neural Networks with Continuous Wavelet Transform

Chinatsu Ukawa,[a] Yoshiyuki Yamashita,[b]

[a]*Department of Food and Energy Systems Science, Graduate School of Bio-Applications Systems Engineering, Tokyo University of Agriculture and Technology, 2-24-16 Naka-cho, Koganei, Tokyo 184-8588, JAPAN*

[b] *Department of Chemical Engineering, Tokyo University of Agriculture and Technology, 2-24-16 Naka-cho, Koganei, Tokyo 184-8588, JAPAN*

Abstract

This study proposed a novel fault detection and diagnosis method using continuous wavelet transform (CWT) and a three-dimensional convolutional neural network (3D-CNN). In particular, multivariate time series data from chemical plants were divided by a time-shifting window and transformed into scalograms using CWT. These scalograms were fed to 3D-CNN to generate outputs indicating the faults that occurred in the process. We applied the proposed method to a Tennessee Eastman process dataset. The proposed method adequately captures the characteristics in the time-frequency domain and exhibited good fault detection and diagnosis performances on the simulation dataset.

Keywords: Fault Detection, Fault Diagnosis, Deep Learning, Wavelet Transformation

1. Introduction

Fault detection and diagnosis (FDD) are crucial steps in process monitoring to maintain high productivity, efficiency, and safety in a chemical plant operation. Data-based FDD methods have attracted attention in recent years due to the dramatic growth of computer and information technology. Among various data-driven methods, machine learning (ML), such as deep learning, is a promising method for capturing process dynamics and nonlinearity. ML have shown great potential in various research fields (Xiong, 2022), where numerous studies have been conducted in the process engineering field (Ge, 2021).

Data preprocessing is a critical phase in developing accurate models using ML. The appropriate selection of preprocessing methods can contribute to achieving ML exhibiting high performance. Signal processing, such as Fourier transform and wavelet transform, are typical examples of preprocessing. They are used for denoising or feature extraction in various fields. Fourier transform has been applied to numerous fields, exhibiting outstanding performance for feature extraction. However, this method misses the time information of the original signal and cannot work well when the original data includes nonlinearity and non-stationarity. In contrast, wavelet transform is a powerful tool to extract the information in the time-frequency domain of the original signal. This method can be applied to nonlinear and nonstationary signals. In particular, wavelet transform has shown great potential in fault detection, including bearing and electrocardiogram. Although ML and wavelet transform have been successfully applied in other research fields, their implementation to obtain the time-frequency information using continuous wavelet transform (CWT) is uncommon for the field of FDD for chemical processes.

Here, we proposed a combined FDD method using CWT and a three-dimensional convolutional neural network (3D-CNN). We obtained 3D inputs called scalograms for 3D-CNN from preprocessing using CWT and classified the result outputs regarding the type of fault that occurred in the chemical process.

2. Method

2.1. Continuous wavelet transform

CWT is a widely mathematically implemented wavelet transform, which can exhibit a complete view of the signal by continuously changing the scale and translation parameters (Ramin, 2022). In particular, CWT has been used for filtering out noise and fault detection and diagnosis in various fields, such as electrocardiogram analysis (Alharbey, 2022), and the development of non-steady process simulation (Hong, 2022). CWT analyses an original signal using a mother wavelet. The equation of the mother wavelet is shown below as Equation 1, where parameters *a* and *b* denoted scaling and time-shifting.

$$W_x(a,b) = \frac{1}{\sqrt{a}} \int_{-\infty}^{\infty} x(t)\,\psi^*\left(\frac{t-b}{a}\right) dt \tag{1}$$

In addition, CWT produces scalograms showing the feature from the original signal in the time-frequency domain. This visualization method is also used in various fields, for instance, fault detection and feature extraction of seismic data (Ali, 2020) and filtering tool for velocity signals of a thunderstorm (Brusco, 2022).

2.2. 3D-CNN

We applied a 3D-CNN in the proposed method for feature extraction and classification to handle multivariable datasets. Various neural network methods have been proposed in recent years, such as Long-Short Term Memory, Recurrent Neural Network, Gaussian Neural Network, and Convolutional Neural Network (CNN). Nevertheless, CNN is an already known method for feature extraction and classification methods by our group. In particular, 2-dimensional CNN is a popular method for image recognition and classification and FDD methods for chemical processes. Moreover, CNN has been widely applied to overcome various problems owing to its great performance. (Kim, 2022)

Furthermore, 3D-CNN is a logical extension of 2D-CNN. The feature extraction scheme is the same as other dimensional CNN. However, 3D-CNN uses 3D filters to convolute 3D data and passes convoluted data to pooling layers before the fully connected layers for classification. This method has been proposed for image recognition with medical 3D scans (Pintelas, 2022), or agricultural hyperspectral images (Diao, 2022). In this study, we selected 3D-CNN to deal with multivariable from a particular chemical process.

2.3. Proposed method

Chemical processes have non-steady and nonlinear multivariable data. These processes can be divided by a time-shifting window. Each divided data can be transformed into a scalogram to show a particular feature of data in the time-frequency domain using CWT.

We decided that the size of a scalogram was 20 x window size. This processing was applied to each process variable to obtain 3D input data (20 x window size x the number of process variables). The generated input data was fed to the 3D-CNN model to obtain outputs representing the chemical process condition.

3. Case study

Tennessee Eastman process (TEP) is one of the most popular benchmarks in chemical process FDD. This process contains five main units: a reactor, a condenser, a stripper, a vapor-liquid separator, and a compressor. The process flow is shown in Figure 1. Several types of TEP simulators and datasets with different types of control loops or various fault scenarios have been developed (Bathelt, 2015, Chiang, 2000D). In this study, we used an extended TEP dataset introduced in 2017 (Rieth, 2017), which includes 20 fault operation modes of all 21 modes and 52 process variables with 41 measured and 11 manipulated variables.

In this study, we set the size of the time shift window for each preprocessing method. In particular, we set 20 for CWT and 128 for Fast Fourier Transform (FFT) and Short-Time Fourier Transform (SFT). The number of selected process values was 33, including 22 measured and 11 manipulated variables.

4. Result

4.1. Metrics for evaluation

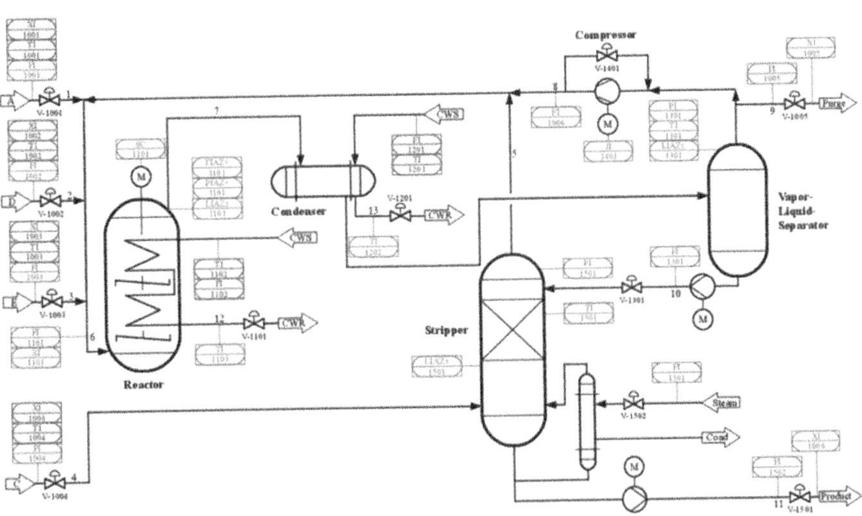

Figure 1 Process flow of TEP

We evaluated the proposed method and compared its performance using the fault detection rate (FDR) and false positive rate (FPR). TP represents true positive, and FN represents false negative.

$$Fault\ Detection\ Rate\ (FDR) = \frac{TP}{TP + FP} \tag{2}$$

$$False\ Positive\ Rate\ (FPR) = \frac{FP}{FP + TP} \tag{3}$$

4.2. Application to TEP

We compared the proposed method with a combination of 3D-CNN methods with FFT and STFT. The results of the comparison are listed in Table 1. The proposed method outperformed the other preprocessing methods.

5. Discussion

The result of the proposed method shows excellent performance on more than half the number of faults compared to FFT and STFT. Figure 2 shows some of the input samples for 3D-CNN. We selected measured variable 1 samples of fault 6 at normal operation and when the fault occurred. We can see the change in the time-frequency domain of this variable, indicating that the proposed model learned the feature from input samples and classifies process conditions based on it.

However, fault-free (fault 00) exhibited the worst FDR result. We verified the prediction output from 3D-CNN, representing a ratio of each fault. The ratio ranged from 0 to 1. The model estimated several fault-free data as fault data of faults 3, 9, and 15. Thus, we have to solve how to divide fault-free and faulty data in future works.

Some FDR of CWT were lower than FFT and STFT. The difference in window size between these methods might be one of the reasons for this result. We compared the effect of window size on the model's ability by setting the window size from 20 to 40. The result showed that expanded window size was valid for fault 21 even though other faults were worse detected. Our next step is establishing a method to decide on the appropriate window size for the original process data.

6. Conclusion

This study proposed a fault detection and diagnosis model for chemical processes using CWT and 3D-CNN. CWT was used as preprocessing method for chemical process data. This method has exhibited great potential for feature extraction for non-steady and nonlinear signals in various research fields. In the proposed method, the chemical process data was first divided into samples with a set length. After that, samples were transformed into scalograms using CWT. The 3D-CNN was applied to extract the feature embedded in the time-frequency domain and perform multivariate analysis with the chemical process variables. The structure of the 3D-CNN comprised convolutional layers, pooling layers, dropout layers, and fully connected layers. The scalograms were fed to 3D-CNN to generate the outputs, representing the faults in the chemical process. The proposed method was applied to the TEP dataset introduced by Rieth et al. in 2017. Our method could capture the characteristics in the time-frequency domain well and showed better performance on more than half the number of faults than FFT and STFT. However, in the case of fault-free, the proposed model should be improved. We will train the proposed model without fault-free data and evaluate its ability to classify faults. Additionally, we will examine the method to decide the window size.

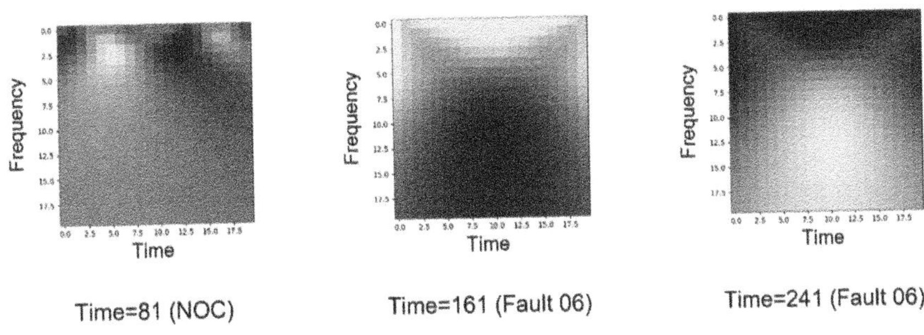

Figure 2 Input samples of fault 06

References

Table 1 Result from the proposed model and other preprocessing methods

Fault	FFT		STFT		CWT			
					window size=20		window size=40	
	FDR	FPR	FDR	FPR	FDR	FPR	FDR	FPR
0	0.063	0.343	0.168	0.767	0.139	0.458	0.138	0.534
1	0.172	0.762	0.673	0.269	0.979	0.030	0.968	0.023
2	0.362	0.889	0.806	0.067	0.972	0.044	0.954	0.049
3	0.112	0.942	0.106	0.859	0.468	0.728	0.399	0.725
4	0.691	0.731	0.716	0.266	0.968	0.024	0.958	0.024
5	0.207	0.877	0.491	0.724	0.959	0.047	0.866	0.076
6	0.836	0.312	0.826	0.342	0.990	0.003	0.975	0.034
7	0.217	0.711	0.734	0.366	0.973	0.040	0.921	0.050
8	0.804	0.268	0.588	0.311	0.865	0.170	0.865	0.218
9	0.107	0.936	0.083	0.862	0.139	0.876	0.090	0.867
10	0.520	0.738	0.292	0.685	0.698	0.579	0.615	0.632
11	0.933	0.084	0.878	0.066	0.733	0.493	0.544	0.554
12	0.712	0.031	0.623	0.047	0.755	0.106	0.697	0.113
13	0.511	0.349	0.571	0.521	0.851	0.136	0.815	0.282
14	0.955	0.019	0.954	0.015	0.864	0.371	0.863	0.355
15	0.299	0.924	0.266	0.839	0.168	0.878	0.131	0.900
16	0.828	0.274	0.740	0.277	0.667	0.356	0.523	0.599
17	0.890	0.034	0.820	0.083	0.922	0.117	0.852	0.226
18	0.624	0.268	0.570	0.311	0.926	0.197	0.875	0.215
19	0.875	0.518	0.732	0.424	0.522	0.651	0.369	0.789
20	0.879	0.203	0.792	0.225	0.906	0.224	0.913	0.158
AVERAGE	0.552	0.486	0.592	0.396	0.736	0.311	0.682	0.353

X. Ge, B. Wang, X. Yang, Y. Pan, B.Liu, B. Liu, 2021, Fault detection and diagnosis for reactive distillation based on convolutional neural network, Computers & Chemical Engineering, Volume 145, 107172

S. Xiong, L. Zhou, Y. Dai, X. Ji, 2022, Attention-based LSTM fully convolutional network for chemical process fault diagnosis, Chinese Journal of Chemical Engineering (In Press)

A. Ali, S. C. Chen, M. Shah, 2020, Continuous wavelet transformation of seismic data for feature extraction. SN Applied Sciences, Volume 2, 1835.

M. Ramin, M. R. Sohrabi, F. Motiee, 2022, Smart spectrophotometric methods for enhancement spectral resolution and rapid simultaneous determination of dapoxetine and ildenafil in commercial tablets using fuzzy inference system and continuous wavelet transform, Chemometrics and Intelligent Laboratory Systems, Volume 230, 104656, ISSN 0169-7439

H. P. Hong, X.Z. Cui, D. 2022, Qiao, Simulating nonstationary non-Gaussian vector process based on continuous wavelet transform, Mechanical Systems and Signal Processing, Volume 165, 108340, ISSN 0888-3270

S. Brusco, G. Buresti, G. Piccardo, 2022, Thunderstorm-induced mean wind velocities and accelerations through the continuous wavelet transform, Journal of Wind Engineering and Industrial Aerodynamics, Volume 221, 104886, ISSN 0167-6105

C. A. Rieth, B. D. Amsel, R. Tran, M. B. Cook, 2017, Additional Tennessee Eastman process simulation data for anomaly detection evaluation", https://doi.org/10.7910/DVN/6C3JR1, Harvard Dataverse, V1

A. Bathelt, N. L. Ricker, M. Jelali, 2015, Revision of the Tennessee Eastman process model, IFAC-PapersOnLine, Volume 48, Issue 8, Pages 309-314, ISSN 2405-8963

L. H. Chiang, E. L. Russell, R. D. Braatz, 2020, Fault diagnosis in chemical processes using Fisher discriminant analysis, discriminant partial least squares, and principal component analysis, Chemometrics and Intelligent Laboratory Systems, Volume 50, Issue 2, Pages 243-252, ISSN 0169-7439

R.A. Alharbey, S. Alsubhi, K. Daqrouq, A. Alkhateeb, 2022, The continuous wavelet transform using for natural ECG signal arrhythmias detection by statistical parameters, Alexandria Engineering Journal, Volume 61, Issue 12, Pages 9243-9248, ISSN 1110-0168

E. Pintelas, P. Pintelas, 2022, A 3D-CAE-CNN model for Deep Representation Learning of 3D images, Engineering Applications of Artificial Intelligence, Volume 113, 104978, ISSN 0952-1976

Z. Diao, J. Yan, Z. He, S. Zhao, P. Guo, 2022, Corn seedling recognition algorithm based on hyperspectral image and lightweight-3D-CNN, Computers and Electronics in Agriculture, Volume 201, 107343, ISSN 0168-1699

M. I. Kim, H. S. Yoon, 2022, Geometric modification for the enhancement of an airfoil performance using deepRa CNN, Ocean Engineering, Volume 266, Part 4, 2022

Antonis Kokossis, Michael C. Georgiadis, Efstratios N. Pistikopoulos (Eds.)
PROCEEDINGS OF THE 33rd European Symposium on Computer Aided Process Engineering (ESCAPE33), June 18-21, 2023, Athens, Greece
 http://dx.doi.org/10.1016/B978-0-443-15274-0.50268-7

Anomaly detection in chemical processes with semantic knowledge graphs: an approach to reduce cause-effect diagrams

Nazanin Hamedi[a], Ilhan Mutlu[b], Fatima Rani[b], Leon Urbas[b]

[a]*DFG Research Training Group 2323 CD-CPPS, Technische Universität Dresden, Georg-Schumann-Straße 7a, 01069 Dresden, Germany*
[b]*Faculty of Electrical and Computer Engineering, Chair of Process Control Systems & Process Systems Engineering Group, Technische Universität Dresden, 01069 Dresden, Germany*
nazanin.hamedi@tu-dresden.de

Abstract

When operators and users face an unexpected problem regarding the process, it would be easier and less time-consuming for them to narrow down the possible causes that result in the final issue. In this study, a pilot modular plant with two Process Equipment Assemblies was considered as the use case. At first, a knowledge graph representing this process was developed in Protégé software, which semantically described not only the equipment type and connectivity but also the behavior of the process. Then, the knowledge graph was imported to Python, where the sensor data were placed in their particular position in the knowledge graph. Afterward, an algorithm was developed to query the knowledge graph and verify if the relevant equipment was functioning correctly or not. Our results indicate that this approach can reduce the cause-effect diagrams in almost all scenarios. Nonetheless, there are situations where further sensor data (e.g., the temperature in a tank) is required for the algorithm to decide.

Keywords: Semantic knowledge graphs; Process behavior, Cause-effect graphs; Decision making.

1. Introduction

The advent of the computer has significantly impacted the control of chemical processes. Indeed, the regulatory control of chemical processes (e.g., opening and closing valves) is entirely automated, resulting in higher product quality and consistency, process safety, and process efficiency (Venkatasubramanian et al., 2003). Despite these advancements, the time-consuming detection of an abnormal situation, identifying its cause, and making the best decision to resolve the problem is still a manual task. Based on the statistics, minor accidents in chemical plants occur daily, leading to workers' injuries and posing huge financial pressure on the industry. Also, it has been reported that 70% of industrial accidents are due to operators' mistakes (Venkatasubramanian et al., 2003). These suggest that developing assistant systems for fault diagnosis is a pressing need.
Notable among different methodologies used for fault diagnosis are qualitative, graph-based approaches, which were comprehensively reviewed by Venkatasubramanian et al. (Venkatasubramanian, Rengaswamy and Kavuri, 2003) and Maurya et al. (Maurya, Rengaswamy and Venkatasubramanian, 2003).

A clear superiority of graph-based methodologies is that they provide a causal structure by which one can track the propagation of an error in a chemical process. However, as the processes get more sophisticated, the graphs become more complex and challenging to navigate (especially because there exist multiple causes for one problem) (Reinartz et al., 2019).
This motivates us to propose an approach for reducing the cause-effect graphs, making them more traceable. In this method, the process has been represented in the form of a knowledge graph in Protégé, in which the equipment type, connectivity, and process behavior are considered. Afterward, by importing this knowledge graph to Python and developing a program that mimics the sensor data, we were able to omit some of the unnecessary branches of the cause-effect graph. It is noteworthy to mention that to examine the applicability of our approach; it has been implemented for a use-case; namely, the Safety Demonstrator (Pelzer et al., 2021).

2. Method

2.1. Cause-effect graph

As mentioned before, the Safety Demonstrator was considered as a use-case in this paper. It is a modular plant with two Process Equipment Assemblies (PEAs). As its name suggests, the Safety Demonstrator aims to investigate a secure operation; and hence, many safety controllers have been taken into account in its design. Therefore, it would be helpful for the operator if he-she received information about why one of the safety control loops is active. In this study, a cause-effect graph is proposed (see figure 1), which is defined as a directed graph in which the nodes are the events (e.g., activation of one controller), and the edges illustrate that the reason for each event is actually the event on the subsequent node. Even though it is a very beneficial tool for identifying a problem in a chemical plant, it can become very complex. Thus, the idea of this research is to automatically reduce the graph based on the specific situation.

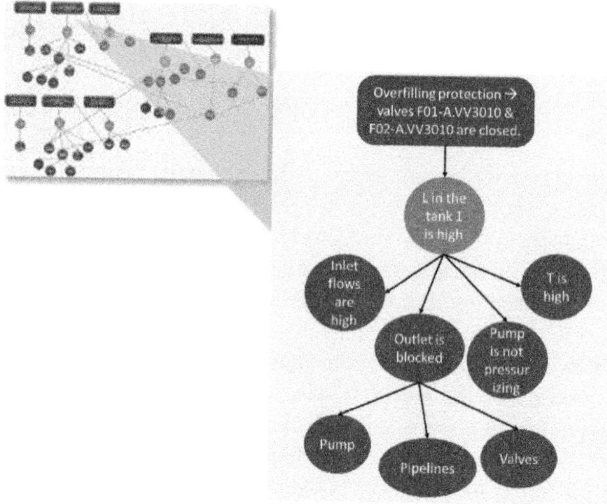

Figure 1- The cause-effect graph of the Safety Demonstrator (Gray nodes are the logics behind the controllers, rectangles show the active control system, green parts are related to PEA 1 and blue parts are related to PEA 2)

2.2. Semantic definition of the process

In this study, Protégé software was used to develop a knowledge graph of the investigated process. As can be seen in figure 2, it consists of 4 major classes: Equipment, Parameter, Chemicals, and Process Function. It is noteworthy to mention that the Process Function class aims to explain each equipment's phenomena semantically. To complete the definition of process function, the knowledge graph was imported from Protégé to Python with the help of owlready (a library in python for ontology descriptions) (Lamy, 2017). Then, as can be seen in figure 3 (A & D), two classes of Mass Transfer and Pressurizing and the related attributes and methods were defined. Indeed, a mass transfer takes place in an equipment when mass enters that equipment and the outlet mass is almost the same as the inlet. Also, an equipment does pressurize when the pressure of its inlet is lower than that of its outlet. These human-understandable facts are defined for the computer, as depicted in figure 3 (B).

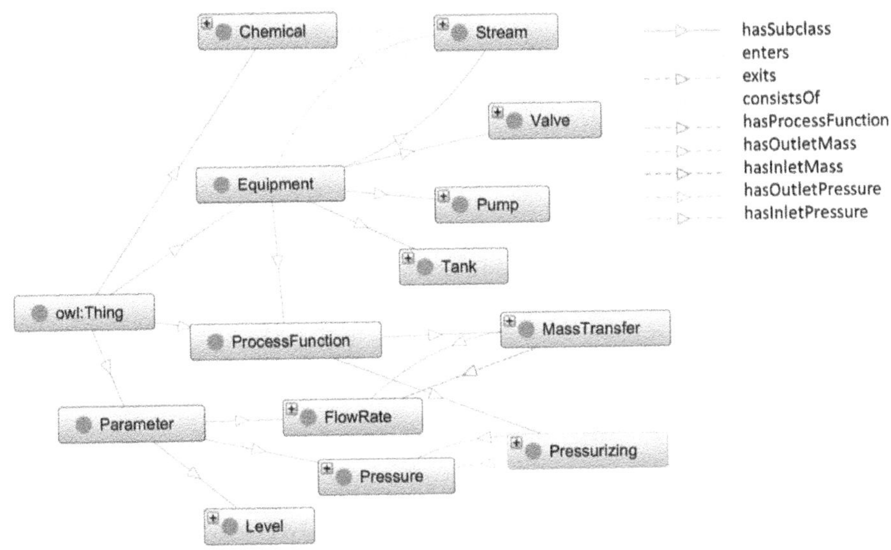

Figure 2 – Knowledge graph of the PEA1 of Safety Demonstrator developed by Protégé.

2.3. Algorithm

After the complete definition of the process in Python (figure 3 (A & B)), the safety control loops were also defined (figure 3 (C)). These controllers are activated when a process parameter (which has already been semantically defined) crosses the considered threshold. For instance, Overfilling-controller needs the value of the level in the tank, which has been already defined in the knowledge graph. Thus, the algorithm is aware that this value is the level of the tank, and the tank is actually connected to pump1 by stream3. The flowchart for drawing the cause-effect graph is shown in figure 3 (D). It should mention that this flowchart only considers the graph for the overfilling-controller, but a similar algorithm can be imagined for other controllers. In short, the algorithm checks whether a safety controller is active or not. If yes, then it searches for the possible reasons (based on the cause-effect graph), and if the cause is valid (based on the semantic definition of the process), it is added to the cause-effect graph.

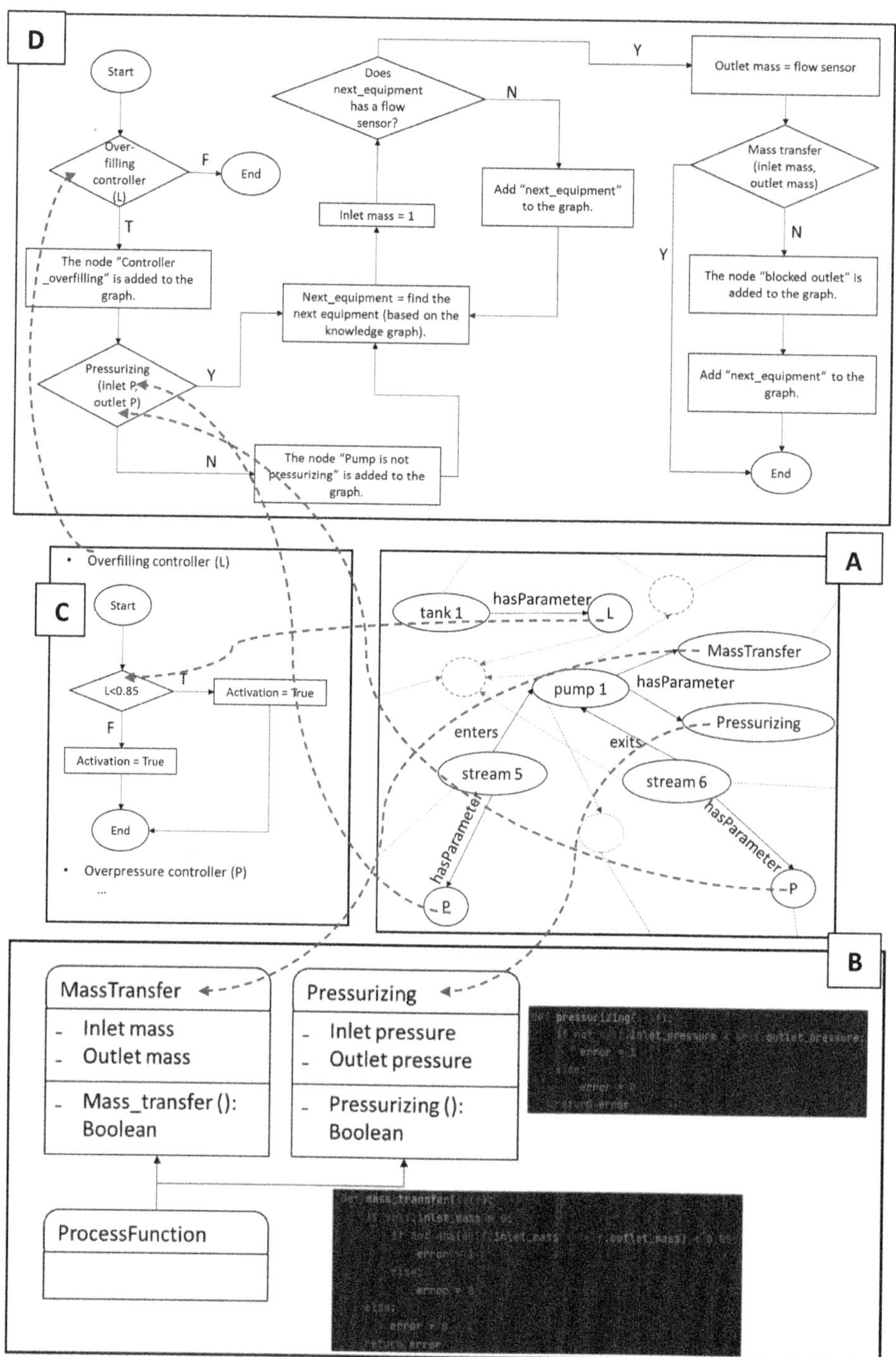

Figure 3 – A) Part of the knowledge graph from protégé (gray nodes and edges indicate that the shown nodes and edges are not the whole graph), B) Definition of process function in Python, C) Definition of the safety controllers, D) Graph development flowchart.

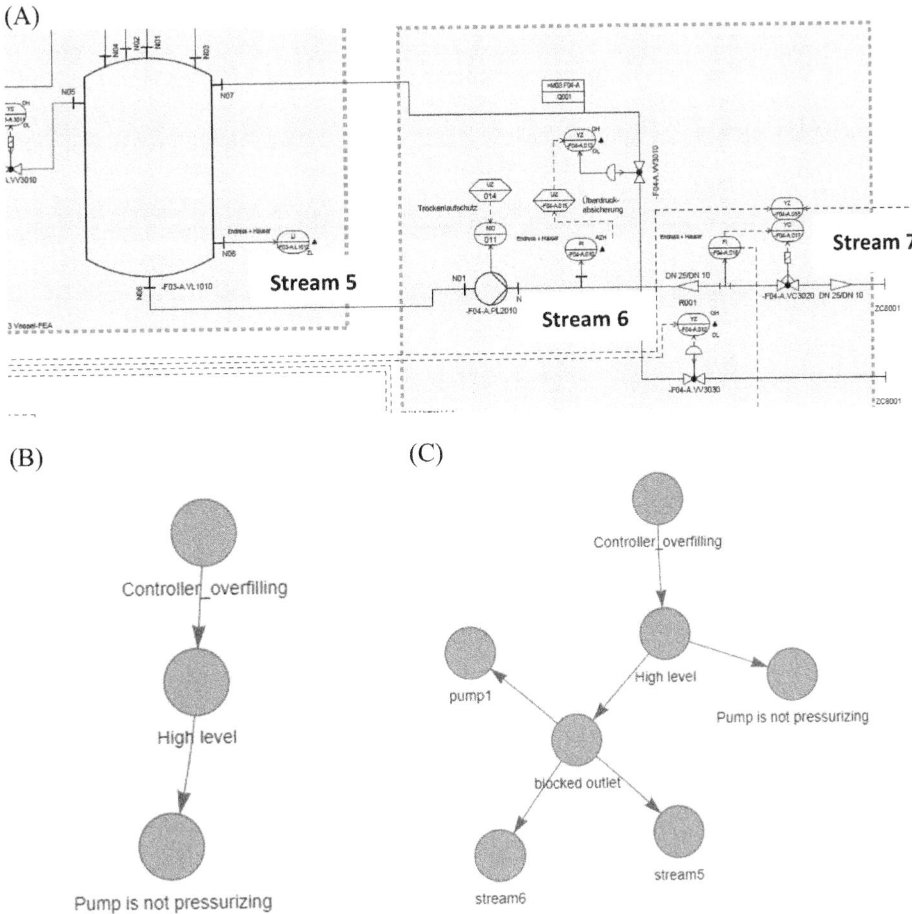

Figure 4 – A) Part of P&ID diagrams of the first PEA of Safety Demonstrator, (B) Reduced graph of a scenario in which pump is not working, (C) Reduced graph of a scenario in which pump is not working and the flow meter shows a blockage in the line.

3. Results and discussion

The above-mentioned method has been examined in different scenarios related to Safety Demonstrator. Because of space limitations, the scenarios investigated here are all related to PEA1 of the Safety Demonstrator (see figure 4 (A)).

- The first scenario is that pump 1 is not pressurizing the flow; i.e., the pressure after the pump (shown by the sensor) is not more than that of stream 5 (calculated from the length of the liquid in the tank), resulting in the activation of the overfilling-controller. The result of this is illustrated in figure 4 (B).
- Second scenario is that the pump is not pressurizing, and at the same time, the flow meter on stream 6 is showing a blockage. The produced graph for this scenario is depicted in figure 4 (C). Since there is no flow sensor in stream 5, our algorithm cannot precisely decide the exact point of blockage; nonetheless, it can identify it in the outlet of the tank and before the flow sensor in stream 6.

4. Conclusion

The purpose of this work is to find a method to reduce cause-effect graphs, facilitating operators' problem-solving ability. To this end, a use case, namely Safety-Demonstrator, was selected, which was first semantically defined in protégé, then imported to Python, where the process behavior was defined entirely. Afterward, an algorithm was developed to communicate with the knowledge graph and, based on the process sensor data, draw the corresponding cause-effect graph. The derived results indicate that this approach is successful in considered scenarios; however, the accuracy of the results would be improved depending on the actual sensor setup on the plant side.

References

J. B. Lamy, 2017, Owlready: Ontology-oriented programming in Python with automatic classification and high level constructs for biomedical ontologies, *Artificial Intelligence in Medicine*, 80, pp. 11–28.

M. R. Maurya, R. Rengaswamy, V. Venkatasubramanian, 2003, A Systematic Framework for the Development and Analysis of Signed Digraphs for Chemical Processes. 1. Algorithms and Analysis, *Industrial & Engineering Chemistry Research*, 42, 20, pp. 4789–4810.

F. Pelzer, *et al.,* 2021, Safety in modular process plants: demonstration of safety concepts, *e & i Elektrotechnik und Informationstechnik*, 138, 7, pp. 462–468.

C. Reinartz, *et al.,* 2019, Generation of Signed Directed Graphs Using Functional Models, *IFAC-PapersOnLine*, 52, 11, pp. 37–42.

V. Venkatasubramanian, *et al.,* 2003, A review of process fault detection and diagnosis: Part I: Quantitative model-based methods, *Computers & Chemical Engineering*, 27, 3, pp. 293–311.

V. Venkatasubramanian, R. Rengaswamy, and S. N. Kavuri, 2003, A review of process fault detection and diagnosis: Part II: Qualitative models and search strategies, *Computers & Chemical Engineering*, 27, 3, pp. 313–326.

Acknowledgments: This work was Funded by the Deutsche Forschungsgemeinschaft (DFG, German Research Foundation) – 319919706 /RTG2323.

Antonis Kokossis, Michael C. Georgiadis, Efstratios N. Pistikopoulos (Eds.)
PROCEEDINGS OF THE 33rd European Symposium on Computer Aided Process Engineering (ESCAPE33), June 18-21, 2023, Athens, Greece
 http://dx.doi.org/10.1016/B978-0-443-15274-0.50269-9

Performance analysis of three-phase fluidized bed absorber for CO_2 capture industrial application

Flavia-Maria Ilea, Ana-Maria Cormos, Simion Dragan, Calin-Cristian Cormos

Babes-Bolyai University, Faculty of Chemistry and Chemical Engineering, Chemical Engineering Department, 11 Arany Janos, RO-400028, Cluj-Napoca, Romania

Abstract

This work aims to analyze the performance achieved using a novel CO_2 capture technology, more specifically three-phase fluidized bed columns, at industrial level. For this purpose, a previously developed detailed mathematical model, validated against experimental data, was adapted to scale-up. It considers the hydrodynamics, mass transfer and chemical reaction that takes place within the process and as such is able to accurately describe the chemical absorption of CO_2 in three phase gas-solid-liquid systems. This system relies on the process intensification which is achieved due to the much higher value of the effective mass transfer area. This is obtained through the continuous renewal of the liquid film that forms on top of the solid particles that are in movement.

A higher scale column was considered with one to three stages, in order to see in which case, the fluidization in more favored. Five different scenarios were chosen that refer to the column configuration. The results show that due to the turbulence within the system, the division of the packing in two stages, with descending static bed height, leads to an 20% increase in carbon capture rate and a 25% decrease in pressure drop.

Keywords: CO_2 capture; Process intensification; Turbulent contact absorbers.

1. Introduction

1.1. Global context and literature overview

Due to the increasing demand of electricity that cannot be fulfilled by sustainable sources, fossil fuels remain the main source of energy in the foreseeable future. As a result, the quantity of carbon dioxide emissions continues to increase over the years, having peaked at an all-time high in 2022, despite the drop that took place in 2021 due to the global pandemic situation (IPCC reports, 2022). Taking into consideration the global warming phenomenon that takes place due to the accumulation of fossil CO_2 in the atmosphere, there are several ongoing European initiatives meant to work against drastic climate changes. Amongst them are some of the UN 2030 sustainability objectives and the goal for EU to be climate neutral by 2050 (UN reports, 2022). The aforementioned reasons, as well as recent events, have led to the price for pollution to increase substantially, having tripled in less than a year. The mid 2022 is represented by a price of over 90 euros per ton of carbon dioxide, which gives more reason to investing in the implementation and research regarding carbon capture technologies (Arcos–Vargas et al, 2022).

Existing capture technologies, involving the post-combustion absorption of CO_2 in amine solutions, still face some important disadvantages that refer not only to the energy penalty, but also to the costs implied by the regeneration and recycling of the solvents (Khosroabadi et al, 2021). As such, the search for new and improved technologies remains a top priority. Turbulent contact absorbers have been used mostly in the biochemical and petroleum fields but have yet to be analyzed regarding their process intensification potential (Ul-Haq, 2012). This equipment unit is based on the use of low-density solid particles that are chemically inert, their purpose being the intensification of the mass transfer between the counter-current flows of gas and liquid.

1.2. Previous work

Previous experimental work (Ilea et al, 2022) shows that the use of three phase fluidization for carbon capture leads to an important increase of 3 – 6 times in the effective mass transfer area, which leads to a proportional increase in the CO_2 transferred flow. This helps to increase the gas treating capacity of the absorber, as well as it minimizes the quantity of solvent needed for the CO_2 capture process. Based on the experimental results, a thorough mathematical model for a pilot scale CO_2 capture column has been developed and validated to describe the process of turbulent contact absorption through its complex hydrodynamics and mass transfer between the liquid and the gas phases.
As relevant innovative aspect, this work presents the adaptation of the previously developed model to industrial scale and analyzes several possible cases in order to find the best possible configuration and parametrization of the system. The solvent used is an aqueous solution of sodium hydroxide (as it was the solvent used in experimental work).

2. Scaled-up three-phase absorption column

Experimental parametrization

The parametrization of the scaled-up column considered was done in accordance with the pilot scale column used for the experimental analysis. The ratio between the column diameter and the particle diameter needs to be above the value of 6 (in the considered cases of this study, the value varies between 10 and 20). The inlet flue gas flows considered are between 1000 and 4000 m^3/h, while for the liquid spray density the values were kept as for the pilot scale column, maintain at the same the packing completely wetted. This was verified by the determination of the minimum liquid flow required to avoid flow at the periphery of the reactor. The density of the particles and the void fraction were determined experimentally using 0.075 m spherical particles. The model parameters are presented in *Table 1*.

Table 1. *Process parameters*

Parameter	Pilot scale column	Scaled-up column
Column diameter [m]	0.14	1
Total static bed height [m]	0.12	1.5
Solid particles diameter [m]	0.01	0.075
Solid particles density [kg/m^3]	170, 210, 330	10 – 100
Liquid flow [m^3/h]	0.1, 0.2	10 – 50
Gas flow [m^3/h]	60 – 120	1000 – 4000

3. Mathematical model

3.1. Mass and energy balance

In order to most accurately describe the process, apart from the aforementioned equations, the model includes mass and energy balance equations, as presented below:

$$F_j^e = F_j^0 \pm w_j \cdot \frac{N_{CO_2} \cdot M_{CO_2}}{\rho_j} \tag{1}$$

$$\frac{dc_{CO_2}^j}{dt} = \frac{F_j^0}{w_j} \cdot C_{CO_2}^0{}^j - \frac{F_j^e}{w_j} \cdot C_{CO_2}{}^j \pm N_{CO_2} - v_r \tag{2}$$

$$\frac{dc_{NaOH}^{l}}{dt} = \frac{F_l^0}{w_l} \cdot C_{NaOH}^{0}{}^{l} - \frac{F_l^e}{w_l} \cdot C_{NaOH}{}^{l} - v_r \tag{3}$$

$$\frac{dT_j}{dt} = \frac{F_j^0}{w_j} \cdot T_j^0 - \frac{F_j^e}{w_j} \cdot T_j - \frac{\Delta H_r \cdot v_r}{\rho_j \cdot cp_j} \mp \frac{h \cdot a_e \cdot (T_l - T_g)}{(\rho_j \cdot cp_j \cdot w_j)} \tag{4}$$

*Note: j subscript refers to the gas and liquid phases respectively; v_r *refers to the reaction velocity*

3.2. Hydrodynamics and mass transfer model

The performance investigation of the carbon capture process using turbulent contact absorbers was done using a previously developed model (Ilea et al, 2022) scaled up to industrial level. The model was updated in terms of the hydrodynamics of the process, including the comprehensive correlations that calculate the liquid holdup, column pressure drops and fluidized bed expansion. The mass transfer model is based on the two-film theory and Billet & Schultes (1993) correlations. In order to describe the performance of turbulent contact absorbers in terms of effective mass transfer area coefficient, an adjustment was performed. These correlations are presented below:

$$\frac{H_S}{H_0} = \left(\frac{D_c}{d_p}\right)^{0.8861} \cdot \left(\frac{\rho_S}{\rho_g}\right)^{-0.3346} \cdot \left(\frac{w_g}{w_{lmf}}\right)^{-0.1172} \cdot \left(\frac{w_l}{w_{lmf}}\right)^{0.2318} \tag{5}$$

$$\Delta P = \rho_l \cdot g \cdot H_0 \cdot \left(\frac{H_S}{H_0}\right)^{-1.7372} \cdot Re_g^{1.2535} \cdot Ga_l^{-0.4632} \cdot Re_l^{0.3897} \cdot We_l^{0.8760} \tag{6}$$

$$h_l = 0.001884 \cdot e^{0.237 \cdot w_g} \cdot (\rho_l \cdot w_l)^{0.616} \cdot H_S^{-0.357} \cdot d_p^{-0.411} \tag{7}$$

$$a_e = a \cdot e^{11.8967} \cdot \left(\frac{w_l^2}{d_e \cdot g}\right)^{-0.0152} \cdot Fr_g^{-0.2466} \cdot Ga_l^{-0.7174} \tag{8}$$

*Note: g and l subscripts refer to the gas and liquid phases respectively

4. Results and discussion

The developed mathematical model was used to analyze the performance of different system configurations. The absorber was divided in either 2 or 3 sections or left undivided. In order to ensure the possibility of the fluidization of the solid phase, decreasing the static bed height with each section was also considered, as a response to the lowering of the gas velocity. The analysis was performed using Matlab/Simulink simulation software.

4.1. Considered cases

Taking into consideration the aforementioned information, 5 different cases were considered in this study. The benchmark case (Case 1) refers to a single stage column, with a static bed height of the packing of 1.5 m. The second case (Case 2) refers to a two-stage column and it is represented in two different configurations: Case 2a – in which the packing static bed heights are the same for both stages (0.75 m) and Case 2b – in which the packing static bed heights descend from the bottom to the top stage (0.9 m and 0.6 m). The third case (Case 3) refers to a three-stage column and it is represented as well in two different configurations: Case 3a – in which the packing static bed heights are the same for all stages (0.5 m) and Case 3b – in which the packing static bed heights decrease from the bottom to the top stage (0.7 m, 0.5 m and 0.3 m).

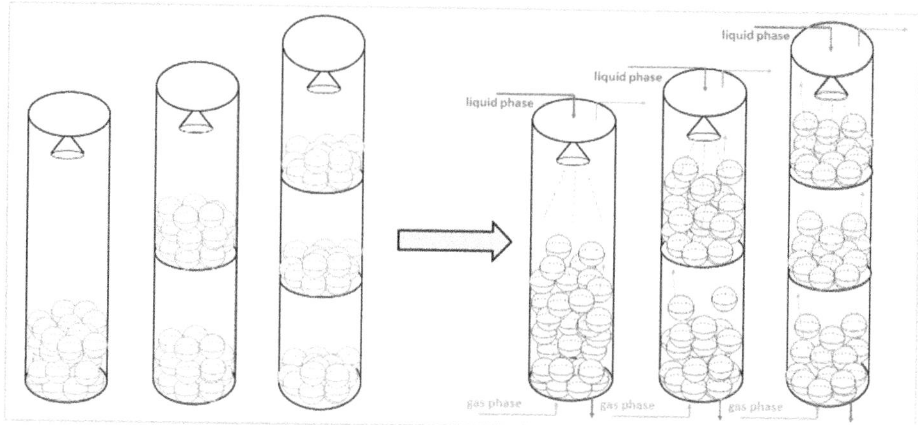

Figure 1. *CO_2 capture column configurations*

4.2. Influence of bed expansion on carbon capture rate and pressure drop

The columns were designed as to have a total static bed height of 1.5 m regardless of the number of stages. In order to find the most suitable configuration of the column, the carbon capture rate was assessed for all the considered cases, dependent on the total fluidized bed expansion rate. This is calculated as the ratio between the sum of the expansions per each stage and the total static bed height. This is presented in *Figure 2*. The pressure drops for each considered case are presented in *Figure 3*.

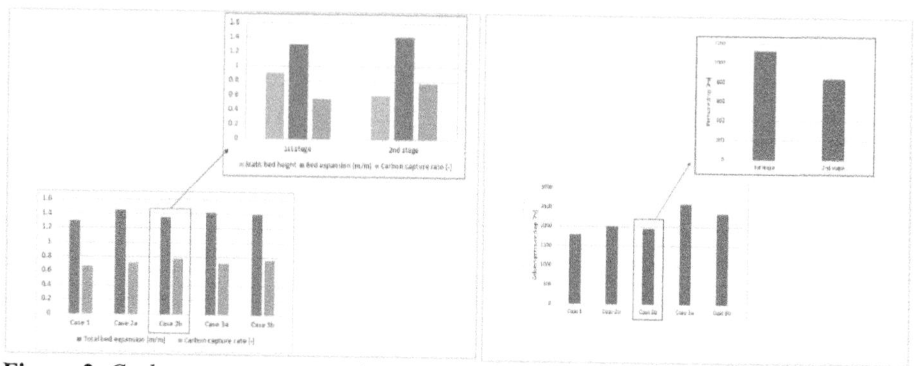

Figure 2. *Carbon capture rate and bed expansion*

Figure 3. *Pressure drops*

As shown in *Figure 2*, the carbon capture rate varies between values of 65 and 80 %. For Case 2b, the highest value of capture rate is obtained (value close to the one obtained in Case 3b). Moreover, for Case 2b, a lower value of bed expansion can be observed. This proves that in order to achieve the intensification of the mass transfer between the liquid and the gas phases, a high gas velocity is not necessary. In fact, a value slightly above the value of the minimum fluidization velocity is enough to reach the turbulence level needed to get the movement of the solid particles and, as such, the renewal of the liquid film. Due to the decreasing in static bed height with the stages and compared to the single stage column, in Case 2b a lower value of the pressure drop is also achieved, it staying at values around 1850 Pa. This is significantly lower than any other packed bed absorption columns, which is an important advantage of using turbulent contact absorbers.

4.3. Influence of inlet CO_2 concentration on carbon capture rate

The inlet carbon dioxide concentration proved to have an important impact on the carbon capture rate. An analysis was made, ranging the concentration values between 10 % and 15 %. The results are presented in *Figure 4.*

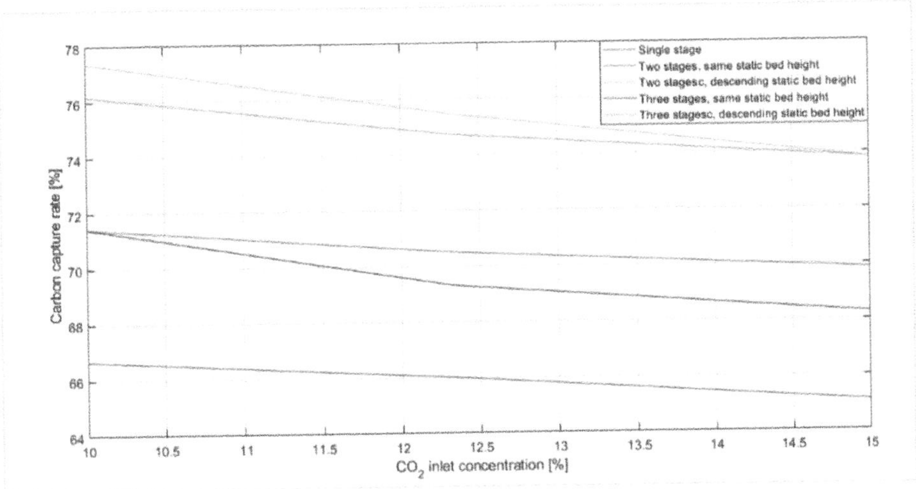

Figure 4. *Influence of inlet CO_2 concentration on carbon capture rate*

It can be observed that the best performance is obtained for Case 2b (two-stage column with descending static bed height). This is due to the fact that in this case, the expansion of the fluidized bed allows for the higher values of turbulence and as such, a higher value of the effective mass transfer area is obtained, leading to the intensification of the process.

4.4. Influence of gas flow on carbon capture rate

The value of the flue gas flow was varied between values of 1000 and 4000 m^3/h. The results in terms of carbon capture are presented in *Figure 5*, for three of the cases.

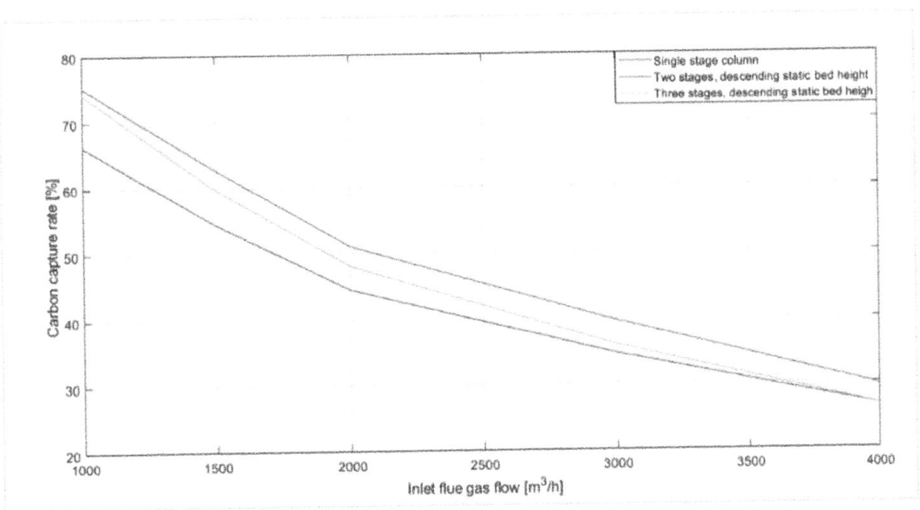

Figure 5. *Influence of flue gas flow on carbon capture rate*

The chosen cases are the one with descending static bed height per stage, as these cases shown better results in the previous analysis. The results show once again that the higher value of the carbon capture rate is obtained in Case 2b (two stage column with descending static bed height (increased by 10 %). They also show a decrease with the increase of gas flow. This is due to the increase in the void fraction of the fluidized bed, which implicitly leads to a decrease of turbulence. As such the liquid film is not as much renewed as in the case of lower gas velocities, velocities closer to the minim fluidization velocity.

5. Conclusions

The presented work analyses the performance of a turbulent contact absorber using a scaled up mathematical model. The absorber is configured as having one to three stages. Five cases are considered that refer to the column having one, two or three stages. The multiple staged ones are configured in two different ways, with the static bed height per stage being equal or descending from the bottom to the top section. The results show that the highest performance is achieved in Case 2b (two stage column with descending static bed height), obtaining a capture rate 20 % higher than in the benchmark case (Case 1 – one stage column). In this case, the pressure drop also decreases with about 25 % compared to the benchmark case. It is concluded that the positive results in this case are due to the high turbulence that is found within the column, which prompts the intensification of the mass transfer between the liquid and the gas phases.

Acknowledgements

This work was supported by a grant of the Ministry of Research, Innovation and Digitization, CNCS/CCCDI – UEFISCDI, project number PN-III-P4-ID-PCE-2020-0632, within PNCDI III.

References

A. Arcos-Vargas, F. Núñez-Hernández, J.A. Ballesteros-Gallardo, 2022, CO_2 price effects on the electricity market and greenhouse gas emissions levels: an application to the Spanish market, Clean Technologies and Environmental Policy.

R. Billet, M. Schultes, 1993, Predicting mass transfer in packed columns, Chemical Engineerig and Technology, 16, 1-9.

F.M. Ilea, A.M. Cormos, S. Dragan, C.C. Cormos, 2022, Assessment of turbulent contact absorber hydrodynamics with application in carbon capture, Chemical Engineering Journal, 449, 137674.

Intergovernmental Panel on Climate Change (IPCC), 2022, Sixth Assessment Report – Mitigations on Climate Change.

F. Khosroabadi, A. Aslani, K. Bekhrad, Z. Zolfaghari, 2021, Analysis of carbon dioxide capturing technologies and their technology developments, Cleaner Engineering and Technology, 5, 100279.

A. Ul-Haq, 2012, Mass Transfer in a Turbulent Contact Absorber, PhD Thesis, Department of Chemical Engineering Pakistan Institute of Engineering and Applied Sciences Islamabad, Pakistan.

United Nations (UN), 2022, Climate Action Reports, https://www.un.org/en/climatechange/reports.

Antonis Kokossis, Michael C. Georgiadis, Efstratios N. Pistikopoulos (Eds.)
PROCEEDINGS OF THE 33rd European Symposium on Computer Aided Process Engineering (ESCAPE33), June 18-21, 2023, Athens, Greece
 http://dx.doi.org/10.1016/B978-0-443-15274-0.50270-5

Decentralized control for optimal operation under changing active constraints

Lucas F. Bernardino,[a] Sigurd Skogestad[a]

[a] *Department of Chemical Engineering, Norwegian University of Science and Technology (NTNU), 7491 Trondheim, Norway*
skoge@ntnu.no

Abstract

Optimal economic operation of chemical plants requires control of active constraints, which may change because of disturbances. In addition, one should ensure optimality with respect to the unconstrained degrees of freedom by driving the reduced cost gradient to zero. One solution is to use on-line optimizing control, but the preferred approach in industry is to use decentralized control and selectors whenever possible. In this paper, we consider a new framework based on identifying the cost gradient projections that can be left uncontrolled when each specific constraint becomes active, leading to a decentralized logic. The proposed framework is applied to the optimal operation of the Williams-Otto reactor, which has two degrees of freedom and two constraints. The projection matrix, which depends on the gradient of the constraints, is assumed constant, resulting in a simple control structure. The approach works well in simulations and it switches between the four active constraint regions.

Keywords: optimal operation, feedback optimizing control, decentralized control, selectors.

1. Introduction

In spite of the potential economic benefits, real-time optimization (RTO) is less used in industry than one may expect. There are several reasons for this, and one is that standard RTO applications operate on a slow time scale, such that for cases with frequent disturbances, optimal operation is not satisfactorily achieved. An appealing solution is to move some of the optimization problem into the control layer design, which operates on a much faster time scale. This means that one should find controlled variables (CVs) that, when controlled to constant setpoints, result in minimal economic loss. This is the idea of self-optimizing control (Skogestad, 2000), which consists of obtaining these CVs as a combination of the available measurements. Another class of feedback optimizing strategies aim to estimate the plant cost gradients in order to control them to zero (Krishnamoorthy et al., 2019).

One of the main challenges that such approaches face is the presence of changing active constraints (Jäschke et al., 2017), which may drastically change the operation mode of the system. For example, if a constraint becomes active due to a disturbance, not taking its control into account leads to infeasible operation. Similarly, if a constraint becomes optimally inactive, the control of such constraint should be given up. For a given set of active constraints, if the cost gradient is measured, it is known that the control of the active constraints together with the control of a projection of the cost gradient over the nullspace of the active constraints' gradient leads to optimal operation (Jäschke and Skogestad, 2012). However, the existence of several active constraint regions may deem necessary the use of several independent control structures, each of them being able to provide near-optimal operation for their respective region of design. The switching between such control structures would also become an issue since the lack of a feedback-based switching strategy could lead to improper operation.

In practical applications of process systems, logic elements have been extensively used for reconfiguring control loops, which is often needed for attaining optimal operation (Reyes-Lúa et al., 2018). In particular, selectors have been successfully applied as a tool for automatic detection of active constraint switching for single input systems (Krishnamoorthy and Skogestad, 2020). In this work, we extend this analysis for multivariable systems, proposing a simple framework for decentralized optimal operation under changing active constraints, with the use of PID controllers and selectors.

2. Methodology

The main idea of this framework consists in identifying the optimal CVs for each region, and consequently proposing control loops for dealing with such CVs, in a way that reconfiguring is minimized. Consider the following convex optimization problem:

$$\begin{aligned} &\min_{u} J(u,d) \\ s.t. \quad & g_i = g_{u,i}^T u + g_{d,i}(d) \leq 0, \qquad i = 1, \dots, n_g \end{aligned} \tag{1}$$

For the purposes of this work, we consider the steady-state cost gradient J_u to be known. Based on that, and assuming that $n_u \geq n_g$ and all vectors $g_{u,i}$ are linearly independent (LI), we can devise a simple control strategy for optimal operation. Firstly, the unconstrained degrees of freedom related to the nullspace of $G_u = [g_{u,1} \quad \cdots \quad g_{u,n_g}]^T$ will always be optimally controlled regardless of how many constraints are active, and therefore $N_0^T J_u$, with N_0 being a basis for the nullspace of G_u, should always be controlled to zero. In addition to that, if all but one constraint is active, an extra unconstrained degree of freedom needs to be considered. The degree of freedom freed by the constraint g_i becoming inactive can be determined by the nullspace of the matrix $G_{u_{-i}}$ comprised of all the remaining rows $g_{u,j}^T$, $j \neq i$. By definition, this nullspace will include the space generated by N_0, and it is sufficient to pick any projection $N_{g,i}$ that is LI from N_0. A unique solution can be obtained by picking $N_{g,i}$ orthogonal to N_0.

With these definitions, we propose the following control strategy for a given active constraint set $\mathcal{A}$:

- if $n_u > n_g$, find N_0 such that $G_u N_0 = 0$ and control $N_0^T J_u = 0$;
- for $i = 1, \dots, n_g$:
  - find $N_{g,i}$ such that $\begin{bmatrix} G_{u_{-i}} \\ N_0^T \end{bmatrix} N_{g,i} = 0$;
  - control $g_i = 0$ if $i \in \mathcal{A}$; else, control $N_{g,i}^T J_u = 0$.

It can be verified that the operating point defined by the forementioned CVs leads to the solution of (1). Furthermore, if decentralized PID control is used for every CV, the decision of controlling $g_i = 0$ or $N_{g,i}^T J_u = 0$ can be performed locally by comparing the corresponding control loops. This problem may be solved with selectors, such that constraint control becomes active when necessary, and the unconstrained degree of freedom is controlled whenever the constraint is not violated. If the system is such that

decentralized PID control can be used for every set of CVs, the framework will lead to optimal operation.
The proposed framework considers linear constraints with respect to the plant inputs, which is a strong assumption that is not accurate for most real systems. We will now demonstrate the effectiveness of the framework in a case study with nonlinear constraints, evaluating the loss that is obtained by the application of the proposed method.

3. Results

3.1. Case study description

We now consider the optimal operation of the Williams-Otto reactor (Williams and Otto, 1960). The optimal operation of the reactor is described by:

$$\begin{aligned} \min_{u} J^{ec} &= p_A F_A + p_B F_B - (F_A + F_B)(p_E x_E + p_P(1 + \Delta p_P)x_P) \\ s.t. \quad & g_1 = x_A - 0.12 \leq 0 \\ & g_2 = x_E - 0.3 \leq 0 \end{aligned} \tag{2}$$

The degrees of freedom for operation are $u = [F_B, T_R]$, with T_R being the reactor temperature, and the considered disturbances are $d = [F_A, \Delta p_P]$. The solution of the above optimization problem as a function of the disturbances leads to the pattern shown in Figure 1, where each region correspond to the set of active constraints at the solution.

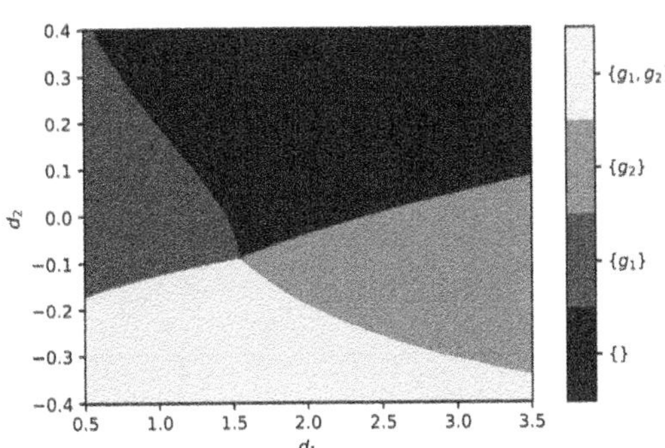

Figure 1: Optimally active constraints as function of the disturbances

3.2. Control structure design

In order to apply the proposed framework, we linearize the constraints at a nominal point. With this, we assume a constant constraint gradient with respect to the inputs, and this ultimately results in constant cost gradient projections to be controlled inside each region. The nominal point was chosen to be the optimal point for $d^* = [1.0 \quad 0.0]$, and the linearization of the constraints at this point results in:

$$\Delta g = \begin{bmatrix} g_{u,1}^T \\ g_{u,2}^T \end{bmatrix} \Delta u + g_d \Delta d = \begin{bmatrix} -0.0492 & 0.0032 \\ -0.0328 & -0.0026 \end{bmatrix} \Delta u + g_d \Delta d \tag{3}$$

Since $n_g = n_u = 2$, there are no unconstrained degrees of freedom that remain active at the fully constrained case, when $\mathcal{A} = \{g_1, g_2\}$. Therefore, no cost gradient projection $N_0^T J_u$ is needed, and all operational degrees of freedom are filled by active constraint control. For $\mathcal{A} = \{g_1\}$, in addition to controlling $g_1 = 0$, we must control one cost gradient projection $N_{g,2}^T J_u = 0$ in order to fill the remaining degree of freedom, where $N_{g,2}$ is chosen such that $G_{u_{-2}} N_{g,2} = g_{u,1}^T N_{g,2} = 0$. Similarly, in the region where $\mathcal{A} = \{g_2\}$, the optimal CVs will be $g_2 = 0$ and $N_{g,1}^T J_u = 0$, where $N_{g,1}$ is chosen such that $G_{u_{-1}} N_{g,1} = g_{u,2}^T N_{g,1} = 0$. In the unconstrained region $\mathcal{A} = \{\}$, the optimal CVs are all the components of the cost gradient $J_u = 0$. However, controlling $N_{g,1}^T J_u = 0$ and $N_{g,2}^T J_u = 0$ simultaneously implies in $J_u = 0$, as the constraints are independent and therefore $[N_{g,1} \quad N_{g,2}]$ is full rank. This means that the same gradient projections used for the partly constrained regions can be used for the fully unconstrained region.

The control structure that results from the application of the proposed methodology is presented in Figure 2, where K_{g_i} and K_{c_i} denote PID controllers related to control of constraints and gradient projections, respectively. It can be seen that control of $N_{g,1}^T J_u$ can be optimally given up when the control of g_1 becomes active, and the same happens with the pair $N_{g,2}^T J_u$ and g_2. This pairing results in optimal operation for all possible active constraint regions, with the switching being performed by max selectors on the controller outputs for this case study.

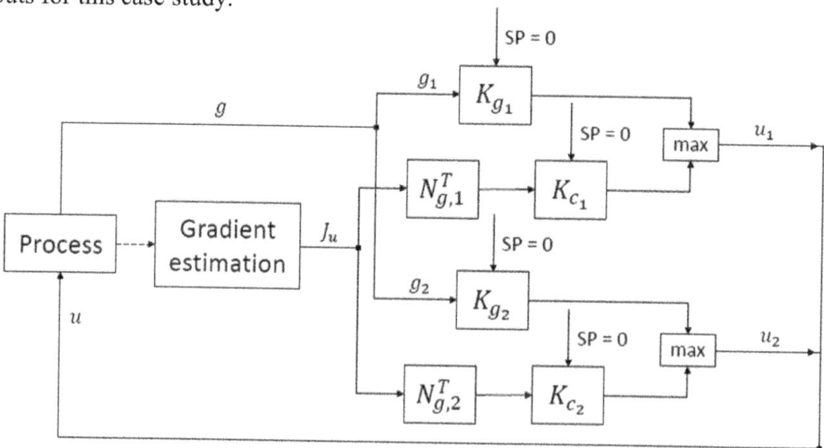

Figure 2: Control structure for optimal operation of case study

3.3. Simulation results

The behavior of the proposed control structure is illustrated by the simulations presented in Figure 3. The disturbance realizations were chosen such that the system operates at all active constraint regions. In the simulations, J_u was obtained using the automatic differentiation tools from CasADi (Andersson et al., 2019) with known disturbances. In practice, an estimator would be needed if the disturbances are not measured.

The system starts at the fully unconstrained region, where perfect optimal operation is possible because the cost gradient, J_u, is known. Once the system moves to the region $\mathcal{A} = \{g_1\}$ at $t = 4$ h, constraint violation is avoided as $g_1 = 0$ becomes the CV chosen by the selector. Even though the inputs are not driven to their optimal value, the choice of CVs is such that low economic loss is achieved. At $t = 6$ h, the disturbance value is equal to the nominal point, where the linearization of the constraints was performed. For this reason, the controlled cost gradient projection $N_{g,2}^T J_u$ corresponds exactly to the optimality conditions, and perfect optimal operation is attained. At $t = 8$ h, the system starts operating at the region $\mathcal{A} = \{g_1, g_2\}$. As all degrees of freedom are associated to active constraint control, perfect optimal operation is achieved. From $t = 12$ h, the system operates at $\mathcal{A} = \{g_2\}$, and near-optimal operation is achieved with low economic loss.

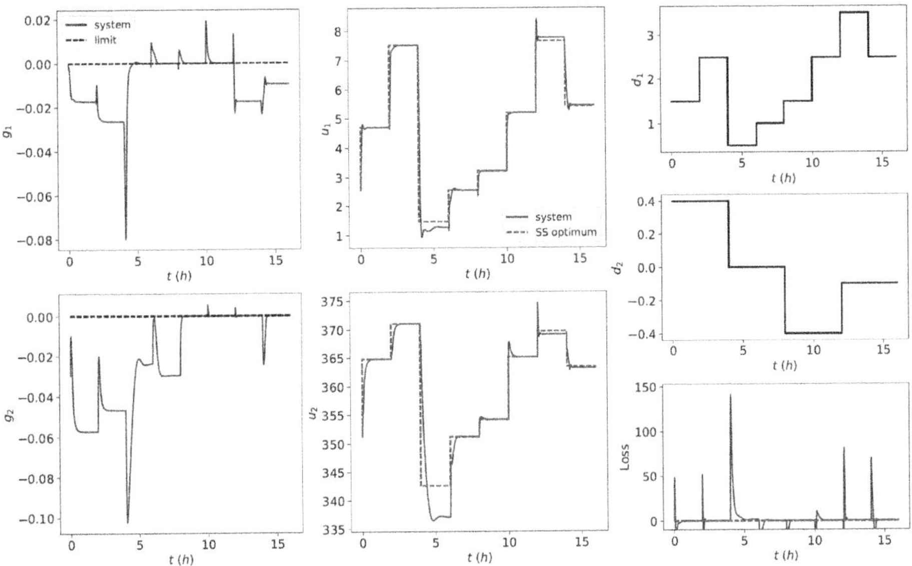

Figure 3: Dynamic simulation of proposed control structure

The steady-state economic loss for the proposed control structure is presented in Figure 4 as a function of the disturbances. Similar to what was observed in the dynamic simulations (Figure 3), the regions with nonzero loss are concentrated at the partly constrained regions. This is because the optimal CVs corresponding to the cost gradient projections change due to the nonlinearity of constraints. In contrast, the fully constrained and fully unconstrained regions mostly present zero operational loss, because the optimal CVs remain constant inside these regions, and the information about J_u is accurate. The linearization of the constraints was performed inside the region $\mathcal{A} = \{g_1\}$, where the operational loss is effectively zero at this reference point, and it increases as the system moves away from it. It can also be seen that optimal switching between regions is subject to errors related to the linearization of the constraints. The largest operational loss is obtained in the switching between regions $\mathcal{A} = \{g_1, g_2\}$ and $\mathcal{A} = \{g_2\}$, where an inaccurate cost gradient projection is tracked, that is, controlling $N_{g,1}^T J_u = 0$ does not lead to exact optimal operation. Therefore, the switching is not performed at the exact optimal boundary, as can be seen in the results, but it nonetheless leads to a reasonable switching policy between regions, and good resulting economic performance.

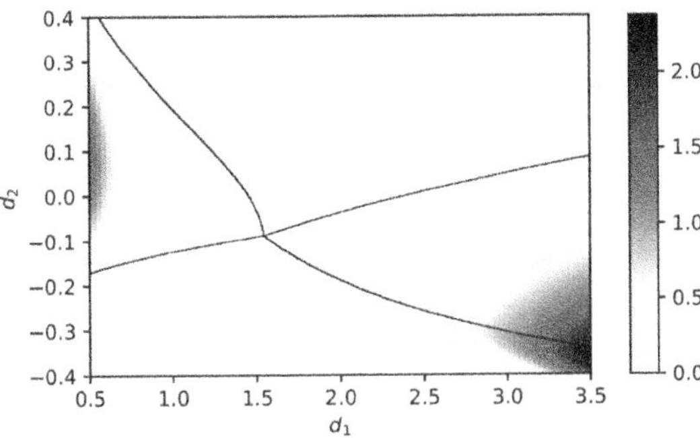

Figure 4: Steady-state economic loss of the proposed control structure as a function of disturbances (black lines represent optimal region switching)

4. Discussion

The control structure resulting from the framework proposed in this work (Figure 2) makes use of selectors as simple elements for switching between operating regions. These elements are frequently used in practice for coordinating conflicting control objectives. In this work, we show how these elements can be used for the optimal operation of systems under changing active constraints in a systematic manner. The min/max nature of the selectors is ultimately related to the nature of the constraint with respect to the input (Krishnamoorthy and Skogestad, 2020).
The approach proposed in this work is based on the analysis of the partly constrained regions, where we find gradient projections that can be optimally controlled. These projected gradients are also the optimal CVs of other regions, which minimizes the number of control loops that are necessary to account for all regions. Moreover, the proposed switching operates independently for each plant input, which means that the detection of each constraint is done independently, and feasible operation is safely achieved. However, the reconfiguring of CVs done by the selectors may significantly change the interactions between loops, and therefore careful tuning of the controllers is necessary, such that a good performance is achieved regardless of which loops are active.

5. Conclusion

In this work, we propose a framework for decentralized optimal operation under changing active constraints, applicable to a class of multivariable problems. Even though the approach is based on the linearization of the constraints, and therefore the quality of the linearization plays a relevant role in the economic performance, the strategy proved to be successful in a nonlinear case study, which encourages its use in other relevant problems of process systems engineering, especially when the gains from the inputs to the constraints do not change greatly in the operating range. The use of adaptive cost gradient projections would also be beneficial for improving economic performance. More theoretical aspects regarding the proposed framework will be expanded in future work.

References

J.A. Andersson, J. Gillis, G. Horn, J.B. Rawlings, M. Diehl, 2019. CasADi: a software framework for nonlinear optimization and optimal control. Mathematical Programming Computation, 11, pp.1-36.

J. Jäschke, Y. Cao, V. Kariwala, 2017. Self-optimizing control–A survey. Annual Reviews in Control, 43, pp.199-223.

J. Jäschke, S. Skogestad, 2012. Optimal controlled variables for polynomial systems. Journal of Process Control, 22, 1, pp.167-179.

D. Krishnamoorthy, E. Jahanshahi, S. Skogestad, 2018. Feedback real-time optimization strategy using a novel steady-state gradient estimate and transient measurements. Industrial & Engineering Chemistry Research, 58, 1, pp.207-216.

D. Krishnamoorthy, S. Skogestad, 2020. Systematic design of active constraint switching using selectors. Computers & Chemical Engineering, 143, p.107106.

A. Reyes-Lúa, C. Zotică, S. Skogestad, 2018. Optimal operation with changing active constraint regions using classical advanced control. IFAC-PapersOnLine, 51, 18, pp.440-445.

S. Skogestad, 2000. Plantwide control: The search for the self-optimizing control structure. Journal of process control, 10, 5, pp.487-507.

T.J. Williams, R.E. Otto, 1960. A generalized chemical processing model for the investigation of computer control. Transactions of the American Institute of Electrical Engineers, Part I: Communication and Electronics, 79, 5, pp.458-473.

Antonis Kokossis, Michael C. Georgiadis, Efstratios N. Pistikopoulos (Eds.)
PROCEEDINGS OF THE 33rd European Symposium on Computer Aided Process Engineering (ESCAPE33), June 18-21, 2023, Athens, Greece
 http://dx.doi.org/10.1016/B978-0-443-15274-0.50271-7

Development of a Deep Learning-based Schedule-Aware Controller: Toward the Integration of Scheduling and Control

M. Abou El Qassime [a,b], A. Shokry[*a], A. Espuña[c], E. Moulines[a,b]

[a]*École polytechnique, Paris, France*
[b]*Emines School of Industrial Management, BenGuerir, Morocco*
[c]*Universitat Politècnica de Catalunya, Barcelona, Spain.*
**ahmed.shokry@polytechnique.edu*

Abstract

Most of the existing approaches for integrating the scheduling and control layers in the process industry suffer from the solution complexity of the resulting Mixed Integer Nonlinear Programming (MINLP) program. This complexity stems from two main obstacles: i) the consideration of nonlinear closed-loop dynamics involved in the control layer and ii) the way of exchanging the information between the control and the scheduling layers. To tackle these two obstacles, this work proposes the use of a schedule-aware controller (SAC) based on deep learning (DL) models, which receives, as input, the scheduling information (i.e., production sequence, targets, and set points) along with the process conditions (values of the state, output, and control variables) to predict the optimal control actions to be applied at the next sampling periods. The method is applied to a benchmark, showing very good performance in terms of the ability to adapt the control actions in accordance with the scheduling decisions, high prediction accuracy, and reduction in the computation cost.

Keywords: Explicit Control- Scheduling- Integration - Machine Learning

1. Introduction

In process industry, the integration of decision making across different layers of the process management hierarchy is of paramount importance. In particular, integrating the scheduling and control layers can lead to significant increase in production profitability, operational reliability, and adaptability to varying environments. Most of the works that have been developed to integrate control and scheduling can be categorized into simultaneous and decoupling approaches. Flores and Grossmann (2006, 2010) proposed a simultaneous approach incorporating the nonlinear process dynamics (involved in the control layer) into the scheduling problem in the form of differential/algebraic constraints, which results in a Mixed Integer Dynamic Optimization (MIDO) problem that handles the control and scheduling simultaneously. They used a discretization method based on orthogonal collocation to transform the MIDO is into a MINLP problem. Zhuge and Ierapetritou (2012) proposed a similar method able to handle online disturbances in the system dynamics. They solve the MIDO problem offline to obtain reference production sequence and control variables, which are initially implemented online. Then, once disturbances occur, the MIDO problem is updated and solved over the remaining time horizon. Conversely, the decoupling approach considers two problems: i) the scheduling one (master), which is formulated as a MINLP and solved based on an economic objective to obtain the mid-term decisions, and ii) the control problem (primal), which is first informed by the scheduling decisions and then solved as dynamic optimization to obtain the short-term decisions. The master-primal problems are solved iteratively until convergence is achieved. Burnak et al. (2018) proposed a decoupling approach based on multi-parametric (MP) programming that consists of i) a MP scheduler, ii) a MP controller based on Model Predictive Control (MPC), and iii) a multiparametric "bridge" that connects these two layers. Martinez and Sandoval (2022) developed a nested approach that solves the scheduling problem to obtain the product switching sequence and the associated set points, which are passed to the MPC to implement them. The method is applied in a closed-loop online mode to minimize perturbations. Although both approaches demonstrated the importance of such integration for economic gain, they still have some drawbacks. In the simultaneous approach, the resulting MIDO problem is usually of large scale and can be intractable, especially in cases where the process model is of high-fidelity and/or highly nonlinear. The decoupling approach leads to more manageable optimization problems, but a significant computational effort is required to iteratively solve the control problem (Burnak et al., 2018; Martinez and Sandoval, 2022). The drawbacks of both approaches result from two main issues: i) handling the nonlinear closed-loop dynamics, and ii) the need to exchange the information between the control and the scheduling layers. To tackle the two obstacles, we propose a novel DL -based SAC that receives, as input, the med-term scheduling information (production targets, sequence, and set points) along with the instantaneous process information (state, output, and control variables) and provides the optimal control actions to be applied to the process in the future sampling period. This novel DL-based SAC represents: a) an accurate and computationally cost-effective controller (tackling the first obstacle) and, b) a controller that accommodates the scheduling decisions, adapts the control actions to pursuit these scheduling targets, and, finally, quantifies the deviation from these targets (e.g., material loss). These deviations can be sent back to the scheduler (tackling the second obstacle). Hence, this DL-based SAC could be integrated seamlessly with a higher-level scheduler.

2. Problem statement

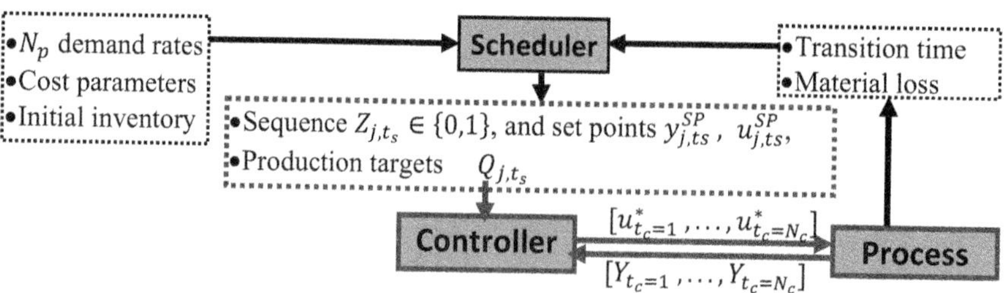

Figure 1. Representation of the framework for integrating scheduling and control.

We consider a framework for integrating the control and scheduling (Figure 1), in which both layers share information. Given N_p products to be manufactured, a demand scenario for each product over of a total time period T_{tot}, and a cost function, the scheduling problem is solved to obtain the best the production sequence $Z_{j,t_s} \in \{0,1\}$ with the associated set points $y^{SP}_{j,ts} \in R^{n_y}$, $u^{SP}_{j,ts} \in R^{n_u}$ and target production rates Q_{j,t_s}, where $j =$

$1,..,N_p$ and $t_s = 1,...,N_s = t_{tot}/\Delta t_s$, and Δt_s is the scheduling time slot. Please note that the solution of the scheduling problem itself is out of the scope of this work, so only the elements highlighted in red in Figure 1are considered.

The scheduling decisions (bold font in the formulation below) are fed to the control layer based on MPC (Eqs. (1):(4)). We denote $t_c = 1,...,N_c$ the control sampling times with $N_c = t_{tot}/\Delta t_c$ and $\Delta t_c \ll \Delta t_s$. At the control level, the scheduling information take constant values within the same scheduling time slot. A nonlinear high-fidelity model, $\mathcal{F}$ and $\mathcal{G}$ (Eq. (2)), is used to calculate the process states $x_{t_c} \in R^{n_x}$ and outputs $y_{t_c} \in R^{n_y}$, and a case-dependent constraint (Eq. (3)) is considered to correlate the control actions $u_{t_c} \in R^{n_u}$ with the production rates advised by the scheduler, Q_{j,t_c}. Denoting P, R and R_1 positive definite matrices, and N_{ph}, M_{ch} the prediction and control horizons, respectively, the control problem can be described as follows:

$$\min_{u_{t_c+1},\dots,u_{t_c+N_{ph}}} J = \sum_{k=1}^{N_{ph}} \left(y_{t_c+k} - \boldsymbol{y}^{SP}_{t_c+k}\right)^T P \left(y_{t_c} - \boldsymbol{y}^{SP}_{t_c+k}\right) + \sum_{k=1}^{M_{ch}} \left(u_{t_c+k} - \boldsymbol{u}^{SP}_{t_c+k}\right)^T R \left(u_{t_c+k} - \boldsymbol{u}^{SP}_{t_c+k}\right) + \sum_{k=1}^{M_{ch}} \Delta u^T_{t_c+k} R_1 \Delta u_{t_c+k} \tag{1}$$

$s.t.$

$$x_{t_c+k+1} = \mathcal{F}\left(x_{t_c+k}, u_{t_c+k}, d^T_{t_c+k}\right),\ y_{t_c+k} = \mathcal{G}\left(x_{t_c+k}, u_{t_c+k}, d^T_{t_c+k}\right) \tag{2}$$

$$u_{t_c+k} \leq \mathcal{H}\left(\boldsymbol{Q}_{j,t_c+k}, x_{t_c+k}, u_{t_c+k}, d^T_{t_c+k}\right) \tag{3}$$

$$x_{min} \leq x_{t_c+k} \leq x_{max},\ y_{min} \leq y_{t_c+k} \leq y_{max},\ u_{min} \leq u_{t_c+k} \leq u_{max} \tag{4}$$

$$\Delta u_{t_c+k} = u_{t_c+k} - u_{t_c+k-1}$$

At each time t_c, the measured values of the state and output variables (x_{t_c} and y_{t_c}) are used as initial conditions for the dynamic optimization problem (Eqs. (1):(4)), which is solved to find the optimal trajectory of the control variables $[u^*_{t_c+1}, ..., u^*_{t_c+N_{ph}}]$. Only the optimal control values corresponding to the first instance, $u^*_{t_c+1}$, are implemented in the process. The final solution of the MPC problem are the closed-loop control profiles along the entire production horizon $[u^*_{t_c=1}, ..., u^*_{t_c=N_c}]$ and the resulting state and output variables $[x_{t_c=0}, ..., x_{t_c=N_c-1}]$ and $[y_{t_c=0}, ..., y_{t_c=N_c-1}]$. Since the solution complexity of this schedule-aware MPC problem could hinder the integration of the scheduling and control layers, we propose a DL-based SAC to cheaply and accurately predict the closed-loop control actions as functions of the scheduling decisions and the process conditions, as in Eq. (5), where l is a time delay and $\mathfrak{f}$ is deep learning model.

$$\hat{u}^*_{t_c+1} = \mathfrak{f}\left(Q_{j,t_c}, .., Q_{j,t_c-l}, y^{SP}_{j,t_c}, ..., y^{SP}_{j,t_c-l}, u^{SP}_{j,t_c}, ..., u^{SP}_{j,t_c-l}, x_{t_c}, ..., x_{t_c-l}, y_{t_c}, ..., y_{t_c-l}, u_{t_c}, ..., u_{t_c-l}\right) \tag{5}$$

3. Methodology steps and techniques

3.1. Sampling of the scheduler outcomes

In the first step, $N_{sc.tr}$ different scheduling scenarios are randomly synthesized over the scheduling slots $t_s = 1, ..., N_s$. For each scenario $i, i = 1, ..., N_{sc.tr}$ a different production sequence $Z_{j,t_s,i} \in \{0,1\}$ (with the associated set points $y^{SP}_{j,ts,i}$, $u^{SP}_{j,ts,i}$) and/or production rates $Q_{j,t_s,i}$ (Figure 2) is defined. The generated scenarios should contain most of the expected transitions between the N_p products and cover the full range of variability in production targets.

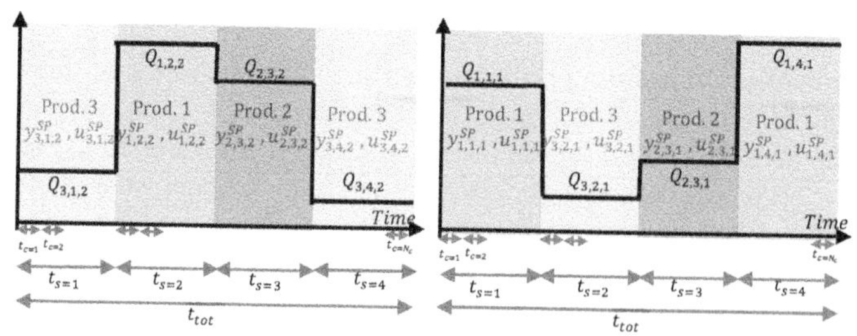

Figure 2. Example of two different scheduling scenarios where $N_p = 3$ and $N_s = 4$.

3.2. Data generation: solving the MPC problem

The scheduling scenarios $[Q_{j,t_s,i}]$, $[y^{SP}_{j,t_s,i},\ u^{SP}_{j,t_s,i}]$ are resampled to the control time scale to obtain the required values at the finer time grid $t_c = 1, \dots, N_c$, i.e., $[Q_{j,t_c=1,i}, \dots, Q_{j,t_c=N_c,i}]$, $[y^{SP}_{j,t_c=1,i}, \dots, y^{SP}_{j,t_c=N_c,i}]$, $[u^{SP}_{j,t_c=1,i}, \dots, u^{SP}_{j,t_c=N_c,i}]$. Then, for each of the $N_{sc.tr}$ scenario, the control problem (Section 2 and Eqs. (1):(4)) is solved to obtain the closed-loop control profiles along the entire production horizon $[u^*_{t_c=1,i}, \dots, u^*_{t_c=N_c,i}]$ and the resulting state variables $[x_{t_c=0,i}, \dots, x_{t_c=N_c-1,i}]$ and output variables $[y_{t_c=0,i}, \dots, y_{t_c=N_c-1,i}]$. The obtained trajectories of the state, output and control variables are unfolded together with the production targets and set points to an input-output data set for training the DL model, depending on the required input lag l. For example, in case of $l = 0$, the input-output dataset will be in the from $[x_{t_c,j}, y_{t_c,j}, u^*_{t_c,j}, y^{SP}_{j,t_c,j}, u^{SP}_{j,t_c,j}, Q_{j,t_c,j}] - [\,u^*_{t_c+1,j}\,]$, $j = 1, \dots, N_{sc.tr} \times N_C$. Following the same steps, a testing dataset with $N_{sc.tst}$ additional scheduling scenarios is also generated.

3.3. Development of DL-based SAC

The unfolded training data is used to fit the DL-SAC in Eq. (5), where f is selected as a feedforward deep neural network (DNN), whose architecture (number of hidden layers, hidden neurons, activation functions), learning parameters, and training algorithm are chosen by a cut and try procedure that balance the DNN complexity and accuracy.

3.4. Testing of the DL-SAC via closed-loop deployment

In the final step, the developed DL-based SAC is used to predict the closed-loop control trajectories corresponding to the scheduling scenarios in the testing set. In each scenario, the DL-based SAC is used in a recursive way as follows:

i. if $t_c = 1$ consider the given initial values of the state, output, and control variables, else-if $t_c > 1$ measure the current values of the state, output, and control variables from the process (high-fidelity model, in our case). Combine them with the target production rates and set points values (scheduling decisions), and use them as input for the SAC to predict the optimal values of the control variables at the next sampling period,
ii. implement the predicted control variables values in the process (i.e., in the model),
iii. if $t_c = N_c$ stop, otherwise back to step i.

The accuracy of DL -based SAC is quantified by comparing the predicted control trajectories and the resulting state and output variables with their exact values. Normalized Root Mean Square Error (NRMSE) is used as accuracy metric.

4. Application: Multi-product CSTR

As a benchmark, we consider an isothermal CSTR (Figure 3-(a)) which is fed by three reactants ($R_i, i = 1,2,3$) and is operated to manufacture three products ($P_j, j = A, B, C$) on a single production line (Flores and Grossmann, 2006; Burnak et al., 2018). In the CSTR, three irreversible reactions take place in parallel, see Figure 3-(b).

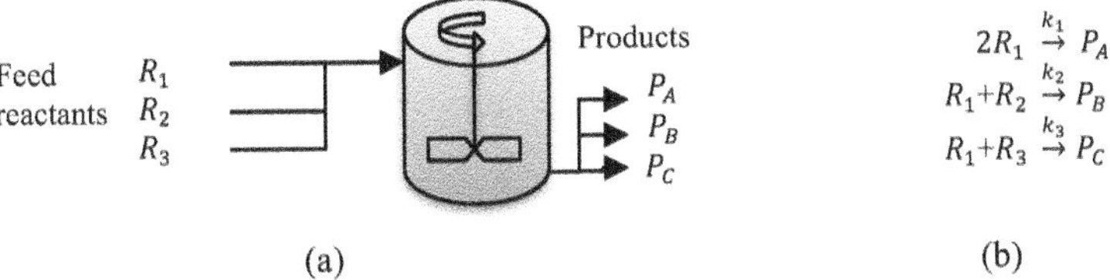

Figure 3: Representation of the multi-product CSTR (a) and its irreversible reactions (b).

The dynamics of the CSTR can be described by six differential equations, as follows:

$$\frac{dC_{R_i}}{dt} = \frac{q_{R_i} C^f_{R_i} - q_{\text{tot}}\, C_{R_i}}{V} + \mathcal{R}_{R_i} \quad and \quad \frac{dC_{P_j}}{dt} = \frac{q_{\text{tot}}\,(C^f_{P_j} - C_{P_j})}{V} + \mathcal{R}_{P_j} \tag{6}$$

Where C_{R_i} (state variables) are the concentrations of the reactants inside the CSTR (mol/L), C_{P_j} (output variables) are the products concentrations (mol/L), and q_{R_i} (control variables) are the reactants feed flowrates (L/min). Notice that $q_{tot} = \sum_{i=1}^{3} q_{R_i}$ is the total feed flowrate, V is the reactor volume (L), while $C^f_{P_j}$ and $C^f_{R_i}$ represent the concentrations in the feed-streams. To manufacture each of the three products, the scheduler determines different set points that the controller must track by manipulating q_{R_i}.The set pints are C^{SP}_{A,t_c}=[0.3,0.0,0.0], C^{SP}_{B,t_c} =[0.0,0.18,0.0], and C^{SP}_{C,t_c} =[0.0,0.0,0.2].

We assume a scheduling scenario which consists of $N_s = 4$ time slots with $\Delta t_s = 120$ min (T_{tot}=480 min). At the control level, $\Delta t_c = 2$ min therefore $t_c = 1, \dots, N_c = 240$ instances. The MPC problem is casted as in Eqs. (1):(4) considering the following parameters: $P = 10^2 \times [100,0,0; 0\ 1,0; 0,0,1]$, R=0 since no set points on the control variables, $R_1 = 10^{-6} \times [1,0,0; 0\ 1,0; 0,0,1]$, $N_{ph} = 6$, and $M_{ch} = 2$. The problem is subjected to: i) a bound constraint on the inlet flowrates $0 \le q_{R_i,t_c} \le 500$ L/min, and ii) an equality constraint $\sum_{i=1}^{3} q_{R_i,t_c} = Q_{tot,t_c}$, which correlates the control variables with the production targets suggested by the scheduler, i.e. equivalent to Eq. (3). A dataset including $N_{sc.tr} = 45$ scenarios is generated (see Sections 3.1 and 3.2) and used for training the DNN-based SAC (Eq. (6)). The DNN consists of three hidden layers with 30, 50, and 5 neurons, and is trained using Bayesian regularization backpropagation algorithm. Another dataset of $N_{sc.tst} = 10$ scenarios is created to test the fitted DNN-based SAC. The time for generating the training data, testing data, and fitting the DNN equals to 269, 199.3, and 15.3 min, respectively.

$$\begin{aligned} &\hat{q}_{1,t_c+1}, \hat{q}_{2,t_c+1}, \hat{q}_{3,t_c+1} = \\ &= \mathcal{f}(C_{R1,t_c}, C_{R2,t_c}, C_{R3,t_c}, C_{A,t_c}, C_{B,t_c}, C_{C,t_c}, q_{1,t_c}, q_{2,t_c}, q_{3,t_c}, Q_{tot,t_c}, C^{SP}_{A,t_c}, C^{SP}_{B,t_c}, C^{SP}_{C,t_c}) \end{aligned} \tag{7}$$

Table 1 reports the performance of the DNN-based SAC over the 10 testing scenarios, in terms of i) low NRMSE of the predicted closed-loop control variables and the resulting state and output variables, and ii) a huge savings in the online computational cost. As example, Figure 4 shows the closed-loop profiles of one inlet flowrate ($\hat{q}_2$), predicted by the DL-based SAC and two of the resulting outputs (C_C and C_B) compared with their exact values obtained by MPC during specific two time slots representing the production transition from product C to product B in scenario 4. The figure further confirms the very good capabilities of the proposed DL-based SAC.

Table 1. Accuracy and online computational time of the DL-based SAC based on the testing set.

Accuracy (NRMSE %)									CPU time *	Saved time %
$\hat{q}_1$	$\hat{q}_2$	$\hat{q}_3$	C_{R1}	C_{R2}	C_{R3}	C_A	C_B	C_C		
0.15	0.23	0.3	0.73	1.8	1.1	0.85	2.26	1.01	0.52	99.7=(199.3-0.52)/1.99.3

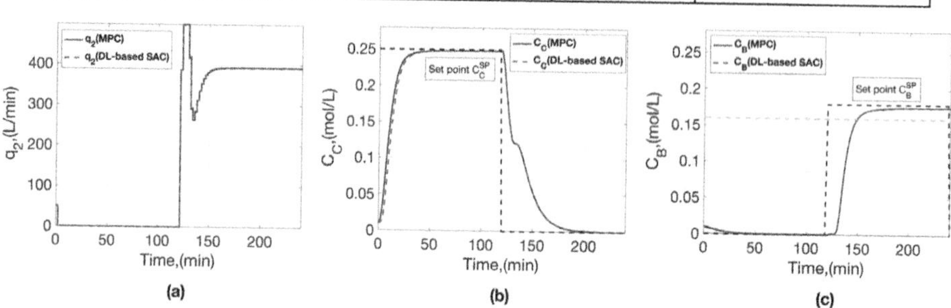

Figure 4. Predicted (red) versus exact (blue) values of the control variable q_2 (a) and the resulting output variables C_C (b) and C_B (c) during transition C$\Rightarrow B$ in the testing scenario no. 4. The green dashed line in subplot (c) represents the minimum quality threshold for product B.

The material loss and transition time during a certain transition can be calculated by comparing the output concentration with a minimum quality threshold. For example, in transition C$\Rightarrow B$, the output concentration C_B is compared to the quality threshold $C_{B,min}=$ 0.16 mol/L (dashed green line in Figure 4-(c)).

5. Conclusions

In this paper, we present a DL-based SAC that is effectively able to receive, as input, the mid-term scheduling decisions along with the process conditions and predict the future control measures that track the scheduling targets. The results of application to a benchmark case study proves the effectiveness of the proposed DL-based SAC in terms of accuracy and very low computation time. A future research line will exploit the DL-based SAC and test its performance when integrated into scheduling problem/layer.

References

A. Flores-Tlacuahuac, I. Grossmann. Simultaneous cyclic scheduling and control of a multiproduct CSTR. Ind. Eng. Chem. Res. 2006, 45, 6698-6712.

A. Flores-Tlacuahuac, I. Grossmann. Simultaneous scheduling and control of multiproduct continuous parallel lines. Ind. Eng. Chem. Res. 2010, 49, 7909-7921.

B. Burnak, J. Katz, N. A. Diangelakis, E. Pistikopoulos, 2018. Simultaneous Process Scheduling and Control: A MP-Based Approach. Ind. Eng. Chem. Res, 57, 3963−3976

J. Zhuge and M. Ierapetritou. Integration of Scheduling and Control with Closed Loop Implementation nd. Eng. Chem. Res. 2012, 51, 8550-8565

O. A.-Martinez, L. R.-Sandoval. A nested online scheduling and nonlinear model predictive control framework for multi-product continuous systems. AICHE journal 68, 2022, 17665

Antonis Kokossis, Michael C. Georgiadis, Efstratios N. Pistikopoulos (Eds.)
PROCEEDINGS OF THE 33rd European Symposium on Computer Aided Process Engineering (ESCAPE33), June 18-21, 2023, Athens, Greece
 http://dx.doi.org/10.1016/B978-0-443-15274-0.50272-9

Robust (explicit) optimization and control via Mixed Integer Programming

Nikolaos A. Diangelakis[a*], Iosif Pappas[b,c], Efstratios N. Pistikopoulos[b,c]

[a]*School of Chemical and Environmental Engineering, Technical University of Crete, Chania, 73100, Greece*

[b]*Artie McFerrin Department of Chemical Engineering, Texas A&M University, College Station, TX, 77845, USA*

[c]*rTexas A&M Energy Institute, Texas A&M University, College Station, TX, 77845, USA*

Abstract

The presence of endogenous and exogenous uncertainty in process systems is one of the key reasons for deviation from set operation policies, resulting in suboptimal or even infeasible operation. In robust rolling horizon optimization (RHO) of discrete-time systems, such as MPC, the derivation of robust optimization policies is mainly characterized by the challenge associated with multiplicative unmeasurable uncertainty. Measurable uncertainty can be treated either "online" with the successive solution of optimization problems or "offline" via multi-parametric programming. We consider the development of closed-loop (explicit) robust rolling horizon solutions that correspond to MPC problems. The unmeasurable multiplicative uncertainty is reformulated into its deterministic mixed integer equivalent form. Furthermore, measurable uncertainty is treated either via online optimization or via multi-parametric programming. With this approach, we show (i) how the need for a multi-stage optimization approach can be alleviated while (ii) guaranteeing the robust nature of the final solution. We present the developments through an example of constrained LQR.

Keywords: robust optimization; robust control; mixed-integer programming; multi-parametric programming

1. Introduction

Model predictive control (MPC) is the established method for the advanced control of multivariable systems, extensively studied and applied by both the academic and industrial communities (D. Q. Mayne, 2014). An MPC problem is most commonly a convex quadratic mathematical optimization problem, whose solution is the optimal vector of inputs to regulate the operation of a system. By utilizing a rolling horizon optimization approach (RHO), the optimal vector of inputs that minimize a performance index while considering (i) the process model, (ii) the constraint set and (iii) a finite prediction horizon is calculated, however only the first control input of the sequence is applied to the system. As soon as the next measurement (or its estimate) is made available, the horizon is rolled over by a single time step, and the optimization problem is resolved. Being model-based, MPC approaches inevitably depend on the accuracy of the underlying model. In practical applications, model uncertainty is prevalent due to the mismatch between the real plant and the model used for the derivation of the control law (endogenous uncertainty) and/or unmeasured disturbances (exogenous uncertainty)

especially when multiple time scales are considered (N.A. Diangelakis et al. (2017), B. Burnak et al. (2020)). Consequently, the closed-loop behavior of the system is not as desired and constraint violations may occur. In an effort to address this challenge strategies that account for additive and multiplicative uncertainty have been developed (e.g. B. Kouvaritakis, et al, (2016) and references therein). Furthermore, robust optimization has received significant attention through the years (e.g. A. Ben-Tal, et. al, 1998, Q. Zhang et al. 2022 and references therein).
Robust explicit MPC aims to derive the explicit solution of the MPC problem by additionally guaranteeing that all sources of uncertainty are taken into account, and at least feasibility is satisfied. In this respect, multiple research efforts have been contributed that tackle the case where the uncertainty source is added to the future prediction (additive uncertainty) and the case where the future prediction is multiplicatively affected by it (multiplicative uncertainty). V. Sakizlis et al. (2004) included constraints in the design phase of the controller that guarantee that for the worst case of the additive uncertainty, the system is feasible. A. Bemporad et al. (2003) proposed a min-max approach where the solution is found for problems with a linear objective function and linear constraints for multiplicative uncertainty. K. Kouramas et al. (2013) tackled the case of explicit MPC problems with a quadratic objective function, linear constraints, and multiplicative uncertainty by employing dynamic programming and robust optimization. More recently, I. Pappas et al. (2022) demonstrated that dynamic programming can be avoided and extended it to hybrid systems by performing projections of the feasible space into the future. An open question in the robust explicit MPC field is how can the robust solution of an explicit MPC problem with a quadratic cost, linear constraints, and multiplicative uncertainty be developed, by using a single multiparametric optimization formulation. This contribution aims to provide a strategy for the later via mixed-integer and multi-parametric quadratic optimization.

2. Problem formulation

Consider a linear discrete time dynamic model of the following form:

$$x_{T+1} = Ax_T + Bu_T \tag{1}$$

where $x_T \in \mathcal{R}^m$ and $u_T \in \mathcal{R}^n$ are the state and control input vectors respectively at time instant T. Instead of considering that the system and input matrices $A \in \mathcal{R}^{m\times m}$ and $B \in \mathcal{R}^{m\times n}$ are constant, in this study we assume that the model is uncertain and described by box uncertainty. Specifically:

$$A = A_0 + \Delta A \qquad (2)$$

$$B = B_0 + \Delta B \qquad (3)$$

$$\Delta A \in \mathcal{A} = \{\Delta A \in \mathcal{R}^{m\times m}: -\epsilon_A|A_0| \leq \Delta A \leq \epsilon_A|A_0|\} \qquad (3)$$

$$\Delta B \in \mathcal{B} = \{\Delta B \in \mathcal{R}^{m\times n}: -\epsilon_B|B_0| \leq \Delta B \leq \epsilon_B|B_0|\} \qquad (4)$$

A_0 and B_0 are the nominal matrices of the model, while ΔA and ΔB are their uncertain component. This element-wise deviation from the nominal matrix value is described by ϵ_A and ϵ_B which are scalars. Here, in a slight abuse of terminology, $|A_0|$ and $|B_0|$ do not represent a norm or determinant of A_0 and B_0, respectively, but their element-wise absolute values, i.e. if the element of A_0 in the ith row and j^{th} column is $a_0^{i,j}$ then the element of $|A_0|$ in the same row and column is $|a_0^{i,j}|$.

Assuming a common prediction and control horizon N, the following linear quadratic regulator problem (LQR) problem can be formulated:

$$\begin{aligned}\min_{u_0,u_1,\dots,u_{N-1}} \quad & x_N^T P x_N + \sum_{k=0}^{N} x_k^T Q x_k + \sum_{k=0}^{N-1} u_k^T R u_k \\ \text{s.t.} \quad & x_{T+1} = A x_T + B u_T, \forall T \in \{0,\dots,N-1\} \\ & \underline{x} \le x_T \le \overline{x}, \forall T \in \{0,\dots,N-1\} \\ & \underline{u} \le u_T \le \overline{u}, \forall T \in \{0,\dots,N-1\} \\ & x_N \in T_s\end{aligned} \tag{5}$$

Its objective is to find the control inputs u_T, that will drive the system to the origin in the presence of the uncertainty. The lower bounds on the inputs and states are implied to be negative and the upper bounds positive. The system and input matrices in eqn. 5 are defined as in eqns. 2-4. The weights on the states and inputs are $Q \in \mathcal{R}^{m\times m}$ and $R \in \mathcal{R}^{n\times n}$ respectively, while $P \in \mathcal{R}^{m\times m}$ is the terminal cost matrix derived from the solution of the discrete-time algebraic Riccati equation for the nominal system and input matrices. The states at the end of the prediction are required to belong to the terminal set T_s (Blanchini, 1999). In the methodology section, the consideration of multiplicative uncertainty within the formulation of eqn. 5 will be shown to provide an optimization based robust counterpart for the LQR.

3. Methodology

The methodology presented in this section aims to ensure feasibility of the system for any value of ΔA and ΔB within their bounds. In other words, it aims to provide a formulation that ensures that for any realization of the multiplicative uncertainty, if the initial values of the states of the system (x_0) lay within a feasible domain, there exists a sequence of u_T that will (a) keep the system feasible for all T and (b) will try to bring the system to the origin. It is understood that through this robustification methodology the feasible domain of x_0 for which (a) can happen will be significantly reduced compared to a nominal case where no multiplicative uncertainty occurs. Conservatism of the approach is not addressed in the current work, only strict feasibility of the system under uncertainty.

The methodology can be broken down into four steps, namely: (i) Robustification, (ii) Reformulation, (iii) Reduction and (iv) Solution. Note that in the following sections the bounds on x_T and u_T are omitted for brevity but are implied and the formulations are valid for $\forall T \in \{0,\dots,N-1\}$.

3.1. Robustification

The Robustification of the problem defined by eqns. 2-5 happens successively starting from the initial state to state N following the approach by (A. Ben Tal, et al. (1998)). Given that the states x_T and inputs u_T of the system are bounded for any time-step T we have from eqns. 2-5:

$$\underline{x} \le x_{T+1} = A x_T + B u_T \le \overline{x} \tag{6}$$

Eqns. 2-4

The robust counterpart for a matrix G of appropriate dimensions and a constraint of the form of eqn. 7 can be expressed as in eqn. 8:

$$G x_{T+1} \le g, \forall T \in \{0,\dots,N-1\} \tag{7}$$

$$G A_0 x_T + \epsilon_A |G||A_0||x_T| + G B_0 u_T + \epsilon_B |G||B_0||u_T| \le g \tag{8}$$

3.2. Reformulation

The main challenge with eqn. (8) is the absolute values of the state and input variables in every time step. We apply a reformulation based on eqn. 8 as follows:

$$GA_0 x_T + \epsilon_A |G||A_0| z_T + GB_0 u_T + \epsilon_B |G||B_0| v_T \leq g \tag{9.1}$$

$$-z_T \leq x_T \leq z_T \tag{9.2}$$

$$0 \leq z_T \leq \max\{|\overline{x}|, |\underline{x}|\} \tag{9.3}$$

$$y_T \underline{u} \leq u_T \leq (1 - y_T)\overline{u}, y_T \in \{0,1\} \tag{9.4}$$

$$w_T = y_T u_T \tag{9.5}$$

$$v_T = u_T - 2w_T \tag{9.6}$$

$$0 \leq u_T - 2w_T \leq \max\{|\overline{u}|, |\underline{u}|\} \tag{9.7}$$

Eqn. (9.2-9.3) corresponds to the standard relaxation of the absolute term of x_T and its bounds, eqn. (9.4) introduces an auxiliary binary variable y_T which is equal to 0 when the input action u_T is positive and 1 when it is negative. A continuous auxiliary variable w_T is introduced in eqn. (9.5) and it represents a bilinear term between the input and auxiliary y_T. Eqn. (9.6) is used to replace v_T in eqn. (9.1) while preserving its bounds by (9.7). The bilinear term introduced can be exactly reformulated with a single McCormick relaxation since it is a product between a binary and a continuous variable, as follows:

$$w_T \geq y_T \underline{u} \tag{9.5.1}$$

$$w_T \geq u_T + y_T \overline{u} - \overline{u} \tag{9.5.2}$$

$$w_T \leq \underline{u} y_T + u_T - \underline{u} \tag{9.5.3}$$

$$w_T \leq \overline{u} y_T \tag{9.5.4}$$

By successively robustifying as in 3.1 for all time step in the output/control horizon, reformulating as in eqn. (9.x), utilizing eqn. (9.5.x) instead of (9.5) and substituting v_T with eqn. (9.6) we derive a robust counterpart of the problem in eqn. (5) that involves the original terms x_0, u_T and the auxiliary terms w_T, z_T and y_T.

3.3. Reduction

The continuous auxiliary variables w_T and z_T can be a source of primal degeneracy as described in E. N. Pistikopoulos, et. al (2020), therefore a procedure is needed to project the problem onto a space where the aforementioned variables do not exist but without affecting the feasible space of the original variables. Here, we utilize a single Fourier-Motzkin elimination step per continuous auxiliary variable. For two linear constraints of the form of eqn. (10), assuming without loss of generality $a_1, a_2 > 0$:

$$a_1 z_T + l_1(x_0, u_T, y_T, w_T) \leq b_1 \Leftrightarrow z_T \leq \frac{1}{a_1}(b_1 - l_1(x_0, u_T, y_T, w_T)) \tag{10.1}$$

$$a_2 z_T + l_2(x_0, u_T, y_T, w_T) \geq b_2 \Leftrightarrow z_T \geq \frac{1}{a_2}(b_2 - l_2(x_0, u_T, y_T, w_T)) \tag{10.2}$$

where l_1 and l_2 are linear functions of their variables we can eliminate z_T by substituting eqn. (10) with:

$$\frac{1}{a_2}(b_2 - l_2(x_0, u_T, y_T, w_T)) \leq \frac{1}{a_1}(b_1 - l_1(x_0, u_T, y_T, w_T)) \tag{11}$$

Note that in the LQR problem, given the proposed reformulation, equations of the form of eqn. (10.x) will always be present for all continuous auxiliary variables as they are bounded by definition. Also note that the procedure is repeated for all z_T and w_T thus creating a problem formulation in u_T, z_0 and y_T, i.e. a mixed-integer program. Lastly, note that the reformulation presented here is based on reformulating the auxiliary variable

v_T of eqn. (9.1). There exists an equivalent reformulation for the term z_T involving binary variables which is more challenging, outside the scope of the current work but within the scope of overall research associated with this approach and currently under development.

3.4. Solution

The reduced problem is symbolically presented below:

$$\begin{aligned} \min_{\mathbf{u},\mathbf{y}} \quad & [\mathbf{u}^T, \mathbf{y}^T]H[\mathbf{u}^T, \mathbf{y}^T]^T + [\mathbf{u}^T, \mathbf{y}^T]Z\mathbf{x_0} \\ \text{s.t.} \quad & Gu + Ey \leq b + F\mathbf{x_0} \\ & CR_A\mathbf{x_0} \leq CR_b \\ & \mathbf{y} \in \mathbf{0}, \mathbf{1}^n, \mathbf{u} \in \mathcal{R}^n, \mathbf{x_0} \in \mathcal{R}^m \end{aligned} \tag{12}$$

The matrix H can be shown to be H a {2n × 2n} positive semi-definite matrix given that the matrices defining the LQR in eqn. (5) are positive definite. Bilinear terms between continuous and binary terms or purely binary terms that may arise due to H can be exactly reformulated similar to eqn. (9.5.x). The program described by eqn. (12) can be solved online as soon as values for x_0 are available with the use of any local MIQP algorithm (i.e. CPLEX, Gurobi, etc.) or explicitly with the use of multi-parametric programming, treating x_0 as uncertain but measurable parameters, as described in E. N. Pistikopoulos, et. al (2020).

4. Uncertain constrained LQR example

Here we consider an LQR of a double integrator system described by eqn. (5) where $Q = \begin{bmatrix} 1 & 0 \\ 0 & 1 \end{bmatrix}$, $R = 0.01$, $P = \begin{bmatrix} 2.6235 & 1.6296 \\ 1.6296 & 2.6457 \end{bmatrix}$, $A_0 = \begin{bmatrix} 1 & 1 \\ 0 & 1 \end{bmatrix}$, $B_0 = \begin{bmatrix} 0 \\ 1 \end{bmatrix}$, $\epsilon_A = \epsilon_B = 0.1$. The system bounds are defined as $\overline{x} = -\underline{x} = [10{,}10]^T$ and $\overline{u} = -\underline{u} = 2$.

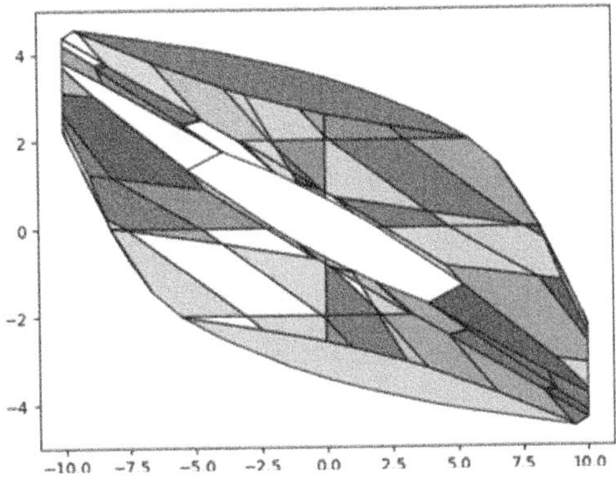

Figure 1: Critical Regions of mpMIQP. The axes represent initial state values

An horizon of 2 time-steps is used. The problem is treated as discussed in section 3 and eqn. 12 is solved via multi-parametric Mixed Integer Quadratic Programming, while preserving the envelopes of solution as described in Pistikopoulos et al. (2020). The solution comprises of 108 Critical Regions (shown in Figure 1) and is able to stir the problem back to the origin from various starting points for a plethora of multiplicative uncertainty realizations. This is presented in Figure 2 where the system returns to the origin for various, random, multiplicative uncertainty realizations and two different starting points for the states, without constraint violations, an indication of successful robustification.

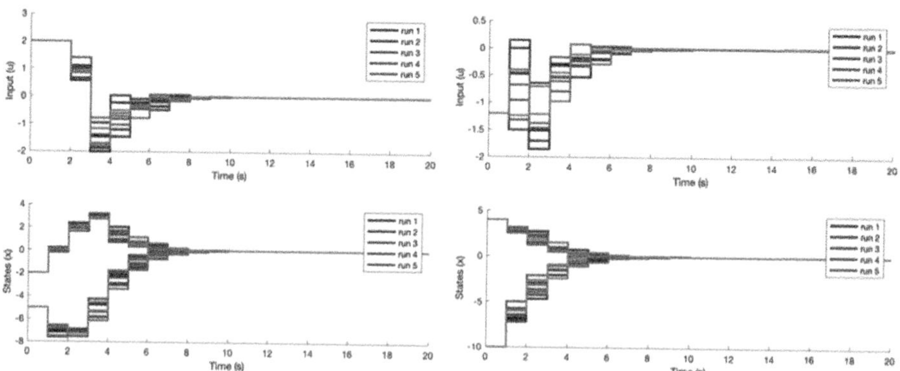

Figure 2: Optimal input and system response of the example for various values of multiplicative uncertainty and two starting points (x0=[-2, -4.4]: left, x0=[-10, 4.4]: right)

5. Concluding remarks

In this work, we presented an algorithm to solve robust explicit RHO problems via mpMIQP. Our approach is based on the (i) robustification, (ii) reformulation and (iii) reduction prior to the (iv) solution via multi-parametric programming. We demonstrated our findings on a constrained LQR case study where we exhibited that the system can be driven to the origin for any arbitrary, bounded value of the multiplicative uncertainty. Our next steps include the extension of step (ii) to other auxiliary variables and the extension to other classes of rolling horizon optimization problems.

References

D. Q. Mayne (2014), Model predictive control: Recent developments and future promise. Automatica, 50 (12), 2967-2986.

N.A. Diangelakis, B. Burnak, J. Katz, E.N. Pistikopoulos (2017), Process design and control optimization: A simultaneous approach by multi-parametric programming. AIChE Journal, 63 (11), 4827-4846.

B. Burnak, N. A. Diangelakis, E. N. Pistikopoulos (2020), Integrated Process Design and Operational Optimization via Multiparametric Programming, Synthesis Lectures on Engineering, Science, and Technology, Morgan & Claypool Publishers.

B.Kouvaritakis, M. Cannon (2016), Model Predictive Control. Switzerland: Springer International Publishing, 38.

D. Q. Mayne, J. B. Rawlings, C. V. Rao, P. O. M. Scokaert (2000) Constrained model predictive control: Stability and optimality. Automatica, 36 (6), 789-814.

A. Ben-Tal, A. Nemirovski (1998), Robust convex optimization, Mathematics of Operations Research, 23 (4), 769-805.

Q. Zhang, C. E. Gounaris (2022), Methodology and applications of robust optimization. Optimization and Engineering 23, 1761-1764

V. Sakizlis, N. M. Kakalis, V. Dua, J. D. Perkins, E. N. Pistikopoulos, (2004), Design of robust model-based controllers via parametric programming. Automatica, 40, 2, 189-201.

A. Bemporad, F. Borelli, M. Morari (2003), Min-max control of constrained uncertain discrete time linear systems. IEEE Transactions on automatic control, 48(9), 1600-1606.

K. Kouramas, C. Panos, N. P. Faísca, E. N. Pistikopoulos (2013), An algorithm for robust explicit/multi-parametric model predictive control. Automatica, 49, 2, 381-389.

I. Pappas, N. A. Diangelakis, R. Oberdieck, E. N. Pistikopoulos (2022), A Robust Optimization Strategy for Explicit Model Predictive Control, CACE, 49, 409-414.

F. Blanchini (1999), Set invariance in control. Automatica, 35 (11), 1747-1767.

E. N. Pistikopoulos, N. A. Diangelakis, R. Oberdieck (2020), Multi-parametric Optimization and Control, John Wiley & Sons, 2020.

Antonis Kokossis, Michael C. Georgiadis, Efstratios N. Pistikopoulos (Eds.)
PROCEEDINGS OF THE 33rd European Symposium on Computer Aided Process Engineering (ESCAPE33), June 18-21, 2023, Athens, Greece
 http://dx.doi.org/10.1016/B978-0-443-15274-0.50273-0

A process data prediction method for chemical process based on the frozen pretrained transformer model

Yiming Bai[a], Jinsong Zhao[a,b]

[a]*Department of Chemical Engineering, Tsinghua University, Beijing 100084, China*
[b]*Beijing Key Laboratory of Industrial Big Data System and Application, Tsinghua University, Beijing, 100084, China*
jinsongzhao@mail.tsinghua.edu.cn

Abstract

With the reform of industrial production digitization and automation, process fault detection and diagnosis (FDD) has been an indispensable technical method to realize the safe and efficient production of chemical process, even if there may be some delay and inaccuracy. Fault prognosis could predict the occurrence of faults in advance, which would give operators more time and reduce the impact of faults. Frozen Pretrained Transformer (FPT) has shown language-pretrained transformers can obtain strong performance on a variety of simple non-language classification tasks. Inspired by the pre-trained models, we proposed a chemical process data prediction method based on FPT models. Specifically, we take a transformer model pretrained on natural language data and finetune some parameters on the chemical process data. And the Tennessee Eastman process (TEP) was used to demonstrate the validity of the method. Finally, we discussed the relevant issues in applying language-pretrained models to process data.

Keywords: Process data prognosis; Chemical process; Transformer model; Frozen pretrained model.

1. Introduction

In recent decades, with the reform of industrial production digitization and automation, various industrial processes have generated massive amounts of data. In order to fully process and apply these data, the computing power of machines has reached an unimaginable level (Rehman et al. 2019). In the modern chemical industry, the development of artificial intelligence technology has made outstanding contributions to industrial intelligence, but the promotion of automation and intelligence has also made modern chemical plants becoming more and more large-scale and highly complex than ever.

As the complexity of chemical engineering systems has increased, chemical process safety has been challenged to a large extent. The faults of chemical processes may cause economic losses and casualties, which pose a great threat to chemical safety (Venkatasubramanian et al. 2003). FDD has been an indispensable technical method to realize the safe and efficient production of chemical processes and has substantial theoretical and application value in the chemical industry (Bi et al. 2022). However, FDD may be not sufficient in some cases, and it is necessary to prognose the fault ahead of time. Fault prognosis mainly aims at potential and future faults, which means the research object of fault prognosis is a future uncertain event (Smith and Randall 2015; Zhang et al. 2011). Bai et al. (Bai and Zhao 2023) explained the relationship between fault prognosis and FDD, which is shown in Fig. 1. In the process monitoring loop, the fault prognosis, located before FDD, attempts to identify that the system is in an abnormal state in advance, which would greatly reduce the probability of abnormal conditions.

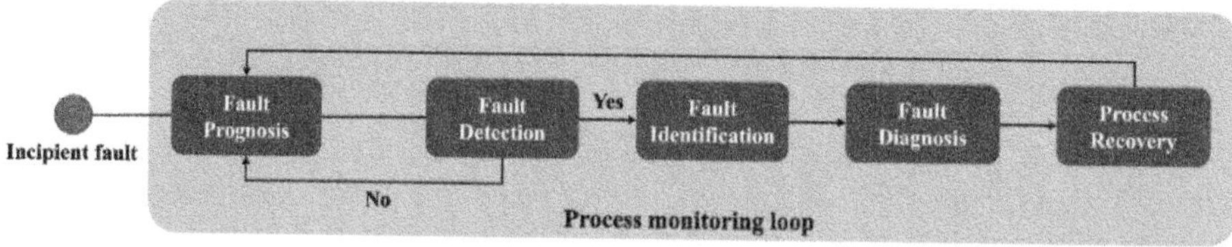

Fig. 1. Relationship between fault prognosis and process monitoring (Bai and Zhao 2023).

Current fault prognosis methods can be generally classified into three types: model-based methods, knowledge-based methods, and data-based methods (Zhong, Han, and Han 2020). Data-based methods determine the status of the system within a certain period by analyzing previously observed data. With the development of automation and digitization, data-based models have been playing an increasing role in the process industry.

Fault prediction methods have evolved with the changing characteristics of industrial data. As data complexity increases, many data-driven methods are flexibly applied in nonlinear scenarios, such as Artificial Neural Networks, autoencoder, Support Vector Machine, Radial Basis Function network, Recurrent Neural Network (RNN) and long short term memory (LSTM) (Yuan et al. 2020). Transformer architecture has widespread applications, particularly in Natural Language Processing (NLP) and computer vision, which also triggered huge interests in the time series data (Wen et al. 2022). It has become common practice to train large models on unsupervised or weakly supervised objectives before finetuning or evaluating zero-shot generalization on a downstream task. Lu et al. found that pre-training in natural language can improve the performance and computational efficiency of non-verbal downstream tasks, and proposed the Frozen Pretrained Transformer (FPT) model (Lu et al. 2021), which inspired the pre-trained models for time series data.

In this paper, we propose a process data prediction method for chemical process based on FPT models. We take a transformer model pretrained on natural language data and finetune some parameters on the chemical process data. And the Tennessee Eastman process (TEP) was used to demonstrate the validity of the method. The results show that the proposed method obtain the strong performance and less consumption on the chemical process fault prognosis task. Additionally, we discussed the relevant issues in applying language-pretrained models to process data, which verified further validity of the proposed method.

The remainder of this paper is organized as follows. Section 2 reviews the basic knowledge of Transformer models and FPT model. Section 3 includes the descriptions of the proposed process data prediction method. The applications and discussions are revealed in Section 4. Section 5 gives summary.

2. Methodology

2.1. Self-attention

The attention mechanism could overcome this weakness by learning the weights of input information at different times. We show a version of a commonly used attention in Fig. 2, which was proposed as self-attention (Vaswani et al. 2017).

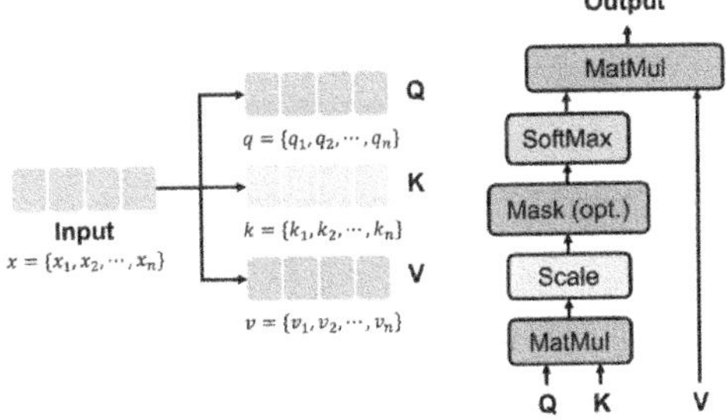

Fig. 2. A schematic diagram of self-attention.

Firstly, given an input sequence $x = \{x_1, x_2, \cdots, x_n\}, x_i \in R^k$, query, key and value vectors can be obtained by mapped with linear transformations and $q_i, k_i, v_i \in R^{d_k}$ are examples to show the calculation process of each element.

$$q_i = W^q x_i, \qquad k_i = W^k x_i, \qquad v_i = W^v x_i, \qquad i = 1, 2, \cdots, n \tag{1}$$

Where $W^q, W^k, W^v \in R^{d_k \times k}$ are learnable matrices.

Then, the similarity score of q_i and k_j, which are chosen as examples to show calculation, is calculated through a scaled dot-product function, and the SoftMax function is then used to normalize the scores and obtain the weights.

$$\alpha_{i,j} = \frac{q_i \cdot k_j}{\sqrt{d_k}} \tag{2}$$

$$\hat{\alpha}_{i,j} = \frac{\exp(\alpha_{i,j})}{\sum_{t=1}^{n} exp(\alpha_{i,t})} \tag{3}$$

Finally, the output is calculated as the weighted sum of values.

$$z_i = \sum_{j=1}^{n} \hat{\alpha}_{i,j} v_j \tag{4}$$

2.2. FPT model

To investigate this hypothesis, we take a transformer model, GPT-2, pretrained on natural language data (Radford et al. 2019). GPT-2 is a transformers model pretrained on a very large corpus of English data in a self-supervised fashion. Crucially, all communication between tokens in the model are frozen. The data in each datapoint is chunked into discrete tokens, and can only reference each other via the frozen attention connections, which are not trained. Our key investigation is to analyze the performance of the model in process data prediction, which is already inherent in the language model. On the basis of the pretrained model, the model is finetuned with a minimal amount of computation on the downstream modality. The FPT model used in process data utilizes the feature extraction part of the pretrained model, and adds two fully connected layers to handle the input process data and the output features. There are five types of parameters in the finetune process, which are input layer, output layer, layer normalization, positional embedding and input embedding.

3. The proposed method

The proposed process data prediction method is shown in the Fig. 3. Firstly, process data can be directly used as model input without embedding. The sequence structure of the process data can reduce the difficulty of position coding, and we can use one-hot matrix to indicate the order of input. Then the input layer expands the dimensionality of input data as needed for GPT-2. And the features of the process data are extracted by the pretrained model by NLP. Finally, the output layer gets the prediction results of the process data by the features. The dimension of the input layer is the sum of the dimension of process variables and the time dimension. The dimension of the output layer is the dimension of process variables. The number of the dimension in the GPT-2 is 762 and the number of frozen attention block is 12.

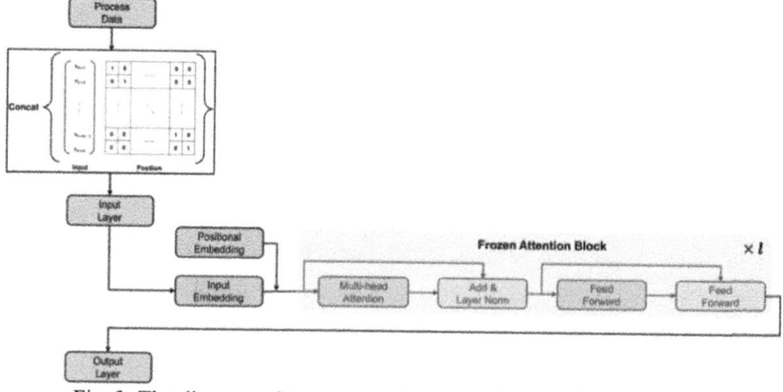

Fig. 3. The diagram of the proposed process data prediction method.

4. Applications and discussions

4.1. Applications on the TEP

TEP is a classical model benchmark for simulating chemical processes, which has been widely used as a study case for FDD and fault prognosis. It is mainly composed of 5 unit operations, including a reactor, a condenser, a recycle compressor, a vapor-liquid separator, and a stripper. The initial version of TEP simulation involves 12 manipulated variables, 41 measured variables and 20 different types of faults. In 2015, Bathelt revised the original TE model and used MATLAB's Simulink to build a new one, adding 8 measurement variables, 24 component variables, and 8 new process disturbances (Bathelt, Ricker, and Jelali 2015). In our work, the applications are based on the revised version of Bathelt, which is shown in the Fig. 4.

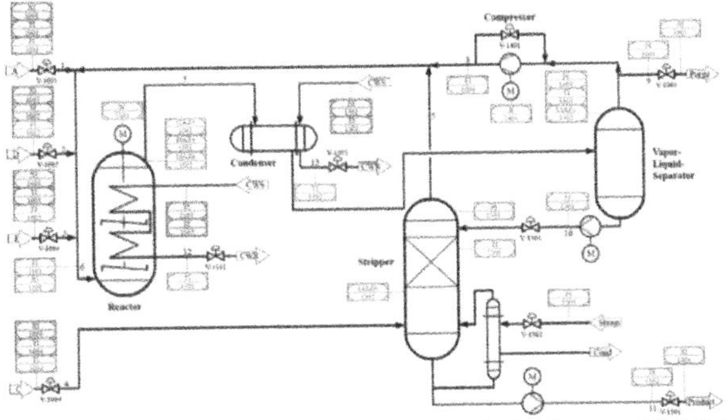

Fig. 4. P&ID of the revised TEP model (Bathelt, Ricker, and Jelali 2015).

Since the sampling frequency of component analysis measurement is much lower, we only choose the process manipulated variables and continuous process measurements except variables related to valves for dataset building. Among the common faults in the TEP, most fault types are step or random changes, which bring significant changes to the system within a short time of the fault occurrence. The process variables prediction and fault prognosis are with little necessarily in the cases, and these cases do little help for validating the effectiveness of the prediction methods in chemical process. According to the application results of the FDD methods in the abnormal state of TEP, we choose fault 13 as the target of the study, which is the fault of reaction kinetics drifting slowly.

A simulation dataset is built to assess the fault detection and identification performance of all the methods. During the simulation, 1 sample is collected every 1 min. The datasets include 10 sets of data for fault 13. Each set of data includes 480 normal samples followed by 600 fault samples, corresponding to 8h of normal data and 10h of fault data. The training dataset includes 8 sets of data and the validation dataset and the testing dataset each includes a set of data. For the selected time window, the goal of the model is to predict the changes in the 30 process variables over a future time window. And MSE of all the process variables in the time window is selected to be the loss function. We chose RNN, LSTM, BiLSTM and full GPT2 to compare the predictive performance. And the full GPT2 means that we trained all the parameters of the GPT2 model in the training stage. The prediction results of different methods with different time window are shown in the following Table. 1.

Table. 1. Prediction results of different methods.

Method	Time widow = 8	Time widow = 16	Time widow =32
RNN	0.826	1.355	**2.648**
LSTM	0.714	1.147	2.883
BiLSTM	0.737	1.157	3.842
FPT	**0.140**	**0.510**	2.774
Full GPT2	0.647	0.999	4.016

The results shown that the FPT had the best performance when the time window is 8 or 16. Theoretically, the result of full GPT2 should be a lower bound of FPT. However, due to the limited training datasets of process variables, the MSE of full GPT2 is larger than that of FPT. And it indicated that to some extent better prediction results could be achieved in the process variables using a pretrained model than a fully trained model. To the time window of 32, FPT didn't get the better performance.

When only the parameters of input layer and output layer were finetuned, the MSE was 7.808. On this basis, we only finetuned some selected parameters to find out which parameters were most important in the GPT2 model for process variables prediction. Firstly, we only finetuned the parameters of the layer normalization, and the MSE reduced to 0.563, which indicated that finetuning the parameters of the layer normalization could improve effectively prediction results. Then we added the parameters of the position embedding to the finetuned parameters, and the MSE reduced to 0.556. Then we added the parameters of the input embedding to the finetuned parameters, and the MSE reduced to 0.551. These two increases in finetuned parameters did not effectively reduce the MSE. The results shown that the paraments of the layer normalization played the most important role in the process variables prediction.

Studies have shown that the number of layers of neural networks for process data is less than that for NLP or CV problems. In our work, the number of layers could be measured by the number of used frozen attention blocks. We finetuned the model with different numbers of frozen attention blocks to learn the most suitable model for the chemical process variables prediction. We found that the number of used frozen attention blocks was selected to be 2 to show the best results in the applications. And the MSE with the time window of 8, 16 and 32 reduced to 0.109, 0.287 and 1.667, which were better than other methods.

5. Summary

In the article, we proposed a process data prediction method for chemical process based on FPT models. In TEP the proposed method shown the best validity with the smallest MSE of process variables. And the FTP method could show the strong performance and less consumption on the chemical process variables prediction task. We found that the paraments of the layer normalization played the most important role in the results. And the number of used frozen attention blocks is selected to be 2, which was shown to have better results in the application of chemical process variable prediction. In the future, the application of pretrained models to process variables prediction and process fault prognosis remains widely open with many challenges left to solve.

References

Bai, Yiming, and Jinsong Zhao. 2023. "A Novel Transformer-Based Multi-Variable Multi-Step Prediction Method for Chemical Process Fault Prognosis." *Process Safety and Environmental Protection* 169 (January): 937–947.

Bathelt, Andreas, N. Lawrence Ricker, and Mohieddine Jelali. 2015. "Revision of the Tennessee Eastman Process Model." *IFAC-PapersOnLine*, 9th IFAC Symposium on Advanced Control of Chemical Processes ADCHEM 2015, 48 (8): 309–314.

Bi, Xiaotian, Ruoshi Qin, Deyang Wu, Shaodong Zheng, and Jinsong Zhao. 2022. "One Step Forward for Smart Chemical Process Fault Detection and Diagnosis." *Computers & Chemical Engineering*, June, 107884.

Lu, Kevin, Aditya Grover, Pieter Abbeel, and Igor Mordatch. 2021. "Pretrained Transformers as Universal Computation Engines." arXiv.

Radford, Alec, Jeffrey Wu, Rewon Child, David Luan, Dario Amodei, Ilya Sutskever, and others. 2019. "Language Models Are Unsupervised Multitask Learners." *OpenAI Blog* 1 (8): 9.

Rehman, Muhammad Habib ur, Ibrar Yaqoob, Khaled Salah, Muhammad Imran, Prem Prakash Jayaraman, and Charith Perera. 2019. "The Role of Big Data Analytics in Industrial Internet of Things." arXiv.

Smith, Wade A., and Robert B. Randall. 2015. "Rolling Element Bearing Diagnostics Using the Case Western Reserve University Data: A Benchmark Study." *Mechanical Systems and Signal Processing* 64–65 (December): 100–131.

Vaswani, Ashish, Noam Shazeer, Niki Parmar, Jakob Uszkoreit, Llion Jones, Aidan N Gomez, Łukasz Kaiser, and Illia Polosukhin. 2017. "Attention Is All You Need." In *Advances in Neural Information Processing Systems*. Vol. 30. Curran Associates, Inc.

Venkatasubramanian, Venkat, Raghunathan Rengaswamy, Kewen Yin, and Surya N. Kavuri. 2003. "A Review of Process Fault Detection and Diagnosis: Part I: Quantitative Model-Based Methods." *Computers & Chemical Engineering* 27 (3): 293–311.

Wen, Qingsong, Tian Zhou, Chaoli Zhang, Weiqi Chen, Ziqing Ma, Junchi Yan, and Liang Sun. 2022. "Transformers in Time Series: A Survey." arXiv. doi:10.48550/arXiv.2202.07125.

Yuan, Xiaofeng, Chen Ou, Yalin Wang, Chunhua Yang, and Weihua Gui. 2020. "A Novel Semi-Supervised Pre-Training Strategy for Deep Networks and Its Application for Quality Variable Prediction in Industrial Processes." *Chemical Engineering Science* 217 (May): 115509.

Zhang, Bin, Chris Sconyers, Carl Byington, Romano Patrick, Marcos E. Orchard, and George Vachtsevanos. 2011. "A Probabilistic Fault Detection Approach: Application to Bearing Fault Detection." *IEEE Transactions on Industrial Electronics* 58 (5): 2011–2018.

Zhong, Kai, Min Han, and Bing Han. 2020. "Data-Driven Based Fault Prognosis for Industrial Systems: A Concise Overview." *IEEE-CAA JOURNAL OF AUTOMATICA SINICA* 7 (2): 330–345.

Antonis Kokossis, Michael C. Georgiadis, Efstratios N. Pistikopoulos (Eds.)
PROCEEDINGS OF THE 33rd European Symposium on Computer Aided Process Engineering (ESCAPE33), June 18-21, 2023, Athens, Greece
 http://dx.doi.org/10.1016/B978-0-443-15274-0.50274-2

Sensor bias detection, isolation, and estimation in a subsea pump system.

Rafael D. Oliveira[a], Halvor A. Krog[a], Galo A. C. Le Roux[b], Johannes Jäschke[a]

[a] *Department of Chemical Engineering, Norwegian University of Science and Technology (NTNU), Trondheim, Norway*

[b] *Department of Chemical Engineering, University of São Paulo, São Paulo, Brazil*

johannes.jaschke@ntnu.no

Abstract

Subsea pump systems are subject to many possible faults and monitoring typically relies on the measurements of key process variables. However, the measurements themselves may be subject to faults like e.g., bias. In this work, a method that combines fault detection and isolation, observability analysis and state estimation is presented and tested on a simulated pump system. The results indicate that a state estimator with correct information about which sensor has drifted gives lower estimation error than a state estimator that does not consider bias. Furthermore, we show that it is impossible to estimate the bias on all sensors simultaneously due to observability issues of the augmented system.

Keywords: Sensor bias, Fault detection, Subsea pump, Fault isolation, Sensor drift.

1. Introduction

Subsea pumping systems operate in harsh environments for long periods, making the maintenance of these pumps challenging and costly. Condition monitoring is crucial to keep these subsea pumps running for long periods and to reduce the possibility of accidents. The pumps are subject to several potential faults, and it is desirable to use measurements to detect if something is going wrong. However, the sensors that provide these measures are also susceptible to faults (e.g., gross errors). These gross errors arise from sensors malfunctioning and can significantly impact the monitoring and controlling of the subsea pump.

Given the long operation times of subsea pumps, the long-term faults (e.g., bias) are more challenging to detect than other abrupt changes. It is desirable to detect and isolate these biases as soon as possible and, if practicable, estimate their magnitude. Approaches to deal with these sensor biases can be classified as passive or active. The passive methods apply robust estimators like M-estimators, robust Kalman Filter (KF), and robust Moving Horizon Estimation (MHE) (Siddhartha et al., 2022). Passive methods efficiently handle gross errors, but no information about the bias's location and size is generated. The location of the gross errors is a valuable information for maintenance and process monitoring. On the other hand, active methods try to estimate these biases, usually representing the bias as a new time-dependent continuous parameter on the state estimation problem (Gatzke & Doyle III, 2002). However, observability problems usually arise (Gatzke & Doyle III, 2002).

In this paper, we propose to combine fault detection and isolation (FDI) methods, observability analysis and state estimators in the context of a subsea pump system. In our methodology, i) FDI is used to identify sensors with bias. ii) Then an observability analysis is performed to check if it is possible to estimate that bias, and iii) then the state estimator is extended with a bias parameter to be estimated. The FDI reduces the number of biases to be estimated, and the observability analysis assures that the bias can be estimated without worsening the state estimator's performance. This illustrates how attempting to estimate sensor bias without any *a priori* information of which sensors are failing is impractical and how combining FDI with state estimators can improve the estimation significantly. Furthermore, a Monte Carlo analysis is performed to show the advantages and limitations of that approach in multiple scenarios.

2. Case study: Subsea pump system

We consider a simple subsea booster pump system, as shown in Figure 1. A mixture of oil and water comes from the reservoir at pressure p_1. The pump runs with a variable speed drive (VSD) at ω rpm, which produces a specific head, H. Given the density of the fluid, ρ, one can calculate the pressure at the outlet of the pump, p_2. The pressure p_3 is determined by downstream facilities (oil platform). The choke valve has a flow rate of q and opening of Z %. The three pressures are measured, and a Venturi flowmeter measures the pressure drop over an internal orifice (not shown).

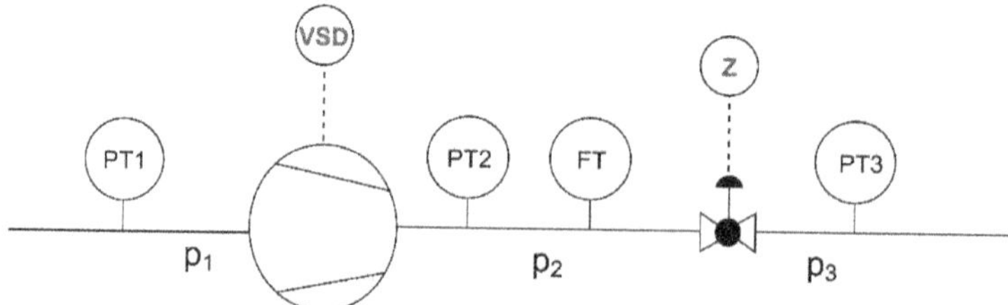

Figure 1: The subsea booster pump with its installed sensors. FT is a Venturi flowmeter.

2.1. Pump model

Sensor bias is a phenomenon that typically takes place in the timescale of months and years. The pressure and density changes in the reservoir are on the same timescale, and we assume the pressure on the oil platform has the same trend as the reservoir pressure. On the other hand, phenomena such as transient flow regimes happen almost instantly in the timescale of seconds. This motivates the use of a quasi-steady state model, where changes in the reservoir and p_3 are modelled dynamically and the pump is modeled in steady-state (and therefore the time-subscript is $k+1$ on both sides of the equality sign in (4)-(6)). It is assumed that p_1 and p_3 decreases at the same rate, hence, they have the same parameter $\theta_{P,1}$. The faultless system is described by:

$$p_{1,k+1} = p_{1,k} + \theta_{P,1}\Delta t + w_{p_1,k} \quad (1)$$

$$\rho_{k+1} = \rho_k + \theta_\rho \Delta t + w_{\rho,k} \quad (2)$$

$$p_{3,k+1} = p_{3,k} + \theta_{P,1}\Delta t + w_{p_3,k} \quad (3)$$

$$p_{2,k+1} = p_{1,k+1} + \rho_{k+1} g H_{k+1} + w_{p_2,k} \quad (4)$$

$$q_{k+1} = C_v Z_{k+1} \sqrt{\frac{p_{2,k+1} - p_{3,k+1}}{\rho_{k+1}/\rho_{water}}} + w_{q,k} \tag{5}$$

where the head at time k, H_k, is found by using the so-called pump curve.

$$H_{k+1} = \theta_{H_0}\omega_{k+1} + \theta_{H_1} q_{k+1}\omega_{k+1} + \theta_{H_2} q_{k+1}^2 \tag{6}$$

The states are $\boldsymbol{x}_k = [p_{1,k}, p_{2,k}, p_{3,k}, \rho_k, q_k]^T$ and the state-propagation equations (1)-(5) can be written in the form of $\boldsymbol{x}_{k+1} = \boldsymbol{f}(\boldsymbol{x}_k, \boldsymbol{u}_{k+1}, \Delta t) + \boldsymbol{w}_k$. The stochastic process noise is defined as $\boldsymbol{w}_k = [w_{p_1,k}, w_{p_2,k}, w_{p_3,k}, w_{\rho,k}, w_{q,k}]^T \sim (\boldsymbol{0}, \boldsymbol{Q}_k)$. The process noise distribution is assumed to be constant at $\boldsymbol{w}_k \sim \mathcal{N}(\boldsymbol{0}, \boldsymbol{Q})$ where $\boldsymbol{Q} = 10^{-5}\boldsymbol{I}$. The parameter values are $\boldsymbol{\theta} = [\theta_{P,1}, \theta_\rho, \theta_{H_0}, \theta_{H_1}, \theta_{H_2}, g, C_v] = [-5.0, 5.0, 3.51 \times 10^{-5}, 3.29 \times 10^{-4}, -0.01, 9.8 m/s^2, 47.50\, m^3/h]^T$,where g is the gravitational constant and C_v is the valve constant. The remaining parameters describe the evolution of pressure and density in time. The initial conditions for all the simulations were $\boldsymbol{x}_0 = [50.0\, bar, 64.5\, bar, 51.0\, bar, 700\, kg/m^3, 136.6\, m^3/h]^T$ and $\boldsymbol{u}_0 = [65\,\%, 3500\, rpm]^T$. When there is no bias in the sensors, the sensor equations are:

$$\boldsymbol{y}_k = \boldsymbol{h}(\boldsymbol{x}_k) + \boldsymbol{v}_k = \left[p_{1,k}, p_{2,k}, p_{3,k}, \frac{\rho_k(1-\beta^4)}{2}\left(\frac{4q_k}{\pi d^2 C\epsilon}\right)^2\right]^T + \boldsymbol{v}_k \tag{7}$$

where $C = 1.01$ is the flow coefficient, $\epsilon = 1$ is the expansibility factor, $d = 0{,}0525mm$ is the Venturi throat diameter, $D = 0{,}154m$ is the internal pipe diameter and $\beta = d/D$. The measurement noise is distributed as $\boldsymbol{v}_k \sim \mathcal{N}(\boldsymbol{0}, \boldsymbol{R})$, where $\boldsymbol{R} = diag([0.1035, 0.1035, 0.1035, R_{venturi}])$.

2.2. Pressure sensor bias

Sensor bias is a constant systematic error in measurement and sensor drift is a time-varying bias in the sensor. Hence, the measurement equation (7) must account for these biases and drift in the following manner, where $\boldsymbol{y}_k^d$ is the bias/drift value at time k and $\boldsymbol{y}'_k$ is the measurement equation with bias/drift:

$$\boldsymbol{y}'_k = \boldsymbol{y}_k + \boldsymbol{y}_k^d = \boldsymbol{h}(\boldsymbol{x}_k) + \boldsymbol{v}_k + \boldsymbol{y}_k^d \tag{8}$$

To estimate the bias/drift, we augment the state vector and model the bias/drift as a random walk. The augmented model is now given by

$$\boldsymbol{x}_{k+1}^a = [\boldsymbol{x}_{k+1}, \boldsymbol{y}_{k+1}^d]^T = [\boldsymbol{f}(\boldsymbol{x}_k, \boldsymbol{u}_{k+1}, \Delta t), \boldsymbol{y}_k^d]^T + [\boldsymbol{w}_k, \boldsymbol{w}_{y^d,k}]^T \tag{9}$$

where $\boldsymbol{w}_{y^d,k} \sim \mathcal{N}(\boldsymbol{0}, \boldsymbol{Q}_y)$ is the distribution describing the steps of the random walk.

3. Combining fault detection and observability analysis to state and bias estimation.

The simultaneous estimation of states and biases in all the sensors simultaneously can make the system non-observable. To avoid these problems, this work combines FDI methods and observability analysis with state estimation. Figure 2 illustrates the methodology. The first step consists of the residual generation. The measurement (y'_k), which may have a bias or drift, is compared to the model's predicted measurement ($\hat{y}_k$). Then this residual is evaluated by a statistical test that is aimed in deciding if the system is fault-free or which sensor has drifted. After that, an observability analysis is performed to evaluate if the identified bias can be estimated. In the positive case, a new term is included on the state estimator as described in section 2.2 to track the bias/ drift on the failing sensor. Otherwise, the standard state estimation is performed without adding additional bias terms.

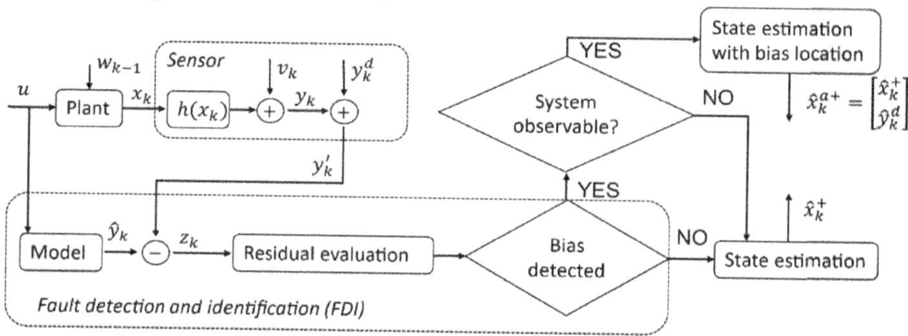

Figure 2: Conceptual flowsheet of how FDI-technique is used to update the structure of the state estimator.

3.1. Residual evaluation for bias detection

Three residuals are made, one for each pressure sensor, which consists of the difference between the model prediction and the measurement. Assuming a perfect model, it is possible to identify a bias in the pressure measurements since these residuals should be zero mean in a normal operating condition and different from zero when a bias occurs. To evaluate if the residuals are different from zero, the statistical method called cumulative sum control chart (CUSUM) was used (Blanke, 2016). A Gaussian distribution was assumed to represent the residual distribution ($\mathcal{N}(\mathbf{0}, \boldsymbol{R})$), and any deviation higher than five times the standard deviation would correspond to a fault.

3.2. State estimation and Observability analysis

We used the Unscented Kalman filter (UKF) to obtain the optimal estimate of the true state and bias (Simon, 2006, ch. 14). The bias terms were included on the UKF as time-dependent continuous parameter as described in section 2.2. The concept of observability tells us if we can uniquely determine the state given some measurements, which is a requirement for state estimators to work. For linear systems, observability can be checked by verifying if the observability matrix has full rank. See Simon (2006, ch. 1.7.2) for a definition of the observability matrix and a discussion about observability for linear systems. In our work, we linearized the model and checked the rank of the observability matrix for the linearized model. If the (augmented) system is not observable, we did *not* include the bias term in the state estimator and used the fault-free estimator as Figure 2 shows.

4. Results

The results section is divided into two parts. First, in section 4.1, an illustrative simulation shows how a previous bias detection and isolation step can improve the state estimation process. Then, in section 4.2, a Monte Carlo analysis is performed to evaluate the method's performance in different scenarios.

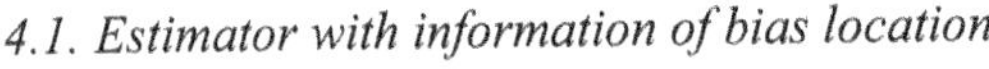

4.1. Estimator with information of bias location

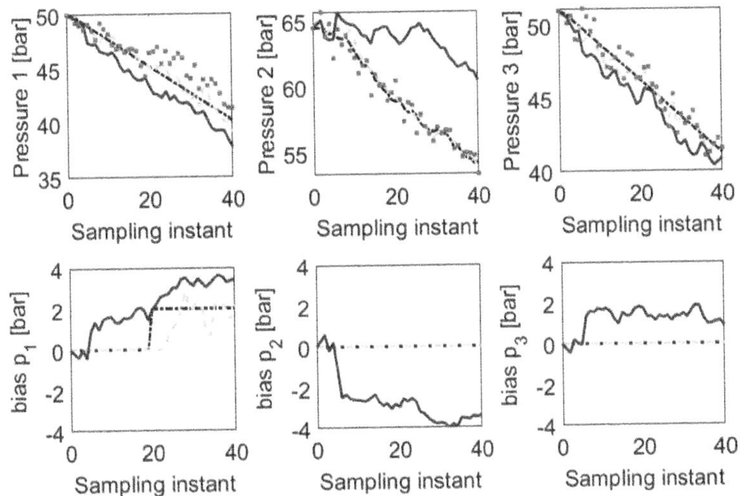

Figure 3- Comparison of true state (black), pressure measurements (red) and the estimates using UKF_b (blue) and UKF_{fdb} (green). Bias was introduced on the measurements of sensor PT1.

The model described in Section 2 was solved for a time window of 40 samplings. After 20 samples, a constant bias of 2 bar is added to the pressure sensor one (PT1). Two methodologies are compared in this scenario; first, the UKF with bias estimation (UKF_b), where the estimators try to estimate the bias on the three sensors since the beginning of the window. Hence, the augmented state dimension is $5 + 3 = 8$ through the entire time.

Second, our proposed method described in Section 3 is applied (UKF_{fdb}). Here, the state dimension is 5 before sampling time 20 and after a successful FDI and observability procedure it is 6. The results of the simulation is presented in Figure 3. The results indicate that the observability problems of adding new bias parameters to be estimated arise even for a small system like the subsea pump. UKF_b fails to estimate the states and bias in all three sensors, even before the bias is added to the measurements on sensor PT1. In fact, the observability analysis showed that the states are non-observable when three bias terms are included on the estimator.

However, when just one bias term is included on the estimator, the states and the bias are observable. This explains why the UKF_{fdb} could obtain a good estimate of the states and the bias on sensor PT1. UKF_{fdb} had a slight delay in detecting the bias on sensor PT1. After that, the observability is verified, and the information on which sensors are failing is sent to the state estimator.

4.2. Monte Carlo analysis

In order to perform a comprehensive analysis of the methodology, a Monte Carlo analysis was performed. This study compared three approaches: UKF_b, UKF_{fdb} and the "standard" UKF without bias estimation. All the simulations started from the same initial condition, with 100 samplings as a time window. Three hundred random samples of different simulation features were used: time to insert the bias (between 10 and 80 samples), the magnitude of the bias (-2 bar to 2 bar), control profile (pump speed and valve opening), sensor location of the bias (sensor 1, 2 or 3) and bias or drift. The results of the simulations are summarized in Table 1.

The results show that the UKF_b has poorer performance than the UKF. The system's observability must be checked before including new bias parameters to be estimated. The standard UKF is the best option in cases of non-observable states. However, when a fault detection and isolation method is applied, there is no need to include a bias term in the UKF for each sensor. Instead, only after the bias is detected a new term can be added to the UKF. The UKF_{fdb} had a better performance in terms of state estimation, with only a slightly larger error on q and H. Besides that, it can also provide a reasonable estimate of the bias in the sensors, which is a valuable information for pump system monitoring.

Table 1: Root-mean-square error states and bias over 300 simulations. * is "sampling instances".

			RMSE							
	Bias correctly Located [%]	Average delay [*]	bias [bar]	p_1 [bar]	p_2 [bar]	p_3 [bar]	ρ [kg/m^3]	q [m^3/h]	H [m]
UKF	-	-	-	0.74	0.77	0.75	24.97	2.54	3.15
UKF_b	-	-	4.89	2.32	3.57	1.92	273.61	2.40	3.08
UKF_{fdb}	86.00	14.96	0.52	0.42	0.41	0.44	10.86	2.58	3.21

5. Conclusion

An approach to monitor and estimate sensor bias on subsea booster pump system was numerically evaluated. The results suggest that estimating sensor bias without any *a priori* information about which sensors are failing is impractical. Therefore, combining fault detection and observability analysis techniques with sensor estimators can improve the estimation significantly and provide helpful information for system monitoring. A more thorough structural and local observability check should be applied in future works. Also, the performance of the methods for multiple simultaneous faults should be evaluated.

Acknowledgement: The authors acknowledge financial support from the Norwegian Research Council through the AutoPRO Project. RDO and GACLR acknowledges financial support from Coordenação de Aperfeiçoamento de Pessoal de Nível Superior - Brasil (CAPES) - Finance Code 001.

References

Blanke, M.; Kinnaert, M.; Lunze, J.; Staroswiecki, M. Diagnosis and Fault-Tolerant Control. Springer Berlin Heidelberg: Berlin, Heidelberg, 2016.

Rangegowda, P. H.; Valluru, J.; Patwardhan, S. C.; Biegler, L. T.; Mukhopadhyay, S. 2022. "Development of a Robust Receding-Horizon Nonlinear Kalman Filter Using M-Estimators." Industrial & Engineering Chemistry Research 61, 1808–29.

Gatzke, E. P.; Doyle III, F. J. 2002 "Use of Multiple Models and Qualitative Knowledge for On-Line Moving Horizon Disturbance Estimation and Fault Diagnosis." Journal of Process Control 12, no. 2: 339–52.

Simon, D. 2006. Optimal state estimation : Kalman, H [infinity] and nonlinear approaches. Hoboken, N.J.: Wiley-Interscience.

Antonis Kokossis, Michael C. Georgiadis, Efstratios N. Pistikopoulos (Eds.)
PROCEEDINGS OF THE 33rd European Symposium on Computer Aided Process Engineering (ESCAPE33), June 18-21, 2023, Athens, Greece
 http://dx.doi.org/10.1016/B978-0-443-15274-0.50275-4

Fusion of pupil and gaze-based features to estimate cognitive workload of control room operators

Mohd Umair Iqbal,[a,b] Babji Srinivasan,[b,c,*] Rajagopalan Srinivasan,[b,c,*]

[a]*Indian Institute of Technology Gandhinagar, Gujarat, 382355, India*
[b]*Indian Institute of Technology Madras, Tamil Nadu, 600036, India*
[c]*American Express Lab for Data Analytics, Risk & Technology, Indian Institute of Technology Madras, Tamil Nadu, 600036, India*
**babji.srinivasan@iitm.ac.in, *raj@iitm.ac.in*

Abstract

Process industries are highly hazardous, and these hazards often lead to accidents. Over 70% of these accidents are attributed to human errors. With the advancements in technology and changing role of operators to the one involving an emphasis on cognitive aspects, most of these errors occur due to limitations in cognitive performance. One of the major constructs to understand cognitive performance is the cognitive workload. An increase in cognitive workload often leads to degradation in performance. Eye tracking has been used in several domains to assess cognitive workload. In this work, we propose a methodology to assess cognitive workload of control room operators during tasks that involve tackling process abnormalities. The methodology employs the fusion of metrics obtained from pupil and gaze data. Our results reveal that fusion of metrics provides better accuracies of classifying cognitive workload at three levels—low, medium and high workload.

Keywords: human errors, operator performance, cognitive workload, eye-tracking, decision trees

1. Introduction

Chemical and process industries are highly hazardous, and these hazards frequently result in disastrous accidents. Even though there have been tremendous increases in safety due to better process design and technical safety approaches, mishaps still happen. A significant majority of these accidents can be traced back to human errors; human errors contribute to around 60-80% of accidents (Jung et al., 2020). Nonetheless, humans form an essential layer of protection, especially in the wake of abnormalities in the process that are often outside the purview of control systems. However, with the advancements in technology in the form of digitalization, and concomitant automation (Khan et al., 2021), the role of human operators has changed more towards monitoring (during regular operation), diagnosis, and prognosis (during abnormal situations). To maintain adequate situational awareness, operators must prioritize, acquire, segregate, and integrate information from multiple sources in distributed control system (DCS). An abundance of information due to digitalization, coupled with the increased complexity of modern plants, makes these tasks cognitively challenging for an operator, especially if they are not highly skilled. Further, the increased automation which often keeps the operator out-of-loop with the system makes it even more challenging for them to develop correct mental models (understanding of the process). As a result, operators' cognitive workload can drastically increase if they are not adequately trained and assessed. An increase in cognitive workload is known to result in the degradation of performance (Iqbal et al., 2020; Islam et al., 2017). There is a strong association between cognitive workload and human errors. While human errors are a result of mismatch between process demands and human capabilities high cognitive workload, at the most general level, is a result of mismatch between mental resources demanded by the process and those supplied by an operator (Parasuraman, 2008).

Traditional approaches of performance assessment are primarily based on process and operator action-based measures such as number of successes and failures, reaction times, deviations from standard operating procedures, and time taken to complete a task (Iqbal et al., 2021). These ignore the cognitive aspects of performance, such as cognitive workload, decision-making, information acquisition pattern, and operators' mental states while they perform a task. Hence, these also fail to unravel the underlying reasons for shortcomings in operator performance. While subjective assessment-based methods such as National Aeronautics Space Administration Task Load Index, or NASA-TLX (Hart, 2006) are employed for assessing cognitive workload, these are often criticized for being prone to subjective-bias. These can also not be used to assess changes in cognitive workload in real time, and thus have limited practical applicability. Therefore, there is a need to assess cognitive workload of control room operators objectively.

Advancements in sensory technology such as eye tracking, electroencephalography (EEG), electrocardiography (ECG) and electrodermal activity (EDA) have made it possible to tap on physiological measures as surrogates of cognitive performance (especially cognitive workload). Physiological measures are intrinsic and provide critical insights into covert factors that shape operator performance. In this paper, we discuss how fusion of gaze and pupil-based metrics obtained using eye-tracking can be employed to better characterize cognitive workload of control room operators while they control process abnormalities.

2. Literature

Researchers in safety critical domains such as aviation, driving and healthcare industries have attempted to understand cognitive workload using eye-tracking due to its non-invasive nature and ease of use. Eye tracking involves tracing the eyes of a person with respect to their head in order to obtain the coordinates of their gaze on an human machine interface (HMI). Eye tracking provides critical insights into visual behavior. Since about 50% of the brain's neural pathways are used for visual processing (Woo, 2019), visual behavior is important to understand cognitive aspects of performance. For instance, during stressful emergency accident situations, expert nuclear power plant operators have a different eye gaze pattern, marked by higher total fixation durations compared to novices (Liu et al., 2019); higher fixation duration reflecting the ability to extract more relevant information. Pupil diameter obtained using eye-tracking also provides crucial information about human cognitive performance. For instance, during simulated surgical tasks, it was found that higher pupil dilations corresponded to increase in complexity of the tasks (Zheng et al., 2015). Experiments conducted to assess air traffic controllers' workload revealed that increase in subjective workload translated into an increase in the amplitude of pupil diameter (Rodríguez et al., 2015). Another study revealed that changes in mental workload can be captured by frequency domain analysis of pupil diameter measurements (Peysakhovich et al., 2015).

Various physiological measures have also been fused for use in machine learning models for classifying operator mental states. For instance, eye gaze and EEG based features have been fused to develop models for differentiating between groups of learners: those who could successfully resolve a problem-solving task involving medical diagnosis via an HMI, and those who could not (Khedher et al., 2019). The study revealed that fusion of features resulted in an improved accuracy of 75% in identifying learners and non-learners. It was observed that using only set of features resulted in poor performance (18% using eye features only and 0% using EEG only) in correctly classifying failure cases (non-learners). Ding et al. (2020) found that fusion of sensors corresponding to ECG and EDA resulted in an accuracy of around 58% for classification of mental workload using neural networks. Similarly, four types of feature sets based on EDA, ECG, respiration, and blood oxygen saturation were used in a study to discriminate between high and low levels of workload corresponding to two different levels of difficulty induced during three different cognitive activities (Game, Numerical test, and Raven's test) (Hirachan et al., 2022). The authors report that respiration and Sp02 based features resulted in individual accuracies of 60% and 56% respectively. There have been a few studies focusing on multi-level classification of workload as well. For instance, McDonald et al., (2020) reported an accuracy in the range of 24% to 56% using several machine learning algorithms for classifying driver distraction based on physiological data. The focus of these studies has primarily been on simpler tasks which do not evolve dynamically.

Process industry operator's role involves dealing with tasks which evolve dynamically and the demands on the operator also vary in response to their actions. The disturbances in chemical processes primarily originate from within the plant rather than because of external environment such as nature of numeric calculations or traffic conditions. This makes the cognitive requirements of control room operators unique (Müller and Oehm, 2019) and makes it challenging to characterize the same. Our previous research on reveals that gaze analysis and pupil diameter can provide critical insights into cognitive aspects of control room operator's performance. For instance, gaze entropy can capture the differences in eye-gaze behaviour between novices and experts (Iqbal et al., 2018). The study found that with learning gaze entropy decreases reflecting increase in expertise owing to decreased cognitive workload. Pupillometry studies indicate that the diameter of the pupil correlates closely with operator's ability to successfully perform the tasks involving bringing process within safe limits (Bhavsar et al., 2015). Likewise, fixation duration and saccade duration can capture changes in operator mental states (decision making) during such tasks (Das et al., 2018). In this paper, we develop a decision tree model that can estimate cognitive workload of control room operators, *at three levels*, using fusion of gaze-based and pupil-based features.

3. Experimental Methodology

We used a simulated ethanol process as an experimental testbed for conducting human subject studies. The HMI via which the operators interact with the process is shown in Fig. 1. Ten participants were involved in the study, each carrying out several repetitive trials. During a typical trial, a participant carried out six tasks (each corresponding to a disturbance scenario). The role of the operator is to bring the process variables within acceptable limits in case of a disturbance which is notified to the operator through an alarm and change in color of the tags (of disturbed variables). During these tasks, participants' eye gaze data and pupil diameter are recorded using an eye-tracker (Tobii 300). In addition, process data, alarm information, and operator actions are also recorded.

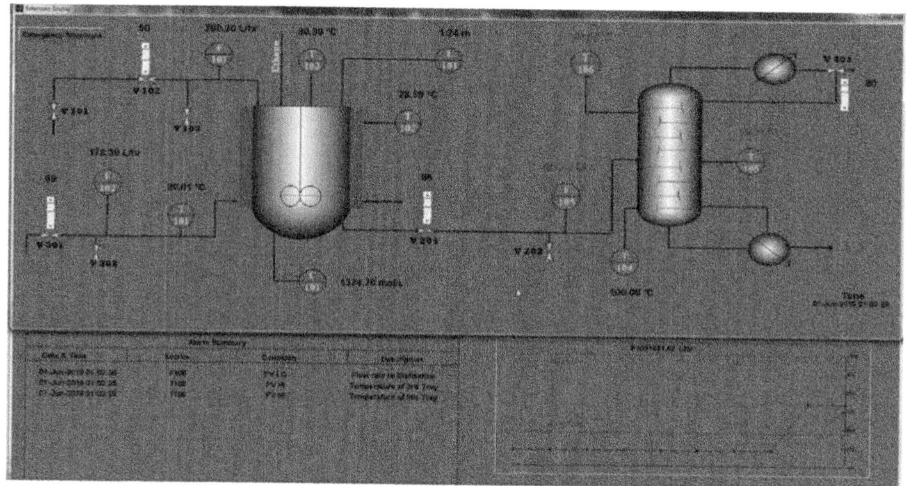

Fig. 1 Human Machine Interface (HMI) of ethanol process

From this information, we calculate integrated absolute abnormality (IAA) which measures the deviation of process variables from acceptable safe limits. We use IAA along with success/failure in a task as a surrogate to classify operators' cognitive workload into three levels- low, medium, and high. Low values of IAA correspond to low workload as these are successfully completed by the operators. On the other hand, if IAA during a task was much higher, operators failed in most of those tasks. Intermediate values of IAA correspond to low workload. Next, pupil and eye gaze-based metrics are extracted for each task (after alarm occurrence) to populate a feature matrix. These metrics for pupil include power spectral densities of pupil diameter and average pupil diameter. For gaze, these include various fixation-based metrics. A summary of these metrics is provided in Table 1. We trained decision tree models based on i) pupil-based features only ii) gaze-based features only and, iii) fusion of pupil and gaze-based features to estimate the cognitive workload of operators as well as assess the impact of fusion.

Table 1. Summary of feature categories

S. NO.	Feature	Feature Description
1	Pupil	P1: PSD of pupil diameter signal in the range 0 – 0.5Hz P2: PSD of pupil diameter signal in the range 0.5 – 1 Hz P3: PSD of pupil diameter signal in the range 1 – 1.5 Hz P4: PSD of pupil diameter signal in the range 1.5 – 2 Hz P5: Average pupil diameter
2	Gaze	G1: Longest Fixation duration/shortest Fixation duration G2: Normalized Fixation Number i.e., No. of fixations/time G3: Normalized Fixation Duration: Total fixation duration/time G4: Standard deviation of fixation duration G5: Mean of fixation duration

4. Results

Consider an operator O1 who performs eight trials of the experiment resulting in a total of 48 tasks. For two of the tasks, some data was missing, and therefore are not considered. Each task is labelled based on the value of integral absolute abnormality (IAA) calculated for post alarm phase of the task as discussed in Section 3.

Table 2. Cognitive workload classification accuracies of trained decision tree model

S. NO.	Feature Combination	Overall Accuracy (%)
1	Pupil only	60.87
2	Gaze only	50.00
3	Pupil + Gaze	65.22

As per the methodology discussed in Section 3, eye tracking based features are obtained for all the tasks that this operator performed. These features are then used as an input to train a decision tree and the model is tested using a 10-fold cross validation technique. It involves randomly dividing the data into 10 parts and then using 9 parts to train the model and test it on the remaining 10%. The procedure is repeated 10 times each time reserving a different one tenth of data for testing. The overall accuracies achieved using different combinations of features for operator O1 are shown in Table 2, and the results clearly highlight the benefits of the fusion of gaze and pupil-based features. Fusion of features results in an increase in classification accuracies to about 65%.

A representative decision tree for O1 trained using fusion of features is shown in Fig. 2(a). The topmost node is the root node and consists of pupil and gaze features for all the 46 tasks that the operator performed i.e., it contains a mix of features which correspond to all the levels of workload (0,1, and 2). The first split is made based on feature P4 (Power spectral density of pupil diameter in the range 1.5 to 2 Hz). This indicates that power spectral density of pupil diameter in range 1.5 to 2Hz is an important predictor of cognitive workload for this participant. For the observations for which P4 is less than 0.55 they are allocated to node 2 (on the left) otherwise to node 3. At node 2, the split is made based on G2 (Normalized Fixation Number i.e., No. of fixations/time). For the observations for which G2 is less than 0.44 they are allocated to leaf node representing workload level of 2 (medium workload). If G2 is greater than 0.22, the observations are allocated to leaf node representing workload level of 3. In a similar way, data at node 3 is split based on the learnt rules till leaf nodes are arrived. Leaf nodes represent the predictions of the target variable which in this case is cognitive workload. The confusion matrix for the decision tree trained using fusion of gaze and pupil-based features is shown in Fig. 2(b). The confusion matrix reveals that decision tree trained on fusion of pupil and gaze-based features can classify high workload levels (class 3) with an accuracy of 75%. Thus, the algorithm performs well in identifying poor performances that can pose safety challenges, and can be used to gauge adequacy of training.

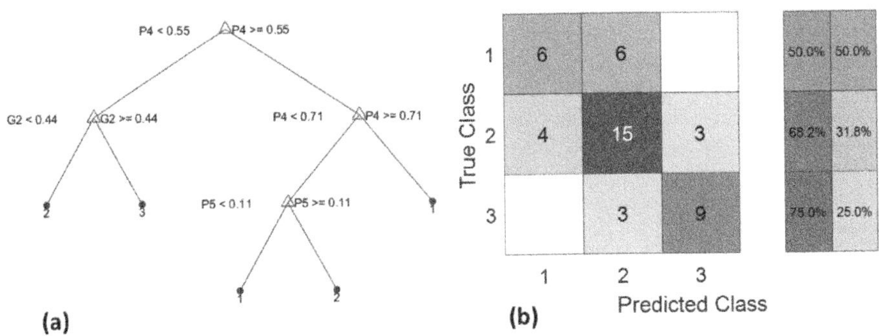

Fig. 2 a) Decision tree trained for operator O1 trained using both pupil and gaze-based features b) Confusion matrix for three levels of cognitive workload

5. Conclusion

In this work, we demonstrated how fusion of different categories of features (gaze and pupil) can improve classification accuracies of control room operator's performance as measured by cognitive workload—a precursor to sub-optimal performance. This is the first work of its kind, as it involves multi-level classification of process industry operator performance. The overall achieved accuracy of the model is 65.22% while it can identify poor operator performance (highest workload) with an accuracy of 75% (critical for ensuring adequacy of training). These accuracies are fairly good given the complexities and dynamic aspects of process industry tasks where operator performance is governed by the correctness of the mental models. Similar accuracies (57% for generic model) have been reported for nuclear operators (Braarud et al., 2021). This work has the potential to evaluate operator's expertise, and is thus crucial for ensuring safety in process industries. It also has the potential to guide the application of human factors by providing objective information about operator's expertise and (indirectly) impact of HMI design. In future, we plan to increase the robustness of our model by employing fusion of eye-tracking with electroencephalography (EEG) based features, and identifying other precursors of suboptimal performance. Our previous studies (Iqbal et al., 2020) reveal that EEG based-metrics have the potential to capture performance of control room operators.

References

Braarud, P. Ø., Bodal, T., Hulsund, J. E., Louka, M. N., Nihlwing, C., Nystad, E., ... & Wingstedt, E. (2021). An investigation of speech features, plant system Alarms, and operator–system interaction for the classification of operator cognitive workload during dynamic work. Human factors, 63(5), 736-756.

Das, L., Iqbal, M. U., Bhavsar, P., Srinivasan, B., & Srinivasan, R. (2018). Toward preventing accidents in process industries by inferring the cognitive state of control room operators through eye tracking. ACS Sustainable Chemistry & Engineering, 6(2), 2517-2528.

Iqbal, M. U., Srinivasan, B., & Srinivasan, R. (2018). Towards obviating human errors in real-time through eye tracking. In Computer Aided Chemical Engineering (Vol. 43, pp. 1189-1194). Elsevier.

Iqbal, M. U., Srinivasan, B., & Srinivasan, R. (2020). Dynamic assessment of control room operator's cognitive workload using Electroencephalography (EEG). Computers & Chemical Engineering, 141, 106726.

Jung, S., Woo, J., & Kang, C. (2020). Analysis of severe industrial accidents caused by hazardous chemicals in South Korea from January 2008 to June 2018. Saf. Sci. 124, 104580

Khan, F., Amyotte, P., & Adedigba, S. (2021). Process safety concerns in process system digitalization. Education for Chemical Engineers, 34, 33-46.

Liu, Z., Zhao, Q., Zhang, L., Zhang, X., Fan, J., Wang, Q., & Wu, P. (2021). Quantitative Evaluation on the Effect of Experience Under Emergency Situations in NPP Main Control Room Based on Multimodal Data. Nuclear Technology, 207(4), 575-581.

Müller, R., & Oehm, L. (2019). Process industries versus discrete processing: How system characteristics affect operator tasks. Cognition, Technology & Work, 21(2), 337-356.

Parasuraman, R., Sheridan, T. B., & Wickens, C. D. (2008). Situation awareness, mental workload, and trust in automation: Viable, empirically supported cognitive engineering constructs. Journal of Cognitive Engineering and Decision Making, 2(2), 140-160.

Peysakhovich, V., Causse, M., Scannella, S. and Dehais, F., 2015. Frequency analysis of a task-evoked pupillary response: Luminance-independent measure of mental effort. International Journal of Psychophysiology, 97(1), pp.30-37.

Rodríguez, S., Sánchez, L., López, P., & Cañas, J. J. (2015, September). Pupillometry to assess Air Traffic Controller workload through the Mental Workload Model. In Proceedings of the 5th International Conference on Application and Theory of Automation in Command and Control Systems (pp. 95-104). ACM.

Antonis Kokossis, Michael C. Georgiadis, Efstratios N. Pistikopoulos (Eds.)
PROCEEDINGS OF THE 33rd European Symposium on Computer Aided Process Engineering (ESCAPE33), June 18-21, 2023, Athens, Greece
 http://dx.doi.org/10.1016/B978-0-443-15274-0.50276-6

Quantification of Process Operability Using Flexibility Analysis

Fei Zhao,[a*] Chentao Mei,[b] Lingyu Zhu,[b] Xi Chen[a]

[a]*State Key Laboratory of Industrial Control Technology, College of Control Science and Engineering, Zhejiang University, Hangzhou, Zhejiang, 310027, China*
[b]*College of Chemical Engineering, Zhejiang University of Technology, Hangzhou, Zhejiang, 310014, China*
zhaofeizju@zju.edu.cn

Abstract

Industrial plants are commonly subject to uncertainties during operation. Operability and flexibility are two key properties for design models with uncertain parameters. In this study, flexibility analysis is first considered, where a derivative-free optimization method based on flexibility index is designed to identify the largest rectangular flexibility space, while the optimal nominal point of uncertain parameters can be obtained. Then, a novel operability analysis method is proposed to quantify the entire system operability. Based on the critical points of uncertainties in the flexibility space, a line projection method is adopted to locate boundary points of the operational space. The volume of the operational space is calculated by the Delaunay Triangulation method, and the ratio to the physical operational ranges can be used to assess the system operability. The case study shows the proposed method is capable of dealing with the nonconvex operational spaces.
Keywords: Process operability; flexibility analysis; flexibility index; derivative-free optimization.

1. Introduction

Uncertainties are process disturbances related to variability of process parameters, which represent the degrees of freedom or manipulated variables in an industrial process, such as feed flow rates and temperatures. The uncertainties can lead to mismatch between a plant and the corresponding design model. Flexibility analysis method was proposed as a quantitative framework for measuring the capability of feasible operations (Floudas, et al., 2001; Zhao, et al., 2022), where the uncertain parameters are measured; the control variables can be manipulated to compensate any variation of uncertain parameters during operation. Flexibility index (Grossmann and Floudas, 1987) is a well-known quantitative indicator, which can define a flexibility space with a specific shape, such as a rectangle.
Apart from the process flexibility, analyzing the process operability (Gazzaneo, et al., 2020) is another indispensable part of process design. Lima et al. (2010) summarized the similarities and differences between the process flexibility and process operability in great detail. The traditional process operability focuses on the examination of whether a process controller might be able to meet all the quality requirements due to limitations of the process design, in other words, whether all quality outputs are operable in the presence of disturbances; by comparison, the process flexibility is concerned with whether the process is feasible for all the desirable quality outputs due to the uncertainties.
Flexibility and operability are two closely linked concepts, which have been recognized to represent two important features in the production process. In a word, with the presence

of uncertainties, the process flexibility quantifies the ability of ensuring the feasibility of quality constraints, and the process operability represents the system implementability of the corresponding flexibility. The process flexibility and operability should be assessed together. The motivation of this study is to quantify the entire system operability.

2. Flexibility space description through the DFO solution strategy

For a design model, the flexibility constraint is defined by the following logic expression.

$$\forall \boldsymbol{\theta} \in FS(\boldsymbol{\theta})\{\exists \boldsymbol{z} \in T_z\{\forall j \in J[g_j(\boldsymbol{\theta},\boldsymbol{z},\boldsymbol{x}) \leq 0], \forall i \in I[h_i(\boldsymbol{\theta},\boldsymbol{z},\boldsymbol{x}) = 0]\}\} \quad (1)$$

where $\boldsymbol{\theta}$ and $\boldsymbol{x}$ represent uncertain parameters and state variables, respectively, and the control variables z represents the degrees of freedom that can be adjusted during operation to offset different realization of $\boldsymbol{\theta}$. The specified set of uncertain parameters and the given physically adjustable ranges of control variables are described by lower and upper bounds. The expected deviations along the negative and positive directions, $\Delta\boldsymbol{\theta}^-$ and $\Delta\boldsymbol{\theta}^+$, can be defined as $(\boldsymbol{\theta}^U - \boldsymbol{\theta}^L)/2$. The flexibility index F can be calculated by the maximum value of δ in the following semi-infinite programming problem.

$$\begin{aligned}
F(\boldsymbol{\theta}^N) &= \max_{\delta\in\mathbb{R}^+} \delta \\
s.t.\ & \max_{\boldsymbol{\theta}} \min_{\boldsymbol{z}} \max_{j\in J} g_j(\boldsymbol{\theta},\boldsymbol{z},\boldsymbol{x}) \leq 0 \\
& h_i(\boldsymbol{\theta},\boldsymbol{z},\boldsymbol{x}) = 0, \ \ \forall i \in I \\
& \boldsymbol{\theta}^N - \delta\Delta\boldsymbol{\theta}^- \leq \boldsymbol{\theta} \leq \boldsymbol{\theta}^N + \delta\Delta\boldsymbol{\theta}^+ \\
& \boldsymbol{z}^L \leq \boldsymbol{z} \leq \boldsymbol{z}^U
\end{aligned} \quad (2)$$

Grossmann and Floudas (1987) proposed an active constraint strategy, where the lower-level problem is reformulated by the Karush-Kuhn-Tucker (KKT) conditions. Hence, the multilevel optimization problem can be converted into a single-level mixed-integer linear or nonlinear programming (MILP/MINLP) problem, which is given by,

$$\begin{aligned}
F(\boldsymbol{\theta}^N) &= \min \delta \\
s.t.\ & g_j(\boldsymbol{\theta},\boldsymbol{z},\boldsymbol{x}) + s_j = 0, \ \ \forall j \in J \\
& h_i(\boldsymbol{\theta},\boldsymbol{z},\boldsymbol{x}) = 0, \ \ \forall i \in I \\
& \textstyle\sum_j \lambda_j \cdot \frac{\partial g_j}{\partial x_k} + \sum_i \mu_i \cdot \frac{\partial h_i}{\partial x_k} = 0, \ \ \forall k \in K \\
& \textstyle\sum_j \lambda_j \cdot \frac{\partial g_j}{\partial z_k} + \sum_i \mu_i \cdot \frac{\partial h_i}{\partial z_k} = 0, \ \ \forall k \in K \\
& \textstyle\sum_j \lambda_j = 1 \\
& \textstyle\sum_j y_j = n_z + 1 \\
& s_j - M(1 - y_j) \leq 0, \ \ \forall j \in J \\
& \lambda_j - y_j \leq 0, \ \ \forall j \in J \\
& \boldsymbol{\theta}^N - \delta\Delta\boldsymbol{\theta}^- \leq \boldsymbol{\theta} \leq \boldsymbol{\theta}^N + \delta\Delta\boldsymbol{\theta}^+ \\
& \boldsymbol{z}^L \leq \boldsymbol{z} \leq \boldsymbol{z}^U \\
& \lambda_j \geq 0, s_j \geq 0, y_j \in \{0,1\}, \forall j \in J
\end{aligned} \quad (3)$$

Moreover, if $\boldsymbol{\theta}^N$ is unknown, the above flexibility index problem can be extended to a design centering problem shown in Equation (4), which indicates that the flexibility index model is applied for candidate nominal points, $\boldsymbol{\theta}^N$, which are regarded as the decision variables of the outer-level optimization problem.

$$\begin{aligned} &\max_{\boldsymbol{\theta}^N} \delta \\ &s.t.\ g_j(\boldsymbol{\theta}^N, \mathbf{z}, \boldsymbol{x}) \le 0,\ \ \forall j \in J \\ &\quad h_i(\boldsymbol{\theta}^N, \mathbf{z}, \boldsymbol{x}) = 0,\ \ \forall i \in I \\ &\quad \boldsymbol{\theta}^L \le \boldsymbol{\theta}^N \le \boldsymbol{\theta}^U \\ &\quad \text{Equation (3)} \end{aligned} \tag{4}$$

The feasibility of $\boldsymbol{\theta}^N$ should be guaranteed during the whole optimization process. Due to the presence of the state variables $\boldsymbol{x}$ and control variables $\mathbf{z}$, the maximum constraint violation (MCV) of all the inequality constraints should be less than or equal to zero, i.e., $MCV(\boldsymbol{\theta}^N) \le 0$.

$$\begin{aligned} &MCV(\boldsymbol{\theta}^N) = \min u \\ &s.t.\ g_j(\boldsymbol{\theta}^N, \mathbf{z}, \boldsymbol{x}) \le u,\ \ \forall j \in J \\ &\quad h_i(\boldsymbol{\theta}^N, \mathbf{z}, \boldsymbol{x}) = 0,\ \ \forall i \in I \\ &\quad \mathbf{z}^L \le \mathbf{z} \le \mathbf{z}^U \end{aligned} \tag{5}$$

After moving $MCV(\boldsymbol{\theta}^N)$ to the objective function and introducing a penalty coefficient M, the DFO formulation of design centering problem can be written as

$$\begin{aligned} &\min_{\boldsymbol{\theta}^N} -F(\boldsymbol{\theta}^N) + M \cdot \max\left(0, MCV(\boldsymbol{\theta}^N)\right) \\ &s.t.\ \boldsymbol{\theta}^L \le \boldsymbol{\theta}^N \le \boldsymbol{\theta}^U \end{aligned} \tag{6}$$

To increase the likelihood of obtaining the largest flexibility space, the Latin hypercube sampling (LHS) strategy is applied to generate a set of points in the uncertain parameter space. Each feasible LHS point will be treated as a starting point for the DFO solution procedure. Finally, the maximum value of δ, i.e., the flexibility index, corresponds to the final rectangular flexibility space, which can be described as

$$\theta_p^- \le \theta_p \le \theta_p^+,\ \ \forall p \in P \tag{7}$$

where θ_p^- and θ_p^+ represent the largest allowable range for each uncertain parameter, $p \in P$. Thus, the vertex set of this rectangular space contains 2^P points, i.e.,

$$\text{Verts} := \left\{\{\theta_1^-, \theta_2^-\} \times \{\theta_1^-, \theta_2^+\} \times \cdots \times \{\theta_{P-1}^+, \theta_P^+\}\right\} \tag{8}$$

The optimal nominal point is $\boldsymbol{\theta}_{op}^N := (\theta_1^N, \ldots, \theta_P^N)$, where $\theta_p^N = \left(\theta_p^- + \theta_p^+\right)/2$. Together with the feasible LHS points sampled in the rectangular flexibility space, denoted as FPs, we can define a critical point set CP, i.e.,

$$CP := \left\{\text{Verts}, \boldsymbol{\theta}_{op}^N, \text{FPs}\right\} \tag{9}$$

These chosen points are representative for reflecting the system flexibility, especially for the vertices, which may reflect the limitation of operations.

3. Quantificational description of operational space

Assuming that the control variables can be adjusted continuously within the physical ranges. For each candidate $\boldsymbol{\theta}_k$ in the critical set CP, the corresponding operational space is the feasible region of the following system with respect to z.

$$\begin{cases} g_j(\boldsymbol{\theta}_k, \mathbf{z}, \boldsymbol{x}) \le 0,\ \ \forall j \in J \\ h_i(\boldsymbol{\theta}_k, \mathbf{z}, \boldsymbol{x}) = 0,\ \ \forall i \in I \\ \mathbf{z}^L \le \mathbf{z} \le \mathbf{z}^U \end{cases} \tag{10}$$

The solution strategy to quantifying the system operability mainly contains three steps:

(1) *Locate boundary points of operational space*

For each point of uncertain parameters in CP, given a sampling number, the LHS method is performed over the physical ranges of control variables. All the feasible LHS points that can satisfy Equation (10) form a set **A**. In this work, we focus on exploring points on the boundary of the operational space. A line projection method is adopted to efficiently locate boundary points. For example, through a pair of points in **A**, e.g., $\mathbf{z}_1$ and $\mathbf{z}_2$, there are two directions, $\mathbf{z}_1 \rightarrow \mathbf{z}_2$ and $\mathbf{z}_2 \rightarrow \mathbf{z}_1$. We can locate two boundary points by solving the maximization problems. The global optimization solver, e.g., GAMS/BARON, is applied to guarantee the solutions located on the boundary. After enumerating the pairs of points in the set **A**, all the obtained boundary points define the set **B**. The more LHS points are specified, the more boundary points we can obtain.

(2) *Delaunay triangulation for the operational space*

The boundary points illustrate the profile of the operational space, and the nonconvex parts also can be exactly captured as long as the boundary points are enough. In this work, in order to provide an effective way to quantify the operational space, a triangle meshing technique using the boundary points is proposed, which is implemented by the Delaunay triangulation method. Quickhull (Barber, 1996) is one of the most efficient methods for high-dimensional triangulations, which was developed based on the convex hull methods. As a numerical implementation of the Quickhull method in SciPy, Qhull is applied to execute the Delaunay triangulation.

(3) *Quantification of the operational space*

Since the Quickhull method creates a simplex mesh of a convex hull for a set of points, some tringles may be generated beyond the nonconvex feasible space. Thus, we should remove these infeasible tringles through checking the feasibility. In $\mathbb{R}^d$, the centroid of the kth simplex, $\bar{\mathbf{z}}^k$, can be simply calculated based on its vertex coordinates.

$$\bar{\mathbf{z}}^k = \frac{1}{n+1}\sum_{i=1}^{n+1} \boldsymbol{x}_i^k \tag{11}$$

where $\boldsymbol{x}_i^k$ is the coordinate vector for vertex i of simplex k, n denotes the dimension. We can check the feasibility of the centroid. Thus, the nonconvex parts can be identified, and we can remove the infeasible meshes. Benefit from the Delaunay triangulation and the feasibility tests, the triangle meshing of the entire operational space is obtained. Then, we can calculate the area or volume of each simplex (triangle or polyhedron).

$$V^k = \frac{1}{n!}\left|det[(x_1^k - x_0^k)^T(x_2^k - x_0^k)^T \ldots (x_n^k - x_0^k)^T]\right| \tag{12}$$

where, k is the index of the simplex, $k \in K$. Each column of the n-by-n matrix is the transpose of difference between two row vectors representing vertex $i(i = 1, 2, \ldots, n)$ and the reference vertex x_0^k, respectively. The reference vertex here defaults to the first point of each mesh. Moreover, the initial operable space is a rectangular space, and the volume can be calculated by the given adjustable ranges of the control variables, i.e,

$$V_g = \prod(\mathbf{z}^U - \mathbf{z}^L) \tag{13}$$

Thus, we can define the system operability by the following operable space ratio.

$$FR = \frac{\sum_k^K V^k}{V_g} \tag{14}$$

4. Case study

Consider the following inequalities,

$$\begin{cases} 2(z_2 - 2)(z_1 - 2) - \theta_1 \leq 0 \\ (z_2 - 2)^2(z_1 - 2)^2 - \theta_2 \leq 0 \\ 0 \leq \theta_1 \leq 2, 0 \leq \theta_2 \leq 2 \\ 0 \leq z_1 \leq 4, 0 \leq z_2 \leq 4 \end{cases}$$

where θ_1 and θ_2 can be regarded as the uncertain parameters, and z_1 and z_2 are the control variables. In order to evaluate the operability, the largest flexibility space of the uncertain parameters should be first identified. According to Equation (6), the flexibility index is calculated as $F = 1$, which means that the flexibility space is identical to the given region, and the optimal nominal point of θ_1 and θ_2 is (1, 1).

Next, the operational space can be evaluated based on this flexibility space. Due to the nonlinearity and nonconvexity, it is generally difficult to describe the complete boundary of the operational space. According to the proposed quantification method, the profile of the operational space can be captured through the line projection approach. To fully assess the overall operability, we need describe the operational space at a critical point set of uncertain parameters. According to Equation (9), four vertices, the nominal point and two feasible LHS points are chosen as *CP*, i.e., {(0, 0), (0, 2), (2, 0), (2, 2), (1.0, 1.0), (1.25089, 0.70663), (0.67603, 1.61001)}. Taking the point (2, 2) in *CP* as an example, as shown in Figure 1(a), the gray and green points are the initial and feasible LHS points in the space of control variable, respectively. Along any pair of feasible points, two boundary points can be located. Figure 1(a) also shows that the nonconvex parts can be identified. Note that the result is greatly influenced by the LHS points. The more LHS points are specified, the more likely we are to find the exact boundary. Figure 1(b) shows the feasible LHS points and all the boundary points that are identified at seven points of uncertainties in $[\mathbf{z}^L, \mathbf{z}^U]$. which indicates that seven operational spaces are overlapping, and also can verify that different uncertain parameters can correspond to different operational spaces. The outermost points depict the boundary of the entire operational space. It is worth noting that, there is only one operational point for (0, 0) and (2, 0), i.e., z_1=2, z_2=2. Because the operational space only contains discrete points, i.e., there is no operating margin, it is very likely to lead the system uncontrollable due to unknown noise. Thus, we need avoid these uncertainties if possible.

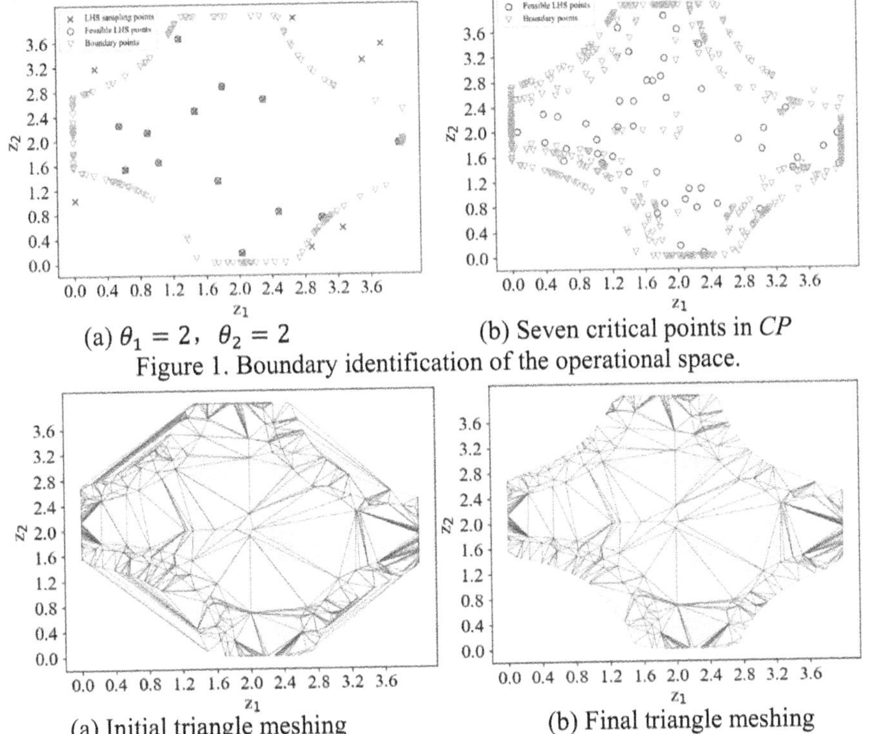

(a) $\theta_1 = 2,\ \theta_2 = 2$ (b) Seven critical points in *CP*

Figure 1. Boundary identification of the operational space.

(a) Initial triangle meshing (b) Final triangle meshing

Figure 2. Delaunay triangulation of the nonconvex operational space.

Then, based on all the boundary points, the convex hull including the entire operational space can be meshed through the Delaunay triangulation method, as shown in Figure 2(a). Due to the nonconvexity, the convex hull contains many infeasible parts. Equation (11) can calculate the centroid for each triangle mesh. A feasibility check can be performed by evaluating Equation (10) at the centroid of each mesh, and the infeasible meshes can be identified and then removed, as shown in Figure 2(b). Finally, the total area of the operational space can be calculated as 10.542 by Equation (12). Based on the given ranges, the area of the given operational space V_g is 16; thus, the operable space ratio FR is 0.659.

5. Conclusions

Variability consistently exists in industrial process plants. Flexibility and operability are two closely linked concepts, which are recognized to represent two important features in the production process. The contribution of this work is to connect the flexibility space, the operability space and all the quality requirements closely.

In this article, a novel process operability analysis method is proposed to quantify the system operability. To assess the process flexibility and operability together, the process flexibility is designed as a prerequisite for operability analysis, which can capture the relationships between uncertain process parameters and the quality attributes, and all the specified quality constraints can be satisfied. Then, a line projection method is proposed to directly identify the boundary of the operational space, where some critical realizations of uncertainties in the rectangular flexibility space are considered. The boundary points are applied to calculate the area or volume of the operational space through the Delaunay triangulation method, and the ratio to the given physical operational ranges can assess the operability of the plant.

Challenges in analyzing the process flexibility and operability simultaneously come from the coupling relationships between the uncertain parameters and control variables. The larger flexibility and operability reflect the better performance of the design model. This work illustrates that the larger operable space ratio, the better operability will be. Thus, the proposed method has potential to guide the improvement of the process design if the operability does not meet the requirements. Once the operational space is known, we not only can provide a robust operating strategy to tackle a wide of uncertainties, but also can analyze the operating limitations. Moreover, benefit from the powerful optimization tools, the proposed method is applicable for any scales of process units or systems; however, currently only steady-state process models can be deal with, and the dynamic models are being researching.

Acknowledgements

The authors gratefully acknowledge the financial support from the National Key Research and Development Program of China (2022YFB3305901), Zhejiang Provincial Natural Science Foundation of China (LZ21B060001), and the Fundamental Research Funds for the Central Universities (226-2022-00038 and 226-2022-00086).

References

C.A. Floudas, Z.H. Gümüş, M.G. Ierapetritou, 2001, Global optimization in design under uncertainty: feasibility test and flexibility index problems, Industrial & Engineering Chemistry Research, 40, 20, 4267-4282.

C.B. Barber, D.P. Dobkin, H. Huhdanpaa, 1996, The quickhull algorithm for convex hulls. ACM Transactions on Mathematical Software, 22, 4, 469-483.

F.V. Lima, Z. Jia, M. Ierapetritou, C. Georgakis, 2010, Similarities and differences between the concepts of operability and flexibility: the steady-state case, AIChE Journal, 56, 3, 702-716.

F. Zhao, M.P. Ochoa, I.E. Grossmann, S. García-Muñoz, S.D. Stamatis, 2022, Novel formulations of flexibility index and design centering for design space definition. Computers and Chemical Engineering, 166, 107969.

I.E. Grossmann, C.A. Floudas, 1987, Active constraint strategy for flexibility analysis in chemical processes. Computers and Chemical Engineering, 11, 6, 675-693.

V. Gazzaneo, J.C. Carrasco, D.R. Vinson, F.V. Lima, 2020, Process operability algorithms: past, present, and future developments. Industrial & Engineering Chemistry Research, 59, 2457-2470.

Antonis Kokossis, Michael C. Georgiadis, Efstratios N. Pistikopoulos (Eds.)
PROCEEDINGS OF THE 33rd European Symposium on Computer Aided Process Engineering (ESCAPE33), June 18-21, 2023, Athens, Greece
 http://dx.doi.org/10.1016/B978-0-443-15274-0.50277-8

Coupling of support vector machine and acoustic models to characterize the droplet size distribution of emulsions using ultrasonic techniques

Samuel V. Saraiva, Carlos A. M. Silva, Darlan Bonetti, Ana M. F. Fileti, e Flávio V. Silva.

School of Chemical Engineering, University of Campinas, Campinas, Brazil.

Abstract

The application of ultrasonic techniques to characterize emulsions concerning the droplet size distribution (DSD) has attracted interest in recent years, mainly due to the advantage of being a non-intrusive technique able to characterize concentrated and opaque emulsions. These characterizations occur by adjusting sound attenuation spectroscopy with the acoustic models. The most used models for this purpose are simplifying more complete models, such as ECAH, mainly due to their mathematical complexity and convergence problems. However, most of these simplified models have restrictions related to the range of applications in the wave propagation regime. Therefore, the objective of this work was to couple the acoustic models to the support vector machines (SVM), which have numerous advantages linked to their successful adaptability to nonlinear data, to acting on this limitation, thus improving the determination of the emulsion DSD. This coupling proved to be satisfactory since the SVM classified all data correctly.

Keywords: support vector machine; acoustic models; droplet size distribution.

1. Introduction

The use of ultrasonic techniques to characterize fluids is useful in a wide range of applications, both on a laboratory and industrial scale. Compared to other techniques, ultrasound has the advantages of being a non-invasive and non-intrusive technique, in addition to being able to be applied in-line for real-time measurements.

Concerning emulsion, Challis et al. (2005) present an extensive review of what has been developed in terms of cell construction and acoustic model development. The acoustic models describe the physical phenomena of ultrasound propagation in emulsions in terms of intrinsic and viscous sound dissipation and through the acoustic scattering of waves incident on the drops. The complete model developed to characterize emulsions via ultrasound was developed by Epstein & Carhart and Allegra & Hawley (ECAH), explained in detail by Challis et al. (2005). The authors also point out that, despite being the most complete, this model has convergence problems.

Silva et al. (2022) studied the application of several simplified models to characterize concentrated water-in-edible oil emulsions, comparing the DSD results obtained using two different techniques, ultrasound, and light scattering. The results have shown a suitable convergence with those obtained from light scattering for situations involving monomodal and bimodal droplet size distributions. However, the choice of the most suitable model for the sample depended on the wave propagation regime.

Among the challenges addressed by the alternative acoustic models to the more complex ECAH model, one main is the wavelength regime identification. It is of fundamental importance for many practical industrial applications. When finding the regime, it is possible to alternate among more suitable models to characterize a particulate system. In this case, intelligent algorithms would be a convenient alternative to mitigate the model's restriction, as suggested by (Challis et al. 2005).

The use of intelligent algorithms based on machine learning models has stood out in many applications (Simon Haykin 2008) and has a good growth perspective for the coming years (Huang et al. 2015). Machine learning models stand out for having suitable adaptability to most of the nonlinear problems in the literature for regression and classification of patterns. Some authors report such machines as universal function approximators. Among these machines, we can emphasize one good performance addressed in the literature, the support vector machine (SVM). The SVM has advantages over other techniques, such as artificial neural networks. The main one is the small number of parameters adjusted in training, implying a smaller data set.

Therefore, the objective of the present work was to develop and investigate the use of empirical models based on SVM to act in the limitations of the acoustic model. The combination of these techniques was not found in the literature. The models based on SVM can assist in wave propagation regime classification to obtain real-time measurements without the prior need to choose specific acoustic models.

2. Theoretical foundations

2.1. Acoustic models for wave propagation in emulsion

Sound attenuation models in the emulsion can be expressed as $\alpha(w, r)$, generally in *Np/m* unit. In which *w* and *r* are the angular frequency of sound

(Hz) and droplet radius of the emulsion, respectively (m). In this context, an important dimensionless number in this analysis is the relative wavelength:

$$ka = 2\pi r/\lambda \tag{1}$$

in which r and λ are the radius of the particle and the wavelength of the ultrasound, respectively. For small particles ($ka \ll 1$), the acoustic scattering can be neglected. This regime is defined as a long-wave regime (LWR). On the other hand, for large particles, values of $ka \geq 1$, the effect of sound absorption is neglected, being defined as a short-wave regime (SWR). According to the results obtained from (Silva et al. 2022), the best model that was used in the present work for LWR was coupled phase model (Evans and Attenborough 1997), while the best model for SWR was the elastic scattering model (Feuillade and Clay 1999).

The experimental data must fit an acoustic model to obtain the DSD of the emulsion, and because of these adjustments, the distribution parameters can be calculated. This adjustment will occur by minimizing the sum of squares of errors (Equation 2).

$$Minimize \;\; J = \sum_{w_{min}}^{w_{max}} \left(\alpha_{experimental} - \int_{r_{min}}^{r_{max}} \alpha(w,r) * P(r)dr \right)^2 \tag{2}$$

P(r) is the droplet radius distribution function. Some of the most used distribution functions are unimodal and bimodal. According to some authors (Silva et al. 2022), these distributions represent most particulate systems processes. Due to greater simplicity (fewer variables to be adjusted), unimodal functions are used more in ultrasonic sensor applications. Therefore, P(r) is the log-normal distribution of the droplet radius.

2.2. Support vector machine

SVM is an intelligent technique used for both classification and regression. Developed by Vapnik (2000), SVM is based on the theory of statistical learning and the principle of minimizing the risk structure, reaching the global solution by transforming the regression or classification problem within quadratic programming. As highlighted in the literature, SVM has a sound theoretical basis and can achieve high performance for complex practical problems (Huang et al. 2015).

In a classification task, the objective of the SVM is to linearly separate the original data in a space with a larger dimensional characteristic. It was then solving a linear separation problem in this space. Therefore, given a data set $T = \{\boldsymbol{x}_i, \; y_i\}_{i=1}^{l}, (X \in Y)^l$, where $\boldsymbol{x}_i \in X = R^N$ is the input vector and $y_i \in$

$Y = R^N$ the corresponding output, and l the total number of data pairs, the SVM classification is defined by the hyperplane given by Equation (3.

$$y = f(x) = \mathbf{w}^T \varphi(\mathbf{x}) + b \tag{3}$$

In which φ is the mapping function, $\mathbf{w}$ and b represent the weight vectors and the term bias, respectively. These parameters are estimated by minimizing the regularized risk function. More details about the algorithm can be found in (Vapnik 2000).

3. Experimental procedure and apparatus

The emulsion used in the analyzes carried out in this study was distilled water in sunflower oil with 5% w/w of PGPR emulsifier. Two sets of emulsions were prepared.

The first set of emulsions was prepared to produce a mono-modal distribution with a droplet radius close to 2 µm (set 1). The second set of emulsions was produced to produce an emulsion with droplets above 10 µm (set 2). Details of the procedure used in both production and the physical-chemical properties of the oil and water used in the study can be found in Silva et al. (2022).

The DSD's quantitative analysis was performed using the light Scattering equipment (Mastersizer 2000 - Malvern Instruments Ltd., Malvern, U.K.) that calculates the distribution using the laser diffraction principle.

The system responsible for the characterization of the emulsion through ultrasound was composed of a Parametrics square wave pulse generator (5077PR model), a pair of 6mm-diameter-Parametrics ultrasonic transducers (model V113) with a 15 MHz center frequency, and a PXI-5105 measurement platform from National Instruments. Furthermore, the cell where the emulsion sample was pumped to flow was constructed by Silva et al. (2022).

4. Implemented routine and SVM coupling algorithm with acoustic models

The SVM architecture developed to classify the wave propagation regime (output) presented the dispersed phase volumetric fraction φ and the area of excess acoustic attenuation (Equation (4)) as inputs. In this case, the input is defined as $\{\phi, Area\}$ and the output is defined as the wavelength regime classification (LWR or SWR). This architecture was chosen because it presented the best performance in relation to several different architectures studied for input variables in the SVM. Emphasizing that there is a clear relationship between the excess of attenuation and the wave propagation regime for different dispersed phase fractions.

$$Area = \int_{f_{min}}^{f_{max}} (\alpha_{emulsion}(f) - \alpha_{oil}(f))df \tag{4}$$

The SVM will act previously identifying the wave propagation regime and then choosing the model that will be applied. Thus, if the SVM classifies the data as LWR, the model that presented the best performance for this regime will be chosen similarly for SWR. The strategies for coupling the SVM to the acoustic models are summarized in the flowchart of the implemented algorithm (Figure 1).

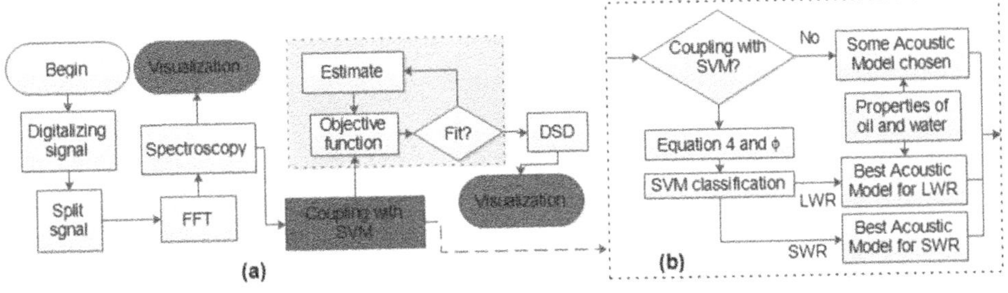

Figure 1 - Flowchart of the strategies of applying SVM.

Following the scheme shown in Figure 1, the ultrasonic signals from the transducers are digitized with the sampling rate and resolution specified previously. The signals are clipped to extract the echoes of interest. The echoes are transformed into the frequency domain, and the spectrum is extracted according to the procedure described by Silva et al. (2020). The spectrum goes to the SVM coupling flow (Figure 1b) and the optimization environment. After the adjustment, the answers are in DSD. The procedure and computational routine were developed in Python.

5. Results

5.1. Wavelength regime classification using SVM

The attenuation and experimental DSD data were subjected to calculations of the excess attenuation area (Equation 4) as a function of the dispersed phase fraction φ. It was observed from Figure 2 that there was a linear separation between the classes of wave propagation regimes. Therefore, the kernel function in the SVM to classify the regime was linear. It can be highlighted that the classification by the SVM of the wave propagation regime proved to be an adequate strategy since the regimes for this data set were separated. The results highlight that the effects of sound energy dissipation in the LRW are more attenuating than in the SWR for a fixed dispersed phase fraction, considering the water-in-oil emulsion.

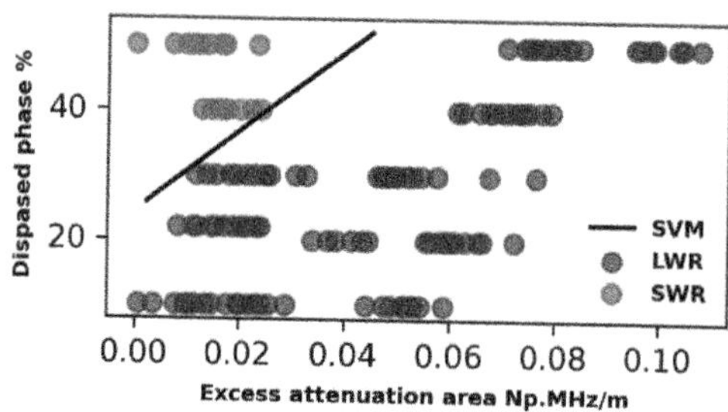

Figure 2 - Results of applying the SVM to classify the emulsions according to the wave propagation regime as a function of the excess attenuation area and the dispersed phase fraction.

5.2. Coupling of acoustic and SVM models

Figure 3 shows the development application. In this figure, three curves are shown. The first curvature refers to using the sum of the phase-coupled and Anderson models without coupling with the SVM. The second curve refers to the use of models with SVM coupling. And finally, the last curvature refers to the analyzes made through the Ligh Scattering.

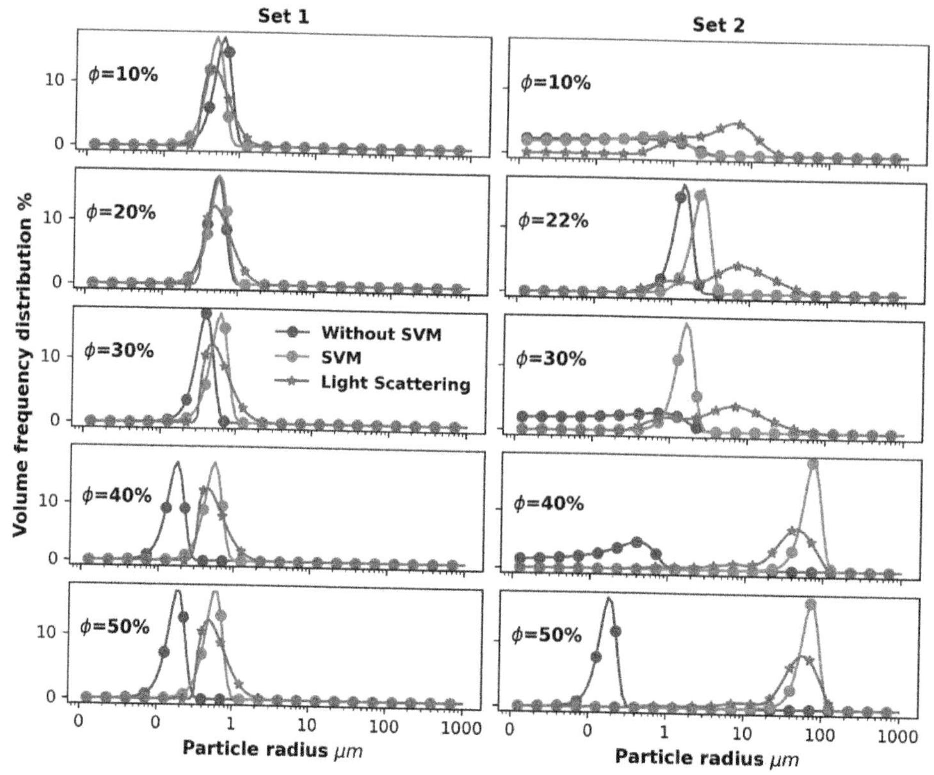

Figure 3 - DSD fitted curves from attenuation spectroscopy, and acoustic models for different phase dispersed concentrations.

6. Conclusion

The present work allowed a distinct and quick procedure for analyzing the ultrasonic signal. Distinct because it does not require prior knowledge of the wave propagation regime since the SVM will act in this limitation. Fast because it requires less computational effort to converge the model in each iteration of the optimization algorithm when compared to the ECAH model. We conclude that these strategies can be implemented in an ultrasonic sensor, provided that the physical parameters, including the volumetric fraction of each emulsion phase, are previously known.

Acknowledgment

We would like to thank Petrobras for the research financial support (Sigitec processes: 2018/00181-9 and 2017/00789-4).

References

Challis, R. E., M. J.W. Povey, M. L. Mather, and A. K. Holmes. 2005. "Ultrasound Techniques for Characterizing Colloidal Dispersions." *Reports on Progress in Physics* 68(7): 1541–1637.

Evans, J. M., and K. Attenborough. 1997. "Coupled Phase Theory for Sound Propagation in Emulsions." *The Journal of the Acoustical Society of America* 102(1): 278–82.

Feuillade, C., and C. S. Clay. 1999. "Anderson (1950) Revisited." *The Journal of the Acoustical Society of America* 106(2): 553–64.

Huang, Gao, Guang-Bin Huang, Shiji Song, and Keyou You. 2015. "Trends in Extreme Learning Machines: A Review." *Neural Networks* 61: 32–48.

Silva, Carlos A.M. et al. 2020. "Application of Acoustic Models for Polydisperse Emulsion Characterization Using Ultrasonic Spectroscopy in the Long Wavelength Regime." *Colloids and Surfaces A: Physicochemical and Engineering Aspects* 602.

———. 2022. "Measurements of Bimodal Droplet Size Distribution of Emulsions Using Ultrasonic Spectroscopy in the Long and Intermediate Wavelength Regimes." *Chemical Engineering Science* 252: 117274.

Simon Haykin. 2008. *Neural Networks and Learning Machines*. 3rd ed. Pearson.

Vapnik, Vladimir N. 2000. *The Nature of Statistical Learning Theory*. New York, NY: Springer New York.

Antonis Kokossis, Michael C. Georgiadis, Efstratios N. Pistikopoulos (Eds.)
PROCEEDINGS OF THE 33rd European Symposium on Computer Aided Process Engineering (ESCAPE33), June 18-21, 2023, Athens, Greece
 http://dx.doi.org/10.1016/B978-0-443-15274-0.50278-X

An Approach to Dependable Hybrid Modeling with Application to an Industrial Distillation Column

Mohamed Elsheikh[a], Yak Ortmanns[b], Felix Hecht[b], Volker Roßmann[b], Stefan Krämer[a,b], Sebastian Engell[a]

[a]*Process Dynamics and Operations Group, TU Dortmund, 44227 Dortmund, Germany.*
[b]*Bayer AG, 51368 Leverkusen, Germany.*
Mohamed.elsheikh@tu-dortmund.de

Abstract

The use of model-based control approaches can improve the operation of many complex processes significantly but necessitates the availability of accurate dynamic process models, and the development of such models is time-consuming which is a major bottleneck for the application of advanced control. In this paper, a hybrid modeling approach is proposed, where a simple mechanistic model is augmented by a machine-learning model to compensate the plant-model mismatch based on observed data. As the data-based models cannot be trusted outside the region in which sufficiently dense training data was available, we propose to adapt the contribution of the data-based model component based on the closeness to its estimated domain of validity. We demonstrate the benefits of the novel hybrid modeling approach via the case study of an industrial distillation column for solvent recovery.

Keywords: hybrid modeling, dynamic modeling, validity domain, distillation column

1. Introduction

The stable and efficient operation of chemical processes is of high importance which is even increasing due to the trend towards more dynamic and flexible operations caused by fluctuating markets and prices and availabilities of raw materials and energy inputs. Model-based control can improve the operation significantly, especially in the case of frequent changes of the operating points. However, for the successful application of model-based control and optimization, accurate models are needed. The development of first-principles-based models requires a significant effort and can lead to large differential-algebraic equations (DAE) models that may be difficult to converge. It is therefore of interest to reduce the modeling effort and the computation times while still accurately describing the processes. An approach to reduce the modeling effort is to adapt model parameters online, e.g. by combined state and parameter estimation, to improve the model fit locally. However this is only feasible for a small number of uncertain parameters. An alternative is the use of hybrid models, i.e. combinations of mechanistic and data-based model elements. One option is to embed surrogate models in the mechanistic model to replace complex calculations e.g. of thermodynamic equilibria, see e.g. Nentwich et al. (2019). Winz and Engell (2022) proposed a systematic method to obtain embedded data-based components of hybrid models from measured data. Another frequently employed approach to hybrid modeling is to add data-based elements to mechanistic models in a parallel structure, see e.g. Bock et al. (2021) and Azadi et al. (2022). An overview of hybrid modeling techniques is given in (Von Stosch et al., 2014). A key problem in using data-based models in online applications is the trustworthiness of such models. In general, data-based models can be trusted only in regions that are dense

in training data, called the domain of validity. Several works proposed to represent the domain of validity as box constraints of the data-based model inputs. However, these approaches tend to overestimate the domain of validity. Other approaches are based on estimating the domain of validity as the convex hull of the training data. This has the shortcoming of the inability to exclude gaps in the training data and to represent nonconvex boundaries in the training data. Schweidtmann et al. (2022) suggested a novel approach to estimate the domain of validity by training a one-class support vector machine (SVM). This approach can overcome the mentioned shortcomings.
In this work, we develop a hybrid dynamic model where we combine a simple mechanistic model and a parallel data-based model that represents the deviations of the observed plant dynamics from the mechanistic model. A one-class SVM is trained on low-dimensional projections of the training data that were obtained by principal component analysis (PCA). The output of the one-class SVM is used to weight the contribution of the data-based model to the overall model, achieving a smooth transition between the full hybrid model and the simplified mechanistic model. We investigate the potential of the novel hybrid modeling approach by the application to modeling of the dynamic behavior of an industrial mother liquor distillation column.

2. Hybrid model with a weighted contribution of the data-based model

We propose a novel approach for obtaining reliable hybrid process models that consist of a simple mechanistic model and a parallel data-driven model for capturing the deficiencies of the mechanistic model. The developed modeling approach weights the contribution of the data-based model based on the domain of validity of the data-based model. Inside the domain of validity, the full hybrid model is used and outside this domain a smooth fading out of the contribution of the data-based element is implemented. The proposed hybrid process model is defined by:

$$x_{k+1} = f_{diff}(x_k, z_k, u_k, p_k), \tag{1}$$

$$\bar{0} = f_{alg}(x_k, z_k, u_k, p_k), \tag{2}$$

$$y_{k+1} = g(x_{k+1}, z_{k+1}), \tag{3}$$

$$e_{k+1} = f_{ML}(h_k), \tag{4}$$

$$\tilde{y}_{k+1} = y_{k+1} + \gamma_{k+1} e_{k+1}, \tag{5}$$

where x_k denotes the state vector of the simplified mechanistic model, z_k represents the vector of the algebraic states of the simplified model, u_k represents the vector of control inputs, p_k the vector of varying parameters, y_k the output vector of the mechanistic model, h_k the input vector of the data-based model, e_k the vector of the mismatches between the outputs of the full model and of the simplified model and $\tilde{y}_k$ the output vector of the hybrid model. γ_k is a scalar time-depending variable that denotes the weighting factor of the data-based model, f_{diff}, f_{alg}, g, and f_{ML} are nonlinear functions.

2.1. The domain of validity of the data-based model component

For the representation of the domain of validity, a one-class SVM is trained on low-dimensional projections of the training data of the input variables of the data-based model. These projections are obtained using principal component analysis (PCA) (Jolliffe and Cadima, 2016). The projection T of the data is defined as $T = XW$, where $T \in \mathbb{R}^{N\times l}$, $X \in \mathbb{R}^{N\times p}$ is the training data, $W \in \mathbb{R}^{p\times l}$ is the loading matrix, N is the number of training samples, p is the number of input variables in training data and l is the number of the used principal components. The training data is centered and scaled before applying PCA.
One-class SVM is a special case of classical SVMs that deals with the case of having only one class of data, and a maximum-margin hyperplane is sought to separate the training

data from the origin in the feature space (Schölkopf et al., 1999). Schweidtmann et al. (2022) proposed to use the decision function $f_{DF}(h)$ of the one-class SVM to determine if a data point h belongs to the domain of validity of the data-based model by checking the inequality $f_{DF}(h) \geq 0$.

2.2. Weighting of the data-based model component

We propose to weight the data-based element of the hybrid model based on the function f_{DF} that indicates the closeness to the domain of validity such that the contribution of the data-based model to the overall model fades out in the regions without sufficient training data. The contribution factor γ_k at time step k is computed according to

$$\gamma_k = \zeta\Gamma(h_k^{s,c}) + (1-\zeta)\gamma_{k-1}, \tag{6}$$

where Γ is a sigmoidal function, $h_k^{s,c} \in \mathbb{R}^{1\times p}$ is the vector of the scaled and centered inputs of the data-based model, ζ is a smoothing factor, and $0< \zeta <1$. In order to ensure smooth transitions, the weighting factor γ_k at each time step k is exponentially smoothed. The weighting function Γ is designed as a sigmoidal function:

$$\Gamma(h_k^{s,c}) = \frac{1}{1+\exp\left(c \times f_{DF}(h_k^{s,c}W) + \ln\left(\frac{\epsilon}{1-\epsilon}\right)\right)}, \quad c = \frac{2\ln\left(\frac{1-\epsilon}{\epsilon}\right)}{z_{cut\,off}}, \tag{7}$$

where $z_{cut\,off} \in \mathbb{R}$ and $\epsilon \in \mathbb{R}$ are the tuning parameters of Γ. The weighting function Γ is chosen to avoid discontinuities in the process model, which can lead to numerical problems as well as abrupt control moves and chattering control behaviour.

3. Case study: Mother liquor distillation column

The considered plant is a mother liquor distillation column that is operated continuously with a varying feed due to the material fed from upstream batch processes. Figure 1 shows a schematic of the distillation column. The feed mixture consists of a low-boiling solvent, a high-boiling component, water, and other impurities. The product streams of the column are the distillate stream that consists of the low-boiling solvent with mass fraction $\boldsymbol{w_{lb}}$ and water, and the side stream that contains water of mass fraction $\boldsymbol{w_w}$ and solvent. $\boldsymbol{w_{lb}}$ and $\boldsymbol{w_w}$ must be above specified values. The sump stream is a waste stream that contains water, the high-boiling component, and impurities. The mass fraction of the high boiling component $\boldsymbol{w_{hb}}$ in the bottom stream should be below a certain limit to avoid fouling in the reboiler and the column sump. Level controllers maintain the levels of the distillate receiver and of the column sump. In addition, flow controllers are used to control the flowrates of the reflux stream, the side stream, and the steam stream. A very detailed model that was developed in Unisim Design (Honeywell, 2016) is available and is used as a virtual plant in this paper.

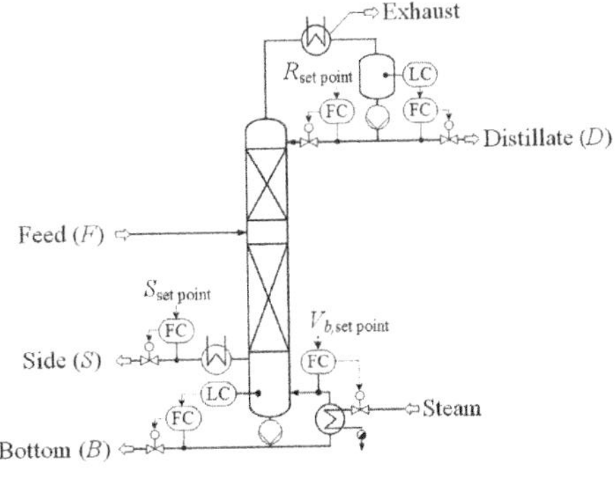

Figure 1: Distillation column schematic

4. Results

The hybrid modeling methodology explained in section 2 is applied to the mother liquor distillation column from section 3. In the simplified mechanistic model we assume that the manipulated variables (the reflux flowrate R, the side flowrate S and the boilup flowrate V_b) are perfectly controlled, so the behavior of the corresponding flow controllers is not included in the model. Similarly, the model does not incorporate the level controllers of the distillate receiver and the column sump. The holdup of each equilibrium stage is assumed to be constant. The model is simplified further by neglecting the energy balances of the equilibrium stages. Thus the simplified model contains only the dynamic mass and component balance equations for each equilibrium stage, the condenser and the reboiler. The vapor phase equilibrium is based on the modified Raoult's law where the liquid activity coefficient of each component is computed by the nonrandom two-liquid (NRTL) model. The saturated vapor pressure of each component at each stage is calculated by the Antoine equation. The equilibrium temperature T_i of each stage i is obtained by solving the summation equation with respect to T_i. This results in a semi-explicit index-1 DAE system, consisting of 108 differential equations and 27 algebraic equations. The DAE system is discretized in time using orthogonal collocation. Due to the simplifications performed in deriving the first-principles-based model, there is a structural mismatch between the true dynamics of the plant and the dynamics of the mechanistic model. A nonlinear autoregressive network with exogenous inputs (NARX) is used to predict and to compensate this plant model mismatch. The feedforward element of the NARX model is a feedforward neural network (FNN) consisting of 2 hidden layers, each formed of 30 neurons. The NARX model has 2 outputs which are the predicted errors of the mass fraction of the low boiling component in the top product and of the mass fraction of the high boiling component in the bottom stream $\{e_{k+1}^{w_{lb}}, e_{k+1}^{w_{hb}}\}$. The mass fraction of water w_w in the side stream is not corrected as the prediction of w_w by the mechanistic model is accurate enough. The NARX model can be represented as

$$e_{k+1} = f_{ML}\left(e_k, \cdots, e_{k-d_e}, \bar{y}_k, \cdots, \bar{y}_{k-d_y}, u_k, \cdots, u_{k-d_u}, F_k, \cdots, F_{k-d_p}\right), \quad (8)$$

where e_k is the vector of the mismatches $\{e_k^{w_{lb}}, e_k^{w_{hb}}\}$, $\bar{y}_k$ is the part of the output vector of the mechanistic model consisting of the mass fractions $\{w_{lb,k}, w_{hb,k}\}$, u_k is the vector of the manipulated variables $\{R_k, S_k, V_{b,k}\}$ and F_k is the feed rate. The numbers of delays d_e, d_y, d_u and d_p are chosen to be equal to 10.

The training data was generated by a reinforcement learning agent to ensure a sufficient excitation of the plant (Ortmanns et al., 2023). With a sampling time of 3 minutes, the outputs of the virtual plant were recorded over 14 (simulated) days. The training data is divided into equally sized batches, and the training objective function is the sum of the mean squared errors of the multi-step ahead predictions and of the single-step predictions over a training batch. Figure 2 shows the results of the multi-step ahead predictions of the hybrid model with the full contribution of the data-based model in the training and validation datasets. The mechanistic model has slower dynamics than the true system

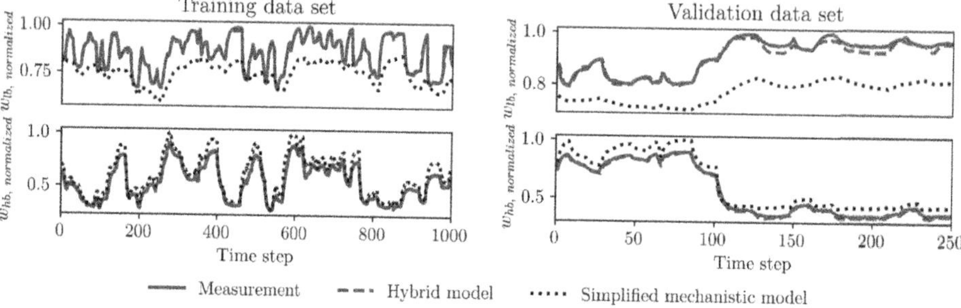

Figure 2: Results of the hybrid model with full contribution of the data_based model.

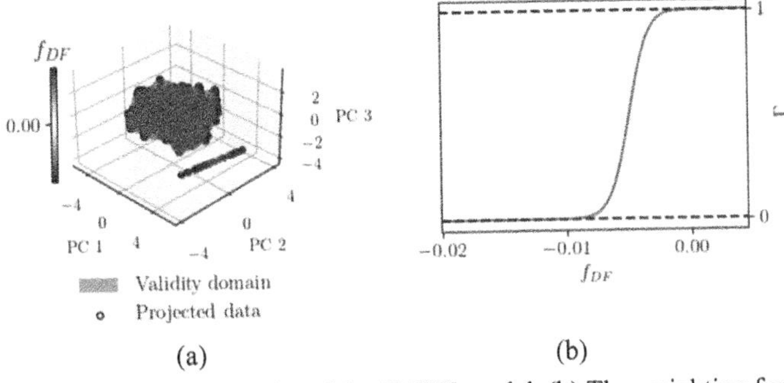

Figure 3: (a) The domain of validity of the NARX model. (b) The weighting function of the NARX model.

especially for the composition for the low boiling solvent in the distillate stream w_{lb}, while the data-based model compensates the mismatches in w_{lb} and w_{hb} very well.

The manipulated variables and the feed rate were found to be the most influential inputs of the data-based model via one-at-a-time sensitivity analysis. The first three principal components explain 95% of the variation in the training data of the influential input variables. A one-class SVM was then trained on the three-dimensional projection of the training data to estimate the domain of validity of the data-based model as shown in Figure 3a. The hyperparameters of the one-class SVM were chosen such that f_{DF} is greater than zero for the projection of the training data. Figure 3a shows the projected training points in red, indicating positive values of f_{DF} according to the colour map. The projected test data points are shown in blue, yielding negative values of f_{DF} as the feed and boilup flow rates in the test data are outside the range of the training data. The relation between the weighting function Γ and the one-class SVM decision function f_{DF} is represented in Figure 3b. The tuning parameters of Γ are $\epsilon = 10^{-3}$ and $z_{cut\,off} = -0.01$, so Γ is equal to 10^{-3} when $f_{DF} = z_{cut\,off}$, and the value of Γ is 0.999 when $f_{DF} = 0$. The smoothing factor ζ was chosen as 0.5.

Figure 4 shows the smooth switching from the full hybrid model to the simplified mechanistic model. As the feed rate changes to the outside of the domain of validity at the 6th time step, the weighting factor γ fades out leading to a better prediction quality of the mechanistic model than that of the full hybrid model without adaptation as the predictions of the data-based component get worse outside the domain of validity.

Based on this modeling approach, Elsheikh et al. (2023) proposed a nonlinear model predictive controller for the considered distillation column where the domain of validity

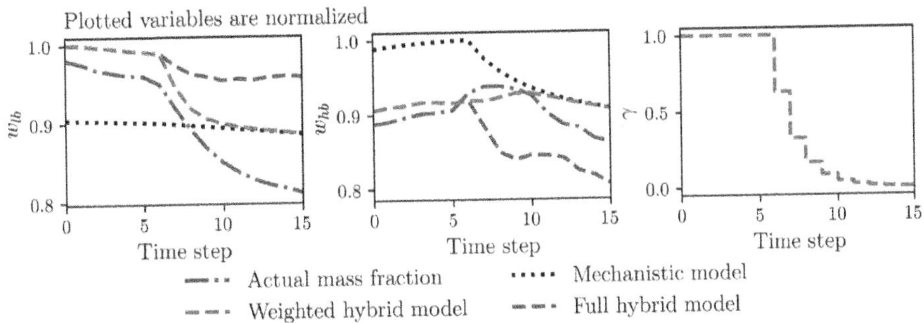

Figure 4: Smooth decay of the contribution of the data-based model.

is in addition extended online based on the observed performance of the data-based component.

5. Conclusion

In this paper, the use of a hybrid model is proposed which is formed of a simple mechanistic model and a data-based model which compensates for the plant dynamics that are not represented by the mechanistic model. A one-class SVM is trained on low-dimensional projections of the training data to characterize the domain of validity of the data-based model component. The contribution of the data-based model is faded out based on the output of the SVM. For the case study of an industrial mother liquor distillation column, it is shown that the data-based component of the hybrid model can accurately compensate the deficiencies of the mechanistic model. The proposed weighting of the data-based model improves the prediction of the hybrid model, as the contribution of the data-based model diminishes smoothly in the areas with insufficient training data.

Acknowledgment

This research has been supported by the project "KI-Inkubator-Labore in der Prozessindustrie - KEEN", funded by the Bundesministerium für Wirtschaft und Klimaschutz (BMWK) under grant number 1MK20014T. This support is gratefully acknowledged.

References

P. Azadi, J. Winz, E. Leo, R. Klock, and S. Engell, 2022, "A hybrid dynamic model for the prediction of molten iron and slag quality indices of a large-scale blast furnace", Computers & Chemical Engineering, 156, 107573.

F. Bock, S. Keller, N. Huber, and B. Klusemann, 2021, "Hybrid Modelling by Machine Learning Corrections of Analytical Model Predictions towards High-Fidelity Simulation Solutions", Materials, 14(8).

M. Elsheikh, Y. Ortmanns, F. Hecht, V. Roßmann, S. Krämer, and S. Engell, 2023, "Model Predictive Control of an Industrial Distillation Column Based on a Hybrid Model: Adapting the Domain of Validity", IFAC- PapersOnLine, Submitted.

Honeywell, 2016, "Unisim design simulation basis", Technical report, Honeywell, London.

I.T. Jollie and J. Cadima, 2016, "Principal component analysis: a review and recent developments", Philosophical Transactions of the Royal Society A: Mathematical, Physical and Engineering Sciences, 374(2065).

C. Nentwich, J. Winz, and S. Engell, 2019, "Surrogate Modeling of Fugacity Coefficients Using Adaptive Sampling", Industrial and Engineering Chemistry Research, 58(40), 18703–18716.

Y. Ortmanns, M. Elsheikh, S. Engell, J. C. Schulze, R. Hille, F. Hecht, V. Rößmann, and S. Krämer, 2023, "Reinforcement Learning for Process Control: Reducing the Training Duration with Pretraining", IFAC- PapersOnLine, Submitted.

B. Schölkopf, R. C. Williamson, A. Smola, J. Shawe-Taylor, and J. Platt, 1999, "Support Vector Method for Novelty Detection", Advances in Neural Information Processing Systems, vol 12.

A. M. Schweidtmann, J. M. Weber, C. Wende, L. Netze, and A. Mitsos, 2022, "Obey validity limits of data-driven models through topological data analysis and one-class classification", Optimization and Engineering, 23(2), 855–876.

M. von Stosch, R. Oliveira, J. Peres, and S. Feyo de Azevedo, 2014, "Hybrid semi-parametric modeling in process systems engineering: Past, present and future", Computers & Chemical Engineering, vol 60, 86–101.

J. Winz and S. Engell, 2022, "A methodology for gray-box modeling of nonlinear ODE systems", 32nd European Symposium on Computer Aided Process Engineering, vol 51, 1483–1488.

Antonis Kokossis, Michael C. Georgiadis, Efstratios N. Pistikopoulos (Eds.)
PROCEEDINGS OF THE 33rd European Symposium on Computer Aided Process Engineering (ESCAPE33), June 18-21, 2023, Athens, Greece
 http://dx.doi.org/10.1016/B978-0-443-15274-0.50279-1

Integrating mineral mining and metallurgical supply chains: a *qualogistics* approach

Mohammed Yaqot, Brenno C. Menezes

Division of Engineering Management and Decision Sciences, College of Science and Engineering, Hamad Bin Khalifa University, Doha, Qatar Foundation, Qatar
bmenezes@hbku.edu.qa

Abstract

The unbalance between supply and demand of raw materials and products worldwide culminates in colossal challenges in global supply chains. Engineering management that relies on lean design and operation is essential for the long-term viability of the value chain to keep enterprises' operations within an acceptable profit margin at the production level in both natural resource exploration and processing transformation sites. In this work, quantity and quality dimensions of mineral raw material feedstocks are considered from a logistics perspective for an integrated mining and metallurgical supply chain. Such a complete producer-processer integration of raw materials, amounts, and properties of the concentrated mineral diet to be fed in metallurgical furnaces, follows a sequence of mined-to-fed resource management with logistics and quality constraints to be met. This includes (1) exploration and processing in the mining sites, (2) shipping of concentrated mineral pellets, and (3) logistics and mixtures of materials before the furnace feeding in metallurgical sites. In this operational decision-making set, mixed-integer solutions for resource management include the following sequence. First, it involves the assignment of equipment to be shared in different quality exploration fields (e.g., trucks and conveyor-belts). Second, conveying the crushed-ore to be mixed-stocked using shuttle-conveyor-belts in simultaneous positioning to match the qualities of the stockpiles to be ground, floated, and processed into concentrated pellets. Finally, shipping of the blended mineral pellets and metallurgical site logistics and blending mixtures before the furnace feeding. In this work, an enterprise-wide optimization of a unified mining and metallurgical of raw materials is proposed for an improved stockpiling and blending in both supply chains. Thus, their integration can be achieved to maximise the ratio of ores (in the mining field) to metals (in the metallurgical site) output while minimising operational and capital costs.

Keywords: Mining fields, metallurgical sites, quality-logistics.

1. Introduction

Most production mining operations aim for a round-the-clock operation, seven days a week, 365 days a year. In practice, semi-autogenous milling facilities run at about 95% capacity (run-time) with an efficient crushing plant running at around 85% (Jenike and Johanson, 2022). These figures should thus be extremely near to, if not greater, the actual percentage of time that the plant works after the start-up phase. However, plant availability is just one factor determining plant run-time; coarse ore's availability (in terms of quantity, but also quality) to feed the semi-autogenous grinding (SAG) milling plant or the fine crushing plant is equally crucial importance.

Stockpile management is a critical concept in the bulk materials handling supply chain at both resource exploration sites and processing plants. The port-based stockyard can represent a mega project investment ($\geq$1 billion USD) with large-scale machinery and automation, e.g., autonomous haul trucks and conveyor belts, etc. Given such a significant investment in infrastructure and technology, the more efficiently raw materials can be processed through the stockyards, the less operational cost and asset installation, the faster the payback, and the higher the operating profits will be. In accordance with the industry-wide aim of optimising asset utilisation, businesses are examining how stockpile management might enhance critical key performance indicators (KPIs) such as utilisation against availability ratios. In order to preserve market dominance, manufacturers are enhancing their stockpiling strategies, taking use of technological advancements in equipment and control systems to renew stockyard infrastructure, and strengthening their capacity to do so through increased material flow and quality control (Morley and Arvidson, 2017). Automation of all or a portion of stockpile operations, replacement of older stacker-reclaimer configurations with newer frameworks, and faster, more accurate stockpile accounting are among the upgrading aims (Menezes et al., 2019). Certain mines with unique processing needs have implemented or are developing bulk ore sorting technology to pre-concentrate low-grade ore from stockpiles, bringing it to a level that is economically viable for milling (Russell, 2018).

The stockyards in both production and metallurgical sites serve as material buffers, reserves, or blendstocks between the mining and processing of the crude-ore or mineral pellets in smelting furnaces. Mining companies are faced with the challenge of balancing the benefits provided by maintaining stockpiles, e.g., guarding against supply fluctuations while also managing costs associated with stockyard management. Ore blending is an essential technique in today's large production mines. It is necessary for the mine's life cycle to be extended and dilute various harmful elements (e.g., sulfur, silica, etc.) found in different ores from the same mine. In this sense, it is also used to combine desired qualities from various ores to provide a high quality and quantity feed. Ore blending reduces input material variability and ensures a consistent flow into the processing unit. It is consequently critical that mines maintain a precise blending strategy that meets the demands of both the mine and the metallurgical processing plants.

In this work, an integrated planning for a unified mining and metallurgical of raw materials by an improved stockpiling and blending design is discussed. Thus, additional safety, stability, and predictability can be achieved to maximise the ratio of ores (in the mining field) to metals (in the metallurgical site) output while minimising the operating and capital costs. The tangible and intangible results of this integration can be applied to fuel and petrochemical industries, food processing, mining-metallurgical parks, etc., that integrates information to be included in strategic, tactical, and operational planning studies between upstream and downstream assets design and operations.

2. Ore blend scheduling at mining processing sites

Kelly and Menezes (2019) have developed a smart conveying and stockpiling flow process that combines process optimisation to leverage control and automation of a cyber-physical system that allows the inventory level control of multiple stockpiles per belt. The model can optimally manage and automate the inbound recipe (by the syncronised mixtures of solid streams) and outbound delivery of mined material including, coal, iron ore, bauxite, etc. Mineral processing facilities are typically complicated systems that must contend with internal and external disturbances, a large number of cyclical loads, and a wide range of interactions. Furthermore, the reducing ore grades as well as the continual variation of raw materials are pushing the boundaries of processing facilities to their limits even further. Through mining supply chain, each process characteristics, which miners are oriented with at each stage, offer opportunities for improvement in terms of reducing costs, energy intensity, and solve the problem of a shortage of qualified and experienced instrumentation technicians, operators, and metallurgists (Rockwell Automation, 2021).

Such control application considers advancements in manufacturing toward the Industry 4.0 (I4.0) mandate involving information and computing technology (ICT) and mechatronics (MEC) evolving together with advances in modelling and solving algorithms (MSA). For such, recent advances in network (flowsheet) optimisation as well as in solving algorithms and computer-aided resources allow fast solutions of complex discrete decision propositions. Moreover, the ICT expansion can provide the demanded velocity of the communication of the online data measurement of complete process networks for the massive volume and variety of information from the plant, which permits the viability of the addressed control strategy. Typically, the objectives of mine planning are to establish a production schedule that culminates in the selection of ore bodies that ensures the timely supply of ore of the required grade quantity and quality to the processing plant. An appropriate ore stockpiling design can significantly reduce milling time, enhance flotation processes, and raise concentrators' production efficiency and economic performance. A complete integrated smart manufacturing system among MSA, ICT, and MEC through sensing, optimising, and actuating cycles can be achieved to precisely control stockpiling process. The I4.0 deployment facilitates a wide range of control strategies for the quantities and qualities of coarse-ore in conveyor belts. This results in a total free of human-being decision, control, calculation, actuation, etc., and moving from one to multiple stockpiles per belt. Figure 1a illustrates a single shuttle-conveyor tripper car apparatus design considering quantity-logic (QL) only, optimised in Kelly and Menezes (2019) and demonstrated in a proof-of-concept demonstration in a laboratory scale in Yaqot et a. (2022). Figure 1b shows two synchronised robotic apparatus considering a blend of both quantity logic quality phenomena (QLQP).

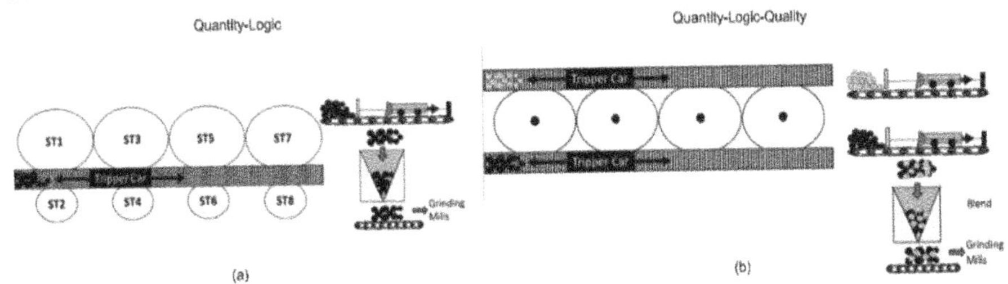

Figure 1. a) Process network with semi-continuous feeding and segregation of solids, from Kelly and Menezes (2019). b) Process network with semi-continuous feeding and blending of solids.

3. Integrating mineral mining and metallurgical supply chains

In light of copper's present and potential future significance in the *electrification of everything*, as an example of one metallurgical industry, there is a need to optimise the scheduling and feed mix blending of concentrates in a copper refinery integrated to the mining operations and controls. The first stage aims to integrate metallurgical planning by means of improving operational and control strategies at the mining site, which in turn potentially yields a lean production by reducing needs of mixtures of streams in the metallurgical sites and, therefore, super dimensioned infrastructure as in today's design. Due to the non-integrated optimisation between supply and demand of raw materials and products in the mining and metallurgical supply chain, besides the super dimensioned infrastructure, remaining blending steps are necessary in metallurgical sites. The flowsheet layout of the copper-refining plant (see Song et al., 2018) is comprised of 2 ships, 6 port piles containing initial inventories or holdups, 1 pre-blender, 3 non-concentrate materials, 6 bins, and 1 smelter, as seen in Figure 2.

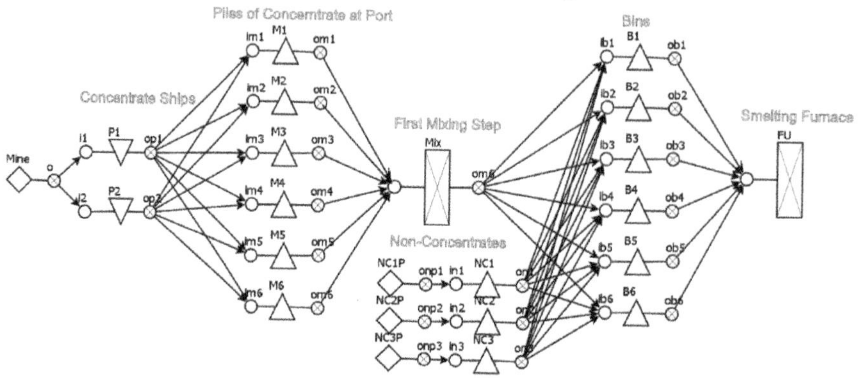

Figure 2. Network sheet for blend-scheduling of a copper-refining plant.

Therefore, quantity and quality constraints of mineral raw material feedstocks must be considered from a *qualogistics* perspective for an integrated mining and metallurgical supply chain. For this producer-processer integration of raw materials, both amounts and properties of the concentrated mineral diet to be fed in metallurgical furnaces follows a sequence of mined-to-fed resource management with logistics and quality constraints to be met. As seen in Figure 3a, this includes: (1) exploration and processing in the mining sites, (2) shipping of concentrated mineral pellets, and (3) logistics and mixtures of materials before the furnace feeding in metallurgical sites. Figure 3b shows the concatenation of solutions to be integrated as proposed, as example, in Menezes et al. (2017) for an enterprise-wide optimization connecting raw materials operations and their processing considering quality throughout the crude-oil supply chain.

In this operational decision-making set integrating mining and metallurgical supply chains, mixed-integer solutions for resource management include the following sequence. First, it involves the assignment of equipment to be shared in different quality exploration fields (e.g., trucks and conveyor-belts). Second, it demands conveying the crushed-ore raw material to be mixed-stocked using shuttle-conveyor-belts in simultaneous positioning (as illustrated in Figure 1a) to match the qualities of the stockpiles to be ground, floated, and processed into concentrated pellets. Finally, it depends in shipping and metallurgical site logistics and blending mixtures before the furnace feeding as illustrated in Figure 2.

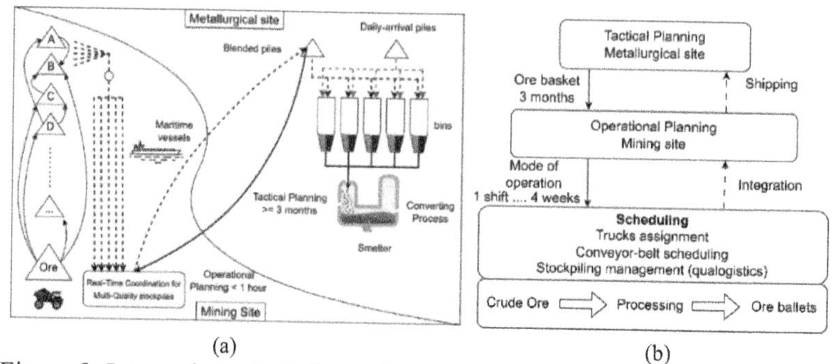

Figure 3. Integrating scheduling of stockpiles in mining-metallurgical sites.

4. Tactical metallurgical and operational mining problems

As proposed in Figure 3a, the tactical planning determines the recipe of segregated crushed-ore raw materials from the mines, whereby it is considered 4 different components (Cu, S, Fe, Rest). In the furnace (FU), there are amounts of quality balances from Kelly et al. (2018) as a linear approximation to find the P1 to P5 (crushed-ore raw materials) and NC1 (non-concentrated or recycled material) that are made of minerals (A to F) or alloys (NC1) in which the results of the FU feed components made by 4 different elements are given as inputs of the operational mining example in Figure 5, whereby the profiles of the level of services of the equipment are shown in Figure 6.

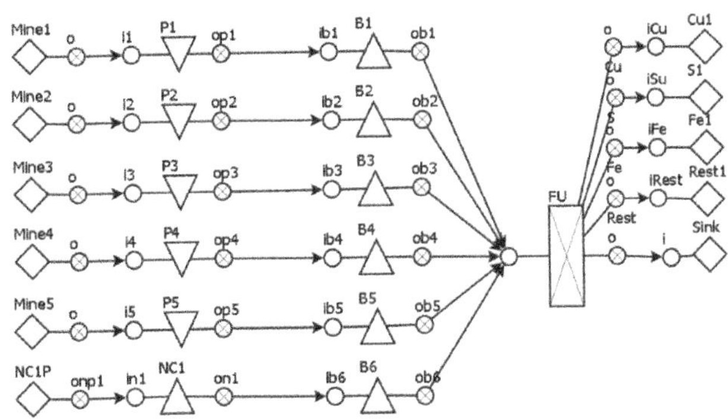

Figure 4. Tactical metallurgical planning to determine FU feed recipes in the furnace.

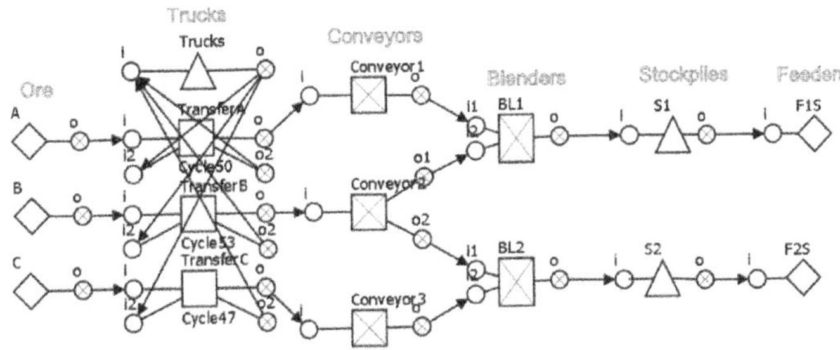

Figure 5. Operational mining scheduling for exploration and process of crushed ore.

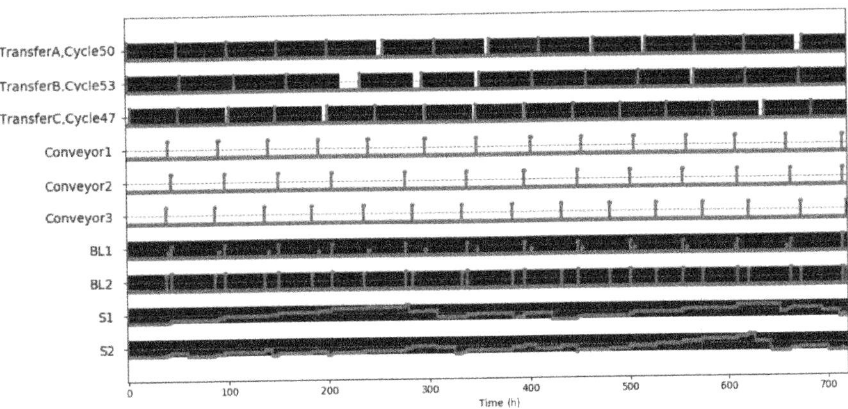

Figure 6. Level of services of the equipment in the operational mining.

5. Conclusion

The disparity between the global supply and demand for raw materials and finished goods creates enormous hurdles for global supply networks. In order to maintain an appropriate profit margin at the production level in both resource exploration and processing transformation sites, it is crucial that businesses use key performance indicators that are based on lean design and operation. As a result of recent developments in the coordination of integrated operations, a comprehensive view of integrated supply and demand throughout the value chain is now attainable, from resource reserves all the way to custom-made feedstocks of blended components intended for final products. This concept, which integrates information to be included in strategic, tactical, and operational planning, scheduling, and control studies between upstream and downstream assets' design and operations, is also used in the fuel and petrochemical sectors, as well as the food processing industry, mining-metallurgical parks, and other similar areas. From a logistics point of view, this study looks at the quantity and quality of mineral raw material feedstocks for an integrated mining and metallurgical supply chain. A mined-to-fed resource management procedure with logistical and quality constraints must be satisfied for such a fully integrated producer-processor raw material, quantity, and property of the concentrated minerals diet to be fed in metallurgical furnaces. Therefore, maximising the ore-to-metal ratio in the mining field and metallurgical site while lowering operational and capital expenses necessitates a higher level of safety, stability, and predictability.

References

Jenike and Johanson, 2022, Importance of coarse-ore stockpile design in mining mega-projects. Bulk Material Engineering. Accessed 20/11/2022 at https://jenike.com/importance-of-coarse-ore-stockpile-design-in-mining-mega-projects/.

Kelly, J.D., Menezes, B.C., 2019, Automating a shuttle-conveyor for multi-stockpile level control, Computer Aided Chemical Engineering, 46, pp. 1153-1158.

Kelly, J.D., 2005, July. The unit-operation-stock superstructure (UOSS) and the quantity-logic-quality paradigm (QLQP) for production scheduling in the process industries. In Multidisciplinary International Scheduling Conference Proceedings: New York, United States. 327, pp. 327-333

Kelly, J.D., Menezes, B.C., Grossmann, I.E., 2018. Successive LP approximation for nonconvex blending in MILP scheduling optimization using factors for qualities in the process industry. Industrial and Engineering Chemistry Research, 57(32), pp.11076-11093.

Menezes, B.C., Grossmann, I.E., Kelly, J.D., 2017, Enterprise-wide optimization for operations of crude-oil refineries: closing the procurement and scheduling gap, Computer Aided Chemical Engineering, 40, pp.1249-1254.

Menezes, B.C., Kelly, J.D., Leal, A.G., 2019. Identification and design of industry 4.0 opportunities in manufacturing: examples from mature industries to laboratory level systems. IFAC-PapersOnLine, 52(13), pp.2494-2500.

Morley, C., Arvidson, H., 2017, Mine value chain reconciliation-demonstrating value through best practice. In Proceedings of the Tenth International Mining Geology Conference, pp. 279-292.

Rockwell Automation, 2021. Model Predictive Control for Mining: Process Optimization Solutions. Accessed 20/11/2022 at https://www.rockwellautomation.com/.

Russell A. Carter, 2018, Staying on Top of Stockpile Management: Smarter, quicker solutions emerge for measuring and controlling stockpile size and quality. Engineering and Mining Journal. Accessed 20/11/2022 at http://www.womp-int.com/story/2018vol12/story026.htm.

Song, Y., Menezes, B.C., Garcia-Herreros, P., Grossmann, I.E., 2018. Scheduling and feed quality optimization of concentrate raw materials in the copper refining industry. Industrial & Engineering Chemistry Research, 57(34), pp.11686-11701.

Yaqot, M., Franzoi, R. E., Islam A., Menezes, B. C., 2022, Cyber-Physical System Demonstration of an Automated Shuttle-Conveyor-Belt Operation for Inventory Control of Multiple Stockpiles: A Proof of Concept, IEEE Access, 10, pp. 127636-127653.

Antonis Kokossis, Michael C. Georgiadis, Efstratios N. Pistikopoulos (Eds.)
PROCEEDINGS OF THE 33rd European Symposium on Computer Aided Process Engineering (ESCAPE33), June 18-21, 2023, Athens, Greece
 http://dx.doi.org/10.1016/B978-0-443-15274-0.50280-8

A simple two-parameter steady-state detection algorithm: Concept and experimental validation

Evren Mert Turan[a], Johannes Jäschke[a],

[a] *Department of Chemical Engineering, Norwegian University of Science and Technology, Sem Sælandsvei 4, 7491, Trondheim, Norway*
johannes.jaschke@ntnu.no

Abstract

Automatic detection of steady state periods is a necessary step for many tasks related to real time operation of a process. Based on the insight that the time series of a controlled process variable at steady state resembles a mean reverting process, we fit a first order auto-regressive model to a window of data and use the Dickey-Fuller test to test for this mean reverting property. We compare the proposed approach on two synthetic data sets and one experimental data set and find that the method performs well in comparison to methods in the literature. As the method is computationally inexpensive, only requires two parameters, and is interpretable, we suggest that this is an effective steady state detection tool.
Keywords: Steady state detection, Process Monitoring.

1. Introduction

The identification of steady state periods in industrial process operations is an essential initial step for many typical tasks, e.g. parameter estimation, abnormal event identification, and real time optimization. Incorrect identification of steady state can severely impact operation, e.g., updating a steady state model with dynamic data can lead to reduced profits and/or unstable operation. Although steady state can be reliably identified by human operators (in low-dimensional data), the automatic identification of steady state, especially with substantial amounts of online data in high dimensions, is still challenging. In this work we use the Dickey-Fuller test to identify steady state time windows, and demonstrate it's effectiveness in comparison to other methods in the literature based on a their perforance on experimental and synthetic data.

Most approaches in for steady state detection in the process engineering literature use 1) a statistical test, or 2) a classification based machine learning method. In this work we focus on the former due to their interpretability, and simple transfer between applications due to their relatively small amount of tuning parameters. Methods relying on statistical tests either tend to perform this test on characteristic values of the system or parameters of a model fitted to online data (Rhinehart, 2013). A key difference between such methods is in the construction of the statistical test, i.e. whether steady state is the null or alternate hypothesis. Although most methods have steady state as the null hypothesis, identification of steady state by the alternate hypothesis is a statistically stronger claim (Rhinehart, 2013).

In this work we fit an auto-regressive (AR) model and use the Dickey-Fuller test to test the null hypothesis that the process is transient. The method is simple to tune, having only two parameters: the size of the time window and the significance level used in the statistical test. In process engineering, the Dickey-Fuller test has been used in fault detection, however to the authors knowledge it has not been applied to steady state detection.

2. Background literature

We briefly outline three methods from the steady state detection literature that make use of a statistical test. In all these methods some quantity based on recent measurements is determined (e.g. a slope), and a statistical test is performed on this quantity (Rhinehart, 2013). Furthermore, in the first two methods we consider a rolling time-window of N recent measurements.

2.1 Method 1: Slope of a line

The simplest approach to steady state detection is to fit a linear model to the data:

$$y_k = mt_k + c \tag{1}$$

where y is a measured process variable, t_k is the kth time point, m is a slope, and c the intercept. After fitting this line by linear least squares, one can check the null-hypothesis of steady state by testing if the slope is zero using a t-test at some significance level. However, this approach violates some assumptions, in particular the y variables at different time points are not independent, i.e. y_{k+1} depends on y_k, and y_{k-1} and so on.

2.2 Method 2: Kelly and Hedengren

In this approach, we test if the process can be described as being at some value, and subjected to independent, identically distributed white noise (Kelly and Hedengren, 2013). As before, we consider the linear model in equation 1. To estimate the parameters averages of the intercept, *c*, and gradient, *m*, are used. Note that if there is gradient is zero then the intercept is the mean value. If the data is sampled uniformly in time by Δt, then these are given by:

$$m = \frac{1}{N\Delta t}\sum_{k=2}^{N} y_k - y_{k-1} \tag{2}$$

$$c = \frac{1}{N}\left(\sum_{k}^{N} y_k - m\Delta t\right) \tag{3}$$

The standard deviation of the noise, σ_n, can then be estimated as:

$$\sigma_n = \sqrt{\frac{\sum_k^N (y_k - mt_k - c)^2}{n-2}} \tag{4}$$

Lastly, steady state identification is performed pointwise by checking if the deviation of the data point from the intercept is within a factor of the standard deviation of the noise. This cut-off point is given by a Student's critical value, t_{crit}, at some significance level and degree of freedom N:

$$s_k = \begin{cases} 1 & \text{if } |y_k - c| \leq t_{crit}\sigma_n \\ 0 & \text{else} \end{cases} \tag{5}$$

$$S = \frac{1}{N}\sum_{k=1}^{N} s_k \tag{6}$$

where S represents the fraction of time in the window "at steady state". If $S \geq 0.5$, then the entire window is at steady state.

2.3 Method 3: Cao and Rhineheart

This method relies on the idea that the ratio of two different estimates of the time series' variance at the same time point should be close to one when the process is at steady state (Cao & Rhinehart, 1995). The first estimate of the variance at t_k, $\sigma_{1,k}^2$, is given by the filtered squared difference between the process measurement, and a filtered value of the measurement, $y_{f,k}$. First the filtered measurement is calculated, and used to calculate a filter of the mean square deviation, v_k^2:

$$y_{f,k} = \lambda_1 y_k + (1 - \lambda_1) y_{f,k-1} \tag{7}$$

$$v_k^2 = \lambda_2 \left(y_k - y_{f,k}\right)^2 + (1 - \lambda_2) v_k^2 \tag{8}$$

Assuming the process is at steady state and the measurements are uncorrelated, then the variance, $\sigma_{1,k}^2$, can be directly estimated as (Cao & Rhinehart, 1995):

$$\sigma_{1,k}^2 = \frac{2 - \lambda_1}{2} v_k^2 \tag{9}$$

The second estimate of the variance, $\sigma_{2,k}^2$, is given by filtering the difference between two consecutive measurements:

$$2\sigma_{2,k}^2 = \lambda_3 (y_k - y_{k-1})^2 + (1 - \lambda_3) 2\sigma_{2,k-1}^2 \tag{10}$$

Lastly, the ratio of these two variances are taken to calculate an R statistic:

$$R_k = \frac{\sigma_{1,k}^2}{\sigma_{2,k}^2} \tag{11}$$

The null hypothesis is that the process is steady, corresponding to a ratio close to 1. If R_k is greater than some critical value, R_{crit} then the null hypothesis is rejected.

3. Proposed method

The proposed method is based on the insight that the time series of a controlled process variable at steady state, subject to stochastic disturbances, resembles that of a *mean reverting process,* i.e. it tends to some mean point despite the stochasticity of the system. In the example of a controlled process unit, this would be due to the relevant controllers rejecting disturbances, or the system settling to a new operation point.

The first step in the appraoch is to find the mean value in the time window, μ, and subtract this from each data point:

$$\tilde{y}_k = y_k - \mu \tag{12}$$

Then we consider the first order auto-regressive (AR) model:

$$\tilde{y}_{k+1} = p\tilde{y}_k + n_k \tag{13}$$

where $\tilde{y}_k$ is the transformed measurement at time t_k, n_k is some random variable with mean zero and finite variance, and p is a variable estimated by linear least squares on the data in the time window. A value of $|p| < 1$ means that the process is mean reverting, as as the absolute disturbance from the mean will deterministically decrease at each iteration. After estimating $\hat{p}$, we then perform a one-sided confidence test on the null-hypothesis that $|p| = 1$, i.e. the process is in a transient state, by calculating the test statistic τ:

$$\tau = (\hat{p} - 1)\sqrt{\frac{(N-2)\sum_{k=2}^{N}\tilde{y}_k^2}{\sum_{k=2}^{N}(\tilde{y}_k - \hat{p}\,\tilde{y}_{k-1})^2}} \tag{14}$$

If the test statistic is less than the critical value from the Dickey-Fuller distribution then the null hypothesis is rejected, i.e. the process is at steady state. The critical values can be found from response surfaces fit to the distribution (MacKinnon 2010). The asymptotic results of this test do not require assumptions of normality or homoscedasticity (MacKinnon 2010). Lastly, note that if the process variable is controlled off-set free to some set-point, then one could use the set-point instead of calculating the mean value.

4. Results

We compare the proposed method (DF) against the line slope, Cao & Rhinehart (CR, 1995) and Kelly & Hedengren (KH, 2013) methods on two synthetic data sets and one experimental data set.. For brevity these acronyms are used throughout this section. For these data sets we label steady state and transient periods based on 1) knowledge of when the system inputs are changing and 2) inspection of the data in the time windows. For the methods with time windows and significance level as hyper-parameters we use a time window of 30 seconds, and significance level of 95%. For the method of Cao & Rhinehart (1995) we use the recommended values of $\lambda_1 = 0.2, \lambda_2 = \lambda_3 = 0.1,\ R_{crit} = 2$.

4.1 Low level of normally distributed noise

We first consider a synthetic example with a low level of Gaussian noise. This is an easier task than the others and therefore serves to give a baseline of the methods performance. The results are summarized in Figure 1. From a visual inspection, the proposed DF method performs the most consistently, with the worst performance by the slope method. For all the methods the primary source of error is due to the delay that occurs when the system goes from transient to steady state (around 180s, 360s, and 540s in Figure 1). In contrast, the transition from steady to transient state is captured without a significant delay. The transition to steady state is harder to detect due to the presence of the transient in the time window, as this heavily influences the fit of the models and the filtered variance. Note that if the DF method is used with a set-point instead of the mean then this delay would be reduced, i.e. the delay for this approach comes from taking the mean of the data.

4.2 Moderate level of t-distributed noise

In this comparison we consider the same process as in section 4.1, but now use t-distributed noise, with a larger variance, instead of Gaussian noise. The t-distribution has heavier tails than the normal distribution leading to a high probability of “high-magnitude” noise, leading to a more challenging classification. The experiment is summarized in Figure 2, with “spikes” due to change in distribution clearly shown in Figure 2a. The aim of using this distribution is 1) to test the methods against higher noise levels, 2) to test against non-normally distributed noise.

Apart from the general decrease in performance, the clearest difference when comparing Figure 2b vs 1b is that DF, KH, and CR have a higher rate of false identification of steady state. This does not occur with slope method, as this approach is extremely sensitive to noise. As in section 4.1 the DF method has the best performance.

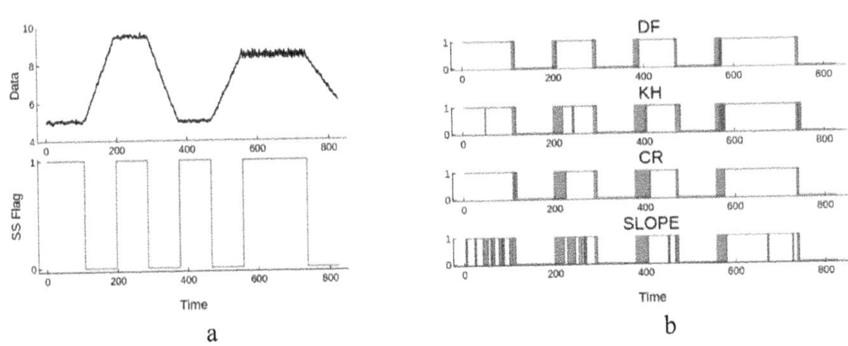

Figure 1: Synthetic data with low levels of Gaussian noise and identified steady state periods (A), and steady state predictions (B). In A and B, a steady state flag of 1 indicates steady state. The shaded red regions in B indicate a mismatch between the predicted and true system state

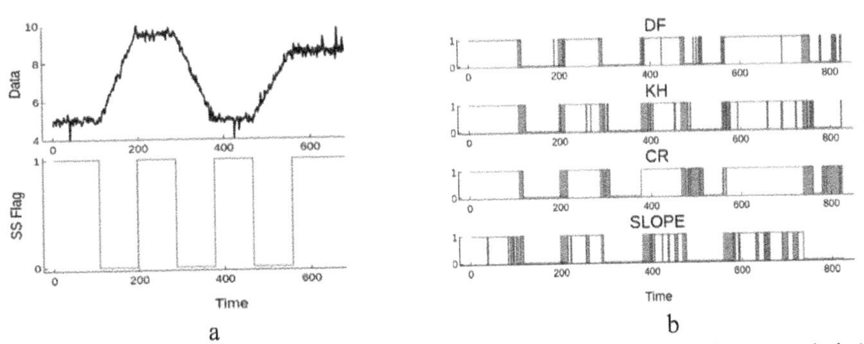

Figure 2: Synthetic data with moderate levels of t-distributed noise and steady state periods (A), and steady state predictions (B).

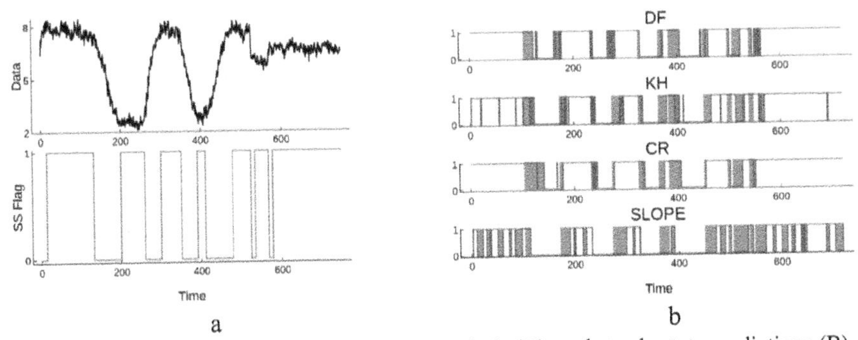

Figure 3: Experimental data with steady state periods (A), and steady state predictions (B)

4.3 Experimental data

Lastly, we consider the steady state algorithms applied to real experimental measurements from a lab rig with the experiment summarized in Figure 3, and Table 1.

Table 1: Summary statistics of the steady state detection algorithms applied to the experimental data, shown in Figure 3. The bolded entries indicate the best method for a statistic

STATISTIC	DF	KH	CR	SLOPE
PRECISION	0.85	0.83	0.81	**0.86**
RECALL	**0.90**	0.78	0.89	0.44
F1 SCORE	**0.87**	0.81	0.85	0.58
ϕ COEFFICIENT	**0.60**	0.46	0.51	0.29

This is a real system with a relatively large amount of noise, leading to worse performance compared to the synthetic results by all the methods. In addition there are two step changes between 500 and 600 seconds, which results poorer performance in this section, as shown in Figure 3. A positive aspect of the CR and DF methods are that they are more consistent as they have much less false flags within the SS and transient periods.

Summary statistics of the methods on this data set are shown in Table 1 and serves as quantitative evidence of the visually better performance of the DF method in Figure 3. Precision is the fraction of correct SS predictions over all SS predictions, while recall is the fraction of correct SS predictions over all true SS periods. The F1 score is the harmonic mean of precision and recall. Lastly the ϕ coefficient is a balanced metric that requires reliable performance in both SS and TS prediction. In general, the proposed DF method performs the best, except for the precision metric where the simple line slope method performs similarly due to rarely identifying steady state, as shown in Figure 3.

5. Conclusions

In conclusion, we present the application of the Dickey-Fuller test for use in steady state detection and compare it with other methods in the literature on a range of examples. Based on summary statistics of these methods, the Dickey-Fuller test performs the best overall. The method is simple to implement and only requires two-hyperparameters. Further work could extend the approach to multivariet system using approaches suggested in the literature (Rhinehart, 2013, Kelly and Hedengren, 2013).

Acknowledgments

The authors acknowledge the support of the Norwegian Research Council through the AutoPRO project.

References

S. Cao and R.R. Rhinehart, 1995, An efficient method for on-line identification of steady state, Journal of Process Control, 5(6), pp.363-374.

D.A. Dickey and W.A. Fuller, 1979, Distribution of the estimators for autoregressive time series with a unit root. Journal of the American statistical association, 74(366a), pp.427-431.

J.D. Kelly, and J.D. Hedengren, 2013, A steady-state detection (SSD) algorithm to detect non-stationary drifts in processes, Journal of Process Control, 23(3), pp.326-331.

J.G. MacKinnon, 2010, Critical values for cointegration tests, Queen's Economics Department Working Paper, No. 1227.

R.R. Rhinehart, 2013, Automated steady and transient state identification in noisy processes, American Control Conference, pp.4477-4493.